COMPREHENSIVE ANALYTICAL CHEMISTRY

Elsevier
Radarweg 29, PO Box 211, 1000 AE Amsterdam, The Netherlands
Linacre House, Jordan Hill, Oxford OX2 8DP, UK

First edition 2007

Library of Congress Cataloging-in-Publication Data
A catalog record for this book is available from the Library of Congress

British Library Cataloguing in Publication Data
A catalogue record for this book is available from the British Library

ISBN: 978-0-444-53053-0
ISBN CD-ROM: 978-0-444-53133-9
ISSN: 0166-526X

For information on all Elsevier publications
visit our website at books.elsevier.com

Printed and bound in The Netherlands

07 08 09 10 11   10 9 8 7 6 5 4 3 2 1

Wilson & Wilson's

COMPREHENSIVE ANALYTICAL CHEMISTRY

Edited by

D. BARCELÓ

Research Professor
Department of Environmental Chemistry
IIQAB-CSIC
Jordi Girona 18-26
08034 Barcelona
Spain

Wilson & Wilson's

COMPREHENSIVE ANALYTICAL CHEMISTRY

VOLUME 49

ELECTROCHEMICAL SENSOR ANALYSIS

Edited by

S. ALEGRET

Autonomous University of Barcelona, Catalonia, Spain

A. MERKOÇI

Catalan Institute of Technology, Catalonia, Spain

ELSEVIER

AMSTERDAM – BOSTON – HEIDELBERG – LONDON – NEW YORK – OXFORD – PARIS
SAN DIEGO – SAN FRANCISCO – SINGAPORE – SYDNEY – TOKYO

Contributors to Volume 49

Patricia Abad-Valle
Departamento de Química Física y Analítica, Universidad de Oviedo, 33006 Asturias, Spain

Hassan Y. Aboul-Enein
Pharmaceutical and Drug Industries Research Division, National Research Center (NRC), Dokki, Cairo 12662, Egypt

Javier Adrian
Applied Molecular Receptors Group (AMRg), IIQAB-CSIC, 08034 Barcelona, Catalonia, Spain

Salvador Alegret
Grup de Sensors i Biosensors, Departament de Química, Universitat Autònoma de Barcelona, 08193 Bellaterra, Catalonia, Spain

Aziz Amine
Faculty of Sciences and Techniques, University Hassan II – Mohammedia, B.P. 146, Mohammedia, Morocco

Jordi Barbé
Unitat de Microbiologia, Departament de Genètica i Microbiologia, Universitat Autònoma de Barcelona, 08193 Bellaterra, Catalonia, Spain

S. Benedetti
Department of Food Science and Technology, University of Milan, Italy

Nina Blomqvist-Kutvonen
Thermo Fisher Scientific, Ratastie 2, FIN-01621 Vantaa, Finland

Johan Bobacka
Process Chemistry Centre, c/o Laboratory of Analytical Chemistry, Åbo Akademi University, Biskopsgatan 8, FIN-20500 Åbo-Turku, Finland

Khiena Z. Brainina
Ural State University of Economics, 8 Marta St., Bld. 62, 620219 Ekaterinburg, Russia

S. Buratti
Department of Food Science and Technology, University of Milan, Italy
Malte Burchardt
Faculty of Mathematics and Natural Sciences, Institute of Chemistry, and Biology of the Marine Environment, Department of Pure and Applied Chemistry, Carl von Ossietzky University of Oldenburg, D-26111 Oldenburg, Germany
Daniel Calvo
Grup de Sensors i Biosensors, Departament de Química, Universitat Autònoma de Barcelona, 08193 Bellaterra, Catalonia, Spain
Luigi Campanella
University of Rome "La Sapienza", Piazzale Aldo Moro, 5 00185 Rome, Italy
M. Campàs
BIOMEM Group, Centre de Phytopharmacie, Université de Perpignan, 52 Avenue Paul Alduy, 66860 Perpignan Cedex, France
Susana Campoy
Centre de Recerca en Sanitat Animal (CReSA), Universitat Autònoma de Barcelona, 08193 Bellaterra, Catalonia, Spain
S. Campuzano
Departamento de Química Analítica, Facultad de Ciencias Químicas, Universidad Complutense de Madrid, Ciudad Universitaria s/n, 28040 Madrid, Spain
Maria Teresa Castañeda
Grup de Sensors i Biosensors, Departament de Química, Universitat Autònoma de Barcelona, 08193 Bellaterra, Catalonia, Spain. On leave from: Departamento de Ciencias, Básicas, Universidad Autònoma Metropolitana-Azcapotzalco, 022000, México, D. F., Mexic
Mario Castaño-Álvarez
Departamento de Química Física y Analítica, Universidad de Oviedo, Asturias, Spain
S. Centi
Università degli Studi di Firenze, Dipartimento di Chimica, Polo Scientifico, Via della Lastruccia 3, 50019 Sesto Fiorentino, Firenze, Italy
Nikos A. Chaniotakis
Department of Chemistry, University of Crete, 71 409 Iraklion, Crete, Greece

Ana M. Chiorcea-Paquim
Departamento de Química, Faculdade de Ciências e Tecnologia, Universidade de Coimbra, 3004-535 Coimbra, Portugal
Edith Chow
School of Chemistry, The University of New South Wales, Sydney, NSW 2052, Australia
D. Compagnone
Department of Food Science, University of Teramo, 64023 Teramo, Italy
Maurice Comtat
Université Paul Sabatier, Laboratoire de Génie Chimique, UMR CNRS 5503, 31062 Toulouse, Cedex 9, France
M.S. Cosio
Department of Food Science and Technology, University of Milan, Italy
Serge Cosnier
Départment de Chimie Moléculaire de Grenoble (UMR-5250, ICMG F-2607), CNRS, Université Joseph Fourier, BP 53, 38041 Grenoble, Cedex 9, France
Agustín Costa-García
Departamento de Química Física y Analítica, Universidad de Oviedo, 33006 Asturias, Spain
Adrian Crew
Centre for Research in Analytical, Materials and Sensors Science, Faculty of Applied Sciences, University of the West of England, Bristol, Frenchay Campus, Coldharbour Lane, Bristol BS16 1QY, UK
Eric Crouch
Centre for Research in Analytical, Materials and Sensors Science, Faculty of Applied Sciences, University of the West of England, Bristol, Frenchay Campus, Coldharbour Lane, Bristol BS16 1QY, UK
Frank Davis
Cranfield Health, Cranfield University, Silsoe, MK45 4DT, UK
Michele Del Carlo
Department of Food Science, University of Teramo, 64023 Teramo, Italy
Manel del Valle
Grup de Sensors i Biosensors, Departament de Química, Universitat Autònoma de Barcelona, 08193 Bellaterra, Catalonia, Spain

Dermot Diamond
Adaptive Sensors Group, National Centre for Sensor Research, School of Chemical Sciences, Dublin City University, Dublin 9, Ireland
Victor C. Diculescu
Departamento de Química, Faculdade de Ciências e Tecnologia, Universidade de Coimbra, 3004-535 Coimbra, Portugal
Arzum Erdem
Ege University, Faculty of Pharmacy, Analytical Chemistry Department, 35100 Bornova, Izmir, Turkey
Orlando Fatibello-Filho
Laboratório de Bioanalítica, Departamento de Química, Universidade Federal de São Carlos, Rod. Washington Luiz, km 235, C.P. 676, 13560-970 São Carlos/SP, Brazil
María Teresa Fernández-Abedul
Departamento de Química Física y Analítica, Universidad de Oviedo, 33006 Asturias, Spain
Didier Fournier
Institut de Pharmacologie et de Biologie Structurale, Université Paul Sabatier, 205, route de Narbonne, 31077 Toulouse, Cedex, France
J. Frederick van Staden
Faculty of Chemistry, University of Bucharest, 4-12 Regina Elisabeta Blvd., 703461 Bucharest-1, Romania
Akira Fujishima
Kanagawa Academy of Science and Technology, 3-2-1 Sakado, Takatsu-ku, Kawasaki-shi, Kanagawa 213-0012, Japan
Roger Galve
Applied Molecular Receptors Group (AMRg), IIQAB-CSIC, 08034 Barcelona, Catalonia, Spain
M. Gamella
Departamento de Química Analítica, Facultad de Ciencias Químicas, Universidad Complutense de Madrid, Ciudad Universitaria s/n, 28040 Madrid, Spain
Tania Gatta
University of Rome "La Sapienza", Piazzale Aldo Moro, 5 00185 Rome, Italy
Tim Gibson
T & D Technology Ltd., Stanley, Wakefield, West Yorkshire WF3 4AA, UK

C. Gondran

Department de Chimie Moléculaire de Grenoble (UMR-5250, JCMG F-2607), CNRS, Université Joseph Fourier, BP 53, 38041 Grenoble, Cedex 9, France

María Begoña. González-García

Departamento de Química Física y Analítica, Universidad de Oviedo, 33006 Asturias, Spain

J. Justin Gooding

School of Chemistry, The University of New South Wales, Sydney, NSW 2052, Australia

John Griffiths

Uniscan Instruments Ltd, Sigma House, Burlow Road, Buxton SK17 9JB, UK

Pierre Gros

Université Paul Sabatier, Laboratoire de Génie Chimique, UMR CNRS 5503, 31062 Toulouse, Cedex 9, France

Coline Guitton

Université Paul Sabatier, Laboratoire de Génie Chimique, UMR CNRS 5503, 31062 Toulouse, Cedex 9, France

B. Gulmez

Analytical Chemistry Department, Faculty of Pharmacy, Ege University, 35100 Bornova, Izmir, Turkey

Manuel Gutiérrez

Grup de Sensors i Biosensors, Departament de Química, Universitat Autònoma de Barcelona, 08193 Bellaterra, Catalonia, Spain

Alan L. Hart

AgResearch Grasslands, Private Bag 11008, Palmerston North, New Zealand

John P. Hart

Centre for Research in Analytical, Materials and Sensors Science, Faculty of Applied Sciences, University of the West of England, Bristol, Frenchay Campus, Coldharbour Lane, Bristol BS16 1QY, UK

David Hernández-Santos

Departamento de Química Física y Analítica, Universidad de Oviedo, 33006 Asturias, Spain

Tibor Hianik

Department of Nuclear Physics and Biophysics, Faculty of Mathematics, Physics and Computer Sciences, Comenius University, Mlynska Dolina F1, 842 48 Bratislava, Slovak Republic

Seamus P.J. Higson
Cranfield Health, Cranfield University, Silsoe, MK45 4DT, UK
Michael Holzinger
Department de Chimie Moléculaire de Grenoble (UMR-5250, JCMG F-2607), CNRS, Université Joseph Fourier, BP 53, 38041 Grenoble, Cedex 9, France
Kevin C. Honeychurch
Centre for Research in Analytical, Materials and Sensors Science, Faculty of Applied Sciences, University of the West of England, Bristol, Frenchay Campus, Coldharbour Lane, Bristol BS16 1QY, UK
R.E. Ionescu
Department of Biotechnology Engineering, Ben-Gurion University of the Negev, P.O. Box 653, Beer-Sheva 84105, Israel
Alla V. Ivanova
Ural State University of Economics, 8 Marta St., Bld. 62, 620219 Ekaterinburg, Russia
Ari Ivaska
Process Chemistry Centre, c/o Laboratory of Analytical Chemistry, Åbo Akademi University, Biskopsgatan 8, FIN-20500 Åbo-Turku, Finland
Graham Johnson
Uniscan Instruments Ltd, Sigma House, Burlow Road, Buxton SK17 9JB, UK
H. Karadeniz
Analytical Chemistry Department, Faculty of Pharmacy, Ege University, 35100 Bornova, Izmir, Turkey
Carolina Nunes Kirchner
Faculty of Mathematics and Natural Sciences, Institute of Chemistry, and Biology of the Marine Environment, Department of Pure and
Applied Chemistry, Carl von Ossietzky University of Oldenburg, D-26111 Oldenburg, Germany
Ulku Anik-Kirgoz
Chemistry Department, Faculty of Art and Science, Muğla University, 48000-Kötekli Muğla, Turkey
Joachim P. Kloock
Aachen University of Applied Sciences, Jülich Campus, Ginsterweg 1, 52428 Jülich, Germany
Alisa N. Kositzina
Ural State University of Economics, 8 Marta St., Bld. 62, 620219 Ekaterinburg, Russia

Pedro José Lamas Ardisana
Grupo de Inmunoelectroanálisis, Departamento de Química-Física y Analítica, Universidad de Oviedo, 33006 Oviedo, Asturias, Spain
Karen A. Law
Cranfield University at Silsoe, Barton Rd, Silsoe MK45 4DT, UK
Viktor Lax
Process Chemistry Centre, c/o Laboratory of Analytical Chemistry, Åbo Akademi University, Biskopsgatan 8, FIN-20500 Åbo-Turku, Finland
Oldair D. Leite
Laboratório de Bioanalítica, Departamento de Química, Universidade Federal de São Carlos, Rod. Washington Luiz, km 235, C.P. 676, 13560-970 São Carlos/SP, Brazil
Anabel Lermo
Grup de Sensors i Biosensors, Departament de Química, Universitat Autònoma de Barcelona, 08193 Bellaterra, Catalonia, Spain
Andrzej Lewenstam
Åbo Akademi University, Process Chemistry Centre, c/o Center for Process Analytical Chemistry and Sensor Technology (ProSens), Biskopsgatan 8, FIN-20500 Turku-Åbo, Finland and Faculty of Material Science and Ceramics, AGH-University of Science and Technology, Al. Mickiewicza 30, PL-30059 Cracow, Poland
Tom Lindfors
Process Chemistry Centre, c/o Laboratory of Analytical Chemistry, Åbo Akademi University, Biskopsgatan 8, FIN-20500 Åbo-Turku, Finland
Karina O. Lupetti
Instituto de Química, Universidade de São Paulo, São Carlos/SP, Brazil
Saverio Mannino
Department of Food Science and Technology, University of Milan, Italy
María Pilar Marco
Applied Molecular Receptors Group (AMRg), IIQAB-CSIC, 08034 Barcelona, Catalonia, Spain
R.S. Marks
Department of Biotechnology Engineering, Ben-Gurion University of the Negev, P.O. Box 653, Beer-Sheva 84105, Israel
G. Marrazza
Dipartimento di Chimica, Università degli Studi di Firenze, Polo Scientifico, Via della Lastruccia 3, 50019 Sesto Fiorentino, Firenze, Italy

Jean-Louis Marty
BIOMEM Group, Centre de Phytopharmacie, Université de Perpignan, 52 Avenue Paul Alduy, 66860 Perpignan Cedex, France
Marcello Mascini
Department of Food Science, University of Teramo, 64023 Teramo, Italy
Arben Merkoçi
Institut Català de Nanotecnologia, Campus UAB, 08193 Bellaterra, Barcelona, Catalonia, Spain
Konstantin Mikhelson
Process Chemistry Centre, c/o Laboratory of Analytical Chemistry, Åbo Akademi University, Biskopsgatan 8, FIN-20500 Åbo-Turku, Finland
Paul Millner
School of Biochemistry and Molecular Biology, University of Leeds, Irene Manton Building, Leeds LS2 9JT, UK
Vladimir M. Mirsky
Institute of Analytical Chemistry, Chemo- and Biosensors, University of Regensburg, D-93040 Regensburg, Germany
Hasna Mohammadi
Faculty of Sciences and Techniques, University Hassan II – Mohammedia, B.P. 146, Mohammedia, Morocco
Patrick Morier
DiagnoSwiss SA, Rte de l'Ile-au-Bois 2, CH-1870 Monthey, Switzerland
Danila Moscone
Dipartimento di Scienze e Tecnologie Chimiche, Università di Roma Tor Vergata, Via della Ricerca Scientifica, 00133 Rome, Italy
Vladimir I. Ogurtsov
Tyndall National Institute, University College, Lee Maltings, Prospect Row, Cork, Ireland
Ana M. Oliveira Brett
Departamento de Química, Faculdade de Ciências e Tecnologia, Universidade de Coimbra, 3004–535 Coimbra, Portugal
Jan Öst
Thermo, Fisher Scientific Ratastie 2, FIN-01621 Vantaa, Finland
M. Ozsoz
Analytical Chemistry Department, Faculty of Pharmacy, Ege University, 35100 Bornova, Izmir, Turkey

Giuseppe Palleschi
Dipartimento di Scienze e Tecnologie Chimiche, Università di Roma Tor Vergata, Via della Ricerca Scientifica, 00133 Rome, Italy
Roy M. Pemberton
Centre for Research in Analytical, Materials and Sensors Science, Faculty of Applied Sciences, University of the West of England, Bristol, Frenchay Campus, Coldharbour Lane, Bristol BS16 1QY, UK
Shane Peper
Advanced Radioanalytical Chemistry Group, Physical and Chemical Sciences Division, Pacific Northwest National Laboratory, Richland, WA 99352, USA
José A.P. Piedade
Departamento de Química, Faculdade de Ciências e Tecnologia, Universidade de Coimbra, 3004-535 Coimbra, Portugal
José Manuel Pingarrón
Departamento de Química Analítica, Facultad de Ciencias Químicas, Universidad Complutense de Madrid, Ciudad Universitaria s/n, 28040 Madrid, Spain
María Isabel Pividori
Grup de Sensors i Biosensors, Departament de Química, Universitat Autònoma de Barcelona, 08193 Bellaterra, Catalonia, Spain
Martin Pumera
ICYS, National Institute for Materials Science, 1-1 Namiki, Tsukuba, Japan
Aleksandar Radu
Adaptive Sensors Group, National Centre for Sensor Research, School of Chemical Sciences, Dublin City University, Dublin 9, Ireland
Ángel Julío Reviejo
Departamento de Química Analítica, Facultad de Ciencias Químicas, Universidad Complutense de Madrid, Ciudad Universitaria s/n, 28040 Madrid, Spain
Frédéric Reymond
DiagnoSwiss S.A., Rte de l'Ile-au-Bois 2, CH-1870 Monthey, Switzerland
Francesco Ricci
Dipartimento di Scienze e Tecnologie Chimiche, Università di Roma Tor Vergata, Via della Ricerca Scientifica, 00133 Rome, Italy
Joël Stéphane Rossier
DiagnoSwiss S.A., Rte de l'Ile-au-Bois 2, CH-1870 Monthey, Switzerland

Maya Yu. Rubtsova
Moscow State University, Chemical Faculty, Leninskiye Gory, 119992 Moscow, Russia
Audrey Ruffien-Ciszak
Université Paul Sabatier, Laboratoire de Génie Chimique, UMR CNRS 5503, 31062 Toulouse, Cedex 9, France
Francisco Sánchez Baeza
Applied Molecular Receptors Group (AMRg), IIQAB-CSIC, 08034 Barcelona, Catalonia, Spain
Svetlana Yu. Saraeva
Ural State University of Economics, 8 Marta St., Bld. 62, 620219 Ekaterinburg, Russia
Matteo Scampicchio
Department of Food Science and Technology, University of Milan, Italy
Michael J. Schöning
Aachen University of Applied Sciences, Institute of Nano- and Biotechnologies, Jülich Campus, Germany
Boris M. Sergeev
Chemical Faculty, Moscow State University, Leninskiye Gory, 119992 Moscow, Russia
Beatriz Serra
Departamento de Química Analítica, Facultad de Ciencias Químicas, Universidad Complutense de Madrid, Ciudad Universitaria s/n, 28040 Madrid, Spain
Sílvia H.P. Serrano
Instituto de Química, Universidade de S. Paulo, Av. Prof. Lineu Prestes, 748 – Bloco 2 Superior, Cidade Universitária, CEP: 05508 – 900, São Paulo, SP, Brazil
Elena N. Sharafutdinova
Ural State University of Economics, 8 Marta St., Bld. 62, 620219 Ekaterinburg, Russia
Michelle A. Sheehan
Tyndall National Institute, University College, Lee Maltings, Prospect Row, Cork, Ireland
Raluca-Ioana Stefan-van Staden
Faculty of Chemistry, University of Bucharest, 4-12 Regina Elisabeta Blvd., 703461 Bucharest-1, Romania
Fredrik Sundfors
Process Chemistry Centre, c/o Laboratory of Analytical Chemistry, Åbo Akademi University, Biskopsgatan 8, FIN-20500 Åbo-Turku, Finland

D. Szydlowska
BIOMEM Group, Centre de Phytopharmacie, Université de Perpignan, 52 Avenue Paul Alduy, 66860 Perpignan Cedex, France
Chiaki Terashima
General Technology Division, Technology Research and Development Department, Central Japan Railway, 1545-33 Ohyama, Komaki City, Aichi 485-0801, Japan
Anthony P.F. Turner
Cranfield University at Silsoe, Barton Rd, Silsoe MK45 4DT, UK
K. van Velzen
Palm Instruments BV, 3992 BZ Houten, The Netherlands
Majlinda Vasjari
Institute of Analytical Chemistry, Chemo- and Biosensors, University of Regensburg, D-93040 Regensburg, Germany
Mercedes Vázquez
Process Chemistry Centre, c/o Laboratory of Analytical Chemistry, Åbo Akademi University, Biskopsgatan 8, FIN-20500 Åbo-Turku, Finland
Iolanda C. Vieira
Departamento de Química, Universidade Federal de Santa Catarina, Florianópolis/SC, Brazil
A. Visconti
Institute of Sciences of Food Production ISPA-CNR, 70126 Bari, Italy
Torsten Wagner
Aachen University of Applied Sciences, Institute of Nano- and Biotechnologies, Jülich Campus, Germany
Joseph Wang
Departments of Chemical and Materials Engineering and Chemistry and Biochemistry, The Biodesign Institute, Arizona State University, Tempe, AZ 85287-5801, USA
Gunther Wittstock
Faculty of Mathematics and Natural Sciences, Institute of Chemistry and Biology of the Marine Environment, Department of Pure and Applied Chemistry, Carl von Ossietzky University of Oldenburg, D-26111 Oldenburg, Germany
Emanuela Zacco
Grup de Sensors i Biosensors, Departament de Química, Universitat Autònoma de Barcelona, 08193 Bellaterra, Catalonia, Spain
Chuan Zhao
Faculty of Mathematics and Natural Sciences, Institute of Chemistry and Biology of the Marine Environment, Department of Pure and Applied Chemistry, Carl von Ossietzky University of Oldenburg, D-26111 Oldenburg, Germany

WILSON AND WILSON'S

COMPREHENSIVE ANALYTICAL CHEMISTRY

VOLUMES IN THE SERIES

Vol. 1A	Analytical Processes
	Gas Analysis
	Inorganic Qualitative Analysis
	Organic Qualitative Analysis
	Inorganic Gravimetric Analysis
Vol. 1B	Inorganic Titrimetric Analysis
	Organic Quantitative Analysis
Vol. 1C	Analytical Chemistry of the Elements
Vol. 2A	Electrochemical Analysis
	Electrodeposition
	Potentiometric Titrations
	Conductometric Titrations
	High-Frequency Titrations
Vol. 2B	Liquid Chromatography in Columns
	Gas Chromatography
	Ion Exchangers
	Distillation
Vol. 2C	Paper and Thin Layer Chromatography
	Radiochemical Methods
	Nuclear Magnetic Resonance and Electron Spin Resonance Methods
	X-Ray Spectrometry
Vol. 2D	Coulometric Analysis
Vol. 3	Elemental Analysis with Minute Sample
	Standards and Standardization
	Separation by Liquid Amalgams
	Vacuum Fusion Analysis of Gases in Metals
	Electroanalysis in Molten Salts
Vol. 4	Instrumentation for Spectroscopy
	Atomic Absorption and Fluorescence Spectroscopy
	Diffuse Reflectane Spectroscopy
Vol. 5	Emission Spectroscopy
	Analytical Microwave Spectroscopy
	Analytical Applications of Electron Microscopy
Vol. 6	Analytical Infrared Spectroscopy
Vol. 7	Thermal Methods in Analytical Chemistry
	Substoichiometric Analytical Methods
Vol. 8	Enzyme Electrodes in Analytical Chemistry
	Molecular Fluorescence Spectroscopy
	Photometric Titrations
	Analytical Applications of Interferometry

Volumes in the series

Vol. 9	Ultraviolet Photoelectron and Photoion Spectroscopy Auger Electron Spectroscopy Plasma Excitation in Spectrochemical Analysis
Vol. 10	Organic Spot Tests Analysis The History of Analytical Chemistry
Vol. 11	The Application of Mathematical Statistics in Analytical Chemistry Mass Spectrometry Ion Selective Electrodes
Vol. 12	Thermal Analysis Part A. Simultaneous Thermoanalytical Examination by Means of the Derivatograph Part B. Biochemical and Clinical Application of Thermometric and Thermal Analysis Part C. Emanation Thermal Analysis and other Radiometric Emanation Methods Part D. Thermophysical Properties of Solids Part E. Pulse Method of Measuring Thermophysical Parameters
Vol. 13	Analysis of Complex Hydrocarbons Part A. Separation Methods Part B. Group Analysis and Detailed Analysis
Vol. 14	Ion-Exchangers in Analytical Chemistry
Vol. 15	Methods of Organic Analysis
Vol. 16	Chemical Microscopy Thermomicroscopy of Organic Compounds
Vol. 17	Gas and Liquid Analysers
Vol. 18	Kinetic Methods in Chemical Analysis Application of Computers in Analytical Chemistry
Vol. 19	Analytical Visible and Ultraviolet Spectrometry
Vol. 20	Photometric Methods in Inorganic Trace Analysis
Vol. 21	New Developments in Conductometric and Oscillometric Analysis
Vol. 22	Titrimetric Analysis in Organic Solvents
Vol. 23	Analytical and Biomedical Applications of Ion-Selective Field-Effect Transistors
Vol. 24	Energy Dispersive X-Ray Fluorescence Analysis
Vol. 25	Preconcentration of Trace Elements
Vol. 26	Radionuclide X-Ray Fluorecence Analysis
Vol. 27	Voltammetry
Vol. 28	Analysis of Substances in the Gaseous Phase
Vol. 29	Chemiluminescence Immunoassay
Vol. 30	Spectrochemical Trace Analysis for Metals and Metalloids
Vol. 31	Surfactants in Analytical Chemistry
Vol. 32	Environmental Analytical Chemistry
Vol. 33	Elemental Speciation – New Approaches for Trace Element Analysis
Vol. 34	Discrete Sample Introduction Techniques for Inductively Coupled Plasma Mass Spectrometry
Vol. 35	Modern Fourier Transform Infrared Spectroscopy
Vol. 36	Chemical Test Methods of Analysis
Vol. 37	Sampling and Sample Preparation for Field and Laboratory
Vol. 38	Countercurrent Chromatography: The Support-Free Liquid Stationary Phase
Vol. 39	Integrated Analytical Systems

Volumes in the series

Vol. 40	Analysis and Fate of Surfactants in the Aquatic Environment
Vol. 41	Sample Preparation for Trace Element Analysis
Vol. 42	Non-destructive Microanalysis of Cultural Heritage Materials
Vol. 43	Chromatographic-mass spectrometric food analysis for trace determination of pesticide residues
Vol. 44	Biosensors and Modern Biospecific Analytical Techniques
Vol. 45	Analysis and Detection by Capillary Electrophoresis
Vol. 46	Proteomics and Peptidomics New Technology Platforms Elucidating Biology
Vol. 47	Modern Instrumental Analysis
Vol. 48	Passive Sampling Techniques in Environmental Monitoring

Contents

NOVEL TRENDS

THICK AND THIN FILM BIOSENSORS

Editors' Preface

There is a continuing demand for the application of electrochemical sensors with a good quality/cost performance not only in comparison to sensors based on other transducer mechanisms but also to some standard analytical methods. *Electrochemical Sensor Analysis* (ECSA) presents novel theoretical considerations along with detailed applications of electrochemical (bio)sensors. The combination of both theoretical and practical aspects provides a comprehensive forum that integrates interdisciplinary research and development, presenting the most recent advances in applications in various important areas related to everyday life. Additionally ECSA reflects that electrochemical sensor analysis is already a well established research and applied area of analytical chemistry.

Our main objective was to prepare a useful reference source for all those involved in the research, teaching, study, and practice of electrochemical (bio)sensor analysis for environmental, clinical and industrial analysis. It covers the entire field of electrochemical (bio)sensor designs and characterisations and encompasses all subjects relevant to their application in real clinical, environment, food and industry related samples as well as for safety and security. The contributors work in a wide diversity of technological and scientific fields.

The book is organized in two parts. The current printed volume represents Part I while Part II is digitally printed as a CD-Rom.

Part I contains general reviews on the theoretical and practical aspects related to the application of (bio)sensors in various fields. Its 38 chapters are grouped in seven sections: 1) Potentiometric sensors, 2) Voltammetric sensors, 3) Gas sensors, 4) Enzyme based sensors, 5) Affinity biosensors, 6) Thick and thin film (bio)sensors and 7) Novel trends.

Each section covers the most important applications of electrochemical (bio)sensor achieved so far along with the recent trends. It provides firstly an overview of the topic (analyte, matrix etc.) of interest as well as a discussion of published data with a selected list of references for further details.

Chapters 1 to 5 deal with ionophore-based potentiometric sensors or ion-selective electrodes (ISEs). Chapters 6 to 11 cover voltammetric sensors and biosensors and their various applications. The third section (Chapter 12) is dedicated to gas analysis. Chapters 13 to 17 deal with enzyme based sensors. Chapters 18 to 22 are dedicated to immunosensors and genosensors. Chapters 23 to 29 cover thick and thin film based sensors and the final section (Chapters 30 to 38) is focused on novel trends in electrochemical sensor technologies based on electronic tongues, micro and nanotechnologies, nanomaterials, etc.

Part II (the CD-ROM) contains 53 procedures related to the design and practical applications of the (bio)sensors mentioned in the chapters of Part I. A detailed list of all the materials, reagents, solutions necessary to carry out each of the proposed procedures is given. The necessary steps to prepare the (bio)sensor including its calibration, measurement sequences followed by sample treatment (if applicable) and analysis are described in detail. Each procedure ends with a brief discussion of the typical results expected and selected recommended literature.

ECSA presents in detail (bio)sensor designs and applications so as to meet, according to analytical circumstances: (1) The relevance of output signal to measurement environment (2) Accuracy and repeatability (3) Sensitivity and resolution (4) Selectivity (or cross selectivity for sensor matrix arrays) (5) Dynamic range (6) Speed of response (7) Insensitivity to temperature (or temperature compensation) (8) Insensitivity to electrical and other environmental interference (9) Testing and calibration (10) Reliability and Self-Checking Capability (11) Physical robustness (12) Service requirements (13) Capital cost (14) Running costs and life (15) Acceptability by user, and (16) Product safety. Nevertheless it will be up to the readers to meet the above objectives in the conditions of their own laboratories while either applying the presented (bio)sensors to real samples or doing further related research.

We sincerely hope that this book will be useful not only to researchers working in the field but also to materials scientists, physicists, biologists, physicians, metallurgists, engineers, and all who are actively engaged in solving analytical problems with electrochemical (bio)sensors in real samples. Those researchers in other fields who require novel and easy to apply electrochemical (bio)sensors will also find much of value here.

Finally we would like to express our gratitude to all the authors for sharing their expertise in the field of electrochemical sensors analysis. The highest quality research presented here will do much to promote real sample analysis with electrochemical (bio)sensors.

Salvador Alegret and Arben Merkoçi
April 2007

Series Editors' Preface

This book on *Electrochemical (Bio)Sensor Analysis*, edited by S. Alegret and A. Merkoçi, is an additional step to advance the field of rapid analysis. It presents advanced sensor developments as well as practical applications of electrochemical (bio)sensors in various fields in a single source. The book contains 38 chapters grouped into seven sections: (a) Potentiometric sensors, (b) Voltammetric (bio)sensors, (c) Gas sensors, (d) Enzyme based sensors, (e) Affinity biosensors, (f) Thick and thin film biosensors, and (g) Novel trends. This interdisciplinary book has contributions from well-known specialists in the field and will be a useful resource for professionals with an interest in the application of electrochemical (bio)sensors.

The book covers the entire field of electrochemical (bio)sensor design and characterization and at the same time gives a comprehensive picture of (bio)sensor applications in real clinical, environmental, food and industry-related samples as well as for citizens' safety/security. In addition to the chapters, this volume offers 53 step-by-step procedures ready to use in the laboratory. This complementary information is offered on a CD-ROM included with the book in order to facilitate hands-on information on the practical use of electrochemical biosensor devices for the interested reader. It is the first time that the *Comprehensive Analytical Chemistry* series offers such complementary information with detailed practical procedures.

Finally, I would like to thank all the contributing authors of this book for their time and efforts in preparing this excellent and useful book on *Electrochemical (Bio)Sensor Analysis* including fundamental aspects, application areas and laboratory procedures. I am especially grateful to my old friend and colleague Salvador Alegret from the Autonomous University of Barcelona and his co-worker Arben Merkoçi from the Catalan Institute of Nanotechnology for their efforts in making possible the publication of this relevant and comprehensive work in our series. This book is the third one on sensors to be published in the *Comprehensive Analytical Chemistry* series, following volume 39 on

Integrated Analytical Systems, edited by S. Alegret, and volume 44 on *Biosenors and Modern Biospecific Analytical Techniques*, edited by L. Gorton. The addition of this volume certainly makes the CAC series a valuable reference in the field of sensors in analytical chemistry.

D. Barceló
Department of Environmental Chemistry,
IIQAB-CSIC, Barcelona, Spain

PART 1: FUNDAMENTALS AND APPLICATIONS

POTENTIOMETRIC SENSORS

Chapter 1

Clinical analysis of blood gases and electrolytes by ion-selective sensors

Andrzej Lewenstam

1.1 INTRODUCTION

Electrochemical ion-selective sensors (ISSs), including potentiometric ion-selective electrodes (ISEs) and potentiometric or amperometric gas-selective sensors (GSSs), attracted the interest of clinical chemistry because they offer fast, reliable, inexpensive analytical results in service-free automated analyzers. In this way, the electrochemical sensors satisfy the present demands of central hospital laboratories and peripheral point-of-care medical service points, such as bedside, emergency or first-contact healthcare centers.

A breakthrough in the clinical application of electrochemical sensors came with blood gas analysis via pH (“glass electrode” [1]), pCO_2 (“Severinghouse gas electrode” [2]) and pO_2 (“Clark sensor” [3]) electrodes, introduced in the 1950s [4]. Another important milestone in this area was the invention of a valinomycin-based potassium ISE in 1970 [5]. The subsequent development of new generations of ISEs and ISSs, gas-selective electrodes and GSSs, supplemented recently with a rapidly growing family of electrochemical biosensors, provided analytical tools that permit the specific and precise determination of a number of species of medical importance.

Today the determination of blood gases (pO_2, pCO_2) is performed using electrochemical GSSs, while major blood electrolytes (Na^+, K^+, Cl^- as well as Ca^{2+}, Mg^{2+}, Li^+ as a therapeutic ion and pH) and main urine electrolytes (Na^+, K^+) are predominantly determined by electrochemical ISSs or ISEs. Some important metabolites (glucose, lactate, urea, creatinine) are often determined with electrochemical biosensors.

The application of ISSs/GSSs in clinical chemistry aroused notable interest in the chemical scientific community marked by dedicated

Comprehensive Analytical Chemistry 49
Alegret and Merkoçi (Eds)
Volume 49 ISSN: 0166-526X DOI: 10.1016/S0166-526X(06)49001-2

reviews [6–11] and monographs [12–14], including a special series of reports devoted exclusively to the application of ISEs in clinical measurements [15]. For clinicians, normalized and traceable application of ISSs/GSSs in routine analysis is a professional obligation. For this reason their activity is supported by the dedicated activity of the International Federation of Clinical Chemistry (IFCC) and its Working Group on Selective Electrodes and Biosensors. Within this group a number of documents devoted to the application of electrochemical sensors in routine clinical chemistry were published [16–25].

1.2 GENERAL CHARACTERISTICS OF CLINICAL ANALYSIS OF ELECTROLYTES AND GASES

1.2.1 Clinical sample, analytical matrix

A clinical sample (whole blood, serum, plasma, urine, gastric juice, bile fluid, sweat, etc.) differs from any other analytical sample because of the presence of heterogeneous organic (e.g., proteins) and organic or inorganic components (e.g., urea or sodium ion), sample changes in time (owing to, e.g., denaturation of proteins, escape of CO_2) and small sample size (even a few tens of microliters).

In the case of frequently measured whole blood samples, anaerobic blood samples (arterial or venous) are typically used for blood gas analysis, while blood plasma (i.e., the liquid component of blood, in which the blood cells are suspended, containing the clotting agent fibrinogen; the clotting process can be blocked by anticoagulants such as heparin) or blood serum (i.e., blood plasma not containing fibrinogen) are used for measurements of electrolytes and soluble molecules (e.g., glucose) dissolved in plasma water. Plasma water comprises approximately 80–96% of the total serum/plasma volume, with the remainder being composed of insoluble proteins, lipids and macromolecules. The normal water content in plasma is ~93%. The major electrolytes contained in plasma water, i.e., sodium, potassium and chlorides, are nearly 100% dissociated. The terms "ionized"—sodium, potassium and chloride—are therefore used to describe the results obtained by ISE measurements in blood. Ionized calcium and magnesium refer to the free (unbound) fraction of total calcium (ca. 50%) and total magnesium (ca. 70%).

In the case of the frequently requested sample urine, the term "urine"—sodium and potassium—is applied. Urine has a different matrix than blood (chemical composition, ionic strength, interferences), which justifies the distinction in measuring terminology, although in both cases the same ISSs/ISEs are used.

The analytical ranges of major parameters in blood and urine are summarized in Table 1.1.

The pathological values higher ("*hyper*") or lower ("*hypo*") are outside of the normal ranges given in Table 1.1, and consequently broader analytical ranges are set for individual sensors. Typical concentration ranges and within-run imprecision values are given in Table 1.2.

TABLE 1.1

Reference ranges for the adults' clinical parameters measured by electrochemical sensors

Fluid and Parameter	Range
Blood gases	*Mixed venous blood*
pH (37°C)	7.33–7.43
pCO_2	5.1–7.2 (kPa)
pO_2	4.7–5.3 (kPa)
	Arterial blood
pH (37°C)	7.35–7.45
pCO_2	4.3–6.4 (kPa)
pO_2	11.0–14.4 (kPa)
Blood serum electrolytes	
pH	7.35–7.45
Na^+	136–145 (mmol/L)
K^+	3.5–5.1 (mmol/L)
Cl^-	98–107 (mmol/L)
Total CO_2[a]	23–29 (mmol/L)
Ionized Ca^{2+}	1.16–1.32 (mmol/L)
Ionized Mg^{2+}	0.50–0.66 (mmol/L)
Li^+ (therapeutic)	0.6–1.2 (mmol/L)
Urine electrolytes	
pH	4.5–8.0
Na^+	40–220 (mmol/24h)
K^+	25–125 (mmol/24h)
Cl^-	110–250 (mmol/24h)
Ca^{2+}	2.5–7.5 (mmol/24h)

[a]HCO_3^- is ~1.2 mmol/L lower.

TABLE 1.2

Analytical ranges and imprecision of measurement in a typical blood gas and electrolyte analyzer

Parameter	Analytical range	Within-run imprecision (SD)
pH	6.50–8.00	0.005
pCO_2 (kPa)	0.5–27	0.15
pO_2 (kPa)	0–107	0.20
Na^+ (mmol/L)	80–200	1.4
K^+ (mmol/L)	2–10	0.05
Cl^- (mmol/L)	50–150	1.0
Total CO_2 (mmol/L)	5–40	0.5
Ionized Ca^{2+} (mmol/L)	0.2–3.0	0.02
Ionized Mg^{2+} (mmol/L)	0.2–2.0	0.02
Na^+ (urine) (mmol/L)	20–300	5
K^+ (urine) (mmol/L)	20–200	5

1.2.2 Types and design of ion-sensors and gas-sensors used in clinical analysis

The demand for reliability of the analyses, a long-life of the electrodes and their readiness to be used any time (for so-called *Stat* samples) directly dictated the design of ISSs/GSSs. Instead of macroelectrodes, a large variety of microelectrodes have been developed [26].

ISSs, as with any potentiometric sensor, can be treated as a galvanic half-cell and represented schematically:

$$\text{metal}|\text{internal contact}|\text{membrane}|\text{sample (blood, urine)} \tag{A.1}$$

Correspondingly the galvanic cell (with ISS) can be represented by the following scheme:

$$\text{metal}|\text{ISS}|\text{sample}|\text{liquid-junction (salt bridge)}|\text{reference electrode}|\text{metal}' \tag{A.2}$$

At present, most clinical measurements are realized in continuous, flow-through assemblies. The overall design of an average electrode, its mechanical arrangement and electrochemical properties conform to the constraints and targets of clinical analysis, even if some firm-to-firm changes can be seen.

Today's design of the ISSs used in clinical measurements is the accumulation of many improvements that have been made over the past years. The changes in the design of ISSs were stimulated by the special character of clinical samples, growing analytical and throughput demands in clinical laboratories, the need to follow international recommendations (e.g., those given by IFCC), the demands of single result and method traceability and, *last but not least*, the developments in sensor technology.

The main challenge in designing clinically useful sensors is definitely the production of the electroactive element, i.e., the sensor membrane. The membrane is the place where the chemical recognition and discrimination processes occur. The membrane dictates, overwhelmingly, the quality of signal and durability of the sensor. Only a restricted number of membranes can be and are used in routine electrolyte and blood gas measurements:

- *glass membranes* are commonly used for pH and sometimes for sodium measurements. Glass membranes are made from melts of Na_2O and SiO_2 in the case of pH, and Na_2O, Al_2O_3 and SiO_2 in the case of Na-ISSs. Glass membranes allow for a stable and reproducible response and a long lifetime. Their main disadvantages are high electrical resistance, mechanical fragility and the liability to adsorb organic macromolecules.
- *polymeric membranes*, sometimes called "plastic membranes", are based mainly on matrices of PVC and its derivatives. Their sensitivity, selectivity and dynamic properties depend on the overall membrane composition, namely on the ionophore (neutral carrier or charged molecules/carriers), modifiers, i.e., plasticizers (typically alkyl-esters or phenyl-ethers) and ionic additives (typically phenyl borates and their derivates) used. The ionophores employed predominantly for sensing sodium, potassium, calcium, magnesium and lithium are neutral carriers, i.e., synthetic neutral molecules able to selectively complex ion of interest (see, e.g., Refs. [27,28]). Most of them are compounds introduced at ETH Zurich, Switzerland, by the research group of the late Prof. Simon (commercialized by Fluka, Switzerland). Occasionally, other neutral carrier molecules, e.g., calixarenes or crown ethers, are employed as well. Charged carriers are currently used as electroactive substances for chloride and bicarbonate ISSs, and also for calcium, magnesium and pH-selective membranes.

  Due to the inherent plasticity of "plastic membranes" and their solubility in organic solvents these membranes can be rather easily placed in different electrode body designs. Moreover, plastic membranes may be casted for all major electrolytes by the same manufacturing technique, which ensures production and service integrity. The lifetime is, as a rule, worse than for glass membranes. The selectivity of these membranes, except for lithium, magnesium and bicarbonates is satisfactory for clinical use (for selectivity patterns see, e.g., Refs. [15,27,28]).

– *solid-state* (crystalline) membranes are rarely used in routine clinical determinations, with the exception of chloride ISSs, based on insoluble silver chloride.

The internal contact of a conventional ISS (see A1) is a liquid contact usually made from a silver chloride reference electrode immersed in the internal (filling) solution, which makes the ISS equivalent to the ISE. A major disadvantage of this design is the evaporation of the filling solution, resulting in a break of the electrical circuit and a need for intervention. A modern way to eliminate this difficulty is achieved by introducing a mediating layer ("solid-contact") between the ionically conducting membrane and the electron-rich contact. This ion-to-electron transducing layer can be achieved in various ways, for instance through the use of conducting polymers [29]. By replacing the internal solution with a solid-contact, conventional ISEs are transformed into solid-contact sensors or simply ISSs. This integration step is increasingly important today because of trends directed toward sensor integration and miniaturization.

GSSs are purposefully designed electrochemical cells, a galvanic cell in the case of carbon dioxide sensors, and an electrolytic cell in the case of the oxygen sensors.

Both cells are equipped with gas-permeable membranes that allow for nearly specific gas transfer from the sample into a thin indicator layer (buffer) of a cell that is in contact with the electrochemical sensing electrode. For the CO_2-GSS, the indicator layer is a flat pH glass membrane, while for the O_2-GSS it is a cathode made of platinum or gold.

Determination of the substance concentration with a potentiometric ISSs and CO_2-GSS is realized through the measurement of the electromotive force (EMF) of the galvanic cell (*potentiometric measurement*). In the case of the O_2-GSS, however, the current is measured (*amperometric measurement*).

1.3 ELECTROCHEMICAL MEASUREMENT IN CLINICAL ANALYSIS

To allow the practical potentiometric measurement of the EMF of a galvanic cell it is ensured that the measurement takes place in the open circuit where the sampling current from the voltammeter is small enough so as not to influence the EMF measured. This condition is realized by using voltammeters with high input ohmic resistance ($R \gg 10^{14}\Omega$) ensured by operational amplifiers with low bias current ($<100\,\text{fA}$). All measurements take place at constant temperature (at least $\Delta T \leqslant \pm 1^\circ\text{C}$, in blood gas analyzers $\Delta T \leqslant \pm 0.1^\circ\text{C}$) and external pressure. This demand is in practice easy to fulfill by use of thermostats while working under atmospheric pressure.

The galvanic cell must be assembled so that each interface is dominated by a well-established, fast and reversible faradaic charge (ion or electron) transfer.

Under these necessary conditions EMF is a function of the ISS potential given by the following equation [10,20]:

$$\text{EMF} = E_{\text{ISS}} - E_{\text{REF}} + E_{\text{D}} \tag{1.1}$$

where E_{ISS} is electrical potential of ISS, E_{REF} is the reference electrode potential and E_{D} is a liquid-junction potential at the reference electrode bridge (all in mV).

If the response of an ISS in the standards and samples is fast, the reference electrode potential and the liquid-junction potential are constant ($-E_{\text{REF}}+E_{\text{D}}$ = constant) and do not change during the calibration and the sample measurement (i.e., residual liquid-junction potential is constant), a simple equation then applies over the course of an analysis with the ISS:

$$\text{EMF} = \text{constant} + E_{\text{ISS}} \tag{1.2}$$

The potential of the ISS is given by the following extended Nikolskii–Eisenman equation:

$$E_{\text{ISS}} = \text{constant}' + s\log[a_i + K_{ij}(a_j)^{z_i/z_j} + L] \tag{1.3}$$

where s is the slope of the electrode (characterizing its sensitivity), a_i is the individual activity of the main ion, a_j is the individual activity of the interfering ion, K_{ij} is the selectivity coefficient (a measure of the preference of the electrode for the interfering ions against the main ions) and z_i, z_j are the charges of the respective ions, and L is the detection limit. The slope has the Nernstian interpretation:

$$s = \frac{2.303RT}{z_i F} \tag{1.4}$$

where R is the gas constant ($R = 8.3145\,\text{J/K}^\circ\text{mol}$), T is the absolute temperature ($T = 273.15 + t(^\circ\text{C})$), and F is the Faraday constant

($F = 96485\,\text{C/mol}$) and s in V. Thus, $s = +0.05916/z_i$ V $= 59.16/z_i$ mV for cations, and $s = -59.16/z_i$ mV for anions at 25°C, and $+61.55/z_i$ mV for cations, and $s = -61.55/z_i$ mV for anions at 37°C.

From Eqs. (1.2) and (1.3) it can be seen that the magnitude of EMF is an explicit function of the activity of ions in the solution given by the following equation:

$$\text{EMF} = \text{constant}'' + s\log[a_i + K_{ij}(a_j)^{z_i/z_j} + L] \tag{1.5}$$

Further, in the ideal case, if $K_{ij} = 0$ and $L = 0$ the electrode response is a specific one, and EMF is a Nernstian function of a main ion activity:

$$\text{EMF} = \text{constant}'' + s\log a_i \tag{1.6}$$

In the case of the potentiometric CO_2 electrode:

$$\text{EMF} = \text{constant}' + s\log p_{CO_2} \tag{1.7}$$

where p_{CO_2} is the partial pressure of blood carbon dioxide. In the case of the amperometric oxygen sensor, the same reasoning as for the ISS applies:

$$I = \text{constant}^* + kp_{O_2} \tag{1.8}$$

where I is the current (A), constant* is the empirical constant, k is the slope of the calibration curve and p_{O_2} is the partial pressure of blood oxygen.

The above equations are targets in the practical application of any ISS/GSS that can be met if special precautions are taken, either by the manufacturer or the user. The task is not straightforward, especially in the clinical application of sensors, and for this reason different precautions have to be taken to block the sources of errors.

1.4 APPLICATION OF SENSORS FROM THE PRODUCER'S POINT OF VIEW

The producer must ensure that the galvanic cell is collected in such a way that the potential-determining ion-transfer processes at the "ISS/GSS | solution" interface is exposed, while all other electric potential drops at the other interfaces are constant, see: A2.

It is preferable that the ISS/GSS is a specific, i.e., when its signal is described by Eqs. (1.6)–(1.8). This target can be achieved whenever the effect(s) of the interfering ion(s) and the influence of the detection limit can be ignored (e.g., when $a_i \gg K_{ij}(a_j)^{z_i/z_j} + L$), and if the sensor slope and "constants" determined in the calibration remain constant during measurements in the samples.

For most ISS/GSSs used in clinical analysis, these conditions are satisfied. In such cases, the galvanic cell can be calibrated easily with a two-point calibration with standards of different but known activities and traceable concentrations of the components.

pH, sodium, potassium, calcium and chloride ISEs, and carbon dioxide and oxygen gas electrodes, are nowadays (rather) specific. Two electrodes, lithium and magnesium, are subject to sodium and calcium ion interferences, respectively. The potential of these electrodes can be, in principle, described by Eq. (1.5) and calibrated with a three-point calibration.

The elements of the galvanic cell should not be influenced by the chemistry and/or number of samples measured. In particular, the sample should maintain its initial physical and chemical properties, even after storage. In other words such processes as the evaporation of plasma water, gas escape, denaturation of proteins, the cell lyses, etc. should be minimized to avoid analyte concentration changes and artifacts.

If this is achieved, then and only then will the overall cell response be fast and reversible (approx. 15 s is the presently required measurement time), the signal be reproducible and stable, and all parasitic processes for sensors, such as adsorption (of, e.g., proteins) or redox processes (for instance induced by variable concentration of dissolved oxygen), be avoided.

Most of these potential sources of error have actually been eliminated, or at least minimized, in commercial analyzers by proper hardware and software implementation, e.g., proper washing and cleansing procedures, signal stability criteria, corrections of the residual potential or simply by declaring "limits", i.e., the types of sample, interfering substances, vial types to be avoided, by limiting the number of reports per hour, adjusting within-day and between-day imprecision and the lifetime of the electrodes.

In the last instance, however, it is the user who has to use, maintain and "understand" the analyzer to avoid the unexpected influences of new drugs on the stability of the "ISE|solution" and "solution|salt bridge" interfaces.

There are certain sources of error which are inherent to potentiometric measurements. The first is the relative error (in %) in a potentiometric measurement of the ith analyte: $\Delta c_i/c_i = 100|z_i|F/RT\Delta E$ (ΔE is an error in the potential readout in V), which, for instance, for a monovalent ion with $\Delta E = 0.05\,\mathrm{mV}$ at 37.0°C leads to an error of ~0.2%. This high resolution is required to satisfy the demand in clinical measurements with potentiometric electrochemical sensors (Tables 1.1 and 1.2).

The list of error sources continues, just to mention a few: the ionic strength of the sample, the liquid-junction and residual liquid-junction potentials, temperature effects, instabilities in the galvanic cell, carry-over effects, improper use of available corrections (e.g., for pH-adjusted ionized calcium or magnesium). An error analysis goes beyond the limited scope of this paper; more details are presented elsewhere [10].

1.5 SENSORS USED IN ROUTINE CLINICAL MEASUREMENTS: A BRIEF OVERVIEW

1.5.1 Hydrogen ISSs (pH electrodes)

The most frequently used hydrogen ISEs are pH electrodes with glass membranes composed of melts of alkaline and silica oxides. pH-selective membranes are mounted into a flow-through body (capillary) or mounted as a miniaturized flat surface. The sensitivity, selectivity and lifetime of a glass electrode imposes no limitations in practical use. However, the natural drying properties of glass and the tendency to be disturbed by protein deposition are an issue. Cleansing solutions containing ammonium difluoride and pepsin dissolved in hydrochloric acid are used for this purpose.

A pH electrode with a plastic membrane can be employed as an alternative to a glass electrode. The active ionophore is a higher order amine, e.g., dodecylamine. The accuracy and reproducibility of this sensor meet all of the requirements for clinical application, although in some instances lipophilic ions (e.g., active component in drugs) may exert an unfavorable effect. The dynamic range of this electrode is inferior to that of the glass type sensor, but still good enough for a practical use.

The design of an integrated, maintenance-free pH-ISS is far simpler with a plastic membrane than with a glass.

1.5.2 Sodium ISSs

Two types of electrodes are used in clinical analyzers, based either on a sodium glass or plastic membrane. As in the case of the pH sensor, a sodium-glass selective membrane is mounted into a flow-through cell with a capillary or as a miniaturized stand-alone electrode. The sensitivity, selectivity and lifetime of a glass electrode impose no limitations on practical use. The glass should be washed and pretreated for the same reasons and in the same way as in the case of the pH electrode.

A plastic sodium membrane is now predominantly based on a neutral carrier (ETH 2120) that ensures sufficient sensitivity, selectivity and lifetime for the sensor. Some other compounds such as neutral carriers ETH 157, 227, 4120, calixarenes, crown ethers and hemispherands have been proposed. Anionic influence observed during measurements in undiluted urine may be circumvented by dilution of the sample.

1.5.3 Potassium ISSs

An electrode with a plastic membrane containing valinomycin as the active carrier is now predominantly used in clinical analyzers. Nearly four decades of experience with this sensor have proven that it fulfils all demands concerning sensitivity, selectivity and lifetime. An anionic interference that can be observed during measurements in undiluted urine may be eliminated by the use of silicone rubber instead of polyvinyl chloride in the membrane or by pre-dilution of urine. Despite some experimental trials, no other ionophore has replaced valinomycin as the active compound in potassium ISEs. This is basically due to the better stability and lipophilicity of this compound in comparison to the others proposed.

1.5.4 Lithium ISSs

At present the plastic membrane with ionophore ETH 2137 is used. The introduction of ionophore ETH 2137 generated a stable selectivity towards sodium ion allowing for numerical correction of sodium influence using the values of independently measured sodium. Thus a complete procedure covering therapeutic concentration range of lithium (which should be <1.2 mmol/L) is available.

1.5.5 Calcium ISSs

A carrier satisfying all of the basic requirements for the determination of calcium in biological fluids—ETH 1001—was invented in the early seventies. Almost all clinical analyzers rely on an electrode with this ionophore in a plastic membrane. For historical reasons, one producer (Radiometer) provided an electrode based on the ion-exchanger (organophosphate) instead. The latter approach induces a need to cover an ISE membrane with a dialysis (cellophane) membrane, which prolongs the response time. Some other ligands have been proposed instead of ETH 1001 but, other than ETH 129, do not seem to be used in clinical analyzers.

1.5.6 Magnesium ISSs

The development of a sensor for ionized magnesium turned out to be one of the most difficult challenges of recent years. Several carriers have been designed for this purpose but none have been satisfactory. The first report of a successful measurement of ionized magnesium in an automated clinical analyzer (Thermo, prev. KONE) was published only in 1990 [30]. The ionophore ETH 5520 was used as the active compound. Two other carriers have been used since then: ETH 7025 (Roche, former AVL), and a derivative of 1,10-phenenthroline (Nova). All of the magnesium sensors are based on a plastic membrane. Numerical compensations of the influence of calcium ion and the ionic strength are used due to insufficient selectivity of the magnesium sensors.

A growing interest in ionized magnesium measurements indicates that further improvements in the analytical properties of the Mg-ISSs, regarding the selectivity as well as the response time and lifetime, are needed, to facilitate wider clinical application of this method [31].

A procedure (Procedure 1) related to the measurement of ionized magnesium in human blood is given in the second part of this book.

1.5.7 Chloride ISSs

The most widely used sensor for chloride ions in clinical analyzers is based on an ion-exchanger, a quaternary alkylammonium chloride, dispersed in a plastic membrane. It is not an ideal sensor due to the interference of lipophilic anions (e.g., salicylates, bromides) and lipophylic cations (e.g., bacteriostatic agents, anesthetics) and a relatively poor selectivity towards hydrogen carbonates (bicarbonates). However, compared to charged anion- and neutral carrier-based membranes that have been tested, it is still the best-suited for automated analyzers.

One exception is a solid-state electrode based on silver chloride with a membrane covered by acetyl cellulose, which is occasionally used in some setups (OCD, prev. Kodak).

1.5.8 Carbon dioxide GSSs

GSSs, often called gas electrodes, belong to a separate class of potentiometric sensors. The carbon-dioxide gas electrode is used for the measurement of the partial pressure of dissolved CO_2. This "double membrane" electrode is a galvanic cell composed of a glass pH half cell made of a flat glass pH-selective membrane, and a reference electrode, both in contact with a thin layer of a bicarbonate buffer solution. The latter is separated from the sample by a CO_2 gas-permeable membrane. CO_2 from the sample comes through the gas-permeable membrane into the buffer layer and the resulting pH shift in the buffer is measured by the glass membrane.

1.5.9 Bicarbonate ion-selective electrodes

A good sensor for the direct measurement of bicarbonates has not yet been invented. However, an electrode selective to carbonate ions, based on triflouroacetophenone derivates, has been used. This electrode is non-selective for a number of ions, and unless substantial improvements or inventions are made, it is unlikely that it will become a commonly used sensor in routine instruments [32].

It is possible to use a carbonate ISE in samples prediluted with base to convert all forms of CO_2 into carbonates or to use the CO_2 gas electrode or pH electrode for the measurement of total CO_2 evolved after addition of acid (e.g., Dade-Behringer prev. Dade).

For all of the above reasons, bicarbonates are most often reported as a calculated value via the use of the Henderson–Hasselbalch equation and measured pH and p_{CO_2}.

1.5.10 Oxygen GSSs

A sensor for dissolved oxygen is based on a similar principle as the CO_2, employing a gas-permeable membrane which selectively allows diffusion of oxygen into a thin layer of—usually—phosphate buffer. However, this sensor typically operates in amperometric mode, i.e., the signal is generated by reduction of oxygen at the platinum or gold cathode with silver chloride used as the anode.

1.5.11 Reference electrodes

The reference electrode (system) and its stability in clinical analyzers is a crucial problem because all typical ISSs (and GSSs) use this electrode. Currently silver chloride electrodes are used. They work in two systems: an open liquid junction or a constraint liquid junction with concentrated KCl ($>2\,M$) or sodium formate ($4\,M$) as the equitransferent "hypertonic" electrolyte bridge. The latter better serves whole blood measurements.

To maintain the condition $E_{REF}+E_D$ = const, numerical corrections are possible, e.g., using the Henderson equation for liquid-junction potential.

In general, restricted hypertonic junctions can introduce EMF errors due to protein contamination (amorphous or precipitate overcoats) and "isotonization" of the junction with a subsequent influence of ionic strength. Open junctions, on the other hand, are apparently affected by the ionic form of proteins. The problem of reference electrode stability is especially important in direct measurements, as discussed above and elsewhere [9,33].

It should again be emphasized that in clinical measurements with every ISS, the reference electrode is an equal partner in the formation of the EMF. This means that any potential error at the reference electrode affects all of the ISSs, but not *vice versa*.

1.6 APPLICATION OF SENSORS: THE USER'S POINT OF VIEW

1.6.1 Direct and indirect measurement of sodium, potassium and chlorides

From the medical point of view, in contrast to the analytical which refers to the mechanism of sensor signal formation, a "direct measurement" is defined as a measurement carried out directly in undiluted sample (whole blood, plasma, serum, urine, etc.), whereas "indirect measurement" employ sample dilution. For an analytical chemist, a direct measurement is more challenging because of small sample volume, interferences and matrix effects, diffusion potential and carry-over effects, and the influence on the sensor lifetime due to high extractability of active components by undiluted samples such as serum or urine. However, this measurement method has one important advantage; it allows the measurement of the activity of analytes "as-they-are". For obvious reasons, blood gases, pH, ionized calcium and magnesium are measured only directly (in the medical sense). Indirect measurements with ISSs are currently applied more frequently for ionized sodium, chloride and potassium for the reasons just listed.

An indirect measurement with an ISS has several advantages: less sample volume is required, the effects of the interferences and matrix are reduced, diffusion potential and carry-over effects become easier to control and the sensor lifetime is extended. The application of diluent itself offers the possibility of imposing the physical and chemical properties of the diluent on the sample, such as pH, ionic strength and even undesirable interferences and influences suppression.

However, in contrast to a direct measurement, where the analyte activity in the native sample is measured, in the indirect measurement (after dilution) this possibility is lost. In the latter case, only total concentration in a sample can be "back-calculated". Additionally, because water plasma in a patient sample is not measured, the indirect method of measuring blood electrolytes in hyperlipemic/hyperproteinemic samples (i.e., in the cases of low plasma water vs. plasma volume) results in a negative bias compared to direct measurement (the bias is bigger with lower water content), while for samples with a high electrolyte content (i.e., high plasma water ionic strength) the indirect measurements will result in a positive bias.

Undoubtedly, while the direct method is more relevant, because the analyte "activity in water plasma" is actually measured, the reporting on blood sodium, potassium and chloride in terms of "concentration in plasma" is preferred by medical professionals, whatever method of measurement is used. This is justified by the fact that before ISEs had been invented, sodium, potassium and chloride were all determined by indirect methods, with flame emission spectroscopy (FES) for Na^+ and K^+, and coulometry for Cl^-.

For the above reasons, the IFCC recommendations on activity coefficients [19] and the measurement of and conventions for reporting sodium and potassium [21] and chlorides [25] by ISEs were developed. At the core of these recommendations is the concept of "the adjusted active substance concentration" (mmol/L), as well as a traceable way to remove the discrepancy between direct and indirect determinations of these electrolytes in normal sera. Extensive studies of sodium and potassium binding to inorganic ligands and proteins, water binding to proteins, liquid-junction effects and the influence of ionic strength have demonstrated that the bias between sodium and potassium reports obtained from an average ISE-based commercial analyzer and FES on patient pooled sera, although statistically significant, is below 1%. For sodium in normal plasma the ratio of sodium concentration in plasma and its activity in plasma water is estimated to be 1.27 [11].

1.6.2 Direct measurement: calcium, magnesium, lithium, pH and blood gases

All direct measurements with electrochemical sensors encounter similar difficulties as those discussed above. Appropriate IFCC recommendations were designed to help the clinical chemist. In the case of ionized calcium, magnesium, pH and blood gases the main idea is to refer the results to water-based or gas-tonometered reference materials (notice: not a "normal serum"). The IFCC recommendations for measuring pH [16], blood gases [17] and CO_2 [23] as well as the reference methods involving well-defined standards, and reference materials were developed. The recommendations on definitions of quantities and conventions related to blood gases and pH have also been proposed [18]. The recommendations for measuring ionized calcium [19,22] and ionized magnesium [24] were recently published.

There are numerous specific calculation procedures that involve the values measured by ISSs/GSSs, just to mention three of them:

(a) the anionic gap (AG), i.e., the difference between the sum of sodium and potassium concentration and the chloride and bicarbonate or total CO_2 concentrations in venous blood, is often used by the medical doctors (especially in US). The normal value of the AG is $\sim$16 mEq/L.
(b) in the case of calcium and magnesium measurements the influence of pH (in the range 7.2–7.6), via calcium- and magnesium-bicarbonate binding, is numerically compensated. Ionized calcium and magnesium "corrected" to $pH = 7.40$ are often reported, especially for separated plasma or serum and aerobic sampling.
(c) in determination of blood gases the system is calibrated with appropriate mixtures of gases and the partial pressure of CO_2 is then measured directly in the samples. Consequently, any other forms of plasma carbonates are then calculated from the Henderson–Hasselbalch equation by use of CO_2 and pH values measured for a particular sample. There is a long and ongoing discussion concerning medical relevance of this procedure, which basically

refers to the question of the reliability of CO_2 measurements and/ or the "stability" of the apparent *overall* first dissociation constant of CO_2 in plasma (typically $pK_1(37^{\circ}C) = 6.08–6.13$ is used).

1.7 THE PROSPECTS FOR USE OF NEW ION OR GAS-SELECTIVE SENSORS

By taking into account the general properties of ISEs (concentration ranges, selectivity, expected life time, etc.) and by listing the ions of physiological importance, it is possible to deduce that some improvements or applications of new ISSs are still feasible. Further amelioration of magnesium electrodes is still likely because existing ones suffer from relatively poor selectivity. The introduction of an electrode for the direct determination of bicarbonates would be most welcome. Inorganic phosphate and sulfate ions are other candidates for direct determination by ISS, while plethora of clinically relevant substances (e.g., different disease markers, drugs, DNA/RNA) are to be covered by indirect electrochemical measurement [34]. Liquid-junction maintenance-free reference electrode appears to be a hot issue [35]. Further simplification and integration of the sensors [36], lowering their detection limits [37], miniaturization (down-to-nanostructure) and applications in smart systems (e.g., lab-on-the-chip or *in situ* medicine administration) is the focus for various research groups. The research activity should in the foreseeable future increase the list of electrochemically measured analytes as well as extend the application of electrochemical sensors beyond well-established routines.

The margin for new scientific ideas, technical inventions and resulting clinical benefits (hopefully, not confusion) is rather well defined and wide. This prospect stimulates the work of academic research and industrial development, making new results inevitable.

1.8 CONCLUSIONS

The use of ISSs and GSSs in clinical analyzers has proven to be an effective answer for the growing clinical demand for specific, fast, inexpensive and fully automated measurements. These electrochemical sensors allow for clinical analysis of the most frequently requested parameters (analytes). They are today a routine element in clinical analyzers and an everyday partner of clinical chemists.

ACKNOWLEDGMENTS

This work is part of the activities at the Åbo Akademi Process Chemistry Group within the Finnish Centre of Excellence Programme (2000–2011) by the Academy of Finland.

Financial support from the Polish State Committee for Scientific Research (KBN), project no 3 T09A 175 267 is gratefully acknowledged. Peter Lingenfelter is acknowledged for suggestions to improve the text.

REFERENCES

1 F. Haber and Z. Klemensiewicz, *Z. Physik. Chem.*, 67 (1909) 385–431.
2 I.W. Severinghaus and A.F.J. Bradley, *Appl. Physiol.*, 13 (1958) 515–520.
3 L.C. Clark, *Trans. Am. Soc. Artif. Intern. Organs.*, 2 (1959) 41–48.
4 A. Astrup and J.W. Severinghous, *The History of Blood Gases, Acids and Bases*, Munskgaard, Copenhagen, 1986.
5 L.A.R. Pioda, W. Simon, H.R. Bosshard and H.C. Curtis, *Clin. Chim. Acta.*, 29 (1970) 289–293.
6 U. Uesch, D. Amman and W. Simon, *Clin. Chem.*, 32 (1986) 1448–1459.
7 T.P. Byrne, *Selective Electrode Rev.*, 10 (1988) 107–124.
8 A. Lewenstam, *Anal. Proc.*, 28 (1991) 106–109.
9 A. Lewenstam, A. Hulanicki and M. Maj-Zurawska, *Electroanalysis*, 3 (1991) 727–734.
10 A. Lewenstam, *Scand. J. Clin. Lab. Inv.*, 54 (1994) 11–20.
11 A. Lewenstam, *Scand. J. Clin. Lab. Inv.*, 56 (1996) 135–140.
12 *Ion Measurements in Physiology and Medicine*. In: M. Kessler, D.K. Harrison, J. Hoper (Eds.), Springer-Verlag, Berlin, 1985.
13 J. Havas, *Ion- and Molecule-Selective Electrodes in Biological Systems*, Akademiai Kiado, Budapest, 1985.
14 U.E. Spichiger-Keller, *Chemical Sensors and Biosensors for Medical and Biological Applications*, Wiley-VCH, Weinheim, 1998.
15 *Methodology and Clinical Applications of Ion-Selective Electrodes*, Vols. 1–13, IFCC, Milan, Italy, 1979–1991.
16 A.H.J. Mass, H.F. Weisberg, R.W. Burnett, O. Mueller-Plathe, P.D. Wimberley, W.G. Zijlstra, R.A. Durst and O. Siggaard-Andersen, *Eur. J. Clin. Chem. Clin. Biochem.*, 25 (1987) 281–289.
17 R.W. Burnett, A.K. Covington, A.J.H. Maas, O. Muller-Plathe, H.F. Weisberg, P.D. Wimberley, W.G. Zijlstra, O. Siggaard-Andersen and R.A. Durst, *Eur. J. Clin. Chem. Clin. Biochem.*, 27 (1989) 403–408.
18 R.W. Burnett, T. Christiansen, A.K. Covington, N.G. Fogh-Andersen, W. Kulpman, A.J.H. Maas, O. Muller-Plathe, A.L. Van Kessel, P.D.

Wimberley, W.G. Zijlstra, O. Siggaard-Andersen and H.F. Weisberg, *Eur. J. Clin. Chem. Clin. Biochem.*, 33 (1995) 399–404.
19 R.W. Burnett, A.K. Covington, N.G. Fogh-Andersen, W. Kulpman, A.J.H. Maas, O. Muller-Plathe, O. Siggaard-Andersen, A.L. VanKessel, P.D. Wimberley and W.G. Zijlstra, *Eur. J. Clin. Chem. Clin. Biochem.*, 35 (1997) 345–349.
20 R.W. Burnett, A.K. Covington, N.G. Fogh-Andersen, W. Kulpman, A. Lewenstam, A.J.H. Maas, O. Muller-Plathe, A.L. VanKessel and W.G. Zijlstra, *Clin. Chem. Lab. Med.*, 38 (2000) 363–370.
21 R.W. Burnett, A.K. Covington, N.G. Fogh-Andersen, W. Kulpman, A. Lewenstam, A.J.H. Maas, O. Muller-Plathe, A.L. VanKessel and W.G. Zijlstra, *Clin. Chem. Lab. Med.*, 38 (2000) 1065–1071.
22 R.W. Burnett, T. Christiansen, A.K. Covington, N.G. Fogh-Andersen, W. Kulpman, A. Lewenstam, A.J.H. Maas, O. Muller-Plathe, C. Sachs, O. Siggaard-Andersen, A.L. Van Kessel and W.G. Zijlstra, *Clin. Chem. Lab. Med.*, 38 (2000) 1301–1314.
23 R.W. Burnett, A.K. Covington, N. Fogh-Andersen, W.R. Külpmann, A. Lewenstam, A.H.J. Maas, A.L. Van Kessel and W.G. Zijlstra, *Clin. Chem. Lab. Med.*, 39 (2001) 283–289.
24 B. Rayana, R.W. Burnett, A.K. Covington, P. D'Orazio, N. Fogh-Andersen, E. Jacobs, R. Kataky, W. Külpmann, K. Kuwa, L. Larsson, A. Lewenstam, A.H.J. Maas, G. Mager, J.W. Naskalski, A.O. Okorodudu, C. Ritter and A. St John, *Clin. Chem. Lab. Med.*, 43 (2005) 564–569.
25 B. Rayana, R.W. Burnett, A.K. Covington, P. D'Orazio, N. Fogh-Andersen, E. Jacobs, R. Kataky, W. Külpmann, K. Kuwa, L. Larsson, A. Lewenstam, A.H.J. Maas, G. Mager, J.W. Naskalski, A.O. Okorodudu, C. Ritter and A. St John, *Clin. Chem. Lab. Med*, 44 (2006) 346–352.
26 E. Bakker, D. Diamond, A. Lewenstam and E. Pretsch, *Anal. Chim. Acta.*, 393 (1999) 11–18.
27 E. Bakker, P. Bühlmann and E. Pretsch, *Chem. Rev.*, 97 (1997) 3083–3132.
28 P. Bühlmann, E. Pretsch and E. Bakker, *Chem. Rev.*, 98 (1998) 1593–1687.
29 A. Cadogan, Z. Gao, A. Lewenstam, A. Ivaska and D. Diamond, *Anal. Chem.*, 64 (1992) 2496–2501.
30 M. Maj-Zurawska and A. Lewenstam, *Anal. Chim. Acta.*, 236 (1990) 331–335.
31 N.-E.L. Saris, E. Mervaala, H. Karppanen, J.A. Khawaja and A. Lewenstam, *Clin. Chim. Acta.*, 294 (2000) 1–26.
32 J. Bobacka, M. Maj-Zurawska and A. Lewenstam, *Biosens. Bioelectron.*, 18 (2003) 245–253.
33 T. Sokalski, M. Maj-Zurawska, A. Hulanicki and A. Lewenstam, *Electroanalysis*, 9 (1999) 632–636.

34 T.G. Drummont, M.G. Hill and J.K. Burton, *Nat. Biotechnol.*, 21 (2003) 1192–1199.
35 T. Blaz, J. Migdalski and A. Lewenstam, *Analyst*, 130 (2005) 637–643.
36 M. Vázquez, J. Bobacka, A. Ivaska and A. Lewenstam, *Talanta*, 62 (2004) 57–63.
37 A. Michalska, *Anal. Bioanal. Chem.*, 384 (2006) 391–406.

Chapter 2

Ion-selective electrodes in trace level analysis of heavy metals: Potentiometry for the XXI century

Aleksandar Radu and Dermot Diamond

2.1 INTRODUCTION: HISTORICAL MILESTONES

The field of potentiometric sensors was established about a century ago, when Cremer in 1906 described the response function of glass to changes in pH [1]. The glass electrode developed by Beckman in 1932 [2], was the first significant commercial event. However, the immense expansion in the field of ion-selective electrodes (ISEs) started with the discovery that some antibiotics are capable of selective binding of particular cations. The group of Simon utilized valinomycin, monensin and nonactin to make potassium-, sodium- and ammonium-selective electrodes, respectively [3–6]. Later, it was realized that other compounds can also be utilized as complexing agents for the desired ions, resulting in the term "ionophores" coined by Pressman [7]. The following years saw vigorous research in the area, and with advances in host–guest chemistry, many ionophores have been synthesized and ISEs for more than 60 ions have been reported [8]. Commercially, the biggest success has been the utilization of ISEs in clinical analysis, with first such analyser being commercialized in 1972. Today, ISEs are the method of choice for clinical analysis of ions of clinical relevance such as K^+, Na^+, Ca^{2+}, Mg^{2+} and Cl^-. In recent years, the field has undergone a renaissance mainly due to the work by the groups of Pretsch and Bakker who introduced the possibility of the application of ISEs in areas requiring significantly lower limits of detection (LODs) than previously thought achievable. By 1997, the research momentum in the area was regained and a new wave of excitement initiated when Bakker reported a new experimental setup for the determination of selectivity coefficients [9], and Sokalski *et al.* in the same year reported a

Comprehensive Analytical Chemistry 49
Alegret and Merkoçi (Eds)
Volume 49 ISSN: 0166-526X DOI: 10.1016/S0166-526X(06)49002-4

Pb^{2+}-selective electrode with LOD in picomolar range—improvements of many orders of magnitude compared to the ones known at the time [10]. Since then, many ISEs with significantly improved LODs have been reported [11–13] and the theory of the response of ISEs has been redefined [14–16]. In 2001, Ceresa *et al.* reported the application of ISEs for trace level analysis and demonstrated the use of a Pb^{2+}-selective electrode for the analysis of Zurich drinking water [17]. Subsequently, this work has been followed by other examples of application of ISEs for the analysis of environmental samples [18,19].

This quiet revolution in the area was noticed and acknowledged in the general analytical chemistry community. For the specialists in the area, various articles and reviews have appeared in recent years [8,9,11–13,20–23]. With this chapter, we target general analytical chemists and non-specialists in the area of ISEs with ultimate goal of spreading this technology as a new analytical tool for various application fields. In addition, we hope that students looking to expand their knowledge will find this text inspiring enough to start studying the field in more detail.

2.2 POTENTIOMETRY AND ITS PLACE AMONG ANALYTICAL TECHNIQUES

Potentiometry differs from all other analytical techniques by its ability to detect an ion's activity in a sample. This unique property makes ISEs particularly useful for bioavailability or speciation studies. The response of an ISE is described by Nernst equation:

$$\mathrm{EMF} = K + \frac{RT}{z_I F} \ln a_I \qquad (2.1)$$

where EMF is the electromotive force (the observed potential of the electrochemical cell at zero current); K symbolizes all constant potential contributions arising at the contact between two phases in the cell including the liquid junction potential at the reference electrode and the concentration of the ion of interest–ionophore complex within the membrane; a_I is the sample activity of the ion I bearing a charge z_I; and R, T and F are the gas constant, absolute temperature and the Faraday constant, respectively. Note that the ion activity describes the so-called free, or uncomplexed, concentration of the analyte and can be orders of magnitude smaller than the total concentration, especially when a complexing agent is present in the sample. The fact that activity rather than the total concentration is the fundamental driving force in chemical and biochemical processes is an important reason for the application of ISEs in clinical analysis. Previously, atomic spectroscopic approaches (see below) had reported the total concentration of the electrolyte of interest rather than on the biochemically important fraction.

Analytical techniques for trace level analysis can be divided in the three principal classes:

(1) *Atomic spectrometric methods*: Here, the entire sample is atomized or ionized either by flame or inductively coupled plasma and transferred into the detector. The most common techniques in this class are flame atomic absorption spectrometry (FAAS) and inductively coupled plasma mass spectrometry (ICPMS). A general characteristic of these methods is the determination of the total concentration of the analyte without the direct possibility of distinguishing its specific forms in the sample.
(2) *Voltammetric sensors*: Here, detection is based on the redox behaviour of the analyte on the electrode. However, when the analyte is bound to suspended particles or present in complexes that are chemically inert, direct determination is generally not possible. Therefore, voltammetric sensors provide information on the species that are chemically available (labile). Uniquely, these sensors typically involve a necessary preconcentration step in which the analyte is usually reduced and accumulated for a certain time at the electrode. This process is followed by its oxidation and stripping from the electrode. Whole family of methods has emerged based on the different potential–current profiles for the stripping step, all having common name "Stripping Analysis" (SA).
(3) *Potentiometric sensors* (*ion-selective electrodes*): These sensors are able to directly determine the activity of the ion of interest in the sample. In the following section, their behaviour is described in more detail.

It is important to note that the information obtained by each of the above techniques is very useful for understanding the composition and behaviour of a particular sample. A particularly appropriate example was the investigation of ground-water pollution by arsenic in the Lower Ganga Plain of West Bengal in India and Bangladesh. At one location, arsenic concentration was clearly well above the safe drinking threshold, while at another in the close proximity it was not toxic [24]. In this case, a detailed knowledge of the interaction of arsenic in its two valence states with local rocks and minerals is crucial for the determination of its speciation and mobility and therefore bioavailability. Only a combination of methods that allow the determination of total and free arsenic concentration can provide a rigorous understanding of the problem.

2.3 THE STATE-OF-THE-ART OF POTENTIOMETRIC SENSORS

Traditionally, potentiometric sensors are distinguished by the membrane material. Glass electrodes are very well established especially in the detection of H^+. However, fine-tuning of the potentiometric response of this type of membrane is chemically difficult. Solid-state membranes such as silver halides or metal sulphides are also well established for a number of cations and anions [25,26]. Their LOD is ideally a direct function of the solubility product of the materials [27], but it is often limited by dissolution of impurities [28–30]. Polymeric membrane-based ISEs are a group of the most versatile and widespread potentiometric sensors. Their versatility is based on the possibility of chemical tuning because the selectivity is based on the extraction of an ion into a polymer and its complexation with a receptor that can be chemically designed. Most research has been done on polymer-based ISEs and the remainder of this work will focus on this sensor type.

These devices are fabricated by doping a polymer membrane with sensing ingredients such as an ionophore which selectively binds a targeted ion and an ion exchanger which attracts a fixed concentration of analyte and preserves charge balance into the membrane phase [8,23,29]. The selectivity of resulting sensor is given by the lipophilicity of the analyte ion relative to an interfering ion and relative strength of the complex formed in the membrane [8,21]. If the charges of analyte (primary ion) and interfering ions are different or the formed complexes are of different stoichiometries, the selectivity can be tuned [8,22,29].

In polymer-based ISEs, electrical contact between the membrane and inner reference electrode is made via an inner filling electrolyte. This type of ISE is the most common and they are usually referred to as liquid contact ISEs or very often simply ISEs. On the other hand, the contact can be obtained by the substitution of the aqueous inner solution with another polymeric material, to produce so-called 'solid-contact' ISEs Table 2.1 provides current achievements in trace level

TABLE 2.1

Potentiometric sensors for trace analysis realized to date with detection limits and a brief description of membrane and inner solution composition

Method/ion	ISE description	ISE	Ref	ICPMS	EPA
Ag^{+}	Polymeric; resin in inner solution	0.1	[16]	0.005	100
	Filled monolithic column	0.2	[84]		
Ca^{2+}	Polymeric; resin in inner solution	0.08	[44]	0.5	
	Polymeric; EDTA in inner solution	0.2	[41]		
	Filled monolithic column	0.2	[84]		
	Polymeric; microparticles in membrane	$\approx$0.004	[54]		
	Polymeric; EDTA in inner solution	$\approx$0.0004	[94]		
Cd^{2+}	Polymeric; NTA in inner solution	0.01	[43]	0.005	5
	Polymeric; Et_4NNO_3 in inner solution	0.01	[83]		
ClO_4^-	Polymeric; resin in inner solution	2	[42]		1*
Cs^{+}	Polymeric; resin in inner solution	1	[80]	0.01	
Cu^{2+}	Polymeric	0.6	[92]	0.005	1300
	Polymeric; hard membrane; Et_4NNO_3 in inner solution	0.01	[18]		
	Solid state; rotating electrode	0.005	[28]		
I^-	polymeric; resin in inner solution	0.25	[42]	1	
K^{+}	polymeric; resin in inner solution	0.2	[44]	0.5	
Na^{+}	filled monolithic column	0.7	[84]	0.05	
NH_4^+	polymeric; resin in inner solution	0.2	[44]		100**

TABLE 2.1 (*continued*)

Method/ion	ISE description	ISE	Ref	ICPMS	EPA
Pb^{2+}	polymeric; NTA in inner solution; optimized for rugged response	0.2	[17]	0.001	30
	polymeric; covalently attached ligand	0.2	[52]		
	polymeric; solid inner contact	0.2	[93]		
	polymeric (plasticizer-free); solid inner contact	0.1	[50]		
	polymeric; EDTA in inner solution	0.016	[41]		
	polymeric; rotating electrode	0.012	[54]		
Vitamin B1	polymeric; very lipophilic ion exchanger	3.3	[95]		

LODs given as μg/l and compared to LODs obtained by ICPMS [91] and action limits set by USA EPA for drinking water if not specified otherwise. [79]
*suggested
**depending on temperature

analysis potentiometric sensors where their LODs are compared to the corresponding values obtained by ICPMS and USA Environmental Protection Agency (EPA) action limits in drinking water.

2.4 POLYMERIC MEMBRANE ISEs WITH LIQUID INNER CONTACT

2.4.1 Problems of the past

Trace level analysis using ISEs was considered impossible only a decade ago. With detection limits in micromolar levels and the discrimination of interfering ions by a factor of 10^3–10^4, researchers at the time did not see any possibility of serious advances in the area and the application of ISEs in new field such as trace metal analysis in environmental samples was regarded as impossible. The potential response of an ISE in the presence of interfering ions is described by the well-known

Nikolskii–Eisenman equation:

$$\mathrm{EMF} = E_I^0 + \frac{RT}{z_I F} \ln\left(a_{I,(\mathrm{aq})} + K_{I,J}^{\mathrm{pot}} a_{J,(\mathrm{aq})}^{z_I/z_J} \right) \tag{2.2}$$

where E_I^0 stands for standard potential of the primary ion I, a_J is the sample activity of the interfering ion J with the charge z_J and $K_{I,J}^{\mathrm{pot}}$ is the selectivity coefficient. This very compact and useful equation is typically used not only for the prediction of the electrochemical response and the determination of unknowns but also in the calculation of selectivity coefficients [9,21]. Historically, researchers had three major methods for the determination of the selectivity coefficients—the separate solution method (SSM), the fixed interference method (FIM) and the matched potential method (MPM) [8]. Please refer to several available review articles about methods for the determination of selectivity coefficients [21,31,32]. The former two were based on Eq. (2.2) and required Nernstian response slopes for both primary and interfering ions. Unfortunately, obtaining Nernstian slopes for interfering ions was in most cases impossible and researchers at the time surprisingly concluded that $K_{I,J}^{\mathrm{pot}}$ is not the real measure of selectivity. In contrast, the MPM method was not bound by any model assumptions and could be applied for cases showing non-Nernstian responses. IUPAC endorsed this method because it seemed to alleviate the problems described above [33]. However, this method suffered from a range of problems. Highly discriminated interfering ions induced responses that generated misleading selectivity factors [32,34]. In addition, the selectivity factor values were highly concentration dependent in the case of primary and interfering ions of different charges [21]. The name "factor" instead at "coefficient" emphasized that the reported values were dependent on experimental conditions [21]. Therefore, in spite of a range of available methods, researchers were left without a method that was capable of providing "true" values of the selectivity coefficients.

It has been widely accepted for many years that the LOD of an ISE in an unbuffered solution is at micromolar level. Interestingly, if a complexing agent is added into the sample and the concentration of the free primary ions is significantly lowered, the LOD is reduced sometimes to subnanomolar levels [35]. In addition, if halide ions are added to samples in which a silver-selective electrode is immersed, the electrode shows a decrease in potential indicating lowering of the activity of a silver at the sample/membrane phase boundary [36]. Moreover, ionophore-based optodes showed picomolar detection limits [37], even though they are based on a very similar approach to their counterpart ISEs [8]. Despite these 'hints' it required the systematical studies led by Pretsch and Baker *et al.* to generate the recent resurgence in research activity.

2.4.2 Frequent bias in the determination of the selectivity coefficient and the limit of detection

ISE membranes contain a rather high concentration of primary ion–ionophore complex, typically in order of 5 mM. The total amount of the complex is in the range of about 30 nmol, and, bearing in mind that a typical membrane is ca. 3 mm in diameter and 0.2 mm in thickness, if there is a mechanism that can expel only 1% of these ions, a significant bias can be created since ISEs are sensitive towards the activity at the phase boundary instead of the bulk concentration [8]. Indeed, Mathison and Bakker showed that primary ions from the inner solution can partition in the membrane together with their counterions [38], and because of the consequent formation of a concentration gradient within the membrane, they are transported into the sample, perturbing the sample/membrane phase boundary [38]. Figure 2.1 left (case A) represents the increase in the membrane/sample phase boundary activity of primary ion compared to the bulk, as a consequence of leaching from the membrane due to transport from the inner solution. It should be noted that the charge balance in the membrane is preserved as there is no charge separation, since both positive and negative charges are transported in the same direction, and therefore, measurements are still performed under zero-current conditions. Typical inner solutions used in these ISEs contained 10^{-3} or 10^{-2} M of primary ion salt solution and the amount of partitioned (coextracted) electrolyte from the inner solution side was around 1% [39]. In spite of the seemingly minor quantities involved, leaching of the excess primary ions from the membrane (so-called outward flux) significantly affects the sample/membrane phase boundary. The experimental detection limit was therefore defined by the amount of ions leaching from the membrane instead of the activity in the bulk as shown in Fig. 2.1 right (curve A). On the other hand, a very dilute inner solution induces significant exchange of primary ions from the membrane by more concentrated interfering ions from the inner solution, lowering the concentration of the primary ion–ionophore complex at the inner membrane phase boundary. In this case, a gradient in the opposite direction (from sample towards inner solution—so-called inward fluxes)

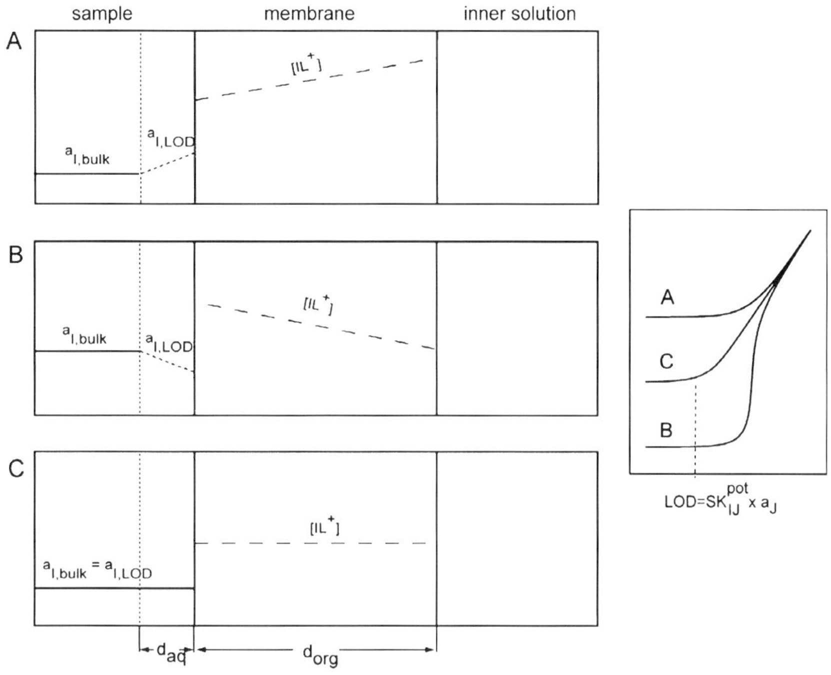

Fig. 2.1. Zero-current ion fluxes in the ion-selective membrane. Left: (A) Concentrated inner solution induces coextraction of electrolyte into the membrane increasing the primary ion–ionophore concentration within the membrane. Consequently, primary ions leach into the sample increasing the activity of primary ions at the membrane/sample phase boundary. (B) Diluted inner solution and ion exchange at the inner solution side decreases the concentration of the complex within the membrane. Primary ions are siphoned-off from the sample, and their activity at the membrane/sample phase boundary is significantly decreased. (C) Ideal case of perfectly symmetric sample and inner solution resulting in no membrane fluxes. Note that fluxes of other species (counterions or interfering ions) are not shown for clarity. Right: potential responses for each case. Ideal LOD is defined by the Nikolskii–Eisenman equation ($\sum K_{I,J}^{\text{pot}} a_J$) and is obtained only in the ideal case (C). Fluxes in either direction significantly affect the LOD.

is formed and primary ions from the sample are siphoned-off into the membrane. Note that at the same time, a flux of interfering ions from the inner solution towards the sample is formed resulting in no net charge transfer and preserving overall charge neutrality. The sample phase boundary concentration of primary ions is far smaller than the bulk concentration, and below the critical activity, the ISE is insensitive towards the bulk concentration. At the critical concentration of the primary ions, a steep EMF change is observed with a so-called super-Nernstian slope that becomes Nernstian in the range of the higher bulk primary ion concentrations. Severe drifts in the region exhibiting super-Nernstian slope and very slow response time mean that these signals are of very little practical use despite the impression of increased sensitivity [40]. Figure 2.1 left (case B) depicts inward fluxes which result in a potential response as shown in Fig. 2.1 right (curve B). It is clear that the true zero-flux condition and therefore true detection limit as predicted by the Nikolskii–Eisenman equation is possible only in the case of perfect symmetry of the sample and the inner solution (see Fig. 2.1 case C). In real-life measurements, however, the system becomes asymmetrical as soon as the sample is varied to some extent compared to inner solution and the concentration gradient is inevitably formed.

The issue of sub-Nernstian responses to interfering ions during the determination of selectivity coefficients by either SSM or FIM originates also via the formation of concentration gradients [21]. Even though primary ions are not added to the sample solution, due to their leaching from the membrane, they are present at the sample/membrane phase boundary. The highly discriminated ions are therefore competing with primary ions and the calibration curve is the representation of a response to a mixed solution rather than of a solution of only interfering ions.

2.4.3 Determination of the true (unbiased) selectivity coefficients

In light of the previous discussion, it is clear that the concentration gradient should be diminished or minimized in order to obtain unbiased selectivity coefficients. Bakker has suggested a slightly modified protocol based on the SSM which avoids exposure of the electrode to the primary ion before the response to interfering ions is obtained [9]. In contrast to previous well-established practise, the electrode is initially conditioned in a solution of an interfering ion. The calibration curves for all interferents are first recorded followed by the calibration curve for the primary ions. Figure 2.2A shows the responses of a silver-selective electrode conditioned in a solution of sodium ions according to the new protocol. Figure 2.2B depicts the response of an electrode based on the same membrane after conditioning in silver solution according to the classical practise. Clearly, Nernstian slopes are readily obtained for all ions in the former case, as opposed to the latter. Therefore, true selectivity coefficients are obtained, $K_{I,J}^{\mathrm{pot}}$ can be regarded as an accurate

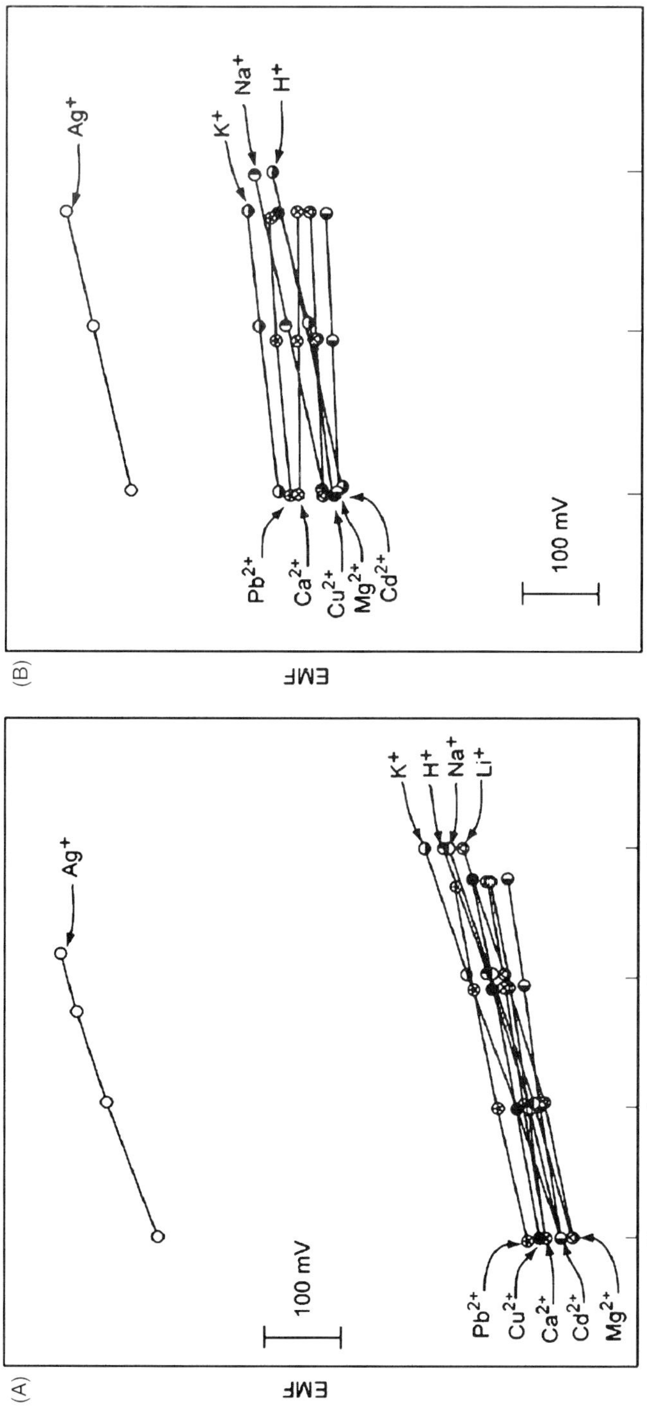

Fig. 2.2. Determination of selectivity coefficient of a silver-selective electrode by SSM for indicated ions. (A) Electrode conditioned in 10^{-3} M $NaNO_3$. (B) Electrode conditioned in 10^{-3} M $AgNO_3$.

TABLE 2.2

Selectivity coefficients obtained by SSM employing modified (membrane conditioned in interfering ion solution (10^{-3} M $NaNO_3$)) and classical (membrane conditioning in primary ion solution (10^{-3} M $AgNO_3$)) experimental setups

Ion	Membrane conditioned in 10^{-3} M $NaNO_3$		Membrane conditioned in 10^{-3} M $AgNO_3$	
	Slope (mV/dec)[a]	$\log K_{I,J}^{pot}$ [a]	Slope (mV/dec)[a]	$\log K_{I,J}^{pot}$ [a]
Ag	62.56±0.95	0.0	54.57±0.35	0.0
Cu	31.96±0.84	−11.48±0.43	38.33±0.66	−6.04±0.09
Li	50.21±0.48	−9.72±0.74	n.d.	n.d.
Na	55.83±0.19	−9.41±0.73	40.77±0.06	−4.98±0.02
K	61.54±0.53	−8.88±0.72	23.67±0.47	−4.09±0.01
H	53.18±0.36	−9.38±0.69	45.60±0.81	−4.49±0.02
Mg	26.36±0.77	−11.60±0.72	9.22±0.90	−6.61±0.06
Ca	28.48±0.49	−11.18±0.69	2.89±0.84	−6.29±0.08
Pb	35.15±2.00	−10.49±0.56	15.62±0.16	−5.71±0.02
Cd	26.46±0.66	−11.32±0.64	8.97±1.00	−6.23±0.04

Silver-selective membrane used is based on *o*-xylylenebis(*N*,*N*-diisobutyldithiocarbamate).
[a]Mean values given with standard deviation ($N \geq 3$).

measure of selectivity. Table 2.2 provides the calculated values of the slopes and the selectivity coefficients based on the both experimental protocols. Perhaps surprisingly, the values obtained by this new protocol are much lower than classical ones. This can be explained as follows—the electrode response is defined by the primary ions that are spontaneously introduced at the membrane/sample phase boundary, in addition to the added interfering ions, thereby inducing mixed response. Pretsch *et al.* have used protocols in which primary ions in the inner solution have been buffered to very low levels, and unbiased selectivity coefficients have been obtained [41]. As a consequence of having no primary ions in the inner solution [9], or their buffering to low levels [41], the outward flux of primary ions as the main source of bias is completely removed and the electrode responds to interfering ions only. Both protocols are widely accepted and used by the research community and improvements of up to 10 orders of magnitude in the value of selectivity coefficients have been reported [9,10,19,32]. It is unfortunate, however, that the dynamics of the advance in the area has not been followed by IUPAC whose last recommendation still recommends the

MPM as the method of choice for determination of selectivity coefficients [33]. However, the number of experimental results demonstrating that these newly described protocols finally remove the problem of biased selectivity suggests that the research community has accepted the new protocols and that IUPAC will soon follow.

2.4.4 Determination of the LOD

This new understanding of the ISE response mechanism generated real excitement, because of dramatic improvements in the LOD. If concentration gradients within the membrane can be minimized, leaching of the primary ions can be significantly reduced. Hence, the perturbation of the sample/membrane phase boundary is minimized and the LOD is improved. The breakthrough in lowering of the detection limit has been achieved by using complexing agents to buffer primary ions in the inner solution [10,16–18,42–44]. By lowering the activity of the primary ions in the inner solution while keeping the concentration of the interfering ion relatively high, the electrolyte coextraction is greatly reduced and the fluxes are minimized. The first successful demonstration of lowering the LOD was by Sokalski *et al.* in which the LOD of a Pb^{2+}-selective ISE was improved by six orders of magnitude compared to the classical one [10]. This work initiated a lot of excitement and was soon followed by many others reporting significantly improved LODs [16,17,19,41–44].

2.4.5 Describing the potential response of ISEs

These significant experimental breakthroughs were very soon followed by improvements in the theoretical treatment of the potentiometric response. A recent review discusses the advantages and disadvantages of approaches that were undertaken to explain these responses and the reader is encouraged to consult it for deeper insight [22]. Initially, a range of treatments were developed that calculated the primary ion–ionophore complex concentration in the membrane and the activity of the primary ion at the sample/membrane interface [14,45]. These treatments considered fundamental coextraction and ion exchange at both phase boundaries assuming steady-state fluxes. Subsequently, a more simplified approach was introduced based on the current selectivity theory, thereby providing the explicit equations that depend only on experimentally available data [16,17]. In this approach, the concentration of the primary ion–ionophore complex in the membrane and the activity of the primary ion at the sample/membrane phase boundary are calculated assuming a linear concentration gradient (steady-state) across the membrane phase and the aqueous diffusion layer (see Fig. 2.1). Considering Fick's first law of diffusion, the problem may be described by the following equation:

$$\frac{a_{I,\mathrm{PB}} - a_{I,\mathrm{bulk}}}{[\mathrm{IL}^+]' - [\mathrm{IL}^+]} = q = \frac{D_M \delta_{\mathrm{aq}}}{D_{\mathrm{aq}} \delta_M} \tag{2.3}$$

where $[\mathrm{IL}^+]'$ and $[\mathrm{IL}^+]$ are the concentration of the primary ion–ionophore complex at the inner and outer boundaries, respectively, $a_{I,\mathrm{PB}}$ and $a_{I,\mathrm{bulk}}$ are the activity of the primary ion at the membrane/sample phase boundary and in the bulk, respectively, and D and δ are the diffusion coefficient and diffusion layer thickness, respectively, for organic and aqueous phases (see Fig. 2.1). As a consequence of flux treatment based on the phase boundary potential model, an extended Nikolskii–Eisenman equation can be developed:

$$\begin{aligned} E = E_I^0 + \frac{RT}{z_I F} \ln\Bigg(\frac{1}{2}\left(a_{I,\mathrm{bulk}} + \sum K_{IJ}^{\mathrm{pot}} a_J + [\mathrm{IL}_n^{z_I^+}]' q - \frac{q R_T}{z_I} \right) \\ + \frac{1}{2}\left(a_{I,\mathrm{bulk}} - \sum K_{IJ}^{\mathrm{pot}} a_J + [\mathrm{IL}_n^{z_I^+}]' q - \frac{q R_T}{z_I} \right)^2 \\ + 4 \sum K_{IJ}^{\mathrm{pot}} a_J \left([\mathrm{IL}_n^{z_I^+}]' q + a_{I,\mathrm{bulk}} \right)^{1/2} \Bigg) \end{aligned} \tag{2.4}$$

where R_T is total amount of a lipophilic salt added to the membrane cocktail in order to preserve electroneutrality [8]. This equation is valid when both primary and interfering ions are of the same charges. Equations for ions of other charges have also been developed and reported in the same work [16]. The most important advantage of this equation compared to others so far reported is that it uses only experimentally available data. The concentration of the primary ion–ionophore complex at the inner side ($[\mathrm{IL}^+]'$) can also be calculated based on a knowledge of the inner solution composition and current selectivity theory [39]. In the simple case of an ideally optimized inner solution and monovalent primary and interfering ions, the complex concentration is given by [39]

$$[\mathrm{IL}^+]' = \frac{a_I' R_T}{K_{I,J}^{\mathrm{pot}} a_J' + a_I'} \tag{2.5}$$

where prime indicates inner solution activities of ions I and J. For more complicated cases please refer to the literature [46].

A convenient equation describing the detection limit in the case of primary and interfering ions having the same charge or monovalent/divalent or divalent/monovalent combination can also be obtained as [16]

$$\log a_{I,(\mathrm{DL})} = \left(\frac{1}{2}+\frac{1}{6}(z_J - z_I)\right) + \log\left(\left(\frac{qR_T}{z_I}\right)^{z_I/z_J} \sum K_{IJ}^{\mathrm{pot}} a_J^{z_I/z_J}\right) \tag{2.6}$$

Figure 2.3 depicts comparison of the theoretical predictions and experimental observations of the potential response of a silver-selective electrode based on *o*-xylylenebis(*N*,*N*-diisobutyldithiocarbamate. Figure 2.3A demonstrates the potential response of an electrode that utilizes a classical experimental setup, i.e. concentrated inner solution (open circles) compared with theoretical prediction based on Eq. (2.2) (full line). The experimentally observed LOD of 10^{-7} M corresponds poorly with the optimistic theoretical prediction of 4×10^{-15} M. On the other hand, after optimization of the inner solution [19], the potential response is extended (Fig. 2.3B closed circles) and the detection limit is improved by almost three orders of magnitude to 3×10^{-10} M. At the same time, an excellent correspondence between experimental observation and theoretical prediction was achieved by employing the extended Nikolskii–Eisenman equation (Eq. (2.4)—full line). This demonstrates the essential role of membrane fluxes in the potential response of ion-selective electrodes. (For all experimental and calculations parameters see the figure caption.)

Time-dependent phenomena that are neglected in steady-state treatments were also a topic of vigorous research, especially during the 1970s and 1980s. The recent improvements in understanding the mechanisms that dictate ISE responses have induced a renewed interest in modelling dynamic responses [15,47]. Recently, a theoretical model involving the solution of the diffusion equation by finite difference in time and finite element in space approximation (based on only experimentally available parameters) demonstrated excellent agreement between predicted and experimental data for time-dependent potentiometric responses [15].

2.4.6 Practical recipes

In the last few years, the area of ISEs has undergone a significant change. Today's understanding of the problems and solutions is greatly different than that only 10 years ago, as the fundamentals of the area

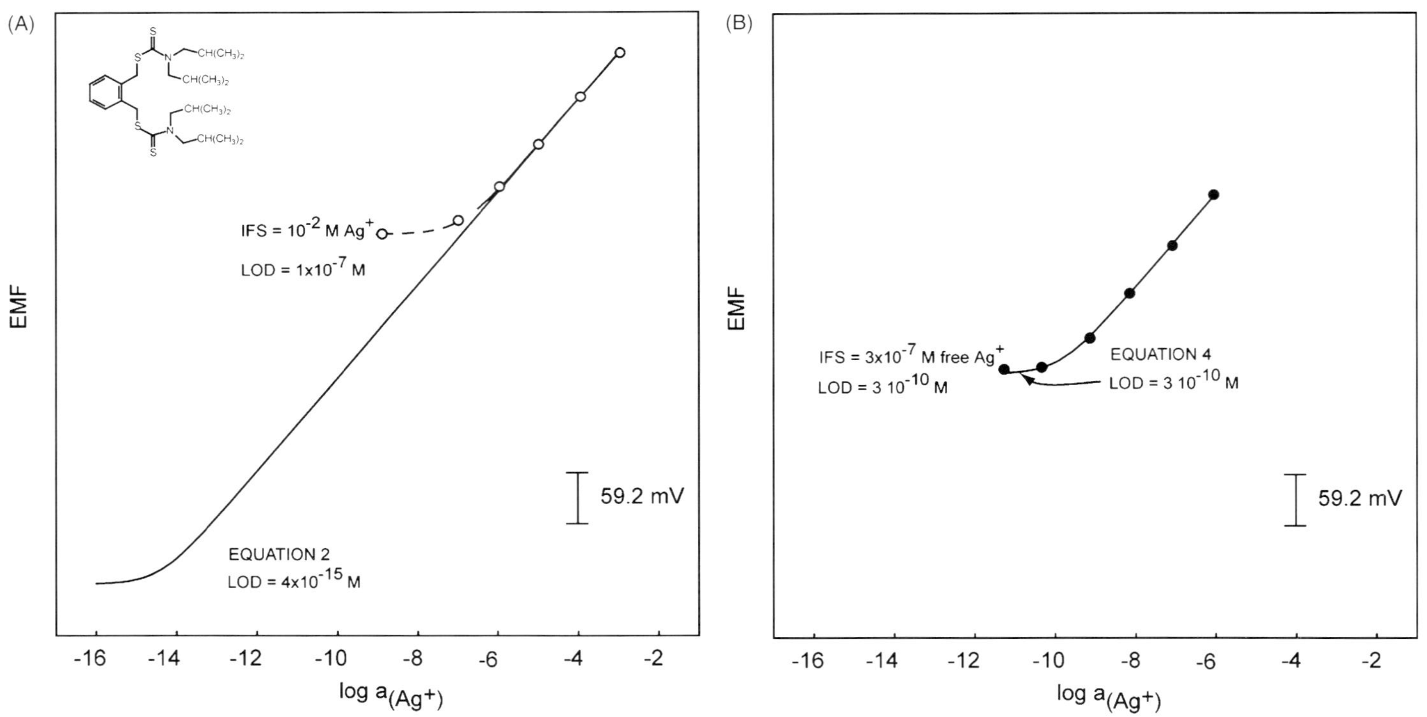

Fig. 2.3. Experimental observation and theoretical predictions of the potential responses of a silver-selective electrode. (A) Open circles: experimental response of electrode having 10^{-2}M $AgNO_3$ in inner solution. Full line: potential response predicted by Eq. (2.2). Inset: *o*-Xylylenebis(*N*,*N*-diisobutyldithiocarbamate). (B) Filled circles: experimental response of electrode having 3×10^{-7}M of free Ag^+ in inner solution buffered by ion exchange resin [19]. Full line: potential response predicted by Eq. (2.4) (used parameters: $R_T = 5.53$ mmol/kg, $\log K^{\mathrm{pot}}_{\mathrm{Ag,Na}} = -9.4$, $c_{\mathrm{Na}} = 10^{-5}$ M, $\delta_M = 200\,\mu$m, $\delta_{\mathrm{aq}} = 0.33\,\mu$m, $D_M = 10^{-8}\,\mathrm{cm}^2$/s, $D_{\mathrm{aq}} = 1.65 \times 10^{-5}\,\mathrm{cm}^2$/s, $a_{\mathrm{Ag,bulk(impurities)}} = 3 \times 10^{-10}$ M) [19].

are better understood and redefined. A researcher, whose interest has wandered away from the field and who has not closely followed these advancements, may find oneself in the situation of rediscovering the area once again, almost as a complete novice. The following section is presented therefore as a summary of routes towards the easy fabrication of low LOD ISEs, both in terms of characterization and optimization of experimental parameters.

2.4.6.1 Selectivity coefficients

With the elimination of bias in the experimental setup, a range of ionophores must be re-characterized, as their reported selectivity coefficients can differ significantly from the unbiased ones [19]. It is very important to obtain true thermodynamic selectivity coefficients as already discussed above. From the three proposed methods that were interchangeably used in the past, a modified SSM has emerged as the most convenient one. Here, the calibration curves for all ions are determined in separate titrations. The modification from the traditional method involves removing the experimental bias by one of the following proposed methods:

(1) Calibration curves for all interfering ions are obtained before the membrane is brought in contact with primary ions [9]. The membrane cocktail must be prepared with ion exchanger that does not contain primary ions as its counterions and the membrane should be conditioned in the solution of an interfering ion. After calibration curves for all interfering ions are recorded, the calibration curve for primary ions is determined.
(2) Inner solutions that induce large super-Nernstian responses can also be used [41], since inward fluxes are created and leaching of the primary ions from the inner solution is prevented.
(3) Unbiased selectivity coefficients can also be obtained by adding a complexing agent for primary ions to the sample [48] as this removed primary ions from the sample/membrane phase boundary.

If the slopes for all ions are Nernstian, the selectivity coefficients are determined using the following equation [9]:

$$\log K_{IJ}^{\mathrm{pot}} = \frac{(E_J - E_I)z_I F}{2.303RT} + \log\left(\frac{a_I}{a_J^{z_I/z_J}}\right) \quad (2.7)$$

Slopes lower than Nernstian indicate a very low preference for the determined ion and the maximal limiting values of $K_{I,J}^{\mathrm{pot}}$ should therefore be determined at the highest activity of the interfering ion. A significantly smaller slope than theoretical for primary ions, or even a slope of opposite sign at higher activities, indicates severe counterion interference [21] and the portion of the curve at lower activities may be used if it is close to Nernstian. If no Nernstian slopes for primary ions are observed, no selectivity coefficients can be calculated.

2.4.6.2 Limit of detection

Accurate estimation of LODs also requires minimization of membrane fluxes. It is unfortunate that fluxes cannot be completely removed and electrodes can only be optimized to generate minimal fluxes. The problem of fluxes is summarized in Eq. (2.3). At the detection limit, $a_{I,\mathrm{bulk}}$ is negligible and $a_{I,\mathrm{PB}}$ is equal to the detection limit DL. Therefore, Eq. (2.3) can be rewritten as

$$\mathrm{DL} = \frac{D_M \delta_{\mathrm{aq}}}{D_{\mathrm{aq}} \delta_M} ([\mathrm{IL}^+]' - [\mathrm{IL}^+]) \tag{2.8}$$

Clearly, the detection limit depends on six different parameters. The part in parenthesis arises from the composition of the inner solution. The concentration of the complex at the inner side ($[\mathrm{IL}^+]'$) increases as the activity of the primary ion in the inner solution increased [39]. It is therefore clear why complexation of primary ions in the inner solution to low levels gave such good initial results [10]. Initially complexing agents such as NTA and EDTA were used [17,41], but later ion exchange resins were introduced [16,42,49]. The advantage of the latter lies in the fact that they are not extracted into the membrane [16] and they bind a range of ions for which there are no chelators available. Therefore, LODs can be significantly improved by simple optimization of the inner solution. While researchers have equations to calculate the optimal detection limit at their disposal [16], the optimization of inner solution is commonly performed empirically due to unexpected problems, such as extraction of the chelator from the inner solutions [43]. A range of the different inner solutions is typically made until a super-Nernstian response is observed implying that the direction of the fluxes is overturned and further dilution of inner solution will lead to stronger super-Nernstian responses [14,40,41]. Recently, an elegant experimental method was proposed which determines the optimization of the inner solution by measuring the so-called stirring effect [39]. This method can also be utilized for the estimation of the impurities in a sample (see caption of Fig. 2.3) [39]. On the other hand, the concentration of the complex at the outer phase boundary ($[\mathrm{IL}^+]$) depends on the total amount of ion exchanger in the membrane (similarly to $[IL^{+}]'$ as expressed in Eq. (2.5)). Therefore, further improvement in the LOD can be achieved by reducing the total amount of ion exchanger in the membrane. Furthermore, from Eq. (2.8) it is clear that reducing the thickness of the aqueous diffusion layer, increasing the thickness of the membrane and lowering the diffusion coefficient in the membrane can consequently lead to further improvement of LOD. Normally, the diffusion coefficient of the aqueous phase cannot be increased. The diffusion coefficient of the membrane phase can be decreased by increasing the viscosity of the membrane (e.g. by decreasing the plasticizer content in the cocktail [18,40]) using other membrane materials [50,51] or by covalent attachment of the ionophore to the polymer matrix [52,53]. A decrease in the aqueous diffusion layer can be achieved by sample stirring [40], using the rotating disc electrode setup [54], or a wall-jet system [55]. These approaches are likely to lead to manufacturing low-limit ISEs which could be easily adopted by the scientists less familiar with the area.

2.5 POLYMERIC MEMBRANE ISEs WITH SOLID INNER CONTACT

The ultimate detection limit as described by the Nikolskii–Eisenman equation (Eq. (2.2)) is defined by the displacement of a fraction of the primary ions with interfering ions at the sample/membrane phase boundary, which amounts to 50% in the case of monovalent primary and interfering ions. Equation (2.8) (the part in parenthesis) implies two obvious solutions: (1) avoiding the bias introduced by the inner solution and (2) reducing the amount of the primary ion–ionophore complex at the sample/membrane phase boundary and hence reducing the absolute amount of released primary ions due to the ion exchange.

Miniaturization of ISEs by completely removing the inner solution was proposed as an elegant solution long time ago, when the concept of so-called coated wire electrodes or solid contact ISEs was introduced [56]. Originally, membranes of the usual composition [8] were directly applied as a coating to a platinum wire. Unfortunately, the electrodes suffered from potential instabilities and surprisingly high detection limits that were comparable to the LODs of classical liquid contact ISEs. These electrodes found limited application as detectors for capillary electrophoresis [57] or in flow-injection analysers [58] for which no long-term stability is required. Note that optodes of very similar construction (without any inner contact) had significantly lower detection limits [37,51].

Originally, the lack of well-defined redox couple at the membrane/metal interface was proposed as a reason for the potential instabilities [59,60]. Several solutions were proposed to alleviate this problem. For example, it was suggested that a redox active component could be added in the membrane [61], or an intermediate redox active polymer layer could be employed [60]. It was also proposed to use a conducting polymer that has both electronic and ionic conductivity either as internal layer [62–64] or as a membrane matrix [65–67]. More recently, interest in lowering the detection limit of solid contact electrodes has intensified. Michalska *et al.* have identified the spontaneous discharge of a conducting polymer serving as an intermediate layer as a possible drawback [68]. The discharge reaction is coupled to a primary ion membrane flux towards the sample which increases the LOD by a similar mechanism as in liquid contact ISEs.

Recently, Fibbioli *et al.* have identified potential instabilities arising from the spontaneous formation of water layer between the membrane and the metal electrode [69]. This layer acts as a reservoir that is re-equilibrated after every change in the sample composition. Theoretical calculations in which the water layer is simulated as inner solution shows good correspondence with experimental observation, implying the existence of membrane fluxes that drastically affect the detection limit. The elimination of this water layer is therefore crucial in order to improve the LOD. Fibbioli *et al.* have suggested the use of redox-active self-assembled monolayers (SAM) as one solution [70]. These layers improve the lipophilicity of the metal surface, thereby blocking the formation of the water layer and improving the establishment of a more stable redox couple at the membrane/metal surface. Moreover, it has been suggested that the polymer matrix plays a significant role in the formation of this water layer, since with the usual membrane composition (PVC and 66% of plasticizer) water layers are normally formed [70]. Plasticizer-free polymer matrices have therefore been identified as a better choice to use in solid contact electrodes. Another important advantage of such polymers is that the number of membrane components capable of leaching from the membrane is reduced [53,71]. For example, Sutter *et al.* have utilized methyl methacrylate–decyl methacrylate (MMA–DMA) copolymer and poly(3-octyltiophene) (POT) as an internal redox layer and reported a Pb^{2+}-selective solid contact electrode with an LOD of 5×10^{-10} M which is currently the lowest detection limit so far for a non-aqueous inner contact Pb^{2+}-selective electrode [50]. It should be noted that these results were obtained by solvent casting of the POT, rather than by its electropolymerization [50].

Clearly, the development of solid contact electrodes is happening at a quite fast pace, and they are becoming highly promising platforms for future work.

2.6 ION-SELECTIVE ELECTRODES IN TRACE LEVEL ANALYSIS

Initial attempts to assess trace levels in natural samples by ISEs were based on solid-state selective electrodes [72–75]. Due to the quite small solubility product for some membrane materials, predicted detection limits were sometimes extremely low, e.g. for Ag_2S membrane electrode, predicted response for free sulphide ions was down to 10^{-17} M and for free silver ions down to 10^{-25} M in the presence of complexes of these respective ions [25]. Unfortunately, such LODs have never been observed due to impurities in the sensing material [30] and the defects in the membrane surface [76], and therefore the predicted improvements for these systems are yet to be realized [28]. Recently, the group of De Marco applied jalpaite membrane electrodes for the determination of Cu^{2+} in seawater samples from San Diego Bay [28]. With the optimized electrode used in rotating disk configuration, they achieved LODs in the nanomolar range. Moreover, chalcogenide glass-based electrodes were used for the determination of Fe^{3+} [77] and Hg^{2+} [78] in seawater. Activities down to 10^{-25} M of Fe^{3+} and 10^{-20} M of Hg^{2+} were detected but in a rather high total concentration of 10^{-3}–10^{-5} M. However, these findings are of great significance as they demonstrate the important role of potentiometry in speciation analysis.

Attempts to optimize polymer-based electrodes for trace analysis have started very recently. Ceresa *et al.* have reported an ion-selective electrode optimized for the determination of Pb^{2+} in drinking water. The detection limit was 0.7 ppb (3×10^{-9} M) which is somewhat poorer than the best LOD reported so far for Pb^{2+}-selective electrodes [10] but the former was optimized for ruggedness and response time rather than LOD. Nevertheless, the obtained LOD was still adequate for the targeted application since it was about 20-fold lower than the 15 ppb action limit for Pb^{2+} in drinking water imposed by the USA EPA [79]. The authors used ICPMS as a reference method and obtained excellent correlation for samples of concentration $\geqslant$3 nM. It was shown that the calibration procedure required ca. 10 min for stable readings in micromolar to nanomolar concentration levels. Moreover, the authors demonstrated that ISEs are a powerful tool for speciation analysis of lead in drinking water. Commonly found anions in water (e.g. carbonates, hydroxides, chlorides, sulphates etc.) form complexes with lead depending on the sample pH. At pH lower than 4, the lead concentration approximately corresponds to free ion activity, while at pH above 8 lead ions are almost completely complexed. Figure 2.4 results obtained with ISEs for the determination of the activity of lead in samples of different pH compared with the calculated lead activity.

In another recent application, a polymer-based ISE was used for the determination of Cu^{2+} in drinking water [18]. The optimal detection limit was reported as 2×10^{-9} M which deteriorated after first week of use by about 0.5 logarithmic units. In the long term, however, it remained at $< 10^{-8}$ M over 55 days. The LOD in water samples containing the most common interferences was 1×10^{-7} M which was much better than the required 2×10^{-6} M (1/10 of EPA action level), so the electrode was applied to four different drinking water samples. The obtained results deviated by $\leqslant 30\%$ from the values obtained by ICPMS as the reference method.

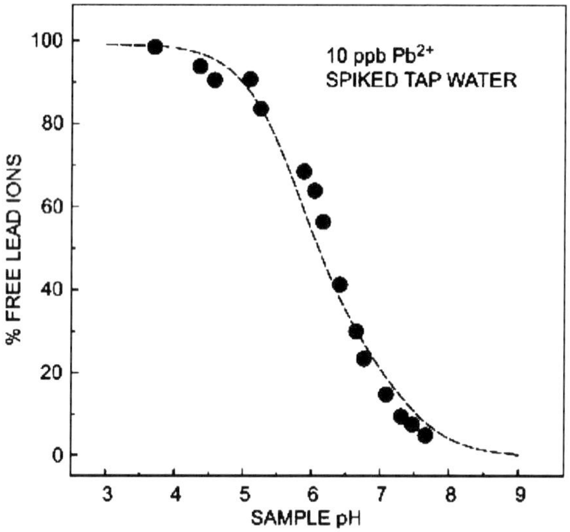

Fig. 2.4. Speciation analysis of drinking water spiked with 10 ppb Pb^{2+} as a function of sample pH performed with Pb^{2+}-selective electrode described in Ceresa *et al.* [17]. The dotted line represents the behaviour calculated on the basis of lead–carbonate complexation [12]. Reprinted from Ref. [12] (p. 205), with permission from Elsevier.

The increasing threat of international terrorism was one motivation for development of ISE for the determination of Cs^+ in environmental samples [80]. In an event such as a Chernobyl-type disaster or the explosion of a "dirty bomb", cesium is one of the most important reaction products and is expected to be the most significant threat to public health [81]. With a detection limit of 10^{-8} M, the developed electrode is sensitive enough for this application and the successful detection of cesium activities in spiked water samples has been demonstrated (see Procedure 2 in CD accompanying this book). In addition, the electrode shows excellent selectivity to cesium in the presence of high levels of strontium, an important interferent originating from nuclear explosions.

Simultaneously with trace level analysis of natural waters, applications of ISEs in metal speciation and bioavailability are attracting significant attention. The same Pb^{2+}-selective electrode described above was used to study the influence of lead speciation and its bioavailability to the fresh-water algae *Chlorella kessleri* in the presence of Suwannee River fulvic acid (SRFA) [82]. At similar concentrations, the authors discovered a discrepancy between Pb bioaccumulation in the presence of synthetic ligands and SRFA. Moreover, they suggested that neither of the currently used steady-state models (free ion activity model and biotic ligand model) are suitable for the prediction of uptake or biological effects. This raises important questions about the predictive capacity and environmental relevance of the simplified models, especially when applied to the determination of metal bioavailability in natural waters.

In a similar application, Plaza *et al.* reported a Cd^{2+}-selective electrode that was used to monitor the uptake of Cd^{2+} by yeast and *Arabidopsis* cell cultures [83]. In the presence of $Ca(NO_3)_2$ and a 0.5 × Murashige and Skoog basal medium for yeast and plant cells, the LODs were 10^{-10} and 10^{-8} M, respectively. Controlled experiments using AAS confirmed that the decrease in Cd^{2+} activities was caused by the uptake of the metal by the cells under investigation.

2.7 FUTURE DIRECTIONS

The last few years have seen significant developments in the area of ISEs. Understanding the mechanisms that dictate LODs and the development of models that successfully describe the potential response have led to new protocols for using ISEs for applications in trace level analysis. Due to the many factors influencing LOD and therefore many options for LOD improvement, for a novice in the area it may seem that the process of optimization is somewhat complicated. Therefore, it is very important to formulate a unified and simplified approach for producing potentiometric sensors with low LODs, short response times, long lifetime and sufficient chemical ruggedness. Recent developments in the area of solid-contact polymeric membrane ISEs suggest that these systems will be very promising platforms for achieving that goal.

In spite of the fact that ISEs for more than 60 ions have been described so far, recent findings imply that these ISEs should be re-characterized and re-optimized for trace level applications [19]. The list of ISEs with low-level LODs needs to be expanded either by re-characterization of existing ionophores or by synthesis of new ones. Important ions for which low LODs have yet to be demonstrated are, for example, mercury, chromium, nickel, arsenate and arsenite ions. Hopefully, synthetic chemists will rise to the challenge and new, selective ionophores will be developed that will achieve this goal.

A very important direction in the development of ISEs is their miniaturization. While microelectrodes have been known for a long time, there is only one report so far describing miniature ISEs with low LODs [84] (see Table 2.1). Because of the relative simplicity of construction and optimization, solid-contact polymer-based membrane ISEs are again very promising platforms for this research direction. Low-cost construction and simple data acquisition [85] facilitate the integration of such platforms into sensor arrays and further into sensor communities based on networks of simple and accurate devices for widely distributed monitoring of, for example, water quality [86]. In addition, microelectrodes can be used for spatial mapping of low chemical concentrations (e.g. in chemical microscopy or the study of ion uptake by the plant roots [87]). Moreover, in contrast to most other analytical techniques, potentiometric sensors ideally do not perturb the sample, and such microelectrodes can therefore be used in very small sample volumes allowing determination of extremely small total quantities.

Instrumental control over the sensitivity of potentiometric sensors will allow controlled ion uptake by the membrane, thereby generating strong super-Nernstian responses. Advances in this direction were recently realized with double- and triple-pulse experiments, where well-defined current and potential pulses were used for accurate control of the otherwise highly transient transport and extraction process [88].

This resulted in the development of robust sensors for the highly charged blood proteins heparin and protamin, which normally have very low theoretical response slopes [89,90].

The list of possible future directions does not end here. Recent advances have shown that ISE research remains very dynamic. This is generating a wealth of knowledge that will lead to the emergence of many new ideas and the expansion of the range of applications in which these electrodes can be used.

ACKNOWLEDGMENT

The authors thank the Science Foundation of Ireland for financial support under the Adoptive information Cluster award (grant 03/IN.3/1361).

REFERENCES

1 M. Cremer, *Z. Biol.*, 47 (1906) 562.
2 M.S. Frant, *Analyst*, 119 (1994) 2293.
3 W.K. Lutz, H.K. Wipf and W. Simon, *Helv. Chim. Acta*, 53 (1970) 1741.
4 L.A.R. Pioda, A.H. Wachter, R.E. Dohner and W. Simon, *Helv. Chim. Acta*, 50 (1967) 1373.
5 Z. Stefanac and W. Simon, *Chimia*, 20 (1966) 436.
6 Z. Stefanac and W. Simon, *Microchem. J.*, 12 (1967) 125.
7 B.C. Pressman, E.J. Harris, W.S. Jagger and J.H. Johnson, *Proc. Natl. Acad. Sci. USA*, 58 (1967) 1949.
8 E. Bakker, P. Buehlmann and E. Pretsch, *Chem. Rev.*, 97 (1997) 3083.
9 E. Bakker, *Anal. Chem.*, 69 (1997) 1061.
10 T. Sokalski, A. Ceresa, T. Zwickl and E. Pretsch, *J. Am. Chem. Soc.*, 119 (1997) 11347.
11 E. Bakker and E. Pretsch, *Anal. Chem.*, 74 (2002) 420A.
12 E. Bakker and E. Pretsch, *Trends Anal. Chem.*, 24 (2005) 199.
13 E. Bakker and E. Pretsch, *Trends Anal. Chem.*, 20 (2001) 11.
14 T. Sokalski, T. Zwickl, E. Bakker and E. Pretsch, *Anal. Chem.*, 71 (1999) 1204.
15 A. Radu, A.J. Meir and E. Bakker, *Anal. Chem.*, 76 (2004) 6402.
16 A. Ceresa, A. Radu, S. Peper, E. Bakker and E. Pretsch, *Anal. Chem.*, 74 (2002) 4027.
17 A. Ceresa, E. Bakker, B. Hattendorf, D. Guenther and E. Pretsch, *Anal. Chem.*, 73 (2001) 343.
18 Z. Szigeti, I. Bitter, K. Toth, C. Latkoczy, D.J. Fliegel, D. Guenther and E. Pretsch, *Anal. Chim. Acta*, 532 (2005) 129.

19 K. Wygladacz, A. Radu, C. Xu, Y. Qin and E. Bakker, *Anal. Chem.*, 77 (2005) 4706.
20 E. Bakker, D. Diamond, A. Lewenstam and E. Pretsch, *Anal. Chim. Acta*, 393 (1999) 11.
21 E. Bakker, E. Pretsch and P. Buehlmann, *Anal. Chem.*, 72 (2000) 1127.
22 E. Bakker, P. Buhlmann and E. Pretsch, *Talanta*, 63 (2004) 3.
23 P. Buehlmann, E. Pretsch and E. Bakker, *Chem. Rev.*, 98 (1998) 1593.
24 D. Chakraborti, S.C. Mukherjee, S. Pati, M.K. Sengupta, M.M. Rahman, U.K. Chowdhury, D. Lodh, C.R. Chanda, A.K. Chakraborti and G.K. Basu, *Environ. Health Perspect.*, 111 (2003) 1194.
25 R.A. Durst (Ed.), *Ion-Selective Electrodes*, National Bureau of Standards, Special Publication No. 314, 1969.
26 M.A. Arnold and R.L. Solsky, *Anal. Chem.*, 58 (1986) 84R.
27 E. Pungor and K. Toth, *Analyst*, 95 (1970) 625.
28 A. Zirino, R. De Marco, I. Rivera and B. Pejcic, *Electroanalysis*, 14 (2002) 493.
29 W.E. Morf, *The Principles of Ion-Selective Electrodes and of Membrane Transport*, Elsevier, New York, 1981.
30 R.P. Buck, *Crit. Rev. Anal. Chem.*, 5 (1975) 323.
31 E. Bakker, *Trends Anal. Chem.*, 16 (1997) 252.
32 M.A. Pineros, J.E. Shaff, L.V. Kochian and E. Bakker, *Electroanalysis*, 10 (1998) 937.
33 Y. Umezawa, K. Umezawa and H. Sato, *Pure Appl. Chem.*, 67 (1995) 507.
34 C. Macca, *Anal. Chim. Acta*, 321 (1996) 1.
35 U. Schefer, D. Ammann, E. Pretsch, U. Oesch and W. Simon, *Anal. Chem.*, 58 (1986) 2282.
36 E. Bakker, *Sens. Actuators B B*, 35 (1996) 20.
37 M. Lerchi, E. Bakker, B. Rusterholz and W. Simon, *Anal. Chem.*, 64 (1992) 1534.
38 S. Mathison and E. Bakker, *Anal. Chem.*, 70 (1998) 303.
39 A. Radu, M. Telting-Diaz and E. Bakker, *Anal. Chem.*, 75 (2003) 6922.
40 A. Ceresa, T. Sokalski and E. Pretsch, *J. Electroanal. Chem.*, 501 (2001) 70.
41 T. Sokalski, A. Ceresa, M. Fibbioli, T. Zwickl, E. Bakker and E. Pretsch, *Anal. Chem.*, 71 (1999) 1210.
42 A. Malon, A. Radu, W. Qin, Y. Qin, A. Ceresa, M. Maj-Zurawska, E. Bakker and E. Pretsch, *Anal. Chem.*, 75 (2003) 3865.
43 A.C. Ion, E. Bakker and E. Pretsch, *Anal. Chim. Acta*, 440 (2001) 71.
44 W. Qin, T. Zwickl and E. Pretsch, *Anal. Chem.*, 72 (2000) 3236.
45 W.E. Morf, M. Badertscher, T. Zwickl, N.F. de Rooij and E. Pretsch, *J. Phys. Chem. B*, 103 (1999) 11346.
46 E. Bakker, R.K. Meruva, E. Pretsch and M.E. Meyerhoff, *Anal. Chem.*, 66 (1994) 3013.
47 T. Sokalski, P. Lingenfelter and A. Lewenstam, *J. Phys. Chem.*, 107 (2003) 2443.

48 T. Sokalski, M. Maj-Zurawska and A. Hulanicki, *Mikrochim. Acta*, 1 (1991) 285.
49 W. Qin, T. Zwickl and E. Pretsch, *Anal. Chem.*, 72 (2000) 3236.
50 J. Sutter, A. Radu, S. Peper, E. Bakker and E. Pretsch, *Anal. Chim. Acta*, 523 (2004) 53.
51 M. Telting-Diaz and E. Bakker, *Anal. Chem.*, 74 (2002) 5251.
52 M. Puntener, T. Vigassy, E. Baier, A. Ceresa and E. Pretsch, *Anal. Chim. Acta*, 503 (2004) 187.
53 Y. Qin, S. Peper, A. Radu, A. Ceresa and E. Bakker, *Anal. Chem.*, 75 (2003) 3038.
54 T. Vigassy, R.E. Gyurcsanyi and E. Pretsch, *Electroanalysis*, 15 (2003) 1270.
55 E. Lindner, R.E. Gyurcsanyi and R.P. Buck, *Electroanalysis*, 11 (1999) 695.
56 R.W. Cattrall and H. Freiser, *Anal. Chem.*, 43 (1971) 1905.
57 P. Schnierle, T. Kappes and P.C. Hauser, *Anal. Chem.*, 70 (1998) 3585.
58 T. Dimitrakopoulos, P.W. Alexander and D.B. Hibbert, *Electroanalysis*, 8 (1996) 438.
59 R.W. Cattrall, D.M. Drew and I.C. Hamilton, *Anal. Chim. Acta*, 76 (1975) 269.
60 P.C. Hauser, D.W.L. Chiang and G.A. Wright, *Anal. Chim. Acta*, 302 (1995) 241.
61 D. Liu, R.K. Meruva, R.B. Brown and M.E. Meyerhoff, *Anal. Chim. Acta*, 321 (1996) 173.
62 A. Cadogan, Z.Q. Gao, A. Lewenstam, A. Ivaska and D. Diamond, *Anal. Chem.*, 64 (1992) 2496.
63 J. Bobacka, M. McCarrick, A. Lewenstam and A. Ivaska, *Analyst*, 119 (1994) 1985.
64 F.Y. Song, J.H. Ha, B. Park, T.H. Kwak, I.T. Kim, H. Nam and G.S. Cha, *Talanta*, 57 (2002) 263.
65 J. Bobacka, T. Lindfors, M. McCarrick, A. Ivaska and A. Lewenstam, *Anal. Chem.*, 67 (1995) 3819.
66 P. Sjoberg, J. Bobacka, A. Lewenstam and A. Ivaska, *Electroanalysis*, 11 (1999) 821.
67 J. Bobacka, A. Ivaska and A. Lewenstam, *Electroanalysis*, 15 (2003) 366.
68 A. Michalska, J. Dumanska and K. Maksymiuk, *Anal. Chem.*, 75 (2003) 4964.
69 M. Fibbioli, W.E. Morf, M. Badertscher, N.F. de Rooij and E. Pretsch, *Electroanalysis*, 12 (2000) 1286.
70 M. Fibbioli, K. Bandyopadhyay, S.G. Liu, L. Echegoyen, O. Enger, F. Diederich, D. Gingery, P. Buhlmann, H. Persson, U.W. Suter and E. Pretsch, *Chem. Mater.*, 14 (2002) 1721.
71 Y. Qin, S. Peper and E. Bakker, *Electroanalysis*, 14 (2002) 1375.
72 M. Trojanowicz, P.W. Alexander and D.B. Hibbert, *Anal. Chim. Acta*, 370 (1998) 267.

73 R. Stella and M.T. Ganzerli-Valentini, *Anal. Chem.*, 51 (1979) 2148.
74 M.J. Smith and S.E. Manahan, *Anal. Chem.*, 45 (1973) 836.
75 R. Jasinski, I. Trachtenberg and D. Andrychuk, *Anal. Chem.*, 46 (1974) 364.
76 W.E. Morf, G. Kahr and W. Simon, *Anal. Chem.*, 46 (1974) 1538.
77 R. De Marco and D.J. Mackey, *Mar. Chem.*, 68 (2000) 283.
78 R. De Marco and J. Shackleton, *Talanta*, 49 (1999) 385.
79 http://www.epa.gov/oerrpage/superfund/resources/remedy/pdf/93-60102-s.pdf.
80 A. Radu, S. Peper, C. Gonczy, W. Runde and D. Diamond, *Electroanalysis*, 13–14 (2006) 1379.
81 P. Weiss, *Sci. News*, 168 (2005) 282.
82 V.I. Slaveykova, K.J. Wilkinson, A. Ceresa and E. Pretsch, *Environ. Sci. Technol.*, 37 (2003) 1114.
83 S. Plaza, Z. Szigeti, M. Geisler, E. Martinoia, B. Aeschlimann, D. Guenther and E. Pretsch, *Anal. Biochem.*, 347 (2005) 10.
84 T. Vigassy, C.G. Huber, R. Wintringer and E. Pretsch, *Anal. Chem.*, 77 (2005) 3966.
85 S. Hudgins, Y. Qin, E. Bakker and C. Shannon, *J. Chem. Educ.*, 80 (2003) 1303.
86 D. Diamond, *Anal. Chem.*, 76 (2004) 278A.
87 M.A. Pineros, J.E. Shaff and L.V. Kochian, *Plant Physiol*, 116 (1998) 1393.
88 S. Makarychev-Mikhailov, A. Shvarev and E. Bakker, *J. Am. Chem. Soc.*, 126 (2004) 10548.
89 A. Shvarev and E. Bakker, *Anal. Chem.*, 77 (2005) 5221.
90 A. Shvarev and E. Bakker, *J. Am. Chem. Soc.*, 125 (2003) 11192.
91 R.A. Meyers (Ed.), *Encyclopedia of Analytical Chemistry: Application, Theory and Instrumentation*, Wiley, Chichester, 2000.
92 S. Kamata, A. Bhale, Y. Fukunaga and H. Murata, *Anal. Chem.*, 60 (1988) 2464.
93 J. Sutter, E. Lindner, R.E. Gyurcsanyi and E. Pretsch, *Anal. Bioanal. Chem.*, 380 (2004) 7.
94 I. Bedlechowicz, M. Maj-Zurawska, T. Sokalski and A. Hulanicki, *J. Electroanal. Chem.*, 537 (2002) 111.
95 G.H. Zhang, T. Imato, Y. Asano, T. Sonoda, H. Kobayashi and N. Ishibashi, *Anal. Chem.*, 62 (1990) 1644.

Chapter 3

Enantioselective, potentiometric membrane electrodes: design, mechanism of potential development and applications for pharmaceutical and biomedical analysis

Raluca-Ioana Stefan-van Staden

3.1 OVERVIEW

Enantioanalysis became increasingly important for pharmaceutical and clinical analysis [1–5] due to the difference between the behaviour of the enantiomers of the same chiral substance in the body. The high enantiopurity for the substances with a chiral centre is essential for the active compounds utilized in pharmaceutical formulations, as in many cases the pharmacological activity and metabolism of two enantiomers of a certain pharmaceutical compound may be different: one of the enantiomers may inhibit the action of the other, it is toxic for the body, is creating unwanted secondary effects by interacting in a different part of the body or is not having any effect on the body [1–4]. It is not enough to determine the total amount of a chiral substance of biological importance in clinical analysis, because each of the enantiomers may be a marker for a different disease, and accordingly each of them should be quantified separately [5].

The methods proposed for enantioanalysis are chromatography [1,3], potentiometry [2] and amperometry [2]. Chromatographic methods may be used mainly for a qualitative determination of enantiomers, because the accuracy of the determination is not always good due to the sample treatment before introducing them into the column. The best results in terms of accuracy, obtained using a chromatographic method, are given by capillary zone electrophoresis, although the limits of detection for this

Comprehensive Analytical Chemistry 49
Alegret and Merkoçi (Eds)
Volume 49 ISSN: 0166-526X DOI: 10.1016/S0166-526X(06)49003-6

technique are not always enough to determine the enantiomers from any type of matrix. Amperometric detection of enantiomers can be performed using amperometric biosensors and immunosensors [6–9] and its selectivity and enantioselectivity is enhanced by the utilization of the biochemical reaction.

Enantioselective, potentiometric membrane electrodes (EPMEs) are proposed for the potentiometric detection of the enantiomers [2,10]. The advantages of utilization of these electrodes over amperometric biosensors and immunosensors are a longer lifetime, a large working concentration range, no dilution required for the samples and possibility of decreasing of limit of detection by utilization of KCl 0.1 mol/L as internal solution [2].

3.2 POTENTIAL DEVELOPMENT FOR EPME

3.2.1 Membrane configuration

The membrane configuration and potential development are similar for ion-selective membrane electrodes and EPME. A sandwich membrane configuration was proposed for EPME by Stefan and Aboul-Enein [11]. The zone free of solution is sandwiched symmetrically between zones impregnated with internal solution and membrane–solution interfaces. The electrode potential is given by the difference between the interfacial potentials. It contains a constant term, which is given by enantiomer transfer between the first zone and the internal solution, and a variable term, which is given by ion transfer between the fifth zone and the analysed solution. Electron transfer is taking place between the second and fourth zones (Fig. 3.1) [11]. The proposed membrane configuration is in agreement with the experimental data obtained through reflection IR spectroscopy as well as by secondary ion mass spectroscopy (SIMS) [12,13].

3.2.2 Mechanism of potential development and chiral recognition

There are two main classes of membrane equilibria [14]: general equilibria and specific equilibria. The general equilibria are ion-exchange equilibria, Donnan equilibria and extraction–re-extraction equilibria. The specific equilibria are, for example, redox equilibria, complexation–de-complexation equilibria, etc. These specific equilibria determine the type of the electrode. For EPMEs, the main equilibria that take place are extraction–re-extraction and complexation–de-complexation.

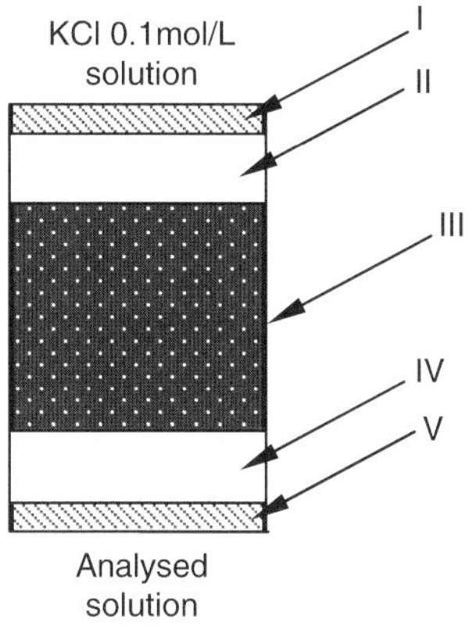

Fig. 3.1. The configuration of the membrane. I and V are membrane–solution interfaces, II and IV are zone impregnated with KCl 0.1 mol/L solution, III is the zone free of solution.

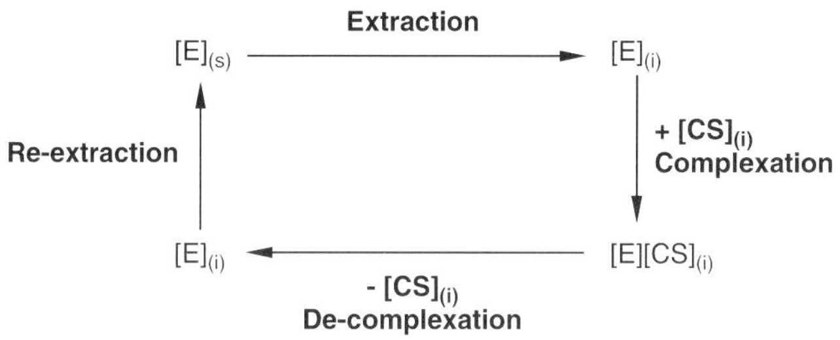

Fig. 3.2. The cycle of membrane equilibria for potential development. [E] is the enantiomer, [CS] is the chiral selector from the membrane, [E-CS] is the complex, (s) and (i) define the bulk solution and the membrane–solution interface, respectively.

Accordingly, the mechanism of potential development is based on membrane equilibria [11] and it can be explained using a cycle of membrane equilibria: extraction, re-extraction, complexation and de-complexation (Fig. 3.2) [14]. When the activity of the main enantiomer is higher in the bulk solution than in the membrane–solution interface, its extraction into the interface is taking place. The second step in the model involves the complexation of the extracted enantiomer at the membrane–solution interface. The de-complexation process will occur when the activity of the main enantiomer in the bulk solution is lower than in the membrane–solution interface. The fourth step is connected with the re-extraction of the main enantiomer from the interface into the bulk solution. The main process that is responsible for the potential development is the complexation one.

The main principle for enantioselective analysis using EPME is to find the lock for the key. The key is the enantiomer that must be analysed and the lock is a substance with a special architecture that can bind enantioselectively with the enantiomer. Accordingly, the principle can be defined as enantioselective binding.

This principle is applied for the potential development in the EPMEs and for obtaining the intensity of the current in amperometric immunosensors. For the enantioselective, potentiometric electrodes, it is necessary to find a molecule with a special architecture that can accommodate the enantiomer. In this regard, cyclodextrins and their derivatives, maltodextrins, antibiotics and fullerenes and their derivatives were proposed [17–52].

The advantage of using antibiotics, cyclodextrins, maltodextrins and fullerenes as chiral selectors is that the enantioselectivity of the molecular interaction takes place in two places: inside the cavity (internal enantioselectivity) and outside the cavity—due to the arrangement, size and type of the radicals, atoms or ions bound on the external chain of the chiral selector (external enantioselectivity) [10]. The thermodynamics of the reaction between the enantiomers and chiral selectors plays the main role in the enantioselectivity of molecular interaction.

If L is the chiral selector and S and R the enantiomers to be determined, the following reactions take place:

$$\mathrm{L} + \mathrm{S} \leftrightarrow \mathrm{LS} \qquad (3.1)$$

$$\mathrm{L} + \mathrm{R} \leftrightarrow \mathrm{LR} \qquad (3.2)$$

The stability constants of the complexes formed between chiral selector and enantiomers are given by the following equations:

$$K_{\mathrm{S}} = \mathrm{e}^{-\Delta G_{\mathrm{S}}/RT} \quad K_{\mathrm{R}} = \mathrm{e}^{-\Delta G_{\mathrm{R}}/RT}$$

where K_{S} and K_{R} are the stability constants of the complexes formed between chiral selector and the S and R enantiomer, respectively, ΔG_{S} and ΔG_{R} are the free energies of the reactions (3.1) and (3.2), respectively, $R = 8.31$ J/Kmol and T is the temperature measured in Kelvin.

The enantioselectivity of the chiral selector is given by the difference between the free energies of the reactions (3.1) and (3.2) [10]:

$$\Delta(\Delta G) = \Delta G_{\mathrm{S}} - \Delta G_{\mathrm{R}}$$

A direct proportionality between log K and ΔG is obvious from the relations between the stability constant and the free energy.

This means that a difference in the free energies of the reactions will result in a difference of the stability of the complexes formed between chiral selector and the S and R enantiomers.

The stability of the complexes is directly correlated with the slope (response) of the EPMEs [14]. Accordingly, a large difference between the free energies of reactions (3.1) and (3.2) will give a large difference between the slopes when the S and R enantiomers will be assayed. Accordingly, the slope is the criterion for molecular recognition of the enantiomer, when the electrode is enantioselective. The minimum value admissible for a 1:n stoichiometry between the enantiomer and chiral selector is 50/n mV/decade of concentration [2].

Mechanisms of interactions between cyclodextrins and enantiomers were reported by Parker and Kataky [15,16]. The complexes have a 1:1 stoichiometry and they all have a host–guest structure.

3.3 DESIGN OF EPME

The main components of the membrane of the enantioselective, potentiometric electrode are chiral selector and matrix. Selection of the chiral selector may be done accordingly with the stability of the complex formed between the enantiomer and chiral selector on certain medium conditions, e.g., when a certain matrix is used or at a certain pH. Accordingly, a combined multivariate regression and neural networks are proposed for the selection of the best chiral selector for the determination of an enantiomer [17]. The most utilized chiral selectors for EPME construction include crown ethers [18–21], cyclodextrins [22–35], maltodextrins [36–42], antibiotics [43–50] and fullerenes [51,52]. The response characteristics of these sensors as well as their enantioselectivity are correlated with the type of matrix used for sensors construction.

Two main types of matrices are described for the design of the sensors: a PVC based matrix [21,35] and a carbon paste based matrix [22–31, 33–52]. A special design was adopted for the construction of the imprinted polymers based sensors [53]. The most reproducible design was proved to be the one based on a carbon paste matrix. The non-reproducibility of PVC-based matrices is due to non-uniformity and non-reproducibility of the repartition of the electroactive material in the matrix [2].

3.3.1 Design of carbon paste based EPMEs

The paraffin oil and graphite powder were mixed in a ratio of 1:4 (w/w) followed by the addition of the solution of chiral selector (10^{-3} mol/L)

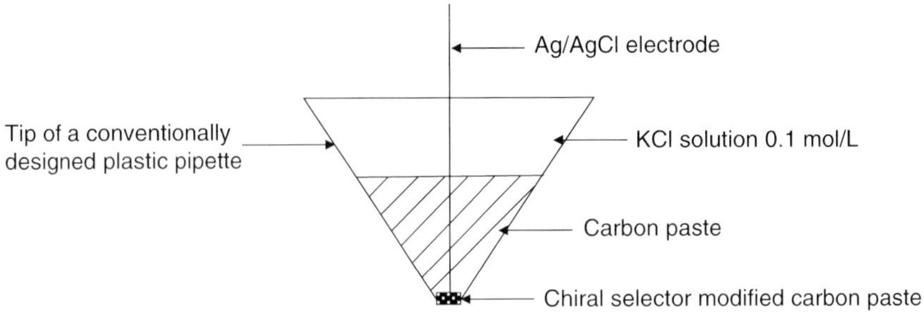

Fig. 3.3. Design of EPMEs based on carbon paste.

(100-μL chiral selector solution to 100-mg carbon paste). The graphite–paraffin oil paste was filled into a plastic pipette peak leaving 3–4 mm empty at the top to be filled with the carbon paste that contains the chiral selector (Fig. 3.3). The diameter of the enantioselective, potentiometric membrane sensor was 3 mm (polishing the electrode did not affect the diameter). Electric contact was made by inserting a silver wire in the carbon paste [15–36]. Before each use, the surface of the electrode was wetted with deionised water and then polished with an alumina paper (polishing strips 30144-001, Orion). When not in use, the electrode was immersed in a 10^{-3} mol/L enantiomer solution under study or analysis.

3.3.2 Design of PVC-based EPMEs

The membrane of these EPMEs contain 1.2% chiral selector, 65.6% *ortho*-nitrophenyl octyl ether (*o*-NPOE) or dioctylsebacate (DOS) (plastifiant), 0.4% tetrakis([3,5-bis(trifluoromethyl)phenyl]borate) (TKB) in a tetrahydrofuran solution containing the PVC (32.4% from the membrane composition) [37,42]. A 7-mm diameter circle was cut from the membrane and inserted into a electrode body along with a inner solution containing the enantiomer to be determined in a certain concentration (usually 10^{-3} mol/L).

3.3.3 Design of molecularly imprinted polymers based enantioselective sensors

Of main importance for this class of sensors is the synthesis of polymer [54]. A reliable synthesis will give reliable response characteristics for the sensor as well as reliable analytical information. Polymers can be prepared by either thermal initiation or photolytic initiation with azobisnitriles (AIBN), or herein by Lucirin LR 8893X as photolytic initiator. In a 50-mL tube, 7800-mg 1,4-butanedienyl diacrylate (BDDA), 3360-mg α-methylacrylic acid (MAA), 165-mg L-phenylalanine and 100-mg photoinitiator ethyl 2,4,6-trimethylbenzophenyphosphinate (LR 8893 X) in 15-mL methanol were thoroughly mixed. Solubilization was achieved by sonication and the mixture was then cooled on ice. After degassing, the tube was sealed under nitrogen and exposed to a UV-laser beam (366 nm) for 18 h. The polymers obtained were ground in a mortar and wet-sieved in water to the desired size distribution. Particles that passed through a 25-μm sieve constituted a fraction of $<25\,\mu$m.

3.4 APPLICATION OF EPMEs IN ENANTIOANALYSIS

3.4.1 EPMEs based on cyclodextrins

Two EPMEs based on α- and γ-cyclodextrins were proposed for the assay of R-baclofen [22]. The slopes of the electrodes were 59.50 and 51.00 mV/pR-baclofen for α- and γ-cyclodextrin-based electrodes, respectively. The detection limits of the proposed electrodes were 7×10^{-9} mol/L for α-cyclodextrin-based electrode and 1.44×10^{-10} mol/L for γ-cyclodextrin-based electrode. The enantioselectivity was determined over S-baclofen. The proposed electrodes can be employed for the assay of R-baclofen raw materials and its pharmaceutical formulation, Norton-Baclofen® tablets.

S-captopril was the first angiotensin-converting enzyme inhibitor tested with this type of electrochemical sensor [23] (see Procedure 3). The EPME based on impregnation of 2-hydroxy-3-trimethylammoniopropyl-β-cyclodextrin (as chloride salt) solution in a carbon paste was successfully used for S-captopril assay in the 10^{-6}–10^{-2} mol/L concentration range (pH range 3.0–6.5), with a near-Nernstian slope of 57.70 mV/decade of concentration, a low limit of detection of 2×10^{-7} mol/L and an average recovery of 99.99% (RSD = 0.05%). Using this sensor, S-captopril can also be reliably assayed in pharmaceutical tablet formulations (25-mg S-captopril/tablet) with an average recovery of 99.69% (RSD = 0.39%). The enantiopurity was tested over R-captopril and D-proline; 10^{-4} magnitude order obtained for potentiometric selectivity coefficients proved its enantioselectivity. The response characteristics of sensor were tested also for R-captopril. A non-Nernstian slope of 27.30 mV/decade of concentration as well as a detection limit of 10^{-6} magnitude order were obtained.

Enantioanalysis of S-cilazapril [24] was performed using an EPME based on impregnation of 2-hydroxy-3-trimethylammoniopropyl-β-cyclodextrin (as chloride salt) solution in a carbon paste, which proved a near-Nernstian slope (56.85 mV/decade of concentration) for the 10^{-5}–10^{-2} mol/L concentration range (pH between 3.0 and 5.5), with a limit of detection of 5.0×10^{-6} mol/L. The enantioselectivity of sensor was checked over D-proline; 10^{-4} magnitude order of potentiometric selectivity coefficient proved the sensors's enantioselectivity. The average recovery of S-cilazapril is 99.36% with an RSD of 0.27%.

S-enalapril assay can be done using the potentiometric electrode based on impregnation of 2-hydroxy-3-trimethylammoniopropyl-β-cyclodextrin (as chloride salt) solution in a carbon paste, in the 3.6×10^{-5}–6.4×10^{-2} mol/L (pH between 3.0 and 6.0) concentration range with a detection limit of 1.0×10^{-5} mol/L [25]. The slope is near-Nernstian: 55.00 mV/decade of concentration. The average recovery of S-enalapril raw material is 99.96% (RSD = 0.098%). The potentiometric selectivity coefficient over D-proline (6.5×10^{-4}) proved the sensor's enantioselectivity. S-enalapril was determined from pharmaceutical tablets with an average recovery of 99.59% (RSD = 0.20%).

(-)-Ephedrine is the pharmacologically active enantiomers of ephedrine. The enantiopurity tests for (-)-ephedrine can be performed using a peroctylated γ-cyclodextrin based plastic membrane electrode [26]. Bis (1-butylpentyl)adipate (BBPAP) was used as plasticizer and 10^{-3} mol/L NH_4Cl as inner solution. The slope of the electrode is 60 mV/decade of concentration, and the potentiometric enantioselectivity coefficient is less than 10^{-4}. The limit of detection is of 10^{-7} mol/L magnitude order.

EPMEs based on immobilization of β-, γ-cyclodextrins or 2-hydroxy-3-trimethylammoniopropyl-β-cyclodextrin (as chloride salt) in carbon paste have been designed for the assay of L- and D-2-hydroxyglutaric acid (2-HGA) [27]. γ-Cyclodextrin or 2-hydroxy-3-trimethylammoniopropyl-β-cyclodextrin based EPMEs can be used in the following concentration ranges: 10^{-8}–10^{-6} and 10^{-7}–10^{-5} mol/L for the assay of L-2-HGA, while γ-cyclodextrin based EPME can be used for the assay of D-2-HGA in the concentration range 10^{-6}–10^{-4} mol/L. The β-cyclodextrin based EPME showed the lowest detection limit (1.0×10^{-9} mol/L) for the detection of L-2-HGA while the detection limit for D-2-HGA was 6.3×10^{-7} mol/L. The enantioselectivity and selectivity over D- or L-2-HGA and creatine, creatinine and different inorganic cations was good. The proposed EPMEs can be reliably used for the assay of L- and D-2-HGA in urine sample, favouring the correct diagnosis of the illness associated with these diseases: L- and D-2-hydroxyglutaric acidurias.

S-pentopril can be determined in the presence of R-pentopril using an EPME based on impregnation of 2-hydroxy-3-trimethylammoniopropyl-β-cyclodextrin (as chloride salt) solution in a carbon paste. This can be done on the 10^{-6}–10^{-2} mol/L (pH = 3.0–5.5) concentration range [24]. The limit of detection is 7.58×10^{-5} mol/L. Slope is near-Nernstian: 58.16 mV/decade of concentration. The potentiometric selectivity coefficient over D-proline is 6.9×10^{-4}. S-pentopril average recovery is 99.79% (RSD = 0.17%).

The EPME based on impregnation of 2-hydroxy-3-trimethylammoniopropyl-β-cyclodextrin (as chloride salt) solution in a carbon paste, can be used for the S-perindopril assay in the 10^{-5}–10^{-2} mol/L (pH = 2.35–6.00) concentration range [28]. The detection limit is 5×10^{-6} mol/L, and the slope is near-Nernstian: 54.23 mV/decade of concentration. The selectivity was checked over R-perindopril and D-proline. The 10^{-4} magnitude orders obtained for potentiometric selectivity coefficients proved enantioselectivity of the sensor. For R-perindopril, the EPME has the following response characteristics: slope is non-Nernstian (38.00 mV/decade of concentration), linear range between 10^{-4} and 10^{-2} mol/L and detection limit of 10^{-6} mol/L magnitude order. S-perindopril can be determined with an average recovery of 99.58% (RSD = 0.33%).

Better results for the enantioanalysis of S-perindopril were obtained when the carbon paste was impregnated with α-, β- and γ-cyclodextrins as chiral selectors [29]. There had been improvements in detection limits (lower with unsubstituted CDs as chiral selectors), enantioselectivity (higher especially when α-CD was used as chiral selector) and the values of the slope of the electrodes (higher especially when α-CD and γ-CD were used as chiral selectors).

An EPME based on impregnation of 2-hydroxy-3-trimethylammoniopropyl-β-cyclodextrin (as chloride salt) solution in a carbon paste was proposed for the assay of L-proline [30]. The linear concentration range for the assay of L-proline is between 5.0×10^{-5} and 1.5×10^{-1} mol/L, with a detection limit of 1.0×10^{-5} mol/L. The average recovery is 99.90% with an RSD of 0.12% ($n = 10$). The electrode is enantioselective over D-proline.

EPMEs based on carbon paste impregnated with α-, β- and γ-cyclodextrins are proposed for the assay of L-proline [31]. Response characteristics showed that the proposed electrodes could be reliably used in the assay of L-proline, with the best enantioselectivity and time-stability exhibited by α-cyclodextrin based EPME. The EPMEs based on the proposed unsubstituted cyclodextrins showed lower detection limits (10^{-10} magnitude order) than the one previously studied, based on β-cyclodextrin derivative [30]. The widest linear concentration range is recorded for the γ-cyclodextrin based EPME (10^{-8}–10^{-3} mol/L). The recovery tests performed for the assay of L-proline in the presence of D-proline (recoveries higher than 99.65%) proved that the proposed electrodes can be reliably used for enantioselectivity tests of proline raw material.

2-Hydroxy-3-trimethylammoniopropyl-β-cyclodextrin was used as chiral selector for the design of EPMEs based on plastic membrane. The electrodes can be used for the determination of S- and R-propranolol [32]. Two plasticizers were used for the design of the electrodes: *o*-NPOE and bis(*t*-butylpentyl)adipate (BBPA). 10^{-3} mol/L R-propranolol and 10^{-3} mol/L NH_4Cl were used as inner filling solutions for electrodes A (designed for the assay of R-propranolol) and B (designed for the assay of S-propranolol). Electrode A has a slope of 59.1 mV/decade of concentration with a detection limit of 1.20×10^{-5} mol/L while electrode B has a slope of 60.0 mV/decade of concentration with a detection limit of 5.24×10^{-6} mol/L.

S-ramipril can be determined as raw material and from its pharmaceutical formulations, using the EPME based on impregnation of 2-hydroxy-3-trimethylammoniopropyl-β-cyclodextrin (as chloride salt) solution in a carbon paste, in the 1.8×10^{-5}–2.3×10^{-1} mol/L (pH between 2.5 and 6.0) concentration range with an average recovery of 99.94% (RSD = 0.030%) and 98.98% (RSD = 0.67%), respectively [33]. The detection limit is of 10^{-5} mol/L magnitude order. The slope is near-Nernstian: 52.00 mV/decade of concentration. Enantioselectivity was proved over D-proline, when a 10^{-4} magnitude order was obtained for potentiometric selectivity coefficient.

The EPME based on impregnation of 2-hydroxy-3-trimethylammoniopropyl-β-cyclodextrin (as chloride salt) solution in a carbon paste can be reliably used for S-trandolapril assay with an average recovery of 99.77% (RSD = 0.22%) [24]. The linear concentration range is 10^{-4}–10^{-2} mol/L on the 2.5–5.5 pH range. The detection limit is of 10^{-5} mol/L magnitude order. The slope is near-Nernstian: 52.45 mV/decade. The sensor enantioselectivity was determined over D-proline, when a 10^{-4} magnitude order was obtained for potentiometric selectivity coefficient.

Four EPMEs based on cyclodextrins (α-, β-, γ- and 2-hydroxy-3-trimethylammoniopropyl-β-cyclodextrin (as chlorine salt) (β-derivative-cyclodextrins)) have been designed for the enantioanalysis of L-vesamicol [34]. The linear concentration ranges for the proposed EPMEs based on cyclodextrins were from 10^{-9} to 10^{-7} mol/L for the electrodes based on α-, β- and β-derivative-cyclodextrins with detection limits of 5.7×10^{-10}, 2.42×10^{-10} and 4.77×10^{-10} mol/L, respectively, while γ-cyclodextrin based electrode is characterized by linear concentration range between 10^{-5} and 10^{-2} mol/L with detection limit of 7.86×10^{-6} mol/L. The electrodes displayed near-Nernstian slopes of 52–59 mV per decade of

L-vesamicol concentration. The electrodes were used for the direct assay of L-vesamicol in serum samples and proved to be reliable, reproducible and stable.

Kataky *et al.* described the design of a PVC membrane based on α-, β- and γ-cyclodextrins as chiral selectors [35]. The enantiomers of aryl ammonium ions of pharmaceutical importance can be discriminated and assayed with low limits of detection and high sensitivities using these electrodes.

3.4.2 EPMEs based on maltodextrins

Maltodextrins (dextrose equivalent (DE) 4.0–7.0, 13.0–17.0 and 16.5–19.5) are proposed as novel chiral selectors for the construction of EPMEs for S-captopril assay [36]. The EPMEs can be used reliably for the assay of S-captopril as raw material and from pharmaceutical formulations as Novocaptopril tablets, using direct potentiometry. The best response was obtained when maltodextrin with higher DE was used for the electrode's construction. The best enantioselectivity and stability in time was achieved for the lower DE maltodextrin. L-Proline was found to be the main interferent for all proposed electrodes. The surface of the electrodes can be regenerated by simply polishing, obtaining a fresh surface ready to be used in a new assay.

Two EPMEs based on maltodextrins with different value of DE (maltodextrin I: DE 4.0–7.0; maltodextrin II: DE 16.5–19.5) were proposed for the assay of baclofen enantiomers in baclofen raw materials and from its pharmaceutical formulation, Norton-Baclofen tablets [37]. The slopes of the electrode function of the proposed electrodes were 55.0 mV/pS-baclofen for maltodextrin I-based electrode and 59.0 mV/pR-baclofen for maltodextrin II-based electrode and the detection limits were 1.34×10^{-6} mol/L (S-baclofen) and 2.52×10^{-10} mol/L (R-baclofen), respectively.

Three EPMEs based on carbon paste impregnated with different maltodextrins (DE 4.0–7.0 (I), 13–17 (II), 16.5–19.5 (III)) were proposed as chiral selectors for the assay of S-flurbiprofen raw materials and from its pharmaceutical formulation, Froben 100® tablets [38]. The best response and enantioselectivity were obtained when maltodextrin with the lowest DE was used for the electrode design. The three EPMEs showed very low detection limits.

D- and L-glyceric acids (D- and L-GA) are human metabolites responsible for two different diseases. Excess excretion of D-GA causes D-glyceric academia/acidurias, while excess excretion of L-GA causes hyperoxaluria type 2, PH II. EPMEs based on maltodextrins I (DE 4.0–7.0), II (DE 13.0–17.0) and III (DE 16.5–19.5) as chiral selectors are proposed for the determination of L- (EPMEs based on maltodextrins I and III) and D-GA (EPME based on maltodextrin II). EPMEs based on maltodextrins I and III can be reliably used for the analyses of L-GA using direct potentiometric method, in the concentration range of 10^{-8}–10^{-6} and 10^{-6}–10^{-3} mol/L, respectively with very low detection limits (1.19×10^{-9} and 1.0×10^{-7} mol/L, respectively). The EPME based on maltodextrin II was successfully used for the enantioanalysis of D-GA in the 10^{-5}–10^{-3} mol/L concentration range with detection limit of 1×10^{-6} mol/L [39]. The enantioselectivity of EPMEs was determined over L- (or D-) GA, and their selectivity over creatine, creatinine and some inorganic cations such as Na^+, K^+ and Ca^{2+}. Simply polishing to obtain a fresh surface ready to be used in a new measurement can regenerate the surface of the electrodes.

Quantitative assay of L-2-HGA is important for the diagnosis of L-2-hydroxyglutaric aciduria. Three EPMEs based on maltodextrins with different DE (DE: 4.0–7.0 (I), 13.0–17.0 (II) and 16.5–19.5 (III)) were designed for the enantioanalysis of L-2-HGA [40]. The EPMEs can be used reliably for enantiopurity assay of L-2-HGA using the direct potentiometric method in the ranges of 10^{-9}–10^{-5}, 10^{-6}–10^{-3} and 10^{-8}–10^{-5} mol/L for the EPMEs based on maltodextrins I, II and III, respectively, with very low detection limits. A high reliability was obtained when the electrodes were used for the assay of L-2-HGA in urine samples.

EPMEs based on maltodextrins of different DE (DE: 4.0–7.0 (I), 13.0–17.0 (II), 16.5–19.5 (III)) as chiral selectors were used for the assay of S-perindopril [41]. The proposed electrodes can be successfully used in the assay of S-perindopril in the presence of R-perindopril, D-proline, PVC and inorganic ions such as Na^+, K^+ and Ca^{2+}. The best results were obtained by using maltodextrins I and II in the design of the EPMEs. Accordingly, these are the electrodes of choice for the enantioselective analysis of S-perindopril. If one compares the results obtained using these maltodextrins-based electrodes and those obtained by using substituted β-CD for the assay of S-perindopril, one can easily observe that there had been improvements in detection limits (lower with maltodextrins as chiral selectors), working concentration ranges, slope and enantioselectivity (higher especially with maltodextrin of lower DE as chiral selector).

L-Pipecolic acid is a marker for peroxisomal disorders. Three EPMEs were used for the enantioanalysis of L-pipecolic acid [42]. These electrodes are based on carbon paste impregnated with different maltodextrins (DE: 4.0–7.0 (I), 13.0–17.0 (II) and 16.5–19.5 (III), respectively) as chiral selectors and they can be used reliably for enantiopurity assay of L-pipecolic acid using the direct potentiometric method in the concentration ranges of 10^{-8}–10^{-3}, 10^{-8}–10^{-5} and 10^{-10}–10^{-6} mol/L for the maltodextrins I, II and III, respectively, based electrodes, with very low detection limits (magnitude orders of 10^{-9} for I and II, respectively, and 10^{-12} mol/L for III). The proposed electrodes can be successfully applied for the enantioanalysis of L-pipecolic acid in serum samples.

EPMEs based on carbon paste impregnated with the following maltodextrins, having dextrose equivalence (DE) 4.0–7.0 (I), 13.0–17.0 (II) and 16.5–19.5 (III) as chiral selectors were proposed for the assay of L-proline [43]. The response characteristics showed that the proposed electrodes could be reliably utilized in the assay of L-proline as raw material, with the best enantioselectivity and time-stability exhibited when the EPME based on maltodextrin I was used. The three EPMEs showed very low detection limits (of 10^{-9} mol/L magnitude order).

3.4.3 EPMEs based on antibiotics

Two EPMEs based on macrocyclic glycopeptide antibiotics—vancomycin and teicoplanin—were designed for the assay of acetyl-L-carnitine [44]. The linear concentration ranges for the proposed electrodes were 10^{-5}–10^{-2} mol/L for the vancomycin-based electrode and 10^{-4}–10^{-2} mol/L for the teicoplanin-based electrode, with slopes of 58.1 and 55.0 mV/p(acetyl-L-carnitine), respectively. The enantioselectivity was determined over D-carnitine.

Two EPMEs were proposed for the assay of R-baclofen. The electrodes were designed using macrocyclic glycopeptide antibiotic, teicoplanin [45]. Acetonitrile was added to the teicoplanin to design a modified teicoplanin-based electrode. The linear concentration ranges for the proposed enantioselective, membrane electrodes were 10^{-7}–10^{-4} mol/L for teicoplanin-based electrode and 10^{-6}–10^{-4} mol/L for the electrode based on teicoplanin modified with acetonitrile. The slopes of the electrodes were 60.0 and 57.2 mV/pR-baclofen for teicoplanin and teicoplanin modified with acetonitrile-based electrodes, respectively. The enantioselectivity was determined over S-baclofen. The proposed electrodes can be employed reliably for the assay of R-baclofen raw material and from its pharmaceutical formulation, Norton-Baclofen® tablets.

In order to determine the enantiopurity of L-carnitine, three EPMEs were proposed for the assay of L-carnitine [46]. The electrodes were designed using macrocyclic glycopeptide antibiotics: vancomycin and teicoplanin. Acetonitrile was added to the teicoplanine to design a modified teicoplanine-based electrode. The linear concentration ranges for the proposed enantioselective membrane electrodes were 10^{-4}–10^{-2} mol/L for electrodes based on vancomycin and teicoplanin and 10^{-5}–10^{-2} mol/L for electrode based on teicoplanin modified with acetonitrile. The slopes of the electrodes were 56.5 , 54.5 and 58.3 mV/pL-carnitine for vancomycin, teicoplanin and teicoplanin modified with acetonitrile-based electrodes, respectively. The enantioselectivity was determined over D-carnitine. The proposed electrodes could be employed reliably for the assay of L-carnitine raw material and its pharmaceutical formulation, Carnilean® capsules.

EPMEs based on carbon paste modified with antibiotics: vancomycin, teicoplanin and teicoplanin modified with acetonitrile are proposed for the determination of D-2-HGA [47]. The proposed electrodes can be used reliably for enantiopurity tests of D-2-HGA using direct potentiometric method of analysis. The linear concentration ranges recorded for EPMEs are 10^{-7}–10^{-3}, 10^{-7}–10^{-2} and 10^{-6}–10^{-2} mol/L with detection limits of 1.00×10^{-8}, 1.00×10^{-8} and 1.00×10^{-7} mol/L for the vancomycin, teicoplanin and teicoplanin modified with acetonitrile-based EPME, respectively. The selectivity was determined over L-2-HGA, creatine, creatinine and some inorganic cations. The proposed EPMEs were applied for the assay of D-2-HGA in urine samples. The duration of one analysis is 10 min, including the calibration of the instrument and the determination of the amount of D-2-HGA in the urine sample.

Three EPMEs based on macrocyclic glycopeptide antibiotics—vancomycin and teicoplanin (modified or not with acetonitrile)—were proposed for the determination of L- and D-enantiomers of methotrexate (Mtx) [48]. The linear concentration ranges for the proposed enantioselective membrane electrodes were between 10^{-6} and 10^{-3} mol/L for L- and D-Mtx. The slopes of the electrodes were 58.00 mV/pL-Mtx for vancomycin-based electrode, 57.60 mV/pD-Mtx for teicoplanin-based electrode and 55.40 mV/pD-Mtx for teicoplanin modified with acetonitrile-based electrode. The detection limits of the proposed electrodes were of 10^{-8} mol/L magnitude order. All proposed electrodes proved to be successful for the determination of the enantiopurity of Mtx as raw material and of its pharmaceutical formulations (tablets and injections).

An EPME based on carbon paste impregnated with macrocyclic antibiotic vancomycin as chiral selector was proposed for the enantioanalysis of S-perindopril raw material and from its pharmaceutical formulation [49]. The proposed electrode was applied for the assay of

S-perindopril raw material and from its pharmaceutical formulation (Coversyl® tablets). The limit of detection for the S-perindopril is 15 nmol/L, with a working concentration range of 5.5×10^{-8}–1.0×10^{-3} mol/L. Unlike the S-perindopril that showed near-Nernstian response (slope = 55.71 mV/decade), the electrode response for R-perindopril was found to be non-Nernstian (slope = –6.20 mV/decade). This result clearly indicates the preference of the electrode for the S-enantiomer. The proposed electrode was stable and reproducible over 2-week test period. The response of the electrode does not depend on the pH in the range 3.0–7.8. The values of pK_{pot} ($pK_{pot} = -\log K_{pot}$, where K_{pot} is the potentiometric selectivity coefficient) are much higher than 5 for both R-perindopril and D-proline and 3.24 and 1.15 for PVP and L-proline, respectively. These indicate that while PVP, R-perindopril and D-proline did not interfere L-proline interfered.

An EPME based on vancomycin was proposed for the assay of D-pipecolic acid [50]. The linear concentration range for the proposed enantioselective membrane electrode is 10^{-9}–10^{-6} mol/L with the slope of electrode function 60.2 mV/p(D-pipecolic acid). The enantioselectivity was determined over L-pipecolic acid. The proposed electrode could be reliably employed for the assay of D-pipecolic acid in serum samples.

EPMEs based on antibiotics are proposed for the enantioanalysis of L-vesamicol [51]. A carbon paste was modified with antibiotics (vancomycin, teicoplanin and teicoplanin modified with acetonitrile), as chiral selectors. The EPMEs based on antibiotics were reliably used for enantiopurity tests of L-vesamicol using the direct potentiometric technique. The following linear concentration ranges: $1.0 \times 10^{-6} - 1.0 \times 10^{-4}$, 1.0×10^{-6}–1×10^{-3} and 1×10^{-7}–1×10^{-2} mol/L; and detection limits: 1.1×10^{-7}, 9.6×10^{-8} and 3.6×10^{-8} mol/L were obtained for vancomycin, teicoplanin and teicoplanin modified with acetonitrile-based EPMEs, respectively. The proposed EPMEs were applied for the enantioanalysis of L-vesamicol in urine samples.

3.4.4 EPMEs based on crown ethers

A crown ether (19-[(10-undecen-1-yl)oxy]-4R,14R-(-)4,14-diphenyl-3,6,9,12,5-pentaoxa-21-azabicyclo[15.3.1]heneicosa-1(21),17,19-triene) was used in a plastic membrane as chiral selector for the design of EPMEs [21]. These electrodes can differentiate between S- and R-1-phenylethylammonium ions. The values of the slopes obtained when different plasticizers were used for EPMEs design were between 51.3 and 60.6 mV/decade of concentration proving the suitability of these electrodes for potentiometric measurements.

3.4.5 EPMEs based on fullerenes

Novel EPMEs based on carbon paste impregnated with (1,2-methanofullerene C60)-61-carboxylic acid (I), diethyl (1,2-methanofullerene C60)-61-61-dicarboxylate (II) and tert-butyl (1,2-methanofullerene C60)-61-carboxylic acid (III) were designed for the assay of S-clenbuterol raw material and in serum samples [52]. All EPMEs showed near-Nernstian responses (56.8, 57.0 and 57.1 mV/decade of concentration) for S-Clen, with correlation coefficients for the equations of calibration of 0.9996 (I), 0.9998 (II) and 0.9998 (III), respectively, and low detection limits (of 10^{-10} and 10^{-11} mol/L magnitude order). R-Clen, on the other hand, showed non-Nernstian response. All electrodes displayed good stability and reproducibility over 1-month test period, when used every day for measurements (RSD $<$ 0.1%).

[5,6] fullerene-C_{70} (I) and diethyl (1,2-methanofullerene C_{70})-71,71-dicarboxylate (II) were proposed as new chiral selectors for the design of EPMEs [53]. L-Proline was selected as model analyte for the study of the behaviour of these electrodes. Both electrodes exhibited a cation-enantioselective behaviour. The best response was obtained when diethyl (1,2-methanofullerene C_{70})-71,71-dicarboxylate was used for the electrode's construction (59.4 mV/decade of concentration). The best enantioselectivity was achieved for the [5,6] fullerene-C_{70} based electrode ($pK_{pot} = 3.85$ over D-proline).

3.5 CONCLUSIONS

The EPMEs are a good alternative for chromatographic methods in enantioanalysis. There are advantages over the amperometric biosensors and immunosensors, e.g., higher lifetime and larger working concentration range.

The selection of chiral selector and matrix for the design of these electrodes are the keys for a successful enantioanalysis using EPMEs. A validation of these electrodes will require a reliable design, because only a reliable design will assure the reliability of the analytical information obtained using these electrodes. So far, the most reliable design was proved to be the one based on carbon paste.

The mechanism of potential development is similar with the one proposed earlier by Stefan and Aboul-Enein, as the equilibrium responsible for the potential development is the same: complexation–de-complexation. The slope of the electrodes is directly dependent on the stability constant of the complex formed at membrane–solution interface. Accordingly, the selection of the best chiral selector may be done using the values of these stability constants.

Enantioanalysis is essential for substances with a chiral moiety from the body and pharmaceutical compounds. Utilization of EPMEs in clinical and pharmaceutical enantioanalysis improves the reliability of the analysis, shortens the time of analysis and minimizes its cost.

REFERENCES

1 H.Y. Aboul-Enein and I. Ali, *Chiral Separations by Liquid Chromatography and Related Technologies*, Marcel Dekker, New York, 2003.
2 R.I. Stefan, J.F. van Staden and H.Y. Aboul-Enein, *Electrochemical Sensors in Bioanalysis*, Marcel Dekker, New York, 2001.
3 H.Y. Aboul-Eein and W. Wainer, *The Impact of Stereochemistry on Drug Development and Use*, Wiley, Germany, 1997.
4 K.M. Rentsch, *J. Biochem. Biophys. Methods*, 54 (2002) 1.
5 A.C. Sewell, M. Heil, F. Podebrad and A. Mosandl, *Eur. J. Pediatr.*, 157 (1998) 185.
6 R.I. Stefan, G.L. Radu, H.Y. Aboul-Enein and G.E. Baiulescu, *Curr. Trends Anal. Chem.*, 1 (1998) 135.
7 R.I. Stefan, J.F. van Staden and H.Y. Aboul-Enein, *Electroanalysis*, 11 (1999) 1233.
8 H.Y. Aboul-Enein, J.F. van Staden and R.I. Stefan, *Anal. Lett.*, 32 (1999) 623.
9 R.I. Stefan and H.Y. Aboul-Enein, *J. Immunoassay Immunochem.*, 23 (2002) 429.
10 R.I. Stefan, H.Y. Aboul-Enein and J.F. van Staden. In: H. Baltes, G.K. Fedder and G. Korvink (Eds.), *Enantioselective Electrochemical Sensors in Sensors Update*, Vol. 10, Wiley-VCH, Weinheim, Germany, 2001.
11 R.I. Stefan and H.Y. Aboul-Enein, *Instrum. Sci. Technol.*, 25 (1997) 169.
12 E. Pungor, *Anal. Method Instrum.*, 1 (1993) 52.
13 H.M. Widmer, *Anal. Method Instrum.*, 1 (1993) 60.
14 R.I. Stefan and H.Y. Aboul-Enein, *Instrum. Sci. Technol.*, 27 (1998) 105.
15 D. Parker and R. Kataky, *Chem. Commun.*, 2 (1997) 141.
16 P.S. Bates, R. Kataky and D. Parker, *J. Chem. Perkin Trans.*, 2 (1994) 669.
17 T.D. Booth, K. Azzaoui and I.W. Wainer, *Anal. Chem.*, 69 (1997) 3879.
18 Y. Yasaka, T. Yamamoto, K. Kimura and T. Shono, *Chem. Lett.*, 20 (1980) 769.

19 T. Shinbo, T. Yamaguchi, K. Sakaki, H. Yanagishita, D. Kitamoto and M. Sugiura, *Chem. Express.*, 7 (1992) 781.
20 W. Bussmann, J.M. Lehn, U. Oesch, P. Plumere and W. Simon, *Helv. Chim. Acta.*, 64 (1981) 657.
21 V. Horvath, T. Takacs, G. Horvai, P. Huszthy, J.S. Bradshaw and R.M. Izatt, *Anal. Lett.*, 30 (1997) 1591.
22 R.I. Stefan-van Staden and A.A. Rat'ko, *Talanta*, 69 (2006) 1049.
23 R.I. Stefan, J.F. van Staden and H.Y. Aboul-Enein, *Talanta*, 48 (1999) 1139.
24 R.I. Stefan, J.F. van Staden and H.Y. Aboul-Enein, *Electroanalysis*, 11 (1999) 192.
25 H.Y. Aboul-Enein, R.I. Stefan and J.F. van Staden, *Analusis*, 27 (1999) 53.
26 R. Kataky, P.S. Bates and D. Parker, *Analyst*, 117 (1992) 1313.
27 R.I. Stefan-van Staden, R.M. Nejem, J.F. van Staden and H.Y. Aboul-Enein, *Anal. Lett.*, 38 (2005) 1847.
28 R.I. Stefan, J.F. van Staden and H.Y. Aboul-Enein, *Chirality*, 11 (1999) 631.
29 K.I. Ozoemena, R.I. Stefan, J.F. van Staden and H.Y. Aboul-Enein, *Sens. Act. B*, 105 (2005) 425.
30 R.I. Stefan, J.F. van Staden and H.Y. Aboul-Enein, *Anal. Lett.*, 31 (1998) 1787.
31 K.I. Ozoemena and R.I. Stefan, *Talanta*, 66 (2005) 501.
32 C.J. Sun, X.X. Sun and H.Y. Aboul-Enein, *Anal. Lett.*, 37 (2004) 2259.
33 R.I. Stefan, J.F. van Staden, G.E. Baiulescu and H.Y. Aboul-Enein, *Chem. Anal.*, 44 (1999) 417.
34 R.I. Stefan-van Staden and R.M. Nejem, *Sens. Act. B*, 117 (2006) 123.
35 R. Kataki, D. Parker and P.M. Kelly, *Scand. J. Clin. Invest.*, 55 (1995) 409.
36 R.I. Stefan, J.F. van Staden and H.Y. Aboul-Enein, *Fresenius J. Anal. Chem.*, 370 (2001) 33.
37 A.A. Rat'ko and R.I. Stefan-van Staden, *Il Farmaco*, 59 (2004) 993.
38 R.I. Stefan-van Staden, R.G. Bokretsion and K.I. Ozoemena, *Anal. Lett.*, 39 (2006) 1065.
39 R.I. Stefan and R.M. Nejem, *Sens. Act. B*, 106 (2005) 736.
40 R.M. Nejem, R.I. Stefan, J.F. van Staden and H.Y. Aboul-Enein, *Talanta*, 65 (2005) 437.
41 K.I. Ozoemena, R.I. Stefan, J.F. van Staden and H.Y. Aboul-Enein, *Talanta*, 62 (2004) 681.
42 R.I. Stefan, R.M. Nejem, J.F. van Staden and H.Y. Aboul-Enein, *Electroanalysis*, 20 (2004) 1730.
43 K.I. Ozoemena and R.I. Stefan, *Sens. Act. B*, 98 (2004) 97.
44 A.A. Rat'ko, R.I. Stefan, J.F. van Staden and H.Y. Aboul-Enein, *Instrum. Sci. Technol.*, 32 (2004) 601.
45 A.A. Rat'ko and R.I. Stefan, *Anal. Lett.*, 37 (2004) 3161.

46 A.A. Rat'ko, R.I. Stefan, J.F. van Staden and H.Y. Aboul-Enein, *Talanta*, 63 (2004) 515.
47 R.I. Stefan, R.M. Nejem, J.F. van Staden and H.Y. Aboul-Enein, *Sens. Act. B*, 106 (2005) 791.
48 A.A. Rat'ko, R.I. Stefan, J.F. van Staden and H.Y. Aboul-Enein, *Talanta*, 64 (2004) 145.
49 K.I. Ozoemena, R.I. Stefan, J.F. van Staden and H.Y. Aboul-Enein, *Instrum. Sci. Technol.*, 32 (2004) 371.
50 A.A. Rat'ko, R.I. Stefan, J.F. van Staden and H.Y. Aboul-Enein, *Sens. Act. B*, 99 (2004) 546.
51 R.I. Stefan-van Staden and R.M. Nejem, *Anal. Lett.*, 39 (2006) 675.
52 R.I. Stefan-van Staden and B. Lal, *Anal. Lett.*, 39 (2006) 1311.
53 R.I. Stefan, *Sens. Lett.*, 1 (2003) 71.
54 S.Z. Zhan, Q. Dai, C.W. Yuan, Z.H. Lu and L. Haeussling, *Anal. Lett.*, 32 (1999) 677.

Chapter 4

Ion sensors with conducting polymers as ion-to-electron transducers

Johan Bobacka and Ari Ivaska

4.1 INTRODUCTION

Potentiometric ion sensors, or ion-selective electrodes (ISEs), are one of the oldest and most successful electrochemical sensors with a large number of applications, especially in clinical analysis [1–3]. The potentiometric measurement technique is attractive for practical applications because it allows the use of small-size, portable, low-energy consumption and relatively low-cost instrumentation. Recent developments of ISEs with extremely low detection limits [4] tend to expand the applications of potentiometry towards trace analysis [5]. Furthermore, the development of solid-state ISEs without internal filling solution gives durable and maintenance-free ion sensors that can easily be miniaturized, if necessary [6]. The discovery and development of conducting polymers, starting from polyacetylene in the mid-1970s, was recognized by the Nobel Prize in Chemistry in the year 2000 [7–10]. Today, conducting polymers have become one of the most important ion-to-electron transducers in solid-state ISEs.

4.1.1 Principle of ion-to-electron transduction in conventional ISEs

An ISE is composed of an ion-selective membrane and an ion-to-electron transducer. In the conventional ISE, the ion-to-electron transduction takes place at the internal reference electrode that is immersed in the inner filling solution. In the case of the commonly used Ag/AgCl reference electrode immersed in an electrolyte solution containing Cl^- anions, the ion-to-electron transduction process can be written as follows:

$$Ag + Cl^- \rightleftharpoons Ag^+Cl^- + e^- \qquad (4.1)$$

Comprehensive Analytical Chemistry 49
Alegret and Merkoçi (Eds)
Volume 49 ISSN: 0166-526X DOI: 10.1016/S0166-526X(06)49004-8

The inner filling solution also contains the ion (primary ion) for which the ion-selective membrane is selective, which assures a reversible ionic equilibrium between the inner filling solution and the ion-selective membrane, resulting in a stable standard potential of the ISE. In order to construct a solid-state ISE with a stable standard potential, the internal reference electrode and inner filling solution should be replaced by a solid material with suitable redox and ion-exchange properties [6]. Conducting polymers belong to this group of materials.

4.1.2 Principle of ion-to-electron transduction in conducting polymer-based ISEs

Conducting polymers based on polymer chains with conjugated double bonds are electroactive materials that have found widespread use also in the field of chemical sensors [11–41]. Oxidation of the conjugated polymer backbone is accompanied by anion insertion or cation expulsion, as follows:

$$P + A^- \rightleftharpoons P^+A^- + e^- \tag{4.2}$$

$$PA^-M^+ \rightleftharpoons P^+A^- + M^+ + e^- \tag{4.3}$$

where P = neutral conducting polymer unit, P^+ = oxidized conducting polymer unit, A^- = anion, M^+ = cation and e^- = electron. Reactions (4.2) and (4.3) describe two limiting cases where either anions or cations are mobile, respectively. The similarity between reactions (4.1) and (4.2) immediately suggests that conducting polymers can work as ion-to-electron transducers in ISEs. Today, conducting polymers are frequently used as ion-to-electron transducers in solid-state ISEs [42].

4.1.3 Design and fabrication of conducting polymer-based ISEs

Most conducting polymer films used as ion-to-electron transducers in ISEs are fabricated by electropolymerization of the monomers on solid electronic conductors with high work function such as platinum, gold or glassy carbon. Ion-recognition sites can be incorporated into the conducting polymer film as counterions during electropolymerization. Alternatively, ion receptors can be covalently bound to the monomer prior to polymerization. Some conducting polymers may be dissolved or dispersed in organic or aqueous solutions, which allows film preparation from solution as well. In this case ion receptors may be dissolved together with the conducting polymer or the ion receptors may be covalently bound to the conducting polymer chains before film casting. A conducting polymer film containing specific ion receptors is an elegant approach towards durable solid-state ISEs [37,42]. Covalent binding of ion-recognition sites to the conducting polymer backbone allows integration of ion-recognition and signal transduction within a single (macro)molecule, which may be of importance for the construction of durable nanosensors. However, since the conducting polymers are electronically conducting, such ISEs usually also show some redox sensitivity, which may limit their practical usefulness in some applications.

The redox sensitivity is effectively eliminated by coating the conducting polymer with a conventional ion-selective membrane. In this case the conducting polymer works only as ion-to-electron transducer and therefore it does not have to be selective to any particular ion, which greatly facilitates the choice of conducting polymer. The conducting polymer is typically coated with a conventional solvent polymeric ion-selective membrane via solution casting. This procedure can be used to fabricate a large variety of solid-contact ISEs [43]. Ion-selective membranes based on plasticized poly(vinyl chloride) (PVC) containing suitable ionophores and additives are most commonly used, but membranes based on acrylic polymers, silicone rubber and polyurethane can be applied as well. The solubility of the conducting polymer in the solvent used to cast the ion-selective membrane determines to what extent the conducting polymer transducer is mixed with the ion-selective membrane. An insoluble conducting polymer results in a solid-contact ISE where the ion-to-electron transducer forms a well-defined intermediate layer between the ion-selective membrane and the electronic conductor so that the ion-selectivity of the electrode is determined solely by the outer ion-selective membrane. On the other hand, complete mixing of the ion-to-electron transducer and the ion-selective membrane results in a so-called single-piece electrode, where the conducting polymer may also influence the selectivity of the ISE [44].

At the interface between the conducting polymer and the ion-selective membrane there is an ionic equilibrium. However, any water present at this interface results in a small volume of internal electrolyte between the conducting polymer and the ion-selective membrane. Variations in the ion concentration of this internal electrolyte can influence the potential of the ISE. The presence of water depends on the conducting polymer and the ion-selective membrane, which should be taken into account in the electrode design.

4.1.4 Conducting polymers used as ion-to-electron transducers

Polypyrrole was the first conducting polymer used as ion-to-electron transducer in solid-state ISEs [43], and is still one of the most frequently used [45–68]. Other conducting polymers that have been applied as ion-to-electron transducers in solid-state ISEs include poly(1-hexyl-3,4-dimethylpyrrole) [69,70], poly(3-octylthiophene) [44,70–74], poly(3,4-ethylenedioxythiophene) [75–86], poly(3-methylthiophene) [87], polyaniline [44,67,73,88–99], polyindole [100,101], poly(α-naphthylamine) [102], poly(*o*-anisidine) [67] and poly(*o*-aminophenol) [103]. The monomer structures are shown in Fig. 4.1.

4.1.5 Conducting polymers used as ion-selective membranes

In some cases, conducting polymer-based ion-selective membranes can substitute the commonly used plasticized PVC-based ion-selective (liquid) membranes. However, there are fundamental differences between the two types of membranes. Conducting polymer membranes are most often non-plasticized and the charge is transported by both

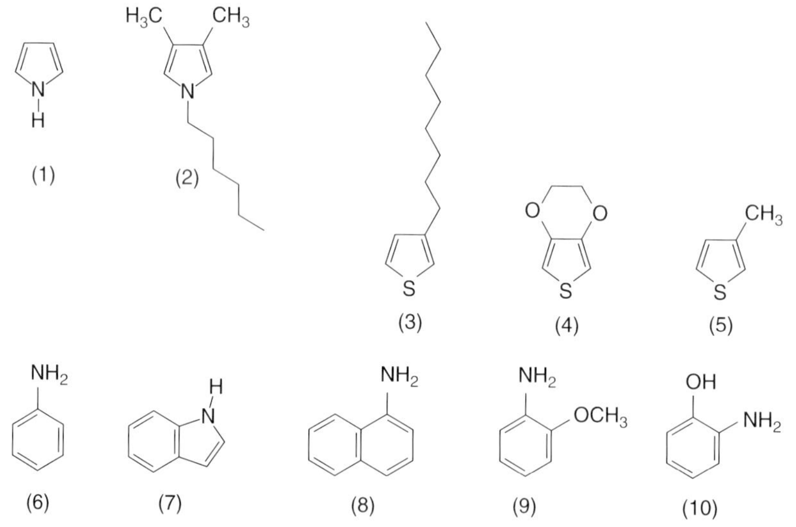

Fig. 4.1. Monomer units of conducting polymers that have been applied as ion-to-electron transducers in solid-state ISEs: (**1**) pyrrole, (**2**) 1-hexyl-3,4-dimethylpyrrole, (**3**) 3-octylthiophene, (**4**) 3,4-ethylenedioxythiophene, (**5**) 3-methylthiophene, (**6**) aniline, (**7**) indole, (**8**) α-naphthylamine, (**9**) *o*-anisidine, (**10**) *o*-aminophenol.

electrons and ions, in contrast to plasticized PVC membranes that are purely ionic conductors. The electronic conductivity ensures reversible electron transfer (electronic equilibrium) between the conducting polymer and the electronic conductor. However, this also means that conducting polymer membranes are redox sensitive, which can have a major influence on the analytical performance of conducting polymer-based potentiometric ion sensors. In the conducting polymer membrane, the ions and ion-recognition sites are embedded in a (charged) polymer network which may restrict the movement of bulky ion-recognition compounds even if they are not covalently bound to the conducting polymer backbone. In the plasticized PVC membrane, on the other hand, the ions and ion-recognition sites are dissolved in the liquid phase (plasticizer) of the membrane and even bulky ionophores are mobile inside the membrane. The mobility of the ionophore (and lipophilic counterions) in the ion-selective membrane may have a significant impact on the ion–ionophore interactions at the polymer/solution interface. Despite such differences, conducting polymers can be applied also as ion-selective membranes, in addition to their use as ion-to-electron transducers.

4.2 APPLICATION

Solid-state ion sensors with conducting polymers as ion-to-electron transducers (and sensing membranes) offer some advantages over conventional liquid-contact ISEs. Solid-state ISEs without internal filling solution are more durable, require less maintenance, are easier to miniaturize, and allow great flexibility in electrode design and fabrication.

4.2.1 Analytical performance of solid-contact ISEs

When the conducting polymer is used as ion-to-electron transducer in the form of an intermediate layer between the electronic conductor and the ion-selective membrane it does not significantly influence the sensitivity and selectivity of the ISE, but it allows high potential stability [75]. For example, microfabricated solid-state K^+-ISEs with polypyrrole as ion-to-electron transducer was found to show even better long-term potential stability than those based on a hydrogel contact [58]. The potential of the polypyrrole-based K^+-ISE was slightly more sensitive to the oxygen concentration of the sample in comparison to the hydrogel-based K^+-ISE. However, both types of solid-state ISEs were found to be suitable for high-precision potentiometric measurements under well-controlled experimental conditions [58]. Furthermore, the sensitivity of this type of solid-contact ISE to dissolved O_2 and CO_2 (pH) can be decreased by replacing polypyrrole with, e.g., poly(3,4-ethylenedioxythiophene) as ion-to-electron transducer [78]. In another example, K^+-selective microelectrodes using polypyrrole as solid contact show similar selectivity and sensitivity as conventional ISEs and were found to be suitable as detectors in micro-flow systems and as potentiometric tips in scanning electrochemical microscopy (SECM) [51]. Flow-injection analysis of K^+ in the concentration range 10^{-5}–10^{-1} M using a solid-state K^+-ISE as detector allow rapid analysis (150 samples/h) with potential applications in food analysis and clinical analysis [50]. Solid-contact Ca^{2+}-ISEs with poly(3,4-ethylenedioxythiophene) as ion-to-electron transducer have proven useful for the determination of Ca^{2+} in wood pulp (see Procedure 4). Further information on recent progress in the area of conducting polymer-based solid-state ISEs is available in the literature [42].

4.2.2 Conducting polymer-based pH sensors

Solid-contact pH sensors can be constructed by using polypyrrole [45,59] or polyaniline [92,96] as ion-to-electron transducer in combination with pH-selective membranes based on plasticized PVC [45,59,92,96]. The dynamic pH range of the sensors depend on the pH ionophore used in the plasticized PVC membranes, as follows: tri-*n*-dodecylamine (pH 2–12) [45], tris(2-phenylethyl)amine (pH 4.5–12.6) [59], tris(3-phenylpropyl)amine (pH 4.6–13.2) [59], tribenzylamine (pH 2.5–11.2) [92,96], dibenzylnaphtalenemethylamine (pH 0.65–10.0) [96], dibenzylpyrenemethyl-amine (pH 0.50–10.2) [96]. Suggested applications include pH measurements in body fluids such as serum [45,96], whole blood [92], and cow milk [59].

The acid–base properties of polyaniline can be utilized to produce solid-state pH sensors where polyaniline works both as the pH-sensitive material and as the ion-to-electron transducer. An excellent example is the electrodeposition of polyaniline on an ion-beam etched carbon fiber with a tip diameter of ca. 100–500 nm resulting in a solid-state pH nanoelectrode with a linear response (slope ca. −60 mV/pH unit) in the pH range of 2.0–12.5 and a working lifetime of 3 weeks [104]. The response time vary from ca. 10 s (around pH 7) to ca. 2 min (at pH 12.5).

The analytical performance of the pH nanoelectrode compares favorably with commercial glass pH electrodes when applied for pH measurements of body fluids (serum, urine) and low ionic strength water samples (rain water, tap water, deionized water) [104].

4.2.3 Polypyrrole-based nitrate sensor

The ionic response of conducting polymers is generally non-selective unless specific ion-recognition sites are included in the conducting polymer membrane [37]. However, there are some exceptions to this rule, such as the polypyrrole-based nitrate sensor [105]. Electrosynthesis of polypyrrole in presence of $NaNO_3$ gives rise to a solid-state NO_3^- sensor that can be applied for the determination of NO_3^- in different water samples, including tap water [106]. The sensor shows an appreciable selectivity to NO_3^- in comparison to several common anions. The logarithm of selectivity coefficients ($\log K^{pot}_{NO_3^-,j}$) determined by fixed interference method are as follows: −0.4 (SCN^-), −1.1 (Br^-), −1.2 (ClO_4^-), −1.2 (Cl^-), −1.3 (I^-), −1.3 (salicylate), −3.0 (PO_4^{3-}), −3.2 (SO_4^{2-}) [105]. The selectivity to nitrate is thought to originate from a process similar to "molecular imprinting" [105]. When the polypyrrole-based nitrate sensors were stored in the light, the useful lifetime was only about 9 days. However, when stored in the dark, the slope of the nitrate sensors remained unchanged for at least 55 days [105].

4.2.4 Sensors for surfactants

Solid-state sensors for anionic surfactants can be constructed by using polyaniline as sensing membrane [107,108], and by using polypyrrole as ion-to-electron transducer in combination with plasticized PVC as sensing membranes [53,66]. The sensors may be applied for the determination of dodecylsulfate in, e.g., mouth-washing solution and tap water [107], and for the determination of dodecylbenzenesulfonate in detergents [66,108]. Solid-state surfactant sensors allow a sample rate of 30 samples/h, when applied in flow-injection analysis [53].

4.2.5 Pharmaceutical analysis

Solid-contact ISEs with conducting polymers as ion-to-electron transducers and plasticized PVC-based sensing membranes may be applied to the analysis of pharmaceutical products, such as clenbuterol [89], salbutamol [90], dimedrol [102], and chlordiazepoxide [103]. In these applications, the ion-to-electron transducers are based on polyaniline [89,90], poly(α-naphthylamine) [102], and poly(*o*-aminophenol) [103].

4.2.6 ISEs for measurements in non-aqueous solvents

The use of ISEs with ion-selective membranes based on plasticized PVC, as well as glass pH electrodes, is limited to the analysis of aqueous solutions. On the other hand, sensors based on conducting polymer membranes are usually insoluble in organic solvents, which extends the range of possible applications. Electrosynthesized polypyrrole doped with calcion works as a Ca^{2+} sensor that can be applied as indicator electrode in the titration of Ca^{2+} with NaF in mixed solvents, such as water–methanol (1:1) and water–ethanol (1:1) [52]. Another example is the use of polyaniline as indicator electrode in order to follow the acid–base precipitation titration of trimeprazine base with tartaric acid in isopropanol solution (see Procedure 5).

4.2.7 Solid-state ion sensors with low detection limits

Solid-state ISEs with conducting polymers are also promising for low-concentration measurements [60,63,74], even below nanomolar concentrations [60,74], which gives rise to optimism concerning future applications of such electrodes. In principle, the detection limit can be improved by reducing the flux of primary ions from the ion-selective membrane (or conducting polymer) to the sample solution, e.g., via complexation of primary ions in the solid-contact material. For example, a solid-state Pb^{2+}-ISEs with poly(3-octylthiophene) as ion-to-electron transducer coated with an ion-selective membrane based on poly(methyl methacrylate)/poly(decyl methacrylate) was found to show detection limits in the subnanomolar range and a faster response at low concentrations than the liquid-contact ISE [74].

4.2.8 Conducting polymer-based reference electrode

In order to fully utilize solid-state ion sensors in potentiometric measurements there is a need for a durable and reliable solid-state reference electrode as well. It was shown recently that it is possible to construct a junctionless solid-state reference electrode based on conducting polymers doped with pH-buffering ligands. The conducting polymer is doped with a large immobile anion that at the same time is the base of a pH buffer pair. After the conducting polymer is soaked in pH buffer solution, the electrode potential is determined mainly by interactions with protons. However, due to the pH buffering effect, the potential of this conducting polymer electrode is practically constant in a certain pH range [109]. The electrode is easy to manufacture, maintenance-free, durable, easy to miniaturize, and inexpensive [109].

4.3 CONCLUSIONS

Conducting polymers have been studied as potentiometric ion sensors for almost two decades and new sensors are continuously developed. The analytical performance of solid-state ion sensors with conducting polymers as ion-to-electron transducer (solid-contact ISEs) has been significantly improved over the last few years. Of particular interest is the large improvement of the detection limit of such solid-contact ISEs down to the nanomolar level. Further optimization of the solid contacts as well as the ion-selective membranes will most certainly extend the range of practical applications.

Solid-state ion sensors with conducting polymers as sensing membranes have also proved useful in some applications. Of particular importance are the pH sensors based on polyaniline that can be also applied in non-aqueous solutions. Polypyrrole-based sensors for nitrate also show great promise for water analysis. However, in addition to these two excellent examples, a large number of functionalized conducting polymers have been synthesized already, and these materials may offer unique possibilities for fabrication of durable, miniaturized ion sensors.

ACKNOWLEDGMENTS

Financial support from the National Technology Agency (TEKES) and the Academy of Finland is gratefully acknowledged. This work is part of the activities at the Åbo Akademi Process Chemistry Group within the Finnish Centre of Excellence Programme (2000–2011) by the Academy of Finland.

REFERENCES

1 E. Bakker, P. Bühlmann and E. Pretsch, *Chem. Rev.*, 97 (1997) 3083–3132.
2 P. Bühlmann, E. Pretsch and E. Bakker, *Chem. Rev.*, 98 (1998) 1593–1687.
3 E. Bakker, D. Diamond, A. Lewenstam and E. Pretsch, *Anal. Chim. Acta*, 393 (1999) 11–18.
4 T. Sokalski, A. Ceresa, T. Zwickl and E. Pretsch, *J. Am. Chem. Soc.*, 119 (1997) 11347–11348.
5 E. Bakker and E. Pretsch, *Trends Anal. Chem.*, 24 (2005) 199–207.
6 B.P. Nikolskii and E.A. Materova, *Ion-Sel. Electrode Rev.*, 7 (1985) 3–39.
7 H. Shirakawa, E.J. Louis, A.G. MacDiarmid, C.K. Chiang and A.J. Heeger, *J. Chem. Soc. Chem. Commun.* (1977) 578–580.
8 H. Shirakawa, *Synth. Met.*, 125 (2002) 3–10.
9 A.J. MacDiarmid, *Synth. Met.*, 125 (2002) 11–22.
10 A.J. Heeger, *Synth. Met.*, 125 (2002) 23–42.
11 A. Ivaska, *Electroanalysis*, 3 (1991) 247–254.
12 M.D. Imisides, R. John, P.J. Riley and G.G. Wallace, *Electroanalysis*, 3 (1991) 879–889.
13 G. Bidan, *Sens. Actuators B*, 6 (1992) 45–56.
14 G. Zotti, *Synth. Met.*, 51 (1992) 373–382.
15 P.N. Bartlett and P.R. Birkin, *Synth. Met.*, 61 (1993) 15–21.
16 P.R. Teasdale and G.G. Wallace, *Analyst*, 118 (1993) 329–334.
17 P.N. Bartlett and J.M. Cooper, *J. Electroanal. Chem.*, 362 (1993) 1–12.
18 M. Josowicz, *Analyst*, 120 (1995) 1019–1024.
19 S.A. Emr and A.M. Yacynych, *Electroanalysis*, 7 (1995) 913–923.
20 O.A. Sadik, *Anal. Methods Instrum.*, 2 (1996) 293–301.
21 J.N. Barisci, C. Conn and G.G. Wallace, *TRIP*, 4 (1996) 307–311.
22 S.B. Adeloju and G.G. Wallace, *Analyst*, 121 (1996) 699–703.
23 L. Rover Jr., G. De Oliveira Neto and L.T. Kubota, *Quim. Nova*, 20 (1997) 519–527 (in Portuguese).
24 W. Göpel and K.-D. Schierbaum. In: H.S. Nalwa (Ed.), *Handbook of Organic Conductive Molecules and Polymers, Vol. 4. Conductive Polymers: Transport, Photophysics and Applications,* Wiley, Chichester, 1997, pp. 621–659.
25 B. Fabre and J. Simonet, *Coord. Chem. Rev.*, 178–180 (1998) 1211–1250.
26 A. Giuseppi-Elie, G.G. Wallace and T. Matsue. In: T.A. Skotheim, R.L. Elsenbaumer and J.R. Reynolds (Eds.), *Handbook of Conducting Polymers*, 2nd ed., Marcel Dekker, New York, 1998, pp. 963–991.
27 T.W. Lewis, G.G. Wallace and M.R. Smyth, *Analyst*, 124 (1999) 213–219.
28 G.G. Wallace, M. Smyth and H. Zhao, *Trends Anal. Chem.*, 18 (1999) 245–251.

29 D.T. McQuade, A.E. Pullen and T.M. Swager, *Chem. Rev.*, 100 (2000) 2537–2574.
30 E. Palmisano, P.G. Zambonin and D. Centonze, *Fresenius J. Anal. Chem.*, 366 (2000) 586–601.
31 L.A.P. Kane-Maguire and G.G. Wallace, *Synth. Met.*, 119 (2001) 39–42.
32 B. Fabre. In: H.S. Nalwa (Ed.), *Handbook of Advanced Electronic and Photonic Materials and Devices, Vol. 8 Conducting Polymers,* Academic Press, San Diego, 2001, pp. 103–129.
33 M. Leclerc. In: H. Baltes, W. Göpel, J. Hesse (Eds.), *Sensors Update*, Vol. 8, Wiley-VCH, Weinheim, 2001, pp. 21–38.
34 A. Ramanaviciene and A. Ramanavicius, *Crit. Rev. Anal. Chem.*, 32 (2002) 245–252.
35 J. Janata and M. Josowicz, *Nat. Mater.*, 2 (2003) 19–24.
36 M. Trojanowicz, *Mikrochim. Acta*, 143 (2003) 75–91.
37 J. Bobacka, A. Ivaska and A. Lewenstam, *Electroanalysis*, 15 (2003) 366–374.
38 J. Bobacka, T. Lindfors, A. Lewenstam and A. Ivaska, *Am. Lab.*, 36 (2004) 13–20.
39 B. Adhikari and S. Majumdar, *Prog. Polym. Sci.*, 29 (2004) 699–766.
40 K.C. Persaud, *Mater. Today*, April (2005) 38–44.
41 J. Bobacka. In: C.A. Grimes, E.C. Dickey and M.V. Pishko (Eds.), *Encyclopedia of Sensors*, Vol. 2, American Scientific Publishers, California, USA, 2006, pp. 279–294.
42 J. Bobacka, *Electroanalysis*, 18 (2006) 7–18.
43 A. Cadogan, Z. Gao, A. Lewenstam, A. Ivaska and D. Diamond, *Anal. Chem.*, 64 (1992) 2496–2501.
44 J. Bobacka, T. Lindfors, M. McCarrick, A. Ivaska and A. Lewenstam, *Anal. Chem.*, 67 (1995) 3819–3823.
45 A. Michalska, A. Hulanicki and A. Lewenstam, *Analyst*, 119 (1994) 2417–2420.
46 A. Hulanicki and A. Michalska, *Electroanalysis*, 7 (1995) 692–693.
47 A. Michalska, A. Hulanicki and A. Lewenstam, *Microchem. J.*, 57 (1997) 59–64.
48 T. Momma, S. Komaba, M. Yamamoto, T. Osaka and S. Yamauchi, *Sens. Actuators B*, 24–25 (1995) 724–728.
49 T. Momma, M. Yamamoto, S. Komaba and T. Osaka, *J. Electroanal. Chem.*, 407 (1996) 91–96.
50 S. Komaba, J. Arakawa, M. Seyama, T. Osaka, I. Satoh and S. Nakamura, *Talanta*, 46 (1998) 1293–1297.
51 R.E. Gyurcsányi, A.-S. Nybäck, K. Tóth, G. Nagy and A. Ivaska, *Analyst*, 123 (1998) 1339–1344.
52 T. Blaz, J. Migdalski and A. Lewenstam, *Talanta*, 52 (2000) 319–328.
53 B. Kovács, B. Csóka, G. Nagy and A. Ivaska, *Anal. Chim. Acta*, 437 (2001) 67–76.

54 H. Kaden, H. Jahn, M. Berthold, K. Jüttner, K.-M. Mangold and S. Schäfer, *Chem. Eng. Technol.*, 24 (2001) 1120–1124.
55 P.C. Pandey, G. Singh and P.K. Srivastava, *Electroanalysis*, 14 (2002) 427–432.
56 R. Zielińska, E. Mulik, A. Michalska, S. Achmatowicz and M. Maj-Żurawska, *Anal. Chim. Acta*, 451 (2002) 243–249.
57 A. Michalska, J. Dumańska and K. Maksymiuk, *Anal. Chem.*, 75 (2003) 4964–4974.
58 R.E. Gyurcsányi, N. Rangisetty, S. Clifton, B.D. Pendley and E. Lindner, *Talanta*, 63 (2004) 89–99.
59 W.-S. Han, S.-J. Yoo, S.-H. Kim, T.-K. Hong and K.-C. Chung, *Anal. Sci.*, 19 (2003) 357–360.
60 A. Konopka, T. Sokalski, A. Michalska, A. Lewenstam and M. Maj-Żurawska, *Chem. Sens. Suppl. B*, 20 (2004) 472–473.
61 A. Konopka, T. Sokalski, A. Michalska, A. Lewenstam and M. Maj-Żurawska, *Anal. Chem.*, 76 (2004) 6410–6418.
62 A. Michalska and K. Maksymiuk, *J. Electroanal. Chem.*, 576 (2005) 339–352.
63 A.J. Michalska, C. Appaih-Kusi, L.Y. Heng, S. Walkiewicz and E.A.H. Hall, *Anal. Chem.*, 76 (2004) 2031–2039.
64 H. Kaden, H. Jahn and M. Berthold, *Solid State Ionics*, 169 (2004) 129–133.
65 W. Vonau, J. Gabel and H. Jahn, *Electrochim. Acta*, 50 (2005) 4981–4987.
66 L. Shafiee-Dastjerdi and N. Alizadeh, *Anal. Chim. Acta*, 505 (2004) 195–200.
67 J.E. Zachara, R. Toczyłowska, R. Pokrop, M. Zagórska, A. Dybko and W. Wróblewski, *Sens. Actuators B*, 101 (2004) 207–212.
68 R. Toczyłowska, R. Pokrop, A. Dybko and W. Wróblewski, *Anal. Chim. Acta*, 540 (2005) 167–172.
69 I.T. Kim, S.W. Lee and R.L. Elsenbaumer, *Synth. Met.*, 141 (2004) 301–306.
70 F. Song, J. Ha, B. Park, T.H. Kwak, I.T. Kim, H. Nam and G.S. Cha, *Talanta*, 57 (2002) 263–270.
71 J. Bobacka, M. McCarrick, A. Lewenstam and A. Ivaska, *Analyst*, 119 (1994) 1985–1991.
72 R. Paciorek, P.D. van der Wal, N.F. de Rooij and M. Maj-Żurawska, *Electroanalysis*, 15 (2003) 1314–1318.
73 R. Paciorek and M. Maj-Zurawska, *Chem. Sens. Suppl. B*, 20 (2004) 556–557.
74 J. Sutter, A. Radu, S. Peper, E. Bakker and E. Pretsch, *Anal. Chim. Acta*, 523 (2004) 53–59.
75 J. Bobacka, *Anal. Chem.*, 71 (1999) 4932–4937.
76 J. Bobacka, T. Lahtinen, J. Nordman, S. Häggström, K. Rissanen, A. Lewenstam and A. Ivaska, *Electroanalysis*, 13 (2001) 723–726.

77 J. Bobacka, A. Lewenstam and A. Ivaska, *J. Electroanal. Chem.*, 509 (2001) 27–30.
78 M. Vázquez, J. Bobacka, A. Ivaska and A. Lewenstam, *Sens. Actuators B*, 82 (2002) 7–13.
79 J. Bobacka, T. Alaviuhkola, V. Hietapelto, H. Koskinen, A. Lewenstam, M. Lämsä, J. Pursiainen and A. Ivaska, *Talanta*, 58 (2002) 341–349.
80 T. Alaviuhkola, J. Bobacka, M. Nissinen, K. Rissanen, A. Ivaska and J. Pursiainen, *Chem. Eur. J.*, 11 (2005) 2071–2080.
81 J. Bobacka, T. Lahtinen, H. Koskinen, K. Rissanen, A. Lewenstam and A. Ivaska, *Electroanalysis*, 14 (2002) 1353–1357.
82 J. Bobacka, V. Väänänen, A. Lewenstam and A. Ivaska, *Talanta*, 63 (2004) 135–138.
83 M. Vázquez, J. Bobacka, A. Ivaska and A. Lewenstam, *Talanta*, 62 (2004) 57–63.
84 M. Vázquez, P. Danielsson, J. Bobacka, A. Lewenstam and A. Ivaska, *Sens. Actuators B*, 97 (2004) 182–189.
85 A. Michalska and K. Maksymiuk, *Anal. Chim. Acta*, 523 (2004) 97–105.
86 A. Michalska, M. Ocypa and K. Maksymiuk, *Electroanalysis*, 17 (2005) 327–333.
87 A. Michalska, A. Konopka and M. Maj-Zurawska, *Anal. Chem.*, 75 (2003) 141–144.
88 G. Cui, J.S. Lee, S.J. Kim, H. Nam, G.S. Cha and H.D. Kim, *Analyst*, 123 (1998) 1855–1859.
89 X.X. Sun and H.Y. Aboul-Enein, *Anal. Lett.*, 32 (1999) 1143–1156.
90 X.X. Sun, L.Z. Sun and H.Y. Aboul-Enein, *Electroanalysis*, 12 (2000) 853–857.
91 W.-S. Han, M.-Y. Park, K.-C. Chung, D.-H. Cho and T.-K. Hong, *Anal. Sci.*, 16 (2000) 1145–1149.
92 W.-S. Han, M.-Y. Park, K.-C. Chung, D.-H. Cho and T.-K. Hong, *Electroanalysis*, 13 (2001) 955–959.
93 W.-S. Han, M.-Y. Park, K.-C. Chung, D.-H. Cho and T.-K. Hong, *Talanta*, 54 (2001) 153–159.
94 T. Lindfors, J. Bobacka, A. Lewenstam and A. Ivaska, *Analyst*, 121 (1996) 1823–1827.
95 T. Lindfors, P. Sjöberg, J. Bobacka, A. Lewenstam and A. Ivaska, *Anal. Chim. Acta*, 385 (1999) 163–173.
96 W.-S. Han, K.-C. Chung, M.-H. Kim, H.-B. Ko, Y.-H. Lee and T.-K. Hong, *Anal. Sci.*, 20 (2004) 1419–1422.
97 T. Lindfors and A. Ivaska, *Anal. Chem.*, 76 (2004) 4387–4394.
98 A.L. Grekovich, N.N. Markuzina, K.N. Mikhelson, M. Bochenska and A. Lewenstam, *Electroanalysis*, 14 (2002) 551–555.
99 T. Lindfors, S. Ervelä and A. Ivaska, *J. Electroanal. Chem.*, 560 (2003) 69–78.
100 P.C. Pandey and R. Prakash, *J. Electrochem. Soc.*, 145 (1998) 4103–4107.

101 P.C. Pandey and R. Prakash, *Sens. Actuators B*, 46 (1998) 61–65.
102 M.V. Kuznetsova, S.S. Ryasenskii and I.P. Gorelov, *Pharm. Chem. J.*, 37 (2003) 599–601.
103 I.P. Gorelov, S.S. Ryasenskii, S.V. Kartamyshev and M.V. Fedorova, *J. Anal. Chem.*, 60 (2005) 65–69.
104 X. Zhang, B. Ogorevc and J. Wang, *Anal. Chim. Acta*, 452 (2002) 1–10.
105 R.S. Hutchins and L.G. Bachas, *Anal. Chem.*, 67 (1995) 1654–1660.
106 T.A. Bendikov and T.C. Harmon, *J. Chem. Educ.*, 82 (2005) 439–441.
107 M.F. Mousavi, M. Shamsipur, S. Riahi and M.S. Rahmanifar, *Anal. Sci.*, 18 (2002) 137–140.
108 H. Karami and M.F. Mousavi, *Talanta*, 63 (2004) 743–749.
109 T. Blaz, J. Migdalski and A. Lewenstam, *Analyst*, 130 (2005) 637–643.

Chapter 5

Light-addressable potentiometric sensors (LAPS): recent trends and applications

Torsten Wagner and Michael J. Schöning

5.1 INTRODUCTION

The light-addressable potentiometric sensor (LAPS) is one of the youngest developments in the family of field-effect-based sensor devices. Based on the early investigations on ion-selective field-effect transistors (ISFETs) [1], researchers developed first the electrolyte–insulator–semiconductor (EIS) capacitance to investigate the complex electrochemical and surface mechanisms of the ISFET-gate region [2–4]. This was a consequential step, and follows the same way as the traditional development of MOSFETs (metal-oxide semiconductor field-effect transistor) and MOS (metal-oxide semiconductor) capacitances. Thus, it was possible to use the same methods, which were introduced for investigating on MOS capacitances [5–8]. Due to the relatively simple manufacturing process and simple encapsulation, (bio-)chemical sensor development recently started to use EIS structures as an individual sensor platform and not only for investigations of ISFET properties [9–12].

Besides the traditional capacitance *versus* voltage (C/V) measurements, which are mainly used for the characterisation of MOS and EIS capacitances, the scanned light pulse technique (SLPT) was introduced by Engström and Alm [13], first for MOS structures. This technique utilises a light source to illuminate a local area of the MOS structure. Thus, a local photo-effect-induced current can be measured, which only depends on the local properties and energy states of the illuminated region of the MOS structure. In 1988, Hafeman *et al.* combined this SLPT method with EIS structures to develop the LAPS [14,15]. This sensor is capable of measuring the surface potential of the electrolyte–transducer interface with a lateral resolution. Hence, the surface

Comprehensive Analytical Chemistry 49
Alegret and Merkoçi (Eds)
Volume 49 ISSN: 0166-526X DOI: 10.1016/S0166-526X(06)49005-X

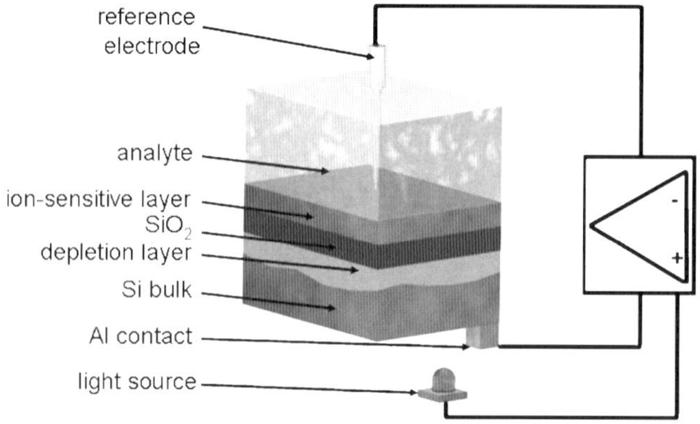

Fig. 5.1. Principal set-up of the light-addressable potentiometric sensor (LAPS).

potential itself depends on the chemical interactions between the transducer surface and the electrolyte solution. Figure 5.1 depicts the principal set-up of such a LAPS system.

This paper will discuss, after a short summary of the theoretical principles, the trends and developments of LAPS-based measurement devices, followed by new developments and future possible tasks and challenges.

5.2 THEORETICAL BACKGROUND

Since LAPS-based devices belong to the family of field-effect-based sensors, they share the basic principles together with EIS and MOS capacitances as well as ISFETs and MOSFETs. The LAPS structure consists, in its simplest form, of a p- or n-doped silicon substrate (e.g., 1–10 Ω cm, 350–400 μm, ⟨100⟩), and a 30–50 nm thick oxide layer, e.g., SiO_2, produced by dry oxidation, and an ohmic rear-side contact (e.g., 300 nm). The top layer depends on the later application, since it provides the sensitivity towards a specific substance. For pH sensing, Si_3N_4 and Ta_2O_5 are known as stable and sensitive transducer materials for EIS sensors and LAPS [16–20]. This manufacturing does not differ from the manufacturing of an EIS or MOS capacitance. Thus, in the absence of illumination, the LAPS behaves like an EIS capacitor. As for an EIS capacitance, a dc bias voltage is applied to the LAPS structure via a reference electrode so that a depletion layer appears at the insulator/semiconductor interface. The width of the depletion layer is a function of the total sum of potentials applied to the sensor surface, e.g., the bias voltage and the surface potential. Under the assumption that the bias voltage remains constant, changes in the depletion-layer width indicate surface-potential changes. Since the space-charge capacitance is a function of the width of the depletion layer, the variation of the field-effect capacitance can be measured to determine surface potential changes. To detect the variation of the capacitance of the depletion layer, the LAPS is, in contrast to EIS-based measurements, illuminated with a modulated light beam, which induces an ac photocurrent to be measured as the sensor signal. In the presence of illumination of the semiconductor with a light source, there is a generation of electron–hole pairs in the semiconductor substrate. In accumulation, nearly all generated electron–hole pairs will recombine after a short diffusion time; there is no total current left, which can be measured. In the depletion region, electron–hole pairs can diffuse, recombine or be separated in the electric field. When the LAPS structure is driven into depletion, electron–hole pairs that have diffused from the bulk semiconductor into the depletion region or that are photogenerated within this region are separated in the electric field. Hence, a total current can be measured, which has to flow to compensate this charge separation (drift current). With longer illumination time, more and more charges are separated by the electric field. Consequently, a diffusion current back towards the silicon bulk followed by a recombination will compensate the charge separation and the external total current tends to zero. This is somewhat similar to the charge of a capacitor. After turning the light source off, the drift current, due to charge separation, tends to zero and the diffusion current dominates now. Hence, it needs to be compensated by an external circuit as before and the same shaped, but negative, current can be measured, similar as discharging of a capacitor. If the time sequence of "on" and "off" of the light source is fast enough, faster than the decay-time constant of the transient currents (i.e., the modulation frequency of the light source is high enough), a sinusoidal response can be measured. The amplitude of this signal depends on the changing of the space-charge region and thus on the surface voltage. Furthermore, this signal form is ideal to be measured by lock-in amplification set-ups and by this way noise and signal disturbances can be eliminated.

The excitation light for LAPS devices may be directed to the semiconductor either through the transparent insulator (front-side illumination) or from the opposite side of the planar structure (back-side illumination) by placing one or more light sources just below the semiconductor. If front-side illumination is used, the injection of excess-charge carriers occurs just below the oxide layer in the space-charge region and the effectiveness of charge generation is high. However, the variation of the optical properties of the analyte under test and the sensor membrane will additionally influence the sensor signal. If, on the other hand, the illumination occurs at the back-side, excess-charge carriers are generated also near the back-side. Only those excess-charge carriers that diffuse through the semiconductor bulk to the space-charge region are able to contribute to the photocurrent. Those photogenerated carriers that recombine either at the back-side surface, in the semiconductor bulk or at interface defects before reaching the space-charge region are not effective in the external total sum current.

To understand the electrical behaviour of the LAPS-based measurement, the LAPS set-up can be represented by an electrical equivalent circuit (see Fig. 5.2). V_{bias} represents the voltage source to apply the dc voltage to the LAPS structure. R_{e} is a simple presentation of the reference electrode and the electrolyte resistance followed by a interface capacitance $C_{\text{interface}}$ (this complex capacitance can be further simulated by different proposed models as they are described, e.g., in Refs. [2,21,22]). In series to the interface capacitance, the insulator capacitance C_{i} will summarise the capacitances of all insulating layers of the LAPS device. The electrical current due to the photogeneration of electron–hole pairs can be modelled as current source I_{P} in parallel to the

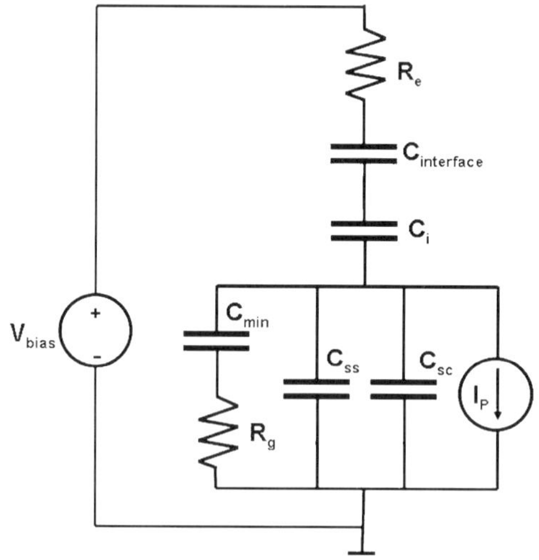

Fig. 5.2. Electrical representation of a LAPS structure.

space-charge capacitance C_{sc}. The capacitance C_{ss} is added in parallel to take surface states between the interface of the semiconductor and the insulator into account. Finally, the capacitance C_{min} in series with the resistance R_g describes the minority charges in the inversion layer, while R_g is the time delay between the generation and recombination process. Figure 5.2 represents a simplified equivalent electric circuit for the LAPS.

Under the assumption that the LAPS sensor is operated in a small linear range with a fixed frequency and amplitude of the light source and under a constant bias voltage, the electrical representation can be simplified. Most capacitances are negligible under these assumptions. Hence, under the condition of strong depletion, the photocurrent I_{photo} of the LAPS can be described by the following simplified equation [23]:

$$I_{photo} = I_P \frac{C_i}{C_i + C_{sc}} \tag{5.1}$$

where I_P is the alternating component of the photogeneration of electron–hole pairs, C_i the capacitance of the insulator materials and C_{sc} the space-charge capacitance. Under the above assumptions, the capacitance of the space-charge region, C_{sc}, depends on the applied bias voltage, whereas C_i remains constant at all time. Thus, the measured photocurrent is a function of the bias voltage applied to the LAPS structure. The LAPS uses this dependence of the photocurrent on the bias voltage to measure chemically sensitive surface potentials on the insulator surface. The steepness of the LAPS photocurrent–voltage curve, which determines the sensitivity of the LAPS, depends on many parameters such as the density of the interface states, the thickness of the insulating layer, the doping level of the semiconductor, etc. [24].

Figure 5.3 depicts a typical current–voltage (I/V) curve. The amplitude of the photocurrent starts at zero for the accumulation region. Since no space-charge region arises, no total sum current can be measured. With increasing bias voltage, the depletion layer increases and hence the amplitude of the photocurrent increases. Finally, the photocurrent saturates as the bias is increased into inversion. The LAPS utilises this dependence of the photocurrent on the bias voltage to detect its interfacial potential change. For, e.g., different pH values, the photocurrent–voltage curve shifts along the voltage axis as can be seen from the figure. By measuring this shift, e.g., the pH value or the ion concentration can be quantitatively determined. The shift of the photocurrent–voltage curve is often quantified by tracking the potential of the inflection point of the curve. At this point, the photocurrent is most

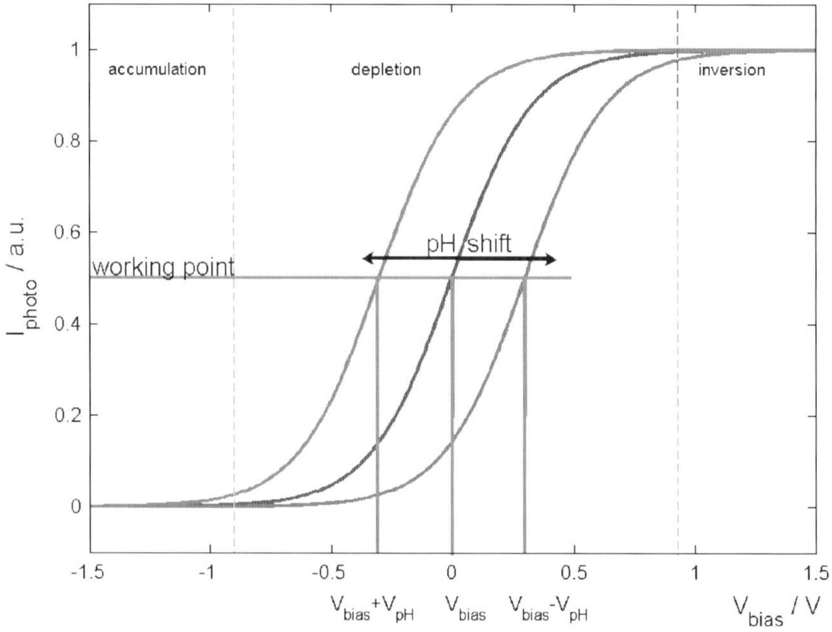

Fig. 5.3. Typical current *versus* voltage curve (I/V) for a p-doped LAPS, with $I_P = 0$ for accumulation, $0 < I_P < I_{\max}$ for depletion and $I_P = I_{\max}$ for the inversion region. Change in the surface potential will shift the I/V curve. A working point for the constant-current (CC) mode can be set at the inflection point.

sensitive to changes in the surface potential. To calculate the analyte sensitivity, several measurements for known analyte concentrations are performed and the bias voltage corresponding to the inflection point of each curve is calculated from the second derivative and plotted *versus* the analyte concentration. This results in a calibration curve for the LAPS.

The determination of the shift of the inflection point can be simplified by the use of an algorithm or a control system. After the specification of the working point (a specific current–voltage pair), a controller will be utilised to keep the current constant. The controller needs to compensate the variation of the surface voltage by varying and operating the bias voltage with the same but inverted amount. For the assumption of a linear behaviour around the working point, the following equations (Eqs. (5.2)–(5.5)) describe the controller actions:

$$I_{\text{photo}} = k(V_{\text{bias}} + V_{\text{surface}}) \tag{5.2}$$

$$I_{\text{photo}} + \Delta I_{\text{photo}} = k(V_{\text{bias}} + V_{\text{surface}} + \Delta V_{\text{surface}}) \tag{5.3}$$

$$\text{for} \quad \Delta V_{\text{surface}} = -V_{\text{controller}} \tag{5.4}$$

$$I_{\text{photo}} = k(V_{\text{bias}} + V_{\text{surface}} + \Delta V_{\text{surface}} - V_{\text{controller}}) \tag{5.5}$$

where, I_{photo} is the resulting photocurrent, k the transimpedance coefficient that depends on the material properties, V_{bias} the applied external bias voltage, V_{surface} the surface voltage due to the interaction with the analyte solution, $\Delta V_{\text{surface}}$ the variation of the surface voltage due to change in the analyte concentration and $V_{\text{controller}}$ the controller voltage which is necessary to compensate $\Delta V_{\text{surface}}$. For this so-called constant-current (CC) measurement mode, the absolute value of the voltage that is adjusted by the controller is equal to the change of the surface voltage and can be directly recorded [25–27]. For that, the photocurrent I_{photo} as described in Eq. (5.1) will be measured continuously. A change in the surface voltage will result in a different photocurrent (Eq. (5.2)) and the controller will calculate the difference, which is necessary to move back to the original photocurrent. To obtain Eq. (5.4), the calculated controller voltage $V_{\text{controller}}$ has to be equal but inverted to the change in the surface voltage V_{surface} (Eq. (5.3)). Thus, the calculated controller voltage can be used to determine the change in the analyte concentration. This method has the benefit that the LAPS system will be operated at a constant working point. A non-linear behaviour of the LAPS material parameters will not affect the measurement as it will be for single I/V measurements, due to different operating parameters. A typical CC curve of a pH-sensitive LAPS measurement in the pH range 5–8 is shown in Fig. 5.4. Every pH step can be identified by a change in the controller voltage. For known pH solutions, e.g., during calibration, the steps width per pH-unit step can be used to calculate the pH sensitivity of the LAPS sensor. Further details can be found in the protocol section.

The benefit of such a LAPS sandwich structure consists in a easy manufacturing process, where no patterning and masking as well as encapsulation of conducting tracks are needed (see Fig. 5.1). In the measurement set-up, the LAPS structure itself will be embedded in a measurement chamber. The flat uniform surface of the LAPS assists a proper sealing by, e.g., an O-ring, silicone or rubber material. Finally, only the sensitive membrane will stay directly in contact with the analyte under test.

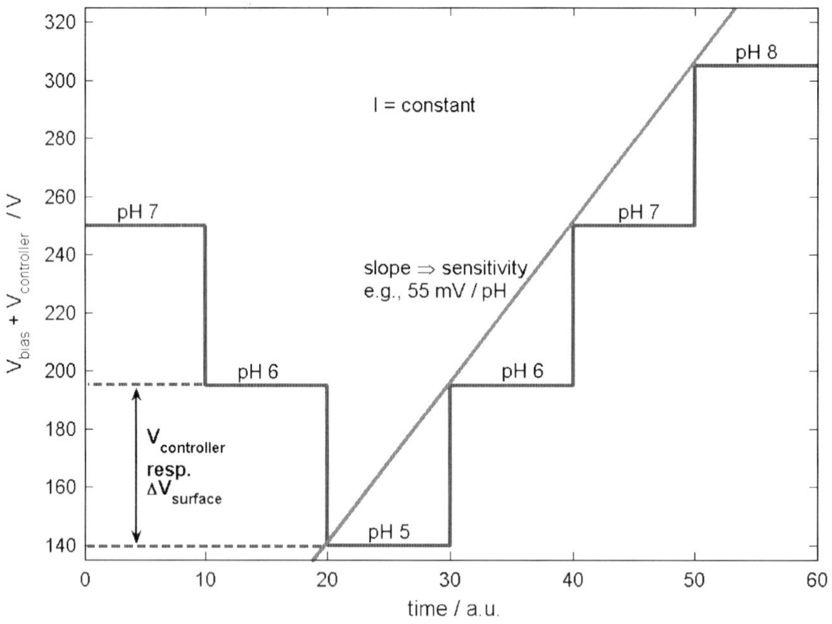

Fig. 5.4. Schematic constant-current (CC) measurement: the controller voltage over time has to be equivalent to the change of the surface voltage, to remain a fixed working point. The change in the pH value can be calculated by a proportional change in the controller voltage.

To allow the illumination of the LAPS from the back-side, it is common to create an opening window in the rear-side contact, thus allowing a direct illumination of the silicon-bulk material. The measurement chain consists (from top to bottom, see Fig. 5.5) of a reference electrode to apply the dc-bias voltage as well as to close the measurement loop, the test sample (analyte) and the LAPS-sandwich structure with an electrical contact at the rear-side as well as a wire to the external measurement circuit. As a light source, either a laser or light-emitting diodes (LEDs) can be utilised. A detailed description of investigations on different light sources for LAPS devices can be found in Section 5.3.1.

The external measurement circuit (see Fig. 5.6) needs to handle three major tasks: the modulation of the light source with a specific frequency; the measurement of the resulting photocurrent with a necessary amplification and signal conditioning; and the calculation and adjustment of the dc-bias voltage to operate the LAPS at a desired working point. The external electronic unit is therefore connected between the rear-side contact and the reference electrode. Furthermore,

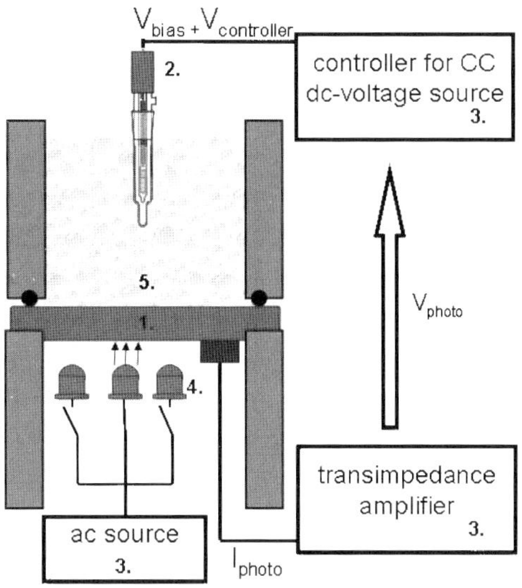

Fig. 5.5. Complete measurement chain for a LAPS. (1) LAPS sensor, (2) reference electrode, (3) electronic unit, (4) light source and (5) analyte.

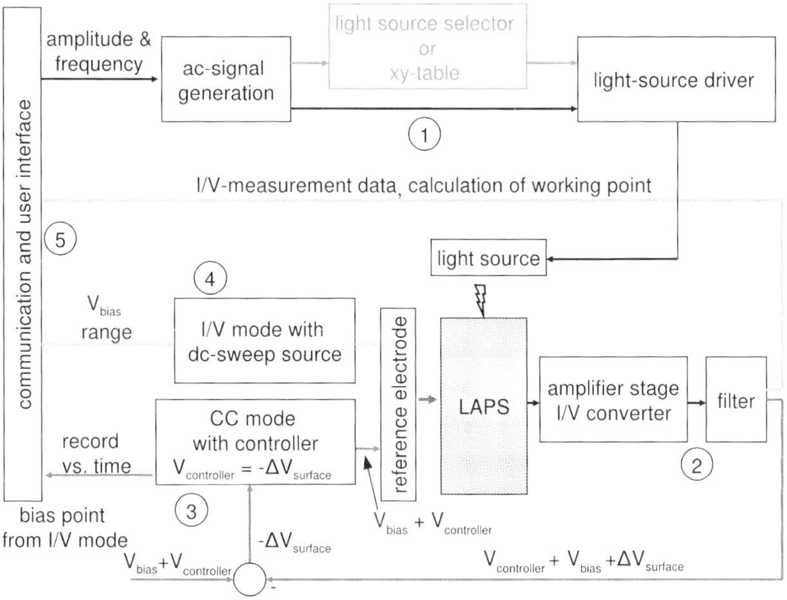

Fig. 5.6. Principal block diagram of the electronic unit. (1) Signal generation, (2) amplification of I_{photo}, (3) controller for CC mode, (4) dc voltage sweep for I/V mode and (5) communication interface.

it is connected to the light source. If required, a potentiostat could be utilised to use a three-electrode configuration, as it was demonstrated in, e.g., Ref. [28].

The LAPS structure provides an additional benefit over EIS and ISFET structures: since electron–hole pairs are only generated in the illuminated region of the LAPS structure, the photocurrent depends only on the local illuminated space-charge capacitance. By moving the light source to another position, one can record a local change in the field-effect capacitance and thus a local difference in the surface voltage. This allows the determination of a particular analyte under test with a spatial resolution. This so-called scanning LAPS device is of great interest and will be described in detail in Section 5.3.1.

5.3 APPLICATIONS OF LAPS

Since the introduction of LAPS in 1988, when Hafeman *et al.* proposed a measurement device for biological applications [15], the first LAPS was mainly developed for biological investigations, e.g., a phospholipid-bilayer membrane-based LAPS [29], a sandwich immunoassay for human chorionic gonadotropin (hCG) [30] and an enzyme-based (urease) microchamber-LAPS device [31]. Starting from these early contributions, LAPS-based devices have been investigated in more detail. It is beneficial to separate the following investigations in mainly three different directions:

- characterisation and investigation of the LAPS principle itself,
- biological applications based on LAPS, and
- alternative applications and materials for LAPS devices.

The following three subsections will discuss the progress and developments in these three different fields.

5.3.1 Characterisation of LAPS

Among the first practical applications, research studies have been interested to fathom possible potentials and limitations of LAPS-based measurements. One distinguishing feature of the LAPS is its spatial resolution. The resolution of a LAPS system is limited by several factors such as the definition of the light source (e.g., wavelength, beam diameter, beam divergence, beam intensity, etc.) and the LAPS geometry of the build-up (layer scheme, layer thickness, material properties, etc.). Therefore, many investigations have been performed to define the highest possible resolution of LAPS.

Nakao *et al.*, for example, described a method to improve the resolution of LAPS (based on a Si_3N_4–SiO_2–Si n-type structure) by the utilisation of thin bulk materials down to 100 μm [32]. He also discussed the infiuence of the diffusion length of the minority carriers on the spatial resolution. The experimental set-up consisted of a laser beam with 635 nm (He–Ne) and the sensor chip was illuminated from the back-side. Thus, the authors already concluded that the penetration depth of the light was only about 2 μm. Hence, the major amount of photo-induced charge carriers was generated close to the rear-side surface of the silicon substrate. The charges had to diffuse along the silicon bulk before contributing to the photocurrent by charge separation. This made the thickness of the bulk material and the diffusion length of the minority carriers inside the bulk material the most limiting factors in terms of resolution. In this experiment, the lateral resolution could be improved from initially 500 μm to better than 100 μm. Furthermore, the authors predicted a resolution of about 50 μm for a combination of 50 μm bulk thickness and a diffusion length of 10 μm, which was demonstrated later by Nakao *et al.* [33]. In 1996, the same method was used to detect yeast *Saccharomyces cerevisiae* and *Escherichia coli* bacteria (see also Section 5.3.2). By thinning the substrate thickness down to 20 μm and with an alternative light source (830 nm), a resolution of 10 μm has been achieved [34]. The resolution was estimated by a test pattern on the sensor surface. The smallest still by LAPS recognisable pattern structure was defined as the resolution.

Parak *et al.* investigated the spatial resolution in 1997 [35] in a similar set-up. Besides the slightly different sensor structures, the main difference was the illumination of the LAPS chip from the front-side. Although a similar light source has been applied as above (wavelength $\lambda = 670$ nm), the generation of the electron–hole pairs took place mostly close to or in the space-charge region. Hence, the thickness of the silicon bulk did not influence the lateral resolution. The authors were able to demonstrate a possible resolution of 100 μm. However, it should be mentioned that the illumination from the front-side has several disadvantages, for example, the light sensitivity of the biological samples or different absorption rates of light for different analyte solutions will adulterate the measurement signal. Moreover, practical reasons, like the size and the geometry of the chamber set-up, avoid often a front-side illumination.

A similar investigation of the resolution of LAPS devices was done in 1998 by applying ultra-thin Si films (0.5 μm) on transparent substrates [36]. Here, the experiments allow to achieve resolutions down to 5 μm, demonstrating the influence of different wavelengths of the light source and comparing the results to a single Si-LAPS structure (thickness 10 μm). Although a back-side illumination set-up was chosen, due to the utilisation of silicon-on-insulator (SOI) and ultra-thin Si films, the experiments approve the results from Nakao *et al.*, where thin silicon layers improved the spatial resolution. The films were prepared with several photolithographic steps together with three layers of wafer employing the silicon direct bonding technique, which is somehow contrary to the original beneficial simplicity of the initial LAPS structures.

All the results were confirmed by investigations with both back- and front-side illumination [37]. In this study, which also describes theoretical aspects, a final resolution of 17 μm for an epi-structure was achieved, which consisted of a thin layer (3 μm, specific resistance of 10 Ω cm) and a thick silicon substrate (0.38 μm, with low specific resistance of 0.005–0.02 Ω cm). Instead of using a pattern grid on the front-side to study the smallest possible resolution, in this work the sensor chip was coated half with a metal layer. This forms on the covered side a metal–insulator–semiconductor (MIS) structure. The light pointer was moved from the metallised area to the uncovered area and the resolution was determined at the borderline by measuring the photocurrent that depends on the diffusion length of the carriers. Since, for the assumption that the diffusion length is the main decisive and critical parameter for the amount of carriers which could reach the metal-covered part of the semiconductor substrate from a specific distance, the calculation of the minimal resolution was experimentally observed.

Starting in 1999, Moritz *et al.* improved the lateral resolution of LAPS by employing GaAs instead of Si [38]. A resolution was achieved as before by a metal pattern to form partial MIS structures and to calculate the possible resolution by evaluating the diffusion length. Due to the fact that the diffusion length of the minority charge carriers is smaller with respect to Si, the authors were able to determine a resolution of 3.1 μm. It is worth mentioning that in these experiments a standard CD-player optic has been used as a light source. Furthermore, the authors propose a so-called scanning photo-induced impedance microscope (SPIM), which utilises the same field-effect principle as the LAPS, and the results obtained regarding the resolution can easily be transferred from one technique to the other. In this SPIM mode, the impedance change at a working point in the inversion region has been investigated. Hence, the laterally resolved impedance behaviour for different polymer films on top of silicon-oxide layers can be characterised [39].

Due to the inability to deposit electrochemically stable gate-insulating materials for GaAs, another approach was developed based on amorphous silicon (a-Si), prepared as a thin layer for LAPS devices on transparent glass substrates [40]. The diffusion length in this material was reported to be as small as 120 nm [41] and a resolution down to 1 μm has been demonstrated, which was mainly limited by the optical set-up. The electrochemical properties of the a-Si-based structures were investigated later with a LAPS device; thus, the above results for SPIM were transferred back and proved for the LAPS, too [42,43].

Another way to improve the lateral resolution of the LAPS was introduced by Krause *et al.* [44], where instead of using one high-energy light source ($E_{photon}E_{bandgap}$), two low-energy photons ($E_{photon} < E_{bandgap}$) are generated. Only at the focused point, where two photons are present at the same time, electron–hole pairs can be generated. This method prevents the generation of electron–hole pairs in the bulk material, where the resolution decreases due to the additional lateral diffusion of the charge carriers. The authors were able to demonstrate an improvement in resolution by 31% from 112 μm with a single light source down to 78 μm for bulk-based structures by using a standard silicone wafer. Furthermore, this principle was also extended to thin films when applying it to SOI- and silicon-on-sapphire (SOS)-based materials, which results in resolutions down to 13 μm for SOI- and even 0.6 μm for SOS-based structures.

Table 5.1 summarises the different research activities to improve the lateral resolution of LAPS. It is clearly demonstrated that the resolution can be extended down to the sub-micron range in the near future. This will broaden the possible application fields of the LAPS principle. For example, biological and chemical sensor arrays will benefit from further investigations and improvements of the lateral resolution of the LAPS. However, the LAPS works with an external light source, i.e., the experimentally applied light source, and its optical pathway will also limit the achievable resolution.

The investigations to obtain high lateral resolutions were accompanied by different simulations and models for LAPS devices. Beside generic electrical representations and SPICE (simulation program with integrated circuit emphasis)-based simulations [45,46],

TABLE 5.1

Investigations and methods to improve the lateral resolution for LAPS-based devices

Method	Resolution (μm)	Verified by	References
Thinning bulk materials down to 100 μm	100	Test pattern	[32]
Thinning bulk materials down to 20 μm	10	Test pattern	[34]
Front-side illumination on bulk material (300–400 μm)	100	Test pattern	[35]
Ultra-thin Si films (0.5 μm) on transparent substrates (SOI)	5	Test pattern	[36]
Front- and back-side illumination of an epi-layer structure	17	Average diffusion length	[37]
Use of GaAs as bulk material	3.1	Average diffusion length	[38]
Use of a-Si as bulk material	1	Average diffusion length	[40]
Two-photon effect on bulk material	78	Average diffusion length	[44]
Use of ultra-thin structures based on SOI and SOS	0.6	Average diffusion length	[44]

comparisons between simulations and measurements have been performed [47]; also numerical models have been discussed and compared with experimentally observed measurement values [48,49]. The different proposed models and simulations have been tailored for particular applications or performances, e.g., the lateral resolution, the electronic interfacing and the chemical interaction [50–56] of LAPS devices.

Other "intrinsic" characteristic parameters of LAPS have been investigated by different research groups such as the chemical response time, the surface-state densities and zeta potential (for Si_3N_4), and the minority carrier diffusion length for resolution estimations. For a more detailed description of these experimental set-ups, see, e.g., Refs. [57–61].

5.3.2 Biological applications based on LAPS

The original idea behind the LAPS was the development of a sensing device for biological applications of living cells on a semiconductor field-effect structure [15]. Therefore, it has not been surprising that the first LAPS applications were created for the detection of biological agents. They can be subdivided into two categories:

- detection of a specific analyte or a group of analytes with the help of a biological "transducer", and
- observation and detection of the biological activity itself.

The first category utilises the biological component on top of the LAPS surface as an additional, integral part of the transducer path to create a two- or more-step reaction chain (see Fig. 5.7a). These biological transducers are very specific to a particular analyte inside the test sample. At the second category, the biological component itself is under investigation. With the help of, e.g., an ion-specific anorganic transducer layer on top of the LAPS structure, the metabolic activity of the biological species under test can be monitored and information can be provided, which helps to understand and control, for instance, biotechnological

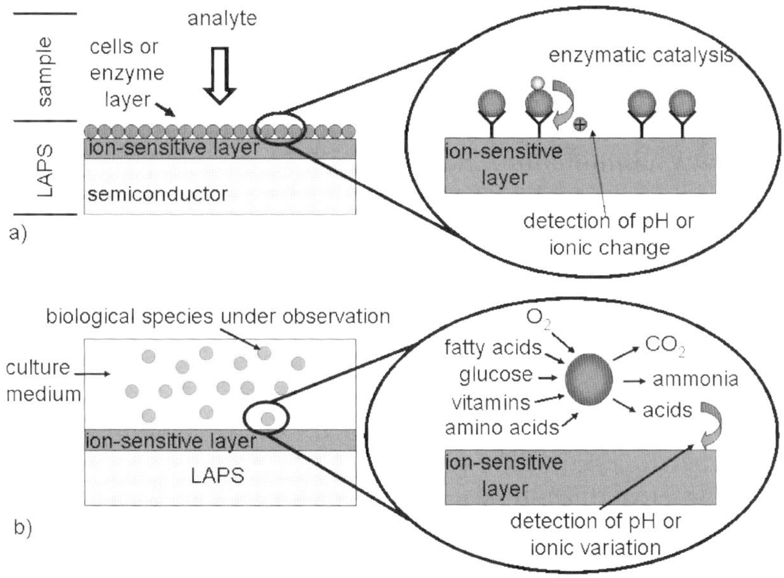

Fig. 5.7. Biological sensors based on the LAPS principle. (a) LAPS set-up with a biological component as integral part of the transducer chain. (b) LAPS set-up for observation and detection of a biological component.

processes, or can be applied for biomedical applications (see Fig. 5.7b). Most of the works presented in this chapter originated from the idea to measure the change of the ion concentration (especially pH) either in the background solution of the biological sample or at the interface between the biological component and the LAPS structure.

The earliest application demonstrated the measurement of the conductance, the capacitance and the trans-membrane potential of phospholipid-bilayer membranes with the help of a LAPS device [29]. This work differs somehow from later works, since it utilised the variation of the overall electrical potential distribution instead of an additional electrochemical potential due to "electrolyte–transducer" interactions. By placing a bilayer membrane on top of the LAPS structure, the additional membrane resistance, capacitance and trans-membrane potential influence the potential distribution and hence, alter the space-charge region layer of the LAPS structure. Thus, under the condition that the surface potentials at the reference electrode and at the LAPS surface remain constant, the resulting photocurrent was directly dependent on the membrane properties only. The presence of ion carriers (valinomycin) or cation carriers (gramicidin D and alamethicin) in the solution changed these properties, and therefore a two-step detection was established: presence of valinomycin at the ion-sensitive membrane ⇒ 1st step—change in the membrane conductance ⇒ 2nd step—change in the photocurrent amplitude.

The first device for the observation of cells based on a LAPS sensor was introduced by Parce *et al.* in 1989 [62]. This device used a flow-through chamber, where adherent cells were placed onto a glass substrate or, in case of non-adherent cells, entrapped by the use of micro-fabricated wells (50 μm × 50 μm in size, 25 μm in height). With a flow speed as low as 10–50 μm/s, this LAPS arrangement was able to measure the metabolic products of about 10^3 cells. To demonstrate possible applications with such a device, mammalian cells as well as human epidermal keratinocytes and tumour cell lines have been cultured on the LAPS surface. All cell types were treated with a particular analyte and the metabolic response was observed. Libby and Wada reported a device based on a LAPS for the detection of pathogenic bacteria (*Yersinia pestis* and *Neisseria meningitidis*) by an immunofiltration method [63]. The results compared with an enzyme-linked immunosorbent assay (ELISA) show to be significantly faster (20 min compared to 2.5 h) and more sensitive (less than 1000 bacterial cells compared to 60,000 for the ELISA). The results have been proven with an hCG immunoassay [30]. Furthermore, the

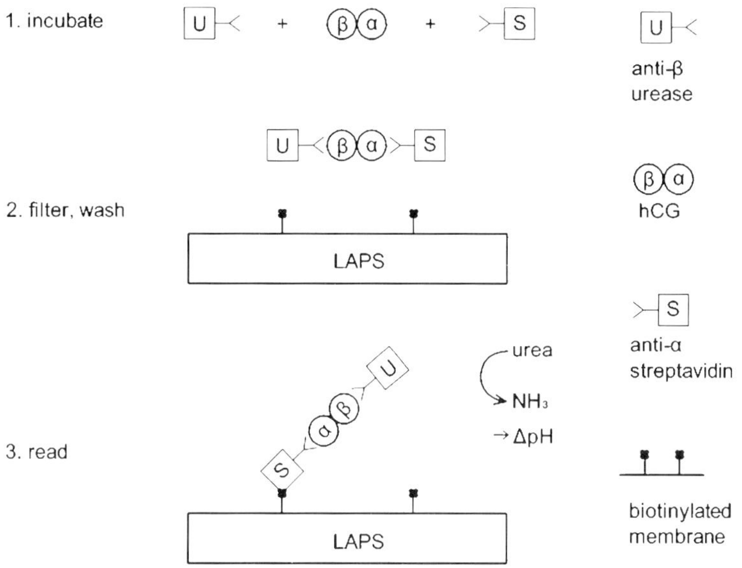

Fig. 5.8. Schematic configuration for an immunoassay based on LAPS.

immobilisation of membranes close above the LAPS surface to entrap cells has been reported. An additional plunger reduced the overall volume under test. This resulted in a small measurement chamber with a low volume of the test sample, which is important in order to detect a metabolism-based acidification (rate) change.

Figure 5.8 depicts a typical immunoassay-based LAPS. The reaction of an antibody to its antigen will immobilise the species and after washing, only those species that have been successfully bound will be left onto the LAPS surface. Here, a conjugated enzyme, e.g., urease, will change the pH value of the test sample by enzymatic catalysis after injection of urea.

Another immunofiltration-based investigation was reported by Hu *et al.* [64] for the detection of the Venezuelan equine encephalitis virus (VEE). Further specifications and a more detailed flow-chamber system are described in Ref. [65], where different kinds of mammalian, human and tumour cells (human keratinocytes, B82, MESS-SA and DX5, respectively) have been applied to demonstrate irritancy determination, detection of receptor-mediated triggering of cells and chemotherapeutic efficacy testing. Owicki *et al.* reported the study of cellular responses of different cells (keratinocytes, CHO and B82 with different expressions) to hormones and neurotransmitters with the same principle as above [66,67]. The determination of a small quantity of enzyme (urease) was presented by Bousse *et al.* including a model to predict the sensitivity of the LAPS system as a function of the chamber volume, specific activity of the enzyme and the amount of pH buffering present [31]. The indirect detection of DNA was demonstrated by using a streptavidin–biotin–SSB–DNA–antiDNA–urease complex when measuring the resulting pH change due to the catalytic process of the urease, with a LAPS sensor [68].

Based on all these early approaches, the Cytosensor® microphysiometer from Molecular Devices was introduced into the market in 1990/1991. This device features four complete individually acting flow-path systems including pumps, debubblers and valves as well as four independently working LAPS sensors and four reference electrodes. The LAPS chip itself is assembled at the bottom of a microchamber that hosts membranes and spacers to encapsulate the cells (see Fig. 5.9). It is beyond the scope of this article to mention all contributions in the field of biological investigations which have utilised the commercial Cytosensor system for their experimental studies. The interested reader might be referred on to Refs. [69–84] for particular applications, based on the Cytosensor microphysiometer.

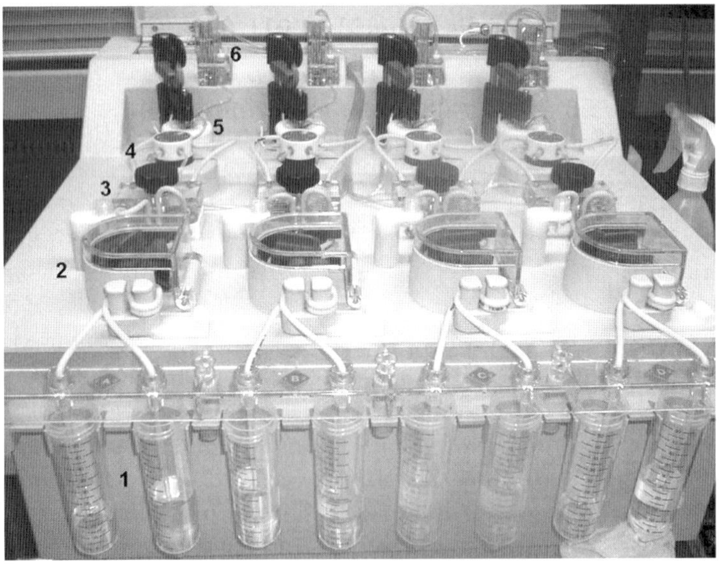

Fig. 5.9. Picture of the commercial Cytosensor microphysiometer from Molecular Devices. Four fluid pathways with (1) analyte container, (2) displacement pump, (3) debubbler, (4) valves, (5) LAPS sensor chamber and (6) reference electrode.

Besides this remarkably fast commercial development, other research groups designed their own laboratory-compatible LAPS systems for similar purposes. The use of, e.g., a nicotinic acetylcholine receptor was reported to create a LAPS-receptor biosensor capable of detecting receptor ligands (acetylcholine, carbamylcholine, succinylcholine, suberyldicholine, nicotine as well as d-tubocurarine, α-bungarotoxin and α-Naja toxin) [85]. Another system quite similar to the Cytosensor set-up was introduced, where mouse fibroblast line 3T6 cells were chosen to demonstrate the determination of metabolic processes of these cells [86].

All the above-mentioned research activities refer on a single light source to measure one spot on the LAPS-sensor chip. The light addressability was not used so far. A development towards a multisensor-LAPS structure with more than one measurement spot on the same sensor chip was introduced by Bousse *et al.*, where a multichannel chip, in which eight fluidic channels were etched into the sensor surface, has been realised. Each channel has its own flow of tissue-culture medium that crosses the measurement spot [87,88]. A further detailed description about the multichannel set-up and a practical measurement procedure based on an exclusion method was reported consequently [89,90]. Aoki *et al.* reported the measurement of three saccharides such as sucrose, maltose and glucose on the same sensor chip for the simultaneous determination of disaccharides [91]. The commercial Cytosensor device was adopted for multisensing by the integration of amperometric-based sensors for the detection of glucose, lactate and oxygen together with the LAPS-based measurement of the pH value for a screening of the cell-culturing process [92,93]. Another modification of the Cytosensor was introduced by the substitution of the single light pointer by four light sources to implement four measurement spots on one sensor chip. Further, the parallel measurement of the metabolic activity of different CHO colonies has been demonstrated [94].

When applying the LAPS as either a single sensing element or for several discrete measurement spots, its spatial resolution can be advantageously combined with a scanning LAPS set-up to obtain an image of a biological process. The use of a urease-based scanning LAPS was reported to measure the local pH changes due to the enzymatic catalysis of urea into carbon dioxide and ammonia [95]. Further, a scanning LAPS set-up with different enzyme pattern immobilised on the sensor surface, for the detection of urea and glucose, was suggested [96].

The last group of LAPS devices for biological applications described in literature deals with the detection of neuronal cell activities. The LAPS-based measurement of extracellular signals of electrically excitable neuronal cells with a spatial resolution was investigated by different research groups. For example, a scanning LAPS set-up was reported to measure, together with standard patch-clamp technique, the extracellular response on electrical stimuli of neural cells obtained from *Aplysia* [97], which was followed by Ismail *et al.* with a similar set-up for a pheochromocytoma cell line (PC-12 cells) and nerve cells of *Lymnaea stagnalis* on poly-L-ornithine and laminin (PLOL)-coated Si_3N_4 [20]. Parak *et al.* demonstrated the use of a spatially resolved LAPS device for the determination of the extracellular potential in chick heart muscle cells [98], and further measurements on the activity of an N1-E115 cell line compared with the results of conventional patch-clamp methods were described by Stein *et al.* [99]. All these investigations demonstrate the beneficial use of the spatial resolution of the LAPS.

On the other hand, another task for LAPS devices and a special group of multi-LAPS systems is represented by differential LAPS systems. These systems use two or more sensing spots to calculate a differential measurement signal, which will reduce the influence of external disturbances such as light and temperature influences as well as sensor drift. The first differential LAPS set-up was build up for the qualitative determination of various "sweet" substances (saccharose, maltose, lactose, levulose) based on different lipid membranes with respect to a non-ion-sensitive reference spot [100]. Furthermore, Adami *et al.* suggested different set-ups and developed a so-called potentiometric alternating biosensor (PAB), which is based on a LAPS (see also Section 5.3.3), to determine enzymes, such as alcohol dehydrogenase (EC 1.1.1.1) from yeast (YADH) [101], redox enzymes such as horse radish peroxidase (HRP), CHO cell populations grown onto a glass substrate [102] and the activity of a monolayer-immobilised urease by a modified Langmuir–Blodgett (LB) technique [103]. The same set-up has been further used to monitor the effects of mitoxantrone on normal "healthy" cells *versus* cancer cells [104] as well as for a redox-reaction-based set-up to monitor the enzymatic activity of HRP, which can be used as a label in an immunoenzymatic application [105]. The detection of terbuthylazine (TBA) in water samples was demonstrated by Mosiello *et al.* with a PAB-based set-up together with a monoclonal antibody (mAb) conjugated with urease [106]. Finally, a comparison between the PAB system and an ISFET-based set-up with respect to the performance for cell-metabolism detection was published in 1996 [107].

More application-orientated LAPS devices were built up for different biological and chemical applications. The detection of cyanide with an *n*-CdSe membrane-based LAPS was reported by Licht [108], and the simultaneous determination of glucose, penicillin and urea by immobilising glucose oxidase, penicillinase and urease, respectively, onto a LAPS structure has been demonstrated by Seki *et al.* [109]. Furthermore, Poghossian *et al.* compared LAPS, EIS and ISFET devices for the detection of penicillin by the catalytic reaction of the enzyme penicillinase [11]. A complete field-portable LAPS system based on an immunoassay-receptor arrangement for the detection of biological warfare agents was demonstrated by Uithoven *et al.* [110], and an organic pollution sensor based on *Trichosporon cutaneum* for determining the biochemical oxygen demand (BOD) was introduced by Murakami *et al.* [111]; a LAPS-based immunoassay for the detection of Staphylococcal enterotoxin B, the Newcastle disease virus and vegetative bacteria (*Brucella melitensis*) was reported by Lee *et al.* [112]. Yicong *et al.* observed the metabolic reaction of nephridium cells for different drugs (chlorpromazine, fluorouracil, insulin and thyroxin) [113,114]. A new immobilisation method of urease and butyrylcholinesterase by photocurable polymeric enzyme membranes for LAPS devices has been elaborated by Mourzina *et al.* [115].

The detection of *E. coli* cells by means of LAPS has been investigated by different research groups. As already mentioned, the measurement of DNA-binding proteins (SSB) from *E. coli* was reported by Kung *et al.* [68]. Nakao *et al.* demonstrated the detection of an extracellular pH distribution of colonies of *E. coli* with the described scanning LAPS set-up [33]. An immunoligand assay (ILA) in conjunction with a LAPS set-up for the rapid detection of *E. coli* bacteria was introduced by Gehring *et al.* [116] and a modified LAPS set-up, for which immunoparamagnetic beads (IMB) and a urease-conjugated anti-fluorescein antibody were combined to label cells captured by the beads, was proposed by Tu *et al.* [117]. The earlier described PAB system has been utilised by Ercole *et al.* to detect *E. coli* strains in water by an immunoassay-based system [28] as well as an antibody-based set-up for the determination of *E. coli* cells in vegetable food [118].

Tables 5.2–5.4 summarise the above-described research activities. Beside the biological agent under test, the measurement principle is given. For more details about the particular experiment, the reader may follow the corresponding references mentioned.

TABLE 5.2

Biological applications based on LAPS devices (Part 1)

Biological agent under test	Principle	References
Phospholipid-bilayer membranes with valinomycin, gramicidin D, alamethicin	Membrane conductance	[29]
Human epidermal keratinocytes, P388DI cells, CHO[a] cells	pH change due to different irritants	[62]
Yersinia pestis, *Neisseria meningitides*	Immunofiltration procedure	[63]
hCG[b]	Immunoassay	[30]
VEE[c]	Immunofiltration procedure	[64]
Human keratinocytes, B82, MESS-SA, DX5	pH change due to different irritants	[65]
Keratinocytes, CHO, B82	pH change for different concentration of EGF[d] and TGF-a[e]	[66,67]
Total single-stranded DNA	Urease-based enzyme-linked immunoassay	[31]
Total single-stranded DNA	DNA-binding proteins and anti-DNA antibody bind to enzymes	[68]
Cholinergic ligands	Nicotinic acetylcholine receptor (nAChR)	[85]
Mouse fibroblast line 3T6	pH shift due to metabolic activity	[86]

[a]Chinese hamster ovary.
[b]Human chorionic gonadotropin.
[c]Venezuelan equine encephalitis virus.
[d]Epidermal growth factor.
[e]Transforming growth factor a.

5.3.3 Alternative applications and materials for LAPS devices

In addition to the development of LAPS devices for biological applications, other research topics based on LAPS have been pursued over the last few years. Nearly all biological applications actually tend to determine the shift of the pH value due to the biological activity. Alternative applications are being focused to realise chemical sensors based on LAPS, which are sensitive to other ions rather than the pH.

TABLE 5.3

Biological applications based on LAPS devices (Part 2)

Biological agent under test	Principle	References
CHO-K1 cells with m1-muscarinic acetylcholine receptor	pH shift with and without carbamylcholine chloride	[87,88]
Lecithin, phosphatidylethanolamine, cholesterol, oleic acid lipid membranes	Response to saccharose, maltose, lactose, levulose	[100]
Alcohol dehydrogenase from yeast (YADH)	pH shift due to enzymatic reaction	[101]
HRP[a], CHO	Immunoenzymatic assays	[102]
Hepatocytes, 3T6, rat hepatocytes, HEPG2	Metabolic reaction on different concentration of mitoxantrone, Ara-C	[104]
Urease	Enzymatic reaction	[103]
HRP	Immunoenzymatic reaction on pesticide	[105]
Cyanide	*n*-CdSe membrane	[108]
Glucose, penicillin and urea	Enzymatic reaction on glucose oxidase, penicillinase and urease	[109]
Penicillin	Enzymatic reaction of penicillinase	[11]
Bacillus subtilis spores	Immunoassay	[110]
Trichosporon cutaneum	BOD test	[111]
Staphylococcal enterotoxin B, Newcastle disease, *Brucella melitensis*	Immunoassay	[112]
Nephridium cells	Metabolic response for chlorpromazine, fluorouracil, insulin and thyroxin	[113,114]
Terbuthylazine (TBA)	Monoclonal antibody (mAb) immunoassay	[106]

[a]Horse radish peroxidase.

Furthermore, different derivations of the LAPS principle have been proposed either to allow a multiple detection of different ions on the same sensor chip or to extend the measurement capabilities of LAPS devices. Another focus for alternative applications is LAPS-based sensors for gas-sensing purposes by making use of catalytic effects of thin

TABLE 5.4

Biological applications based on LAPS devices (Part 3)

Biological agent under test	Principle	References
DNA streptavidin–biotin–SSB–DNA–antiDNA–urease	Catalytic reaction of [68]	
Saccharomyces cerevisiae (yeast), *Escherichia coli*	pH change with spatial resolution	[33]
Escherichia coli	Detection by ILA	[116]
Escherichia coli	Detection by immunoassay with immunomagnetic bead	[117]
Escherichia coli	pH shift by anti-*E. coli* IgG coupled to urease	[28,118]
Urea and urease	Local pH change due to enzymatic reaction	[95]
Urea and glucose	Local pH change due to enzymatic reaction	[96]
CHO	Metabolic activity	[94]
Neural cells from *Aplysia*	Extracellular response on electrical stimulus	[97]
Pheochromocytoma (PC-12) cells nerve cells of *Lymnaea stagnalis*	Extracellular response on electrical stimulus	[20]
Chick heart muscle cells	Extracellular response on electrical stimulus	[98]
N1-E115 cell	Extracellular response on electrical stimulus	[99]
Disaccharides	Enzymatic reaction by invertase, alpha-D-glucosidase, glucokinase	[91]

metal films as transducer materials. In this section, selected examples covering these "alternative" applications and materials for LAPS devices will be presented.

Engström and Carlsson already introduced in 1983 an SLPT [119] for the characterisation of MIS structures, which was extended to chemical gas sensors by Lundström *et al.* [26]. Both SLPT and LAPS base upon the same technique and principle. However, due to the different fields of applications in history, one refers to LAPS for chemical sensors in electrolyte solutions and for biosensors, and the SLPT for gas sensors. A description of the development of a hydrogen sensor based on catalytic field-effect devices including the SLP technique can be found, e.g., in Refs. [120,121]. The SPLT consists of a metal surface as sensitive material which is heated by, for instance, underlying resistive heaters to a specific working-point temperature, and a prober tip replaces the reference electrode (see Fig. 5.10).

The electrical changes of the thin discontinuous metal film, due to chemical reactions of the gas ambient on the film, will influence the photocurrent in the same manner as for the LAPS principle. Thus, imaging of the potential changes occurring on the thin metal surface upon gas exposure can be performed [27]. Lundström *et al.* demonstrated the laterally resolved measurement of the hydrogen concentration over a large palladium surface by using a hydrogen-sensitive palladium-oxide-semiconductor device. Furthermore, a mathematical model compared to the measured results with model-based approximations was demonstrated [122]. The spillover effect of hydrogen was reported in the above works as a critical parameter, since it influences the measurement signal, especially near the lateral metal-oxide tran-

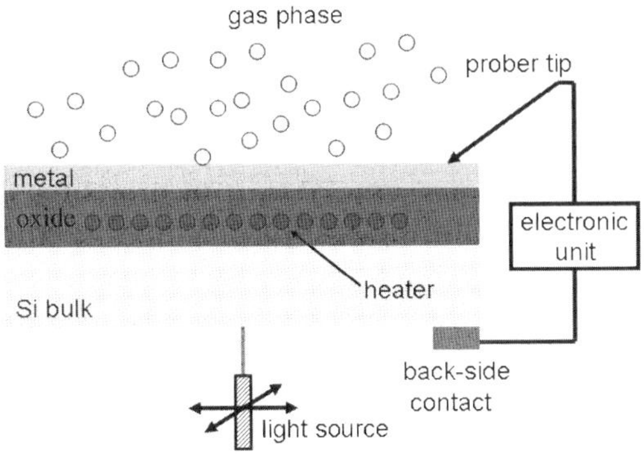

Fig. 5.10. Principal set-up of the scanning light pulse technique for gas-sensing LAPS.

sition. The effect of H_2 gas with SLPT and, further, the hydrogen spillover effect between the metal layer and the insulator [123] as well as the differences of the spillover effect between hydrophilic and hydrophobic sensing structures [124] have been investigated. The use of this effect to create hydrogen sensors with higher sensitivity was demonstrated by Filippini and Lundström [125].

An application-oriented multisensing approach was demonstrated by Lundström *et al.*, where different metal layers were used together with the SLP technique for the detection of ethanol and hydrogen in the gas ambient [126]. The first practical application here was the determination of diabetes by imaging the breathing air with the SLP technique [127]. Different catalytic metals (Pt, Pd or Ir) with a gradient in thickness (5 and 50 nm) have been deposited, in order to obtain characteristic images for different gas mixtures (H_2, NH_3 and C_2H_5OH) [128]. Filippini and Lundström reported a two-dimensional chemical imaging SLPT for hydrogen, ammonia and nitrogen dioxide including the spillover effect on SiO_2 with the aid of a porous metal layer [129]. The porous metal layer made it difficult to apply the necessary bias voltage. Therefore, underlying metal contacts [130] and a metal grid pattern [131] were introduced, to ensure a proper biasing of the surface; the complete measurement procedure for chemical gas sensing allows an unsupervised operation [132]. A summary of the experiments related to the gas/odour flow visualisation based on SLPT can be found, e.g., in Ref. [133].

With the SLPT devices, different practical attempts have been performed to obtain a high spatial resolution for the chemical imaging in electrolyte solutions [38,39,134]. However, in addition to the resolution itself, the scanning time to generate a two-dimensional image is a critical parameter. Too long measurement times can tend to time discrepancies between different measurement "pixels" and hence, a distortion of the "correct" chemical image. Therefore, different methods to speed up the scanning procedure have been proposed. The measurement time was successfully decreased by applying digital data processing [135]. In the first step, the measurement data were captured directly by an A/D converter and the final signal processing was done in digital form parallel to the measurement. By adjusting the ratio of the sampling frequency to the light-modulation frequency, Yoshinobu *et al.* were able to reduce the number of necessary measurement samples, consequently obtaining a distinct speed improvement [51].

Furthermore, the scanning LAPS was used to generate images of chemical processes. The spatial distribution and temporal evolution of redox potentials in chemical reaction systems were reported by Oba and Yoshinobu [136], followed by the visualisation of electrolytic processes [137,138]. The pH distribution around a cathode during the electrochemical deposition of calcium phosphate with the help of a scanning LAPS device [139] and the diffusion of protons in porous media were observed and evaluated [140].

Different approaches of the LAPS method have been investigated to enhance their performance. Adami *et al.* described a bi-LAPS set-up, where n- and p-doped silicon served as substrates to create a differential LAPS set-up. Since both LAPS devices react contrariwise, different disturbing effects (e.g., external light, analyte conductance and offset errors of the bias voltage) compensated (partially) each other and a more stable measurement signal was obtained [141]. Furthermore, the application of LAPS devices in an extended gate configuration was suggested by Adami *et al.* [142]. Another differential LAPS method was introduced by Ismail *et al.* based on two light sources with the same modulation frequency but with 90° phase shift to each other [143]. Due to the addition of the two out-of-phase photocurrents, low-frequency components were compensated. Yicong *et al.* and Qintao *et al.* investigated the use of a multi-LAPS (MLAPS) set-up. Instead of a single modulated light source, several light sources with different modulation frequencies were used simultaneously. The overall sum signal was then separated in the frequency domain [113,144]. K^+- and Ca^{2+}-sensitive measurement spots based on PVC membranes as well as a pH-sensitive spot (Si_3N_4) were placed on a single chip to measure the metabolic reaction of cells during drug influence (see also Section 5.3.2) [114]. Recently, the MLAPS was combined with stripping voltammetry (SV) to detect heavy metal ions (Fe, Cr, Mn and As) in water samples based on chalcogenide-glass membranes [145].

Many LAPS set-ups comprise the disadvantage to use a commercially expensive external amplifier stage for the signal read-out of the sensor chip. This avoids the use of equipment of the LAPS principle for "low-budget" applications, like for the point-of-care market and in-field applications. Yoshinobu *et al.* described the use of a miniaturised set-up with integrated electronics to create a hand-held LAPS device within the size of a standard glass electrode. The measurement with four different sensor spots sensitive to Li^+, K^+, Ca^{2+} and the pH value was demonstrated [146]. Schöning *et al.* extended this pen-shaped LAPS device to finally 16 sensor spots, including the complete electronic unit and a PC-based software package to perform the complete measurement procedure [147,148]. Another drawback of some LAPS systems is the difficult assembling of the LAPS chip into the measurement chamber containing the test sample. To avoid a long time of complex and difficult disassembling and assembling (mounting of the sensor chip), which might include breakage of the LAPS chip, recently a LAPS-card based on a standard plastic card and a suitable reader has been introduced [149,150] (see Fig. 5.11).

Finally, researchers became interested to work on alternative materials for LAPS devices to extend the range of possible ions that can be detected by LAPS arrangements. Beside the investigations of surface modifications with regard to the pH sensitivity [151] and to improve the biological cell adhesion [19], different ion-selective materials have been used together with LAPS devices: for the detection of K^+, Ca^{2+} and Mg^{2+} ions [152,153] as well as for anions (NO_3^- and SO_4^-) [154], sensitive membranes based on ionophores entrapped in a PVC matrix on top of the LAPS surface are reported. Another gas-sensing device with an LaF_3 membrane for F^- sensing based on SLPT and

Fig. 5.11. Miniaturised complete LAPS device with integrated electronic unit. (1) LAPS-card with integrated LAPS sensor, (2) electronic unit, (3) light sources, (4) measurement chamber and (4) reference electrode holder.

LAPS was introduced by Sato *et al.* [155]. The detection of heavy metal ions by thin films of chalcogenide-glass membranes using the pulsed laser deposition method (PLD) was reported by Mourzina *et al.* [156]. The PLD technique was also introduced to evaporate Al_2O_3 as a pH-sensitive material for LAPS devices [157]. The first practical application of the above-described LAPS card was demonstrated by Kloock *et al.* for a comparative study of Cd-sensitive chalcogenide glasses for ISFETs, LAPS and μISEs (ion-selective electrodes) [158].

Details on the fabrication of a LAPS for the detection of the pH value and the cadmium-ion concentration in aqueous solutions are given in Procedure 6 (see CD accompanying this book).

5.4 CONCLUSIONS

The LAPS is already successful in (laboratory) use for a wide range of applications. However, commercial devices are not available at the moment. The manufacturing of the Cytosensor from Molecular Devices was stopped in 1999. The commercial sensor market for chemical and especially biosensors is known to be conservative for new sensor trends: a typical example for this is the ISFET with more than 35 years of development (i.e., 20 years more compared to the LAPS); ISFET-based devices are still used only for some specialised applications in laboratories rather than becoming a mass-market product. Some reasons that LAPS devices are not yet commercially available might be found in the historical background. The first and only worldwide commercially available device was the Cytosensor. Developed as an explicit tool for biological research, it was delivered as a complete measurement environment, including computer and software (based on an Apple Macintosh). As usual in biological research, it included many disposable parts (e.g., membranes, tubes, sample containers). This tends to be an expensive and time-consuming task to operate and maintain the Cytosensor device, which only specialists were able to afford. However, it was used for a large series of mainly biological research activities (see Section 5.3.2).

The development of alternative approaches based on the LAPS principle started just around 10 years ago with the scanning LAPS device for chemical imaging of electrolyte solutions [32] and for gas-sensing applications [26]. These scanning LAPS devices demonstrate a great potential for a fast screening of analyte mixtures. The somehow fuzzy image can be used with intelligent signal processing (e.g., by artificial neuronal networks (ANN), fuzzy technique, principal component analysis (PCA) and clustering) to create an individual "finger print" of the test sample. The present scanning systems are relatively huge in size and fragile and the measurement time is high (in the order of several minutes up to hours). Thus, future developments have to investigate the possibilities to create fast, portable and robust LAPS "scanners" for a quick determination of analyte mixtures in "point-of-care" and in-field applications.

Besides further development for biological applications, LAPS platforms might be used to detect other ions in gases or liquids. With the possibility to design several independent measurement spots onto one single chip area, LAPS devices represent an ideal platform to develop and test new transducer materials. Either by scanning LAPS or by multiple discrete light sources, one can investigate transducer materials with different configurations at the same time under the same conditions in a "gradient-type" sensor configuration.

Another promising technique is the use of several light sources at the same time but with different modulation frequencies. The separation of the sum signal by digital signal processing allows the development of a high-speed LAPS for time-critical applications. Furthermore, the signals of different measurement spots can be compared at exactly the same time "window".

Until now, all in the literature proposed LAPS devices are complete autarkic measurement systems. Further applications can be found by the integration of LAPS devices into existing analytic fields. This requires the development of inexpensive integrated electronic units to operate the LAPS and to provide a standardised communication with higher process levels. The LAPS devices need to be easy in use to allow the operation in commercial environments. Due to the simple structure of the LAPS, the integration into micro-electro-mechanical systems (MEMS), "lab-on-chip" and micro-total analysis systems (μ-TAS) might be of special interest in the near future.

With more than 150 publications on LAPS devices, a wide range of possible applications for LAPS has been demonstrated in the past. However, since LAPS devices belong to the family of field-effect-based sensors, results and studies on, e.g., ISFETs and EIS sensors can be easily adapted to LAPS. This includes especially the transfer of alternative sensor materials, which were initially developed for ISFET and EIS devices and will further extend the range of possible applications for LAPS.

REFERENCES

1 P. Bergveld, Development of an ion-sensitive solid-state device for neurophysiological measurements, *IEEE Trans. Biomed. Eng.* BME-17(1) (1970) 70–71.

2 W.M. Siu and R.S. Cobbold, Basic properties of the electrolyte–SiO_2–Si system: physical and theoretical aspects, *IEEE Trans. Electron Devices*, ED-26(11) (1979) 1805–1815.

3 C.D.J. Fung, Ph.D. thesis: Characterization and Theory of Electrolyte–Insulator–Semiconductor Field-Effect Transistor, Case Western Reserve University, Cleveland, OH, March 1980.

4 L. Bousse, N.D. Rooij and P. Bergveld, Operation of chemically sensitive field-effect sensors as a function of the insulator–electrolyte interface, *IEEE Trans. Electron Devices*, ED-30(10) (1983) 1263–1270.

5 M. Kuhn, A quasi-static technique for MOS C-V and surface state measurements, *Solid-State Electron*, 13(6) (1970) 873–885.

6 F.H. Hielscher and H.M. Preier, Non-equilibrium C-V and I-V characteristics of metal–insulator–semiconductor capacitors, *Solid-State Electron.*, 12(7) (1969) 527–538.

7 D.R. Lamb, Some electrical properties of the silicon–silicon dioxide system, *Thin Solid Films*, 5(4) (1970) 247–276.

8 J.R. Brews and A.D. Lopez, A test for lateral nonuniformities in MOS devices using only capacitance curves, *Solid-State Electron.*, 16(11) (1973) 1267–1277.

9 M. Schöning, D. Brinkmann, D. Rolka, C. Demuth and A. Poghossian, CIP (cleaning-in-place) suitable ''non-glass'' pH sensor based on a Ta_2O_5-gate EIS structure, *Sens. Actuators B Chem.*, 111 (2005) 423–429.

10 M. Schöning, N. Näther, V. Auger, A. Poghossian and M. Koudelka-Hep, Miniaturised flow-through cell with integrated capacitive EIS sensor fabricated at wafer level using Si and SU-8 technologies, *Sens. Actuators B Chem.*, 108(1) (2005) 986–992.

11 A. Poghossian, T. Yoshinobu, A. Simonis, H. Ecken, H. Lüth, and M.J. Schöning, Penicillin detection by means of field-effect based sensors: EnFET, capacitive EIS sensor or LAPS? *Sens. Actuators B Chem.*, 78 (1–3) (2001) 237–242.

12 A. Poghossian, M. Thust, M. Schöning, M. Müller-Veggian, P. Kordos and H. Lüth, Cross-sensitivity of a capacitive penicillin sensor combined with a diffusion barrier, *Sens. Actuators B Chem.*, 68(1) (2000) 260–265.

13 O. Engström and A. Alm, Energy concepts of insulator–semiconductor interface traps, *J. Appl. Phys.*, 54(9) (1983) 5240–5244.

14 H. Mc Connell, J. Parce and D. Hafeman, Light addressable potentiometric biosensor, *J. Electrochem. Soc.*, 134(8B) (1987) C523.

15 D. Hafeman, J. Parce and H. McConnell, Light-addressable potentiometric sensor for biochemical systems, *Science*, 240(4856) (1988) 1182–1185.
16 U. Teravaninthorn, Y. Miyahara and T. Moriizumi, The suitability of Ta_2O_5 as a solid-state ion-sensitive membrane, *Jpn. J. Appl. Phys.*, 26(12) (1987) 2116–2120.
17 P. Gimmel, K.D. Schierbaum, W. Göpel, H.H. van den Vlekkert and N.F. de Rooij, Microstructured solid-state ion-sensitive membranes by thermal oxidation of Ta, *Sens. Actuators B Chem.*, 1(1–6) (1990) 345–349.
18 L. Bousse, S. Mostarshed, B. van der Schoot and N.F. de Rooij, Comparison of the hysteresis of Ta_2O_5 and Si_3N_4 pH-sensing insulators, *Sens. Actuators B Chem.*, 17(2) (1994) 157–164.
19 I. Hirata, H. Iwata, A.B.M. Ismail, H. Iwasaki, T. Yukimasa and H. Sugihara, Surface modification of Si_3N_4-coated silicon plate for investigation of living cells, *Jpn. J. Appl. Phys. Pt. 1: Regular Pap. Short Notes Rev. Pap.*, 39(11) (2000) 6441–6442.
20 A.B.M. Ismail, T. Yoshinobu, H. Iwasaki, H. Sugihara, T. Yukimasa, I. Hirata and H. Iwata, Investigation on light-addressable potentiometric sensor as a possible cell–semiconductor hybrid, *Biosens. Bioelectron.*, 18(12) (2003) 1509–1514.
21 S. Lamperski, Semi-empirical modelling of the properties of the electrode/electrolyte interface in the presence of specific anion adsorption, *Electrochim. Acta*, 41(14) (1996) 2089–2095.
22 D.E. Yates, S. Levine and T.W. Healy, Site-binding model of electrical double-layer at oxide–water interface, *J. Chem. Soc. Faraday Trans. I*, 70 (1974) 1807–1818.
23 A. Poghossian and M.J. Schöning, Silicon-based chemical and biological field-effect sensors. In: C.A. Grimes, E.C. Dickey and M.V. Pishko (Eds.), *Encyclopedia of Sensors*, Vol. X, American Scientific Publishers (USA), 2006, pp. 463–534.
24 M.Y. Doghish and F.D. Ho, A comprehensive analytical model for metal–insulator semiconductor (MIS) devices, *IEEE Trans. Electron Devices*, 39(12) (1992) 2771–2780.
25 T. Yoshinobu, H. Ecken, A. Poghossian, A. Simonis, H. Iwasaki, H. Lüth and M.J. Schöning, Constant-current-mode LAPS (CLAPS) for the detection of penicillin, *Electroanalysis*, 13(8–9) (2001) 733–736.
26 I. Lundström, R. Erlandsson, U. Frykman, E. Hedborg, A. Spetz, H. Sundgren, S. Welin and F. Winquist, Artificial 'olfactory' images from a chemical sensor using a light-pulse technique, *Nature*, 352(6330) (1991) 47–50.
27 F. Winquist, H. Sundgren, E. Hedborg, A. Spetz and I. Lundström, Visual images of gas mixtures produced with field-effect structures, *Sens. Actuators B Chem.*, 6(1–3) (1992) 157–161.

28 C. Ercole, M.D. Gallo, M. Pantalone, S. Santucci, L. Mosiello, C. Laconi and A. Lepidi, A biosensor for *E. coli* based on a potentiometric alternating biosensing (PAB) transducer, *Sens. Actuators B Chem.*, 83(1–3) (2002) 48–52.
29 G.B. Sigal, D.G. Hafeman, J.W. Parce and H.M. Mcconnell, Electrical-properties of phospholipid-bilayer membranes measured with a light addressable potentiometric sensor, *ACS Symp. Ser.*, 403 (1989) 46–64.
30 J.D. Olson, P.R. Panfili, R. Armenta, M.B. Femmel, H. Merrick, J. Gumperz, M. Goltz and R.F. Zuk, A silicon sensor-based filtration immunoassay using biotin-mediated capture, *J. Immunol. Methods*, 134(1) (1990) 71–79.
31 L. Bousse, G. Kirk and G. Sigal, Biosensors for detection of enzymes immobilized in microvolume reaction chambers, *Sens. Actuators B Chem.*, 1(1–6) (1990) 555–560.
32 M. Nakao, T. Yoshinobu and H. Iwasaki, Improvement of spatial resolution of a laser-scanning pH-imaging sensor, *Jpn. J. Appl. Phys. Pt. 2: Lett.*, 33 (3A) (1994) L394–L397.
33 M. Nakao, S. Inoue, R. Oishi, T. Yoshinobu and H. Iwasaki, Observation of microorganism colonies using a scanning-laser-beam pH-sensing microscope, *J. Ferment. Bioeng.*, 79(2) (1995) 163–166.
34 M. Nakao, S. Inoue, T. Yoshinobu and H. Iwasaki, High-resolution pH imaging sensor for microscopic observation of microorganisms, *Sens. Actuators B Chem.*, 34(1–3) (1996) 234–239.
35 W.J. Parak, U.G. Hofmann, H.E. Gaub and J.C. Owicki, Lateral resolution of light-addressable potentiometric sensors: an experimental and theoretical investigation, *Sens. Actuators A Phys.*, 63(1) (1997) 47–57.
36 Y. Ito, High-spatial resolution LAPS, *Sens. Actuators B Chem.*, 52(1–2) (1998) 107–111.
37 M. George, W.J. Parak, I. Gerhardt, W. Moritz, F. Kaesen, H. Geiger, I. Eisele and H.E. Gaub, Investigation of the spatial resolution of the light-addressable potentiometric sensor, *Sens. Actuators A Phys.*, 86(3) (2000) 187–196.
38 W. Moritz, I. Gerhardt, D. Roden, M. Xu and S. Krause, Photocurrent measurements for laterally resolved interface characterization, *Fresenius J. Anal. Chem.*, 367(4) (2000) 329–333.
39 S. Krause, H. Talabani, M. Xu, W. Moritz and J. Griffiths, Scanning photo-induced impedance microscopy—an impedance based imaging technique, *Electrochim. Acta*, 47(13–14) (2002) 2143–2148.
40 W. Moritz, T. Yoshinobu, F. Finger, S. Krause, M. Martin-Fernandez and M.J. Schöning, High resolution LAPS using amorphous silicon as the semiconductor material, *Sens. Actuators B Chem.*, 103 (2004) 436–441.
41 J.C. van den Heuvel, R.C. van Oort and M.J. Geerts, Diffusion length measurements of thin amorphous silicon layers, *Solid State Commun.*, 69(8) (1989) 807–810.

42 T. Yoshinobu, M.J. Schöning, F. Finger, W. Moritz and H. Iwasaki, Fabrication of thin-film LAPS with amorphous silicon, *Sensors*, 4 (2004) 163–169.

43 T. Yoshinobu, W. Moritz, F. Finger and S. Michael, Application of thin-film armorphous silicon to chemical imaging. In: *Material Research Society Symposium Proceeding*, Vol. 910, Material Research Society, 2006.

44 S. Krause, W. Moritz, H. Talabani, M. Xu, A. Sabot and G. Ensell, Scanning photo-induced impedance microscopy—resolution studies and polymer characterization, *Electrochim. Acta*, 51(8–9) (2006) 1423–1430.

45 M. Grattarola, S. Martinoia, G. Massobrio, A. Cambiaso, R. Rosichini and M. Tetti, Computer simulations of the responses of passive and active integrated microbiosensors to cell activity, *Sens. Actuators B Chem.*, 4(3–4) (1991) 261–265.

46 G. Massobrio, S. Martinoia and M. Grattarola, Light-addressable chemical sensors: Modelling and computer simulations, *Sens. Actuators B Chem.*, 7(1–3) (1992) 484–487.

47 M. Sartore, M. Adami and C. Nicolini, Computer simulation and optimization of a light addressable potentiometric sensor, *Biosens. Bioelectron.*, 7(1) (1992) 57–64.

48 G. Verzellesi, L. Colalongo, D. Passeri, B. Margesin, M. Rudan, G. Soncini and P. Ciampolini, Numerical analysis of ISFET and LAPS devices, *Sens. Actuators B Chem.*, 44(1–3) (1997) 402–408.

49 L. Colalongo, G. Verzellesi, D. Passeri, A. Lui, P. Ciampolini and M.V. Rudan, Modeling of light-addressable potentiometric sensors, *IEEE Trans. Electron Devices*, 44(11) (1997) 2083–2090.

50 Q. Zhang, Theoretical analysis and design of submicron-LAPS, *Sens. Actuators B Chem.*, 105 (2005) 304–311.

51 T. Yoshinobu, N. Oba, H. Tanaka and H. Iwasaki, High-speed and high-precision chemical-imaging sensor, *Sens. Actuators A Phys.*, 51(2–3) (1996) 231–235.

52 W. Bracke, P. Merken, R. Puers and C. Hoof, On the optimization of ultra low power front-end interfaces for capacitive sensors, *Sens. Actuators A Phys.*, 117(2) (2005) 273–285.

53 P. Temple-Boyer, J. Launay, G. Sarrabayrouse and A. Martinez, Amplifying structure for the development of field-effect capacitive sensors, *Sens. Actuators B Chem.*, 86(2–3) (2002) 111–121.

54 A.L. Kukla and Y.M. Shirshov, Computer simulation of transport processes in biosensor microreactors, *Sens. Actuators B Chem.*, 48(1–3) (1998) 461–466.

55 G. Ferri, P.D. Laurentiis, A. D'Amico and C.D. Natale, A low-voltage integrated CMOS analog lock-in amplifier prototype for LAPS applications, *Sens. Actuators A Phys.*, 92(1–3) (2001) 263–272.

56 M. George, W.J. Parak and H.E. Gaub, Highly integrated surface potential sensors, *Sens. Actuators B Chem.*, 69(3) (2000) 266–275.
57 L. Bousse, D. Hafeman and N. Tran, Time-dependence of the chemical response of silicon nitride surfaces, *Sens. Actuators B Chem.*, B1(1–6) (1990) 361–367.
58 M. Adami, D. Alliata, C.D. Carlo, M. Martini, L. Piras, M. Sartore and C. Nicolini, Characterization of silicon transducers with Si_3N_4 sensing surfaces by an AFM and a PAB system, *Sens. Actuators B Chem.*, 25(1–3) (1995) 889–893.
59 L.J. Bousse, S. Mostarshed and D. Hafeman, Combined measurement of surface potential and zeta potential at insulator/electrolyte interfaces, *Sens. Actuators B Chem.*, 10(1) (1992) 67–71.
60 R. Raiteri, B. Margesin and M. Grattarola, An atomic force microscope estimation of the point of zero charge of silicon insulators, *Sens. Actuators B Chem.*, 46(2) (1998) 126–132.
61 M. Sartore, M. Adami, C. Nicolini, L. Bousse, S. Mostarshed and D. Hafeman, Minority carrier diffusion length effects on light-addressable potentiometric sensor (LAPS) devices, *Sens. Actuators A Phys.*, 32(1–3) (1992) 431–436.
62 J.W. Parce, J.C. Owicki, K.M. Kercso, G.B. Sigal, H.G. Wada, V.C. Muir, L.J. Bousse, K.L. Ross, B.I. Sikic and H.M. McConnell, Detection of cell-affecting agents with a silicon biosensor, *Science*, 246(4927) (1989) 243–247.
63 J.M. Libby and H.G. Wada, Detection of *Neisseria meningitidis* and *Yersinia pestis* with a novel silicon-based sensor, *J. Clin. Microbiol.*, 27(7) (1989) 1456–1459.
64 W.-G. Hu, H.G. Thompson, A.Z. Alvi, L.P. Nagata, M.R. Suresh and R.E. Fulton, Development of immunofiltration assay by light addressable potentiometric sensor with genetically biotinylated recombinant antibody for rapid identification of Venezuelan equine encephalitis virus, *J. Immunol. Methods*, 289(1–2) (2004) 27–35.
65 J.W. Parce, J.C. Owicki and K.M. Kercso, Biosensors for directly measuring cell affecting agents, *Ann. Biol. Clin. (Paris)*, 48(9) (1990) 639–641.
66 J.C. Owicki and J.W. Parce, Bioassays with a microphysiometer, *Nature*, 344(6263) (1990) 271–272.
67 J. Owicki, J. Parce, K. Kercso, G. Sigal, V. Muir, J. Venter, C. Fraser and H. McConnell, Continuous monitoring of receptor-mediated changes in the metabolic rates of living cells, *PNAS*, 87(10) (1990) 4007–4011.
68 V.T. Kung, P.R. Panfili, E.L. Sheldon, R.S. King, P.A. Nagainis, B. Gomez, D.A. Ross, J. Briggs and R.F. Zuk, Picogram quantitation of total DNA using DNA-binding proteins in a silicon sensor-based system, *Anal. Biochem.*, 187(2) (1990) 220–227.
69 F. Hafner, Cytosensor® microphysiometer: technology and recent applications, *Biosens. Bioelectron.*, 15(3–4) (2000) 149–158.

70 J.C. Owicki and J.W. Parce, Biosensors based on the energy metabolism of living cells: the physical chemistry and cell biology of extracellular acidification, *Biosens. Bioelectron.*, 7(4) (1992) 255–272.
71 H.M. McConnell, J. Owicki, J. Parce, D. Miller, G. Baxter, H. Wada and S. Pitchford, The Cytosensor microphysiometer: biological applications of silicon technology, *Science*, 257(5078) (1992) 1906–1912.
72 K. Neve, M. Kozlowski and M. Rosser, Dopamine D2 receptor stimulation of Na^+/H^+ exchange assessed by quantification of extracellular acidification, *J. Biol. Chem.*, 267(36) (1992) 25748–25753.
73 K.M. Raley-Susman, K.R. Miller, J.C. Owicki and R.M. Sapolsky, Effects of excitotoxin exposure on metabolic rate of primary hippocampal cultures: application of silicon microphysiometry to neurobiology, *J. Neurosci.*, 12(3) (1992) 773–780.
74 D.L. Miller, J.C. Olson, J.W. Parce and J.C. Owicki, Cholinergic stimulation of the Na^+/K^+ adenosine triphosphatase as revealed by microphysiometry, *Biophys. J.*, 64(3) (1993) 813–823.
75 J.C. Fernando, K.R. Rogers, N.A. Anis, J.J. Valdes, R.G. Thompson, A.T. Eldefrawi and M.E. Eldefrawi, Rapid detection of anticholinesterase insecticides by a reusable light addressable potentiometric biosensor, *J. Agric. Food Chem.*, 41(3) (1993) 511–516.
76 J.T. Colston, P. Kumar, E.D. Rael, A.T.C. Tsin, J.J. Valdes and J.P. Chambers, Detection of sub-nanogram quantities of Mojave toxin via enzyme immunoassay with light addressable potentiometric detector, *Biosens. Bioelectron.*, 8(2) (1993) 117–121.
77 P. Kumar, R.C. Willson, J.J. Valdes and J.P. Chambers, Monitoring of oligonucleotide hybridization using light-addressable potentiometric and evanescent wave fluorescence sensing, *Mater. Sci. Eng. C*, 1(3) (1994) 187–192.
78 J.C. Owicki, L.J. Bousse, D.G. Hafeman, G.L. Kirk, J.D. Olson, H.G. Wada and J.W. Parce, The light-addressable potentiometric sensor: principles and biological applications, *Annu. Rev. Biophys. Biomol. Struct.*, 23 (1994) 87–113.
79 M.N. Garnovskaya, T.W. Gettys, T. van Biesen, V. Prpic, J.K. Chuprun and J.R. Raymond, 5-HT_{1A} receptor activates Na^+/H^+ exchange in CHO-K1 cells through $G_{i\alpha2}$ and $G_{i\alpha3}$, *J. Biol. Chem.*, 272(12) (1997) 7770–7776.
80 K. Dill, J.H. Song, J.A. Blomdahl and J.D. Olson, Rapid, sensitive and specific detection of whole cells and spores using the light-addressable potentiometric sensor, *J. Biochem. Biophys. Methods*, 34(2) (1997) 161–166.
81 K. Dill, L.H. Stanker and C.R. Young, Detection of salmonella in poultry using a silicon chip-based biosensor, *J. Biochem. Biophys. Methods*, 41(1) (1999) 61–67.
82 H. Fischer, A. Seelig, N. Beier, P. Raddatz and J. Seelig, New drugs for the Na^+/H^+ exchanger. Influence of Na^+ concentration and

determination of inhibition constants with a microphysiometer, *J. Membr. Biol.*, 168(1) (1999) 39–45.
83 Y. Okada, T. Taniguchi, Y. Akagi and I. Muramatsu, Two-phase response of acid extrusion triggered by purinoceptor in Chinese hamster ovary cells, *Eur. J. Pharmacol.*, 455(1) (2002) 19–25.
84 D. Bironaite, L. Gera and J.M. Stewart, Characterization of the B_2 receptor and activity of bradykinin analogs in SHP-77 cell line by Cytosensor microphysiometer, *Chem. Biol. Interact.*, 150(3) (2004) 283–293.
85 K.R. Rogers, J.C. Fernando, R.G. Thompson, J.J. Valdes and M.E. Eldefrawi, Detection of nicotinic receptor ligands with a light addressable potentiometric sensor, *Anal. Biochem.*, 202(1) (1992) 111–116.
86 P. Gavazzo, S. Paddeu, M. Sartore and C. Nicolini, Study of the relationship between extracellular acidification and cell viability by a silicon-based sensor, *Sens. Actuators B Chem.*, 19(1–3) (1994) 368–372.
87 L. Bousse and W. Parce, Applying silicon micromachining to cellular-metabolism, *IEEE Eng. Med. Biol. Mag.*, 13(3) (1994) 396–401.
88 L. Bousse, R.J. McReynolds, G. Kirk, T. Dawes, P. Lam, W.R. Bemiss and J.W. Parce, Micromachined multichannel systems for the measurement of cellular metabolism, *Sens. Actuators B Chem.*, 20(2–3) (1994) 145–150.
89 G.T. Baxter, L.J. Bousse, T.D. Dawes, J.M. Libby, D.N. Modlin, J.C. Owicki and J.W. Parce, Microfabrication in silicon microphysiometry, *Clin. Chem.*, 40(9) (1994) 1800–1804.
90 L. Bousse, Whole cell biosensors, *Sens. Actuators B Chem.*, 34(1–3) (1996) 270–275.
91 K. Aoki, H. Uchida, T. Katsube, Y. Ishimaru and T. Iida, Integration of bienzymatic disaccharide sensors for simultaneous determination of disaccharides by means of light addressable potentiometric sensor, *Anal. Chim. Acta*, 471(1) (2002) 3–12.
92 S.E. Eklund, D. Taylor, E. Kozlov, A. Prokop and D.E. Cliffel, A microphysiometer for simultaneous measurement of changes in extracellular glucose, lactate, oxygen, and acidification rate, *Anal. Chem.*, 76(3) (2004) 519–527.
93 S.E. Eklund, R.M. Snider, J. Wikswo, F. Baudenbacher, A. Prokop and D.E. Cliffel, Multianalyte microphysiometry as a tool in metabolomics and systems biology, *J. Electroanal. Chem.*, 587(2) (2006) 333–339.
94 B. Stein, M. George, H.E. Gaub, J.C. Behrends and W.J. Parak, Spatially resolved monitoring of cellular metabolic activity with a semiconductor-based biosensor, *Biosens. Bioelectron.*, 18(1) (2003) 31–41.
95 S. Inoue, M. Nakao, T. Yoshinobu and H. Iwasaki, Chemical-imaging sensor using enzyme, *Sens. Actuators B Chem.*, 32(1) (1996) 23–26.
96 M. Shimizu, Y. Kanai, H. Uchida and T. Katsube, Integrated biosensor employing a surface photovoltage technique, *Sens. Actuators B Chem.*, 20(2–3) (1994) 187–192.

97 H. Tanaka, T. Yoshinobu and H. Iwasaki, Application of the chemical imaging sensor to electrophysiological measurement of a neural cell, *Sens. Actuators B Chem.*, 59(1) (1999) 21–25.

98 W. Parak, M. George, J. Domke, M. Radmacher, J. Behrends, M. Denyer, and H. Gaub, Can the light-addressable potentiometric sensor (LAPS) detect extracellular potentials of cardiac myocytes? *IEEE Trans. Biomed. Eng.*, 47 (2000) 1106–1113.

99 B. Stein, M. George, H.E. Gaub and W.J. Parak, Extracellular measurements of averaged ionic currents with the light-addressable potentiometric sensor (LAPS), *Sens. Actuators B Chem.*, 98(2–3) (2004) 299–304.

100 Y. Sasaki, Y. Kanai, H. Uchida and T. Katsube, Highly sensitive taste sensor with a new differential LAPS method, *Sens. Actuators B Chem.*, 25(1–3) (1995) 819–822.

101 M. Adami, L. Piras, M. Lanzi, A. Fanigliulo, S. Vakula and C. Nicoli, Monitoring of enzymatic activity and quantitative measurements of substrates by means of a newly designed silicon-based potentiometric sensor, *Sens. Actuators B Chem.*, 18(1–3) (1994) 178–182.

102 M. Adami, M. Sartore and C. Nicolini, PAB: a newly designed potentiometric alternating biosensor system, *Biosens. Bioelectron.*, 10(1–2) (1995) 155–167.

103 S. Paddeu, A. Fanigliulo, M. Lanzi, T. Dubrovsky and C. Nicolini, LB-based PAB immunosystem: activity of an immobilized urease monolayer, *Sens. Actuators B Chem.*, 25(1–3) (1995) 876–882.

104 C. Nicolini, M. Lanzi, P. Accossato, A. Fanigliulo, F. Mattioli and A. Martelli, A silicon-based biosensor for real-time toxicity testing in normal versus cancer liver cells, *Biosens. Bioelectron.*, 10(8) (1995) 723–733.

105 L. Piras, M. Adami, S. Fenu, M. Dovis and C. Nicolini, Immunoenzymatic application of a redox potential biosensor, *Anal. Chim. Acta*, 335(1–2) (1996) 127–135.

106 L. Mosiello, C. Laconi, M. Del Gallo, C. Ercole and A. Lepidi, Development of a monoclonal antibody based potentiometric biosensor for terbuthylazine detection, *Sens. Actuators B Chem.*, 95(1–3) (2003) 315–320.

107 A. Fanigliulo, P. Accossato, M. Adami, M. Lanzi, S. Martinoia, S. Paddeu, M.T. Parodi, A. Rossi, M. Sartore, M. Grattarola and C. Nicolini, Comparison between a LAPS and an FET-based sensor for cell-metabolism detection, *Sens. Actuators B Chem.*, 32(1) (1996) 41–48.

108 S. Licht, Developments in photoelectrochemistry: light-addressable photoelectrochemical cyanide sensors, *Colloids Surf. A Physicochem. Eng. Aspects*, 134(1–2) (1998) 231–239.

109 A. Seki, S. Ikeda, I. Kubo and I. Karube, Biosensors based on light-addressable potentiometric sensors for urea, penicillin and glucose, *Anal. Chim. Acta*, 373(1) (1998) 9–13.

110 K.A. Uithoven, J.C. Schmidt and M.E. Ballman, Rapid identification of biological warfare agents using an instrument employing a light addressable potentiometric sensor and a flow-through immunofiltration-enzyme assay system, *Biosens. Bioelectron.*, 14(10–11) (2000) 761–770.
111 Y. Murakami, T. Kikuchi, A. Yamamura, T. Sakaguchi, K. Yokoyama, Y. Ito, M. Takiue, H. Uchida, T. Katsube and E. Tamiya, An organic pollution sensor based on surface photovoltage, *Sens. Actuators B Chem.*, 53 (1998) 163–172.
112 W.E. Lee, H.G. Thompson, J.G. Hall and D.E. Bader, Rapid detection and identification of biological and chemical agents by immunoassay, gene probe assay and enzyme inhibition using a silicon-based biosensor, *Biosens. Bioelectron.*, 14(10–11) (2000) 795–804.
113 W. Yicong, W. Ping, Y. Xuesong, Z. Qingtao, L. Rong, Y. Weimin and Z. Xiaoxiang, A novel microphysiometer based on mLAPS for drugs screening, *Biosens. Bioelectron.*, 16(4–5) (2001) 277–286.
114 W. Yicong, W. Ping, Y. Xuesong, Z. Gaoyan, H. Huiqi, Y. Weimin, Z. Xiaoxiang, H. Jinghong and C. Dafu, Drug evaluations using a novel microphysiometer based on cell-based biosensors, *Sens. Actuators B Chem.*, 80(3) (2001) 215–221.
115 I.G. Mourzina, T. Yoshinobu, Y.E. Ermolenko, Y.G. Vlasov, M.J. Schöning and H. Iwasaki, Immobilization of urease and cholinesterase on the surface of semiconductor transducer for the development of light-addressable potentiometric sensors, *Microchim. Acta*, 144(1–3) (2004) 41–50.
116 A.G. Gehring, D.L. Patterson, and S.-I. Tu, Use of a light-addressable potentiometric sensor for the detection of *E. coli* O157:H7, Anal. Biochem., 258(2) (1998) 293–298.
117 S.I. Tu, J. Uknalis and A. Gehring, Detection of immunomagnetic bead captured *E. coli* O157:H7 by light addressable potentiometric sensor, *J. Rapid Methods Autom. Microbiol.*, 7(2) (1999) 69–79.
118 C. Ercole, M.D. Gallo, L. Mosiello, S. Baccella and A. Lepidi, *E. coli* detection in vegetable food by a potentiometric biosensor, *Sens. Actuators B Chem.*, 91(1–3) (2003) 163–168.
119 O. Engström and A. Carlsson, Scanned light pulse technique for the investigation of insulator–semiconductor interfaces, *J. Appl. Phys.*, 54(9) (1983) 5245–5251.
120 I. Lundström, C. Svensson, A. Spetz, H. Sundgren and F. Winquist, From hydrogen sensors to olfactory images—twenty years with catalytic field-effect devices, *Sens. Actuators B Chem.*, 13(1–3) (1993) 16–23.
121 I. Lundström, Why bother about gas-sensitive field-effect devices? *Sens. Actuators A Phys.*, 56(1–2) (1996) 75–82.
122 M.L.I. Lundström, Monitoring of hydrogen consumption along a palladium surface by using a scanning light pulse technique, *J. Appl. Phys.*, 86(2) (1999) 1106–1113.

123 M. Holmberg and I. Lundström, A new method for the detection of hydrogen spillover, *Appl. Surf. Sci.*, 93(1) (1996) 67–76.
124 E. Hedborg, F. Winquist, H. Sundgren and I. Lundström, Charge migration on hydrophobic and hydrophilic silicon dioxide, *Thin Solid Films*, 340(1–2) (1999) 250–256.
125 D. Filippini and I. Lundström, Hydrogen detection on bare SiO_2 between metal gates, *J. Appl. Phys.*, 91(6) (2002) 3896–3903.
126 I. Lundström, H. Sundgren and F. Winquist, Generation of response maps of gas mixtures, *J. Appl. Phys.*, 74(11) (1993) 6953–6961.
127 Q. Zhang, P. Wang, J. Li and X. Gao, Diagnosis of diabetes by image detection of breath using gas-sensitive LAPS, *Biosens. Bioelectron.*, 15(5–6) (2000) 249–256.
128 M. Löfdahl, M. Eriksson and I. Lundström, Chemical images, *Sens. Actuators B Chem.*, 70(1–3) (2000) 77–82.
129 D. Filippini and I. Lundström, Distinctive photocurrent chemical images on bare SiO_2 between continuous metal gates, *Sens. Actuators B Chem.*, 95(1–3) (2003) 116–122.
130 D. Filippini, I. Lundström and H. Uchida, Gap-gate field effect gas sensing device for chemical image generation, *Appl. Phys. Lett.*, 84(15) (2004) 2946–2948.
131 D. Filippini, J. Gunnarsson and I. Lundström, Chemical image generation with a grid-gate device, *J. Appl. Phys.*, 96(12) (2004) 7583–7590.
132 H. Uchida, D. Filippini and I. Lundström, Unsupervised scanning light pulse technique for chemical sensing, *Sens. Actuators B Chem.*, 103(1–2) (2004) 225–232.
133 J. Mizsei, Chemical imaging by direct methods, *Thin Solid Films*, 436(1) (2003) 25–33.
134 M. Nakao, T. Yoshinobu and H. Iwasaki, Scanning-laser-beam semiconductor pH-imaging sensor, *Sens. Actuators B Chem.*, 20(2–3) (1994) 119–123.
135 H. Uchida, W.Y. Zhang and T. Katsube, High speed chemical image sensor with digital LAPS system, *Sens. Actuators B Chem.*, 34(1–3) (1996) 446–449.
136 N. Oba, T. Yoshinobu and H. Iwasaki, Redox potential imaging sensor, *Jpn. J. Appl. Phys. Pt. 2 – Lett.* 35 (4A) (1996) L460–L463.
137 T. Yoshinobu, H. Iwasaki, M. Nakao, S. Nomura, T. Nakanishi, S. Takamatsu and K. Tomita, Application of chemical imaging sensor to electro generated pH distribution, *Jpn. J. Appl. Phys. Pt. 2 – Lett. Express Lett.* 37 (3B) (1998) L353–L355.
138 T. Yoshinobu, T. Harada and H. Iwasaki, Application of the pH-imaging sensor to determining the diffusion coefficients of ions in electrolytic solutions, *Jpn. J. Appl. Phys. Pt. 2 – Lett.*, 39 (4A) (2000) L318–L320.

139 S. Ban and S. Maruno, pH distribution around the electrode during electrochemical deposition process for producing bioactive apatite, *Jpn. J. Appl. Phys. Pt. 2 -Lett.* 38 (5A) (1999) L537–L539.
140 S. Nomura, Y. Yang, C. Inoue and T. Chida, Observation and evaluation of proton diffusion in porous media by the pH-imaging microscopy using a flat semiconductor pH sensor, *Anal. Sci.*, 18 (2002) 1081–1084.
141 M. Adami, M. Sartore, E. Baldini, A. Rossi and C. Nicolini, New measuring principle for LAPS devices, *Sens. Actuators B Chem.*, 9(1) (1992) 25–31.
142 M. Adami, M. Sartore, A. Rapallo and C. Nicolini, Possible developments of a potentiometric biosensor, *Sens. Actuators B Chem.*, 7(1–3) (1992) 343–346.
143 A.B.M. Ismail, H. Sugihara, T. Yoshinobu and H. Iwasaki, A novel low-noise measurement principle for LAPS and its application to faster measurement of pH, *Sens. Actuators B Chem.*, 74(1–3) (2001) 112–116.
144 Z. Qintao, W. Ping, W.J. Parak, M. George and G. Zhang, A novel design of multi-light LAPS based on digital compensation of frequency domain, *Sens. Actuators B Chem.*, 73(2–3) (2001) 152–156.
145 H. Men, S. Zou, Y. Li, Y. Wang, X. Ye and P. Wang, A novel electronic tongue combined MLAPS with stripping voltammetry for environmental detection, *Sens. Actuators B Chem.*, 110(2) (2005) 350–357.
146 T. Yoshinobu, M.J. Schöning, R. Otto, K. Furuichi, Y. Mourzina, Y. Ermolenko and H. Iwasaki, Portable light-addressable potentiometric sensor (LAPS) for multisensor applications, *Sens. Actuators B Chem.*, 95(1–3) (2003) 352–356.
147 M. Schöning, T. Wagner, C. Wang, R. Otto and T. Yoshinobu, Development of a handheld 16 channel pen-type LAPS for electrochemical sensing, *Sens. Actuators B Chem.*, 108(1–2) (2005) 808–814.
148 T. Yoshinobu, H. Iwasaki, Y. Ui, K. Furuichi, Y. Ermolenko, Y. Mourzina, T. Wagner, N. Näther and M. Schöning, The light-addressable potentiometric sensor for multi-ion sensing and imaging, *Methods*, 37(1) (2005) 94–102.
149 T. Wagner, T. Yoshinobu, C. Rao, R. Otto and M.J. Schöning, All-in-one'' solid-state device based on a light-addressable potentiometric sensor platform, *Sens. Actuators B Chem.*, 117(2) (2006) 472–479.
150 T. Wagner, C. Rao, J.P. Kloock, T. Yoshinobu, R. Otto, M. Keusgen and M.J. Schöning, ''LAPS Card''—A novel chip card-based light-addressable potentiometric sensor (LAPS), *Sens. Actuators B Chem.*, 118 (2006) 33–40.
151 Y. Yoshimi, T. Matsuda, Y. Itoh, F. Ogata and T. Katsube, Surface modifications of functional electrodes of a light addressable potentiometric sensor (LAPS): non-dependency of pH sensitivity on the surface functional group, *Mater. Sci. Eng. C*, 5(2) (1997) 131–139.

152 A. Seki, K. Motoya, S. Watanabe and I. Kubo, Novel sensors for potassium calcium and magnesium ions based on a silicon transducer as a light-addressable potentiometric sensor, *Anal. Chim. Acta*, 382(1–2) (1999) 131–136.
153 Y. Ermolenko, T. Yoshinobu, Y. Mourzina, K. Furuichi, S. Levichev, M.J. Schöning, Y. Vlasov and H. Iwasaki, The double K^+/Ca^{2+} sensor based on laser scanned silicon transducer (LSST) for multi-component analysis, *Talanta*, 59(4) (2003) 785–795.
154 Y.G. Mourzina, Y.E. Ermolenko, T. Yoshinobu, Y. Vlasov, H. Iwasaki and M.J. Schöning, Anion-selective light-addressable potentiometric sensors (LAPS) for the determination of nitrate and sulphate ions, *Sens. Actuators B Chem.*, 91(1–3) (2003) 32–38.
155 T. Sato, M. Shimizu, H. Uchida and T. Katsube, Light-addressable suspended-gate gas sensor, *Sens. Actuators B Chem.*, 20(2–3) (1994) 213–216.
156 Y. Mourzina, T. Yoshinobu, J. Schubert, H. Lüth, H. Iwasaki and M.J. Schöning, Ion-selective light-addressable potentiometric sensor (LAPS) with chalcogenide thin film prepared by pulsed laser deposition, *Sens. Actuators B Chem.*, 80(2) (2001) 136–140.
157 A.B.M. Ismail, T. Harada, T. Yoshinobu, H. Iwasaki, M.J. Schöning and H. Lüth, Investigation of pulsed laser-deposited Al_2O_3 as a high pH-sensitive layer for LAPS-based biosensing applications, *Sens. Actuators B Chem.*, 71(3) (2000) 169–172.
158 J.P. Kloock, L. Moreno, A. Bratov, S. Huachupoma, J. Xu, T. Wagner, T. Yoshinobu, Y. Ermolenko, Y.G. Vlasov and M.J. Schöning, PLD-prepared cadmium sensors based on chalcogenide glasses—ISFET, LAPS and μISE semiconductor structures, *Sens. Actuators B Chem.*, 118 (2006) 149–155.

VOLTAMMETRIC (BIO)SENSORS

Chapter 6

Stripping-based electrochemical metal sensors for environmental monitoring

Joseph Wang

6.1 INTRODUCTION

Contamination of heavy metals (such as lead, arsenic, cadmium or mercury) is widespread around the world. Major concerns regarding the toxic effect of heavy metal have led to increasing needs to monitor trace metals in a variety of matrices. Traditionally, such trace-metal measurements have been carried out in the central laboratory, in connection with time-consuming sampling, transportation and storage steps, and a bulky atomic spectroscopy instrumentation. *In situ* and on-site monitoring of trace metals is preferable for most practical situations, since it offers various advantages. For example, in water quality control, a continuous surveillance must be kept to maintain the level of toxic metals within legally defined limits. In industrial process control, rapid measurements of metals are essential if a prompt corrective action has to be taken. *In situ* measurements also obviate errors due to contamination, losses, and other (equilibrium) changes, associated with the sample collection, and can greatly reduce the cost of metal analysis. Atomic spectroscopic techniques, commonly used for measuring trace metals in the central laboratory, are not suitable for the task of on-site assays.

Electrochemical stripping analysis has always been recognized as a powerful tool for measuring trace metals [1–3]. Its remarkably high sensitivity is attributed to the 'built-in' accumulation step, during which the target metals are preconcentrated onto the working electrode. The inherent miniaturization of electrochemical instruments and their minimal power requirements satisfy many of the requirements for on-site and *in situ* measurements of toxic metals. Stripping analysis is uniquely suited for speciation studies because of its

Comprehensive Analytical Chemistry 49
Alegret and Merkoçi (Eds)
Volume 49 ISSN: 0166-526X DOI: 10.1016/S0166-526X(06)49006-1

sensitivity both to low metal concentrations and to the chemical form of metals in solution. Recent technological advances have successfully addressed previous obstacles for such field applications. The consequence of these developments is that major considerations are now given to decentralized electrochemical testing for trace metals.

The present chapter reviews recent research efforts aimed at developing new devices for *in situ* and on-site electrochemical stripping analysis of trace metals. It is not a comprehensive review, but rather focuses on new tools for decentralized metal testing, including remotely deployed submersible stripping probes, hand-held metal analyzers coupled with disposable microfabricated strips, and newly developed 'green' bismuth film sensors.

6.2 PRINCIPLES

Stripping analysis is a two-step technique. The first or deposition step commonly involves the electrolytic deposition of a small portion of the metal ions in solution into the mercury electrode to preconcentrate the metals. Non-electrolytic (adsorptive) accumulation schemes have also been developed for expanding the scope of stripping analysis to trace metals that cannot be electrodeposited. The preconcentration step is followed by the stripping (measurement) step, which involves the dissolution (stripping) of the deposit. Different versions of stripping analysis can be employed, depending upon the nature of the deposition and measurement steps.

6.2.1 Anodic stripping voltammetry

Anodic stripping voltammetry (ASV) is the oldest, and still the most widely used version of stripping analysis [3]. The technique is applicable to metal ions that can be readily deposited at the working electrode, and particularly for those metals that dissolve in mercury. In this case, the metals are being preconcentrated by electrodeposition into a small-volume mercury electrode (a thin mercury film or a hanging mercury drop). The preconcentration is done by cathodic deposition at a controlled potential and time. The deposition potential is usually ca. 0.3 V more negative than E° for the least easily reduced metal ion to be determined. The metal ions reach the mercury electrode by diffusion and convection, where they are reduced and concentrated as amalgams:

$$M^{n+} + ne^{-} + Hg \rightarrow M(Hg) \quad (6.1)$$

Since the sensitivity of the stripping operation is dependent on the deposition time, the latter should be selected according to the concentration of the target metals (from around 2 min at the 10^{-8} M level to 15 min for 10^{-9}–10^{-10} M concentrations). The deposition step is usually facilitated by convective transport of the analyte to the surface of the working electrode. Quiescent solutions can be used in connection to ultramicroelectrodes. Only a small, and yet reproducible, fraction of the metal in the solution is being deposited (Fig. 6.1).

Following the preselected deposition period, the forced convection is stopped, and an anodic potential scan is initiated, during which the amalgamated metals are reoxidized and 'stripped' away from the surface:

$$M(Hg) \rightarrow M^{n+} + ne^{-} + Hg \tag{6.2}$$

The stripping voltammogram, recorded during this measurement step, consists of multiple current peaks, corresponding to the reoxidation of the amalgamated metals, and their 'stripping' out of the electrode. Such output provides the qualitative and quantitative information through measurements of the peak potential and current,

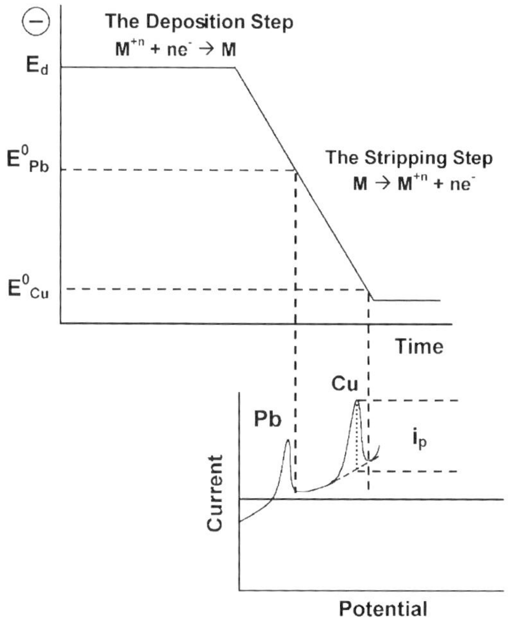

Fig. 6.1. The operation and response of anodic stripping voltammetry: the potential-time waveform (top), along with the resulting stripping voltammogram (bottom).

respectively. The ASV operation thus allows the simultaneous measurements of four to six trace metals.

6.2.2 Potentiometric stripping analysis

Potentiometric stripping analysis (PSA) is another attractive version of stripping analysis [4,5]. PSA offers several important advantages (over analogous ASV protocols), including reliable operation in nondeaerated samples and reduced susceptibility to adsorption of organic surfactants. Similar to other stripping operations, PSA consists of two steps. The first (accumulation) step, in which the metals are electrodeposited onto the working electrode (usually a mercury film), is followed by a stripping step in which the amalgamated metals are brought back to the solution with the aid of an oxidant (e.g., oxygen, mercury ion) or a constant anodic current. The resulting potentiogram (E vs. t plot) consists of stripping plateaus, as in a redox titration curve, with sharp potential changes accompanying the depletion of each metal from the surface. Such potentiogram provides both qualitative and quantitative analytical information. Qualitative identification relies on potential measurements (in accordance with the Nernst equation for the amalgamated metal). The transition time needed for the oxidation of a given metal, t_M, is a quantitative measure of the sample concentration of the metal:

$$\frac{t_M \propto C_{M^{n+}} t_{dep}}{C_{ox}} \tag{6.3}$$

where C_{ox} is the concentration of the oxidant. Modern PSA instruments use microcomputers to register fast stripping events and to convert the wave-shaped signal to a more convenient peak over a flat baseline (i.e., to dt/dE vs. E signals). The peak potential thus providing the qualitative identification with the peak area is proportional to the bulk concentration. Because of the short (ms) transition times, a fast rate of data acquisition (kHz) is required for obtaining a sufficient number of 'counts' for a defined peak.

6.2.3 Adsorptive stripping voltammetry

Adsorptive stripping analysis greatly expands the scope of stripping measurements towards numerous trace metals (e.g., Cr, Al, Fe, V, Mo and U) that are not electrodeposited readily [6,7]. This protocol involves the formation of a surface-active complex of the target metal (in the presence of an appropriate chelating agent), followed by its adsorptive accumulation (usually at a positive potential), and reduction by a negative-going scan. Both voltammetric and potentiometric stripping schemes, with a negative-sweeping potential scan or constant cathodic current, respectively, can be employed for measuring the adsorbed metal–ligand complex.

6.3 WORKING ELECTRODES FOR STRIPPING ANALYSIS: FROM MERCURY ELECTRODES TO DISPOSABLE STRIPS

Table 6.1 summarizes the development of sensing electrodes for electrochemical stripping analysis of trace metals. Traditionally, laboratory-based stripping measurements have relied on relatively bulky, expensive and toxic electrodes, such as the static mercury drop electrode (SMDE) or a mercury-coated rotating disk electrode (with a large mercury reservoir and heavy motor, respectively). Such bulky electrodes were combined with large cells (of 25–50 mL solution volume), the operation of which required careful cleaning, a prolonged oxygen removal (by bubbling nitrogen), solution stirring during the preconcentration, standard additions and replacement with a new solution.

In order to address the needs of decentralized (field) testing, it is necessary to move away from such cumbersome electrodes and operation. The exploitation of advanced microfabrication techniques allows the replacement of traditional ('beaker-type') electrochemical cells with

TABLE 6.1

Development of sensing electrodes for electrochemical stripping analysis of trace metals

Year	Development	Reference
1958	Introduction of the hanging mercury drop electrode	
1965	Introduction of mercury film electrodes	
1970	Introduction of *in situ* plated mercury films	
1978	Introduction of on-line metal detectors	[16]
1992	Introduction of disposable stripping sensors	[8]
1994	Introduction of hand-held stripping analyzers	
1997	Introduction of remote *in situ* metal sensors	[18,23]
2000	Introduction of bismuth film electrodes	[12]

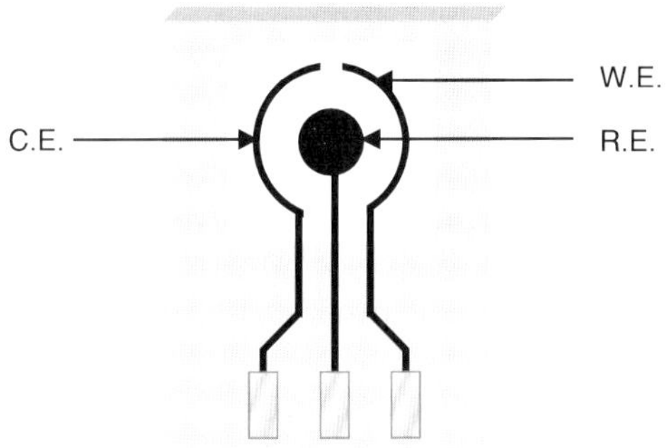

Fig. 6.2. Screen-printed (thick-film) sensors for decentralized metal testing. WE: working reference; RE: reference electrode; CE: counter electrode.

easy-to-use disposable sensor strips [8,9], similar to those used for the management of diabetes.

Such strips rely on planar working and reference electrodes that are screen-printed on a plastic or ceramic substrate (Fig. 6.2). These strips can be considered as self-contained electrochemical cells onto which the sample droplet is placed. The screen-printing technology, adapted from the microelectronic industry, offers high-volume production of extremely inexpensive, and yet highly reproducible single-use metal sensors. As desired for routine field applications, the performance of these disposable metal sensors is not compromised in comparison with traditional glassy-carbon-based electrodes. With their single-use character, such strips eliminate the need for surface polishing, solution replacement and cell cleaning (and related contamination and carry-over problems). Each experiment is thus performed on a fresh electrode surface in connection to a mercury or bismuth film.

It is also possible to use standard silicon-based lithographic technologies for fabricating planar sensor strips for stripping analysis of trace metals [10]. The higher resolution of such thin-film microfabrication route is particularly attractive for designing microelectrode arrays. For example, Buffle's group employed an array of 64 individually addressable gel-integrated mercury-plated iridium microelectrodes (of 5 μm diameter) for high-resolution concentration profiling measurements [11].

6.4 BISMUTH-BASED METAL SENSORS

A major limitation of field-based stripping sensing is the toxicity of mercury (used as the working electrode). Intensive research efforts have thus been devoted to the development of alternative stripping electrodes, with a performance approaching that of mercury-based ones. Since the year 2000 [12], bismuth film electrodes (BFEs) have become an attractive topic as a potential replacement for mercury electrodes [13,14]. Bismuth is an environmentally friendly element, with very low toxicity, and a widespread pharmaceutical use. Bismuth films have been deposited on carbon substrates, particularly glassy carbon, carbon fiber, carbon paste or screen-printed carbon strips. Such bismuth films can be prepared *ex situ* (preplated) or *in situ* (by adding 0.25–1.0 ppm bismuth(III) directly to the sample solution, and simultaneously depositing the target heavy metals and bismuth). The resulting BFEs offer a well-defined, undistorted and highly reproducible response, excellent resolution of neighboring stripping peaks, high hydrogen evolution, with signal-to-background characteristics comparable to those of common mercury electrodes (e.g., Fig. 6.3). The coupling of BFEs with screen-printed carbon substrates [15] is

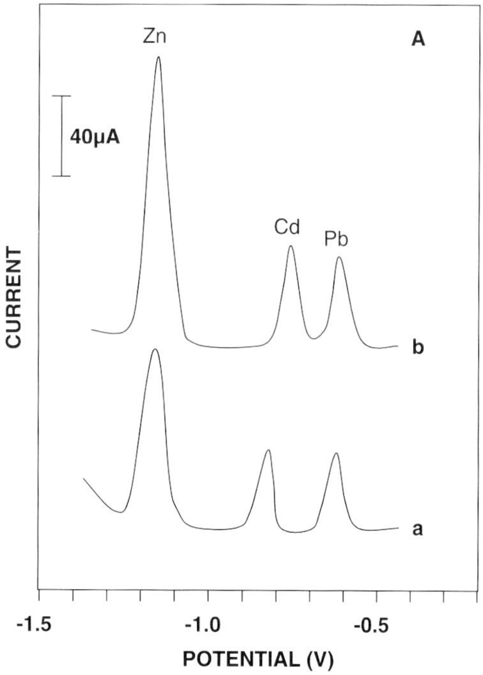

Fig. 6.3. Stripping voltammograms for 50 ppb Zn, Cd and Pb at glassy-carbon electrodes coated with bismuth (a) and mercury (b) films. Two-min deposition at –1.4 V. (Based on Ref. [12], with permission obtained from The American Chemical Society).

particularly attractive for meeting the demands of decentralized field metal testing. Such use of bismuth electrodes in electrochemical stripping analysis has taken off rapidly and will surely continue to expand.

6.5 *IN SITU* METAL SENSORS

The portable instrumentation and low power demands of stripping analysis satisfy many of the requirements for on-site and *in situ* measurements of trace metals. Stripping-based automated flow analyzers were developed for continuous on-line monitoring of trace metals since the mid-1970s [16,17]. These flow systems involve an electrochemical flow detector based on a wall-jet or thin-layer configuration along with a mercury-coated working electrode, and downstream reference and counter electrodes.

6.6 REMOTE METAL SENSORS

The need for continuous monitoring of trace metals in a variety of matrices has led to the development of submersible sensors based on electrochemical stripping analysis [18–22]. By providing a fast return of the analytical information in a timely, safe and cost effective fashion, such remotely deployed metal sensors offer direct and reliable assessment of the fate and gradient of contaminant sites, while greatly reducing the huge analytical costs. Submersible probes circumvent the need for solution pumping and offer greater simplification and miniaturization. Such remote metal monitoring has been realized by eliminating the needs for mercury electrodes, oxygen removal, forced convection or supporting electrolyte (which previously prevented the direct immersion of stripping electrodes into sample streams) [18,19]. This was accomplished through the judicious coupling of potentiometric stripping operation, the development of non-mercury electrodes, and the use of advanced ultramicroelectrode technology. Compatibility with field operations was achieved by connecting the three-electrode housing (including a gold fiber working electrode, in the PVC tube), via environmentally sealed three-pin connectors, to a 25 m long shielded cable. Convenient and simultaneous quantitation of several trace metal levels (e.g., Cu, Pb, Hg and Ag) has thus been realized in connection to measurement frequencies of 20–30/h (based on typical 1–2 min deposition periods).

The *in situ* monitoring capability of the remote metal sensor was documented in studies of the distribution of labile copper in San Diego Bay (CA, USA) [20]. For this purpose, the probe was floating on the side of a small US Navy vessel. The resulting map of copper distribution reflected the circulation pattern and metal discharge in the bay. A similar study, conducted in collaboration with Prof. Daniele, involved the use of a submersible gold-fiber microelectrode for assessing the distribution of metal contaminants (Hg and Cu) in the canals and lagoon of Venice, Italy [21].

Buffle's group has also made pioneering contributions to *in situ* measurements of trace metals [2,11,22,23]. It described a submersible and robust voltammetric *in situ* profiling (VIP) system (based on advanced microprocessor and telemetry technology) for autonomous water profiling down to 500 m [16] and a novel 64-microelectrode array for real-time, high-spatial resolution, ASV monitoring of the fluxes of metals across the sediment–water interface [11]. Both devices employ an effective protective agar coating, covering the mercury-coated iridium electrode. Such a thick gel layer acts as a dialysis membrane that excludes co-existing surface-active macromolecules (fouling components) and offers the surface protection (anti-fouling capability) necessary for achieving the system robustness (and a stable long-term *in situ* monitoring) [2,23]. Solid-state gold–mercury electrodes have been used by Luther's group for *in situ* voltammetric measurements of trace iron and manganese in marine environments [24].

The extension of remote stripping electrodes to additional metals that cannot be deposited electrolytically (e.g., Cr, Ni and U) relies on the adaptation of adsorptive stripping protocol for a submersible operation [25,26]. Remote adsorptive stripping sensors rely on a different probe design based on an internal solution chemistry. This includes continuous delivery of the ligand, its complexation reaction with the metal 'collected' in a semipermeable microdialysis sampling tube, and transport of the complex to the working electrode compartment. Such dialysis sampling also offers extension of the linear range and protection against surface fouling (due to its dilution and filtration actions).

6.7 CONCLUSIONS

The power and scope of stripping analysis have grown dramatically during the past decade. Our rethinking has paved the way to conceptionally new approaches for performing stripping experiments. We are no longer bound by the traditional use of bulky electrodes and cumbersome operation, but rather rely on faster, smaller and better easy-to-use sensor strips, hand-held metal analyzers, remotely deployed stripping probes, or 'green' electrode materials. While the development of these metal-sensing systems is still at infancy, they offer great promise for faster, simpler and cheaper trace metal testing.

ACKNOWLEDGMENTS

The work was supported by grants from the US DOE (EMSP Program) and the US EPA (STAR Program).

REFERENCES

1 J. Wang, *Stripping Analysis*, VCH Publishers, New York, 1985.
2 J. Buffle and M.L. Tercier-Waeber, *Trends Anal. Chem.*, 24 (2005) 172.
3 J. Wang, *Analytical Electrochemistry*, 3rd ed, Wiley, New York, 2006.
4 D. Jagner, *Trends. Anal. Chem.*, 2(3) (1983) 53.
5 D. Jagner, *Analyst*, 107 (1982) 593.
6 C.M.G. van den Berg, *Anal. Chim. Acta*, 250 (1991) 265.
7 M. Paneli and A. Voulgaropoulos, *Electroanalysis*, 5 (1993) 355.
8 J. Wang and B. Tian, *Anal. Chem.*, 64 (1992) 1706.
9 J. Wang, J. Lu, B. Tian and C. Yarnitzky, *J. Electroanal. Chem.*, 361 (1993) 77.
10 A. Uhlig, U. Schnakemberg and R. Hintsche, *Electroanalysis*, 9 (1997) 125.
11 M.L. Tercier-Waeber, J. Pei, J. Buffle, G. Fiaccabrino, M. Koudelka-Hep, G. Riccardi, F. Confalonieri, A. Sina and F. Graziottin, *Electroanalysis*, 12 (2000) 27.
12 J. Wang, J. Lu, S. Hocevar, P. Farias and B. Ogorevc, *Anal. Chem.*, 72 (2000) 3218.
13 A. Economou, *Trends Anal. Chem.*, 24 (2005) 334.
14 J. Wang, *Electroanalysis*, 17 (2005) 1341.
15 J. Wang, J.M. Lu, S.B. Hočevar and B. Ogorevc, *Electroanalysis*, 13 (2001) 13.
16 A. Zirino, S. Lieberman and C. Clavell, *Environ. Sci. Technol.*, 12 (1978) 73.
17 J. Wang and M. Ariel, *J. Electroanal. Chem.*, 83 (1977) 217.
18 J. Wang, D. Larson, N. Foster, S. Armalis, J. Lu, X. Rongrong, K. Olsen and A. Zirino, *Anal. Chem.*, 67 (1995) 1481.
19 J. Wang, *Trends Anal. Chem.*, 16 (1997) 84.

20 J. Wang, N. Foster, S. Armalis, D. Larson, A. Zirino and K. Olsen, *Anal. Chim. Acta*, 310 (1995) 223.
21 S. Daniele, C. Bragato, M. Baldo, J. Wang and J. Lu, *Analyst*, 125 (2000) 731.
22 M.L. Tercier and J. Buffle, *Electroanalysis*, 5 (1993) 187.
23 M.L. Tercier, J. Buffle and F. Graziottin, *Electroanalysis*, 10 (1998) 355.
24 M. Taillefert, G.W. Luther and D.B. Nuzzio, *Electroanalysis*, 12 (2000) 401.
25 J. Wang, J. Lu, D. Luo, J. Wang, B. Tian and K. Olsen, *Anal. Chem.*, 69 (1997) 2640.
26 J. Wang, J. Wang, J. Lu, B. Tian, D. MacDonald and K. Olsen, *Analyst*, 124 (1999) 349.

Chapter 7

Graphite-epoxy electrodes for stripping analysis

Arben Merkoçi and Salvador Alegret

7.1 INTRODUCTION

The electrochemical stripping analysis is strongly affected by the working electrode material. An ideal working electrode should possess low ohmic resistance, chemical and electrochemical inertness over a broad range of potentials, high hydrogen and oxygen overvoltage resulting in a wide potential window, low residual current, ease of reproduction of the electrode surface, maximum versatility, low cost, and no toxicity.

Various materials have been used as working electrodes for detection of heavy metals, the most popular being mercury-based electrodes. The usefulness of mercury electrodes for the determination of metal ions is due to their ability to form amalgams, allowing for preconcentration of the metal ions prior to their determination by voltammetric stripping methods. Another advantage of using mercury in working electrodes is associated with the high overpotential of hydrogen evolution on such electrodes [1].

A common stripping procedure involves the use of mercury solutions added to the electrochemical cell and the posterior electrochemical generation of the mercury film onto the surface of the graphite working electrode. Mercury-modified electrodes coupled with stripping techniques have been recognized as the most-sensitive methods for determination of heavy metals, especially the detection of lead. The use of hanging mercury drop electrode (the most common mercury electrode) and mercury-film electrode has allowed a sub-ppb lead determination [2]. However, these techniques require tedious experimental precautions regarding the stability and recovery of mercury drop after each experiment or careful manipulation of mercury solutions for film deposition.

Comprehensive Analytical Chemistry 49
Alegret and Merkoçi (Eds)
Volume 49 ISSN: 0166-526X DOI: 10.1016/S0166-526X(06)49007-3

7.1.1 *In Situ* generation of mercury

The potential danger associated with mercury has led to the development of other strategies that avoid the use of a mercury solution. These strategies use glassy carbon electrodes (GCE) coated with a mercury film modified with Nafion [3,4], cellulose acetate [5], naphthol derivative [6], etc., where mercury is generated *in situ* and this way avoiding the manipulation of mercury solutions as done previously. Composite electrode containing HgO as a built-in mercury precursor, which supply mercury-film formation, has even been reported to avoid the use of mercury solution [7].

7.1.2 Mercury-free modified electrodes

Although there are advantages brought by the above strategies a growing interest is appearing in the use of mercury-free electrodes. A mercury-free voltammetric sensor for detection of copper based on chemical accumulation of the trace metal onto the surface of GCE modified with tetraphenylporphyrin has been reported [8]. Similar efforts have resulted in the use of electrodes modified with PAN/Nafion [9] or *N*-*p*-chlorophenylcinnamohydroxamic acid [10]. Disposable screen-printed electrodes for lead determination based on carbon inks mixed with dithizone, Nafion, or ionophore have been also developed [11]. Unmodified electrodes like bare carbon, gold, or iridium as possible alternative to mercury have been also used [12–15]. Gold electrodes from recordable CDs [16] or silver-plated rotograved carbon electrodes [2] have been also reported.

7.1.3 Bismuth electrodes

Bismuth-film electrodes (BiFEs), consisting of a thin bismuth-film deposited on a suitable substrate, have been shown to offer comparable performance to MFEs in ASV heavy metals determination [17]. The remarkable stripping performance of BiFE can be due to the binary and multi-component ‘‘fusing’’ alloys formation of bismuth with metals like lead and cadmium [18]. Besides the attractive characteristics of BiFE, the low toxicity of bismuth makes it an alternative material to mercury in terms of trace-metal determination. Various substrates for bismuth-film formation are reported. Bismuth film was prepared by electrodeposition onto the micro disc by applying an *in situ* electroplating procedure [19]. Bismuth deposition onto gold [20], carbon paste [21], or glassy carbon [22–24] electrodes have been reported to display an attractive stripping voltammetric behavior. *In situ* or *ex situ* preparation [25] of the BiFEs including bismuth precursor salt and a variety of substrate surface (platinum, gold, glassy carbon, carbon paste, carbon fiber) [26] for bismuth plating were carefully examined for their effects in the preconcentration and stripping steps including the constant-current potentiometric stripping [27].

7.1.4 Composite electrodes

Composite represents one of the most-interesting materials for the preparation of working electrodes to be applied in electrochemical analysis. A composite results from the combination or integration of two or more dissimilar materials. Each individual component maintains its original characteristics while giving the composite distinctive chemical, mechanical, and physical qualities. These qualities are different from those shown by the individual elements of the composite [28].

Usually the composites are classified according to the nature of the conducting material (platinum, gold, carbon, etc.) and the arrangement of its particles (i.e., whether the conducting particles are dispersed in the polymer matrix or if they are grouped randomly in clearly defined conducting zones and insulating zones). Furthermore, using a given conducting material we may use different types of polymers thus establishing novel composites such as epoxy composites, methacrylate composites, silicone composites, etc. Moreover, composites can be classified according to their rigidity as rigid composites or soft composites (known also as pastes or inks).

Conducting (bio)composites are interesting alternatives for the construction of electrochemical (bio)sensors. The capability of incorporating several materials within the composite result in enhanced sensitivity and selectivity. This incorporation is possible to be performed either through a previous modification of one of the components of the composite before its preparation or through physical incorporation into the composite matrix. Composite and biocomposite electrodes offer many potential advantages [29–32] compared to more traditional electrodes consisting of a surface-modified single-conducting phase. Composite electrodes can often be fabricated with great flexibility in size and shape permitting easy adaptation to a variety of electrode configurations. Even more they can be smoothed or polished to provide fresh active material ready to be used in a new assay. Composite electrodes have higher signal-to-noise (S/N) ratio, compared to the corresponding pure conductors, resulting in an improved (lower) detection limit.

Conducting (bio)composites applied to electrochemical analysis have opened a new range of possibilities for the construction of electrochemical (bio)sensors for several applications [29–32].

7.2 CONSTRUCTION AND SURFACE CHARACTERIZATION OF COMPOSITE ELECTRODES

7.2.1 Construction

Various designs based on graphite-epoxy composite electrodes (GECE) for electrochemical stripping analysis of metals have been used in our laboratories. Figure 7.1 summarizes the main strategies used for that purpose: Fig. 7.1A is a schematic of the GECE without any modification which have been firstly applied in metal analysis [33–35]. Figure 7.1B represents a Bi-GECE which is a GECE without modification used for metal detection in the presence of bismuth in the measuring solution [36]. The third sensor design is the $Bi(NO_3)_3$-GECE [37] that represents a GECE modified internally with bismuth nitrate salt.

The GECE were prepared using graphite powder with a particle size of 50 µm (BDH, UK) and Epotek H77 (epoxy resin) and hardener (both from Epoxy Technology, USA). Graphite powder and epoxy resin (mixed with hardener) were hand-mixed in a ratio of 1:4 (w/w) as

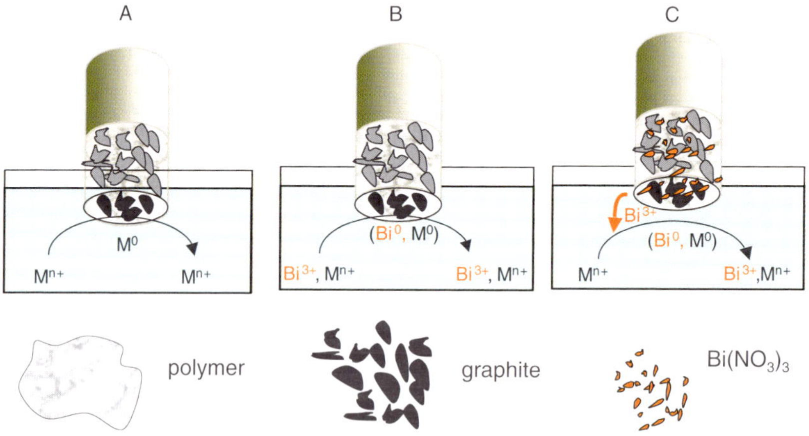

Fig. 7.1. Schematics of the GECEs used for stripping analysis of metals. (A) Sensing based on GECE sensors without modifications. (B) Sensing based on Bi-GECE. It represents GECE without modification but in the presence of bismuth in the measuring solution. (C) Sensing based on $Bi(NO_3)_3$-GECE. It represents GEC modified internally with bismuth nitrate salt.

described in a previous work [38,39]. Bismuth nitrate salt was added in the GECE paste before curing so as to prepare the $Bi(NO_3)_3$-GECE.

The resulting pastes, for all cases, were placed into a PVC cylindrical sleeve body. The conducting composite material glued to the copper contact was cured at 40°C during a week. Before each use, the surface of the electrode was wet with doubly distilled water and then thoroughly smoothed, first with abrasive paper and then with alumina paper (see more details on the preparation of GECE in Procedure 7).

7.2.2 Surface characterization

Scanning electron microscopy (SEM) and profilometry were used to characterize the surface of GECE. Figure 7.2 represents the obtained results. The SEM image of GECE (Fig. 7.2, left image) appears to have clusters of material gathered in random areas. Topographically speaking, these appear to be of varying heights due to their apparent depth while the SEM image of the GCE for purposes of comparison (not shown) was characterized by a smooth surface and the absence of any clusters. The presence of clusters in the GECE surface results in an increase in its surface area compared to that of the GCE and consequently in the increase of the lead uptake due to an increased physical adsorption.

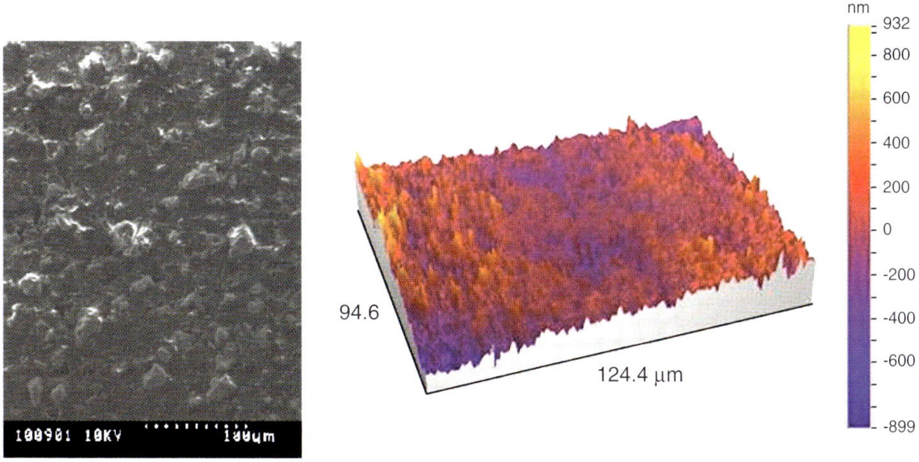

Fig. 7.2. Images of GECE obtained by using SEM (left; accelerated voltage, 10 kV, and resolution, 100 μm, were used) and white light interferometric profilometry (right; at 50 × magnification). The surfaces have been polished as explained in the text (see Procedure 7).

Figure 7.2 (right image) shows the 3D topographic map of the surface of the GECE. The GECE surface is characterized by a very rough surface (R.M.S roughness = 151 nm, max. peak to valley height = 1.83 μm), covered with "peaks" and "valleys" which create sub-micron sized "wells" in agreement with the SEM image. The surface of the GCE (not shown) was much smoother. Therefore there is strong evidence that these "wells" were going to play an important role in the physical adherence of the reduced metals on the surface of the GECE.

7.3 STRIPPING ANALYSIS WITH NON-MODIFIED COMPOSITES

7.3.1 Detection in the absence of bismuth

7.3.1.1 Stripping potentiometry

Potentiometric stripping analysis (PSA) with two modes: (a) constant current and (b) chemical oxidation has been performed [33]. The constant-current mode consisted of two steps. In the first step (accumulation), the GECE as working electrode, the Ag/AgCl as reference electrode, and the platinum as auxiliary electrode were immersed in a stirred 25 mL of 0.1 M sodium acetate solution (pH = 3.76) and a constant potential (E_{acc}) of –0.9 V (vs. Ag/AgCl) was applied during a fixed interval time (τ_{acc}) 1–30 min. In the second step (stripping), a constant current (I_{strip}) of 1 μA was applied while the potential is recorded until a limit of −0.2 V during a measurement time of 300 s (in unstirred solution). The same procedure, without the removal of the electrodes, was repeated after an addition of a known quantity of heavy metal standard solution to obtain a calibration curve. All experiments were carried out without removal of oxygen.

In PSA with chemical oxidation mode the procedure was the same except that the stripping step was performed by dissolved oxygen in equilibrium with the atmosphere.

Responses in standard solutions

Figure 7.3 represents the stripping curves and the corresponding calibration plot (inset) for lead obtained for a τ_{acc} of 60 s and a I_{strip} of 1 μA. Each point in the calibration curve corresponds to the mean of three parallel measurements performed consecutively in the same cell without polishing the electrode. The error bars are the standard deviations of these measurements. Possible changes of graphite-epoxy electrode surface in contact with solution should have the effect of repeatability of the response. A DL of approximately 200 ppb of lead was determined

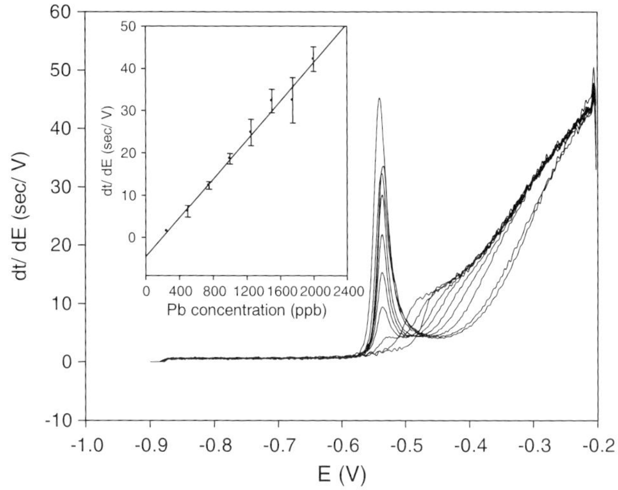

Fig. 7.3. PSA curves and the calibration plot (inset) using non-modified GECE. Constant-current method for different lead concentrations The composition of the cell was: 25 mL of 0.1 M acetate buffer pH = 3.76 with 0.5 ppm of Pb. The other electrodes were: Auxiliary: platinum electrode; reference electrode: Ag/AgCl; Deposition potential: −0.9 V; Accumulation time: 60 s; stripping current: 1 μA; potential limit: −0.2 V; Max. time of measurement: 300 s. Adapted from Ref. [33].

as the concentration corresponding to three times of the PSA background signal. Taking into consideration that a τ_{acc} of 60 s is used in this work, it would be expected that an increase in τ_{acc} would result in a lower DL comparable with those reported for modified GCE (12–15 ppb for a τ_{acc} of 600 s).

A mixture containing lead with copper and cadmium was also checked using PSA with chemical oxidation mode (results not shown). It was shown that GECE was more sensitive for the analysis of copper than for lead and cadmium.

7.3.1.2 Stripping voltammetry

In the accumulation step the three electrodes—GECE as working electrode, the Ag/AgCl as reference electrode, and the platinum as auxiliary electrode—were immersed in a stirred 25 mL of 0.1 M HCl solution. A conditioning potential of 1 V during 30 s was applied to clean the electrode from the previous deposited metals; after that an accumulation potential (E_{acc}) of −1.4 V (vs. Ag/AgCl) was applied for a fixed interval time (τ_{acc}) ranging from 1 to 30 min. The second step was the stripping step. In this step the potential was changed within the range of −0.7 to

−0.1 V, using a potential step of 0.0024 V. Modulation time was 0.05 s and interval time of applied pulses was 0.2 s. During the stripping step the current is recorded in quiescent solution. To obtain a calibration curve a known quantity of heavy-metal solution was successively added and the above accumulation and stripping procedures were applied without the removal of the electrodes. All experiments were carried out without the removal of oxygen.

Flow-through measurements with DPASV were also performed. The working parameters were the same as in the batch measurements except that the metal accumulation step is performed in a flowing stream while the stripping step in a stop mode.

Responses in standard solutions

Figure 7.4 shows the voltammograms for a mixture containing lead, copper, and cadmium using an accumulation time of 60 s [35]. The peaks for each metal are well separated. The corresponding calibration curves (inset Fig. 7.4) are also presented. It can be seen that the sensitivity (change of I_{str} per unit of metal concentration change) for lead is higher than copper with cadmium being the lowest. The detection limits (DL,

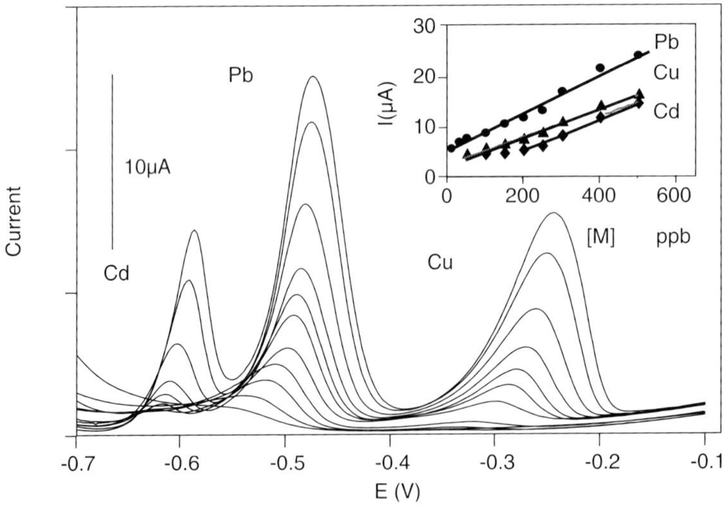

Fig. 7.4. DPASV curves obtained with a graphite-epoxy composite electrode for increasing concentration of Cd, Pb, and Cu along with the corresponding calibration curves (inset). The cell composition was: 25 mL 0.1 N HCl; the reference electrode: Ag/AgCl; counter electrode: Pt; accumulation potential: −1.4 V; τ_{acc}: 60 s; step potential: 0.0024 V; modulation time: 0.05 s; interval time: 0.2 s. Adapted from Ref. [35].

evaluated as the concentration corresponding to 3σ of the DPASV blank signal) were 100 ppb for Cd, 10 ppb for lead, and 50 ppb for copper.

A simple flow-through system that permits the constant flow of lead solution in 0.1 M HCl was also used. The electrochemical cell used permitted the integration of the same three-electrode configuration as in the batch measurements. The detection limit was similar to batch measurements. The stability of the system for 16 runs of a 500 ppb lead solution showed a 4% RSD.

Real samples

The GECE sensors were used for lead determination in real water samples suspected to be contaminated with lead obtained from water suppliers. The same samples were previously measured by three other methods: a potentiometric FIA system with a lead ion-selective-electrode as detector (Pb-ISE); graphite furnace atomic absorption spectrophotometry (AAS); inductively coupled plasma spectroscopy (ICP). The results obtained for lead determination are presented in Table 7.1. The accumulation times are given for each measured sample in the case of DPASV. Calibration plots were used to determine the lead concentration. GEC electrode results were compared with each of the above methods by using paired *t*-Test. The results obtained show that the differences between the results of GECE compared to other methods were not significant. The improvement of the reproducibility of the methods is one of the most important issues in the future research of these materials.

TABLE 7.1

Results obtained for real water samples using non-modified GECE

Sample	[Pb] ($\mu g\ L^{-1}$)			
	DPASV/τ_{acc} (min)	Pb-ISE	AAS	ICP
1	–/60	17	<6	<5
2	17/6	28	44	36
3	89/2	64	88	80
4	50/2	122	176	141
5	18/10	45	41.3	37
6	263/1	175	240	194
7	14/20	18	18.1	15
8	121/1	158	192	174
9	404/1	344	495	398

Experimental conditions as in Fig. 7.4.

7.3.2 Detection in the presence of bismuth in solution

The analytical use of GECE modified *in situ* by using bismuth solution for square wave anodic stripping voltammetry (SWASV) of heavy metals is also studied [36]. The use of this novel format is a simpler alternative to the use of mercury for analysis of trace levels of heavy metals. The applicability of these new surface-modified GECE to real samples (tap water and soil samples) is presented.

Square wave voltammetry stripping measurements were performed by *in situ* deposition of the bismuth and the target metals in the presence of dissolved oxygen. The three electrodes were immersed into 25 mL electrochemical cell containing 0.1 M acetate buffer (pH 4.5) and 400 $\mu g L^{-1}$ of bismuth. The deposition potential of –1.3 V was applied to the GECE while the solution was stirred. Following the 120 s deposition step, the stirring was stopped and after 15 s the voltammogram was recorded by applying a square-wave potential scan between −1.3 and −0.3 V (with a frequency of 50 Hz, amplitude of 20 mV, and potential step of 20 mV). Aliquots of the target metal standard solution were introduced after recording the background voltammograms. A 60 s conditioning step at +0.6 V (with solution stirring) was used to remove the target metals and the bismuth, prior to the next cycle.

Responses in standard solutions were tested for lead, cadmium, and zinc (see Fig. 7.5). The results obtained show well-defined and single peaks for all of the metals. Sharper peaks were obtained for lead and cadmium compared to zinc. Detection limits of 23.1, 2.2, and 600 $\mu g L^{-1}$ were estimated for lead, cadmium, and zinc, respectively, based on the signal-to-noise characteristics of these data ($S/N = 3$). The reproducibility of the Bi-GECE was also tested and found to be 2.99%, 1.56%, and 2.19% for lead, cadmium, and zinc, respectively. The difference in peak shapes (sharper for lead and cadmium) and in detection limits of these heavy metals can be explained by the binary and multi-component "fusing" alloys formation of lead and cadmium with bismuth [40]. According to these results, it was deduced that zinc competes with bismuth for the surface site rather than involving an alloy formation with this metal.

The most attractive property of the Bi-GECE can be observed at the peak potentials of heavy metals. Compared to BiFE on glassy carbon and carbon fiber substrates [18], the approximate positive shifts of peak potential are of 125, 150, and 305 mV for lead, cadmium, and zinc, respectively. This shifted peak potential can be attributed

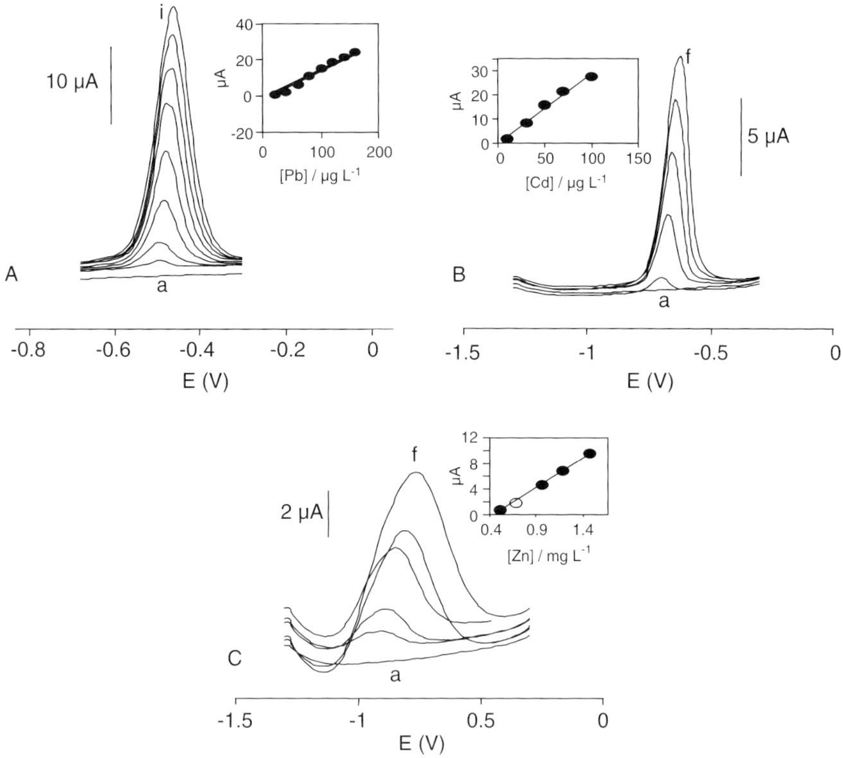

Fig. 7.5. Square-wave stripping voltammograms for increasing concentration of lead (A) in 20 mg L^{-1} steps (b–i), cadmium (B) in 10 mg L^{-1} steps (b–f), and zinc (C) in concentration of 500, 700, 1000, 1200, and 1500 mg L^{-1} (b–e). Also shown are the corresponding blank voltammograms (a) and as inset the calibration plots over the ranges 10–100 mg L^{-1} cadmium, 20–160 mg L^{-1} lead, and calibration plot of the mentioned zinc concentrations. Solutions 0.1 M acetate buffer (pH 4.5) containing 400 mg mL^{-1} bismuth. Square-wave voltammetric scan with a frequency of 50 Hz, potential step of 20 mV, and amplitude of 25 mV. Adapted from Ref. [36].

to the more homogenous and uniform film formation due to the novel supporting material. The rich microstructure of GECE, composed of a mixture of carbon microparticles forming internal microarrays might have a profound effect upon the bismuth-film structural features. The obtained peak shifts may be useful for improving resolutions overall taking into consideration metals like Bi and Cu. The obtained peak widths of 230, 360, and 430 mV for lead, cadmium, and zinc, respectively, were similar to other BiFEs reported previously.

The simultaneous measurement of lead and cadmium with Bi-GECE was also performed (see Procedure 7). Detection limits of around 30 $\mu g L^{-1}$ have been estimated for lead and cadmium, respectively, based on the signal-to-noise characteristics of these data ($S/N = 3$).

The stripping performances of Bi film on glassy carbon or carbon fiber substrates were examined very carefully by Wang *et al.* [17]. In addition to these materials, GECE (combined with bismuth film), a very easy to prepare and low-cost electrode, can also be used successfully for simultaneous stripping analysis of cadmium and lead. Zinc was also tried to be detected simultaneously with lead and cadmium but it was not possible to obtain undistorted and linearly increased peaks. The poor response to zinc can be probably attributed to the preferable accumulation of Bi on GECE rather than of Zn which is a result of the competition of these two metals for the GECE surface sites as also observed in other works [18].

The performance of Bi-GECE was also tested for measuring lead and cadmium in tap water and acetic acid extracted soil sample and the results are shown in Procedure 7.

7.4 STRIPPING ANALYSIS WITH GRAPHITE-EPOXY ELECTRODES MODIFIED WITH BISMUTH NITRATE

SWASV measurements were carried out for Pb^{2+}, Cd^{2+} and Zn^{2+}, using the $Bi(NO_3)_3$-GECE as working, Ag/AgCl as reference, and platinum as auxiliary electrode. The measurements were carried out in a stirred 25 mL of 0.1 M acetate buffer (pH 4.5). The deposition potential of −1.3 V was applied to the $Bi(NO_3)_3$-GECE while the solution was stirred. Following the 120 s deposition step, the stirring was stopped and after 15 s equilibration, the voltammogram was recorded by applying a square-wave potential scan between −1.3 and −0.3 V (with a frequency of 50 Hz, amplitude of 20 mV, and potential step of 20 mV). Aliquots of the target metal standard solution were introduced after recording the background voltammograms. A 60 s conditioning step at +0.6 V (with solution stirring) was used to remove the target metals and the reduced bismuth, prior to next cycle. The electrodes were washed thoroughly with deionized water between each test.

SWASV measurements were performed using 0.1 and 0.5 M HCl solutions as electrolytic medium in calibrations for lower concentrations of Pb^{2+} (from 1 to 10 ppb), the experimental conditions being the same as for the measurements in acetate buffer.

7.4.1 Responses in standard solutions

The stripping performance of $Bi(NO_3)_3$-GECE was tested for lead and cadmium and the resulting voltammograms were given in Fig. 7.6.

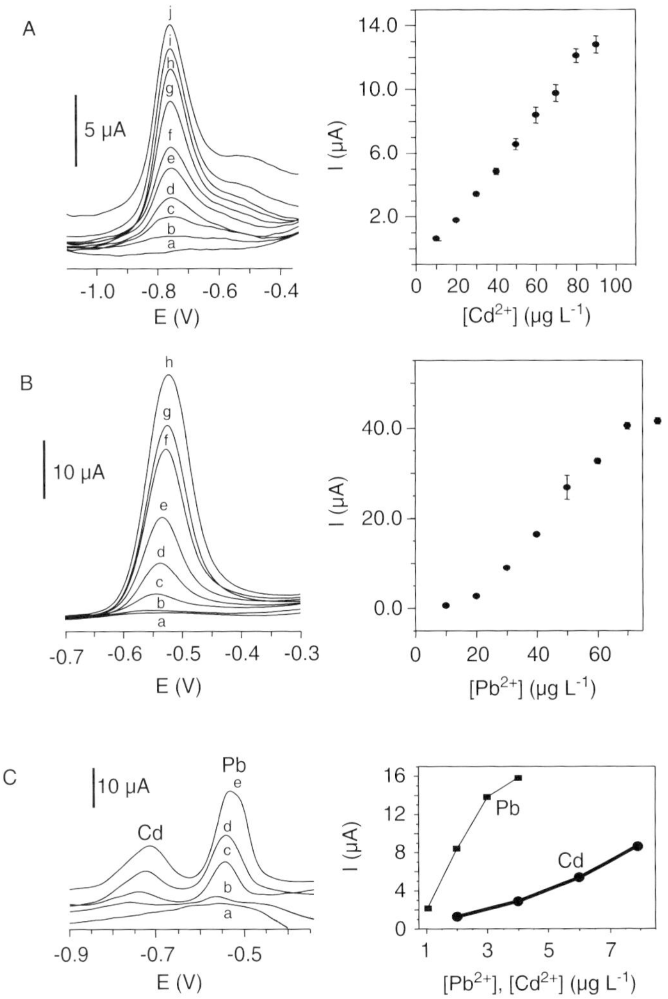

Fig. 7.6. Determination of cadmium (A) and lead (B) for increasing concentrations in $10\,\mu g\,L^{-1}$ steps (b–e); concentration ranges of 10–40 (Cd) and 20–80 (Pb) $\mu g\,L^{-1}$. Simultaneous determination of cadmium and lead (C) for increasing concentrations in $10\,\mu g\,L^{-1}$ steps (Pb) and $20\,\mu g\,L^{-1}$steps (Cd). Also shown is the blank (a) and the corresponding calibration plots. Solutions 0.1 M acetate buffer (pH 4.5). Square-wave voltammetric scan with a frequency of 50 Hz, potential step of 20 mV, and amplitude of 25 mV. Deposition potential of −1.3 V during 120 s.

Figure 7.6 demonstrates the square wave stripping voltammograms for increasing concentration of cadmium (A) in 10 μg L^{-1} steps (b–j) and lead (B) in 10 μg L^{-1} steps (b–h). Also shown are the corresponding blank voltammograms (a) and the calibration plots (right) over the ranges 10–100 μg L^{-1} cadmium and 10–70 μg L^{-1} lead. The $Bi(NO_3)_3$-GECE displays well-defined and single peaks for cadmium (Ep = −0.76 V) and lead (Ep = −0.54 V). Detection limits of 7.23 and 11.81 μg L^{-1} can be estimated for cadmium and lead, respectively, based on the upper limit approach (ULA) [41], which utilizes the one-sided confidence band around the calibration line. Lower detection limits are expected in connection with longer deposition periods. Also in the concentration ranges mentioned above, the calibration plots (right) were linear exhibiting the R values of 0.9968 and 0.9953 for cadmium and lead, respectively.

The difference in peak shapes (sharper for lead and cadmium) and in detection limits of these heavy metals can be explained by the binary and multi-component "fusing" alloys formation of lead and cadmium with bismuth.

The SWASV for zinc was also checked but the results obtained were not satisfactory. According to these results, it can be deduced that zinc competes with bismuth for the surface site rather than involving an alloy formation with this metal as also observed for Bi-GECE studied previously [36].

As in the case of Bi-GECE the bismuth-film formation onto $Bi(NO_3)_3$-GECE is shown to be a homogenous and uniform one due to the novel supporting material. The rich microstructure of $Bi(NO_3)_3$-GECE, composed of a mixture of carbon microparticles forming internal microarrays, might have a profound effect upon the bismuth-film structural features. This novel stripping platform may be useful for improving resolutions overall taking into consideration metals like Bi and Cu. The obtained peak widths of 20 mV for lead and cadmium, respectively, were similar to other BiFEs reported previously.

The simultaneous measurement of lead and cadmium with $Bi(NO_3)_3$-GECE was also performed as shown in Fig. 7.6C. This figure displays square wave stripping voltammograms for cadmium (Ep = −0.72 V) and lead (Ep = −0.54 V) for increasing concentrations in steps of 10 μg L^{-1} (Pb) and 20 μg L^{-1} (Cd) (b–e). The well-resolved peaks increase linearly with the metal concentration. The voltammogram clearly indicates that these metals can be measured simultaneously following a short deposition time of 2 min. In the concentration range of 10–40 μg Cd L^{-1} and 20–80 μg Cd L^{-1} the stripping signals remained undistorted and the resulting calibrating plots of this concentration range is linear exhibiting *R* values of 0.9562 and 0.9762 for lead and cadmium. Detection limits of around 19.1 and 35.8 μg L^{-1} can be estimated for lead and cadmium, respectively, based on the same method [41].

A more sensitive measurement was observed for lead at 0.5 M HCl as measuring solution. Figure 7.7 represents typical subtractive square-wave stripping voltammograms (removing blanks) for increasing concentration of lead ranging from 1 to 10 μg L^{-1} steps (a–h). Also shown is the calibration plot (right) over the studied range. This highly sensitive response in HCl medium, as expected also from the study of the pH effect is probably related with an improved bismuth release and alloy formation in this medium.

The stability of the $Bi(NO_3)_3$-GECEs in 10 consecutive measurements for 50 ppb cadmium in 0.1 M acetate buffer of pH 4.5 and using the same surface was tested. The relative standard deviation of this measurement was 9.33%.

Although the $Bi(NO_3)_3$ particles were not uniform in size they were expected to be exposed in a reproducible way onto the freshly obtained $Bi(NO_3)_3$-GECE surfaces after each mechanical polishing procedure.

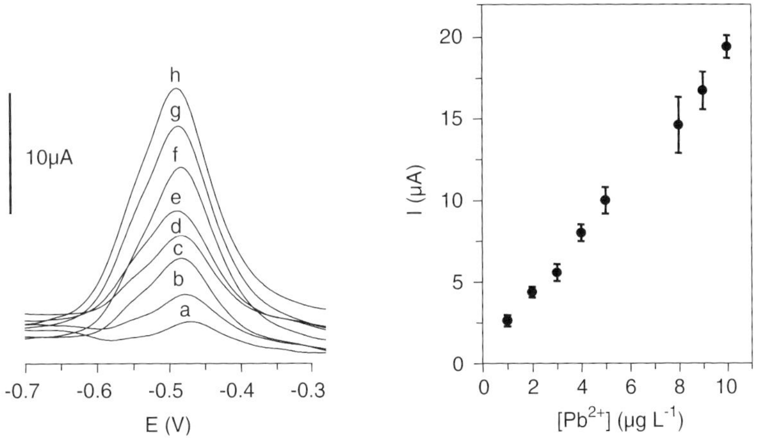

Fig. 7.7. Square-wave stripping voltammograms for increasing concentration of lead: (a) 1, (b) 2, (c) 3, (d) 4, (e) 5, (f) 8, (g) 9, (h) 10 μg L^{-1}. Also shown is the corresponding calibration plot (right) over the range 1–10 μg L^{-1} lead. The measuring solution was 0.5 M HCl. Other experimental conditions are as in Fig. 7.6.

This was confirmed by checking the reproducibility of the measurements for a series of 10 different surfaces of the same $Bi(NO_3)_3$-GECE. The relative standard deviation of these measurements performed in the same experimental conditions as for the stability study was 10.69 for cadmium measurements.

7.5 CONCLUSIONS

Several free-mercury sensors for stripping analysis of heavy metals have been developed. The first strategy, graphite-epoxy composite electrodes (GECE) without any modification brings several advantages. Possibly the greatest advantage to be offered by the proposed sensors is the avoidance of the use of harmful mercury or other time-consuming procedures to modify glassy carbon or other electrodes. Additionally the metal stripping can be performed without oxygen removal. Graphite-epoxy electrodes are cheaper and easy to be prepared in the laboratory. Although, a higher detection limit compared to the mercury-film electrode has been found, this kind of electrode can be envisioned as an attractive alternative for mercury-free detection of heavy metals.

The second strategy, the use of bismuth solution to form bismuth-film GECE (Bi-GECE) for the determination of cadmium, lead, and zinc is also demonstrated. This strategy combines the GECE with bismuth-film formation *in situ* during the stripping analysis of metals. The coupling of GECE with bismuth film results in sensitive, well-defined, and undistorted peaks especially for cadmium and lead.

The third strategy, a novel GECE that incorporates $Bi(NO_3)_3$ salt in the sensing matrix is also developed. The resultant $Bi(NO_3)_3$-GECE is compatible with BiFEs for use in stripping analysis of heavy metals. The built-in bismuth property is the distinctive feature of this $Bi(NO_3)_3$ modified GECE which can be utilized for the generation of bismuth adjacent to the electrode surface. The developed $Bi(NO_3)_3$-GECE is related with *in situ* bismuth ion generation and film formation without the necessity of external addition of bismuth in the measuring solution. This sensor is also related with a higher sensitivity and improved detection limits during measurements in HCl medium.

The developed bismuth-based GECE sensors compared to unmodified GECE show good stability which owing to the unique surface morphology results in enhanced contact between the GECE matrix and the electrochemically reduced bismuth. Additionally in the case of $Bi(NO_3)_3$-GECE, the avoidance of external bismuth addition improved the sensor integration and consequently its utility.

The possible use of graphite-epoxy material by screen-printing technology opens the possibility of mass production of disposable sensors for heavy-metal analysis using stripping techniques. The utilization of these sensors for an extensive application in real heavy-metal samples is underway in our laboratories.

ACKNOWLEDGMENTS

This work was financially supported by Ministry of Education and Culture (MEC) of Spain (Projects MAT2005-03553, BIO2004-02776), the Spanish foundation Ramón Areces (project 'Bionanosensores') and EC FP6 project WARMER (Reference: FP6-034472-2005-IST-5).

REFERENCES

1 Z. Galus. In: P.T. Kissinger and W. Heineman (Eds.), *Laboratory Techniques in Electroanalytical techniques*, 2nd ed., Marcel Dekker, New York, 1996.
2 P.R.M. Silva, M.A. El Khakani, M. Chaker, A. Dufrense and F. Courchesne, *Sens. Actuators B*, 76 (2001) 250.
3 A. Merkoçi, M. Vasjari, E. Fabregas and S. Alegret, *Mikrochim. Acta*, 135 (2000) 29.
4 J. Vidal, R. Viñao and J. Castillo, *Electroanalysis*, 4 (1992) 653.
5 J. Wang and L.D. Hutchins-Kumar, *Anal. Chem.*, 58 (1986) 402.
6 A.A. Esnafi and M.A. Naeeni, *Anal. Lett.*, 33 (2000) 1591.
7 K. Seo, S. Kim and J. Park, *Anal. Chem.*, 70 (1998) 2936.
8 H.H. Frey, C.J. McNeil, R.W. Keay and J.V. Bannister, *Electroanalysis*, 10 (1998) 480.
9 Zh. Hu, C.J. Seliskar and W.R. Heineman, *Anal. Chim. Acta*, 369 (1998) 93.
10 T.H. Degefa, B.S. Chandravanshi and H. Alemu, *Electroanalysis*, 11 (1999) 1305.
11 I. Palchetti, C. Upjohn, A.P.F. Turner and M. Mascini, *Anal. Lett.*, 33 (2000) 1231.
12 J. Wang, *Stripping Analysis*, VCH Publishers, Deerfield Beach, 1985.
13 E.P. Achterberg and C. Braungardt, *Anal. Chim. Acta*, 400 (1999) 381.
14 J. Wang and B. Tian, *Anal. Chem.*, 65 (1993) 1529.

15 M.A. Nolan and S.P. Kounaves, *Anal. Chem.*, 71 (1999) 3567.
16 L. Angnes, E.M. Richter, M.A. Augelli and G.H. Kume, *Anal. Chem.*, 72 (2000) 5503.
17 J. Wang, J. Lu, S.B. Hocevar, P.A. Farias and B. Ogorevc, *Anal. Chem.*, 72 (2000) 3218.
18 J. Wang, J. Lu, Ü.A. Kirgöz, S.B. Hocevar and B. Ogorevc, *Anal. Chim. Acta*, 434 (2001) 29.
19 M.A. Baldo and S. Daniele, *Anal. Lett.*, 37 (2004) 995.
20 H.P. Chang and D.C. Johnson, *Anal. Chim. Acta*, 248 (1991) 85.
21 G.U. Flechsig, O. Korbout, S.B. Hocevar, S. Thongngamdee, B. Ogorevc, P. Grundler and J. Wang, *Electroanalysis*, 14 (2002) 192.
22 E.A. Hutton, B. Ogorevc, S.B. Hocevar, F. Weldon, M.R. Smyth and J. Wang, *Electrochem. Commun.*, 3 (2001) 707.
23 J. Wang and J. Lu, *Electrochem. Commun.*, 2 (2000) 390.
24 G. Kefala, A. Economou, A. Voulgaropoulos and M. Sofoniou, *Talanta*, 61 (2003) 603.
25 E.A. Hutton, J.T. Van Elteren, B.I. Ogorevc and M.R. Smyth, *Talanta*, 63 (2004) 849.
26 S.B. Hocevar, B. Ogorevc, J. Wang and B. Pihlar, *Electroanalysis*, 14 (2002) 1707.
27 S.B. Hocevar, J. Wang, R.P. Deo and O. Ogorevc, *Electroanalysis*, 14 (2002) 112.
28 G. Ruschan, R.E. Newnham, J. Runt and E. Smith, *Sens. Actuators*, 269 (1989) 20.
29 S. Alegret, *Analyst*, 121 (1996) 1751.
30 F. Céspedes, E. Fàbregas and S. Alegret, *Trends Anal. Chem.*, 296 (1996) 15.
31 F. Céspedes and S. Alegret, *Trends Anal. Chem.*, 276 (2000) 19.
32 S. Alegret, E. Fàbregas, F. Céspedes, A. Merkoçi, S. Solé, M. Albareda and M.I. Pividori, *Quím. Anal.*, 23 (1999) 18.
33 M. Serradell, S. Izquierdo, L. Moreno, A. Merkoçi and S. Alegret, *Electroanalysis*, 14 (2002) 1281.
34 L. Moreno, A. Merkoçi and S. Alegret, *Electrochim. Acta*, 48 (2003) 2599.
35 S. Carrégalo, A. Merkoçi and S. Alegret, *Microchim. Acta*, 147 (2004) 245.
36 Ü.A. Kırgöz, S. Marín, M. Pumera, A. Merkoçi and S. Alegret, *Electroanalysis*, 17 (2005) 881.
37 M.T. Castañeda, B. Pérez, M. Pumera, M. Del Valle, A. Merkoçi and S. Alegret, *Analyst*, 130 (2005) 971.
38 M. Santandreu, F. Céspedes, S. Alegret and E. Martínez-Fàbregas, *Anal. Chem.*, 69 (1997) 2080.

39 A. Merkoçi, S. Braga, E. Fàbregas and S. Alegret, *Anal. Chim. Acta*, 391 (1999) 65.
40 G.G. Gong, L.D. Freedman and G.O. Doak, Bismuth and bismuth alloys. In: M. Grayson (Ed.), *Encyclopedia of Chemical Technology*, Vol. 3, Wiley, New York, 1978, pp. 912–937.
41 J. Mocak, A.M. Bond, S. Mitchell and G. Scollary, *Pure Appl. Chem.*, 69 (1997) 297.

Chapter 8

Voltammetric sensors for the determination of antioxidant properties in dermatology and cosmetics

Coline Guitton, Audrey Ruffien-Ciszak, Pierre Gros and Maurice Comtat

8.1 VOLTAMMETRIC METHODS

For about 90 years, we assist to the emergence and the development of analytical chemistry, in which voltammetry occupies a central place. During this period, many electrochemical methods have been proposed to improve the performances, and simple and effective equipments have been designed, developed and marketed. The milestones of these progresses are summarized in Table 8.1. Lastly, theoretical treatments allowed to establish the general laws, giving the equations that connect various experimental parameters to the concentrations.

The first voltammetric methods met are stationary voltammetries performed on a dropping mercury electrode (polarography) or on a solid rotating disk electrode. The limiting current measured is directly proportional to the concentration of the electroactive species in the solution. Experimental potential scan rate is lower than $10\,\mathrm{mV\,s^{-1}}$.

Then appears linear sweep rate voltammetry in which the electrode potential is a linear function of time. The current–potential curve shows a peak whose intensity is directly proportional to the concentration of electroactive species. If the potential sweep takes place in two directions, the method is named cyclic voltammetry. This method is one of the most frequently used electrochemical methods for more than three decades. The reason is its relative simplicity and its high information content. It is very useful in elucidating the mechanisms of electrochemical reactions in the case where electron transfer is coupled

Comprehensive Analytical Chemistry 49
Alegret and Merkoçi (Eds)
Volume 49 ISSN: 0166-526X DOI: 10.1016/S0166-526X(06)49008-5

TABLE 8.1
Some milestones about the evolution of dynamic electrochemistry

Period	Main authors	Proposed topics
1920s	Heyrovsky and Kucera	Electrocapillary measurements with the dropping mercury electrode
	Heyrovsky and Shikata	Performed $I(E)$ curves—polarographies
1940s	Kolthoff and Laitinen	Introduction of the term voltammetry—solid electrode
	Heyrovsky and Forejt	Oscillographic polarography
1950s	Sargent	Automatic polarography
	Barker	Pulse polarography
	Breyer	Alternating current polarography
	Barker	Square wave polarography
	Rogers and Shain	Anodic stripping voltammetry
1960s		Apparition of operational amplifiers for constructing electrochemical instrumentation
	Levich	Rotating disk electrode—rotating ring-disk electrode
	Nicholson, Shain, Savéant and Vianello	Linear scan and cyclic voltammetry
	Geske and Maki	Electron spin resonance and electrochemistry
	Murray, Heineman and Kuwana	Spectroelectrochemistry—transparent electrodes
	Feldberg	Digital simulation methods in electrochemistry
	Hubbard, Murray and Kuwana	Modified electrodes by covalently fixed monolayers
1970s	Miller, Bard and Murray	Electrochemical generation of polymer layers
1980s	Fleischman	Ultramicroelectrodes
	Gonon and Adams	Determination of catecholamines in vivo
	Wightman	Ultramicroelectrodes, submicrosecond electrochemistry Electrochemistry in low-conductivity media Electrochemistry under time-independent conditions

TABLE 8.1 (*continued*)

Period	Main authors	Proposed topics
	Manz and Simon	HPLC coupled to electrochemical detection
	He and Faukner	Software to perform several voltammetric experiments
		Modification of surfaces with clays, zeolites, inorganic crystals, enzyme layers, organic metals, composites
	Wallingford and Ewing	Electrochemical detection in capillary electrophoresis

with chemical reactions occurring before or after the electrode reaction. Scan rates generally used are higher than $15\,mV\,s^{-1}$.

The main improvements of these techniques are essentially gathered with:

- *Selectivity*: New electrode materials were developed such as the boron-doped diamond electrodes, carbon nanotubes electrodes, edge plane pyrolytic graphite electrodes, etc. [1–3]. Another approach deals with the conventional electrodes' surface modification by various species acting as oxido-reduction mediator or capable of changing the adsorption of the analyzed species [4]. Electrodes are indifferently modified by covalent grafting of chemical, biochemical and biological catalysts, or by strong adsorption of chemicals or biochemicals [5,6].
- *Sensibility*: The first strategy to lower the detection limits aims at minimizing the capacitive versus the faradic current intensities. In the second strategy, a preliminary step of electroactive species preconcentration in or on the electrode is realized [7,8].

The introduction of ultramicroelectrodes in the field of voltammetric analysis offers access to cyclic voltammetry experiments that are impossible with conventionally sized macroelectrodes. In addition to analyses in small volumes or at microscopic locations, microelectrodes allow measurements in resistive media and make it possible to perform high scan rate voltammetry [9,10].

- *Minimization of time scale measurements*: Ultrafast undistorted cyclic voltammetry may be performed at ultramicroelectrodes using an ultrafast potentiostat allowing on-line ohmic drop compensation.

In such cases, results may be obtained with scan rates in the range of the megavolts per second [11]. So, voltammetry appears as a kinetic method on a nine order of magnitude of time, from second to nanosecond.

- *Electronic tongues*: An electronic tongue is a device that uses an array of nonselective electrochemical sensors coupled with chemometric methods for the recognition and discrimination of molecules in liquids.

Voltammetry appears as an analytical method with high precision accuracy, sensitivity and wide linear range with relatively low-cost instrumentation. Among all these methods, cyclic voltammetry is not usually performed to analyze samples because of its low sensibility compared to the others. However, in some media, it appears as an appropriate tool, namely in complex media where chemical reactions are coupled with electron transfer reactions.

8.2 OXIDATIVE STRESS AND ANTIOXIDANT DEFENSE SYSTEMS: A RAPID SURVEY

8.2.1 Oxygen reactive species

Oxygen is a vital molecule for organism's life, acting as an acceptor for the electrons released during biologic oxidations; its reduction to water is also essential for energy production in mitochondria. About 2% of oxygen consumed in respiratory chain is lost in the form of intermediates, leading to the production of highly reactive oxygen species (ROS) like superoxide anion $O_2\bullet^-$, hydroxyl radical $OH\bullet$, nitric oxide $NO\bullet$ or peroxynitrite $ONOO^-$ [12].

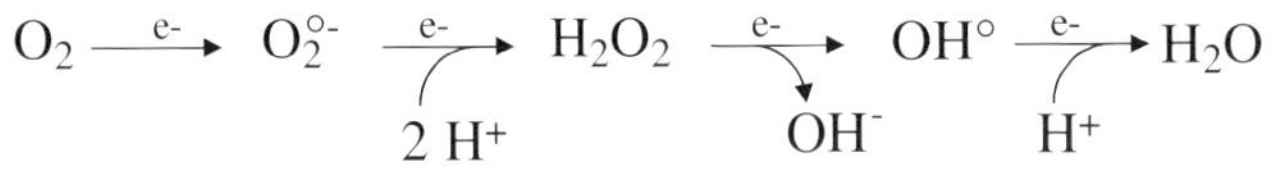

Other enzymatic reactions are responsible for the production of ROS, such as immune cells' NADPH oxidase or neuron's NO synthase:

$$2O_2 + NADPH + H^+ \xrightarrow{\text{NADPH oxidase}} 2O_2\bullet^- + NADP^+ + 2H^+$$

$$\text{Arginine} + 2O_2 \xrightarrow{\text{NOS/NADPH}} \text{Citrulline} + NO\bullet + 2H_2O$$

Moreover, metallic ions like Fe^{2+} are pro-oxidants as they catalyze Fenton's reaction, inducing formation of strongly reactive hydroxyl radical OH• [13,14]:

$$H_2O_2 + Fe^{2+} = Fe^{3+} + OH\bullet + OH^-$$

Naturally produced by the organism, these reactive metabolites play a role in many physiological processes: cell respiration, immune reactions, neurotransmission or blood vessel dilatation. Their concentration has to be finely regulated because these species are unstable and aggressive. They usually display very short life-times, in the range of 10^{-6}–10^{-10} s, and show an important reactivity towards cellular constituents. For instance, the reaction rate constants of hydroxyl radicals with lipids, proteins or nucleic acids are between 10^8 and 10^{10} L mol^{-1} s^{-1}. However, their production can be accentuated during a physical effort, an intense immune reaction or under the influence of pro-oxidant external factors: ionizing radiations (UV, X or γ), chemical pollutants, ozone, metals or tobacco smoke [15].

An excessive production of ROS induces the so-called "oxidative stress" and contributes to cellular dysfunctions and death. Oxidative stress is involved in aging process and contributes to the development of several diseases: cataract, cardiovascular and degenerative diseases (Parkinson, Alzheimer), cancers, etc. [16].

8.2.2 Antioxidant protection system

To reduce the injurious effect of oxidative stress, cells are equipped with two major antioxidant defense systems. The first concerns numerous enzymes, which catalyze ROS degradation, such as superoxide dismutase (SOD), catalase or glutathione peroxidase (GPx):

$$2O_2\bullet^- + 2H^+ \xrightarrow{SOD} O_2 + H_2O_2$$

$$ROOH + 2GSH \xrightarrow{GPx} GSSG + H_2O + ROH$$

The second system involves numerous hydrophilic or lipophilic molecules like ascorbic acid (AH^-) (vitamin C) and α-tocopherol (αT-OH) (vitamin E). These low-molecular-weight antioxidants (LMWA) reduce ROS by oxido-reduction reactions:

$$AH^- + OH\bullet = A\bullet + H_2O$$

$$\alpha T - OH + ROO\bullet = \alpha T - O\bullet + ROOH$$

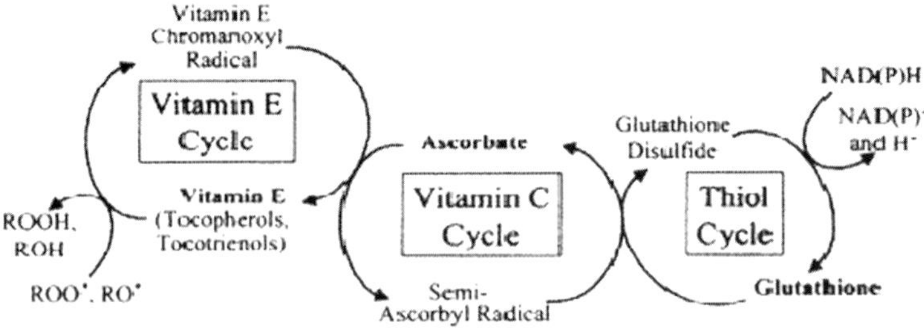

Fig. 8.1. Interactions between principal antioxidants [18].

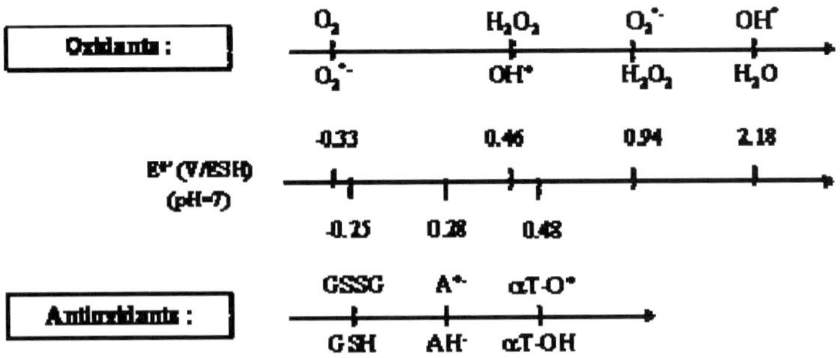

Fig. 8.2. Apparent standard redox potentials of oxidant and antioxidant species (GSH: glutathione, AH^-: ascorbate, αT-OH: α-tocopherol).

The large number of LMWA species, their possible interactions and their reactivity with enzymatic antioxidants are essential for the protection against ROS [17]. Synthesized by the cell, glutathione (GSH) and NADPH are key molecules of the antioxidant system. They react directly with ROS and allow the regeneration of the main antioxidants provided by nutrition: ascorbic acid and α-tocopherol. Figure 8.1 shows the interactions between these antioxidants [18].

Other antioxidant species are synthesized by cells like uric acid, ubiquinol or thiols (cystein, homocystein, etc.). In addition, many compounds found in food display antioxidant properties: retinol (vitamin A) and its precursor β-carotene, and polyphenols (flavonoids, etc.). Figure 8.2 shows the apparent standard potential of some LMWA and ROS explaining the spontaneous oxido-reduction reactions at the origin of the antioxidant protection system.

8.2.3 Voltammetry for the evaluation of antioxidant global capacity

The early diagnosis of diseases, the study of premature aging, the determination of the biomolecular mechanisms taking place and the evaluation of the therapy efficiency need the development of simple, rapid and reliable techniques of detection and quantification of oxidative stress. Numerous analytical methods are available to evaluate oxidative stress, namely electron spin resonance, chromatography, spectroscopy or mass spectrometry. They allow the detection of ROS, the assay of the lipids, proteins and nucleic acids oxidation products, and the determination of enzymatic and nonenzymatic antioxidant species [19]. Nevertheless, all these techniques require expensive materials, involve complex protocols, often need tissue biopsy and provide only delayed results.

Electrochemistry has been considered in the last few years as a promising alternative approach. It represents a suitable method to study oxido-reduction reactions and to determine the species properties. Cyclic voltammetry was particularly involved in evaluating the global antioxidant capacity of real samples like wine, biological fluids and animal or vegetal tissues. The anodic part of the voltammogram provided information concerning the ability of LMWA to act as reducing agents [20–22]. However, in most of the cases, the analytical process was either invasive since it involved tissue homogenates, or performed indirectly, i.e. the sample was diluted in an electrolytic solution.

The new protocols presented here deal with the direct evaluation of antioxidant global capacity applied to skin and dermo-cosmetic creams. The analyses are performed directly in the medium without pre-treatment of the samples. This allows a simple and noninvasive measurement without introducing any possible interference. Moreover, cyclic voltammetry provides results in quasi-real time, allowing the direct monitoring of reaction kinetics.

8.3 ELECTROCHEMISTRY FOR THE STUDY OF SKIN AND COSMETICS ANTIOXIDANT PROPERTIES

8.3.1 State-of-the-art

Skin constitutes the interface between the human body and the environment. It represents a major target of oxidative stress since it is exposed to external oxidant aggressions like UV radiation, ozone, chemicals or pollution. Continuous exposure to such damaging effects and/or deficiency of the antioxidant protection systems result in skin premature aging and contribute to the development of cutaneous diseases and cancers [23]. Electrochemical studies dealing with the effect of oxidative stress on skin and the antioxidant capacity of cutaneous surface are very scarce. Kohen group recently applied this technique successfully to *stratum corneum* analysis; nevertheless, the method adopted suffered from two major drawbacks. First, the protocol was either invasive because it involved skin homogenates, or it was indirect as it was performed in an electrolytic solution in contact with the skin surface. Second, the measurements used macroelectrodes with surface areas in the range of a square centimeter; they presented rather low sensitivity and did not allow localized measurements.

A lot of dermo-cosmetic creams provide antioxidants to skin, in order to protect it from external aggressions. The choice of the active species, the determination of their concentration and the study of their possible interactions require the development of analytical techniques. Electrochemical studies dealing with the analysis of creams systematically involve modification of the sample by expensive operations: dilution, extraction, filtration, heating, etc. [24,25]. These protocols do not allow the study of the overall interactions between the components. Moreover, the evaluation of the variation of the properties of the cream is difficult because of the low resolution time of the measurements.

The research activities in dermatology and cutaneous biology have undergone an important expansion in the last few years [26]. The improvement in our comprehension of cutaneous physiology and diseases mechanisms leads to the identification of the reactions involved in skin submitted to chemical aggressions, light or other oxidations. This knowledge allows to define the nature of the effective principles of dermo-cosmetic products. Therefore, these researches represent a significant stake not only in terms of health but also in terms of market for dermo-cosmetic industries.

8.3.2 Cyclic voltammetry in cosmetic creams and on skin surface

As described in Procedure 9, direct measurements were performed in the bulk of cream, just taken out of its container; the product was not modified by any other operation such as dilution. The cyclic voltammograms were reproducible, as presented in Figure 8.3 (see also Table 9.2 in Procedure 9). Reproducibility varies from one cream to another probably because adsorption phenomena vary with the cream composition.

As described in Procedure 8, direct measurements were performed on skin surface without adding water or gel. Although macroelectrodes

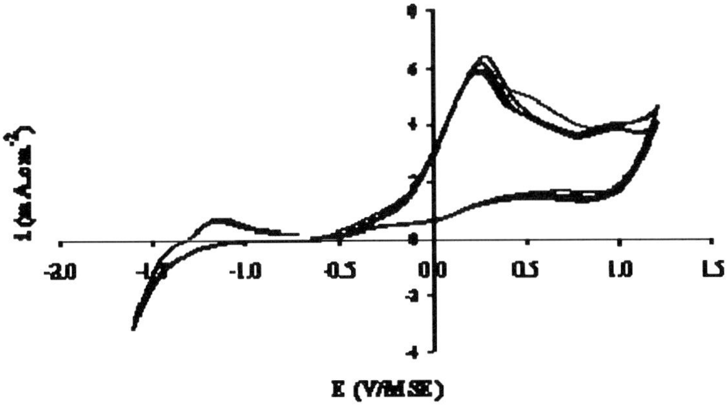

Fig. 8.3. Cyclic voltammograms performed directly in the depilatory cream Klorane with a 2 mm diameter platinum disk. Potential scan rate: $50\,mV\,s^{-1}$.

TABLE 8.2

Charge density on platinum, gold and vitreous carbon in seven commercial dermo-cosmetic creams

Cream	q_{moy} ($mC\,cm^{-2}$)		
	Platinum	Gold	Vitreous carbon
Depigmenting emulsion Trio D (Laboratoires d'Evolution Dermatologique)	7 (25%)	18 (10%)	3 (48%)
After sun balm repair Uriage (Laboratoires Dermatologiques Uriage)	7.6 (9%)	27 (32%)	6.2 (11%)
Depilatory cream (Klorane)	116 (3%)	78 (7%)	103 (9%)
Restructuring care NIVEA Vital (NIVEA)	70 (16%)	66 (3%)	60 (41%)
Emulsion Ystheal+(Laboratoires Dermatologiques Avène)	11 (16%)	4 (44%)	9 (46%)
Epithelial cream A Derma (Laboratoires Dermatologiques Ducray)	4.3 (5%)	1.8 (29%)	1.2 (28%)
Antiwrinkles and radiance day cream Daylift+Day Onagrine (Eucerin)	20 (14%)	15.0 (5%)	8.6 (7%)

allowed study of cosmetic creams, they were not suitable for skin analysis and it was necessary to use 50 μm diameter microelectrodes. In fact, using a 1 mm diameter disk platinum working electrode on skin, the corresponding cyclic voltammogram displayed a current with high resistive and capacitive components (Fig. 8.4a), leading to unusable data. On the contrary, the miniaturized size of the working microelectrode improved the sensitivity of the response owing to the hemispherical diffusion field induced by edge effects (Fig. 8.4b). This voltammogram showed well-defined anodic and cathodic waves. No ohmic drop between the working and the reference electrodes was generated. Comparison of these results demonstrates that microelectrodes allow to realize cyclic voltammetry measurements in resistive media [9].

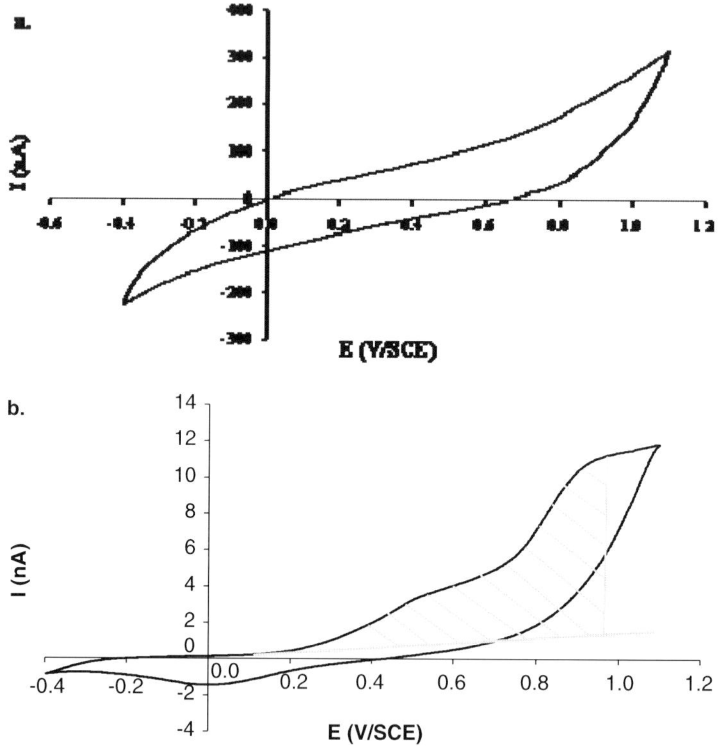

Fig. 8.4. Cyclic voltammograms performed directly on the skin with a (a) 1 mm and (b) 50 μm diameter platinum electrode. Potential scan rate: 50 mV s^{-1}.

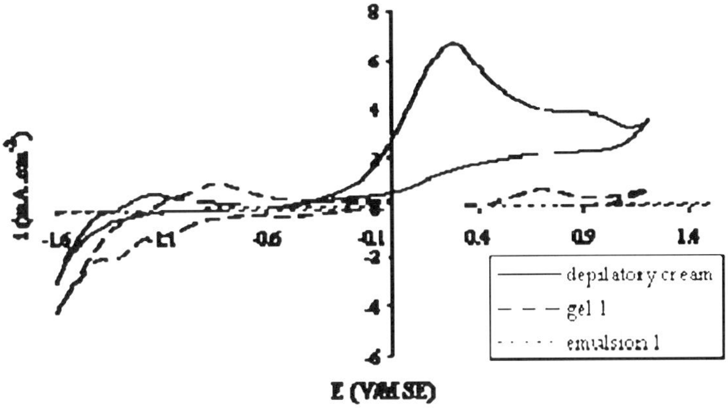

Fig. 8.5. Cyclic voltammograms performed at 50 mV s^{-1} with a 2 mm diameter platinum electrode in a commercial product (depilatory cream Klorane) and in two home-made formulas containing no antioxidant species (gel 1 and emulsion 1).

8.3.3 Evaluation of global antioxidant capacity

Figure 8.5 shows cyclic voltammograms obtained with a 2 mm diameter platinum electrode in a depilatory cream Klorane containing butylhydroxyanisole (BHA) and thiolactic acid and in two home-made formulas containing no antioxidant species. Figure 8.4b shows cyclic voltammograms performed with a 50 µm diameter platinum electrode directly on the skin surface. The anodic parts of these voltammograms were related to the presence of reduced species, in particular antioxidants.

Comparison of cyclic voltammograms recorded in home-made creams without antioxidant (gel 1, gel 2 and emulsion 1) and commercial creams containing antioxidants illustrated the close relationship between the important anodic response and the antioxidant properties of the product (Fig. 8.5). Other results presented in Table 9.2 in Procedure 9 confirmed this relationship. Moreover, current and charge densities increase in the presence of antioxidant(s), proportionally with the antioxidant concentration (see Procedure 9). The overall anodic charge density was chosen as the indicator to estimate the antioxidant global capacity of the creams.

By analogy, *stratum corneum* global antioxidant capacity was estimated from the anodic current recorded at around 0.9 V/SCE or from the anodic charge recorded between 0 and around 0.9 V/SCE, after correction of the residual current estimated by the extrapolation of the tangent at 0 V/SCE (Fig. 8.4b).

8.3.4 Influence of electrode materials

The particular affinity and the interactions between the different species and materials provide the first step of selective electrochemical detection. Also, the use of several electrode materials allows to collect complementary information, in particular in these cases of complex media.

Three electrode materials were tested in nine different commercial creams: platinum, gold and vitreous carbon. On the one hand, in some creams, the shape of the cyclic voltammograms remained nearly the same, whatever the material involved (Fig. 8.6a). On the other hand, in

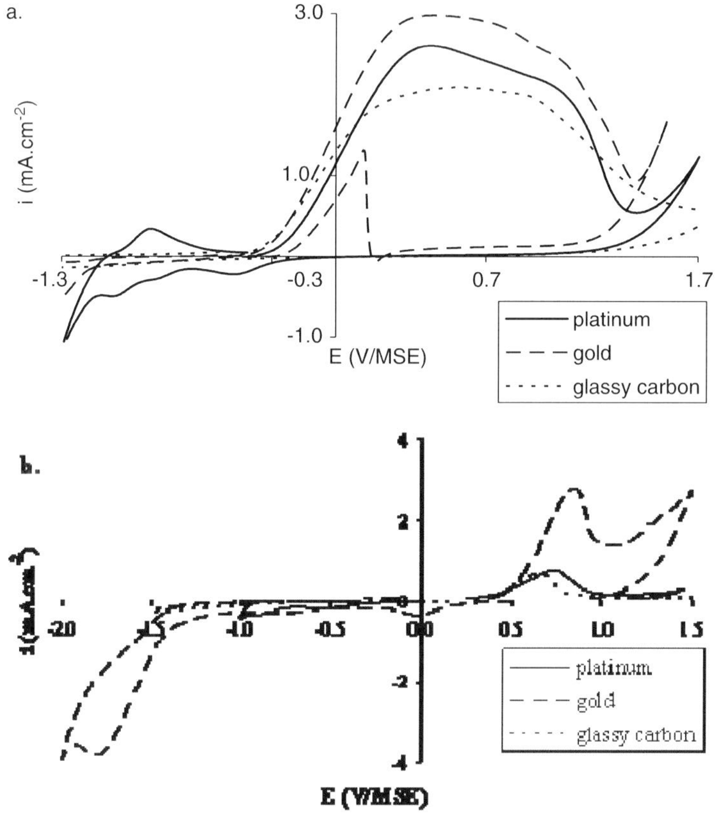

Fig. 8.6. Influence of electrode materials on cyclic voltammograms performed at 50 mV s^{-1} (a) in the restructuring care NIVEA Vital (NIVEA) and (b) in the after sun balm repair Uriage.

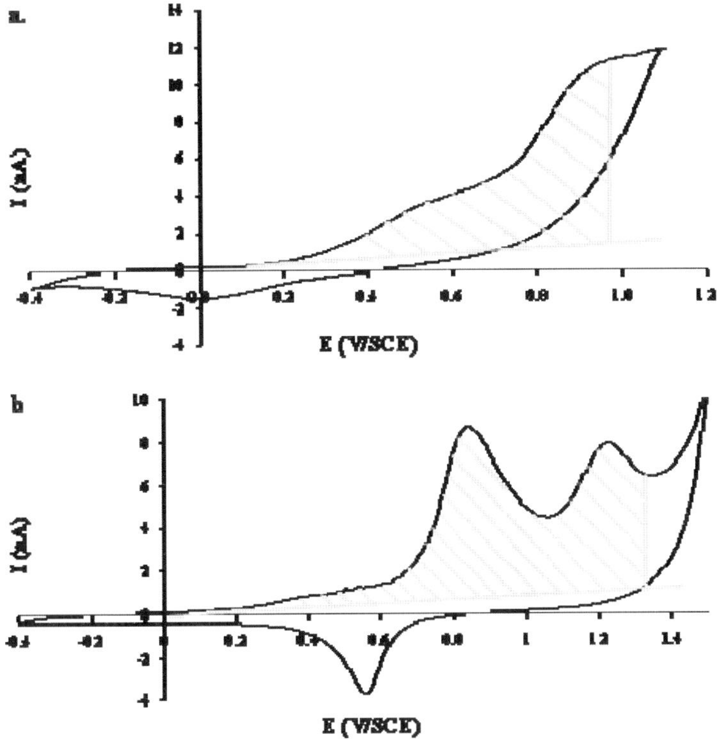

Fig. 8.7. Influence of electrode materials on cyclic voltammograms performed on skin surface at 50 mV s^{-1} (a) with a 50 μm diameter platinum electrode and (b) with a 50 μm diameter gold electrode.

other products, the orders of magnitude of charge density were quite different, as were the shapes of the curves (Fig. 8.6b). Table 8.2 presents charge density recorded in seven commercial creams. Evaluation of antioxidant global capacity differed from one material to another.

Figures 8.7a and b show cyclic voltammograms obtained on the skin with a platinum microelectrode and with a gold microelectrode, respectively. The shape of the current–potential curves was different.

These first results show that platinum was the most pertinent material to estimate the reductive capacity of the medium as charge densities were the most important indicator, in particular in the case of creams. But gold was more interesting to identify components because more peaks were observed. Adsorption phenomena, different on each material, could be responsible of these differences.

8.3.5 Principal antioxidants involved in anodic responses

Identification of the main electroactive antioxidants involved in the global anodic signal recorded on the *stratum corneum* or in the cream is difficult due to the wide diversity of components. Indeed, species can react according to a heterogeneous way directly at the electrode or according to a homogeneous coupled reaction.

In creams, the important viscosity of the media disabled to identify the constituents involved in the electrochemical signal of a commercial product by comparison of the electrochemical characteristic of the species in aqueous media [27]. Moreover, mixing and adding an amount of solution in the cream changes its structure and quantification with an internal calibration is not conceivable. Identification and quantification of species implied to make home-made creams with different compositions.

The analyses of cutaneous removals have been performed using HPLC. The results confirmed the presence of antioxidant species on epidermis surface, i.e. ascorbic and uric acids and glutathione (Table 8.3).

In other respects, voltammograms were performed in solution containing different antioxidants. Figure 8.8a shows cyclic voltammograms obtained with a platinum microelectrode in a deaerated ascorbic acid 1 mmol L^{-1} and uric acid 1 mmol L^{-1} solution, and Figure 8.8b shows cyclic voltammograms obtained with a gold microelectrode in a deaerated glutathione 1 mmol L^{-1} solution. The shape of these voltammograms was similar to those obtained on skin surface with the same electrode material.

Ascorbic and uric acids were the principal antioxidants involved in the anodic response obtained with a platinum electrode. Glutathione was the principal antioxidant involved in the anodic peak observed at around 1.2 V/SCE with a gold electrode.

TABLE 8.3

Chromatographic analysis of cutaneous samples (expressed in nmol L^{-1} because they cannot be weighted to the protein quantity)

	Vitamin C (nmol L^{-1})	Uric acid (nmol L^{-1})	Glutathione (nmol L^{-1})
Subject 1	128	502	4512
Subject 2	95	374	5655
Subject 3	42	535	3618
Subject 4	69	1002	5789

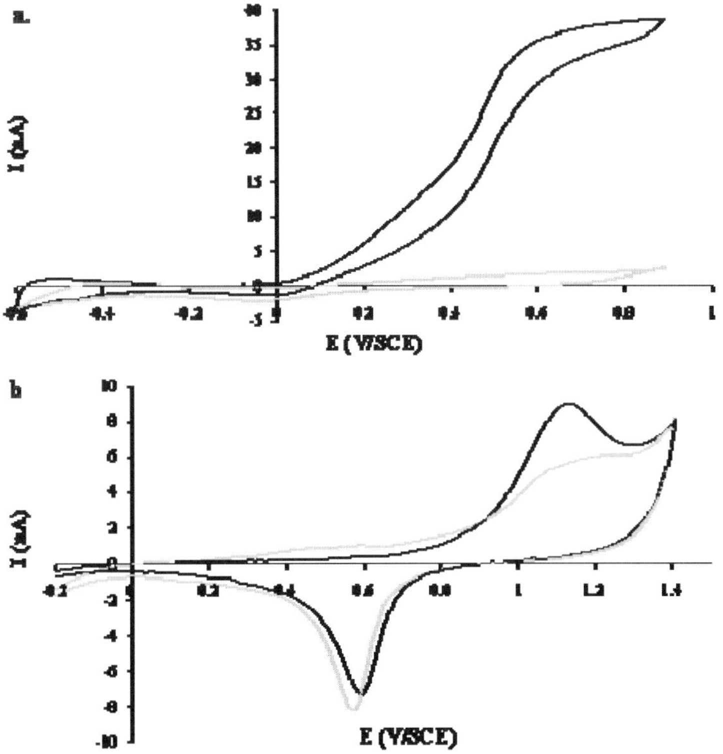

Fig. 8.8. Cyclic voltammograms obtained (a) with a platinum microelectrode in a deaerated phosphate buffer solution (—) or in a deaerated ascorbic acid 1 mmol L^{-1} and uric acid 1 mmol L^{-1} solution (—) and (b) with a gold microelectrode in a deaerated phosphate buffer solution (—) or in a deaerated glutathione 1 mmol L^{-1} solution (—). Potential scan rate: 50 mV s^{-1}.

8.3.6 Evolution of the antioxidant global properties as a function of time

As classical analytical methods only give delayed results, no evolution of the global antioxidant capacity of these media with time can be examined. For the first time, owing to fast response of cyclic voltammetry, the presented direct electrochemical measurements give results in real time, thus allowing the monitoring of reaction kinetics.

In this way, cyclic voltammograms were performed with platinum electrodes at regular time intervals of about 8 h. The antioxidant global capacity of creams changed as a function of the time when the product was exposed to light or air (see Fig. 9.5 in Procedure 9). However, the variation of the oxido-reductive properties was especially located at the surface of the product as it was the most exposed part (data not shown): in the bulk of the product, the loss of antioxidant properties was of 60% whereas at surface it was 95%.

Fine layers of cream were made in order to study these variations and to simulate the product evolution as it was spread on skin. Fluctuations of ingredients concentrations were observed (Fig. 8.9), explained by coupled chemical reactions.

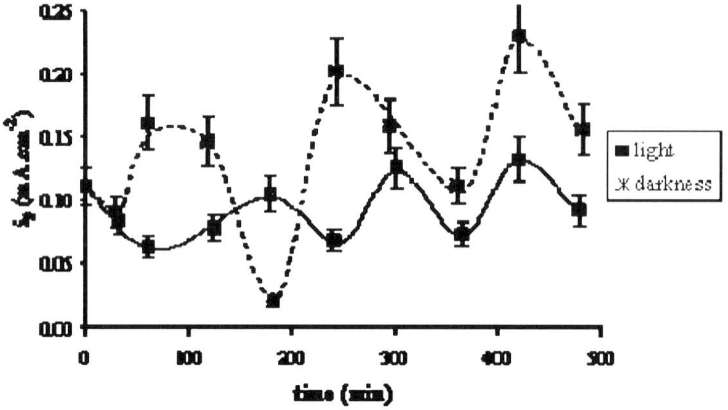

Fig. 8.9. Variation of the antioxidant properties of the restructuring care NIVEA Vital (NIVEA) disposed in a 40 μm layer: peak current density (corresponding to ascorbate oxidation) as a function of the time.

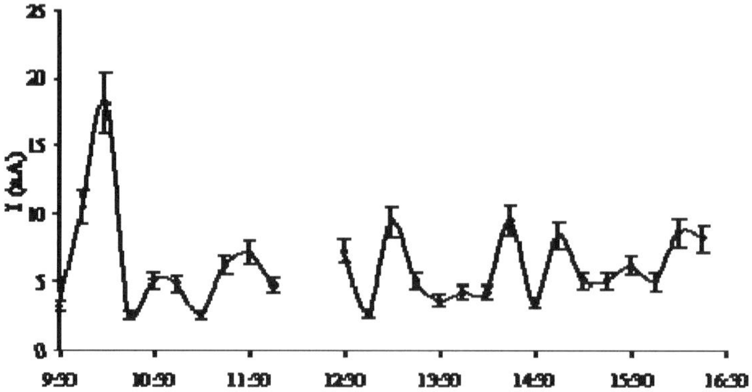

Fig. 8.10. Variation of the current at 0.9 V/SCE recorded with a 50 μm diameter platinum microelectrode on forearm skin with time. Error bars correspond to the mean accuracy of the measurement (see Procedure 8).

Cyclic voltammograms were also performed with platinum microelectrodes on skin surface at regular time intervals of about 7 h. Figure 8.10 shows the typical curve giving the evolution of the anodic current at 0.9 V/SCE as a function of time. A sinusoidal evolution was observed for the nine volunteers. Current values as well as the amplitude and the period of the variations were different for each subject. It has been verified that the amplitude of the current variations was significantly higher than the accuracy of the amperometric response. Consequently, the variation of the anodic current was actually due to a variation of the oxido-reductive properties of the skin and was not an artifact of the measurements.

8.4 CONCLUSIONS

Cyclic voltammetry proved to be a convenient method to reveal the oxido-reduction properties of the skin and of dermo-cosmetic creams. On the one hand, using microelectrodes, it was for the first time possible to evaluate the antioxidant properties on the skin surface. This simple protocol allowed to study in real time the global oxido-reductive state and to determine several antioxidant species. On the other hand, results showed the effect of oxidative stress on the evolution of the antioxidant properties of dermo-cosmetic products in time.

The development of these studies contributes to a better understanding of cutaneous biochemistry and constitutes the foundation for a strict elaboration of creams based on scientific principles. This information makes it possible to evaluate the efficiency of dermo-cosmetic products applied on skin by electrochemical analyses both on the cutaneous surface and in the cosmetic products.

REFERENCES

1 J. Koehne, J. Li, A.M. Cassel, H. Chen, Q. Ye, H.T. Ng, J. Han and M. Meyyappan, *J. Mater. Chem.*, 14 (2004) 676–684.
2 N.S. Lawrence, M. Pagels, A. Meredith, T.G.J. Jones, C.E. Hall, C.S.J. Pickles, H.P. Godfried, C.E. Banks, R.G. Compton and L. Jiang, *Talanta* [Available online 9 December 2005].
3 C.E. Banks and R.G. Compton, *Anal. Sci.*, 21 (2005) 1263–1268.
4 A.A. Karyakin, *Electroanalysis*, 13 (2001) 813–819.
5 S. Casili, M. de Luca, C. Apetrei, V. Parra, A.A. Arrieta, L. Valli, J. Jiang, M.L. Rodriguez-Mendez and J.A. de Saja, *Appl. Surf. Sci.*, 246 (2005) 304–312.

6 K.C. Lin and S.M. Chen, *J. Electroanal. Chem.*, 578 (2005) 213–222.
7 L. Qiong, W. Lirong, X. Danli and L. Guanghan, *Food Chem.*, 97 (2006) 176–180.
8 D. Giovanelli, N.S. Lawrence, S.J. Wilkins, L. Jiang, T.G.J. Jones and R.G. Compton, *Talanta*, 61 (2003) 211–220.
9 R.M. Wightman and D.O. Wipf, Voltammetry at ultramicroelectrodes. In: V. Allen and J. Bard (Eds.), *Electroanalytical Chemistry*, Marcel Dekker, New York, 1989, pp. 267–353.
10 K. Stulik, C. Amatore, K. Holub, V. Marecek and W. Kutner, *Pure Appl. Chem.*, 72 (2000) 1483–1492.
11 C. Amatore, Y. Bouret, E. Maisonhaute, H.D. Abruna and J.I. Goldsmith, *Comptes Rendus Acad. Sci. Chim.*, 6 (2003) 99–115.
12 B. Halliwell, *Am. J. Med.*, 91 (1991) 14–22.
13 H.J.H. Fenton, *J. Chem. Soc. Proc.*, 10 (1894) 157–158.
14 B. PalYu, *Physiol. Rev.*, 74 (1994) 139–162.
15 H. Sies, *Am. J. Med.*, 91 (1991) 31–38.
16 B.N. Ames, M.K. Shigenaga and T.M. Hagen, *Proc. Natl. Acad. Sci.*, 90 (1993) 7915–7922.
17 H. Sies, *Eur. J. Biochem.*, 215 (1993) 213–219.
18 J.J. Thiele, F. Dreher and L. Packer, *J. Toxicol. Cutan. Ocul. Toxicol.*, 21 (2002) 119–160.
19 R.L. Prior and G. Cao, *Free Radic. Biol. Med.*, 27 (1999) 1173–1181.
20 S. Chevion, M.A. Roberts and M. Chevion, *Free Radic. Biol. Med.*, 28 (2000) 860–870.
21 P.A. Kilmartin, *Antioxid. Redox Signal.*, 3 (2001) 941–955.
22 W.R. Sousa, C. da Rocha, C.L. Cardoso, D.H.S. Silva and M.V.B. Zanoni, *J. Food Compos. Anal.*, 17 (2004) 619–633.
23 K.J. Trouba, H.K. Hamadeh, R.P. Amin and D.R. Germolec, *Antioxid. Redox Signal.*, 4 (2002) 665–673.
24 J.M. Irache, I. Ezpeleta and F.A. Vega, *Chromatographia*, 35 (1993) 232–236.
25 L. Galgiardi, D.D. Orsi, L. Manna and D. Tonelli, *J. Liq. Chromatogr. Rel. Technol.*, 20 (1997) 1797–1808.
26 M.L. Wilgus, P.A. Adcock and A. Takashima, *J. Dermatol. Sci.*, 37 (2005) 125–136.
27 W. Miao, Z. Ding and A.J. Bard, *J. Phys. Chem. B*, 106 (2002) 1392–1398.

Chapter 9

Sensoristic approach to the evaluation of integral environmental toxicity

Luigi Campanella and Tania Gatta

9.1 BIOSENSORS OF INTEGRAL TOXICITY

For some years now, the toxicity of many chemical substances has been studied by means of living organisms used as biological indicators. Higher organisms are generally used as the guinea pig, as the results they produce are often extremely reliable, even though the response times may be lengthy (ranging from a few days to several months, or even years in the testing for chronic toxicity). Nevertheless, by exploiting unicellular organisms, in particular yeasts, it is possible to reduce testing time considerably. Furthermore, by using suitable biological systems, such as immobilized yeast colonies, it is possible to implement the method even *in situ*, with the advantage of obtaining a value of integral toxicity as due neither to this nor to that compound but to all together with the eventual antagonisms and synergisms too.

This is one of the most-interesting results of modern biosensor research. For several years, our laboratory has been engaged in this type of research. As a result, a number of integral toxicity biosensors have been developed using immobilized *Saccharomyces cerevisiae* yeast cells and different types of electrochemical transducers [1]. One of the crucial aspects of this research was the development of a suitable method for immobilizing the yeast cells. For this purpose, two classical immobilization methods were adopted. The first method involved the immobilization of yeast cells in a dialysis membrane and the second the entrapment in a membrane of cellulose triacetate. Unfortunately, the results obtained using such methods were not encouraging. It was consequently tempted to develop a new immobilization method involving the entrapment of *Saccharomyces cerevisiae* cells in agar gel containing culture medium [2,3].

Comprehensive Analytical Chemistry 49
Alegret and Merkoçi (Eds)
Volume 49 ISSN: 0166-526X DOI: 10.1016/S0166-526X(06)49009-7

The results obtained using this method were immediately seen to be very interesting: the agarized medium layer on a Petri dish, containing the entrapped yeast cells, can be stored at $+5^{\circ}C$ for more than one month. During this period, whenever the toxicity biosensor has to be used to make a measurement, all that needs to be done is to cut a 0.5 cm diameter disk out of the agarized layer and place it on the selected indicating electrode using a suitable net or dialysis membrane. With this aim the following indicating electrodes have been tested: gas-diffusion amperometric electrode for oxygen, gas-diffusion potentiometric electrode for carbon dioxide, glass electrode for measuring pH and H_3O^+ sensitive FET.

The method is based on the perturbation of the respiratory activity of yeast, *Saccharomyces cerevisiae*, immobilized on an agar gel containing the culture medium (i.e., ''agarized medium''), by the toxic test substance. Glucose is used as substrate while the substances tested consist of several metallic ions, phenols and cationic or anionic surfactants, pesticides and other toxics.

In recent years, the unicellular nature of planktonic algae has been exploited for the construction of whole-cell based biosensors capable of real-time response on critical change of the aquatic ecosystems caused by pollutant emissions. Most of the proposed devices are based on the electrochemical detection of the inhibiting effect on the photosynthetic activity of algae and cyanobacteria exerted by some toxicants.

This approach is quiet interesting and may result as an effective way for improving environmental quality of waterbodies. At present, the impact of xenobiotics on the aquatic environment is evaluated using bioassays (including algal tests) carried out in laboratories under static conditions, which produce a delayed response on water toxicity. The increasing need for automatic devices allowing real-time detection and on-line monitoring may find a successful solution in algal biosensors. However, up to now, few advances have been made in this research field so that the application of biosensors in water-quality monitoring is not yet common. Moreover, no attempt has been made to develop alert systems for toxicity assessment of seawater in estuaries and enclosed bays, which are the marine ecosystems more, affected by pollutant emissions in the sea.

The analytical device is represented by an amperometric biosensor able to monitor the evolution of photosynthetic O_2, obtained by coupling a suited algal bioreceptor to a Clark electrode. The selected algal species is the marine filamentous cyanobacterium *Spirulina subsalsa* (Campanella [4]). Further details on the application of biosensors for integral toxicity measurements can be found in Procedure 10 (in CD accompanying this book).

9.2 PHOTOSENSORS OF ENVIRONMENTAL PERMANENCE

This is a preliminary approach to the use of a new generation of solid-state sensors based on the capacity of the sensor element to catalyze the photodegradation of various kinds of organic compounds and to recognize their structure on the basis of the type of process catalyzed. The electron holes present in the TiO_2 structure are able to promote the oxidative process of substances present in the environment, in particular the ones easily adsorbed on it. Titanium dioxide is a well-known photocatalyst [5–13]. Less famous are its characteristics as sensor material [14–18] of the ability of the organic molecules to be completely degraded, that is mineralized.

During mineralization acidification occurs [19,20]. Since TiO_2 is a photocatalyst of the process and is also able to behave as pH sensor, it is in a position to activate a process and to monitor its proceeding. The time needed in order to record a pH shift can be assumed as a delay time proportional to the recalcitrancy of a compound.

This parameter could be of great help in the case of an unknown or not characterized compound: a White Book of European Community invites the scientific community to make the most of efforts in order to set up chemical tests able to give information—especially alarm advices—in real- or almost real-time about the toxicity of a compound.

By reacting with TiO_2 these substances produce potential variations corresponding to values specific to the above-mentioned photo-oxidative processes. In this first attempt, TiO_2 in the prevailing anatase form was used as sensor, and phenols and wine as known compounds and matrices. Several experimental details were thoroughly studied and optimized: they include nanoparticulated titanium dioxide in suspension, the presence in the water used as solvent of carbonate and organic substances (which affect sensor response), the buffering properties of the phosphate over the entire linear range of detector response, the verification using alternative techniques of the results obtained, the rigorous analysis of the data (kinetic curves, uncertainty, and equation describing the phenomenon), and lastly, the size (owing to the growing demand for "portable" instruments), ease of use and the materials characterizing the sensor.

Titanium dioxide has also found profitable use as a component of an electronic nose or tongue: by different types of treatments useful for the sensing-film preparation, different chemical species can be determined. The fingerprint so obtained can be interpreted as diagnostic on the basis of chemometric treatment of data.

Further details on the application of biosensors for environmental permanence measurements can be found in Procedure 11 (in CD accompanying this book).

9.3 BIOSENSORS FOR THE DETERMINATION OF RADICALS

To the question about the contribution of Chemistry to the evaluation of the human and environmental risk from chemicals and production, also a radical approach answer can be given. When we produce or use energy some of the energy remains segregated within secondary pathway molecules called radicals that are highly energetic and reactive due to the presence of unpaired electrons which can be so or lost or coupled in their electronic structure. As a consequence the radical acts as a reducing or oxidizing agent. Radicals are able to attack any substrate or tissue and so behave as a very dangerous chemical species.

When a free radical reacts with a non-radical, a new radical results and chain reactions can be set up:

1. A radical ($X\bullet$) may add on to another molecule. The adduct must still have an unpaired electron

   $X\bullet + Y \rightarrow [X - Y]$

   Example: $OH\bullet$ adds to guanine in DNA; the initial product is an 8–hydroxyguanine radical.

2. A radical may be a reducing agent, donating a single electron to a non-radical.
   The recipient then has an unpaired electron

   $X\bullet + Y \rightarrow X^+ + Y\bullet^-$

   Example: $CO_2\bullet^-$ reduces Cu^+ to Cu

   $CO_2\bullet^- + Cu^+ \rightarrow CO_2 + Cu$

3. A radical may be an oxidizing agent, accepting a single electron from a non-radical. The non-radical must then have an unpaired electron left behind.
   Example: Hydroxyl radical oxidizes the sedative drug promethazine to the radical cation
   $PR + OH\bullet \rightarrow PR\bullet^+ + OH^-$

4. A radical may abstract a hydrogen atom a C–H bond. As the hydrogen atom has only one electron, an unpaired electron must be left on the carbon.
   Example: Hydroxyl radical abstracts hydrogen from a hydrocarbon side-chain of a fatty acid to initiate lipid peroxidation

   $$CH + OH\bullet \rightarrow C\bullet + H_2O$$

Radicals particularly attack proteins and produce oxidative stress.

All amino-acid residues of proteins are potential targets for attack by reactive oxygen species (ROS) produced in the radiolysis of water; however, in only a few cases have the oxidation products been fully characterized. Moreover, under most physiological conditions, cysteine, methionine, arginine, lysine, proline, histidine, and the aromatic amino acids are primary targets for ROS-mediated oxidation.

Exposure of proteins to ROS can lead to the formation of protein–protein cross-linkages

$$2PHS + \text{``O''} \rightarrow P - S - S - P + H_2O$$

$$PNH_2 + PCHO \rightarrow PN = CHP + H_2O$$

$$2PNH_2 + CH_2(CHO)_2 \rightarrow PN = CHCH_2CH = NP + 2H_2O$$

$$2PCHR \rightarrow \begin{array}{c} PCHR \\ | \\ PCHR \end{array}$$

The more they are present in the environment, the more their concentration within Roman urban air, monitored by us by means of opportunely designed and built sensors (see below), is proportional to traffic intensity and inhabitants number of considered area.

Our research group recently approached the problem of radical determination starting from the determination of oxygen free radicals, in particular superoxide radical, and assembling several new kinds of electrochemical sensors and biosensors suitable for this purpose [21–24]. Firstly, a voltammetric system based on the detection of reduced cytochrome c: this system was also applied to develop a suitable amperometric carbon-paste electrode. Secondly, two potentiometric sensors (one classical selective-membrane sensor and the other a solid-state field effect transistor sensor) based on selective membrane [25–32] entrapping benzylidenephenylnitrone with potentiometric detection. More recently we studied two different kinds of biosensors to determine superoxide radicals obtained by coupling a transducer consisting of an amperometric gas-diffusion electrode for the oxygen, or an other amperometric electrode for hydrogen peroxide, with superoxide dismutase enzyme immobilized in kappa carrageenan gel. Both the sensors showed suitable response to the superoxide radical. We consider that the second type of biosensor is now mature from both an engineering and an operative point of view and employ this biosensor to check antioxidant properties of several compounds comparing the response of the biosensor both in the presence and in the absence of the considered scavenger compound. Radical reactions as said are involved in the processes leading to the breakdown of the lipids contained in most foodstuffs and their deterioration and are responsible for ageing processes and cause of numerous diseases. This is the cause of considerable economic loss and potential problems for health itself.

As a consequence of the increasing number of biochemical problems involving free radicals, the use of scavenging and antioxidant compounds in the health, drugs, and food field has constantly grown. Different screening tests have been developed to determine the antioxidant properties of natural and synthetic antioxidant compounds.

To this end, several studies have been reported involving organic solvents, micelles and liposomes, sometimes using the rate of oxygen uptake via the pressure transducer method, or measured by an oxygen electrode. Successively, methods based on conjugated diene formation, which have a higher sensitivity than oxygen uptake and allow spectrophotometric or fluorimetric measurements, are proposed such as RANSOD, in which the superoxide radical reacts with a derivative of phenyltetrazolium chloride to form a red formazan dye, or the ones that measure the absorbance of the reduced nitro blue tetrazolium compound. Several *in vitro* tests have recently been proposed for measuring the antioxidant properties of food products, based on the inhibition of human low-density lipoprotein oxidation, or by using the oxygen radical absorbance capacity (ORAC), used to determine the total antioxidant activity.

REFERENCES

1 L. Campanella, G. Favero, D. Mastrofini, L. Piccinelli, M. Tomassetti and A. Torresi, *Ann. Chim.*, 88 (1998) 687.

2 L. Campanella, G. Favero and M. Tomassetti, *Sci. Tot. Environ.*, 171 (1995) 227.

3 L. Campanella, G. Favero, D. Mastrofini and M. Tomassetti, *Sens. Actuators B*, 44 (1997) 279.

4 L. Campanella, F. Cubadda and M.P. Sammartino, *Oriental J. Chem.*, 13 (1997) 131–136.

5 V. Vamathevan, R. Amal, D. Beydoun, G. Low and S. McEvoy, *J. Photochem. Photobiol. A: Chem.*, 148 (2002) 233–245.

6 M.R. Dhananjeyan, R. Annapoorani and R. Renganathan, *J. Photochem. Photobiol. A: Chem.*, 109 (1997) 147–153.

7 D. Dvoranovà, V. Brezovà, M. Mazùr and M. Malati, *Appl. Catal. B: Environ.*, 37 (2002) 91–105.

8 L.R. Skubal and N.K. Meshlov, *J. Photochem. Photobiol. A: Chem.*, 149 (2002) 211–214.

9 K.H. Lee, Y.-C. Kim, H. Suzuki, K. Ikebukuro, K. Hashimoto and I. Karube, *Electroanalysis*, 12 (2000) 1334–1338.

10 R.-A. Doong, C.-H. Chen, R.A. Maithreepala and S.-M. Cgang, *Wat. Res.*, 35 (2001) 2873–2880.

11 Y. Mishima, J. Motonaka, K. Maruyama and S. Ikeda, *Anal. Chim. Acta*, 358 (1998) 291–296.

12 F. Oliva, L. Avalle, E. Santos and O. Càmara, *J. Photochem. Photobiol. A: Chem.*, 146 (2002) 175–188.

13 G. Epling and C. Lin, *Chemosphere*, 46 (2002) 561–570.

14 V. Maurino, C. Minero, E. Pelizzetti, P. Piccinini, N. Serpone and H. Hidaka, *J. Photochem. Photobiol. A: Chem.*, 109 (1997) 171–176.

15 S. Glab, A. Hulanicki, G. Edwall and F. Ingman, *Anal. Chem.*, 21 (1989) 29–34.

16 J.A. Byrne, A. Davidson, P.S.M. Dunlop and B.R. Egginns, *J. Photochem. Photobiol. A: Chem.*, 148 (2002) 365–374.

17 M.F.S. Teixeira, F. Moraes, O. Fatibello-Filho, L.C. Ferracin, R.C. Rocha-Filho and N. Bocchi, *Sens. Actuators B*, 56 (1999) 169–174.

18 P. Shuk, K.V. Ramanujachary and M. Geenblatt, *Solid State Ionics*, 85 (1996) 257–263.

19 C. Ravichandran, C.J. Kennady, S. Chellammal, S. Thangevolu and P.N. Anantharaman, *J. Appl. Electrochem.*, 21 (1991) 60–63.

20 A. Fujishima, R.X. Cai, J. Otsuki, K. Hashimoto, K. Itoh, T. Yamashita and Y. Kubota, *Electrochim. Acta*, 38 (1993) 153–157.

21 L. Campanella, M.P. Sammartino, M. Tomassetti and S. Zanella, *Sens. Actuators B: Chem.*, 76 (2001) 158–165.

22 L. Campanella, G. Favero, L. Persi, M.P. Sammartino, M. Tomassetti and G. Visco, *Anal. Chim. Acta*, 426 (2001) 235–247.
23 L. Campanella, G. Favero, L. Persi and M. Tomasetti, *J. Pharm. Biomed. Anal.*, 24 (2001) 1055–1064.
24 L. Campanella, S. De Luca, G. Favero, L. Persi and M. Tomassetti, *Fresenius J. Anal. Chem.*, 369 (2001) 594–600.
25 L. Campanella, G. Favero and M. Tomassetti, *Sens. Actuators B*, 44 (1997) 559–565.
26 L. Campanella, G. Favero and M. Tomassetti, A biosensor for determination of free radicals. In: R. Puers (Ed.), *Proceedings of Eurosensors X, Vol. 3, The 10th European Conference on Solid-State Transducers*, Leuven, 1996, pp. 917–918.
27 L. Campanella, G. Favero and M. Tomassetti, *Anal. Lett.*, 32 (1999) 2559–2581.
28 L. Campanella, M.P. Sammartino, M. Tomassetti and S. Zanella, *Sens. Actuators B*, 76 (2001) 158–165.
29 L. Campanella, S. De Luca, G. Favero, L. Persi and M. Tomassetti, *Fresenius J. Anal. Chem.*, 369 (2001) 594–600.
30 L. Campanella, G. Favero, L. Persi and M. Tomassetti, *J. Pharm. Biomed. Anal.*, 23 (2000) 69–76.
31 L. Campanella, G. Favero, L. Persi and M. Tomassetti, *J. Pharm. Biomed. Anal.*, 24 (2001) 1055–1064.
32 L. Campanella, G. Favero, G. Furlani, L. Persi, M.P. Sammartino and M. Tomassetti, *Proceeding of the 4th Croating Congress of Food Technologists, Biotechnologists and Nutritionists Central European Meeting*, 2001, pp. 373–383.

Chapter 10

Peptide-modified electrodes for detecting metal ions

J. Justin Gooding

10.1 INTRODUCTION

The binding of metals to biological molecules is ubiquitous in nature. In the case of living systems, such metal binding can have a deleterious effect, which is the reason for the toxicity of many metal ions [1]. The toxicity of metal ions is partly related to non-specific binding of metals to proteins and DNA with a concomitant alteration or cessation of their function. Conversely, however, the ingestion of trace amounts of metal ions can be a biological necessity as specific binding of certain metals in proteins is required to form the active centres of many enzymes. The ability of enzymes to bind metal ions with a high degree of selectivity provides a clear sign post that peptide motifs could form the basis of biomimetic ligands for the development of sensors for the detection of metal ion. It is exploiting the ability of peptides to selectively bind metal ions for the development of metal ion sensors, as well as the fact that different peptides allow sensors for different metals to be developed using the same generic method of fabricating the sensor, that is the subject of this chapter.

10.1.1 The binding of metals by amino acids and peptides

The binding of metal ions to peptides and proteins is a consequence of these molecules containing a great number of potential donor atoms through both the peptide backbone and amino acid side chains. The complexes formed exist in a variety of conformations that are sensitive to the pH environment of the complex [2,3]. With at least 20 amino acid combinations available, some with coordinating side chains, which can be linked in any particular order and length, the number of ligands that

Comprehensive Analytical Chemistry 49
Alegret and Merkoçi (Eds)
Volume 49 ISSN: 0166-526X DOI: 10.1016/S0166-526X(06)49010-3

can be synthesized using simple amino acids is practically infinite. In a single amino acid, with a non-coordinating side chain within a peptide sequence, there are two donor atoms that complex the metal, the terminal amine and carbonyl oxygen or amide nitrogen as shown in Scheme 10.1.

With a peptide, the potential donor atoms are extended to the amide nitrogen in the peptide backbone or the carbonyl oxygen. A significantly stronger binding is achieved when the amide nitrogen is involved [2] and therefore, again in the absence of coordinating side chains, the complexation of metal ions occurs as depicted in Scheme 10.2 for the tetrapeptide Gly-Gly-Gly-Gly (the standard three-letter abbreviations for amino acids will be used throughout this chapter).

In view of these binding modes, it is clear that to coordinate strongly to the amide group the metal ions must be capable of substituting for the nitrogen-bound amide hydrogen. Therefore, the number of peptide nitrogens involved in the bonding is dependent on the pH. The ability of

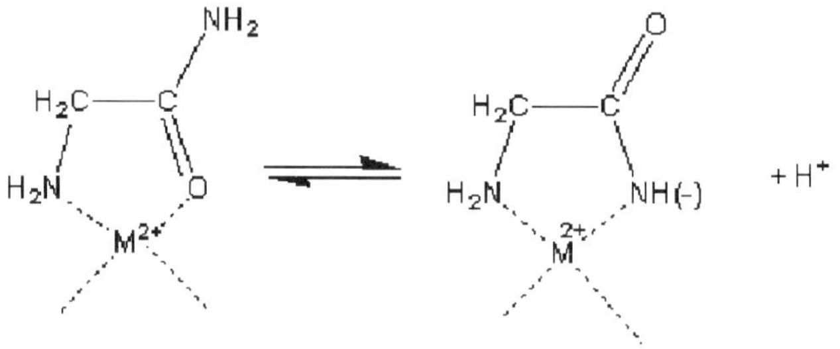

Scheme 10.1.

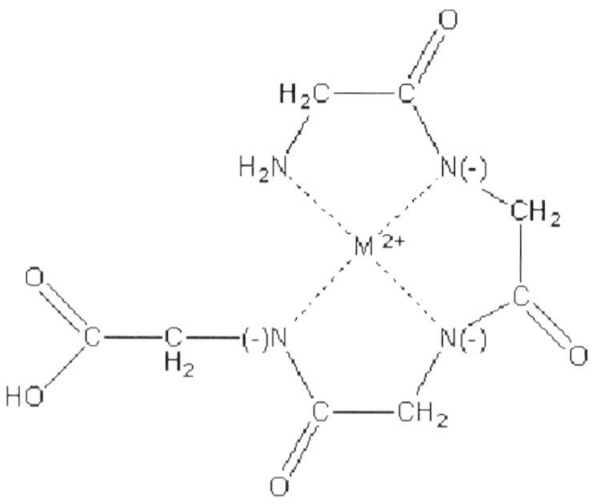

Scheme 10.2.

the metal ion to promote deprotonation of the amide nitrogens plays a key role in the final stability of the resultant complex. Metal-induced deprotonation of the amide nitrogens is important because it increases the number of coordination points between the metal and the ligand. The order of peptide hydrogen displacement by metal ions, with representative pK_a value in brackets for short peptides, is given by Pd^{2+} (2) > Cu^{2+} (4) > Ni^{2+} (8) > Co^{2+} (10) [2]. Note that the order relates to reasonably hard metal ions as they interact with the hard amide ligands [4].

The selectivity of peptide motifs for certain metals comes from the coordinating contribution from amino acid side chains, the common coordination number of the metal, hardness/softness of the metal ion, ligand field stabilisation effects and the hardness/softness of any coordinating side chains of the amino acid sequence. An example of the influence of side chains and the importance of the position of the side chain comes from the tripeptides Gly-Gly-His, also known as copper binding peptide. The side chain imidazole ring of the His residue has a very efficient nitrogen donor (the imidazole N), which can form a tetradentate chelate ring for coordination as in Scheme 10.3.

This coordination geometry has been verified in a number of ways including using mass spectrometry [5,6]. The position of the His residue, however, plays a key role in the coordination properties of the peptide. If the His residue is the first or second amino acid in the tripeptide (His-Gly-Gly or Gly-His-Gly) there is a greater than 10-fold reduction in complex stability [2].

The importance of the peptide ligand having soft ligands for soft metals and hard ligands for hard metals is illustrated by the tripeptide

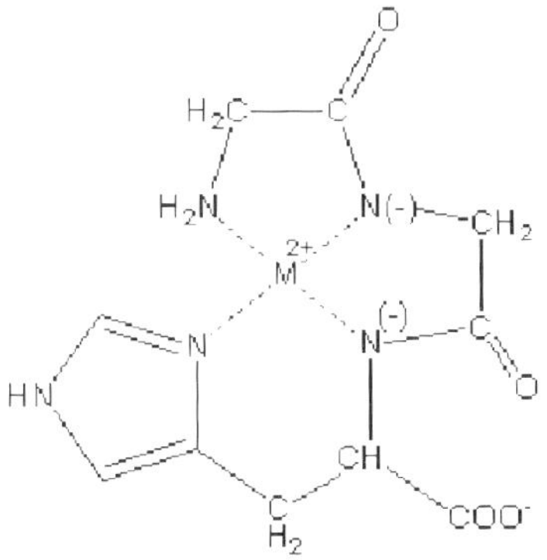

Scheme 10.3.

glutathione. Glutathione has the amino acid sequence γ-Glu-Cys-Gly. Glutathione has eight possible coordination sites, which fall into three classes. There are two carboxyl groups (one on the Gly at the –OOH end and one on the side chain of the glutamic acid) and one amino acid at the glutamic acid end, the soft thiol side chain of the cysteine and the two pairs of carbonyl and amide nitrogen donors associated with the peptide bonds. Hard metal donors interact primarily with the glutamic acid moiety while the primary anchor for soft metals is through the thiol [7].

The above discussion is designed to highlight that the amino acid building blocks provide a myriad of peptide ligands that will have a broad spectrum of affinities for different metal ions. The fact that the synthesis of different peptides essentially involves the same chemistry regardless of the sequence, and is reasonably simple, makes peptide ligands a highly attractive and under-exploited class of ligands for the development of solid-state metal ion sensors. This attractiveness is perhaps enhanced by the development of generic immobilisation chemistries so that arrays of different peptide-based metal ion sensors can be fabricated [8–12] relatively easily.

10.1.2 Using peptides to detect metal ions

The complexation of metal ions with peptides in electroanalysis dates back to the 1980s [13] for the selective detection of peptides with the metal ions used as a redox marker. The first reports of peptides [14] and amino acids [15] for the enhancement of cathodic stripping analysis of metals also commenced around the same time. Even with a simple amino acid such as cysteine, very low detection limits (as low as 0.6 nM Cu(II)) can be achieved with cathodic stripping analysis at a mercury electrode [15]. Similarly, Fogg and co-workers used hanging mercury drop electrodes in combination with poly-L-histidine to detect Cu(II) in the range of 5–400 nM with accumulation of the metal in 2 min [16]. Mercury electrodes have also been used to explore speciation of metal ions such as Cd(II) and Zn(II) to peptides such as glutathione and fragments of metallothioneins using multivariate analysis [17–21]. Designing peptide ligands with high selectivity for a given metal ion was first demonstrated using fluorescence-based sensing by Imperiali and co-workers for the detection of first Zn(II) [22] and then Cu(II) [23]. These fluorescence-based metal ions employ environmentally sensitive fluorophores to give a change in fluorescence upon binding of the metal ion with the peptide. The work of Imperiali and co-workers is significant for the use of peptides in developing electrochemical metal ion sensors for three important reasons. Firstly, the synthetic peptides used were derived from peptide motifs used in nature to selectively bind metals, zinc fingers in the case of the detecting Zn(II) [22], secondly, it was demonstrated that simply by changing one of two amino acids in the modular ligand, the selectivity and binding constant of the ligand–metal complex was altered, and finally, the dissociation constants for these complexes were in the pM to nM range, thus suggesting that very low detection limits could be achieved. It was this work that was an inspiration for us to use peptide-modified electrodes as sensors for the detection of metal ions [24].

10.1.3 Peptide-modified electrodes as metal ion selective sensors

The immobilisation of peptides onto solid-state electrodes to give a sensor has almost exclusively been achieved using self-assembled monolayers (SAMs). The monolayer system used has most frequently been alkanethiols on gold but more recently has also been diazonium salt derived monolayers on carbon electrodes [25–27]. Possibly, the first example of modifying an electrode with a peptide to give an electrochemical metal ion sensor was by Takehara *et al.* [28] where a gold electrode was modified with glutathione (Glu-Cys-Gly) as a sensor for detecting lanthanide metals. The modification process involved self-assembly of glutathione onto the gold electrode via the thiol side chain on the cysteine. In the absence of metal ions, and at a pH greater than 5.7 the monolayer has a net negative charge that repels the redox active species ferricyanide, which is in solution. Upon binding of any of the lanthanides La^{3+}, Eu^{3+} and Lu^{3+} the ferricyanide can penetrate the monolayer and give an electrochemical signal at the electrode. The limitation of the approach is one of selectivity; all three lanthanides give a response and later workers have used glutathione-modified electrodes to detect Cd^{2+} [10] and Cu^{2+} [29]. The work is important, however, in demonstrating the immobilisation of peptides on electrodes by self-assembly.

This idea of using self-assembly to form peptide-modified electrodes was further followed by Zeng *et al.* [29] who used a mixed SAM of glutathione and 3-mercaptopropionic acid on gold as an amperometric sensor for detecting copper in the range of 10–100 μM without preconcentration via the reduction current. Even lower detection limits (down to 0.4 nM) for Cu(II) were achieved by modifying the electrode with cysteine alone and preconcentrating the copper [30,31] where each

Cu(II) has been shown to be bound by two cysteines [32]. The main limitation of the use of cysteine or glutathione for detecting metal ions is the thiol side chain which having a high affinity for soft metals is being employed to immobilise the peptide and hence is not available for coordination with the metal ion.

A generic method of modifying carbon electrode materials has very recently been developed by Compton and co-workers [25,26] for the purpose of copper sequestration. One approach [25] involves modifying graphite powders with a 4-nitrobenzenediazonium tetratfluoroborate to give a nitrophenyl terminates surface. Reduction of the nitrogroup into an amine allows subsequently coupling of a peptide from the carboxylic acid end (a C–N attachment). In this example, poly-L-cysteine was attached to the graphite powder that was dispersed into the sample to complex any free Cd(II). In the later work [26], glassy carbon microspheres were treated with thionyl chloride to give a surface susceptible to nucleophilic attack by amines such as from the amino terminus of L-cysteine methyl ester. The latter approach would generally be preferred as it gives an N–C attachment of the peptide. In the work by Compton and co-workers the direction of the peptide attachment is not important as the complexation of the metal is dominated by the thiol side chain of the cysteine. However, in some cases the direction of immobilisation is exceedingly important as shown by Tao and co-workers [33] where gold surfaces were modified with SAMs that were terminated in either a carboxylic acid group to allow N–C attachment of the peptide or an amine terminated surface to give C–N attachment. Attachment of the tripeptide Gly-Gly-His to each surface resulted in quite different copper binding behaviour as determined by differential surface plasmon resonance (SPR). The N–C attachment resulted in much lower concentration range in which the sensor was sensitive to changes in copper concentration than the C–N case that reflects a higher binding constant. The different binding affinity can be understood from Scheme 10.4, which shows the two surface constructs. In the case of the N–C immobilisation, the imidazole ring of the histidine is much more accessible for complexing with the copper to give a complex as depicted in Scheme 10.4 than in the C–N case.

Our work has been motivated towards making the use of peptide-modified electrodes for detecting metal ions as generic as possible. Essentially, this means developing a generic approach to modifying the electrode so that any peptide can be attached in a manner to provide optimal metal binding capability.

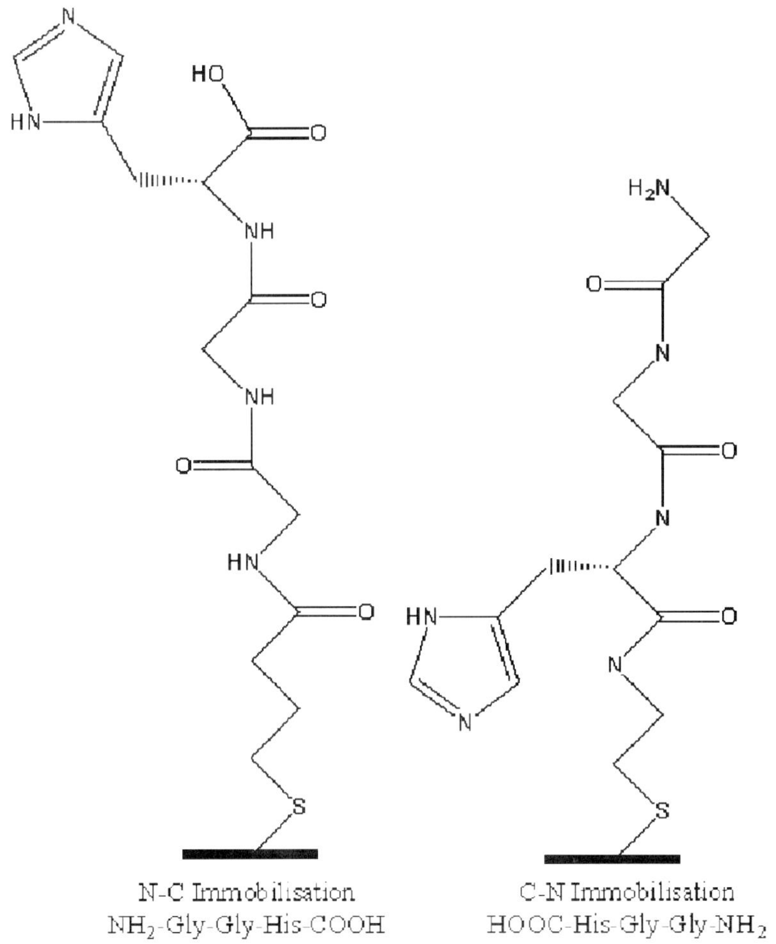

Scheme 10.4.

10.2 APPLICATION

10.2.1 Interfacial design

The generic approach to modifying electrodes with peptides for the detection of metal ions is depicted in Fig. 10.1 and outlined in Procedure 13 (in CD accompanying this book). The essential feature is an electrode modified with an SAM, which contains a carboxylic acid moiety at the distal end. The carboxylic acid moiety is activated using carbodiimides, typically 1-ethyl-3-(3-dimethylaminopropyl)

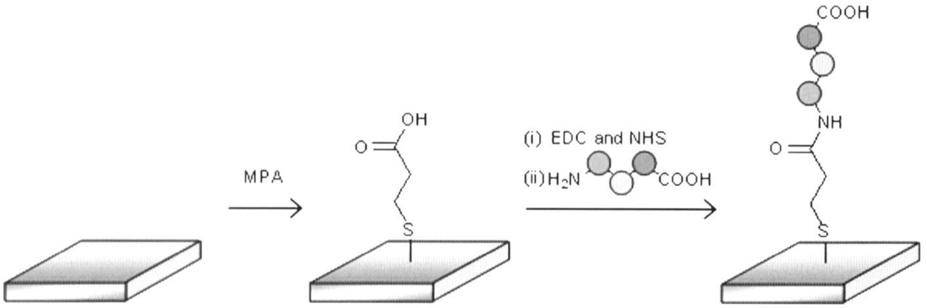

Fig. 10.1. Schematic of the generic interface used to immobilise peptides on an electrode surfaces for metal ion sensing.

carbodiimide hydrochloride (EDC) and *N*-hydroxysuccinimide (NHS) in aqueous solution or 1,3-dicyclohexylcarbodiimide (DCC) in organic solvents are employed. The activation converts the carboxylic acid into an active intermediate that is susceptible to nucleophilic attack from amines, such as those found on the N-ends of peptides.

In most cases, the monolayer forming molecule is an alkanethiol on gold electrodes, most frequently, 3-mercaptopropioic acid (MPA) [5,6,8–10,12,24,32,34,35] but thioctic acid has also been used [11]. The choice of a short chain alkanethiol is to ensure that the underlying electrode is still electrochemically accessible. However, short chain alkanethiols form significantly less stable SAMs than longer alkanethiols due to less stabilisation of the SAM by van der Waals forces between alkyl chains. The consequences of the lower stability are (1) the gold–thiolate bond can easily be oxidised in the case of MPA [36,37] and (2) even prior to oxidation short chain alkanethiols give a narrower potential window in which the modified electrode can operate without reductive or oxidative desorption of the SAM. To extend the range of potentials in which alkanethiol chemistry can effectively be employed we used thioctic acid [11]. Greater stability is achieved with this molecule as there are two thiolate bonds formed with the gold surface per molecule. This was necessary in the case of a sensor for lead because the current maximum for the reduction of lead is approximately −0.4 V (vs. Ag/AgCl) at an Angiotensin I modified electrode and at this potential MPA reductive desorption of the gold–thiolate bond has commenced. Thiotic acid does not begin to desorb until −0.5 V or below [11,38] and hence allows detection of lead without loss of the SAM. The drawbacks of thiotic acid are that the potential range is still limited and the longer alkyl chain results in a slower rate of electron transfer between the metal ion and the electrode, relative to MPA [11], which results in a slightly lower sensitivity. To give even more stable SAMs, carboxyphenyl diazonium salts have been employed to modify carbon [39] and gold electrodes [27]. This modification is not as simple as the alkanethiol modification as molecules must be custom synthesized and the assembly process is performed under potentiostatic control. However, the monolayers are significantly more stable than alkanethiol monolayers and this opens the potential window in which peptide-modified electrodes can be employed for detecting metal ions.

The coupling of the N-terminus of a peptide gives an N to C immobilisation of the peptide which, as shown by Tao and co-workers [33] can be important with regards to the affinity of a peptide for a specific metal. A carboxylic acid terminated surface is also advantageous with regards to electrode fabrication. The activation of the surface bound carboxylic acid moieties by the carbodiimide means the activated surface can be removed from the carbodiimide solution and rinsed prior to the exposure of the surface to peptide. This stepwise fabrication greatly simplifies the preparation of modified electrodes as avoiding the carbodiimide and peptide coexisting in the same solution obviates the need for protection of any carboxylic acid moieties on the peptide. If the carbodiimide and peptide were in the same solution and no protection of the carboxylic acids was employed the peptide would cross-link to itself. Note *t*-Butoxycarbonyl (Boc) or similar protection strategies are still required if the peptide contains amino acids, such as lysine, with amine side chains [9].

The modification of electrode surfaces with monolayer forming molecules allows precise positioning and density of immobilised peptide ligands. The density of ligands can be controlled by forming mixed monolayers where the peptide ligand can only be attached to one of the components. Hence by varying the number of molecules to which the peptides can be attached the ligand density can be controlled. This approach has been exploited using a mixed monolayer of MPA and mercaptopropane (MP) to explore the effect of the density of the tripeptide Gly-Gly-His on the sensitivity of the modified electrode for the detection of copper (see Fig. 10.2) [6]. As the mole fraction of MPA increased relative to MP the amount of copper complexed increased until a maximum was reached at 0.5 mole fraction of MPA. Further increases in MPA mole fraction resulted in no further change in copper current (see Fig. 10.2). The insensitivity of the copper current to the density of coupling points above a mole fraction of 0.5 was attractive for fabrication purposes as minor changes in the monolayer composition had no effect on the sensor response. The insensitivity was attributed to

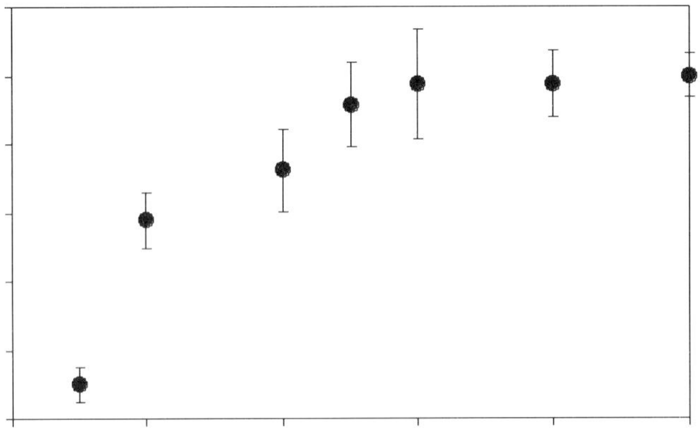

Fig. 10.2. Cu(II) coverage (Γ) of Gly-Gly-His modified gold electrodes on mixed SAMs of MPA and MP, determined by integration of CV peaks. Mixed SAMs comprising MPA and MP were prepared by immersing the gold-coated substrates in solutions of mixtures of MPA and MP of a given fraction. In all cases, Cu(II) was accumulated at the Gly-Gly-His modified electrode at open circuit for 10 min in a 0.05 M ammonia acetate buffer solution (pH 7.0) containing 0.1 μM copper nitrate, removed, rinsed and then placed in a copper-free ammonium acetate buffer solution. Scan rate: 100 mV s^{-1}. Reproduced with permission of The Royal Society of Chemistry from Ref. [6], Copyright, Royal Society of Chemistry (2003).

steric effects preventing more Gly-Gly-His ligands being immobilised. Additional evidence for such a conclusion was derived from surfaces where the tripeptide was synthesized directly on the electrode surface from the individual amino acids with a view to demonstrating new peptide ligands for a given metal could be identified by combinatorial synthesis directly on the electrode surface [40]. Different compositions of mixed monolayers were systematically evaluated for the composition that gave the best yield of complete ligands on the electrode surface as determined by electrochemistry and mass spectrometry [8]. The optimal density of coupling points between MP and MPA was again identified as a mole fraction of 0.5. At higher mole fractions of MPA the yield of the successfully synthesized peptide on the surface decreased. These studies illustrate that the key to optimal sensor performance comes from appropriate interfacial design.

10.2.2 The detection of copper ions at Gly-Gly-His-modified electrodes

The tripeptide glycine glycine histidine is known as copper binding peptide and is used for treating of copper disorders such as Wilson's disease [2]. Immobilisation of the peptide onto an MPA-modified gold electrode via attachment at the N-terminus of the peptide gives the final sensor as depicted in Scheme 10.4. The surface construct was confirmed by mass spectroscopy [6] and X-ray photoelectron spectroscopy [12]. Cyclic voltammograms (CVs) and Osteryoung square wave voltammograms (OSWV) of a freshly prepared MPA-Gly-Gly-His modified electrode prior to and after copper accumulation are illustrated in Fig. 10.3. CVs were used for characterisation of the electrode interface whilst OSWVs were employed for analytical measurements due to the higher sensitivity of OSWV. The voltammograms typically display no electrochemistry between +0.4 and −0.2 V prior to exposure to copper. Occasionally, small peaks due to copper are evident due to contamination of the electrodes in the preparation process. When such contamination was detected, copper was removed from the electrode by applying a potential of +0.5 V in 0.1 M $HClO_4$ prior to use. Figure 10.3a(ii) shows the CV after incubating the electrode in 46 nM Cu^{2+} for 10 min, followed by rinsing with copper-free ammonium acetate buffer, and electrochemical measurement in copper-free ammonium acetate buffer (60 mM CH_3COONH_4, 50 mM NaCl, pH 7.0). Copper redox peaks at $E_{1/2}$ of +0.14 V appear, which are stable over several cycles in the voltammogram. The electrochemistry is attributed to Cu^{2+}/Cu^0 where copper is underpotential deposited (UPD) on gold when electrochemically reduced [12]. Some early papers have attributed the electrochemistry to Cu^{2+}/Cu^{1+}; however the $E_{1/2}$ is more consistent with the E° for Cu^{2+}/Cu^0 rather than that for Cu^{2+}/Cu^{1+}. UPD of copper on SAM-modified surfaces has also been reported previously [41–43]. Importantly, on glassy carbon surfaces modified with Gly-Gly-His, where UPD of copper could not occur, the $E_{1/2}$ for reduction of Cu^{2+} was −0.06 V [39], a value more consistent with Cu^{2+}/Cu^{1+}. The stability of the electrochemistry in Fig. 10.3 indicates that Cu^0 being oxidised to Cu^{2+} must be recaptured at the electrode at the scan rate of 100 mV employed. At slower scan rates the electrochemistry was observed to decay with multiple scans as, at slower scan rates, the Cu^{2+} formed in the oxidation reaction has sufficient time to diffuse away from the electrode prior to the cathodic cycle reducing any oxidised metal.

The binding of copper by Gly-Gly-His as depicted in Scheme 10.3 was confirmed using mass spectroscopy [40] and semi-quantitative molecular modelling [6]. The amount of copper bound, and hence the current signal observed was shown to be a function of the accumulation time [12] with a Langmuir-like variation in amount of copper accumulated with time [12]. Other parameters that were shown to influence the

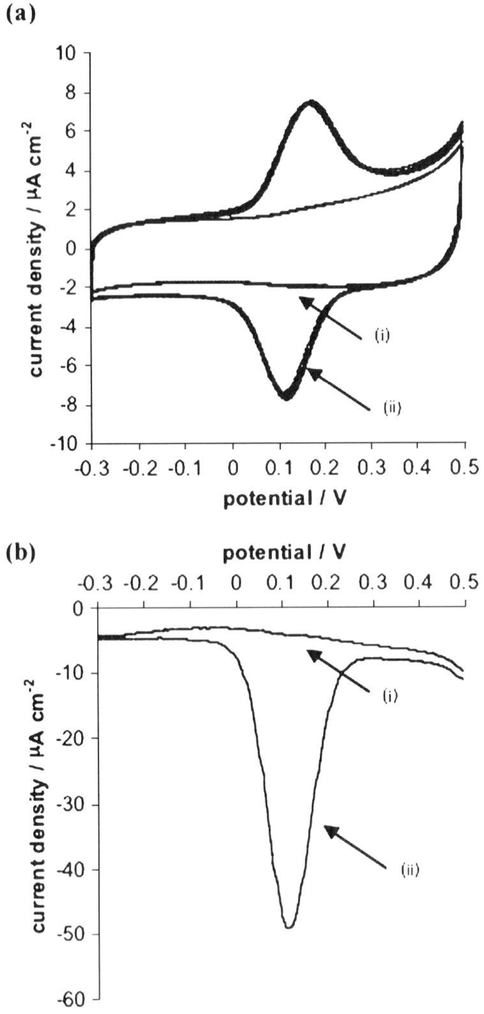

Fig. 10.3. General electrochemical performance of MPA-Gly-Gly-His modified electrodes for the detection of Cu^{2+} ions. Cu^{2+} ions are complexed to Gly-Gly-His in the accumulation process and are electrochemically reduced to Cu(0) to give UPD Cu. (a) Cyclic voltammograms of MPA-Gly-Gly-His modified electrodes in 50 mM ammonium acetate (pH 7.0) and 50 mM NaCl at 25°C at a scan rate of 100 mV s^{-1} (i) before accumulation of metal ions and (ii) after accumulation in 46 nM Cu^{2+} in 50 mM ammonium acetate (pH 7.0) for 10 min. Multiple cycles in the copper voltammogram illustrate stable electrochemistry. (b) Cathodic Osteryoung square wave voltammograms of MPA-Gly-Gly-His modified gold electrodes in 50 mM ammonium acetate (pH 7.0) and 50 mM NaCl (i) before accumulation of metal ions and (ii) after accumulation in 46 nM Cu^{2+} in 50 mM ammonium acetate (pH 7.0) for 10 min. Reprinted from Ref. [12]. Copyright (2005) with permission from Elsevier.

sensors sensitivity were pH, the concentration of background electrolyte and temperature [39]. The highest sensitivity was observed at pH between 7 and 9, with minimal background electrolyte during copper accumulation and higher temperatures. A calibration curve under the optimal conditions is shown in Fig. 10.4. Each point in the calibration curve is obtained by incubating the electrode in the appropriate copper standard for 10 min, followed by removal, rinsing and then electrochemical measurement in copper-free ammonium acetate buffer (60 mM CH_3COONH_4 and 50 mM NaCl, pH 7.0). Before the next measurement, the electrode was regenerated in $HClO_4$ as described in Procedure 13 (in CD accompanying this book). Detection limits of 0.3 nM were routinely recorded [39] although on some occasions copper concentrations down to 3 pM (0.2 ppt) were detected [5]. These exquisitely low detection limits are difficult to verify because the concentration of standards are below the detection limit of ICP-MS and therefore are nominal concentration. The low detection limits are, however, realistic as the affinity constant for Gly-Gly-His for Cu^{2+} in solution is

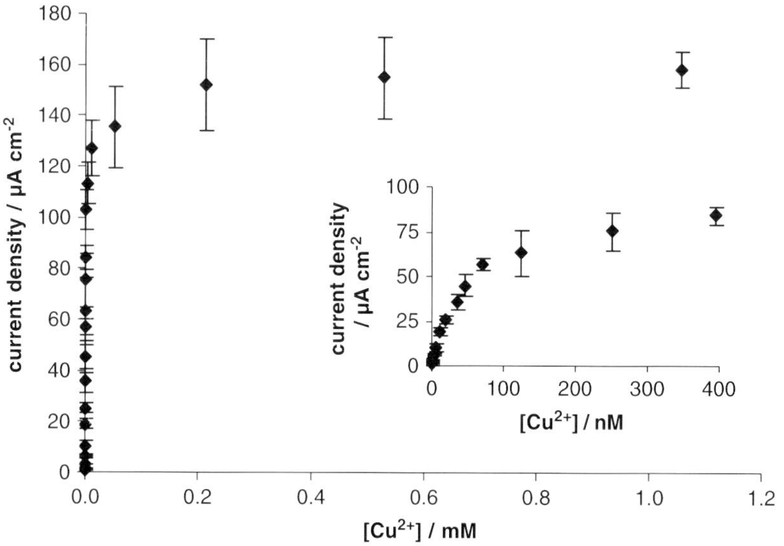

Fig. 10.4. Cathodic OSWV peak current densities for the reduction of Cu^{2+} at MPA-Gly-Gly-His modified electrodes as a function of Cu^{2+} concentration. Error bars are ± 1 standard deviation of the current densities of four individual electrodes. Inset shows the peak current density in the region between 0 and 400 nM Cu^{2+}. OSWVs were measured in 50 mM ammonium acetate (pH 7.0) and 50 mM NaCl at a pulse amplitude of 0.025 V, a step of 0.004 V and frequency of 25 Hz.

$1.3 \times 10^{11} M^{-1}$ at 25°C [44] whilst the binding constant for the surface bound Gly-Gly-His with Cu^{2+} was $8 \times 10^{10} M^{-1}$ for the most sensitive electrodes prepared. The lower affinity of the surface bound Gly-Gly-His is reasonable considering the loss of configurational freedom associated with the immobilisation of the peptide. Furthermore, the affinity constant for the immobilised peptide relative to the peptide in solution indicates the low detection limits reported from the nominal concentrations are conceivable.

The Gly-Gly-His modified electrodes exhibited exquisite sensitivity but the next important issue was the selectivity of the sensor. The selectivity of the MPA-Gly-Gly-His modified electrode for Cu^{2+} was investigated in the presence of common metal ion interferents: Cd^{2+}, Pb^{2+}, Ni^{2+}, Zn^{2+}, Ba^{2+} and Cr^{3+}. A study of the interferences was carried out by a Plackett–Burman experimental design consisting of eight experimental runs with each run containing 46 nM Cu^{2+}, plus a low level (nominal 200 nM) or high level (nominal 5 μM) of each of the metal ion interferents Zn^{2+}, Pb^{2+}, Ni^{2+}, Cd^{2+}, Ba^{2+} and Cr^{3+} (see Procedure 13). A variable in which no change to the solution is made, known as the dummy variable, was introduced as an estimate of the measurement uncertainty. The main effect of a particular metal ion (i.e. the average change in current density as that metal concentration goes from the low value to the high value) is determined by summing the copper current density of each experimental run multiplied by the metal's contrast coefficient, +1 or −1, then dividing by half the number of runs. The significance of an effect is tested by a Rankit plot (Fig. 13.33 in Procedure 13 in CD accompanying this book). If the data are normally distributed, that is, the effects are zero and any changes in current density may be ascribed to random measurement error, then all effects should fall on a straight line through zero. Effects that fall off the line are considered as arising from significantly interfering metal ions. The insignificant effects are Zn^{2+}, Cd^{2+}, Ba^{2+}, Ni^{2+} and the dummy variable. The significant interfering ions are Cr^{3+} and Pb^{2+}, which lead to a decrease in the copper current since they compete for binding sites but do not provide an electrochemical signal at the applied potential.

The stability of the peptide-modified electrodes with repeated use is the final issue in the evaluation of their performance. Between measurements, the electrodes were regenerated by poising the peptide-modified electrode at +0.5 V for 30 s in 0.1 M $HClO_4$. With repeated accumulation–measurement–regeneration cycles (where the copper concentration in each accumulation cycle was 23 nM) the degradation between measurements was 1–2% regardless of whether the measurement was once per day (stored at 4°C in ammonium acetate buffer between measurements) or continually repeated [39]. This degradation was attributed to loss of the capability of the peptide to complex the metal rather than loss of the alkanethiol SAM from the gold surface because identical stability was recorded for Gly-Gly-His modified glassy carbon electrodes. The monolayer on the glassy carbon surfaces derived from diazonium salts are exceedingly stable and, unlike alkanethiols, are difficult to remove and do not oxidise [39].

When applied to the analysis of real samples the Gly-Gly-His modified electrodes, regardless of whether gold or glassy carbon is used as the electrode material gave lower copper concentrations to those derived using ICP-MS (see Table 13.2 in Procedure 13). In contrast, recovery of copper from spiked pure water samples is close to 100%. These differences between the copper concentrations measured in real samples by ICP-MS and the modified electrodes can be attributed to the peptide-modified electrodes not measuring total metal ion concentrations but the concentration of free or loosely bound copper. Using different peptide ligands with different affinity constants for a given water sample the concentration of copper detected by the peptide-modified electrodes increased with increasing affinity constant for the metal [45]. The ability of the peptide-modified electrodes to measure loosely bound copper, and the fact that the amount of recovered copper depends on the affinity constant of the peptide for the metal in question, indicates that peptide-modified electrodes could be used to gain an indication of the bioavailability of metal ions in a sample for a specific species. For example, the uptake of copper by some fish is achieved through specific peptide motifs in their gills. Electrodes modified with the same peptide motifs could then be used to gain a measure of the concentration copper in a water sample that the fish can take up.

10.2.3 Peptide-modified electrodes for other metals: towards multianalyte arrays

The detailed study of Gly-Gly-His modified electrodes has been used to provide the basic knowledge of what are the important parameters in fabricating and using peptide-modified electrodes. The real power of the approach to making metal ion sensors though is that a single interfacial design can be used to fabricate a number of different metal ion sensors simply by changing the peptide attached to the SAM-modified electrode. Peptide sequences selective for certain metals can be identified from protein databases, using phage display libraries of peptides [46] or using a variety of traditional instrumental techniques for characterising peptide–metal interactions such as mass spectrometry [47,48] or nuclear magnetic resonance spectroscopy [49,50]. Metal ion sensors for cadmium [9,10], lead [11] and silver [51] have been fabricated by attaching selective peptides for these metals to MPA or thioctic-modified gold electrodes.

The detection of cadmium was achieved using two different peptides. In the first instance, glutathione (GSH with sequence γ-Glu-Cys-Gly) was employed [10]. The electrode exhibited a dynamic range of two orders of magnitude with a lowest detected concentration of Cd^{2+} of 5 nM, which was well below the WHO guidelines for drinking water of 27 nM [52]; however, the electrode did suffer from negative interference from Cu^{2+} (meaning the copper binds to the GSH without contributing to the electrochemical signal) and positive interference from Pb^{2+}. The positive interference from Pb^{2+} is due to Pb^{2+} having a redox potential that overlaps with Cd^{2+} as well as being able to bind to the peptide ligand. The interference from Cu^{2+} was perhaps not surprising considering glutathione-modified electrodes have also been used for the detection of copper [29] and also considering the cysteine in the peptide leaves a free sulfhydryl group that will have a strong affinity to any soft metals. Superior selectivity for Cd^{2+} was achieved using the hexapeptide His-Ser-Gln-Lys-Val-Phe rather than GSH. The lysine amino acid, with a free amine in the hexapeptide, meant that side chain protection was required to ensure the peptide attached to the carboxylate surface via the amine end of the peptide. His-Ser-Gln-Lys-Val-Phe with a *t*-butoxycarbonyl (Boc) protection group on the lysine side chain was attached to an MPA-modified gold electrode followed by deprotection in 1% trifluoroacetic acid, 2.5% triisopropylsilane and 2% water in dichloromethane. The resultant electrode has a lower detection limit (lowest detected cadmium concentration of 0.9 nM) and a higher sensitivity than when GSH was employed with the only interfering ion being Pb^{2+} [9].

For the detection of Pb^{2+}, human angiotensin I (Asp-Arg-Val-Tyr-Ile-His-Pro-Phe-His-Leu) was attached to a thioctic acid modified electrode. Thioctic acid was used to widen the potential window over which the SAM would remain stable on the electrode surface [11]. The electrode showed no significant interferences to the complexation of the peptide with the metal although, as with the other peptides, mercury and silver caused some degradation of the SAM at high concentrations (500 nM) [11]. The detection limit for Pb^{2+} was reported to be 1.9 nM with a dynamic range over two orders of magnitude.

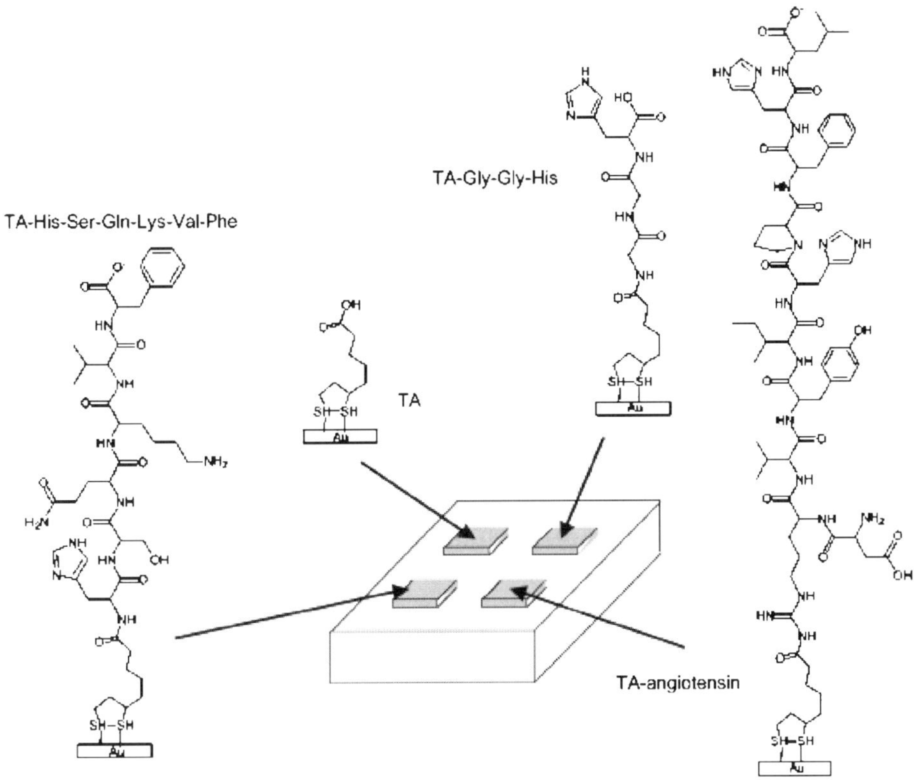

Fig. 10.5. A schematic of the four elements electrode array where elements were modified with no peptide (that is thioctic acid alone), Gly-Gly-His for Cu^{2+}, GSH for Cd^{2+} and human angiotensin I for Pb^{2+}.

The ability to detect Cu^{2+}, Cd^{2+} and Pb^{2+} with different electrodes modified in the same way but with different peptides selective for different metals opens the door to fabricating an electrode array for more than one metal ion. To explore that possibility an electrode array with four elements was employed where each electrode was modified with thioctic acid. The four elements were electrodes modified with (a) no peptide (that is thioctic acid alone), (b) Gly-Gly-His, (c) GSH and (d) human angiotensin I (see Fig. 10.5). The electrode array was calibrated by preparing 16 different cocktails of Cu^{2+}, Cd^{2+} and Pb^{2+} according to a central composite design [53]. OSWV for each electrode for one of the calibration solutions (0.100 μM Cu^{2+}, 3.00 μM Cd^{2+} and 0.600 μM Pb^{2+} in 50 mM ammonium acetate at pH 7.0) are shown in Fig. 10.6. The two features apparent from Fig. 10.6 are that there are only two main peaks due to the overlap of the electrochemistry from

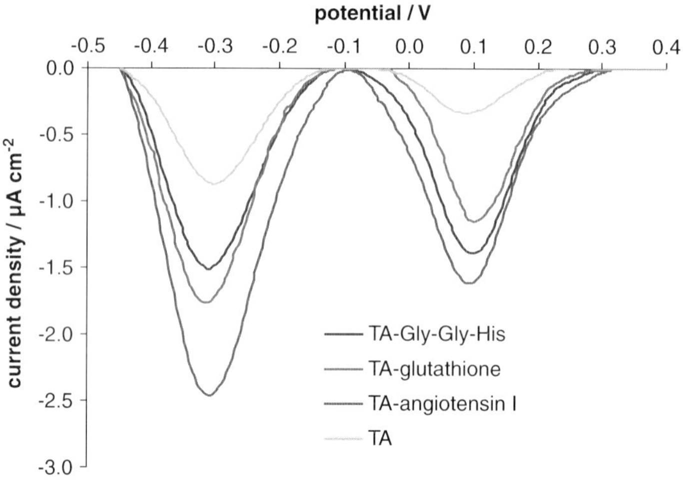

Fig. 10.6. Baseline subtracted OSW voltammograms of the difference between the voltammograms obtained before and after metal ion accumulation in test sample number 1 for 10 min at 25°C. Test sample 1 contained 0.100 µM Cu^{2+}, 3.00 µM Cd^{2+} and 0.600 µM Pb^{2+} in 50 mM ammonium acetate (pH 7.0). The electrodes were modified with TA-Gly-Gly-His, TA-glutathione, TA-angiotensin I and TA. OSW voltammograms were measured in 50 mM ammonium acetate (pH 7.0) and 50 mM NaCl at 25°C at a pulse amplitude of 0.025 V, a step of 0.004 V and frequency of 25 Hz.

TABLE 10.1

Predictions of Cu^{2+}, Cd^{2+} and Pb^{2+} in test samples using *N*-PLS using two, five and seven principal components respectively

	Analyte	Concentration (actual) (µM)	Concentration (predicted) (µM)
Test sample 1	Cu^{2+}	0.10	0.19
	Cd^{2+}	3.00	3.6
	Pb^{2+}	0.60	0.47
Test sample 2	Cu^{2+}	0.40	0.63
	Cd^{2+}	5.0	6.5
	Pb^{2+}	0.10	0.22
Test sample 3	Cu^{2+}	0.60	0.64
	Cd^{2+}	4.0	5.6
	Pb^{2+}	0.70	0.53

Cd^{2+} and the Pb^{2+} and all electrodes display the two peaks but to different extents. Despite the peak overlap, the electrode array can be calibrated for each metal ion using a three-way partial least squares regression (*N*-PLS) [53]. The electrode array was employed to analyse three test samples of known concentration of Cu^{2+}, Cd^{2+} and Pb^{2+} and the concentrations of each analyte predicted by the calibrated electrode array are shown in Table 10.1. As can be seen from Table 10.1 there is reasonable agreement between the actual and predicted values despite the fact that all electrodes respond to all analytes and that the electrochemical responses to lead and cadmium overlap. Further improvements would be expected if the calibrations were performed with a box experimental design, which encompassed the linear range of all the sensors.

10.3 CONCLUSIONS

The binding of metal ions by peptide motifs, both nonselectively and with high selectivity, is ubiquitous in nature. As a consequence, metal ion sensors where peptides are used as the recognition molecule have begun to attract considerable interest. Peptide-modified electrodes have been shown to provide a generic technology for the fabrication of sensors for the detection of metal ions. By modifying an electrode with an SAM with a short alkyl chain and a carboxylic acid terminus, simple carbodiimide coupling allows the attachment of amino acids and short peptide sequences via the N-terminus of the peptide. The advantages of using SAMs are (1) the short alkyl chain ensures the metal ion is maintained close to the electrode surface such that it is electrochemically accessible and (2) the control over the interfacial design afforded by SAMs enables control over the accessibility of the peptide for the metal ion. Such control has been exploited to allow the synthesis of peptide ligands directly on electrode surfaces, which have similar binding capacity for the metal ion in question as the attachment of presynthesized and purified peptides. The final sensor possesses two layers of selectivity, the selectivity of the peptide for the metal as well as the different redox potentials of the different metal ions.

Electrodes modified with the tripeptide Gly-Gly-His have been shown to be able to selectively detect copper with detection limits that are unrivalled. Optimal response of the Gly-Gly-His electrode is obtained at higher temperature, lower salt concentration and longer accumulation time in a pH range of 7–9. The stability of such electrodes to repeated use was limited by the stability of the peptide with a 1–2% degradation in signal with repeated use. The analysis of real samples revealed that peptide-modified electrodes accumulate only loosely bound copper as distinct from total copper. The amount of copper in a given water sample detected with peptide-modified electrodes was dependent on the affinity of the peptide for the metal ion. This latter observation suggests the peptide with which an electrode was modified could be tuned to match peptides involved in the uptake of metals by a particular species, thus providing a sensor that gives an indication of bioavailable metal concentrations for that species.

The generic nature of the method of fabricating peptide-modified electrodes enables sensors for different metal ions to be developed. Peptide-modified electrodes for cadmium, lead and silver have also been prepared all of which show low detection limits and good selectivity for the metal ion in question. Arrays of these sensors were fabricated and used to analyse solutions containing a cocktail of Cu^{2+}, Cd^{2+} and Pb^{2+}. To calibrate the sensor array, a partial least squares regression analysis approach was used. The final sensor array was shown to give reasonable agreement between the concentration of the three metals predicted from the measured responses and the calibration model with the assigned values of each metal. Thus single metal ion sensors can be developed that will detect a suite of metal ions in a sample, in the field in less than 20 min.

ACKNOWLEDGMENTS

I would like to acknowledge the Australian Research Council and the CASS Foundation for funding aspects of this work. I would also like to thank my colleagues with whom I have collaborated on the use of peptide-modified electrodes for the detection of metal ions namely, Professor Brynn Hibbert, Dr. Wenrong Yang and Dr Edith Chow, Quynh Nguyen, Guozhen Liu, Associate Professor Gary Willett and Diako Ebrahimi.

REFERENCES

1 J.J. Fausto da Silva and R.J.P. Williams, *The Biological Chemistry of the Elements*, Oxford University Press, Oxford, 1991.
2 H. Sigel and R.B. Martin, *Chem. Rev.*, 82 (1982) 385–426.
3 H. Kozlowski, W. Bal, M. Dyba and T. Kowalik-Jankowska, *Coord. Chem. Rev.*, 184 (1999) 319–346.

4 K.M. Mackay and R.A. Mackay, *Introduction to Modern Inorganic Chemistry*, International Textbook Company, London, 1981.
5 W. Yang, D. Jaramillo, J.J. Gooding, D.B. Hibbert, R. Zhang, G.D. Willett and K.J. Fisher, *Chem. Commun.* (2001) 1982–1983.
6 W.R. Yang, E. Chow, G.D. Willett, D.B. Hibbert and J.J. Gooding, *Analyst*, 128 (2003) 712–718.
7 A. Krezel and W. Bal, *Acta Biochim. Pol.*, 46 (1999) 567–580.
8 W.R. Yang, R. Zhang, G.D. Willett, D.B. Hibbert and J.J. Gooding, *Anal. Chem.*, 75 (2003) 6741–6744.
9 E. Chow, D.B. Hibbert and J.J. Gooding, *Electrochem. Commun.*, 7 (2005) 101–106.
10 E. Chow, D.B. Hibbert and J.J. Gooding, *Analyst*, 130 (2005) 831–837.
11 E. Chow, D.B. Hibbert and J.J. Gooding, *Anal. Chim. Acta*, 543 (2005) 167–176.
12 E. Chow, E.L.S. Wong, T. Böcking, Q.T. Nguyen, D.B. Hibbert and J.J. Gooding, *Sens. Actuators B*, 111 (2005) 540–548.
13 R. Bilewicz, *J. Electroanal. Chem.*, 267 (1989) 231–241.
14 T. Shimizu and Y. Tanaka, *Ion-Sensitive Cyclic Octapeptides*, Vol., Agency of Industrial Sciences and Technology, Japan, 1986, p. 6.
15 S. Tanaka and H. Yoshida, *J. Electroanal. Chem.*, 137 (1982) 261–270.
16 J.C. Moreira, R. Zhao and A.G. Fogg, *Analyst*, 115 (1990) 1561–1564.
17 J. Mendieta, M.S. Diaz-Cruz, R. Tauler and M. Esteban, *Anal. Biochem.*, 240 (1996) 134–141.
18 J. Mendieta and A.R. Rodriguez, *Electroanalysis*, 8 (1996) 473–479.
19 J. Mendieta, M.S. Diaz-Cruz, A. Monjonell, R. Tauler and M. Esteban, *Anal. Chim. Acta*, 390 (1999) 15–25.
20 M.S. Diaz-Cruz, J.M. Diaz-Cruz, J. Mendieta, R. Tauler and M. Esteban, *Anal. Biochem.*, 279 (2000) 189–201.
21 M.S. Diaz-Cruz, M. Esteban and A.R. Rodriguez, *Anal. Chim. Acta*, 428 (2001) 285–299.
22 G.K. Walkup and B. Imperiali, *J. Am. Chem. Soc.*, 119 (1997) 3443–3450.
23 A. Torrado, G.K. Walkup and B. Imperiali, *J. Am. Chem. Soc.*, 120 (1998) 609–610.
24 J.J. Gooding, D.B. Hibbert and W. Yang, *Sensors*, 1 (2001) 75–90.
25 G.G. Wildgoose, H.C. Leventis, I.J. Davies, A. Crossley, L.N.S. Lawrence, L. Jiang, T.G.J. Jones and R.G. Compton, *J. Mater. Chem.*, 15 (2005) 2375–2382.
26 B. Sljukic, G.G. Wildgoose, A. Crossley, J.H. Jones, L. Jiang, T.G.J. Jones and R.G. Compton, *J. Mater. Chem.*, 16 (2006) 970–976.
27 G.Z. Liu, T. Böcking and J.J. Gooding, *J. Electroanal. Chem.*, 600 (2006) 335–344.
28 K. Takehara, M. Aihara and N. Ueda, *Electroanalysis*, 6 (1994) 1083–1086.
29 B.Z. Zeng, X.G. Ding and F.Q. Zhao, *Electroanalysis*, 14 (2002) 651–656.

30 D.W.M. Arrigan and L. Le Bihan, *Analyst*, 124 (1999) 1645–1649.
31 A.-C. Liu, D.-C. Chen, C.-C. Lin, H.-H. Chou and C.-H. Chen, *Anal. Chem.*, 71 (1999) 1549–1552.
32 W. Yang, J.J. Gooding and D.B. Hibbert, *J. Electroanal. Chem.*, 516 (2001) 10–16.
33 E.S. Forzani, H.Q. Zhang, W. Chen and N.J. Tao, *Environ. Sci. Technol.*, 39 (2005) 1257–1262.
34 W. Yang, J.J. Gooding and D.B. Hibbert, *Analyst*, 126 (2001) 1573–1577.
35 J.J. Gooding, E. Chow and R. Finlayson, *Aust. J. Chem.*, 56 (2003) 159–162.
36 N.J. Brewer, R.E. Rawsterne, S. Kothari and G.J. Leggett, *J. Am. Chem. Soc.*, 123 (2001) 4089–4090.
37 J.J. Gooding, A. Chou, F.J. Mearns, E. Wong and K.L. Jericho, *Chem. Commun.* (2003) 1938–1939.
38 M. Akram, M.C. Stuart and D.K.Y. Wong, *Anal. Chim. Acta*, 504 (2004) 243–251.
39 G.Z. Liu, Q.T. Nguyen, E. Chow, T. Böcking, D.B. Hibbert and J.J. Gooding, *Electroanalysis*, 18 (2006) 1141–1151.
40 W.R. Yang, D.B. Hibbert, R. Zhang, G.D. Willett and J.J. Gooding, *Langmuir*, 21 (2005) 260–265.
41 D. Oyamatsu, M. Nishizawa, S. Kuwabata and H. Yoneyama, *Langmuir*, 14 (1998) 3298.
42 D. Oyamatsu, S. Kuwabata and H. Yoneyama, *J. Electroanal. Chem.*, 473 (1999) 59.
43 M. Nishizawa, T. Sunagawa and H. Yoneyama, *Langmuir*, 13 (1997) 5215.
44 J. Masuoka, J. Hegenauer, B.R. van Dyke and P. Saltman, *J. Biol. Chem.*, 268 (1993) 21533–21537.
45 Q.T. Nguyen, Thesis: *Gly-Gly-His Biosensors for the Detection of Copper Ions in the Environment*, Vol. Honours Thesis, The University of New South Wales, Sydney, 2004.
46 U. Kriplani and B.K. Kay, *Curr. Opin. Biotechnol.*, 16 (2005) 470–475.
47 H.B. Li, K.W.M. Siu, R. Guevremont and J.C.Y. LeBlanc, *J. Am. Soc. Mass Spectrom.*, 8 (1997) 781–792.
48 T.C. Rohner and H.H. Girault, *Rapid Commun. Mass Spectrom.*, 19 (2005) 1183–1190.
49 S. Fujiwara, G. Formicka-Kozlowska and H. Kozlowski, *Bull. Chem. Soc. Jpn.*, 50 (1977) 3131–3135.
50 L.A.P. Kane-Maguire and P.J. Riley, *J. Coord. Chem.*, 28 (1993) 105–120.
51 E. Chow, *Peptide-Modified Electrochemical Sensors for the Detection of Heavy Metal Ions*, Vol, The University of New South Wales, Sydney, 2006.
52 Lenntech—*WHO/EU Drinking Water Standards Comparative Table*, Vol.
53 E. Chow, D. Ebrahimi, L. Hejazi, J.J. Gooding and D.B. Hibbert, *Analyst*, (2006) submitted for publication.

Chapter 11

Reproducible electrochemical analysis of phenolic compounds by high-pressure liquid chromatography with oxygen-terminated diamond sensor

Chiaki Terashima and Akira Fujishima

11.1 INTRODUCTION

Phenol and substituted phenols are important industrial products, which are absolutely essential for manufacturing phenolic resin used in the plywood, automotive, and appliance industries. Phenols are also used for manufacturing caprolactam and bisphenol A, which are intermediates in the production of nylon and epoxy resin. Therefore, some of the larger and more common sources of phenols are paper mills, petrochemical refineries, glue manufacturers, coke plants, and municipal waste incinerators [1–4]. The leachate might contain phenolic compounds such as chlorinated phenols (CPs) and aromatic phenols. Forms of important environmental contamination involving CPs result from the fact that phenols react with chlorine during the water treatment process to produce CPs [5]. In addition, when household garbage is incinerated with vinyl chloride, CPs, chlorinated benzenes, and polychlorinated dibenzo-*p*-dioxins and dibenzofurans (PCDD/F) are known to be present in the flue gases discharged from municipal waste incinerators [6–9]. Due to the harmful nature and persistency of CP pollutants, the U.S. Environmental Protection Agency (EPA) recognized that CPs are priority pollutants, and should be monitored periodically because they exert an extensive influence on the public health [10,11]. Moreover, CPs in flue gases are considered to be a precursor of PCDD/F [8,12], and need continuous monitoring on a simple alternative monitoring index.

The EPA method [10,11] of gas chromatography (GC) with either electron capture detector or mass spectrometry after liquid–liquid

Comprehensive Analytical Chemistry 49
Alegret and Merkoçi (Eds)
Volume 49 ISSN: 0166-526X DOI: 10.1016/S0166-526X(06)49011-5

extraction is recommended for the determination of phenols and CPs, and this method enables precise analysis to be conducted in the laboratory. However, it is not considered appropriate for on-site field analysis. The pK_a values of, for example, dichlorophenols (DCPs) are within the range of 4.5–8.0 [13], and the pH of environmental water can easily vary between 6 and 8. This fact indicates that most of phenols exist in ionic form in environmental water. Therefore, as CPs are soluble in environmental water, it is necessary for GC analysis to treat samples for analysis preliminarily with liquid–liquid and solid-phase extractions in order to convert the CPs into organic solvents, except when the purpose is for concentration and purification. Consequently, high-pressure liquid chromatography (HPLC), which can treat a liquid sample directly, is employed for the separation and detection of environmental CPs including complex matrices. Although, as HPLC detectors and spectroscopic, photochemical, and enzymatic detection methods have been developed for phenol analysis, they are not very successful due to their complexity and limitation to only a few CPs [14]. On the other hand, electrochemical methods offer both sensitivity and selectivity without the need for having expensive instruments or using complicated derivatization procedures [15,16]. Therefore, HPLC with electrochemical detection (HPLC-ECD) has been applied in the determination of phenols in environmental water, and can be adapted for the analysis of CPs found in flue gas discharged from municipal waste incinerators. The crucial issue for environmental analysis is the capability of a consecutive and robust method to be accomplished with high sensitivity and high selectivity. A combination of three technologies, (i) reversed-phase HPLC that enables water sample to be directly injected, (ii) column-switching HPLC that allows a large-volume injection for higher sensitivity and purification of dirty samples [17,18], and (iii) amperometric detection based on redox reaction, satisfied the above-mentioned requisite. Although the clinical assay of catecholamines and their metabolites has routinely been performed in an electrochemical detection with carbon electrodes of glassy carbon and graphite, at which relatively low potentials around 0.5 V vs. Ag/AgCl are applied [19], the detection of phenols, however, seems to differ.

It is well known that the oxidation of phenolic compounds at solid electrodes produces phenoxy radicals, which couple to form a passivating polymeric film on the electrode surfaces [20,21]. The anodic reaction proceeds through an initial one-electron step to form phenoxy radicals, which subsequently can undergo either polymerization or further oxidation with the transfer of oxygen from hydroxyl radicals at the electrode surface to produce *o*-benzoquinone and *p*-benzoquinone. Surface fouling, due to the building of adsorbed reaction products, causes problems in the electroanalysis of phenols. The electrochemically generated passive film is strongly adherent and continuous over the electrode surface. Washing with organic solvents is usually insufficient to regenerate the electrode activity. The film must be mechanically removed by, for example, polishing with diamond powder and alumina powder slurried in ultrapure water on felt pads. Wang and Lin [22] reported that repetitive electrochemical treatment can be used *in situ* to prevent solid electrode fouling in the presence of various deactivating compounds. The optimum parameters for electrochemical treatment of glassy carbon, however, are dependent on the electroactive reactants. In the case of electrochemical treatment of wastewater, PbO_2 anodes were investigated to minimize electrode corrosion during the oxidation of organic compounds, including phenols [23–25]. Kawagoe *et al.* demonstrated the utility of bismuth-doped lead dioxide electrodes (Bi-PbO_2), which exhibit relatively low overpotentials for anodic O_2 evolution and thus facilitate the mechanical removal of the passivating layer by means of bubble evolution between the film and the substrate. Modification of electrodes with an electroactive polymer layer prior to their use in electroanalysis is another approach to stabilize the electrode response in the detection of phenols [26–28]. Although such modified electrodes exhibited improved performance in the detection of phenols, they could not be reactivated on-line. The regeneration procedure of solid electrodes has not been reported for practical applications.

It is worthwhile searching for a new electrode material capable of overcoming the above problems to some extent. Highly boron-doped diamond electrodes (BDD electrodes), which have recently emerged, appear to be promising in this regard [29–34]. These electrodes have recently received a great deal of attention, particularly for electroanalysis, owing to their unique electrochemical properties, e.g., (i) wide electrochemical potential window in aqueous solutions, (ii) very low background current, and (iii) long-term stability of the response. Anodically pretreated diamond electrodes have recently been shown to be useful in the selective detection of dopamine and uric acid in the presence of ascorbic acid [35,36]. These electrodes have also exhibited excellent stability and high reproducibility. Also, diamond, like other dimensionally stable oxide electrodes (e.g., PbO_2, SnO_2, and Ru-TiO_2), has a high overpotential for oxygen production [37–39]. The diamond electrode is reported to be promising for the treatment of phenols in wastewater [40,41]. In the potential region of water discharge on diamond, the production of highly oxidizing hydroxyl radicals was proposed, which would cause the complete oxidation of organic compounds in solution as well as passive organic films [42,43].

In this chapter, the analytical performance of diamond electrodes, which have been oxygen-terminated by anodic oxidation, was studied using cyclic voltammetry, flow injection analysis, and liquid chromatography. On-line regeneration of the slightly deactivated diamond electrodes during an accelerated test for fouling resistance was also found to extend to the continuous monitoring of real samples, although anodically treated diamond electrodes exhibited excellent resistance to fouling in nature. We have paid particular attention to the analysis of dichlorinated and trichlorinated phenols in phenolic compounds, because of the consideration that polychlorophenols have been involved in producing the most toxic tetrachlorinated dibenzo-*p*-dioxins. Analytical conditions of chromatographic separation and detection potential were investigated to enable reproducible analysis of target phenols. Finally, we demonstrated the column-switching HPLC analysis of 11 dichlorophenols and trichlorophenols in drain-water condensed from the flue gas discharged by waste incinerators.

11.2 APPLICATIONS

11.2.1 Electrode characterization

Highly boron-doped diamond electrodes were deposited on Si (100) wafers using a microwave plasma-assisted chemical vapor deposition (CVD) system (ASTeX Corporation, Woburn, MA, USA). The details of the preparation are described in Procedure 14 (see in CD accompanying this book). Figure 11.1 indicates a contact mode AFM image taken from the BDD film, which was deposited for 8 h under conditions described in Procedure 14 (see in CD accompanying this book). The AFM observations showed that the film was a well-faceted, polycrystalline, and continuous morphology. Using image section analysis, the roughness factor, *Ra*, was estimated to be 183 nm in the range of 10 μm × 10 μm. Figure 11.2 is a cross-section SEM image showing the morphology of the columnar diamond grains and a thin film of 19 μm thickness without pinholes. Over the entire diamond thickness, the typical boron concentration obtained under present conditions was ca. 2.1×10^{21} atoms cm^{-3}, estimated by secondary ion mass spectrometry. This boron doping level involves the diamond film showing a metallic behavior [44]. In fact, the film resistivity, which was measured by the four-point probe

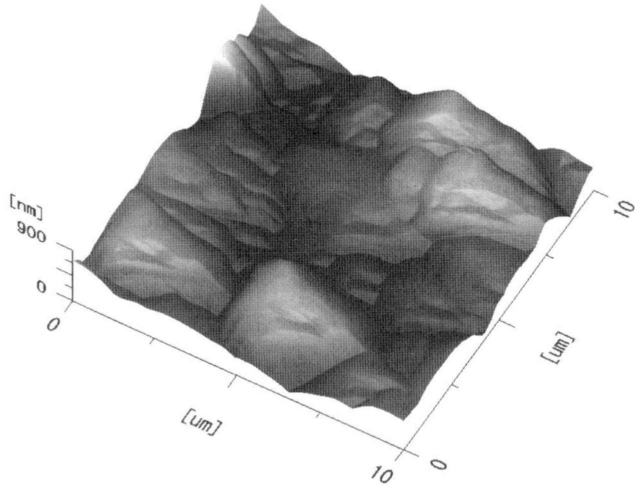

Fig. 11.1. Topographical AFM image of boron-doped polycrystalline diamond grown for 8 h. Scan size was 10 μm × 10 μm.

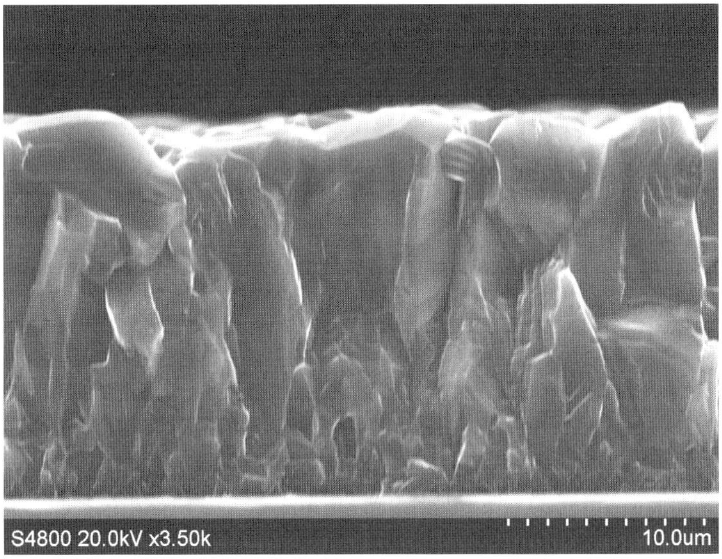

Fig. 11.2. Cross-section SEM image of boron-doped polycrystalline diamond grown for 8 h.

method, was on the order of $10^{-3}\,\Omega\,\mathrm{cm}$. The film quality was also confirmed by Raman spectroscopy (Renishaw System 2000). From the Raman spectrum shown in Fig. 11.3 diamond film of high quality and doped by boron was verified. It is well known that the strong

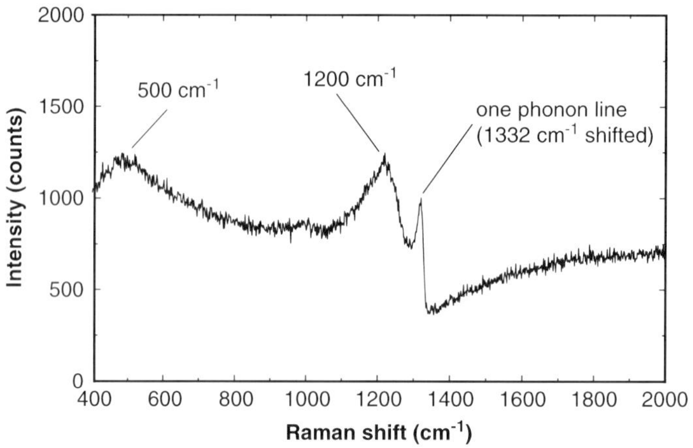

Fig. 11.3. Raman spectrum of boron-doped polycrystalline diamond showing metallic conductivity. Excitation line from an Ar^+ laser was 514.5 nm.

characteristic peak at 1332 cm^{-1} decreases in intensity and shifts to lower wave numbers, and the intensity of the broad peaks at 500 and 1200 cm^{-1} increases when the boron concentration in the film increases [45]. Moreover, no additional peaks due to sp^2 carbon are observed around 1500 cm^{-1}, indicating the high quality of these films.

11.2.2 Electrochemical oxidation of 2,4-dichlorophenol

Basic electrochemical reaction of 2,4-dichlorophenol (2,4-DCP), which was taken as representative of CPs, was first examined at glassy carbon, as-deposited BDD (AD-BDD), and anodically oxidized BDD (AO-BDD) electrodes. Figure 11.4 depicts the voltammograms for 0.3 mM 2,4-DCP obtained at a 0.1 V s^{-1} sweep rate. The background voltammograms obtained at diamond electrodes before and after anodic oxidation show that those electrodes have a very low residual current property and that anodic treatment has no effect on damage of the diamond surface. The peaks appeared in the reversed scan in the first cycle and the anodic peaks in the fifth cycle were experimentally confirmed to be chlorinated benzoquinones. The main anodic peaks in the first cycle were due to 2,4-DCP oxidation, and the peak potentials at glassy carbon, AD-BDD, and AO-BDD were different. The observed anodic shift at the AO-BDD electrode is probably because the surface dipoles on the oxidized diamond electrostatically repel both anions and neutral molecules. Recently, our group has confirmed experimentally the presence of $C=O$ and OH groups on the oxidized polycrystalline diamond surface by

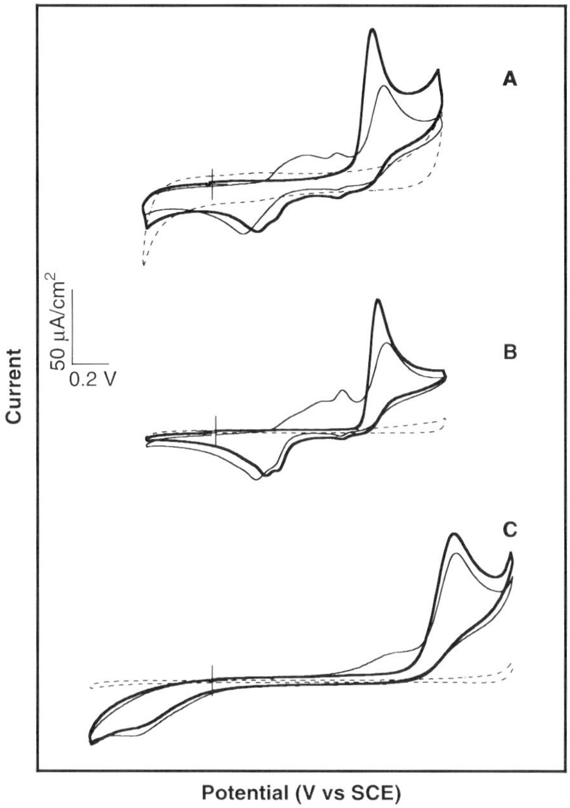

Fig. 11.4. Cyclic voltammograms for 0.3 mM 2,4-DCP in Britton–Robinson buffer (pH 2) at (A) glassy carbon, (B) as-deposited diamond, and (C) anodically oxidized diamond electrodes. Dotted lines show the residual voltammograms. Bold lines and thin lines show the first cycle and fifth cycle, respectively. The sweep rate was $0.1\,V\,s^{-1}$.

surface modification of DNPH and APTES, respectively [46,47]. Martin and Morrison have also demonstrated the presence of OH functional groups on the diamond surface after anodic polarization, while they could not see the C = O feature which may be hidden underneath the H_2O vapor bands [48]. The possibility of diamond surface oxide dipoles was further tested with other redox analytes, which are positively charged (Fe^{3+}/Fe^{2+}), negatively charged ($Fe(CN)_6^{4-}/Fe(CN)_6^{3-}$), and uncharged (ascorbic acid in acidic pH). The results (see Procedure 14 in CD accompanying this book) obtained with these species are well in agreement with our previous reports and other's reports related to ion–dipole and dipole–dipole interactions [46,47,49,50].

The voltammogram for phenol at carbon electrodes is known to exhibit a linear dependence of the anodic peak potential on pH, with a slope of 58 mV/pH under pH 10 [51]. We examined the pH dependence for 2,4-DCP oxidation at BDD electrodes. It is known that the pK_a values of the phenols decrease as the number of chlorine atoms substituted on the ring increases [13]. Figure 11.5 shows the pH dependence of peak potential (E_p) for 0.1 mM 2,4-DCP at AD-BDD and AO-BDD electrodes. The E_p vs. pH plots in both cases are linear below pH 8, the value being in good agreement with the pK_a of 2,4-DCP (7.85) [52]. Below pH 8, the slope of the E_p vs. pH plots at AD-BDD and AO-BDD electrodes were estimated to be 66 and 87 mV/pH, respectively. The slope at AD-BDD is reasonably close to 59 mV/pH, indicating the participation of equal numbers of protons and electrons in the oxidation reaction. A possible reason for the higher slope (87 mV/pH) at AO-BDD could be associated with the higher positive potential required for 2,4-DCP oxidation at AO-BDD, because it is likely that, in this potential range, the hydroxyl radicals produced may involve other side reactions. Although there is no experimental evidence in this state, the existence of surface hydroxyl groups could also be the reason for such a behavior. Above pH 8, all compounds are in the phenolate forms, which are oxidized more easily than the neutral phenols.

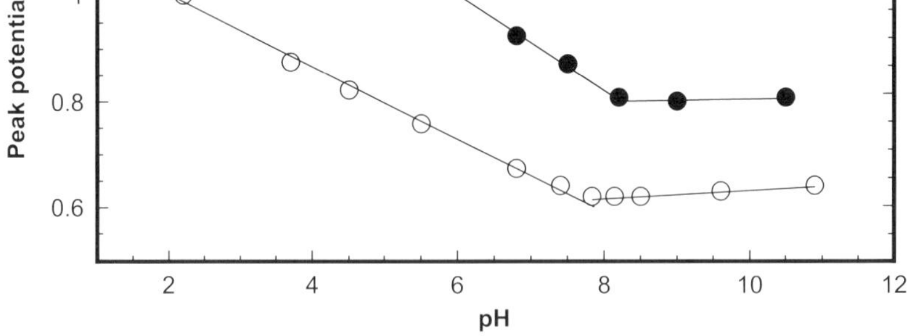

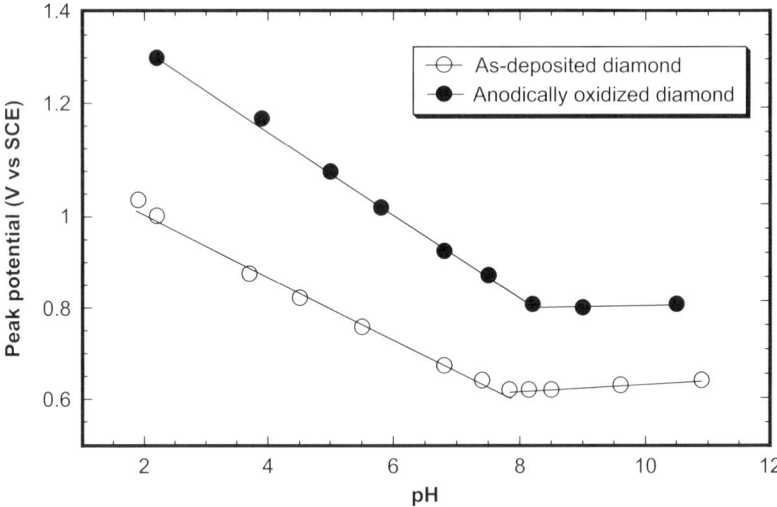

Fig. 11.5. pH dependence of the peak potential for 2,4-DCP at (○) as-deposited diamond and (●) anodically oxidized diamond electrodes in Britton–Robinson buffer.

11.2.3 Deactivation of diamond electrodes

Other than from the standpoint of analytical chemistry, phenols can often be used as model compounds to examine electrode stability because of their well-known electrode fouling properties [53–56]. Unmodified glassy carbon electrodes undergo rapid deactivation during the oxidation of phenols [26,57]. In order to undertake successful electroanalysis of phenols at the diamond electrode, we must ensure its stability by examining repetitive measurements. We first examined the durability of diamond electrodes at high 2,4-DCP concentration, because it has been observed previously that fouling of the electrode is pronounced at higher concentrations [26]. Figure 11.6A compares the first cycle and the fifth cycle of cyclic voltammograms for 5 mM 2,4-DCP at an AD-BDD electrode. Even in the case of AD-BDD, which is known to be less adsorptive material [58,59], the peak current disappeared by the fifth cycle due to electrode fouling. In this state, no recovery of the peak current could be accomplished after a few minutes of recovery time, even after vigorous stirring by bubbling with nitrogen gas in the vicinity of the electrode surface.

11.2.4 Regeneration of diamond electrodes

Figure 11.6A(b), which shows the fouled AD-BDD electrode, has drastically changed to Fig. 11.6B(a) by the anodic oxidation treated in a solution at 2.64 V vs. SCE for 4 min. This phenomenon is very interesting and should be paid attention to in the following three respects. First, the deactivation of the AD-BDD electrode is regenerated and the peak current is observed. At high anodic potentials, diamond is expected to produce hydroxyl radicals, which can completely oxidize the passive layer on the diamond film. Panizza *et al.* demonstrated the effect of such treatment to recover the activity of diamond in the detection of 2-naphthol [42]. The treatment potential of 2.64 V vs. SCE is in the water discharge region for the diamond electrode but, according to Panizza *et al.*, also involves the production of hydroxyl radicals. Furthermore, the electrogeneration of hydroxyl radicals at a BDD electrode was experimentally demonstrated by electron spin resonance (ESR) in presence of the spin trapping, compared with the spectrum obtained using the Fenton reaction [60]. The current density of 0.1 mA cm^{-2} in the ESR experiment was extremely lower than our anodic treatment, 3 mA cm^{-2} at 2.64 V vs. SCE. Thus, the anodic oxidation in the present conditions was expected to produce hydroxyl radical effectively. The second observation in this study

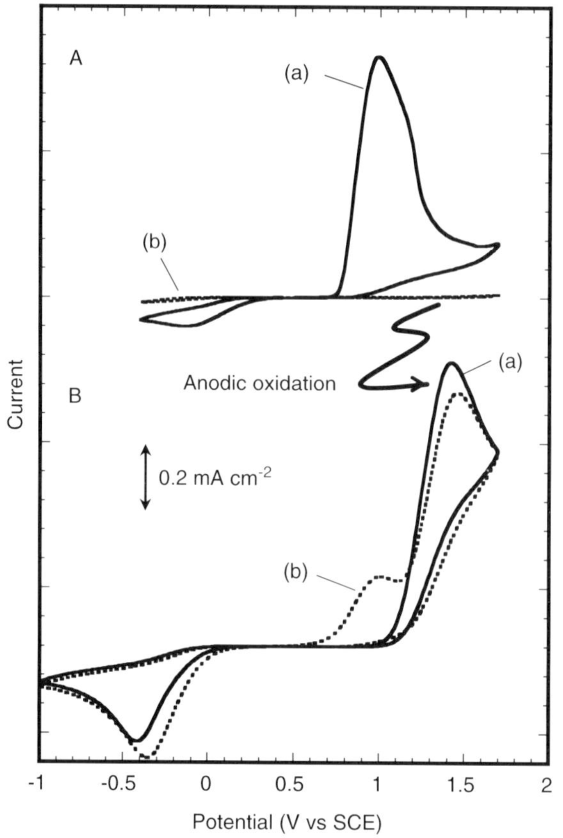

Fig. 11.6. Cyclic voltammograms for 5 mM 2,4-DCP in Britton–Robinson buffer (pH 2) at diamond electrodes. (A) At the as-deposited diamond electrodes, (a) the first cyclic voltammogram changed to (b) the fifth cyclic one by deactivation due to the high concentration. (B) After anodic oxidation of the fouled electrode, i.e., (a) the first cycle and (b) the fifth cycle at the anodically oxidized diamond electrodes.

is that the voltammetric peak potential has shifted dramatically to a more positive potential, moving from 1.0 to 1.4 V due to anodic oxidation. No further shift in the peak potential was observed after repeated treatments. In a word, the fouled AD-BDD electrode changed to an AO-BDD electrode during the anodic treatment for the electrode regeneration. A detailed explanation of the peak shift is described in Section 11.2.2. Third, it is very interesting and surprising that the AO-BDD electrodes appear to be less prone to fouling than AD-BDD electrodes, as evident from a well-defined voltammogram obtained in the fifth cycle (Fig. 11.6B). Gattrell and Kirk have concluded that phenoxy radicals play an important role in producing these types of passive films on the electrode surface [55,56]. We believe that the type of functional group and its distribution over the surface are the factors that control adsorption behavior of oxidized diamond. Although polar molecules are expected to be adsorbative on an oxygen-covered surface containing polar oxygen functional groups, the situation in case of diamond seems to be different. We believe that oxygen groups such as the carbonyl or hydroxyl groups formed on the diamond surface are well aligned in the facets of microcrystals, which are expected to form a negative surface dipolar field [61]. This field probably repels the phenols or their reaction products due to ion–dipole and dipole–dipole interactions [46,47,49,50], resulting in very low adsorption and an anodic shift in the peak potential. These features were specific in diamond electrodes, and not observed at glassy carbon.

11.2.5 Flow injection analysis

To investigate the performance of the BDD electrode in flow conditions, the electrode stabilities and the optimum detection potentials for CPs were examined in a flow injection analysis (FIA) mode. Figure 11.7 shows the FIA results for 100 repetitive injections of 5 mM 2,4-DCP at AO-BDD and glassy carbon electrodes. The detection potential was 1.4 V for AO-BDD and 1.2 V for glassy carbon. The mobile phase was 60% methanol–water containing 0.5% phosphoric acid. Although high methanol content slightly increased the polymerization reaction, our results, which are explained in a later section, indicated that this mobile phase is very suitable for the separation of DCPs and trichlorophenols (TCPs). A highly stable detection signal was observed when AO-BDD was used with a relative standard deviation (RSD) of 2.3% and with a decrement of ~10% in relative peak intensity (Fig. 11.7). As mentioned above, phenol oxidation produces polymer film on the electrode. In the FIA, since hydrodynamic mass transport is caused by the solution flowing through a flow cell, the mobile phase carries electrogenerated product away from the electrode surface depending on the flow rate. This phenomenon helps in minimizing the deactivation of the electrode in FIA compared to a bulk-solution experiment such as cyclic voltammetry. Nevertheless, repetitive injections of highly concentrated 2,4-DCP at glassy carbon resulted in a 70% reduction of the peak height. The value of RSD in the peak height ranges from 30 to 40%. After this durability experiment, a visible passivating layer could be observed on the glassy carbon surface. Under flow conditions, fouling of the electrode is not rapid but there is a gradual

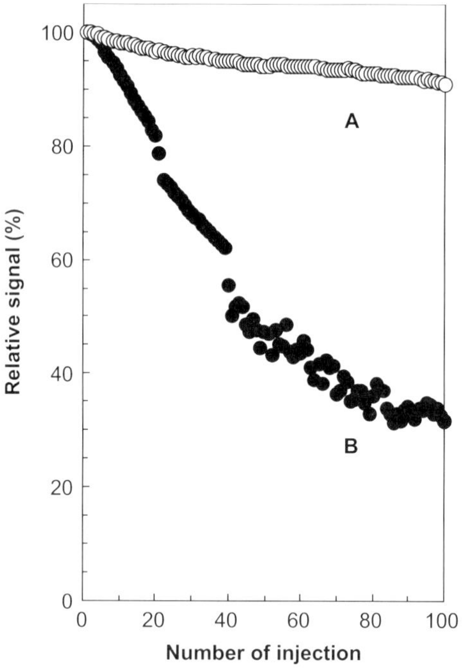

Fig. 11.7. Plots of the relative signal vs. injection number at (A) anodically oxidized diamond and (B) glassy carbon electrodes in flow injection analysis. 10 μL of 5 mM 2,4-DCP was injected at intervals of 2 min.

deactivation due to continuous deposition of polymer on the glassy carbon electrode surface.

In addition, we verified not only the stability of the response for repetitive injection measurements but also an on-line regeneration of the fouled electrode by anodic treatment. Continuous operation causes deactivation after several measurements even in the case of the AO-BDD electrode. Figure 11.8 indicates that the AO-BDD electrode was regenerated by applying a high positive potential (2.64 V, 10 min) after 100 injections. This effect was consecutively confirmed at 260 and 520 min in Fig. 11.8. The FIA conditions were same as Fig. 11.7. A complete recovery of the peak current was observed after this treatment. Although such an anodic treatment is very effective, the stabilization time after switching to the operation potential is also an important factor in electroanalysis. Figure 11.9 shows a comparison of the background current stabilization time between AO-BDD and glassy carbon electrodes. In case of glassy carbon electrode, the background current was monitored at the polished glassy carbon, which was

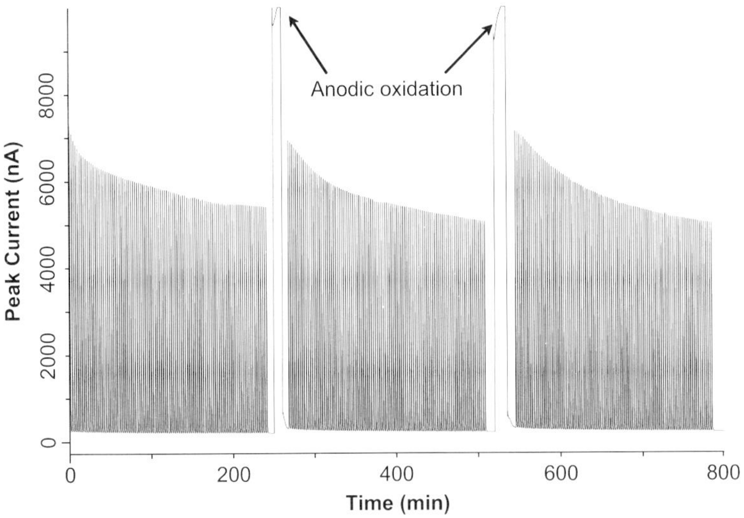

Fig. 11.8. Long-term stability test of anodically oxidized diamond electrode. The conditions of flow injection analysis were same as Fig. 11.7. After 100 injections, the electrode was regenerated on-line by anodic oxidation.

mechanically polished to regenerate the electrode activity. The background current of AO-BDD electrode stabilizes immediately after switching on the operation potential, while glassy carbon required more than 60 min to stabilize and at a substantially higher background current. Thus, AO-BDD electrode works as a dimensionally stable anode. This is another unique property of the AO-BDD electrode.

11.2.6 Hydrodynamic voltammetry

Figure 11.10A and B show hydrodynamic voltammograms at the AO-BDD electrode for DCPs and TCPs, respectively. A general observation is that TCPs appear to oxidize at lower potentials than DCPs, which is in contrast to expectations, because the higher CPs have more difficulty in undergoing oxidation. However, this apparent discrepancy is due to the fact that the currents for TCPs fall at relatively negative potentials without attaining a limiting value, due to the formation of an insoluble deposit on the electrode surface [62]. It is also important to note that the current for 2,6-DCP, for example, is higher than that for 2,3,6-TCP due to the same reason. The hydrodynamic voltammograms shown in Fig. 11.10 also indicate that *meta*-chlorinated phenol, for example 3,5-DCP, has more difficulty in undergoing oxidization than *ortho*-chlorinated 2,6-DCP. The potential for amperometric detection was set at 1.2 V vs. Ag/

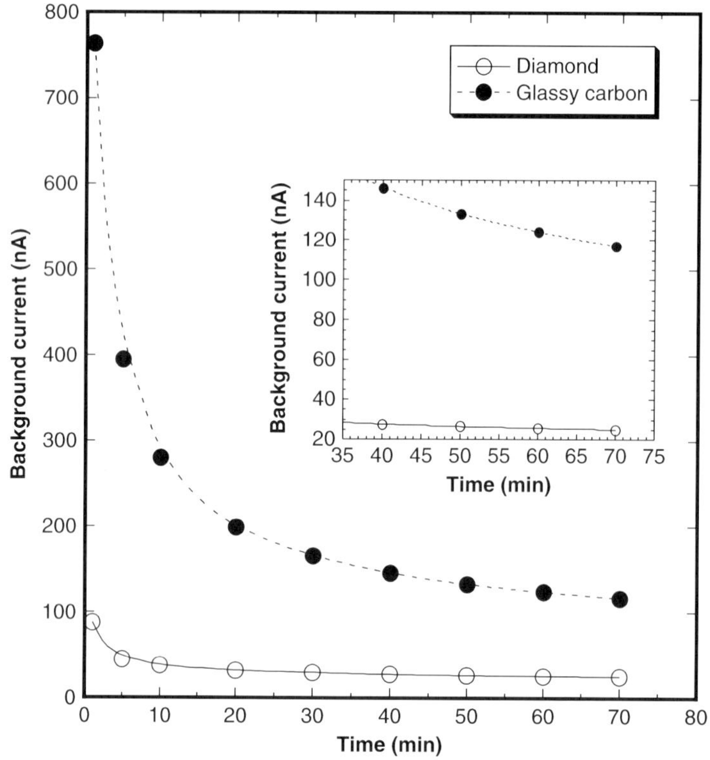

Fig. 11.9. Background current vs. time profiles for (○) anodically oxidized diamond and (●) glassy carbon electrodes. The applied potential was set at 1.4 V for diamond and 1.2 V for glassy carbon. The mobile phase was 60% methanol–water containing 0.5% phosphoric acid at a 0.2 mL min^{-1} flow rate.

AgCl, at which interferences from other impurities in real samples are diminished and all CPs produce significant currents.

11.2.7 Chromatographic separation

The optimum separation conditions for 11 CPs of DCPs and TCPs were examined in an isocratic elution. The separation column used was octadecyl silica, Inertsil ODS-3, which was provided from GL Sciences Inc. (Tokyo, Japan). Figure 11.11 indicates the relationship of capacity factor, k', and pH for different organic solvents in (A) 50% acetonitrile buffer and (B) 60% methanol buffer. The value of k' represents alternatively the elution time. In the case of acetonitrile solvent shown in Fig. 11.11A, the separation of the selected 11 CPs was not observed completely in any pH region. On the contrary, the mobile phase of the 60% methanol buffer

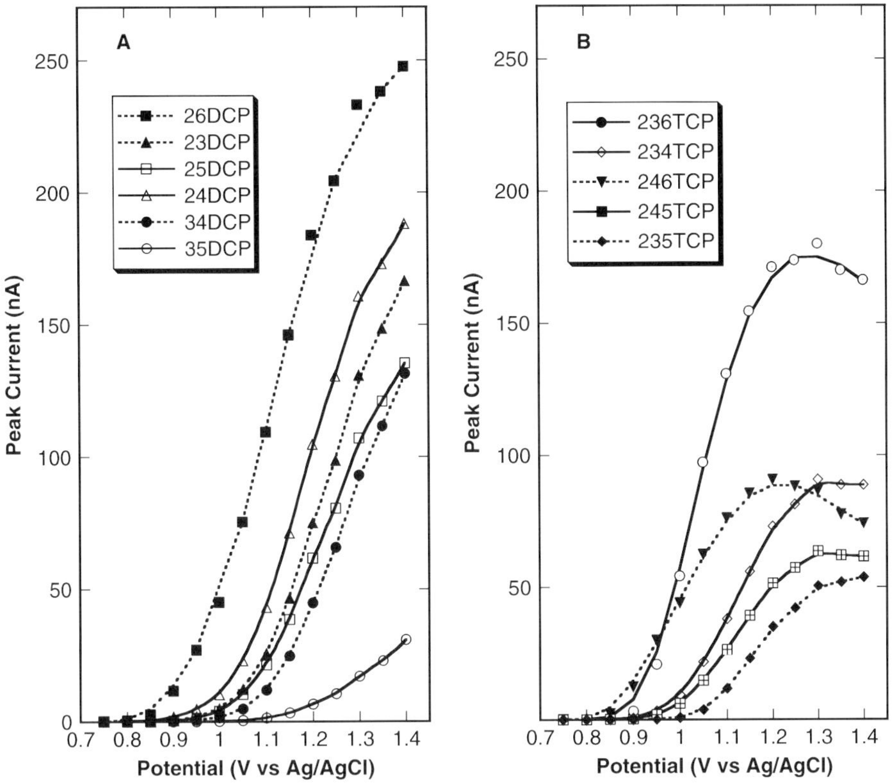

Fig. 11.10. Hydrodynamic voltammograms for (A) DCPs (6.2 μM each) and (B) TCPs (5.1 μM each) at anodically oxidized diamond electrodes.

below pH 5 accomplished perfect separation in less than 30 min. The mobile phase of 60% methanol–water containing 0.5% phosphoric acid was as a result selected due to the robust and easy-to-adjust solvent characteristics.

11.2.8 Column-switching HPLC

Advantages of the column-switching technique provided an ultrasensitive and short-time assay for biological and environmental samples [63]. Additionally, miniaturizing the separation column diameter enhanced mass sensitivity because of the less dilution in the HPLC mobile phase at a low flow rate [64]. For the purpose of routine analysis of CPs in environmental water, column-switching HPLC was performed. Figure 11.12 depicts a schematic diagram of an automated HPLC system. The time program of this system is shown in Table 11.1. HPLC columns were

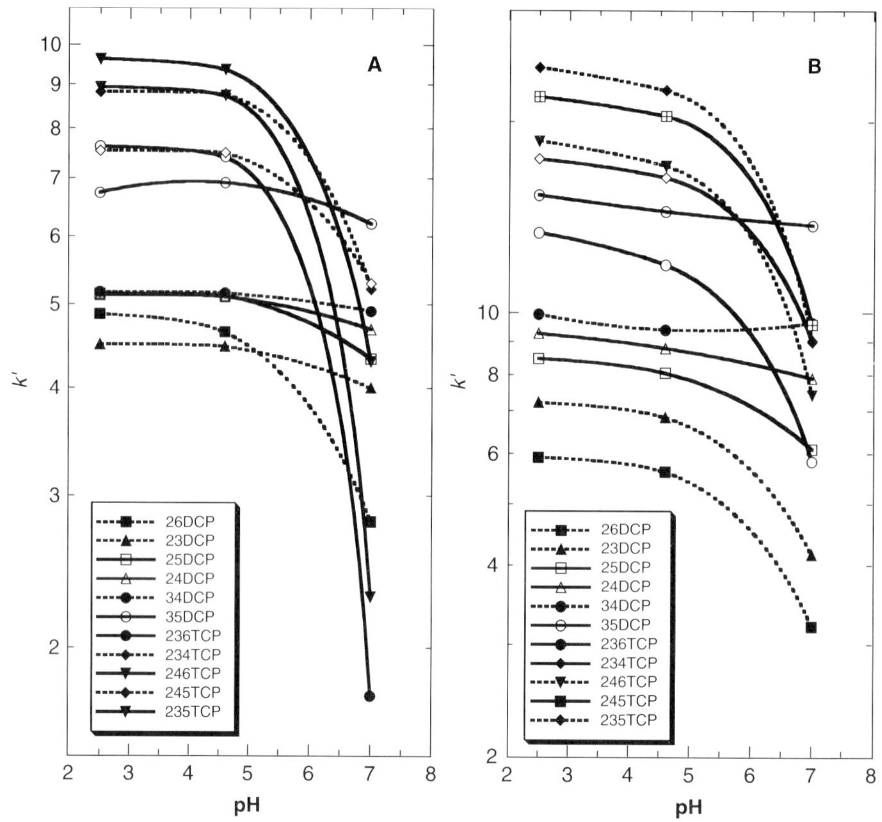

Fig. 11.11. Relationship of the capacity factor with pH in (A) 50% acetonitrile buffer and (B) 60% methanol buffer. An Inertsil ODS-3 column was used in an isocratic elution.

octadecyl silica (ODS), and the dimensions were 50 mm × 2.1 mm i.d. for Column-1 and 150 mm × 2.1 mm i.d. for Column-2. The mobile phases consisted of 60% methanol–water containing 0.5% phosphoric acid, and the flow rates were set at 0.2 mL min^{-1}. At first, the sample loop of 500 μL is fulfilled with a standard sample or a real sample, which can be switched by valve V-1. At 4 min, the valve V-2 starts to introduce the sample onto the precolumn Column-1. Weakly polar compounds and ions such as Cl^- and Br^- contained in the samples introduced are eluted from a precolumn prior to the elution of the polar target CPs. Then, at 9.5 min, the six-port valve V-3 is switched to allow a fraction of the sample to flow into an analytical column Column-2 for separation and detection. To shorten the analysis time, valve V-3 returns to the initial position and precolumn with remaining strong polar compounds is

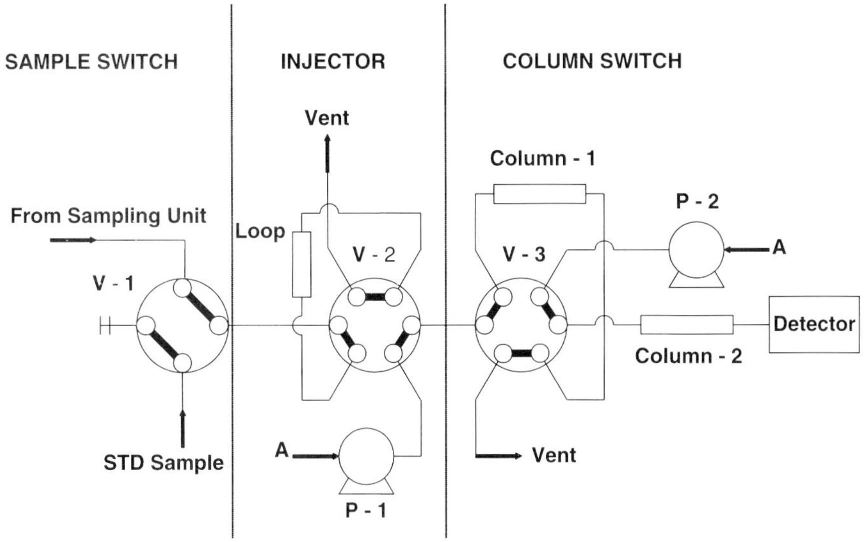

Fig. 11.12. Schematic diagram of a column-switching HPLC system. V-1, V-2, and V-3: switching valves. The position of ○—○ means position 1, and ○ ○ is position 2. P-1 and P-2: pumping system at 0.2 mL min^{-1} flow rate. Detector: electrochemical detector with diamond electrodes. Column-1 and Column-2: Inertsil ODS-3. Loop: 500 μL. A: Mobile phase of 60% methanol–water containing 0.5% phosphoric acid.

TABLE 11.1

Time program for column-switching HPLC system

Time (min)	Valve			Event
	V-1[a]	V-2[a]	V-3[a]	
0.0	1	1	1	Fulfills the loop with sample
4.0	1	2	1	Introduces the sample to Column-1
9.5	1	2	2	Starts analysis
23.0	1	2	1	Washes the loop and Column-1
40.0	1	1	1	End

[a]V-1, V-2, and V-3 are six-port valves. Position 1 is ○—○ and 2 is ○ ○ shown in Fig. 11.12, respectively refers to positions in Fig. 11.12.

cleaned during analysis. This technique is very attractive for sample clean-up and routine analysis, which allows analysis to be completed in a shorter time. Also, for analyses that require ultrahigh sensitivity, large-volume injection of purely aqueous samples adjusted to low pH was

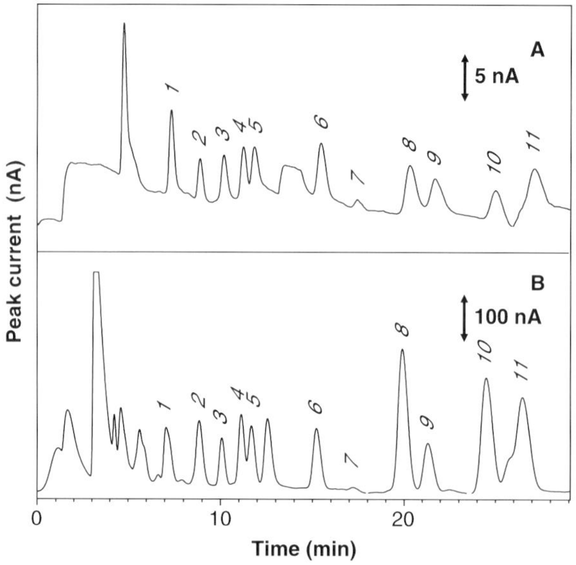

Fig. 11.13. Typical chromatograms of the column-switching HPLC for (A) standard chlorophenols (1 ng mL^{-1} each) and (B) drain-water sample from a municipal waste incinerator without any pretreatment. The applied potential at anodically oxidized diamond electrode was 1.2 V vs. Ag/AgCl. The peak assignments are (1) 26DCP, (2) 23DCP, (3) 25DCP, (4) 24DCP, (5) 34DCP, (6) 236TCP, (7) 35DCP, (8) 234TCP, (9) 246TCP, (10) 245TCP, and (11) 235TCP.

adopted in reversed-phase chromatography in order to achieve an effective preconcentration effect.

A typical chromatogram of a standard sample mixture containing each CP at the 1 ng mL^{-1} (ppb) level is shown in Fig. 11.13A. This chromatogram was obtained automatically by using the column-switching technique. The detection limits were estimated to be in the range of 0.038 ng mL^{-1} (0.23 nM) for 2,6-DCP to 0.361 ng mL^{-1} (2.23 nM) for 3,5-DCP (S/N = 3). In addition, Fig. 11.13B shows a chromatogram for a drain-water sample from a municipal waste incinerator. Table 11.2 summarizes the detection limits of 11 CPs in this method and the detected concentrations of CPs contained in the drain-water sample.

11.3 CONCLUSIONS

Anodically oxidized BDD thin-film electrodes have been shown to be best suited for the electrochemical detection of CPs. Anodic pretreatment of diamond is crucial for achieving high stability and reproducibility of the

TABLE 11.2

Limits of detection for chlorophenols estimated from Fig. 11.13A and the concentrations detected in Fig. 11.13B

Compound	LOD[a] (ng mL^{-1})	Detected concentration (ng mL^{-1})	Compound	LOD[a] (ng mL^{-1})	Detected concentration (ng mL^{-1})
26DCP	0.038	14.8	236TCP	0.059	27.0
23DCP	0.082	39.6	234 TCP	0.072	76.7
25DCP	0.076	26.9	246 TCP	0.098	34.4
24DCP	0.065	27.9	245 TCP	0.102	85.1
34DCP	0.073	23.6	235 TCP	0.092	61.6
35DCP	0.361	9.1			

[a]S/N = 3.

electrode performance. Another advantage of the AO-BDD electrode is that electrochemical reactivation of diamond is also possible by applying a high positive potential (2.64 V vs. SCE). In this way, a complete recovery of the current response is possible due, presumably, to the destruction of passive film by electrogenerated hydroxyl radicals. Such a treatment is not effective for glassy carbon, for which the treatment resulted in a substantial increase in background current. The outstanding stability of the AO-BDD electrodes was demonstrated by coupling with an FIA system, where the response variability for a concentrated (5 mM) 2,4-DCP solution was only 2.3% after 100 injections. When coupled with chromatography, the sample volume injected was very large owing to the reversed-phase column-switching technique, so the detection limit of 2,4-DCP was lowered to 0.4 nM. Using HPLC with column-switching technique and coupling with a diamond sensor, we were able to detect 11 CPs (Table 11.2) in water samples obtained from a municipal incineration plant. The present system can be applied for several phenols in water samples from a wide range of sources, just by modifying HPLC conditions such as column materials, mobile phases, and detection potential.

REFERENCES

1 N. Kristiansen, M. Froshaug, K. Aune and G. Becher, *Environ. Sci. Technol.*, 28 (1994) 1669.
2 M. Galceran and O. Jauregui, *Anal. Chim. Acta*, 304 (1995) 75.
3 M.J. Christophersen and T.J. Cardwell, *Anal. Chim. Acta*, 323 (1996) 39.
4 K. Rogers and C. Gerlach, *Environ. Sci. Technol.*, 30 (1996) 486A.

5 R.C.C. Wegman and A.W.M. Hofstee, *Water Res.*, 13 (1979) 651.
6 S.B. Ghorishi and E.R. Altwicker, *Chemosphere*, 32 (1996) 133.
7 E.R. Altwicker, *Chemosphere*, 33 (1996) 1897.
8 R. Weber and H. Hagenmaier, *Chemosphere*, 38 (1999) 529.
9 K. Hell, E.R. Altwicker, L. Stieglitz and R. Addink, *Chemosphere*, 40 (2000) 995.
10 U.S. EPA Method Study 14, Method 604 Phenols, Government Reports Announcements & Index (GRA&I), 1984.
11 U.S. EPA Method Study 30, Method 625 Base Neutrals, Acids and Pesticides, Government Reports Announcements & Index (GRA&I), 1984.
12 W.M. Shaub and W. Tsang, *Environ. Sci. Technol.*, 17 (1983) 721.
13 W. Lu and R.M. Cassidy, *Anal. Chem.*, 66 (1994) 200.
14 A.P. Durand, R.G. Brown, D. Worrall and F. Wilkinson, *J. Photochem. Photobiol. A: Chem.*, 96 (1996) 35.
15 T. Galeano-Diaz, A. Guiberteau-Cabanillas, N. Mora-Diez, P. Parrilla-Vazquez and F. Salinas-Lopez, *J. Agric. Food Chem.*, 48 (2000) 4508.
16 N.A. Lacher, K.E. Garrison and S.M. Lunte, *Electrophoresis*, 22 (2001) 2526.
17 P. Kuban and H. Flowers, *Anal. Chim. Acta*, 437 (2001) 115.
18 G. Mitulovic, M. Smoluch, J.-P. Chervet, I. Steinmacher, A. Kungl and K. Mechtler, *Anal. Biochem.*, 376 (2003) 946.
19 M.A. Raggi, C. Sabbioni, G. Casamenti, G. Gerra, N. Calonghi and L. Masotti, *J. Chromatogr. B*, 730 (1999) 201.
20 M. Gattrell and D.W. Kirk, *J. Electrochem. Soc.*, 140 (1993) 903.
21 S.K. Johnson, L.L. Houk, J. Feng, R.S. Houk and D.C. Johnson, *Environ. Sci. Technol.*, 33 (1999) 2638.
22 J. Wang and M.S. Lin, *Anal. Chem.*, 60 (1988) 499.
23 V. DeScure and A. Watkinson, *Can. J. Chem. Eng.*, 59 (1981) 852.
24 B. Fleszar and J. Ploszynska, *Electrochim. Acta*, 30 (1985) 31.
25 K.T. Kawagoe and D.C. Johnson, *J. Electrochem. Soc.*, 141 (1994) 3404.
26 J. Wang and R. Li, *Anal. Chem.*, 61 (1989) 2809.
27 V.C. Dall'Orto, C. Danilowicz, S. Sobral, A.L. Balbo and I. Rezzano, *Anal. Chim. Acta*, 336 (1996) 195.
28 T. Mafatle and T. Nyokong, *Anal. Chim. Acta*, 354 (1997) 307.
29 T.N. Rao and A. Fujishima, *Diamond Relat. Mater.*, 9 (2000) 384.
30 D. Shin, D.A. Tryk, A. Fujishima, A. Merkoci and J. Wang, *Electroanalysis*, 17 (2005) 305.
31 G.W. Muna, V. Guaiserova-Mocko and G.M. Swain, *Anal. Chem.*, 77 (2005) 6542.
32 I. Duo, P.A. Michaud, W. Haenni, A. Perret and Ch. Comninellis, *Electrochem. Solid-State Lett.*, 3 (2000) 325.
33 M. Pagels, C.E. Hall, N.S. Lawrence, A. Meredith, T.G. Jones, H.P. Godfried, C.S.J. Pickles, J. Wilman, C.E. Banks, R.G. Compton and L. Jiang, *Anal. Chem.*, 77 (2005) 3705.

34 V.A. Pedrosa, L. Codognoto, S.A.S. Machado and L.A. Avaca, *J. Electroanal. Chem.*, 573 (2004) 11.
35 E. Popa, H. Notsu, T. Miwa, D.A. Tryk and A. Fujishima, *Electrochem. Solid-State Lett.*, 2 (1999) 49.
36 E. Popa, Y. Kubota, D.A. Tryk and A. Fujishima, *Anal. Chem.*, 72 (2000) 1724.
37 J.-F. Zhi, H.-B. Wang, T. Nakashima, T.N. Rao and A. Fujishima, *J. Phys. Chem. B*, 107 (2003) 13389.
38 P. Canizares, J. Garcia-Gomez, C. Saez and M.A. Rodrigo, *J. Appl. Electrochem.*, 34 (2004) 87.
39 X. Chen, F. Gao and G. Chen, *J. Appl. Electrochem.*, 35 (2005) 185.
40 M. Fryda, D. Herrmann, L. Schafer, C. Klages, A. Perret, W. Haenni, Ch. Comninellis and D. Gandini, *New Diamond Front. Carbon Technol.*, 9 (1999) 229.
41 P.L. Hagans, P.M. Natishan, B.R. Stoner and W.E. O'Grady, *J. Electrochem. Soc.*, 148 (2001) E298.
42 M. Panizza, I. Duo, P.A. Michaud, G. Cerisola and Ch. Comninellis, *Electrochem. Solid-State Lett.*, 3 (2000) 429.
43 G. Foti, D. Gandini, Ch. Comninellis, A. Perret and W. Haenni, *Electrochem. Solid-State Lett.*, 2 (1999) 228.
44 A. Fujishima, Y. Einaga, T.N. Rao and D.A. Tryk, *Diamond Electrochemistry*, BKC and Elsevier, Tokyo, 2005.
45 S. Prawer and R.J. Nemanich, *Philos. Trans. R. Soc. Lond. A*, 362 (2004) 2537.
46 H. Notsu, I. Yagi, T. Tatsuma, D.A. Tryk and A. Fujishima, *J. Electroanal. Chem.*, 492 (2000) 31.
47 H. Notsu, T. Fukazawa, T. Tatsuma, D.A. Tryk and A. Fujishima, *Electrochem. Solid-State Lett.*, 4 (2001) H1.
48 H.B. Martin and P.W. Morrison, *Electrochem. Solid-State Lett.*, 4 (2001) E17.
49 H. Notsu, I. Yagi, T. Tatsuma, D.A. Tryk and A. Fujishima, *Electrochem. Solid-State Lett.*, 2 (1999) 522.
50 M.C. Granger and G.M. Swain, *J. Electrochem. Soc.*, 146 (1999) 4551.
51 V.F.P. Bub, K. Wisser, W.J. Lorenz and W. Heimann, *Ber. Bunsenges. Phys. Chem.*, 77 (1973) 823.
52 *Kirk-Othmer Encyclopedia of Chemical Technology*, Vol. 6, 4th ed., Wiley, New York, 1993, p. 156.
53 J. Wang and T. Martinez, *J. Electroanal. Chem.*, 313 (1991) 129.
54 J. Wang, M. Jiang and F. Lu, *J. Electroanal. Chem.*, 444 (1998) 127.
55 M. Gattrell and D.W. Kirk, *J. Electrochem. Soc.*, 139 (1992) 2736.
56 M. Gattrell and D.W. Kirk, *J. Electrochem. Soc.*, 140 (1993) 1534.
57 O. Jauregui and M.T. Galceran, *Anal. Chim. Acta*, 340 (1997) 191.
58 J. Xu, Q. Chen and G.M. Swain, *Anal. Chem.*, 70 (1998) 3146.

59 T.N. Rao, I. Yagi, T. Miwa, D.A. Tryk and A. Fujishima, *Anal. Chem.*, 71 (1999) 2506.
60 B. Marselli, J. Garcia-Gomez, P.-A. Michaud, M.A. Rodrigo and Ch. Comninellis, *J. Electrochem. Soc.*, 150 (2003) D79.
61 T.N. Rao, T.A. Ivandini, C. Terashima, B.V. Saraada and A. Fujishima, *New Diamond Front. Carbon Technol.*, 13 (2003) 79.
62 J.D. Rodgers, W. Jedral and N.J. Bunce, *Environ. Sci. Technol.*, 33 (1999) 1453.
63 S.E. Nielsen, R. Freese, C. Cornett and L.O. Dragsted, *Anal. Chem.*, 72 (2000) 1503.
64 R. Grimm, M. Serwe and J.-P. Chervet, *LC–GC*, 15 (1997) 960.

GAS SENSORS

Chapter 12

Chemical sensors for mercury vapour

Vladimir M. Mirsky and Majlinda Vasjari

12.1 INTRODUCTION

Mercury is a toxic substance that, through human and natural activities, cycles through the atmosphere, hydrosphere, and ecosphere affecting the health of both humans and wildlife. It enters the environment naturally through erosion, fire, and volcanic processes, as well as a result of human industrial practices. The human activities such as combustion, smelting, and mining have elevated global mercury levels to approximately three times those found before industrialization. Once released, mercury persists in the environment where it circulates between air, water, sediments, and biota in various forms. Mercury is present everywhere in the environment. The level of Hg in air varies from $0.5\,ng/m^3$ to $10\,\mu g/m^3$.

Mercury may be present in air in different chemical states such as the elemental form (as a vapour or adsorbed on particular matter) or in the form of volatile mercury compounds (mercury chloride, methylmercuric chloride, and dimethyl mercury). Although elemental mercury is only one of the mercury forms which is not as toxic as its organic or ionic forms, analytical determination of elemental mercury is of special importance. Such analysis is used not only for determination of elemental mercury in environment, but also as a method for determination of other forms of mercury after reductive treatment.

A number of analytical methods were developed for determination of elemental mercury. The methods are reviewed in Refs. [1–4]. They include traditional analytical techniques, such as atomic adsorption spectroscopy (AAS), atomic fluorescence spectroscopy (AFS), and atomic emission spectroscopy (AES). The AAS is based on measurements of optical adsorption at 253.7 or 184.9 nm. Typical value of the detection limit without pre-concentration step is over $1\,\mu g/l$. The AEF is much more sensitive and allows one to detect less than $0.1\,ng/l$ of mercury

Comprehensive Analytical Chemistry 49
Alegret and Merkoçi (Eds)
Volume 49 ISSN: 0166-526X DOI: 10.1016/S0166-526X(06)49012-7

vapour without pre-concentration. However, because of quenching of mercury fluorescence by oxygen and other species, this approach is essentially less selective. Depending on particular application, the detection limit of AEF without pre-concentration step can be from 15 μg/l to 5 ng/l [1,2]. Other analytical methods include neutron-activation analysis and laser photo-acoustic spectroscopy (detection limit about 5 ng/l). To increase sensitivity and selectivity of these methods, a pre-concentration of mercury is widely performed. This can be realized by exploring high affinity of mercury to gold. Mercury and its volatile compounds can easily be adsorbed on gold at normal temperatures and can be released quantitatively upon heating. Hence, gold has been used as an adsorbent for the accumulation of mercury to a concentration sufficient for further spectroscopic analyses. A common disadvantage of all these detection techniques is that they require relatively complicated technical solutions; typical costs of commercial devices are over $50,000 [5].

The requirement to develop inexpensive but highly sensitive and highly selective sensor for mercury vapour was a motivation to apply non-traditional analytical approaches based on the concept of chemical-affinity sensors. According to the classical definition, a chemical sensor should include receptor and transducer. The receptor should possess high affinity and selectivity towards molecules to be detected, while the transducer provides electrical or optical signal characterizing the amount of bound molecules. This approach was realized by using a thin gold layer as the receptor. Mercury adsorption onto this receptor was indicated by decreasing electrical conductance of thin gold layer or by changes of mechanoacoustic or optical properties [6–11].

12.2 MERCURY–GOLD INTERACTION

High affinity of mercury to gold is the base for development of both pre-concentration methods and chemical sensors for mercury. An initial phase model was suggested to describe the interaction [4,12–14]. According to this model, adsorbed mercury atoms form Hg–Au amalgam in the surface gold layer. As long as the amount of Hg adsorbed is lower than the maximum surface concentration value, the adsorption occurs with a sticking probability close to unity. Diffusion of mercury through gold is much slower than the adsorption on the surface and uptake from the surface (Fig. 12.1). This leads to saturation of the surface gold layer with mercury and blocking of further gold adsorption. Such an effect

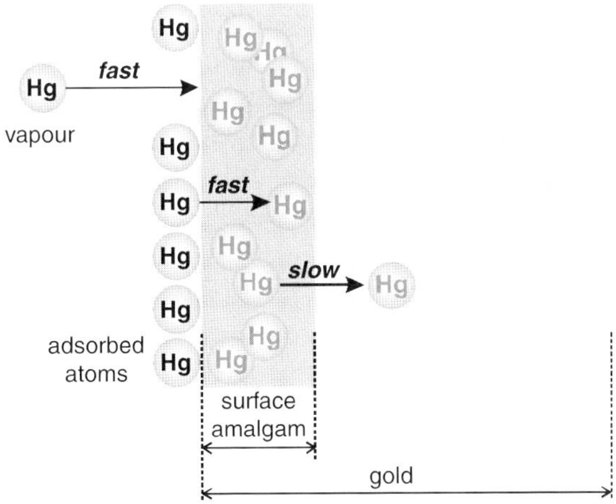

Fig. 12.1. Stages of mercury absorption by gold.

occurs when about 50% of the gold surface is saturated with mercury [12,14,15].

This process has been investigated recently by means of simultaneous surface plasmon resonance spectroscopy and electrochemistry [16,17]. Comparison of the surface plasmon resonance (SPR) angle dependencies with the theoretical ones has demonstrated that the two-phase model based on pure gold layer and pure mercury layer can be excluded. Better results were obtained with the assumption that the whole layer is transformed into the gold amalgam. Assuming additive contributions of both metals to optical properties, the mercury content in the amalgam can be estimated to be about 5–15%. However, the best fitting was obtained by using the two-phase model consisting of the pure gold phase and gold–mercury amalgam (Fig. 12.1). The thickness of this gold amalgam is about 8 nm. This result is close to the experimental data obtained by secondary ion mass spectroscopy [13], demonstrating that Hg penetrates into the gold not deeper than 5–6 nm.

Literature data on the surface capacitance of gold for mercury vary from 0.7 to 1.5 μg Hg(0)/cm^2 [15,18,19]; the mean value corresponds to about 0.3 nm^2 per mercury atom. This density is about twice less than the density of gold atoms in the crystal lattice, i.e., it corresponds to about a half of the monolayer. An increasing of temperature decreases the amount of the adsorbed mercury [14].

An increase of the time of mercury exposure up to hours leads to the formation of mercury aggregates in the form of islands or three-dimensional dendritic structures on/in the gold layer [13,14]. Such structures were detected by scanning electron microscopy and scanning tunnelling microscopy. These irreversible transformations of the film morphology at high exposure time or at very high mercury concentrations are considered as the main limitation in the implementation of thin-film gold sensors for Hg monitoring.

12.3 TRANSDUCERS FOR MERCURY SENSORS BASED ON THIN GOLD LAYERS

High affinity of gold to mercury provides a simple way for the development of chemical sensors for mercury vapour. A number of transducers can be used for detection of mercury binding to gold; all of them are based on exploitation of thin gold layers with a thickness from about 10 to about 100 nm.

Mercury binding leads to an increase of mass of the gold layer which can be detected by electro-acoustic transducers based on quartz microbalance (QMB; the abbreviation QCM = quartz crystal microbalance is also widely used), surface acoustic waves (SAW)—devices [20] or microcantilevers [21,22]. Adsorption of mercury vapour increases resonance frequency of shear vibrations of piezoelectric quartz crystals (Fig. 12.2). This process can be described by Sauerbrey equation [23]. For typical AT-cut quartz, this equation is

$$\Delta f = -\alpha f_0^2 \frac{\Delta m}{A}$$

where $\alpha = 2.3 \times 10^{-6}\,\mathrm{cm^2\,Hz\,g}$, f_0 is the quartz resonance frequency, and $\Delta m/A$ is the adsorbed mass in g/cm^2. The square dependence of the sensor sensitivity on resonance frequency makes favourable an increase of the resonance frequency; however, it requires a decrease of the quartz thickness which is accompanied by decreasing of its mechanic stability. Therefore, quartz with resonance frequency between 5 and 20 MHz are usually used.

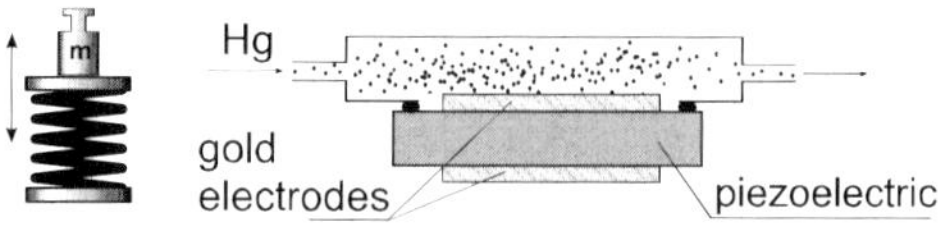

Fig. 12.2. An increase of mass leads to an increase of the resonance frequency of elastic mechanic oscillator (left). Similar principle is used for detection of mercury vapour adsorption onto gold electrode of a quartz resonator (right).

Chemical sensors for mercury vapour, based on QMB, provide sensitivity of 0.7–5 ng/l [24]. Reversibility is reached by heating the sensor; a microheater can be integrated on the sensor surface. This detection principle was also used by several other groups, for example [7,9,25]. Another realization of this transducing principle is based on exploitation of gold-coated microcantilevers for mercury detection [21,22].

Sensors based on SAWs operate at much higher frequencies and therefore provide higher mass sensitivity. SAW device consists of two pairs of interdigital electrodes forming correspondingly transmitter (generator) and receiver of SAWs. Time delay required for the wave propagation between transmitter and receiver depends on the length and physico-chemical properties of surface between transmitter and receiver (delay line). Such devices are widely used in electronics as selective filters and phase-shifting elements. Deposition of selective coating on the delay line makes this device sensitive towards definitive species; in the case of mercury sensors, thin gold layer is used as receptor (Fig. 12.3). The electrodes can be prepared by photolithography. Temperature fluctuations are compensated by using differential measurements; in this case a pair of sensitive (with receptor) and reference (without receptor) SAWs is formed on one crystal. The SAWs devices can be included into the electronic circuit to measure shift of resonance frequency caused by binding of analytes. However, a typical approach is based on the detection of difference in the phase shift caused by changes in the acoustic velocities through delay lines of the sensitive and the reference SAWs. This technology was used for mercury detection not only in air but also in aqueous solutions [8]; employment of a shear-horizontal acoustic plate mode led to minimization of signal losses, which are typical for the devices with vertically polarized displacements. The sensor operated at 91 MHz. The sensor detected about 10 ng of mercury in 120 μl aqueous solution; the detection limit for mercury vapour was not reported [8].

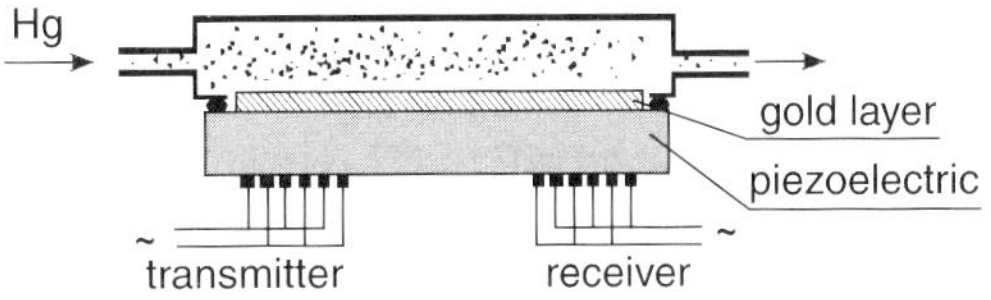

Fig. 12.3. Mercury sensor based on surface acoustic waves (SAW) with shear-horizontal acoustic plate mode. This approach was tested in Ref. [8].

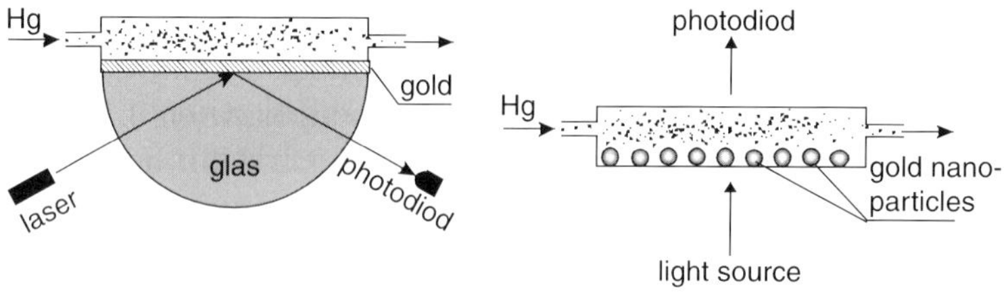

Fig. 12.4. Suggestive set-ups for detection of mercury vapour by surface plasmon resonance (left) and localized plasmon resonance (right). The simplest configurations are shown; a higher sensitivity can be obtained by using double wavelength technique [28], distributed referencing [29], bimetallic resonant mediums [26], or other approaches.

One can also use an SPR reflectometry to detect interaction of mercury with gold. The method is based on measurements of monochromatic p-polarized light reflected by thin (about 50 nm) planar gold layer under conditions of total internal reflection. Resonance of electron plasma in the gold layer, observed as a loss of the reflected light, depends on the complex refractive index of this and neighbouring layers. Adsorption of mercury leads to changes of the refractive index of the gold layer and to the corresponding shift of the resonance curve (Fig. 12.4, left) [16]. Several ways to improve sensitivity of SPR technique were suggested. For example, one can use bimetallic Ag/Au coating: a resonance curve on silver is sharper while a thin gold layer protects silver against oxidation (and also works as a receptor in the case of mercury sensors) [26]. An application of this approach for mercury vapour was tested in Ref. [27]. A shift of the resonance curve was observed; however, the reported data do not provide information on the detection limit. Other possibilities to increase sensitivity of SPR transducers can be based on the use of double-wavelength technique [28] or distributed referencing [29].

Another plasmon resonance approach for detection of mercury vapour is based on localized plasmon resonance in gold nanoparticles deposited on transparent support (Fig. 12.4, right). Changes of the refractive index of gold nanoparticles due to adsorption of mercury should lead to modification of the gold plasmon band of optical adsorption spectra. This approach has been applied successfully for investigation of interaction of biomolecules; however, to our knowledge there is still no report on its applications for detection of mercury vapour.

Adsorption of mercury atoms onto thin gold film leads to scattering of conduction electrons at the gold surface resulting in an increase of surface electrical resistance. This phenomenon is used for preparation of conductometric transducers (chemoresistors) for mercury detection [6,10,11,25]. The detection limit of this sensor is below 1 ng/l; response time is in the minutes time scale. Temperature fluctuations are compensated by using differential measurement set-up. Desorption of adsorbed mercury required for sensor reversibility is performed by heating to about 110°–150°, a microheater can be integrated into the sensor. Typical design of the simplest resistive sensor for mercury vapour (without reference channel and microheater) is shown in Fig. 15.1 of Procedure 15 (see in CD accompanying this book).

The sensor signal is caused by changes of surface electrical resistance (ΔR_s); this resistance is shunted by bulk gold resistance (R_b). The measurable resistance of the parallel connected R_s and R_b is $R_sR_b/R_s + R_b$. Small changes of the surface resistance R_s for the value ΔR_s (where $\Delta R_s \ll R_s$) lead to changes of the measurable resistance ΔR:

$$\Delta R = \frac{R_sR_b}{R_s+R_b} - \frac{(R_s+\Delta R_s)R_b}{(R_s+\Delta R_s)+R_b} \approx \Delta R_s \frac{R_b}{R_s+R_b} \approx \Delta R_s \frac{R_b}{R_s+R_b}$$

This is the reason for decreasing of the sensor sensitivity with an increase of thickness of the gold layer (Fig. 12.5). However, the minimal thickness of electrically conductive gold layer is limited by its surface relief (Fig. 12.5b). Depending on the gold-deposition technique, the surface roughness can be in the range between 5 and 15 nm. The experimental results demonstrating influence of thickness of the gold layer on sensitivity of conductometric chemical sensors are shown in Fig. 12.5c.

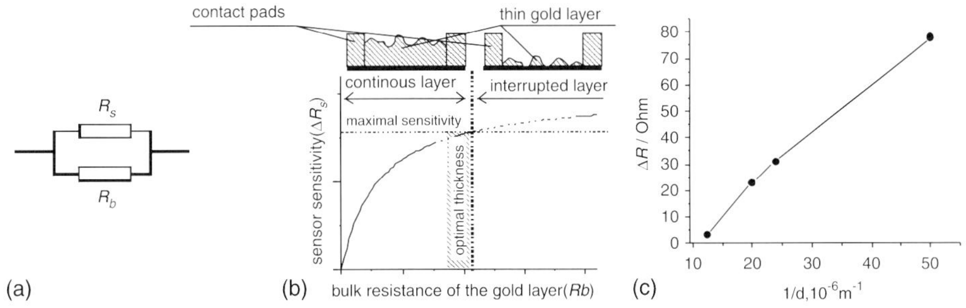

Fig. 12.5. Resistance change of thin gold layer ΔR due to increase of its surface resistance R_s, ΔR_s, depends on the ratio $R_b/R_s + R_b$, where R_b is the bulk resistance of the gold layer (a). However, an increase of R_b by decreasing the thickness of the gold layer is limited by surface roughness (b). Changes in the resistance (ΔR) of thin gold layers due to exposure to 10 ng/l of mercury vapour on reciprocal thickness of these layers (1/d) are shown in (c) (adapted from Ref. [25]). With kind permission of Springer Science and Business Media.

12.4 SELECTIVITY IMPROVEMENT

The main disadvantage of mercury sensors based on bare gold layers is their poor selectivity. This is illustrated in Fig. 12.6: an incubation at 100% humidity (Fig. 12.6a), with saturated vapour of sulphuric acid (Fig. 12.6), volatile sulphides or thiols (10 μg/l of 1-butanethiol vapour, Fig. 12.6c), or halogens (10 μg/l of iodine vapour, Fig. 12.6d), results in conductivity changes of the same magnitude as an incubation with 10–20 ng/l of mercury vapour. This interference with widely spread substances is a serious problem in applications of such sensors for real probes and makes necessary a pre-treatment of probes.

The pre-treatment can be performed by several ways. The first approach is the same as in the AAS/AFS techniques and is based on selective mercury accumulation and subsequent injection into measurement cell. The second approach is based on selective extraction of

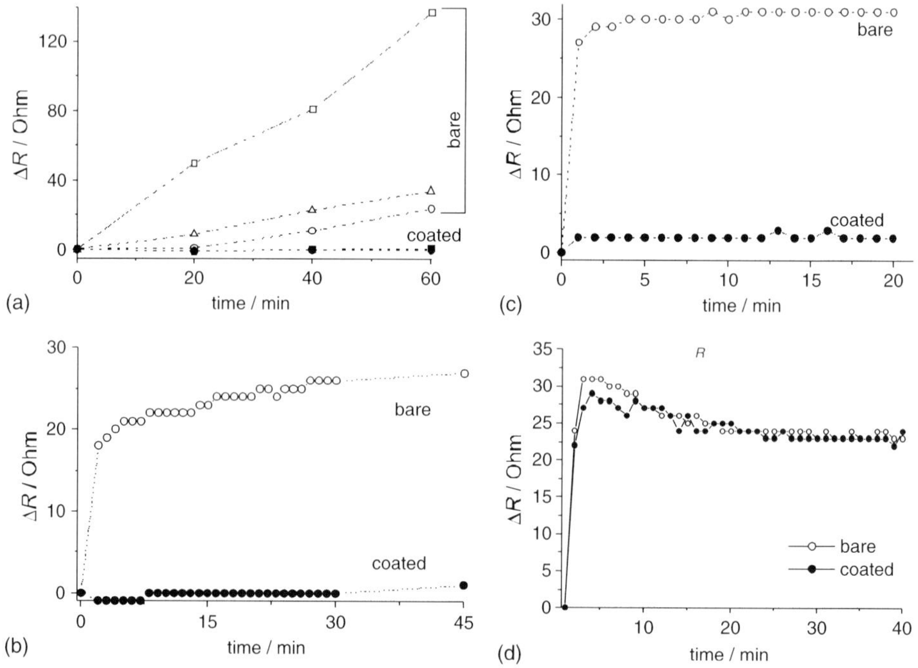

Fig. 12.6. Influence of typical interferents on mercury sensors based on changes of lateral resistance of thin gold layers. Sensors: bare gold layers (open symbols), gold layers coated by self-assembled monolayers of 1-hexadecanethiol (filled symbols). Interferents: 100% humidity (a), saturated vapour of sulphuric acid (b), 10 μg/l of 1-butanethiol (c), 10 μg/l of iodine vapour (d), Thickness of gold layers: 42 nm [25].

typical interferents from the probe. Both techniques increase the weight and size of the sensor device at least hundred times; however, one can expect future miniaturization of these pre-treatment devices by using microsystem technology. Another approach is based on preparation of ultrathin filtering coatings for gold layers operating as mercury receptors.

A perspective solution is the coating of gold by self-assembled monolayers of alkylthiols. In 1987, a blocking effect of self-assembled monolayers of alkanethiols on Faraday's processes was described [30,31]. This was a strong motivation for intensive investigation of this system and a basis for its numerous analytical and technical applications [31–36]. Later, a blocking effect of such self-assembled monolayers for a number of neutral molecules was observed [25]. It was found [25,37] that the mercury can easily penetrate such monolayers. Incubation of thin gold film with mercury vapour leads to an increase of its lateral conductance (Fig. 12.7a, curve 2). Compared to the uncoated sensor, changes in the resistance of coated sensors were lower by a factor of ~2. However, even in this case the detection limit remains about 50 times less than the maximal mercury vapour concentration allowed at workplaces (100 $\mu g/m^3$). The response of the coated as well as uncoated sensors was monotonously dependent on mercury concentration (Fig. 12.7b); at low mercury concentrations this dependence was linear (Fig. 15.3 of Procedure 15; see in CD accompanying this book). The concentration dependence of the initial rate of the resistance increase was also linear at low concentrations of mercury vapour.

Mercury interaction with gold electrodes coated by hexadecanethiol was confirmed by measurements with a quartz crystal microbalance (Fig. 12.7b): an incubation of quartz with hexadecanethiol-coated gold electrodes in the presence of mercury vapour resulted in a decrease of the resonance frequency, thus indicating an increase of the electrode mass. The adsorbed amount of mercury can be estimated from Sauerbrey equations; the effect observed corresponds to adsorption of 0.5 μg of mercury per cm^2 of the gold surface.

While conductometric responses of coated and uncoated gold layers on exposure to mercury vapour are similar, a strong difference in their responses to other compounds is observed (Fig. 12.6). No response of the coated sensors to saturated water vapour was detected (Fig. 12.6a). Also in contrast to the uncoated sensors, practically no effect of either vapour of sulphuric acid (Fig. 12.6b) or butanethiol (Fig. 12.6c) on the coated sensors was observed. The effect of iodine vapour was not changed.

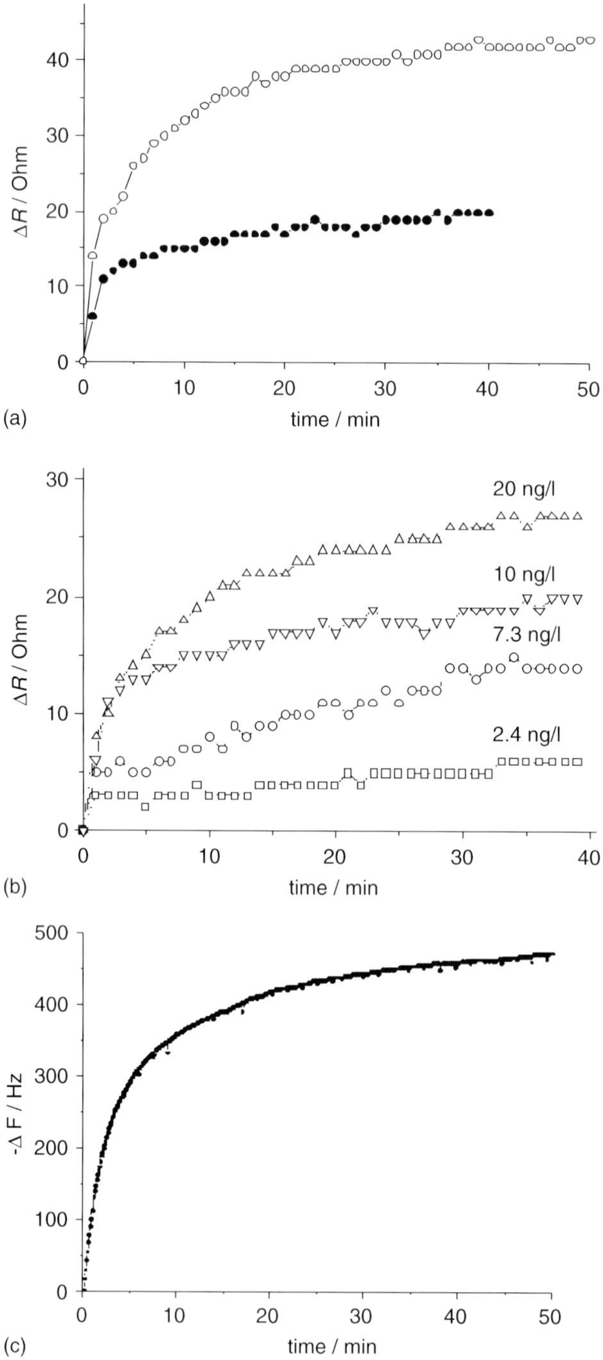
ΔR / Ohm
0
10
20
30
40
0
10
20
30
40
50
(a)
time / min
ΔR / Ohm
20 ng/l
10 ng/l
7.3 ng/l
2.4 ng/l
0
10
20
30
0
10
20
30
40
(b)
time / min
-Δ F / Hz
0
100
200
300
400
500
0
10
20
30
40
50
(c)
time / min

The observed selective permeability of alkanethiol monolayers can be caused by different factors. Certainly, a transfer of water, sulphuric acid, and other polar molecules through highly ordered hydrophobic monolayers requires considerable activation energy. Additionally, low surface energy formed by methyl groups of the alkanethiol monolayer leads to formation of droplets of hydrophilic compounds on this surface, thus decreasing the number of molecules interacting with the surface. The discrimination based on difference in hydrophobic/hydrophilic properties of species also explains a high permeability of iodine through a monolayer of hexadecanethiol (Fig. 12.6d). In the case of organo-sulphuric compounds, the effect can be also diminished by an occupation of sulphur-binding sites on the gold surface by thiol groups of alkanethiols. The finding that mercury vapour can penetrate a monolayer of alkanethiols on gold, while most of interferents cannot, allows one to design highly selective sensors for mercury vapour without additional macroscopic pre-treatment modules. Thus, a simple integrated analytical system including a filter and a sensor, both scaled down almost until the technical limit (monomolecular layer as the filter and tens or hundreds of atomic layers as the sensor) can be prepared. The permeability of alkanethiol monolayers can be probably controlled by its chemical modification, by formation of multilayers with laterally organized nanostructures or subsequently deposited polymer layers, or by variation of a physical state of the monolayer. One can expect a further extension of this principle including a combination with other analytical techniques, especially with electrochemical reduction and subsequent conductometric detection [38].

12.5 CALIBRATION

Mercury sensors based on thin gold layers require regular calibrations. In ideal cases, the calibration should be performed before each measurement. Therefore, a calibration technique should be compatible with design and concept of the mercury vapour sensor. Several approaches were suggested.

Fig. 12.7. Interaction of mercury vapour with thin gold films coated by self-assembled monolayer of 1-hexadecanethiol: (a) comparison of the kinetics of resistive response for bare (open symbols) and coated (filled symbols) gold films on exposure to 10 ng/l mercury vapour; (b) influence of different mercury vapour concentrations on the resistance of coated electrodes; (c) kinetics of changes of the resonance frequency of a 1-hexadecanethiol-coated piezo-quartz due to exposure to 8.3 ng/l of mercury vapour [25].

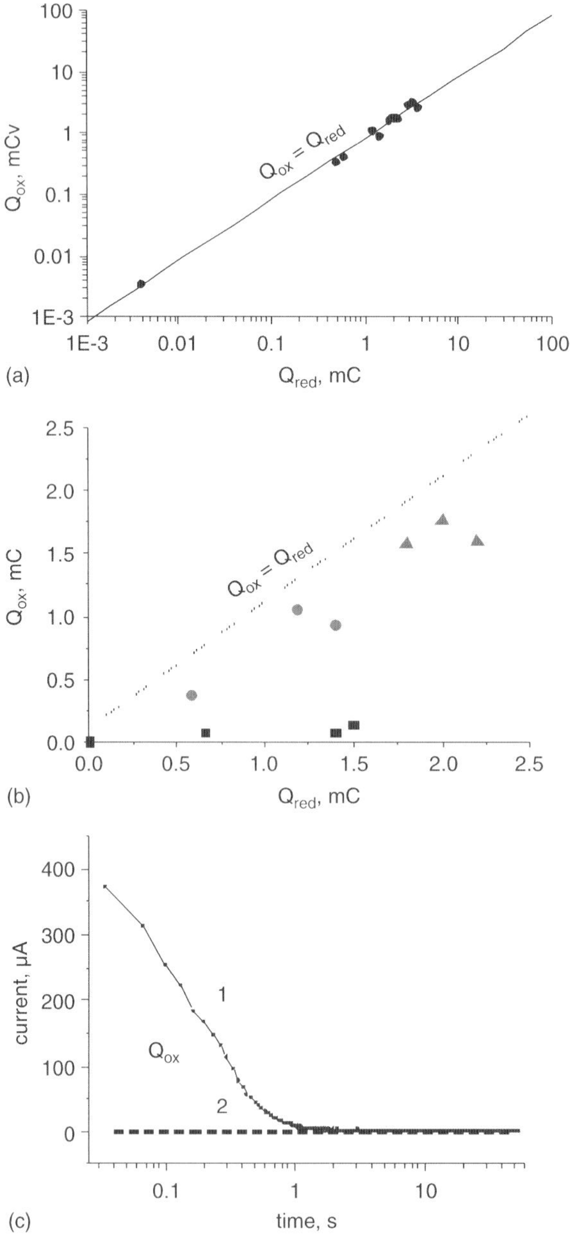
100
10
1
0.1
0.01
1E-3
Qox, mCv
1E-3
0.01
0.1
1
10
100
Qox = Qred
(a)
Qred, mC
2.5
2.0
1.5
1.0
0.5
0.0
Qox, mC
0.0
0.5
1.0
1.5
2.0
2.5
Qox = Qred
(b)
Qred, mC
400
300
200
100
0
current, μA
1
Qox
2
0.1
1
10
(c)
time, s

The most frequently used calibration procedure is based on temperature dependence of pressure of saturated mercury vapour [19,39–41]. At 25°C this pressure is of 0.0018 mm Hg height; it corresponds to the vapour density of 20 μg/l. To get in the measurement cell a mercury concentration of about 10 ng/l, the saturated vapour should be strongly diluted. Instead of dilution, a lower temperature can be used; however, the density of saturated vapour of 10 ng/l corresponds to the temperature of less than −40°C. Both dilution and temperature decrease can be realized easily in laboratory conditions but their incorporation into a miniaturized chemical sensor is rather complicated. An attempt to develop such a device is reported in Ref. [41]. An additional problem in application of these techniques in portable sensing devices with integrated calibration is the necessity to have a reservoir with mercury in the device; it complicates recycling of these devices and does not correspond to modern trends in technology.

Another calibration approach is based on generation of defined amount of mercury vapour from aqueous solution of mercury chloride by addition of reductive agents [42], for example tin chloride or sodium borohydride [43,44]. A disadvantage of this approach is that the generated mercury is contaminated with water and other reagents. Also, under usual laboratory conditions this approach can hardly be used for calibration of sensitive mercury sensors operating in nanogram scale. However, an application of modern microfluidics technology can solve this problem and make possible realization of this technique with negligibly small amount of toxic materials which is acceptable to be stored in portable devices.

The third approach is based on thermoinjection of electrochemically generated mercury [45]. Gold wire electrodes with diameter of 50 or 100 μm were used. Metallic mercury was formed on the gold wires by electrochemical reduction from water-soluble mercury salt.

The amount of mercury formed on the gold electrode under different reduction conditions or that remained on the electrode after different storage conditions was measured by subsequent electrochemical oxidation (Fig. 12.8). First, the relation of the reduction (Q_{red}) and

Fig. 12.8. Test of reproducibility of electrochemical oxidation/reduction and of completeness of thermoinjections. Dependence of the oxidative charge on the reduction charge is shown in (a, b). The oxidation was performed in pure water immediately after reduction (all data in (b) and ◆ in (c)), after 72 h storage at room temperature (■) and after 96 h storage at −20°C (●). Kinetics of the limiting oxidation current of wire gold electrode with reduced mercury before (1) and after (2) thermoinjection is shown in (c) [45].

oxidative (Q_{ox}) charges was investigated in the range from 5 μC to 5 mC, corresponding to the range of amounts of metallic mercury on the electrode from 10 ng to 10 μg (Fig. 12.8a). If the mercury oxidation was performed immediately after mercury reduction, the charge of oxidation over the whole investigated range was proportional to the charge of reduction, linear correlation coefficient being 0.9738. However, the ratio Q_{ox}/Q_{red} was somewhat below 100%, typically between 70% for nanogram amounts to 100% for microgram amounts of reduced mercury. Therefore, at least the main part of mercury can be reduced and oxidized reversibly. The deviation from complete reversibility can be explained by the high volatility of metallic mercury and by red/ox impurities in the electrochemical cells. However, the deviation of the ratio Q_{ox}/Q_{red} from 100% is caused by mercury loss during both metallic mercury formation and electrochemical analysis of this mercury quantity; therefore, a real discrepancy between the amount of metallic mercury on the wire electrode and the amount calculated from the Q_{red} is less than this deviation. Therefore, the deviation observed can be considered as an overestimated error of the suggested technique.

The effect of storage conditions on the amount of metallic mercury retained on gold electrodes was studied (Fig. 12.8b). It was found that mercury completely evaporated from gold electrodes during 72 h storage at room temperature, but no loss of mercury was observed even during 96 h storage at -20°C. Therefore, thermoinjection should be made immediately after mercury reduction, otherwise a storage at low temperature or in closed volumes is required.

Thermoinjection is performed by heating of the wire electrode placed in mercury measurement chamber; it was realized by application of electrical current through the wire. The residual amount of mercury on the gold electrode after thermoinjection, which is determined by electrochemical oxidation, has demonstrated that heating results in complete evaporation of mercury from the wire gold electrode (Fig. 12.8c). Therefore, the technique of thermoinjection [45] is simple and reproducible. One can imagine integration of this technique into sensing device. In the case of portable devices, small single-use ampoules loaded with defined amounts of mercury wire electrodes can be used.

12.6 CONCLUSION

Although the first chemical sensor for detection of mercury vapour was suggested more than 30 years ago, there is still a high potential for its further development. The main directions of this progress will probably include a design of new selective filters, an integration of calibration system into sensor devices, and an evaluation of new detection principles. These purposes are strongly supported by current trends in the development of new technologies, such as an exploitation of combinatorial and high-throughput techniques [46,47] to design selective coatings, an application of microsystem technology and microfluidics for development of integrated sensing systems, or an application of nanotechnology and nanophotonics for development of new detection principles. Procedure 15 (see in CD accompanying this book) describes preparation of simple conductometric sensors for mercury vapour.

ACKNOWLEDGMENT

The authors are grateful to Prof. O. S. Wolfbeis for helpful discussions.

REFERENCES

1 T.G. Laperdina, *Mercury Determination in Natural Waters*, Nauka, Novosibirsk, 2000 (in Russian).
2 C. Luca, I. Tanase and A.F. Dänet, *Rev. Anal. Chem.*, 9 (1987) 1.
3 H.B. Cooper, R.S. Foote and G.D. Rawlings, *Adv. Instr.*, 28 (1973) 731.
4 W.H. Shroeder, M.C. Hamilton and R.S. Stobart, *Rev. Anal. Chem.*, 8 (1985) 3.
5 N.B. French, S.J. Priebe and W.J. Haas. Jr., *Anal. Chem. News Features*, 71 (1999) 470A.
6 J.J. McNerney, P.R. Buseck and R.C. Hanson, *Science*, 178 (1972) 611.
7 A.N. Mogilevski, A.D. Mayorov, N.S. Stroganova, D.B. Stavrovski, I.P. Galkina, D.B. Stavrovski, I.P. Galkina, L. Spassov, D. Mikhailov and R. Zaharieva, *Sens. Actuators A*, 28 (1991) 35.
8 M.G. Schweyer, J.C. Andle, D.J. McAllister and J.F. Vetelino, *Sens. Actuators B*, 170 (1996) 35.
9 M.H. Ho, G.G. Guilbaut and E.P. Scheide, *Anal. Chim. Acta*, 130 (1981) 141.
10 G. Braeker, G. Wiegleb and L. Winter, *Technische Überwachung*, 9 (1996) 21 (in German).
11 J.J. McNerney and P.R. Buseck, *Gold Bull.*, 6 (1973) 106.
12 C.R. Brundle and M.W. Roberts, *Chem. Phys. Lett.*, 18 (1973) 380.
13 C. Battistoni, E. Bemporad, A. Galdikas, S. Kaciulis, G. Mattorno, S. Mickevicius and V. Olevano, *Appl. Surf. Sci.*, 103 (1996) 107.

14 M. Levlin, E. Ikävalko and T. Laitinen, *Fresenius J. Anal. Chem.*, 365 (1999) 577.
15 R.W. Joyner and M.W. Roberts, *Chem. Soc. Faraday Trans. 1*, 69 (1973) 1242.
16 M. Vasjari, Yu.M. Shirshov, A.V. Samoylov and V.M. Mirsky, *J. Electroanal Chem.* (2007), in press.
17 Yu.M. Shirshov, R. Khristosenko, M. Vasjari, A.V. Samoylov and V.M. Mirsky, in preparation.
18 S. Lythgoe, P.J. Robinson and R.D. Sedgwick, *J. Phys. E*, 3 (1970) 401.
19 F. Slemr, W. Seiles, C. Eberling and P. Roggendorf, *Anal. Chim. Acta*, 110 (1979) 35.
20 J. Janata, *Principles of Chemical Sensors*, Plenum Press, New York, 1989, pp. 55–80.
21 T. Thundat, E.A. Wachter, S.L. Sharp and R.J. Warmack, *Appl. Phys. Lett.*, 66 (1995) 1695.
22 B. Rogers, L. Manning, M. Jones, T. Sulchek, K. Murray, B. Beneschott, J.D. Adams, Z. Hu, T. Thundat, H. Cavazos and S.C. Minne, *Rev. Sci. Instrum.*, 74 (2003) 4899.
23 G. Sauerbrey, *Z. Phys.*, 155 (1959) 206.
24 Q. Bristow, *J. Geochim. Explor.*, 1 (1972) 55.
25 V.M. Mirsky, M. Vasjari, I. Novotny, V. Rehacek, V. Tvarozek and O.S. Wolfbeis, *Nanotechnology*, 13 (2002) 175.
26 S.A. Zynio, A.V. Samoylov, E.R. Surovtseva, Y.M. Shirshov and V.M. Mirsky, *Sensors*, 2 (2002) 62–70.
27 T. Morris and G. Szulczewski, *Langmuir*, 18 (2002) 5823.
28 A. Zybin, Ch. Grunwald, V.M. Mirsky, J. Kuhlmann, O.S. Wolfbeis and K. Niemax, *Anal. Chem.*, 77 (2005) 2393–2399.
29 V.M. Mirsky, A. Zybin, D. Boecker, K. Niemax and O.S. Wolfbeis, *Opt(r)odes-2006, Abstract book*.
30 H.O. Finklea, S. Avery, M. Lynch and T. Furtsch, *Langmuir*, 3 (1987) 409–413.
31 H.O. Finklea, *Electroanal. Chem.*, 19 (1996) 109.
32 V.M. Mirsky, *Trends Anal. Chem.*, 21 (2002) 439.
33 J.J. Gooding and D.B. Hibbert, *Trends Anal. Chem.*, 18 (1999) 525.
34 R.M. Crooks and A.J. Ricco, *Acc. Chem. Res.*, 31 (1998) 219.
35 S. Flink, F.C.J.M. van Veggel and D.N. Reinhoudt, *Adv. Mater.*, 12 (2000) 1315.
36 A.N. Shipway, E. Katz and I. Willner, *Chem. Phys. Chem.*, 1 (2000) 18.
37 J. Thome, M. Himmelhaus, M. Zharnikov and M. Grunze, *Langmuir*, 14 (1998) 7435.
38 M.J. Schoening, "Voltohmmetry"—a new transducer principle for electrochemical sensors. In: V.M. Mirsky (Ed.), *Ultrathin Electrochemical Chemo- and Biosensors. Technology and Performance*, Springer, Berlin, 2004.

39 R. Dumarey, R. Heindryckx, R. Dams and J. Hoste, *Anal. Chim. Acta*, 107 (1979) 159.
40 W.F. Fitzgerald and G.A. Gill, *Anal. Chem.*, 51 (1979) 1714.
41 F.R. Lanziillotta and G. Chiti, *Environ. Technol.*, 21 (2000) 713.
42 Z. Yoshida and K. Motojama, *Anal. Chim. Acta*, 106 (1979) 405.
43 P. Schierling and K.H. Schaller, *At. Spectrosc.*, 2 (1981) 91.
44 G. Tuncel and O.Y. Ataman, *At. Spectrosc.*, 1 (1980) 126.
45 M. Vasjari and V.M. Mirsky, *Fresenius J. Anal. Chem.*, 368 (2000) 727.
46 R. Potyrailo, *Angew. Chem.*, 45 (2006) 702.
47 V.M. Mirsky and V. Kulikov, Combinatorial electropolymerization: concept, equipment and applications, in: *High Throughput Analysis: A Tool for Combinatorial Materials Science*, Kluwer Academic/Plenum Publishers, 2003, 431–446.

ENZYME BASED SENSORS

Chapter 13

Application of electrochemical enzyme biosensors for food quality control

Beatriz Serra, Ángel Julio Reviejo and José Manuel Pingarrón

13.1 INTRODUCTION

The development and innovation of the agricultural and food industry rely generically on two fundamental axes: food safety and food quality. The more and more complexity of the food chain demands the development of efficient traceability systems which guarantee the firmness of all chain links. In this context, a clear priority is defined in the development of molecular methods of detection, analysis and diagnosis. These methods should be rapid, of high sensitivity and should permit automated screening.

Biosensor technologies provide powerful analysis tools with numerous applications in agricultural and food chemistry. The more remarkable characteristics of biosensing devices which convert them in unique attractive options to compete with other technologies in the agricultural and food market are: high selectivity, high sensitivity, short time of analysis, ability to be included in integrated systems, automation easiness, capability of response in real time, versatility allowing the design of "*à la carte*" devices and low cost. Among the three large areas of biosensors applications in the food and agricultural industry: food safety, food quality and processes control, in this chapter the latter two will be addressed with some specific examples.

13.2 FOOD QUALITY CONTROL

The term "food quality" can be envisaged as a set of factors capable of differentiating food products according to their organoleptic characteristics, composition and functional properties. The increase of regulatory action and a heightened consumer demand have led to extensive

Comprehensive Analytical Chemistry 49
Alegret and Merkoçi (Eds)
Volume 49 ISSN: 0166-526X DOI: 10.1016/S0166-526X(06)49013-9

labelling of all the major and minor constituents as well as to a scientific evaluation of product freshness. Table 13.1 gathers most of the target compounds to be analysed for food quality assessment.

TABLE 13.1

Food quality substances

Target compound	Enzymes	Food product	Significance
Glucose	Glucose oxidase, PQQ-glucose dehydrogenase,	Fruits, honey, juices, soft drinks, milk, musts	Maturity and ripening index, taste attributes
	NAD(P)-glucose dehydrogenase	Yoghurt, wines	Fermentation process control, affects quality of the final product
Fructose	Fructose oxidase, PQQ-fructose dehydrogenase NAD(P)-fructose dehydrogenase	Milk, honey, gelatine, juice, artificial sweeteners	Taste, sweetening ability, ripening indicator
Lactose	Lactase, β-galactosidase	Milk	Products destined to lactose intolerant persons Cow and milk diagnosis
Lactulose	Galactosidase	Milk	Distinction between UHT and in-container sterilised milk
Sucrose	Invertase, sucrose phosphorylase	Milk, sugar cane juices	Maturity and ripening, taste
Lactate	Lactate oxidase, NAD-lactate dehydrogenase	Milk, dairy products, wines, ciders	Detection of microbial contamination; fermentation process control
Starch	Amyloglucosidase	Maize	Quality of the products
Amino acids	Amino acid oxidase (generic) specific amino acid oxidase	Milk, fruits, teas	Quality of the product, nutritive value, aging
	Specific amino acid dehydrogenase Specific amino acid decarboxylase	Wines, vinegars, beers	Fermentation process control: D- and L-ratio: indicative of quality of the technological process, adulteration, contamination; quality evaluation of the initial product; estimation of bacteria activity
L-Glutamate	Glutamate dehydrogenase, L-glutamate decarboxylase, L-glutamate oxidase	Fruits, juices, wines, ciders	Flavour enhancer
L-Lysine	L-Lysine α-oxidase	Milk, pasta	Nutritional value; assessment of food processing techniques
Ethanol	Alcohol oxidase, alcohol dehydrogenase (ADH), alcohol dehydrogenase (PQQ-ADH)	Alcoholic beverages	Fermentation process control, tax regulations
		Fruits	Degradation index

TABLE 13.1 (*continued*)

Target compound	Enzymes	Food product	Significance
Glycerol	Glycerol dehydrogenase (PQQ-GlyDH), glycerol dehyrogenase (NAD^+)	Wines	Contributes to the smoothness and viscosity of a wine; taste; fermentation process control
Catechol	Tyrosinase, laccase, peroxidase	Beer	Antioxidant behaviour
Cholesterol	Cholesterol oxidase	Lard, egg, butter	Levels relevant to health
Citric acid	Citrate lyase	Fruit, juices, beverages	Flavour indicator
Acetic acid	Acetate kinase	Wines, vinegar, soy sauces	Taste and flavour
Malic acid	Malate oxidase, $NAD(P)^+$-malate dehydrogenase	Wines	Flavour blender, prevents turbidity, fermentation process control, affects acidity
Lecithin	Phospholipase D	Egg yolk, flour, soy oil	Healthy effects
Isocitrate	Isocitric dehydrogenase	Fruits	Flavour and ripeness indicator
Polyphenols	Tyrosinase, laccase, peroxidase	Olive oil, greens, red wine, tea, coffee, cacao	Dietary value due to their chain-breaking antioxidant activity
Catechins	Peroxidase	Teas	Antiviral and antioxidant activity
Tannins	Peroxidase, laccase	Teas, wines	Astringents, bitter taste, colour formation
Glucosinolates	Myrosinase	Vegetable seeds	Positive or negative character, e.g. antinutritional, toxic and off-flavour effects, as well as anticarcinogenic properties
Short chain fatty acids	Lipase	Milk, dairy products	Rancidification evaluation
Trimethylamine, ornithine, amines, histamine, hypoxanthine	Trimethylamine dehydrogenase, ornithine carbamyl transferase, nucleoside phosphorylase, adenosine deaminase, xanthine oxidase, etc.	Fish, meat, fruits, greens	Freshness index
2,4,6-Trichloroanisole		Wines	Aging
Aldehydes	Alcohol dehydrogenase	Cereals, honey, coffee, fruits, spirits	Indicators of quality deterioration, temperature overheating, microbial fermentation and off-flavour
Oxalate	Oxalate oxidase	Vegetables	Possibly related to kidney stones formation
Phosphate		Drinking water	Toxicity
Sulfite	Sulfite oxidase, sulfite dehydrogenase	Wines	Antioxidant agent

TABLE 13.2

Fundamental approaches for enzyme electrochemical biosensors applied to food analysis

Type of enzyme	Recognition reaction	Detection reaction(s)	Electrochemical technique/ applied potential	Characteristics
Oxidase	Analyte+O_2→ oxidase→ oxidised analyte+H_2O_2	O_2 monitoring [1–4]	Amperometry/−600 mV	Advantages: no electrochemical interference from other sample constituents. Disadvantages: low response, dependence on oxygen can reduce the accuracy and reproducibility. Due to high background signal, the lowest detectable concentration is typically in the ppm range.
		H_2O_2 monitoring [5–9]	Amperometry/+600 mV	Advantages: relative ease of manufacturing, possibility of constructing in small sizes. Usually characterised by high upper linearity and wide linear range. Disadvantages: high potential necessary to oxidise hydrogen peroxide→ electrochemical interference, due to the presence of reducing compounds in real samples (e.g. ascorbic acid, uric acid, bilirubin and acetaminophen), slower responses.
		Only when an amino acid is the substrate: NH_3 is also generated by the enzymatic reaction: NH_3 monitoring [10]	Potentiometry/NH_4^+-selective electrodes	Advantages: very selective, large linear range. Disadvantages: slow responses.
		Use of electrocatalyst to reduce the potential for oxidation or reduction of H_2O_2: e.g. Prussian Blue, rhodium, cobalt [11–13]	Amperometry/close to 0 V	Advantages: reduction of the applied potential, interferences are reduced.
		Use of mediators: e.g. ferrocenes, potassium ferrocyanide osmium complexes [14–16]	Amperometry/close to 0 V	Advantages: interferences are reduced. Detection is speeded up.
		Coupling of a second enzyme reaction: H_2O_2+peroxidase→ H_2O+oxidised peroxidase: Monitoring of reduction of oxidised peroxidase [17]	Amperometry/0 to +200 mV	Advantages: reduction of the applied potential, interferences are reduced.
		Coupling of a second enzyme reaction: H_2O_2+peroxidase→ H_2O+oxidised peroxidase: Mediator for peroxidase reaction: H_2O_2+mediator→ peroxidase→ H_2O+oxidised mediator→Monitoring of reduction of oxidised mediator [18]	Amperometry/0 to +200 mV	Advantages: electron transfer between the electrode and HRP is much faster, thereby increasing the sensitivity and lowering the detection limit
		Coupling of a second enzyme reaction: H_2O_2+peroxidase→ H_2O+oxidised peroxidase: H_2O_2+*p*-fluoroaniline→ peroxidase→H_2O+F^- [19]	Potentiometry/fluoride-sensitive electrode	Advantages: very selective, large linear range. Disadvantages: slow responses.

		Direct electron transfer between the active site and the electrode surface [20]	Amperometry/close to 0 V	Third generation biosensors: very few cases reported. Active site usually deeply buried within the protein, thus enzyme engineering is required for protein modification to allow electrical wiring between enzyme and electrode. Advantages: interferences could be completely eliminated.
Nicotinamide adenine dinucleotide-dependent dehydrogenases	Analyte+$NAD(P)^+$→ dehydrogenase→oxidised analyte+NAD(P)H+H^+	NAD(P)H monitoring [21]	Amperometry/+600 to 700 mV	Disadvantages: large amount of coenzyme required, low stability of the enzyme and relatively high applied potential for amperometric detection.
		Use of electrocatalyst: conducting poly-(*o*-phenylenediamine) film, NAD^+ oxidation products (can act as electrocatalysts for NADH oxidation) [22]	Amperometry/close to 0 mV	Still require the addition of the coenzyme.
		Use of mediators: e.g. Meldola's Blue, Prussian Blue, hexacyanoferrate [24,25]	Amperometry/close to 0 mV	Still require the addition of the coenzyme.
		Coupling of a second enzyme reaction: NAD(P)H+O_2→NADH oxidase→ $NAD(P)^+$+H_2O_2→H_2O_2 monitoring [26] (For the particular case of lactate dehydrogenase, where pyruvate is one of the reaction products, pyruvate oxidase or pyruvate transaminase has been coupled to the main enzyme reaction) [23,27]	Amperometry/close to 0 mV with appropriate electrocatalyst or mediator	Advantages: reduction of the applied potential, interferences are reduced.
		Monitoring of the oxidation of the enzyme cofactor [28]	Amperometry/+500–600 V	Advantages: cofactor is attached to the enzyme, thus no addition of coenzyme is required. Disadvantages: high potential values required.
Pyrroloquinoline-quinone (QQP)-dependent dehydrogenases	Analyte+oxidised dehydrogenase→oxidised analyte+reduced dehydrogenase	Use of mediators: e.g. Meldola's Blue, sodium ferrocyanide, ferrocene derivatives, osmium complex-modified conducting polymer, $Os(bpy)_2Cl_2$, tetrathiafulvalene, tetrathiafulvalene-tetracyanoquinodimethane (TTF-TCNQ) [29–31]	Amperometry/close to 0 mV	Advantages: reduction of the applied potential, interferences are reduced.
Decarboxylases	Analyte+H_2O→ decarboxylase→ product+CO_2	Monitoring of CO_2 [32]	Potentiometric/CO_3^{2-}-selective electrodes	Advantages: very selective, large linear range. Disadvantages: slow responses.
Ammonia lyases	Amino acids→ammonia lyase→product+NH_3	Monitoring of NH_3 [33]	Potentiometric/NH_4^+-selective electrodes	Advantages: very selective, large linear range. Disadvantages: slow responses.

Other types of enzymes: When no oxidase or dehydrogenase is available for a target analyte, other types of enzymes have been used for biospecific recognition: e.g. for citric acid detection, citrate lyase, and amperometric detection was possible by coupling to two more enzymatic reactions: oxaloacetate decarboxylase and pyruvate oxidase, which convert citric acid into H_2O_2 with the latter being monitored amperometrically with an H_2O_2 probe. For detection of acetic acid, acetate kinase is used, coupled to pyruvate kinase and pyruvate oxidase [34,35].

The biosensors technology is being more and more applied to control—in a rapid, effective, reliable, easy-to-use and low-cost manner—such target compounds for the assessment of food quality, and enzyme biosensors and, in particular, electrochemical enzyme biosensors play a crucial role in this area. As it can be seen in Table 13.1, specific enzymes can be used for the selective recognition of the targeted compounds of interest in food quality control. The nature and selectivity of their catalytic activity, and the ability of a single enzyme molecule to catalyse the reaction of numerous substrate molecules, providing an amplification effect, make enzymes excellent tools for the analysis of such compounds. A further advantage is that most enzyme-catalysed reactions can be followed by simple, widely available electrochemical methods. Typical enzymes used for biosensor construction belong to the oxidoreductase group (oxidases, dehydrogenases) and the hydrolase and lyase groups. The oxidoreductase group presents the advantage of the possibility of direct electric coupling with the electrochemical transducer, allowing direct control over the enzyme reaction and real-time monitoring.

Table 13.2 summarises the different approaches used to construct enzyme electrochemical biosensors for application to food analysis based on the different types of enzymes available. Generally, the main problems of many of the proposed amperometric devices have been poor selectivity due to high potential values required to monitor the enzyme reaction, and poor sensitivity. Typical interferences in food samples are reducing compounds, such as ascorbic acid, uric acid, bilirubin and acetaminophen. Electrocatalysts, redox mediators or a second enzyme coupled reaction have been used to overcome these problems (see Table 13.2), in order to achieve the required specifications in terms of selectivity and sensitivity.

Despite these improvements, there are other important biosensor limitations related to stability and reproducibility that have to be addressed. In this context, enzyme immobilisation is a critical factor for optimal biosensor design. Typical immobilisation methods are direct adsorption of the catalytic protein on the electrode surface, or covalent binding. The first method leads to unstable sensors, and the second one presents the drawback of reducing enzyme activity to a great extent. A commonly used procedure, due to its simplicity and easy implementation, is the immobilisation of the enzyme on a membrane. The simplest way is to sandwich the enzyme between the membrane and the electrode. Higher activity and greater stability can be achieved if the enzyme is previously cross-linked with a bi-functional reagent.

Membranes can be used as well as a supporting material for immobilisation (e.g. polycarbonate membranes with created amino groups on the surface that allow covalent binding with glutaraldehyde). Entrapment of enzymes on electrode surfaces can be carried out with polymeric membranes such as polyacrylamide and gelatine, or by electropolymerisation of small monomers (*o*-phenylenediamine). Enzyme encapsulation within a sol–gel matrix has also been reported.

Membranes can also act as a protective layer, preventing electrochemically interfering compounds from reaching the electrode surface. This exclusion effect can be achieved by the presence of charged groups on the surface of the membrane or by size exclusion. Polycarbonate, nitrocellulose acetate and Nafion membranes have been used for these purposes. Moreover, a membrane on the surface of an electrode can also be useful to create a diffusion barrier between the enzyme and the substrate, avoiding swamping effect of a high substrate concentration and allowing a linear response to the concentration.

Another interesting approach for enzyme immobilisation is the entrapment into the bulk of a composite matrix. Some examples of composite matrices used in the development of biosensors applied to food analysis are carbon paste, graphite–ceramic, solid binding compounds, graphite-epoxy resin and graphite–Teflon. The use of bulk-modified bioelectrodes offers several advantages: (i) the close proximity of the biocatalytic and sensing sites; (ii) the possibility of incorporation of other components, e.g. cofactors or redox mediators; (iii) an easy renewability of the surface; (iv) the ease and economy of fabrication; (v) a high stability of the incorporated biocatalysts. Enzymes inside the composite matrices are protected against environmental deactivating factors, such as oxygen, humidity and biological contaminants. Besides, they are more protected against undesired conformational stretching which can lead to deactivation.

Finally, self-assembled monolayers (SAMs) on gold electrodes constitute electrochemical interfaces of supramolecular structures that efficiently connect catalytic reactions, substrate and product diffusion and heterogeneous electron transfer step when enzymes are immobilised on them. Resulting enzyme–SAM electrodes have demonstrated to exhibit good performance and long-term enzyme stability.

A selected revision of enzyme electrochemical biosensors applied to food quality control in the last years is presented in Tables 13.3–13.12.

TABLE 13.3
Glucose biosensors

Ref.	Matrix	Enzyme/immobilisation method	Electrode configuration/ applied potential	Mediator
Niculescu *et al.* [31]	Wine	Glucose dehydrogenase (PQQ-GDH)/integrated in redox hydrogel poly(ethyleneglycol)-diglycidyl ether (PEGDGE) as the cross-linking agent	Spectroscopic graphite/ +200 mV vs. Ag/AgCl	Os-complex-modified nonconducting polymer
Razumiene *et al.* [36]	Lemonade	Glucose dehydrogenase (PQQ-GDH)	Screen-printed carbon electrode/+0.4 V vs. Ag/ AgCl	4-Ferrocenylphenol (FP)
Llopis *et al.* [37]	Cola beverage	Glucose oxidase (GOD)/ entrapped together with the mediator in the bulk of the biocomposite material	Graphite and nonconducting epoxy resin composite electrode/ +0.3 V vs. Ag/AgCl	Tetrathiafulvalene–tetracyanoquino-dimethane (TTF···TCNQ)
Badea *et al.* [38]	Fruit juices Lemon tea	Glucose oxidase (GOx)/on 2,6-dihydroxynaphthalene (2,6-DHN) copolymerised with 2-(4-aminophenyl)-ethylamine (AP-EA) films	Pt electrodes/+650 mV vs. Ag/AgCl	–
Campuzano *et al.* [39]	Soft drinks	Glucose oxidase (GOx)/ cross-linked with glutaraldehyde	Mercaptopropionic acid (MPA)-SAM-modified gold electrode/+0.2 V vs. Ag/AgCl	Tetrathiafulvalene (TTF) (also cross-linked with glutaraldehyde)
Turkusic *et al.* [40]	Beer	Glucose oxidase/mixed with MnO_2-carbon paste	Screen-printed carbon electrode/0.48 V vs. Ag/ AgCl	MnO_2
Ricci *et al.* [13]	Beverages	Glucose oxidase (GOx)/ with glutaraldehyde and Nafion	Prussian Blue (PB)-modified glassy carbon (PB-GC) particles screen-printed electrodes/–0.05 V vs. Ag/AgCl	Prussian Blue
Del Cerro *et al.* [18]	Must Wine	Glucose oxidase and peroxidase/physically included in the composite matrix together with the mediator	Graphite–Teflon composite electrode/0 V vs. Ag/AgCl	Ferrocene (included in the composite matrix)
Jawaheer *et al.* [12]	Fruits (maturity and quality)	Glucose oxidase, mutarotase, invertase/ pectin matrix and covered with a cellulose acetate (CA) membrane	Screen-printed rhodinised carbon electrodes/ +350 mV vs. Ag/AgCl	
Luiz de Mattos *et al.* [41]	Coffee	Glucose oxidase/covered with a Nafion layer	Glassy carbon electrode/–50 mV vs. Ag/ AgCl	Prussian Blue electrodeposited on the electrode
Liu *et al.*, 1998 [106]	Milk	Glucose oxidase (GOD)/ cross-linked with β-cyclodextrin polymer. Coated with Nafion film	Glassy carbon electrode/ +0.35 V vs. Ag/AgCl	Ferrocene (also included in the polymer)
Ivekovic *et al.* [42]	Fruit juice, yoghurt drink	Glucose oxidase (GOx)/ into palladium hexacyanoferrate (PdHCF) hydrogel	Nickel hexacyanoferrate (NiHCF) electrodeposited onto graphite electrode/–0.075 V vs. SCE	Nickel hexacyanoferrate

Analytical characteristics				Observations
Dynamic range	Sensitivity	Repeatability	Response speed	
20–800 μM	87 mA M^{-1} cm^{-2}, LOD: 9 μM		Within seconds	– Flow injection analysis. – Sample dilution required. – Operational stability: 60% response after continuous operation for 20 h.
Up to 10 mM	30 nA mM^{-1}		2–3 min	2,4,7-Trinitro-9-fluorenone (TNF) was electrochemically polymerised on the carbon surface to avoid interferences. Flow-measurements.
	322 ± 19 nA mg^{-1} mL	RSD = 1.9% (15 injections 1 mM glucose solution)	≈ 3 min	– Easy to prepare. – FIA system. – Oxygen has to be removed using a nitrogen flow.
0.1–10 mM	LOD = 0.07 mM	RSD = 1.88% (15 injections 5 mM glucose)	25 s (90% of the signal)	– FIA. – 4.5% of the enzymatic activity lost after 10 h.
1.0×10^{-5}–6.0×10^{-3} M	LOD: 3.5 × 10^{-6} M	RSD = 5.2% (n = 10, repeatability) RSD = 4.5% (n = 10, between biosensors)		No significant changes in the slope value for approximately 5 days, with a decrease in sensitivity of 10% and 17% for 6 and 7 days, respectively.
11–13,900 μmol L^{-1}	LOD: 1 μmol L^{-1}	RSD = 7% (n = 5)	150 s	
Up to 0.5 mM	LOD: 4×10^{-6} mol L^{-1}		15 s	A storage stability up to 20 weeks (at RT) and an operational stability of 1 day with 11 calibration curves performed.
1×10^{-5}–8×10^{-4} M	LOD: 1.9×10^{-6} M	RSD = 4.7% (n = 10, same surface) RSD = 8.0% (n = 10, after polishing the electrode surface)	20 s	The same biosensor could be used with no need of surface regeneration for 4–5 days of use, after that, it was possible to recover the 100% of the initial response by polishing. Both batch and FIA modes.
Up to 30 mM (β-D-glucose)				Pectin: matrix to enhance stabilisation. CA membrane stopped the interferents and extended the linear range.
0.15–2.50 mM	LOD: 0.03 mM	RSD < 1.5%		60 samples per hour.
50 μM–13.5 mM	LOD: 10 μM	RSD = 6.5% (n = 9, real samples)	40 s (95% of the response)	Proteins in milk were previously precipitated with trichloroacetic acid.
0.05–1.0 mM (in batch analysis mode); 0–7.0 mM (in FIA)			30 s	During the 32 days testing period, no significant decrease in the sensor sensitivity.

TABLE 13.4
Other sugars biosensors

Ref.	Analyte	Matrix	Enzyme/immobilisation method	Electrode configuration/ applied potential	Mediator
Gloria *et al.* [43]	Starch	Maize	Amyloglucosidase hydrolysis and glucose oxidase		
Sharma *et al.* [44]	Lactose	Milk and dairy products	Lactase and galactose oxidase (GaO)/in Langmuir–Blodgett (LB) films of poly(3-hexyl thiophene) (P3HT)/stearic acid (SA)	Indium-tin-oxide coated glass plates/0.4 V vs. Pt reference electrode	–
Eshkenazi *et al.* [46]	Lactose	Fresh raw milk	β-Galactosidase, glucose oxidase, and horseradish peroxidase/with PEI and glutaraldehyde	Glassy carbon electrode/ 0 V vs. SCE	5-Aminosalicylic acid (in solution)
Tkac *et al.* [47]	Lactose	Dairy products	β-Galactosidase, galactose oxidase and peroxidase	Graphite rod with preadsorbed ferrocene electrode	Ferrocene
Sharma *et al.* [45]	Galactose	Milk	Galactose oxidase/with poly(3-hexyl thiophene)/ stearic acid (P3HT/SA) onto ITO coated glass plates using Langmuir–Blodgett (LB) film deposition technique	Indium tin oxide (ITO) coated glass plate/0.4 V vs. Pt electrode	–
Amárita *et al.* [48]	Lactose	Milk	Lactases (Lactozym and β-galactosidase of *Enterobacter agglomerans*) adsorbed onto diethylaminoethyl cellulose (DE-52), and confined between the surface of a gas-permeable membrane of a CO_2 electrode and a dialysis membrane	CO_2 potentiometric electrode	–
Moscone *et al.* [49]	Lactulose	Milk	β-Galactosidase immobilised in a reactor and fructose dehydrogenase on the electrode	Pt electrode	$K_3[Fe(CN)_6]$
Paredes *et al.* [15]	Fructose	Honey Fruit juices Soft drinks	D-PQQ-fructose dehydrogenase/physically included in the electrode matrix, also containing mediator and polyethyleneimine (PEI)	Carbon paste electrode/ 0.1 V vs. Ag/AgCl	$Os(bpy)_2Cl_2$ (included in the electrode matrix)
Sekine and Hall [50]	Lactulose	Milk	PQQ-Fructose dehydrogenase (FDH) and β-galactosidase (β-gal)/by covering the electrode surface with a dialysis membrane	Carbon paste ring band electrode/200 mV vs. Ag/ AgCl	Tetrathiafulvalene–tetracyanoquinodimethane (TTF-TCNQ) salt/physically packed
Campuzano *et al.* [29]	Fructose	Honey Pear juice	Fructose dehydrogenase (FDH)/with glutaraldehyde	Self-assembled monolayer (SAM) modified gold electrode/ +0.2 V vs. Ag/AgCl	Tetrathiafulvalene (TTF)/ with glutaraldehyde
Garcia *et al.*, 1998 [107]	Fructose	Dietetic products	Fructose 5-dehydrogenase/in a polypyrrole (PPY) film	Platinum electrode/ +0.25 mV vs. Ag/AgCl	Sodium ferricyanide
Maestre *et al.* [51]	Sucrose	Fruit juices	Sucrose phosphorylase, phosphoglucomutase and glucose-6-phosphate 1-dehydrogenase/in carbon paste together with PEI, $NADP^+$ and the mediator	Carbon paste electrodes/ +0.15 V vs. Ag/AgCl	Os(4,4′-dimethyl-2,2-bypyridine)$_2$ (1,10-phenanthroline-5,6-dione)
Jawaheer *et al.* [12]	β-D-glucose, total D-glucose, sucrose and ascorbic acid	Fruits (maturity and quality)	Glucose oxidase, mutarotase, invertase/pectin matrix and covered with a cellulose acetate (CA) membrane	Screen-printed rhodinised carbon electrodes/+350 mV vs. Ag/AgCl	
Campuzano *et al.* [52]	Lactulose	Milk	β-Galactosidase (GAL) and fructose dehydrogenase (FDH)/cross-linked with glutaraldehyde	MPA-SAM-modified gold electrode/+0.1 V vs. Ag/ AgCl	Tetrathiafulvalene (TTF) (also cross-linked with glutaraldehyde)

Analytical characteristics				Observations
Dynamic range	Sensitivity	Repeatability	Speed of response	
				YSI Bioanalyser.
1–6 g dL^{-1}		5%		Shelf life more than 120 days. The reusability of electrode was found ten times with 3% loss in current response.
10–340 mg L^{-1}				After 140 days, the sensor still retained 40% of its initial activity.
	LOD: 44 μM	2.02%	75 s (90% of the steady-state value)	Addition of DEAE-dextran and inositol to the enzyme layer improved the half-life more than 16-fold.
1–4 g dL^{-1}	LOD: 1 g dL^{-1}		60 s	Shelf life of more than 90 days.
1–10%			≈ 1 h	2 weeks
1×10^{-6}–5×10^{-3} mol L^{-1} (fructose) 1×10^{-5}–5×10^{-3} mol L^{-1} (lactulose)	LOD: 5×10^{-7} mol L^{-1} (fructose)	RSD = 2%		
0.2–20 mM	LOD: 35 μM	RSD < 3% (n = 26, FIA)	30 s (FIA peak)	100% of the response within 4 h of continuous operation. After this period, a sharp decrease in the response is observed 100% of the response was retained for 1 week of dry storage.
Up to 50 μM (fructose) p to 30 μM (lactulose)	LOD: 1.0 μM (lactulose)	RSD = 4.45% (real samples)		The sensor could be used for 2 days with recalibration, since the sensor keeps 80% of the initial response, and could still determine 1.0 μM lactulose after 2 days.
1.0×10^{-5}–1.0×10^{-3} M	LOD: 2.4×10^{-6} M	RSD = 6.7% (slope values, n = 10) RSD = 5.8% (values from five different electrodes fabricated in the same way)	40 s	Useful lifetime: 30 days.
0.1–0.8 mmol L^{-1}		RSD = 0.68% (n = 7)	15 s	200 analyses in 2 weeks of continuous use. Operational stability: 7 days.
1–15 mM	LOD: 1 mM	RSD = 10% (for five different electrodes)	5 min	Partial loss of the activity within the first 2 h, then, the response stayed stable for the next 10 h.
Up to 30 mM (β-D-glucose) Up to 30 Mm (sucrose)				Pectin: matrix to enhance enzyme entrapment and stabilisation. CA membrane stopped the interferents and extended the linear range.
3.0×10^{-5}–1.0×10^{-3} M	LOD: 9.6×10^{-6} M	RSD = 6.6% (n = 10)	100 s	Useful lifetime of one single biosensor: 81 days.

TABLE 13.5
Ethanol biosensors

Ref.	Matrix	Enzyme/immobilisation method	Electrode configuration/applied potential	Mediator
Niculescu *et al.* [31]	Wine	Alcohol dehydrogenase (PQQ-ADH)/integrated in redox hydrogel poly(ethyleneglycol)-diglycidyl ether (PEGDGE) as the cross-linking agent	Spectroscopic graphite/+200 mV vs. Ag/AgCl	Os-complex-modified nonconducting polymer
Razumiene *et al.*, 2003b [108]	Wine	Alcohol dehydrogenase (PQQ-ADH)/with aq. solution of glutaric aldehyde and covered with semi-permeable terylene film	Screen-printed carbon electrode/ 0.4 V vs. Ag/AgCl	*N*-(4-hydroxybenzylidene)-4-ferrocenylaniline (HBFA)
Yao *et al.* [53]	Wine Beer Soy sauce	Alcohol dehydrogenase (ADH)/entrapped in carbon paste	Carbon paste: (NAD^+, polyethylene glycol-modified alcohol dehydrogenase and an oil soluble mediator)/+0.3 V vs. Ag/AgCl	7-Dimethylamine-2-methyl-3-β-naphtamido-phenothiazinium chloride (3-NTB)
Lobo-Castañón *et al.*, 1997 [22]	Cider Wine Whisky	Alcohol dehydrogenase (ADH)/entrapped in carbon paste and covered with poly-(*o*-phenylenediamine)	ADH-NADH-modified carbon paste electrode with poly-(*o*-phenylenediamine) electropolymerised on the surface/ +0.15 V vs. Ag/AgCl	
Boujtita *et al.* [11]	Beer	Alcohol oxidase (AOx)	Screen-printed carbon electrode doped with 5% cobalt phthalocyanine (CoPC-SPCE), and coated with AOx; a perm-selective membrane on the surface acts as a barrier to interferents/+400 mV vs. Ag/AgCl	Cobalt phthalocyanine
Katrlík *et al.* [25]	Wine	Alcohol dehydrogenase (ADH) and diaphorase (DP)/added either to the bulk composite or to the surface	Graphite-solid binding matrix composite modified with NAD^+ and the mediator/+300 mV vs. SCE	Organic dyes, vitamin K_3, hexacyanoferrate(III), ferrocene
Leca and Marty [26]	Cider Wine Whisky	Co-immobilisation of alcohol dehydrogenase (ADH), NADH oxidase and high molecular weight NAD (NAD-dextran) in a poly(vinyl alcohol) bearing stryrylpyridinium groups (PVA-SbQ) matrix	Pt/+550 vs. Pt	
Morales *et al.* [54]	Wine Brandy	Alcohol oxidase (AOx)/ entrapped in the composite matrix	Graphite powder and epoxy resin composite/+1100 mV vs. Ag/AgCl	–
Akyilmaz and Dinçkaya [55]	Wine Beer Liquor	Alcohol oxidase in *C. tropicalis* cells/cells over a Teflon membrane and immobilised with glutaraldehyde	Clark type O_2 electrode	–
Rotariu *et al.* [56]	Alcoholic beverages	yeast cells (*Saccharomyces ellipsoideus*)/micro-organisms were placed on the oxygen membrane and covered with a dialysis membrane (DM) and fixed with a rubber ring	Potentiometric oxygen electrode	–
Patel *et al.* [9]	Wine	Alcohol oxidase (AOx)/with poly(carbamoyl)sulfonate (PCS) hydrogel containing PEI	Screen-printed platinum electrode/ +600 mV vs. Ag/AgCl	–
Peña *et al.* [57]	Beer	Alcohol oxidase, peroxidase/ physically entrapped in the electrode matrix	Reticulated vitreous carbon (RVC)–epoxy resin electrode/0 V vs. Ag/AgCl	Ferrocene (physically entrapped in the electrode matrix)
Guzmán-Vázquez de Prada *et al.* [58]	Beer Wine Liquor	Alcohol oxidase and peroxidase/physically included in the composite matrix together with the mediator	Graphite–Teflon composite electrode/0 V vs. Ag/AgCl	Ferrocene (included in the composite matrix)

Analytical characteristics				Observations
Dynamic range	Sensitivity	Repeatability	Speed of response	
2.5–250 μM	220 mA M^{-1} cm^{-2}, LOD: 1.2 μM		Within seconds	– Flow injection analysis – Sample dilution required – Operational stability: 90% response was kept after continuous operation for 100 h
	28 μA $(cm^2\ mM)^{-1}$			Measurements in a flow system. The response decreases by ca. 50% in 2 days.
Up to 1 mM	LOD: 50 μM		1 min	Stable response during 2 weeks
3×10^{-8}–3×10^{-6} M	3.1×10^6 nA M^{-1} LOD: 2×10^{-8} M		20 s.	
		RSD = 0.64%		Disposable
0.2–4.0 mM		RSD = 2–2.5%	5 min	Stable for at least 3 months
3×10^{-7}–3×10^{-4} M	~64 mA L mol^{-1} cm^{-2}	RSD = 2.5%	<2 min	More than 80 reproducible measurements without requiring cofactor addition
0.5–29 mg mL^{-1}	22 nA mg^{-1} mL	RSD = 0.5%	65 s (95% of the signal)	– The biocomposite is robust, polishable and easily mechanised. – The enzyme is stable in the biocomposite matrix for more than 3 months. – Interferences from other matrix substances.
0.5–7.5 Mm			2 min	50% of its initial activity lost after 7 h
0.02–50 mM			7 min for the steady-state and 2 min for kinetic measurements	– Ethanol assimilation by the yeast cell is accompanied by oxygen consumption in the respiratory process. – After 5 days, the signal has decreased to 70% of its initial value, and reached 40% after 8 days.
0.01–3 mM	30 nA mM^{-1} LOD: 0.01 mM	<3%	12 s (90% of the signal)	– Operational stability: 95% after 12 h. – Storage stability: 60% after 8 days.
Methanol: $(0.1–6.0) \times 10^{-5}$ M Ethanol: $(0.4–25) \times 10^{-5}$ M	LOD: Methanol: 4.2×10^{-7} M Ethanol: 1.5×10^{-6} M	RSD = 6.4% ($n = 10$, same surface) RSD = 6.3% ($n = 10$, after polishing the electrode surface) RSD = 6.0% ($n = 10$, different biosensors)	1.5 min	Methanol and ethanol as analytes. Possibility of regeneration of the electrode surface by polishing. Lifetime of a single biosensor: 20 days.
Methanol: $(0.2–15) \times 10^{-5}$ M Ethanol: $(2–200) \times 10^{-5}$ M	LOD: Methanol: 0.54×10^{-6} M Ethanol: 5.3×10^{-6} M	RSD = 6.1% ($n = 10$, same surface) RSD = 8.3% ($n = 10$, after polishing the electrode surface) RSD = 6.5% ($n = 9$, different biosensors)	40 s	Methanol and ethanol as analytes. The useful lifetime of a single AOD–HRP–ferrocene composite electrode was of approximately 15 days with no need of enzyme stabilisers in the electrode matrix. Both batch and FIA modes.

TABLE 13.6

Lactic acid and other organic acids biosensors

Ref.	Analyte	Matrix	Enzyme/immobilisation method	Electrode configuration/ applied potential	Mediator
Choi [59]	L-Lactate	Dairy products	L-Lactate oxidase/with chitosan in an eggshell membrane	Oxygen electrode	–
Avramescu *et al.* [24]	D-Lactate	Wines	D-Lactate dehydrogenase/with NAD^+ deposited onto the surface of the electrode and covered with polyethyleneimine–Nafion membrane	Carbon electrode, modified with an insoluble salt of Meldola's Blue/–50 mV vs. Ag/AgCl	Meldola's Blue
Kwan *et al.* [27]	L-Lactate	Yoghurt milk, soda, sport drinks, and healthy supplement	Salicylate hydroxylase (SHL), L-lactate dehydrogenase (LDH), and pyruvate oxidase (PyOD)/entrapped by a poly(carbamoyl) sulfonate (PCS) hydrogel on a Teflon membrane	Clark-type oxygen electrode	–
Lobo-Castañón *et al.* [23]	L-Lactate	Cider	Glutamic pyruvic transaminase (GPT) and L-lactate dehydrogenase (LDH) together with NAD^+/by an electropolymerised poly-(*o*-phenylenediamine) (PPD) film	Carbon paste modified with PPD and poly(*o*-aminophenol)/0 V vs. Ag/AgCl	–
Serra *et al.* [60]	L-Lactate	Yoghurt Wine	L-Lactate oxidase (LOx) and peroxidase (HRP)/physically entrapped in the composite electrode matrix together with the mediator	Graphite–Teflon composite electrode/0 V vs. Ag/AgCl	Ferrocene
Torriero *et al.* [61]	L-Lactate	Milk	L-Lactate oxidase (LOx) and peroxidase (HRP)/LOx on 3-aminopropyl-modified controlled-pore glass reactor HRP and Os-PAA were covalently immobilized on the electrode surface	Glassy carbon electrode/0 vs. Ag/AgCl	Redox polymer
Kriz *et al.* [62]	L-Lactate	Tomato paste baby food	L-Lactate oxidase/injected in solution from a small reservoir	Pt electrode/+650 mV vs. Ag/AgCl	–
Katrlík *et al.* [63]	L-Malate, L-lactate	Wines	L-Malate dehydrogenase or L-lactate dehydrogenase and diaphorase/covered by a dialysis membrane	NAD^+-modified graphite–2-hexadecanone composite electrode	Hexacyanoferrate(III)
Gajovic *et al.* [64]	L-malate	Fruits, fruit juices, ciders and wines	$NAD(P)^+$-dependent L-malate dehydrogenase oxaloacetate decarboxylating with salicylate hydroxylase (SHL)/in gelatine membrane sandwiched between a dialysis membrane and a PET membrane	Clark-electrode	–

Analytical characteristics				Observations
Dynamic range	Sensitivity	Repeatability	Speed of response	
Up to 0.5 mM	LOD: 8.6 μM	RSD = 5%, 0.10 mM (n = 10)	60 s (95% of the signal)	Long shelf-life of at least 1 year. It maintained its activity above 80% after being kept at 4 °C for a year.
0.1–1 mM	LOD: 0.05 mM	RSD = 10.8% ($n = 30$)	150 s	After 2 weeks of storage at 4 °C, the biosensor showed 75% of its initial response.
10–400 μM	LOD: 4.3 μM		100 s	The trienzyme system allows the determination of lactate with high reliability, sensitivity and reproducibility.
6×10^{-7}–8.5×10^{-5} M	LOD: 3×10^{-8} M	RSD = 4.0% (reproducibility among electrodes fabricated from the same batch of carbon paste)	80 s	Equilibrium displaced towards the products by coupling the LDH reaction with the GPT reaction. Interferences minimisation with poly(*o*-aminophenol) film.
5.0×10^{-6}–1.0×10^{-4} mol L^{-1}	LOD: 1.4 μM (batch mode) and 0.9 μM (FIA mode)	RSD = 5.6% ($n = 10$), 2.5×10^{-4} mol L^{-1} RSD = 6.8% (after renewing of electrode surface by polishing) RSD = 6.6% (responses from eight different electrodes)	60 s	Renewability of the biosensor surface. Useful lifetime of one single biosensor: 60 days. No significant loss of the enzyme activity occurred after 6 months of storage at 4 °C.
0.010–2.50 mM	LOD: 5 nmol L^{-1}	RSD = 5% ($n = 6$), 0.100 mM	3 min	After 30 min, the system reached a stable response which lasted for at least 3 h. In the FIA system, there is practically no decay in the catalytic current after 20 milk samples.
0–0.1 mM		RSD = 2.5% ($n = 15$)		The whole analysis is performed in 3 min.
Up to 1.1 mM for L-malate and 1.3 mM for L-lactate	LOD: 10 μM for both	From 1.1% to 3.4% ($n = 6$)	From 3 to 6 min	After 5 months storage at room temperature, the L-malate sensor exhibited almost 100% and L-lactate 90% of the initial sensitivity.
0.01–1.2 mmol L^{-1}	18.5 μA L cm^{-2} $mmol^{-1}$		40 s (90% response)	45 tests per hour. Working stability more than 30 days. The reagent and enzyme costs per test were ca. 2 US cents.

TABLE 13.6 (*continued*)

Ref.	Analyte	Matrix	Enzyme/immobilisation method	Electrode configuration/ applied potential	Mediator
Albareda-Sirvent and Hart [65]	L-Malate, L-lactate	Wines	Lactate oxidase or malate dehydrogenase/in sol–gel matrix	Sol–gel thick-film printed graphite electrode (contained NAD^+ for malate sensors)/+ 350 mV vs. Ag/AgCl for lactate sensor and −125 mV vs. Ag/AgCl for malate	Meldola's Blue (mediator for the dehydrogenase)
Lupu *et al.* [66]	L-Malate Glucose	Wine-making process	Glucose oxidase or $NAD(P)^+$-dependent L-malate dehydrogenase oxaloacetate decarboxylating/in a polyethyleneimine–glutaraldehyde cross-linking membrane	Screen-printed electrodes/0 V vs. Ag/AgCl (glucose) and 0.2 V vs. Ag/AgCl (malic acid)	Meldola's Blue (malic acid) Prussian Blue (glucose)
Prodromidis *et al.* [35]	Citric acid	Fruits, juices and sport drinks	Citrate lyase (in solution), oxaloacetate decarboxylase and pyruvate oxidase/ sandwiched between a cellulose acetate membrane and a protective polycarbonate membrane	H_2O_2 probe (Pt electrode)	–
Kim and Kim [67]	Isocitrate	Apple, grape, strawberry and tomato juices	Isocitric dehydrogenase/in aminopropyl glass beads with glutaraldehyde and packed in a Teflon tube	Potentiometric CO_3^{2-}-selective electrode	–
Mizutani *et al.* [34]	Acetic acid	Wines	Acetate kinase (AK), pyruvate kinase (PK) and pyruvate oxidase (PyOx)/on a photo-cross-linkable poly(vinyl alcohol) bearing stilbazolium group membrane	Platinum electrode coated with poly(dimethylsiloxane)/−0.4 V vs. Ag/AgCl	–

Analytical characteristics				Observations
Dynamic range	Sensitivity	Repeatability	Speed of response	
Up to 1 mM for lactate and up to 15 mM for malate				Interferences from the sample matrix were detected.
Malate: Up to 10^{-3} M Glucose: Up to 5×10^{-3} M	LOD: 10^{-5} M for malic acid and 10^{-6} M for glucose	Malate: RSD = 9.3% (five electrodes, 0.05 mM); RSD = 5.8% (five measurements, 0.05 mM). Glucose: RSD = 5.9% (five electrodes, 0.05 mM); RSD = 2.5% (five measurements, 0.05 mM).		Malate biosensor: After 15–20 measurements, the initial response of the sensor is decreased by 25%.
0.015–0.5 mM		RSD = 1.0% ($n = 8$)		Multi-membrane system, consisting of a cellulose acetate membrane for elimination of interferents, an enzymic membrane and a protective polycarbonate membrane placed on a Pt electrode. Used with a fully automated flow injection manifold. An 8–10% loss of the initial activity of the sensor was observed after 100–120 injections.
10^{-3}–10^{-1} M		RSD: <2%		Enzyme reactor with an FIA system. The interference effect of major sugars and organic acids on the sensor system was less than 5%.
0.05–20 mM	LOD: 0.05 mM	RSD = 1.7% ($n = 10$, 5 mM)	3 min	FIA system with a sampling rate of 20 h^{-1}, and stable for a month. The addition of expensive reagent such as ATP, PEP and FAD is required, but the cost of these reagents was ca. 5 cents per assay.

TABLE 13.7
Amino acids biosensors

Ref.	Analyte	Matrix	Enzyme/immobilisation method	Electrode configuration/ applied potential	Mediator
Kelly *et al.* [68]	L-Lysine	Protein samples from milk and pasta	L-Lysine α-oxidase/with BSA and glutaraldehyde	Ruthenium/rhodium coated glassy carbon electrode covered with 1,2-diaminobenzene polymer/ +100 mV vs. an Ag/AgCl	–
Oliveira *et al.* [32]	L-Glutamate	Commercial soups	L-Glutamate decarboxylase/ in glass beads	Potentiometric ionic sensor	–
Alvarez-Crespo *et al.* [69]	L-Glutamate	Chicken bouillon cubes	L-Glutamate dehydrogenase/ entrapped in carbon paste	NAD^+ modified carbon paste electrode/+0.15 V vs. Ag/ AgCl	*o*-Phenylenediamine polymerised on the electrode surface
Almeida and Mulchandani [70]	L-Glutamate	Food samples	L-Glutamate oxidase/on the electrode surface cross-linked with glutaraldehyde	Carbon paste electrode/ +0.15 V vs. Ag/AgCl	Tetrathiafulvalene (TTF)
Domínguez *et al.* [71]	L-Amino acids D-Amino acids	Grapes	L-Amino acid oxidase or D-amino acid oxidase and peroxidase/entrapped in electrode composite matrix by physical inclusion	Graphite–Teflon–ferrocene composite electrode/0 V vs. Ag/AgCl	Ferrocene
Váradi *et al.* [72]	L-Amino acids D-Amino acids	Beers (brewing process)	L-Amino acid oxidase and D-amino acid oxidase/in a Plexi-cell thin-layer on natural protein membrane	Pt wire electrode/+100 mV vs. Ag/AgCl	–
Sarkar *et al.* [73]	L-Amino acids, D-amino acids	Milk (ageing effects)	L- and/or D-Amino acid oxidase	Screen-printed rhodinised carbon electrode	Rh
Inaba *et al.* [2]	D-Alanine	Fish sauces	D-Amino acid oxidase (D-AAOx) and pyruvate oxidase (PyOx)/PyOx is immobilised on a membrane and sandwiched between dialysis membrane and Teflon membrane of the oxygen electrode	Oxygen electrode	–

TABLE 13.8
Glycerol biosensors

Ref.	Analyte	Matrix	Enzyme/immobilisation method	Electrode configuration/ applied potential	Mediator
Niculescu *et al.* [31]	Glycerol	Wine	Glycerol dehydrogenase (PQQ-GlyDH)/integrated in redox hydrogel poly(ethyleneglycol)-diglycidyl ether (PEGDGE) as the cross-linking agent	Spectroscopic graphite/ +200 mV vs. Ag/AgCl	Os-complex-modified nonconducting polymer
Eftekhari [30]	Glycerol	Grape juice	Glycerol dehyrogenase (NAD^+ dependent)/into polyaniline film	Aluminium electrode/ +0.35 V vs. NHE	$Fe(CN)_6^{3-}$
Alvarez-Gonzalez *et al.* [74]	Glycerol	Plant-extract syrup	Glycerol dehyrogenase (NAD^+)/by electropolymerised layer of nonconducting poly-(*o*-phenylenediamine) (PPD)	Carbon paste electrodes modified with the oxidation products of NAD^+/0 mV vs. Ag/AgCl	–

Analytical characteristics				Observations
Dynamic range	Sensitivity	Repeatability	Speed of response	
2–125 μM	LOD: 2 μM	RSD = 2.3% (40 replicate injections of 1 mM)	2 min	Microwave hydrolysis reduces the time to hydrolyse proteins into their constituent amino acids.
2.5–75 mmol L^{-1}		RSD < 5%		Enzymatic reactors for automatic continuous flow systems. Useful life time of 78 days or 638 determinations.
5.0×10^{-6}–7.8×10^{-5} M	LOD: 3.8×10^{-6} M	RSD = 4.8%	120 s. (98% of the signal, 2×10^{-5} M)	Lifetime: 3 days with recalibration. Storage time: 1 month.
Up to 0.8 Mm	LOD: 2.6 μM		2 min	Retains more than 90% of its original activity over a period of 3 weeks.
(1.0–25) × 10^{-5} M (L-tryptophan) (1.0–140) × 10^{-5} M, (D-methionine)	LOD: 2.3×10^{-6} M (L-tryptophan) LOD: 3.2×10^{-6} M (D-methionine)	RSD = 5.0% (4.0×10^{-4} M L-tryptophan) RSD = 5.3% (2.0×10^{-4} M D-methionine) Reproducibility between biosensors: RSD: 4.9% (L-AAOx biosensor), 6.7% (D-AAOx biosensor)	2 min	Batch and FIA modes. Renewability of the biosensor surface. Useful lifetime of a single biosensor: 30 days (L-amino acid oxidase biosensor) and 15 days (D-amino acid biosensor).
0.1–3 mM (L-methionine) 0.2–3 mM (D-methionine)		(L- and D-Methionine, 1 mM, $N = 25$) RSD = 2% and 2.7%		Enzyme reactors. The LAO reactor could be used for ca. 900 measurements, while the DAO reactor could be used for ca. 1000 measurements.
	LOD: 0.47, 0.15 and 0.20 mM, L-leucine, L-glycine and L-phenylalanine, respectively			Stability over a 56 days test period.
0.1–1 mM	LOD: 0.05 mM	RSD = 4.9% ($n = 5$) for 0.5 mM		D-Amino acid oxidase (D-AAOx) reactor. Enzyme immobilised on controlled pore glass beads. Difference in signal between two oxygen electrodes.

Analytical characteristics				Observations
Dynamic range	Sensitivity	Repeatability	Speed of response	
1–200 μM	32 mA M^{-1} cm^{-2}, LOD: 1 μM		Within seconds	– Flow injection analysis. – Sample dilution required. – Operational stability: 80% response after continuous operation for 20 h.
5×10^{-6}–2×10^{-3} M	LOD = 1×10^{-6} M	RSD = 1.7% (0.1 mM glycerol, $n = 10$)	60–80 s	Long-term stability: 75% of the response was retained after 100 days of storage at 5 C.
10^{-6}–10^{-4} M	LOD: 4.3×10^{-7} M			The amperometric response remains stable for at least 3 days.

TABLE 13.9
Polyphenols biosensors

Ref.	Analyte	Matrix	Enzyme/immobilisation method	Electrode configuration/ applied potential	Mediator
Campanella *et al.* [75]	Polyphenols, hydrogen peroxide, KO_2, lecithin	Olive oil and other vegetable oils Egg yolk Ground soya seed oil	Tyrosinase or catalase or superoxide dismutase or phospholipase D/choline oxidase were entrapped in kappa-carrageenan gel	Oxygen electrode	–
Campanella *et al.* [1]	Superoxide radical polyphenols Sulfite Ascorbic acid	Wine	Superoxide dismutase (SOD), or tyrosinase, or sulfite oxidase, or ascorbate oxidase/in a gel-like kappa-carrageenan membrane sandwiched between a cellulose acetate membrane and a dialysis membrane	Amperometric Pt electrode for hydrogen peroxide (for superoxide determination)/ +650 mV vs. Ag/AgCl/ Amperometric oxygen electrode (for the rest)/–650 mV vs. Ag/AgCl	–
Mello *et al.* [76]	Polyphenols	Tea	Peroxidase (HRP)/ adsorption on silica–titanium and cross-linking with glutaraldehyde, then mixed with carbon paste	Carbon paste electrode/–50 mV vs. Ag/ AgCl	–
Mello *et al.* [77]	Polyphenols	Vegetable extract	Peroxidase (HRP)/ adsorption on silica–titanium and cross-linking with glutaraldehyde, then mixed with carbon paste	Carbon paste electrode/–50 mV vs. Ag/ AgCl	–
Capannesi *et al.* [78]	Polyphenols	Olive oil	Tyrosinase/with glutaraldehyde onto a preactivated membrane (Immobilon) sandwiched between the gas permeable membrane of the electrode and a dialysis membrane	Amperometric oxygen probe	–
Gomes *et al.* [79]	Polyphenols	Wine	Laccase/on derivatised polyethersulfone membranes	Pt electrode	–
Carralero Sanz *et al.* [80]	Polyphenols	Wine	Tyrosinase/by cross-linking with glutaraldehyde	Gold nanoparticles-modified glassy carbon electrode/–0.10 V vs. Ag/ AgCl	–
Eggins *et al.* [81]	Flavanols	Beer	Tyrosinase from banana tissue	Carbon paste electrode	–

Analytical characteristics				Observations
Dynamic range	Sensitivity	Repeatability	Speed of response	
For phenol: 1–37 μM For H_2O_2: 30–130 μM For lecithin: 2.0–48.7 mg L^{-1}				Working media: *n*-hexane for tyrosinase electrode; chloroform for catalase electrode; chloroform/hexane (50% v/v) containing 1% by volume of methanol for lecithin electrode.
Superoxide radical: 0.02–2.0 mM		<5%		Antioxidant capacity of wines was evaluated in terms of polyphenol content, sulfite and ascorbic acid, and compared to superoxide radical scavenging
Chlorogenic acid (CGA) polyphenol compound: 1–50 μmol L^{-1}	181 μmol^{-1} L nA cm^{-2}, LOD:0.7 μmol L^{-1}	RSD = 5.6% ($n = 10$), 30 μmol L^{-1}	3 s (90% of the response)	Decrease in the current density about 20% after 60 successive measurements. Decrease in sensitivity about 50% after 8 days of storage, at 4°C and pH 7.0 After this period the electrode signal was constant for 30 days.
Chlorogenic acid (CGA) polyphenol compound: 1–50 μmol L^{-1}	181 μmol^{-1} L nA cm^{-2}, LOD: 0.7 μmol L^{-1}	RSD = 5.6% ($n = 10$), 30 μmol L^{-1}	3 s (90% of the response)	
Phenol: 0.5–6 ppm	Phenol: LOD = 4.0 ppm	RSD = 2%		FIA. Measurements in *n*-hexane, direct measurement of the oil, with no extraction.
2.0–4.0 × 10^{-6} M, for equimolar mixed solutions of catechin and caffeic acid	0.0566 mA M^{-1}, LOD = 1.0 × 10^{-6} M for equimolar mixed solutions of catechin and caffeic acid	RSD< 10%		Previous solid-phase extraction for polyphenols enrichment.
Phenol: 0.1–4 μM	Phenol: 0.21 μM	RSD = 3.6% ($n = 6$)		The useful lifetime of one single biosensor was of at least 18 days, and an RSD of 4.8% was obtained for the slope values of catechol calibration plots obtained with five different biosensors. Error due to flavanols present in plant tissues.

TABLE 13.10
Freshness index biosensors

Ref.	Analyte	Matrix	Enzyme/immobilisation method	Electrode configuration/ applied potential	Mediator
Loechel *et al.* [20]	Trimethylamine	Fish	Trimethylamine dehydrogenase (TMADH)		(Dimethylamine) methylene ferrocene (DMAMFe)
Cayuela *et al.* [82]	Hypoxanthine	Sardines	Xanthine oxidase and peroxidase/physically entrapped in the composite matrix, together with the mediator	Graphite–Teflon composite electrode/0 V vs. Ag/AgCl	Ferrocene (physically entrapped in the composite matrix)
Agüí *et al.* [83]	Hypoxanthine	Sardines Chicken	Xanthine oxidase/cross-linked with glutaraldehyde and BSA	Carbon paste electrode modified with electrodeposited gold nanoparticles/0 V vs. Ag/AgCl	–
Hu *et al.* [84]	Hypoxanthine	Fish	Xanthine oxidase/within a polyaniline film on the electrode surface by electropolymerisation	Sodium montmorillonite-methyl viologen carbon paste modified electrode/–0.72 V vs. Ag/AgCl	Methyl viologen
Basu *et al.* [85]	Hypoxanthine	Fish Meat	Xanthine oxidase (XO)/ cross-linked with glutaraldehyde in the presence of BSA. Covered with gelatine and a membrane of cellulose acetate		
Niu and Lee [86]	Hypoxanthine	Carp	Xanthine oxidase (XO)/ bulk modified	Sol–gel graphite electrode/–0.95 V	Benzyl viologen
Okuma and Watanabe [3]	*K*-value ($K = \{(HxR+Hx)/(ATP+ADP+AMP+IMP+HxR+Hx)\} \times 100$), ATP, ADP, AMP, IMP, HxR and Hx are adenosine triphosphate, adenosine diphosphate, adenosine monophosphate, inosine monophosphate, inosine and hypoxanthine, respectively	Tuna Scallop	Enzyme reactor I: nucleoside phosphorylase and xanthine oxidase immobilised simultaneously on chitosan beads Enzyme reactor II: nucleoside phosphorylase, xanthine oxidase, phosphatase alkaline and adenosine deaminase immobilised simultaneously on chitosan beads	Oxygen electrode	–
Ghosh *et al.* [14]	Hypoxanthine (HX), inosine (INO) and inosine monophosphate (IMP)	Fish Catla-Catla	Xanthine oxidase, nucleoside phosphorylase and nucleotidase	Conducting polypyrrole enzyme Pt electrodes/0.7 V vs. SCE	Ferrocene carboxylic acid (FCA)

Analytical characteristics				Observations
Dynamic range	Sensitivity	Repeatability	Speed of response	
	LOD: 2 mg TMA-N per 100 g wet fish muscle			
5×10^{-7}–1×10^{-5} M	LOD: 9.0×10^{-8} mol L^{-1}	RSD = 4.5% ($n = 10$, same electrode surface) RSD = 6.0% ($n = 10$, after renewing electrode surface by polishing) RSD = 5.4% ($n = 5$, different electrodes)	10 s	Measurements at 30 ± 1 °C. More than 6 months of storage at 4 °C without loss of enzyme activity.
0.5–10 μM	LOD: 2.2×10^{-7} M	RSD = 3.9% ($n = 10$), 5.0×10^{-6} M	30 s	Lifetime: at least 15 days.
1–400 μM	LOD: 0.8 μM			Sodium montmorillonite-methyl viologen in the carbon paste has excellently performed catalytic activity for the oxygen reduction. 14 days of stable signal. Approximately 60% of the initial enzyme activity after 5 weeks.
0.05–2 mM				
1.2–1100 μM	LOD: 0.38 μM	RSD = 5% ($n = 10$, different electrode surfaces) RSD = 8.6% ($n = 6$, different electrodes)		The surface can be renewed by mechanical polishing. Stable response for 20 days.
Very good correlation between *K*-value and electrode output. Response was linear between 0% and 100% *K*-value		RSD = 2.3%	5 min per assay.	In repeated run over 25 h (1500 assays), the reactor (II) was found to retain 40% of its initial activity.
Up to 2 mM of HX Up to 1 mM of IMP Up to 1.5 mM of INO				

TABLE 13.10 (*continued*)

Ref.	Analyte	Matrix	Enzyme/immobilisation method	Electrode configuration/ applied potential	Mediator
Carsol and Mascini [87]	*K*-value	Fish	Two reactors in series: one packed with nucleoside phosphorylase (Np) and the other with xanthine oxidase (XO)/immobilised on aminopropyl glass	Screen-printed graphite electrode/450 mV vs. Ag/AgCl	–
Sato *et al.* [4]	Ornithine	Prawns	Ornithine carbamyl transferase and pyruvate oxidase in a reactor	Oxygen electrode	–
Shin *et al.* [88]	Ornithine	Kuruma prawn	Ornithine carbamyl transferase, nucleoside phosphorylase and xanthine oxidase/in membranes prepared from cellulose triacetate, glutaraldehyde and 1,8-diamino-4-aminomethyl octane	Oxygen electrode	–
Draisci *et al.* [7]	Biogenic amines: putrescine, cadaverine, histamine, tyramine, spermidine, spermine, tryptamine	Anchovies	Diamine oxidase (DAO)/on a nylon-net membrane, using glutaraldehyde	Platinum electrode/ +0.650 V vs. Ag/AgCl	–
Mitsubayashi *et al.* [89]	Trimethylamine	Mackerel	Flavin-containing monooxygenase type-3 (FMO3)/in polyvinyl alcohol membrane containing stilbazolium groups. Covered with a Nylon membrane	Oxygen electrode	–
Shin *et al.* [90]	Octopine	Scallop	Octopine dehydrogenase (ODH)/ODH-bonded beads were packed into a polypropylene reactor tube/ pyruvate oxidase/in a cellulose triacetate, glutaraldehyde and 1,8-diamino-4-aminomethyl octane membrane	Oxygen electrode	–
Campanella *et al.* [91]	Hydroperoxide	Extra virgin olive oil (rancidification process)	Catalase/in kappa-carrageenan gel	Oxygen electrode	–

Analytical characteristics				Observations
Dynamic range	Sensitivity	Repeatability	Speed of response	
		RSD = 2–3% ($n = 4$) for the assay	Each assay could be done in 5 min	Sample treated by alkaline phosphatase (AlP) for 5–10 min at 45 C, to measure together Hx, HxR and IMP. The immobilised enzymes were fairly stable for at least 3 months at 4 C. More than 200–300 samples could be analysed in about 1 month by using these enzyme reactors, provided the disposable screen-printed electrode is changed every 30–40 real samples.
Up to 3 mM	LOD: 0.05 mM	RSD = 4.9% ($n = 15$, 1.1 mM) RSD = 3.9% ($n = 15$, 3.0 mM)	One assay was completed within 4 min	The immobilised enzymes were stable for 2 months at 4 C and more than 150 samples could be continually determined.
Up to 40 mM	LOD: 0.05 mM		10 s	The current decrease was stable ($p > 0.003$) for more than 80 assays. 15 min for sample preparation and one assay was completed within 5 min.
1×10^{-6}–5×10^{-5} M	LOD: 5×10^{-7} M			
1.0–50.0 mmol L^{-1}		RSD = 4.39%, $n = 5$		
Up to 40 mM			2 min	The current was reproducible for more than 50 assays. When the enzyme membrane was stored at 4 C, no appreciable decrease of the current was observed for more than 30 days. Each assay could be performed within 5 min.
0.03–0.2 mM (H_2O_2, in toluene)		RSD ≤ 10 ($n = 5$)	15 min	The biosensor operates in organic media, such as *n*-decane, toluene or chloroform. Lifetime ≈ 15 days.

TABLE 13.11

Other analytes biosensors

Ref.	Analyte	Matrix	Enzyme/immobilisation method	Electrode configuration/ applied potential	Mediator
Campanella *et al.* [92]	Lecithin	Soya seed flour, soya seed oil, ground soya seed oil, ground maize germ oil, milk chocolate, soya lecithin in gel capsules, soya lecithin in tablets	Phospholipase D and choline oxidase/both immobilised in kappa-carrageenan gel or in dialysis membrane	Amperometric gas diffusion electrode for oxygen	–
Pati *et al.* [93]	Choline	Milk, milk powder and soy lecithin hydrolysates	Choline oxidase/ immobilised onto an electropolymerised polypyrrole film	Pt electrode	–
Wu *et al.* [94]	Glucosinolates	Seeds of commonly consumed vegetables	Myrosinase and glucose oxidase/onto an eggshell membrane	Oxygen electrode	–
Peña *et al.* [95]	Free and total cholesterol	Butter Lard Egg yolk	Cholesterol oxidase and horseradish peroxidase/ physically entrapped in the composite electrode matrix together with the mediator	Graphite–70% Teflon electrode/+0.1 V vs. Ag/ AgCl	Potassium ferrocyanide

Analytical characteristics				Observations
Dynamic range	Sensitivity	Repeatability	Speed of response	
1.1–66.1 mg L^{-1}	LOD: 0.55 mg L^{-1}	7.6%		– Operate in a chloroform–hexane–methanol (1+1+0.02 by volume) mixture: for sample dissolving. – Lifetime: 3 days.
	LOD: 0.12 μM			– Flow injection
2.5×10^{-5}–7.5×10–4 mol L^{-1} glucosinolates		0.50 mmol L^{-1} sinigrin standard (RSD = 4.15%, $n = 6$)	100 s	Common matrix interferents such as tartaric acid, glycine, oxalic acid, succinic acid, DL-α-alanine, adipic acid, DL-cysteine, calcium chloride and sodium chloride did not show significant interference. 85% of its initial value over 3 months of storage.
1.0×10^{-5}–3.0×10^{-3} mol L^{-1}	LOD: 6.2×10^{-6} mol L^{-1}	RSD = 5.0% ($n = 10$, same surface) RSD = 6.4% ($n = 10$, after polishing the electrode surface) RSD = 9.7% ($n = 10$, different biosensors)	5 min	Compatibility with predominantly nonaqueous media. Reversed micelles working medium: ethyl acetate as continuous phase (in which cholesterol is soluble), a 4% of 0.05 mol L^{-1} phosphate buffer solution, pH 7.4, as dispersed phase, and 0.1 mol L^{-1} AOT as emulsifying agent. Lifetime of one biosensor: 1 month.

TABLE 13.12
Multidetection biosensors

Ref.	Analyte	Matrix	Enzyme/immobilisation method	Electrode configuration/ applied potential	Mediator
Guzman-Vázquez de Prada *et al.* [96]	Methanol Ethanol Glucose Lactate (in the same assay)	Wine	Glucose oxidase (GOx), lactate oxidase (LOx), alcohol oxidase (AOx) and peroxidase (HRP)/ physically entrapped in the graphite–Teflon composite matrix	Ferrocene–graphite–Teflon composite electrode/0 V vs. Ag/AgCl	Ferrocene
Palmisano *et al.* [97]	Glucose L-Lactate	Tomato juice	Lactate oxidase or glucose oxidase/entrapped in gel, obtained by glutaraldehyde co-cross-linking with bovine serum albumin. This was deposited on an electrosynthesised overoxidised polypyrrole (PPYox) anti-interference membrane	Pt-modified dual disc electrode (parallel configuration)	–
Miertus *et al.* [98]	Glucose Fructose Ethanol Lactate Malate Sulfite	Wine	(1) Glucose/fructose/ ethanol multibiosensor: glucose oxidase, fructose dehydrogenase and alcohol dehydrogenase surface-modified enzyme electrode. (2) L-Lactate/L-malate/ sulfite multibiosensor: L-lactate dehydrogenase/L-malate dehydrogenase/ sulfite oxidase surface-modified enzyme electrodes/enzymes were deposited on the composite electrodes and covered with a dialysis membrane	Carbon paste electrodes (modified with mediator or NAD^+ cofactor)/+0.20 V vs. Ag/AgCl	With ferrocene for oxidase electrodes
Jawaheer *et al.* [12]	β-D-Glucose α-D-Glucose Sucrose	Pineapple	Glucose oxidase (GOx) invertase, mutarotase/with CA membrane	Thick-film screen-printed rhodinised carbon electrode/+350 mV vs. Ag/ AgCl	–
Campuzano *et al.* [99]	Glucose Fructose	Honey, cola soft drink, apple juice	Glucose oxidase (GOD) and fructose dehydrogenase (FDH)/with glutaraldehyde	3-Mercaptopropionic acid (MPA) self-assembled monolayer (SAM) modified gold disc electrode/+0.1 V vs. Ag/AgCl	Tetrathiafulvalene (TTF)/with glutaraldehyde
Menzel *et al.* [19]	Glucose, maltose, maltotriose and other oligosaccharide	Sugar syrup	Co-immobilised β-amylase, amyloglucosidase, mutarotase, glucose oxidase and peroxidase	Potentiometric, fluoride sensitive (pF), buffer capacity insensitive electrolyte isolator semiconductor capacitor (pF-EIS-CAP) chip	
Lvova *et al.* [100]	Caffeine, catechines, sugar, amino acid L-arginine	Natural coffee, black tea and different sorts of green teas	Glucose oxidase/with glutaraldehyde and BSA, then covered with an aromatic polyurethane membrane	Carbon paste screen-printed electrodes/ potentiometric (different active components on each of 30 sensors to construct an array)	Iron hexacyanoferrate Prussian Blue, $Fe^{III}_4[Fe^{II}(CN)_6]_3$

Analytical characteristics				Observations
Dynamic range	Sensitivity	Repeatability	Speed of response	
Methanol: $(0.5–8) \times 10^{-4}$ mol L^{-1} Ethanol: $(2.0–128) \times 10^{-4}$ mol L^{-1} Glucose: $(1.0 \pm 32) \times 10^{-4}$ mol L^{-1} Lactate: $(0.05–1.0) \times 10^{-4}$ mol L^{-1}	LOD: Methanol: 4.0×10^{-5} mol L^{-1} Ethanol: 4.0×10^{-5} mol L^{-1} Glucose: 5.0×10^{-5} mol L^{-1} Lactate: 1.4×10^{-6} mol L^{-1}	RSD: 0.8–3.8%	2 min	Enzymes were co-immobilised in the same electrode and the assay was performed after HPLC separation, or enzymes were immobilised in different biosensors placed in parallel, and detection was performed simultaneously with no separation step. Surface regeneration by polishing.
Glucose: up to 100 mM Lactate: up to 20 mM		RSD = 5.8% (n = 15) for glucose RSD = 5.3% (n = 15) for lactate		High stability of the biosensors. After 2 days of continuous use the responses were around 75% and 86% of the initial ones, for glucose and lactate, respectively. Response remained constant for 60 days of consecutive use.
Glucose: 0.03–15 mM Fructose: 0.01–10 mM Ethanol: 0.014–4 mM Lactate: 0.011–1.5 mM Malate: 0.015–1.5 mM Sulfite: 0.01–0.1 mM	LOD: Glucose: 0.03 mM Fructose: 0.01 mM Ethanol: 0.014 mM Lactate: 0.011 mM Malate: 0.015 mM Sulfite: 0.01 mM		60–120 s	Two hybrid three-channel multibiosensors: disposable biosensors.
Up to 35 mM glucose				– Ascorbate oxidase is added to the sensor for ascorbic acid interference elimination. – Pectin is also added for enzyme stabilisation. – Only preliminary work with real samples
1.0×10^{-5}–1.0×10^{-2} M (glucose) 1.0×10^{-5}–1.0×10^{-3} M (fructose)	LOD: 5.7×10^{-6} M (glucose) 2.5×10^{-6} M (fructose)	Responses from different electrodes: RSD = 9.1% (glucose), 9.2% (fructose)		Simultaneous determination of glucose and fructose after HPLC separation. Operational stability: 388 h (approximately 16 days) for fructose, and 96 h (4 days) for glucose
Glucose: 10^{-4}–10^{-2} M				All-solid-state "electronic tongue" microsystem.

TABLE 13.13

Process control applications of enzyme electrochemical biosensors

Ref.	Analyte	Matrix	Enzyme/immobilisation method	Electrode configuration/ applied potential	Mediator
Niculescu *et al.* [101]	Ethanol	Wine production	Alcohol dehydrogenase (QH-ADH)/integrated in redox hydrogel poly(ethyleneglycol)-diglycidyl ether (PEGDGE) as the cross-linking agent	Screen-printed (SP) electrodes/+300 mV vs. Ag/AgCl	Os-complex-modified poly(vinylimidazole) redox polymer
Tkac *et al.* [16]	Ethanol	Off-line monitoring of *S. cerevisiae* batch fermentation	Alcohol oxidase (AOx) in *Gluconobacter oxydans* cells/covered with a cellulose acetate (CA) membrane	Glassy carbon electrode/ +300 mV vs. SCE	Ferricyanide
Çökeliler and Mutlu [6]	Ethanol	Beer	Alcohol oxidase (AOx)/on polycarbonate membrane prepared by plasma polymerisation technique	Pt electrode/+650 mV vs. Ag/AgCl	–
Menzel *et al.* [19]	Glucose Ethanol Phosphate	*Saccharomyces cerevisiae*, *Acremonium chrysogenum* and recombinant baby hamster kidney (rBHK) cell cultivations	Co-immobilised glucose oxidase and peroxidase Co-immobilised alcohol oxidase and peroxidase Co-immobilised nucleoside phosphorylase, xanthin oxidase and peroxidase	Potentiometric, fluoride sensitive (pF), buffer capacity insensitive electrolyte isolator semiconductor capacitor (pF-EIS-CAP) chip	–
Compagnone *et al.* [5]	Glycerol	Alcoholic fermentation	Glycerokinase (GK) and glycerol-3-phosphate oxidase (GPO)/ covalently immobilised (Bioreactor of GK and bioelectrode of GPO)	Pt-based hydrogen peroxide probe/+650 mV vs. Ag/AgCl	–

Analytical characteristics				Observations
Dynamic range	Sensitivity	Repeatability	Speed of response	
1–250 μM	0.336 ± 0.025 mA M^{-1} cm^{-2}, LOD: 1 μM			Integration into an automated sequential-injection analyser. Successful on-line monitoring of ethanol during wine fermentation processes. No signal decrease over an analysis time of > 100 h under continuous injection of a 500 μM ethanol standard solution.
2–270 μM	3.5 μA mM^{-1}, LOD: 0.85 μM	RSD = 1.79% (10 μM ethanol solution, $n = 30$)	13 s (90% of the signal)	– Size exclusion effect of cellulose acetate membrane – Monitoring of ethanol concentration during 10 h (the sensitivity remained unchanged for 8.5 h of continuous use) – 50% of initial sensitivity after 8 days storage at 20 C
Up to 2 mM	5.6 nA mM^{-1}		50 s	Beer samples were degassed
Glucose: 0.1–2.0 g L^{-1}			12–20 s	The (Bio-pF-EIS-CAP) sensors were integrated in flow injection analysis systems. The medium components were monitored on-line. Satisfactory agreement between on- and off-line measurements (within 10% of mean relative deviation). Long range stability (many weeks to months depending on number of assays).
2×10^{-6}–10^{-3} mol L^{-1}	LOD: 5×10^{-7} mol L^{-1}	RSD = 1–2% (3 mM with ATP and Mg^{2+} in the working buffer)		– FIA system. – Immobilisation of glycerokinase in a glass beads reactor coupled with glycerol-3-phosphate oxidase on a preactivated Immobilon AV membrane kept at the electrode surface. – Lifetime of the glycerol-3-phosphate membrane extended up to 1 month by storage in the working buffer containing 1% DEAE-dextran and 5% lactitol. More than 350 samples can be assayed.

TABLE 13.13 (*continued*)

Ref.	Analyte	Matrix	Enzyme/immobilisation method	Electrode configuration/ applied potential	Mediator
Esti *et al.* [8]	L-Lactate L-Malate	Micro-malolactic fermentation in red wine	Lactate oxidase/in a nylon membrane with glutaraldehyde or $NAD(P)^+$-dependent L-malate dehydrogenase oxaloacetate decarboxylating/ immobilised in an aminopropyl glass beads reactor	Platinum electrode/ +650 mV vs. Ag/AgCl	Phenazine methosulfate (for malate sensor)
Chen and Su [102]	L-Glutamate	L-Glutamic acid fermentation	L-Glutamate oxidase/ onto a cellulose triacetate membrane	Oxygen electrode	–
Ferreira *et al.* [103]	Lactose	Yeast *Kluyveromyces marxianus* fermentation	β-Galactosidase and glucose oxidase	Oxygen electrode	–
Zaydan *et al.* [104]	L-Lactate	Milk *Streptococcus thermophilus* fermentation	L-Lactate oxidase and peroxidase	Carbon paste/–100 mV vs. Ag/AgCl	Ferrocene
Min *et al.* [17]	Glucose L-Lactate	*Lactococcus lactis* (ATCC 19435) fermentation	Glucose oxidase (GOx) or L-lactate oxidase (LOx) with horseradish peroxidase (HRP)/in a carbon paste matrix with polyethyleneimine and a polyester sulfonic acid cation exchanger (Eastman AQ-29D)	Carbon paste electrode/–50 mV vs. Ag/ AgCl	–
Serra *et al.* [105]	Alkaline phosphatase	Milk (process evaluation: pasteurisation)	Tyrosinase/physically included in the composite matrix	Graphite–Teflon composite electrode/–0.10 V vs. Ag/ AgCl	–

Analytical characteristics				Observations
Dynamic range	Sensitivity	Repeatability	Speed of response	
Lactate: 5×10^{-6}–10^{-3} M Malate: 1×10^{-5}–10^{-4} M	LOD: Lactate: 2×10^{-6} M Malate: 3×10^{-6} M	2.6–3.9%	3 min (lactic acid) 3.5 min (malic acid)	Off-line FIA measurements. For malate sensor $NADP^+$ and mediator are added in solution. For L-malic acid the response of the probe decreased by about 10% after 150 sample injections, for L-lactic acid it decreased by 35%.
0.12–0.84 mM			<3 min	The sensor retained 95% of its original activity after 400 assays over a period of 3 weeks.
Up to 30 g lactose L^{-1}				A glucose sensor was also used for subtracting glucose interference.
0.01–0.1 g L^{-1} for glucose 0.01–0.05 g L^{-1} for L-lactic acid				Microdialysis sampling. Sampling frequency of 15 h^{-1}. Around 20% decrease in the signal in 20 h of continuous operation for both the glucose and L-lactate electrodes. During the on-line monitoring of the fermentation process standard injections were made intermittently to check the decrease in the responses of the electrodes. The decrease in the responses were taken into account.
2.0×10^{-13}–2.5×10^{-11} M	LOD: 6.7×10^{-14} M	RSD = 8.4% (n = 10, 2.0×10^{-13}) RSD = 5.6% (n = 10, 2.5×10^{-11} M)	5 min	Phenyl phosphate is added to the working solution.

13.3 PROCESS CONTROL APPLICATIONS

Systems of continuous monitoring of industrial processes allow real-time detection of possible errors in the chain production to be rectified in an immediate manner. Besides the continuous control of parameters such as pH, temperature, pressure, etc., biosensors for in-line or at least on-line analysis may also find several applications in the food industry to facilitate the quality control of products during processing. In process control applications, biosensors offer two main advantages, high selectivity and rapid response in comparison with other analytical tools. For example, fermentation processes are widely employed in numerous applications. They are dynamic events in which substances are consumed by microorganisms so that new products are continuously produced. Monitoring of these changes provides important information on media composition, and helps understanding of the dynamic behaviour of the process, thus enabling on-line control and optimisation of such a process which will be translated to costs reduction and pollution impact minimisation in the food industry. However, constrictions of some biosensor technologies associated with stability, useful lifetime under continuous operation and robustness of biosensing devices have limited, up to date, to a few cases their usefulness for process monitoring. Some selected examples of enzyme electrochemical biosensors in this field are summarised in Table 13.13.

13.4 CONCLUSIONS AND SOME REMARKS FROM THE COMMERCIAL POINT OF VIEW

Even though the huge literature on the topic few biosensors are being currently used for food quality control in the food industry. This fact can be explained by several reasons. First of all, the food market, although very large, has, in general, low profit margins and huge competition. Consequently, this industry cannot afford to invest in modern analytical methods at the same level as other high-technology sectors, such as clinical or pharmaceutical. Moreover, biosensors in food industry still have to compete with other analytical methods in terms of cost, performance and reliability.

In general, the commercially available biosensors in the food industry are based on a very similar technology, either an oxygen electrode or a hydrogen peroxide electrode in conjunction with an immobilised oxidase. They are available in several forms, such as autoanalysers, manual laboratory instruments and portable devices. Nevertheless, screen-printed technology has been one of the most revolutionary improvements in electrochemical biosensors, allowing inexpensive mass fabrication, which lends itself to an increasingly price competitive industry, such as the fresh products sector. Given this fact, it is perhaps surprising that this biosensor fabrication technique has not been transferred to the measurement of important target analytes in the food industry yet.

However, taking into account all the considerations mentioned above, biosensors in food industry still require to overcome several technical limitations. For example, biosensors cannot be sterilised and function only within a limited concentration range. In addition, recalibration and/or replacement of their biocomponents must be done frequently, and sensor fouling still represents the main drawback for using biosensors *in situ* or for real sample analysis. Nonetheless, microdialysis has been coupled to biosensors as an interesting improvement of their capabilities of use *in situ*. New approaches have to be developed benefiting the incipient protein engineering technology, reaching the third generation biosensor design in which direct wiring between enzyme and electrode can be obtained.

As a conclusion it can be said that, although some research is still necessary to substantially improve their robustness, biosensors are powerful analytical tools that probably will introduce important changes in the analysis of foodstuffs in a short period of time. Applications of biosensor to food industry comprise the whole food chain, from the primary production to the final distribution to the consumer, which implies an enormous potential of application to food traceability. Finally, it should be remarked that biosensors are not a substitution technology of the conventional analytical techniques, but they represent alternatives for specific demands as a response to specific problems. This specificity of the biosensor purpose allows the "*à la carte*" design of devices, by combining different technologies involved in the recognition elements, the immobilisation methodologies and the signal transduction systems.

REFERENCES

1 L. Campanella, A. Bonanni, E. Finotti and M. Tomassetti, Biosensors for determination of total and natural antioxidant capacity of red and white wines: comparison with other spectrophotometric and fluorimetric methods, *Biosens. Bioelectron.*, 19(7) (2004) 641–651.

2 Y. Inaba, K. Mizukami, N. Hamada-Sato, T. Kobayashi, C. Imada and E. Watanabe, Development of a D-alanine sensor for the monitoring of a fermentation using the improved selectivity by the combination of D-amino acid oxidase and pyruvate oxidase, *Biosens. Bioelectron.*, 19(5) (2003) 423–431.
3 H. Okuma and E. Watanabe, Flow system for fish freshness determination based on double multi-enzyme reactor electrodes, *Biosens. Bioelectron.*, 17(5) (2002) 367–372.
4 N. Sato, K. Usui and H. Okuma, Development of a bienzyme reactor sensor system for the determination of ornithine, *Anal. Chim. Acta*, 456(2) (2002) 219–226.
5 D. Compagnone, M. Esti, M.C. Messia, E. Peluso and G. Palleschi, Development of a biosensor for monitoring of glycerol during alcoholic fermentation, *Biosens. Bioelectron.*, 13(7–8) (1998) 875–880.
6 D. Çökeliler and M. Mutlu, Performance of amperometric alcohol electrodes prepared by plasma polymerization technique, *Anal. Chim. Acta*, 469(2) (2002) 217–223.
7 R. Draisci, G. Volpe, L. Lucentini, A. Cecilia, R. Federico and G. Palleschi, Determination of biogenic amines with an electrochemical biosensor and its application to salted anchovies, *Food Chem.*, 62(2) (1998) 225–232.
8 M. Esti, G. Volpe, L. Micheli, E. Delibato, D. Compagnone, D. Moscone and G. Palleschi, Electrochemical biosensors for monitoring malolactic fermentation in red wine using two strains of *Oenococcus oeni*, *Anal. Chim. Acta*, 513(1) (2004) 357–364.
9 N.G. Patel, S. Meier, K. Cammann and G.-C. Chemnitius, Screen-printed biosensors using different alcohol oxidases, *Sens. Actuators B Chem.*, 75(1–2) (2001) 101–110.
10 N. Garcia-Villar, J. Saurina and S. Hernandez-Cassou, Potentiometric sensor array for the determination of lysine in feed samples using multivariate calibration methods, *Fresenius J. Anal. Chem.*, 371(7) (2001) 1001–1008.
11 M. Boujtita, J.P. Hart and R. Pittson, Development of a disposable ethanol biosensor based on a chemically modified screen-printed electrode coated with alcohol oxidase for the analysis of beer, *Biosens. Bioelectron.*, 15(5–6) (2000) 257–263.
12 S. Jawaheer, S.F. White, S.D.D.V. Rughooputh and D.C. Cullen, Development of a common biosensor format for an enzyme based biosensor array to monitor fruit quality, *Biosens. Bioelectron.*, 18(12) (2003) 1429–1437.
13 F. Ricci, A. Amine, C.S. Tuta, A.A. Ciucu, F. Lucarelli, G. Palleschi and D. Moscone, Prussian Blue and enzyme bulk-modified screen-printed electrodes for hydrogen peroxide and glucose determination with improved storage and operational stability, *Anal. Chim. Acta*, 485(1) (2003) 111–120.

14 S. Ghosh (Hazra), D. Sarker and T.N. Misra, Development of an amperometric enzyme electrode biosensor for fish freshness detection, *Sens. Actuators B Chem.*, 53(1–2) (1998) 58–62.
15 P.A. Paredes, J. Parellada, V.M. Fernández, I. Katakis and E. Domínguez, Amperometric mediated carbon paste biosensor based on D-fructose dehydrogenase for the determination of fructose in food analysis, *Biosens. Bioelectron.*, 12(12) (1998) 1233–1243.
16 J. Tkac, I. Vostiar, L. Gorton, P. Gemeiner and E. Sturdik, Improved selectivity of microbial biosensor using membrane coating. Application to the analysis of ethanol during fermentation, *Biosens. Bioelectron.*, 18(9) (2003) 1125–1134.
17 R.W. Min, V. Rajendran, N. Larsson, L. Gorton, J. Planas and B. Hahn-Hägerdal, Simultaneous monitoring of glucose and L-lactic acid during a fermentation process in an aqueous two-phase system by on-line FIA with microdialysis sampling and dual biosensor detection, *Anal. Chim. Acta*, 366(1–3) (1998) 127–135.
18 M.A. Del Cerro, G. Cayuela, A.J. Reviejo, J.M. Pingarrón and J. Wang, Graphite-teflon-peroxidase composite electrodes. Application to the direct determination of glucose in musts and wines, *Electroanalysis*, 9(14) (1997) 1113–1119.
19 C. Menzel, T. Lerch, K. Schneider, R. Weidemann, C. Tollnick, G. Kretzmer, T. Scheper and K. Schüger, Application of biosensors with an electrolyte isolator semiconductor capacitor (EIS-CAP) transducer for process monitoring, *Process Biochem.*, 33(2) (1998) 175–180.
20 C. Loechel, A. Basran, J. Basran, N. S. Scrutton and E. A. Hall, Using trimethylamine dehydrogenase in an enzyme linked amperometric electrode. Part 1. Wild-type enzyme redox mediation, *Analyst*, 128(2) (2003) 166–172; Part 2. Rational design engineering of a ''wired'' mutant, *Analyst*, 128(7) (2003) 889–898.
21 J. Wang, E. González-Romero and M. Ozsoz, Renewable alcohol biosensors based on alcohol-dehydrogenase/nicotinamide-adenine-dinucleotide graphite epoxy electrodes, *Electroanalysis*, 4 (1992) 539–544.
22 M.J. Lobo Castañón, A.J. Miranda Ordieres and P. Tuñón Blanco, Amperometric detection of ethanol with poly-(*o*-phenylenediamine)-modified enzyme electrodes, *Biosens. Bioelectron.*, 12(6) (1997) 511–520.
23 M.J. Lobo-Castañón, A.J. Miranda-Ordieres and P. Tuñón-Blanco, A bienzyme-poly-(*o*-phenylenediamine)-modified carbon paste electrode for the amperometric detection of L-lactate, *Anal. Chim. Acta*, 346(2) (1997) 165–174.
24 A. Avramescu, T. Noguer, V. Magearu and J.-L. Marty, Chronoamperometric determination of D-lactate using screen-printed enzyme electrodes, *Anal. Chim. Acta*, 433(1) (2001) 81–88.

25 J. Katrlík, J. Svorc, M. Stred'anský and S. Miertus, Composite alcohol biosensors based on solid binding matrix, *Biosens. Bioelectron.*, 13(2) (1998) 181–191.
26 B. Leca and J.-L. Marty, Reagentless ethanol sensor based on a NAD-dependent dehydrogenase, *Biosens. Bioelectron.*, 12(11) (1997) 1083–1088.
27 R.C.H. Kwan, P.Y.T. Hon, K.K.W. Mak and R. Renneberg, Amperometric determination of lactate with novel trienzyme/poly(carbamoyl) sulfonate hydrogel-based sensor, *Biosens. Bioelectron.*, 19(12) (2004) 1745–1752.
28 J. Razumiene, M. Niculescu, A. Ramanavicius, V. Laurinavicius and E. Csöregi, Direct bioelectrocatalysis at carbon electrodes modified with quinohemoprotein alcohol dehydrogenase from *Gluconobacter* sp. 33, *Electroanalysis*, 14 (2002) 43–49.
29 S. Campuzano, R. Galvez, M. Pedrero, F.J. Manuel de Villena and J.M. Pingarron, An integrated electrochemical fructose biosensor based on tetrathiafulvalene-modified self-assembled monolayers on gold electrodes, *Anal. Bioanal. Chem.*, 377(4) (2003) 600–607.
30 A. Eftekhari, Glycerol biosensor based on glycerol dehydrogenase incorporated into polyaniline modified aluminum electrode using hexacyanoferrate as mediator, *Sens. Actuators B Chem.*, 80(3) (2001) 283–289.
31 M. Niculescu, R. Mieliauskiene, V. Laurinavicius and E. Csöregi, Simultaneous detection of ethanol, glucose and glycerol in wines using pyrroloquinoline quinone-dependent dehydrogenases based biosensors, *Food Chem.*, 82(3) (2003) 481–489.
32 M.I.P. Oliveira, M.C. Pimentel, M.C.B.S.M. Montenegro, A.N. Araújo, M.F. Pimentel and V.L. da Silva, L-Glutamate determination in food samples by flow-injection analysis, *Anal. Chim. Acta*, 448(1–2) (2001) 207–213.
33 R.R. Walters, P.A. Johnson and R.P. Buck, Histidine ammonia-lyase enzyme electrode for determination of L-histidine, *Anal. Chem.*, 52 (1980) 1684–1690.
34 F. Mizutani, Y. Hirata, S. Yabuki and S. Iijima, Flow injection analysis of acetic acid in food samples by using trienzyme/poly(dimethylsiloxane)-bilayer membrane-based electrode as the detector, *Sens. Actuators B Chem.*, 91(1–3) (2003) 195–198.
35 M.I. Prodromidis, S.M. Tzouwara-Karayanni, M.I. Karayannis and P.M. Vadgama, Bioelectrochemical determination of citric acid in real samples using a fully automated flow injection manifold, *Analyst*, 122(10) (1997) 1101–1106.
36 J. Razumiene, V. Gureviciene, V. Laurinavicius and J.V. Grazulevicius, Amperometric detection of glucose and ethanol in beverages using flow cell and immobilised on screen-printed carbon electrode PQQ-dependent

glucose or alcohol dehydrogenases, *Sens. Actuators B Chem.*, 78(1–3) (2001) 243–248.
37 X. Llopis, A. Merkoçi, M. del Valle and S. Alegret, Integration of a glucose biosensor based on an epoxy-graphite-TTF·TCNQ-GOD biocomposite into a FIA system, *Sens. Actuators B Chem.*, 107(2) (2005) 742–748.
38 M. Badea, A. Curulli and G. Palleschi, Oxidase enzyme immobilisation through electropolymerised films to assemble biosensors for batch and flow injection analysis, *Biosens. Bioelectron.*, 18(5–6) (2003) 689–698.
39 S. Campuzano, R. Gálvez, M. Pedrero, F.J. Manuel de Villena and J.M. Pingarrón, Preparation, characterization and application of alkanethiol self-assembled monolayers modified with tetrathiafulvalene and glucose oxidase at a gold disk electrode, *J. Electroanal. Chem.*, 526(1–2) (2002) 92–100.
40 E. Turkusic, J. Kalcher, E. Kahrovic, N.W. Beyene, H. Moderegger, E. Sofic, S. Begic and K. Kalcher, Amperometric determination of bonded glucose with an MnO_2 and glucose oxidase bulk-modified screen-printed electrode using flow-injection analysis, *Talanta*, 65(2) (2005) 559–564.
41 I. Luiz de Mattos and M. Carneiro da Cunha Areias, Automated determination of glucose in soluble coffee using Prussian Blue–glucose oxidase–Nafion® modified electrode, *Talanta*, 66(5) (2005) 1281–1286.
42 D. Ivekovic, S. Milardovic and B.S. Grabaric, Palladium hexacyanoferrate hydrogel as a novel and simple enzyme immobilization matrix for amperometric biosensors, *Biosens. Bioelectron.*, 20(4) (2004) 872–878.
43 E.M. Gloria, C.F. Ciacco, J.F. Lopes Filho, C. Ericsson and S.S. Zochi, Influence of low and high levels of grain defects on maize wet milling, *J. Food Eng.*, 55(4) (2002) 359–365.
44 S.K. Sharma, R. Singhal, B.D. Malhotra, N. Sehgal and A. Kumar, Lactose biosensor based on Langmuir–Blodgett films of poly(3-hexyl thiophene), *Biosens. Bioelectron.*, 20(3) (2004a) 651–657.
45 S.K. Sharma, R. Singhal, B.D. Malhotra, N. Sehgal and A. Kumar, Langmuir–Blodgett film based biosensor for estimation of galactose in milk, *Electrochim. Acta*, 49(15) (2004b) 2479–2485.
46 I. Eshkenazi, E. Maltz, B. Zion and J. Rishpon, A three-cascaded-enzymes biosensor to determine lactose concentration in raw milk, *J. Dairy Sci.*, 83(9) (2000) 1939–1945.
47 J. Tkac, E. Sturdik and P. Gemeiner, Novel glucose non-interference biosensor for lactose detection based on galactose oxidase-peroxidase with and without co-immobilised beta-galactosidase, *Analyst*, 125(7) (2000) 1285–1289.
48 F. Amárita, C. Rodríguez Fernández and F. Alkorta, Hybrid biosensors to estimate lactose in milk, *Anal. Chim. Acta*, 349(1–3) (1997) 153–158.

49 D. Moscone, R.A. Bernardo, E. Marconi, A. Amine and G. Palleschi, Rapid determination of lactulose in milk by microdialysis and biosensors, *Analyst*, 124(3) (1999) 325–329.
50 Y. Sekine and E.A.H. Hall, A lactulose sensor based on coupled enzyme reactions with a ring electrode fabricated from tetrathiafulvalen–tetracyanoquinodimetane, *Biosens. Bioelectron.*, 13(9) (1998) 995–1005.
51 E. Maestre, I. Katakis and E. Domínguez, Amperometric flow-injection determination of sucrose with a mediated tri-enzyme electrode based on sucrose phosphorylase and electrocatalytic oxidation of NADH, *Biosens. Bioelectron.*, 16(1–2) (2001) 61–68.
52 S. Campuzano, M. Pedrero, F.J. Manuel de Villena and J.M. Pingarrón, A lactulose bienzyme biosensor based on self-assembled monolayer modified electrodes, *Electroanalysis*, 16(17) (2004) 1385–1392.
53 Q. Yao, S. Yabuki and F. Mizutani, Preparation of a carbon paste/alcohol dehydrogenase electrode using polyethylene glycol-modified enzyme and oil-soluble mediator, *Sens. Actuators B Chem.*, 65(1–3) (2000) 147–149.
54 A. Morales, F. Céspedes, E. Martínez-Fàbregas and S. Alegret, Ethanol amperometric biosensor based on an alcohol oxidase-graphite-polymer biocomposite, *Electrochim. Acta*, 43(23) (1998) 3575–3579.
55 E. Akyilmaz and E. Dinçkaya, An amperometric microbial biosensor development based on *Candida tropicalis* yeast cells for sensitive determination of ethanol, *Biosens. Bioelectron.*, 20(7) (2005) 1263–1269.
56 L. Rotariu, C. Bala and V. Magearu, New potentiometric microbial biosensor for ethanol determination in alcoholic beverages, *Anal. Chim. Acta*, 513(1) (2004) 119–123.
57 N. Peña, R. Tárrega, A.J. Reviejo and J.M. Pingarrón, Reticulated vitreous carbon-based composite bienzyme electrodes for the determination of alcohols in beer samples, *Anal. Lett.*, 35(12) (2002) 1931–1944.
58 A. Guzmán-Vázquez de Prada, N. Peña, M.L. Mena, A.J. Reviejo and J.M. Pingarrón, Graphite–Teflon composite bienzyme amperometric biosensors for monitoring of alcohols, *Biosens. Bioelectron.*, 18(10) (2003) 1279–1288.
59 M.M.F. Choi, Application of a long shelf-life biosensor for the analysis of L-lactate in dairy products and serum samples, *Food Chem.*, 92(3) (2005) 575–581.
60 B. Serra, A.J. Reviejo, C. Parrado and J.M. Pingarrón, Graphite–Teflon composite bienzyme electrodes for the determination of L-lactate: application to food samples, *Biosens. Bioelectron.*, 14(5) (1999) 505–513.
61 A.A.J. Torriero, E. Salinas, F. Battaglini and J. Raba, Milk lactate determination with a rotating bioreactor based on an electron transfer mediated by osmium complexes incorporating a continuous-flow/stopped-flow system, *Anal. Chim. Acta*, 498(1–2) (2003) 155–163.
62 K. Kriz, L. Kraft, M. Krook and D. Kriz, Amperometric determination of L-lactate based on entrapment of lactate oxidase on a transducer surface

with a semi-permeable membrane using a SIRE technology based biosensor. Application: tomato paste and baby food, *J. Agric. Food Chem.*, 50(12) (2002) 3419–3424.
63 J. Katrlík, A. Pizzariello, V. Mastihuba, J. Svorc, M. Stred'anský and S. Miertus, Biosensors for L-malate and L-lactate based on solid binding matrix, *Anal. Chim. Acta*, 379(1–2) (1999) 193–200.
64 N. Gajovic, A. Warsinke and F.W. Scheller, A bienzyme electrode for L-malate based on a novel and general design, *J. Biotechnol.*, 61(2) (1998) 129–133.
65 M. Albareda-Sirvent and A.L. Hart, Preliminary estimates of lactic and malic acid in wine using electrodes printed from inks containing sol–gel precursors, *Sens. Actuators B Chem.*, 87(1) (2002) 73–81.
66 A. Lupu, D. Compagnone and G. Palleschi, Screen-printed enzyme electrodes for the detection of marker analytes during winemaking, *Anal. Chim. Acta*, 513(1) (2004) 67–72.
67 M. Kim and M.-J. Kim, Isocitrate analysis using a potentiometric biosensor with immobilized enzyme in a FIA system, *Food Res. Int.*, 36(3) (2003) 223–230.
68 S.C. Kelly, P.J. O'Connell, C.K. O'Sullivan and G.G. Guilbault, Development of an interferent free amperometric biosensor for determination of L-lysine in food, *Anal. Chim. Acta*, 412(1–2) (2000) 111–119.
69 S.L. Alvarez-Crespo, M.J. Lobo-Castañón, A.J. Miranda-Ordieres and P. Tuñón-Blanco, Amperometric glutamate biosensor based on poly(*o*-phenylenediamine) film electrogenerated onto modified carbon paste electrodes, *Biosens. Bioelectron.*, 12(8) (1997) 739–747.
70 N.F. Almeida and A.K. Mulchandani, A mediated amperometric enzyme electrode using tetrathiafulvalene and L-glutamate oxidase for the determination of L-glutamic acid, *Anal. Chim. Acta*, 282(2) (1993) 353–361.
71 R. Domínguez, B. Serra, A.J. Reviejo and J.M. Pingarrón, Chiral analysis of amino acids using electrochemical composite bienzyme biosensors, *Anal. Biochem.*, 298(2) (2001) 275–282.
72 M. Váradi, N. Adányi, E.E. Szabó and N. Trummer, Determination of the ratio of D- and L-amino acids in brewing by an immobilised amino acid oxidase enzyme reactor coupled to amperometric detection, *Biosens. Bioelectron.*, 14(3) (1999) 15335–15340.
73 P. Sarkar, I.E. Tothill, S.J. Setford and A.P. Turner, Screen-printed amperometric biosensors for the rapid measurement of L- and D-amino acids, *Analyst*, 124(6) (1999) 865–870.
74 M.I. Alvarez-Gonzalez, S.B. Saidman, M.J. Lobo-Castanon, A.J. Miranda-Ordieres and P. Tunon-Blanco, Electrocatalytic detection of NADH and glycerol by NAD(+)-modified carbon electrodes, *Anal. Chem.*, 72(3) (2000) 520–527.

75 L. Campanella, G. Favero, M.P. Sammartino and M. Tomassetti, Analysis of several real matrices using new mono-, bi-enzymatic, or inhibition organic phase enzyme electrodes, *Anal. Chim. Acta*, 393(1–3) (1999) 109–120.
76 L.D. Mello, A.A. Alves, D.V. Macedo and L.T. Kubota, Peroxidase-based biosensor as a tool for a fast evaluation of antioxidant capacity of tea, *Food Chem.*, 92(3) (2005) 515–519.
77 L.D. Mello, M.P. Taboada Sotomayor and L.T. Kubota, HRP-based amperometric biosensor for the polyphenols determination in vegetables extract, *Sens. Actuators B Chem.*, 96(3) (2003) 636–645.
78 C. Capannesi, I. Palchetti, M. Mascini and A. Parenti, Electrochemical sensor and biosensor for polyphenols detection in olive oils, *Food Chem.*, 71(4) (2000) 553–562.
79 S.A.S.S. Gomes, J.M.F. Nogueira and M.J.F. Rebelo, An amperometric biosensor for polyphenolic compounds in red wine, *Biosens. Bioelectron.*, 20(6) (2004) 1211–1216.
80 V. Carralero Sanz, M.L. Mena, A. González-Cortés, P. Yáñez-Sedeño and J.M. Pingarrón, Development of a tyrosinase biosensor based on gold nanoparticles-modified glassy carbon electrodes. Application to the measurement of a bioelectrochemical polyphenols index in wines, *Anal. Chim. Acta*, 528(1) (2005) 1–8.
81 B.R. Eggins, C. Hickey, S.A. Toft and D. Min Zhou, Determination of flavanols in beers with tissue biosensors, *Anal. Chim. Acta*, 347(3) (1997) 11281–11288.
82 G. Cayuela, N. Peña, A.J. Reviejo and J.M. Pingarrón, Development of a bienzymic graphite–Teflon composite electrode for the determination of hypoxanthine in fish, *Analyst*, 123(2) (1998) 371–377.
83 L. Agüí, J. Manso, P. Yáñez-Sedeño and J.M. Pingarrón, Amperometric biosensor for hypoxanthine based on immobilized xanthine oxidase on nanocrystal gold-carbon paste electrodes, *Sens. Actuators B*, 113(1) (2006) 272–280.
84 S. Hu, C. Xu, J. Luo, J. Luo and D. Cui, Biosensor for detection of hypoxanthine based on xanthine oxidase immobilized on chemically modified carbon paste electrode, *Anal. Chim. Acta*, 412(1–2) (2000) 55–61.
85 A.K. Basu, P. Chattopadhyay, U.R. Choudhury and R. Chakraborty, Development of an amperometric hypoxanthine biosensor for determination of hypoxanthine in fish and meat tissue, *Indian J. Exp. Biol.*, 43(7) (2005) 646–653.
86 J. Niu and J.Y. Lee, Bulk-modified amperometric biosensors for hypoxanthine based on sol–gel technique, *Sens. Actuators B Chem.*, 62(3) (2000) 190–198.
87 M.-A. Carsol and M. Mascini, Development of a system with enzyme reactors for the determination of fish freshness, *Talanta*, 47(2) (1998) 335–342.

88 S.J. Shin, H. Yamanaka, H. Endo and E. Watanabe, Development of ornithine biosensor and application to estimation of prawn freshness, *Anal. Chim. Acta*, 364(1–3) (1998) 159–164.
89 K. Mitsubayashi, Y. Kubotera, K. Yano, Y. Hashimoto, T. Kon, S. Nakakura, Y. Nishi and H. Endo, Trimethylamine biosensor with flavin-containing monooxygenase type 3 (FMO3) for fish-freshness analysis, *Sens. Actuators B Chem.*, 103(1–2) (2004) 463–467.
90 S.J. Shin, H. Yamanaka, H. Endo and E. Watanabe, Development of an octopine biosensor and its application to the estimation of scallop freshness, *Enzyme Microbial Technol.*, 23(1–2) (1998) 10–13.
91 L. Campanella, M.P. Sammartino, M. Tomassetti and S. Zannella, Hydroperoxide determination by a catalase OPEE: application to the study of extra virgin olive oil rancidification process, *Sens. Actuators B Chem.*, 76(1–3) (2001) 158–165.
92 L. Campanella, F. Pacifici, M.P. Sammartino and M. Tomassetti, A new organic phase bienzymatic electrode for lecithin analysis in food products, *Bioelectrochem. Bioenerg.*, 47(1) (1998) 25–38.
93 S. Pati, M. Quinto, F. Palmisano and P.G. Zambonin, Determination of choline in milk, milk powder, and soy lecithin hydrolysates by flow injection analysis and amperometric detection with a choline oxidase based biosensor, *J. Agric. Food Chem.*, 52(15) (2004) 4638–4642.
94 B. Wu, G. Zhang, S. Shuang, C. Dong, M.M.F. Choi and A.W.M. Lee, A biosensor with myrosinase and glucose oxidase bienzyme system for determination of glucosinolates in seeds of commonly consumed vegetables, *Sens. Actuators B Chem.*, 106(2) (2005) 700–707.
95 N. Peña, G. Ruiz, A.J. Reviejo and J.M. Pingarrón, Graphite-teflon composite bienzyme electrodes for the determination of cholesterol in reversed micelles. Application to food samples, *Anal. Chem.*, 73(6) (2001) 1190–1195.
96 A. Guzman-Vázquez de Prada, N. Peña, C. Parrado, A.J. Reviejo and J.M. Pingarrón, Amperometric multidetection with composite enzyme electrodes, *Talanta*, 62(5) (2003) 896–903.
97 F. Palmisano, R. Rizzi, D. Centonze and P.G. Zambonin, Simultaneous monitoring of glucose and lactate by an interference and cross-talk free dual electrode amperometric biosensor based on electropolymerized thin films, *Biosens. Bioelectron.*, 15(9–10) (2000) 531–539.
98 S. Miertus, J. Katrlík, A. Pizzariello, M. Stred'anský, J. Svitel and J. Svorc, Amperometric biosensors based on solid binding matrices applied in food quality monitoring, *Biosens. Bioelectron.*, 13(7–8) (1998) 911–923.
99 S. Campuzano, Ó.A. Loaiza, M. Pedrero, F.J. Manuel de Villena and J.M. Pingarrón, An integrated bienzyme glucose oxidase–fructose dehydrogenase–tetrathiafulvalene-3-mercaptopropionic acid–gold electrode for the simultaneous determination of glucose and fructose, *Bioelectrochemistry*, 63(1–2) (2004) 199–206.

100 L. Lvova, A. Legin, Y. Vlasov, G.S. Cha and H. Nam, Multicomponent analysis of Korean green tea by means of disposable all-solid-state potentiometric electronic tongue microsystem, *Sens. Actuators B Chem.*, 95(1–3) (2003) 391–399.
101 M. Niculescu, T. Erichsen, V. Sukharev, Z. Kerenyi, E. Csöregi and W. Schuhmann, Quinohemoprotein alcohol dehydrogenase-based reagentless amperometric biosensor for ethanol monitoring during wine fermentation, *Anal. Chim. Acta*, 463(1) (2002) 39–51.
102 C.-Y. Chen and Y.-C. Su, Amperometric L-glutamate sensor using a novel L-glutamate oxidase from *Streptomyces platensis* NTU 3304, *Anal. Chim. Acta*, 243 (1991) 9–15.
103 L.S. Ferreira, M.B. De Souza Jr., J.O. Trierweiler, O. Broxtermann, R.O.M. Folly and B. Hitzmann, Aspects concerning the use of biosensors for process control: experimental and simulation investigations, *Comput. Chem. Eng.*, 27(8–9) (2003) 1165–1173.
104 R. Zaydan, M. Dion and M. Boujtita, Development of a new method, based on a bioreactor coupled with an L-lactate biosensor, toward the determination of a nonspecific inhibition of L-lactic acid production during milk fermentation, *J. Agric. Food Chem.*, 52(1) (2004) 8–14.
105 B. Serra, M.D. Morales, A.J. Reviejo, E.H. Hall and J.M. Pingarrón, Rapid and highly sensitive electrochemical determination of alkaline phosphatase using a composite tyrosinase biosensor, *Anal. Biochem.*, 336(2) (2004) 289–294.
106 H. Liu, H. Li, T. Ying, K. Sun, Y. Qin and D. Qi, Amperometric biosensor sensitive to glucose and lactose based on co-immobilization of ferrocene, glucose oxidase, β-galactosidase and mutarotase in *β*-cyclodextrin polymer, *Anal. Chim. Acta.*, 358(2) (1998) 137–144.
107 C.A.B. Garcia, G. de Oliveira Neto and L.T. Kubota, New fructose biosensors utilizing a polypyrrole film and D-fructose 5-dehydrogenase immobilized by different processes, *Anal. Chim. Acta.*, 374(2-3) (1998) 201–208.
108 J. Razumiene, A. Vilkanauskyte, V. Gureviciene, V. Laurinavicius, N.V. Roznyatovskaya, Y.V. Ageeva, M.D. Reshetova and A.D. Ryabov, New bioorganometallic ferrocene derivatives as efficient mediators for glucose and ethanol biosensors based on PQQ-dependent dehydrogenases, *J. Organometal. Chem.*, 668(1-2) (2003b) 83–90.

Chapter 14

Electrochemical biosensors for heavy metals based on enzyme inhibition

Aziz Amine and Hasna Mohammadi

14.1 INTRODUCTION

Heavy metals are major contributors to the environmental pollution because of their involvement in many natural and industrial processes. During recent decades, there has been growing interest in the development of analytical devices for the detection, quantification, and monitoring of these compounds [1–7]. The determination of heavy metals is usually performed with techniques such as atomic absorption spectrometry, stripping voltammetry, and inductively coupled plasma spectrometry, which require sample pre-treatment, expensive instrumentation, and skilled operators. Moreover, metal measurement in field is often hampered by complexity of the devices. In this context, biosensors appear as suitable alternative or complementary tools. Several biosensors based on electrochemical, thermal, and optical detection for the determination of heavy metals have been reported [1,8–18]. Luque de Castro and Herrera have reported inhibition-based biosensors and biosensing systems [8]. In this review, Luque de Castro and Herrera have discussed the application of a huge number of the biosensors for heavy-metal determinations in standard solution and different matrices. They concluded that the enzymatic-inhibition method must be very carefully controlled particularly when in the presence of suspect interfering species. Rodrigues-Mozaz *et al.* reviewed the trends in biosensors for environmental applications [19]. Some examples of biosensors in advanced stage of development, which have been applied to real samples, as well as of commercial devices, are given. Biosensor designs for measurement of either specific chemicals or their biological effects, such as toxicity biosensors and endocrine-effect biosensors, are discussed [19]. Enzymatic methods are commonly

Comprehensive Analytical Chemistry 49
Alegret and Merkoçi (Eds)
Volume 49 ISSN: 0166-526X DOI: 10.1016/S0166-526X(06)49014-0

used for metal ion determinations, as these can be based on the use of a wide range of enzymes that are specifically inhibited by low concentrations of certain metal ions [20]. The inhibition effect of heavy metals ions on the activity of more specific enzymes has been studied and used for the construction of calibration curve and determination of the pollutants in samples.

According to literature, the inhibition of the activity of certain enzymes seems to involve the –SH group of these proteins owing to covalent binding with mercury [21,22].

Mercury and their compounds are potentially the most toxic elements for the environment. They can be found in all the environmental compartments and can be accumulated in the trophic chain [23]. Nowadays it is recognized worldwide that the toxicity of an element (e.g., Hg) is determined by its particular species occurring in the sample. In general terms, the organic forms of metals, especially methyl mercury, which is more hydrophobic than mercury goes through the biological membranes quite easily as compared to inorganic forms [22]. Therefore, recently we are witnessing a growing interest in analytical speciation of trace elements. The speciation of mercury ions and organic mercury compounds by using different techniques has been reported [2–4,24,25]. In general, this technique involves three essential steps before analysis such as acid digestion, solvent extraction, and preconcentration using a solid-phase extraction column. When the final detection is carried out by the gas chromatographic methods, back-extraction with an aqueous phase and further re-extraction with the organic solvent is recommended to clean up the extracts. For the clean-up step different complex agents have been used [2,4,24].

Concerning the detection of the inhibitors with biosensors, this is generally performed in aqueous solution. However, these compounds are generally characterized by a low solubility in water and a high solubility in organic solvent [18]. Extraction and concentration of heavy metal or pesticides from solid matrices (fruits, vegetables, fish, etc.) are thus commonly carried out in such solvents. Depending on the nature and the amount of organic solvent, the enzyme can be strongly inactivated when experiments are performed in these media [18,26]. Thus, the choice of the organic solvent should be studied prior to the development of any method in order to avoid any inactivation. Until now, a large part of research was applied for the determination of pesticides in organic phase [27–30].

A new procedure for determination of inhibitors (pesticides and methylmercury) based on extraction of the inhibitor by the enzyme (acethylcholinesterase or invertase) using an organic/aqueous mixture has been developed by Amine *et al.* [26]. The enzyme in aqueous phase selectively extracts the irreversible inhibitor dissolved in organic solvent at the interface of organic/aqueous phases. The best results were obtained with toluene and invertase [31] and with hexane for AChE [32]. Using the same approach, the enzyme invertase in presence of the organic phase combined with a glucose oxidase biosensor was used for methyl mercury determination in fish samples. The new approach will be discussed in our protocol (see Procedure 20 in CD accompanying this book).

An overview concerning the enzyme-based electrochemical biosensors published during the last 5 years for heavy-metal determinations is reported. Their sensitivity and selectivity toward inhibitors and the factors (pH, buffer, enzyme, and inhibitor concentrations) affecting the analytical characteristics of these biosensors are also discussed.

14.2 PARAMETERS AFFECTING THE ENZYME INHIBITION SYSTEM

There have been some factors affecting the response of the biosensors based on enzyme inhibition such as pH, enzyme, and substrate concentrations, method of enzyme immobilization, incubation time, and the mechanism of enzyme inhibition [18]. Since the performance of a biosensor device is strongly dependent on its configuration and the experimental conditions, the study and optimization of these conditions were the essential steps before the inhibitor determination. For these reasons there is confusion in the literature on what type of responses or quantitative relationships are expected from immobilized enzyme inhibition biosensors. A variety of linear, non-linear, logarithmic, and other relationships between inhibition percentage and either inhibitor concentration and/or incubation time have been reported. Depending on the exact experimental conditions under which the devices are used, a variety of responses may be obtained. The inhibition of the enzyme-based biosensors can be either reversible or result in an irreversible inactivation of the enzyme [18]. The reversible inhibition is independent of incubation time and enzyme concentration; in this case, the initial activity of the enzyme biosensor can be regenerated by simple washing with a buffer solution. Moreover in the case of the native enzyme some chemical agents can be used for restoration of the initial activity. On the other hand, the irreversible inhibitors depend strongly on the enzyme concentration and incubation time. In this case, the term irreversible means that the decomposition of the enzyme–inhibitor complex results in the destruction of enzyme. A variety of the heavy-metal biosensors have been described in the literature [18,20,33,34]. Recently it has been reported that the immobilization under a negatively charged polymer induces an increase in the inhibitory effect of heavy metals due to the cations accumulation in the polymeric matrix [13,14].

14.3 ANALYTICAL CHARACTERIZATION OF BIOSENSORS-BASED ENZYME INHIBITION

There has been a rather phenomenal growth in the field of biosensors with application in a variety of disciplines, including environment monitoring. Demands of sensitivity, specificity, speed, and accuracy of analytical measurements have stimulated considerable interest in developing bio-sensing probes as diagnostic tools in technology. A large number of methods based on enzyme-inhibition-based biosensors can be found in the literature, some of which are summarized in Table 14.1 [10,14,20,35–43]. The first column of this table shows the heavy metals analyte, which is, naturally present in an inorganic form (80%). The mercury compounds represent 60% with regard to the other metal analysis. The second column refers to the inhibited biocatalysis, which are most of the times urease, followed by glucose oxidase. The third column summarizes the methods used for enzyme immobilization.

Alexander and Rechnitz investigated the determination of mercury (II) using glucose oxidase enzyme in an amperometric technique [10]. In this study, the inhibition of glucose oxidase immobilized in polyvinylpyridine (PVP) in presence of 2-aminoethanethiol mediator appears reversible. Employing the amperometric glucose oxidase biosensor Malitesta and Guascito confirmed the reversible inhibition of glucose oxidase by mercury (II). In this work they described an original application of biosensors based on enzyme immobilized by electro polymerization in poly-*o*-phenylenediamine for heavy metals determination [38]. In both studies, after stabilization of glucose response and injection of mercury in the electrochemical cell a rapid decrease of enzymatic activity occurs, reaching approximately a steady state within 100 s. However the biosensor based on glucose oxidase immobilized in a PVP in presence of 2-aminoethanethiol mediator appears more sensitive toward mercury determination with a limit of detection

TABLE 14.1

Survey of enzyme inhibition based biosensors for heavy metals for the last 5 years

Inhibitors	Enzymes	Immobilization matrix	Techniques	Incubation time	Working range/LOD	Nature of inhibition	Sample	Ref.
Ag^+	Urease	Deposition onto SPE	Potentiometric	30 min	0.01–1000 μM	–	–	[35]
Cu^{2+}								
Chromium (VI)	GOx	Cross-linking with GA and covering with aniline membrane	Amperometric	–	$0.49\,\mu g\,l^{-1}$–$8.05\,mg\,l^{-1}$	–	Soil samples	[36]
					LOD = $0.49\,\mu g\,l^{-1}$			
Ag^+	Urease	Deposition onto electrode area and covering with poly(4-vinylpyridine and Nafion)	Potentiometric	–	LOD = $3.5 \times 10^{-8}\,mol\,l^{-1}$	Irreversible	–	[13]
Ni^{2+}			pH-SFET		LOD = $7 \times 10^{-5}\,mol\,l^{-1}$	The use of the negatively charged polymer/regeneration with EDTA		
Cu^{2+}					LOD = $2 \times 10^{-6}\,mol\,l^{-1}$			
Cu^{2+}	AChE	Cross-linking with GA vapor onto SPE	Amperometric	–	0.05 to $4.0\,mmol\,l^{-1}$	Reversible inhibition	–	[37]
Hg^{2+}	Glucose oxidase	Electropolymerization in PPD	Amperometric	<2 min	$2.5\,\mu mol\,l^{-1}$ to $0.2\,mmol\,l^{-1}$	Reversible rapid regeneration with EDTA	–	[38]
Cu^{2+}	(GOx)				$2.5\,\mu mol\,l^{-1}$ to $0.2\,mmol\,l^{-1}$			
Hg^{2+}	Urease	Entrapment in alginate gel and adsorption on the SPE in a nafion film	Amperometric	–	LOD = $64\,\mu g\,l^{-1}$	–	Water and soil samples	[14]
Cu^{2+}					LOD = $55\,\mu g\,l^{-1}$			
Hg^{2+}	AChE		Amperometric	<3 min	1×10^{-10}–1×10^{-5}	Reversible	–	[39]
					LOD = $1 \times 10^{-10}\,mol\,l^{-1}$			
$Hg(NO_3)_2$	Urease	Entrapped in PVC polymer at the surface of pH-electrode	Potentiometric	30 min	0.05–1 μM	Irreversible	Water samples	[20]
$HgCl_2$					0.05–1 μM	Total regeneration EDTA and thioacetamide after less than 10 min		
$Hg_2(NO_3)_2$					0.01–1.0 μM			
$PhHg\bullet^{-2}$					0.1–4 μM			

TABLE 14.1 (*continued*)

Inhibitors	Enzymes	Immobilization matrix	Techniques	Incubation time	Working range/LOD	Nature of inhibition	Sample	Ref.
Hg^{2+}	HRP	Entrapment in β-cyclodextrin polymer	Amperometric	8 s and 1–8 min	LOD = 0.1 ng ml^{-1}	Reversible in less than 8 s and irreversible in 1–8 min	–	[40]
Hg^{1+} MeHg Hg-glutathione complex					LOD = 0.1 ng ml^{-1} LOD = 1.7 ng ml^{-1}			
$HgCl_2$	Invertase	Cross-linkage with GA and deposition on laponite modified electrode	Amperometric	20 min	I_{50} = 0.27 ppm	Irreversible reactivation with cysteine	–	[41]
$Hg(NO_3)_2$ Hg_2Cl_2 $MeHg^{2+}$ $PhHg^{2+}$					I_{50} = 0.032 ppm I_{50} = 0.27 ppm I_{50} = 0.34 ppm I_{50} = 0.12 ppm			
Hg^{2+}	GOx inhibition of invertase in solution and detection of product with glucose biosensor	Cross linking with GA and BSA	Amperometric	10 min	2.5–12 ng ml^{-1}	–	Spiked water	[42]
Hg^{2+}	GOx	Immobilized in polyvinylpyridine (PVP) in presence of 2-aminoethanethiol mediator	Amperometric	<4 min	LOD = 1 ng ml^{-1} 1–100 ppb	Reversible	–	[10]
Hg^{2+}	Urease	Adsorption on gold nanoparticles in presence of PVC-NH_2	Potentiometric	<1 min	LOD = 0.2 ppb 0.09–1.99 μM	Reversible	Waste water	[43]

equal to 0.2 ppb. In the other approach, mercury (II) and mercury (I) were determined at 0.1 ppb level using horseradish peroxidase (HRP) entrapped in β-cyclodextrin polymer [40]. Employing the methylene blue mediated (HRP) enzyme electrode Han *et al.* studied the kinetics of the inhibition of HRP immobilized with mercury [40]. It appears that the inhibition of this enzyme was apparently reversible and non-competitive in the presence of $HgCl_2$ in less than 8 s but irreversibly inactivated when incubated with different concentrations of $HgCl_2$ for 1–8 min.

Using the negatively charged nafion film, Rodriguez *et al.* have improved the sensitivity of urease immobilized reaching a limit of 64 and 55 $\mu g\,l^{-1}$ of mercury and copper, respectively, instead of 2.9 and 29.8 $mg\,l^{-1}$ using the alginate gel alone [14].

The feasibility of amperometric sucrose and mercury biosensors based on the immobilization of invertase, glucose oxidase, and mutarotase entrapped in a clay matrix (laponite) was investigated by Mohammadi *et al.* [31]. In this work, the effect of pH of a tri-enzymatic biosensor in which the optimum pH of the three enzymes is different (Invertase, pH 4.5; Glucose oxidase, pH 5.5; and Mutarotase, pH 7.4) [41] was studied. The pH effect on the biosensor response was analyzed between pH 4 and 8 and the highest activity was found at pH 6.0. In order to improve the selectivity of the invertase toward mercury and to avoid silver interference, a medium exchange technique was carried out. The biosensor was exposed to mercury in an acetate buffer solution at pH 4 while the residual activity was evaluated with phosphate buffer solution at pH 6 [41].

Recently, a renewable potentiometric urease inhibition biosensor based on self-assembled gold nanoparticles has been developed by Yunhui *et al.* for the determination of mercury ions [43]. The advantages of self-assembled immobilization are low detection limit (0.05 μM), fast response, and relatively easy regeneration of the biosensors. The assembled gold nanoparticles and inactivated enzyme layers denatured by Hg^{2+} can be rinsed out via a saline solution with acid and alkali successively.

It was shown that the biosensors can be assembled using different matrices and modes of exposure of immobilization (simple deposition, cross-linking, electro polymerization, or by entrapment way) on different electrode materials.

There are two general strategies for analysis with enzyme inhibitors. Such biosensors can work as single-use (in the case of irreversible inhibition) or as reusable analytical devices (in the case of reversible inhibition). The latter way is some times difficult in analytical practice as a regeneration of the enzyme layer after inhibitor recognition is not convenient and is a time-consuming process. Single-use devices seem to be an attractive alternative. The most-important factors are the high reproducibility, low-cost and simplicity of manufacturing, possibility of mass-production, and long-term storage stability. All these requirements are fulfilled by the screen-printed technology. Ogonezyk *et al.* reported how to prepare, by means of screen-printing, very simple, cheap, and reproducible urease-based biosensors and demonstrated their utility for heavy-metal ions detection [35]. The bio-sensing thick film is prepared using bio-composite screen-printable material composed of ruthenium dioxide, urease, graphite, and the organic polymer [35].

It is seen that the amperometric technique is more used for heavy-metal determination in comparison to potentiometric techniques. The research published during the last 5 years concerning heavy-metal determination has shown the applications especially for water samples (Table 14.1).

14.4 CONCLUSIONS

Biosensors based on enzyme inhibition are still limited in analytical applications since these sensor technologies are not usually able to discriminate various toxic compounds in the same sample.

However, these biosensors can be used as alarm systems; they would provide either quantification of one contaminant when this analyte is present alone or an indication of total contamination of particular samples. They could also be used as complementary techniques to classical methods. On the other hand, a particular advantage of these biosensing systems is that they offer the possibility of analysis in both batch and flow mode, allowing the use of these sensors for analysis of a large number of samples in a reasonable time interval. These methods can be recommended also for a single use with screen-printed electrodes (SPEs), to avoid problems related to fouling of the electrode which generally involves a chemical or electrochemical deactivation of the working electrode surface. SPEs offer several additional advantages including low cost, simple handling, mass production, and suitability for instrument miniaturization for *in loco* analysis. Also, the single use of these sensors avoids the need of reactivation.

Current research studies involve numerous efforts in improving the analytical performance of the biosensing systems in order to be able to monitor a wide range of pollutants in environmental and food samples. Effectively, artificial neural networks (ANN) have already been shown to be useful for interpretation of experimental data in analysis of binary mixtures [44].

The use of genetically modified enzymes for the design of biosensors can be an effective way to improve the sensor sensitivity.

14.5 FUTURE PERSPECTIVES

Engineered variants of enzymes could be another approach in biosensor design for the discrimination and detection of various enzyme-inhibiting compounds when used in combination with chemometric data analysis using ANN. The crucial issues that should be addressed in the development of new analytical methods are the possibility of simultaneous and discriminative monitoring of several contaminants in a multi-component sample and the conversion of the biosensing systems to marketable devices suitable for large-scale environmental and food applications.

REFERENCES

1 S.V. Dzyadevych, A.P. Soldatkin, Y.I. Korpan, V.N. Arkhypova, A.V. El'skaya, J.-M. Chovelon, C. Martelet and N. Jaffrezic-Renault, Biosensors based on enzyme field-effect transistors for determination of some substrates and inhibitors, *Anal. Bioanal. Chem.*, 377 (2003) 496–506.

2 M. Zenki, K. Minamisawa and T. Yokoyama, Clean analytical methodology for the determination of lead with Arsenazo III by cyclic flow-injection analysis, *Talanta*, 68 (2005) 281–286.

3 J. Potedniok and F. Buhl, Speciation of vanadium in soil, *Talanta*, 59 (2003) 1–8.

4 C.F. Harrington, The speciation of mercury and organomercury compounds by using high-performance liquid chromatography, *Trends Anal. Chem.*, 12 (2000) 167–178.

5 Y. Bonfil, M. Brand and E. Kirowa-Eisner, Trace determination of mercury by anodic stripping voltammetry at the rotating gold electrode, *Anal. Chim. Acta*, 424 (2000) 65–76.

6 O. Cankur, D. Korkmaz and O.Y. Ataman, Flow injection–hydride generation–infrared sepectrophotometric determination of Pb, *Talanta*, 66 (2005) 789–793.

7 S. Demirci Cekic, H. Filik and R. Apak, Use of an *o*-aminobenzec acid-functionalized XAD-4 copolymer resin for the separation and preconcentration of heavy metal (II) ions, *Anal. Chim. Acta*, 505 (2004) 15–24.

8 M.D. Luque de Castro and M.C. Herrera, Enzyme inhibition-based biosensors and biosensing systems: questionable analytical devices, *Biosens. Bioelectron.*, 18 (2003) 279–294.
9 D. Compagnone, A.S. Lupu, A. Ciucu, V. Magearu, C. Cremisini and G. Palleschi, Fast amperometric FIA procedure for heavy metal detection using enzyme inhibition, *Anal. Lett.*, 34 (2001) 17–27.
10 P.W. Alexander and G.A. Rechnitz, Enzyme inhibition assays with an amperometric glucose biosensor based on thiolate self-assembled monolayer, *Electroanalysis*, 12 (2000) 343–350.
11 M.M. Lee and D.A. Russel, Novel determination of cadmium ions using an enzyme self assembled monolayer with surface plasmon resonance, *Anal. Chim. Acta*, 500 (2003) 119–125.
12 D. Shan, C. Mousty and S. Cosnier, Subnanomolar cyanide detection at polyphenol oxidase/clay biosensors, *Anal. Chem.*, 76 (2004) 178–183.
13 A.P. Soldatkin, V. Volotovsky, A.V. El'skaya, N. Jaffrezic-Renault and C. Martelet, Improvement of urease based biosensor characteristics using additional layers of charged polymers, *Anal. Chim. Acta*, 403 (2000) 25–29.
14 B.B. Rodriguez, J.A. Bolbot and I.E. Tothill, Urease-glutamic dehydrogenase biosensor for screening heavy metals in water and soil samples, *Anal. Bioanal. Chem.*, 380 (2004) 284–292.
15 B. Kuswandi, Simple optical fibre biosensor based on immobilized enzyme for monitoring of trace heavy metal ions, *Anal. Bioanal. Chem.*, 376 (2003) 1104–1110.
16 H.-C. Tsai and R.-a. Doong, Simultaneous determination of pH, urea, acetylcholine and heavy metals using array-based enzymatic optical biosensor, *Biosens. Bioelectron.*, 20 (2005) 1796–1804.
17 S. Pirvutoiu, I. Surugiu, E.S. Dey, A. Ciucu, V. Magearu and B. Danielsson, Flow injection analysis of mercury (II) based on enzyme inhibition and thermometric detection, *Analyst*, 126 (2001) 1612–1616.
18 A. Amine, H. Mohammadi, I. Bourais and G. Palleschi, Enzyme inhibition based biosensors for food safety and environmental monitoring (review), *Biosens. Bioelectron.*, 21 (2005) 1405–1423.
19 S. Rodrigues-Mozaz, M.j. Lopes de Alda, M.-P. Macro and D. Barcelo, Biosensors for environmental monitoring: a global perspective, *Talanta*, 65 (2005) 291–297.
20 T.K. Krawezynski, M. Moszezynska and M. Trojanowiez, Inhibitive determination of mercury and other metal ions by potentiometric urease biosensor, *Biosens. Bioelectron.*, 15 (2000) 681–691.
21 J.L. Webb, *Enzyme and Metabolic Inhibitors*, Vol. 2, Academic Press, New York, 1966, pp. 635–653.
22 E.M. Faustman, R.A. Ponce, Y.C. Ou, M.A.C. Mendoza, T. Lewandowski and T. Kavanagh, Investigation of methylmercury-induced alterations in neurogenesis, *Environ. Health Perspect.*, 110 (2002) 859–864.

23 B. Gagnaire, H. Thomas-Guyon and T. Renault, *In vitro* effects of cadmium and mercury on Pacific oyster, *Crassostrea gigas* (Thunberg), Haemocytes, *Fish Shellfish*, 16 (2004) 501–512.
24 G.G. Pandit, S.K. Jha, R.M. Tripathi and T.M. Krishnamoorthy, Intake of methyl mercury by the population of Mumbai, India, *Sci. Total Environ.*, 205 (1997) 267–270.
25 M. Hempel, H. Hintelmann and R.D. Wilken, Determination of organic mercury species in soil by high performance chromatography with ultraviolet detection, *Analyst*, 117 (1992) 669.
26 A. Amine, H. Mohammadi, F. Arduini, F. Ricci, D. Moscone and G. Palleschi, Extraction of enzyme inhibitors using a mixture of organic solvent and aqueous solution and their detection with electrochemical biosensors. The Eighth World Congress on Biosensors. P 3.7.64, Abstract Book. Granada, Spain, 2004.
27 S. Andreescu, T. Noguer, V. Magearu and J.-L. Marty, Screen-printed electrode based on AChE for the detection of pesticides in presence of organic solvents, *Talanta*, 57 (2002) 169–176.
28 M. Del Carlo, M. Mascini, A. Pepe, D. Compagnone and M. Mascini, Electrochemical bioassay for the investigation of chlorpyrifos-methyl in vine samples, *J. Agric. Food Chem.*, 50 (2002) 7206–7210.
29 A. Ciucu, C. Ciucu and R.B. Baldwin, Organic phase potentiometric biosensor for detection of pesticides, *Roum. Biotechnol. Lett.*, 7 (2002) 625–630.
30 E. Wilkins, M. Carter, J. Voss and D. Ivnitski, A quantitative determination of organophosphate pesticides in organic solvents, *Electrochem. Commun.*, 2 (2000) 786–790.
31 H. Mohammadi, A. Amine, A. Ouarzane and M. El Rhazi, Screening of fish tissue for methyl mercury using the enzyme invertase in a solvent interface, *Microchim. Acta*, 149 (2005) 251–257.
32 F. Arduini, F. Ricci, I. Bourais, A. amine, D. Moscone and G. Palleschi, Extraction and detection of pesticides by cholinesterase inhibition in a two-phase system: a strategy to avoid heavy metal interference, *Anal. Lett.*, 48 (2005) 17003–17019.
33 M.D. Trevan, *Immobilized Enzyme: An Introduction and Application in Biotechnology*, Wiley, Chichester, 1980.
34 S.V. Dzyadevych, V.N. Arkhypova, A.P. Soldatkin, A.V. El'skaya and C. Martelet, N. Jaffrezic-Renault, Enzyme biosensor for tomatine detection in tomatoes, *Anal. Lett.*, 37 (2004) 1611–1624.
35 D. Ogonezyk, L. Tymecki, I. Wyzkiewiez, R. Koncki and S. Glab, Screen-printed disposable urease-based biosensors for inhibitive detection of heavy metal ions, *Sens. Actuators B*, 106 (2005) 450–454.
36 Z. Guang-Ming, T. Lin, S. Guo-Li, H. Guo-He and N. Cheng-Gang, Determination of trace chromium (VI) by an inhibition-based enzyme biosensor incorporating an electro polymerized aniline membrane and

ferrocene as electron transfer mediator, *Int. J. Environ. Sci. Eng.*, 84 (2004) 761–774.
37 G.A. Evtyugin, I.I. Stoikov, C.K. Budnikov and E.E. Stoikova, A cholinesterase sensor based on a graphite electrode modified with 1,3-disubstituted calixarenes, *J. Anal. Chem.*, 58 (2003) 1151–1156.
38 C. Malitesta and M.R. Guascito, Heavy metal determination by biosensors based on enzyme immobilised by electropolymerisation, *Biosens. Bioelectron.*, 20 (2005) 1643–1647.
39 M. Stoytcheva and V. Sharkova, Kinetics of the inhibition of immobilized acetylcholinesterase with Hg (II), *Electroanalysis*, 14 (2002) 1007–1010.
40 S. Han, M. Zhu, Z. Yuan and X. Li, A methylene blue-mediated enzyme electrode for the determination of trace mercury (II), mercury (I), methylmercury, and mercury-glutathione complex, *Biosens. Bioelectr.*, 16 (2001) 9–16.
41 H. Mohammadi, A. Amine, S. Cosnier and C. Mousty, Mercury-enzyme inhibition assays with an amperometric sucrose biosensor based on a trienzymatic-clay matrix, *Anal. Chim. Acta*, 543 (2005) 143–149.
42 H. Mohammadi, M. El Razi, A. Amine, A.M.O. Brett and C.M.A. Brett, Determination of mercury (II) by invertase enzyme inhibition coupled with batch injection analysis, *Analyst*, 27 (2002) 1088–1093.
43 Y. Yang, Z. Wang, M. Yang, M. Guo, Z. Wu, G. Shen and R. Yu, Inhibitive determination of mercury ion using a renewable urease biosensor based on self-assembled gold nanoparticles, *Sens. Actuators B*, 114 (2006) 1–8.
44 T.T. Bachmann, B. Leaca, F. Vilatte, J.-L. Marty, D. Fournier and D.R. Schmid, Improved multianalyte detection of organophosphates and carbamates with disposable multielectrode biosensors using recombinat mutants of *Drosophila* acethylcholinesterase and artificial neural networks, *Biosens. Bioelectr.*, 15 (2000) 193–201.

Chapter 15

Ultra-sensitive determination of pesticides via cholinesterase-based sensors for environmental analysis

Frank Davis, Karen A. Law, Nikos A. Chaniotakis, Didier Fournier, Tim Gibson, Paul Millner, Jean-Louis Marty, Michelle A. Sheehan, Vladimir I. Ogurtsov, Graham Johnson, John Griffiths, Anthony P.F. Turner and Seamus P.J. Higson

15.1 INTRODUCTION

While pesticides are used extensively within modern agricultural techniques to control insect infestation, increasing concern is being shown towards their indiscriminate use and the long-term effects they may cause to the environment, livestock and human health [1,2]. A significant proportion of the pesticides used within agriculture become washed off or are otherwise lost from the large areas of agricultural-land-treated surfaces—and for this reason an excess of active ingredient is commonly applied [3]. The problem is compounded by the fact that many pesticides such as DDT have very long lifetimes under environmental conditions. Although organophosphate pesticides (OPs) are now commonly used instead of the organochlorine pesticides due to their lower persistence in the environment whilst still remaining effective, they are, however, neurotoxins and therefore present a serious risk to human health. These compounds may still find their way into our food and water supplies, which necessitates the use of analytical approaches for the reliable detection of pesticides for environmental protection and food safety purposes. Legislation has now been passed to help restrict pesticides within water supplies; European Commission: EU Water Framework Directive 2000/60/EC, European Commission: Drinking Water Directive 98/83/EC, which recommends levels within water supplies of 0.1 $mg\,l^{-1}$ for individual pesticides and 0.5 $mg\,l^{-1}$ for total pesticide. It is likely that with the widespread concerns about

Comprehensive Analytical Chemistry 49
Alegret and Merkoçi (Eds)
Volume 49 ISSN: 0166-526X DOI: 10.1016/S0166-526X(06)49015-2

these materials that these levels could come down. More recently, the area of biodefense is receiving much attention, with organophosphate-based nerve agents also needing to be analysed.

Contemporary methods for environmental determination and/or the monitoring of pesticides include gas and liquid chromatography and various spectroscopic techniques [4]. Each of these approaches suffers from several disadvantages such as being costly, time consuming, not sufficiently sensitive and/or requiring complex sample preparation [1,5,6]. Continuous monitoring moreover is not possible with any of these methods and it follows that a simplified analytical approach would prove highly beneficial.

A potential solution to this problem is the utilisation of biosensor technology. Biosensors generically offer simplified reagentless analyses for a range of biomedical and industrial applications and for this reason this area has continued to develop into an ever expanding and multi-disciplinary field during the last couple of decades. Electrochemical techniques are amongst the easiest and most inexpensive methods for detection of binding events and many groups have previously demonstrated the fabrication of enzymatic and affinity-based sensors that lend themselves to interrogation by either (i) amperometric or (ii) impedimetric approaches.

Much of the work described was carried out within a collaborative project between a number of academic and industrial groups under the remit of the SAFEGARD consortium, an EU funded Framework 5 research contract ref QLRT-1999-30481. The various expertises available from the academic and industrial collaborators made this project feasible.

The detection of many pesticides at extremely low levels can be best achieved not by direct detection of the pesticide itself but rather by detection of its inhibitory effects on enzyme reactions. An enzyme-electrode is first constructed and its response when exposed to a suitable concentration of its substrate determined. When an electrode is then exposed to a dilute pesticide solution, the pesticide interacts with the enzyme and diminishes (or completely destroys) its activity. This inhibition can then be easily quantified by further exposure to the initial substrate concentration and comparison with the response prior to pesticide exposure.

The detection of organophosphate and other pesticides based on the inhibition of the enzyme acetylcholinesterase by these compounds has received considerable attention primarily due to high specificity and sensitivity [1,7–16]. Cholinesterases, such as acetylcholinesterase, catalyse the hydrolysis of choline esters to the corresponding carboxylic acid and choline:

$$\text{Acetylcholine} + H_2O \xrightarrow{\text{acetylcholinesterase}} \text{choline} + \text{acetic acid} \tag{15.1}$$

The use of electrochemical techniques combined with biological molecules has been extensively reviewed [17] and will not be discussed in detail here. The most widely used method for the AChE-containing electrodes is via the simple amperometric detection of the product of the ester hydrolysis enzyme catalysed reaction [17].

A typical approach is to utilise a substrate which when hydrolysed by the enzyme gives rise to a product which can be easily detected electrochemically. Thiocholine can be easily detected using screen-printed carbon electrodes doped with cobalt phthalocyanine (CoPC) [18,19], which acts as an electrocatalyst for the oxidation of thiocholine at a lowered working potential of approximately +100 mV (vs. Ag/AgCl) [18,19], thereby minimising interference from other electroactive compounds:

$$\begin{aligned}&\text{acetylthiocholine chloride} + H_2O\\&\xrightarrow{\text{acetylcholinesterase}} \text{thiocholine}_{(\text{red})} + \text{acetic acid} + Cl^-\\&2\ \text{thiocholine}_{(\text{red})}\\&\xrightarrow{+100\text{mV versus Ag/AgCl}} \text{thiocholine}_{(\text{ox})} + 2e^- + 2H^+\end{aligned} \tag{15.2}$$

A similar approach utilises *p*-aminophenyl acetate [20].

There are problems with this approach since enzymes isolated from natural sources such as the electric organ of electric eels often display low sensitivity and selectivity to the wide range of potential pesticide targets [21]. A possible solution to this is the development of a multisensor array where a variety of genetically modified acetylcholinesterases are immobilised on an array of electrochemical sensors and the responses from these are then processed via a neural network.

A wide variety of methods exist for the immobilisation of enzymes on a sensor surface. Screen-printed carbon electrodes are often the favourite base for these sensors due to their inexpensiveness and ease of mass production. Methods used for the construction of AChE-containing electrodes include: simple adsorption from solution [22], entrapment within a photo-crosslinkable polymer [20,23], adsorption from solution onto microporous carbon and incorporation into a hydroxyethyl cellulose membrane [24], binding to a carbon electrode via Concanavalin A affinity [25,26] and entrapment within conducting electrodeposited polymers [27].

15.2 APPLICATION

15.2.1 Synthesis of the acetylcholinesterase

Earlier work in this field [28] indicated that acetylcholinesterase enzymes would be suitable biomolecules for the purpose of pesticide detection, however, it was found that the sensitivity of the method varied with the type and source of cholinesterase used. Therefore the initial thrust of this work was the development of a range of enzymes via selective mutations of the *Drosophila melanogaster* acetylcholinesterase *Dm.* AChE. For example mutations of the (*Dm.* AChE) were made by site-directed mutagenesis expressed within baculovirus [29]. The acetylcholinesterases were then purified by affinity chromatography [30]. Different strategies were used to obtain these mutants, namely: (i) substitution of amino acids at positions found mutated in AChE from insects resistant to insecticide, (ii) mutations of amino acids at positions suggested by 3-D structural analysis of the active site, (iii) Ala-scan analysis of amino acids lining the active site gorge, (iv) mutagenesis at positions detected as important for sensitivity in the Ala-scan analysis and (v) combination of mutations which independently enhance sensitivity. The activity of the enzymes was determined photometrically at 412 nm using the Ellman method [31]. The use of these different strategies allowed the development of sensitive enzymes, for a mutant which was 300-fold more sensitive to dichlorvos than the *Drosophila* wild-type enzyme and 288,000-fold more sensitive than the electric eel enzyme which is commonly used to detect organophosphates. The most-effective method of increasing the sensitivity to the pesticide peared to be via the incorporation of hydrophobic amino acids at the rim of the active site of the enzyme [24,32]. For example, in solution the sensitivity to methamidophos of a genetically modified *Drosophila* AChE was one order of magnitude higher than a commercial electric eel AChE [32]. When immobilised in a biosensor, this led to a useful working analytical range for the *Drosophila* AChE electrode of 0.5–100 ppb methamidophos as against 0.05–24 ppm for the commercial AChE electrode.

15.2.2 Immobilisation of the enzymes

Several different approaches have been developed within the group of the SAFEGARD consortium to immobilise the engineered AChEs. The simplest of these was reported by Bonnet *et al.* [22] in which a screen-printed graphite electrode was exposed to a solution of commercial electric eel AChE in phosphate buffer. The resultant enzyme electrodes were then used to detect acetylthiocholine chloride (10^{-3} M) which gave currents in the range of 225 nA. Inhibition studies with chlorpyrifos ethyl oxon (exposure time 10 min) were performed and gave a detection limit of 1.2 ng l^{-1} with good operational stability. The absence of diffusion barriers, however, gave a high level of sensitivity, although there was high variability in response between electrodes. In the work of Andreescu *et al.* [33], a comparative study where the enzyme was immobilised using the following techniques was made. This included

(a) a mixing of AChE with graphite, tetracyanoquinodimethanide (TCNQ, used as a mediator), hydroxyethylcellulose (HEC) and a methyltrimethoxysilane based sol–gel which was then deposited on a screen-printed working electrode surface as a paste. This was then allowed to dry.
(b) Screen-printing a graphite/TCNQ/HEC composite electrode, then printing a layer consisting of the enzyme and a 30% solution of a photopolymerisable poly(vinyl alcohol)/styryl pyridinium copolymer on top of the electrode and finally irradiation with UV light to photocrosslink the polymer.
(c) Screen-printing a graphite/TCNQ/HEC composite electrode and then printing a layer containing a nickel compound attached to a silica support. This was then exposed to a solution of a histidine$_6$-tagged AChE in phosphate buffer, with the histidine tag binding strongly to the immobilised nickel compound.

The three types of electrode were exposed to solutions of acetylthiocholine chloride and the resulting current recorded. These gave slightly different calibration curves over the concentration range 0–2.5×10^{-3} M of substrate, with the nickel containing composite being the least sensitive. The nickel composite also gave the poorest storage performance, with the other two electrodes being stable up to 12 days, and the poorest reproducibility also being observed for the nickel-binding method.

Inhibition studies were made with chlorpyrifos ethyl oxon with the sol–gel method giving the largest linear range (0–6×10^{-8} M), although the nickel-binding method gave an electrode which was more sensitive at lower concentrations. The sol–gel electrode also gave good response behaviour to paraoxon and dichlorvos.

Further work [34] also compared enzyme electrodes formulated using the photocrosslinking technique above—with the electrodes being treated by simple immobilisation of AChE inside a matrix of bovine albumin crosslinked by glutaraldehyde. A variety of experimental conditions were utilised. The glutaraldehyde crosslinking technique has the advantage of simplicity and gives electrodes which have fast response times while being robust and reproducible. They did require a far higher enzyme content (80 mUA), however, to give similar responses to those of the photopolymerised electrodes containing 0.7–1.0 mUA of enzyme. It should also not be forgotten that this greatly increases the expense of such systems.

Microporous carbon was also studied as a potential substrate for binding of AChE [24,35]. Discs cut from a commercial porous carbon rod were cleaned and then exposed to a solution of AChE in phosphate buffer for 20 h to allow for simple physisorption and chemisorption of the enzyme. Initial tests using electric eel AChE [35] gave linear detection of dichlorvos in the range 10^{-6}–10^{-12} M. The sensitivity of this method was increased still further by utilisation of the genetically engineered AChE mentioned earlier, with the detection limit of these systems being extended down to 10^{-17} M [35].

Instability of the mutant AChE can be a problem with up to 50% of its activity in solution being lost in 10 days. This led to a study in which the enzyme was immobilised in porous silica (pore size 10 nm) or porous carbon (<70 nm) beads [36]. The AChE is known to be approximately 6 nm in size and therefore it is thought that entrapment within the pores could well inhibit unfolding of the enzyme, so enhancing its stability.

Following immobilisation, the beads were dispersed in an aqueous solution of HEC and cast onto Pt electrodes. Activity tests showed that leaching of immobilised enzyme was 2.5 times slower than that of free enzyme dispersed in HEC. Comparisons of activity to acetylthiocholine after 72 h constant operation showed a large stability enhancement for enzymes immobilised on both silica and carbon when compared to dispersion in HEC [36].

Affinity binding was also used to bind AChE to a working-electrode surface [25]. Amino-grafted silica beads were used as the starting point and reacted with glutaraldhyde. The resultant beads containing active aldehyde groups on the surface were then treated with Concanavalin A, a lectin type protein with binding affinity for mannose, a sugar which is present at the surface of AChE. Finally the protein-grafted silica beads were treated with a solution of commercial electric eel AChE [25].

Treatment with divinyl sulfone followed by a disaccharide was used as alternative activating step before Concanavalin A adsorption. Monitoring of enzymatic activity showed binding of the AChE only for systems containing sugar/Concanavalin A affinity links, indicating that unspecific adsorption did not lead to immobilisation of the enzyme. The beads were then mixed with graphite/TCNQ composite and cast onto a screen-printed working electrode.

Amperometric activity of the electrodes in thiocholine before and following exposure to solutions of pesticides was measured. Sample to sample reproducibility was found to be favourable (RSD 6.6%), as was stability with electrodes being shown to be capable of being stored for up to 2 months at $-18^{\circ}C$. Linear detection of chlorpyrifos methyl oxon by inhibition was obtained between 1×10^{-8} and 5×10^{-8} M by this approach.

A similar method was used [26] to directly immobilise AChE on the electrode. A screen-printed carbon electrode was treated with a Nafion/heptylamine mixture. The amino groups were then activated with divinyl sulfone and then treated with a disaccharide. This was then used to bind first Concanavalin A and then electric eel AChE via affinity binding. The resultant electrodes had similar reproducibility to the silica-containing analogues with no enzyme leakage occurring upon storage for 2 h in buffer. Bovine albumin was used in this instance to block non-specific binding. The electrode activity was completely inhibited upon exposure to chlorpyrifos methyl oxon (10^{-5} M), but could be completely regenerated simply by exposing the electrode to fresh AChE solution; this behaviour was observed for three inhibition/regeneration cycles.

15.2.3 Use of microelectrodes

Another potential method for immobilising AChE is to entrap the enzyme within a conducting polymer such as polyaniline. The entrapment of biological molecules within conducting polymers has been widely studied and extensively reviewed elsewhere [37]. All the methods described so far in this paper have been related to the production of planar electrodes. Microelectrodes offer several advantages over conventional larger working electrodes within biosensors, since they experience hemispherical solute diffusional profiles, and it is this phenomenon that can impart stir independence to sensor responses whilst also offering lowered limits of detection.

Individual microelectrodes offer very small responses and one approach for overcoming this problem is to use many microelectrodes together in the form of an array to allow a cumulative and so larger response to be measured. Microelectrode arrays may be fabricated by a number of approaches although techniques such as photolithography or laser ablation have to date proved cost prohibitive for the mass production of disposable sensor strips. We have previously described a novel sonochemical fabrication approach [38,39] for the production of microelectrodes, that lends itself to the mass production of sensor arrays.

The method of producing microelectrodes will be described in more detail within the protocol (Procedure 24 in CD accompanying this book). The method is as shown schematically in Fig. 15.1. A conducting surface, for example formed by screen-printed carbon can be insulated by deposition of poly(*o*-phenylene diamine). Sonochemical ablation has been shown to form pores in this insulating surface [39] with population densities of up to 2×10^5 pores cm^{-1}. Electrochemical deposition of conducting polyaniline at these pores can be performed and used to grow protrusions of polyaniline at the surface [38] and if AChE is included in the deposition solution, the enzyme may be entrapped within a conducting polyaniline matrix [27]. These arrays of polyaniline protrusions can be visualised by scanning electron microscopy [40] (Fig. 15.2) and display the typical stir-independent behaviour of microelectrodes [27].

In this way a sonochemically fabricated microelectrode array was used to form an array of conducting microelectrodes [27] containing a genetically modified AChE which had been modified to maximise pesticide sensitivity. Use of an I^{125}-labelled AChE meant that the amount of enzyme deposited could be measured and in this instance corresponded to 0.15 units activity. Measuring the amperometric

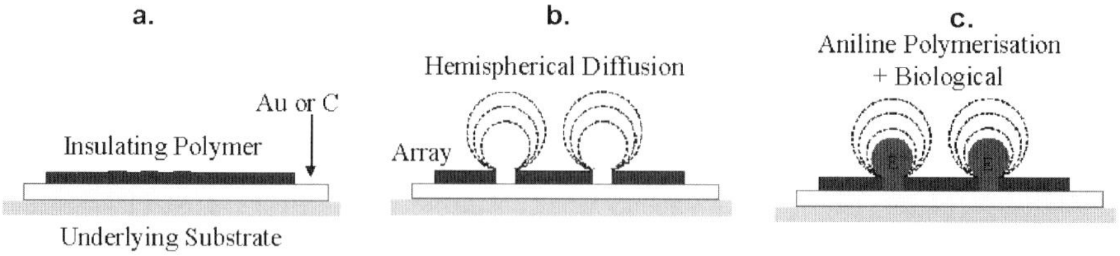

Fig. 15.1. Schematic of sonochemical microelectrode formation: (a) formation of the insulating layer on the electrode surface, (b) sonochemical ablation leading to formation of microelectrode pores, (c) electropolymerisation of aniline and AChE at the pores to form enzyme microelectrodes.

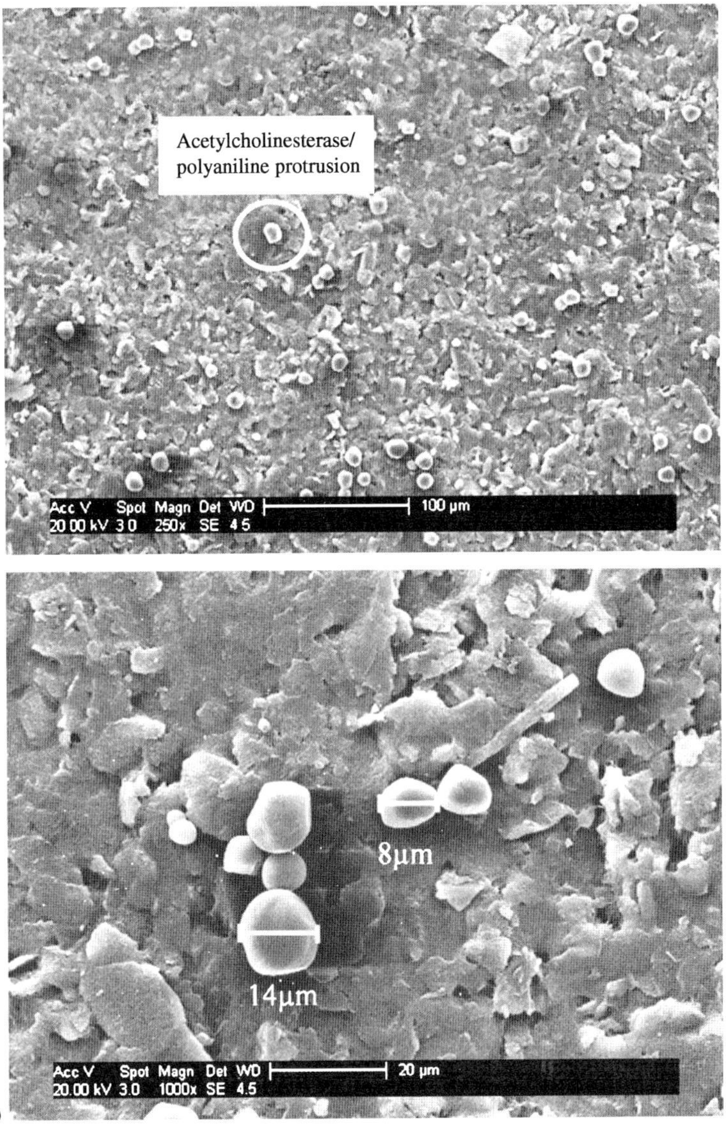

(a)

(b)

Fig. 15.2. Scanning electron micrographs of enzyme containing polyaniline protrusions at (a) 250 ×, (b) 1000 ×, (c) side view at an angle of 50° 1000 ×, (d) side view 5000 ×.

response of the electrode in acetylthiocholine before and following exposure to paraoxon solutions allowed a measurement of the inhibition of enzyme activity. Levels as low as 10^{-17} M paraoxon could be reproducibly detected [27]. Although very low, these levels are

(c)

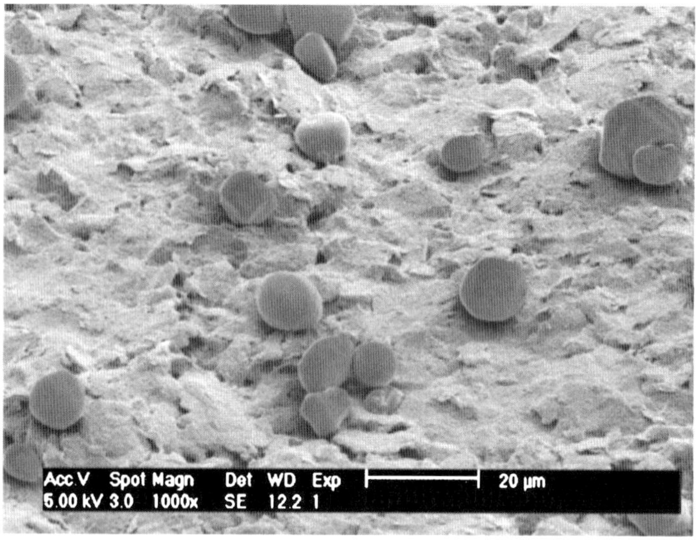

(d)

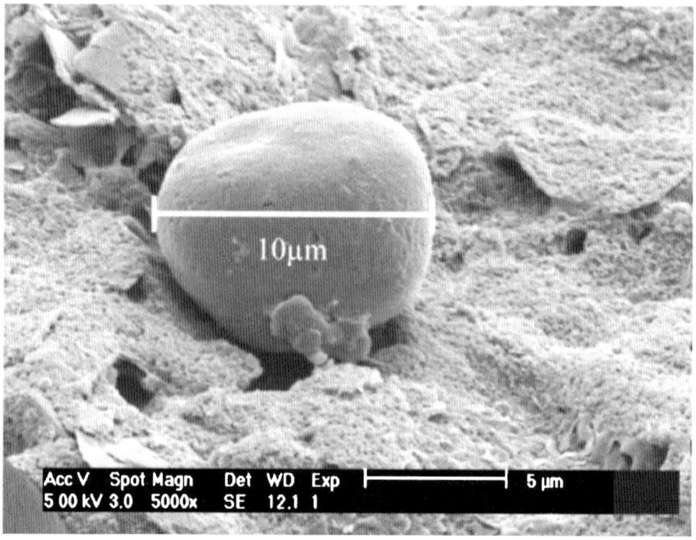

Fig. 15.2. (*Continued*)

comparable to those determined using acetylcholineesterase immobilised on microporous conductive carbon [24].

15.2.4 Multiple pesticide detection

One major problem with determining pesticides in real samples is that one or several of a range of pesticides could be present. Therefore we need a sensor that can interrogate the sample and determine not only which pesticides are present but at what levels. One possible method is to manufacture a sensor, usually by screen-printing, containing multiple working electrodes with each containing a different AChE. Pattern-recognition software can then be used to monitor the varying inhibitory response pesticides and mixtures of pesticides. Alternatively a range of single AChE electrodes can be manufactured and then incorporated into a flow injection system so that they are all simultaneously exposed to the sample, with responses being monitored and pattern-recognition software used as before.

A series of multielectrode sensors were developed based on *Drosophila* mutant AChE immobilised via photocrosslinking onto screen-printed carbon electrodes [8]. Four different mutant and wild-type AChE were evaluated for their sensitivity to the organophosphate paraoxon and the carbamate pesticide carbofuran. The response of the electrodes in thiocholine before and following a 15-min exposure to solutions of the pesticides was compared. The data was then processed using a feed-forward neural network generated with NEMO 1.15.02 as previously described [8,9]. Networks with the smallest errors were selected and further refined. This approach together with varying the AChE led to the construction of a sensor with capability to analyse the binary pesticide mixtures.

When solutions of individual pesticides were used, concentrations as low as $0.5\,\mu g\,l^{-1}$ (10^{-9} M) could be determined. When binary mixtures with pesticide levels from 0 to $5\,\mu g\,l^{-1}$ were measured, the concentration of each pesticide could be determined within the range with errors of $0.4\,\mu g\,l^{-1}$ for paraoxon and $0.5\,\mu g\,l^{-1}$ for carbofuran. Similar levels were obtained when river water samples spiked with pesticide were used but with a higher degree of inaccuracy. When different mutant AChEs were utilised, binary mixtures of the very similar pesticides paraoxon and malaoxon could be analysed in the range $0–5\,\mu g\,l^{-1}$, with resolution of the two components with accuracies of the order of $1\,\mu g\,l^{-1}$. The use of more sensitive and selective mutant enzymes together with the addition of extraction and concentration steps to the assay could greatly enhance the methods range and accuracy.

A flow injection system combined with an enzyme electrode was used to detect and quantify a variety of pesticides [41]. Photocrosslinkable poly(vinyl alcohol) was used to immobilise AChE onto platinum wire working electrodes. These were then placed inside a flow injection cell and the electrochemical response to injections of thiocholine measured. A series of measurements were made before and following the injection of a pesticide solution. Under constant flow, the sensors were found to be stable for several days. The inhibition of the current after exposure to various pesticide solutions was measured with detection limits using mutant AChEs being found to be as low $1.1\,\mu g\,l^{-1}$ (Chlorpyrifos oxon), $30\,\mu g\,l^{-1}$ (paraoxon) and $25\,\mu g\,l^{-1}$ (malaoxon). What makes this system of interest is that it could potentially be used for multiple tests with the sample electrode, since injection of and incubation of the electrode with pyridine-2-aldoxime methochloride reversed the inhibition effect of the pesticide. Detection of pesticides in spiked river water samples was also achieved.

Some of the work described previously showed diminution of the biosensor performance when pesticide solutions using river water rather than laboratory water were used. This is thought to be partially due to other compounds present within river water affecting electrode performance. A system containing triple enzyme electrodes within a flow injection system was developed in an attempt to combat this [33]. Three different AChE variants were immobilised on screen-printed electrodes by photocrosslinking, one a wild-type *Drosophila*, the second a mutant with extremely high sensitivity to pesticides and the third a wild-type electric eel AChE which is relatively resistant to pesticide. However, any matrix interference would affect both electrodes equally and therefore can be subtracted, allowing us to distinguish inhibition due to the presence of non-pesticide inhibitors, e.g., Hg from specific interactions which occur only if pesticides are present. Limits of detection for the pesticide omethoate were found to be 2×10^{-6} M for the wild-type *Drosophila* and 10^{-7} M for the mutant, levels which caused only minimal inhibition of the electric eel AChE control.

Heavy metals and hypochlorite can both inhibit AChE [33] and so similar tests for pesticides were repeated in solutions containing either $20\,mg\,l^{-1}$ Hg^{2+} or $0.1\,mg\,l^{-1}$ $NaClO_4$. In both cases, large inhibition effects were noted for both the enzyme electrodes, not just the mutant, so indicating the presence of a non-specific interferent. When river water was introduced to the system, no inhibition effects were observed, however, when omethoate spiked river water samples were used, then inhibition effects could be measured for the mutant with similar levels of sensitivity to when pure water was used as the matrix.

As an alternative to simple AChE electrodes, a bienzyme system containing AChE and tyrosinase which utilised phenyl acetate as a substrate has been developed [42]. The AChE hydrolyses the phenyl acetate to phenol which the tyrosinase enzyme oxidises to *p*-quinone which can in turn be detected electrochemically. The bienzyme system was found to be less sensitive than the AChE electrode, although it did display a large tolerance for hexane.

Microelectrode arrays containing AChE were also utilised within a flow injection system [40]. A system was developed where a sample was separated and flushed simultaneously through eight cells, each containing a screen-printed electrode and fitted with a separate bespoke mini-potentiostat (Fig. 15.3). This allowed multiple measurements to be made on a single water sample using multiple electrodes, each specific for a different pesticide due to inclusions of different AChE mutants in each of the electrodes. Pattern-recognition software could then be utilised to deduce the pesticide levels in a potentially complex sample.

Early results indicate a high sensitivity for pesticide detection, with the system being capable of detecting dichlorvos at concentrations as low as 1×10^{-17} M and parathion and azinphos both at concentrations as low as 1×10^{-16} M [40].

15.2.5 Signal processing for pesticide detection

The development of user-friendly automated instrumentation able for identification and quantitative detection of pesticides is needed for a

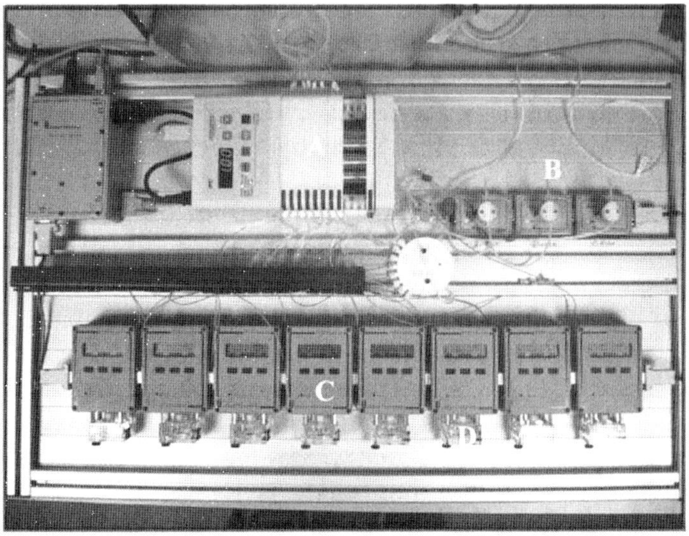

Fig. 15.3. Flow-injection analyser. (A) Pump set at 1 ml min^{-1}, (B) injection valves for substrate and pesticide samples, (C) one of eight potentiostats and (D) flow cell which comprises one sensor.

wide variety of application areas. For the identification and quantification of the pesticide type the multisensor approach combined with pattern-recognition software is highly promising. To enable pesticide quantification, special algorithms for the signal processing of the biosensor response have been developed.

These algorithms can be divided into two groups. The first group consists of algorithms related to the signal processing of a separate sensor response on a pesticide injection. They decrease the influence of noise on the measurements by increasing the signal-to-noise ratio thereby providing a lowering of detection limit and increasing the sensitivity and reproducibility of the instrumentation.

The second group consists of algorithms associated with the pesticide concentration quantification. In this case, the initial data is the processed sensor response for an unknown pesticide concentration and the parameters of the calibration curve (which is derived from preliminary experimental calibration measurements for a range of standard pesticide concentrations) or alternatively, a set of sensor responses obtained by addition of known amounts of pesticide to the analysed sample. This group of algorithms allows the automation of the pesticide quantification, thereby enabling the use of the instrumentation by unskilled personal. This removes the sensing platform from specialised laboratories to the realm of the end-users.

The structure of the algorithm for the developed software is presented in Fig. 15.4. It integrates biosensor signal processing together with pesticide quantification algorithms and includes:

- Preliminary biosensor signal processing.
- Analytical signal extraction.
- Analytical signal processing.

Preliminary biosensor signal processing combines the analysis of the background signal and biosensor response and smoothing/filtration of the biosensor response upon pesticide injection. Its purpose is to increase the signal-to-noise ratio of the biosensor response resulting from pesticide injection by using an optimal smoothing/filtration procedures, the parameters of which are defined by the analysis of background signal and the biosensor response after pesticide injection in the time and frequency domains. Application of this approach to the electric eel AChE electrode demonstrated that the biosensor background signal in the time domain represents Gaussian noise with non-zero medium. The correlation time was equal to 17.47 s which defined the lower limit of

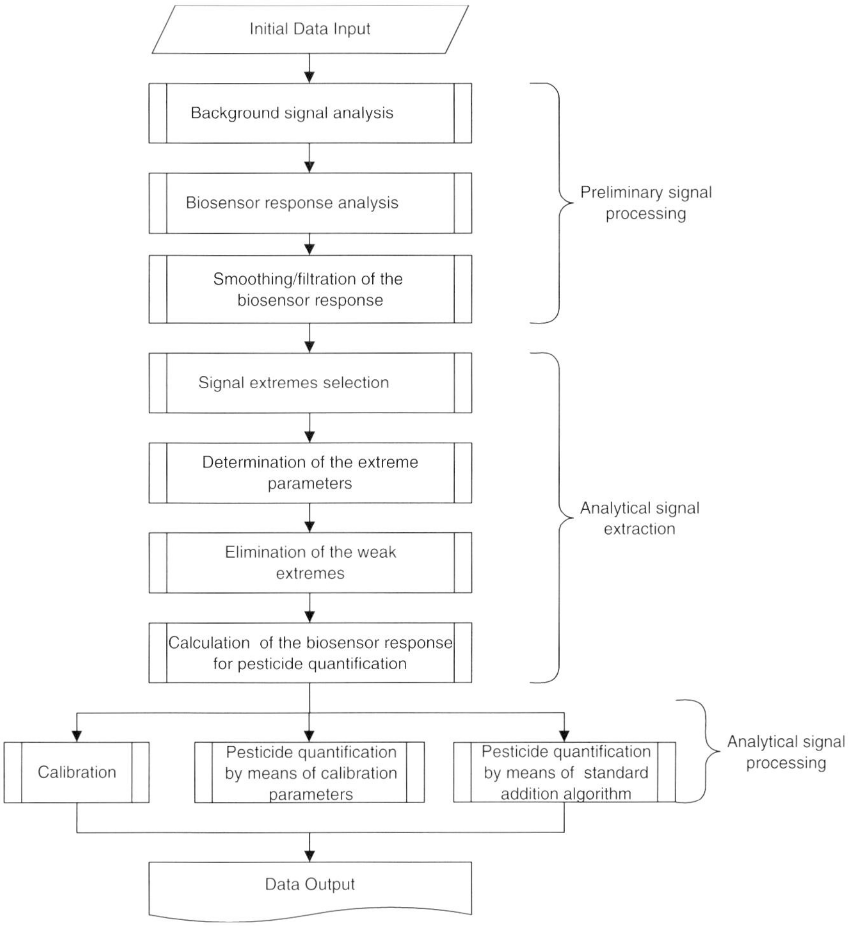

Fig. 15.4. Flow-chart of software for pesticide quantification.

the integration time for noise filtration. In the frequency domain, the background signal presented mainly uniform distributed noise with a small region below 8 mHz with flicker noise type of frequency dependence ($1/f$ function).

Analysis of biosensor response on the pesticide injections with different acteylthiocholine concentrations in the range of 1–50 mM displayed a shape of normalised sensor response that only slightly depended on the pesticide concentration. Signal time (the time interval containing 90% of signal energy), which gives the upper limit of the integration time for the signal filtration, decreased only by 16% from 100 s if the concentration increased by two orders of magnitude. In the frequency domain the biosensor responses presented a bell-shaped profile where the frequency band of the signal (the band containing 90% of the signal energy) slightly increased with increasing pesticide concentration. The filter band equal to 5 mHz can be selected as a lower limit for the frequency filtration approach.

Examination of the different algorithms for filtration/smoothing of the biosensor response, which included low frequency filtration, Gaussian kernel and running median smoothing, demonstrated that the running median smoothing method could be recommended. This is due to its good adaptability to fast signal variations, which are typical for the biosensor response on a pesticide injection.

The analytical signal extraction block is based on determination of the maximum signal response and includes the following sections: selection of response extremes, determination of the extreme parameters, elimination of the weak extremes and calculation of the analytical signal for pesticide quantification. The differences between the maxima in the biosensor response and baselines were used for the calculation of analytical signals, where the line drawn between two nearby minima within the limits of each injection was taken to be the base line.

This analytical signal of the biosensor was used for:

- calculation of the biosensor calibration parameters (slope and intercept of the calibration line) by statistical processing of the biosensor responses following pesticide injections with known concentrations or
- pesticide quantification in the sample by means of calibration parameters or, in case of need more accurate data, by means of the standard addition software analogous to the algorithm described in Ref. [43].

15.3 CONCLUSIONS

This chapter describes the wide range of research undertaken by several groups during the course of the SAFEGARD European Commission funded Framework V project. A wide range of mutant acetylcholinesterase enzymes have been obtained with some being determined to have sensitivities to selected pesticides orders of magnitude greater than wild-type enzymes. A wide range of immobilisation techniques have been studied to develop sensitive and selective enzyme electrodes which can measure concentrations of a range of pesticides down to levels hitherto undetectable (1×10^{-17} M).

Other techniques such as use of multiple electrodes, pattern recognition software and flow-injection techniques have enabled the subtraction of matrix effects such as heavy metals from the system as well as the determination of pesticides in systems containing more than one compound. The signal processing algorithms allow automation of the pesticide quantification enabling use of the instrumentation by unskilled personal, thereby removing this sensing platform from specialised laboratories and making it available to the end-users. Thus this application could conceivably be utilised in the field as well as under laboratory conditions. The relative low cost of electrochemical technology compared with many of the other technologies used makes it an attractive alternative, especially if the enzyme electrodes can be inexpensively mass-produced using screen-printing to allow single-shot use.

REFERENCES

1 G. Jeanty, Ch. Ghommidh and J.L. Marty, Automated detection of chlorpyrifos and its metabolites by a continuous flow system-based enzyme sensor, *Anal. Chim. Acta*, 436(1) (2001) 119–128.

2 I. Palchetti, A. Cagnini, M. Del Carlo, C. Coppi, M. Mascini and A.P.F. Turner, Determination of anticholinesterase pesticides in real samples using a disposable biosensor, *Anal. Chim. Acta*, 337 (1997) 315–321.

3 A.N. Ivanov, G.A. Evtugyn, R.E. Gyurcsányi, K. Tóth and H.C. Budnikov, Comparative investigation of electrochemical cholinesterase biosensors for pesticide determination, *Anal. Chim. Acta*, 404 (2000) 55–65.

4 C. Aprea, C. Colosio, T. Mammone, C. Minoia and M. Maroni, Biological monitoring of pesticide exposure: a review of analytical methods, *J. Chromatogr. B. Anal. Tech. Biomed. Life Sci.*, 769 (2002) 191–219.

5 S. Gupta, S.K. Handa and K.K. Sharma, A new spray reagent for the detection of synthetic pyrethoids containing a nitrile group on thin-layer plates, *Talanta*, 45 (1998) 1111–1114.

6 J. Diehl-Faxon, A.L. Ghindilis, P. Atanasov and E. Wilkins, Direct electron transfer-based tri-enzyme electrode for monitoring of organophosphorus pesticides, *Sens. Actuat. B. Chem.*, 36 (1996) 448–457.

7 T. Montesinos, S. Pérez-Munguia, F. Valdez and J.L. Marty, Disposable cholinesterase biosensor for the detection of pesticides in water-miscible organic solvents, *Anal. Chim. Acta*, 431 (2001) 231–237.

8 T.T. Bachmann, B. Leca, F. Vilatte, J.-L. Marty, D. Fournier and R.D. Schmid, Improved multianalyte detection of organophosphates and carbamates with disposable multielectrode biosensors using recombinant

mutants of *Drosophila* acetylcholinesterase and artificial neural networks, *Biosens. Bioelectron*, 15 (2000) 193–201.
9 T.T. Bachmann and R.D. Schmid, A disposable multielectrode biosensor for rapid simultaneous detection of the insecticides paraoxon and carbofuran at high resolution, *Anal. Chim. Acta*, 401 (1999) 95–103.
10 J.M. Abad, F. Pariente, L. Hernandez, H.D. Abruna and E. Lorenzo, Determination of organophosphorus and carbamate pesticides using a piezoelectric biosensor, *Anal. Chem.*, 70 (1998) 2848–2855.
11 J.J. Rippeth, T.D. Gibson, J.P. Hart, I.C. Hartley and G. Nelson, Flow-injection detector incorporating a screen-printed disposable amperometric biosensor for monitoring organophosphorus pesticides, *Analyst*, 122 (1997) 1425–1429.
12 G.A. Evtugyn, H.C. Budnikov and E.B. Nikolskaya, Influence of surface-active compounds on the response of cholinesterase biosensors for inhibitor determination, *Analyst*, 121 (1996) 1911–1915.
13 A.L. Ghindilis, T.G. Morzunova, A.V. Barmin and I.N. Kurochkin, Potentiometric biosensors for cholinesterase inhibitor analysis based on mediatorless bioelectrocatalysis, *Biosens. Bioelectron*, 11 (1996) 873–880.
14 N. Mionetto, J.-L. Marty and I. Karube, Acetylcholinesterase in organic solvents for the detection of pesticides: biosensor application, *Biosens. Bioelectron*, 9 (1994) 463–470.
15 P. Skladal and M. Mascini, Sensitive detection of pesticides using amperometric sensors based on cobalt phthalocyanine-modified composite electrodes and immobilised cholinesterase, *Biosens. Bioelectron*, 7 (1992) 335–343.
16 G. Palleschi, M. Bernabei, C. Cremisini and M. Mascini, Determination of organophosphorus insecticides with a choline electrochemical biosensor, *Sens. Actuators B*, 7 (1992) 513–517.
17 S. Andreescu and O.A. Sadik, Trends and challenges in biochemical sensors for clinical and environmental monitoring, *Pure Appl. Chem.*, 76 (2004) 861–878.
18 A.L. Hart, W.A. Collier and D. Janssen, The response of screenprinted enzyme electrodes containing cholinesterases to organophosphates in solution and from commercial formulations, *Biosens. Bioelectron*, 12 (1997) 645–654.
19 J.P. Hart and R.M. Pemberton, Application Note 121: A review of screen-printed electrochemical sensors. Perkin–Elmer Instruments, Princeton Applied Research, 1999.
20 S. Andreescu, T. Noguer, V. Magearu and J.-L. Marty, Screen-printed electrode based on AChE for the detection of pesticides in the presence of organic solvents, *Talanta*, 57 (2002) 169–176.
21 H. Schulze, S. Vorlova, F. Vilatte, T.T. Bachmann and R.D. Schmid, Design of acetylcholinesterases for biosensor applications, *Biosens. Bioelectron*, 18 (2003) 201–209.

22 C. Bonnet, S. Andreescu and J.-L. Marty, Adsorption and easy and efficient immobilisation of acetylcholinesterase on screen-printed electrodes, *Anal. Chim. Acta*, 481 (2003) 209–211.

23 B. Bucur, M. Dondoi, A. Danet and J.-L. Marty, Insecticide identification using a flow injection analysis system with biosensors based on various cholinesterases, *Anal. Chim. Acta*, 539 (2005) 195–201.

24 S. Sotiropoulou, D. Fournier and N.A. Chaniotakis, Genetically engineered acetylcholinesterase-based biosensor for attomolar detection of dichlorvos, *Biosens. Bioelectron*, 20 (2005) 2347–2352.

25 B. Bucur, A.F. Danet and J.-L. Marty, Versatile method of cholinesterase immobilisation via affinity bonds using Concanavalin A applied to the construction of a screen-printed biosensor, *Biosens. Bioelectron*, 20 (2004) 217–225.

26 B. Bucur, A.F. Danet and J.-L. Marty, Cholinesterase immobilisation on the surface of screen-printed electrodes based on Concanavalin A affinity, *Anal. Chim. Acta*, 530 (2005) 1–6.

27 J. Pritchard, L.A. Law, A. Vakurov, P. Millner and S.P.J. Higson, Sonochemically fabricated enzyme microelectrode arrays for the environmental monitoring of pesticides, *Biosens. Bioelectron*, 20 (2004) 765–772.

28 J.-L. Marty, D. Garcia and R. Rouillon, Biosensors: potential in pesticides detection, *Trends. Anal. Chem.*, 14 (1995) 3329–3333.

29 Y. Boublik, P. Saint-Aguet, A. Lougarre, M. Arnaud, F. Villatte, S. Estrada-Mondaca and D. Fournier, Acetylcholinesterase engineering for the detection of insecticide residues, *Prot. Eng.*, 15 (2002) 43–50.

30 S. Estrada-Mondaca and D. Fournier, Stabilization of recombinant *Drosophila* acetylcholinesterase, *Prot. Exp. Purif.*, 12 (1998) 166–172.

31 G.L. Ellman, K.D. Courtney and R.M. Featherstone, A new and rapid colorimetric determination of acetylcholinesterase activity, *Biochem. Pharmacol.*, 7 (1961) 88–95.

32 P.R.B.O. Marques, G.S. Nunes, T.C.R. dos Santos, S. Andreescu and J.-L. Marty, Comparative investigation between acetylcholinesterase obtained from commercial sources and genetically modified *Drosophila melanogaster*. Application in amperometric detectors for methamidophos pesticide detection, *Biosens. Bioelectron*, 20 (2004) 825–832.

33 S. Andreescu, L. Barthelmebs and J.-L. Marty, Immobilisation of acetylcholinesterase on screen-printed electrodes: comparative study between three immobilisation methods and applications to the detection of organophosphorus pesticides, *Anal. Chim. Acta*, 464 (2002) 171–180.

34 G.S. Nunes, G. Jeanty and J.-L. Marty, Enzyme immobilisation procedures on screen-printed electrodes used for the detection of anticholinesterase pesticides. Compartive study, *Anal. Chim. Acta*, 523 (2004) 107–115.

35 S. Sotiropoulou and N.A. Chaniotakis, Lowering the detection limit of the acetylcholinesterase biosensor using a nanoporous carbon matrix, *Anal. Chim. Acta*, 530 (2004) 199–204.
36 S. Sotiropoulou, V. Vamvakaki and N.A. Chaniotakis, Stabilization of enzymes in nanoporous materials for biosensor applications, *Biosens. Bioelectron*, 20 (2005) 1674–1679.
37 M. Gerard, A. Chaubey and B.D. Malholtra, Application of conducting polymers to biosensors, *Biosens. Bioelectron*, 17 (2002) 345–359.
38 A.C. Barton, S.D. Collyer, F. Davis, D.D. Gornall, K.A. Law, E.C.D. Lawrence, D.W. Mills, S. Myler, J.A. Pritchard, M. Thompson and S.P.J. Higson, Sonochemically fabricated micro-electrode arrays for biosensors offering widespread applicability, *Part I, Biosens. Bioelectron*, 20 (2004) 328–337.
39 S.P.J. Higson, Sensor International Patent: PCT/GB96/00922, continuation of UK patent 9507991—filed 19th November 1996. Patents granted to date in Europe, US, Canada, Japan and Australia.
40 K.A. Law and S.P.J. Higson, Sonochemically fabricated acetylcholinesterase micro-electrode arrays within a flow injection analyser for the determination of organophosphate pesticides, *Biosens. Bioelectron*, 20 (2005) 1914–1924.
41 G. Jeanty, A. Wojciechowska and J.-L. Marty, Flow-injection amperometric determination of pesticides on the basis of their inhibition of immobilised acetylcholinesterases of different origin, *Anal. Bioanal. Chem.*, 373 (2002) 691–695.
42 S. Andreescu, A. Avramescu, C. Bala, V. Magearu and J.-L. Marty, Detection of organophosphorus insecticides with immobilised acetylcholinesterase—comparative study of two enzyme sensors, *Anal. Bioanal. Chem.*, 374 (2002) 39–45.
43 V. Beni, V.I. Ogurtsov, N.V. Bakounine, D.W.M. Arrigan and M. Hill, Development of a portable electroanalytical system for the stripping voltammetry of metals: determination of copper in acetic acid soil extracts, *Anal. Chim. Acta*, 552 (2005) 190–200.

Chapter 16

Amperometric enzyme sensors for the detection of cyanobacterial toxins in environmental samples

M. Campàs and J.-L. Marty

16.1 INTRODUCTION

16.1.1 Cyanobacterial toxins: the overview

Cyanobacteria are also known as blue-green algae due to their chlorophyll-*a* and other accessory pigments. They are one of the oldest life forms on earth and their presence has been observed in fresh, brackish and marine water all over the world.

Certain strains of cyanobacteria produce toxins. These cyanobacterial toxins can be classified according to their chemical structure or their toxicity. Table 16.1 summarises the characteristics of the main cyanobacterial toxins. Depending on the chemical structure, there are cyclic peptides, alkaloids and lipopolysaccharides. According to the toxic effects, they are classified as:

(a) *Hepatotoxins*: They are the most commonly found cyanobacterial toxins. They include two groups: the cyclic heptapeptide microcystins and the cyclic pentapeptide nodularins. Microcystins have been found in species from the genera *Microcystis*, *Anabaena*, *Oscillatoria* (*Planktothrix*), *Nostoc* and *Anabaenopsis* [1–4]. Nodularins have been isolated only from *Nodularia spumigena* [3,5]. Hepatotoxins are implicated in the inhibition of serine/threonine protein phosphatases type 2A (PP2A) and 1 (PP1) [6–11], which results in hyperphosphorylation of cytoskeletal filaments, hepatocyte deformation, tumour promotion and liver damage [8].

(b) *Neurotoxins*: They are low-molecular-weight alkaloids. They include two groups: the anatoxins and the saxitoxins. Anatoxin-a

Comprehensive Analytical Chemistry 49
Alegret and Merkoçi (Eds)
Volume 49 ISSN: 0166-526X DOI: 10.1016/S0166-526X(06)49016-4

TABLE 16.1
General features of cyanotoxins

Toxin	Structure	Pathology	Organism	Country
Microcystin	Cyclic heptapeptide	Hepatotoxin	*Microcystis, Anabaena, Oscillatoria, Nostoc, Anabaenopsis, Hapalosiphon*	Australia, Belgium, Brazil, Canada, China, Czech Republic, Denmark, Egypt, England, Finland, France, Germany, Greece, Japan, Korea, Latvia, Norway, Philippines, Portugal, Slovenia, Switzerland, Taiwan, Thailand, USA
Nodularin	Cyclic pentapeptide	Hepatotoxin	*Nodularia*	Australia, Baltic Sea, New Zealand, Tasmania
Anatoxin-a	Alkaloid	Neurotoxin	*Anabaena, Oscillatoria, Aphanizomenon, Cylindrospermum*	Canada, Chine, Finland, Germany, Ireland, Italy, Japan, Scotland, USA

Homoanatoxin-a	Alkaloid	Neurotoxin	*Oscillatoria*	Norway
Anatoxin-a(s)	Alkaloid	Neurotoxin	*Anabaena, Aphanizomenon*	Canada, Denmark, USA
Saxitoxin	Alkaloid	Neurotoxin	*Anabaena, Aphanizomenon, Oscillatoria, Lyngbya, Cylindrospermopsis, Alexandrium, Gymnodinium, Pyrodinium*	Australia, Brazil, Portugal, USA
Cylindrospermopsin	Alkaloid	Cytotoxin	*Aphanizomenon, Umezakia, Cylindrospermopsis*	Australia, Brazil, Hungary, Israel, Japan, USA
Lyngbyatoxin	Alkaloid	Dermatoxin	*Lyngbya*	Australia, Japan, USA
Aplysiatoxin	Alkaloid	Dermatoxin	*Lyngbya, Oscillatoria, Schizothrix*	Australia, Japan, USA
Debromoaplysiatoxin	Alkaloid	Dermatoxin	*Oscillatoria, Schizothrix*	Australia, Japan, USA
Lipopolysaccharide endotoxins (LPS)	Lipopolysaccharides	Irritant toxins	All	All

See Refs. [85,97,98] and text for references.

and homoanatoxin-a act as postsynaptic cholinergic agonists, binding to neuronal nicotinic acetylcholine receptors. But unlike natural acetylcholinesterase substrates, they cannot be broken down by this enzyme and cause overstimulation of the muscle cells [12]. They have been characterised from *Anabaena*, *Oscillatoria* (*Planktothrix*), *Aphanizomenon* and *Cylindrospermum* [13–16]. Anatoxin-a(s) has been found in *Anabaena* and *Aphanizomenon* [17,18] and is implicated in the inhibition of acetylcholinesterase, acting similarly to some organophosphate insecticides [19]. Saxitoxins have been isolated from *Anabaena*, *Aphanizomenon*, *Lyngbya* and *Cylindrospermopsis raciborskii* cyanobacteria and from marine *Alexandrium tamarense*, *Gymnodinium catenatum* and *Pyrodinium bahamense* dinoflagellates [5,20–24]. They are carbamate alkaloids that block sodium channels, inhibiting nerve conduction [25].

(c) *Cytotoxins*: Cylindrospermopsin is a cyclic guanidine alkaloid produced by *Aphanizomenon ovalisporum*, *Umezakia natans* and *C. raciborskii* [26–28]. It is implicated in the inhibition of glutathione and protein synthesis and is responsible for necrotic and genetic damage [29–32].

(d) *Dermatoxins*: Lyngbyatoxin, aplysiatoxin and debromoaplysiatoxin are alkaloids produced by benthic marine cyanobacteria *Lyngbya*, *Oscillatoria* and *Schizothrix* [24]. They are protein kinase C activators and cause skin irritation, tumour formation and, sometimes, gastrointestinal inflammation [33].

(e) *Irritant toxins*: Lipopolysaccharide endotoxins (LPS) have been isolated from *Anacystis nidulans* [34], although in fact are constituents of the gram-negative cell wall, and are therefore common to all cyanobacteria. They are related to gastrointestinal inflammatory incidents [35,36].

This chapter is focused on microcystins and anatoxin-a(s), for which our research group has developed the corresponding electrochemical biosensors. As previously mentioned, microcystins are monocyclic heptapeptides with the general structure cyclo(D-Ala^1-L-X^2-D-MeAsp^3-L-Y^4-Adda^5-D-Glu^6-Mdha^7), containing five constant amino acids and two variable ones (X and Y), with a molecular weight between 900 and 1100 Da (Fig. 16.1) [37]. They contain two novel amino acids: the Mdha (*N*-methyl-dehydroalanine) and the hydrophobic Adda (3-amino-9-methoxy-2,6,8-trimethyl-10-phenyldeca-4,6-dienoic acid). This chain allows the toxins to penetrate the hepatocytes [38], where they irreversibly

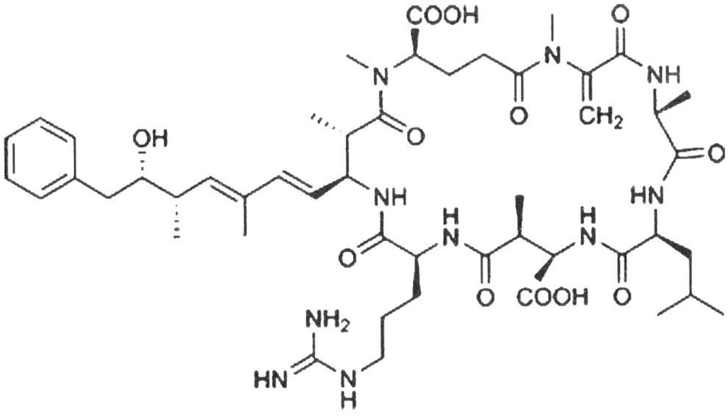

Fig. 16.1. Chemical structure of microcystin-LR. This image is licensed under the "http://www.gnu.org/copyleft/fdl.html" GNU Free Documentation License. It uses material from the "http://en.wikipedia.org/wiki/Cyanotoxin" Wikipedia article.

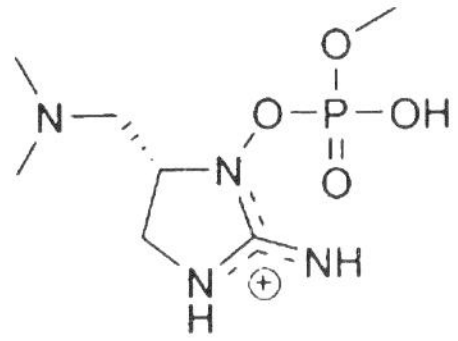

Fig. 16.2. Chemical structure of anatoxin-a(s). This image is licensed under the "http://www.gnu.org/copyleft/fdl.html" GNU Free Documentation License. It uses material from the "http://en.wikipedia.org/wiki/Cyanotoxin" Wikipedia article.

inhibit the protein phosphatases by the formation of covalent bonds. Thanks to their cyclic structure, microcystins are extremely stable in water and highly resistant to heat, hydrolysis and oxidation, characteristics that explain the difficulty to degrade or remove them. To date, nearly 80 variants [5] of microcystins have been identified, each one of them with different polarity, lipophilicity and toxicity. Among them, microcystin-LR, with leucine (L) and arginine (R) as variable amino acids, is the most frequent and most toxic congener.

Anatoxin-a(s) is a phosphate ester of a cyclic *N*-hydroxyguanidine (Fig. 16.2) [5]. It is the only natural organophosphate known and, as the synthetic parathion and malathion, irreversibly inhibits acetylcholinesterase. When this enzyme is inhibited, acetylcholine is no longer hydrolysed, the postsynaptic membrane cannot be repolarised, the nerve influx is blocked and the muscle cannot be contracted.

16.1.2 Cyanobacteria growth and blooms – toxin release

Cyanobacteria growth is associated with several environmental factors: (a) high nutrient concentrations, intensified by agricultural and industrial activities near aquatic areas; (b) low turbulence flow regimes; (c) high light intensity and quality, required for the photosynthesis; (d) warm temperatures, which explain why most blooms occur in summer; and (e) the presence of trace metals, like iron [39,40]. The characteristic buoyancy of cyanobacteria allows them to regulate their position in the water column to achieve the optimal nutrient and light conditions [41]. Despite the age of cyanobacteria and their toxins, the conditions that cause toxin production, the variations in toxin production observed between strains and along years, and the physiological function of the toxins within or outside the cell are still not completely understood. All these mysteries make necessary the combination of tools from the molecular ecology, taxonomy, genomics, proteomics, toxicology and analytical chemistry fields for an interdisciplinary comprehension of cyanobacteria and cyanotoxins.

16.1.3 Analytical methods for microcystin and anatoxin-a(s) detection

In order to guarantee the water and food quality and to preserve the human health, several detection and quantification methods, each one of them having advantages and disadvantages, have been developed for microcystins and anatoxin-a(s).

With respect to microcystins, the mouse bioassay screening method [42,43] is being replaced by other methods, more sensitive and reliable and less problematic from an ethical point of view. The two most common methods, usually used in parallel, are the protein phosphatase inhibition assay and the high-performance liquid chromatography (HPLC). The enzymatic inhibition assay, usually colorimetric although also radiometric and fluorimetric, informs about the toxicity of the sample [10,44–49], whereas HPLC, usually coupled to a UV detector, allows the microcystin variant identification [50,51]. Nevertheless, these techniques present some drawbacks: the protein phosphatase may also be inhibited by other toxins, which makes the inhibition assay not specific, and the HPLC technique requires trained personnel, expensive equipment and sample pre-treatment, involving long analysis times. Enzyme-linked immunosorbent assays (ELISAs) are based on monoclonal or polyclonal antibodies. They have the advantage of being highly specific but usually present problems of cross-reactivity among microcystin and nodularin congenerers [52–54]. Novel technologies are being exploited, such as liquid chromatography/mass spectrometry (LC/MS) [51,55], thin-layer chromatography (TLC) [56], capillary electrophoresis (CE) [57,58] and gas chromatography/mass spectrometry (GC/MS) [59]. The choice of the detection method depends on the purpose and, since they have different and complementary information, the combination of them provides the complete identification.

Anatoxin-a(s) was first detected by the mouse bioassay. However, the technique most commonly used for its detection is HPLC, coupled to MS detection. The irreversible inhibitory power of this toxin towards acetylcholinesterase has been described [60] and the corresponding colorimetric inhibition assay has also been developed [61–63]. To date, no antibodies towards anatoxin-a(s) have been produced.

Our research group is working on the development of electrochemical biosensors for the detection of microcystin and anatoxin-a(s), based on the inhibition of protein phosphatase and acetylcholinesterase, respectively. These enzyme biosensors represent useful bioanalytical tools, suitable to be used as screening techniques for the preliminary yes/no detection of the toxicity of a sample. Additionally, due to the versatility of the electrochemical approach, the strategy can be applied to the detection of other cyanobacterial toxins.

16.1.4 Environmental and health effects – guideline values

Cyanobacterial toxins produce adverse health effects on animals and humans, ranging from mild to fatal [6,64–74]. Although oral ingestion by humans is difficult, due to the ability of cyanobacterial algae to discolour the water, accumulate as blooms, scums, mats and biofilms, and sometimes produce odour compounds, these evidences may slip past animals. Moreover, the absence of cyanobacteria does not guarantee an absence of toxins, as they can occur in the water after the release [50]. Other human exposure routes are accidental ingestion or dermal contact with poisoned water in recreational areas, inhalation of aqueous aerosols [75,76], consumption of toxic cyanobacterial dietary supplements [77], vegetables irrigated by contaminated water [78] and poisoned animals [79–81] and the particular case of intravenous uptake via contaminated haemodialysis water [82–84].

The potential risk for the public health has led the World Health Organisation (WHO) to define a provisional guideline value of $1\,\mu g\,L^{-1}$

for microcystin-LR in drinking water that, in principle, should be enough to protect the consumer [85]. This value is derived from the typical daily intake of water, the individual's body weight and the toxicity, and it is provisional because is subjected to the appearance of new data. This current recommendation is integrated into several action plans for the management of water resources. In fact, although the WHO guidance value has not been yet incorporated into the EC legislation, it has been used, sometimes with slight modifications, in the establishment of national legislations for water and food quality in Spain, Poland, France and Canada. To date, no guideline values for other cyanobacterial toxins have been established due to the lack of data to derive them. In fact, anatoxin-a(s) is currently not monitored in drinking water. Consequently, powerful analytical tools are becoming necessary to obtain reliable and meaningful experimental data for the implementation of water quality assessment programs.

16.2 APPLICATION

16.2.1 Protein phosphatase-based biosensor for electrochemical microcystin detection

16.2.1.1 Enzyme substrates

Protein phosphatase inhibition is usually detected by colorimetric methods, but the development of a biosensor requires the search of other transduction techniques. Electrochemistry has been widely used in biosensors because of the simplicity, easy to use, portability, disposability and cost-effectiveness of the devices. As protein phosphatase is not an oxidoreductase enzyme, our work has been devoted to the investigation of novel enzymatic substrates, electrochemically active only after their dephosphorylation by the protein phosphatase. Nevertheless, colorimetric assays have been used for the optimisation of several experimental parameters.

Two different protein phosphatases were used: the one from Upstate Biotechnology (New York, USA), from human red blood cells, and the one from GTP Technology (Toulouse, France), isolated from SF9 insect cells infected by baculovirus. The enzymatic activity of these two enzymes towards several substrates was investigated by cyclic voltammetry and steady-state chronoamperometry (see experimental details in Refs. [86,87]). First, commercial substrates were tested. Ascorbic acid 2-phosphate and phenyl phosphate were not recognised by the protein phosphatases. Anyway, the experiments performed with the alkaline phosphatase model enzyme showed that ascorbic acid requires high working potentials to be detected and phenol blocks the electrode surface. α-Naphthyl phosphate, *p*-aminophenyl phosphate and 4-methylumbelliferyl phosphate were recognised only by the protein phosphatase from Upstate. However, the fouling effect, the instability of the product, and the high working potential required, respectively, forced us to attempt the synthesis of two novel substrates: 4-methoxyphenyl phosphate and catechyl phosphate. The former was not recognised by the enzymes and, in any case, it would also cause a fouling effect. Nevertheless, the latter was recognised by the enzyme from Upstate (Fig. 16.3) and showed less electrode fouling than other substrates. Catechyl phosphate allows us to work at +450 mV (vs. Ag/AgCl), oxidation peak corresponding to the produced catechol (Fig. 16.4). The partial electrode fouling due to the phenolic radicals is responsible for the shift of the oxidation peak to higher potentials, when compared to the cyclic voltammogram of freshly prepared catechol. None of the trials with the enzyme from GTP provided satisfactory results. It seems that this enzyme is only active towards the colorimetric *p*-nitrophenyl phosphate substrate.

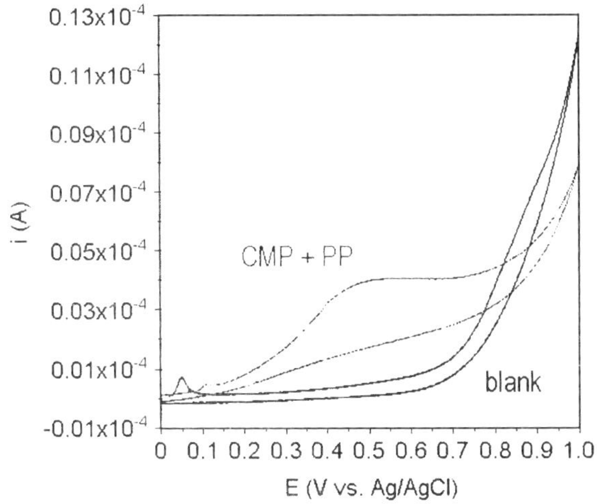

Fig. 16.3. Cyclic voltammogram of 13.6 U PP2A from Upstate (blank) and 0.3 mM catechol monophosphate+13.6 U PP2A from Upstate (CMP+PP) at 100 mV s^{-1} using a single-drop configuration on a horizontally supported screen-printed two-electrode system, with graphite as working and Ag/AgCl as reference/auxiliary electrode. Reprinted from Campàs *et al.* [86], with permission from Elsevier.

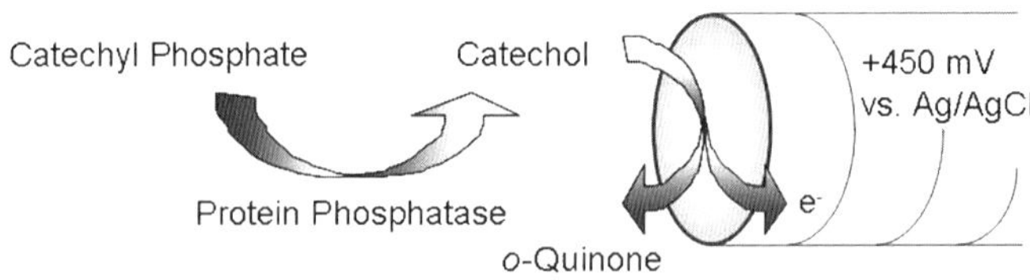

Fig. 16.4. Schematic representation of the enzymatic reaction between protein phosphatase and catechyl phosphate and the subsequent catechol detection on the electrode surface.

16.2.1.2 Enzyme immobilisation

One of the key factors in biosensor design is the immobilisation technique used to attach the biorecognition molecule to the transducer surface so as to render it in a stable and functional form. The challenge is to have a stable layer (or layers) of biorecognition molecules that do not desorb from the surface and that retain their activity. Entrapment or encapsulation techniques avoid the chemical changes that usually change the structure of the enzymes and modify their recognition capacity.

The GTP protein phosphatase was immobilised with poly(vinyl alcohol) bearing styrylpyridinium groups, also named Poly(Vinyl Alcohol)-Stilbazolium Quaternary (PVA-SbQ), on Maxisorp microtitre wells, UltraBind modified polyethersulfone membranes and screen-printed graphite electrodes, by simply mixing the two components (enzyme and polymer), spreading the mixture on the surfaces and letting it polymerise under neon light (see details in Ref. [86]). Colorimetric enzymatic assays were performed to characterise the immobilisation yield and to optimise the drying time after photocrosslinking. The highest immobilisation yields were obtained when the enzyme was immobilised on microtitre wells. Compared to the electrodes, wells have a higher surface area, which favours the spreading of the enzyme:polymer mixture and the enzyme accessibility. Compared to the membranes, whereas in the immobilisation on wells only physical entrapment is present, in the immobilisation on membranes there is also a covalent contribution due to the reaction of their active aldehyde groups with the free amino groups of the enzyme. These covalent bonds may be restricting the enzyme functionality.

Regarding the drying time, the enzyme-modified electrodes dried for 30 min provided higher absorbance values than the ones dried for 22 h. Hence, short drying times were enough to immobilise the enzyme and longer times only accelerated the enzyme inactivation. At this stage it is important to mention the fast inactivation rate of the enzyme. The best storage conditions for the enzyme are in a 50% glycerol solution at −18°C. Its simple manipulation for the biosensor construction involves a problem, as glycerol-free enzyme solutions (required for the immobilisation) do not show any enzymatic activity after some hours at 4°C. Nevertheless, the absorbance values obtained from the enzyme-modified electrodes demonstrated that the PVA entrapment technique not only immobilises the enzyme but also retains its stability.

The Upstate protein phosphatase was immobilised with Poly(Vinyl Alcohol) Azide-unit pendant Water-soluble Photopolymer (PVA-AWP) on gold and platinum foils, and screen-printed graphite electrodes (see experimental details in Ref. [87] and Procedure 21 in CD accompanying this book). Theoretically, this polymer has higher photoreactivity and provides lower swelling than PVA-SbQ. Colorimetric enzymatic assays were performed to characterise the immobilisation yield and to optimise the enzyme:polymer ratio. With the purpose of improving the electrode reproducibility, gold and platinum foils were used. However, the deposited PVA polymer was removed with rinsing. The lack of adsorption of the polymer on these materials forced us to continue with the screen-printed graphite electrodes. Colorimetric assays demonstrated that after 30-min incubation of the enzyme-modified electrodes in buffer (time required for the microcystin incubation), the absorbance values decreased to 62% with respect to non-incubated electrodes. The decrease in the activity could be due to the partial enzyme inactivation and/or to the leakage of the enzyme from the net during the 30-min incubation step. Despite this phenomenon, the response was high enough to continue with the electrochemical biosensor development.

Regarding the enzyme:polymer ratio, high polymer amounts may provide higher immobilisation yields but may restrict the enzyme flexibility and functionality, and high enzyme amounts may provide higher responses but may limit the sensitivity of the biosensor. The dilemma has appeared and the choice will depend on the own particular interests. The 1:2 ratio provided higher absorbance values (0.259; coefficient of variation (CV) = 17%) than those obtained with the 2:1 (0.041; CV = 9%) and the 1:3 (0.189; CV = 15%) ratios. In the same way, the 1:2 ratio also provided the highest immobilisation yields after 30-min incubation in buffer (62% as compared to 16% and 41% for the 2:1 and the 1:3 ratios, respectively). In our case, the results clearly demonstrate that the 1:2 enzyme:polymer ratio is the optimum one for the biosensor construction.

16.2.1.3 Microcystin detection

In order to demonstrate the viability of the approach, protein phosphatase inhibition was first performed with the enzyme in solution and detected by colorimetric methods. Two microcystin variants, microcystin-LR and microcystin-RR, were used. Both enzymes were inhibited by these toxins, although to a different extent. The 50% inhibition coefficients (IC_{50}) towards microcystin-LR were 0.50 and 1.40 $\mu g\,L^{-1}$ (concentrations in the microtitre well) for the Upstate and the GTP enzymes, respectively. Hence, the Upstate enzyme was more sensitive. The IC_{50} towards microcystin-RR were 0.95 and 2.15 $\mu g\,L^{-1}$ for the Upstate and the GTP enzymes, respectively. As expected, microcystin-LR was demonstrated to be a more potent inhibitor.

Microcystin detection with immobilised protein phosphatase from GTP was first characterised by colorimetric methods. Figure 16.5 plots the colorimetric standard curves for the inhibition of this enzyme immobilised on different supports by microcystin-LR and microcystin-RR. The immobilised enzyme also recognised the microcystin presence, although the net limited the mass transport of both the toxin and the enzyme substrate towards the immobilised protein phosphatase. This phenomenon decreased the sensitivity of the assay, as the higher limits of detection, IC_{50} and IC_{100}, show. The diffusion barrier also decreased the reproducibility between electrodes, the CVs increasing from 15 (with the enzyme in solution) to 25%. Nevertheless, results were satisfactory enough to continue with the biosensor development.

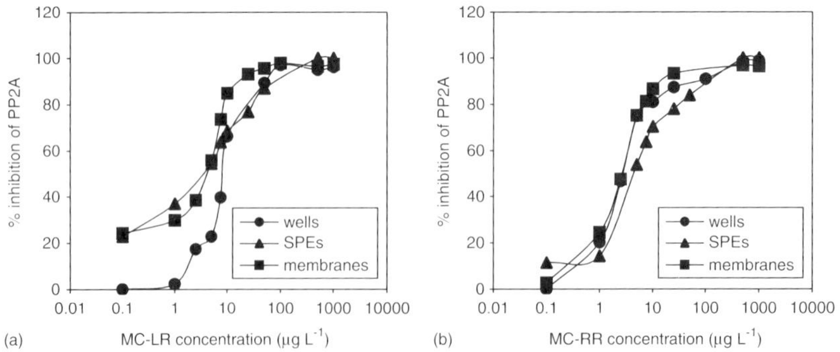

Fig. 16.5. Colorimetric standard curves for the inhibition of PP2A from GTP Technology immobilised on different supports by (a) microcystin-LR (MC-LR) and (b) microcystin-RR (MC-RR). Inhibition is expressed as percentage of the control (no microcystin). Concentrations refer to those in the well. Reprinted from Campàs *et al.* [86], with permission from Elsevier.

After the colorimetric assays, microcystin detection with immobilised protein phosphatase from Upstate was characterised by electrochemical methods. Catechyl phosphate was used as substrate and, after 20-min incubation with the enzyme, the catechol detection was carried out by applying a potential of +450 mV vs. Ag/AgCl for 1 min. The electrochemical standard curve for the inhibition of this enzyme immobilised on screen-printed graphite electrodes by microcystin-LR (see Fig. 21.2 of Procedure 21 in CD accompanying this book) showed an IC_{50} of 83.08 $\mu g L^{-1}$ and a limit of detection between 10 and 20 $\mu g L^{-1}$, with 100% inhibition at about 1000 $\mu g L^{-1}$. Despite the lower sensitivity with respect to the colorimetric approach with the enzyme in solution (IC_{50} of 3.06 $\mu g L^{-1}$, limit of detection between 0.5 and 1 $\mu g L^{-1}$, and 100% inhibition at about 10 $\mu g L^{-1}$), the amperometric biosensor is still useful as screening tool to discriminate between toxic and non-toxic samples. Work is in progress to improve the reproducibility and the sensitivity of the electrochemical strategy by the use of novel electrode materials, signal amplification systems and molecularly engineered supersensitive enzymes.

16.2.1.4 Analysis of real samples

The developed biosensor should be, in principle, applicable to any sample suspicious of containing microcystins. This includes water samples, cyanobacterial cells and infected organisms. Whereas water samples could be directly applied to the biosensor, specific extraction protocols are required for the detection of the toxin in cyanobacteria and other organisms. In all the cases, matrix effects should be evaluated.

The developed biosensor was applied to the analysis of cyanobacterial bloom samples from the Tarn Region (Midi-Pyrénées, France) collected in 2002. That summer, 26 dogs, which had gone to quench their thirst at the Tarn River, fell ill and 20 of them died. It was hypothesised that cyanobacterial toxins could be responsible for that intoxication. This theory was confirmed the following years: whereas no cases were counted in 2004, year characterised by the intense rains, the scorching heat and the drought of 2003 and 2005 resulted in 15 poisoned dogs (nine of them died). The high temperatures contributed not only to the cyanobacterial growth but also to the toxin release.

Results obtained with the electrochemical biosensor were compared to those obtained from the colorimetric PPI assay with the enzyme in solution and by HPLC (see Table 21.1 of Procedure 21 in CD accompanying this book). All real samples contained microcystin at levels detectable by the amperometric biosensor and the colorimetric PPI assay. However, when HPLC was used, the microcystin presence was detected in only three samples. This fact can be explained taking into account that only microcystin-LR, microcystin-RR and microcystin-LF were used as standards in HPLC, whereas protein phosphatase can be inhibited by more toxin variants. The two enzymatic methods correlate well ($y = 4.7991x - 7.3555$; $R^2 = 0.9921$), but higher toxin levels are always obtained when samples are analysed with the biosensor. Since when microcystin standard solutions were used, the colorimetry assay showed lower limits of detection, it seems that the electrochemical biosensor gives a slight overestimation of the toxin content. This effect could be due to the fouling of the electrode by some of the cell extract components, which would decrease the steady-state oxidation currents, simulating an enzymatic inhibition. Nevertheless, results clearly justify the use of the amperometric biosensor as screening tool. In positive samples, the use in parallel to other analytical techniques, such as LC combined with UV MS detection, would accurate the microcystin determination and would enable the variant identification.

16.2.2 Acetylcholinesterase-based biosensor for electrochemical anatoxin-a(s) detection

16.2.2.1 Enzyme choice

Acetylcholinesterase inhibition has been widely used for pesticide detection [88–94], but less exploited than protein phosphatase inhibition for cyanobacterial toxin detection. Nevertheless, the anatoxin-a(s) has more inhibition power than most insecticides, as demonstrated by the higher inhibition rates [95]. In order to detect toxin concentrations smaller than usually, mutant enzymes with increased sensitivity were obtained by genetic engineering strategies: residue replacement, deletion, insertion and combination of mutations. Modifications close to the active site, located at the bottom of a narrow gorge, made the entrance of the toxin easier and enhanced the sensitivity of the enzyme.

The main drawback of acetylcholinesterase-based biosensors is the lack of selectivity because, as we mentioned, this enzyme is inhibited not only by anatoxin-a(s) but also by insecticides such as organophosphorates and carbamates. This problem can be overcome by the choice of specific mutant enzymes. The combined use of mutants highly sensitive to anatoxin-a(s) and resistant to most insecticides and vice versa allows us to unambiguously discriminate between the cyanobacterial toxin and insecticides.

It is also important to mention the use of the reactivation of the acetylcholinesterase by pyridine-2-aldoxime methochloride to discriminate between the toxin and potential insecticides [96]. Once phosphorylated, the active site serine of the enzyme can be reactivated by powerful nucleophilic agents such as oximes. However, this reactivation is not possible if attempted too late due to the stable adduct formed by the dealkylation (aging) of the inhibitor's remaining group. When acetylcholinesterase is inhibited by anatoxin-a(s), it shows immediately the characteristics of an aged enzyme and cannot be reactivated. In this way, it is possible to distinguish between the inhibition caused by anatoxin-a(s) and the one provoked by other insecticides.

16.2.2.2 *Enzyme immobilisation*

Acetylcholinesterase was immobilised by entrapment into a PVA-SbQ matrix (see experimental details in Refs. [88,95]). The need of polymer hydration slightly increases the response times, when compared to other immobilisation techniques. Nevertheless, the entrapment presents the advantage of providing biosensors with longer lifetimes due to the protective effect of the polymer matrix.

The enzyme immobilisation was carried out on 7,7,8,8-tetracyanoquinodimethane (TCNQ)-modified screen-printed electrodes. TCNQ allows the electrochemical oxidation of thiocholine, the product of the reaction between acetylthiocholine and the enzyme, at +100 mV (vs. Ag/AgCl) (Fig. 16.6). The enzymatic activity of the acetylcholinesterase can thus be monitored by electrochemical methods.

16.2.2.3 *Anatoxin-a(s) detection*

Acetylcholinesterase inhibition was first performed with the enzyme in solution and detected by colorimetric methods. As the inhibition is irreversible and stoichiometric, it is possible to calculate the toxin concentration by the percentage of the enzyme that has not been inhibited and remains active. The use of several enzyme concentrations and the

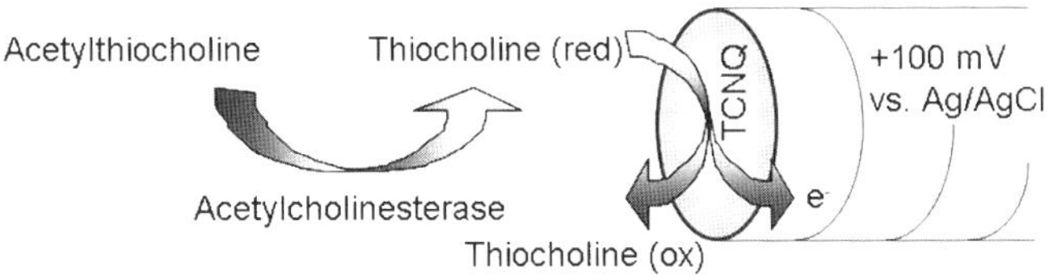

Fig. 16.6. Schematic representation of the enzymatic reaction between acetylcholinesterase and acetylthiocholine and the subsequent thiocholine detection on the TCNQ-modified electrode surface.

multiple non-linear data fitting allowed us to confirm and refine the toxin concentration.

When the amperometric biosensor was used, results correlated with those obtained by colorimetric methods. The two sensitive enzymes allowed the detection of 0.5 nM anatoxin-a(s) and the limits of detection obtained with the two insensitive mutants were 16- and 50-fold higher than those obtained with the sensitive ones.

16.2.2.4 Analysis of real samples

The developed biosensor was applied to the analysis of cyanobacterial bloom samples from freshwater lakes of Spain, Greece, France, Scotland and Denmark. Two samples from Scotland and one from Denmark irreversibly inhibit the acetylcholinesterase. The estimated concentrations were between 1.5 and 30 nmol/g of dry weight, values extremely high when compared to the intraperitoneal 50% lethal dose of anatoxin-a(s) in mice (121 nmol/kg).

16.3 CONCLUSIONS

The prompt detection of cyanotoxin in water reservoirs and recreational areas is the only way to guarantee the water quality and to assure the public health. As a consequence, vigilance programs are being implemented around the world, most of them based on the continuous monitoring of water resources and supplies. The method par excellence for cyanobacterial toxin analysis is HPLC, but it suffers from long analysis times and lack of commercially available standards. Consequently, the development of biotools for fast, simple and reliable screening of environmental samples is becoming a need.

On the one hand, protein phosphatase and acetylcholinesterase inhibition assays for microcystin and anatoxin-a(s) detection, respectively, are excellent methods for toxin analysis because of the low limits of detection that can be achieved. On the other hand, electrochemical techniques are characterised by the inherent high sensitivities. Moreover, the cost effectiveness and portability of the electrochemical devices make attractive their use in *in situ* analysis. The combination of enzyme inhibition and electrochemistry results in amperometric biosensors, promising as biotools for routine analysis.

Real samples of cyanobacterial cells around the world have been analysed with the developed enzyme sensors and the comparison with other analytical techniques has shown the viability of the approach.

However, problems associated with the reproducibility between electrodes derived from the screen-printing process and the partial electrode fouling have compromised the sensitivity of the biosensors. Work is in progress to improve both the reproducibility and the limits of detection by the use of new types of electrodes. The toxin overestimation observed with the amperometric biosensor, in the case of the microcystin analysis, suggests the use in parallel to other analytical techniques in order to minimise the risk of false-positive results. Nevertheless, the electrochemical strategy is appropriate to discriminate between toxic/non-toxic samples.

Although at present these biosensors cannot be considered as an accurate quantification technique, the applicability as tools for a first and extremely useful screening of the toxicity of real environmental samples is demonstrated. The simplicity of both the biosensor construction and the electrochemical measurement, together with the electrode disposability and the sufficient sensitivity, make the amperometric biosensors attractive for routine analysis, even at home.

Finally, it is necessary to underline the versatility of the approach. By selecting the appropriate enzyme, biosensors for other toxins can be developed. In this direction, our group has started the development of an enzyme sensor for okadaic acid, a marine toxin produced by toxigenic dinoflagellates and implicated in the diarrheic shellfish poisoning (DSP).

ACKNOWLEDGMENTS

Dr Campàs acknowledges the "Secretaría de Estado de Educación y Universidades", the "Fondo Social Europeo" and the project "ACI, Nouvelles méthodologies analytiques et capteurs", CNRS, for financial support. The authors are grateful to GTP Technology for the PP2A, Toyo Gosei Kogyo Co. for the PVA-AWP polymer and CRITT Bio-Industries for the real cyanobacterial samples.

REFERENCES

1 D.P. Botes, A.A. Tuinmann, P.L. Wessels, C.C. Viljoen, H. Kruger, D.H. Williams, S. Santikarn, R.J. Smith and S.J. Hammond, The structure of cyanoginosin-LA, a cyclic heptapeptide toxin from the cyanobacterium *Microcystis aeruginosa*, *J. Chem. Soc. Perkin Trans.*, 1 (1984) 2311–2318.
2 W.W. Carmichael, The cyanotoxins, *Adv. Bot. Res.*, 27 (1997) 212–256.

3 K.L. Rinehart, G.R. Shaw and G.K. Eaglesham, Structure and biosynthesis of toxins from blue-green algae (cyanobacteria), *J. Appl. Phycol.*, 6 (1994) 159–176.
4 O.M. Skulberg, W.W. Carmichael, G.A. Codd and R. Skulberg, Taxonomy of toxic Cyanophyceae (cyanobacteria). In: I.R. Falconer (Ed.), *Algal Toxins in Seafood and Drinking Water*, Academic Press, London, 1993, pp. 145–164.
5 K. Sivonen and G. Jones, Cyanobacterial toxins. In: I. Chorus and J. Bartram (Eds.), *Toxic Cyanobacteria in Water—A Guide to Their Public Health Consequences, Monitoring and Management*, WHO, E & FP Spon, London, 1999, pp. 41–111.
6 W.W. Carmichael, Cyanobacteria secondary metabolites—the cyanotoxins, *J. Appl. Bacteriol.*, 72 (1992) 445–459.
7 R.M. Dawson, The toxicology of microcystins, *Toxicon*, 36 (1998) 953–962.
8 J.E. Eriksson, D. Toivela, J.A.O. Meriluto, H. Karaki, Y.G. Han and D. Harstshorne, Hepatocyte deformation induced by cyanobacterial toxins reflects inhibition of protein phosphatases, *Biochem. Biophys. Res. Commun.*, 173 (1990) 1347–1353.
9 T. Kuiper-Goodman, I. Falconer and J. Fitzgerald, Human health aspects. In: I. Chorus and J. Bartram (Eds.), *Toxic Cyanobacteria in Water—A Guide to Their Public Health Consequences, Monitoring and Management*, WHO, E & FP Spon, London, 1999, pp. 113–153.
10 C. MacKintosh, K.A. Beattie, S. Klumpp, P. Cohen and G.A. Codd, Cyanobacterial microcystin-LR is a potent and specific inhibitor of protein phosphatases 1 and 2A from both mammals and higher plants, *FEBS Lett.*, 264 (1990) 187–192.
11 S. Yoshizawa, R. Matsushima, M.F. Watanabe, K. Hard, A. Ichihara, W.W. Carmichael and H. Fujiki, Inhibition of protein phosphatase by microcystin and nodularin associated with hepatotoxicity, *J. Cancer Res. Clin. Oncol.*, 116 (1990) 609–614.
12 L. Soliakov, T. Gallagher and S. Wonnacott, Anatoxin-a-evoked ^{3}H dopamine release from rat striatal synaptosomes, *Neuropharmacology*, 34 (1995) 1535–1541.
13 W.W. Carmichael, D.F. Biggs and P.R. Gorham, Toxicology and pharmacological action of *Anabaena flos-aquae* toxin, *Science*, 187 (1975) 542–544.
14 C. Edwards, K.A. Beattie, C.M. Scrimgeour and G.A. Codd, Identification of anatoxin-a in benthic cyanobacteria (blue-green algae) and in associated dog poisonings at Loch Insh, Scotland, *Toxicon*, 30 (1992) 1165–1175.
15 K. Sivonen, K. Himberg, R. Luukkainen, S.I. Niemelä, G.K. Poon and G.A. Codd, Preliminary characterization of neurotoxic blooms and strains from Finland, *Toxicity Assess.*, 4 (1989) 339–352.
16 O.M. Skulberg, W.W. Carmichael, R.A. Anderson, S. Matsunaga, R.E. Moore and R. Skulberg, Investigations of a neurotoxic Oscillatorialean

strain (Cyanophyceae) and its toxin. Isolation and characterization of homoanatoxin-a, *Environ. Toxicol. Chem.*, 11 (1992) 321–329.
17 P. Henriksen, W.W. Carmichael, J.S. An and Ø. Moestrup, Detection of an anatoxin-a(s)-like anticholinesterase in natural blooms and cultures of cyanobacteria/blue-green algae from Danish lakes and in the stomach contents of poisoned birds, *Toxicon*, 35 (1997) 901–913.
18 H. Onodera, Y. Oshima, P. Henriksen and T. Yasumoto, Confirmation of anatoxin-a(s) in the cyanobacterium *Anabaena lemmermannii* as the cause of bird kills in Danish lakes, *Toxicon*, 35 (1997) 1645–1648.
19 W.W. Carmichael, The toxins of cyanobacteria, *Sci. Am.*, 270 (1994) 78–86.
20 W.W. Carmichael, W.R. Evans, Q.Q. Yin, P. Bell and E. Mocauklowski, Evidence for paralytic shellfish poisons in the freshwater cyanobacterium *Lyngbya wollei* (Farlow ex Gomont) comb. nov., *Appl. Environ. Microbiol.*, 63 (1997) 3104–3110.
21 A.R. Humpage, J. Rositano, A.H. Bretag, R. Brown, P.D. Baler, B.C. Nicholson and D.A. Steffensen, Paralytic shellfish poisons from Australian cyanobacterial blooms, *Aust. J. Mar. Freshwater Res.*, 45 (1994) 761–771.
22 N. Lagos, J.L. Liberona, D. Andrinolo, P.A. Zagatto, R.M. Soares and S.M.F.O. Azevedo, First evidence of paralytic shellfish toxins in the freshwater cyanobacterium *Cylindrospermopsis raciborskii* isolated from Brazil, *Abstract in the VIII International Conference on Harmful Algae*, Vigo, Spain, June 25–29, 1997, p. 115.
23 N.A. Mahmood and W.W. Carmichael, Paralytic shellfish poisons produced by the freshwater cyanobacterium *Aphanizomenon flos-aquae* NH-5, *Toxicon*, 24 (1986) 175–186.
24 H. Onodera, M. Satake, Y. Oshima, T. Yasumoto and W.W. Carmichael, *Nat. Toxins*, 5 (1997) 146–151.
25 C.Y. Kao, Paralytic shellfish poisoning. In: I.R. Falconer (Ed.), *Algal Toxins in Seafood and Drinking Water*, Academic Press, London, 1993, pp. 75–86.
26 P.D. Banker, S. Carmeli, O. Hadas, B. Teltsch, R. Porat and A. Sukenik, Identification of cylindrospermopsin in *Aphanizomenon ovalisporum* (Cyanophyceae) isolated from Lake Kinneret, Israel, *J. Phycol.*, 33 (1997) 613–616.
27 K.I. Harada, I. Ohtani, K. Iwamoto, M. Suzuki, M.F. Watanabe, M. Watanabe and K. Terao, Isolation of cylindrospermopsin from a cyanobacterium *Umezakia natans* and its screening method, *Toxicon*, 32 (1994) 73–84.
28 P.R. Hawkins, N.R. Chandrasena, G.J. Jones, A.R. Humpage and I.R. Falconer, Isolation and toxicity of *Cylindrospermopsis raciborskii* from an ornamental lake, *Toxicon*, 35 (1997) 341–346.
29 A.R. Humpage, M. Fenech, P. Thomas and I.R. Falconer, Micronucleous induction and chromosome loss in transformed human white cells

indicate clastogenic and aneugenic action of the cyanobacterial toxin, cylindrospermopsin, *Mutat. Res.*, 472 (2000) 155–164.
30 M. Runnegar, K. Shou-Ming, Z. Ya-Zhen and C. Shelly, Inhibition of reduced glutathione synthesis by cyanobacterial alkaloid cylindrospermopsin in cultured rat hepatocytes, *Biochem. Pharmacol.*, 49 (1995) 219–255.
31 X. Shen, P.K.S. Lam, G.R. Shaw and W. Wickramasingue, Genotoxicity investigation of cyanobacterial toxin, cylindrospermopsin, *Toxicon*, 40 (2002) 1499–1501.
32 K. Terao, S. Ohmori, K. Igarashi, I. Ohtani, M. Watanabe, K. Harada, E. Ito and M. Watanabe, Electron microscopic studies on experimental poisoning in mice induced by cylindrospermopsin isolated from blue-green algae *Umezakia Natans*, *Toxicon*, 32 (1994) 833–843.
33 J.S. Mynderse, R.E. Moore, M. Kashiwagi and T.R. Norton, Antileukemia activity in the Oscillatoriaceae, isolation of debromoaplysiatoxin from *Lyngbya*, *Science*, 196 (1977) 538–540.
34 G. Weise, G. Drews, B. Jann and K. Jann, Identification and analysis of lipopolysaccharide in cell walls of the blue-green algae *Anacystis nidulans*, *Arch. Microbiol.*, 71 (1970) 89–98.
35 E.C. Lippy and J. Erb, Gastrointestinal illness at Sewickley, PA, *J. Am. Water Works Assoc.*, 68 (1976) 606–610.
36 G. Keleti and J.L. Sykora, Production and properties of cyanobacterial endotoxins, *Appl. Environ. Microbiol.*, 43 (1982) 104–109.
37 W.W. Carmichael, V. Beasley, D.L. Bunner, J.N. Eloff, I.R. Falconer, P. Gorham, K.-I. Harada, T. Krishnamurthy, Y. Min-Juan, R.E. Moore, K. Rinehart, M. Runnegar, O.M. Skulberg and M. Watanabe, Naming of cyclic heptapeptide toxins of cyanobacteria (blue-green algae), *Toxicon*, 26 (1988) 971–973.
38 K.I. Harada, K. Matsuura, M. Suzuki, M.F. Watanabe, S. Oishi, A.M. Dahlem, V.R. Beasley and W.W. Carmichael, Isolation and characterization of the minor components associated with microcystins LR and RR in the cyanobacterium (blue-green algae), *Toxicon*, 28 (1990) 55–64.
39 M.J. Pearson, A.J.D. Ferguson, G.A. Codd, C.S. Reynolds, J.K. Fawell, R.M. Hamilton, S.R. Howard and M.R. Attwood, *Toxic Blue-Green Algae*, Report of the National Rivers Authority, Water Quality, Series no. 2., London, 1990.
40 R. Ressom, F.S. Soong, J. Fitzgerald, L. Turczynowicz, O. El Saadi, D. Roder, T. Maynard and I.R. Falconer, *Health Effects of Toxic Cyanobacteria (Blue-Green Algae)*, National Health and Medical Council, Australian Government Publishing Service, Camberra, 1994.
41 R.L. Oliver and G.G. Ganf, Freshwater blooms. In: B.A. Whiton and M. Potts (Eds.), *The Ecology of Cyanobacteria*, Kluwer Academic Publishers, Dordrecht, 2000, pp. 149–194.
42 D.L. Campbell, L.A. Lawton, K.A. Beattie and G.A. Codd, Comparative assessment of the specificity of the brine shrimp and Microtox assays to

hepatotoxic (microcystin-LR containing) cyanobacteria, *Environ. Toxicol. Water Qual.*, 9 (1994) 71–77.
43 I.R. Falconer, Measurement of toxins from blue-green algae in water and foodstuffs. In: I.R. Falconer (Ed.), *Algal Toxins in Seafood and Drinking Water*, Academic Press, London, 1993, pp. 165–175.
44 J.S. An and W.W. Carmichael, Use of a colorimetric protein phosphatase inhibition assay and enzyme-linked immunosorbent assay for the study of microcystins and nodularins, *Toxicon*, 32 (1994) 1495–1507.
45 N. Bouaïcha, I. Maatouk, G. Vincent and Y. Levi, A colorimetric and fluorometric microplate assay for the detection of microcystin-LR in drinking water without preconcentration, *Food Chem. Toxicol.*, 40 (2002) 1677–1683.
46 T. Heresztyn and B.C. Nicholson, Determination of cyanobacterial hepatotoxins directly in water using a protein phosphatase inhibition assay, *Water Res.*, 35 (2001) 3049–3056.
47 C. Rivasseau, P. Racaud, A. Deguin and M.-C. Henion, Development of a bioanalytical phosphatase inhibition test for the monitoring of microcystins in environmental samples, *Anal. Chim. Acta*, 394 (1999) 243–257.
48 B.S.F. Wang, P.K.S. Lam, L. Xu, Y. Zhang and B.J. Richardson, A colorimetric assay for screening microcystin class compounds in aquatic systems, *Chemosphere*, 38 (1999) 1113–1122.
49 C.J. Ward, K.A. Beattie, E.Y.C. Lee and G.A. Codd, Colorimetric protein phosphatase inhibition assay of laboratory strains and natural blooms of cyanobacteria: comparisons with high-performance liquid chromatography analysis for microcystins, *FEMS Microbiol. Lett.*, 153 (1997) 465–473.
50 L.A. Lawton, C. Edwards and G.A. Codd, Extraction and high-performance liquid chromatography method for the determination of microcystins in raw and treated waters, *Analyst*, 119 (1994) 1525–1530.
51 K. Tsuji, S. Naito, F. Kondo, M.F. Watanabe, S. Suzuki, H. Nakazawa, M. Suzuki, T. Shimada and K.-I. Harada, A clean-up method for analysis of trace amounts of microcystins in lake waters, *Toxicon*, 32 (1994) 1251–1259.
52 W.P. Brooks and G.A. Codd, Immunoassay of hepatotoxic cultures and water blooms of cyanobacteria using *Microcystis areuginosa* peptide toxin polyclonal antibodies, *Environ. Technol. Lett.*, 9 (1988) 1343–1348.
53 F.S. Chu, X. Huang and R.D. Wei, Enzyme-linked immunosorbent assay for microcystins in blue-green algal blooms, *J. Assoc. Off. Anal. Chem.*, 73 (1990) 451–456.
54 S. Nagata, H. Soutome, T. Tsutsumi, A. Hasegawa, M. Sekijima, M. Sugamata, K.-I. Harada, M. Suganuma and Y. Ueno, Novel monoclonal antibodies against microcystin and their protective activity for hepatotoxicity, *Nat. Toxins*, 3 (1995) 78–86.
55 C. Edwards, L.A. Lawton, K.A. Beattie, G.A. Codd, S. Pleasance and G.J. Dear, Analysis of microcystins from cyanobacteria by liquid

chromatography with mass spectroscopy using atmospheric-pressing ionisation, *Rapid Commun. Mass Spectrom.*, 7 (1993) 714–721.
56 A. Pelander, I. Ojanperä, K. Lahti, K. Niinivaara and E. Vuori, Visual detection of cyanobacterial hepatotoxins by thin-layer chromatography and application to water analysis, *Water Res.*, 10 (2000) 2643–2652.
57 G. Vasas, D. Szydlowska, A. Gáspár, M. Welker, M. Trojanowicz and G. Borbély, Determination of microcystins in environmental samples using capillary electrophoresis, *J. Biochem. Biophys. Methods*, 66 (2006) 87–97.
58 H. Sirèn, M. Jussila, H. Liu, S. Peltoniemi, K. Sivonen and M.-L. Riekkola, Separation, purity testing, and identification of cyanobacterial hepatotoxins with capillary electrophoresis and electrospray mass spectrometry, *J. Chromatogr. A*, 839 (1999) 203–215.
59 K. Tsuji, H. Masui, H. Uemura, Y. Mori and K.-I. Harada, Analysis of microcystins in sediments using MMPB method, *Toxicon*, 39 (2001) 687–692.
60 N.A. Mahmood and W.W. Carmichael, Anatoxin-a(s), an anticholinesterase from the cyanobacterium *Anabaena flos-aquae* NRC-525-17, *Toxicon*, 25 (1987) 1221–1227.
61 P. Henriksen, W.W. Carmichael, J. An and O. Moestrup, Detection of anatoxin-a(s)-like anticholinesterase in natural blooms and cultures of cyanobacteria/blue-green algae from Danish lakes and in stomach contents of poisoned birds, *Toxicon*, 25 (1997) 901–903.
62 L.E. Llewellyn, A.P. Negri, J. Doyle, P.D. Baker, E.C. Beltran and B.A. Neilan, Radioreceptor assays for sensitive detection and quantitation of saxitoxin and its analogues from strains of the freshwater cyanobacterium, *Anabaena circinalis, Environ. Sci. Technol.*, 35 (2001) 1445–1451.
63 J.R. Molica, E.J.A. Oliveira, P.V.V.C. Carvalho, A.N.S.F. Costa, M.C.C. Cunha, G.L. Melo and S.M.F.O. Azevedo, Occurrence of saxitoxins and an anatoxin-a(s)-like anticholinesterase in a Brazilian drinking water supply, *Harmful Algae*, 4 (2005) 743–753.
64 I. Chorus, I.R. Falconer, H.J. Salas and J. Bartram, Health risks caused by fresh cyanobacteria in recreational waters, *J. Toxicol. Environ. Health, Part B*, 3 (2000) 323–347.
65 G.A. Codd, W.P. Brooks, L.A. Lawton and K.A. Beattie, Cyanobacterial toxins in European waters: occurrence, properties, problems and requirements. In: D. Wheeler, M.L. Richardson and J. Bridges (Eds.), *Watershed'89. The Future for Water Quality in Europe*, Vol. II, Pergamon, Oxford, 1989, pp. 211–220.
66 G.A. Codd, Cyanobacterial toxins, the perception of water quality and the prioritization of eutrophication control, *Ecol. Eng.*, 16 (2000) 51–60.
67 W.O. Cook, V.R. Beasley, R.A. Lovell, A.M. Dahlem, S.B. Hooser, N.A. Mahmood and W.W. Carmichael, Consistent inhibition of peripheral cholinesterases by neurotoxins from the freshwater cyanobacterium

Anabena flos-aquae: studies of duck, swine, mice and a steer, *Environ. Toxicol. Chem.*, 8 (1989) 915–922.

68 I.R. Falconer, Potential impact on human health by cyanobacteria, *Phycologia*, 35 (1996) 6–11.

69 I.R. Falconer, Algal toxins and human health. In: J. Hrubec (Ed.), *The Handbook of Environmental Chemistry, Vol. 5 (Part C), Quality and Treatment of Drinking Water II*, Springer-Verlag, Berlin, 1998, pp. 54–72.

70 B.C. Hitzfeld, S.J. Hoger and D.R. Dietrich, Cyanobacterial toxins: removal drinking water treatment and human risk assessment, *Environ. Health Perspect.*, 108 (2000) 113–122.

71 N.A. Mahmood, W.W. Carmichael and D. Pfahler, Anticholinesterase poisonings in dogs from a cyanobacterial (blue-green algae) bloom dominated by *Anabaena flos-aguae*, *Am. J. Vet. Res.*, 49 (1988) 500–503.

72 L.S. Pilotto, R.M. Douglas, M.D. Burch, S. Cameron, M. Beers, G.R. Rouch, P. Robinson, M. Kirk, C.T. Cowie, S. Hardiman, C. Moore and R.G. Attewell, Health effects of recreational exposure to cyanobacteria (blue-green algae) during recreational water-related activities, *Aust. N. Z. J. Public Health*, 21 (1997) 562–566.

73 S.-Z. Yu, Drinking water and primary liver cancer. In: Z.Y. Tang, M.C. Wu and S.S. Xia (Eds.), *Primary Liver Cancer*, China Academic Publishers, Beijing, Spring-Verlag, Berlin, 1989, pp. 30–37.

74 S.-Z. Yu, Primary prevention of hepatocellular carcinoma, *J. Gastroenterol. Hepatol.*, 10 (1995) 674–682.

75 R.B. Fitzgeorge, S.A. Clark and C.W. Keevil, Routes of intoxication. In: G.A. Codd, T.M. Jefferies, C.W. Keevil and E. Potter (Eds.), *Detection Methods for Cyanobacterial Toxins*, The Royal Society of Chemistry, Cambridge, 1994, pp. 69–74.

76 T.W. Lambert, C.F.B. Holmes and S.E. Hrudey, Microcystin class of toxins: health effects and safety of drinking water supplies, *Environ. Rev.*, 2 (1994) 167–186.

77 D.J. Gilroy, K.W. Kauffman, R.A. Hall, X. Huang and F.S. Chu, Assessing potential health risks from microcystin toxins in blue-green algae dietary supplements, *Environ. Health Perspect.*, 108 (2000) 435–439.

78 G.A. Codd, J.S. Metcalf and A.B. Kenneth, Retention of *Microcystis aeruginosa* and microcystin by salad lettuce (*Lactuca sativa*) after spray irrigation with water containing cyanobacteria, *Toxicon*, 37 (1999) 1181–1185.

79 I.R. Falconer, M. Dornbusch, G. Moran and S.K. Yeung, Effect of the cyanobacterial (blue-green algal) toxins from *Microcystis aeruginosa* on isolated enterocytes from the chicken small intestine, *Toxicon*, 30 (1992) 790–793.

80 I.R. Falconer, M.D. Burch, D.A. Steffensen, M. Choice and O.R. Coverdale, Toxicity of the blue-green alga (cyanobacterium) *Microcystis aeruginosa* in drinking water to growing pigs, as an animal model for

human injury and risk assessment, *Environ. Toxicol. Water Qual.*, 9 (1994) 131–139.

81 E.E. Prepas, E.G. Kotak, L.M. Campbell, J.C. Evans, S.E. Hrudey and C.F.B. Holmes, Accumulation and elimination of cyanobacterial hepatotoxins by the freshwater clam *Anodonta grandis simpsoniana*, *Can. J. Fish. Aquat. Sci.*, 54 (1997) 41–46.

82 W.W. Carmichael, S.M.F.O. Azevedo, J.S. An, R.J.R. Molica, E.M. Jochimsen, S. Lau, K.L. Rinehart, G.R. Shaw and G.K. Eaglesham, Human fatalities from cyanobacteria: chemical and biological evidence for cyanotoxins, *Environ. Health Perspect.*, 109 (2001) 663–668.

83 E.M. Jochimsen, W.W. Carmichael, J.S. An, D.M. Cardo, S.T. Cookson, C.E.M. Aholmes, M.B. Antunes, D.A. de Melo, T.M. Lyra, V.T. Spinelli, S.M.F.O. Azevedo and M.D. Jarvis, Liver failure and death after exposure to microcystins at a hemodialysis center in Brazil, *N. Engl. J. Med.*, 338 (1998) 873–878.

84 S. Pouria, A. de Andrade, J. Barbosa, R.L. Cavalcanti, V.T.S. Barreto, C.J. Ward, W. Preiser, G.K. Poon, G.H. Neild and G.A. Codd, Fatal microcystin intoxication in haemodialysis unit in Caruaru, Brazil, *Lancet*, 352 (1998) 21–26.

85 WHO, *Guidelines for Drinking-Water Quality*, 2nd ed., addendum to Vol. 1, Recommendations, WHO, Geneva, 1998.

86 M. Campàs, D. Szydlowska, M. Trojanowicz and J.-L. Marty, Towards the protein phosphatase-based biosensor for microcystin detection, *Biosens. Bioelectron.*, 20 (2005) 1520–1530.

87 M. Campàs, D. Szydlowska, M. Trojanowicz and J.-L. Marty, Enzyme inhibition-based biosensor for the electrochemical detection of microcystins in natural blooms of cyanobacteria (2006), submitted for publication.

88 S. Andreescu, L. Barthelmebs and J.-L. Marty, Immobilisation of acetylcholinesterase on screen-printed electrodes: comparative study between three immobilisation methods and applications to the detection of organophosphorus insecticides, *Anal. Chim. Acta*, 464 (2002) 171–180.

89 S. Andreescu, T. Noguer, V. Magearu and J.-L. Marty, Screen-printed electrode based on AChE for the detection of pesticides of organic solvents, *Talanta*, 57 (2002) 169–176.

90 T.T. Bachmann, B. Leca, F. Vilatte, J.-L. Marty, D. Fournier and R.D. Schmid, Improved multianalyte detection of organophosphates and carbamates with disposable multielectrode biosensors using recombinant mutants of *Drosophila* acetylcholinesterase and artificial neural networks, *Biosens. Bioelectron.*, 15 (2000) 193–201.

91 P.R. Brasil de Oliveira Marques, G. Silva Nunes, T.C. Rodrigues dos Santos, S. Andreescu and J.-L. Marty, Comparative investigation between acetylcholinesterase obtained from commercial sources and genetically modified *Drosophila melanogaster*: application in amperometric

biosensors for methamidophos pesticide detection, *Biosens. Bioelectron.*, 20 (2004) 825–832.
92 B. Bucur, M. Dondoi, A. Danet and J.-L. Marty, Insecticide identification using a flow injection analysis system with biosensors based on various cholinesterases, *Anal. Chim. Acta*, 539 (2005) 195–201.
93 G. Jeanty and J.-L. Marty, Detection of paraoxon by continuous flow system based enzyme sensor, *Biosens. Bioelectron.*, 13 (1998) 213–218.
94 G. Jeanty, Ch. Ghommidh and J.-L. Marty, Automated detection of chlorpyrifos and its metabolites by a continuous flow system-based enzyme sensor, *Anal. Chim. Acta*, 436 (2001) 119–128.
95 E. Devic, D. Li, A. Dauta, P. Henriksen, G.A. Codd, J.-L. Marty and D. Fournier, Detection of anatoxin-a(s) in environmental samples of cyanobacteria by using a biosensor with engineered acetylcholinesterases, *Appl. Environ. Microbiol.*, 68 (2002) 4102–4106.
96 F. Villate, H. Schulze, R.D. Schmid and T.T. Bachmann, A disposable acetylcholinesterase-based electrode biosensor to detect anatoxin-a(s) in water, *Anal. Bioanal. Chem.*, 372 (2002) 322–326.
97 WHO, *Toxic Cyanobacteria in Water—A Guide to Their Public Health Consequences, Monitoring and Management*, WHO, E & FP Spon, London, 1999.
98 D.R. de Figueiredo, U.M. Azeiteiro, S.M. Esteves, F.J.M. Gonçalves and M.J. Pereira, Microcystin-producing blooms—a serious global public health issue, *Ecotoxicol. Environ. Saf.*, 59 (2004) 151–163.

Chapter 17

Electrochemical biosensors based on vegetable tissues and crude extracts for environmental, food and pharmaceutical analysis

Orlando Fatibello-Filho, Karina O. Lupetti, Oldair D. Leite and Iolanda C. Vieira

17.1 INTRODUCTION

The first tissue biosensor employing a slice of a plant material was proposed by Kuriyama and Rechnitz [1] for determining L-glutamate. The biosensor was constructed by placing a slice of yellow squash tissue at the surface of a carbon dioxide gas-sensing electrode. The high concentration of glutamate decarboxylase present in yellow squash catalyzes the following reaction:

$$\text{L-glutamate} \xrightarrow{\text{glutamate decarboxylase}} \text{4-aminobutyrate} + CO_2 \qquad (17.1)$$

By coupling this biocatalytic activity with a CO_2 probe, L-glutamate was successfully determined potentiometrically in the concentration range from 4.4×10^{-4} to $4.7 \times 10^{-3}\,mol\,l^{-1}$ with a limit of detection (LD) of $2.0 \times 10^{-4}\,mol\,l^{-1}$ and a slope of 48 mV per concentration decade.

In recent years, there has been an increased preference for using vegetable or plant tissue instead of purified enzymes for preparation of electrochemical biosensors [2,3]. Vegetable or plant tissues may be used directly with minimal preparation. This class of biocatalytic materials maintains the enzyme of interest in its natural environment, often resulting in better stability (increased lifetime), lower cost and higher enzyme activity, compared with those biosensors using the corresponding purified enzyme. Other advantages comprise simplicity of biosensor construction, and the required cofactors may already be present in the vegetable cell and may not need to be separately immobilized. On the

Comprehensive Analytical Chemistry 49
Alegret and Merkoçi (Eds)
Volume 49 ISSN: 0166-526X DOI: 10.1016/S0166-526X(06)49017-6

other hand, tissues generally contain a multiplicity of enzymes and thus may not be as selective as purified enzyme biosensors. For these cases, a selective enhancement strategy should be developed.

The response times of biosensors based on slices of plant tissues depend on the thickness of the slices employed, as the substrates and reaction products need to diffuse through the tissue. An alternative approach that results in very short response times is the use of carbon paste biosensors [4] and, more recently, biocomposites [5] containing powdered plant tissue (or lyophilized crude extract) immobilized in a rigid carbon–polymer matrix. Additional advantages of such biosensors are long lifetime, high rigidity and mechanical stability and a sensing surface that can be renewed by a simple polishing procedure. Furthermore, as pointed out by Céspedes and Alegret [6], rigid conducting carbon–polymer biocomposites are ideal for the construction of electrochemical electrodes. The plastic nature of these materials makes them modifiable, permitting the incorporation of the fillers before they are cured. Also, the proximity of the redox center of the biological material and the conducting sites on the electrode surface favors the transfer of electrons between electroactive species.

In this chapter, we describe the methods of preparation and applications of electrochemical biosensors based on vegetable tissues and crude extracts (homogenates) for environmental, food and pharmaceutical analysis.

17.1.1 Electrochemical biosensors based on vegetable tissues

Vegetable tissue based electrochemical sensors can be divided into two groups according to their principle of operation: potentiometric and amperometric. Such devices are usually prepared in a manner similar to that of conventional enzyme electrodes, with the detection of an electroactive species that is consumed or produced by the enzyme present in the vegetable tissue.

17.1.1.1 Potentiometric biosensors

The potentiometric biosensor is a combination of an ion-selective electrode (ISE) base sensor with a vegetable tissue (the source of enzyme), which provides a highly selective and sensitive method for the determination of a given substrate. Advantages of such potentiometric biosensors are simplicity of instrumentation (only a pH meter is needed), low cost, availability of a large number of goods, reliable ISEs and easy production. Potentiometric biosensors generate an electric potential proportional to the logarithm of substrate concentration (Nernst equation) in the sample to be analyzed. Table 17.1 summarizes some potentiometric biosensors [1,5,7–10] based on vegetable tissues and related systems.

17.1.1.2 Amperometric biosensors

An amperometric plant tissue biosensor can be defined as a device incorporating a vegetable or plant tissue (containing the sensing element) into the electrochemical transducer, and the electrical signal obtained can be correlated with the concentration of the substrate to be detected. In a biosensor, the function of the tissue is to generate (or to consume) an electroactive species in a stoichiometric relationship with its substrate or target analyte. The amperometric transducer allows the electrochemical reaction (oxidation or reduction) to proceed at the electrode surface, giving rise to a current proportional to the bulk substrate concentration.

Several examples of tissue-based amperometric biosensors are given below.

17.1.1.3 Amperometric biosensors based on oxygen consumption

Given an enzymatic reaction that consumes O_2, a biosensor may be devised where the tissue-based electrode is assembled with a Clark-type oxygen electrode. The optimum thickness of the tissue slice reflects a compromise between mechanical stability and the response time, which tends to be slow because of the long diffusion path between the sample solution and the oxygen electrode surface. Some typical amperometric biosensors [11–16] where the oxygen uptake was measured with an O_2 electrode are listed in Table 17.2, together with their response characteristics. The principle involves the relationship between substrate and decreasing oxygen concentrations, i.e., as the substrate reacts according to the following equation:

$$\text{Substrate} + O_2 \rightarrow \text{product(s)} \tag{17.2}$$

the flow of oxygen traversing the electrode membrane decreases. Oxygen diffuses through the membrane (usually made of Teflon, silicone or polyethylene), and is reduced at a platinum cathode at about $-600\,\text{mV}$ with a cathodic current proportional to the oxygen concentration or inversely proportional to the substrate concentration.

TABLE 17.1

Configuration and some characteristics of potentiometric tissue based electrodes

Analyte	Sample matrix	Tissue	Sensor	Linear range ($mol\,l^{-1}$)	Limit of detection	Response time	Lifetime	Ref.
L-Asparagine	–	Chrysanthemum flower receptacle	NH_3	1.0×10^{-5}–2.0×10^{-4}		10 min	18 days	[7]
L-Cysteine	–	Cucumber leaf	NH_3 or H_2S	10^{-5}–10^{-3}			4 weeks	[8]
L-Glutamate	–	Yellow squash	CO_2	4.4×10^{-4}–4.7×10^{-3}	2.0×10^{-4}	10 min	7 days	[1]
Pyruvate	–	Corn kernel	CO_2	2.5×10^{-4}–3.0×10^{-3}	8.0×10^{-5}	10–25 min	7 days	[9]
Urea	–	Jack bean meal	NH_3	3.4×10^{-5}–1.5×10^{-3}	2.1×10^{-6}	1–5 min	3 months	[10]
Urea	Fertilizers	Soybean	MnO_2 composite	1.0×10^{-4}–1.0×10^{-1}	6.0×10^{-5}	40 s	5 months	[5]

TABLE 17.2

Configuration and some characteristics of amperometric tissue based electrodes

Analyte	Sample matrix	Tissue	Sensing element	Linear range (mol l^{-1})	Limit of detection	Response time	Lifetime	Ref.
Ascorbic acid	Fruit juices and vitamin tablets	Cucumber	O_2	2.0×10^{-5}–5.7×10^{-4}	–	1–1.5 min	Up 60 assays	[11]
Atrazine	Natural water	Potato	O_2	2.0×10^{-5}–1.3×10^{-6}	1×10^{-5}	20 min	8–14 days	[12]
Catechol	Pharmaceutical products	Potato	O_2	2.5×10^{-5}–2.3×10^{-4}	–	–	3 months	[13]
Catechol	–	Eggplant	O_2	5.0×10^{-6}–2.5×10^{-5}	–	–	3 months	[14]
Catechol	–	Banana and mushroom	Carbon paste	2.0×10^{-5}–3.0×10^{-2}	–	–	1 month	[23]
Dopamine	–	Banana	O_2	2.0×10^{-4}–1.2×10^{-3}		1–3 min	2 weeks	[15]
Dopamine	–	Banana	Carbon paste	5.0×10^{-7}–2.5×10^{-5}	1.3×10^{-8}	12 s	30 days	[18]
Dopamine	–	Spinach	Carbon paste	1.7×10^{-6}–1.6×10^{-4}	7.1×10^{-7}	–	10 days	[24]
Flavanols	Beers	Banana	Carbon paste	2.0×10^{-5}–1.0×10^{-6}	–	–	>1 month	[25]
		Apple		2.0×10^{-6}–1.2×10^{-5}			>3 months	
	Food	Apple	Carbon paste	4.1×10^{-6}– 7.3×10^{-4}	–	2 min	3 months	[26]
Fluoride	Pharmaceutical products	Asparagus	Carbon paste	2.6×10^{-6}–5.2×10^{-4}	1.3×10^{-5}	–	24 days	[27]
Glycolic acid	–	Spinach	Carbon paste	Up 5.0×10^{-4}	1.0×10^{-6}	1 min	24 days	[28]
Hydroquinone	Cosmetic creams	Sweet potato	Carbon paste	6.2×10^{-5}–1.5×10^{-3}	8.5×10^{-6}	–	7 months	[29]
Hydroquinone	Cosmetic creams	Sweet potato	Carbon paste	7.5×10^{-5}–1.6×10^{-3}	8.1×10^{-6}	–	7 months	[30]
Hydrogen peroxide	–	Tobacco callus	Carbon paste	5.0×10^{-6}–1.1×10^{-4}	7.5×10^{-7}	–	5 months	[33]
Hydrogen peroxide	Milk	Asparagus	Carbon paste	5.0×10^{-6}–7.0×10^{-5}	4.0×10^{-7}	2 min	1 month	[34]
Hydrogen peroxide	–	Turnip	Carbon paste	6.0×10^{-5}–8.5×10^{-4}	–	–	1 month	[35]
Paracetamol	Pharmaceutical products	Avocado	Carbon paste	1.2×10^{-4}–5.8×10^{-3}	8.8×10^{-5}	2 min	3 months	[31]
Pectin	–	Orange peel	Carbon paste	0.1–0.9 g l^{-1}	–	–	2 weeks	[32]
Phenol	–	Artichoke	O_2	2.0×10^{-6}–1.0×10^{-5}	–	10 min	–	[16]

17.1.1.4 Amperometric biosensors based on carbon paste electrode
Carbon paste electrode, composed of a matrix of carbon or graphite particles with water-immiscible non-conducting liquid (mineral oil, silicone, liquid paraffin, etc.) has been employed over the past four decades in electroanalytical chemistry [17]. The approach for preparing electrochemical biosensor, based on a "mixed plant tissue-carbon paste electrode" was reported by Wang and Lin in 1988 [18]. The tissue carbon paste electrode was prepared by crushing 0.11 g of banana pulp and mixing with 0.9 g of mineral oil. Subsequently, 1.1 g of graphite powder was added to the banana–oil slurry and thoroughly mixed and a portion of this modified carbon paste was packed into the end of a 4 mm i.d. glass tube. Since then, the modification of these electrodes with vegetable or plant tissues has received considerable interest and attention [2,19]. Several monographs dedicated to carbon paste electrodes modified with tissues and cells have been published [2,19–22], where the prospects and the progress of this area were evaluated and it was seen that the amperometric biosensors, as opposed to potentiometric biosensors, had been the most successful and have the most promising future. Examples of carbon paste electrodes modified with vegetable tissues [18,23–35] are presented in Table 17.2.

The amperometric biosensor based on carbon paste electrode ensures proximity at the molecular level between the catalytic and electrochemical sites because the carbon electrode is both the biocatalytic phase and the electrode sensor (Table 17.2). The tissue containing carbon paste can be incorporated in various electrode configurations and these have very rapid response times, extended lifetimes, high rigidity, mechanical stability and very low cost.

17.1.2 Electrochemical biosensors based on crude extracts (homogenates)

Uchiyama *et al.* [36] used cucumber juice (source of ascorbate oxidase) for the first time as carrier in a flow injection system for the determination of L-ascorbic acid. In another work, the same researchers used banana pulp and spinach leaf solution as a source of polyphenol oxidase (PPO) in a flow injection system for the determination of polyphenols [37]. However, the first biosensor based on vegetable crude extract (homogenate) was constructed by Signori and Fatibello-Filho [38]. In this study, an amperometric biosensor for the determination of phenols was proposed using a crude extract of yam (*Alocasia macrohiza*) as an enzymatic source of PPO. The biosensor was constructed by immobilization of this crude extract with glutaraldehyde and bovine serum albumin onto an oxygen membrane and was used for the determination of phenols in industrial wastewaters in the concentration ranges of 2.5×10^{-5}–$8.0 \times 10^{-5}\,\mathrm{mol\,l^{-1}}$; 1.0×10^{-5}–$8.5 \times 10^{-5}\,\mathrm{mol\,l^{-1}}$; 1.0×10^{-5}–$9.0 \times 10^{-5}\,\mathrm{mol\,l^{-1}}$ and 1.0×10^{-5}–$1.0 \times 10^{-4}\,\mathrm{mol\,l^{-1}}$ for pyrogallol, catechol, phenol and *p*-cresol, respectively. The response time ranged from 1 to 4 min, depending on the concentration and/or the substrate used. The crude extract electrode was stable for at least 2 weeks (over 300 determinations for each enzymatic membrane), without significant loss in sensitivity during this period (decrease of only 8–10%).

Plants produce a variety of phenolic compounds, which often interfere in the isolation of their enzymes. Careful separation procedures were devised to maximize the enzymatic activity in cell-free extracts, generally using insoluble forms of polyvinylpyrrolidone (PVP), a compound with great affinity for natural phenolic compounds and low solubility that can be removed by filtration or by centrifugation. By using insoluble polymers, several authors [39–42] have separated natural phenolic compounds that form strong H-bonded complexes (i.e., those with isolated hydroxyl groups) from many crude extracts, obtaining very active soluble enzymes (Fig. 17.1) [2,40,41].

The activity and total protein extracted may vary according to the extraction procedure and medium used. The buffer-to-tissue ratio was an important factor in the extraction of PPO from sweet potato root. PPO was extracted using ratios of 2–6:1 v/m and the highest specific activity was obtained at a ratio of 4:1 v/m [43]. It was also found that PPO could be extracted with a phosphate buffer solution of low

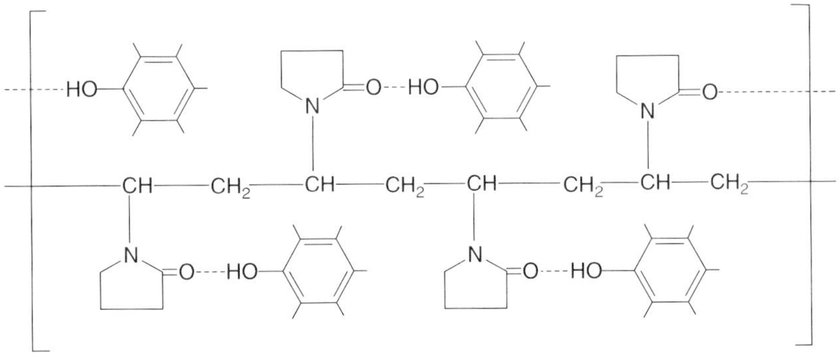

Fig. 17.1. Hydrogen bonding of plant phenol to polyvinylpyrrolidone.

concentration such as 0.05–0.4 mol l^{-1} with the maximum yield achieved at 0.1 mol l^{-1}. The effect of the buffer pH on the extraction of PPO was also investigated in the pH range 6.0–8.0. The highest enzymatic activity was reached at pH 7.0. In addition, it is well known that the enzymatic browning tendency of sweet potato is related to the presence of natural phenolic compounds (substrates), particularly chlorogenic and isochlorogenic acids that comprise at least 80% of the total phenolic content in the root. This process and the oxidation by atmospheric oxygen are responsible for the decrease in the PPO activity in the crude extract. In order to minimize this effect, several protective agents and/or stabilizers have been investigated such as Polyclar K-30, L-cysteine, L-cysteine+Amberlite CG-400 ion-exchange resin, Amberlite CG-400 ion-exchange resin+EDTA+Polyclar AT, Amberlite CG-400 ion-exchange resin, Polyclar AT, Polyclar R and Polyclar SB-100. Table 17.3 presents the activity (U ml^{-1}), total protein (mg ml^{-1}) and specific activity (U (mg of protein)$^{-1}$) of the crude extracts obtained in triplicate using 0.1 mol l^{-1} phosphate buffer solution (pH 7) and these compounds. As can be seen, Polyclar SB-100 in the concentration ratio of 2.5:25.0 m/m was the best compound among those studied. The inhibition of PPO by compounds containing -SH and other reducing groups is well documented [44]. The L-cysteine used in this work probably

TABLE 17.3

Effect of protective agents and/or stabilizers on the extraction of PPO from sweet potato root at 25°C

Protective agent and/or stabilizer	Activity (U ml^{-1})	Total protein (mg ml^{-1})	Specific activity (U (mg protein)$^{-1}$)
Without[a]	1515	4.17	363
Polyclar K-30	1482	3.51	422
L-Cysteine	1605	3.68	436
L-Cysteine +Amberlite CG-400	1570	3.25	483
Amberlite CG-400+EDTA+Polyclar A.T.	1930	3.80	508
Amberlite CG-400	2270	3.75	605
Polyclar A.T.	2440	3.80	642
Polyclar R	2795	3.85	726
Polyclar SB-100	2997	3.30	908

[a]Extraction in 0.1 mol l^{-1} phosphate buffer solution.

inhibited the enzyme by interaction with the enzyme's copper, leading to a decrease in the crude extract activity (see Table 17.3). The same effect was observed in a previous study [38] and in that of Lourenço *et al.* [45] in the preparation of crude extracts of yam (*Alocasia marcrohiza*) and sweet potato root (*Ipomoea batatas (L.) Lam.)*, respectively.

Using the optimum extraction conditions described above (Polyclar SB-100), the enzyme activity of the crude extract did not vary for at least 5 months when the extract was stored in a refrigerator at 4°C. Cysteine plus Amberlite CG-400 ion-exchange resin also provided a good enzymatic stabilization, but not cysteine alone, as can be seen in Fig. 17.2.

Crude extracts of vegetables and fruits such as apple (*Pirus malus*) [38], apple-flavored dessert banana (*Musa paradisiaca*) [38], artichoke (*Cynara scolymus L.*) [2], Cavendish banana (*Musa acuminata*) [38], cara (*Dioscorea bulbifera*) [46], eggplant (*Solanum melongena*) [38], gilo (*Solanum gilo*) [47], ginger (*Zingiber officinales* Rosc.) [48], jack fruit (*Artocarpus integrifolia L.*) [49], manioc (*Manihot utilissima*) [2,38], peach (*Prunus persica*) [38], pear (*Pirus communis*) [2,38], potato (*Solanum tuberosum*) [38], sweet potato (*Ipomoea batatas* (*L.*) *Lam.*) [50], yam (*Alocasia macrorhiza*) [38] and zucchini (*Cucurbita pepo*) [43,51–54] have been successfully employed by our group as the

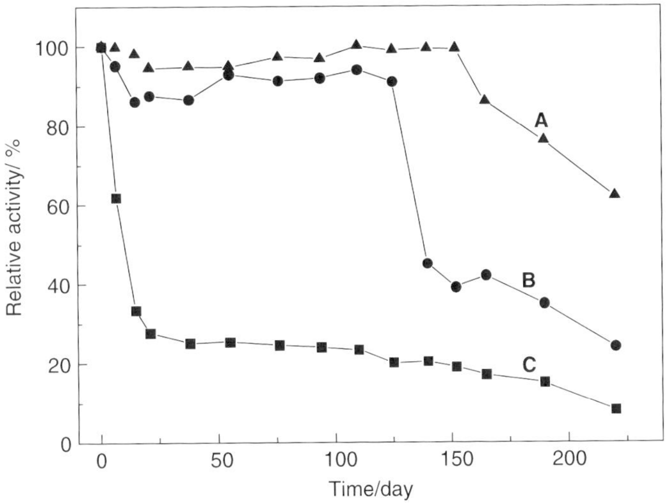

Fig. 17.2. Effect of the extraction medium (A, Polyclar SB-100; B, L-cysteine+ Amberlite CG-400 and C, L-cysteine) on the storage time and/or stability of the crude extract of sweet potato root.

source of peroxidase and PPO for the construction and application of several biosensors.

Table 17.4 presents the configuration and analytical characteristics of some electrochemical biosensors based on crude extracts (homogenates).

17.2 APPLICATION

17.2.1 Preparation of a typical electrode based on a slice of plant material

Take the base electrode (chosen from Tables 17.1 and/or 17.2) and turn it upside down. Fix a thin slice of a plant material (chrysanthemum flower receptacle, cucumber leaf, yellow squash, corn kernel, apple, potato or banana) at the end of a base electrode (ammonia gas-sensing electrode, carbon dioxide gas-sensing electrode or oxygen gas-sensing electrode). Immobilize a section of plant material using BSA-glutaraldehyde matrix containing an appropriate tissue preservative [58]. Then let it dry and cover with a piece of dialysis membrane (20–25 μm thick cellophane) about twice the diameter of the electrode sensor. Place a rubber O-ring that fits the electrode body snugly around the cellophane membrane, and gently push the O-ring onto the electrode body, so that tissue forms a good uniform layer on top of the electrode surface. Place the electrode in a buffer solution that is optimum for the enzyme system being used for 2–4 h before use and store it in the same buffer solution between uses.

17.2.2 Preparation of typical carbon paste electrode based on tissue powder

Wash, hand-peel, chop and lyophilize each plant tissue material for 3–4 h at 25°C (e.g., soybean (Table 17.1), potato, banana, eggplant, sweet potato and artichoke (Table 17.2). Grind the dry tissue to obtain a fine powder and select the particle size using sieves of lower than 200 μm mesh. Store the dry powder in a desiccator at 25°C and use it as the enzymatic source of PPO or peroxidase in the preparation of biosensor. Determine enzymatic activity and total protein content as described in Procedure 22 (in CD accompanying this book). Prepare the carbon paste electrode with dry tissue as described in the same procedure.

TABLE 17.4

Configuration and some characteristics of amperometric crude extract based electrodes

Analyte	Sample matrix	Crude extract	Sensor	Linear range ($mol\,l^{-1}$)	Limit of detection	Response time	Lifetime	Ref.
Ascorbic acid	Pharmaceutical products	Zucchini	Carbon paste	2.0×10^{-4}–5.5×10^{-3}	2.0×10^{-5}		3 months	[51]
Catecholamines	Pharmaceutical products	Zucchini (peroxidase) and fungi (laccase)	Carbon paste	6.1×10^{-6}–1.0×10^{-4}	2.5×10^{-8}		2 months	[52]
Dopamine	Pharmaceutical products	Zucchini	Carbon paste	5.0×10^{-4}–3.0×10^{-3}	2.6×10^{-5}			[53]
Dopamine and epinefrine	Pharmaceutical products	Cara root	Carbon paste	2.0×10^{-3}– 8.0×10^{-3}				[46]
				2.0×10^{-4}–1.2×10^{-3}				
Hydroquinone	Wastewaters of photographic and X-ray processes	Gilo	Carbon paste	2.5×10^{-4}–5.5×10^{-3}	2.0×10^{-6}			[47]
	Wastewaters of photographic and X-ray processes	Ginger	Carbon paste	2.5×10^{-4}–2.4×10^{-3}	2.0×10^{-5}			[48]
	Photographic developers	Zucchini	Carbon paste	6.2×10^{-5}–8.9×10^{-3}	8.3×10^{-6}		7 months	[54]
Hydrogen peroxide	Milk	Artichoke	O_2	5.0×10^{-5}–5.0×10^{-6}		3 min		[55]
Paracetamol	Pharmaceutical products	Zucchini	Carbon paste	1.2×10^{-4}–2.5×10^{-3}	6.9×10^{-5}	–	7 months	[43]
Phenolic compounds	Wastewaters	Yam	O_2	2.5×10^{-5}–8.0×10^{-5}	–	1–4 min		[38]
				1.1×10^{-5}–8.5×10^{-5}				
	Wastewaters	Jack fruit	Carbon paste	5.2×10^{-6}–7.8×10^{-4}	2.0×10^{-7}		6 months	[49]
	Wastewaters	Sweet potato	O_2	2.4×10^{-5}– 4.5×10^{-4}				[50]
		Mushroom	O_2	2.0×10^{-4}–2.0×10^{-2}				[56]
Sulfite	Food products	*Malva vulgaris*	O_2	2.0×10^{-4}–1.8×10^{-3}	2.0×10^{-4}	–	–	[57]

17.2.3 Preparation of typical carbon paste electrode based on tissue crude extract

Wash, hand-peel, chop and homogenize 25 g of each tissue material chosen from Table 17.4 for 2 min at 4–6°C in a blender with 100 ml of 0.1 mol l^{-1} phosphate buffer solution (pH 7.0) and 2.5 g of Polyclar SB-100. Filter the homogenate through four layers of gauze and centrifuge at 15,000 rpm for 15 min at 4°C. Store the resulting supernatant (homogenate) at this temperature in a refrigerator and use it as the enzymatic source after the determination of the peroxidase activity and total protein as described in Procedure 23 (Part II of this book). Prepare the carbon paste electrode containing vegetable crude extract as described in the same procedure.

17.2.4 Operational properties of electrodes

Tables 17.1, 17.2 and 17.4 give a list of some vegetable tissue based and crude extract electrodes that have been prepared for analysis of several substrates in environmental, food and pharmaceutical samples. These tables also present the sample matrix, the tissue or crude extract used, the sensor, the range of determinable concentration, the LD, the response time and the stability and/or lifetime of the biosensor.

In several cases, different base sensors could be used. For example, L-cysteine is hydrolyzed in the presence of the enzyme L-cysteine desulfhydrolase according to the following equation:

$$\text{L-cysteine} + H_2O \xrightarrow{\text{L-cysteine desulfhydrolase}} \text{pyruvate} + H_2S + NH_3 \qquad (17.3)$$

yielding both NH_3 and H_2S. Thus, either an ammonia gas-sensing electrode or an H_2S gas-sensing electrode could be used to determine the L-cysteine concentration. This biosensor gives a response to L-cysteine in phosphate buffer solution (pH 7.6) in the concentration range from 10^{-5} to 10^{-3} mol l^{-1} with a slope of about 35 mV per decade. The relatively poor slope and long response times between 5 and 15 min of this biosensor could be due to the cucumber leaf disc fitting the NH_3 gas-sensor tip.

As a second example, a biosensor for urea could be constructed using an ammonia gas-sensing electrode [10] or a biocomposite containing soybean powder, polymer matrix and manganese dioxide as pH-sensing electrode [5].

The range of applicability for potentiometric tissue based electrodes shown in Table 17.1 is from 1.0×10^{-5} to $2.0 \times 10^{-4}\,mol\,l^{-1}$ for L-asparagine [7], from 1.0×10^{-5} to $1.0 \times 10^{-3}\,mol\,l^{-1}$ for L-cysteine [8], from 4.4×10^{-4} to $4.7 \times 10^{-3}\,mol\,l^{-1}$ for L-glutamate [1], from 2.5×10^{-4} to $3.0 \times 10^{-3}\,mol\,l^{-1}$ for pyruvate [9], from 3.4×10^{-5} to 1.5×10^{-3} and from 1.0×10^{-4} to $1.0 \times 10^{-1}\,mol\,l^{-1}$ for urea using jack bean meal [10] or soybean [5], respectively. The ranges presented depend on the solubility of the substrate in the aqueous solution and/or the detection limit of the base sensor used in the construction of the biosensor.

The stability of enzyme electrodes is difficult to define because an enzyme can lose some of its activity. Deterioration of immobilized enzyme in the potentiometric electrodes can be seen by three changes in the response characteristics: (a) with age the upper limit will decrease (e.g., from 10^{-2} to $10^{-3}\,mol\,l^{-1}$), (b) the slope of the analytical (calibration) curve of potential vs. log [analyte] decrease from 59.2 mV per decade (Nernstian response) to lower value, and (c) the response time of the biosensor will become longer as the enzyme ages [59]. The overall lifetime of the biosensor depends on the frequency with which the biosensor is used and the stability depends on the type of entrapment used, the concentration of enzyme in the tissue or crude extract, the optimum conditions of enzyme, the leaching out of loosely bound cofactor from the active site, a cofactor that is needed for the enzymatic activity and the stability of the base sensor.

The response time is often governed by the immobilized biocatalyst layer rather than by the sensor response, and there are many factors that affect the biosensor response time, as shown in Table 17.5.

TABLE 17.5

Factors affecting the response time of an enzyme electrode [59,60]

Chemical factors	Physical factors	Instrumental factor
Enzyme concentration	Temperature	Sensor response time
Substrate concentration	Thickness of slice of tissue or crude extract layer	
pH	Stirring rate	
Availability of cofactors	Protective membrane: thickness and permeability	
Activators		
Inhibitors		

To obtain a response, the substrate must (a) diffuse through solution to the membrane surface, (b) diffuse through the membrane and react at the enzyme active site, and (c) the products formed must then diffuse to the electrode surface where they are measured.

Biosensors that were constructed with a slice of the vegetable (or plant) tissue (e.g., L-asparagine, L-cysteine, L-glutamate and Pyruvate in Table 17.1) have long response times. On the other hand, the urea minibiosensor using soybean (*Glycine max L. Merr.*) powder tissue [5] has a shorter response time and a longer durability when compared to biosensors constructed with a slice of tissue. The additional advantages of the flexible potentiometric minibiosensor for urea based on MnO_2–soybean tissue composite [5] include a good reproducibility after polishing, the high chemical and mechanical resistances, the high biocatalytic activity, the uniformity of the conductivity of biomaterial and the ease with which the biocomposite can be modeled into several shapes and sizes.

According to the Michaelis–Menten equation, the rate of the reaction depends on the activity of the enzyme, enzyme concentration, and upon several other factors such as the pH, the temperature, availability of cofactors, activators and/or inhibitors, and the concentration of the substrate (analyte). Moreover, the reaction rate depends on the thickness of the slice of tissue or crude extract, and on the size of the dialysis membrane, when used. By contrast, the steady-state potential obtained is dependent only on the substrate concentration and the temperature.

The final factor that affects the speed of the response is the electrode sensor itself and how fast it reaches a potential (potentiometric biosensor) or current (amperometric/voltammetric biosensor) proportional to the concentration of the products or reactants it has detected. A consideration of each of these factors affecting the response time is discussed in detail elsewhere [59,60].

For the first biosensor shown in Table 17.2, L-ascorbic acid has been determined in fruit juices and vitamin tablet, by immobilizing a slice of cucumber (*Cucumis sativus*) onto an O_2 electrode in the pH range 6.0–6.5 (phosphate buffer solution) [11]:

$$2\ \text{L-ascorbic acid} + O_2 \xrightarrow{\text{L-ascorbate oxidase}} 2\ \text{dehydroascorbic acid} + 2\ H_2O \tag{17.4}$$

The calibration graph based on the decrease in steady-state current was rectilinear in the L-ascorbic acid concentration range from

2.0×10^{-5} to $5.7 \times 10^{-4}\,mol\,l^{-1}$, with a response time of 1.0–1.5 min and a lifetime of about 60 assays.

The tissue-based amperometric electrodes for the determination of catechol [13,14], dopamine [15] and phenol [16] shown in Table 17.2 operate on the same principle as the biosensor for L-ascorbic acid [11].

A biosensor containing a thin slice of potato (*Solanum tuberosum*) tissue, which contains high levels of the enzyme PPO, coupled to a commercial O_2-selective Clark electrode, responds to catechol concentration. When atrazine [12] is added to a catechol solution, the inhibition of the enzymatic activity, after an incubation time of 20 min, is proportional to its concentration in the range from 20 to 130 $\mu mol\,l^{-1}$. The performance of this biosensor is not affected by the presence of organophosphorous or carbamic pesticides.

As can be seen in Tables 17.2 and 17.4, a wide variety of different enzymes present in vegetable tissue or in the crude extracts have been immobilized onto Clark-type oxygen electrode membranes or in carbon paste electrodes. To facilitate understanding of the operation of electrode containing PPO or peroxidase in combination with amperometric biosensors a brief discussion is provided below.

PPO and peroxidase are widely distributed in the plant kingdom and have been detected in most known fruits and vegetables. The concentration of these enzymes in plants depends on the species, cultivation, age and maturity [61–64].

Polyphenol oxidase (EC 1.10.3.1) (PPO) is also known as tyrosinase, catechol oxidase, cresolase and phenolase. Tyrosinase was the first name used because tyrosine was first used as a substrate. PPO is a copper-containing enzyme capable of catalyzing two different types of reactions in the presence of molecular oxygen (Fig. 17.3, Eqs. (17.5) and (17.6)); the hydroxylation of monophenols to *o*-diphenols (Fig. 17.3, Eq. (17.5)) and the dehydrogenation of *o*-diphenols to *o*-quinones (Fig. 17.3, Eq. (17.6)) [61]. In Eqs. (17.5) and (17.6), E-2Cu^{+} and E-2Cu^{2+} represent the PPO in the reduced and oxidized forms, respectively. The *o*-quinone is electrochemically active and can be very efficiently reduced close to or below 0 mV (vs. SCE) meaning these biosensors can operate within a potential range essentially free of interfering reactions (Fig. 17.3, Eq. (17.7)). Another advantage of these systems is that the species electrochemically produced (*o*-diphenol) is active as substrate for the enzyme, thus resulting in an amplification of the analytical signal and detection around $\mu mol\,l^{-1}$ has been reported [19]. The other possibility discussed in Section 17.1.1.3 comprises the electrochemical detection of molecular oxygen using a Clark-type

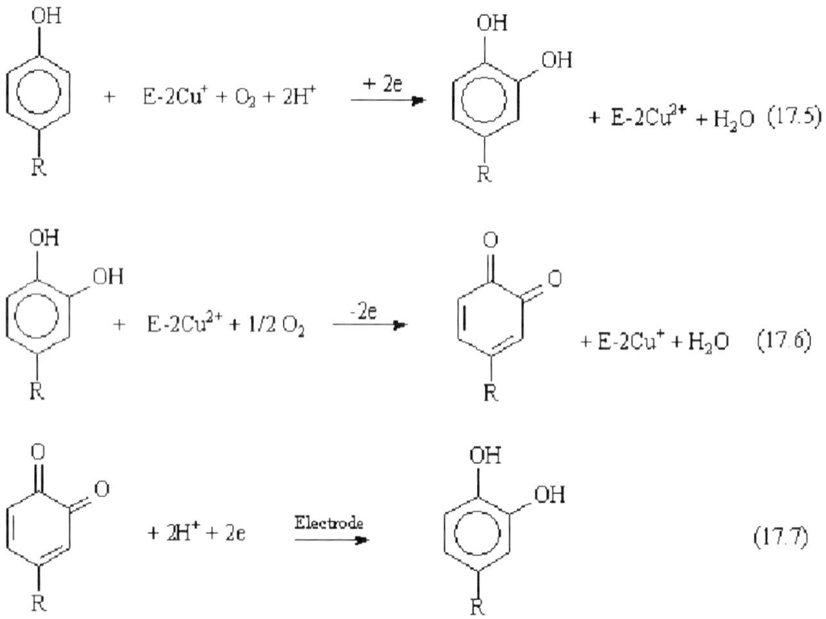

Fig. 17.3. Reactions catalyzed by PPO: (17.5) Hydroxylation of monophenol to *o*-diphenol and (17.6) Dehydrogenation of *o*-diphenol to *o*-quinone. Reaction (17.7) is the electrochemical reduction of *o*-quinone to *o*-diphenol.

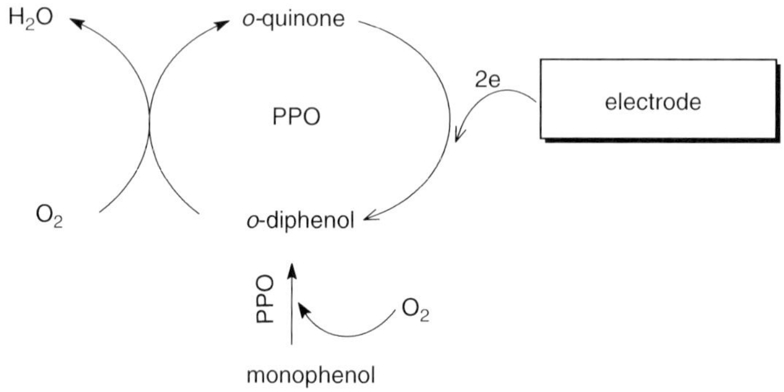

Fig. 17.4. Operation principles of a biosensor based on enzymatic oxidation of monophenol and/or *o*-diphenol by PPO and electrochemical detection by determining molecular oxygen or the oxidation product derived from monophenol and/or *o*-diphenol.

oxygen electrode. The operation principles of the most popular biosensors for phenolic compounds analysis are shown schematically in Fig. 17.4. Due to enzyme's broad selectivity, PPO-based electrodes have been used for the determination of several phenolic compounds as can be seen in Tables 17.2 and 17.4.

Peroxidase (EC 1.11.1.7) (HRP) is recognized as being one of the most heat-stable enzymes present in higher plants that catalyzes the oxidation of organic and inorganic substrates by hydrogen peroxide or organic peroxides. Most peroxidases are heme proteins and contain iron (III) protoporphyrin IX (ferriprotoporphyrin IX) as the prosthetic group [62–64]. The catalytic cycle involving native peroxidase (Fe^{3+}), hydrogen peroxide and donor substrate (SH) can be explained by the following three steps [19,29]:

$$HRP + H_2O_2 \rightarrow CpdI + H_2O \tag{17.8}$$

$$CpdI + SH \rightarrow CpdII + S\bullet \tag{17.9}$$

$$CpdII + SH \rightarrow HRP + S\bullet + H_2O \tag{17.10}$$

The first step (Eq. (17.8)) comprises a two-electron oxidation of the ferriheme prosthetic group of the native peroxidase, HRP, by hydrogen peroxide with formation of an unstable intermediate, two oxidizing equivalents above ferric state, consisting of ferryl iron and porphyrin π cation radical (Cpd I) and water. In the next step (Eq. (17.9)), compound-I (Cpd I) loses one oxidizing equivalent by one-electron reduction of the first electron donor SH forming compound-II (Cpd II, iron with an oxidation state +4) and S• free radical. Then, in step 3 (Eq. (17.10)), Cpd II receives an additional electron from SH, whereby the enzyme is returned to its native state (Fe^{3+}). Also, the peroxidase immobilized in the electrode can be oxidized by hydrogen peroxide according to step 1 (Eq. (17.8)) and then subsequently reduced by electrons provided by the electrode, according to the following reaction:

$$CpdI + 2e + 2H^{+} \rightarrow HRP + H_2O \tag{17.11}$$

As pointed out by Gorton and co-workers [19,65], this process is usually referred to a direct electron transfer (DET), that is, when an electrode substitutes the electron donor substrates in a common peroxidase reaction cycle (Eqs. (17.8)–(17.10)). When an electron donor (SH), such as phenolic compounds and the catecholamines shown in Tables 17.2 and 17.4, is present in a peroxidase-electrode system, both processes can occur simultaneously [29,30,43,48,52–54] and the oxidized donor S• is reduced electrochemically by the electrode as shown in the following reaction:

$$S\bullet + e + H^{+} \rightarrow SH \tag{7.12}$$

The reaction sequence (Eqs. (17.8)–(17.10) and (17.12)) [19,65] is known as mediated electron transfer (ET). In an amperometric electrode, both these approaches result in a reduction current, which may be correlated to the concentration of peroxide [33–35] and/or phenolic compounds and catecholamines in the solution [29,30,43,48,52–54].

17.3 CONCLUSIONS

Electrochemical biosensors based on vegetable tissues have been employed for environmental, food and pharmaceutical analysis over the past two decades and those employing crude extract (homogenates) over the last decade. These biocatalytic materials maintain the enzyme in its natural environment, a characteristic that results, in many cases, in a longer lifetime, a lower cost and a higher enzyme activity. The analytical performance of these electrochemical biosensors has been significantly improved with the immobilization of powdered plant tissue or lyophilized crude extracts in a rigid carbon–polymer matrix (biocomposite) or in a carbon paste. The additional advantages of such biosensors are the ease with which they can be incorporated in various electrode configurations, the simplicity of construction, reliability and the proximity at the molecular level between the electrochemical and catalytic sites that favors the transfer of electrons tremendously. Finally, the great variety of vegetables and plants enables us to foresee many possibilities for the construction of other biosensors.

ACKNOWLEDGMENTS

The authors gratefully acknowledge financial support from Brazilian foundations (FAPESP, CNPq and CAPES). Profs. Drs. A. Moura and E. Dockal are thanked for critical comments.

REFERENCES

1 S. Kuriyama and G.A. Rechnitz, *Anal. Chim. Acta*, 131 (1981) 91–96.
2 O. Fatibello-Filho and I.C. Vieira, *Quim. Nova*, 25 (2002) 455–464.
3 M.K. Sezgintürk and E. Dinçkaya, *Talanta*, 65 (2005) 998–1002.
4 I.C. Vieira, K.O. Lupetti and O. Fatibello-Filho, *Anal. Lett.*, 35 (2002) 2221–2231.
5 O. Fatibello-Filho, R.R.I. Portugal and L.A. Ramos, *A Flexible Potentiometric Minibiosensor for Urea Based on MnO_2–Soybean (Glycine Max L. Merr.) Tissue Composite*, 54th Annual Meeting of the International Society of Electrochemistry, São Pedro, Brazil, 2003, p. 329.

6 F. Céspedes and S. Alegret, *Trends Anal. Chem.*, 19 (2000) 276–284.
7 S. Uchiyama and G.A. Rechnitz, *Anal. Lett.*, 20 (1987) 451–470.
8 N. Smit and G.A. Rechnitz, *Biotechnol. Lett.*, 6 (1984) 209–214.
9 S. Kuriyama, M.A. Arnold and G.A. Rechnitz, *J. Membr. Sci.*, 12 (1983) 269–278.
10 F. Schubert, U. Wollenberg and F. Scheller, *Biotechnol. Lett.*, 5 (1983) 239–242.
11 L. Macholán and B. Chmelíková, *Anal. Chim. Acta*, 185 (1986) 187–193.
12 F. Mazzei, F. Botrè, G. Lorenti, G. Simonetti, F. Porcelli, G. Scibona and C. Botrè, *Anal. Chim. Acta*, 316 (1995) 79–82.
13 F. Botrè, F. Mazzei, M. Lanzi, G. Lorenti and C. Botrè, *Anal. Chim. Acta*, 255(1) (1991) 59–62.
14 E. Dinckaya, E. Akyilmaz, A. Telefoncu and S. Ahgol, *Indian J. Biochem. Biophys.*, 36(1) (1999) 36–38.
15 Y. Chen and T.C. Tan, *Sens. Actuators B*, 28 (1995) 39–48.
16 D. Odaci, S. Timur and A. Telefoncu, *Artif. Cells Blood Substit. Immobil. Biotechnol.*, 32(2) (2004) 315–324.
17 R.N. Adams, *Anal. Chem.*, 30 (1958) 1576–1576.
18 J. Wang and M.S. Lin, *Anal. Chem.*, 60 (1988) 1545–1548.
19 L. Gorton, *Electroanalysis*, 7 (1995) 23–45.
20 L. Macholán. In: D.L. Wise (Ed.), *Bioinstrumentation and Biosensors*, Marcel Dekker, New York, 1991, pp. 329–377.
21 B.R. Eggins, *Biosensors: An Introduction*, Wiley/Teubner, Chichester/Stuttgart, 1996.
22 D.C. Wijesuriya and G.A. Rechnitz, *Biosens. Bioelectron.*, 8(3–4) (1993) 155–160.
23 T.Y. Feng, *Anal. Lett.*, 26(8) (1993) 1557–1566.
24 L. Zhihong, Q. Wenjean and W. Meng, *Anal. Lett.*, 25(7) (1992) 1171–1181.
25 B.R. Eggins, C. Hickey, S.A. Toft and D.M. Zhou, *Anal. Chim. Acta*, 347 (1997) 281–288.
26 E.A. Cummings, P. Milley, S. Linquette-Mailley, B.R. Eggins, E.T. McAdams and S. McFadden, *Analyst*, 123(10) (1998) 1975–1980.
27 S. Liawruangrath, W. Oungpipat, S. Watanesk, B. Liawruangrath, C. Dongduen and P. Purachat, *Anal. Chim. Acta*, 448 (2001) 37–46.
28 W. Ougpipat and P.W. Alexander, *Anal. Chim. Acta*, 295(1–2) (1994) 37–46.
29 O. Fatibello-Filho and I.C. Vieira, *Fresenius J. Anal. Chem.*, 368(4) (2000) 338–343.
30 I.C. Vieira and O. Fatibello-Filho, *Talanta*, 52 (2000) 681–689.
31 O. Fatibello-Filho, K.O. Lupetti and I.C. Vieira, *Talanta*, 55 (2001) 685–692.
32 H. Horie and G.A. Rechnitz, *Anal. Chim. Acta*, 306 (1995) 123–127.
33 A. Navaratne and G.A. Rechnitz, *Anal. Chim. Acta*, 257(1) (1992) 59–66.
34 W. Oungpipat, P.W. Alexander and P. Sounthwell-Keely, *Anal. Chim. Acta*, 309 (1995) 35–45.

35 S.L. Chut, J. Li and S. Ngin-Tan, *Anal. Lett.*, 30(11) (1997) 1993–1998.
36 S. Uchiyama, Y. Tofuku and S. Suzuki, *Anal. Chim. Acta*, 208 (1988) 291–294.
37 S. Uchiyama and S. Suzuki, *Anal. Chim. Acta*, 261 (1992) 361–365.
38 C.A. Signori and O. Fatibello-Filho, *Quim. Nova*, 17 (1994) 38–42.
39 W.D. Loomis and J. Battaile, *Phytochemistry*, 5 (1966) 423–438.
40 A.F. Hsu, C.E. Thomas and D. Brauer, *J. Food Sci.*, 53 (1988) 1743–1745.
41 W.M. Walter and A.E. Purcell, *J. Agric. Food Chem.*, 27 (1979) 942–946.
42 R.A. Andersen and J.A. Sowers, *Phytochemistry*, 7 (1968) 293–302.
43 I.C. Vieira, K.O. Lupetti and O. Fatibello-Filho, *Quim. Nova*, 26(1) (2003) 39–43.
44 A.M. Mayer and E. Harel, *Phytochemistry*, 18 (1979) 193–215.
45 E.J. Lourenço, V.A. Neves and M.A. Da-Silva, *J. Agric. Food Chem.*, 40 (1992) 2369–2373.
46 C.S. Caruso, I.C. Vieira and O. Fatibello-Filho, *Anal. Lett.*, 32 (1999) 39–50.
47 I.R.W.Z. Oliveira and I.C. Vieira, *Enzyme Microb. Technol.*, 38 (2006) 449–456.
48 I.R.W.Z. Oliveira, I.C. Vieira, K.O. Lupetti, O. Fatibello-Filho, V.T. Fàvere and M.C.M. Laranjeira, *Anal. Lett.*, 37 (2004) 3111–3127.
49 K.O. Lupetti, I.C. Vieira and O. Fatibello-Filho, *Anal. Lett.*, 37 (2004) 1833–1846.
50 I.C. Vieira and O. Fatibello-Filho, *Anal. Lett.*, 30 (1997) 895–907.
51 O. Fatibello-Filho and I.C. Vieira, *J. Braz. Chem. Soc.*, 11 (2000) 412–418.
52 O.D. Leite, K.O. Lupetti, O. Fatibello-Filho, I.C. Vieira and A.M. Barbosa, *Talanta*, 59 (2003) 889–896.
53 K.O. Lupetti, L.A. Ramos, I.C. Vieira and O. Fatibello-Filho, *Il Fármaco*, 60 (2005) 179–183.
54 I.C. Vieira, O. Fatibello-Filho and L. Angnes, *Anal. Chim. Acta*, 398 (1999) 145–151.
55 G. Ozturk, F.N. Ertas, E. Akyilmaz, E. Dinçkaya and H. Tural, *Artif. Cells Blood Substit. Biotechnol.*, 32 (2004) 637–645.
56 S. Topçu, M.K. Sezgintürk and E. Dinçkaya, *Biosens. Bioelectron.*, 20 (2004) 592–597.
57 M.K. Sezgintürk and E. Dinçkaya, *Talanta*, 65 (2005) 998–1002.
58 M.A. Arnold and A. Rechnitz. In: A.P.F. Turner, I. Karube and G.S. Wilson (Eds.), *Biosensors Fundamentals and Applications*, Oxford University Press, New York, 1987, pp. 30–59.
59 G.G. Guilbault, *Analytical Uses of Immobilized Enzymes*, Marcel Dekker, New York, 1984.
60 G.G. Guilbault, A.A Suleiman, O. Fatibello-Filho and M.A. Nabirahni. In: D.L. Wise (Ed.), *Bioinstrumentation and Biosensors*, Marcel Deckker, New York, 1990, pp. 659–692.

61 J. Zawistowski, C.G. Biliaderis and N.A.M. Eskin. In: D.S. Robinson and N.A.M. Eskin (Eds.), *Oxidative Enzymes in Foods*, Elsevier, New York, 1991, pp. 217–273.
62 J. Zawistowski, C.G. Biliaderis and N.A.M. Eskin. In: D.S. Robinson and N.A.M. Eskin (Eds.), *Oxidative Enzymes in Foods*, Elsevier, New York, 1991, pp. 1–47.
63 J.R. Whitaker, *Principles of Enzymology for the Food Sciences*, Marcel Dekker, New York, 1985.
64 R. Yoruk and M.R. Marshall, *J. Food Biochem.*, 27(5) (2003) 361–422.
65 T. Ruzgas, J. Emnés, L. Gorton and G. Marko-Varga, *Anal. Chim. Acta*, 311 (1995) 245–253.

AFFINITY BIOSENSORS

Chapter 18

Immunosensors for clinical and environmental applications based on electropolymerized films: analysis of cholera toxin and hepatitis C virus antibodies in water and serum

Serge Cosnier and Michael Holzinger

18.1 INTRODUCTION

Rapid detection and monitoring in clinical and food diagnostics, environmental and bio-defense monitoring has paved the way for the elaboration of alternative, state-of-the-art analytical devices, known as biosensors, microarrays or biochips. A biosensor will couple an immobilized bio-specific recognition component such as protein (enzymes, antibodies and antigens), DNA, cells or tissue slices to the surface of a transducer, which "transduces" a molecular recognition event into a measurable electrical or spectroscopic signal, pinpointing the presence of the target analyte [1–3]. The generated electrical or fluorescent signal is often directly proportional to the analyte concentration. In a time when dangerous viruses (such as AIDS, SARS, West Nile virus (WNV)) are afflicting more and more humans, early detection of pathogenic germs is believed to be an essential tool to help to fight them. Fast and reliable detection of these viruses, based on antibody–antigen interactions, are continuously gaining importance due to fast diffusion of diseases. Nanostructured immuno-interfaces that will detect and quantify antibody–antigen interactions have minimal limitation in the number of analytes they can detect. The specificity remains the property of the antibody, and since many thousands of antibodies can be prepared to detect their respective antigens, immunosensors will be able to be made for any number of analytes required. Since the analyte scope is relatively unlimited, the application areas are also very diverse.

Comprehensive Analytical Chemistry 49
Alegret and Merkoçi (Eds)
Volume 49 ISSN: 0166-526X DOI: 10.1016/S0166-526X(06)49018-8

Wherever an antibody assay is used now, there will be the prospect of producing an immunosensor for the same analyte. Therefore, the development of immunosensors is directed to obtain sensors with fast response, low detection limits, little preparation efforts, low price availability and high specificity. Thanks to the possibility of fabricating fast portable sensors, immunosensors now constitute a potential alternative to centralized and sophisticated bioanalytical systems.

The principle of an immunosensor is based on antibody–antigen interactions, where the specific receptor is immobilized on a transducer surface, ready to form a complex with the analyte. The immunoreaction triggers directly or indirectly a signal via the physical transducer allowing the determination of the analyte target. Beside field effect transistors, transducers are mainly based on metal surfaces or conducting materials for obtaining electronic signals while glass fibers were used for optical detection.

For the fabrication of an immunosensor, a layer of antibodies or antigens with optimum density and adjusted orientation has to be formed on the transducer surface. It should be noted that the quality of the immobilized biolayer on which the molecular recognition process takes place is of extreme importance in the achievement of suitable sensitivity. As a consequence, the procedure of biomolecule immobilization on transducer surfaces remains a key step for the performance of the resulting bioanalytical devices. The ideal immobilization of a biological component leading to the biocoating should fulfill the following conditions: the immobilization protocol of the biomolecule involves solely a single attachment point preserving an excellent accessibility to each immobilized biomolecule. In addition, the latter may induce an appropriate orientation propitious to affinity interactions. Besides its stability, this binding should also fully retain the biological recognition properties of the immobilized target. Taking into account the optimal conditions for generating a recognition signal, the immobilized biocoating should be compact, ordered and homogeneous as well as thin and free of defects. Moreover, the methodologies for the biomolecule immobilization should have the capacity to be transferred to industrial mass production scale where the deposition of biomolecules is secured in controlled spatial resolution. Standard procedures are physical adsorption, cross-linking, covalent binding, attachment on self-assembled monolayer and entrapment in gels or membranes [4–19]. Among these conventional approaches of protein immobilization, the electrogeneration of polymers constitutes one of the few methods that have the potential for automated mass production and provide an easy control over the properties of the polymeric biocoating such as morphology and thickness. In addition, the polymeric films are stable in organic and aqueous solvents. Conducting and insulating polymers can be electrogenerated using monomers like thiophenes, phenols, anilines and pyrroles [20–24]. The latter are particularly appropriate for the fabrication of biosensors since pyrrole derivatives can be polymerized in water at neutral pH under mild conditions as well as in organic solvents. The physical and chemical properties of the resulting films can be modulated with N-substituted pyrrole derivatives that can be easily synthesized in large scale leading to electropolymerized films bearing specific functionalities.

A straightforward procedure of biomolecule immobilization was based on the entrapment of biomolecules in polymer films during their electrochemical growth on electrode surface [25,26]. This simple one-step entrapment procedure occurs without chemical reaction between the formed polymers and the proteins, thus preserving their biological activity. However, the protein immobilization by entrapment obviously suffers from the poor accessibility of the entrapped proteins that is conflicting with the development of immunoreactions. Besides the conventional protein entrapment, the main strategies used for protein immobilization encompass the chemical or photochemical grafting onto the functionalized polymers and the anchoring by affinity interactions with the underlying film [27–29]. The electropolymerized functional groups used for the immobilization of the antibody or antigen receptors are *N*-hydroxysuccinimide, *N*-hydroxyphthalimide or pentafluorophenyl esters or carboxylic or amino groups or photoactive entities for the covalent attachment and biotin derivatives or nitrilotriacetic acid for affinity binding via the formation of biotin–avidin complexes or Ni(II) chelation [30,31].

The direct immobilization of the immunochemical reagents on the transducer surface allows various detection methods. The conventional detection methods of biological interactions such as antigen–antibody are mainly based on sophisticated fluorescent-detection technologies [32,33]. These indirect approaches, however, require a previous labeling step of the protein target by an enzyme marker or a fluorescent molecule. Concerning the development of direct detection strategies for immunosensors, the principal attempts were based on gravimetric and electrochemical transductions. The latter involve conductimetric, impedimetric and potentiometric methods that may use the unusual electronic or redox properties of conducting polymers, thus conferring an active role to polymer coatings in the transduction step. For instance, the characterization of the interaction between a human serum albumin and an anti-human serum albumin was carried out by cyclic voltammetry and impedance spectroscopy [34]. Nevertheless, amperometric methods were also developed that required previous labeling step of the analyte. Labels are usually enzymes, like oxidases, which catalyze the formation of an electrochemically active product. The main advantage of electrochemical detections lies in the possibility for miniaturization and hence portability.

Optical detection is possible when photoactive reagents were activated with the help of the enzyme label catalysis. For example, horseradish peroxidase could be used for chemiluminescence [32,33,35].

Quartz crystal microbalances (QCMs) can be modified to serve as gravimetric immunosensors. Antigens or antibodies are immobilized at the surface of one of the two electrodes covering the quartz crystal. The intrinsic vibration frequency of the support changes after the immunochemical recognition reaction, caused by the mass-increase of the crystal. Drastic improvement of the noise-signal ratio was realized by labeling the analyte with weight-carrying compounds like latex beads [36] or gold nanoparticles [37].

Due to the vast achievements in immunosensor development, only few highlights of novel immobilization techniques and investigative detection principles based on electrogenerated polypyrrole films are presented with cholera toxin and hepatitis C Virus (HCV) as exemplary target analyte.

18.2 IMMOBILIZATION TECHNIQUES

18.2.1 Affinity immobilization based on biotin–avidin interactions

A very famous example of the immobilization of biomolecules is the biotin–avidin complex attached to an electrochemical-generated polypyrrole or polyphenol film [27–29]. The avidin–biotin strategy has some notable advantages compared to other immobilization techniques. The extremely specific and high-affinity interactions between four biotins, a vitamin and the glycoprotein avidin (association constant $K_a = 10^{15}$) [38] lead to strong associations similar to covalent bonds (Fig. 18.1A). The use of these remarkable strong affinity interactions was applied to the successful elaboration of nanostructured bioarchitectures on polymers by an affinity-driven immobilization (Fig. 18.1B). The anchoring of protein or oligonucleotide monolayer was thus performed by the formation of avidin bridges between electropolymerized films bearing biotin groups (Fig. 18.1C) and biotinylated enzymes, antibodies, bacteria or oligonucleotides [30,39–41]. The successive protein layers were elaborated by simple incubation of the biotinylated polymer with the corresponding aqueous solution of protein.

The avidin–biotin immobilization method maintains the biological activity of the receptor molecules. Besides the biocompatibility of the procedure, the surface geometry of these films provides high accessibility of the immobilized biomolecules. In addition, the avidin molecules form a passivation layer on the transducer surface that prevents nonspecific adsorption of proteins on the surface. In contrast with conventional grafting or affinity binding, this step-by-step approach can also be applied to the preparation of assemblies containing multilayers of biological molecules [42].

The biotin monomers are usually electropolymerized on gold, platinum or glassy carbon electrode surfaces providing pure biotinylated homopolymer or biotinylated copolymers by co-polymerizations with regular pyrrole, a pyrrole ammonium or a polypyridinyl complex of ruthenium(II) containing three pyrrole groups [27,28,43]. A particularly exciting achievement has been the deposition of a transparent indium tin oxide (ITO) layer onto a glass fiber tip which was used as electrode for the electrochemical formation of polypyrrole films [44]. The formation of the ITO layer on the glass fiber was performed by DC-Magnetron sputtering with an ITO target. The fiber tips were exposed to the sputter beam with an angle of 45°. This allowed the ITO coating of the fiber tip surface, the area of the electropolymerization, as well as a part of the glass cylinder of the fiber for the electric contact. The resulting ITO layer has an approximate thickness of 200 nm.

Another advantage of the avidin–biotin procedure is the fact that the surface of the transducer, after the recognition event, can be regenerated. It has been demonstrated that biotinylated conducting polypyrroles, used for the fabrication of DNA chips showed the capability to be regenerated after "denaturation" of the biotin/avidin links by surfactants [45]. The recovered biotinylated polypyrrole presents the same ability to form biological architectures based on avidin–biotin bridges opening the way to a reusable sensor.

Due to the importance of cholera as an endemic scourge or infrequently as a pandemic in a number of countries in Asia, Africa and Latin America, the possibility to develop an immunosensor for this disease was examined with anti-cholera toxin (anti-CT) antibody. Indeed, rapid identification of cholera is essential for a healthy society

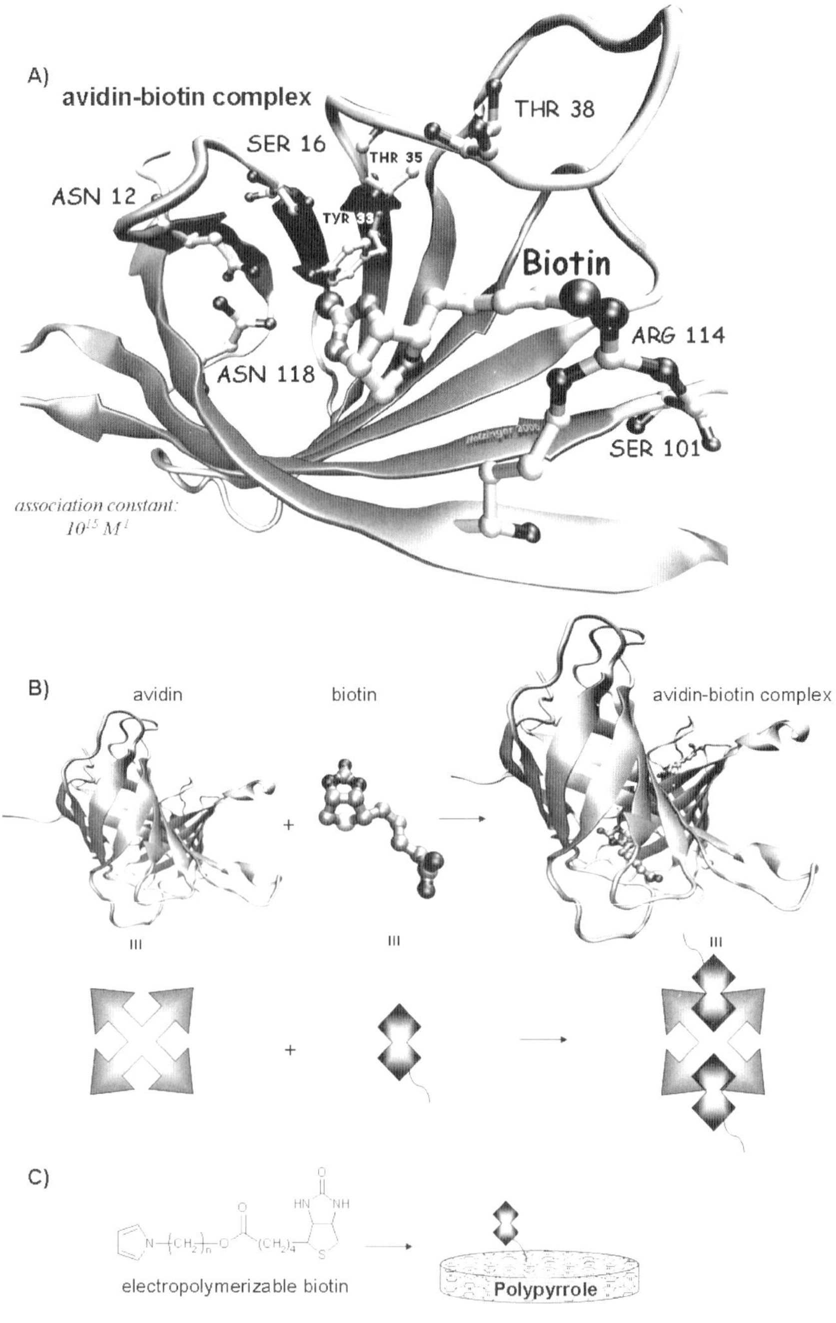
A)
avidin-biotin complex
THR 38
SER 16
THR 35
ASN 12
TYR 33
Biotin
ARG 114
ASN 118
SER 101
association constant:
10^{15} M^{-1}
B)
avidin
biotin
avidin-biotin complex
C)
electropolymerizable biotin
Polypyrrole

since cholera still remains a serious threat in developing countries where sanitation is poor, health care limited and drinking water unsafe. The immobilization and detection of rabbit anti-CT antibody was achieved using a sandwich immunoassay [46] (Fig. 18.2). The first step required the electrochemical deposition of a thin transparent poly(pyrrole–biotin) film onto an ITO-coated optical fiber tip with an approximate thickness of 200 nm [44].This affinity film allowed by successive incubation steps the conjugation of avidin and the subsequent binding of the biotin-labeled cholera toxin B subunit. The analyte, rabbit anti-CT antibody, thereafter bound the corresponding immobilized cholera toxin B subunit epitopes. The presence of the immobilized antibody was determined and quantified by a chemiluminescence reaction after its labeling by an enzyme.

18.2.2 Photoaffinity immobilization

Among the conventional immobilization methods, photopatterning has been widely used for the fabrication of bioanalytical devices. Actually, immobilization of protein by light is topically addressable and compatible with biological functions. In addition, this reagentless approach based on a light-induced reaction between a photoreactive group and C–H bonds, is easily applicable to a wide variety of proteins [47–51]. However, the immobilization of a protein by a photochemical linkage previously requires the homogeneous fixation of photoreactive compounds on transducer surfaces. An innovative approach consists of combining the advantages of the preceding method (affinity immobilization via electropolymerized films) with those of the photochemical grafting [52]. This approach was first based on the electrogeneration of a biotinylated polypyrrole followed by its incubation with avidin to form a compact avidin layer. The binding sites of the latter were then saturated with photobiotin. Since the commercially available photobiotin was widely employed for the light-driven biotinylation of biological

Fig. 18.1. (A) Schematic presentation of the biotin–avidin complex with detailed configuration of biotin and surrounding polar residues. (B) Detailed (top) and simplified (bottom) representations of the complication possibilities of biotin and avidin. (C) Structure of an electropolymerizable biotin derivative and schematic presentation of an electropolymerized polypyrrole film bearing biotin groups on its surface. All three-dimensional representations of avidin have been generated using the program VMD [70] and subsequently rendered with PovRay.

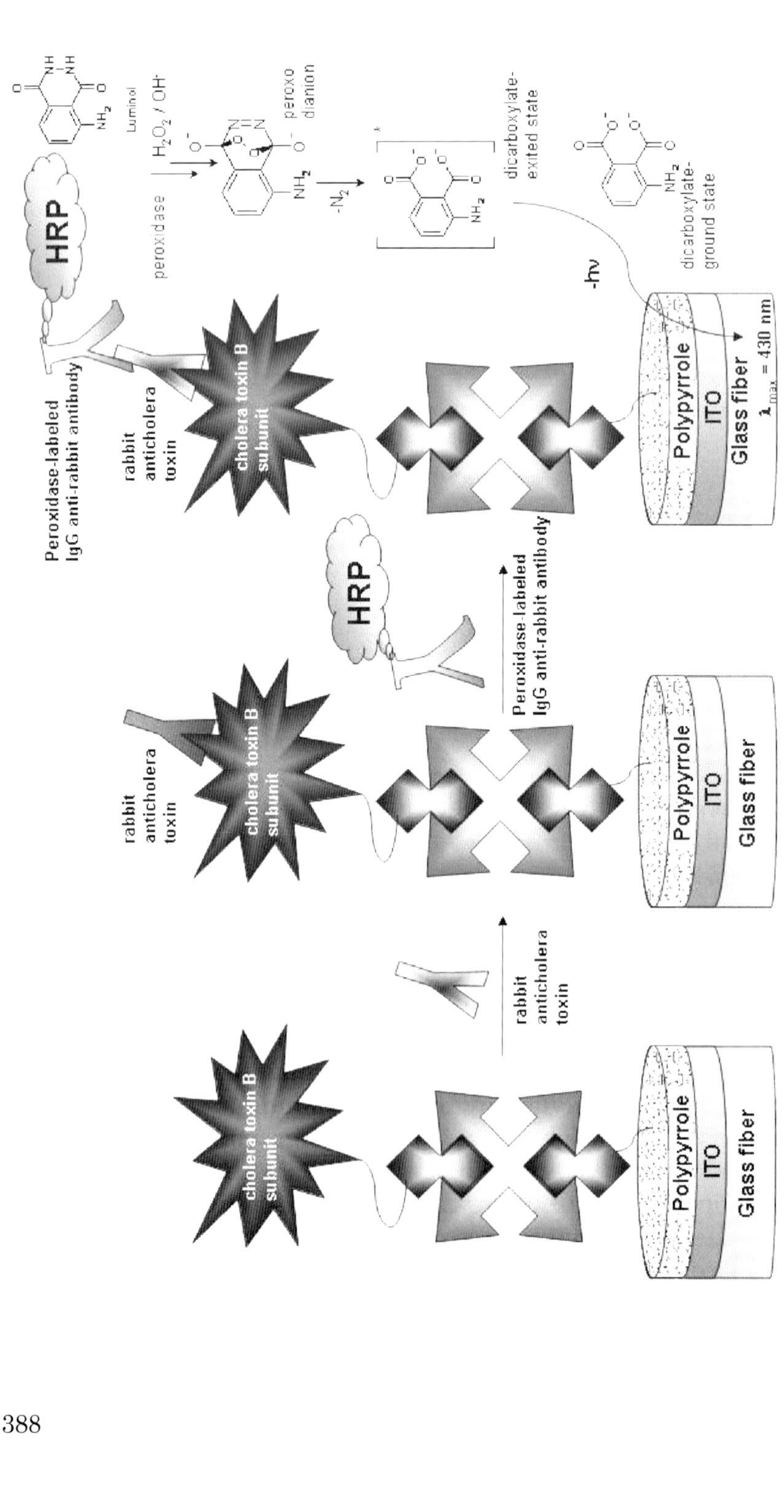

Fig. 18.2. Biosensor scheme describing the various steps involved in the immunoassay using ITO-poly(pyrrole–biotin)-coated optical fibers for the detection of anti-cholera toxin B subunit using the biotin–avidin immobilization technique.

macromolecules, this photoreactive compound fixed on the avidin layer, should allow, under irradiation, the covalent binding of proteins onto the electropolymerized film (Fig. 18.3).

For instance, the immobilization of cholera toxin (anti-CT) was carried out through the light activation of photobiotin, previously fixed on a biotinylated polypyrrole film by affinity interactions. The advantage of this immobilization method is the attachment of the antibody to the transducer surface with only one single covalent bond generated by a light-induced insertion reaction. The accessibility of the anti-CT for further immunoreactions is therefore preserved. In order to demonstrate the presence of the photografted antibody, the modified electrode was incubated with a marker of the cholera toxin antibody: a secondary antibody labeled with a horseradish peroxidase. The resulting intensity of the enzymatic activity of the electrode constitutes a signal proportional to the amount of immobilized anti-CT. To verify the efficiency of this photografting process, parallel experiments have been performed by using standard biotin instead of photobiotin. Keeping identical conditions for the immunosensor fabrication, the binding of anti-CT by irradiation led to only a partial coverage of the surface by the antibody via a non-covalent adsorption phenomenon. This has been proven by measurements of the enzymatic activity of the resulting electrodes where the immunosensor, fabricated with photobiotin, gives a three times higher signal (31.6 mU/cm^2) than the system with non-covalently adsorbed anti-CT (10 mU/cm^2). Furthermore, the spatial homogeneous immobilization of photografted anti-CT was confirmed by identical enzymatic activity values recorded for the photografted antibody and for a monolayer of biotinylated cholera toxin saturated with the target: anti-CT [52].

18.2.3 Photochemical immobilization

Although the positioning of the protein by affinity interactions or photografting at the polymer–solution interface seems to constitute a key parameter of the recognition process for immunosensors, the presence of an intermediate avidin layer as building block may have a detrimental effect on the sensitivity of the transduction step [53]. A recent simplified approach in protein deposition by photografting lies in the direct covalent binding of proteins by irradiation of electrogenerated photoactive polymer films. The strengths and weaknesses are the same as those reported for the covalent binding via activated ester groups

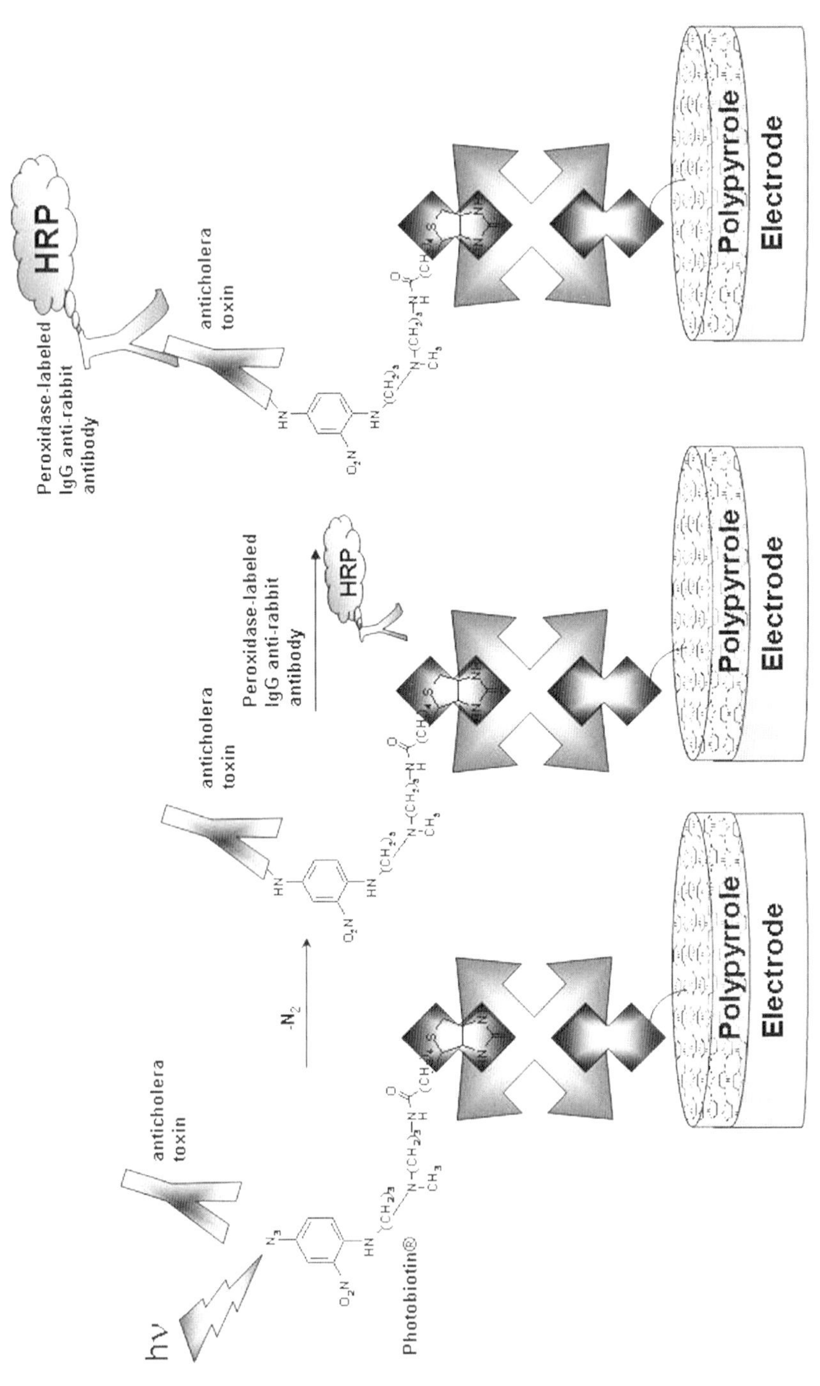

Fig. 18.3. Schematic representation of the photoaffinity immobilization of the antibody cholera toxin (anti-CT). After irradiation of the photobiotin® layered electrode in presence of the antibody, a covalent bond has been formed between the electrode and the antibody keeping its accessibility for immunoreactions with HRP-IgG anti-rabbit antibody.

except that the procedure was faster and spatially addressable by light [24]. This photografting procedure was successfully applied to the elaboration of enzyme electrodes and optical immunosensors [54–57]. Photochemical immobilization of protein was achieved by the electropolymerization of a benzophenone derivative, a photoactive component modified by an electropolymerizable pyrrole group and irradiation of the resulting polypyrrole film in the presence of proteins. Benzophenone and most of its derivatives form biradical transition states at the carbonyl group when exited at around 345 nm.This activated benzophenone reacts preferably with acidic C–H bonds in an insertion-like reaction accompanied by hydrogen migration to the carbonyl oxygen atom forming a hydroxy group. In biological systems, the most effective position for this kind of reaction is the NH_2-methylene group in α-amino acids [58] (Fig. 18.4).

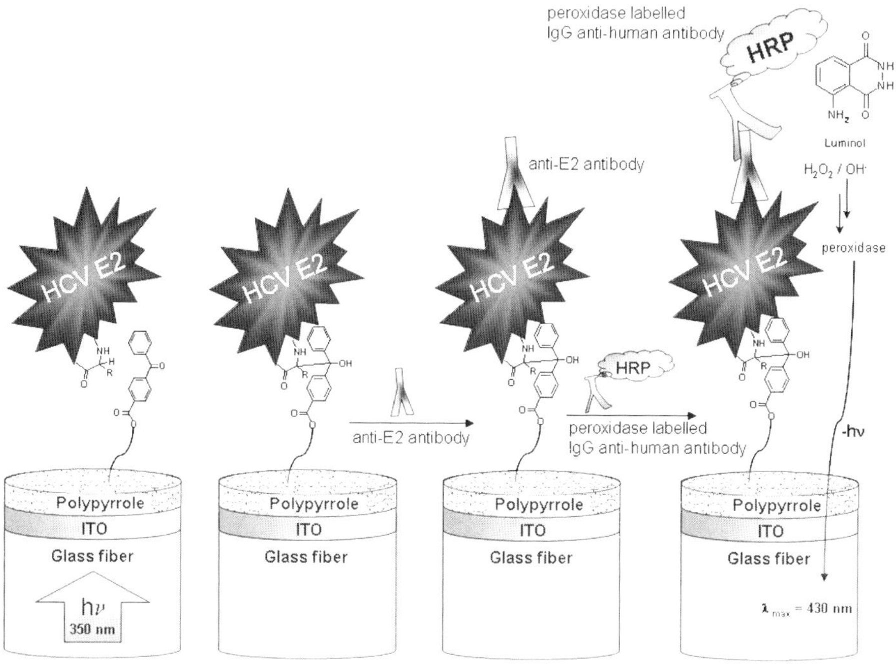

Fig. 18.4. Mechanism for the photochemical reaction of a benzophenone-coated ITO-modified optical fiber with a C–H bond of an amino acid side chain of the hepatitis C virus. The virus can act as an antigen for the detection of the anti-HCV antibody. After the immunoreaction, the subsequent binding of a marker, a secondary peroxidase-labeled antibody, allowed to catalyze a chemiluminescence reaction for the receipt of an optical signal.

The development of optical immunosensors for the detection of antibody of HCV was performed by electrogeneration of the photoactive polypyrrole on ITO-modified optical fibers followed by internal light irradiation in the presence of the virus. The resulting layer of HCV can act as an antigen for the detection of the anti-HCV antibody. After the immunoreaction, the subsequent binding of a marker, a secondary peroxidase-labeled antibody, allowed to catalyze a chemiluminescence reaction in the presence of luminol and H_2O_2 [59].

18.3 DETECTION TECHNIQUES FOR IMMOBILIZED ANALYTES

Besides groundbreaking immobilization methods of antibodies, innovative detection strategies developed for high-performance immunosensors were illustrated with antibodies of cholera toxin or HCV as target. Enzymes that catalyze light-emitting processes or the formation of redox active components labeled the recognized target. Besides labeling for signal receipt, an example is highlighted where the immobilized antibody (anti-CT) is directly detected [60].

18.3.1 Chemiluminescence catalyzed by enzyme labeling

The possibility to electropolymerize pyrrole derivatives on ITO-coated fiber tips opened a new access to highly sensitive optical immunosensors. An optical immunosensor for the detection of rabbit anti-CT antibody was achieved by immobilization of a cholera toxin via biotin–avidin interactions onto a biotinylated polypyrrole formed on an optical fiber. The optical detection scheme based on a sandwich enzymatic immunoassay is illustrated in Fig. 18.5. After immunoreaction and hence immobilization of the target (anti-CT antibody), the subsequent binding of a marker peroxidase-labeled IgG anti-rabbit antibody allowed then to carry out a chemiluminescence reaction by adding peroxidase substrates, namely luminol and H_2O_2 [59]. The emitted light is collected through the polypyrrole–ITO films by the optical fiber and transferred to the measurement setup. Fig. 18.5 shows the calibration curve for the cholera toxin antibody obtained by collecting data in triplicate at antibody dilutions (titers) ranging from 1:30 to 1:1,200,000 [46].

The detection of anti-CT dilution up to 1:1,200,000 constitutes a clear improvement in detection limit compared to similar

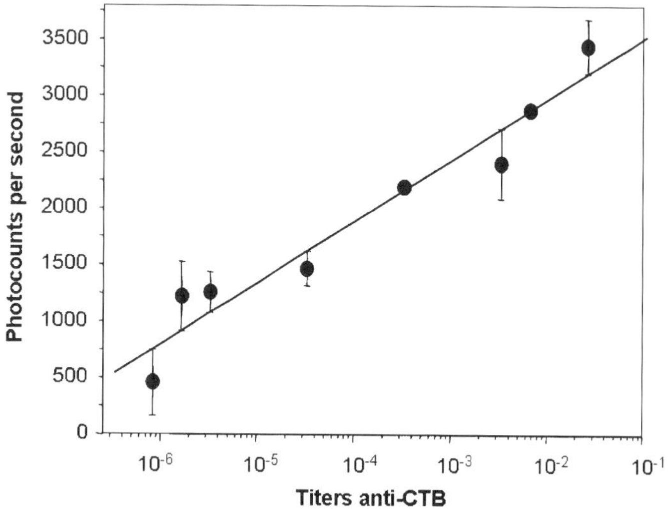

Fig. 18.5. Calibration curve obtained by the immunoassay procedure for the detection of anti-CTB using the ITO-PPB-coated optical fibers. The curve was fitted according to the equation $y = A+B\ \ln(x)$, where x is the anti-CTB dilution value and y is the chemiluminescence response. The obtained correlation coefficient is $R^2 = 0.95$.

immunosensors. For example, when the biotinylated polypyrrole coating on the optical fiber tip is generated by wet-chemical polymerization, the lower detection limit of anti-CTB dilution has been 1:300,000 [61]. Such improvements are attributed to the electrochemical procedure that assures more uniform film coatings with well-controlled thicknesses [62].

Another example of optical transduction was developed with an immunosensor for the detection of antibody of the HCV [53]. The optical fiber was modified by the HCV E2 envelope protein photoimmobilized on polypyrrole–benzophenone film as mentioned above. After the immunoreaction, the subsequent binding of a marker, peroxidase-labeled IgG anti-human antibody, allowed to catalyze a chemiluminescence reaction in the presence of luminol and H_2O_2. The immunosensor response in term of photocounts/s, was recorded in triplicate with antibody titer dilutions ranging from 1:500 up to 1:1,024,000. Fig. 18.6 shows a linear calibration curve for antibody titer dilutions in the range of 1:64,000–1:1,024,000. At higher concentrations, the curve levels off with a response saturation observed from titers 1:4000 and more. The target (anti-E2 antibody) was thus detected at the extremely low titer of 1:1,024,000 in real serum tests exhibiting higher sensitivity than

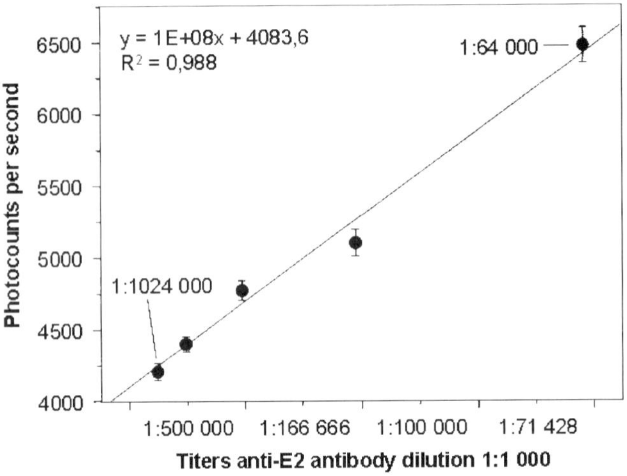

Fig. 18.6. Calibration curve obtained by the immuno-assay procedure in the detection of anti-E2 using the ITO-Poly (pyrrole-benzophenone) coated optical fibers. The linear range of the calibration curve was obtained for titer 1:64,000 and lower. The curve was fitted according to the equation $y = A+B(x)$, where x is the human sera (anti-E2 antibodies) dilution value and y is the chemiluminescence response. The obtained correlation coefficient was $R^2 = 0.988$.

Western blot and ELISA tests [63,64]. Besides the higher sensitivity, an improved selectivity has also been observed using the optical fiber immunosensor. Anti-E2 antibodies in all sera positive for anti-HCV (anti-core, NS3 and NS5) and HCV RNA (100% sensitivity) could be detected. Using Western blot tests, only 69% of anti-HCV (anti-core, NS3 and NS5) positive samples could be detected [63,64]. Furthermore, standard test for anti-HCV antibodies requires the use the three cited antigens (core, NS3 and NS5) because of its limited sensitivity. The high sensitivity of the optical fiber immunosensor, containing only the HVC E2 envelope protein, allows the identification of all RNA positive patients.

18.3.2 Amperometric detection via enzyme labeling

The amperometric transduction of an immunoreaction consists in the post-labeling of the detected target by an enzyme able to catalyze the production or the consumption of electroactive species. The electrochemical oxidation or reduction of the latter at a constant potential applied to the immunosensor provides thus a current, whose intensity is proportional to the amount of immobilized target.

For amperometric immunosensors based on electropolymerized films, the enzymically generated electroactive species indicating the immunoreaction must diffuse through the polymer coating to be detected at the electrode surface. As a consequence, the immunosensor sensitivity is strongly related to the permeability of the underlying polymer used for the immobilization of antibody or antigen. The electropolymerization of pyrrole–biotin alone provides a compact and fully active layer of biotin. However, its hydrophobic character prevents the polymer from swelling in water and passivates the electrode surface. It has been reported that composites of polypyrrole films with highly hydrophilic hydrogels were designed to improve the hydration level of the polymer coating and hence its permeability in aqueous solutions [65,66]. In the same vein, we recently reported the co-polymerization of pyrrole–biotin with a hydrophilic pyrrole derivative (lactobionamide–pyrrole) for the fabrication of permeable biotinylated films [67] (Fig. 18.7). The introduction of hydrophilic

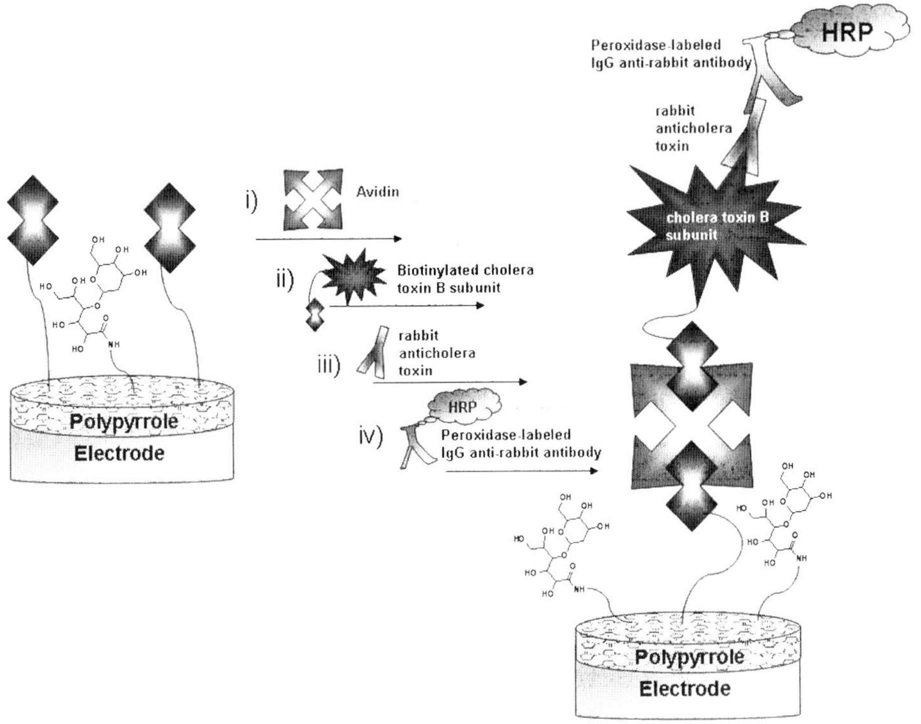

Fig. 18.7. Schematic representation of the immobilization of the biotinylated cholera toxin B subunit using the biotin–avidin bridging system. The hydrophilic co-polymer on the transducer surface improves the permeation of the enzymatically generated quinone.

groups to biotinylated, electropolymerized copolymers has no significant influence on the formation of a homogeneous avidin layer on the modified electrode. QCM experiments have shown that the amount of complexed avidin on the biotin–lactobionamide pyrrole copolymer decreases less than ten percent compared to a pure biotinylated polypyrrole surface [68].

An amperometric immunosensor for the detection of rabbit anti-CT antibody was achieved by immobilization of a cholera toxin by an avidin bridge onto the preceding biotinylated copolymer formed on electrode surface. The transduction step involved the recognition of the captured target (anti-CT) by a secondary antibody marker labeled by a peroxidase or a biotin group. The enzyme markers, biotinylated glucose oxidase and biotinylated polyphenol oxidase, were then attached to the secondary antibody by an avidin bridge. Different amperometric detection configurations involving various substrates (glucose, catechol, hydroquinone, ferrocyanide, ferrocene di-carboxilic acid), carbon or platinum electrodes, an electro-reduction or electro-oxidation step depending on the enzyme marker, were developed and tested for the amperometric detection of anti-CT. The comparison of the electro-enzymatic performances of these three configurations with different substrates clearly shows that the more sensitive amperometric immunosensor was based on the simplest enzymatic configuration based on a secondary antibody conjugated with a peroxidase and the electroactive hydroquinone/H_2O_2 system [67]. The polymerized lactobionamide groups must counterbalance the hydrophobic character of the biotinylated polypyrrole film and thus improve the permeation of the enzymatically generated quinine (for details, see Procedure 26 in CD accompanying this book). The peroxidase-amperometric immunosensor thus exhibited a very sensitive limit of detection of 50 ng/mL anti-CT with a response time for the enzyme product within the range of 5–30 s (Fig. 18.8). It should be noted that this detection limit is more sensitive than those reported for a conventional spectrometric detection using the ELISA test (100 ng/mL)[69]. Moreover, operational stability tests demonstrate that the immunosensors show a very low (5%) decay of the initial amperometric response after 4 h.

18.3.3 Photoelectrochemical detection without labeling step

The conventional detection methods of antigen–antibody interactions were mainly based on sophisticated detection technologies requiring

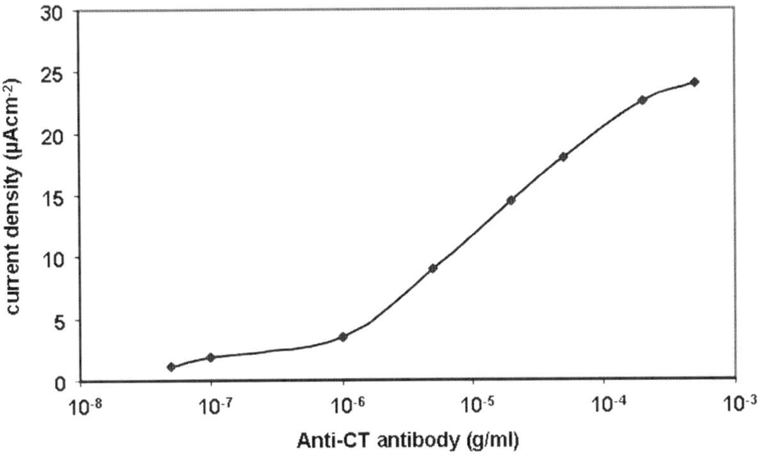

Fig. 18.8. Current response of the immunosensors as a function of anti-CT antibody concentrations ranging from 5×10^{-8} to 5×10^{-3} g/mL with peroxidase as marker and hydroquinone/H_2O_2 as electrochemical system. Applied potential: -0.1 V vs. SCE.

a previous labeling step of the protein target by an enzyme marker or a fluorescent molecule. The avidin–biotin technique is a powerful tool for the immobilization of bio-receptors keeping their recognition abilities. However, the examples of high-performance immunosensors based on this construction method, also requires a labeling procedure by enzymes that catalyze reactions forming redox- or photoactive species. Concerning the development of direct detection strategies for immunosensors, a clear step forward has been the development of a new photoelectrochemical concept for the direct detection of an immunoreaction without labeling step of the antibody target [60]. This strategy consists of conferring photoelectrochemical properties to the probe layer covering the transducer, the recognition event triggering a change of these properties. The innovative point has been the first synthesis of an electropolymerizable, photosensitive, biotinylated ruthenium complex (Fig. 18.9). After the electrogeneration of a polypyrrole film of Ru(II), the immobilization of a cholera toxin as antigen was successfully performed by the formation of avidin bridges between the biotinylated film and the biotinylated toxin. The analyte, anti-CT antibody, was anchored by immunoreaction on the antigen-coated electrode. The oxidation of the polymerized Ru(II) into Ru(III) is induced by the irradiation of the immunosensor in presence of pentaaminechlorocobalt(III) chloride as oxidative quencher. The subsequent

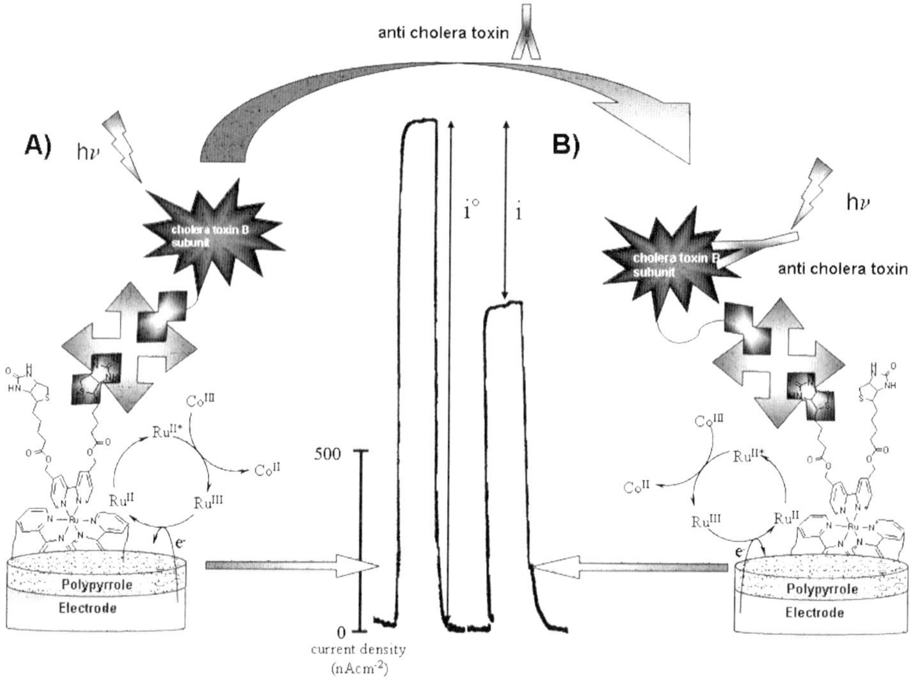

Fig. 18.9. (A) Schematic presentation of biotinylated polypyrrole-ruthenium electrode in presence of an oxidative quencher and the resulting photocurrent response by switching the excitation light on and off. (B) Reduced photocurrent response and schematic presentation of the polypyrrole-ruthenium electrode after incubation in anti-cholera toxin solution.

reduction of the polymerized complex by the underlying electrode potentiostated at 0.5 V vs. SCE completes the circuit, thus generating a cathodic photocurrent. Fig. 18.9A shows the photocurrent response of the initial biotinylated Ru complex film when the excitation light was turned on and off. The regenerated photocurrent keeps its intensity 20 times over 50 min without noticeable decrease illustrating the mechanical and photo physical stability of the film. The successive binding of avidin and biotinylated antigen on the biotinylated film results in a decrease of the photocurrent response. This effect is due to the hindered diffusion of quencher molecules to the ruthenium film. The final immunoreaction with anti-CT results in a more considerable photocurrent decrease, reflecting the antibody anchoring (Fig. 18.9B). The photocurrent produced by irradiation of this film in the presence of a Co(III) complex can be thus exploited for directly detecting the immunoreaction without labeling step of the cholera toxin antibody. A calibration curve based on the decrease of the photocurrent intensity as a function of the antibody target concentrations in the range 0–200 μg/mL, illustrates the validity of this new detection concept of protein interactions, the detection limit being 0.5 μg/mL. This immunosensor setup has passed several control experiments like the examination for non-specific binding such as simple antibody adsorption. This original photoelectrochemical approach that is characterized by a good spatial and temporal resolution may constitute an attractive alternative to fluorescence and other detection systems that require tagged analyte.

18.4 CONCLUSION

The presented examples of recent achievements in immunosensor development show the tremendous progress in this field. Detection limits lower continuously, miniaturized setups by keeping the possibility of automated production progressed, and procedures for detection are constantly simplified. Direct immobilization of the receptor antibody by photochemical reactions enables the production of ready to use immunosensors in few steps. With photovoltaic components, time consuming and expensive labeling procedures of the analyte can be avoided.

REFERENCES

1 J. Wang, *Anal. Chim. Acta*, 469 (2002) 63–71.
2 E. Katz and I. Willner, *Electroanalysis*, 15 (2003) 913–947.
3 S. Rodriguez-Mozaz, M.J.L.d. Alda, M.-P. Marco and D. Barcelo, *Talanta*, 65 (2005) 291–297.
4 X.J. Liu, W. Farmerie, S. Schuster and W.H. Tan, *Anal. Biochem.*, 283 (2000) 56–63.
5 A. Navas-Di'az and M.C. Ramos-Peinado, *Sens. Actuators B*, 39 (1997) 426–431.
6 B. Polyak, E. Bassis, A. Novodvorets, S. Belkin and R.S. Marks, *Water Sci. Technol.*, 42 (2000) 305–311.
7 B. Polyak, E. Bassis, A. Novodvorets, S. Belkin and R.S. Marks, *Sens. Actuators B*, 74 (2001) 18–26.
8 L. Henke, P.A.E. Piuno, A.C. McClure and U.J. Krull, *Anal. Chim. Acta*, 344 (1997) 201–213.

9 J. Cordek, X. Wang and W. Tan, *Anal. Chem.*, 71 (1999) 1529–1533.
10 D.R. Walt, *Acc. Chem. Res.*, 31 (1998) 267–278.
11 F.J. Steemers and D.R. Walt, *Mikrochim. Acta*, 131 (1999) 99–105.
12 M. Lee and D.R. Walt, *Anal. Biochem.*, 282 (2000) 142–146.
13 S. Szunerits and D.R. Walt, *Anal. Chem.*, 74 (2002) 1718–1723.
14 M.Y. Rubtsova, G.V. Kovba and A.M. Egorov, *Biosens. Bioelectron.*, 13 (1998) 75–85.
15 B. Lu, M.R. Smyth and R. O'Kennedy, *Analyst*, 121 (1996) 29R–32R.
16 P.L. Domen, J.R. Nevens, A.K. Mallia, G.T. Hermanson and D.C. Klenk, *J. Chromatogr.*, 510 (1990) 293–302.
17 P. Pantano and W.G. Kuhr, *Anal. Chem.*, 65 (1993) 623–630.
18 E. Brynda, M. Houska, J. Kvor and J.J. Ramsden, *Biosens. Bioelectron*, 13 (1998) 165–172.
19 H. Gao, M. Sanger, R. Luginbuhl and H. Sigrist, *Biosens. Bioelectron*, 10 (1995) 317–328.
20 M. Gerard, A. Chaubey and B.D. Malhotra, *Biosens. Bioelectron*, 17 (2002) 345–359.
21 W. Schuhmann, *Rev. Mol. Biotech.*, 82 (2002) 425–441.
22 S. Cosnier, *Anal. Bioanal. Chem.*, 377 (2003) 507–520.
23 J.C. Vidal, E. Garcia-Ruiz and J.R. Castillo, *Mikrochim. Acta*, 143 (2003) 93–111.
24 S. Cosnier, *Electroanalysis*, 17 (2005) 1701–1715.
25 M. Umana and J. Waller, *Anal. Chem.*, 58 (1986) 2979–2983.
26 P.N. Bartlett, R. Whitaker, M.J. Green and J. Frew, *J. Chem. Soc. Chem. Commun.*, 20 (1987) 1603–1604.
27 S. Cosnier, B. Galland, C. Gondran and A. LePellec, *Electroanalysis*, 10 (1998) 808.
28 L.M. Torres-Rodriguez, A. Roget, M. Billon, T. Livache and G. Bidan, *Chem. Soc. Chem. Commun.*, 18 (1998) 1993–1994.
29 S.T. Yang, A. Witkowski, R.S. Hutchins, D.L. Scott and L.G. Bachas, *Electroanalysis*, 10 (1998) 58.
30 S. Cosnier, M. Stoytcheva, A. Senillou, H. Perrot, R.P.M. Furriel and F.A. Leone, *Anal. Chem.*, 71 (1999) 3692–3697.
31 N. Haddour, S. Cosnier and C. Gondran, *J. Am. Chem. Soc.*, 127 (2005) 5752–5753.
32 K. Ramanathan, A. Dzgoev, J. Svitel, B.R. Jonsson and B. Danielsson, *Biosens. Bioelectr.*, 17 (2002) 283–288.
33 A.E. Botchkareva, F. Fini, S. Eremin, J.V. Mercader, A. Montoya and S. Girroti, *Anal. Chim. Acta*, 453 (2002) 43–52.
34 A. Sargent, T. Loi, S. Gal and O.A. Sadik, *J. Electroanal. Chem.*, 470 (1999) 144–156.
35 M.A. Gonzalez-Martinez, R. Puchades and A. Maquieira, *Anal. Chem.*, 73 (2001) 4326–4332.
36 H. Aizawa, S. Kurosawa, M. Tozuka, J.-W. Park, K. Kobayashi and H. Tanaka, *Biosens. Bioelectr.*, 18 (2003) 765–771.

37 Z. Ma, J. Wu, T. Zhou, Z. Chen, Y. Dong, J. Tang and S.-F. Sui, *New J. Chem.*, 26 (2002) 1795–1798.
38 P.C. Weber, D.H. Ohlendorf, J.J. Wendoloski and F.R. Salemme, *Science*, 243 (1989) 85–88.
39 S. DaSilva, L. Grosjean, N. Ternan, P. Mailley, T. Livache and S. Cosnier, *Bioelectrochemistry*, 63 (2004) 297–301.
40 O. Ouerghi, A. Touhami, N. Jaffrezic-Renault, C. Martelet, H.B. Ouada and S. Cosnier, *Bioelectrochemistry*, 56 (2002) 131–133.
41 L.M. Torrez-Rodrigez, M. Billon, A. Roget and G. Bidan, *J. Electroanal. Chem.*, 523 (2002) 70–78.
42 J. Anzai, H. Takeshita, Y. Kobayashi, T. Osa and T. Hoshi, *Anal. Chem.*, 67 (1995) 811–817.
43 S. Cosnier and A. LePellec, *Electrochim. Acta*, 44 (1999) 1833–1836.
44 R.S. Marks, A. Novoa, T. Konry, R. Krais and S. Cosnier, *Mater. Sci. Eng. C*, 21 (2002) 189–194.
45 A. Dupont-Filliard, A. Roget, T. Livache and M. Billon, *Anal. Chim. Acta*, 449 (2001) 45–50.
46 T. Konry, A. Novoa, S. Cosnier and R.S. Marks, *Anal. Chem.*, 75 (2003) 2633–2639.
47 L. Rozsnyai, D.R. Benson, S.P.A. Fodor and P.G. Schultz, *Angew. Chem. Int. Ed.*, 6 (1992) 759.
48 H. Sigrist, A. Collioud, J.-F. Clémence, H. Gao, R. Luginbühl, M. Sänger and G. Sundarababu, *Opt. Eng.*, 34 (1995) 2339–2348.
49 G. Sundarababu, G. Gao and H. Sigrist, *Photochem. Photobiol.*, 61 (1995) 540.
50 O. Prucker, A. Naumann, J. Rûhe, W. Knoll and W. Curtis, *J. Am. Chem. Soc.*, 121 (1999) 8766.
51 T. Panasyuk, V.M. Mirskyand and O.S. Wolfbeis, *Electroanalysis*, 14 (2002) 221.
52 S. Cosnier, C. Molins, C. Mousty, B. Galland and A. LePellec, *Mater. Sci. Eng. C*, 26 (2006) 436–441.
53 T. Konry, A. Novoa, Y. Shemer-Avni, N. Hanuka, S. Cosnier, A. LePellec and R.S. Marks, *Anal. Chem.*, 77 (2005) 1771–1779.
54 S. Cosnier. In *PCT/FR00/02267*; WO 0112699 A1 20010222: France, 2001.
55 S. Cosnier and A. Senillou, *Chem. Commun.*, 3 (2003) 414–415.
56 G. Herzog, K. Gorgy, T. Gulon and S. Cosnier, *Electrochem. Commun.*, 7 (2005) 808–814.
57 T. Konry, A. Novoa, Y. Shemer-Avni, N. Hanuka, S. Cosnier, A. Lepellec and R.S. Marks, *Anal. Chem.*, 77 (2005) 1771–1779.
58 G.D. Prestwich, G. Dorman, J.T. Elliott, D.M. Marecak and A. Chaudhary, *Photochem. Photobiol.*, 65 (1997) 222–234.
59 T.M. Freeman and W.R. Seltz, *Anal. Chem.*, 50 (1978) 1242–1246.
60 N. Haddour, S. Cosnier and C. Gondran, *Chem. Commun.*, 21 (2004) 2472–2473.

61 R.S. Marks, A. Novoa, D. Thomassey and S. Cosnier, *Anal. Bioanal. Chem.*, 374 (2002) 1056–1063.
62 D.T. McQuade, A.E. Pullen and T.M. Swager, *Chem. Rev.*, 100 (2000) 2537–2574.
63 N. Hanuka, E. Sikuler, D. Tovbin, L. Neville, O. Nussbaum, M. Mostoslavsky, M. Orgel, A. Yaari, S. Manor, S. Dagan, N. Hilzenrat and Y. Shemer-Avni, *J. Med. Virol.*, 9999 (2004) 1–7.
64 N. Hanuka, E. Sikuler, D. Tovbin, M. Mostoslavsky, M. Hausman, M. Orgel, A. Yaari and Y. Shemer-Avni, *J. Viral Hepatitis*, 9 (2002) 141–145.
65 C.J. Small, C.O. Too and G.G. Wallace, *Polym. Gels Networks*, 5 (1997) 251–265.
66 S. Brahim and A. Guiseppi-Elie, *Electroanalysis*, 17 (2005) 556–570.
67 R.E. Ionescu, C. Gondran, S. Cosnier, L.A. Gheber and R.S. Marks, *Talanta*, 66 (2005) 15–20.
68 R.E. Ionescu, C. Gondran, L.A. Gheber, S. Cosnier and R.S. Marks, *Anal. Chem.*, 76 (2004) 6808.
69 B. Leshem, G. Sarfati, A. Novoa, I. Breslav and R.S. Marks, *Luminescence*, 19 (2004) 69–77.
70 W. Humphrey, A. Dalke and K. Schulten, *J. Mol. Graphics*, 14.1 (1996) 33–38.

Chapter 19

Genosensor technology for electrochemical sensing of nucleic acids by using different transducers

Arzum Erdem

19.1 INTRODUCTION

The electroactivity in nucleic acids was described by Palecek [1] in the beginning of 1960s before many electrochemical approaches had been performed and had proven useful for analyzing or quantification of nucleic acids. There has been an increase in the use of nucleic acids as tools in the recognition and monitoring of hybridization of analytical interest [2–7]. Nucleic acid layers combined with electrochemical or optical transducers produce a new kind of affinity biosensors as DNA biosensor (genosensor) for low-molecular-weight molecules. A biosensor associates a bioactive sensing layer with any suitable transducer and gives an output signal. Biomolecular sensing can be defined as the possibility of detecting analytes of biological interest, like metabolites, drugs and toxins, by using an affinity layer, like enzymes, receptors, antibody or nucleic acids. The affinity layer can be a natural system or an artificial one, capable of recognizing a target molecule in a complex medium among thousands of others. Various combinations of biological material associated with different types of transducers are an attractive subject of research.

A genosensor, or gene-based biosensor/DNA biosensor, normally employs immobilized DNA probes as the recognition element and measures specific binding processes such as the formation of DNA–DNA and DNA–RNA hybrids, and the interactions between proteins or ligand molecules and DNA at the sensor surface [5].

Typically, the design of a genosensor involves the following steps [7]: (1) modification of the sensor surface to create an activated layer for the attachment of the DNA probe; (2) immobilization of the probe

Comprehensive Analytical Chemistry 49
Alegret and Merkoçi (Eds)
Volume 49 ISSN: 0166-526X DOI: 10.1016/S0166-526X(06)49019-X

molecules onto the surface, preferably with controlled packing density and orientation; and (3) detection of target gene sequence by DNA hybridization at the sensor–liquid interface.

DNA biosensors based on electrochemical transduction of hybridization couple the high specificity of hybridization reactions with the excellent sensitivity and portability of electrochemical transducers. The ultimate goal of all researches is to design DNA biosensors for preparing a basis for the future DNA microarray system.

Recent developments in genosensor design with the advances in nanotechnology provide new tools in order to develop new techniques to monitor biorecognition and interaction events on solid surfaces and also in solution phase. Typical applications include environmental monitoring and control, and chemical measurements in the agriculture, food, and drug industries [8–17].

In aspect of chip-based technology, electrochemical genosensors based on different materials and transducers have been recently developed in response to clinical demand of giving promising results [18–25]. Different sensor technologies provide a unique platform in order to immobilize molecular receptors by adsorption, crosslinking or entrapment, complexation, covalent attachment, and other related methods on nanomaterials [5,7,26].

An overview on the genosensor technologies for detection of nucleic acids (NA) immobilized onto different transducers by adsorption, cross-linking, complexation and covalent attachment is briefly summarized in Table 19.1. The applications of electrochemical genosensor technology are discussed in the following section.

19.2 APPLICATIONS OF ELECTROCHEMICAL GENOSENSOR TECHNOLOGIES

The reported strategies utilized in DNA sensing include (1) sequence-specific hybridization processes based on the oxidation signal of most electroactive DNA bases, guanine and adenine [13,24] or (2) quasi-specific detection of small molecules capable of binding by intercalation or complexation with DNA, such as metal coordination complexes, antibiotics, pesticides, pollutants, etc. [17,18] or in the presence of some metal tags such as gold, silver nanoparticles, etc. [23,50,51].

In recent years, electrochemical genosensors developed on the principle of nanotechnology have become one of the most exciting forefront fields in analytical chemistry due to the recent advances in

TABLE 19.1

An overview on the development of genosensor technology for electrochemical sensing of nucleic acids by using various transducers: carbon paste electrode (CPE), pencil graphite electrode (PGE), screen-printed carbon and gold electrode (SCPE and AuSCPE), hanging mercury drop electrode (HMDE) and plane pyrolytic graphite electrode (PyGE) graphite epoxy composite electrode (GECE), carbon fiber microelectrode (CFME) or metal electrodes such as gold electrode (AuE), platinum electrode (PtE), etc.

Immobilization method	Technique	Response	Electrode	References
Adsorption				
Electrochemical adsorption of DNA by applying a potential at +0.5 V or at open circuit for a changing period of time	DPV, SWV, PSA	The oxidation signal of guanine/adenine	CPE, GCE, SCPE, PGE	[16,27–32]
Electrochemical adsorption of DNA by applying a potential at −0.1 V for 2 min	PSA, DPV	The oxidation signal of guanine	CPE	[33,34]
Passive adsorption of DNA	DPV, SWV, PSA, CV	The oxidation signal of guanine/adenine or the oxidation/reduction signal of intercalators	CPE, GCE, GECE, PyGE, SCPE, PGE, AuE	[16,35–39]
		The reduction signal of guanine/cytosine and adenine	HMDE	[40,41]

TABLE 19.1 (*continued*)

Immobilization method	Technique	Response	Electrode	References
Physical adsorption of DNA by using conductive polymers, e.g. polypyrrole	CV	The oxidation/reduction signal of intercalators	GCE	[42]
Crosslinking or entrapment				
Entrapment in dextran matrix, polyacrylamide or DNA crosslinked by using some agents such as glutaraldehyde				[43]
Complexation				
Streptavidin–biotin or avidin–biotin complexation	CV, SWV, DPV	The oxidation signal of guanine or the oxidation/reduction signal of intercalators	CPE, GCE, GECE, SCPE, PGE, CFME, AuE	[30,44,45]
Covalent attachment				
By using EDC/NHS	DPV, CV	The oxidation/reduction signal of intercalators	CPE, GCE	[46,47]
By using AET/EDC or MCH	CV		AuE	[7,18,19,48]
By using MSi/CDS	CV		PtE	[49]

Electrochemical measurements have been developed by using different electrochemical techniques (differential pulse voltammetry (DPV), cyclic voltametry (CV), potentiometric stripping analysis (PSA), square wave voltammetry (SWV), adsorptive stripping transfer voltammetry (ASTV), etc.). The abbreviations given in covalent attachment of DNA onto different transducers are water soluble carbodimide 1-(3-dimethyaminopropyl)-3-ethyl-carbodimide (EDC), *N*-hydroxysuccimide (NHS), mercaptohexanol (MCH), aminoethanethiol (AET), mercaptosilane (MSi), and *N*-cyclohexyl-*N′*-[2-(N-methylmorpholino)-ethyl]carbodimide-4-tolune sulfonate (CDS).

nanomaterials such as magnetic particles/nanoparticles [23,50–55], nanotubes [56–59], and nanowires [60–62].

The use of magnetic particles/nanoparticles can bring novel capabilities to bioaffinity assays and sensors. A biomagnetic assay of DNA sequences related to *BRCA1* breast cancer gene was reported by Wang *et al.* [52]. This study was performed using biotinylated inosine-substituted probes and streptavidin-coated magnetic beads in connection with potentiometric stripping analysis (PSA) measurements at a renewable pencil graphite electrode (PGE). Another study based on enzyme-linked immunoassay coupled with magnetic particles for the detection of the DNA hybridization by using linear square voltammetry (LSV) technique and pyrolytic graphite electrode (PyGE) was reported by Palecek *et al.* [53]. Recently, there have been studies reported by Erdem *et al.* [24,54] that represent the electrochemical detection of DNA hybridization related to specific sequences using different transducers. A label-free genomagnetic assay for the electrochemical detection of *Salmonella* spp. sequence was presented by using graphite–epoxy composite electrode (GECE) and magneto-GEC electrodes as electrochemical transducers [54]. In another study performed by Erdem *et al.* [24], hepatitis B virus (HBV) detection was performed successfully in PCR amplicons by using genomagnetic assay in combination with disposable sensor, PGE.

Some magnetic assays connected by using metal nanoparticles were reported in the literature [55]. Two different particle-based assays for monitoring DNA hybridization based on PSA detection of an iron tracer were reported [55]. The probes labelled with gold-coated iron core–shell nanoparticles were used and thus, the captured iron-containing particles were dissolved following hybridization step and the released iron was quantified by cathodic-stripping voltammetry by using hanging mercury drop electrode (HMDE), in the presence of the 1-nitroso-2-naphthol ligand and a bromate catalyst.

The modification of electrodes with carbon nanotubes has recently attracted considerable attention in the field of electrochemical genosensing technology. In one of the earlier studies, an enhanced guanine oxidation signal at multiwalled carbon nanotubes (MWCNTs)-modified glassy carbon electrode (GCE) was shown in 100 fmol detection limit [56]. Additionally, the use of carbon nanotubes was reported for enzyme amplification of electrochemical DNA sensing and also the transduction events by Wang's group [57]. The fabrication of carbon nanotube paste electrode was represented recently, and it was used for adsorption and electrochemical oxidation of nucleic acids by Pedano and Rivas [58].

A nanoelectrode array based on vertically aligned MWCNTs, with controlled density, embedded in an SiO_2 matrix was reported by Li's group to be useful for detecting DNA hybridization [59]. Oligonucleotide probes were selectively functionalized to the open ends of the MWCNTs and thus, DNA targets could be detected by combining the nanoelectrode array with ruthenium bipyridine-mediated guanine oxidation.

The specific properties of novel nanomaterials as nanowires offer an excellent prospect for biological recognition surfaces in order to develop a more selective and sensitive biosensor technology [60–62].

Li *et al.* [61] reported a novel method using a sequence-specific label-free DNA sensors based on silicon nanowires (Si-NWs) by measuring the change in the conductance. Kelley's group [62] developed a gold nanowire array (Au-NW) in 15–20 nm diameter range and this array was used for electrochemical DNA detection with the help of the electrocatalytic reporter systems, $Ru(NH_3)_6^{3+}$ and $Fe(CN)_6^{3-}$.

19.3 CONCLUSION

Material science has a key importance because it can offer the possibilities of how to apply the new materials prepared from micro to nanoscales into optical, electrical, magnetic, chemical, and biological applications. In aspect of chip technology, nanomaterial based electrochemical genosensors have been recently developed using different transducers in response to clinical demands of giving promising results for detection of interaction between drugs or small molecules, pollutants, etc. and DNA, and also, of the detection of specific DNA sequences related to inherited and genetic diseases, such as hepatitis B, factor-V leiden mutation, breast cancer *BRCA1* gene, etc. It is hoped that continued development through combined efforts in micro/nanoelectronics, surface chemistry, molecular biology, and analytical chemistry will lead to the establishment of genosensor technology based on different transducers and nanomaterials. Immediate applications will include directly quantifying DNA samples for use in sequencing or polymerase chain reactions (PCR), or pharmaceutical testing and quality control. Eventually, they could be applied for producing credit card-sized sensor arrays [63] for clinical applications such as detection of pathogenic bacteria, tumors and genetic disease, drug–protein interactions, or for forensics.

Procedure 27 (see in CD accompanying this book), related to this chapter, gives some details on the use of disposable graphite sensors for DNA detection.

ACKNOWLEDGMENTS

A.E. acknowledges the Turkish Academy of Sciences in the framework of the Young Scientist Award Program (KAE/TUBA-GEBIP/2001-2-8) and Ege University, Faculty of Pharmacy, project coordination (Project No. 04/ECZ/003) for their financial support.

REFERENCES

1 E. Palecek, *Nature*, 188 (1960) 656–657.
2 J. Wang, *Chem. Eur. J.*, 6 (1999) 1681–1685.
3 J. Wang, *Nucleic Acids Res.*, 28 (2000) 3011–3016.
4 E. Palecek and M. Fojta, *Anal. Chem.*, 73 (2000) 75A–83A.
5 M.I. Pividori, A. Merkoci and S. Alegret, *Biosens. Bioelectron.*, 15 (2000) 291–303.
6 A. Erdem and M. Ozsoz, *Electroanalysis*, 14 (2002) 965–974.
7 M. Yang, M.E. McGovern and M. Thompson, *Anal. Chim. Acta*, 346 (1997) 259–275.
8 W.G. Kuhr, *Nat. Biotechnol.*, 18 (2000) 1042–1043.
9 J. Wang, *Electroanalysis*, 17 (2005) 7–14.
10 J. Wang, *Analyst*, 130 (2005) 421–426.
11 S. Zhang, G. Wright and Y. Yang, *Biosens. Bioelectron.*, 15 (2000) 273–282.
12 J. Wang, *Anal. Chim. Acta*, 469 (2002) 63–71.
13 A. Erdem, M. Pividori, M. Del Valle and S. Alegret, *J. Electroanal. Chem.*, 567 (2004) 29–37.
14 K. Besteman, J.O. Lee, F.G. Wiertz, H.A. Heering and C. Dekker, *Nano Lett.*, 3 (2003) 727–730.
15 L.A. Terry, S.F. White and L.J. Tigwell, *J. Agric. Food Chem.*, 53 (2005) 1309–1316.
16 H. Karadeniz, B. Gulmez, F. Sahinci, A. Erdem, G.I. Kaya, N. Unver, B. Kivcak and M. Ozsoz, *J. Pharm. Biomed. Anal.*, 35 (2003) 298–302.
17 A. Erdem, B. Kosmider, R. Osiecka, E. Zyner, J. Ochocki and M. Ozsoz, *J. Pharm. Biomed. Anal.*, 38(4) (2005) 645–652.
18 S.O. Kelley, E.M. Boon, J.K. Barton, N.M. Jackson and M.G. Hill, *Nucleic Acids Res.*, 27 (1999) 4830–4837.
19 E.M. Boon, D.M. Ceres, T.G. Drummond, M.G. Hill and J.K. Barton, *Nat. Biotechnol.*, 18 (2000) 1096–1100.
20 J. Wang, A.N. Kawde, A. Erdem and M. Salazar, *Analyst*, 126(11) (2001) 2020–2024.
21 J. Wang, D. Xu, A. Erdem, R. Polsky and M. Salazar, *Talanta*, 56 (2002) 931–938.
22 F. Jelen, A. Erdem and E. Palecek, *Bioelectrochemistry*, 55 (2002) 165–167.

23 M. Ozsoz, A. Erdem, K. Kerman, D. Ozkan, B. Tugrul, N. Topcuoglu, H. Erken and M. Taylan, *Anal. Chem.*, 75(9) (2003) 2181–2187.

24 A. Erdem, D. Ozkan Ariksoysal, H. Karadeniz, P. Kara, A. Sengonul, A.A. Sayiner and M. Ozsoz, *Electrochem. Commun.*, 7 (2005) 815–820.

25 J.E. Koehne, H. Chen, A.M. Cassell, Q. Ye, J. Han, M. Meyyapan and J. Li, *Clin. Chem.*, 50 (2004) 1886–1893.

26 S.W. Metzger, M. Natesan, C. Yanavich, J. Schneider and G.U. Lee, *J. Vac. Sci. Technol. A*, 17(5) (1999) 2623–2628.

27 J. Wang, X. Cai, C. Johnsoon and M. Balakrishnan, *Electroanalysis*, 8 (1996) 20–24.

28 J. Wang, X. Cai, G. Rivas and H. Shiraishi, *Anal. Chim. Acta*, 326 (1996) 141–147.

29 J. Wang, X. Cai, G. Rivas, H. Shiraishi and N. Dontha, *Biosens. Bioelectron.*, 12 (1997) 587–599.

30 G. Marazza, L. Chinella and M. Mascini, *Biosens. Bioelectron.*, 14 (1999) 43–51.

31 A. Erdem, K. Kerman, B. Meric, U.S. Akarca and M. Ozsoz, *Anal. Chim. Acta*, 422(2) (2000) 139–149.

32 A. Erdem and M. Ozsoz, *Anal. Chim. Acta*, 437 (2001) 107–114.

33 J. Wang, E. Palecek, P.E. Nielsen, G. Rivas, X. Cai, H. Shiraishi, N. Dontha, D. Luo and P.A.M. Farias, *J. Am. Chem. Soc.*, 118 (1996) 7667–7670.

34 K. Kerman, D. Ozkan, P. Kara, A. Erdem, B. Meric, P.E. Nielsen and M. Ozsoz, *Electroanalysis*, 15 (2003) 667–670.

35 A.M. Oliveira Brett, S.H.P. Serrano, I. Gutz, M.A. La-Scalea and M.L. Cruz, *Electroanalysis*, 9 (1997) 1132–1137.

36 L. Labuda, M. Buckova, M. Vanickova, L. Mattusch and R. Wennrich, *Electroanalysis*, 11 (1999) 101–107.

37 M.I. Pividori, A. Merkoçi, J. Barbé and S. Alegret, *Electroanalysis*, 15 (2003) 1815–1823.

38 M.I. Pividori and S. Alegret, *Anal. Lett.*, 36 (2003) 1669–1695.

39 D.-W. Pang and H.D. Abruna, *Anal. Chem.*, 70 (1998) 3162–3169.

40 E. Palecek, P. Boublikova and F. Jelen, *Anal. Chim. Acta*, 187 (1986) 99–107.

41 E. Palecek, *Electroanalysis*, 8 (1996) 7–14.

42 L. Li, M. Wang, S. Dong and E. Wang, *Anal. Sci.*, 13 (1997) 305–310.

43 B. Johnsson, S. Lofas and G. Lindquist, *Anal. Biochem.*, 198 (1991) 268–273.

44 P. Pantano, T.H. Morton and W.G. Kuhr, *J. Am. Chem. Soc.*, 113 (1991) 1832–1833.

45 S. Cosnier, B. Galland, C. Gondran and A. Le Pellec, *Electroanalysis*, 10 (1998) 808–813.

46 K.M. Millan, A.L. Spurmanis and S.K. Mikkelsen, *Electroanalysis*, 4 (1992) 929–932.

47 K.M. Millan and S.K. Mikkelsen, *Anal. Chem.*, 65 (1993) 2317–2323.
48 H. Karadeniz, B. Gulmez, A. Erdem, F. Jelen, M. Ozsoz and E. Palecek, *Front. Biosci.*, 11 (2005) 1870–1877.
49 I. Moser, T. Schalkhammer, R. Pittner and G. Urban, *Biosens. Bioelectron.*, 12 (1997) 729–737.
50 L. Authier, C. Grossiord, P. Brossier and B. Limoges, *Anal. Chem.*, 73 (2001) 4450–4456.
51 J. Wang, D. Xu, A.-N. Kawde and R. Polsky, *Anal. Chem.*, 73 (2001) 5576–5581.
52 J. Wang, G.U. Flechsig, A. Erdem, O. Korbut and P. Grundler, *Electroanalysis*, 16(11) (2004) 928–931.
53 E. Palecek, R. Kizek, L. Havran, S. Billova and M. Fojta, *Anal. Chim. Acta*, 469 (2002) 73–83.
54 A. Erdem, M.I. Pividori, A. Lermo, A. Bonanni, M. del Valle and S. Alegret, *Sens. Actuators B Chem.*, 114 (2006) 591–598.
55 J. Wang, L. Guodong and A. Merkoci, *Anal. Chim. Acta*, 482 (2003) 149–155.
56 J. Wang, A.N. Kawde and M. Musameh, *Analyst*, 128 (2003) 912–916.
57 J. Wang, G.D. Liu and M.R. Jan, *J. Am. Chem. Soc.*, 126 (2004) 3010–3011.
58 M.L. Pedano and G.A. Rivas, *Electrochem. Commun.*, 6 (2004) 10–16.
59 J. Li, H.T. Ng, A. Cassell, W. Fan, H. Chen, Q. Ye, J. Koehne, J. Han and M. Meyyappan, *Nano Lett.*, 3 (2003) 597–602.
60 F. Patolsky and C.M. Lieber, *Mater. Today*, April (2005) 20–28.
61 Z. Li, Y. Chen, X. Li, T.I. Kamins, K. Nauka and R.S. Williams, *Nano Lett.*, 4(2) (2004) 245–247.
62 M.A. Lapierre-Devlin, C.L. Asher, B.J. Taft, R. Gasparac, M.A. Roberts and S.O. Kelley, *Nano Lett.*, 5(6) (2005) 1051–1055.
63 W.G. Kuhr, *Nat. Biotechnol.*, 18 (2000) 1042–1043.

Chapter 20

DNA-electrochemical biosensors for investigating DNA damage

Ana M. Oliveira Brett, Victor C. Diculescu, Ana M. Chiorcea-Paquim and Sílvia H.P. Serrano

20.1 INTRODUCTION

Electrochemical research on DNA is of great relevance to explain many biological mechanisms. The interaction of many chemical compounds, including water, some metal ions and their complexes, small organic molecules and proteins, with DNA is reversible and stabilizes DNA conformations. However, hazard compounds such as drugs and carcinogens interact with DNA causing irreversible damage and these interactions have to be carefully studied. DNA-electrochemical biosensors are a very good model for simulating nucleic acid interactions with cell membranes, potential environmental carcinogenic compounds and clarifying the mechanisms of interaction with drugs and chemotherapeutic agents.

The electrochemical behaviour and the adsorption of nucleic acid molecules and DNA constituents have been extensively studied over recent decades [1–6]. Electrochemical studies demonstrated that all DNA bases can be electrochemically oxidized on carbon electrodes [7–13], following a pH-dependent mechanism. The purines, guanine (G) and adenine (A), are oxidized at much lower positive potentials than the pyrimidines, cytosine (C) and thymine (T), the oxidation of which occurs only at very high positive potentials near the potential corresponding to oxygen evolution, and consequently are more difficult to detect. Also, for the same concentrations, the oxidation currents observed for pyrimidine bases are much smaller than those observed for the purine bases. Consequently, the electrochemical detection of oxidative changes occurring in DNA has been based on the detection of purine base oxidation peaks or of the major

Comprehensive Analytical Chemistry 49
Alegret and Merkoçi (Eds)
Volume 49 ISSN: 0166-526X DOI: 10.1016/S0166-526X(06)49020-6

oxidation product, 8-oxoguanine (8-oxoGua) [14,15], which is a biomarker of oxidative stress.

Electrochemical oxidation of natural and synthetic DNA performed at pyrolytic graphite [16] and glassy carbon [3–6,17,18] electrodes showed that at pH 4.5 only the oxidation of the purine residues in polynucleotide chains is observed. Using differential pulse voltammetry, the less positive peak corresponds to the oxidation of guanine residues and the peak at more positive potentials is due to the oxidation of adenine residues.

Large differences in the currents obtained at carbon electrodes for dsDNA and ssDNA oxidation were observed. The greater difficulty for the transition of electrons from the inside of the rigid helix form of dsDNA to the electrode surface than from the flexible ssDNA, where guanine and adenine residues can reach the surface, leads to much higher peak currents for ssDNA. Thus, the oxidation currents of guanine and adenine residues in DNA can be used to probe individual adenine–thymine (A–T) and guanine–cytosine (G–C) pairs in dsDNA. In this way, the irreversible DNA damages caused by health hazardous compounds and oxidizing substances can be monitored electrochemically either by using the changes in the oxidation peaks of the purinic DNA bases, guanine and adenine or by the appearance of 8-oxoGua characteristic peaks due to DNA oxidative stress.

When natural or synthetic DNA molecules interact with electrode surfaces adsorption occurs. The knowledge about the adsorption of nucleic acids onto the electrode surface leads to the development of DNA-modified electrodes, also called electrochemical DNA biosensors [3–6,19–24]. An electrochemical DNA biosensor is an integrated receptor–transducer device that uses DNA as the biomolecular recognition element to measure specific binding processes with DNA, using electrochemical transduction.

This chapter describes different methodologies used in the design of DNA-electrochemical biosensors, their surface morphological characterization as well as their application to DNA–drug interaction studies.

20.2 AFM IMAGES OF DNA-ELECTROCHEMICAL BIOSENSORS

A DNA-electrochemical biosensor is formed by a DNA film, which constitutes the molecular recognition element (the probe), directly immobilized on the electrochemical transducer. The performance, sensitivity and reliability of the DNA biosensor and the electrochemical response are dictated by the DNA immobilization procedure. The DNA biophysical properties, such as flexibility, and DNA–drug interactions, are influenced by the adsorbed DNA structure (ssDNA or dsDNA), concentration, pH and supporting electrolyte [25–28]. Therefore, a full understanding of the surface morphology of the DNA-electrochemical biosensor is necessary to guarantee the correct interpretation of the experimental results.

Magnetic AC atomic force microscopy (MAC Mode AFM) has proved to be a powerful surface analysis technique to investigate the interfacial and conformational properties of biological samples softly bound to the electrode surface and can be used as an important tool to characterize DNA-electrochemical biosensor surfaces [25,27].

Carbon electrodes such as glassy carbon, carbon fibres, graphite or carbon black have wide electrochemical applications but they do not represent a good substrate to obtain AFM images. Highly oriented pyrolytic graphite (HOPG), which is easy to clean, inert in air and has extremely smooth terraces on its basal plane (Fig. 20.1A) can be used with success for imaging biological molecules. Thus the MAC Mode AFM surface characterization of the nanoscale DNA adsorbed films was performed using HOPG electrodes.

Depending on the required application, three different methods of preparation of DNA films were developed. The optimized experimental conditions used to obtain AFM images are described in Procedure 28 (see in CD accompanying this book).

The AFM image of the HOPG substrate modified by a thin layer of dsDNA, when small DNA concentrations are used, is shown in Fig. 20.1B. The dsDNA molecules adsorbed on the HOPG surface formed two-dimensional lattices with uniform coverage of the electrode.

DNA films grown in acid buffer solutions presented greater DNA surface coverage, due to overlaying and superposition of DNA molecules, in relation to films formed in neutral buffer solutions [26]. The DNA network patterns define nanoelectrode systems with different active surface areas on the graphite substrate and form a biomaterial matrix to attach and study interactions with molecules such as drugs [27].

The thin dsDNA films do not completely cover the HOPG electrode surface and the network structure has holes exposing the electrode underneath. The drug molecules from the bulk solution will also diffuse and adsorb non-specifically on the electrode's uncovered regions. If the drugs are electroactive, this leads to two different contributions to the electrochemical signal, one from the simple adsorbed drug and the other due to the damage caused to immobilized DNA, being difficult to distinguish between them.

A complete coverage of the electrode surface is obtained using the multi-layer and thick dsDNA films described in Procedure 28 (see in CD

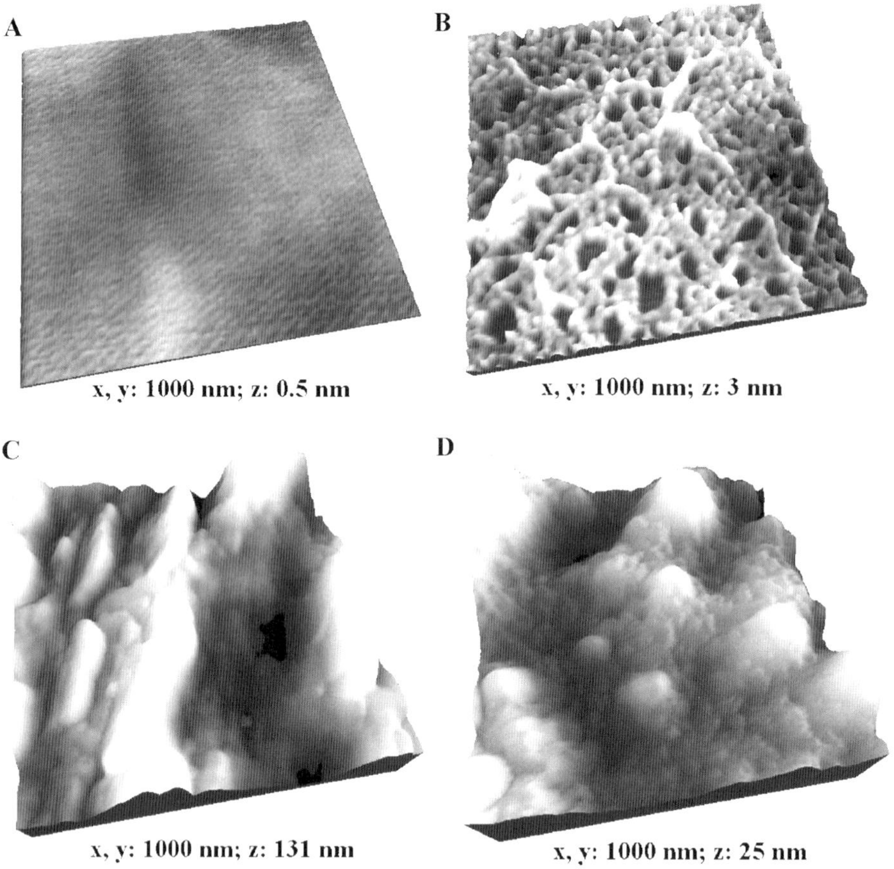

Fig. 20.1. MAC Mode AFM three-dimensional images in air of (A) clean HOPG electrode; (B) thin-film dsDNA-biosensor surface, prepared onto HOPG by 3 min free adsorption from 60 μg/mL dsDNA in pH 4.5 0.1 M acetate buffer; (C) multi-layer film dsDNA biosensor, prepared onto HOPG by evaporation of three consecutive drops each containing 5 μL of 50 μg/mL dsDNA in pH 4.5 0.1 M acetate buffer; (D) thick-film dsDNA biosensor, prepared onto HOPG by evaporation from 37.5 mg/mL dsDNA in pH 4.5 0.1 M acetate buffer. With permission from Refs. [28,29].

accompanying this book). The AFM images of both multi-layer and thick-layer films show uniform and complete coverage of the electrode, with regularly dispersed peaks and valleys (Figs. 20.1(C) and (D)). The DNA–electrode surface interactions are stronger and these DNA films are more stable on the HOPG surface. The big advantage of the dsDNA thick film is that the HOPG surface is completely covered by dsDNA and consequently the undesired binding of molecules to the electrode surface is not possible. As a result, the multi-layer and thick-film DNA biosensors [27,29] are useful tools to study, *in situ*, the interactions between DNA and health hazardous compounds and drugs with therapeutic activity.

20.3 DNA-ELECTROCHEMICAL BIOSENSORS FOR DETECTION OF DNA DAMAGE

The aim of developing DNA-modified electrodes is to study the interaction of DNA immobilized on the electrode surface with analytes in solution and to use the DNA biosensor to evaluate and to predict DNA interactions and damage by health hazardous compounds based on their ability to bind to nucleic acids. In this way, DNA acts as a promoter between the electrode and the biological molecule under study.

The DNA-biosensing design usually employs electrochemical, optical and mechanical transduction techniques [5]. Electrochemical methods have the advantage of being rapid, sensitive and cost effective. Nevertheless, the most important advantage in using electrochemical DNA biosensors is the possibility of *in situ* generation of reactive intermediates and the detection of their interaction with DNA. Comprehensive descriptions of research on DNA and DNA sensing [1–6,19–22,30–34] show the great possibilities of using electrochemical transduction in DNA diagnostics.

The electrochemical sensor for DNA damage consists of an electrode with DNA immobilized on the surface. The interactions of the surface-immobilized DNA (either by electrostatic adsorption or by evaporation) with the damaging agent are converted, via changes in electrochemical properties of the DNA recognition layer, into measurable electrical signals corresponding to the oxidation of DNA purine bases [35]. The double-stranded DNA structure makes access of the bases to the electrode surface difficult, hindering their oxidation. The occurrence of DNA damage causes the unwinding of the double helix. As the double helix unwinds, closer access of the bases to the surface is possible, leading to the possibility of voltammetric detection of DNA damage. This biosensor has been successfully applied to study the interaction of several substances with dsDNA and the interpretation of the results has contributed to the elucidation of the mechanisms by which DNA is damaged by health hazardous compounds [5,6,35–37].

Although several different DNA adsorption methods have been used on different types of electrodes [19,25], the immobilization of dsDNA to electrode surfaces can be attained very easily by evaporation or adsorption and no reagents or DNA modifications will occur since the immobilization does not involve formation of covalent bonds with the surface. Therefore, DNA-electrochemical biosensors have been prepared using a glassy carbon electrode (GCE) instead of HOPG as the conducting transducer substrate for DNA adsorption (see Procedure 29 in CD accompanying this book).

Differential pulse (DP) voltammetry, a voltammetric technique with high sensitivity, is normally performed and the equipment as well as the electrochemical procedures used for the voltammetric studies of DNA–drug interaction are described (see Procedure 29 in CD accompanying this book).

20.4 DNA DAMAGE PRODUCED BY REACTIVE OXYGEN SPECIES (ROS)

Free radicals are produced in living cells by normal metabolism and by exogenic sources such as carcinogenic compounds and ionizing radiation. Several drugs, such as nitroimidazoles, show biological activity after *in vivo* reduction to produce free radicals, hydroxylamine or nitroso derivatives [38], which react with biological molecules in the cell, including DNA. The result of these interactions usually involves DNA damage giving rise to the so-called oxidative stress, which is the main cause of mutagenesis, carcinogenesis and ageing [39].

By analogy, the DNA immobilization on a conducting solid support provides an interface that models the processes occurring in the living cell where DNA interacts with charged surfaces. Therefore, by controlling (conditioning) the electrode/biosensor potential, oxidation or reduction of different compounds previously linked to DNA can occur. These redox reactions cause the *in situ* formation of reactive intermediates, such as free radicals, and their action on DNA is detected by electrical measurable signals in a voltammogram using the DNA-electrochemical biosensor.

Next, some typical examples will be presented of how a DNA-electrochemical biosensor is appropriate to investigate the DNA damage caused by different types of substances, such as the antioxidant agent quercetin (Scheme 20.1), an anticancer drug adriamycin (Scheme 20.2) and nitric oxide. In all cases, the dsDNA damage is detected by changes in the electrochemical behaviour of the immobilized dsDNA, specifically through modifications of the purinic base oxidation peak current [3,5,40].

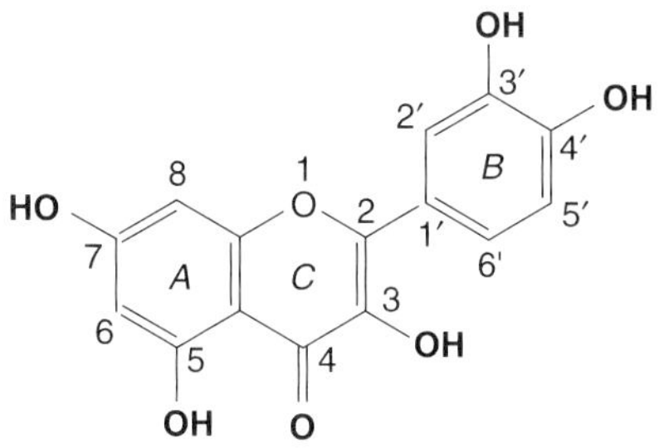

Scheme 20.1. Chemical structure of quercetin.

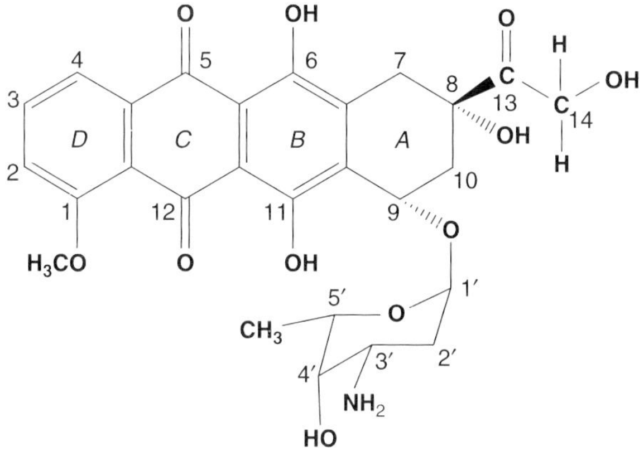

Scheme 20.2. Chemical structure of adriamycin.

20.4.1 Quercetin

Flavonoids, compounds found in rich abundance in all land plants, owing to their polyphenolic nature often exhibit strong antioxidant properties [41]. They are considered as potential chemopreventive agents against certain carcinogens since it was demonstrated that the intake of a large quantity of flavonoid inhibited the incidence of ROS producing damage to DNA. However, in contrast with this commonly accepted role, there is also evidence that flavonoids themselves are mutagenic and have DNA damaging ability [42,43].

Quercetin (Scheme 20.1) a major flavonoid in human diet, acts as a pro-oxidant [42–44] under certain circumstances. Although the mechanism of interaction of quercetin with DNA is not yet fully understood, there is considerable evidence that quercetin-induced DNA damage occurs via reaction with Cu(II). Quercetin can directly reduce transition metals, thus providing all the elements necessary to generate the highly oxidizing hydroxyl radical (•OH). On the other hand, there is experimental support that the formation of quercetin radicals via auto-oxidation of the catechol ring leads to the generation of superoxide radicals. Therefore, it was proposed that quercetin can promote oxidative damage to DNA through the generation of ROS.

DNA–Cu(II)–quercetin interactions can be followed electrochemically using a DNA-electrochemical biosensor [29,35]. This knowledge about the electrochemical behaviour of the dsDNA incubated with quercetin–Cu(II) complexes at GC electrode [45] is an important feature to understand quercetin–DNA interactions at a DNA-electrochemical biosensor. The preparation of the solutions and the quercetin–Cu(II) complex used during the characterization of *in situ* electrochemical DNA damage promoted by the quercetin–Cu(II) complex using a DNA biosensor is described (see Procedure 29 in CD accompanying this book).

20.4.1.1 Quercetin–dsDNA interaction at thick DNA biosensors

In Fig. 20.2 are shown DP voltammograms obtained with a thick-layer DNA biosensor previously immersed into a quercetin–$CuSO_4$ solution for 30 min and for 6 h. During incubation of the DNA biosensor, the solution containing quercetin and Cu(II) was continuously stirred. For comparison, a DP voltammogram obtained with the dsDNA-modified GCE in acetate buffer is also presented in Fig. 20.2.

After 30 min of incubation of the thick-layer dsDNA biosensor in a solution of quercetin–Cu(II), a typical quercetin oxidation peak 1 is observed followed by a small peak at about +0.45 V. Increasing the incubation time to 6 h led to the total disappearance of quercetin peak 1 and the appearance of a larger peak at +0.45 V, and big changes occurred in the dsDNA layer. Two new anodic peaks that could be identified with oxidation of the deoxyguanosine (dGuo) and deoxyadenosine (dAdo) residues appeared and their currents increased with incubation time. This can be explained, since quercetin interacts with DNA especially at pyrimidinic residues oxidizing them. The thymine and cytosine oxidation products will not be able to form hydrogen bonds with adenine and guanine residues, which now become more accessible to the electrode surface leading to an increase in their oxidation peak currents.

Using the thick-layer dsDNA-modified GCE, long periods of incubation with quercetin–Cu(II) complex are necessary for the detection of dsDNA interaction. The necessity of a faster response led to the construction of different types of biosensors such as the thin-layer dsDNA biosensors obtained by electrostatic adsorption of dsDNA strands at the

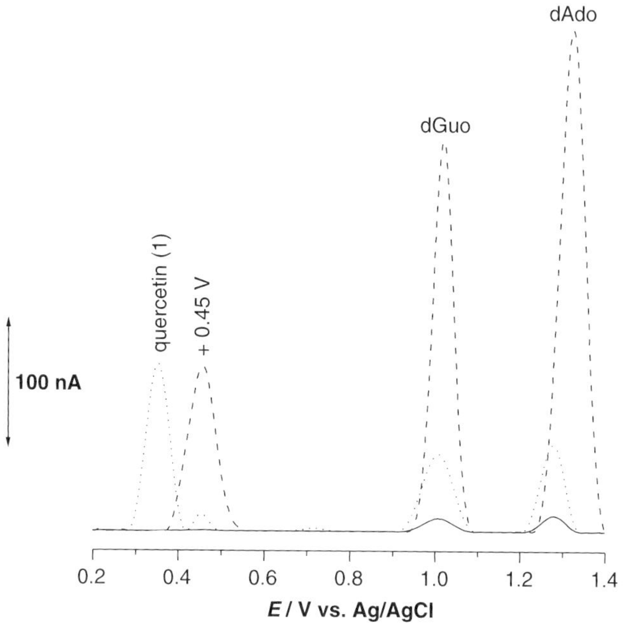

Fig. 20.2. DP voltammograms obtained in pH 4.3 0.1 M acetate buffer with a thick-layer dsDNA-modified GCE after (—) 0 min, (····) 30 min and (---) 6 h incubation in a mixture of 100 μM quercetin with 50 μM $CuSO_4$. With permission from Ref. [35].

GCE surface. Such kinds of devices have been shown to be inappropriate since they do not ensure a complete coverage of the GCE surface allowing the non-specific adsorption of the compound. However, a new type of biosensor-multi-layer dsDNA-electrochemical biosensor obtained by successive additions of small quantities of dsDNA on the GCE surface has been developed (see Procedure 29 in CD accompanying this book) and further used to study the interaction between dsDNA and the quercetin–Cu(II) complex.

20.4.1.2 Quercetin–dsDNA interaction at a multi-layer DNA biosensor

To study the interaction between dsDNA and the quercetin–Cu(II) ion complex, a newly prepared multi-layer dsDNA biosensor was kept for 10 min in a mixture of quercetin and Cu(II) ions. The electrode was transferred to supporting electrolyte and a DP voltammogram recorded. The quercetin oxidation peak 1 occurs followed by another peak at +0.45 V (Fig. 20.3). Comparing with the voltammogram obtained before incubation, the peaks due to dGuo and dAdo base oxidation are several

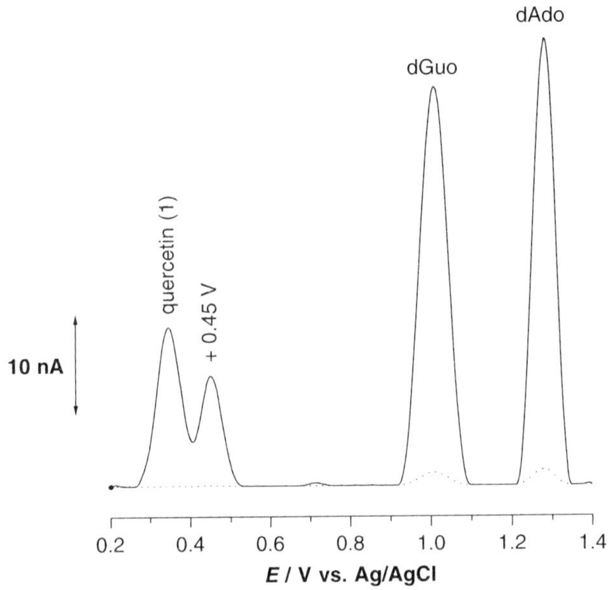

Fig. 20.3. DP voltammograms obtained in pH 4.3 0.1 M acetate buffer with a multi-layer dsDNA-modified GCE (••••) before and (▬) after 10 min of incubation in a mixture of 100 μM quercetin with 50 μM $CuSO_4$. With permission from Ref. [35].

times higher. This means that large modifications in the dsDNA have occurred after interaction with the quercetin–Cu(II) complex, and the presence of the peak at a potential of +0.45 V indicates the formation of a new product after DNA interaction with quercetin–Cu(II).

The same quercetin–DNA interaction occurs either at a thick or multi-layer dsDNA-electrochemical biosensor, but at the multi-layer DNA sensor it occurs more rapidly. Besides, the construction of a multi-layer dsDNA biosensor is faster than that of the thick-layer biosensor, it consumes less reagents, and so was further used to study the interaction between dsDNA and ROS produced by redox activation of quercetin.

20.4.1.3 The role of ROS in the dsDNA damage promoted by quercetin
Quercetin–Cu(II) complexes bind to dsDNA causing DNA oxidative damage. Several studies have shown that the hydroxyl groups of the catechol ring of quercetin is important for the Cu(II) ions chelation [45–47] and this reaction produces ROS, which can attack the dsDNA, disrupting the helix and leading to the formation of 8-oxodeoxyguanosine (8-oxodGuo). It is known that the electrochemical oxidation of quercetin occurs first at the hydroxyl groups of the catechol ring and

the quercetin radicals formed react with molecular oxygen thus producing ROS. Thus, it can be proposed that the oxidation of quercetin bonded to DNA could cause oxidative damage, and an electrochemical multi-layer biosensor was used to confirm the dsDNA damage produced by oxidation of quercetin.

A newly prepared biosensor was incubated for 10 min in a solution of quercetin; the electrode was washed with deionized water in order to remove the unbound quercetin molecules and transferred to electrolyte solution. The DP voltammogram recorded (Fig. 20.4) shows the quercetin peak 1 due to the oxidation of the hydroxyl groups of the catechol ring followed by small peaks of dGuo and dAdo residues confirming that quercetin interacts with dsDNA and even after interaction quercetin can still undergo oxidation.

The incubation procedure was repeated using a new biosensor and the electrode was transferred to pure buffer solution where a potential of +0.40 V was applied for 5 min. During this conditioning period, the quercetin molecules bound to DNA are oxidized leading to formation of ROS. The radical damages dsDNA, detected by the occurrence of high oxidation peaks of dGuo and dAdo residues (Fig. 20.4). Moreover, the DP voltammogram obtained in these conditions shows a peak at +0.45 V, confirming the formation of a new product.

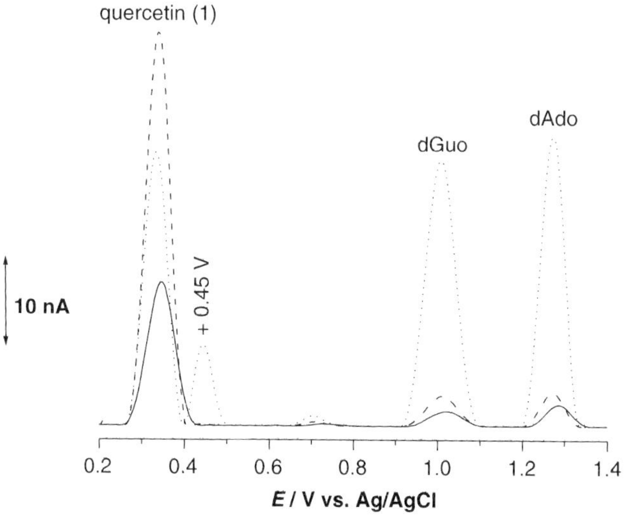

Fig. 20.4. DP voltammograms in pH 4.3 0.1 M acetate buffer obtained with a multi-layer dsDNA biosensor incubated for 10 min in 100 μM quercetin (- - -) before and after applying +0.40 V for 300 s (—) with and (····) without bubbling N_2 in the solution. With permission from Ref. [35].

In order to obtain information about the origin of the peak at +0.45 V, the GCE surface was modified with DNA-like sequences (polyguanylic and polyadenylic acids) that contain or not guanine residues [35]. These new types of biosensors were incubated in a quercetin solution and then conditioned (see Procedure 29 in CD accompanying this book). In this way, it was shown that the peak at +0.45 V is directly related with the presence of guanine residues in the polynucleotidic chain and that it is due to the formation of 8-oxodGuo.

To prove the involvement of ROS in the process of DNA damage during quercetin oxidation, another experiment was carried out after bubbling nitrogen into the buffer electrolyte solution. After O_2 removal from the solution, the quercetin radicals formed during the oxidation of quercetin could not react with oxygen and no ROS were formed to damage the DNA film. The DP voltammogram obtained (Fig. 20.4) showed only a small oxidation peak of dGuo and dAdo proving that no DNA damage had occurred and no additional peak, at +0.45 V due to 8-oxodGuo, was observed although quercetin peak 1 occurred with a smaller current due to oxidation of quercetin during the conditioning procedure. It was demonstrated that the presence of oxygen is fundamental for extensive DNA damage, promoted by the highly reactive oxygen radicals. The interpretation of the results obtained enabled the understanding of the quercetin–dsDNA interaction mechanism [35].

20.4.2 Adriamycin

Adriamycin is an antibiotic of the family of anthracyclines with a wide spectrum of chemotherapeutic applications and anti-neoplasic action but it causes cardiotoxicity ranging from delayed and insidious cardiomyopathy to irreversible heart failure [48–52].

There was experimental evidence that adriamycin can promote oxidative damage to DNA in cancerous cells through the generation of ROS [50–52] and high levels of 8-oxoguanine (8-oxoGua), a known biomarker of oxidative stress, were detected in *in vitro* studies [53]. The generation of this main product of guanine oxidation within DNA is strongly mutagenic and can contribute to cell dysfunction [54]. There was ample evidence that adriamycin interacts with DNA through intercalation [55] but less was known as to whether it could directly oxidize DNA after intercalation has occurred.

Electrochemical-dsDNA biosensors were used to detect *in situ* adriamycin DNA oxidative damage [37]. The experimental conditions used during the investigation of *in situ* electrochemical DNA damage caused by adriamycin are described in Procedure 29 (see in CD accompanying this book).

20.4.2.1 *Adriamycin–DNA interaction at a thick-dsDNA electrochemical biosensor*

The oxidation of adriamycin at the thick-layer DNA-electrochemical biosensor was investigated and the effect of the immersion time of the thick-layer DNA biosensor in 1 μM adriamycin solution (Fig. 20.5) was compared with the results obtained with a bare GCE [37,56,57]. Using the thick-layer DNA-electrochemical biosensor it was possible to pre-concentrate adriamycin on the thick layer of DNA and the adriamycin oxidation peak current was found to increase with immersion time and to reach saturation after 1 h of immersion (see insert, Fig. 20.5A).

However, if a potential of −0.60 V was applied to the DNA biosensor during 120 s, big changes occurred inside the DNA layer (Fig. 20.6). Two new oxidation peaks appeared, which can be identified [2]: the first at +0.80 V, as guanine oxidation, and the second at +1.05 V, as adenine oxidation. Nevertheless, the oxidation peak potentials for guanosine and adenine are very close, making their identification difficult.

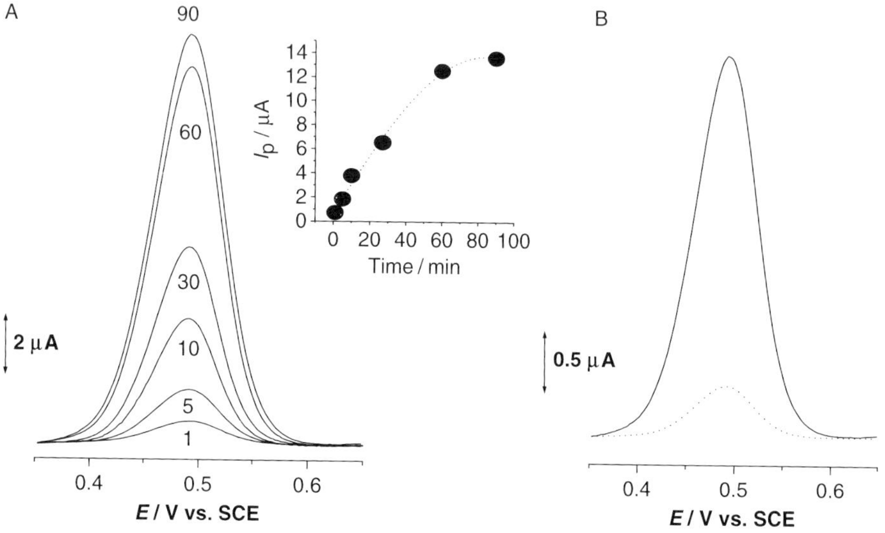

Fig. 20.5. Background-subtracted DP voltammograms of 1 μM adriamycin in 0.1 M pH 4.5 acetate buffer obtained with a thick-layer dsDNA-electrochemical biosensor: (A) Effect of immersion time, insert I_p vs. t; (B) Current decrease in successive differential pulse voltammograms: (▬) First voltammogram after 5 min immersion and (····) Fifth voltammogram. With permission from Ref. [37].

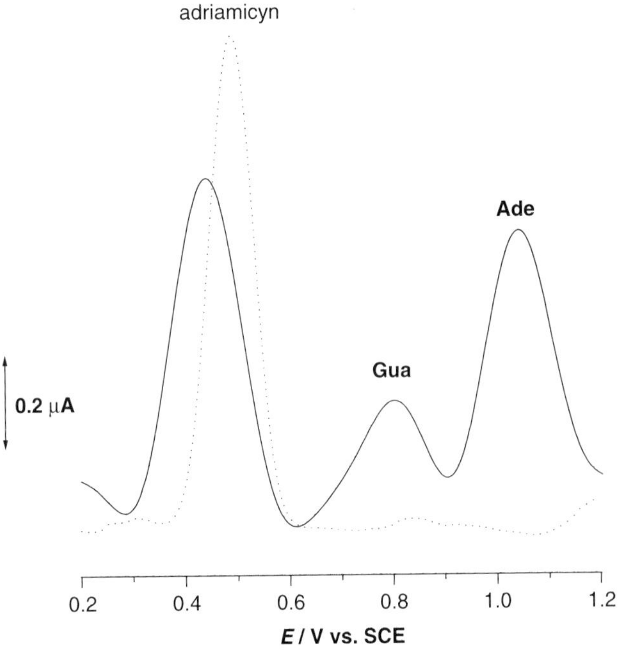

Fig. 20.6. Background-subtracted DP voltammograms in pH 4.5 0.1 M acetate buffer obtained with a thick-layer dsDNA-electrochemical biosensor GCE after being immersed during 10 min in a 1 μM adriamycin solution and rinsed with water before the experiment in buffer: (····) without applied potential and (▬) subsequent scan after applying a potential of –0.6 V during 120 s. With permission from Ref. [37].

20.4.2.2 Adriamycin–dsDNA interaction at a thin-dsDNA electrochemical biosensor

The thin-layer DNA biosensor was immersed during 3 min in an adriamycin solution, rinsed with water, and later transferred to buffer, where a DP voltammogram was recorded. The peak for adriamycin oxidation occurs at +0.50 V, and only after applying the potential of −0.60 V during 60 s, the oxidation peak for guanine, at +0.84 V, and the oxidation peak for 8-oxoGua, at +0.38 V (Fig. 20.7) appeared.

These results are in agreement with those obtained with the thick-layer dsDNA-electrochemical biosensor. The clear separation of the adriamycin and 8-oxoGua peaks can be explained by the non-uniform coverage of the electrode surface by DNA and the adsorption of adriamycin [37] on the uncovered glassy carbon. This peak separation is

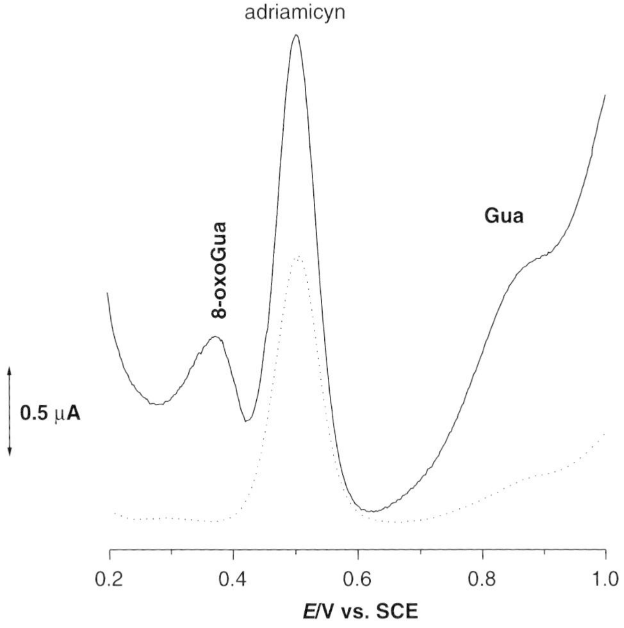

Fig. 20.7. DP voltammograms in pH 4.5 0.1 M acetate buffer obtained with a thin-layer dsDNA-electrochemical biosensor after being immersed in a 5 μM adriamycin solution during 3 min and rinsed with water before the experiment in buffer: (····) without applied potential; (—) applying a potential of −0.6 V during 60 s. First scans with permission from Ref. [37].

very important as it enables the use of the less positive 8-oxoGua oxidation peak to detect adriamycin damage to dsDNA.

Adriamycin intercalates within double helix DNA and can undergo a reaction beginning with the transfer of a single electron to the quinone portion of the adriamycin ring system, generating a free radical [57], which can interact with DNA *in situ* with the products of this interaction being retained in the DNA layer.

Adriamycin electroactive functional groups (the oxidizable hydroquinone group in ring B, and the reducible quinone function in ring C) are intercalated between the base pairs in the dsDNA [37]. The reducible quinone group in ring C protrudes slightly into the major groove, and this enables *in situ* (*in helix*) generation of an adriamycin radical within the double helix [58]. Therefore, a redox reaction between adriamycin and guanine residues inside the double helix of DNA can occur and 8-oxoGua is the main product of guanine oxidation. The mechanistic pathway involving two electrons and two protons [14] depends on the chemical environment surrounding guanine.

20.4.3 Nitric oxide

Nitric oxide (NO), the nitrogen monoxide, is a physiologically active molecule regulating numerous biological processes including vasodilatation, neurotransmission and cell-dependent immunity [59,60]. However, it was shown that in certain conditions (infected or inflamed tissues), large amounts of NO produce genotoxic effects that have been associated with carcinogenesis.

The whole spectra of DNA damage by NO and its derivatives include nitrosative deamination of DNA bases causing transition mutations during DNA replication, strand breakage and both oxidation and nitration of the bases [61]. For example, it was demonstrated that peroxynitrite ($ONOO^-$) formed during the interaction of NO and superoxide radicals ($O_2^{-\bullet}$) causes DNA cleavage at every nucleotide with a slight predominance for guanine sites and high amounts of 8-oxoGua (Scheme 20.3A) were measured [62]. There are conflicting reports concerning the formation of 8-oxoGua, this compound being a target for further oxidation by $ONOO^-$ giving rise to oxaluric acid; and other products could be generated by $ONOO^-$-mediated guanine modification. After interaction of $ONOO^-$ with guanine residues, 8-nitroguanine (8-nitroGua) (Scheme 20.3B) and 4,5-dihydro-5-hydroxy-4-(nitrosooxy)-2′-deoxyguanosine were also separated and characterized [63,64].

Due to the importance of nitric oxide in regulating various cell functions, an electrochemical study of the NO–DNA interaction has been undertaken (see Procedure 29 in CD accompanying this book).

The source of nitric oxide was diethylenetriamine/nitric oxide, DETA/NO (Scheme 20.4) a compound that has been used in studies of the cytostatic, vasodilatory and other pharmacological properties of NO [65–67]. DETA/NO is a 1-substituted diazen-1-ium-1,2-diolate containing the $[N(O)NO]^-$ functional group that has been proved to be useful for the reliable generation of nitric oxide in homogenous solutions [68]. When dissolved in blood, cell culture medium or buffer this compound dissociates to generate NO leaving the nucleophilic structure as a by-product.

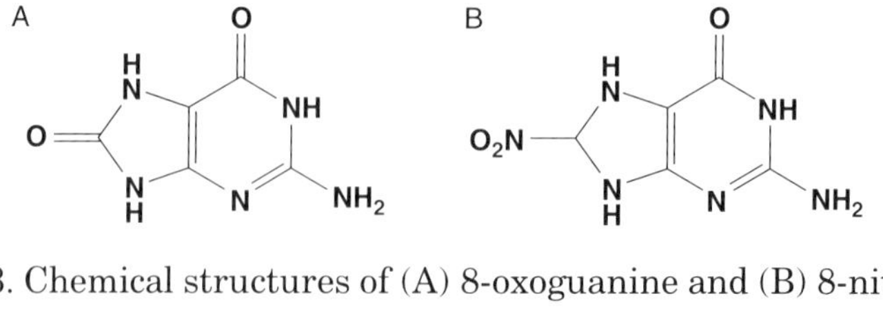

Scheme 20.3. Chemical structures of (A) 8-oxoguanine and (B) 8-nitroguanine.

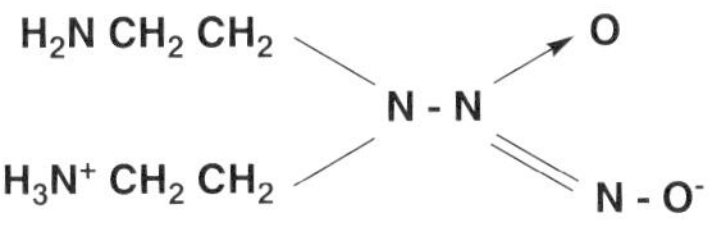

Scheme 20.4. Chemical structure of DETA/NO.

The electrochemical behaviour of DETA/NO was studied using a GCE [40]. One main oxidation peak at $E_{pa} = +0.80\,V$ independent of pH was observed. This potential corresponds to the oxidation of NO. The peak current is constant with time, meaning that the NO is dissolved in the solution after being liberated from the nucleophilic DETA/NO [69].

20.4.3.1 NO–DNA interaction at a thick-layer dsDNA biosensor

The DP voltammogram obtained in supporting electrolyte with the thick-layer dsDNA biosensor after incubation for 5 min in a solution of 100 μM DETA/NO (Fig. 20.8) shows the anodic peak specific to NO oxidation ($E_{pa} = +0.80\,V$) followed by the two small peaks of dGuo and dAdo, comparable with those obtained with the biosensor before incubation, meaning that the NO radicals themselves do not damage DNA.

After incubation in DETA/NO, a newly prepared biosensor was transferred to buffer and a potential of $-0.60\,V$ was applied during 3 min, causing the electrochemical reduction of oxygen and generation of $O_2^{-\bullet}$. The superoxide reacts with NO pre-concentrated into the thick DNA film to form peroxynitrite ($ONOO^-$), a highly reactive species that can cause oxidative damage to DNA [70]. The DP voltammogram (Fig. 20.8) shows increase of dGuo and dAdo peaks due to the cleavage of the dsDNA helix. The NO oxidation peak still appears but with a smaller current because of having been consumed, due to its reaction with superoxide radicals.

In order to produce the peroxynitrite radicals that damage DNA, the thick-layer biosensor has to be first incubated in the solution of DETA/NO and then conditioned at $-0.60\,V$ such that $O_2^{-\bullet}$ reacts with NO pre-concentrated into the thick DNA film. On the contrary, the production of peroxynitrite at the multi-layer dsDNA biosensor is carried out in a single step.

20.4.3.2 NO–DNA interaction at a multi-layer dsDNA biosensor

A multi-layer dsDNA biosensor was incubated in DETA/NO at $-0.60\,V$, and the $O_2^{-\bullet}$ radicals produced at the electrode surface reacted with the NO molecules that diffuse from solution towards the biosensor surface, giving rise to $ONOO^-$ that damage DNA. The DP voltammogram

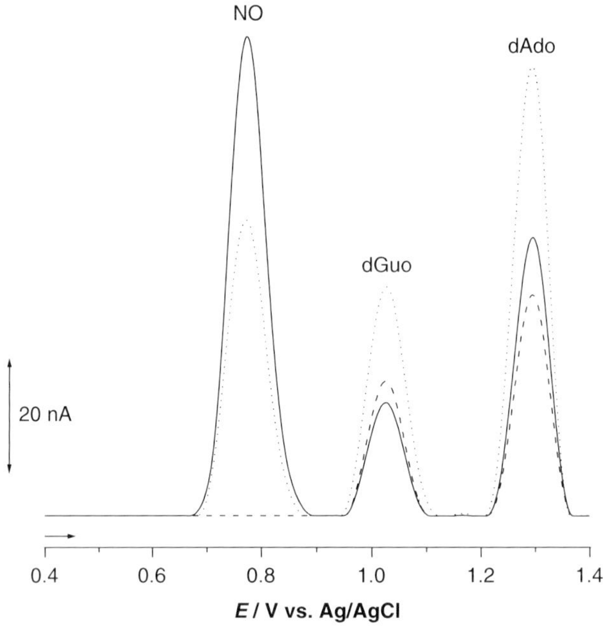

Fig. 20.8. DP voltammograms obtained in pH 4.5 0.1 M acetate buffer with a thick-layer DNA biosensor (----) before and after incubation during 5 min in 100 μM DETA/NO (—) without or (·····) with application of −0.60 V during 3 min. With permission from Ref. [40].

obtained after the transfer of the biosensor into buffer electrolyte (Fig. 20.9) shows the variation with time of the cleavage of DNA recognizable by the higher oxidation peaks of dGuo and dAdo.

For higher DETA/NO concentrations, the DP voltammogram showed two large oxidation peaks for dGuo and dAdo plus a third peak at a lower potential, $E_{pa} = +0.77\,V$ (Fig. 20.10). This peak is a consequence of the interaction between peroxynitrite radicals and DNA, corresponding to guanine (Gua) oxidation.

This result is in agreement with the fact that peroxynitrite induces DNA cleavage predominantly at the 5′-G of GG and GGG sequences [62]. Hence, the 5′ terminal guanine residues will be in direct contact with the electrode surface.

For longer incubation periods, 3 min at −0.60 V in concentrated DETA/NO, the dAdo peak current remained unchanged, and the dGuo peak became smaller whereas the Gua oxidation current increases (Fig. 20.10). This experiment showed that for longer incubation times, more $ONOO^-$ is available to damage DNA and, as a consequence, more guanine residues are accessible to the electrode surface.

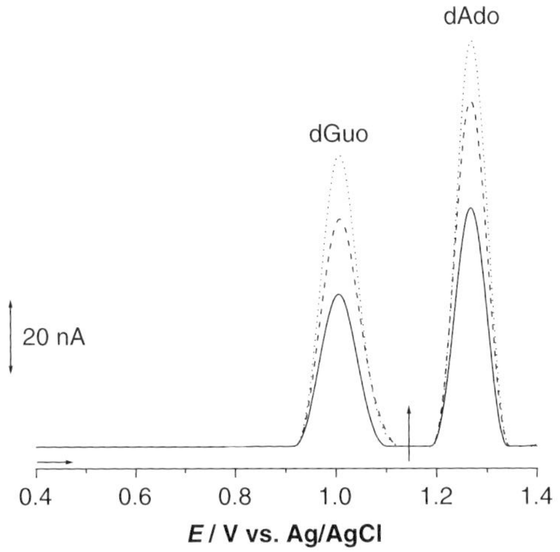

Fig. 20.9. DP voltammograms obtained in pH 4.5 0.1 M acetate buffer with a thin multi-layer DNA biosensor (▬) before and after incubation in 100 μM DETA/NO at −0.60 V during (----) 3 or (·····) 5 min. With permission from Ref. [40].

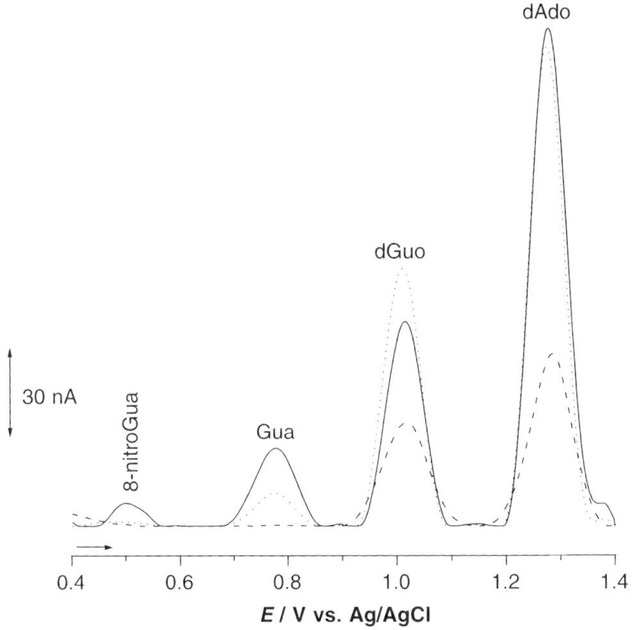

Fig. 20.10. DP voltammograms obtained in pH 4.5 0.1 M acetate buffer with a thin multi-layer dsDNA biosensor previously incubated in 1.5 mM DETA: (----) during 3 min without applying any potential or during (·····) 2 and (▬) 3 min applying −0.60 V. With permission from Ref. [40].

Also, a new peak is observed at $E_{pa} = +0.49\,V$ (Fig. 20.10). This new peak is due to oxidative damage to immobilized DNA after interaction with peroxynitrite. Since guanine is the most easily oxidized base in DNA, it is expected that $ONOO^-$ readily interacts with guanine. Although 8-oxoGua is the main guanine oxidation product, literature reports 8-nitroGua as the principal oxidative damage to DNA treated with peroxynitrite. 8-nitroGua has a structure very similar to 8-oxoGua (Scheme 20.3) and it is expected that their oxidation potentials are very close to each other. Both products are highly mutagenic and can contribute to cell dysfunction.

20.5 CONCLUSION

The development of the electrochemical DNA biosensors has opened a wide perspective using particularly sensitive and selective methods for the detection of specific DNA damaging interactions. Electrochemical voltammetric *in situ* sensing of DNA oxidative damage caused by different compounds with or without therapeutic activity is possible using different types (thin, multi and thick layer) of electrochemical dsDNA biosensors. The choice of the best approach to be used depends on the drug and the time necessary to cause DNA damage. However, previous knowledge of the electrochemical DNA biosensor surface morphology and of the electrochemical behaviour of the drug at a bare electrode is most important to avoid possible misinterpretations.

The understanding of DNA interaction with molecules or ions using voltammetric techniques for *in situ* generation of reactive intermediates is a complementary tool for the study of these biomolecular interaction mechanisms. The electrochemical DNA biosensor enables the preconcentration of compounds on the biosensor surface. Controlling the potential, the *in situ* electrochemical generation of radical intermediates is possible. Monitoring the changes of dGuo and dAdo oxidation peak currents or the appearance of new redox peaks it is possible to conclude about the damaging and potential toxic effect of different compounds.

The understanding of the mechanism of action of health hazard compounds that interact with DNA can aid in explaining the differences in reactivity between similar moieties. This knowledge can and should be used as an important parameter for quantitative structure–activity relationships (QSAR) and/or molecular modelling studies, as a contribution to the design of new structure-specific DNA-binding drugs, and for the possibility of pre-screening the damage they may cause to DNA integrity.

REFERENCES

1 A.M. Oliveira Brett and S.H.P. Serrano, Development of DNA-based biosensors for carcinogens. Biosensors. In: P. Frangopol, D.P. Nikolelis and U.J. Krull (Eds.), *Current Topics in Biophysics*, Vol. 2, Al. I. Cuza University Press, Iaşi, Romania, 1997, pp. 223–238 Chap. 10.
2 A.M. Oliveira Brett, S.H.P. Serrano and J.A.P. Piedade, Electrochemistry of DNA. In: R.G. Compton and G. Hancock (Eds.), *Comprehensive Chemical Kinetics, Applications of Kinetic Modelling*, Vol. 37, Elsevier, Oxford, UK, 1999, pp. 91–119 Chap. 3.
3 E. Palecek, Past, present and future of nucleic acids electrochemistry, *Talanta*, 56 (2002) 809.
4 E. Paleček, M. Fojta, F. Jelen and V. Vetterl, Electrochemical analysis of nucleic acids. *Bioelectrochemistry*. In: A.J. Bard and M. Stratmann (Eds.), *The Encyclopedia of Electrochemistry*, Vol. 9, Wiley-VCH, Weinheim, Germany, 2002, pp. 365–429 Chap. 12, and references therein.
5 A.M. Oliveira Brett, DNA-based biosensor. In: L. Gorton (Ed.), *Comprehensive Analytical Chemistry, Biosensors and Modern Specific Analytical Techniques*, Vol. XLIV, Elsevier, Amsterdam, 2005, pp. 179–208 Chap. 4.
6 A.M. Oliveira Brett, Electrochemistry for probing DNA damage. In: C.A. Grimes, E.C. Dickey and M.V. Pishko (Eds.), *Encyclopaedia of Sensors*, Vol. 3, American Scientific Publishers, USA, 2007, pp. 301–314.
7 G. Dryhurst and P.J. Elving, Electrochemical oxidation of adenine: reaction products and mechanisms, *J. Electrochem. Soc.*, 5 (1968) 1014–1022.
8 G. Dryhurst and P.J. Elving, Electrochemical oxidation–reduction paths for pyrimidine, cytosine, purine and adenine. Correlation and application, *Talanta*, 16 (1969) 855–874.
9 G. Dryhurst, Adsorption of guanine and guanosine at the pyrolytic graphite electrode, *Anal. Chim. Acta*, 57 (1971) 137–149.
10 E. Paleček, F. Jelen, M.A. Hung and J. Lasovsky, Reaction of the purine and pyrimidine derivatives with the electrode mercury, *Bioelectrochem. Bioenerg.*, 8 (1981) 621–631.
11 A.M. Oliveira Brett and F.-M. Matysik, Voltammetric and sonovoltammetric studies on the oxidation of thymine and cytosine at a glassy carbon electrode, *J. Electroanal. Chem.*, 429 (1997) 95–99.
12 A.M. Oliveira Brett and F.-M. Matysik, Sonovoltammetric studies of guanine and guanosine, *Bioelectrochem. Bioenerg.*, 42 (1997) 111–116.
13 A.M. Oliveira-Brett, J.A.P. Piedade, L.A. Silva and V.C. Diculescu, Voltammetric determination of all DNA nucleotides, *Anal. Biochem.*, 332 (2004) 321–329.
14 A.M. Oliveira Brett, J.A.P. Piedade and S.H.P. Serrano, Electrochemical oxidation of 8-oxoguanine, *Electroanalysis*, 13 (2001) 199–203.

15 I. Rebelo, J.A.P. Piedade and A.M. Oliveira Brett, Development of an HPLC method with electrochemical detection of femtomoles of 8-oxo-7,8-dihydro-2′ deoxyguanosine in the presence of uric acid, *Talanta*, 63 (2004) 323–331.
16 V. Brabec and G. Dryhurst, Electrochemical behaviour of natural and biosynthetic polynucleotides at the pyrolytic graphite electrode. A new probe for studies of polynucleotide structure and reactions, *J. Electroanal. Chem.*, 89 (1978) 161–173.
17 T. Tao, T. Wasa and S. Mursha, The anodic voltammetry of desoxyribonucleid acid at a glassy carbon electrode, *Bull. Chem. Soc. Jpn.*, 51 (1978) 1235–1236.
18 C.M.A. Brett, A.M. Oliveira Brett and S.H.P. Serrano, On the adsorption and electrochemical oxidation of DNA at glassy carbon electrodes, *J. Electroanal. Chem.*, 366 (1994) 225–231.
19 M.I. Pividori, A. Merkoci and S. Alegret, Electrochemical genosensor design: immobilisation of oligonucleotides onto transducer surfaces and detection methods, *Biosens. Bioelectron.*, 15 (2000) 291–303.
20 D. Ivnitski, I. Abdel-Hamid, P. Atanasov, E. Wilsins and S. Stricker, Application of electrochemical biosensors for detection of food pathogenic bacteria, *Electroanalysis*, 12 (2000) 317–325.
21 J. Wang, Survey and summary: from DNA biosensors to gene chips, *Nucleic Acids Res.*, 28 (2000) 3011–3016.
22 M. Mascini, I. Palchetti and G. Marrazza, DNA electrochemical biosensors, *Fresenius J. Anal. Chem.*, 369 (2001) 15–22.
23 K. Kerman, B. Meric, D. Ozkan, P. Kara, A. Erdem and M. Ozsoz, Electrochemical DNA biosensor for the determination of benzo[*a*]pyrene–DNA adducts, *Anal. Chim. Acta*, 450 (2001) 45–52.
24 A.M. Oliveira Brett and L.A. Silva, A DNA-electrochemical biosensor for screening environmental damage caused by s-triazine derivatives, *Anal. Bioanal. Chem.*, 373 (2002) 717–723.
25 A.M. Oliveira Brett and A.-M. Chiorcea, Atomic force microscopy of DNA immobilized onto a highly oriented pyrolytic graphite electrode surface, *Langmuir*, 19 (2003) 3830–3839.
26 A.M. Oliveira Brett and A.-M. Chiorcea, Effect of pH and applied potential on the adsorption of DNA on highly oriented pyrolytic graphite electrodes. Atomic force microscopy surface characterisation, *Electrochem. Commun.*, 5 (2003) 178–183.
27 A.M. Oliveira Brett and A.-M. Chiorcea Paquim, Atomic force microscopy characterization of an electrochemical DNA-biosensor, *Bioelectrochemistry*, 63 (2004) 229–232.
28 A.M. Oliveira Brett and A.-M. Chiorcea Paquim, DNA imaged on a HOPG electrode surface by AFM with controlled potential, *Bioelectrochemistry*, 66 (2005) 117–124.

29 A.M. Oliveira-Brett, A.-M. Chiorcea Paquim, V.C. Diculescu and J.A.P. Piedade, Electrochemistry of nanoscale DNA surface films on carbon, *Med. Eng. Phys.*, 28 (2006) 963–970.
30 S.R. Mikkelsen, Electrochemical biosensors for DNA sequence detection, *Electroanalysis*, 8 (1996) 15–19.
31 J. Wang, G. Rivas, X. Cai, E. Palecek, P. Nielsen, H. Shiraishi, N. Dontha, D. Luo, C. Parrado, M. Chicharro, P.A.M. Farias, F.S. Valera, D.H. Grant, M. Ozsoz and M.N. Flair, DNA electrochemical biosensors for environmental monitoring, *Anal. Chim. Acta*, 347 (1997) 1–8.
32 J. Wang, X. Cai, G. Rivas, H. Shiraishi and N. Dontha, Nucleic-acid immobilization, recognition and detection at chronopotentiometric DNA chips, *Biosens. Bioelectron.*, 12 (1997) 587–599.
33 M. Yang, M.E. McGovem and M. Thompson, Genosensor technology and the detection of interfacial nucleic acid chemistry, *Anal. Chim. Acta*, 346 (1997) 259–275.
34 E. Palecek and M. Fojta, Detecting DNA hybridisation and damage, *Anal. Chem.*, 73 (2001) 74A–83A.
35 A.M. Oliveira-Brett and V.C. Diculescu, Electrochemical study of quercetin–DNA interactions. Part II—*In situ* sensing with DNA-biosensors, *Bioelectrochemistry*, 64 (2004) 143–150.
36 A.M. Oliveira Brett, T.R.A. Macedo, D. Raimundo, M.H. Marques and S.H.P. Serrano, Voltammetric behaviour of mitoxantrone at DNA-biosensor, *Biosens. Bioelectron.*, 13 (1998) 861–867.
37 A.M. Oliveira Brett, M. Vivan, I.R. Fernandes and J.A.P. Piedade, Electrochemical detection of *in situ* adriamycin oxidative DNA damage to DNA, *Talanta*, 56 (2002) 959–970.
38 A.M. Oliveira Brett, S.H.P. Serrano, I. Gutz, M.A. La-Scalea and M.L. Cruz, Voltammetric behaviour of nitroimidazoles at a DNA-biosensor, *Electroanalysis*, 9 (1997) 1132–1137.
39 B. Halliwell and J.M.C. Gutteridge, Oxidative stress: adaptation, damage, repair and death, *Free Radicals in Biology and Medicine*, 3rd ed., Oxford University Press, New York, 1993, pp. 247–349.
40 V.C. Diculescu, R.M. Barbosa and A.M. Oliveira–Brett, *In situ* sensing of DNA damage by a nitric oxide-releasing compound, *Anal. Lett.*, 38 (2005) 2525–2540.
41 C.A. Rice-Evans, N.J. Miller and G. Paganga, Structure-antioxidant activity relationships of flavonoids and phenolic acids, *Free Radic. Biol. Med.*, 20 (1996) 933–956.
42 H. Ohshima, Y. Yoshie, S. Auriol and I. Gilibert, Antioxidant and pro-oxidant actions of flavonoids: effects on DNA damage induced by nitric oxide, peroxynitrite and nitroxyl anion, *Free Radic. Biol. Med.*, 25 (1998) 1057–1065.
43 M.K. Johnson and G. Loo, Effects of egpigallocatechin gallate and quercetin on oxidative damage to cellular DNA, *Mutat. Res.*, 459 (2000) 211–218.

44 S.J. Duthie, W. Johnson and V.L. Dobson, The effect of dietary flavonoids on DNA damage (strand breaks and oxidised pyrimidines) and growth in human cells, *Mutat. Res.*, 390 (1997) 141–151.
45 A.M. Oliveira-Brett and V.C. Diculescu, Electrochemical study of quercetin-DNA interactions. Part I—Analysis in incubated solutions, *Bioelectrochemistry*, 64 (2004) 133–141.
46 M. Hollstein, D. Sidransky, B. Vogelstein and C.C. Harris, p53 mutations in human cancers, *Science*, 253 (1991) 49–53.
47 J.E. Brown, H. Khodr, R.C. Hider and C.A. Rice-Evans, Structural dependence of flavonoid interactions with Cu^{n+} ions: implications of their antioxidant properties, *Biochem. J.*, 330 (1998) 1173–1178.
48 G. Minotti, G. Cairo and E. Monti, Role of iron in anthracycline cardiotoxicity: new tunes for an old song?, *FASEB J.*, 13 (1999) 199–212.
49 E.L. de Beer, A.E. Bottone and E.E. Voest, Doxorubicin and mechanical performance of cardiac trabeculae after acute and chronic treatment: a review, *Eur. J. Pharmacol.*, 415 (2001) 1–11.
50 S. Zhou, A. Starkov, M.K. Froberg, R.L. Leino and K.B. Wallace, Cumulative and irreversible cardiac mitochondrial dysfunction induced by doxorubicin, *Cancer Res.*, 61 (2001) 771–777.
51 E.L. Kostoryz and D.M. Yourtee, Oxidative mutagenesis of doxorubicin–Fe(III) complex, *Mutat. Res.*, 490 (2001) 131–139.
52 K. Kiyomiya, S. Matsuo and M. Kuruebe, Differences in intracellular sites of action of adriamycin in neoplastic and normal differentiated cell, *Cancer Chemother. Pharmacol.*, 47 (2001) 51–56.
53 I. Muller, A. Jenner, G. Bruchelt, D. Niethammer and B. Halliwell, Effect of concentration on the cytotoxic mechanism of doxorubicin—apoptosis and oxidative DNA damage, *Biochem. Biophys. Res. Commun.*, 230 (1997) 254–257.
54 S.S. David and S.D. Williams, Chemistry of glycosylases and endonucleases involved in base-excision repair, *Chem. Rev.*, 98 (1998) 1221–1261.
55 H. Berg, G. Horn, U. Luthardt and W. Ihn, Interaction of anthracycline antibiotics with biopolymers: Part V—Polarographic behavior and complexes with DNA, *Bioelectrochem. Bioenerg.*, 8 (1981) 537–553.
56 A.M. Oliveira-Brett, J.A.P. Piedade and A.-M. Chiorcea, Anodic voltammetry and AFM imaging of picomoles of adriamycin adsorbed onto carbon surfaces, *J. Electroanal. Chem.*, 538–539 (2002) 267–276.
57 J.A. Piedade, I.R. Fernandes and A.M. Oliveira Brett, Electrochemical sensing of DNA–adriamycin interactions, *Bioelectrochemistry*, 56 (2002) 81–83.
58 M.C. Perry, *The Chemotherapy Source Book*, Williams & Wilkins, Baltimore, USA, 1996.
59 S. Burney, J.L. Caulfield, J.C. Niles, J.S. Wishnok and S.R. Tannenbaum, The chemistry of DNA damage from nitric oxide and peroxynitrite, *Mutat. Res.*, 424 (1999) 37–49.

60 A. Martinez, A. Urios, V. Felipo and M. Blanco, Mutagenicity of nitric oxide-releasing compounds in *Escherichia coli*: effect of superoxide generation and evidence for two mutagenic mechanisms, *Mutat. Res.*, 47 (2001) 159–167.
61 J.C. Beckman, T.W. Beckman, J. Chen, P.A. Marshall and B.A. Freeman, Apparent hydroxyl radical formation by peroxinitrite: implications from endothelial injury from nitric oxide and superoxide, *Proc. Natl. Acad. Sci. USA*, 87 (1990) 1620–1624.
62 S. Kawanishi, Y. Hiraku and S. Oikawa, Mechanism of guanine-specific DNA damage by oxidative stress and its role in carcinogenesis and aging, *Mutat. Res.*, 488 (2001) 65–76.
63 V. Yermilov, J. Rubio and H. Oshima, Formation of 8-nitroguanine in DNA treated with peroxinitrite *in vitro* and its rapid removal from DNA by depurination, *FEBS Lett.*, 376 (1995) 207–210.
64 T. Douki, J. Cadet and B.N. Ames, An adduct between peroxinitrite and 2′-deoxiguanosine: 4,5-dihydro-5-hydroxy-4-(nitosooxy)-2′-deoxyguanosine, *Chem. Res. Toxicol.*, 9 (1996) 3–7.
65 J.B. Salom, M.D. Barbera, J.M. Centeno, M. Orti, G. Torregrosa and E. Alborch, Relaxant effects of sodium nitroprusside and NONOates in rabbit basilar artery, *Pharmacology*, 57 (1998) 79–87.
66 G. Lemaire, F.-J. Alvarez-Pachon, C. Beuneu, M. Lepoivre and J.-F. Petit, Differential cytostatic effects of NO donors and NO producing cells, *Free Radic. Biol. Med.*, 26 (1999) 1274–1283.
67 D.D. Kindler, C. Thiffault, N.J. Solenski, J. Dennis, V. Kostecki, R. Jenkins, P.M. Keeney and J.P.Rr. Bennett, Neurotoxic nitric oxide rapidly depolarizes and permeabilizes mitochondria by dynamically opening the mitochondrial transition pore, *Mol. Cell. Neurosci.*, 23 (2003) 559–573.
68 N.R. Ferreira, A. Ledo, J.G. Frade, G.A. Gerhardt, J. Laranjinha and R.M. Barbosa, Electrochemical measurement of endogenously produced nitric oxide in brain slices using Nafion/*o*-phenylenediamine modified carbon fiber microelectrodes, *Anal. Chim. Acta*, 535 (2005) 1–7.
69 I. Zacharia and W. Deen, Diffusivity and solubility of nitric oxide in water and saline, *Ann. Biomed. Eng.*, 33 (2005) 214–222.
70 S. Inoue and S. Kawanishi, Oxidative DNA damage induced by simultaneous generation of nitric oxide and superoxide, *FEBS Lett.*, 371 (1995) 86–88.

Chapter 21

Electrochemical genosensing of food pathogens based on graphite–epoxy composite

María Isabel Pividori and Salvador Alegret

21.1 INTRODUCTION

21.1.1 Contamination of food, food pathogens and food safety

The control of food quality has become of growing interest for both consumer and food industry since the increasing incidence of food poisoning is a significant public health concern for customers worldwide [1].

Contaminants in foods can be grouped according to their origin and nature. Essentially, these are microbes (bacteria, viruses, parasites), exogenous matter (biological, chemical, physical), natural toxins (seafood toxins, mycotoxins), other chemical compounds (such as pesticides, toxic metals, veterinary drug residues, undesirable fermentation products) and packaging materials. Most of the agents found in food are natural contaminants from environmental sources, but some are deliberate additives. While additives were at one time a major concern, nowadays microbiological issues are the greatest, followed by pesticide and animal drug residues and antimicrobial drug resistance. Although the majority of microorganisms carry out essential activities in nature and many are closely associated with plants or animals in beneficial relations, certain potentially harmful microorganisms result in numerous foodborne diseases with profound effects on animals and humans. Bacteria that cause foodborne diseases occur worldwide, the most common being *Salmonella*, *Staphylococcus aureus*, *Clostridium perfringens* and *Bacillus cereus* [2]. These bacteria can multiply rapidly in moist, warm, protein-rich foods, such as meat, poultry, fish, shellfish, milk, eggs, and most foods after they have been processed. Infectious organisms such as *Salmonella* and *C. perfingens* can multiply in the digestive

Comprehensive Analytical Chemistry 49
Alegret and Merkoçi (Eds)
Volume 49 ISSN: 0166-526X DOI: 10.1016/S0166-526X(06)49021-8

tract and cause illness by invasion of the cell lining, production of toxins, or both. Other microorganisms are able to produce enterotoxins, such as *S. aureus* and *B. cereus*, or neurotoxins, such as *Clostridium botulinum*, in the food during their growth and metabolism.

In the last decade, *Salmonella enteritidis* has been the source of many outbreaks involving egg products through transovarian infection from hen to egg. In the last few years, *Salmonella typhimurium* DT 104 and other antibiotic-resistant salmonellae have also recently become a concern [3].

Many factors have contributed to recent food emergencies. The food production chain is becoming increasingly complex because of mass production. Modern farming practices are intensive and can result in microorganisms contaminating a large number of crops or infecting a large number of animals. Healthy animals may carry pathogens that cause disease in humans. Examples include *Salmonella* spp., *Listeria* spp. and some strains of *Escherichia coli* and *Campylobacter*. Animals may be infected from feed, other animals or the environment.

Food regulatory agencies have thus established control programs in order to avoid food pathogens from entering the food supply. Food quality and safety can only be ensured through the application of quality-control systems throughout the entire food chain. They should be implemented at farm level with the application of good agricultural practices and good veterinary practices at production, good manufacturing practice at processing and good hygiene practices at retail and catering levels. One of the most effective ways for the food sector to protect public health is to base their food management programs on Hazard Analysis and Critical Control Point (HACCP). HACCP consists of seven principles that outline how to establish, implement and maintain a quality assurance plan for a food establishment.

21.1.2 Food pathogen detection and culture methods

Conventional bacterial identification methods usually include a morphological evaluation of the microorganism as well as tests for the organism's ability to grow in various media under a variety of conditions, which involve the following basic steps: (i) preenrichment, (ii) selective enrichment, (iii) biochemical screening and (iv) serological confirmation [4]. Although standard microbiological techniques allow the detection of single bacteria, amplification of the signal is required through growth of a single cell into a colony in order to detect pathogens which typically occur in low numbers in food or water. In the case of *Salmonella*, as well as for other pathogens, it is essential that initial incubation in a non-selective medium at 37°C, such as buffered peptone water (BPW) and lactose broth, is first performed [5]. This step allows the organism not only to overcome the effects of sub-lethal injury but also to multiply. The period of recovery or repair is often referred to as "preenrichment". Incubation during preenrichment step is usually performed for 18–24 h. After preenrichment, selective isolation is usually performed in two stages: growth in an enrichment broth followed by plating on a selective agar. Selective media make use of either dyes or chemicals. Malachite green, in combination with high concentrations of magnesium chloride and a medium pH of 5.0–5.2 in soya peptone broth (Rappaport–Vassiliadis broth, RV), has been shown to be effective for selective culture following preenrichment of *Salmonella*. Further plating in selective agars is usually performed in bismuth sulfite, xylose lysine deoxycholate (XLD) and brilliant green agar (BGA). After a further growing for 24 h, the confirmation of '*Salmonella*-like' colonies is mandatory. Biochemical confirmation test relies on assessment of urease and lysine decarboxylase activity, fermentation of dulcitol, indole production, growth in the presence of potassium cyanide, utilization of sodium malonate and, more recently, pyrrolindonylarylamidase (PYR) activity. These reactions, in combination with serological tests, are usually sufficient for identification, but additional tests may sometimes be necessary. As a laboratory routine, the use of commercial kits may be cost-effective. Antibodies against somatic (O) and flagella (H) antigens are used to confirm/identify *Salmonella*-like isolates.

These growing and enrichment steps are relatively time-consuming, having a total assay time of up to 1 week in certain food pathogens [6]. Although classical culture methods can be sensitive, they are greatly restricted by the assay time.

As food regulatory agencies have established strict control programs in order to avoid food pathogens from entering the food supply, official laboratories should be able to efficiently process a high number of samples. As a result, routine, rapid and efficient food control becomes mandatory. According to these requirements, the development of rapid, inexpensive, sensitive and high sample throughput and on-site analytical strategies, which can be used as an "alarm" to rapidly detect the risk of contamination by food pathogens in wide variety of food matrixes, is thus a priority, since the standard methods do not meet these requirements.

21.1.3 Rapid detection methods to detect food pathogens

The time necessary to obtain a negative result can take up to 96 h when the whole culture method has to be applied, while presumptive positive results may take up to 48–96 h. Many alternative methods have consequently been introduced in recent years to reduce analytical time and also save staff time and media requirements. These rapid methods are designed to obviate the need for selective culture and serological/biochemical identification.

Many new instrumental methods have been developed using various principles of detection, and these include chromatography, infrared or fluorescence spectroscopy, bioluminescence, flow cytometry, impedimetry and many others. Among these methods, techniques based on electrical conductance and impedance are in use in many laboratories. When microbial growth and metabolism take place in a culture medium, changes in conductance of the medium occur, but the reliability of the method largely depends on the performance of the selective medium used for the assay. Moreover, these methods are centralized in large stationary laboratories because complex instrumentation are used, and require highly qualified technical staff. Furthermore, the capital cost involved in these instrumental analyses is high which restricts its use [6].

During the past decade, immunological detection of bacteria, cells, spores, viruses and toxins has become more sensitive, specific, reproducible and reliable with many commercial immunoassays (IAs) available for the detection of a wide variety of microbes and their products in food. Advances in antibody production have stimulated this technology, since polyclonal antibodies can be now quickly and cheaply obtained, and do not require the time or expertise associated with the production of monoclonal antibodies. However, polyclonal antibodies are limited both in terms of their specificity and abundance. Many test kits are available, including immunodiffusion, enzyme-linked immunosorbent assays (ELISA) and the use of specific antibodies to capture and concentrate the organism [7]. Moreover, IAs have shown the ability to detect not only contaminating organisms but also their biotoxins. However, IAs give total bacterial load rather than just the number of viable cells.

Nucleic acid-based detection may be more specific and sensitive than immunological-based detection. Furthermore, the polymerase chain reaction (PCR) can be easily coupled to enhance the sensitivity of nucleic acid-based assays. Target nucleic segments of defined length and sequence are amplified by PCR by repetitive cycles of strand denaturation, annealing and extension of oligonucleotide primers by the thermostable DNA polymerase, *Thermus aquaticus* (Taq) DNA polymerase. Nucleic acid-based detection coupled with PCR has distinct advantages over culture and other standard methods for the detection of microbial pathogens, such as specificity, sensitivity, rapidity, accuracy and capacity to detect small amounts of target nucleic acid in a sample [8]. Moreover, multiple primers can be used to detect different pathogens in one multiplex reaction [9]. This kind of method requires an enrichment period, which delays results but sensitivity can be of the order of 3 CFU per 25 g of food [5]. This new approach cannot be applied directly to samples because of the fact that DNA is heat-stable. Thus, intact DNA will be present in processed foods. DNA in both a free form and within dead cells of *Salmonella* can also survive for considerable periods in seawater. One possible way to overcome this problem is to include a culture step as a means of detecting viable cells, this step being not sufficient to allow the growth of more severely damaged cells.

The use of nucleic acids recognition layers represents a new and exciting area in analytical chemistry which requires an extensive study. Besides classical methodologies to detect DNA, novel approaches have been designed, such as the DNA chips [10–12] and lab-on-a-chips based on microfluidic techniques [13]. However, these technologies are still out of the scope of food industry, since it requires simple, cheap and user-friendly analytical devices.

21.1.4 Biosensing as a novel strategy for the rapid detection of food pathogens

Biosensors offer an exciting alternative to the more traditional methods, allowing rapid "real-time" and multiple analyses that are essential for the detection of bacteria in food [9]. Biosensors use a combination of biological receptor compounds (antibody, enzyme, nucleic acid) and the physical or physico-chemical transducer producing, in most cases, "real-time" observation of a specific biological event (e.g. antibody–antigen interaction) [14]. They allow the detection of a broad spectrum of analytes in complex sample matrices, and have shown great promise in areas such as clinical diagnostics, food analysis, bioprocess and environmental monitoring [15,16]. HACCP system, which is generally accepted as the most effective system to ensure food safety, can utilize biosensor to verify that the process is under control [17]. The sensitivity of each of the sensor systems may vary depending on the transducer's properties, and the biological recognizing elements. An ideal biosensing device for the rapid detection of microorganisms should be fully automated, inexpensive and routinely used both in the field and the laboratory. Biosensors may be divided into four basic groups, depending on the signal transduction method: optical, mass, electrochemical and thermal sensors [18]. Optical transducers are particularly attractive as they can allow direct "label-free" and "real-time" detection of bacteria, but they lack sensitivity. The phenomenon of surface plasmon resonance (SPR) has shown good biosensing potential and many commercial SPR systems are now available (e.g. BIAcore^{TM}) [18]. Electrochemically based transduction devices are more robust, easy to use, portable and inexpensive analytical systems [19,20]. Furthermore, electrochemical biosensors can operate in turbid media and offer comparable instrumental sensitivity. Regarding the molecular recognition of food pathogens, mainly immunological [21,22] reagents as well as DNA probes [23] can be used as molecular receptor in order to obtain a useful signal after a biological reaction on the transducer element (antigen/antibody and hybridization, respectively) [19,24,25].

Electrochemical DNA biosensors can meet the demands of food control, offering considerable promise for obtaining sequence-specific information in a faster, simpler and cheaper manner compared to traditional hybridization assays. Such devices possess great potential for numerous applications, ranging from decentralized clinical testing to environmental monitoring, food safety and forensic investigations. The following section is focused on electrochemical DNA biosensing.

21.2 DNA ELECTROCHEMICAL BIOSENSORS

21.2.1 Immobilization of DNA and detection features

The design of a DNA electrochemical biosensor involves the immobilization of a DNA probe—usually 20 mer and complementary in their bases to the DNA pathogen genome—on a proper electrochemical transducer.

The most successful immobilization techniques for DNA or oligonucleotides appear to be those involving multi-site attachment—either electrochemical or physical adsorption—or single-point attachment—mainly covalent immobilization or strept(avidin)/biotin linkage [26]. As an example, single-point covalent immobilization can be performed on different surface-modified electrochemical transducers, such as glassy carbon [27,28], carbon paste [29], gold [30] or platinum [31], through carbodiimide chemistry. In the case of gold electrodes, the controlled-single-point immobilization of mercapto-modified DNA is also possible [32,33]. Single-point attachment is beneficial to hybridization kinetics, especially if a spacer arm is used. However, among different DNA immobilization methodologies reported, multi-site adsorption is the simplest and most easily automated procedure, avoiding the use of pre-treatment procedures based on previous activation/modification of the surface transducer and subsequent DNA immobilization. Such pre-treatment steps are known to be tedious, expensive and time-consuming. Furthermore, the adsorption properties of DNA on various supports (i.e. nylon, nitrocellulose) have been known for a long time [34].

Electrochemical detection of successful DNA hybridization event should be also considered. A typical DNA hybridization sensor involves a conversion of the hybridization event into useful electrical signal. The DNA recognition event for electrochemical genosensing is based mostly on external electrochemical markers such as electroactive indicators [35,36], metal nanoparticles or quantum dots [37], or enzymes. Enzyme labelling has been transferred from non-isotopic DNA classical methods to electrochemical genosensing. The enzyme labelling relies on the reaction between a biotinylated or digoxigenin-modified DNA probe and the streptavidin or anti-digoxigenin enzyme conjugate, respectively. Although a second incubation step is normally required for labelling, higher sensitivity and specificity have been reported for the enzyme labelling method compared with the other reported methods [38,39].

The direct electrochemical detection of DNA was initially proposed by Paleček [40,41] who recognized the capability of both DNA and RNA to yield reduction and oxidation signals after being adsorbed. The exploitation of the intrinsic DNA oxidation signal coming from the guanine base in electrochemical genosensing requires multi-site attachment—such as adsorption—as immobilization technique. The oxidation of DNA was shown to be strongly dependent on both the DNA adsorption conditions as well as the substrate whereon DNA is being absorbed, thus requiring a meticulous control of the DNA adsorbed layer. While immobilization and detection are important features, the choice of a suitable electrochemical substrate is also of great importance in determining the overall performance of the analytical electrochemical-based device, especially regarding the immobilization efficiency of

DNA. Next section is focused on the most important transducers features for DNA electrochemical genosensing.

21.2.2 Transducer materials for electrochemical DNA biosensing

The development of new transducing materials for DNA analysis is a key issue in the current research efforts of electrochemical-based DNA analytical devices. Carbonaceous materials, such as carbon paste [42], glassy carbon [43] and pyrolitic graphite [44], are the most popular choice of electrodes used in electrochemical DNA biosensing design. However, the use of platinum [31], gold [45], indium tin oxide [46], copper solid amalgam [47], mercury [48] and other continuous conducting metal substrates has been reported [26]. Most DNA biosensor transducers require pre-treatment of their surfaces. In some incidences, transducing materials have to be modified [49,50]. This modification process can be tedious, expensive (because they involve many reagents), time-consuming and difficult to reproduce. The improved electrochemical properties of carbon nanotubes have been lately demonstrated for DNA biosensing [51]. Although almost all these transducing materials have been successfully used for electrochemical genosensing, this chapter is focused on rigid carbon composites and their properties for the immobilization of DNA.

Carbon composites result from the combination of carbon with one or more dissimilar materials. Each individual component maintains its original characteristics while giving the composite distinctive chemical, mechanical and physical properties. The capability of integrating various materials is one of their main advantages. Some components which are incorporated within the composite result in enhanced sensitivity and selectivity. The best composite compounds will give the resulting material with improved chemical, physical and mechanical properties. As such, it is possible to choose between different binders and polymeric matrix in order to obtain better signal-to-noise (S/N) ratio, lower non-specific adsorption and improved electrochemical properties (electron transfer rate and electrocatalytic behaviour). Powdered carbon is frequently used as the conductive phase in composite electrodes due to its high chemical inertness, wide range of working potentials, low electrical resistance and a crystal structure responsible for low residual currents. A key property of polycrystalline graphite is porosity. The term "graphite" is used to designate materials that have been subjected to high temperatures, and thus have aligned the sp^2 planes parallel to each other.

Regarding their mechanical properties, carbon composites can be thus classified as rigid composites [52,53] or soft composites—the carbon pastes [54]. The composites are also classified by the arrangement of their particles, which can be (i) dispersed or (ii) randomly grouped in clearly defined conducting zones within the insulating zones.

The inherent electrical properties of the composite depend on the nature of each of the components, their relative quantities and their distribution. The electrical resistance is determined by the connectivity of the conducting particles inside the non-conducting matrix; therefore, the relative amount of each composite component has to be assessed to achieve optimal composition. Carbon composites show improved electrochemical performance, similar to an array of carbon fibres separated by an insulating matrix and connected in parallel. The signal produced by this macroelectrode formed by a carbon fibre ensemble is the sum of the signals of the individual microelectrodes. Composite electrodes thus showed a higher signal-to-noise (S/N) ratio, compared to the corresponding pure conductors, that accompanies an improved (lower) detection limit.

Rigid composites are obtained by mixing graphite powder with a non-conducting polymeric matrix into a soft paste that becomes rigid after a curing step [52,53]. They could be classified according to the nature of the binder or the polymeric matrix into epoxy composites, methacrylate composites or silicone composites.

Rigid conducting graphite–polymer composites (GEC) and biocomposites (GEB) have been extensively used in our laboratories and shown to be a suitable material for electrochemical (bio)sensing due to their unique physical and electrochemical properties [52,53]. In particular, we have used graphite–epoxy composite (GEC) made by mixing the non-conducting epoxy resin (Epo-Tek, Epoxy Technology, Billerica, MA, USA) with graphite powder (particle size below 50 μm). An ideal material for electrochemical genosensing should allow an effective immobilization of the probe on its surface, a robust hybridization of the target with the probe, a negligible non-specific adsorption of the label and a sensitive detection of the hybridization event. GECs fulfil all these requirements.

GECs present numerous advantages over more traditional carbon-based materials: higher sensitivity, robustness and rigidity. Additionally, the surface of GEC can be regenerated by a simple polishing procedure. Unlike carbon paste, the rigidity of GEC permits the design of different configurations, and these materials are compatible with non-aqueous solvents. In the following section, GEC materials for the development of electrochemical DNA biosensors are reviewed. Different graphite–epoxy platforms for electrochemical DNA biosensing, as well as strategies for detecting DNA hybridization, are also described. The advantages of these new graphite–epoxy platforms for electrochemical DNA biosensing are discussed and compared with the current state-of-the-art in DNA sensing techniques.

21.2.3 Electrochemical genosensing of food pathogens based on DNA dry-adsorption on GEC as electrochemical transducer

Adsorption is an easy way to attach nucleic acids to solid surfaces, since no reagents or modified-DNA are required. These features have promoted extensive use of adsorption as immobilization methodology in genetic analysis. The mainly claimed disadvantages of adsorption with respect to covalent immobilization are that (i) nucleic acids may be readily desorbed from the substrate and (ii) base moieties may be unavailable for hybridization if they are bonded to the substrate in multiple sites [55]. However, the electrochemical detection strategy based on the intrinsic oxidation of DNA requires the DNA to be adsorbed in close contact with the electrochemical substrate by multi-site attachment. This multi-site attachment of DNA can be thus detrimental for its hybridization but is crucial for the detection based on its oxidation signals.

The common method for the multi-site physical adsorption of DNA on carbonaceous-based material can be classified into dry or wet adsorptions. Dry adsorption relies on leaving DNA to dry on the carbonaceous surface. It can be assisted by light treatment (except UV that is able to induce changes in DNA molecule) or heated until 100°C. DNA can adopt a variety of conformations depending on the degree of hydration. As an example, the most familiar double helix DNA—called "B-DNA"—can become into the "A-DNA" form if it is strongly dehydrated. A structural alteration occurs due to a greater electrostatic interaction between the phosphate groups, leading to A-DNA. When the DNA solution is evaporated to dryness, the bases of DNA which have been dehydrated are exposed, thus the hydrophobic bases are strongly adsorbed flat on the electrode surfaces. Once it is adsorbed, DNA is difficult to re-hydrate. Hence, DNA is not desorbed, no matter how long the adsorbed-DNA is soaked in water, characteristic of irreversible adsorption. The "irreversible" behaviour of the dry-adsorbed DNA layer has been previously reported on glassy carbon electrodes [56]. DNA can be tightly and irreversible immobilized on GEC by both dry and wet adsorption procedures under static conditions [57]. The dual nature of GEC composed by islands of conducting material within the non-conducting and hydrophobic epoxy resin could play an important role in stabilizing the dehydrated A-form of DNA adsorbed on GEC.

Once immobilized on GEC, DNA preserves its unique hybridization properties, which can be revealed using different strategies based on both enzymatic labelling and the intrinsic signal coming from the DNA oxidation.

The DNA immobilization on GEC surface by dry-adsorption was performed by covering the GEC surface with a small drop of DNA in 10 × SSC, and allowing the electrode to dry at 80°C for 45 min in upright position (Fig. 21.1(A1)) [58]. The DNA electrochemical detection was then achieved by an enzymatic labelling step, as previously reported [59,60]. Briefly, the procedure consists of the following steps: (i) *DNA target immobilization*; (ii) *hybridization* with the complementary probe modified with either biotin or digoxigenin; (iii) *enzyme labelling* of the DNA duplex using streptavidin-HRP or anti-DIG-HRP; and (iv) *amperometric determination* based on the enzyme activity by adding H_2O_2 and using hydroquinone as a mediator. Further experimental details can be found in Procedure 30 in CD accompanying this book.

If the PCR product—or any other double-stranded DNA—is directly adsorbed on GEC transducer, a denaturing alkaline procedure after the DNA dry-adsorption is mandatory to break the hydrogen bonds linking the complementary DNA strands in order to ensure proper hybridization [61].

Although a compact thick ssDNA layer can be achieved by dry adsorption, DNA preserves its unique hybridization properties, which can be monitored using different strategies, suggesting that the DNA bases are not fully committed in the adsorption mechanism. DNA bases are mostly available for hybridization, taking into account the differences in signal compared with the non-specific adsorption [60,61]. This strategy was able to electrochemically detect the PCR amplicon coming from *Salmonella* spp. in a very simple and cheap way [61].

Besides this strategy in which the DNA target can be easily attached and detected by its complementary DNA signalling probe, a sandwich assay in which the DNA target is in solution can be easily performed by a double hybridization with a capture and a signalling probe [58]. This strategy was demonstrated to be useful for the detection of a novel

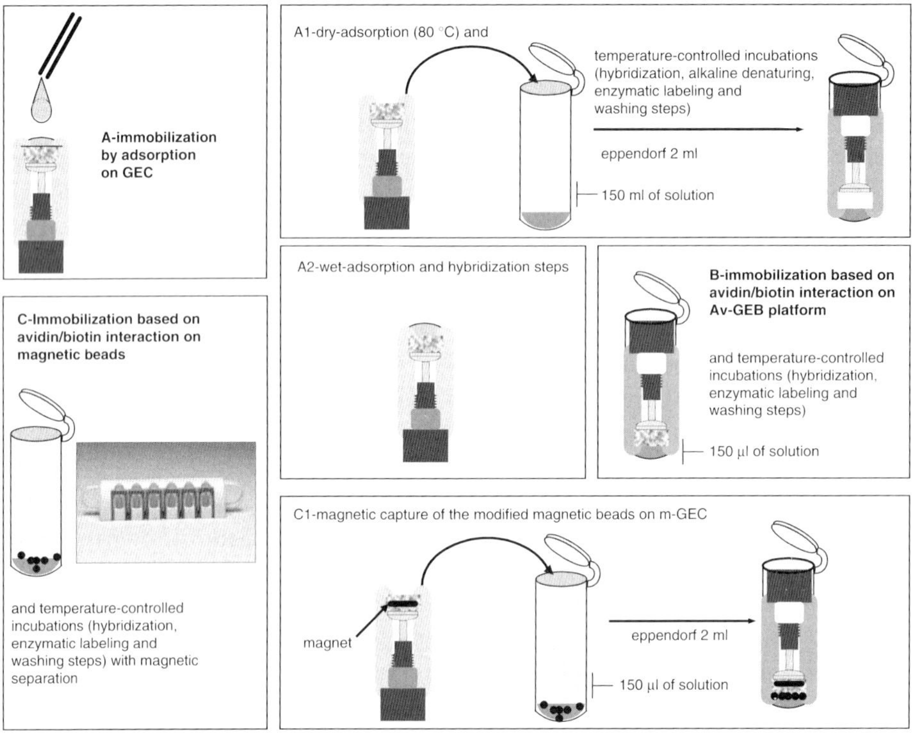

Fig. 21.1. Schematic representation of the manipulation of DNA biosensors using different strategies. (A) DNA biosensors modified with DNA by (A1) dry-adsorption and (A2) wet-adsorption on GEC platform. (B) DNA biosensors based on the single-point immobilization of biotinylated DNA on Av-GEB universal affinity platform. (C) Immobilization of DNA on magnetic beads followed by (C1) the capture of the modified beads on m-GEC electrode (more details in Pividori and Alegret [58]).

determinant of β-lactamase resistance in *S. aureus* using one- and two-step capture format. According to the results, the one-step capture format is more convenient, as a higher sensitivity was achieved [58]. When compared with other reported genosensor designs using a similar capture format [62] and the same labelling system, the genosensor design based on dry adsorption is simpler and cheaper, showing detection limits of the same order of magnitude. The procedures based on previous activation/modification of the surface transducer and subsequent immobilization, as well as some blocking and washing steps that are tedious, expensive and time-consuming, were avoided using GEC as electrochemical platform. These are the principal advantages of GEC platform with respect to other reported devices [38,50,62].

Moreover, by easily controlling the concentration of the DNA solution being dried on GEC platform, a thick or a thin layer of DNA can be formed on the GEC surface by dry adsorption [59]. Depending on the application of the DNA-modified substrate, a thick or thin DNA layer would be necessary. If a stringency control of non-specific DNA adsorption issues is required, a thick DNA layer is more convenient. However, the yield in hybridization is better on a thin DNA layer [59].

Furthermore, GEC has shown unique and selective adsorption behaviour. While DNA is firmly adsorbed under dry conditions, the wet adsorption of non-specific DNA, proteins, enzymes or other biomolecules is negligible under stirring or convection conditions in solution [58,60,61]. The DNA-modified GEC surface does not require blocking steps to minimize the non-specific adsorption on the free sites of the surface [59] since the non-specific adsorption is very low and similar to the instrumental background noise. Moreover, no blocking reagents are required during hybridization to reduce the non-specific adsorption. It was previously demonstrated that the hybridization signals (as well as the non-specific adsorption signals) were essentially the same when performing the hybridization without blocking reagents and using different blocking commercial solution [58].

21.2.4 Electrochemical genosensing of food pathogens based on DNA wet-adsorption on GEC as electrochemical transducer

Wet adsorption relies on leaving DNA to interact with the carbonaceous surface through physical forces in the presence of water. During wet adsorption, the stabilization of B-DNA is expected to occur on the carbonaceous surface, by keeping the hydration water of the DNA molecule. In this case, the hydrated B-DNA form is stabilized over the GEC surface by weaker forces: as the water is kept on the DNA adsorbed molecule, it can be easily desorbed from the GEC surface if soaked in aqueous solutions.

DNA can be easily immobilized on GEC by simple wet-adsorption onto GEC surface (Fig. 21.1(A2)). A small drop of DNA probe in acetate saline solution pH 4.8 [58] is put onto the surface of a GEC electrode in upright position. The immobilization of the probe was allowed to proceed for 15 min without applying any potential under static conditions. After the inosine-modified DNA probe immobilization, the DNA target was detected by the intrinsic DNA oxidation signal coming from the guanine moieties. Briefly, the procedure consists of the following four steps [63]: (i) *electrochemical pre-treatment of the GEC transducer*; (ii) *inosine-substituted probe immobilization by wet-adsorption on GEC transducer*; (iii) *hybridization with the target*; and (iv) *electrochemical determination based on differential pulse voltammetry (DPV)*, in which the oxidation signal of guanine (or adenine) was measured by scanning from +0.30 to +1.20 V at a pulse amplitude of 100 mV and a scan rate of 15 mV/s. This procedure was demonstrated to be useful for the detection of IS200 element specific for *Salmonella* spp. [63].

Although a thick or a thin layer of DNA can be attached on the surface during dry adsorption by controlling the concentration of the DNA solution being dried, the wet adsorption normally yields a thin DNA layer. Less compact DNA layers with wider gaps exposing free-GEC surface are normally obtained during wet adsorption. As a consequence, the thin-layer DNA/GEC surface required blocking treatment to avoid non-specific adsorption. During wet adsorption, the substrate is progressively modified with negative charges coming from the DNA being adsorbed; thus, rejecting the successive DNA molecules that are approaching the substrate. Wet adsorption thus leads to a ''self-control'' surface coverage and is less stringent than dry adsorption [58].

21.2.5 Electrochemical genosensing of food pathogens based on Av-GEB biocomposite as electrochemical transducer

A rigid and renewable transducing material for electrochemical biosensing, based on avidin bulk-modified graphite–epoxy biocomposite (Av-GEB), can be easily prepared by adding a 2% avidin (or streptavidin) in the formulation of the composite and using dry chemistry techniques, avoiding tedious, expensive and time-consuming immobilization procedures. The rigid conducting biocomposite acts not only as a transducer but also as a reservoir for avidin.

Confocal laser scanning fluorescence microscopy was used to study the exposure of the avidin-specific binding sites in the Av-GEB platform by the immobilization of a small and flexible biotinylated fluorescein molecule as a fluorescence marker. Fluorescence microscopy thus confirms that Av-GEB platform exposes active binding sites for biotin, acting as affinity matrix (Fig. 21.2B). After use, the electrode surface can be renewed by a simple polishing procedure for further uses, highlighting a clear advantage of this new material with respect to surface-modified approaches such as classical biosensors and other common

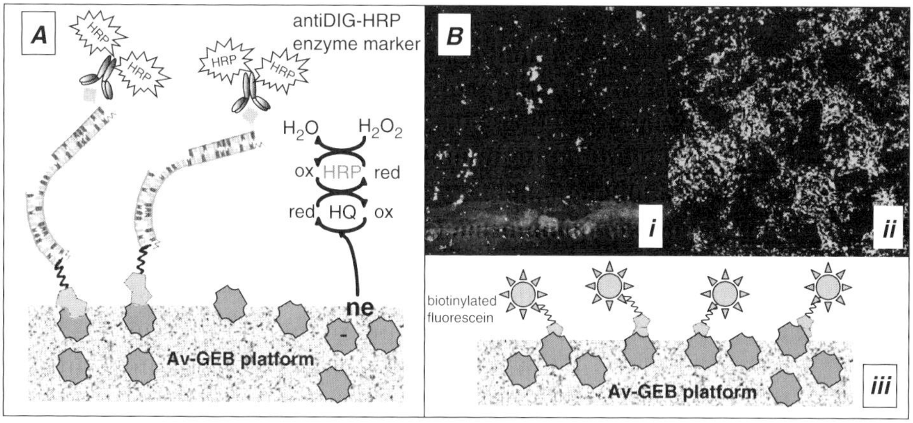

Fig. 21.2. (A) Schematic representation of the electrochemical DNA biosensing procedures based on Av-GEB. (B) Confocal laser scanning fluorescence microphotograph of Av-GEB transducers submitted to (i) non-biotinylated fluorescein (background adsorption) and (ii) 80 pmol of biotinylated fluorescein. Laser excitation was at 568 nm. Voltage: 352 V (more details in Zacco *et al.*, [65]).

biological assays. DNA probe can be easily immobilized on the surface of the avidin-modified transducer through the avidin–biotin reaction, since both nucleic acids as well as short oligonucleotides can be readily linked to biotin without serious effects on their biological, chemical or physical properties. The knowledge about the avidin–biotin interaction has advanced significantly and offers an extremely versatile tool. Moreover, this interaction presents a variety of specific advantages over other single-point immobilization techniques. In particular, the extremely specific and high affinity reaction between biotin and the glycoprotein avidin (association constant $K_a = 10^{15}$ M) leads to strong associations similar to the formation of a covalent bonding. This interaction is highly resistant to a wide range of chemical (detergents, protein denaturants), pH range variations and high temperatures [64]. In addition, much progress has been done in the modification of biomolecules with biotin. Moreover, the strept(avidin) could be considered as a universal affinity biomolecule because it is able to link not only biotinylated DNA or ODNs but also enzymes or antibodies [65].

Biotinylated DNA can be firmly single-point attached in Av-GEB. In this case, a capture format was used in which the immobilization of the biotinylated probe together with the hybridization was performed in a one-step procedure (Figs. 21.1B and 21.2A) [66]. Briefly, the three-step experimental procedure consists of: (i) *one-step immobilization/*

hybridization procedure in which the biotin-labelled capture probe is immobilized onto the electrode surface through a biotin–avidin interaction, while the hybridization with the target and with a second complementary probe—in this case labelled with digoxigenin—is occurring at the same time; (ii) *enzymatic labelling* using as enzyme label the antibody anti-DIG-HRP; and (iii) *amperometric determination* based on the enzyme activity by adding H_2O_2 and using hydroquinone as a mediator.

The utility of Av-GEB platform was demonstrated for the determination of the mecA DNA sequence related with methicillin-resistant *S. aureus* (MRSA) [67] in a simpler and specific manner with respect to previous DNA biosensing devices [58,65,66].

The genosensor design based on Av-GEB not only is able to successfully immobilize onto the electrode surface the mecA biotin-labelled capture probe, while the hybridization with the mecA target and the mecA digoxigenin-labelled probe is occurring at the same time, but also is capable of distinguishing SNPs [58,65].

Compared to genosensors based on GEC, the novelty of this approach is in part attributed to the simplicity of its design, combining the hybridization and the immobilization of DNA in one analytical step. The optimum time for the one-step immobilization/hybridization procedure was found to be 60 min [66]. The proposed DNA biosensor design has proven to be successful in using a simple bulk modification step, hence, overcoming the complicated pre-treatment steps associated with other DNA biosensor designs. Additionally, the use of a one-step immobilization and hybridization procedure reduces the experimental time. Stability studies conducted demonstrate the capability of the same electrode to be used for a 12-week period [66].

21.2.6 Electrochemical genosensing of food pathogens based on magnetic beads and m-GEC electrochemical transducer

One of the most promising materials, which have been developed, is biologically modified magnetic beads [68] based on the concept of magnetic bioseparations. Magnetic beads offer some new attractive possibilities in biomedicine and bioanalysis since their size is comparable to those of cells, proteins or genes [69,70]. Moreover, they can be coated with biological molecules and they can also be manipulated by an external magnetic field gradient. As such, the biomaterial, specific cells, proteins or DNA, can be selectively bound to the magnetic beads and then separated from its biological matrix by using an external magnetic field. Moreover, magnetic beads of a variety of materials and sizes, and modified with a wide variety of surface functional groups, are now commercially available. They have brought novel capabilities to electrochemical immunosensing [71–73]. The magnetic beads have also been used in novel electrochemical genosensing protocols [74,75]. These approaches using magnetic beads for detection of DNA hybridization have been combined with different strategies for the electrochemical detection, such as label-free genosensing [76–78] or different external labels, such as enzymes [39,79], electrochemical indicators [80] or metal tags, e.g. gold or silver nanoparticles [81–83], and using different electrochemical techniques, such as DPV, potentiometric stripping analysis (PSA) or square wave voltammetry (SWV).

Instead of the direct modification of the electrode surface, the biological reactions (as immobilization, hybridization, enzymatic labelling or affinity reactions) and the washing steps can be successfully performed on magnetic beads (Fig. 21.1C). After the modifications, the magnetic beads can be easily captured by magnetic forces onto the surface of GEC electrodes holding a small magnet inside (m-GEC) (Fig. 21.1(C1)). Once immobilized on m-GEC, the hybridization performed on the magnetic beads can be electrochemically revealed using different strategies based on both enzymatic labelling (Fig. 21.3A) and the intrinsic signal coming from the DNA oxidation [84]. In the case of using magnetic beads, a single-point immobilization procedure based on streptavidin–biotin interaction is performed. Biotinylated DNA can be firmly attached on streptavidin-modified magnetic beads in that way. When the electrochemical detection is based on enzymatic activity determination, a capture format was used in which the immobilization of the biotinylated probe together with the hybridization was performed in a one-step procedure. The procedure consists briefly of the following steps: (i) *one-step immobilization/hybridization procedure* in which the biotin-labelled capture probe is immobilized on the streptavidin magnetic beads, while the hybridization with the target and with a second complementary probe—in this case labelled with digoxigenin—is occurring at the same time (30 min at 42°C); (ii) *enzymatic labelling* using as enzyme label the antibody anti-DIG-HRP; (iii) *magnetic capture of the modified magnetic particles*; and (iv) *amperometric determination* based on the enzyme activity by adding H_2O_2 and using hydroquinone as a mediator.

When the DNA target immobilized on the magnetic beads was detected by the intrinsic DNA oxidation signal coming from the guanine

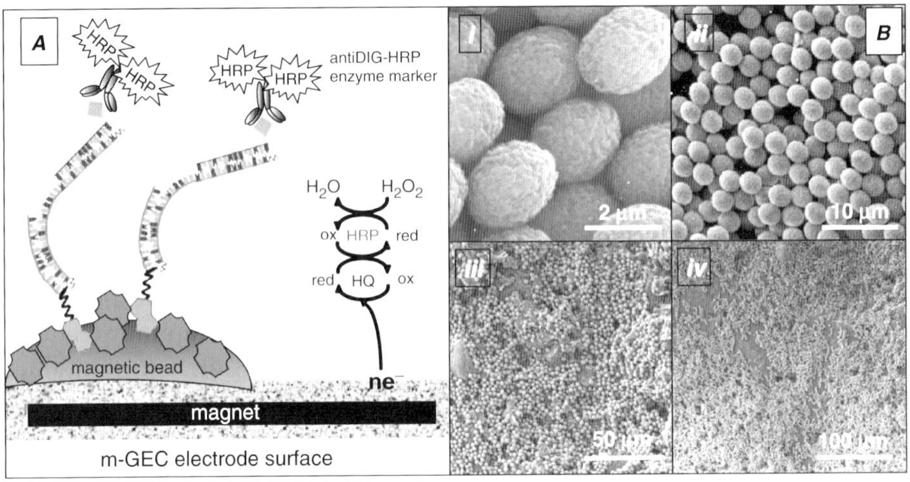

Fig. 21.3. (A) Schematic representation of the electrochemical DNA biosensing procedures based on magnetic beads and m-GEB. (B) Scanning electron microphotographs showing the captured magnetic beads on the surface of m-GEC magneto sensor. Resolution: (i) 2 µm, (ii) 10 µm, (iii) 50 µm and (iv) 100 µm. Acceleration voltage: 10 kV. Number of magnetic beads: 1.6×10^6.

moieties, the procedure consists of the following steps, as previously described in detail [84]: (i) *electrochemical pre-treatment of the m-GEC transducer*; (ii) *biotinylated inosine-substituted probe immobilization on streptavidin magnetic beads*; (iii) *hybridization with the target*; (iv) *magnetic capture of the modified magnetic particles, followed by dry adsorption, was performed during 45 min at 80°C*; (v) *electrochemical determination based on DPV.*

The procedure for electrochemical DNA biosensing based on magnetic beads was also used for the detection of IS200 element specific for *Salmonella* spp [85].

This new electrochemical genomagnetic strategy using magneto electrodes in connection with magnetic particles offers many potential advantages compared to more traditional strategies for detecting DNA. This new strategy takes advantages of working with magnetic particles, such as improved and more effective biological reactions, washing steps and magnetic separation after each step. This electrochemical genomagnetic assay provides much sensitive, rapid and cheaper detection than other assays previously reported. This sensitivity of the GEC with respect to other electrochemical transducer and selectivity conferred by the magnetic separation were also used for the detection of PCR amplicons coming from real samples.

The rapid electrochemical verification of the amplicon coming from the *Salmonella* genome is performed by double-labelling the amplicon during PCR with a set of two labelled PCR primers—one of them with biotin and the other one with digoxigenin (Fig. 21.4, step 1A). During PCR, not only the amplification of the *Salmonella* genome is achieved but also the double-labelling of the amplicon ends with (i) the biotinylated capture primer to achieve the immobilization on the streptavidin-modified magnetic bead and (ii) the digoxigenin signalling primer to achieve the electrochemical detection [86]. Beside this double-labelling

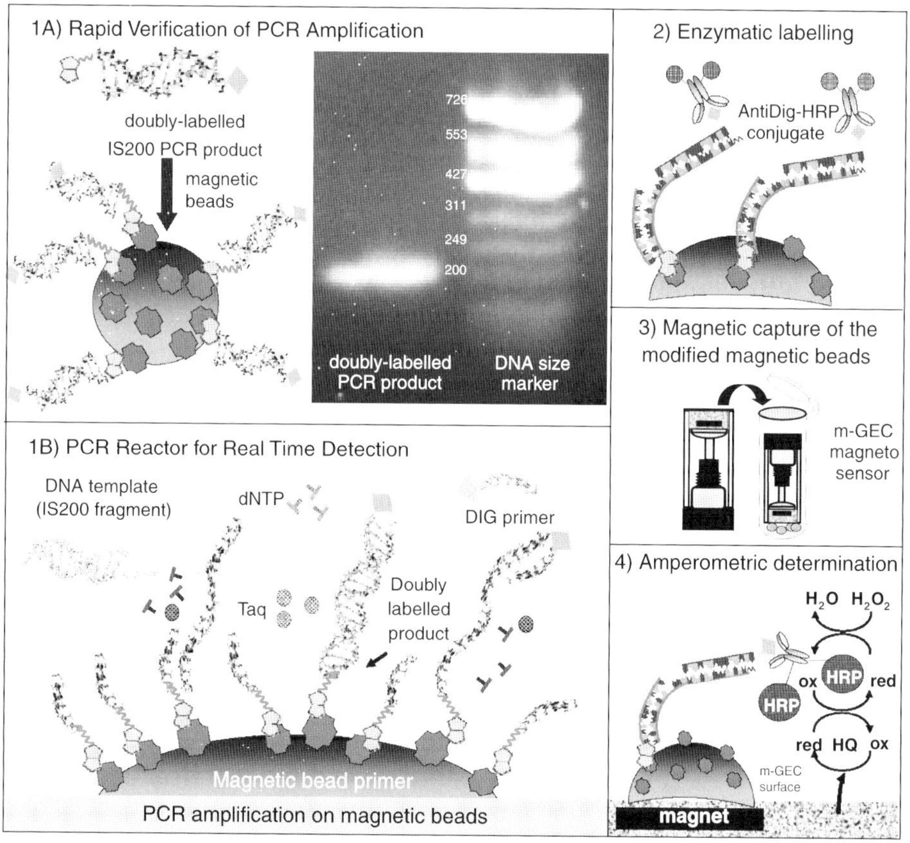

Fig. 21.4. Schematic representation for the detection of *Salmonella* through (1A) a rapid verification of PCR amplification based on the doubly labelled PCR product detection and (1B) real-time PCR reactor based on PCR amplification with magnetic bead primers on streptavidin-modified magnetic beads. (2) Enzymatic labelling; (3) magnetic capture of the modified magnetic beads by the magneto electrode (m-GEC); and (4) chronoamperometric determination are common steps for all of these strategies (1A, 1B).

PCR strategy, a single-labelling PCR strategy with a further confirmation of the amplicon by its hybridization was achieved by performing the PCR with the biotin primer and a further hybridization step with a digoxigenin probe. The procedure consists briefly of the following steps: (i) DNA amplification and double-labelling of *Salmonella* IS200 insertion sequence; (ii) immobilization of the doubly-labelled amplicon in which the biotin extreme of the dsDNA amplicon was immobilized on the streptavidin magnetic beads; (iii) enzymatic labelling using as enzyme label the antibody anti-DIG-HRP capable of bonding the other labelled extreme of the dsDNA amplicon; (iv) magnetic capture of the modified magnetic particles; and (v) amperometric determination (Fig. 21.4, steps 1A, 2, 3 and 4). Further experimental details can be found in Procedure 31 in CD accompanying this book.

Moreover, a PCR reactor for real-time electrochemical detection was also developed. In this case, the amplification and double-labelling is performed directly on the streptavidin magnetic beads by using 'magnetic bead primers' (Fig. 21.4, step 1B). The procedure consists briefly of the following steps: (i) *in situ* DNA amplification and double-labelling of *Salmonella* IS200 insertion sequence on streptavidin-modified magnetic beads by using a magnetic bead primer; (ii) enzymatic labelling using as enzyme label the antibody anti-DIG-HRP capable of bonding the other labelled extreme of the dsDNA amplicon; (iii) magnetic capture of the modified magnetic particles; and (iv) amperometric determination (Fig. 21.4, steps 1B, 2, 3 and 4). Further experimental details can be found in Procedure 32 in CD accompanying this book.

The rapid and sensitive verification of the PCR amplicon related with *Salmonella* can be achieved with 2.8 fmol of amplified product [86]. This strategy can be used for the electrochemical real-time quantification of amplicon since a linear relationship with the amount of amplified product was obtained. Moreover, this strategy is useful only when a unique and specific band is observed by gel electrophoresis, because of the high specificity of the set of primer being used in the PCR for the amplification of the *Salmonella* genome (Fig. 21.4, step 1A). On the contrary, if the set of primers amplifies not only the sequence of interest but also other non-specific fragments, it is necessary to confirm the internal sequence of the amplicon by a second hybridization with a digoxigenin signalling probe. In this case, a single labelling with biotin during PCR was performed followed by a further selective hybridization with a digoxigenin signalling probe. Moreover, magnetic beads primers were used for *in situ* amplification on magnetic beads of the *Salmonella* genome and for further electrochemical detection of the amplified product. The DNA amplification on magnetic beads by using the magnetic bead primer with electrochemical detection of the amplified product demonstrated to be an alternative strategy to the classic detection systems.

21.3 CONCLUSIONS

GEC-based platforms are useful and versatile transducer materials for electrochemical DNA genosensing of amplicon food pathogens.

DNA can be attached directly onto a GEC surface by simple adsorption (the simplest immobilization method and the easiest to automate), avoiding the use of procedures based on previous activation/modification of the surface transducer and subsequent immobilization, which are tedious, expensive and time-consuming. This procedure implies multi-site attachment. Although DNA has been widely attached onto carbonaceous materials, the underlying mechanism of adsorption has not been fully clarified. Adsorption is a complex interplay between the chemical properties, structure and porosity of the substrate surface with the molecule being adsorbed. DNA is a structurally polymorphic macromolecule which, depending on nucleotide sequence and environmental conditions, can adopt a variety of conformations. As a highly negatively charged molecule, dsDNA is considered a hydrophilic molecule. While dsDNA only partially shows its hydrophobic domain through its major and minor grooves or through those sites where dsDNA is open and exposing DNA bases, ssDNA has the hydrophobic bases freely available for their interactions with hydrophobic surfaces. These structural and chemical differences between ssDNA and dsDNA are reflected in different adsorption patterns for both the molecules. The greater size and the more rigid shape of dsDNA with respect to ssDNA are other parameters affecting their adsorption. Beside the DNA molecule and the solid support, the solvent (normally water), in particular the ionic strength, pH and the nature of the solutes play an important role in the adsorption process, mainly in the stabilization of the adsorbed molecule on the substrate.

The hybridization event can be detected both with label-free or enzymatic labelling procedures. The single-point attachment of DNA can be achieved by the immobilization of biotinylated DNA on Av-GEB platform. In this case, a one-step immobilization/hybridization procedure is achieved. The capability of surface regeneration of the biocomposite electrodes allows repeated analyses with the same electrode as the Av-GEB platform can be considered not only the transducer but also the reservoir for the avidin.

The same immobilization strategy based on avidin/biotin linkage can be achieved on magnetic beads. After their efficient modification, the magnetic beads can be easily captured on an m-GEC transducer for the electrochemical determination of the hybridization event.

The sensibility conferred by the m-GEC electrode in connection with the use of magnetic beads and enzymatic labelling results in a rapid, cheap, robust and environmental-friendly device which allows the detection of pathogenic species on food, environmental and clinical samples.

GEC materials present a low non-specific adsorption either for DNA probes or enzyme labels. They do not require blocking steps to minimize the non-specific adsorption on the free sites of the transducer. Although the non-specific adsorption issues can be easily controlled when DNA is simply adsorbed on GEC, stringent conditions can be used for both hybridization and washing when biotin-DNA is immobilized on Av-GEB over longer period of times, due to the stronger avidin/biotin linkage.

DNA biosensors based on GEC meet the demands of genetic analysis, especially in food, biotechnology and pharmaceutical industries, while also generating new possibilities for the development of DNA biosensors that are sensitive, robust, low cost and easily produced.

For all the aforementioned reasons, it is possible to conclude that GECs-based platforms are very suitable for DNA analysis of food pathogens.

However, the nucleic acid-based assays for the detection of food pathogens show problems regarding the sensitivity of the polymerase enzyme to environmental contaminants, difficulties in quantification, the generation of false-positives through the detection of naked nucleic acids, non-viable microorganisms or contamination of samples in the laboratory, and may limit the use of PCR for the direct detection of microbial contamination.

In addition, nucleic acid-based assays are limited in that they can only indicate the genetic potential of a microorganism to produce toxin or to express virulence and do not give any information on toxins in foods or environmental samples. Use of selective enrichment media overcomes the difficulties that may arise in direct PCR by increasing the amount of target DNA and decreasing the amount of inhibitors of PCR that may be present in samples.

With this rapid detection test based on electrochemical DNA biosensor, time is saved because selective plating and biochemical and/or serological confirmatory tests are not necessary. However, a proper sample handling and an initial preenrichment culture are mandatory for the recovery and growth of injured organisms.

ACKNOWLEDGMENTS

M.I.P. and S.A. thank the following former Ph.D. students for valuable work on biosensor research and development involving composite materials, which is partially reported in this chapter: E. Zacco (UAB, Bellaterra), A. Lermo (UAB, Bellaterra), A. Bonanni (UAB, Bellaterra) and Edna Williams (Dublin City University, Ireland). Dr Arzum Erdem (Faculty of Pharmacy, Izmir, Turkey), Dr Susana Campoy and Prof Jordi Barbé (*Unitat de Microbiologia, Departament de Genètica i Microbiologia, Universitat Autònoma de Barcelona*) are also acknowledged for their collaboration. M.I.P. also acknowledges the Universidad Nacional del Litoral (Argentina).

REFERENCES

1 D. Roberts, Food poisoning. Classification,. In: B. Caballero, L. Trugo and P.M. Finglas (Eds.), *Encyclopedia of Food Science and Nutrition*, Academic Press, 2003, pp. 2654–2658.
2 E. Todd, Contamination of food. In: B. Caballero, L. Trugo and P.M. Finglas (Eds.), *Encyclopedia of Food Science and Nutrition*, Academic Press, 2003, pp. 1593–1600.
3 C. Wray, *Salmonella*. Properties and occurrence. In: B. Caballero, L. Trugo and P.M. Finglas (Eds.), *Encyclopedia of Food Science and Nutrition*, Academic Press, 2003, pp. 5074–5079.
4 M. Tietjen and D.Y.C. Fung, *Salmonella* and food safety, *Crit. Rev. Microb.*, 21 (1995) 53–83.
5 T. Humphrey and P. Stephens, *Salmonella* detection. In: B. Caballero, L. Trugo and P.M. Finglas (Eds.), *Encyclopedia of Food Science and Nutrition*, Academic Press, 2003, pp. 5079–5084.
6 D. Ivnitski, I. Abdel-Hamid, P. Atanasov and E. Wilkins, Biosensors for detection of pathogenic bacteria, *Biosens. Bioelectron.*, 14 (1999) 599–624.
7 J.M. Luk, U. Kongmuang, R.S.W. Tsang and A.A. Lindberg, An enzyme-linked immunoadsorbent assay to detect PCR products of the rfbS gene from serogroup D salmonellae: a rapid screening prototype, *J. Clin. Microbiol.*, 35 (1997) 714–718.
8 J. Wan, K. King, H. Craven, C. Mcauley, S.E. Tan and M.J. Coventry, Probelia^{TM} PCR system for rapid detection of *Salmonella* in milk powder and ricotta cheese, *Lett. Appl. Microbiol.*, 30 (2000) 267–271.

9 P. Leonard, P. Hearty, J. Brennan, L. Dunne, J. Quinn, T. Chakraborty and R. O'Kennedy, Advances in biosensors for detection of pathogens in food and water, *Enzyme Microb. Technol.*, 32 (2003) 3–13.
10 D.D.L. Bowtell, Options available—from start to finish—for obtaining expression data by microarray, *Nat. Genet. Suppl.*, 21 (1999) 25–32.
11 F.S. Collins, Microarrays and macroconsequences, *Nat. Genet. Suppl.*, 21 (1999) 2.
12 E.S. Lander, Array of hope, *Nat. Genet. Suppl.*, 21 (1999) 3–4.
13 G.H.W. Sanders and A. Manz, Chip-based microsystems for genomic and proteomic analysis, *Trends Anal. Chem.*, 19 (2000) 364–378.
14 A.K. Deisingh and M. Thompson, Biosensors for the detection of bacteria, *Can. J. Microbiol.*, 50 (2004) 69–77.
15 M.N. Velasco-Garcia and T. Mottram, Biosensors for detection of pathogenic bacteria, *Biosyst. Eng.*, 84 (2003) 1–12.
16 P.D. Patel, (Bio)sensors for measurements of analytes implicated in food safety: a review, *TRAC*, 21 (2002) 96–115.
17 L.D. Mello and L.T. Kubota, Review of the use of biosensors as analytical tools in the food and drink industries, *Food Chem.*, 77 (2002) 237–256.
18 L.A. Terry, S.F. White and L.J. Tigwell, The application of biosensors to fresh produce and the wider food industry, *J. Agric. Food Chem.*, 53 (2005) 1309–1316.
19 D. Ivnitski, I. Abdel-Hamid, P. Atanasov, E. Wilkins and S. Stricker, Application of electrochemical biosensors for detection of food pathogenic bacteria, *Electroanalysis*, 12 (2000) 317–325.
20 M. Mehervar and M. Abdi, Recent development, characteristics, and potential applications of electrochemcial biosensors, *Anal. Sci.*, 20 (2004) 1113–1126.
21 A.G. Gehring, C.G. Crawford, R.S. Mazenko, L.J. Van Houten and J.D. Brewster, Enzyme-linked immunomagnetic electrochemical detection of *Salmonella typhimurium*, *J. Immunol. Methods*, 195 (1996) 15–25.
22 L. Croci, E. Delibato, G. Volpe and G. Palleschi, A rapid electrochemical ELISA for the detection of *Salmonella* in meat samples, *Anal. Lett.*, 34 (2001) 2597–2607.
23 L. Croci, E. Delibato, G. Volpe, D. De Medici and G. Palleschi, Comparison of PCR, electrochemical enzyme-linked immunosorbent assays, and culture method for detecting *Salmonella* in meat products, *Appl. Environ. Microbiol.*, 70 (2004) 1393–1396.
24 A.J. Baeumner, Biosensors for environmental pollutants and food contaminants, *Anal. Bioanal. Chem.*, 377 (2003) 434–445.
25 H. Nakamura and I. Karube, Current research activity in biosensors, *Anal. Bioanal. Chem.*, 377 (2003) 146–168.
26 M.I. Pividori, A. Merkoçi and S. Alegret, Electrochemical genosensor design: immobilization of oligonucleotides onto transducer surfaces and detection methods, *Biosens. Bioelectron.*, 15 (2000) 291–303.

27 K.M. Millan, A.L. Spurmanis and S.K. Mikkelsen, Covalent immobilization of DNA onto glassy carbon electrodes, *Electroanalysis*, 4 (1992) 929–932.
28 K.M. Millan and S.K. Mikkelsen, Sequence-selective biosensor for DNA based on electroactive hybridization indicators, *Anal. Chem.*, 65 (1993) 2317–2323.
29 K.M. Millan, A. Saraullo and S.K. Mikkelsen, Voltammetric DNA biosensor for cystic fibrosis based on a modified carbon paste electrode, *Anal. Chem.*, 66 (1994) 2943–2948.
30 X. Sun, P. He, S. Liu, L. Ye and Y. Fang, Immobilization of single-stranded deoxyribonucleic acid on gold electrode with self-assembled aminoethanethiol monolayer for DNA electrochemical sensor applications, *Talanta*, 47 (1998) 487–495.
31 I. Moser, T. Schalkhammer, F. Pittner and G. Urban, Surface techniques for an electrochemical DNA biosensor, *Biosens. Bioelectron.*, 12 (1997) 729–737.
32 R.G. Nuzzo and D.L. Allara, Adsorption of bifunctional organic disulfides on gold surfaces, *J. Am. Chem. Soc.*, 105 (1983) 4481–4483.
33 T.M. Herne and M.J. Tarlov, Characterization of DNA probes immobilized on gold surfaces, *J. Am. Chem. Soc.*, 119 (1997) 8916–8920.
34 E. Southern, K. Mir and M. Shchepinov, Molecular interactions on microarrays, *Nat. Genet. Suppl.*, 21 (1999) 5–9.
35 M.T. Carter, M. Rodriguez and A.L. Bard, Voltammetric studies of the interaction of metal chelates with DNA. 2. Tris-chelated complexes of cobalt(III) and iron(II) with 1,10-phenanthroline and 2,2′-bipyridine, *J. Am. Chem. Soc.*, 111 (1989) 8901–8911.
36 A. Erdem, K. Kerman, B. Meric and M. Ozsoz, Methylene blue as a novel electrochemical hybridization indicator, *Electroanalysis*, 13 (2001) 219–223.
37 J. Wang, D. Xu, A.-N. Kawde and R. Polsky, Metal nanoparticle-based electrochemical stripping potentiometric detection of DNA hybridization, *Anal. Chem.*, 73 (2001) 5576–5581.
38 L. Alfonta, A.K. Singh and I. Willner, Liposomes labelled with biotin and horseradish peroxidase: a probe for the enhanced amplification of antigen–antibody or oligonucleotide–DNA sensing processes by the precipitation of an insoluble product on electrodes, *Anal. Chem.*, 73 (2001) 91–102.
39 E. Paleček, R. Kizek, L. Havran, S. Billova and M. Fotja, Electrochemical enzyme-linked immunoassay in a DNA hybridization sensor, *Anal. Chim. Acta*, 469 (2002) 73–83.
40 E. Paleček, Oscillographic polarography of nucleic acids and their building blocks, *Naturwiss*, 45 (1958) 186–187.
41 E. Paleček, Oscillographic polarography of highly polymerized deoxyribonucleic acid, *Nature*, 188 (1960) 656–657.

42 J. Wang, X. Cai, C. Jonsson and M. Balakrishnan, Adsorptive stripping potentiometry of DNA at electrochemically pretreated carbon paste electrodes, *Electroanalysis*, 8 (1996) 20–24.
43 A.M. Oliveira Brett, S.H.P. Serrano, I. Gutz, M.A. La-Scalea and M.L. Cruz, Voltammetric behavior of nitroimidazoles at a DNA-biosensor, *Electroanalysis*, 9 (1997) 1132–1137.
44 K. Hashimoto, K. Ito and Y. Ishimori, Novel DNA sensor for electrochemical gene detection, *Anal. Chim. Acta*, 286 (1994) 219–224.
45 D.W. Pang and H.D. Abruña, Micromethod for the investigation of the interactions between DNA and redox-active molecules, *Anal. Chem.*, 70 (1998) 3162–3169.
46 P.M. Armistead and H.H. Thorp, Modification of indium tin oxide electrodes with nucleic acids: detection of attomole quantities of immobilized DNA by electrocatalysis, *Anal. Chem.*, 72 (2000) 3764–3770.
47 F. Jelen, B. Yosypchuk, A. Kourilová, L. Novotný and E. Paleček, Label-free determination of picogram quantities of DNA by stripping voltammetry with solid copper amalgam mercury in the presence of copper, *Anal. Chem.*, 74 (2002) 4788–4793.
48 M. Fojta and E. Paleček, Supercoiled DNA modified mercury electrode: a highly sensitive tool for the detection of DNA damage, *Anal. Chim. Acta*, 342 (1997) 1–12.
49 Z. Li, H. Wang, S. Dong and E. Wang, Electrochemical investigation of DNA adsorbed on conducting polymer modified electrode, *Anal. Sci.*, 13 (1997) 305–310.
50 T. Lumley-Woodyear, C.N. Campbell, E. Freeman, A. Freeman, G. Georgiou and A. Heller, Rapid amperometric verification of PCR amplification of DNA, *Anal. Chem.*, 71 (1999) 535–538.
51 M.D. Rubianes and G.A. Rivas, Carbon nanotubes paste electrode, *Electrochem. Commun.*, 5 (2003) 689–694.
52 S. Alegret, Rigid carbon–polymer biocomposites for electrochemical sensing, *Analyst*, 121 (1996) 1751–1758.
53 F. Céspedes, E. Martínez-Fàbregas and S. Alegret, New materials for electrochemical sensing. I. Rigid conducting composites, *Trends Anal. Chem.*, 15 (1996) 296–304.
54 R.N. Adams, Carbon paste electrodes, *Anal. Chem.*, 30 (1958) 1576.
55 S.R. Rasmussen, M.R. Larsen and S.E. Rasmussen, Covalent immobilization of DNA onto polystyrene microwells: the molecules are only bound at the 5′ end, *Anal. Biochem.*, 198 (1991) 138–142.
56 D.W. Pang, M. Zhang, Z.L. Wang, Y.P. Qi, J.K. Cheng and Z.Y. Liu, Modification of glassy carbon and gold electrodes with DNA, *J. Electroanal. Chem.*, 403 (1996) 183–188.
57 M.I. Pividori and S. Alegret, DNA adsorption on carbonaceous materials. In: C. Wittman (Ed.), *Immobilization of DNA on Chips I, Topics in Current Chemistry*, Vol. 260, Springer Verlag, Heidelberg, Berlin, 2005, pp. 1–36.

58 M.I. Pividori and S. Alegret, Electrochemical genosensing based on rigid carbon composites. A review, *Anal. Lett.*, 38 (2005) 2541–2565.
59 M.I. Pividori and S. Alegret, Graphite–epoxy platforms for electrochemical genosensing, *Anal. Lett.*, 36 (2003) 1669–1695.
60 M.I. Pividori, A. Merkoçi and S. Alegret, Graphite–epoxy composites as a new transducing material for electrochemical genosensing, *Biosens. Bioelectron.*, 19 (2003) 473–484.
61 M.I. Pividori, A. Merkoçi, J. Barbé and S. Alegret, PCR-genosensor rapid test for detecting *Salmonella*, *Electroanalysis*, 15 (2003) 1815–1823.
62 C.N. Campbell, D. Gai, N. Cristler, C. Banditrat and A. Heller, Enzyme-amplified amperometric sandwich test for RNA and DNA, *Anal. Chem.*, 74 (2002) 158–162.
63 A. Erdem, M.I. Pividori, M. Del Valle and S. Alegret, Rigid carbon composites: a new transducing material for label-free electrochemical genosensing, *J. Electroanal. Chem.*, 567 (2004) 29–37.
64 M.L. Jones and G.P. Kurzban, Noncooperativity of biotin binding to tetrameric streptavidin, *Biochemistry*, 34 (1995) 11750–11756.
65 E. Zacco, M.I. Pividori and S. Alegret, Electrochemical biosensing based on universal affinity biocomposite platforms, *Biosens. Bioelectron.*, 21 (2006) 1291–1301.
66 E. Williams, M.I. Pividori, A. Merkoçi, R.J. Forster and S. Alegret, Rapid electrochemical genosensor assay using a streptavidin carbon polymer biocomposite electrode, *Biosens. Bioelectron.*, 19 (2003) 165–175.
67 Y. Katayama, T. Ito and K. Hiramatsu, A new class of genetic element, staphylococcus cassette chromosome mec, encodes methicillin resistance in *Staphylococcus aureus*, *Antimicrob. Agents Chemother.*, 44 (2000) 1549–1555.
68 B.I. Haukanes and C. Kvam, Application of magnetic beads in bioassays, *Biotechnology*, 11 (1993) 60–63.
69 K. Larsson, K. Kriz and D. Kriz, Magnetic transducers in biosensors and bioassays, *Analusis*, 27 (1999) 617–621.
70 S. Solé, A. Merkoci and S. Alegret, New materials for electrochemical sensing III. Beads, *Trends Anal. Chem.*, 20 (2001) 102–110.
71 E. Zacco, M.I. Pividori, S. Alegret, R. Galve and M.-P. Marco, Electrochemical magnetoimmunosensing strategy for the detection of pesticides residues, *Anal. Chem.*, 78 (2006) 1780–1788.
72 A.G. Gehring, J.D. Brewster, P.L. Irwin, S.I. Tu and L.J. Van Houten, 1-Naphthyl phosphate as an enzymatic substrate for enzyme-linked immunomagnetic electrochemistry, *J. Electroanal. Chem.*, 469 (1999) 27–33.
73 M. Dequaire, C. Degrand and B. Limoges, An immunomagnetic electrochemical sensor based on a perfluorosulfonate-coated screen-printed electrode for the determination of 2,4-dichlorophenoxyacetic acid, *Anal. Chem.*, 71 (1999) 2571–2577.

74 E. Palecek and F. Jelen, Electrochemistry of nucleic acids and development of DNA sensors, *Crit. Rev. Anal. Chem.*, 32(3) (2002) 261–270.
75 J. Wang and A. Erdem, An overview of magnetic beads used in electrochemical DNA biosensors. In: Y.G. Gogotsi and I.V. Uvarova (Eds.), *NATO-ARW Book, Nanostructured Materials and Coatings for Biomedical and Sensor Applications*, 4–8 August 2002, Kluwer Academic Publication, Kiev, Ukraine, 2003, pp. 307–314.
76 J. Wang, A.-N. Kawde, A. Erdem and M. Salazar, Magnetic bead-based label-free electrochemical detection of DNA hybridization, *Analyst*, 126 (2001) 2020–2024.
77 J. Wang and A.-N. Kawde, Magnetic-field stimulated DNA oxidation, *Electrochem. Commun.*, 4 (2002) 349–352.
78 J. Wang, G.-U. Flechsig, A. Erdem, O. Korbut and P. Grundler, Labelfree DNA hybridization based on coupling of a heated carbon paste electrode with magnetic separations, *Electroanalysis*, 16(11) (2004) 928–931.
79 J. Wang, D. Xu, A. Erdem, R. Polsky and M. Salazar, Genomagnetic electrochemical assays of DNA hybridization, *Talanta*, 56 (2002) 931–938.
80 E. Palecek, S. Billova, L. Havran, R. Kizek, A. Miculkova and F. Jelen, DNA hybridization at microbeads with cathodic stripping voltammetric detection, *Talanta*, 56 (2002) 919–930.
81 J. Wang, D. Xu and R. Polsky, Magnetically-induced solid-state electrochemical detection of DNA hybridization, *J. Am. Chem. Soc.*, 124 (2002) 4208–4209.
82 J. Wang, L. Guodong, R. Polsky and A. Merkoci, Electrochemical stripping detection of DNA hybridization based on cadmium sulfide nanoparticle tags, *Electrochem. Commun.*, 4 (2002) 722–726.
83 J. Wang, L. Guodong and A. Merkoci, Particle-based detection of DNA hybridization using electrochemical stripping measurements of an iron tracer, *Anal. Chim. Acta*, 482 (2003) 149–155.
84 A. Erdem, M.I. Pividori, A. Lermo, A. Bonanni, M. del Valle and S. Alegret, Genomagnetic assay based on label-free electrochemical detection using magneto-composite electrodes, *Sens. Actuators B*, 114 (2006) 591–598.
85 M.I. Pividori, A. Lermo, S. Hernandez, J. Barbé, S. Alegret and S. Campoy, Rapid electrochemical DNA biosensing strategy for the detection of food pathogens based on enzyme-DNA-magnetic bead conjugate, *Afinidad*, 62 (2006) 13–18.
86 A. Lermo, S. Campoy, J. Barbé, S. Hernández, S. Alegret and M.I. Pividori, *In situ* DNA amplification with magnetic primers for the electrochemical detection of food pathogens, *Biosensors and Bioelectronics*, 22 (2007) 2010–2017.

Chapter 22

Electrochemical immunosensing of food residues by affinity biosensors and magneto sensors

María Isabel Pividori and Salvador Alegret

22.1 INTRODUCTION

22.1.1 Contamination of food, food residues, and food safety

Contaminated food is one of the most widespread public health problems of the contemporary world and causes considerable mortality [1]. In recent years, a number of high-profile food-safety emergencies have shaken consumer confidence in the production of food and have focused attention on the way food is produced, processed, and marketed. Contaminants in foods can be grouped according to their origin and nature. Most of the agents found in food are natural contaminants from environmental sources, but some are deliberate additives [2]. While the term "contaminant" covers harmful substances or microorganisms that are not intentionally added to food, chemicals are also added during food processing in the form of "additives". Contaminants may enter the food accidentally during growth, cultivation, or preparation, accumulate in food during storage, form in the food through the interaction of chemical components, or may be concentrated from the natural components of the food. Additives were at one time a major concern, but today, microbiological issues are the greatest [3], followed by pesticide and animal-drug residues and antimicrobial drug resistance [4]. Although microbiological issues are the greatest concern, many consumers are worried about the long-term impact of mixtures of chemicals additives (such as pesticides, toxic metals, veterinary drug residues, flavorings, and colors) and chronic as well as acute effects on vulnerable groups [5].

Comprehensive Analytical Chemistry 49
Alegret and Merkoçi (Eds)
Volume 49 ISSN: 0166-526X DOI: 10.1016/S0166-526X(06)49022-X

Pesticides are designed to prevent, destroy, repel, or reduce pests (animal, plant, and microbial). Pesticides are categorized according to their mode of action and include insecticides, herbicides, fungicides, acaricides, nematicides, and rodenticides. Pesticides are also used as plant growth regulators and for public-health purposes. Some are selective, impacting only target organisms, whereas others have a broad-range toxicity. Some residues may remain in both fresh produce and processed foods. Atrazine, a triazine herbicide used to control broadleaf and grassy weeds in corn, sorghum, soybeans, sugarcane, pineapple, and other crops, has a low toxicity, but animal studies indicate its potential for endocrine disruption and carcinogenicity [4].

Concerning the veterinary drug residues, antibiotics are added to reduce disease and improve the growth of farm animals and aquaculture fish; these are of less concern for their chemical effect and more for their ability to increase antimicrobial resistance in strains that might subsequently infect humans. There is evidence that antimicrobial resistance is increasing worldwide but particularly in developing countries. The human effect of consuming foods with these drugs is still being debated, but many countries refuse to accept products derived from animals given these drugs. The withdrawal times of such drugs are critical to keep the residues in food as low as possible.

Food regulatory agencies have thus established control programs in order to avoid drug residues to enter the food supply. Food quality and safety can only be insured through the application of quality-control systems throughout the entire food chain. They should be implemented at farm level with the application of good agricultural practices and good veterinary practices at production, good manufacturing practice at processing, and good hygiene practices at retail and catering levels. One of the most effective ways for the food sector to protect public health is to base their food management programs on Hazard Analysis Critical Control Point (HACCP). HACCP consists of a systematic approach to the control of potential hazards in a food operation and aims to identify problems before they occur [5].

Application of the HACCP program is recommended for all food production processes, from small catering procedures to large-scale manufacturing. Once the sources of contamination (the hazards) and the important points at which, and means by which, they can be controlled (critical control points) are identified, and controls introduced, monitored, and verified, the food manufacturers can have greater assurance that they are producing a safe product [2].

22.1.2 Pesticides and drug residues detection methods

As pesticides and drug residues produce undesirable residues, various national and international authorities regulate their use and set maximum residue levels (MRLs) in food [1]. An MRL is the maximum concentration of a food residue and/or its toxic metabolites legally permitted in food commodities and animal feeds. As an example, if pesticides are properly applied at the recommended rates, and crops are only harvested after the appropriate time intervals have elapsed, residue levels are not expected to exceed MRLs. In the EU, the regulation of the agrochemical industry, and the setting of MRLs is currently being harmonized across all member states by the EC. Authorities have also introduced a definition for residues of veterinary medicinal drugs, which, according to EU council regulation 2377/90, is: '*All pharmacologically active substances whether active principles, excipients or degradation products, and their metabolites which remain in food stuffs obtained from animals to which the veterinary medicinal product in question has been administered*'.

The monitoring of the residues in foods is often at the microgram per kilogram level or lower and has to be supported by strict analytical quality-control standards, so that the analysis produces unequivocal, precise, and accurate residue data. An analytical method to be used in food residues determination should accomplish an adequate specificity, sensitivity, linearity, accuracy, and precision at the relevant residue concentration and in appropriate food matrices.

Multiresidue analysis has been carried out using conventional chromatographic methods, such as HPLC, gas chromatography (GC), and capillary electrophoresis. In the special case of veterinary drug residues, conventional microbiological methods can also be performed based on the inhibition of growing of bacteria promoted by the antibiotic.

A classical food residue analysis based on chromatography include the following steps: (i) sampling; (ii) sample preparation/subsampling; (iii) extraction; (iv) clean-up; (v) chromatographic separation and (vi) instrumental determination.

Regarding the sampling in food residues, a representative sample consists of a large number of randomly collected units. Monitoring of pesticide residues for MRL compliance involves analysis of a composite sample, made up of a number of individual units [1]. The storage of a laboratory sample of homogenated or intact tissue may influence the final determined residue content dramatically. Samples should be analyzed without any delay, as some pesticide residues may degrade rapidly.

As an example, some pesticides are known to degrade during the processing of fruit and vegetable samples at ambient temperature. Milk and diluted honey, but also urine or serum used to trace contamination of the animal, may be analyzed in some cases without extensive sample processing. However, solid matrices, including fat, hair, meat, kidney, retina, and skin, however necessitate extraction of residues prior to analysis. Specific solvents are mixed with the homogenized matrix to facilitate solubilization of the residues. Sample extracts not only contain the target analyte(s) but may also contain coextractives, such as plant pigments, proteins, and lipids. These coextractives may have to be removed prior to instrumental analysis to avoid possible contamination of instruments and to eliminate compounds that interfere during the determination step. Clean-up of samples is therefore necessary in most cases to overcome any interference of the analysis method. Adsorption chromatography is used in many residue laboratories for the clean-up of sample extracts.

The final stage of the residue analysis procedures involves the chromatographic separation and instrumental determination. Where chromatographic properties of some food residues are affected by sample matrix, calibration solutions should be prepared in sample matrix. The choice of instrument depends on the physicochemical properties of the analyte(s) and the sensitivity required. As the majority of residues are relatively volatile, GC has proved to be an excellent technique for pesticides and drug residues determination and is by far the most widely used. Thermal conductivity, flame ionization, and, in certain applications, electron capture and nitrogen phosphorus detectors (NPD) were popular in GC analysis. In current residue GC methods, the universality, selectivity, and specificity of the mass spectrometer (MS) in combination with electron-impact ionization (EI) is by far preferred.

High-performance liquid chromatography (HPLC) is increasingly being used for the determination of pesticide and drug residues, as it is especially suited to the analysis of nonvolatile, polar, and thermally labile residues that are difficult to analyze using GC. The resolution achieved on HPLC can be comparatively low, and therefore, the use of selective detection systems may be necessary for reliable residue analysis. Ultraviolet (UV) spectroscopy is the most common choice for detection of residues. Although UV detection is not a very selective technique, it is commonly used for screening purposes due to its low cost, simplicity, and wide application range. Elimination of interferences and optimized chromatography are essential prior to detection in order to enhance the selectivity of UV-based methods. The use of diode array detectors can further enhance the selectivity of UV-detection procedures. Fluorescence detection offers a greater selectivity and sensitivity than UV. The electrochemical detectors are used for a number of residues in relatively clean samples. The on-line combination of HPLC and mass spectroscopy (HPLC-MS) offers a high sensitivity and specificity. There are a number of ionization techniques used to interface HPLC with MS analyzers, of which the most widely used are electrospray and atmospheric pressure chemical ionization.

Some residues require derivatization to enhance the extractability, clean-up, or subsequent chromatographic resolution and determination steps. Instead of chromatography, capillary electrophoresis with a high resolving power may be considered as well.

For regulatory purposes, it is essential that pesticide residues are unequivocally confirmed using MS. However, if an MS method is not available, the sample extract is reanalyzed using a different chromatographic column and/or a different detection system to confirm the initial results [1].

Besides physicochemical methods, the use of microbiological growth-inhibition assays to test meat and milk for the presence of antibiotics residues is popular over a long period of time. These tests use antibiotic-sensitive bacterial reporter strains, such as *Bacillus subtilis* and *Bacillus stearothermophilus* var. *calidolactis*. These bacteria are inoculated under optimal conditions with and without sample. After culturing, results are read from visible inhibition zones or from the color change of the bacterial suspension in agar gels [6].

22.1.3 Rapid detection methods for the food residues. Immunochemical methods

Immunochemical methods [1] are used for rapid screening of an individual or a group of closely related residues. These methods require little or no sample clean-up, require no expensive instrumentation, and are suitable for field use.

The development of immunoassays for the detection of food components and contaminants has progressed rapidly in the last few years [7]. Antibodies against almost all the important food residues compounds are currently available. Classical immunochemical methods such as immunodiffusion and agglutination methods for food analyses generally involve no labeled antigen or antibody. Concentration of the antigen–antibody complex is estimated from the secondary reaction that leads to precipitation or agglutination. These methods are not sensitive, are subject to nonspecific interference, and can only be used for analysis of high-molecular-weight proteins. However, the development of radioimmunoassay (RIA) has widened the scope of immunoassays. This method combines the unique properties of specific antibody–antigen interaction and the use of a radioactive labeled marker to monitor complex formation. Thus, RIA provides specificity, sensitivity, and simplicity, and can be used for analysis of both antigen and haptens. RIA involves the use of a radioactive marker which competes with an analyte in the sample for binding to an antibody. For RIA of high-molecular-weight antigen, either the antigen or antibody molecules can be labeled. It is also common to use a radiolabeled second antibody, i.e., an antibody against the primary antibody. In contrast, labeled hapten is typically used in RIA for low-molecular-weight substances. Although RIA is simple and sensitive, it is limited by the need for a marker with high specific radioactivity, instruments for measuring radioisotopes, and licenses for using radioactive materials and disposal of radioactive materials. Using different nonradioactive labeled markers, a variety of immunoassays, including fluorescence immunoassay (FIA), time-resolved FIA, FIA polarization immunoassay, enzyme immunoassay (EIA), luminescent immunoassay (LIA), metalloimmunoassay (MIA), and viroimmunoassay (VIA) have been developed. Since no radioactive substances are used, the assays avoid the problems encountered in handling radioactivity.

EIA is a general term for immunoassays involving use of an enzyme as a marker for the detection of immunocomplex formation. Enzyme labeling can be achieved by conjugation of the enzyme to an antigen or antibody via periodate oxidation with subsequent reductive alkylation method, cross-linking using glutaraldehyde, or others. Some of the methods used in the conjugation of hapten to proteins can also be used. However, to avoid nonspecific interaction, methods for coupling of protein/hapten to enzyme should be different from the one that had been used for conjugating the hapten to protein for immunization purpose. Although horseradish peroxidase and alkaline phosphatase are the two enzymes most commonly used, others, such as glucose-6-phosphate dehydrogenase, coupled with oxidoreductase and luciferase, glucose oxidase, beta-galactosidase and urease have also been employed. Depending on whether or not the immunocomplex is separated from the free antigen, EIAs are further divided into two types. One type is a homogeneous system, which is called enzyme multiplied immunoassay (EMIT), is based on the modification of enzyme activity occurring when antibody binds with the enzyme-labeled antigen/hapten in solution. Because modification of enzymatic activity is generally not significant, this system is not very sensitive and has not been widely used in food analysis. The other is a heterogeneous system involving separation of free and bound antigen–antibody. In this system, either antigen or antibody is covalently or noncovalently bound to the solid matrix. Nonreacted antibody or antigen is simply removed by washing or centrifugation. The term "enzyme-linked immunosorbent assay" (ELISA) is used for this type of assay. Solid phases such as microtiter plates, cellulose, nylon beads/tubes, nitrocellulose membrane, polystyrene tubes/balls, and modified magnetic beads have been used. In some cases, staphylococcal protein A or protein G is coated on the solid surface, entrapping the antibody for subsequent analysis.

This method is further divided into two major types. One type is competitive ELISA, which can be used for the analysis of both hapten and macromolecule; the other is noncompetitive sandwich-type ELISA, which is only used for divalent and multivalent antigens. Two major types, i.e., direct competitive ELISA (dC-ELISA) and indirect competitive ELISA (inC-ELISA), are used most commonly in food analysis.

In the direct competitive ELISA, the analyte specific antibodies are first coated on a solid phase. The sample or standard solution of analyte is generally incubated simultaneously with the analyte enzyme conjugate or incubated separately in two steps. The amount of enzyme bound to the plate is then determined by incubation with a chromogenic substrate solution. The resulting color/fluorescence, which is inversely proportional to the analyte concentration present in the sample, is then measured instrumentally or by visual comparison with the standards.

A number of quick screening tests based on the ELISA principle described above have been developed. For example, microtiter plate ELISA assay can be completed in less than 20 min. Other approaches involve immobilizing the antibody on a paper disk or other membrane which is mounted either in a plastic card or in a plastic cup, or on the top of a plastic tube. In the "dipstick" assay, antibody or antigen is coated on to a stick, which is then dipped in various reagents for subsequent reactions. Substrates leading to the formation of a water-insoluble chromogenic product are used in these assays. Most of these screening tests are very simple and easy to perform and take 10–15 min to complete. They are designed to provide semiquantitative information at certain cut-off concentrations for the substance in which one is interested. The immunoscreening tests have gained wide application for monitoring residues in foods and versatile assay kits are commercially available.

Immunoassays are available for almost all the important antibiotic residues that might be present in foods. For example, β-lactone antibiotics such as ampicillin, cloxacillin, and penicillin G could easily be measured in milk. Likewise, immunoassays for many pesticides are also available, including 2,4-D, aldicarb, carbendazim, thiabendazole, chlopyrifos, diazinon, endosulfan, and metalaxyl [1]. The triazine immunoassay is now available commercially, and, like most immunoassays, it is specific, sensitive, rapid, and cost-effective [8].

22.1.4 Antibodies as immunological reagents. Immobilization strategies

The antibodies are a group of glycoproteins present in the serum and tissue fluids of all mammals. Five distinct classes of immunoglobulin molecule are recognized in higher mammals, namely, IgG, IgA, IgM, IgD, and IgE. They differ in size, charge, amino acid composition, and carbohydrate content. In addition to the difference between classes, the immunoglobulins within each class are also very heterogeneous. Electrophoretically the immunoglobulins show a unique range of heterogeneity which extends from the γ to the α fractions of normal serum.

Each immunoglobulin molecule is bifunctional. One region of the molecule is concerned with binding to antigen while a different region mediates so-called effector functions. The effector functions are biological activities such as complement activation and cell binding which are related to sites that are distant from the antigen-binding sites (mostly in the Fc region). Effector functions include binding of the immunoglobulin to host tissues, to various cells of the immune system, to some phagocytic cells, and to the first component (C1q) of the classical complement system. Moreover, the region related to the effector function is also involved in the immobilization of antibodies to solid supports for analytical purposes.

The basic structure of all immunoglobulin molecules is a unit consisting of two identical light polypeptide chains and two identical heavy polypeptide chains, as shown in Fig. 22.1. These are linked together by disulphide bonds. The class and subclass of an immunoglobulin molecule are determined by its heavy chain type. The C-terminal half of the light chain (approximately 107 amino acid residues) is constant except for certain allotypic and isotypic variations and is called the CL (constant light chain) region, whereas the N-terminal half of the chain shows much sequence variability and is known as the VL (variable light chain) region. Many proteolytic enzymes cleave antibody molecules into three fragments—two identical Fab (antigen binding) fragments and one Fc (crystallizable) fragment.

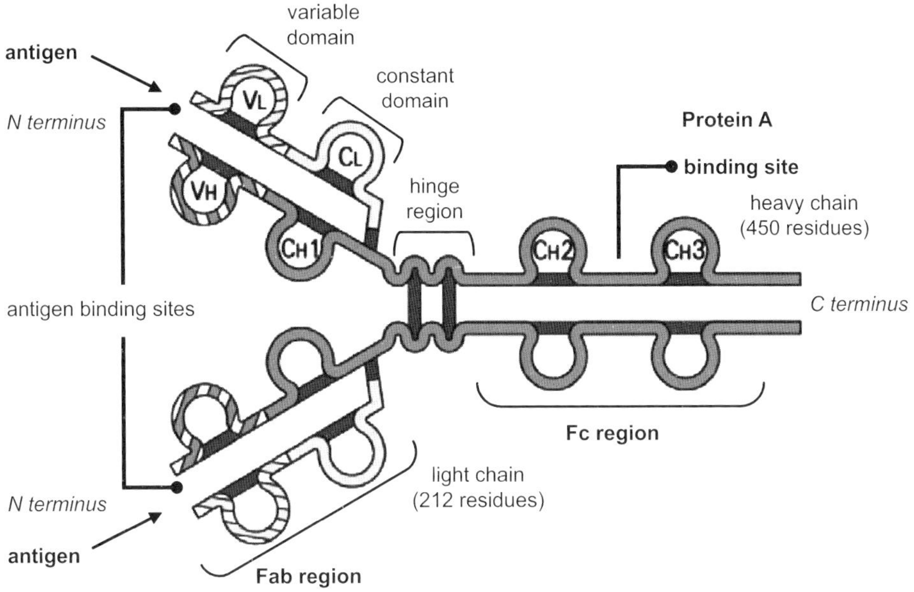

Fig. 22.1. Schematic representation of an IgG antibody. The N-terminal end of IgG_1 is characterized by sequence variability (*V*) in both the heavy and light chains, referred to as the V_H and V_L regions, respectively. The rest of the molecule has a relatively constant (C) structure. The constant portion of the light chain is termed the C_L region. The constant portion of the heavy chain is further divided into three structurally discrete regions: CH_1, CH_2, and CH_3. These globular regions, which are stabilized by intra chain disulphide bonds, are referred to as 'domains'. The sites at which the antibody binds antigen are located in the variable domains. The hinge region is a segment of heavy chain between the CH_1 and CH_2 domains. Flexibility in this area permits the two antigen-binding sites to operate independently. Carbohydrate moieties are attached to the CH_2 domains.

The design of an immunological-based assay involves for the most part the immobilization of an antibody on a proper solid support. The most successful immobilization techniques for antibodies appear to be those involving multisite attachment—physical adsorption—or single-point attachment—mainly (i) covalent immobilization; (ii) protein A interaction; or (iii) strept(avidin)/biotin linkage.

As with other kinds of glycoproteins, physical adsorption is an easy way to attach the antibodies to solid surfaces, since no reagents or modification are required. These features have promoted extensive use of adsorption as immobilization methodology in immunological analysis. The adsorption has been performed on different surface, mainly polystyrene or other membrane materials such as nylon and nitrocellulose. The mainly claimed disadvantages of adsorption with respect to covalent immobilization are that (i) antibodies may be readily desorbed from the substrate, (ii) antigen binding site of antibodies may be unavailable for the antigen since they are bonded to the substrate in multiple sites and thus not oriented away the solid support.

Numerous coupling strategies have been specially developed for the immobilization of antibodies on different surfaces through the formation of defined linkages. Single-point attachment is beneficial to the immunological reaction kinetics, since the antigen-binding site is oriented away from the solid phase. Covalent immobilization can be performed on different surface-modified substrates, through carbodiimide chemistry, by the modification of the NH_2– residues of the basic amino acid moieties. However, by using the common coupling strategies it is not possible to ensure totally free specific binding sites of the immobilized antibody. Some spatial orientations on the surface may prohibit the formation of an antibody–antigen complex. Several approaches have been developed in order to achieve an improved antibody orientation, which leads to a better binding capacity.

Better orientations are achieved with SH– modification of the antibody for the subsequent immobilization on gold-covered substrates. On the other hand, one of the most valuable strategies for the effective immobilization of biomaterial on different substrates is based on the avidin–biotin affinity reaction. Nowadays, the knowledge about this interaction has advanced significantly and offers an extremely versatile tool. The avidin–biotin system has gained great importance over the years as a tool for general application in the biological assays of an almost unlimited number of biological molecules [9–11]. This powerful approach has facilitated the localization, identification, and immobilization of an almost unlimited number of biological molecules [12]. Historically, the development and extensive application of the avidin–biotin system was a definitive breakthrough in the area of nonradioactive labeling and detection of biologically active molecules. The avidin–biotin reaction as an immobilization strategy for biomolecules presents a variety of specific advantages over other single-point immobilization techniques. In particular, ‘‘chicken egg white’’ avidin—a glycosylated and positively charged ($pI \sim 10.5$) protein—and its bacterial analogue streptavidin, share a similarly high affinity ($K_a \sim 10^{15}\,M^{-1}$) for the vitamin biotin [13] similar to the formation of a covalent bonding. Despite the relatively modest sequence homology ~30% identity and 40% overall similarity), the two proteins—avidin and streptavidin—share the same tertiary fold, similar tetrameric quaternary structures, and a nearly identical arrangement of amino acid residues within the respective binding pockets. The interaction with biotin is highly resistant to a wide range of chemical (detergents, protein denaturants), pH range variations, and high temperatures [14]. In addition, the avidin–biotin-based immobilization method maintains the biological activity of the biomolecule being immobilized more successfully than other commonly used methods [15–17]. Much progress has been done in the modification of biomolecules with biotin. A wide range of macromolecules including proteins [18]—enzymes or antibodies—polysaccharides and nucleic acids or short oligonucleotides can be readily linked to biotin without serious effects on their biological, chemical, or physical properties. As such, avidin should be considered as a universal affinity molecule capable to attach different biotinylated biomolecules [19].

Another immobilization strategy is based on the antibody bonding through Fc fragment to protein A or G (Fig. 22.1). The bond strength between Protein A (or G) and an antibody is greatly affected by the antibody classes and subclasses [20–29]. The affinity constant can be varied from strong to weak. The interaction between Protein A of *Staphylococcus aureus* and the Fc region of IgG has also been mapped in some detail. The data suggest a binding site spanning the Cγ2–Cγ3 junction in the Fc region. The binding involves the formation of multiple noncovalent bonds between the protein A and amino acids of the binding Fc site. Considered individually, the attractive forces (hydrogen and electrostatic bonds, Van der Waals, and hydrophobic forces) are weak in comparison with covalent bonds. However, the large number of interactions results in a large total binding energy in the case of IgG. These immobilization strategies allow the binding sites of the antibodies to be oriented away from the solid phase. As a difference with avidin, protein A is able to link the Fc region of many immunoglobulins, thus it is not necessary to have the antibody modified with biotin.

22.1.5 Biosensing as a novel strategy for the rapid detection of food residues

Traditional microbial screening methods have insufficient sensitivities to meet new regulations and classical physicochemical techniques, such as chromatographic methods and mass spectrometry which are often precluded due to the level of experience, skills, and cost involved. Moreover, these methods require laborious extraction and clean up steps that increase analysis time and the risk of errors.

The development of biosensors is a growing area, in response to the demand for rapid 'real-time', simple, selective, and low-cost techniques for food residues. They allow the detection of a broad spectrum of analytes in complex sample matrices, and have shown great promise in areas such as clinical diagnostics, food analysis, bioprocess, and environmental monitoring [30]. A biosensor is a compact analytical device, incorporating a biological sensing element, either closely connected to, or integrated within, a transducer system. Depending on the method of signal transduction, biosensors can also be divided into different groups: electrochemical, optical, thermometric, piezoelectric, or magnetic [31]. HACCP systems which is generally accepted as the most effective system to ensure food safety, can utilize biosensor to verify that the process is under control [32]. The sensitivity of each of the sensor systems may vary depending on the transducer's properties, and the biological recognizing elements. An ideal biosensing device for the rapid detection of food residues is fully automated, inexpensive, and can be routinely used both in the field and the laboratory. Optical transducers are particularly attractive as they can allow direct 'label-free' and 'real-time' detection of food residues. The phenomenon of surface plasmon resonance (SPR) has shown good biosensing potential and many commercial SPR systems are now available. The Pharmacia BIAcore^{TM} (a commercial SPR system) is so far the most-reported method for biosensing of food residues in food and it is based on optical transducing. As an example, the SPR biosensor was compared with existing methods (microbial inhibitor assays, microbial receptor assays, ELISA, HPLC) for detection of sulfamethazine residues in milk by an immunological reaction [33,34]. BIAcore has indicated the occurrence of sulfamethazine at a concentration below the detection limit of HPLC and offered sufficient advantages (no sample preparation, high sensitivity, and rapid and full analysis in real time) to be an alternative for the control of residues and contaminants in food. A similar commercial transducer system was also reported for the determination of β-lactam antibiotics [35–37], multisulfonamide residues [38], and chloramphenicol residues [39]. In the case of pesticides in food, they are mainly based on both the biosensing of the enzyme inhibition by the pesticide [40,41] as well as the immunosensing of the pesticide performed with the specific antibody with SPR transducer [42–44].

However, electrochemically based transduction devices are more robust, easy to use, portable, and inexpensive analytical systems [45]. Furthermore, electrochemical biosensors can operate in turbid media and offer comparable instrumental sensitivity. Many electrochemical sensing and biosensing devices were reported [46–48].

However, regarding the molecular recognition of food residues, immunological reagents are mainly used as molecular receptor in order to obtain a useful signal after the immunological reaction on the transducer element [49,50].

Electrochemical immunosensors can meet the demands of food control, offering considerable promise for obtaining information in a faster, simpler, and cheaper manner compared to traditional methods. Such devices possess great potential for numerous applications, ranging from decentralized clinical testing to environmental monitoring, food safety, and forensic investigations. The following section is focused on electrochemical immunosensing for food residues.

22.2 ELECTROCHEMICAL BIOSENSING OF FOOD RESIDUES BASED ON UNIVERSAL AFFINITY BIOCOMPOSITE PLATFORMS

22.2.1 Transducer materials for electrochemical immunoassay and analytical strategies

Carbonaceous materials such as carbon paste, glassy carbon, and pyrolitic graphite are the most popular choice of electrodes used in electrochemical biosensing design. However, the use of platinum, gold, and other continuous conducting metal substrates has been also reported. Most biosensor transducers require pretreatment of their surfaces. Moreover, the transducing materials have to be surface-modified with the biological molecule. This modification process can be tedious, expensive (because they involve many reagents), time-consuming, and difficult to reproduce. Although almost all these transducing materials may be successfully used for electrochemical biosensing, this chapter is focused on rigid carbon composite and biocomposite and their properties for the immobilization of antibodies.

Carbon composites result from the combination of carbon with one or more dissimilar materials. Each individual component maintains its original characteristics while giving the composite distinctive chemical, mechanical, and physical properties. The capability of integrating various materials is one of their main advantages. The main features of carbon composites are explained in Chapter 21 [3]. Regarding their mechanical properties carbon composites can be classified as rigid [51,52] or soft composites—the carbon pastes [53]. The composites are also classified by the arrangement of its particles, which can be (i) dispersed or (ii) randomly grouped in clearly defined conducting zones within the insulating zones. Rigid composites are obtained by mixing graphite powder with a nonconducting polymeric matrix obtaining a soft paste that becomes rigid after a curing step [51,52]. They could be classified according to the nature of the binder or the polymeric matrix, in epoxy composites, methacrylate composites, or silicone composites.

Rigid conducting graphite-polymer composites (GEC) and biocomposites (GEB) have been extensively used in our laboratories and shown to be a suitable material for electrochemical (bio)sensing due to their unique physical and electrochemical properties. In particular, we have used graphite-epoxy composite (GEC) made by mixing the nonconducting epoxy resin (Epo-Tek, Epoxy Technology, Billerica, MA, USA) with graphite powder (particle size below 50 μm). An ideal material for electrochemical immunosensing should allow an effective immobilization on its surface, a robust immunological reaction, a negligible nonspecific adsorption, and a sensitive detection of the immunological event. GEC fulfills all these requirements. GECs present numerous advantages over more traditional carbon-based materials: higher sensitivity, robustness, and rigidity. Additionally, the surface of GEC can be regenerated by a simple polishing procedure. Unlike carbon paste, the rigidity of GEC permits the design of different configurations, and these materials are compatible with nonaqueous solvents.

A rigid and renewable transducing material for electrochemical immunosensing of pesticide residues, based on bulk-modified GEB, can be easily prepared by adding 2% of avidin, Protein A, or anti-atrazine antibodies in the formulation of the composite to obtain Av-GEB, ProtA-GEB, and Ab-GEB, respectively, and using dry chemistry techniques, avoiding tedious, expensive, and time-consuming immobilization procedures. The rigid conducting biocomposite acts not only as a transducer, but also as a reservoir for the biomolecule.

After use, the electrode surface can be renewed by a simple polishing procedure for further uses, highlighting a clear advantage of this new material with respect to surface-modified approaches such as classical biosensors and other common biological assays. Biotinylated antiatrazine antibodies can be easily immobilized on the surface of the avidin-modified transducer through the avidin–biotin reaction since antibodies can be readily linked to biotin without serious effects on their biological, chemical, or physical properties. Moreover, antiatrazine antibodies can be easily immobilized on the surface of the Protein A-modified transducer without any modification of the antibodies.

In these cases, a competitive immunological assay was used. Briefly, the three-step experimental procedure consists of: (i) immobilization

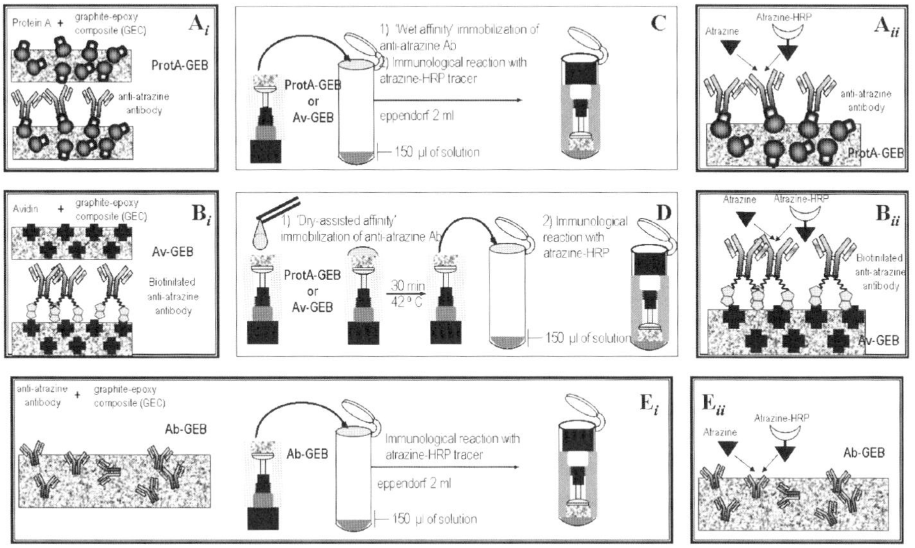

Fig. 22.2. The different biocomposite materials (protA-GEB, Av-GEB, Ab-GEB) and the different strategies for the immobilization of anti-atrazine antibodies on the surface of the electrochemical transducer for the electrochemical biosensing of atrazine. Further experimental details can be found in the text (more details are given in Zacco *et al.*).

procedure in which the antibody or the biotin-labeled antibody is immobilized onto the electrode surface through the Fc-Protein A or biotin–avidin interaction; (ii) competitive immunological reaction between the atrazine and an atrazine-HRP enzymatic tracer; (iii) amperometric determination based on the enzyme activity by adding H_2O_2 and using hydroquinone as a mediator.

Three different strategies for the immobilization of the antibodies on the biosensor were used, as shown in Fig. 22.2. The first one relies on 'wet-affinity' immobilization (Fig. 22.2C). It was performed by dipping the electrodes (ProtA-GEB or Av-GEB) in Eppendorf tubes containing 150 μL of the nonmodified or the biotinylated antibodies, respectively, in order to achieve the immobilization based on the affinity reaction on the surface of ProtA-GEB (Fig. 22.2Aii) or Av-GEB (Fig. 22.2Bii) electrodes, respectively, with slight agitation. The second strategy relies on 'dry-assisted affinity' immobilization (Fig. 22.2D), and it was performed by adding 20 μL of the nonmodified or the biotinylated antibodies to the surface of ProtA-GEB (Fig. 22.2Aii) or Av-GEB (Fig. 22.2Bii) electrodes, at 40°C, until dryness (approximately for 30 min) under static conditions. The third strategy is based on the use of Ab(antiatrazine)-GEB

(Fig. 22.2Ei) as immunosensing transducer. In this case it is not necessary to proceed to the surface immobilization of the antibodies since they were retained on the bulk of the immunocomposite (Fig. 22.2Eii). Briefly, the two-step experimental procedure consist of (i) competitive immunological reaction between the atrazine and the atrazine-HRP enzymatic tracer; (ii) amperometric determination based on the enzyme activity by adding H_2O_2 and using hydroquinone as a mediator. The performance of the electrochemical immunosensing strategy based on universal affinity biocomposite platforms was successfully evaluated using spiked real orange juice samples.

The competitive electrochemical immunosensing strategy can easily reach the required LOD for potable water orange juice (MRL 0.1 $\mu g L^{-1}$) established by the European Community directives with very simple sample pretreatments. The orange juice samples spiked with atrazine was first diluted five times with PBST pH 7.5 buffer and then filtered through 0.22 μm filter before measurement. The features of these approaches are described in detail in the following section.

22.2.1.1 Protein A graphite-epoxy biocomposite

One of the most important features in the immunosensor design is the proper choice of the immobilization method for keeping the affinity of the antibodies. As was previously demonstrated for Protein A, when the antibodies are immobilized through their Fc fragment to Protein A (or G), their F_{ab} binding sites are mostly oriented away from the solid phase. As Protein A is able to link the Fc region of different antibodies, there is no need to modify the antibody with biotin. As an antecedent, we have previously demonstrated the utility of Protein A biocomposite (ProtA-GEB) for the universal attachment of antibodies with different specificities [54].

The antiatrazine antibodies can be easily immobilized by 'wet affinity' immobilization on ProtA-GEB while the nonspecific adsorption of the antibody on the polymeric matrix is negligible [55]. However, a better immobilization yield is achieved if 'dry-assisted' affinity immobilization of antiatrazine antibodies on ProtA(2%)-GEB is performed. It can be explained because of the nature of the bond between Protein A and the Fc part of the antibody. Among the different kinds of binding forces (hydrogen bonding, electrostatic forces, Van der Waals, and hydrophobic bonds), the latter may contribute up to half the total strength of the bond. Hydrophobic bonds rely on the association of nonpolar, hydrophobic groups so that contact with water molecules is minimized. Comparing 'wet-affinity' immobilization, i.e., the affinity immobilization in the presence of water, and 'dry-assisted affinity' immobilization, i.e., the affinity immobilization in the absence of water in order to favor hydrophobic bonds, the latter has shown better immobilization performance. The results confirm that the absence of water (i.e., 'dry-assisted' affinity immobilization) favors the hydrophobic bonds, and renders the binding capacity of the biocomposite higher than in the case of 'wet-affinity' immobilization.

Another way to attach the specific antibody is dry-adsorption on GEC. Like many other biomolecules that were previously dry-adsorbed on GEC (Protein G, Protein A, IgG, DNA, Avidin), antiatrazine antibodies can be easily adsorbed on GEC by dry-adsorption while keeping its specific binding capacity [54]. However, dry adsorption on GEC does not provide a well-oriented layer of specific antibodies while 'dry-assisted' affinity immobilization on ProtA-GEB provides a compact and well-oriented layer of specific antibodies.

22.2.1.2 Avidin graphite-epoxy biocomposite

The Av-GEB could be considered as a universal immobilization platform where biotinylated antibodies can be captured by means of the high-affinity (strept)avidin–biotin reaction ($K_a \sim 10^{15}\,M^{-1}$). The immobilization of antiatrazine biotinylated antibody can be performed either through 'wet-affinity' immobilization as well as 'dry-assisted affinity immobilization'. In contrast with ProtA-GEB, although a higher signal was obtained with dry-assisted affinity immobilization, the performance of both strategies for immobilizing biotinylated antibodies on Av-GEB is almost the same. This could be explained because of the high K_a between avidin and biotin. Although a great influence of water in the immobilization efficiency of antibodies through the Protein A and the constant fragment of the antibodies was found, such effect was not observed when the immobilization was performed through the avidin–biotin interaction. The nature of the bonding depending on avidin showed to be not as dependent on hydrophobic bonds as in the case of Protein A.

22.2.1.3 Antiatrazine graphite-epoxy biocomposite

The antiatrazine antibodies can be directly included in the formulation of the biocomposite (Fig. 22.2Ei). In this approach, the proper orientation of the antibody can be an issue. It was found the antiatrazine antibody is available for the immunological reaction when it was included in the formulation of the biocomposite. However, the biocomposite surface showed to be different after each surface renewal step, according to the reproducibility of the assay. As a result, the availability of the antibodies binding site on the surface is not always reproducible in the immunocomposite, suggesting that the antibodies are randomly oriented, and not always available in the same rate for the immunological reaction.

22.3 ELECTROCHEMICAL BIOSENSING OF FOOD RESIDUES BASED ON MAGNETIC BEADS AND M-GEC ELECTROCHEMICAL TRANSDUCER

22.3.1 Transducer material for electrochemical immunoassay, magnetic beads, and analytical strategies

One of the most promising materials, which have been developed, is biologically modified magnetic beads [56] based on the concept of magnetic bioseparation. Magnetic beads offer some new attractive possibilities in biomedicine and bioanalysis since their size is comparable to those of cells, proteins, or genes [57,58]. Moreover, they can be coated with biological molecules and they can also be manipulated by an external magnetic field gradient. As such, the biomaterial, specific cells, proteins, or DNA, can be selectively bound to the magnetic beads and then separated from its biological matrix by using an external magnetic field. Moreover, magnetic beads of a variety of materials and sizes, and modified with a wide variety of surface functional groups are now commercially available. They have brought novel capabilities to electrochemical immunosensing [59–61].

Instead of the direct modification of the electrode surface, the biological reactions (as immobilization, immunological, enzymatic labeling, or affinity reactions) and the washing steps can be successfully performed on magnetic beads (Fig. 22.3A). After the modifications, the magnetic beads can be easily captured by magnetic forces onto the surface of GEC electrodes holding a small magnet inside (m-GEC) (Fig. 22.3Ai). Once immobilized on m-GEC, the immunological reaction performed on the magnetic beads (Fig. 22.3Aiii) can be electrochemically revealed using enzymatic labeling.

The immunological determination of atrazine in orange juice samples was performed, by using antiatrazine-specific antibodies immobilized on magnetic beads, and an atrazine-HRP tracer for the electrochemical detection.

Different strategies for the attachment of specific antiatrazine antibodies on different magnetic beads have been performed: (i) covalent attachment based on MP-COOH magnetic particles; (ii) covalent

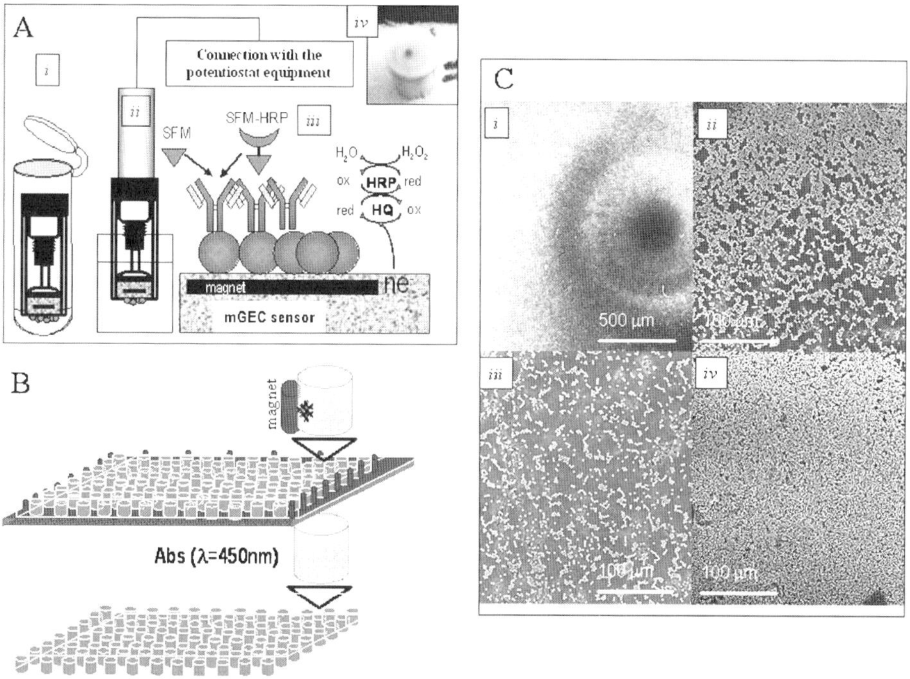

Fig. 22.3. (A) Schematic representation of the electrochemical magneto immunosensing strategy for the detection of sulfonamides (SFM). After the immunoreaction, the antibody-modified magnetic beads were captured for the m-GEC electrode (i). Chemical reactions occurring at the m-GEC surface polarized at −0.150 V (vs. Ag/AgCl) upon the addition of H_2O_2 in the presence of mediator (HQ) (ii and iii). Appearance of the m-GEC electrode covered with the magnetic beads (iv). Schematic representation of the magneto ELISA strategy for the detection of SFM performed on magnetic microtiter plates (B). Scanning electron microphotographs of tosylated magnetic beads on the surface of m-GEC sensors taken at 0.5 µm (i) and 100 µm (ii, iii, iv, for different surface region) of resolution. Identical acceleration voltage (15 kV) was used in all cases.

attachment based on MB-Tosyl magnetic beads; (iii) affinity interaction with MB-ProtA magnetic beads. Although all the strategies for the immobilization of antiatrazine antibodies were found to be useful, the better immobilization performance was achieved with tosyl-modified magnetic beads [59].

The protocol consists briefly of the following steps: (i) competitive immunological reaction between atrazine and HRP-atrazine marker for the antiatrazine antibody immobilized on the magnetic beads, performed with gentle shaking for 30 min at room temperature; (ii) magnetic capture of the modified magnetic particles; (iii) amperometric determination based on the enzyme activity by adding H_2O_2 and using hydroquinone as a mediator [59].

The orange juice samples spiked with atrazine was diluted and pH adjusted to 7.5, with PBST, pH = 4.5 and filtered through 0.22 μm filter before measurement.

As the European legislation fixes the MRL of atrazine at a value of 0.1 $\mu g L^{-1}$ (ppb), the detection limit of the novel electrochemical magneto immunosensing strategy allows to measure orange juices and potable water samples according to the EC legislation.

A similar strategy was performed for the immunological determination of sulfonamides (SFM) in raw full cream, as well as in all varieties of ultra high temperature (UHT) milk, such as full cream (about 3.25% fat), semiskimmed (about 1.5–1.8% fat), and skimmed (0.1% fat), by using a class specific antisulfonamides (anti-SFM) antibodies immobilized on magnetic beads, and an SFM-HRP tracer for the electrochemical detection [62].

Different strategies for the attachment of specific anti-SFM antibodies on different magnetic beads have been performed: (i) covalent attachment based on MP-COOH magnetic particles; (ii) covalent attachment based on MB-Tosyl magnetic beads; (iii) affinity interaction with MB-ProtA magnetic beads. Although all the strategies for the immobilization of anti-SFM antibodies were found to be useful, the better immobilization performance was achieved with tosyl-modified magnetic beads, as in the case of atrazine determination.

The protocol consists briefly of the following steps: (i) competitive immunological reaction between SFM and HRP-SFM marker for the anti-SFM antibody immobilized on the magnetic beads, performed with gentle shaking for 30 min at room temperature; (ii) magnetic capture of the modified magnetic particles; (iii) amperometric determination based on the enzyme activity by adding H_2O_2 and using hydroquinone as a mediator [62].

The raw full-cream samples spiked with sulfonamide were diluted five times with PBST, while all the UHT samples were processed without any treatment.

As the European legislation fix the MRL of sulfonamide to a value of 100 $\mu g\,kg^{-1}$ (ppb), the detection limit of the novel electrochemical magneto immunosensing strategy allows to measure any kind of milk according to the EC legislation [62].

This new electrochemical genomagnetic strategy using magneto electrodes in connection with magnetic particles offer many potential advantages compared to more traditional strategies. This new strategy takes advantages of working with magnetic particles, such as improved and more effective biological reactions, washing steps, and magnetic separation after each step. This electrochemical genomagnetic assay provides much sensitive, rapid, and cheaper detection than other assays previously reported.

22.4 CONCLUSIONS

GEC based platforms are useful and versatile transducer materials for electrochemical immunosensing of food residues. Graphite-epoxy composites (GEC) and biocomposites (GEB), widely studied by our research group, show excellent transducing features for the detection of atrazine and sulfo-drugs based on electrochemical immunosensing.

Although the specific antibodies can be directly attached onto a GEC surface by simple dry-adsorption (the simplest immobilization method and the easiest to automate), this procedure implies multisite attachment, and improved immobilization yields have been obtained with single-point attachment strategies.

Antiatrazine specific antibodies have been successfully surface-immobilized on both biocomposite materials: avidin and protein A modified GEC (Av-GEB and protA-GEB, respectively). Moreover, the antiatrazine specific antibodies have been bulk-immobilized on a specific immunocomposite for the detection of atrazine. In all cases, using the different biocomposite materials (ProtA-GEB, Av-GEB, and Ab-GEB) followed by a competitive immunological assay for the detection of atrazine, excellent detection limits were achieved. However, the electrochemical signal is greater using the ProtA-GEB due to the higher exposure of the antibody-binding site. Moreover, using the ProtA-GEB as transducer, it is not necessary to perform the previous biotinylation of the antiatrazine antibodies. It was found that the binding capacity of the biocomposite showed to be higher when a 'dry-assisted' affinity strategy was performed. In this case, the hydrophobic bonds—between protein A and the Fc of the antiatrazine antibodies—that rely on the association of nonpolar groups seemed to be favored.

Because of the simplicity of the immunochemical procedure based on the biocomposite affinity platform, this strategy can be suitable for fast semiquantitative and quantitative on-site analysis for the presence of atrazine (or atrazine immunoreactive herbicides) in real samples. The fabrication of the biocomposite-based biosensor can be easily transferred to industrial scale. The fact that the same biocomposite material could be used for the immobilization of many antibodies is an important practical feature to be considered for the massive fabrications of electrochemical biosensing devices. Moreover, this material can be easily prepared through dry chemistry using procedures that can be transferred to mass fabrication establishing a clear advantage for the development of *biokits*. Additionally, the biosensor design based on ProtA-GEB fulfills the requirements desired for these devices: ease of preparation, robustness, sensitivity, low cost of production, ease of miniaturization, and simple use and fast response.

This new material coupled with its compatibility with miniaturization and mass fabrication technologies (screen-printing techniques), and versatility, makes them very attractive as user-friendly devices for quick and on-site analyses in industrial and environmental applications.

Similar detection limits for atrazine were obtained with an electrochemical magneto immunosensing strategy based on the use of tosyl-activated magnetic beads. Magnetic beads show a minimized matrix effect since improved washing steps are achieved. Moreover, the use of magnetic beads is known to greatly improve the performance of the immunological reaction, due to (i) increased surface area and (ii) faster assay kinetics.

The competitive electrochemical immunosensing strategy can easily reach the required LOD for potable water orange juice (MRL 0.1 $\mu g\,L^{-1}$) established by the European Community directives with very simple sample pretreatments.

A similar strategy based on the electrochemical magneto immunosensing is able to detect antibiotic residues in milk at very low concentration and without any sample pretreatment. Covalent coupling of antisulfonamide specific antibodies have been successfully performed on tosyl-modified magnetic beads, achieving an excellent antibody coupling efficiency. The competitive electrochemical magneto immunosensing strategy based on m-GEC electrode can easily reach the required LOD for food samples (0.1 $mg\,L^{-1}$) established by the European Community directives without any or very simple sample pretreatments, such as just a dilution in PBST buffer. Both raw and UHT full cream, semiskimmed, and skimmed milk can be easily determined by the electrochemical magneto immunosensing strategy. The magnetic beads provide an easy way for completely removing the matrix effect and improving the nonspecific adsorption. Moreover, the use of magnetic beads provides improved features regarding sensitivity and selectivity of the assay.

The sensibility conferred by the m-GEC electrode in connection with the use of magnetic beads and enzymatic labeling result in a rapid, cheap, robust, and environment-friendly device which allows the detection of food residues and environmental samples. Because of the simplicity of the immunochemical procedure, these strategies can be suitable for fast semiquantitative and quantitative on-site analysis of the presence of food residues in all kinds of samples, in a simple, low-cost, and user-friendly manner.

For all the aforementioned reasons, it is possible to conclude that GEC-based platforms are very suitable for the immunological analysis of food residues according to the food management programs on Hazard Analysis Critical Control Point (HACCP). Procedures 33 and 34 (see in CD accompanying this book) give experimental details on the detection of pesticide atrazine and sulfonamide residues in food samples.

ACKNOWLEDGMENTS

M.I.P. and S.A. thank the PhD students and E Zacco for her valuable work on biosensor research and development involving composite materials. M.-P Marco is also acknowledged for her collaboration. M.I.P. also acknowledges the support of Universidad Nacional del Litoral (Argentina).

REFERENCES

1 S. Nawaz, Pesticides and herbicides, Residue determination. In: B. Caballero, L. Trugo and P.M. Finglas (Eds.), *Encyclopedia of Food Science and Nutrition*, Academic Press, 2003, pp. 4487–4493.
2 D. Roberts, Food Poisoning. In: B. Caballero, L. Trugo and P.M. Finglas (Eds.), *Encyclopedia of Food Science and Nutrition*, Academic Press, 2003, pp. 2654–2658.
3 M.I. Pividori and S Alegret, Electrochemical genosensing of food pathogens based on graphite epoxy-composite. In: S. Alegret and A. Merkoçi (Eds.), *Comprehensive Analytical Chemistry*, Vol. 49, Elsevier, 2007, pp. 437–464.
4 E. Todd, Contamination of food. In: B. Caballero, L. Trugo and P.M. Finglas (Eds.), *Encyclopedia of Food Science and Nutrition*, Academic Press, 2003, pp. 1593–1600.
5 R. Rooney and P.G. Wall, Food safety. In: B. Caballero, L. Trugo and P.M. Finglas (Eds.), *Encyclopedia of Food Science and Nutrition*, Academic Press, 2003, pp. 2682–2688.
6 A.A. Bergwerff and J. Schloesser, Antibiotics and drugs. Residue determination. In: B. Caballero, L. Trugo and P.M. Finglas (Eds.), *Encyclopedia of Food Science and Nutrition*, Academic Press, 2003, pp. 254–261.

7 F.S. Chu, Immunoassays. Radioimmunoassay and enzyme immunoassay. In: B. Caballero, L. Trugo and P.M. Finglas (Eds.), *Encyclopedia of Food Science and Nutrition*, Academic Press, 2003, pp. 3248–3255.
8 A.M. Au, Pesticides and herbicides. Types, uses, and determination of herbicides. In: B. Caballero, L. Trugo and P.M. Finglas (Eds.), *Encyclopedia of Food Science and Nutrition*, Academic Press, 2003, pp. 4483–4487.
9 M. Wilchek and E.A. Bayer, The avidin biotin complex in bioanalytical applications, *Anal. Biochem.*, 171 (1988) 1–32.
10 M. Wilchek and E.A. Bayer, Applications of avidin–biotin technology. Literature survey, *Methods Enzymol.*, 184 (1990) 14–45.
11 M. Wilchek, E.A. Bayer and O. Livnah, Essentials of biorecognition: the (strept)avidin–biotin system as a model for protein–protein and protein–ligand interaction, *Immunol. Lett.*, 103 (2006) 27–32.
12 T. Huberman, Y. Eisenberg-Domovich, G. Gitlin, T. Kulik, E.A. Bayer, M. Wilchek and O. Livnah, Chicken avidin exhibits pseudo-catalytic properties. Biochemical, structural, and electrostatic consequences, *J. Biol. Chem.*, 276 (2001) 32031–32039.
13 N.M. Green, Avidin and streptavidin, *Methods Enzymol.*, 184 (1990) 51–67.
14 M.L. Jones and G.P. Kurzban, Noncooperativity of biotin binding to tetrameric streptavidin, *Biochemistry*, 34 (1995) 11750–11756.
15 F. Darain, S.-U. Park and Y.-B. Shim, Disposable amperometric immunosensor system for rabbit IgG using a conducting polymer modified screen-printed electrode, *Biosens. Bioelectron.*, 18 (2003) 773–780.
16 B. Limoges, J.-M. Savéant and D. Yazidi, Quantitative analysis of catalysis and inhibition at horseradish peroxidase monolayers immobilized on an electrode surface, *J. Am. Chem. Soc.*, 125 (2003) 9192–9203.
17 S. Da Silva, L. Grosjean, N. Ternan, P. Mailley, T. Livache and S. Cosnier, Biotinylated polypyrrole films: an easy electrochemical approach for the reagentless immobilization of bacteria on electrode surfaces, *Bioelectrochemistry*, 63 (2004) 297–301.
18 M. Snejdarkova, M. Rehak and M. Otto, Design of a glucose minisensor based on streptavidin glucose–oxidase complex coupling with self-assembled biotinylated phospholipid membrane on solid support, *Anal. Chem.*, 65 (1993) 665–668.
19 E. Zacco, M.I. Pividori and S. Alegret, Electrochemical biosensing based on universal affinity biocomposite platforms, *Biosens. Bioelectron.*, 21 (2006) 1291–1301.
20 J. Sjoquist, H. Hjelm, I.B. Johansso and J. Movitz, Localization of protein-A in bacteria, *Eur. J. Biochem.*, 30 (1972) 190.
21 J. Sjoquist, B. Meloun and H. Hjelm, Protein A isolated from *Staphylococcus aureus* after digestion with lysostaphin, *Eur. J. Biochem.*, 29 (1972) 572–578.

22 B. Akerstrom, T. Brodin, K. Reis and L. Bjorck, Protein-G. A powerful tool for binding and detection of monoclonal and polyclonal antibodies, *J. Immunol.*, 135 (1985) 2589–2592.
23 B.J. Compton, M. Lewis, F. Whigham, J.S. Gerald and G.E. Countryman, Analytical potential of protein-A for affinity—chromatography of polyclonal and monoclonal-antibodies, *Anal. Chem.*, 61 (1989) 1314–1317.
24 L.J. Janis and F.E. Regnier, Dual-column immunoassays using protein G affinity chromatography, *Anal. Chem.*, 61 (1989) 1901–1906.
25 A. Larsson, An ELISA procedure for the determination of protein G-binding antibodies, *J. Immunol. Methods*, 135 (1990) 273–275.
26 B. Lu, M.R. Smyth and R. O'Kennedy, Immunological activities of IgG antibody on pre-coated Fc receptor surfaces, *Anal. Chim. Acta*, 331 (1996) 97–102.
27 P.F. Zatta, A new bioluminescent assay for studies of protein G and protein A binding to IgG and IgM, *J. Biochem. Biophys. Methods*, 32 (1996) 7–13.
28 M. Reinecke and T. Scheper, Fast online flow injection analysis system for IgG monitoring in bioprocesses, *J. Biotechnol.*, 59 (1997) 145–153.
29 C. Valat, B. Limoges, D. Huet and J.-L. Romette, A disposable protein A-based immunosensor for flow-injection assay with electrochemical detection, *Anal. Chim. Acta*, 404 (2000) 187–194.
30 P.D. Patel, (Bio)sensors for measurements of analytes implicated in food safety: A review, *TRAC*, 21 (2002) 96–115.
31 L.A. Terry, S.F. White and L.J. Tigwell, The application of biosensors to fresh produce and the wider food industry, *J. Agric. Food Chem.*, 53 (2005) 1309–1316.
32 L.D. Mello and L.T. Kubota, Review of the use of biosensors as analytical tools in the food and drink industries, *Food Chem.*, 77 (2002) 237–256.
33 C. Mellgren, A. Sternesjo, P. Hammer, G. Suhren, L. Bjorck and W. Heeschen, Comparison of biosensor, microbiological, immunochemical and physical methods for detection of sulfamethazine residues in raw milk, *J. Food Prot.*, 59(11) (1996) 1223–1226.
34 A. Sternesjo, C. Mellgren and L. Bjorck, Determination of sulphametazine residues in milk by a surface resonance based biosensors assay, *Anal. Biochem.*, 226 (1995) 175–181.
35 E. Gustavsson, P. Bjurling and Å. Sternesjö, Biosensor analysis of penicillin G in milk based on the inhibition of carboxypeptidase activity, *Anal. Chim. Acta*, 468 (2002) 153–159.
36 G. Cacciatore, M. Petz, S. Rachid, R. Hakenbeck and A. Bergwerff, Development of an optical biosensor assay for detection of β-lactam antibiotics in milk using the penicillin-binding protein 2 ×, *Anal. Chim. Acta*, 520 (2004) 105–115.
37 V. Gaudin, J. Fontaine and P. Maris, Screening of penicillin residues in milk by a surface plasmon resonance-based biosensor assay: comparison

of chemical and enzymatic sample pre-treatment, *Anal. Chim. Acta*, 436 (2001) 191–198.
38 W. Haasnoot, M. Bienenmann-Ploum, U. Lamminmäki, M. Swanenburg and H. van Rhijn, Application of a multi-sulfonamide biosensor immunoassay for the detection of sulfadiazine and sulfamethoxazole residues in broiler serum and its use as a predictor of the levels in edible tissue, *Anal. Chim. Acta*, 552 (2005) 87–95.
39 J. Ferguson, A. Baxter, P. Young, G. Kennedy, C. Elliott, S. Weigel, R. Gatermann, H. Ashwin, S. Stead and M. Sharman, Detection of chloramphenicol and chloramphenicol glucuronide residues in poultry muscle, honey, prawn and milk using a surface plasmon resonance biosensor and Qflex® kit chloramphenicol, *Anal. Chim. Acta*, 529 (2005) 109–113.
40 A. Vakurov, C.E. Simpson, C.L. Daly, T.D. Gibson and P.A. Millner, Acetylecholinesterase-based biosensor electrodes for organophosphate pesticide detection: II. Immobilization and stabilization of acetylecholinesterase, *Biosens. Bioelectron.*, 20 (2005) 2324–2329.
41 S. Andreescu and J.-L. Marty, Twenty years research in cholinesterase biosensors: From basic research to practical applications, *Biomol. Eng.*, 23 (2006) 1–15.
42 M. Shimomura, Y. Nomura, W. Zhang, M. Sakino, K.-H. Lee, K. Ikebukuro and I. Karube, Simple and rapid detection method using surface plasmon resonance for dioxins, polychlorinated biphenylx and atrazine, *Anal. Chim. Acta*, 434 (2001) 223–230.
43 W.M. Mullett, E.P.C. Lai and J.M. Yeung, Surface plasmon resonance-based immunoassays, *Methods*, 22 (2000) 77–91.
44 E. Mauriz, A. Calle, L.M. Lechuga, J. Quintana, A. Montoya and J.J. Manclús, Real-time detection of chlorpyrifos at part per trillion levels in ground, surface and drinking water samples by a portable surface plasmon resonance immunosensor, *Anal. Chim. Acta*, 561 (2006) 40–47.
45 M. Mehervar and M. Abdi, Recent development, characteristics, and potential applications of electrochemical biosensors, *Anal. Sci.*, 20 (2004) 1113–1126.
46 G.E. Pellegrini, G. Carpico and E. Coni, Electrochemical sensor for the detection and presumptive identification of quinolone and tetracycline residues in milk, *Anal. Chim. Acta*, 520 (2004) 13–18.
47 S.J. Setford, R.M. Van Es, Y.J. Blankwater and S. Kröger, Receptor binding protein amperometric affinity sensor for rapid β-lactam quantification in milk, *Anal. Chim. Acta*, 398 (1999) 13–22.
48 L. Agüí, A. Guzmán, P. Yáñez-Sedeño and J.M. Pingarrón, Voltammetric determination of chloramphenicol in milk at electrochemically activated carbon fibre microelectrodes, *Anal. Chim. Acta*, 461 (2002) 65–73.
49 A.J. Baeumner, Biosensors for environmental pollutants and food contaminants, *Anal. Bioanal. Chem.*, 377 (2003) 434–445.

50 H. Nakamura and I. Karube, Current research activity in biosensors, *Anal. Bioanal. Chem.*, 377 (2003) 146–168.
51 S. Alegret, Rigid carbon-polymer biocomposites for electrochemical sensing, *Analyst*, 121 (1996) 1751–1758.
52 F. Céspedes, E. Martínez-Fàbregas and S. Alegret, New materials for electrochemical sensing. I. Rigid conducting composites, *Trends Anal. Chem.*, 15 (1996) 296–304.
53 R.N. Adams, Carbon paste electrodes, *Anal. Chem.*, 30 (1958) 1576.
54 E. Zacco, M.I. Pividori, X. Llopis, M. del Valle and S. Alegret, *J. Immunol. Methods*, 286 (2004) 35–46.
55 E. Zacco, R. Galve, M.P. Marco, S. Alegret and M.I. Pividori, Electrochemical biosensing of pesticides residues based on affinity biocomposite platforms, *Biosens. Bioelectron.*, 22 (2007) 1707–1715.
56 B.I. Haukanes and C. Kvam, Application of magnetic beads in bioassays, *Biotechnology*, 11 (1993) 60–63.
57 K. Larsson, K. Kriz and D. Kriz, Magnetic transducers in biosensors and bioassays, *Analysis*, 27 (1999) 617–621.
58 S. Solé, A. Merkoci and S. Alegret, New materials for electrochemical sensing III. Beads, *Trends Anal. Chem.*, 20 (2001) 102–110.
59 E. Zacco, M.I. Pividori, S. Alegret, R. Galve and M.-P. Marco, Electrochemical magnetoimmunosensing strategy for the detection of pesticides residues, *Anal. Chem.*, 78 (2006) 1780–1788.
60 A.G. Gehring, J.D. Brewster, P.L. Irwin, S.I. Tu and L.J. Van Houten, 1-Naphthyl phosphate as an enzymatic substrate for enzyme-linked immunomagnetic electrochemistry, *J. Electroanal. Chem.*, 469 (1999) 27–33.
61 M. Dequaire, C. Degrand and B. Limoges, An immunomagnetic electrochemical sensor based on a perfluorosulfonate-coated screen-printed electrode for the determination of 2,4-dichlorophenoxyacetic acid, *Anal. Chem.*, 71 (1999) 2571–2577.
62 E. Zacco, J. Adrian, R. Galve, M.-P. Marco, S. Alegret and M.I. Pividori, Electrochemical magneto immunosensing of antibiotic residues in milk, *Biosens. Bioelectron.*, 22 (2007) 2184–2191.

THICK AND THIN FILM BIOSENSORS

Chapter 23

Screen-printed electrochemical (bio)sensors in biomedical, environmental and industrial applications

John P. Hart, Adrian Crew, Eric Crouch, Kevin C. Honeychurch and Roy M. Pemberton

23.1 INTRODUCTION

Electrochemical sensors and biosensors offer the achievable opportunity of simplifying the analyses of complex matrices, outside of the laboratory, by suitable modification of appropriate electrode materials [1–5]. One of the most attractive methods for the fabrication of such devices involves the use of screen-printing technology. This allows the (bio)sensors to be manufactured in a wide range of geometries at low cost, particularly when carbon is used; therefore, this allows the devices to become disposable [1,2]. A typical screen-printed electrode design commonly used in our laboratories for prototype investigations is shown in Fig. 23.1.

Although numerous papers have been published on biosensors for a vast range of analytes, comparatively few have been marketed as commercial systems. In a recent review, Malhotra and Chaubey [6] list 14 companies producing glucose biosensor systems, five systems for lactate, two for uric acid and one for cholesterol. The Yellow Springs Instrument Company also produces systems for ethanol, methanol, starch and a variety of sugars. Problems in commercialising biosensor systems arise from many factors, such as difficulty in stabilising and storing the biological materials, lack of methods for mass-production and problems with *in vivo* compatibility [7,8].

This chapter reviews publications which describe the design and development of screen-printed electrochemical (bio)sensors that have

Comprehensive Analytical Chemistry 49
Alegret and Merkoçi (Eds)
Volume 49 ISSN: 0166-526X DOI: 10.1016/S0166-526X(06)49023-1

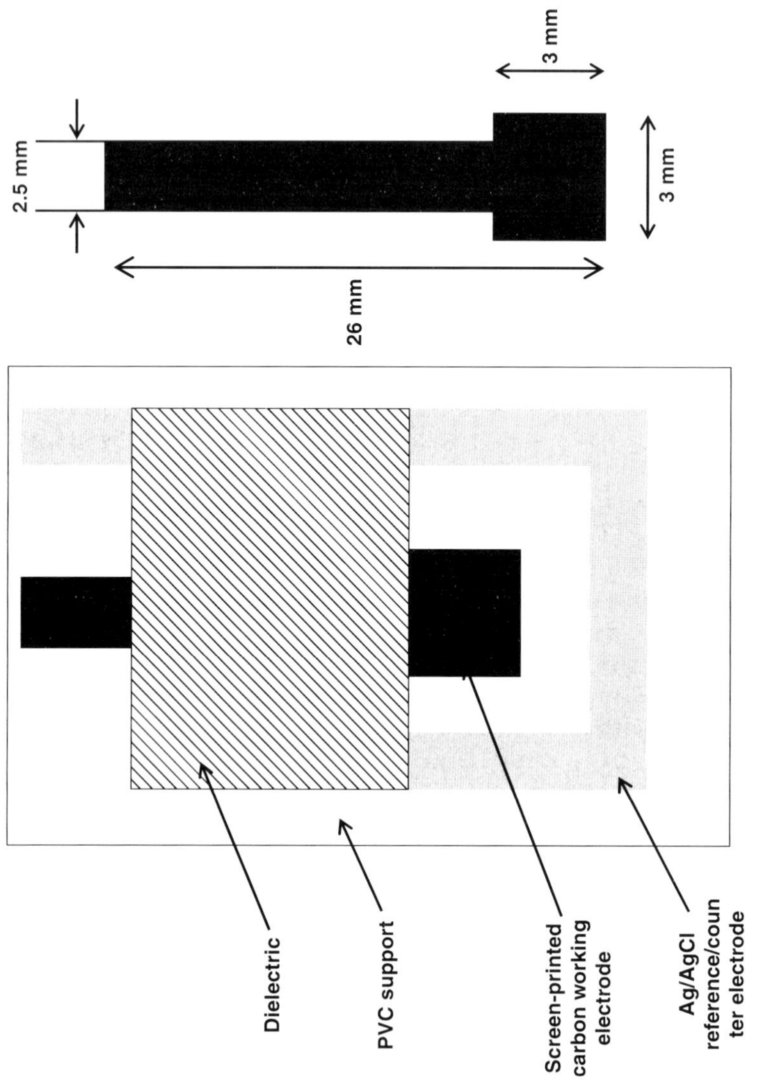

Fig. 23.1. Example of a screen-printed electrode.

mainly been published since 1997; however, a few key papers from earlier reports are also included. The chapter is divided into sections according to applications selected from the fields of biomedical, environmental and industrial analyses as these have seen the largest growth in the development of prototype devices. In order to give a comprehensive survey of earlier publications with new examples, we have used tables where appropriate to group related areas and (bio)sensors; alternatively, where applications have not been previously reviewed by the present authors these have been discussed in the text.

23.2 BIOMEDICAL

23.2.1 Glucose

The determination of blood glucose levels is an indispensable test for the exact diagnosis and therapy of many disorders, the most common of which is diabetes mellitus; a condition that is estimated to affect over 170 million people worldwide, with prevalence rising rapidly [9].

Biosensors for the determination of blood glucose have enjoyed widespread commercial success since the introduction of the pen-sized 30 s blood glucose meter [10]. However, researchers have continued to devise novel approaches in the development of amperometric biosensors based on screen-printing technology; Table 23.1 summarises some examples of these approaches together with their performance characteristics.

23.2.2 Mediators for H_2O_2

The first 15 examples given in Table 23.1 [18–32] have an H_2O_2 mediator, as well as glucose oxidase (GOD), incorporated into the device. The purpose of this type of mediator is to allow electron transfer from H_2O_2 to the working electrode to occur at potentials significantly lower than required for direct oxidation of this product. Table 23.1 shows that potentials of over +1 V are required for the oxidation of H_2O_2 at non-mediated biosensors [46,47]; this can lead to interference problems as many other species present in the samples are also oxidised at these high potentials.

Iron hexacyanoferrate (Prussian blue) [11,21–26] is inexpensive, and can effectively reduce the required operating potential to around 0 V, resulting in a minimal risk of interference. The main disadvantages of

TABLE 23.1

Some recent glucose biosensor systems utilising screen-printed carbon electrodes

GOD immobilisation technique	Mediator	Assay time (s)	Lower linear range (μM)	Upper linear range (mM)	Applied potential (mV)	Storage stability (weeks)	Reference
Entrapment in liposomes	CoPC	–	–	20	400	–	Memoli *et al.* [18]
Bulk incorporation into water-based ink	CoPC	<3[a]	25	2	500[†]	⩾78	Crouch *et al.* [19]
Bulk incorporation into water-based ink	CoPC	100	200	5	500	–	Crouch *et al.* [20]
Drop-coated within Nafion[R] membrane	Prussian blue	–	220	4.6	0[*]	–	O'Halloran *et al.* [21]
Bulk ink incorporation followed by glutaraldehyde cross-linking	Prussian blue	–	200	4	0[*]	–	Pravda *et al.* [22]
Cross-linked with BSA and Nafion[R] using glutaraldehyde	Prussian blue	90	2	0.1	−50	–	Ricci *et al.* [23]
Bulk ink incorporation	Prussian blue	135	30	6	−50	20	Ricci *et al.* [24]
Polyethylenimine/glutaraldehyde immobilization	Prussian blue	60[a]	1	5	0	–	Lupu *et al.* [25]
Cross-linked with glutaraldehyde and Nafion[R]	Prussian blue	–	25	1	−50	⩾52	Ricci *et al.* [26]
Bulk ink incorporation	Cupric hexacyanoferrate	30	–	100	−100[*]	3	Wang and Zhang [27]
Drop-coated	Rhodinised carbon	120	–	1	300[*]	⩾52	Kröger *et al.* [28]
Bulk ink incorporation	Manganese dioxide	<100	11.1	13.9	480[*]	–	Turkušić *et al.* [29]
Drop-coated within Nafion[R] membrane	Ruthenium dioxide	FIA	5.6	2.8	480[*]	–	Kotzian *et al.* [30]
Cross-linked with BSA using glutaraldehyde	Copper-plated SPCE	–	–	26.7	Peak in linear sweep voltammogram at 150 mV[*]	–	Kumar and Zen [31]
Drop-coated within Nafion[R] membrane	MWCNT	10[a]	–	6	100[*]	–	Wang and Musameh [32]
Drop-coated	MWCNT and ferricyanide	–	–	30	300	–	Guan *et al.* [33]
Dispersed in membrane	Alumina nanoparticles and diffusional polymeric mediator	5	278	33.3	300	⩾26	Gao *et al.* [34]

Nitrocellulose strip	Hexamine ruthenium chloride	30	–	27	0	–	Cui *et al.* [35]
Cross-linked to mediator-incorporating membrane	Osmium complex	90	100	10	500*	⩾4	Zhang *et al.* [36]
Contained within capillary-fill device	Osmium complex	60	100	25	550	4	Zhang *et al.* [37]
Drop-coated with carboxymethyl cellulose	Bacterial pyocyanin	–	1000	20	−200	4	Ohfuji *et al.* [38]
Drop-coated GOD with amyloglucosidase for maltose	Ferrocene derivative	60	–	40	300	16	Ge *et al.* [39]
Drop-coated GOD with mutorotatase and invertase for sucrose	Ferrocene derivative	60	–	15	400	–	Hu *et al.* [40]
Drop-coated genetically modified GOD	Ferrocene derivative	30	–	45	450	10	Chen *et al.* [41]
Impregnated into porous carbon with sodium alginate	Ferrocene derivative	23	⩽2.3	⩾24.7	400	⩾78	Forrow and Bayliff [42]
Trapped in photocurable polymeric film	Ferrocene and (vinylferrocene)	–	0.004 (0.005)	17 (42)	350 (400)†	–	Bean *et al.* [43]
Drop-coated GDH	NAD^+ and Meldola's Blue	–	200	1	0	–	Hart *et al.* [44]
GDH cross-linked with glutaraldehyde	PQQ and ferrocene derivative	–	–	2.4	350*	–	Razumiene *et al.* [45]
Bulk ink incorporation	None	–	50	2.5	1150*	8	Galán-Vidal *et al.* [46]
Bulk ink incorporation	None	25[a]	50	2.5	1150*	–	Galán-Vidal *et al.* [47]

All potentials vs. screen-printed Ag/AgCl pseudo-reference, except values marked with asterisk (*), which are vs. Ag/3 M AgCl double-junction reference electrode, and values marked with dagger (†), which are vs. saturated calomel. Abbreviations: CoPC: cobalt phthalocyanine, SPCE: screen-printed carbon electrode, GOD: glucose oxidase, MWCNT: multi-walled carbon nanotubes, NAD: nicotinamide adenine dinucleotide, PQQ: pyrroloquinoline quinone, FIA: flow injection analysis.

[a]The response time from a steady state in a continuous amperometric technique.

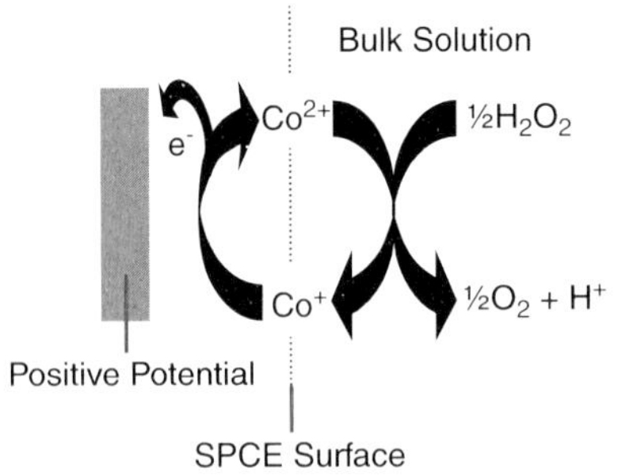

Fig. 23.2. Operating principle of CoPC-modified SPCE used to detect H_2O_2 [12].

this system are the water-soluble nature of the mediator, which may be leached from the electrode surface, and the low pH needed to stabilise the mediator, which may detrimentally affect enzyme stability and operation.

Cobalt phthalocyanine [12–20] (CoPC) requires a slightly higher potential (around +400 mV) in order to catalyse H_2O_2 oxidation. However, this mediator is insoluble in all aqueous and most organic solvents, so the risk of diffusion into the sample is minimised. The principle of operation for this system is shown in Fig. 23.2.

Nanosized CoPC particles were prepared by precipitation into a cationic surfactant solution, and subsequently GOD molecules were adsorbed onto the resulting nanoparticles via electrostatic attraction [13]. A glucose biosensor was prepared by depositing the CoPC/GOD biocomposite particles (mean particle size 187 nm) onto a pyrolytic graphite electrode and casting a Nafion® membrane over the surface. The resulting biosensor performed well in both standard solutions and biological samples, indicating promise for this new form of nanoscaled CoPC mediator. Other groups have also used nanosized mediators for H_2O_2; in this case multi-walled carbon nanotubes (MWCNTs) were employed [32]. The MWCNT formed the bulk of the ink, providing both electrical conductivity and H_2O_2 mediation. The authors also observed catalytic activity towards the cofactor NADH, implying that the MWCNT would be of additional use in dehydrogenase-based biosensors. In this paper, they concluded that the analytical signal resulted only from the electrocatalytic oxidation of H_2O_2, although other groups have demonstrated direct electron transfer between carbon nanotubes (CNTs) and GOD, when the nanotubes are deposited onto glassy carbon [14], packed into a microcavity [15], or regularly aligned on a cysteamine-modified gold electrode [16].

23.2.3 Mediators capable of direct electron transfer with GOD

In biosensor systems employing this type of mediator, the natural substrate for GOD (molecular oxygen) is replaced by a compound which is carefully chosen to be capable of carrying out the same role. The mediator serves to re-oxidise the reduced form of GOD (Eq. (23.2)), following its interaction with glucose (Eq. (23.1)). Consequently, these systems can be operated at the potential of the mediator couple (Eq. (23.3)) and O_2 concentration at the electrode/solution interface does not become a limiting factor in the analysis; this is particularly important in samples where O_2 concentrations are low, such as neonatal blood or microbial fermentations.

$$\mathrm{GOD_{ox}} + \mathrm{Glucose} \rightarrow \mathrm{Gluconolactone} + \mathrm{GOD_{red}} \tag{23.1}$$

$$\mathrm{GOD_{red}} + \mathrm{Mediator_{ox}} \rightarrow \mathrm{GOD_{ox}} + \mathrm{Mediator_{red}} \tag{23.2}$$

$$\mathrm{Mediator_{red}} \rightleftarrows \mathrm{Mediator_{ox}} + n\mathrm{e}^- \tag{23.3}$$

Ferrocene was one of the earliest mediators used [10] but is somewhat hydrophobic so derivatives of the molecule are often employed [39–43]. Ferricyanide can also be used, and the use of MWCNT with this mediator was shown to enhance its effectiveness [33]. Other groups have studied a wide diversity of novel mediator systems such as poly(vinylferrocene-co-acrylamide) dispersed within an alumina nanoparticle membrane [34], ruthenium [35] and osmium [36,37] complexes, and the phenazine pigment pyocyanin, which is produced by the bacteria *Pseudomonas aeruginosa* [38].

23.2.4 The use of glucose dehydrogenase (GDH)

An alternative biosensor system has been developed by Hart *et al.* [44] which involves the use of the NAD^+-dependent GDH enzyme. The first step of the reaction scheme involves the enzymatic reduction of NAD^+ to NADH, which is bought about by the action of GDH on glucose. The analytical signal arises from the electrocatalytic oxidation of NADH back to NAD^+ in the presence of the electrocatalyst Meldola's Blue (MB), at a potential of only 0 V. Biosensors utilising this mediator have been reviewed elsewhere [1,17]. Razumiene *et al.* [45] employed a similar system using both GDH and alcohol dehydrogenase with the cofactor pyrroloquinoline quinone (PQQ), the oxidation of which was mediated by a ferrocene derivative.

23.2.5 Cholesterol

The determination of cholesterol is important for the diagnosis and prevention of a number of clinical disorders such as hypertension, arteriosclerosis, cerebral thrombosis and coronary heart disease. As the majority of cholesterol in human blood is present in an esterified form, a separate saponification step is required to obtain a total cholesterol analysis; early methods for this involved caustic and toxic reagents, long analysis times and a relatively large sample volume. Free cholesterol can be determined chromatographically, although this requires cumbersome and expensive laboratory-based equipment. Modern methods use the enzyme cholesterol esterase to release esterified cholesterol which is then oxidised by a second enzyme, cholesterol oxidase (ChOx, Fig. 23.3) [48].

The production of H_2O_2 from this step can be monitored spectrophotometrically by the formation of a dye in the presence of peroxidase [49]; disadvantages of this system include the use of carcinogenic dyes, lengthy incubation times and the need for laboratory-based equipment. As discussed in the previous section, SPCEs can be modified with mediators to produce effective H_2O_2 transducers; when combined with immobilised enzymes, disposable biosensors ideal for de-centralised clinical analysis can be fabricated. The cholesterol molecule is non-polar and thus provides an additional challenge to the development of

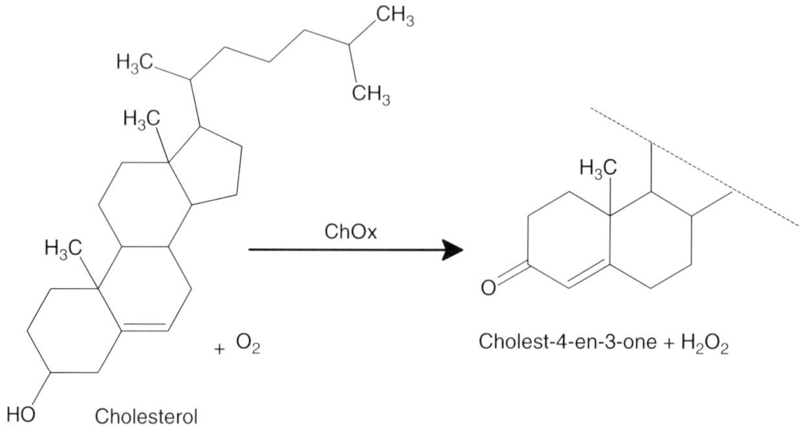

Fig. 23.3. The structure of cholesterol, and the action of ChOx on its functional groups. The dashed line represents the remainder of the molecule, which is unchanged.

electrochemical biosensors for its analysis, as most biosensors operate in aqueous systems. Solubilisation of the analyte is usually achieved by the addition of a non-ionic surfactant such as Triton X-100® to the test solution.

An SPCE modified with CoPC was employed as an H_2O_2 transducer in a cholesterol biosensor fabricated by the drop-coating of ChOx, followed by a cellulose acetate membrane [50]. The resulting cholesterol biosensors were operated in stirred solutions using an applied potential of +400 mV vs. screen-printed Ag/AgCl, and displayed a linear range of 0.06–5 mM.

MWCNTs have also found use in elaborate three-enzyme biosensors for total cholesterol [51]. In this system, carboxyl-modified MWCNTs, together with ferrocyanide, mediate peroxidase which reduces the H_2O_2 produced by ChOx. Cholesterol esterase was immobilised onto the biosensor with a stabiliser in a separate layer to provide a one-step measurement. The biosensors were operated in the chronoamperometric mode with an applied potential of +300 mV vs. screen-printed carbon, and were shown to give a linear response over the range of at least 2.6–15.5 mM. The biosensors required only 2 μl of blood, gave a result in 2 min, and were in good agreement with a standard hospital technique. Tan *et al.* [52] utilised MWCNTs to immobilise ChOx within a chitosan/silica sol–gel matrix which was drop-coated onto a Prussian blue-coated glassy carbon electrode. The authors attributed the desirable performance characteristics of the resulting biosensor to the properties of the MWCNTs, which appeared to effectively promote electron transfer between the enzyme and the surface of the electrode.

A patented device for the determination of total cholesterol that also relies on the peroxidase/ferricyanide mediation system has been described [53]. This device was incorporated with cholesterol esterase into a lateral flow system and was operated at an applied potential of −200 mV vs. screen-printed Ag/AgCl, following a 7.5 min incubation step. Chronoamperometric currents were recorded 3.5 min later and total cholesterol concentrations between 2.81 and 13 mM compared well with the reference method. A novel method has been reported for electrochemical cholesterol detection, using the flavocytochrome P450scc (in place of ChOx) as the biological recognition element [54]. Riboflavin was used in conjunction with rhodinised graphite as a direct electron transfer mediator system. The applied potential was –600 mV vs. screen-printed Ag/AgCl and a detection limit for free cholesterol of 155 μM could be achieved in 2 min.

23.2.6 Lactate

Lactate is an important marker compound in many biomedical, food and beverage applications. Blood lactate is monitored in exercise control and sports medicine, and D-lactate is an indicator of bacterial activity in wine fermentation [59,60,63]. Lactate produced from anaerobic respiration of cattle muscle subsequent to slaughter has been investigated as an indicator of meat freshness [62].

Two enzymes are commonly used for amperometric biosensors, namely lactate oxidase (LOD) and lactate dehydrogenase (LDH). It should be noted that, in this instance, LOD refers to the enzyme which catalyses the reaction shown in Fig. 23.4, in which the products are pyruvate and H_2O_2. This type of enzyme was formerly assigned the E.C. number 1.1.3.2, but this was confused with lactate monooxygenase (E.C. 1.13.12.4), which is also commonly referred to as 'type I lactate oxidase' [55] or simply 'lactate oxidase' [56] whose products are acetate, CO_2 and H_2O_2. The LOD which catalyses the reaction shown in Fig. 23.4 has also been referred to as 'type II lactate oxidase' [55]; following clarification of this point in a published letter [57], current publications refer to this enzyme as E.C. 1.1.3.x.

Table 23.2 describes some recent publications on carbon-based amperometric screen-printed biosensors for lactate. Systems utilising LOD rely on the oxidation of enzymatically produced H_2O_2 (Fig. 23.4), as described in the glucose section. The electrochemical mediator described in several recent reports [62–64] was platinised carbon. This mediator is able to greatly reduce the operating potential, although it does have the disadvantage that it is expensive and has a high electrocatalytic activity for many other substances, including reducing sugars [58]. This drawback necessitates the addition of a perm-selective membrane, and thus complicates the fabrication procedure. Rohm *et al.* [65] described an ultraviolet-polymerisable water-based paste containing LOD and graphite. The biocomposite layer was printed over a preformed platinum electrode that had been fired at 850°C. The resulting

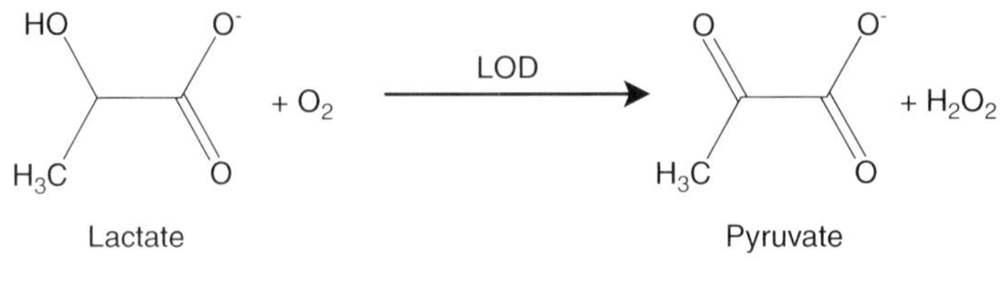

Fig. 23.4. Action of type II LOD (E.C. 1.1.3.x) on the lactate anion.

TABLE 23.2

Some recent lactate biosensor systems utilising screen-printed carbon electrodes

Enzyme and immobilisation technique	Mediator	Assay Time (s)	Lower linear range (μM)	Upper linear range (mM)	Applied potential (mV)	Storage	stability (weeks)
Reference							
LDH drop-coated with NAD^+ and Nafion[R]	NAD^+ and insoluble Meldola's Blue salt	150	100	1	−50	1	Avramescu *et al.* [59]
LDH entrapped in PVA-SbQ polymer	NAD^+ and insoluble Meldola's Blue salt	150	75	1	−150	< 2	Avramescu *et al.* [60]
LDH entrapped in electropolymerised Meldola's Blue	NAD^+ and poly(Meldola's Blue)	100	1000	12	0	2	Litescu *et al.* [61]
LOD incorporated in bulk-modified ink	Platinised carbon	–	–	10	350	32	Hart *et al.* [62]
LOD entrapped within sol–gel ink	Platinised carbon	–	–	0.3	350*	⩾20	Albareda-Sirvent and Hart [63]
LOD immobilised in bulk ink	Platinised carbon	10 s (once stabilised)	–	5	350*	–	Collier *et al.* [64]
LOD and stabilizers entrapped within UV-polymerisable ink	Underlying Pt working electrode	Online FIA	0.02/550[a]	1/50[a]	600‡	–	Rohm *et al.* [65]

All potentials vs. screen-printed pseudo Ag/AgCl reference, except values marked with double-dagger (‡), which are vs. screen-printed Ag/Pd, and those marked with asterisk (*), which are vs. Ag/3 M AgCl double-junction reference electrode. Abbreviations: LDH: lactate dehydrogenase, LOD: lactate oxidase, PVA-SbQ: styrlpyridinium-modified poly(vinyl alcohol), FIA: flow injection analysis.

[a]Linear ranges using a dialysis system.

biosensors were used in a flow injection system without the need for a membrane and still performed well. The same biosensors were also employed in a dialysis-based flow injection system for the analysis of a complex fermentation broth, and gave a close correlation to the reference method. However, the separate screen-printed layers required for the conducting tracks also led to quite a complex fabrication procedure, and a large overpotential was required (+600 mV.)

Biosensors employing the NAD^+-dependent LDH enzyme [59–61] require an immobilisation procedure for the deposition of the mediator and water-soluble cofactor. The MB mediator was rendered water-insoluble by the use of its Reinecke salt in two studies [59,60], and by electropolymerisation with the enzyme in an another study [61]; both of these approaches required an operating potential close to 0 V.

23.2.7 Ascorbic acid, steroid and protein hormones

Table 23.3 lists recent publications concerning SPCE-based approaches towards the analysis of the naturally occurring compounds ascorbate, steroid hormones (including progesterone and estradiol) and protein hormones (hCG and insulin).

23.2.8 Immunoglobulins

The immunoglobulins have been the subject of biosensor analysis, both as model analytes for generic sensor development and in their own right as antibodies of clinical interest. A selection of recent papers is given in Table 23.4.

23.2.9 Miscellaneous proteins, peptides and amino acids

Table 23.5 summarises the main features of papers which detail experimental approaches to the analysis of amino acids, peptides and proteins (other than hormones) using electrochemical biosensor-based methodologies involving SPCEs.

23.2.10 Nucleic acids and purines

Several different approaches have been used to monitor DNA, RNA and purines at SPCEs (Table 23.6).

TABLE 23.3

Reports of biosensors for naturally occurring compounds ascorbate, steroids and protein hormones

Analyte	Reaction	Detection	Applied potential vs. Ag/AgCl	Linear/dynamic range	Detection limit	Reference
Ascorbic acid	Mediated oxidation at a silica sol–gel glass coated ferricyanide-doped Tosflex-modified SPCE	FIA	+0.3 V	46 nM to 300 μM	46 nM	Zen *et al.* [66]
L-Ascorbic acid	MnO_2-modified SPCE	Amperometric	+0.6 V	50–250 mg l^{-1}	0.2 mg l^{-1}	Turkušić *et al.* [67]
Ascorbic acid	Electrocatalytic oxidation at a nickel hexacyanoferrate-modified SPCE	Cyclic voltammetry	+0.4 V	5×10^{-5} to 1.5×10^{-3} M	5×10^{-5} M	Lin *et al.* [68]
Progesterone in buffer	Competitive immunoassay with ALP-labelled progesterone on SPCE bearing surface-immobilised antibody	Chronoamperometric detection of phenol	+0.7 V	1×10^{-9} to 5×10^{-6} M 0.5–50 ng ml^{-1}	1×10^{-9} M 7 ng ml^{-1}	Hart *et al.* [69] Pemberton *et al.* [70]
Progesterone in milk	Competitive immunoassay with ALP-labelled progesterone on SPCE bearing surface-immobilised antibody	Chronoamperometric detection of 4-aminophenol	+0.2 V	2.5–500 ng ml^{-1} 5–50 ng ml^{-1}	10 ng ml^{-1} 20 ng ml^{-1}	Pemberton *et al.* [70] Pemberton *et al.* [71]
Progesterone in milk	See Procedure 35 in CD accompanying this book	Chronoamperometric detection of 1-naphthol	+0.3 V	5–50 ng ml^{-1}	7 ng ml^{-1}	Pemberton *et al.* [71]
Progesterone in milk	See Procedure 35 in CD accompanying this book	Amperometric detection of 1-naphthol	+0.3 V	0–25 ng ml^{-1}	2.5 ng ml^{-1}	Pemberton *et al.* [72]
Progesterone in buffer	Competitive immunoassay with HRP-labelled progesterone on SPCE bearing surface-immobilised antibody	Chronoamperometric detection of TMB substrate	+450 mV	0–25 ng ml^{-1}	5 ng ml^{-1}	Hart *et al.* [44]
Progesterone in milk	Competitive immunoassay using SPCE with immobilised prog-BSA conjugate	Amperometric or DPV detection of ALP-generated *p*-aminophenol	+350 mV or scan from +100 to +500 mV	16–256 ng ml^{-1}	3 ng ml^{-1}	Kreuzer *et al.* [73]
Estradiol (E_2) in serum	Competitive immunoassay with ALP-labelled E_2 on SPCE bearing surface-immobilised antibody	DPV of enzyme-generated 1-naphthol	Scan from 0 to +0.3 V	25–500 pg ml^{-1}	50 pg ml^{-1}	Pemberton *et al.* [74]

TABLE 23.3 (*continued*)

Analyte	Reaction	Detection	Applied potential vs. Ag/AgCl	Linear/dynamic range	Detection limit	Reference
Human chorionic gonadotrophin (hCG)	Competitive immunoassay on Nafion-coated SPCEs in microwells	DPV oxidation of accumulated cationic product of substrate hydrolysis by ALP enzyme	Scan from 0 to +0.7 V	200–10,000 mIU ml^{-1}	100 mIU (1 mg = 9825 IU)	Bagel *et al.* [75]
Insulin	Oxidation of tyrosine residues at pre-treated (+1.7 V, 30 s) microfabricated SPCEs	Potentiometric stripping analysis (2 min accumulation at −0.2 V)	Stripping current of 3 μA	0–600 nM	20 nM	Wang *et al.* [76]

TABLE 23.4

Reports of biosensors for immunoglobulins

Analyte	Reaction	Detection	Applied potential vs. Ag/AgCl	Linear range	Detection limit	Reference
IgE in plasma	Competitive or displacement immunoassay using surface-immobilised anti-IgE and ALP-labelled second antibody	Amperometric detection of 4-aminophenol	+300 mV	63–1000 ng ml^{-1} (competitive); 100–1500 ng ml^{-1} (displacement)	0.09 ng ml^{-1}	Kreuzer *et al.* [77]
IgG (rabbit)	Competitive immunoassay: capture of analyte by sol–gel carbon ink incorporating IgG	Amperometric detection of ALP-generated α-naphthol	+400 mV	50–5000 ng ml^{-1}	5 ng ml^{-1} (32 pM)	Wang *et al.* [78]
IgG (rabbit)	Homogeneous immunoassay: competition between free and GOD-labelled IgG for surface-bound antibody	Amperometric detection of H_2O_2 produced from glucose; channelled through surface conducting polymer-bonded HRP	−0.35 V	0.5–2.0 μg ml^{-1}	0.33 μg ml^{-1}	Darain *et al.* [79]
IgG (goat)	Microwell-based sandwich immunoassay using gold-labelled second antibody	Oxidative gold metal dissolution, then anodic stripping voltammetry at a bare SPCE	5 min accumulation at −0.3 V, then scan from 0 to +0.8 V	0.5–100 ng ml^{-1}	0.5 ng ml^{-1} (3×10^{-12} M)	Dequaire *et al.* [80]
IgG (mouse)	Sandwich immunoassay using SPCE-surface-immobilised anti-IgG antibody	FIA: amperometric, reduction of enzymatically generated iodine	0.0 V	30–700 ng ml^{-1}	3 ng ml^{-1}	Gao *et al.* [81]
IgG (mouse)	Electrophoretic separation of free analyte/ferrocene-labelled anti-IgG	Amperometric, oxidation of ferrocene tracer at gold-plated SPCE	+0.6 V	7.5×10^{-11} to 1×10^{-6} g ml^{-1}	2.5×10^{-12} g ml^{-1}	Wang *et al.* [82]

TABLE 23.4 (*continued*)

Analyte	Reaction	Detection	Applied potential vs. Ag/AgCl	Linear range	Detection limit	Reference
IgG (rabbit)	Competitive immunoassay using ALP-labelled antibody at streptavidin-modified SPCEs bearing capture antibody	Cyclic voltammetry of indigo carmine generated from 3-IP substrate	−0.15 V (scan from −0.3 to +0.2 V)	5×10^{-11} to 1×10^{-9} M	5×10^{-11} M (7.0 ng ml^{-1})	Diaz-Gonzalez *et al.* [83]
IgG (mouse)	Sandwich immunoassay on Os redox polymer-modified SPCEs	Cyclic voltammetry of HRP-mediated electrocatalytic reduction of H_2O_2	+0.2 V (scan from +0.5 to −0.2 V)	4.4–440 pM	0.63 pM	Gao *et al.* [84]
IgG (rabbit)	Sandwich immunoassay upon Os redox hydrogel/ avidin-modified SPCE	Amperometric: HRP-mediated electrocatalytic reduction of H_2O_2	+0.2 V	1 pg ml^{-1} to 10 ng ml^{-1}	7 pg ml^{-1}	Zhang and Heller [85]
Anti-DNA antibody in serum from SLE or bronchial asthma patients	Suppression of enzyme activity due to formation of DNA–antibody adducts on graphite SPCEs	Amperometric detection of cholinesterase or peroxidase activity	+680 mV (cholinesterase) or −150 mV (peroxidase)	nd[a]	nd[a]	Evtugyn *et al.* [86]

[a]nd: not determined.

TABLE 23.5

Reports of biosensors for proteins/peptides other than hormones and immunoglobulins

Analyte	Reaction	Detection	Applied potential vs. Ag/AgCl	Linear range	Detection limit	Reference
Alpha-1-fetoprotein (AFP)	One-step sandwich assay using anti-AFP antibody on Prussian blue-modified SPCEs, with antibody–GOD conjugate	Flow injection, oxidation of H_2O_2 produced in presence of glucose	0.0 V	5–500 ng ml^{-1}	10 ng ml^{-1}	Guan *et al.* [87]
AFP	Inhibition of HRP activity when AFP binds to SPCE-surface-immobilised HRP-labelled AFP antibody/chitosan	Decrease in DPV current response for reduction of oxidised thionine (catalyst = H_2O_2)	−350 mV	0–20 and 20–150 ng ml^{-1}	0.74 ng ml^{-1}	Yu *et al.* [88]
L- or D-amino acids	H_2O_2 production via amino acid oxidase (AAO)	Amperometric, H_2O_2 at rhodinised carbon	+400 mV	0.1–1.0 mM	0.15–0.47 mM	Sarkar *et al.* [89]
20 underivatised amino acids	Photoelectrochemical deposition of Cu nanoparticles: formation of reversible 1:1 $Cu^{II}O$ nanoparticle–amino acid complex	Amperometric, flow injection in PBS carrier solution	0.0 V	5–500 μM	24 nM to 2.7 μM (depending upon amino acid)	Zen *et al.* [90]
Cysteine	Direct electrochemical oxidation at bare SPCEs	Amperometry in stirred solution	+0.6 V	5×10^{-5} to 5×10^{-4} M	5×10^{-5} M	Vasjari *et al.* [91]
Tyrosine	Direct electrochemical oxidation at bare SPCEs	Amperometry in stirred solution	+0.8 V	5×10^{-5} to 5×10^{-4} M	5×10^{-5} M	Vasjari *et al.* [91]
BSA	Selective binding to anti-BSA antibody in electropolymerised polypyrrole film on SPCE surface	Impedance	+500 to +100 mV at $f = 1$ kHz	0–75 ppm	20 ppm	Grant *et al.* [92]
BSA, cytochrome-*c*, haemoglobin, Casilan-90	Membrane-immobilised protease+AAO	Amperometric, H_2O_2 at rhodinised carbon	+400 mV	0.1–1 mg ml^{-1}	0.2 mg ml^{-1}	Sarkar [93]
Single proteins (Casilan-90)	Protease digestion, then H_2O_2 production via L-AAO	Amperometric, H_2O_2 at rhodinised carbon	+400 mV	1.7–1000 μg ml^{-1}	170 μg ml^{-1}	Setford *et al.* [94]
Amino acids and proteins	Consumption of Br_2, produced by electrolysis of KBr electrolyte (dual potential method)	Bromine, at rhodinised carbon	+600 mV (after a primary potential of +1.2 V)	0–2 mg ml^{-1}	<0.5 mg ml^{-1}	Sarkar and Turner [95]

TABLE 23.5 (*continued*)

Analyte	Reaction	Detection	Applied potential vs. Ag/AgCl	Linear range	Detection limit	Reference
Prostate specific antigen (PSA)	Sandwich or enzyme-channelling competitive immunoassay with HRP-labelled antibody and GOX enzyme	Amperometric, dissociation of H_2O_2 in presence of HRP and KI at rhodinised carbon	−70 mV	1.5–15 ng ml^{-1}	0.25 or 0.72 ng ml^{-1}	Sarkar *et al.* [96]
Glutathione	Sol–gel graphite composite SPCEs incorporating surfactant, CoPC/diphenylthiocarbazone	Cyclic voltammetry, electrocatalytic oxidation of GSH	+0.35 V (scan from −0.1 to +1.0 V)	1.98–15 mM	1.98 mM	Guo and Guadalupe [97]
Glutathione	Electrocatalytic oxidation of GSH in presence of naphthoquinone	Chronoamperometry	+0.2 V	15–60 mM	< 15 mM	Stone *et al.* [98]
Glutathione	Electrocatalytic Ellman reaction using cystamine, at Prussian blue-modified SPCE	Amperometric	+0.2 V	2×10^{-6} to 5×10^{-4} M	2×10^{-6} mol l^{-1}	Ricci *et al.* [99,100]
Granulocyte-macrophage-colony stimulating factor (GM-CSF)	Competitive immunoassay with ALP-labelled antigen	*p*-Aminophenol, amperometric	+300 mV	1.1–30 μg ml^{-1}	0.10 μg ml^{-1}	Crowley *et al.* [101]
Human heart fatty-acid binding protein (H-FABP)	Direct sandwich assay with ALP-labelled antibody	*p*-Aminophenol, amperometric	+300 mV	4–250 ng ml^{-1}	4 ng ml^{-1}	O'Regan *et al.* [102]
Interleukin 6 (IL-6)	Sandwich immunoassay using HRP-labelled antibody with TMB substrate	FIA, amperometry	+0.1 V	3.12–300 pg ml^{-1}	3.1 pg ml^{-1}	Fanjul-Bolado *et al.* [103]
Mycobacterium tuberculosis antigens (Ag 231 and Ag 360)	Sandwich immunoassay based on ALP enzyme, using either streptavidin-coated SPCEs or antibody-coated SPCEs	Cyclic voltammetry or SWV of indigo carmine generated from 3-IP substrate	Scan from −0.25 (CV) or −0.3 (SWV) to +0.2 V	5–100 ng ml^{-1} (Ag 231); 100–5000 ng ml^{-1} (Ag 360)	1.0 ng ml^{-1} (Ag 231); 60 ng ml^{-1} (Ag 360)	Diaz-Gonzalez *et al.* [104]
Pneumolysin	Sandwich immunoassay with ALP or HRP labelled antibody using 3-indoxyl phosphate (3-IP) as substrate	Cyclic voltammetric detection of indigo carmine ($E_{1/2} = -0.15$ V)	Scan from −0.4 to +0.1 V	1.25 (AP) or 2.5 (HRP) to 50 ng ml^{-1}	0.6 (AP) or 2.1 (HRP) ng ml^{-1}	Fanjul-Bolado *et al.* [105,106]

Vitellogenin (Vtg)	Competitive immunoassay via covalently attached anti-Vtg antibody and HRP to polymer-modified SPCE	Amperometric detection of H_2O_2 produced from glucose by GOD-labelled Vtg	−0.3 V	0.25–7.8 ng ml^{-1}	0.09 ng ml^{-1}	Darain *et al.* [107]
Secreted placental alkaline phosphatase (SPAP)	Hydrolysis of 2-naphthyl phosphate by SPAP in cell supernatant	2-Naphthol, Osteryoung SWV	Scan from +0.1 to +0.6 V	nd[a]	< 10 pfu/cell herpes simplex virus or < 20 μM PMA	Kelso *et al.* [108]
Alkaline phosphatase	Enzymatic conversion of anionic substrate into cationic product, P^+	Cyclic voltammetric, oxidation of P^+ accumulated at Nafion-SPCE	Scan from 0 to +0.8 V	4×10^{-16} to 1×10^{-13} M	4×10^{-16} M	Bagel *et al.* [109]
Alkaline phosphatase	Enzymatic production of phenolic products	Cyclic voltammetric, oxidation of product at cation-exchanger-doped SPCE	Scan from 0 to +0.8 V	nd[a]	1×10^{-14} M	Authier *et al.* [110]
N-Acetyl-β-D-glucosamine (NAGase)	Enzyme assay: hydrolysis of 1-naphthyl-*N*-acetyl-β-D-glucosaminide	1-Naphthol, amperometric	+650 mV	3.1–108 mU ml^{-1}	3.1 mU ml^{-1}	Pemberton *et al.* [111]

[a]nd: not determined.

TABLE 23.6

Reports of biosensors for nucleic acids and purines

Analyte	Reaction	Detection	Applied potential/ applied current	Linear/working range	Detection limit	Reference
SsDNA	SPCE with polycationic redox electropolymer/20-base ssDNA probe	Hybridisation, then binding of HRP-labelled sequence; electrocatalytic reduction of H_2O_2 to water	+0.1 V	nd[a]	Mismatched sequence at 200 pM in 25 µl	Dequaire and Heller [112]
SsDNA	Hybridisation at Au electrode, loading of Ag clusters, then dissolution	Chronopotentiometric stripping of Ag ions at bare SPCE	+3.0 µA	500–2500 ng ml^{-1}	100 ng ml^{-1}	Wang *et al.* [113]
SsDNA	Intercalation of daunomycin into hybridised dsDNA on SPCE surface	Chronopotentiometric stripping analysis	+1.0 µA	1–3 µg ml^{-1}	2 µg ml^{-1}	Mascini *et al.* [114–116]
SsDNA	Target ssDNA binds to probe on SWCNT-modified SPCEs and prevents binding of single-strand binding protein	Simultaneous monitoring of guanine oxidation and Tyr, Trp oxidation signals by DPV	+0.6 V (Tyr, Trp); +1.0 V (guanine)	5–25 µg ml^{-1}	0.15 µg ml^{-1}	Kerman *et al.* [117]
Chlamydia trachomatis DNA	Target ssDNA binds to probe on SWCNT-modified SPCEs and prevents binding of single-strand binding protein	Simultaneous monitoring of guanine oxidation and Tyr, Trp oxidation signals by DPV	+0.6 V (Tyr, Trp); +1.0 V (guanine)	0.2–3.0 mg l^{-1}	0.2 mg l^{-1}	Marrazza *et al.* [118]
Apolipoprotein-E DNA	Target ssDNA binds to probe on SWCNT-modified SPCEs and prevents binding of single-strand binding protein	Simultaneous monitoring of guanine oxidation and Tyr, Trp oxidation signals by DPV	+0.6 V (Tyr, Trp); +1.0 V (guanine)	1–2 mg l^{-1}	Mismatched strands at $\geq$1 mg l^{-1}	Marrazza *et al.* [119]
SsDNA/tRNA	Oxidation of guanine/ adenine residues at MWCNT-modified SPCEs	Cyclic voltammetry	+610 mV (guanine); +920 mV (adenine)	17.0–345 µg ml^{-1} (DNA); 8.2 µg ml^{-1} to 4.1 mg ml^{-1} (tRNA)	2.0 µg ml^{-1} (DNA)	Ye and Ju [120]
Breast cancer E908X-WT DNA	Hybridisation of biotinylated target strand to oligonucleotide probe-coated magnetic beads, then addition of streptavidin-Au particles	Dissolution of Au tag, and its chronopotentiometric stripping analysis at a single-use SPCE	+5.0 µA	0.3–30 ppm	100 ng ml^{-1}	Wang *et al.* [121]

BRCA1 (breast cancer) gene DNA	Enzyme-linked sandwich assay using magnetic particle-labelled probe hybridising to a biotinylated DNA target that captures streptavidin-AP	1-Naphthol, produced from 1-NP. DPV at bare SPCE	+0.1 V	0.25–8.0 ppm	10 ppb	Wang *et al.* [122]
Cryptosporidium parvum DNA	Adsorptive hybridisation to immobilised probe at SPCE at +0.5 V, then addition of $Co(phen)_3^{3+}$ indicator	Constant current chronopotentiometric stripping analysis	+0.5 V with constant negative current	50–750 ng ml^{-1}	50 ng ml^{-1}	Wang *et al.* [123]
M. tuberculosis DNA	Adsorptive hybridisation to immobilised probe at SPCE at +0.5 V, then addition of $Co(phen)_3^{3+}$ indicator	Constant current chronopotentiometric stripping analysis	+0.5 V with constant negative current	nd[a]	1 μg ml^{-1}	Wang *et al.* [124]
Streptococcus pneumoniae virulence genes	Hybridisation of FITC-labelled target DNA to biotinylated probes on streptavidin-modified SPCEs. Addition of anti-FITC-ALP conjugate	Voltammetric detection of indigo carmine formed by ALP from 3-indoxyl phosphate	Scan from −0.25 to +0.2 V	0.1–5.0 pg μl^{-1} (ply gene)	0.1 pg μl^{-1} (ply)	Hernandez-Santos *et al.* [125]
Streptococcus pneumoniae virulence genes	Hybridisation of Pt(II) complexed target DNA to biotinylated probes on streptavidin-modified SPCEs	Chronoamperometric detection of hydrogen evolution catalysed by Pt	+1.35 V (1 min), then −1.40 V (5 min)	5.0–100 pg μl^{-1}	5 pg μl^{-1}	Hernandez-Santos *et al.* [126]
HCMV DNA	Adsorbed DNA from sample probed using biotinylated complimentary DNA, then SA-HRP	DPV reduction of 2,2′-diaminoazobenzene product	−0.154 V	50–2000 amol ml^{-1}	3.6×10^{-5} copies ml^{-1} amplified DNA (0.6 amol ml^{-1})	Azek *et al.* [127]
HCMV DNA	Addition of PCR-amplified virus DNA to Nafion-coated SPCEs in microwells	Biotinylated probe, then DPV oxidation of accumulated cationic product of substrate hydrolysis by ALP enzyme	+0.5 V	10–10,000 amol ml^{-1}	10 amol ml^{-1}	Bagel *et al.* [75]
Heat shock transcription factor 1 (HSF1) gene	Microfluidic electrophoretic separation, then oxidation at poly-TTCA-modified SPCE	Amperometric detection of catalytically oxidised DNA	+0.8 V	1–100 pg μl^{-1}	584 fg μl^{-1}	Shiddiky *et al.* [128]

TABLE 23.6 (*continued*)

Analyte	Reaction	Detection	Applied potential/applied current	Linear/working range	Detection limit	Reference
P53 gene DNA	Hybridisation to surface covalent-linked ssDNA probe on multi-SPCE network	Impedance measurement	+400 mV dc, 10 mV ac at 10 kHz to 80 Hz	1–200 nM	1 nM	Marquette *et al.* [129]
Hypoxanthine (Hx), xanthine (X), uric acid (UA)	Electrochemical oxidation at nontronite-coated SPCEs	SWV	+1.0 V (Hx), +0.75 V (X), +0.45 V (UA)	Up to 30 or 40 μM	0.34 (Hx), 0.07 (X) and 0.427 μM (UA)	Zen *et al.* [130]
Uric acid in blood	Direct electrochemical oxidation at bare SPCE	SWV	Scan from −0.2 to +1.0 V	200–1000 μM	<300 μM	Chen *et al.* [131]

[a]nd: not determined.

23.2.11 Pharmaceuticals

Table 23.7 presents an overview of SPCE-based sensors and biosensors developed before 2003 for the determination of pharmaceutical compounds.

Ecstasy, 3,4-methylenedioxymethamphetamine (MDMA) (**I**) and its analogues, 3,4-methylenedioxyamphetamine (MDA) (**II**) and 3,4-methylenedioxyethylamphetamine (MDEA) (**III**), are frequently abused for their stimulant and euphoric effects. As a result of this, there is a growing demand for methods which are cable of determining low levels of these drugs in a number of complicated matrices. Recently, an amperometric SPCE-based immunosensor for detection of MDA, MDMA and MDEA in saliva and urine has been reported by Butler *et al.* [136]. The assay was based on a competition between free analyte and horseradish peroxidase-labelled species for binding to an immobilised polyclonal antibody. Amperometric detection was performed at +100 mV vs. Ag/AgCl, using tetramethylbenzidine/H_2O_2 as a substrate. The total assay time was 45 min and a linear range of 0.61–400 ng ml^{-1} with corresponding limit of detection of 0.36 and 0.042 ng ml^{-1} for saliva and urine, respectively, were reported. The cross-reactivity pattern of the analytes was determined, which showed that the order of sensitivity increases with increase in alkyl chain length (MDA < MDMA < MDEA).

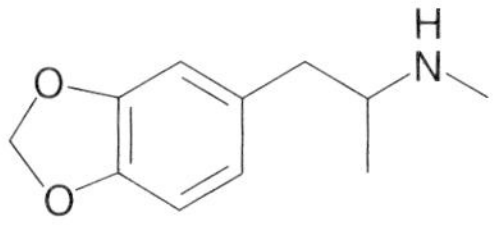

(I)

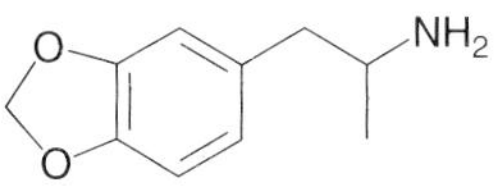

(II)

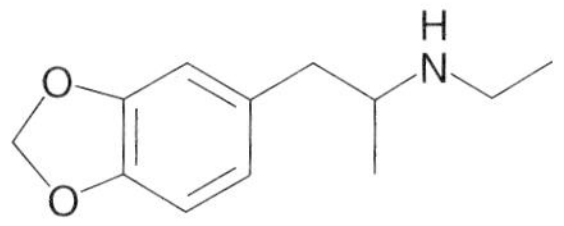

(III)

Levodopa (2-amino-3-(3,4-dihydroxyphenyl)-propanoic acid) (**IV**) is administered to Parkinson's sufferers to alleviate the reduced levels of dopamine commonly seen in such patients. Dopamine itself is not administered, as this compound cannot cross the blood–brain barrier unlike levodopa which is then converted in the body to dopamine [137]. Bergamini *et al.* [138] have utilised a gold screen-printed electrode as a sensor for monitoring levodopa both in stationary solution and in flow systems. This was achieved by using the oxidation of levodopa at +0.63 V in acetate buffer pH 3.0. A linear response was reported from 9.9×10^{-5} to 1.2×10^{-3} M and an associated detection limit of 6.8×10^{-5} M. The sensor was successfully used to detect the levodopa

TABLE 23.7

Reports of SPCE-based sensors and biosensors for pharmaceuticals, prior to 2003

Analyte	SPCE type	Lower linear range	Upper linear range	Analytical application (% recovery)	Detection limit	Remarks	Reference
Codeine	Nontronite clay-modified SPCE	0.625 μM	15 μM	Human urine (96.8–105.9) by square-wave stripping voltammetry	20 nM	Flow injection analysis and square-wave stripping voltammetry	Shih *et al.* [132]
Tricyclic anti-depressive drugs	Cyclodextrin-modified SPCE	0.196 μM	2.75 μM	Serum (86–98)	0.04–0.5 μM	Differential pulse voltammetry	Ferancová *et al.* [133]
Methamphetamine	SPCE modified with monoclonal anti-methamphetamine antibody	200 ng ml^{-1}	1500 ng ml^{-1}	Urine (91.5–100.4)	200 ng ml^{-1}	Amperometric immunosensor	Luangaram *et al.* [134]
Triclosan	Unmodified SPCE	1.2 μM	1.0 mM	Toothpaste and mouthrinse products	1.2 μM	Differential pulse voltammetry	Pemberton *et al.* [135]

concentrations present in two commercial pharmaceutical formulations (Sinemet® and Prolopa®) using a method of multiple standard addition.

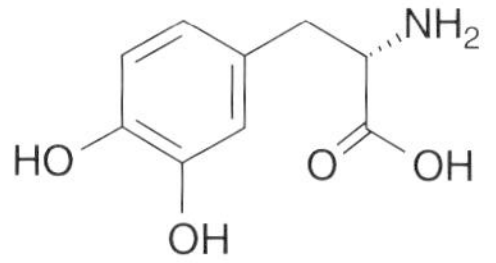

(IV)

23.3 ENVIRONMENTAL

Table 23.8 highlights some of the previously reported environmental applications of SPCEs. These have been dealt with in further detail in our previous review [2].

23.3.1 Biosensors for metal ion detection

Recently, reports [149–151] have been made on the use of the enzyme urease as a bio-recognition element for the detection of metal ions at an SPCE. Urease catalyses the hydrolysis of urea and the formation of NH_3 and can be determined using an NADH–glutamate dehydrogenase coupled reaction system. The NADH consumption can then be monitored correlating to the urease activity. The presence of metal ions is known to inhibit urease activity, resulting in a lowering of the NH_3 (Eq. (23.5)) production and hence a decrease in the consumption of NADH (Eq. (23.4)), thereby giving a negative correlation between NADH signal and metal ion concentration.

$$\text{Urea} + H_2O \xrightarrow{\text{Urease}} CO_2 + 2NH_3 \qquad (23.4)$$

$$NH_3 + \alpha\text{-ketoglutarate} + \text{NADH} + H^+ \xrightarrow{\text{GLDH}} \text{L} - \text{glutamate} + \text{NAD}^+ \qquad (23.5)$$

NADH can be readily monitored electrochemically, and can be used as a simple and effective method to monitor metal ion concentrations. Such an approach has been recently utilised by Rodriguez *et al.* [149] for an SPCE-based biosensor for the amperometric detection of Hg^{2+}, Cu^{2+}, Cd^{2+}, Zn^{+} and Pb^{2+}. Devices used in this study were printed onto 250 μm thick polyester sheet. The working electrode (planar area: $0.16\,cm^2$) was fabricated from a commercially available carbon powder containing 5% rhodium plus promoters, which was made into a screen-printable paste by mixing 1:4 in 2.5% (w/v) hydroxyethyl cellulose in water. The reference electrode ink contained 15% silver chloride in silver paste. The counter electrode and basal tracks were fabricated

TABLE 23.8

Environmental SPCE applications prior to 2003

Analyte	SPCE type	Lower linear range	Upper linear range	Analytical application (% recovery)	Detection limit	Remarks	Reference
Polyaromatic compounds							
Phenanthrene	SPCE immunosensor	5.4 ng ml^{-1}	62.5 ng ml^{-1}	River (98–121) and tap water (70–99)	0.8 ng ml^{-1}	Amperometric	Fähnrich *et al.* [139]
1-Hydroxypyrene	Molecularly imprinted polymer-modified-SPCE	0.1 mM	5 mM	–	–	Linear sweep voltammetry	Kirsch *et al.* [140]
Phenolic compounds							
Phenolic compounds	Horseradish peroxidase modified-SPCE, coupled with immobilised tyrosinase	Phenol 0.025 μM, catechol 0.120 μM, *p*-cresol 0.035 μM	Phenol 45 μM, catechol 43 μM, cresol 38 μM	Lake and river water	2.5 nM for phenol, 10 nM for catechol and 5 nM for *p*-cresol	Amperometric	Cheol Chang *et al.* [141]
Phenolic compounds in beer	Tyrosinase, immobilised SPCE	(–)-Epicatechin 2.5 μM	(–)-Epicatechin 440 μM	Beer	–	Amperometric flow injection analysis	Cummings *et al.* [142]
Pentachlorophenol	CoPC-SPCE competitive inhibition of lactate dehydrogenase using NADPH as a co-factor	–	20	400	26.6 ng ml^{-1}	Amperometric	Young *et al.* [143]
2,4-Dichlorophenoxyacetic acid	MIP-modified SPCE	1 μM	1000 μM	–	1 μM	Differential pulse voltammetry	Kröger *et al.* [144]
Miscellaneous							
Ethanol	SPCE doped with 5% CoPC, coated with alcohol oxidase	0.12 mM	2.00 mM	Beer		Chronoamperometry	Boujtita *et al.* [145]
Sulfite and SO_2	Sulfite oxidase- and cytochrome-modified SPCE	4 ppm	50 ppm	Water and gas streams	4.0 ppm	Amperometric	Abass *et al.* [146], Hart *et al.* [147]
Sulfide	SPCE modified with electrochemically generated nickel oxide layer	20 μM	80 μM	–	5 μM	Linear sweep voltammetry	Giovanelli *et al.* [148]

from 145R carbon ink. The electrodes were then heat-treated at 125°C for 2 h, in order to cure the epoxy resin and to stabilise the three electrodes to allow prolonged use in aqueous solution.

The assay was undertaken by measuring the urease activity in the absence and presence of metal ions. The reaction was undertaken by the addition of 0.1 ml of urea (0.3 M in 0.1 M KCl solution) to an assay cocktail containing 2.5 ml of Tris–HCl buffer (50 mM, pH 8, 0.1 M KCl), 0.1 ml of NADH (8.5 mM in phosphate–KCl buffer), 0.1 ml of α-ketoglutaric acid (25 mM, 0.1 M KCl solution), 0.1 ml of GLDH ($250\,U\,ml^{-1}$ in phosphate–KCl buffer) and 0.1 ml urease ($10\,U\,ml^{-1}$ in Tris–HCl/KCl buffer). A 0.1 ml aliquot was immediately removed and placed onto the electrode surface, and the NADH concentration measured at +0.3 V.

A linear range for Hg^{2+} and Cu^{2+} was obtained between 10 and $100\,\mu g\,l^{-1}$ with detection limits of 7.2 and $8.5\,\mu g\,l^{-1}$, respectively. Cd^{2+} and Zn^{2+} produced enzyme inhibition in the range 1–30 $mg\,l^{-1}$, with limits of detection of $0.3\,mg\,l^{-1}$ for Cd^{2+} and $0.2\,mg\,l^{-1}$ for Zn^{2+}. Lead was found not to inactivate the urease enzyme significantly at the studied range (up to $50\,mg\,l^{-1}$). Coefficients of variation of 6–9% in all cases were reported for each ion studied. The analytical application of the assay was examined on leachate samples obtained from an arsenic production mine (Salsigne, France) and was found to give a good correlation with that found by atomic absorption spectrometry and inductively coupled plasma atomic emission spectroscopy analysis.

Recently, Ogonczyk *et al.* [151] have developed a biosensor for the determination of Cu and Ag, also based on the inhibition of urease at an SPCE, however utilising the co-printing of the enzyme in the sensor manufacture step. The screen-printed sensor was manufactured by hand-mixing 40% of graphite paste, 59.5% of RuO_2 and 0.5% of urease. Potentiometric measurements were performed in stirred, 5.0 mM phosphate buffer in 0.1 M KCl. The developed biosensors were tested as single-use devices, gaining a standard deviation of better than 6% ($n = 64$). Investigations were also made into the possible reuse of the biosensor, and after five cycles of washing and detection, negligible decay of the signal was reported. Storage studies showed that the developed biosensors exhibited over 95% of initial sensitivity after 4 months under ambient conditions (dry, without any special control of humidity and temperature). Interestingly, it was found that the measured signals and their reproducibility were independent of the time between exposure to the sample and the subsequent measurement step. This allowed for the possibility of exposing the sensor to the sample step in the field without need of portable instrumentation and reagents, and then undertaking the subsequent measurement step at some later point in the laboratory. The possible interferences of Pb^{2+}, Sb^{3+}, Cd^{2+}, Co^{2+} and Ni^{2+} were investigated, and under optimised conditions only weak signals were seen for lead, antimony and cadmium ions at concentrations higher than 1 mM. Signals for cobalt and nickel ions were not observed even at concentrations of 10 mM. Calibration graphs for these ions were sigmoidal, and determination of these inhibitors was possible in the ppm range. Copper ions were found to give increased inhibition compared to Ag^{+}; this was concluded to be the result of effects of possible complexation caused by materials used for the enzyme immobilisation.

23.3.2 Voltammetric sensors for metal ion detection

The technique of voltammetric stripping analysis is one of the most sensitive techniques available for the determination of metal ions in complex sample matrices. Traditionally such techniques have been undertaken at Hg electrodes, but in recent years, a large number of reports have focused on the use of SPCEs in this area. Reports prior to 2003 have been reviewed recently by the present authors [3].

A common approach used in a number of investigations is to modify the SPCE with a thin film of Hg. Recently, Palchetti *et al.* [152] have reported on the use of an SPCE modified with a thin film of Hg and cellulose (Methocel) for the determination of Pb^{2+}, Cd^{2+} and Cu^{2+}, using square-wave anodic stripping voltammetry. Detection limits of 0.3, 1 and 0.5 $\mu g\,l^{-1}$ were obtained for Pb^{2+}, Cd^{2+} and Cu^{2+}, respectively; deoxygenation of sample solution was found unnecessary. In all cases, linearity ranging from 1 $\mu g\,l^{-1}$ to 1 $mg\,l^{-1}$ and correlations of 0.9998 for Cd and Pb, and 0.9987 for Cu were obtained.

The SPCE used in this study consisted of a graphite working electrode (diameter 3 mm), a graphite counter electrode and a silver pseudo-reference electrode. These devices were manufactured by depositing ink onto a polyester substrate (thickness 350 μm) in a film of controlled pattern and thickness to obtain overlapping layers. At first the silver tracks were printed, and then a graphite pad was positioned over part of the silver track, to obtain the working and the counter electrodes. An insulating layer with openings that allowed the electrical contact with the potentiostat at one end and the analyte solution at the other end was deposited. Before use the screen-printed graphite working electrode was modified using a solution of cellulose and mercury acetate. Each mercury-coated electrode was pre-treated, before using, by applying −1.1 V for 300 s to convert the Hg acetate to Hg metal. Square-wave voltammetric scans were then carried out until a low and stable background was obtained. Hydrochloric acid was used as supporting electrolyte as the authors suggested that a constant level of chloride ions was needed to maintain the screen-printed Ag pseudo-reference electrode. The analytical application of the device was investigated for a number of different sample types, including water, sediment and soil obtained from Aznalcollar (Spain). A number of different sample preparation steps were investigated and once isolated, sample extracts were simply diluted with electrolyte (0.1 M HCl); a good correlation with results obtained by inductively coupled plasma mass spectrometry (ICP-MS) was reported. Kadara *et al.* [153] have utilised a similar approach for the determination of Cu^{2+} in the acetic acid soluble fraction of soil samples.

Yang *et al.* [154] have reported on the use of a mercury-plated pre-anodised SPCE, used with a thallium internal standard, for the determination of human blood Pb levels. Tl^{3+} was chosen as the internal standard in this research due to its reportedly low endogenous blood concentration. The developed assay was used to investigate the Pb levels of employees in the battery manufacturing industry. The electrochemical pre-treatment of the SPCE was found to enhance the Pb stripping response of the Hg-coated SPCE, in comparison to both bare Hg-free SPCE and Hg-coated SPCEs without prior treatment. The SPCE was first electrochemically pre-treated in 0.1 M H_2SO_4 by continuously cycling in the window of −1.0 to +1.0 V at a scan rate of 100 mV s^{-1} until a stable background current was achieved. The SPCE was then activated by anodising at +2.2 V vs. Ag/AgCl for 10 s in the same solution. An Hg film was deposited by setting the potential at −1.0 V for 180 s. Investigations using ac impedance and chronocoulometry showed that the double-layer charge increases enormously during pre-anodisation due to the reported creation of quinone and other functional groups on the carbon surface. Mercury deposition was found to cause a substantial decrease in the resistance value over those of bare SPCEs and electrochemical pre-treated SPCEs. In contrast, the impedance responses of Hg-coated and Hg-coated electrochemically treated SPCEs were nearly the same. The authors concluded that this was indicative of the lack of any notable interaction of the deposited Hg with the underlying surface functional groups. Investigations were undertaken to optimise the Hg plating time, deposition potential and the concentration of the Tl^{3+} used. Under the optimised conditions, five replicates of 20 ppb Pb^{2+} in the presence of 20 ppb Tl^{3+} internal standard gave a coefficient of variation of 3.16% at the Hg-modified SPCE. The response was found to be linear from 1 to 300 ppb with a slope and a correlation coefficient of 0.039 and 0.999, respectively. Good overall agreement with results gained by graphite furnace atomic absorption spectroscopy was obtained, with an R^2 value of 0.981 being found.

Due to the possible toxic effects of Hg, alternative electrodes have been sought, such as bismuth, which has recently been used by Kadara and Tothill [155,156] for the determination of Pb^{2+} and Cd^{2+} at an SPCE. In this investigation, stripping chronopotentiometric measurements were carried out by depositing a metallic film of bismuth *in situ* with the target metal ions (lead and cadmium). A deposition potential was applied to preconcentrate the analytes, after which, a constant current was applied to strip the preconcentrated analytes until a limit of -0.2 V. The concentrations for Pb^{2+} and Cd^{2+} in the wastewater samples and acetic acid extracts of soils were quantified by the use of a multiple standard addition method.

Cu^{2+} was found to inhibit the stripping response of Pb^{2+} and Cd^{2+} on the bismuth-modified SPCE. The decrease in the Pb^{2+} stripping response (at -0.58 V) obtained in the presence of Cu^{2+} was due to copper altering the character of the electrode by formation of a mixed layer of lead and copper during the deposition step. Hence, the enhancing effect of the bismuth film plated *in situ* on the electrode surface is negated. Recently, Kadara and Tothill [156] have sought to overcome this interference problem by the simple addition of a complexing agent to mask the interfering effect of Cu on the stripping response of Pb. In this study, they have shown that the simple addition of 0.1 M ferricyanide to the test solution effectively achieves this. Good correlations between soil samples determined using this approach and that obtained by ICP-MS were reported.

A gold-based SPCE-based sensor has been described by Laschi *et al.* [157] for the determination of Pb^{2+} and Cu^{2+} in river water using square-wave anodic stripping voltammetry. The working surface of the SPCE was made using a thermoplastic ink containing Au particles. The assay was optimised in terms of accumulation time, supporting electrolyte concentration, deposition time and potential. The electrochemical cells were planar three-electrode strips, based on a gold working electrode, a carbon counter electrode and a silver pseudo-reference electrode printed onto a polyester substrate. Hydrochloric acid was chosen as the supporting electrolyte, for the same reason stated earlier [152]. However, elevated chloride ion concentrations were found to give a decrease in sensitivity, concluded to result from chloride ions reducing the bond strength between metal adatom and the substrate. The assay was found to be linear over 0–50 ng ml^{-1} using a deposition time of 120 s, with a percentage coefficient of variation of 3% for a 20 ng ml^{-1} solution after 50 repetitive measurements. The analytical performance characteristics of the method were evaluated by undertaking three replicate determinations of a river water sample fortified with 100 ng ml^{-1} Pb^{2+} and Cu^{2+}. Recoveries of 93% and 97%, respectively, were obtained, which were in close agreement to those obtained using an Hg-film-based electrode.

Modifications of the SPCE surface with suitable chelating agents or ligands to increase both the selectivity and the sensitivity has become an important area of research in recent years [3]. Recently, Yantasee *et al.* [158] have used such an approach for the determination of Pb^{2+}. In this study, they have modified an Hg-free SPCE with an acetamide phosphoric acid-based chelating agent, formed as a self-assembled monolayer on mesoporous silica. The SPCE was modified by printing a mixture of the chelating agent in ink (5 or 10%, w/w) onto a commercially available SPCE. Pb was accumulated at open circuit and using a medium exchange technique, measured via ASV using a potential of −1.0 V for 120 s, followed by a scan from −0.7 to −0.49 V using DPV. The assay was linear from 1 to 500 ng ml^{-1} when an accumulation time of 3–5 min was used. A detection limit, using a 5 min accumulation time, based on a signal-to-noise ratio of 3 was calculated to be 0.91 ng ml^{-1}. A coefficient of variation of 5% was found for a single sensor. Further studies showed the possibility of detecting both Cd^{2+} and Cu^{2+} alongside Pb^{2+} at the same sensor.

23.3.3 Molecularly imprinted polymer-modified SPCEs

Studies by Hart and co-workers [159,160] have been undertaken to investigate the possibility of using the molecularly imprinted polymers (MIPs) to improve the selectivity of SPCEs for the determination of 1-hydroxypyrene (1-OHP). An initial voltammetric study was undertaken to ascertain the electrochemical behaviour of this molecule at an SPCE [159]. Cyclic voltammograms exhibited two quasi-reversible pairs of peaks generated from the oxidation of the parent molecule. The nature of the redox reactions was further investigated by observing the effect of scan rate and pH on the voltammetric behaviour; the results suggested that the oxidation reactions giving rise to two peaks involved electrochemical–chemical–electrochemical (ECE) and EECE processes.

Additional studies were carried out by HPLC and GC/MS and the results indicated that the oxidation products were a mixture of several pyrene–quinones. In addition, LC/MS investigations showed that 1-OHP had undergone oxidation reactions to produce several dimers.

Further investigations were made to develop a two-step non-competitive affinity method for the trace determination of 1-OHP using a disposable MIP-modified screen-printed carbon electrode (MIP-SPCE) [160]. The MIP was synthesised according to a novel strategy, which is described, and was capable of rebinding the phenolic analyte, 1-OHP, from high pH aqueous organic media, via ionic interactions. Figure 23.5 describes the outline of the method employed. In the first step of the method, 1-OHP was accumulated at the MIP-SPCE from 35% aqueous methanol containing 0.014 M NaOH and 0.14 M NaCl, at open circuit. In the second step, the resulting SPCE with accumulated 1-OHP was then transferred to fresh, clean phosphate-buffered aqueous methanol, and subjected to cyclic voltammetry (CV) or DPV. The latter technique proved to be more sensitive at detecting 1-OHP, with a limit of detection of 182 nM and a linear range up to 125 μM on unmodified electrodes. The possible effects of interference by related phenolic compounds in the MIP-SPCE of 1-OHP were investigated. The method

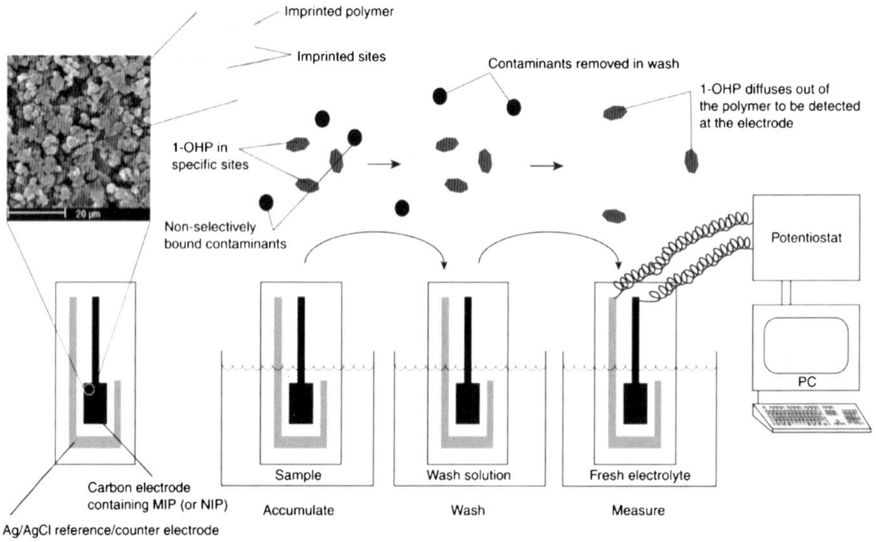

Fig. 23.5. Schematic representation of the MIP-SPCE sensor device and the measurement steps. NIP: non-imprinted polymer (control). Reproduced from Ref. [160] by permission of Wiley-VCH Verlag GmbH & Co.

was evaluated by carrying out 1-OHP determinations on spiked human urine samples; the recovery of 1-OHP was 79.4% and the coefficient of variation was found to be 7.7% ($n = 4$) using a separate MIP-SPCE for each determination.

23.3.4 Organophosphates (OPs)

Standard analytical techniques used for the analysis of OP pesticides have been based on chromatographic techniques [161]. These techniques are time-consuming and expensive when applied to the volumes of samples generated in monitoring OPs in soil, water and agricultural produce. Immunosensors have proved difficult and complex to develop and manufacture [162], especially the problems associated with matrix effects and the regeneration of the immunosurface [163]. Therefore, development of SPCE biosensors for the detection of OP and carbamate pesticides has mainly involved enzymatic systems. Such sensor systems have been based on the enzymes organophosphorus hydrolase (OPH) and acetylcholinesterase (AChE). The development of SPCEs for the production of enzymatic biosensors forms a reliable, rapid and economic method for analysing the presence of OPs in a wide variety of media [2]. SPCE biosensors have the potential to provide a range of applications, from the general preliminary evaluation of a sample to indicate the presence of OPs prior to a more detailed identification and quantification of specific OPs. Comparisons between enzyme-based biosensors and chromatographic techniques have shown comparable sensitivity in measuring individual OPs [164].

OPH-based biosensors have been fully discussed in previous reviews [2,165]. AChE-based biosensors are based on the principle that OP pesticides have an inhibitory effect on the activity of AChE that may be permanent or partially reversible. The extent of the inhibition is directly related to the concentration of the pesticide and therefore enzyme activity may be used as a measure of the inhibition [166]. The amperometric measurement of AChE activity can be based on the measurement of any of the following three mechanisms [167]: (1) production of hydrogen peroxide from choline, (2) oxygen consumption during the enzyme reaction or (3) production of electroactive compounds directly from the oxidation of acetylthiocholine chloride such as thiocholine. The measurement of hydrogen peroxide and oxygen consumption has been described in more details in other reviews [167].

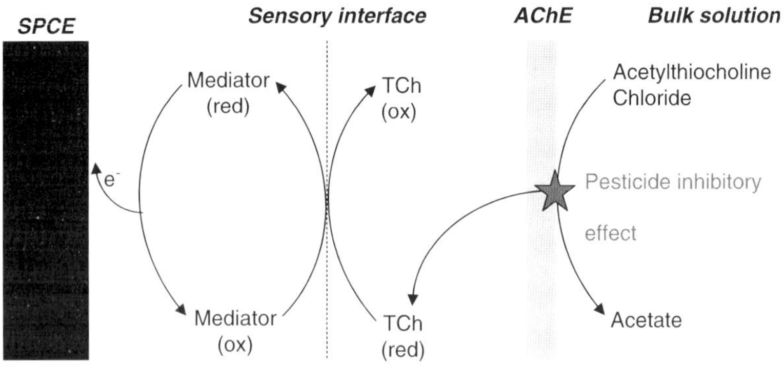

Fig. 23.6. Sequence of reactions occurring at a pesticide biosensor based on an SPCE modified with AChE with a reversible mediator.

Mediators such as 7,7,8,8-tetracyanoquinodimethane (TCNQ) [166], hexacyanoferrate(III) [168] and CoPC [169], which have been applied to a carbon paste or ink, are reduced by the reaction with thiocholine and then reoxidised at the carbon electrode (Fig. 23.6). A second type of mediator has involved the addition of Prussian blue (ferric hexanocyanoferrate) into an AChE and choline oxidase (ChO) bienzyme biosensor. Prussian blue mediates the reduction of hydrogen peroxide produced by the conversion of choline to betaine by ChO [170].

Several methods for immobilising AChE onto the electrode surface have been proposed. SPCE systems have successfully used carbon electrodes by cross-linking AChE with glutaraldehyde [171] or photocross-linkable polyvinyl alcohol (PVA) [172]. Investigations into the immobilisation of AChE onto the electrode surface have more recently examined methods with no physical entrapment of the AChE or covalent binding between the enzyme and the electrode surface. Such affinity methods have been based either on metal ions reversibly binding histidine [173] or cysteine-tagged AChE, or on linkage between the AChE molecule and the lectin Concanavalin A where sugar and divinyl sulfone are used to link to amino groups derivatised onto the electrode surface [174]. Studies by Bonnet *et al.* [175] showed that the direct adsorption of the enzyme onto the electrode surface was possible, resulting in a detection limit for chlorpyrifos-ethyl-oxon in water of $1.2\,ng\,l^{-1}$. However, this method used large quantities of enzyme and leaching of the enzyme into the substrate occurred, leading to high variability between biosensors.

The incorporation of CNTs into OP biosensors may improve the amperometric response and the reversibility of the mediator used in the carbon SPCE in AChE- and OPH-based systems. Deo *et al.* [176] incorporated CNTs into OPH-based biosensors resulting in a 10-fold increase in sensitivity to methyl parathion and paraoxon with limits of detection of 0.8 and 0.15 μM, respectively. Lin *et al.* [177] immobilised AChE and ChO enzymes on CNTs on the surface of the working electrode to produce an SPCE biosensor. Following optimisation, these biosensors achieved a limit of detection for methyl parathion of 0.05 μM and a linear range up to 200 μM.

At present, the ability of AChE-based systems has been demonstrated to operate effectively with a wide range of substrates and matrices. Del Carlo *et al.* [178] used electric eel AChE-based assays to detect parathion-methyl and carbaryl in a variety of biological matrices including egg, meat, milk and honey with a negligible matrix effect. In this study, parathion-methyl and carbaryl were detected at concentrations of 10 ng ml^{-1} in solvent extracts. Using biosensors based on electric eel AChE with a Prussian blue mediator, Del Carlo *et al.* [179] obtained sample detection limits of 5 mg kg^{-1} (65–133 ng ml^{-1} in extracts) for pirimiphos-methyl in durum wheat. This method included an oxidation step to produce the oxon form of the OP and the inclusion of the oxidant (*N*-bromosuccinimide) significantly increased the detection limits. A bienzyme system (AChE and ChO) with a Prussian blue mediator developed by Longobardi *et al.* [180] was a multi-step analysis of dichlorvos in wheat with a limit of detection of 0.05 μg g^{-1} in spiked wheat extracts. The results had a good correlation with GC measurements made on the same extracts. Similarly, Schulze *et al.* [181] found good correlation with GC analysis in the use of TCNQ-mediated electric eel and recombinant human AChE in a wide-ranging analysis of OPs in babyfood. In the absence of discriminative data, the method produced evidence of potential endocrine disruptors above maximum residue levels in 3 of the 23 babyfoods tested rather than identifying OPs present. The results of this study highlighted the importance of developing biosensor assays for qualitative testing that can discriminate between OPs for improved risk assessment.

A strategy for producing AChE biosensors for quantitative analysis of substrates and for a qualitative assessment of the OPs present has concentrated on the modification of the enzyme active site. Possible differences in the inhibition mechanism have been further identified in genetically modified AChE, principally from *Drosophila melanogaster* [182]; structural changes and the replacement of specific amino acids within the AChE have resulted in the greater sensitivity of particular mutant AChE to specific OPs. For example, Boublik *et al.* [183]

discovered that a combination of *D. melanogaster* mutant AChEs DmE69Y and DmY71D was 300-fold more sensitive to dichlorvos than the wild-type. Nunes *et al.* [184] identified that the *D. melanogaster* mutant AChE B03 was capable of detecting 1.4 ppb methamidophos in a standard solution compared to 4.8 ppb for the wild-type, and Avramescu *et al.* [185] observed an order of magnitude increase in biosensor sensitivity to dichlorvos on replacing wild-type AChE with a modified strain ($(His)_6$). Schulze *et al.* [186] described a similar procedure by engineering 10 single- and 4 double-point mutations to the AChE peptide in *Nippostrongylus brasiliensis*. Incorporating three of these mutated enzymes and the wild-type into a multi-enzyme array system allowed the improved detection of 11 out of 14 OPs tested to concentrations below $10\,\mu g\,kg^{-1}$. The engineered mutations caused a variety of changes to the active site of the AChE resulting in a greater specificity of the active site to individual OPs. Bachmann *et al.* [172] developed a multi-electrode system for water analysis incorporating two multi-sensors, each containing four mutant AChEs. This system made it possible to distinguish between paraoxon and carbofuran and between malaoxon and paraoxon, within the concentration range of 0–5 ppb in a total assay time of 40 min.Crew *et al.* [187] described a multi-enzyme array comprising five genetically modified strains of AChE from *D. melanogaster* and wild-type, which was produced for the analysis of five OPs detected in food at a concentration range of 1–0.01 mM. The results illustrated that differing responses between mutant AChEs and OPs that can improve the detectability of OPs can also form the basis of a recognition system to identify individual OPs. Several recent reviews have discussed the potential of such biosensors for the detection of pesticides in different matrices and for qualitative and quantitative measurements [188–190].

23.3.5 Polychlorinated biphenyls

Polychlorinated biphenyls (PCBs) have been known for sometime as persistent pollutants, which can be readily bioaccumulated through the food chain causing well-documented toxic effects in number of species including humans [191]. Consequently, PCBs are commonly routinely monitored as potential industrial pollutants. Due to their environmental persistence and toxicity, detection limits in the ng ml^{-1} region are generally required. Consequently, such work has generally required solvent or solid extraction and concentration steps prior to separation by GC in conjunction with electron capture detection, or mass spectroscopy. However, such techniques require highly trained personnel and a well-equipped laboratory for their implementation. One attractive alternative method is that of immunoassay, which is quite often less demanding of laboratory equipment and personnel.

Recently, Centi *et al.* [192] have used such technology to develop a disposable antibody-coated magnetic bead SPCE-based immunoassay for the detection of PCBs. The electrochemical cells used were planar three-electrode strips, based on a carbon working electrode, a carbon counter electrode and a silver pseudo-reference electrode. The magnetic beads were first coated with IgG anti-PCB28, and an aliquot of these was introduced to the sample in the presence of PCB28-AP conjugate solution. After an incubation time of 20 min, the beads were separated magnetically from the supernatant. The isolated beads were then washed and re-suspended in the electrolyte before an aliquot of the resulting mixture was transferred to the SPCE surface. To better localise the beads onto the electrode, a magnet-holding block was placed on the bottom part of the electrode. The enzymatic substrate (α-naphthyl phosphate) was then added to the dried mixture on the SPCE surface, and after 5 min the resulting enzymatic product was determined by DPV.

Both marine sediment and soil samples were investigated by the developed assay. These were first either Soxhlet extracted or sonicated to isolate the PCB fraction. An aliquot of the resulting extract was then taken and the associated solvent allowed to evaporate before examination by the new assay. Detection limits for the Aroclor 1248, 1242 and 1254 mixtures were reported to be 0.4, 0.3 and 0.5 $ng\,ml^{-1}$, respectively.

23.3.6 Explosives and their residues

We have previously reviewed in more detail SPCE-based sensors for the determination of this class of compounds [2]. Table 23.9 summaries the salient details for these devices.

More recently, a polyphenol-coated SPCE was used for the voltammetric measurement of the 2,4,6-trinitrotoluene (TNT) explosive in the presence of surface-active substances [195]. SPCEs were modified with a polyphenol film to exclude large surface-active macromolecules present in the sample matrix, while allowing the relatively small TNT molecules access to the electrochemically active surface of the SPCE. Electrochemical polymerisation of the polyphenol film on the SPCE was achieved by scanning the potential between 0.0 and +0.8 V (vs. Ag/AgCl reference electrode) at a rate of 100 $mV\,s^{-1}$ for three cycles

TABLE 23.9

Reports of SPCE-based sensors and biosensors for explosives and their residues, prior to 2003

Analyte	SPCE type	Lower linear range	Upper linear range	Analytical application (% recovery)	Detection limit	Remarks	Reference
2,6 Dinitrotoluene	Unmodified SPCE	$161\,ng\,ml^{-1}$	$137\,\mu g\,ml^{-1}$	Potable water (102.4), dust wipes (73.4) and saliva (47.5)	$161\,ng\,ml^{-1}$	Linear sweep voltammetry	Honeychurch *et al.* [193]
2,4,6-Trinitrotoluene	Unmodified SPCE	–	–	Potable water and river water	$100\,ng\,ml^{-1}$[a]	SWV	Wang *et al.* [194]

[a]Using computerized background-correction.

in a 0.5 M NaCl solution, containing 0.1 M phenol. The number of cycles employed was found to be an important factor as beyond nine cycles complete disappearance of the TNT was observed. Good resistance to fouling by high concentrations of the various compounds such as humic acid, SDS and gelatine on the voltammetric response of TNT was reported. The bare carbon electrode displayed a rapid attenuation of the TNT peak with increasing surfactant concentration. The stability of the sensor was investigated over a 3 h experiment involving 60 repetitive measurements of 6 ppm TNT in the presence of 50 ppm humic acid. No apparent loss in the sensitivity of the coated electrode, or change in its peak potential or width, was reported. Square-wave voltammograms were recorded at room temperature by using a 0.5 M NaCl supporting electrolyte. A detection limit of around 1 ppm TNT was reported for the polyphenol-modified SPCEs with a coefficient of variation of 8.2% for a 6 ppm concentration of TNT in the presence of 50 ppm humic acid. Under the same conditions, a coefficient of variation of 42% was obtained for unmodified SPCEs, demonstrating the improvements obtained using this form of modification.

23.3.7 Food toxins

23.3.7.1 Domoic acid

Domoic acid (**V**) is present in a number of different types of shellfish and crabs as a result of their ingestion of this algal toxin, a product of the diatom *Pseudonitzschia pungens*. In sufficient amounts, domoic acid is known to cause amnesic shellfish poisoning in humans, which causes a number of rather unpleasant effects [196]. A number of SPCE-based sensors have been reported for the determination of domoic acid [197,198]. Kania *et al.* [197] have raised polyclonal antibodies against this food toxin: the antiserum produced against a keyhole haemocyanin conjugate displayed a high affinity for the free domoic acid and showed little cross-reactivity with similar structured analytes (kainic acid, aspartic acid, glutamic acid, geranic acid and 2-methyl-3-butenoic acid). The identified polyclonal antibody was then used in the optimisation of an SPCE based on the amperometric detection of alkaline phosphatase-generated *p*-aminophenol. Detection limits were reported in the ng ml^{-1} range, and the linear range of the sensor was comparable with that generally reported for food poisoning incidents. The developed SPCE assay was found to be within $\pm 12\%$ of that found by a previously developed ELISA assay.

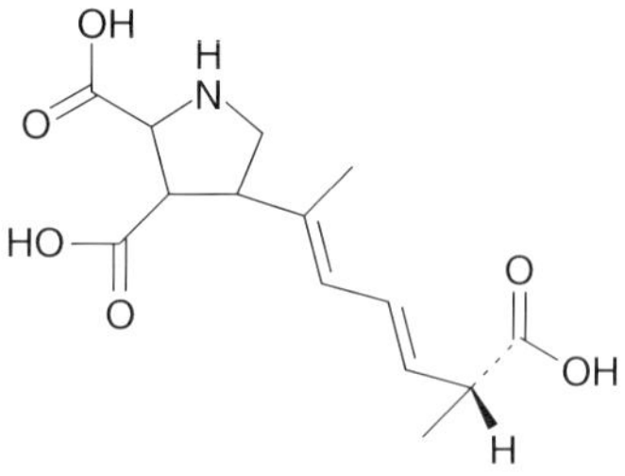

(V)

Micheli *et al.* [198] have reported the use of an electrochemical immunosensor coupled to differential pulse voltammetry for the detection of domoic acid. Domoic acid conjugated to bovine serum albumin (BSA-DA) was coated onto the working electrode of the SPCE, followed by incubation with sample and anti-domoic acid antibody. An anti-goat IgG-alkaline phosphatase conjugate was used for signal generation. The analytical application of the assay was investigated for mussels spiked with domoic acid. Levels as low as 20 $\mu g\,g^{-1}$ domoic acid were discernable, which represents the maximum acceptable limit defined by the Food and Drug Administration. A detection limit of 5 $ng\,ml^{-1}$ of domoic acid was reported.

23.3.7.2 β-ODAP sensor

Recently, an SPCE-based biosensor as part of a flow injection system has been described for the determination of *β*-*N*-oxalyl-L-*α*,*β*-diaminopropionic acid (*β*-ODAP) (**VI**) [199,200], a non-essential amino acid responsible for neurolathyrism, a disorder caused by eating excessive amounts of the seeds of grass pea, *Lathyrus sativus*. In the recent study undertaken by Beyene *et al.* [199], the developed biosensor was based on amperometric measurement of hydrogen peroxide at an MnO_2 bulk-modified SPCE, the H_2O_2 being generated as an intermediate in the oxidation of *β*-ODAP by glutamate oxidase. The biosensors were manufactured by simply mixing the MnO_2 with the screen-printing ink, and then immobilising the enzyme on the surface in Nafion. Both the enzymatic and electrochemical reactions were found to be dependent on pH, with the optimum response being obtained at pH 8. Both pH and the applied potential were found to be important in the assay. At higher pH values, the chemical reduction of H_2O_2 was found to occur, competing with the electrochemical re-oxidation of MnO_2. Similarly, at elevated pH values *β*-ODAP also underwent hydrolysis. At potentials lower than +440 mV, decreasing responses were observed and the background current was reductive, which reportedly can lead to a gradual leaching of the modifier from the electrode due to the formation of a soluble Mn^{2+} species. At potentials above this value, the background current was found to increase.

The analytical application of the biosensor system was investigated on a number of pea grass samples. The natural levels of glutamate present in the samples were first removed via the addition of glutamate dehydrogenase, the addition of which was shown to have no effect on the overall β-ODAP present in the sample. The assay was linear from 195 to 1950 μM with a detection limit of 111.0 μM β-ODAP. The flow injection SPCE biosensor assay for pea grass samples spiked at $50\,mg\,l^{-1}$ β-ODAP was found to give a 98.6% recovery, with real samples only varying between ±0.02% and 0.07%.

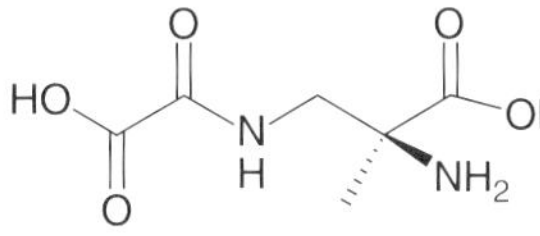

(VI)

23.3.7.3 Mycotoxins

Alarcon *et al.* [201] have reported an immunosensor for Ochratoxin A (OTA) (**VII**) based on a competitive immunoassay using an alkaline phosphatase–OTA conjugate taking place on the surface of SPCEs bearing surface-bound anti-OTA antibody. By using 1-naphthyl phosphate (1-NP) as enzyme substrate and detection of the product, 1-naphthol, by DPV, the immunosensor was able to measure OTA in buffer over the dynamic range $0.25–250\,ng\,ml^{-1}$, with a detection limit of $180\,pg\,ml^{-1}$. In red wine, the working range was $250\,pg\,ml^{-1}$ to $250\,ng\,ml^{-1}$ with a detection limit for OTA contamination of $900\,pg\,ml^{-1}$.

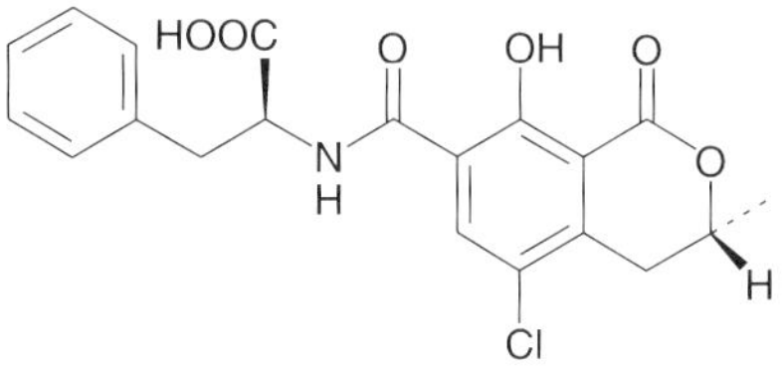

(VII)

Studies towards the development of a sensor array for detecting mycotoxins in grain have been described [202]; an immunosensor for aflatoxin B_1 (AFB_1) (**VIII**) is described as the first example sensor of this array. The sensor is based on an SPCE bearing surface-absorbed anti-AFB_1 antiserum. Competition between biotinylated and free AFB_1 is followed by the addition of streptavidin-labelled alkaline phosphatase enzyme. Finally, addition of 1-NP substrate yields 1-naphthol which is detected at the SPCE surface by linear sweep voltammetry. The use of a biotin–toxin conjugate avoids the need to synthesise an enzyme–toxin conjugate and the SPCEs are designed so that all of the experimental steps take place in 96-well microtitre plates. A calibration plot for AFB_1 in buffer is reported over the concentration range 0.15–2.5 $ng\,ml^{-1}$, with a detection limit of around 0.15 $ng\,ml^{-1}$.

A disposable immunosensor for AFB_1 determination in barley has been reported [203], this time using an SPCE bearing surface-immobilised AFB_1 antigen conjugated to BSA. After incubation of the sensor with the sample and an anti-AFB_1 monoclonal antibody, the addition of a secondary antibody labelled with alkaline phosphatase, enabled the generation of 1-naphthol from 1-NP, and its detection using DPV. The assay had a reported detection limit of 30 $pg\,ml^{-1}$ in barley.

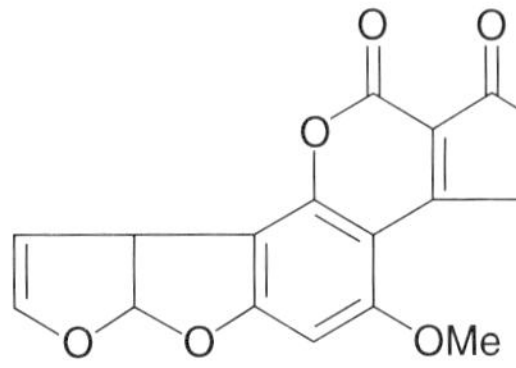

(VIII)

Most recently [204], SPCEs have been tailored using surface-adsorbed rabbit IgG and monoclonal anti-aflatoxin M1 (anti-AFM1) antibody, for the determination of AFM1 (**IX**) in milk. Competition between free AFM1 and an AFM1–HRP enzyme conjugate was followed by the addition of a 3,3′,5,5′-tetramethylbezidine dihydrochloride substrate, then electrochemical reduction of the resulting enzymatic product by chronoamperometry at –100 mV. The assay showed a working range in milk of between 30 and 160 ppt, with a detection limit of 25 ppt.

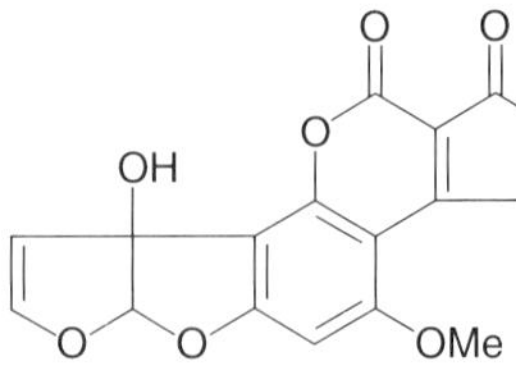

(IX)

23.3.7.4 Porphyrin compounds

The technique of stripping voltammetry has been applied to the detection of chlorophyll a (Chl a) at SPCEs [205]. By exploiting the adsorption of Chl a to the electrode surface, an accumulation step at pH 7, followed by medium exchange and an anodic voltammetric scan, was seen to give a peak response for Chl a at +400 mV vs. Ag/AgCl. Using this approach, a working range of 0.014–2.24 mM was obtained for the detection of Chl a in phosphate buffer. This approach was demonstrated to be capable of quantifying the Chl a content of cows' faeces and therefore of being applicable to the determination of the faecal content in water. Such an approach may be of value in the dairy industry for monitoring the cleanliness of cows' udders prior to milking.

23.3.7.5 Phosphate

An important parameter in a number of fields is the study of inorganic phosphate. Recently, Kwan *et al.* [206,207] have reported on a screen-printed phosphate biosensor based on immobilised pyruvate oxidase (PyOD) for monitoring phosphate concentrations in a sequencing batch reactor system [206] and in human saliva [207]. The enzyme was immobilised by drop-coating a Nafion solution onto the working electrode surface; this was then covered by a poly(carbamoyl) sulfonate (PCS) hydrogel membrane.

Phosphate is detected via the immobilised PyOD which converts it in the presence of pyruvate and oxygen to hydrogen peroxide, carbon dioxide and acetylphosphate (Eq. (23.6)). The enzymatically generated hydrogen peroxide can then be monitored at +420 mV vs. Ag/AgCl:

$$\text{Phosphate} + \text{Pyruvate} + O_2 \xrightarrow{\text{PyOD}} \text{Acetylphosphate} + H_2O_2 + CO_2 \qquad (23.6)$$

The time required for one measurement using this phosphate biosensor was found to be 4 min, which compares favourably with the time required using a commercial phosphate testing kit at 10 min. The biosensor had a reported linear range from 7.5 to 625 μM phosphate with an associated detection limit of 3.6 μM. A good correlation ($R^2 = 0.9848$) between the commercial phosphate testing kit and the developed phosphate biosensor for the measurement of synthetic wastewater in a sequencing batch reactor system was found. This sensor maintained a high working stability (>85%) after 12 h of operation and involved a simple operation procedure.

23.3.7.6 Ammonia

Recently, a disposable biosensor for ammonium ions in sewage effluent has been developed and is based on an SPCE coated with the enzyme glutamate dehydrogenase, 2-oxoglutarate and NADH [208]. This rapid, and selective, amperometric biosensor is capable of detecting NH_4^+ in sewage effluent at concentrations in the range 1–10 ppm. The amperometric biosensor was based on a transducer fabricated by screen-printing a proprietary carbon ink formulation, containing the electrocatalyst MB, onto a plastic substrate. This was further modified by coating the surface with a proprietary solution containing GLDH, stabiliser, NADH and 2-oxoglutarate. An Ag/AgCl electrode was screen-printed alongside the biosensor and served as the reference/counter electrode. The device was completed by covering the active area with a mesh which served to protect the surface from damage, and allowed automatic fixed volume filling by a simple dipping procedure. Sewage effluent determinations were performed using a PC-controlled automated hand-held instrument, in conjunction with the biosensor shown in Fig. 23.7. The programme is initiated so that the heating cycle is started, meanwhile the biosensor is inserted into the effluent sample so that the mesh area is submerged to allow automatic filling to a fixed volume. When the temperature of the heater has reached 37°C, the exposed biosensor is inserted into the heating compartment of the instrument and the lid closed. An incubation period is undertaken, and the measurement step

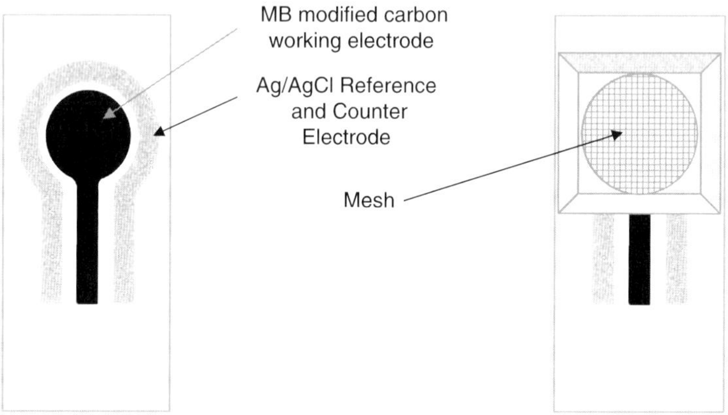

Fig. 23.7. Commercially available (AET, Ltd., UK) screen-printed NH_4^+ amperometric biosensors. The diagram on the left shows the shape of the counter/reference electrode and working electrode (area 0.28 cm^2); the diagram on the right shows the complete biosensor with the mesh in position (After Ref [208]).

is then performed using chronoamperometry with an operating potential of +50 mV vs. the screen-printed Ag/AgCl reference electrode. The instrument delivers an output directly in concentration units of ammonium ion, as an electronic calibration plot is provided with each batch of amperometric biosensors. The total time for the assay is 5 min and 20 s, and the results give a good correlation with those produced by the standard spectroscopic method ($R^2 = 0.9869$). The biosensors were found to be stable for at least 396 days at a temperature of 37°C (equivalent to 665 days at room temperature).

23.4 CONCLUSIONS

This chapter has described a number of approaches for the fabrication of prototype screen-printed electrochemical (bio)sensors together with selected applications in the areas of biomedical, environmental and industrial analyses. Novel methods of immobilising the bio-recognition elements of the biosensors have been emphasised and it is clear that researchers are designing simpler methods of manufacture. One example of this has been the one-step fabrication of a glucose biosensor using a water-based carbon ink incorporating both mediator and enzyme [19,20]. The authors believe that portable electrochemical instruments will become more readily available allowing measurements to be made outside of the laboratory. In one example, we have collaborated with several organisations to develop an automatic instrument for use with a biosensor to determine NH_4^+ in sewage effluent [208].

However, it should be mentioned that there is a flexible hand-held electrochemical instrument on the market, which can be programmed to be used in a variety of voltammetric/amperometric modes in the field [209]. Although the majority of biosensor applications described in this review were for single analyte detection, it is very likely that future directions will involve development of biosensor arrays for multi-analyte determinations. One example of this approach has been described in an earlier section, where five OPs could be monitored with an array of biosensors based on mutant forms of AChE from *D. melanogaster* [187]. This array has considerable potential for monitoring the quality of food, such as wheat and fruit. Developments and applications of biosensors in the area of food analysis are expected to grow as consumer demand for improved quality and safety increases. Another area where biosensor developments are likely to increase significantly is in the field of environmental analysis, particularly with respect to the defence of public safety. In the biomedical field, the development of biosensors for early warning of cancer and heart problems would be of major benefit to the individual as well as to the economy.

Screen-printing is a proven technology, readily adaptable to the manufacture of diverse sensor/biosensor devices for a wide range of applications, and has great potential for new devices, particularly where simplicity of fabrication, and operations at low cost, are important factors. Future devices are likely to incorporate nanoparticles where enhanced sensitivity with miniaturisation may be required.

ACKNOWLEDGMENTS

The authors wish to thank HEFCE for funding. They would also like to thank fellow researchers whose work has been included in this review.

REFERENCES

1 J.P. Hart and S.A. Wring, Recent developments in the design and application of screen-printed electrochemical sensors for biomedical, environmental and industrial analyses, *TrAC*, 16 (1997) 89–103.
2 J.P. Hart, A. Crew, E. Crouch, K.C. Honeychurch and R.M. Pemberton, Some recent designs and developments of screen-printed carbon electrochemical sensors/biosensors for biomedical, environmental and industrial analyses, *Anal. Lett.*, 37 (2004) 789–830.
3 K.C. Honeychurch and J.P. Hart, Screen-printed electrochemical sensors for monitoring metal pollutants, *TrAC*, 7/8 (2003) 456–469.
4 C.A. Galán-Vidal, J. Muñoz, C. Domínguez and S. Alegret, Chemical sensors, biosensors and thick-film technology, *TrAC*, 14 (1995) 225–231.
5 M. Alvarez-Icaza and U. Bilitewski, Mass-production of biosensors, *Anal. Chem.*, 65 (1993) A525–A533.
6 B.D. Malhotra and A. Chaubey, Biosensors for clinical diagnostics industry, *Sens. Actuators B Chem.*, 91 (2003) 117–127.
7 H.H. Weetall, Chemical sensors and biosensors, update, what, where, when and how, *Biosens. Bioelectron.*, 14 (1999) 237–242.
8 P.T. Kissinger, Electrochemical biosensors—promise vs. reality, *Quim. Anal.*, 19(Suppl. 1) (2000) 5–7.
9 WHO, *Diabetes Action Now: An Initiative of the World Health Organisation and the International Diabetes Federation*, WHO, Geneva, Switzerland, 2004.
10 D.R. Matthews, E. Bown, A. Watson, R.R. Holman, J. Steemson, S. Hughs and D. Scott, Pen-sized digital 30-second blood-glucose meter, *Lancet*, 1 (1987) 778–779.

11 F. Ricci and G. Palleschi, Sensor and biosensor preparation, optimisation and applications of Prussian Blue modified electrodes, *Biosens. Bioelectron.*, 21 (2005) 389–407.
12 M.A.T. Gilmartin, R.J. Ewen, J.P. Hart and C.L. Honeybourne, Voltammetric and photoelectron spectral elucidation of the electrocatalytic oxidation of hydrogen-peroxide at screen-printed carbon electrodes chemically-modified with cobalt phthalocyanine, *Electroanalysis*, 7 (1995) 547–555.
13 K. Wang, J.-J. Xu and H.-Y. Chen, A novel glucose biosensor based on the nanoscaled cobalt phthalocyanine–glucose oxidase biocomposite, *Biosens. Bioelectron.*, 20 (2005) 1388–1396.
14 J.Z. Xu, J.J. Zhu, Q. Wu, Z. Hu and H.Y. Chen, Direct electron transfer between glucose oxidase and multi-walled carbon nanotubes, *Chin. J. Chem.*, 21 (2003) 1088–1091.
15 Y.D. Zhao, W.D. Zhang, H. Chen and Q.M. Luo, Direct electron transfer of glucose oxidase molecules adsorbed onto carbon nanotube powder microelectrode, *Anal. Sci.*, 18 (2002) 939–941.
16 J.Q. Liu, A. Chou, W. Rahmat, M.N. Paddon-Row and J.J. Gooding, Achieving direct electrical connection to glucose oxidase using aligned single walled carbon nanotube arrays, *Electroanalysis*, 17 (2005) 38–46.
17 R. Wedge, R.M. Pemberton, J.P. Hart and R. Luxton, Recent developments towards disposable screen-printed biosensors incorporating a carbon ink modified with the redox mediator, Meldola's Blue, *Analysis*, 27 (1999) 570–577.
18 A. Memoli, M.C. Annesini, M. Mascini, S. Papale and S. Petralito, A comparison between different immobilised glucoseoxidase-based electrodes, *J. Pharm. Biomed. Anal.*, 29 (2002) 1045–1052.
19 E. Crouch, D.C. Cowell, S. Hoskins, R.W. Pittson and J.P. Hart, A novel, disposable, screen-printed amperometric biosensor for glucose in serum fabricated using a water-based carbon ink, *Biosens. Bioelectron.*, 21 (2005) 712–718.
20 E. Crouch, D.C. Cowell, S. Hoskins, R.W. Pittson and J.P. Hart, Amperometric, screen-printed, glucose biosensor for analysis of human plasma samples using a biocomposite water-based carbon ink incorporating glucose oxidase, *Anal. Biochem.*, 347 (2005) 17–23.
21 M.P. O'Halloran, M. Pravda and G.G. Guilbault, Prussian Blue bulk modified screen-printed electrodes for H_2O_2 detection and for biosensors, *Talanta*, 55 (2001) 605–611.
22 M. Pravda, M.P. O'Halloran, M.P. Kreuzer and G.G. Guilbault, Composite glucose biosensor based on screen-printed electrodes bulk modified with Prussian blue and glucose oxidase, *Anal. Lett.*, 35 (2002) 959–970.
23 F. Ricci, A. Amine, G. Palleschi and D. Moscone, Prussian Blue based screen printed biosensors with improved characteristics of long-term lifetime and pH stability, *Biosens. Bioelectron.*, 18 (2003) 165–174.

24 F. Ricci, A. Amine, C.S. Tuta, A.A. Ciucu, F. Lucarelli, G. Palleschi and D. Moscone, Prussian Blue and enzyme bulk-modified screen-printed electrodes for hydrogen peroxide and glucose determination with improved storage and operational stability, *Anal. Chim. Acta*, 485 (2003) 111–120.
25 A. Lupu, D. Compagnone and G. Palleschi, Screen-printed enzyme electrodes for the detection of marker analytes during winemaking, *Anal. Chim. Acta*, 513 (2004) 67–72.
26 F. Ricci, D. Moscone, C.S. Tuta, G. Palleschi, A. Amine, A. Poscia, F. Valgimigli and D. Messeri, Novel planar glucose biosensors for continuous monitoring use, *Biosens. Bioelectron.*, 20 (2005) 1993–2000.
27 J. Wang and X. Zhang, Screen printed cupric-hexacyanoferrate modified carbon enzyme electrode for single-use glucose measurements, *Anal. Lett.*, 32 (1999) 1739–1749.
28 S. Kröger, S.J. Setford and A.P.F. Turner, Assessment of glucose oxidase behaviour in alcoholic solutions using disposable electrodes, *Anal. Chim. Acta*, 368 (1998) 219–231.
29 E. Turkušić, K. Kalcher, K. Schachl, A. Komersova, M. Bartos, H. Moderegger, I. Svancara and K. Vytras, Amperometric determination of glucose with an MnO_2 and glucose oxidase bulk-modified screen-printed carbon ink biosensor, *Anal. Lett.*, 34 (2001) 2633–2647.
30 P. Kotzian, P. Brazdilova, K. Kalcher and K. Vytras, Determination of hydrogen peroxide, glucose and hypoxanthine using (bio)sensors based on ruthenium dioxide-modified screen-printed electrode, *Anal. Lett.*, 38 (2005) 1099–1113.
31 A.S. Kumar and J.M. Zen, Electrochemical investigation of glucose sensor fabricated at copper-plated screen-printed carbon electrodes, *Electroanalysis*, 14 (2002) 671–678.
32 J. Wang and M. Musameh, Carbon nanotube screen-printed electrochemical sensors, *Analyst*, 129 (2004) 1–2.
33 W.-J. Guan, Y. Li, Y.-Q. Chen, X.-B. Zhang and G.-Q. Hu, Glucose biosensor based on multi-wall carbon nanotubes and screen printed carbon electrodes, *Biosens. Bioelectron.*, 21 (2005) 508–512.
34 Z. Gao, F. Xie, M. Shariff, M. Arshad and J.Y. Ying, A disposable glucose biosensor based on diffusional mediator dispersed in nanoparticulate membrane on screen-printed carbon electrode, *Sens. Actuators B Chem.*, 111–112 (2005) 339–346.
35 G. Cui, J.H. Yoo, B.W. Woo, S.S. Kim, G.S. Cha and H. Nam, Disposable amperometric glucose sensor electrode with enzyme-immobilized nitrocellulose strip, *Talanta*, 54 (2001) 1105–1111.
36 C.X. Zhang, Q. Gao and M. Aizawa, Flow injection analytical system for glucose with screen-printed enzyme biosensor incorporating Os-complex mediator, *Anal. Chim. Acta*, 426 (2001) 33–41.
37 C. Zhang, T. Haruyama, E. Kobatake and M. Aizawa, Disposable electrochemical capillary-fill device for glucose sensing incorporating a

water-soluble enzyme/mediator layer, *Anal. Chim. Acta*, 442 (2001) 257–265.

38 K. Ohfuji, N. Sato, N. Hamada-Sato, T. Kobayashi, C. Imada, H. Okuma and E. Watanabe, Construction of a glucose sensor based on a screen-printed electrode and a novel mediator pyocyanin from *Pseudomonas aeruginosa*, *Biosens. Bioelectron.*, 19 (2004) 1237–1244.

39 F. Ge, X.E. Zhang, Z.P. Zhang and X.M. Zhang, Simultaneous determination of maltose and glucose using a screen-printed electrode system, *Biosens. Bioelectron.*, 13 (1998) 333–339.

40 T. Hu, X.E. Zhang, Z.P. Zhang and L.Q. Chen, A screen-printed disposable enzyme electrode system for simultaneous determination of sucrose and glucose, *Electroanalysis*, 12 (2000) 868–870.

41 L.Q. Chen, X.E. Zhang, W.H. Xie, Y.F. Zhou, Z.P. Zhang and A.E.G. Cass, Genetic modification of glucose oxidase for improving performance of an amperometric glucose biosensor, *Biosens. Bioelectron.*, 17 (2002) 851–857.

42 N.J. Forrow and S.W. Bayliff, A commercial whole blood glucose biosensor with a low sensitivity to hematocrit based on an impregnated porous carbon electrode, *Biosens. Bioelectron.*, 21 (2005) 581–587.

43 L.S. Bean, L.Y. Heng, B.M. Yamin and M. Ahmad, Photocurable ferrocene-containing poly(2-hydroxyl ethyl methacrylate) films for mediated amperometric glucose biosensor, *Thin Solid Films*, 477 (2005) 104–110.

44 J.P. Hart, A.K. Abass, K.C. Honeychurch, R.M. Pemberton, S.L. Ryan and R. Wedge, Sensors/biosensors, based on screen-printing technology for biomedical applications, *Indian J. Chem., Sect. A*, 42 (2003) 709–718.

45 J. Razumiene, V. Gureviien, A. Vilkanauskyt, L. Marcinkeviien, I. Bachmatova, R. Mekys and V. Laurinaviius, Improvement of screen-printed carbon electrodes by modification with ferrocene derivative, *Sens. Actuators B Chem.*, 95 (2003) 378–383.

46 C.A. Galán-Vidal, J. Muñoz, C. Domínguez and A. Alegret, Glucose biosensor based on a reagentless graphite–epoxy screen-printable biocomposite, *Sens. Actuators B Chem.*, 45 (1997) 55–62.

47 C.A. Galán-Vidal, J. Muñoz, C. Dominguez and S. Alegret, Glucose biosensor strip in a three electrode configuration based on composite and biocomposite materials applied by planar thick film technology, *Sens. Actuators B Chem.*, 52 (1998) 257–263.

48 J. MacLachlan, A.T. Wotherspoon, R.O. Ansell and C.J. Brooks, Cholesterol oxidase: sources, physical properties and analytical applications, *J. Steroid Biochem. Mol. Biol.*, 72 (2000) 169–195.

49 C.C. Allain, L.S. Poon, C.S. Chan, W. Richmond and P.C. Fu, Enzymatic determination of total serum cholesterol, *Clin. Chem.*, 20 (1974) 470–475.

50 M.A.T. Gilmartin and J.P. Hart, Fabrication and characterization of a screen-printed, disposable, amperometric cholesterol biosensor, *Analyst*, 119 (1994) 2331–2336.
51 G. Li, J.M. Liao, G.Q. Hu, N.Z. Ma and P.J. Wu, Study of carbon nanotube modified biosensor for monitoring total cholesterol in blood, *Biosens. Bioelectron.*, 20 (2005) 2140–2144.
52 X. Tan, M. Li, P. Cai, L. Luo and X. Zou, An amperometric cholesterol biosensor based on multiwalled carbon nanotubes and organically modified sol–gel/chitosan hybrid composite film, *Anal. Biochem.*, 337 (2005) 111–120.
53 R. Foster, J. Cassidy and E. O'Donoghue, Electrochemical diagnostic strip device for total cholesterol and its subfractions, *Electroanalysis*, 12 (2000) 716–721.
54 V. Shumyantseva, G. Deluca, T. Bulko, S. Carrara, C. Nicolini, S.A. Usanov and A. Archakov, Cholesterol amperometric biosensor based on cytochrome P450scc, *Biosens. Bioelectron.*, 19 (2004) 971–976.
55 P. Xu, T. Yano, K. Yamamoto, H. Suzuki and H. Kumagai, Screening for bacterial strains producing lactate oxidase, *J. Ferm. Bioeng.*, 81 (1996) 357–359.
56 G.L. Zubay, W.W. Parson and D.E. Vance, *Principles of Biochemistry*, Wm. C. Brown, Dubuque, US, 1995.
57 K. Naka, Enzyme confusion, *Clin. Chem.*, 39 (1993) 2351.
58 J.D. Newman, S.F. White, I.E. Tothill and A.P. Turner, Catalytic materials, membranes, and fabrication technologies suitable for the construction of amperometric biosensors, *Anal. Chem.*, 67 (1995) 4594–4599.
59 A. Avramescu, T. Noguer, V. Magearu and J.L. Marty, Chronoamperometric determination of D-lactate using screen-printed enzyme electrodes, *Anal. Chim. Acta*, 433 (2001) 81–88.
60 A. Avramescu, T. Noguer, M. Avramescu and J.L. Marty, Screen-printed biosensors for the control of wine quality based on lactate and acetaldehyde determination, *Anal. Chim. Acta*, 458 (2002) 203–213.
61 S.C. Litescu, L. Rotariu and C. Bala, Immobilisation of lactate dehydrogenase on electro-polymerised Meldola blue matrix, *Rev. Chim.*, 56 (2005) 57–60.
62 A.L. Hart, C. Matthews and W.A. Collier, Estimation of lactate in meat extracts by screen-printed sensors, *Anal. Chim. Acta*, 386 (1999) 7–12.
63 M. Albareda-Sirvent and A.L. Hart, Preliminary estimates of lactic and malic acid in wine using electrodes printed from inks containing sol–gel precursors, *Sens. Actuators B Chem.*, 87 (2002) 73–81.
64 W.A. Collier, P. Lovejoy and A.L. Hart, Estimation of soluble L-lactate in dairy products using screen-printed sensors in a flow injection analyser, *Biosens. Bioelectron.*, 13 (1998) 219–225.
65 I. Rohm, M. Genrich, W. Collier and U. Bilitewski, Development of ultraviolet-polymerizable enzyme pastes: bioprocess applications of screen-printed L-lactate sensors, *Analyst*, 121 (1996) 877–881.

66 J.-M. Zen, D.-M. Tsai and A.S. Kumar, Flow injection analysis of ascorbic acid in real samples using a highly stable chemically modified screen-printed electrode, *Electroanalysis*, 15 (2003) 1171–1176.
67 E. Turkušić, V. Milicevic, H. Tahmiscija, M. Vehabovic, S. Basic and V. Amidzic, Amperometric sensor for L-ascorbic acid determination based on MnO_2 bulk modified screen printed electrode, *Fresenius J. Anal. Chem.*, 368 (2000) 466–470.
68 J. Lin, D.M. Zhou, S.B. Hocevar, E.T. McAdams, B. Ogorevc and X.J. Zhang, Nickel hexacyanoferrate modified screen-printed carbon electrode for sensitive detection of ascorbic acid and hydrogen peroxide, *Front. Biosci.*, 10 (2005) 483–491.
69 J.P. Hart, R.M. Pemberton, R. Luxton and R. Wedge, Studies towards a disposable screen-printed amperometric biosensor for progesterone, *Biosens. Bioelectron.*, 12 (1997) 1113–1121.
70 R.M. Pemberton, J.P. Hart and J.A. Foulkes, Development of a sensitive, selective electrochemical immunoassay for progesterone in cow's milk based on a disposable screen-printed amperometric biosensor, *Electrochim. Acta*, 43 (1998) 3567–3574.
71 R.M. Pemberton, J.P. Hart, P. Stoddard and J.A. Foulkes, A comparison of 1-naphthyl phosphate and 4-aminophenyl phosphate as enzyme substrates for use with a screen-printed amperometric immunosensor for progesterone in cows' milk, *Biosens. Bioelectron.*, 14 (1999) 495–503.
72 R.M. Pemberton, J.P. Hart and T.T. Mottram, An electrochemical immunosensor for milk progesterone using a continuous flow system, *Biosens. Bioelectron.*, 16 (2001) 715–723.
73 M.P. Kreuzer, R. McCarthy, M. Pravda and G.G. Guilbault, Development of electrochemical immunosensor for progesterone analysis in milk, *Anal. Lett.*, 37 (2004) 943–956.
74 R.M. Pemberton, T.T. Mottram and J.P. Hart, Development of a screen-printed carbon electrochemical immunosensor for picomolar concentrations of estradiol in human serum extracts, *J. Biochem. Biophys. Methods*, 63 (2005) 201–212.
75 O. Bagel, C. Degrans, B. Limoges, M. Joannes, F. Azek and P. Brossier, Enzyme affinity assays involving a single-use electrochemical sensor. Applications to the enzyme immunoassay of human chorionic gonadotrophin hormone and nucleic acid hybridization of human cytomegalovirus DNA, *Electroanalysis*, 12 (2000) 1447–1452.
76 J. Wang, J.W. Mo and A. Erdem, Single-use thick-film electrochemical sensor for insulin, *Electroanalysis*, 14 (2002) 1365–1368.
77 M.P. Kreuzer, C.K. O'Sullivan, M. Pravda and G.G. Guilbault, Development of an immunosensor for the determination of allergy antibody (IgE) in blood samples, *Anal. Chim. Acta*, 442 (2001) 45–53.
78 J. Wang, P.V.A. Pamidi and K.R. Rogers, Sol–gel-derived thick-film amperometric immunosensors, *Anal. Chem.*, 70 (1998) 1171–1175.

79 F. Darain, S.-U. Park and Y.-B. Shim, Disposable amperometric immunosensor system for rabbit IgG using a conducting polymer modified screen-printed electrode, *Biosens. Bioelectron.*, 18 (2003) 773–780.
80 M. Dequaire, C. Degrand and B. Limoges, An electrochemical metalloimmunoassay based on a colloidal gold label, *Anal. Chem.*, 72 (2000) 5521–5528.
81 Q. Gao, Y. Ma, Z.L. Cheng, W.D. Wang and M.R. Yang, Flow injection electrochemical enzyme immunoassay based on the use of an immunoelectrode strip integrate immunosorbent layer and a screen-printed carbon electrode, *Anal. Chim. Acta*, 488 (2003) 61–70.
82 J. Wang, A. Ibanez and M.P. Chatrathi, Microchip-based amperometric immunoassays using redox tracers, *Electrophoresis*, 23 (2002) 3744–3749.
83 M. Diaz-Gonzalez, D. Hernandez-Santos, M.B. Gonzalez-Garcia and A. Costa-Garcia, Development of an immunosensor for the determination of rabbit IgG using streptavidin modified screen-printed carbon electrodes, *Talanta*, 65 (2005) 565–573.
84 Q. Gao, B. Qi, Y. Sha and X. Yang, Electrodepositing redox polymer on sandwich complex for the improvement of sensitivity in sandwich enzyme-linked immunoassay, *Chem. Lett.*, 33 (2004) 1198–1199.
85 Y. Zhang and A. Heller, Reduction of the non-specific binding of a target antibody and of its enzyme-labeled detection probe enabling electrochemical immunoassay of an antibody through the 7 pg/mL–100 ng/mL (40 fM–400 pM) range, *Anal. Chem.*, 77 (2005) 7758–7762.
86 G. Evtugyn, A. Mingaleva, H. Budnikov, E. Stoikova, V. Vinter and S. Eremin, Affinity biosensors based on disposable screen-printed electrodes modified with DNA, *Anal. Chim. Acta*, 479 (2003) 125–134.
87 J.G. Guan, Y.Q. Miao and J.R. Chen, Prussian blue modified amperometric FIA biosensor: one-step immunoassay for alpha-fetoprotein, *Biosens. Bioelectron.*, 19 (2004) 789–794.
88 H. Yu, F. Yan, Z. Dai and H.X. Ju, A disposable amperometric immunosensor for alpha-1-fetoprotein based on enzyme-labeled antibody/chitosan-membrane-modified screen-printed carbon electrodes, *Anal. Biochem.*, 331 (2004) 98–105.
89 P. Sarkar, I.E. Tothill, S.J. Setford and A.P.F. Turner, Screen-printed amperometric biosensors for the rapid measurement of L- and D-amino acids, *Analyst*, 124 (1999) 865–870.
90 J.M.N. Zen, C.T. Hsu, A.S. Kumar, H.J. Lyuu and K.Y. Lin, Amino acid analysis using disposable copper nanoparticle-plated electrodes, *Analyst*, 129 (2004) 841–845.
91 M. Vasjari, A. Merkoci, J.P. Hart and S. Alegret, Amino acid determination using screen-printed electrochemical sensors, *Microchim. Acta*, 150 (2005) 233–238.

92 S. Grant, F. Davis, K. Law, A.C. Barton, S.D. Collyer, S.P.J. Higson and T.D. Gibson, Label-free and reversible immunosensor based upon an ac impedance interrogation protocol, *Anal. Chim. Acta*, 537 (2005) 163–168.

93 P. Sarkar, One-step separation-free amperometric biosensor for the detection of protein, *Microchem. J.*, 64 (2000) 283–290.

94 S.J. Setford, S.F. White and J.A. Bolbot, Measurement of protein using an electrochemical bi-enzyme sensor, *Biosens. Bioelectron.*, 17 (2002) 79–86.

95 P. Sarkar and A.P.F. Turner, Application of dual-step potential on single screen-printed modified carbon paste electrodes for detection of amino acids and proteins, *Fresenius J. Anal. Chem.*, 364 (1999) 154–159.

96 P. Sarkar, P.S. Pal, D. Ghosh, S.J. Setford and I.E. Tothill, Amperometric biosensors for detection of the prostate cancer marker (PSA), *Int. J. Pharm.*, 238 (2002) 1–9.

97 Y. Guo and A.R. Guadalupe, Screen-printable surfactant-induced sol–gel graphite composites for electrochemical sensors, *Sens. Actuators B Chem.*, 46 (1998) 213–219.

98 C.G. Stone, M.F. Cardosi and J. Davis, A mechanistic evaluation of the amperometric response of reduced thiols in quinone mediated systems, *Anal. Chim. Acta*, 491 (2003) 203–210.

99 F. Ricci, F. Arduini, A. Amine, D. Moscone and G. Palleschi, Characterisation of Prussian blue modified screen-printed electrodes for thiol detection, *J. Electroanal. Chem.*, 563 (2004) 229–237.

100 F. Ricci, F. Arduini, C.S. Tuta, U. Sozzo, D. Moscone, A. Amine and P. Palleschi, Glutathione amperometric detection based on a thiol-disulfide exchange reaction, *Anal. Chim. Acta*, 558 (2006) 164–170.

101 E. Crowley, C. O'Sullivan and G.G. Guilbault, Amperometric immunosensor for granulocyte-macrophage colony-stimulating factor using screen-printed electrodes, *Anal. Chim. Acta*, 389 (1999) 171–178.

102 T.M. O'Regan, M. Pravda, C.K. O'Sullivan and G.G. Guilbault, Development of a disposable immunosensor for the detection of human heart fatty-acid binding protein in human whole blood using screen-printed carbon electrodes, *Talanta*, 57 (2002) 501–510.

103 P. Fanjul-Bolado, M.B. Gonzalez-Garia and A. Costa-Garcia, Amperometric detection in TMB/HRP-based assays, *Anal. Bioanal. Chem.*, 382 (2005) 297–302.

104 M. Diaz-Gonzalez, M.B. Gonzalez-Garcia and A. Costa-Garcia, Immunosensor for *Mycobacterium tuberculosis* on screen-printed carbon electrodes, *Biosens. Bioelectron.*, 20 (2005) 2035–2043.

105 P. Fanjul-Bolado, M.B. Gonzalez-Garcia and A. Costa-Garcia, 3-Indoxyl phosphate as an electrochemical substrate for horseradish peroxidase, *Electroanalysis*, 16 (2003) 988–993.

106 P. Fanjul-Bolado, M.B. Gonzalez-Garcia and A. Costa-Garcia, Voltammetric determination of alkaline phosphatase and horseradish

peroxidase activity using 3-indoxyl phosphate as substrate. Application to immunoassay, *Talanta*, 64 (2004) 452–457.
107 F. Darain, D.S. Park, J.S. Park, S.C. Chang and Y.B. Shim, A separation-free amperometric immunosensor for vitellogenin based on screen-printed carbon arrays modified with a conductive polymer, *Biosens. Bioelectron.*, 20 (2005) 1780–1787.
108 E. Kelso, J. McLean and M.F. Cardosi, Electrochemical detection of secreted alkaline phosphatase: implications to cell based assays, *Electroanalysis*, 12 (2000) 490–494.
109 O. Bagel, B. Limoges, B. Schollhorn and C. Degrand, Subfemtomolar determination of alkaline phosphatase at a disposable screen-printed electrode modified with a perfluorosulfonated ionomer film, *Anal. Chem.*, 69 (1997) 4688–4694.
110 L. Authier, B. Schollhorn and B. Limoges, Detection of cationic phenolic derivatives at a surfactant-doped screen-printed electrode for the sensitive indirect determination of alkaline phosphatase, *Electroanalysis*, 10 (1998) 1255–1259.
111 R.M. Pemberton, J.P. Hart and T.T. Mottram, An assay for the enzyme *N*-acetyl-β-D-glucosaminidase (NAGase) based on electrochemical detection using screen-printed carbon electrodes (SPCEs), *Analyst*, 126 (2001) 1866–1871.
112 M. Dequaire and A. Heller, Screen-printing of nucleic acid detecting carbon electrodes, *Anal. Chem.*, 74 (2002) 4370–4377.
113 J. Wang, O. Rincon, R. Polsky and E. Dominguez, Electrochemical detection of DNA hybridization based on DNA-templated assembly of silver cluster, *Electrochem. Commun.*, 5 (2003) 83–86.
114 G. Marrazza, I. Chianella and M. Mascini, Disposable DNA electrochemical sensor for hybridization detection, *Biosens. Bioelectron.*, 14 (1999) 43–51.
115 M. Mascini, I. Palchetti and G. Marrazza, DNA electrochemical biosensors, *Fresenius J. Anal. Chem.*, 369 (2001) 15–22.
116 M. Minunni, S. Tombelli, E. Mariotti, M. Mascini and M. Mascini, Biosensors as new analytical tools for detection of Genetically Modified Organisms (GMOs), *Fresenius J. Anal. Chem.*, 369 (2001) 589–593.
117 K. Kerman, Y. Morita, Y. Takamura and E. Tamiya, *Escherichia coli* single-strand binding protein–DNA interactions on carbon nanotube-modified electrodes form a label-free electrochemical hybridization sensor, *Anal. Bioanal. Chem.*, 381 (2005) 1114–1121.
118 G. Marrazza, I. Chianella and M. Mascini, Disposable DNA electrochemical biosensors for environmental monitoring, *Anal. Chim. Acta*, 387 (1999) 297–307.
119 G. Marrazza, G. Chiti, M. Mascini and M. Anichini, Detection of human apolipoprotein E genotypes by DNA electrochemical biosensor coupled with PCR, *Clin. Chem.*, 46 (2000) 31–37.

120 Y. Ye and H. Ju, Rapid detection of ssDNA and RNA using multi-walled carbon nanotubes modified screen-printed carbon electrode, *Biosens. Bioelectron.*, 21 (2005) 735–741.
121 J. Wang, D.K. Xu, A.N. Kawade and R. Polsky, Metal nanoparticle-based electrochemical stripping potentiometric detection of DNA hybridization, *Anal. Chem.*, 73 (2001) 5576–5581.
122 J. Wang, D. Xu, A. Erdem, R. Polsky and M.A. Salazar, Genomagnetic electrochemical assays of DNA hybridization, *Talanta*, 56 (2002) 931–938.
123 J. Wang, G. Rivas, C. Parrado, X. Cai and M.N. Flair, Electrochemical biosensor for detecting DNA sequences from the pathogenic protozoan *Cryptosporidium parvum*, *Talanta*, 44 (1997) 2003–2010.
124 J. Wang, X. Cai, G. Rivas, H. Shiraishi and N. Dontha, Nucleic-acid immobilization recognition and detection at chronopotentiometric DNA chips, *Biosens. Bioelectron.*, 12 (1997) 587–599.
125 D. Hernandez-Santos, M. Diaz-Gonzalez, M.B. Gonzalez-Garcia and A. Costa-Garcia, Enzymatic genosensor on streptavidin-modified screen-printed carbon electrodes, *Anal. Chem.*, 76 (2004) 6887–6893.
126 D. Hernandez-Santos, M.B. Gonzalez-Garcia and A. Costa-Garcia, Genosensor based on a platinum (II) complex as electrocatalytic label, *Anal. Chem.*, 77 (2005) 2868–2874.
127 F. Azek, C. Grossiord, M. Joannes, B. Limoges and P. Brossier, Hybridization assay at a disposable electrochemical biosensor for the attomole detection of amplified human cytomegalovirus DNA, *Anal. Biochem.*, 284 (2000) 107–113.
128 M.J.A. Shiddiky, D.-S. Park and Y.-B. Shim, Detection of polymerise chain reaction fragments using a conducting polymer-modified screen-printed electrode in a microfluidic device, *Electrophoresis*, 26 (2005) 4656–4663.
129 C.A. Marquette, M.F. Lawrence and L.J. Blum, DNA covalent immobilization onto screen-printed electrode networks for direct label-free hybridization detection of p53 sequences, *Anal. Chem.*, 78 (2006) 959–964.
130 J.-M. Zen, Y.-Y. Lai, H.-H. Yang and A.S. Kumar, Multianalyte sensor for the simultaneous determination of hypoxanthine, xanthine and uric acid based on a preanodized nontronite-coated screen-printed electrode, *Sens. Actuators B Chem.*, 84 (2002) 237–244.
131 J.-C. Chen, H.-H. Chung, C.-T. Hsu, D.-M. Tsai, A.S. Kumar and J.-M. Zen, A disposable single-use electrochemical sensor for the detection of uric acid in human whole blood, *Sens. Actuators B Chem.*, 110 (2005) 364–369.
132 Y. Shih, J.-M. Zen and H.-H. Yang, Determination of codeine in urine and drug formulations using a clay-modified screen-printed carbon electrode, *J. Pharm. Biomed. Anal.*, 29 (2002) 827–833.
133 A. Ferancová, E. Korgová, T. Buzinkaiová, W. Kutner, I. Stěpánek and J. Labuda, Electrochemical sensors using screen-printed carbon

electrode assemblies modified with the β-cyclodextrin or carboxymethylated β-cyclodextrin polymer films for determination of tricyclic antidepressive drugs, *Anal. Chim. Acta*, 447 (2001) 47–54.
134 K. Luangaram, D. Boonsua, S. Soontornchai and C. Promptmas, Development of an amperometric immunosensor for the determination of methamphetamine in urine, *Biocatal. Biotransform.*, 20 (2002) 397–403.
135 R.M. Pemberton and J.P. Hart, Electrochemical behaviour of triclosan at a screen-printed carbon electrode and its voltammetric determination in toothpaste and mouthrinse products, *Anal. Chim. Acta*, 390 (1999) 107–115.
136 D. Butler, M. Pravda and G.G. Guilbault, Development of a disposable amperometric immunosensor for the detection of ecstasy and its analogues using screen-printed electrodes, *Anal. Chim. Acta*, 556 (2006) 333–339.
137 H.V. Barnes, *Clinical Medicine*, Year Book Medical Publisher, New York, 1988, pp. 745–750.
138 M.F. Bergamini, A.L. Santos, N.R. Stradiotto and M.V.B. Zanoni, A disposable electrochemical sensor for the rapid determination of levodopa, *J. Pharm. Biomed. Anal.*, 39 (2005) 54–59.
139 K.A. Fähnrich, M. Pravda and G.G. Guilbault, Disposable amperometric immunosensor for the detection of polycyclic aromatic hydrocarbons (PAHs) using screen-printed electrodes, *Biosens. Bioelectron.*, 18 (2003) 73–82.
140 N. Kirsch, J.P. Hart, D.J. Bird, R.W. Luxton and D.V. McCalley, Towards the development of molecularly imprinted polymer based screen-printed sensors for metabolites of PAHs, *Analyst*, 126 (2001) 1936–1941.
141 S. Cheol Chang, K. Rawson and C.J. McNeil, Disposable tyrosinase-peroxidase bi-enzyme sensor for amperometric detection of phenols, *Biosens. Bioelectron.*, 17 (2002) 1015–1023.
142 E.A. Cummings, S. Linquette-Mailley, P. Mailley, S. Cosnier, B.R. Eggins and E.T. McAdams, A comparison of amperometric screen-printed, carbon electrodes and their application to the analysis of phenolic compounds present in beers, *Talanta*, 55 (2001) 1015–1027.
143 S.J. Young, J.P. Hart, A.A. Dowman and D.C. Cowell, The non-specific inhibition of enzymes by environmental pollutants: a study of a model system towards the development of electrochemical biosensor arrays, *Biosens. Bioelectron.*, 16 (2001) 887–894.
144 S. Kröger, A.P.F. Turner, K. Mosbach and K. Haupt, Imprinted polymer based sensor system for herbicides using differential-pulse voltammetry on screen-printed electrodes, *Anal. Chem.*, 71 (1999) 3698–3702.
145 M. Boujtita, J.P. Hart and R. Pittson, Development of a disposable ethanol biosensor based on a chemically modified screen-printed electrode coated with alcohol oxidase for the analysis of beer, *Biosens. Bioelectron.*, 15 (2000) 257–263.

146 A.K. Abass, J.P. Hart and D. Cowell, Development of an amperometric sulphite biosensor based on sulphite oxidase with cytochrome *c*, as electron acceptor, and a screen-printed transducer, *Sens. Actuators B Chem.*, 62 (2000) 148–153.
147 J.P. Hart, A.K. Abass and D. Cowell, Development of disposable amperometric sulphur dioxide biosensors based on screen-printed electrodes, *Biosens. Bioelectron.*, 17 (2002) 389–394.
148 D. Giovanelli, N.S. Lawrence, S.J. Wilkins, L. Jiang, T.G.J. Jones and R.G. Compton, Anodic stripping voltammetry of sulphide at a nickel film: towards the development of a reagentless sensor, *Talanta*, 61 (2003) 211–220.
149 B.B. Rodriguez, J.A. Bolbot and I.E. Tothill, Development of urease and glutamic dehydrogenase amperometric assay for heavy metals screening in polluted samples, *Biosens. Bioelectron.*, 19 (2004) 1157–1167.
150 B.B. Rodriguez, J.A. Bolbot and I.E. Tothill, Urease–glutamic dehydrogenase biosensor for screening heavy metals in water and soil samples, *Anal. Bioanal. Chem.*, 380(2) (2004) 284–292.
151 D. Ogonczyk, Ł. Tymecki, I. Wzykiewicz, R. Koncki and S. Galb, Screen-printed disposable urease-based biosensors for inhibitive detection of heavy metal ions, *Sens. Actuators B Chem.*, 106 (2005) 450–454.
152 I. Palchetti, S. Laschi and M. Mascini, Miniaturised stripping-based carbon modified sensor for in field analysis of heavy metals, *Anal. Chim. Acta*, 530 (2005) 61–67.
153 R.O. Kadara, J.D. Newman and I.E. Tothill, Stripping chronopotentiometric detection of copper using screen-printed three-electrode system—application to acetic-acid bioavailable fraction from soil samples, *Anal. Chim. Acta*, 493 (2003) 95–104.
154 C.-C. Yang, A.S. Kumar and J.-M. Zen, Precise blood lead analysis using a combined internal standard and standard addition approach with disposable screen-printed electrodes, *Anal. Biochem.*, 338 (2005) 278–283.
155 R.O. Kadara and I.E. Tothill, Stripping chronopotentiometric measurements of lead(II) and cadmium(II) in soils extracts and wastewaters using a bismuth film screen-printed electrode assembly, *Anal. Bioanal. Chem.*, 378 (2004) 770–775.
156 R.O. Kadara and I.E. Tothill, Resolving the copper interference effect on the stripping chronopotentiometric response of lead (II) obtained at bismuth film screen-printed electrode, *Talanta*, 66 (2005) 1089–1093.
157 S. Laschi, I. Palchetti and M. Mascini, Gold-based screen-printed sensor for detection of trace lead, *Sens. Actuators B Chem.*, 114 (2006) 460–465.
158 W. Yantasee, L.A. Deibler, G.E. Fryxell, C. Timchalk and Y. Lin, Screen-printed electrodes modified with functionalized mesoporous silica for voltammetric analysis of toxic metal ions, *Electrochem. Commun.*, 7 (2005) 1170–1176.

159 K.C. Honeychurch, J.P. Hart and N. Kirsch, Voltammetric, chromatographic and mass spectral elucidation of the redox reactions of 1-hydroxypyrene occurring at a screen-printed carbon electrode, *Electrochim. Acta*, 49 (2004) 1141–1149.
160 N. Kirsch, K.C. Honeychurch, J.P. Hart and M.J. Whitcombe, Voltammetric determination of urinary 1-hydroxypyrene using molecularly imprinted polymer-modified screen-printed carbon electrodes, *Electroanalysis*, 17 (2005) 571–578.
161 J. Sherma, Pesticides, *Anal. Chem.*, 67 (1995) 1R–20R.
162 E.M. Garrido, C. Delerue-Matos, J.L.F.C. Lima and A.M.O. Brett, Electrochemical methods in pesticides control, *Anal. Lett.*, 37 (2004) 1755–1791.
163 S. Rodriguez-Mozaz, M.-P. Marco, M.J. Lopez de Alda and D. Barceló, Biosensors for environmental monitoring of endocrine disruptors: a review article, *Anal. Bioanal. Chem.*, 378 (2004) 588–598.
164 Z. Grosmanová, J. Krejčí, J. Týnek, P. Cuhra and S. Baršová, Comparison of biosensoric and chromatographic methods for the detection of pesticides, *Int. J. Environ. Anal. Chem.*, 85 (2005) 885–893.
165 A. Mulchandani, W. Chen, P. Mulchandani, J. Wang and K.R. Rogers, Biosensors for direct determination of organophosphate pesticides, *Biosens. Bioelectron.*, 16 (2001) 225–230.
166 T. Montesinos, S. Pérez-Munguia, F. Valdez and J.-L. Marty, Disposable cholinesterase biosensor for the detection of pesticides in water-miscible organic solvents, *Anal. Chim. Acta*, 431 (2001) 231–237.
167 J.P. Hart and S.A. Wring, Recent developments in the design and application of screen-printed electrochemical sensors for biomedical, environmental and industrial analyses, *TrAC*, 16 (1997) 89–103.
168 A. Ciucu and C. Ciucu, Organic phase amperometric biosensor for detection of pesticides, *Roum. Biotechnol. Lett.*, 7 (2002) 667–676.
169 J.P. Hart and I.C. Hartley, Voltammetric and amperometric studies of thiocholine at a screen-printed carbon electrode chemically modified with cobalt phthalocyanine: studies towards a pesticide sensor, *Analyst*, 119 (1994) 259–263.
170 E. Suprun, G. Evtugyn, H. Budnikov, F. Ricci, D. Moscone and G. Palleschi, Acetylcholinesterase sensor based on screen-printed carbon electrode modified with Prussian blue, *Anal. Bioanal. Chem.*, 383 (2005) 597–604.
171 E.V. Gogol, G.A. Evtugyn, J.-L. Marty, H.C. Budnikov and V.G. Winter, Amperometric biosensors based on Nafion coated screen-printed electrodes for the determination of cholinesterase inhibitors, *Talanta*, 53 (2000) 379–389.
172 T.T. Bachmann, B. Leca, F. Vilatte, J.-F. Marty, D. Fournier and R.F. Schmid, Improved multianalyte detection of organophosphates and carbamates with disposable multielectrode biosensors using recombinant

mutants of *Drosophila* acetylcholinesterase and artificial neural networks, *Biosens. Bioelectron.*, 15 (2000) 193–201.
173 S. Andreescu, D. Fournier and J.-L. Marty, Development of highly sensitive sensor based on bioengineered acetylcholinesterase immobilized by affinity method, *Anal. Lett.*, 36 (2003) 1865–1885.
174 B. Bucur, S. Andreescu and J.-L. Marty, Affinity methods to immobilize acetylcholinesterases for manufacturing biosensors, *Anal. Lett.*, 37 (2004) 1571–1588.
175 C. Bonnet, S. Andreescu and J.-L. Marty, Adsorption: an easy and efficient immobilisation of acetylcholinesterase on screen-printed electrodes, *Anal. Chim. Acta*, 481 (2003) 209–211.
176 R.P. Deo, J. Wang, I. Block, A. Mulchandani, K.A. Joshi, M. Trojanowicz, F. Scholz, W. Chen and Y. Lin, Determination of organophosphate pesticides at a carbon nanotube/organophosphorus hydrolase electrochemical biosensor, *Anal. Chim. Acta*, 530 (2005) 185–189.
177 Y. Lin, F. Lu and J. Wang, Disposable carbon nanotube modified screen-printed biosensor for amperometric detection of organophosphorus pesticides and nerve agents, *Electroanalysis*, 16 (2004) 145–149.
178 M. Del Carlo, M. Mascini, A. Pepe, G. Diletti and D. Compagnone, Screening of food samples for carbamate and organophosphate pesticides using an electrochemical bioassay, *Food Chem.*, 84 (2004) 651–656.
179 M. Del Carlo, A. Pepe, M. Mascini, M. De Gregorio, A. Visconti and D. Compagnone, Determining pirimiphos-methyl in durum wheat samples using an acetylcholinesterase inhibition assay, *Anal. Bioanal. Chem.*, 381 (2005) 1367–1372.
180 F. Longobardi, M. Solfrizzo, D. Compagnone, M. Del Carlo and A. Visconti, Use of electrochemical biosensor and gas chromatography for determination of dichlorvos in wheat, *J. Agric. Food Chem.*, 53 (2005) 9389–9394.
181 H. Schulze, E. Scherbaum, M. Anastassiades, S. Vorlova, R.D. Schmid and T.T. Bachmann, Development, validation, and application of an acetylcholinesterase-biosensor test for the direct detection of insecticide residues in infant food, *Biosens. Bioelectron.*, 17 (2002) 1095–1105.
182 H. Schulze, S. Vorlová, F. Villatte, T.T. Bachmann and R.D. Schmid, Design of acetylcholinesterases for biosensor applications, *Biosens. Bioelectron.*, 18 (2003) 201–209.
183 Y. Boublik, P. Saint-Aguet, A. Lougarre, M. Arnaud, F. Villatte, S. Estrada-Mondaca and D. Fournier, Acetylcholinesterase engineering for detection of insecticide residues, *Protein Eng.*, 15 (2002) 43–50.
184 G.S. Nunes, T. Montesinos, P.B.O. Marques, D. Fournier and J.-F. Marty, Acetylcholine enzyme sensor for determining methamidophos insecticide: evaluation of some genetically modified acetylcholinesterases from *Drosophila melanogaster*, *Anal. Chim. Acta*, 434 (2001) 1–8.

185 A. Avramescu, S. Andreescu, T. Noguer, C. Bala, D. Andreescu and J.-L. Marty, Biosensors designed for environmental and food quality control based on screen-printed graphite electrodes with different configurations, *Anal. Bioanal. Chem.*, 374 (2002) 25–32.
186 H. Schulze, S.B. Muench, F. Villatte, R.D. Schmid and T.T. Bachmann, Insecticide detection through protein engineering of *Nippostrongylus brasiliensis* acetylcholinesterase B, *Anal. Chem.*, 77 (2005) 5823–5830.
187 A. Crew, J.P. Hart, R. Wedge and J.-L. Marty, A screen-printed, amperometric, biosensor array for the detection of organophosphate pesticides based on inhibition of wild type, and mutant acetylcholinesterases, from *Drosophila melanogaster*, *Anal. Lett.*, 37 (2004) 1601–1610.
188 R. Luxton and J.P. Hart, The rapid detection of pesticides in food. In: I.E. Tothill (Ed.), *Rapid and On-Line Instrumentation for Food Quality Assurance*, Woodhead Publishing Limited/CRC Press, Cambridge, 2003, pp. 55–74.
189 R. Luxton and J.P. Hart, The rapid detection of pesticide residues. In: D.H. Watson (Ed.), *Pesticide, Veterinary and Other Residues in Food*, Woodhead Publishing Limited/CRC Press, Cambridge, 2004, pp. 294–313.
190 R. Luxton and J.P. Hart, The rapid detection of pesticide residues. In: W. Jongen (Ed.), *Improving the Safety of Fresh Fruit and Vegetables*, Woodhead Publishing Limited/CRC Press, Cambridge, 2005, pp. 156–176.
191 G. Ross, The public health implications of polychlorinated biphenyls (PCBs) in the environment, *Ecotoxicol. Environ. Saf.*, 59 (2004) 275–291.
192 S. Centi, S. Laschi and M. Fránekand M. Mascini, A disposable immunomagnetic electrochemical sensor based on functionalised magnetic beads and carbon-based screen-printed electrodes (SPCEs) for the detection of polychlorinated biphenyls (PCBs), *Anal. Chim. Acta*, 538 (2005) 205–212.
193 K.C. Honeychurch, J.P. Hart, P.R.J. Pritchard, S.J. Hawkins and N.M. Ratcliffe, Development of an electrochemical assay for 2,6-dinitrotoluene, based on a screen-printed carbon electrode, and its potential application in bioanalysis, occupational and public health, *Biosens. Bioelectron.*, 19 (2003) 305–312.
194 J. Wang, F. Lu, D. MacDonald, J. Lu, M.E.S. Ozsoz and K.R. Rogers, Screen-printed voltammetric sensor for TNT, *Talanta*, 46 (1998) 1405–1412.
195 J. Wang, S. Thongngamdee and A. Kumar, Highly stable voltammetric detection of nitroaromatic explosives in the presence of organic surfactants at a polyphenol-coated carbon electrode, *Electroanalysis*, 16 (2004) 1232–1235.
196 E.C.D. Todd, Domoic acid and amnesic shellfish poisoning—a review, *J. Food Prot.*, 56 (1993) 69–83.

197 M. Kania, M. Kreuzer, E. Moore, M.B. Pravda, B. Hock and G. Guillbault, Development of polyclonal antibodies against domoic acid for their use in electrochemical biosensors, *Anal. Lett.*, 36(9) (2003) 1851–1863.
198 L. Micheli, A. Radoi, R. Guarrina, R. Massaud, C. Bala, D. Moscone and G. Palleschi, Disposable immunosensor for the determination of domoic acid in shellfish, *Biosens. Bioelectron.*, 20 (2004) 190–196.
199 N.W. Beyene, H. Moderegger and K. Kalcher, Development of an amperometric biosensor for β-*N*-oxalyl-L-α,β-diaminopropionic acid (β-ODAP), *Electroanalysis*, 16(4) (2004) 268–274.
200 G. Moges, N. Wodajo, L. Gorton, Y. Yigzaw, K. Kalcher, A. Belay, G. Akalu, B.M. Nair and T. Solomon, Glutamate oxidase advances the selective bioanalytical detection of the neurotoxic amino acid beta-ODAP in grass pea: a decade of progress, *Pure Appl. Chem.*, 76 (2004) 765–775.
201 S.H. Alarcon, L. Micheli, G. Palleschi and D. Compagnone, Development of an electrochemical immunosensor for Ochratoxin A, *Anal. Lett.*, 37 (2004) 1545–1558.
202 R.M. Pemberton, R. Pittson, N. Biddle, G.A. Drago and J.P. Hart, Studies towards the development of a screen-printed carbon electrochemical immunosensor for mycotoxins: a sensor for Aflatoxin B_1, *Anal. Lett.*, 39 (2006) 1573–1586.
203 N.H.S. Ammida, L. Micheli and G. Palleschi, Electrochemical immunosensor for determination of aflatoxin B_1 in barley, *Anal. Chim. Acta*, 520 (2004) 159–164.
204 L. Micheli, R. Grecco, M. Badea, D. Moscone and G. Palleschi, An electrochemical immunosensor for aflatoxin M1 determination in milk using screen-printed electrodes, *Biosens. Bioelectron.*, 21 (2005) 588–596.
205 R.M. Pemberton, A. Amine and J.P. Hart, Voltammetric behaviour of chlorophyll a at a screen-printed carbon electrode and its potential role as a biomarker for monitoring fecal contamination, *Anal. Lett.*, 37 (2004) 1625–1643.
206 R.C.H. Kwan, H.F. Leung, P.Y.T. Hon, J.P. Barford and R. Renneberg, A screen-printed biosensor using pyruvate oxidase for rapid determination of phosphate in synthetic wastewater, *Appl. Microbiol. Biotechnol.*, 66 (2005) 377–383.
207 R.C.H. Kwan, H.F. Leung, P.Y.T. Hon, H.C.F. Cheung, K. Hirota and R. Renneberg, Amperometric biosensor for determining human salivary phosphate, *Anal. Biochem.*, 343 (2005) 263–267.
208 J.P. Hart, S. Serban, L.J. Jones, N. Biddle, R. Pittson and G.A. Drago, Selective and rapid biosensor integrated into a commercial hand-held instrument for the measurement of ammonium ion in sewage effluent, *Anal. Lett.*, 39 (2006) 1657–1667.
209 http://www.uniscan.co.uk/potentiostat.html [accessed 6/03/06].

Chapter 24

Mediated enzyme screen-printed electrode probes for clinical, environmental and food analysis

Francesco Ricci, Danila Moscone and Giuseppe Palleschi

24.1 INTRODUCTION

Since the appearance of the first amperometric biosensors based on the coupling of oxidase enzymes and relying on the final detection of H_2O_2, it was clear that one of the major problems related to this kind of configuration was due to the high overpotential needed for H_2O_2 oxidation (ca. 0.7 V vs. Ag/AgCl). At this potential in fact, many electroactive substances (i.e. ascorbic acid, uric acid etc.), usually present in real samples, could also be oxidised to give interfering signals.

The same problem was also present when other amperometric biosensors, based on different class of enzymes, were assembled. For example, the amperometric detection of nicotinamide adenine dinucleotide (NADH) has been a matter of investigation for many years in the biosensor field [1,2].

The problem associated with amperometric detection of NADH is similar to that of H_2O_2, with a very high overpotential required [3,4] and with electrode fouling due to the presence of radical intermediates produced during NADH oxidation that then interfere with the measurement [5].

Another class of enzymes that has found wide application in the biosensor field in the last decades is that of the cholinesterases which have been mainly used for the detection of pesticides. For the amperometric detection of cholinesterase activity, both the substrates acetylcholine and acetylthiocholine have been extensively used [6–9], the latter being preferred because this avoids the use of another enzyme, choline oxidase, which is usually coupled with acetylcholinesterase. However, the amperometric measurement of thiocholine, produced by

Comprehensive Analytical Chemistry 49
Alegret and Merkoçi (Eds)
Volume 49 ISSN: 0166-526X DOI: 10.1016/S0166-526X(06)49024-3

the enzymatically catalysed hydrolysis of acetylthiocholine, is difficult to achieve at classic electrode surfaces due to the high overpotential needed and possible problems of surface passivation [9–11].

For these reasons, the drawbacks of electrochemical interferences are the primary subject of many research groups involved in the biosensor field, and strategies to overcome them have become a major goal.

First approaches to solving this problem were based on the use of selective membranes [12], which however still pose some problems of response time, of time-consuming procedures during the assembling, decrease in sensor sensitivity and detection limit. Research activity was then devoted to obtaining a sensor capable of detecting the analyte (H_2O_2, NADH, thiols etc.) in a potential range where electrochemical interferences could be avoided or greatly minimised. It is well known that this potential range, which could be defined as "optimum", is usually located between 0.0 and −0.2 V vs. Ag/AgCl [13].

To work in this optimum range, different kinds of approaches were proposed depending on the analyte and on the electrode surface.

First examples of the amperometric detection of H_2O_2 accomplished in such a range were based on the use of an enzyme, namely horseradish peroxidase (HRP), a prototypical hemeprotein peroxidase, which catalyses the reduction of H_2O_2 and due to its peculiar structure, allows direct electron transfer between its active site and the electrode surface at low applied potential [14–17]. This approach, although it shows good sensitivity and accuracy, suffers from some important shortcomings such as low stability and the limited binding of HRP to solid surfaces.

For this reason, in the last decade, inorganic electrochemical mediators, which catalyse the oxidation or reduction of H_2O_2 have been preferred to HRP and have been used for the assembling of oxidase-based biosensors [18–20]. This results in a decrease of the applied potential and the consequent avoidance of electrochemical interferences. Many electrochemical mediators have been used and many of them have found broad application, especially in glucose biosensors for diabetes control. However, due to the solubility of the mediator, they are generally employed in a single-use sensor and present some problems due to the low operative stability.

24.1.1 Prussian blue as electrochemical mediator

In this perspective, hexacyanoferrates and in particular Prussian blue (ferric hexacyanoferrate—PB) have found a wide use [21–23]. It was in

1994 that Karyakin made the claim for a PB-modified electrode as a powerful tool for hydrogen peroxide detection at low applied potential [24]. He demonstrated the possibility of the effective electrochemical deposition of a PB layer onto a glassy carbon electrode providing an efficient and selective catalytic activity towards hydrogen peroxide reduction [24,25]. PB-modified glassy carbon electrodes were used at an applied potential of 0.0 V vs. Ag/AgCl with a sensitivity in the micromolar range [26] and they showed, under the optimised conditions (thin layer of PB), a bimolecular rate constant for the reduction of H_2O_2 of $3 \times 10^3\,mol^{-1}\,l\,s^{-1}$, which is very similar to that measured for the peroxidase enzyme ($2 \times 10^4\,mol^{-1}\,l\,s^{-1}$) [27]. This high catalytic activity and selectivity led Karyakin to consider PB as an "artificial peroxidase". The operating potential (i.e. 0.0 V vs. Ag/AgCl) was low enough to avoid, or greatly reduce, the contribution from all the most common interferents (ascorbic acid, paracetamol, uric acid) [25,26], rendering the PB-modified electrode selective for hydrogen peroxide. However, the low applied potential does not by itself explain the high selectivity of PB. This feature has been accounted for in terms of the structure of PB which enables low molecular weight molecules to penetrate the cubic lattice and to be reduced while excluding molecules with higher molecular weight. This is probably the main advantage of using PB as mediator for H_2O_2 reduction. The selectivity and activity achieved are comparable to that of a biological binding component (HRP) but with all the advantages of an inorganic species (low cost, high stability at certain conditions, ease of electrode surface modification and no saturation effect for substrate) [21]. For this reason, much attention has been devoted to PB-modified electrodes; and their use has increased in recent years. PB-modified glassy carbon [26,28,29], graphite [30–32], carbon paste [33] and platinum [34] electrodes have been studied, leading to the construction of biosensors for glucose [25,35], lactate [36], glutamate [37], aminoacid [30] and alcohol [26] detection. Recently, some reviews on the analytical applications of PB have appeared in the literature [21–23,38]. Among the classic electrode materials, the screen-printing (thick-film) technology applied to sensor and biosensor construction has been considerably improved during recent years and a large number of papers, and recently some reviews, have been published on this subject [39–41]. Screen-printed electrodes (SPEs) are in fact inexpensive, simple to prepare, rapid and versatile and this technology also appears to be the most economical means for large-scale production and for the assembling of spot tests for clinical and environmental analysis. The use of screen-printed electrodes coupled with a redox mediator is dependent on the procedure adopted for its deposition on the electrode surface. This was initially one of the drawbacks for the coupling of the screen-printing technology and PB. Almost all the procedures adopted for PB deposition are in fact based on an electrochemical step which employs a constant applied potential (galvanostatic) in a solution of ferricyanide and ferric chloride [26,29,35,37,42,43] or potential cycling in the same solutions [42]. The galvanostatic strategy is usually followed by (1) a series of cyclic voltammetry (CV) (15–25) which enables a sort of activation of the PB layer and (2) by a heating step (100°C for 1–1.5 h) for the stabilisation of the same layer. All these procedures, however, present two major drawbacks. The first one is related to the fact that any procedure for surface modification which involves the use of an electrochemical step is always difficult to adapt to a large-scale production of sensors. All the electrochemical procedures adopted for PB modification are thus not suitable for the modification of a large number of screen-printed electrodes in a limited time. Another reason which makes the electrochemical procedure less than optimal for PB deposition is the low operative stability at alkaline pH of the PB layer formed [33,42,44]. This fact limited the choice of the oxidase enzymes that could be coupled with PB-modified electrodes to those having a pH optimum in the acid range. The reason for this limited stability is probably to be ascribed to the strong interaction between ferric ions and hydroxyl ions (OH^-) which form $Fe(OH)_3$ at pH higher than 6.4 [45] thus leading to the destruction of the Fe–CN–Fe bond, and hence solubilising PB [46]. Its leakage from the electrode surface results in a decreased signal. For many years, this low stability has represented the main drawback to the use of PB-modified electrodes, especially when they are coupled with an enzyme having its optimum pH in the basic range [42,47].

Our research in this field, which is summarised in this chapter, has been directed at obtaining a sensor modified with PB as electrochemical mediator which could avoid electrochemical interferences and could also couple the advantages of the screen-printed electrodes. For this purpose, an in-depth study of the modification procedure for PB deposition on the electrode surface was first conducted and then when an optimised procedure capable of providing an efficient and stable PB layer was obtained, it was applied with screen-printed electrodes in real analytical systems. Thus, our main goal has been not only to obtain a PB modification procedure suitable for a mass production of modified screen-printed electrodes, as already pointed out above, but also to achieve a stable PB layer in terms of operative and storage stability.

Moreover, the use of PB will not only be limited to the detection of H_2O_2 and its use in conjunction with oxidase enzymes. A recent disclosure of the electrocatalytic activity of PB towards the oxidation of thiols will also be discussed and an application with acetylcholinesterase enzymes for pesticide detection reported.

24.2 APPLICATION

Screen-printed electrodes used for the PB modification were home produced. A detailed description of the electrodes used and of the procedure adopted for PB modification is found in Procedure 17 (in CD accompanying this book). The most important thing to note about this procedure is that it does not involve any electrochemical step and, for this reason, it has been designed as "chemical deposition". This procedure is also very easy to perform and could be adapted to mass production of modified electrodes (see Procedure 17 in CD accompanying this book). The suitability of the proposed deposition procedure was carefully evaluated with different electrochemical techniques and its application in real samples has been summarised and discussed here.

24.2.1 Prussian blue modified screen-printed electrodes as sensitive and stable probes for H_2O_2 and thiol measurements

PB-modified electrodes have been first tested as H_2O_2 probes for which the response time, detection limit, linear range, sensitivity and reproducibility were studied.

The analytical parameters with respect to H_2O_2 amperometric measurement were evaluated in both batch and continuous flow mode and their stability, especially at basic pHs, was studied.

All these results were obtained using screen-printed electrodes with batch amperometric or continuous flow techniques (for more details see Procedure 17 in CD accompanying this book).

The catalytic activity of PB towards H_2O_2 and thiocholine is demonstrated by CV experiments and summarised in Fig. 24.1, which shows the behaviour of the PB oxidation and reduction peaks in response to the injection of different concentrations of H_2O_2 and thiocholine in a buffer solution. In the presence of H_2O_2, the increase of the reduction peak together with the decrease of the oxidation peak clearly indicates the activity of the reduced form of PB towards the reduction of H_2O_2.

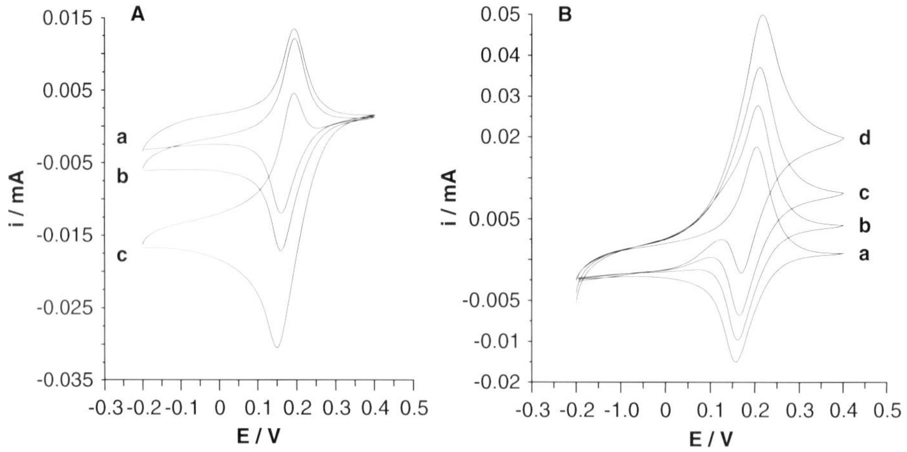

Fig. 24.1. Cyclic voltammograms revealing the catalytic oxidation of H_2O_2 (A) and thiocholine (B) at a PB-modified SPE. (A) CVs obtained in the absence (a) and presence of 0.5 (b) and 2.0 mmol l^{-1} (c) of H_2O_2. (B) CVs obtained in the absence (line a) and presence of 3 (line b), 6 (line c) and 15 mmol l^{-1} (line d) of thiocholine. Scan rate: 10 mV s^{-1}. 0.05 mol l^{-1} phosphate buffer +0.1 mol l^{-1} KCl, pH 7.4.

In the case of thiocholine, on the other hand, the oxidation peak is increased while the reduction one is diminished. This demonstrates that PB acts as electrocatalyst for the oxidation of thiocholine, decreasing its overpotential.

24.2.1.1 H_2O_2 measurement in batch amperometry

PB-modified screen-printed electrodes were first tested in batch amperometry for H_2O_2 detection. The sensors showed a good linearity in the range between 0.1 and 100 μmol l^{-1} with a detection limit of 0.1 μmol l^{-1}. The regression equation of the linear part of the curve was $y = 22.90x - 0.013$, where y represents the current in μA and x the H_2O_2 concentration in mmol l^{-1}; the R^2 value was 0.9980. The sensitivity was 324 μA mmol l^{-1} cm^{-2} with an RSD% of 5% (for all the concentrations tested by six different electrodes) (see Table 24.1).

The interference signal due to the most common electrochemical interfering species was also evaluated by the comparison of the signal for a fixed concentration of H_2O_2 with the current value obtained in the presence of the same concentration (0.05 mmol l^{-1}) of the interfering species. The relative current signal gives a measure of the sensor selectivity. Results showed a high selectivity of the H_2O_2 sensor. The

TABLE 24.1

Analytical parameters of PB-modified SPEs for H_2O_2 and thiocholine

Analyte	Applied potential (mV vs. Ag/AgCl)	Amperometric method	Detection limit ($\mu mol\, l^{-1}$)	Noise ($nA\, cm^{-2}$)	Linearity range ($mol\, l^{-1}$)	Sensitivity ($mA\, mol^{-1}\, l\, cm^{-2}$)	RSD % ($n = 5$)	Time to reach stable background(s)	Time to reach 90% of the signal (s)
H_2O_2	−50	Batch amperometry	0.1	2	0.1×10^{-6}–0.05×10^{-3}	324	5	60	5
H_2O_2	−50	Continuous flow ($10\, \mu l\, min^{-1}$)	1.0	20	1.0×10^{-6}–1.0×10^{-3}	213	5	120	30
Thiocholine	+200	Batch amperometry	5.0	57	5.0×10^{-6}–0.5×10^{-3}	143	7	60	5

interference due to ascorbic acid is around 2%, while there was no signal due to uric acid and catechol.

24.2.1.2 H_2O_2 measurement in continuous flow mode
The magnitude of the hydrogen peroxide signal and electrochemical interferences as well as operational and storage stability were also evaluated in a continuous flow mode. These results were utilised to determine whether these sensors could be used in flow mode for continuous glucose monitoring after coupling the sensor with glucose oxidase.

The detection limit for H_2O_2 (s/n = 3) obtained was 10^{-6} mol l^{-1} while the linear range was between 10^{-6} and 10^{-3} mol l^{-1}, the noise current in the range 15–20 nA and the sensitivity towards hydrogen peroxide was 213 mA mol^{-1} l^{-1} cm^2. The time needed to reach a stable current background was around 2 min, while 30 s were sufficient to reach 90% of the final signal after the injection of the hydrogen peroxide (i.e. 10^{-4} mol l^{-1}). Reproducibility of these sensors was 5% ($n = 5$).

The effect of the substances that usually interfere in the electrochemical determination of glucose was also examined with this technique. The signal due to ascorbic acid, uric acid, 4-acetaminophenol was evaluated using a fixed concentration of 0.5 mmol l^{-1}. No response was observed for these electrochemical interferents except for ascorbic acid, which gave a signal equal to 3% of the hydrogen peroxide. These results demonstrated that neither the sensitivity nor the selectivity of the PB-modified sensors changed when using different amperometric techniques.

24.2.1.3 Stability of Prussian blue modified screen-printed electrodes
The operational stability of all the PB-modified sensors is a critical point, especially at neutral and alkaline pH. A possible explanation for reduced stability could be the presence of hydroxyl ions at the electrode surface as a product of the H_2O_2 reduction. Hydroxyl ions are known to be able to break the Fe–CN–Fe bond, hence solubilising the PB [21]. An increased stability of PB at alkaline pH was first observed by our group after adopting a chemical deposition method for the modification of graphite particles with PB for the assembling of carbon paste electrodes [48].

A similar stability was obtained with our PB-modified screen-printed electrodes. The operational stability was studied using both CV and amperometry. In the first case, the sensors were cycled 250 times at different pHs and the decrease of the oxidation and reduction peaks was measured as a function of time.

The modified electrode showed a very good stability even at pH 9. Although at this pH a decrease in the redox peaks in CV is observed after 250 cycles, the PB layer is still highly electroactive and sufficient to catalyse the H_2O_2 reduction. In fact, stirred batch amperometric measurements of H_2O_2 (10 μmol l^{-1}) were carried out before and after the continuous cycling at pH 9, and the decrease of the signal was only 10% of the initial value.

When tested in stirred batch amperometry, the PB-modified screen-printed electrodes showed no loss of signal for H_2O_2 (20 μmol l^{-1}) after 50 h at pH 7.4 in batch amperometry. Other experiments performed at pH 9 confirmed the high pH stability of the PB-modified SPEs produced. At pH 9 there was still 80% of the residual activity of PB recorded after 16 h of continuous work in a solution of H_2O_2 (20 μmol l^{-1}) (Table 24.2).

In continuous flow mode, the stability experiments were carried out for a total period of 48 and 100 h. Two different solutions (0.1 and 0.2 mmol l^{-1}) of hydrogen peroxide were continuously passed in the wall-jet cell and the signal was first recorded for a total of 48 h. A decrease of around 10% and 15%, respectively, was detected at the end of 48 h of monitoring for 0.1 and 0.2 mM. Under these conditions (presence of hydrogen peroxide and an applied potential), PB is forced to continuously change its oxidative state according to the following equation (Eq. (24.1)) [28]:

$$(PB_{red})K_4Fe_4^{2+}[Fe^{2+}(CN)_6]_3 \rightleftarrows (PB_{ox})Fe_4^{3+}[Fe^{2+}(CN)_6]_3 + 4K^+ + 4e^- \qquad (24.1)$$

From the results obtained in this study, it seems that the crucial point concerning the stability of the PB is represented by this reaction. When conditions are such that the PB is not forced to go into the oxidised form (less stable), the stability of the PB layer is very high. This is probably the reason why, in the absence of an applied potential and of hydrogen peroxide, the PB retained 100% of its activity even after 100 h of continuous flow of buffer (data not shown), while in the case of the applied potential and in the presence of hydrogen peroxide a decrease was observed. Moreover, it should be pointed out that even taking into consideration the slight decrease, the stability of the PB layer is extremely good and this is quite new for PB-based sensors, which have always been affected by a low stability [25].

TABLE 24.2

Operational amperometric pH stability: % of residual activity of PB-modified screen-printed electrode in continuous flow mode at different conditions

Analyte	Applied potential (mV vs. Ag/AgCl)	Amperometric method	Time of operation	pH	$[H_2O_2]$	Residual activity
H_2O_2	−50	Batch amperometry	50 h	7.4	20 μmol l^{-1}	100
H_2O_2	−50	Continuous flow amperometry (10 μl min^{-1})	48 h	7.4	0.1 mmol l^{-1}	90
H_2O_2	−50	Continuous flow amperometry (10 μl min^{-1})	100 h	7.4	0.2 mmol l^{-1}	90
H_2O_2	−50	Cyclic voltammetry+batch amperometry	250 cycles	9.0	10 μmol l^{-1}	90
H_2O_2	−50	Batch amperometry	16 h	9.0	20 μmol l^{-1}	80

All the values are the average of three electrodes.

As already pointed out in our previous papers [48–50], the high stability is probably the result of the newly developed chemical modification procedure which may lead to a stronger adsorption of the PB particles on the electrode surface. In contrast to the PB layer obtained with the more commonly used electrochemical procedures, these modified electrodes are in fact more stable at basic pH and their continuous use is possible with a minimal loss of activity after several hours. Moreover, with respect to the electrochemical procedure, our chemical deposition is much more suitable for mass production since no electrochemical steps are required and a highly automated process could be adopted (see Procedure 17 in CD accompanying this book).

A test of the operational stability has also been carried out using electrodes after storage at room temperature for 6 months. The electrodes showed the same behaviour during 100 h of monitoring. As shown in Fig. 24.2, the stability was very good with a loss of ca. 15% observed after 100 h. This demonstrates that the stability of the PB layer does not decrease even after a long period of storage.

Finally, the PB electrodes have been tested in terms of their storage stability. To do this, the electrodes were left dry at room temperature in the dark after the surface modification. One year after their production, the PB electrodes tested for H_2O_2 reduction, retained $90 \pm 10\%$ ($n = 10$) of their initial signal response to hydrogen peroxide.

24.2.1.4 Thiol measurements with PB-modified screen-printed electrodes

Our group has recently investigated the possibility of using PB as catalyst also for the oxidation of thiols and, for the first time to our knowledge, an extensive study of the electrocatalytic response of PB-modified SPEs to many thiol compounds was performed to better understand the parameters affecting the catalytic mechanism [9].

Thiocholine was studied first as a standard thiol compound because of its importance in environmental analysis. Determination of thiocholine is in fact usually carried out to evaluate the degree of inhibition of acetylcholinesterase, a target enzyme of pesticides, as a means for indirect measurements of organophosphorous and carbamate pesticides.

The effect of PB modification on thiocholine oxidation at the SPE surface has already been demonstrated with voltammetric experiments (Fig. 24.1) [9]. In the case of CVs recorded with PB-modified electrodes in the presence of thiocholine, the current due to PB oxidation starts to increase (with respect to that obtained in buffer solution) at ca. 100 mV

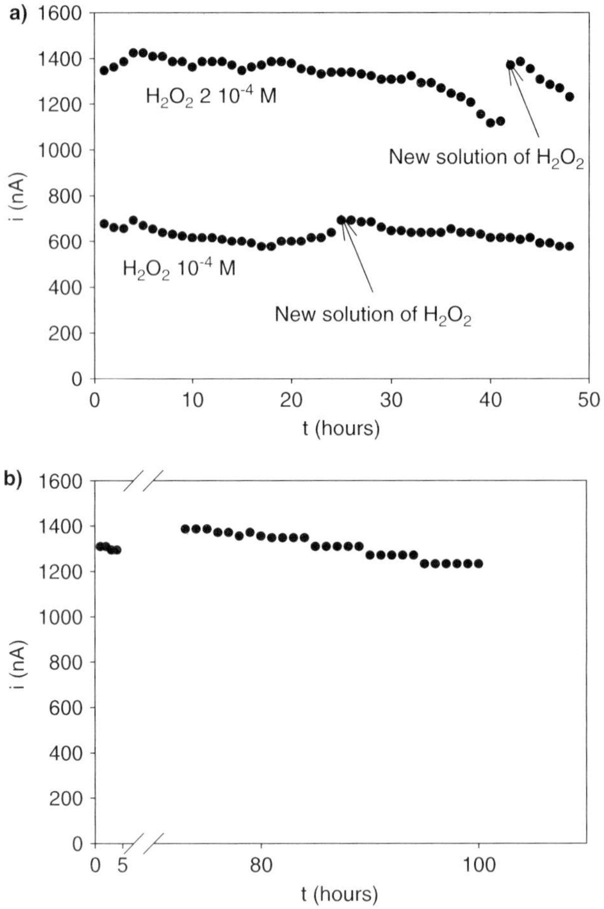

Fig. 24.2. Operational stability of hydrogen peroxide sensors (PB-modified sensors). Applied potential −50 mV vs. int. ref. Continuous monitoring of current in continuous flow mode (10 μl min^{-1}). Arrows indicate where solution of hydrogen peroxide was renewed. (a) Two sensors tested with 10^{-4} and 2×10^{-4} mol l^{-1} of hydrogen peroxide. (b) Six months old sensor tested with a solution of 2×10^{-4} mol l^{-1} of hydrogen peroxide. Reprinted from Ref. [59] with permission from Elsevier.

vs. Ag/AgCl while, in the reverse scan, a decrease of the reduction current is observed. The oxidation current observed in the presence of thiocholine seems, in fact, to appear in correspondence with the formation of the oxidised form of PB ($Fe^{III}Fe^{II}(CN)_6$). This result seems to demonstrate that PB has an electrocatalytic activity towards the oxidation of thiocholine. The generic reactions that occur on the electrode surface could be postulated as follows:

$$\begin{aligned} &PB_{ox} + RSH \rightleftarrows PB_{red} + RSSR \\ &PB_{red} + \text{electrode} \rightleftarrows PB_{ox} + \text{electrode} \end{aligned} \quad (24.2)$$

According to Eq. (24.2), the injection of RSH causes an increase in the concentration of the PB_{red} in the proximity of the electrode, resulting in an increase of the anodic peak current. By contrast, the cathodic peak is proportional to the PB_{ox} concentration that is diminished at the electrode surface by the reaction with thiol (RSH). On the basis of CV experiments and after performing a hydrodynamic voltammogram, a potential of 200 mV to be applied at the PB-modified sensor for thiocholine detection was chosen.

A detection limit (s/n = 3) of 5×10^{-6} mol l^{-1} together with a linear range up to 5×10^{-4} mol l^{-1} has been reached with a batch amperometric system using thiocholine standard solutions. Reproducibility has been evaluated studying the response to 10^{-4} mol l^{-1} of thiocholine for five different sensors giving similar results to those obtained with H_2O_2. Other analytical parameters are summarised in Table 24.1 and demonstrate the suitability of PB-modified sensors for thiocholine measurement.

24.2.2 Biosensor applications of PB-modified screen-printed electrodes

After demonstrating that the PB sensors were highly electroactive and sensitive towards both H_2O_2 and thiocholine, their coupling with enzymes was studied in order to assemble electrochemical biosensors to be applied in real samples. In this perspective, two applications will be discussed here and are illustrated in Scheme 1. The first one is directed towards glucose detection and is based on the final measurement of H_2O_2. The enzyme used is glucose oxidase and the biosensor is intended to be used for the continuous monitoring of glucose in diabetic patients. In the second application, the response due to thiocholine at the PB-modified electrode is used to assemble a pesticide sensor for organophosphorous and carbamate pesticides.

24.2.2.1 Glucose biosensors

In recent years, the use of a microdialysis probe coupled on-line with a glucose biosensor has provided very good results for the continuous monitoring of glucose in diabetic patients [51–54].

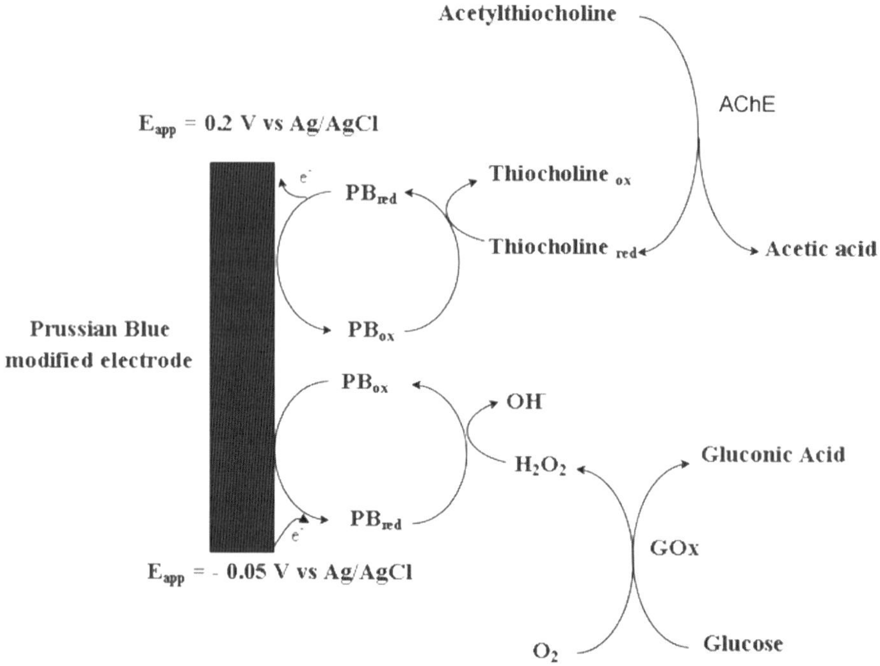

Scheme 1. Schematic representation of the system adopted for glucose and pesticide detection. In the upper part of the scheme is shown the reaction chain for the detection of acetylthiocholine giving a measure of acetylcholinesterase (AChE) activity which can be related to pesticide content. In the lower part of the scheme is shown the classic reaction utilised in the case of an oxidase enzyme (glucose oxidase—GOx) for the detection of glucose. In the first case, the final product is thiocholine and in the second, H_2O_2, both are measured at the Prussian blue modified electrode at an applied potential of 0.2 V vs. Ag/AgCl and −0.05 V vs. Ag/AgCl, respectively.

We report here the results obtained by the use of a screen-printed electrode as electrochemical probe to be coupled with a microdialysis fibre for continuous glucose monitoring. The most significant advance is represented by the introduction of a mediator (PB) as the principal factor for hydrogen peroxide measurement. The improved operational stability observed with the PB-modified screen-printed electrodes has demonstrated that these sensors could serve as tool to be applied for the continuous monitoring of many analytes. The application to diabetic care seems to be the most promising and advantageous area in which to test these sensors.

PB-modified electrodes were used as support for glucose oxidase immobilisation (see Procedure 17 in CD accompanying this book for details) to assemble biosensors that could be used for the continuous monitoring of glucose. The performance of the glucose biosensors was first tested in terms of the glucose signal and the most important analytical parameters were evaluated. A detection limit of around $2 \times 10^{-5}\,mol\,l^{-1}$ was obtained together with a linear range up to $1\,mmol\,l^{-1}$. The reproducibility of these biosensors was 7% ($n = 5$). The sensitivity of the biosensors was $54\,mA\,mol^{-1}\,l^{-1}\,cm^2$ and the current noise signal was almost the same as that obtained with the PB-modified electrodes (15–20 nA). The sensitivity of the biosensors towards glucose was almost 25% with respect to that obtained for hydrogen peroxide. This result is probably related to the composition of the enzymatic membrane. The presence of Nafion, a polyanionic membrane, could in fact have a shielding effect which results in the lower signal of glucose relative to that of hydrogen peroxide.

A very important parameter for a system designed to be used for continuous monitoring of glucose is the response time, which should be as short as possible in order to detect any changes of the analyte concentration in real time. The response time is dependent on the geometry and internal volume of the cell and on the flow rate ($10\,\mu l\,min^{-1}$). To reach 90% of the final signal for glucose ($0.5\,mmol\,l^{-1}$) starting from the background signal (buffer solution), 2 min were required. Another important characteristic to be taken into account is the time needed to reach the stable baseline. In this case, a time of 3–5 min is necessary to reach a stable current with buffer solution.

24.2.2.2 Operational stability of the glucose biosensors

Glucose biosensors were then tested in terms of operational stability. In this case, they were inserted into a wall-jet cell and when the baseline due to the buffer solution was reached, the glucose solution ($0.5\,mmol\,l^{-1}$) was passed into the cell. The signal due to glucose was recorded continuously for 50 and out to 72 h. Results are reported in Fig. 24.3. From these data, it could be concluded that the enzyme immobilisation procedure adopted provides a high operational stability under these operative conditions. The decrease of the signal is more pronounced during the first 12 h where an average loss of the signal of 20% is observed. After this initial period, the signal from the control solution of glucose was highly stabilised and a further decrease of 10–15% was observed during the next 40–50 h. In Fig. 24.3 the continuous monitoring (one point every 3 min) of some of the tested electrodes is shown. As already pointed out, the stability is very high and all the electrodes showed a similar trend. By comparing these results with

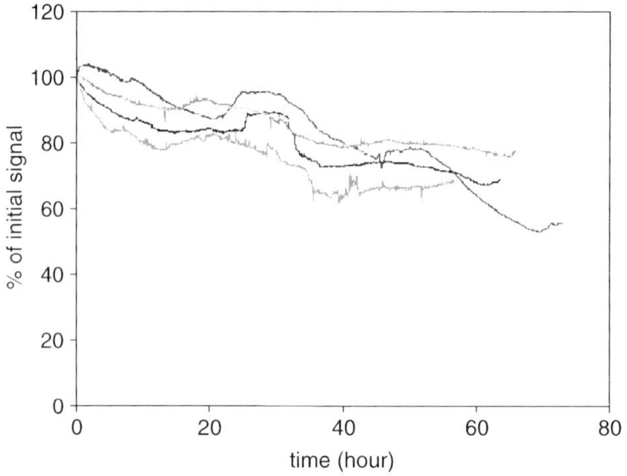

Fig. 24.3. Operative stability of six different glucose biosensors from a minimum of 50 and a maximum of 72 h. Continuous flow mode: 10 μl min^{-1}. Glucose concentration: 0.5 mmol l^{-1}. Applied potential: −50 mV vs. int. ref. The current was continuously recorded for the whole experiment. Reprinted from Ref. [59] with permission from Elsevier.

those obtained with PB-modified electrodes it is possible to say that the decrease of the signal can be almost completely ascribed to enzyme inactivation. The initial drift of the signal in the first 12 h is also probably due to the loss of part of the enzyme which is not strongly bound to the membrane.

Results of the operational stability are extremely interesting for the future application of these probes to continuous monitoring of glucose using a microdialysis probe.

A characteristic that should presently be improved is the limited linear range of the biosensors which could cause some problem in the cases of hyperglycemic levels. Studies are in progress to solve this problem by finding a suitable microdialysis probe which would be able to recover the subcutaneous glucose in the desired range of concentrations.

It should also be pointed out that in the case of an in vivo measurement, the microdialysis probe will be able to recover not only glucose but also many other biological compounds with low molecular weight from the subcutaneous tissue. The electrochemical interferents are greatly reduced by the use of PB at a low applied potential. However, other biological compounds could negatively affect the stability of the enzymatic membrane. Also, it is possible to have a sort of passivation or fouling of the electrode surface due to the absorption of biological compounds on the enzymatic membrane. In this perspective, the slow flow rate could represent a disadvantage since it will not be able to rapidly remove these compounds from the volume surrounding the electrode. An inhibition of the enzyme by certain compounds present in the subcutaneous tissue could also lead to an underestimation of the glucose present in the blood.

For this reason, experiments to test the operational stability of glucose sensors have been performed with dialysed biological samples. Microdialysis probes were inserted into a biological solution (i.e. human serum with glucose and a preservative added) and the signal due to glucose was continuously monitored for ca. 24 h. A control solution of glucose ($5\,mmol\,l^{-1}$) was also used to estimate the stability of the sensors. From the results shown in Fig. 24.4, it seems that the presence of biological compounds in the solution does not contribute to a lower stability of the probe since the signal due to the control solution of glucose is almost unchanged after ca. 20 h. This is an important result because even in the absence of any cut-off membrane placed on the

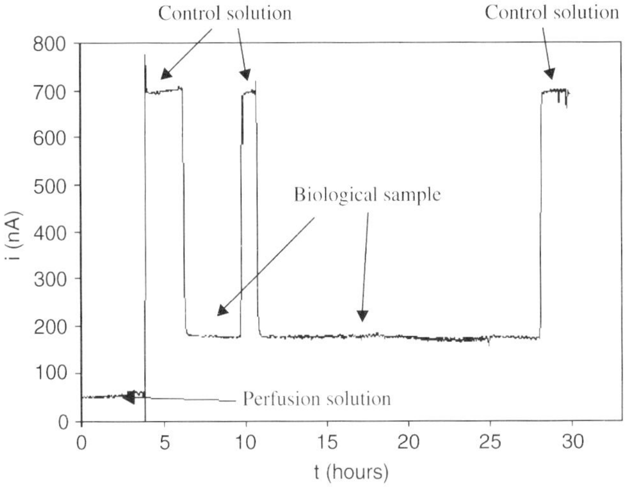

Fig. 24.4. Study of the biosensor stability with biological sample. Arrows indicate where (i.e. biological sample) a solution obtained by dialysing a human serum with the microdialysis probe was flowed in the biosensor cell. At the beginning a perfusion solution and control solution (glucose $5\,mmol\,l^{-1}$) were used instead of the serum to test the biosensor response. Control solution of glucose was also used during and at the end of the experiment to evaluate the stability of the biosensor. Continuous flow mode: $10\,\mu l\,min^{-1}$). Applied potential: $-50\,mV$ vs. int. ref. Reprinted from Ref. [59] with permission from Elsevier.

electrode surface, a high stability could be achieved in the presence of complex fluids.

24.2.2.3 Storage stability of glucose biosensors

To evaluate the shelf life of the glucose biosensors, a series of sensors were produced and stored at RT (i.e. 25°C and dry) and at 45°C in an oven (dry). Three biosensors were then tested for glucose (0.5 mmol l^{-1}) response to determine their residual activity. Taking into account the good reproducibility of these biosensors (ca. 7%) it is possible to evaluate the residual activity by looking only at the signal of the electrodes after the study of each storage interval and reporting, as reference signal for the initial activity, an average value obtained with six new biosensors. The results are shown in Fig. 24.5. As can be seen, the storage stability of these biosensors is extremely good and can be ascribed to the well-known stability of the glucose oxidase enzyme. These results also confirm the high stability of the PB layer during storage. Moreover, looking at the results obtained at 45°C (an average decrease response of ca. 10% after 60 days), it could be concluded that the shelf life of these biosensors is another encouraging point relative to their use in continuous glucose monitoring.

24.2.2.4 Acetylcholinesterase biosensors

The measurement of thiocholine discussed in Section 24.2.1.4 was used for the detection of pesticides by the study of the residual activity of

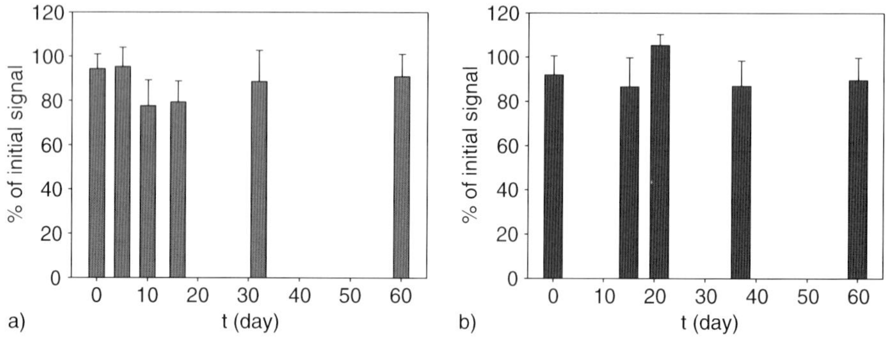

Fig. 24.5. Storage stability studies. Percentage of initial activity of different glucose biosensors after the storage time. (a) Biosensors stored at RT (i.e. 25°C and dry). (b) Biosensors stored at 45°C (dry). For each period, three biosensors were tested. See text for details. Reprinted from Ref. [59] with permission from Elsevier.

acetylcholinesterase [55] according to the following reaction:

$$\text{Acetylthiocholine} \xrightarrow{H_2O,\ AChE} \text{Thiocholine} + \text{Acetic Ac.} \tag{24.3}$$

Acetylcholinesterase (AChE, from electric eel) was immobilised on the PB-modified screen-printed electrode and inhibition measurements were first carried out using standard solutions of an organophosphorous (i.e. Paraoxon) and a carbammic (i.e. Aldicarb) pesticide.

The inhibition and the subsequent signal detection were performed in two different solutions. First the pesticide solution was added and then after 10 min (incubation time) the sensor was moved into a new buffer solution where the substrate (5 mmol l^{-1} acetylthiocholine) was injected and the signal measured. This procedure is particularly suitable when a complex matrix, which could pose problems for the direct measurement of thiocholine oxidation, is used. The analytical characteristics of pesticide determination in standard solutions were then evaluated. Detection limits, defined in this work as the concentrations giving an inhibition of 20%, were 30 and 10 ppb for aldicarb and paraoxon, respectively. By increasing the incubation time up to 30 min, an increase in the degree of inhibition could be observed and lower detection limits both for Aldicarb (5 ppb) and Paraoxon (3 ppb) were achieved.

ChE/PB sensors showed a high reproducibility resulting in a relative standard deviation for the pesticide determination of 7% ($n = 5$). The results were obtained with single-use ChE/PB sensors.

These sensors have then been adopted for the quantification of pesticides in grape samples. Anticholinesterase pesticides are in fact widely used in the production of table and wine grapes [56]. The maximum permissible levels of pesticides in wine grapes during harvesting are 0.1–0.5 mg kg^{-1}. Although more than 80% of pesticides decompose during grape fermentation, the influence of pesticides on fermentation and wine quality is a matter of concern. Anticholinesterase pesticides are spontaneously hydrolysed in the environment to their non-toxic products. However, the rate of this process is strongly affected by matrix characteristics. For this reason, a study of the fate of two standard pesticides during wine fermentation was carried out.

Either Aldicarb or Paraoxon was added to grape juice at the beginning of grape fermentation.

The amount of pesticide was evaluated during the initial days of fermentation. The sensor was incubated for 10 min with must, then washed and immersed into the standard buffer solution for signal

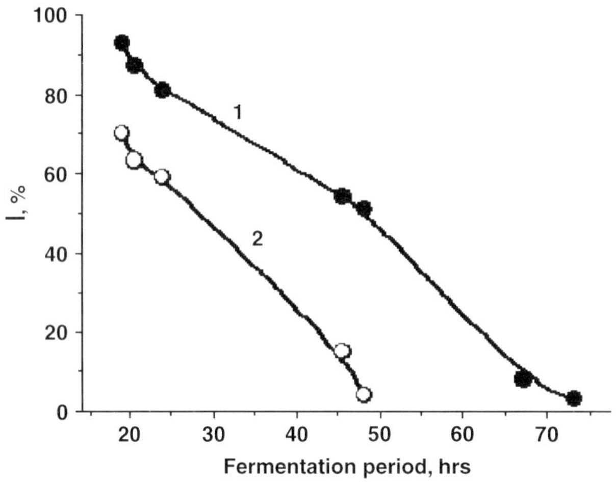

Fig. 24.6. The degree of inhibition measured in fermentation process of white grape juice spiked with (1) 500 ppb of Paraoxon and (2) 500 ppb of Aldicarb. Reprinted from Ref. [55] with kind permission of Springer Science.

measurement. Blank and spiked samples of juice under fermentation were tested without any treatment step. The results of the monitoring of pesticide degradation are shown in Fig. 24.6. The Paraoxon and Aldicarb life-time was estimated as 3 and 2 days, respectively. The decrease of the inhibition during the fermentation can be also due to interaction of pesticides with matrix components; in any case, this trend corresponds to the reported HPLC estimation of pesticide residues in wine and in fermentation media [56–58]. In accordance with their results, we found that about 90% of initial amounts of pesticides tested decomposed during the first 3 days of fermentation.

24.3 CONCLUSIONS

The need for electrochemical sensors which are able to monitor important analytes at low concentrations and with little or low interference effect and which could also allow mass production of probes and an in situ application has become more and more urgent during recent years.

The modified sensors for H_2O_2 and thiocholine detection presented in this chapter seem to be an adequate response to this demand. The use of PB-modified sensors has in fact been demonstrated to be extremely useful for H_2O_2 amperometric detection at low applied potentials with a number of possible biosensor applications based on oxidase enzymes. Moreover, the demonstration of the electrocatalytic effect of PB towards thiocholine oxidation represents an important finding for the construction of acetylcholinesterase biosensors for pesticide detection.

Finally, the optimisation of a procedure suitable for the deposition of a PB layer on a large number of screen-printed electrodes in a time-effective way makes the use of the sensors presented in this chapter particularly advantageous.

The analytical parameters of the PB-modified sensors towards H_2O_2 and thiocholine were carefully evaluated at an applied potential of −50 and +200 mV vs. Ag/AgCl, respectively. A detection limit of 10^{-7} and 10^{-6} mol l^{-1} was obtained for H_2O_2 using a stirred batch amperometry and a continuous flow mode, respectively. In both cases, the level of electrochemical interference was very low and equal to 2% for ascorbic acid, while no detectable signals were observed for uric acid and catechol (0.05 mol l^{-1}). In the case of thiocholine, a lower signal was observed with a detection limit of 5.0×10^{-6} mol l^{-1}. PB-modified sensors have also demonstrated a very high operative stability even at alkaline pH, which makes possible the use of these probes in continuous mode for up to 50 h at pH 7.4 and for 20 h at pH 9.0.

Two applications of the PB-modified screen-printed electrodes were reported. The first one is based on the coupling of the modified sensor with glucose oxidase for assembling a glucose biosensor to be applied in a continuous glucose monitoring. Analytical parameters such as stability of the biosensor were found to be satisfactory. The operational stability was found to be suitable for clinical application. Moreover, studies with biological samples demonstrated that the biosensor signal was not affected by the presence of interferents. Storage stability was found to be very high, with a loss of activity of only 10% after 2 months and at 45°C under dry conditions, and no loss after 1 year of storage at room temperature.

In the second application, the use of PB-modified sensors as powerful tools for thiocholine detection has been proposed for pesticide quantification in grape samples during wine fermentation. Two standard pesticides were used, one belonging to the class of carbamates (aldicarb) and another to the class of organophosphates (paraoxon). In both cases, the detection limits were found to be in the ppb range. The detection of anticholinesterase pesticides was demonstrated to be accurate even in a complicated matrix such as grape juice during fermentation. The limits of detection are low enough to detect the trace amounts of pesticides at the level of their admissible concentrations in wine grapes or soil. Taking into account the accuracy of the probe signal and the variation of the matrix content, ChE/PB sensors can be used for semi-quantitative estimation of pesticide traces in grapes and must and for preliminary tests of contaminant levels in the field.

REFERENCES

1 L. Gorton, Chemically modified electrodes for the electrocatalytic oxidation of nicotinamide coenzymes, J. Chem. Soc., *Faraday Trans*, 82 (1986) 1245–1258.

2 P.N. Bartlett, P. Tebbutt and R.P. Whitaker, Kinetic aspects of the use of modified electrodes and mediators in bioelectrochemistry, *Progr. React. Kinet.*, 16 (1991) 55–156.

3 F.D. Munteanu, D. Dicu, I.C. Popescu and L. Gorton, NADH oxidation using carbonaceous electrodes modified with dibenzo-dithia-diazapentacene, *Electroanalysis*, 15 (2003) 383–391.

4 C.O. Schmakel, K.S.V. Santhanam and P.J. Elving, Nicotinamide adenine dinucleotide (NAD+) and related compounds. Electrochemical redox pattern and allied chemical behavior, *J. Am. Chem. Soc.*, 97(18) (1975) 5083–5092.

5 J. Moiroux and P.J. Elving, Adsorption phenomena in the NAD+/NADH system at glassy carbon electrodes, *J. Electroanal. Chem.*, 102 (1979) 93–108.

6 S. Andreescu, L. Barthelmebs and J.-L. Marty, Immobilization of acetylcholinesterase on screen-printed electrodes: comparative study between three immobilization methods and applications to the detection of organophosphorus insecticides, *Anal. Chim. Acta*, 464(2) (2002) 171–180.

7 G. Jeanty and J.L. Marty, Detection of paraoxon by continuous flow system based enzyme sensor, *Biosens. Bioelectron.*, 13(2) (1998) 213–218.

8 L. Doretti, D. Ferrara, S. Lora, F. Schiavon and F.M. Veronese, Acetylcholine biosensor involving entrapment of acetylcholinesterase and poly(ethylene glycol)-modified choline oxidase in a poly(vinyl alcohol) cryogel membrane, *Enzyme Microb. Technol.*, 27 (2000) 279–285.

9 F. Ricci, F. Arduini, A. Amine, D. Moscone and G. Palleschi, Characterisation of Prussian blue modified screen-printed electrodes for thiol detection, *J. Electroanal. Chem.*, 563 (2004) 229–237.

10 J.-L. Marty, N. Mionetto, T. Noguer, F. Ortega and C. Roux, Enzyme sensors for the detection of pesticides, *Biosens. Bioelectron.*, 8 (1993) 273–280.

11 J.L. Marty, N. Mionetto, S. Lacorte and D. Barcelo, Validation of an enzymic biosensor with various liquid-chromatographic techniques for

determining organophosphorus pesticides and carbaryl in freeze-dried waters, *Anal. Chim. Acta*, 311 (1995) 265–271.

12 G. Palleschi, M.A. Nabi Rahni, G.J. Lubrano, J.N. Nwainbi and G.G. Guilbault, A study of interferences in glucose measurements in blood by hydrogen peroxide based glucose probes, *Anal. Biochem.*, 159 (1986) 114–121.

13 L. Gorton, Carbon paste electrodes modified with enzymes, tissues, and cells, *Electroanalysis*, 7 (1995) 23–45.

14 A. Lindgren, T. Ruzgas, L. Gorton, E. Csoregi, G. Bautista Ardila, I.Y. Sakharov and I.G. Gazaryan, Biosensors based on novel peroxidases with improved properties in direct and mediated electron transfer, *Biosens. Bioelectron.*, 15 (2000) 491–497.

15 S. Razola Serradilla, B. Ruiz Lopez, N. Diez Mora, H.B. Mark and J.M. Kauffmann, Hydrogen peroxide sensitive amperometric biosensor based on horseradish peroxidase entrapped in a polypyrrole electrode, *Biosens. Bioelectron.*, 17(11–12) (2002) 921–928.

16 E. Csoregi, L. Gorton, G. Marko-Varga, A.J. Tudos and W.T. Kok, Peroxidase-modified carbon fiber microelectrodes in flow-through detection of hydrogen peroxide and organic peroxides, *Anal. Chem.*, 66 (1994) 3604.

17 T. Ruzgas, E. Csoregi, J. Emmeus, L. Gorton and G. Marko-Varga, Peroxidase-modified electrodes: fundamentals and application, *Anal. Chim. Acta*, 220 (1996) 123–138.

18 J.D. Newman, S.F. White, I.E. Tothill and A.P.F. Turner, Catalytic materials, membranes, and fabrication technologies suitable for the construction of amperometric biosensors, *Anal. Chem.*, 67 (1995) 4594–4607.

19 A. Chaubey and B.D. Malhorta, Mediated Biosensors, *Biosens. Bioelectron.*, 17 (2002) 441–456.

20 J.M. Zen, A.S. Kumar and D.M. Tsai, Recent updates of chemically modified electrodes in analytical chemistry, *Electroanalysis*, 15(13) (2003) 1073–1087.

21 A.A. Karyakin, Prussian blue and its analogues: electrochemistry and analytical applications, *Electroanalysis*, 13 (2001) 813–819.

22 R. Koncki, Chemical sensors and biosensors based on Prussian blues, *Crit. Rev. Anal. Chem.*, 32(1) (2002) 79–96.

23 F. Ricci and G. Palleschi, Sensor and biosensor preparation, optimisation and applications of Prussian Blue modified electrodes, *Biosens. Bioelectron.*, 21 (2005) 389–407.

24 A.A. Karyakin, O. Gitelmacher and E.E. Karyakina, A high-sensitive glucose amperometric biosensor based on Prussian blue modified electrodes, *Anal. Lett.*, 27(15) (1994) 2861–2869.

25 A.A. Karyakin, O.V. Gitelmacher and E.E. Karyakina, Prussian blue-based first-generation biosensor. A sensitive amperometric electrode for glucose, *Anal. Chem.*, 67 (1995) 2419–2423.

26 A.A. Karyakin, E.E. Karyakina and L. Gorton, Prussian-blue based amperometric biosensors in flow-injection analysis, *Talanta*, 43 (1996) 1597–1606.
27 B.B. Hasinoff and H.B. Dunford, Kinetics of the oxidation of ferrocyanide by horseradish peroxidase compounds I and II, *Biochemistry*, 9 (1970) 4930–4932.
28 K. Itaya, N. Shoji and I. Uchida, Catalysis of the reduction of molecular oxygen to water at Prussian blue modified electrodes, *J. Am. Chem. Soc.*, 106 (1984) 3423–3429.
29 A.A. Karyakin, E.E. Karyakina and L. Gorton, The electrocatalytic activity of Prussian blue in hydrogen peroxide reduction studied using a wall-jet cell with continuous flow, *J. Electroanal. Chem.*, 456 (1998) 97–104.
30 Q. Chi and S. Dong, Amperometric biosensors based on the immobilization of oxidases in a Prussian blue film by electrochemical codeposition, *Anal. Chim. Acta*, 310 (1995) 429–436.
31 S.A. Jaffari and A.P.F. Turner, Novel hexacyanoferrate (III) modified graphite disc electrodes and their application in enzyme electrodes (Part I), *Biosens. Bioelectron.*, 12 (1997) 1–9.
32 Q. Deng, B. Li and S. Dong, Self-gelatinizable copolymer immobilized glucose biosensor based on Prussian blue modified graphite electrode, *Analyst*, 123 (1998) 1995–1999.
33 R. Garjonyte and A. Malinauskas, Electrocatalytic reactions of hydrogen peroxide at carbon paste electrodes modified by some metal hexacyanoferrates, *Sens. Actuators B*, 46 (1998) 236–241.
34 K. Itaya, H. Akahoshi and S. Toshima, Electrochemistry of Prussian blue modified electrodes: an electrochemical preparation method, *J. Electrochem. Soc.*, 129(7) (1982) 1498–1500.
35 R. Garjonyte and A. Malinauskas, Amperometric glucose biosensors based on Prussian blue and polyaniline glucose oxidase modified electrodes, *Biosens. Bioelectron.*, 15 (2000) 445–451.
36 R. Garjonyte, Y. Yigzaw, R. Meskys, A. Malinauskas and L. Gorton, Prussian-blue and lactate oxidase-based amperometric biosensor for lactic acid, *Sens. Actuators B*, 79 (2001) 33–38.
37 A.A. Karyakin, E.E. Karyakina and L. Gorton, Amperometric biosensor for glutamate using Prussian blue-based 'artificial peroxidase' as a transducer for hydrogen peroxide, *Anal. Chem.*, 72 (2000) 1720–1723.
38 N.R. de Tacconi, K. Rajeshwar and R.O. Lezna, Metal hexacyanoferrate: electrosynthesis, in situ characterization and applications, *Chem. Mater.*, 15 (2003) 3046–3062.
39 J.P. Hart and S.A. Wring, Recent developments in the design and application of screen-printed electrochemical sensors for biomedical, environmental and industrial analyses, *TRAC*, 16 (1997) 89–103.

40 M. Albareda-Silvert, A. Merkoci and S. Alegret, Configurations used in the design of screen printed enzymatic biosensors. A review, *Sens. Actuators B*, 69 (2000) 153–163.
41 J.P. Hart, A. Crew, E. Crouch, K. Honeychurch and R. Pemberton, Some recent design and developments of screen-printed carbon electrochemical sensors/biosensors for biomedical, environmental and industrial analyses, *Anal. Lett.*, 37(5) (2004) 789–830.
42 R. Garjonyte and A. Malinauskas, Operational stability of amperometric hydrogen peroxide sensors, based on ferrous and copper hexacyanoferrates, *Sens. Actuators B*, 56 (1999) 93–97.
43 R. Garjonyte and A. Malinauskas, Glucose biosensor based on glucose oxidase immobilized in electropolymerized polypyrrole and poly(o-phenylenediammine) films on a Prussian blue-modified electrode, *Sens. Actuators B*, 63 (2000) 122–128.
44 X. Zhang, J. Wang, B. Ogorevc and U.S. Spichiger, Glucose nanosensor based on Prussian-blue modified carbon-fiber cone nanoelectrode and an integrated reference electrode, *Electroanalysis*, 11 (1999) 945–949.
45 B.J. Feldman and R.W. Murray, Electron diffusion in wet and dry Prussian blue films on interdigitated array electrodes, *Inorg. Chem.*, 26 (1987) 1702–1708.
46 A.A. Karyakin, E.E. Karyakina and L. Gorton, On the mechanism of H_2O_2 reduction at Prussian Blue modified electrodes, *Electrochem. Commun.*, 1 (1999) 78–82.
47 A. Malinauskas, R. Araminaite, G. Mickeviciute and R. Garjonyte, Evaluation of operational stability of Prussian blue and cobalt hexacyanoferrate-based amperometric hydrogen peroxide sensors for biosensing application, *Mater. Sci. Eng. C*, 24 (2004) 513–519.
48 D. Moscone, D. D'Ottavi, D. Compagnone, G. Palleschi and A. Amine, Construction and analytical characterization of Prussian blue based carbon paste electrodes and their assembly as oxidase enzyme sensors, *Anal. Chem.*, 73 (2001) 2529–2535.
49 F. Ricci, A. Amine, G. Palleschi and D. Moscone, Prussian blue based screen printed biosensors with improved characteristics of long-term lifetime and pH stability, *Biosens. Bioelectron.*, 18 (2003) 165–174.
50 F. Ricci, A. Amine, C.S. Tuta, A.A. Ciucu, F. Lucarelli, G. Palleschi and D. Moscone, Prussian blue and enzyme bulk modified screen-printed electrodes for hydrogen peroxide and glucose determination with improved storage and operational stability, *Anal. Chim. Acta*, 485 (2003) 111–120.
51 D. Moscone and M. Mascini, Microdialysis coupled with glucose biosensor for subcutaneous monitoring, *Analysis*, 21 (1993) M40–M42.
52 C. Meyeroff, F. Bischof, F. Sternberg, H. Zier and E.F. Pfeiffer, On line continuous monitoring of subcutaneous tissue glucose in men by combining portable glucose sensor with microdialysis, *Diabetologia*, 35 (1992) 224–230.

53 D. Moscone, M. Pasini and M. Mascini, Subcutaneous microdialysis probe coupled with glucose biosensor for in vivo continuous monitoring, *Talanta*, 39(8) (1992) 1039–1044.
54 D. Moscone and M. Mascini, Microdialysis and glucose biosensor for in vivo monitoring, *Ann. Biol. Clin.*, 21 (1993) M40–M42.
55 E. Suprun, G. Evtugyn, H. Budnikov, F. Ricci, D. Moscone and G. Palleschi, Acetylcholinesterase sensor based on screen-printed carbon electrode modified with Prussian blue, *Anal. Bioanal. Chem.*, 383 (2005) 597–604.
56 FAO, *Submission and Evaluation of Pesticide Residues Data for the Estimation of Maximum Residue Levels in Food and Feed*, 1st ed., FAO, Rome, 2002.
57 J.R. Corbett, *The Biochemical Mode of Action of Pesticides*, Academic Press, London, 1974.
58 P. Cabras and A. Angioni, Pesticide residues in grapes, wine, and their processing products, *J. Agric. Food Chem.*, 48 (2000) 967–973.
59 F. Ricci, D. Moscone, C.S. Tuta, G. Palleschi, A. Amine, A. Poscia, F. Valgimigli and D. Messeri, Novel planar glucose biosensors for continuous monitoring use, *Biosens. Bioelectron.*, 20(10) (2005) 1993–2000.

Chapter 25

Coupling of screen-printed electrodes and magnetic beads for rapid and sensitive immunodetection: polychlorinated biphenyls analysis in environmental samples

S. Centi, G. Marrazza and Marco Mascini

25.1 INTRODUCTION

25.1.1 Polychlorinated biphenyls compounds

Polychlorinated biphenyls (PCBs) are among the 16 chemicals designated as persistent organic pollutants (POPs) that are the subject of negotiations on a global agreement for their control. POPs are highly stable compounds that persist in the environment, accumulate in the fatty tissues of most living organisms since they are lipophilic, and are toxic to humans and wildlife [1].

PCBs were once used in industrial applications, particularly as electrical insulating fluids and as heat-exchange fluids, until concern over possible adverse effects on the environment and on human health resulted in the cessation of PCBs production and an ultimate ban on manufacture in most countries. PCBs were discovered over 100 years ago and their production and commercial use began in 1929. Because of their remarkable electrical insulating properties and their flame resistance, they soon gained widespread use as insulators and coolants in transformers and other electrical equipment where these properties are essential. For several decades, PCBs were also routinely used in the manufacture of a wide variety of common products such as plastics, adhesives, paints, varnishes and carbonless copying paper [2]. Despite their ban almost a quarter of a century ago, these pollutants are largely diffused in the environment. Their presence is mainly due to their

Comprehensive Analytical Chemistry 49
Alegret and Merkoçi (Eds)
Volume 49 ISSN: 0166-526X DOI: 10.1016/S0166-526X(06)49025-5

physical and chemical properties such as low inflammability, chemical stability and solubility in most organic solvents. Because of their characteristics, all degradation mechanisms are difficult and environmental and metabolic degradation is generally very slow. Moreover, PCBs can be unintentionally produced as by-products in a wide variety of chemical processes that contain chlorine and hydrocarbon sources, during water chlorination and by thermal degradation of other chlorinated organics [3].

PCBs levels in the polluted environmental samples change according to the matrices considered; for example in soil their concentrations can be in the range of several ng/g to mg/g (this last value is for highly contaminated sites), whereas in sediments the upper limit is generally hundreds of μg/g. Very high levels can also be reached in living organisms, as a result of bioaccumulation processes [4].

PCBs are a group of 209 structurally related chemical compounds, consisting of two connected benzene rings and 1–10 chlorine atoms (Fig. 25.1). The positions of the chlorine substituents on the rings are denoted by numbers assigned to each of the carbon atoms, with the carbons supporting the bond between the rings being designated 1 and 1′.

Any single chemical compound in the PCB category is called a "congener". The name of a congener specifies the total number of chlorine substituents and the position of each chlorine. The 209 congeners are shown in Table 25.1. Although the physical and chemical properties vary widely across the congeners, PCBs have low water solubilities and low vapour pressure. The molecular weight of PCBs varies from 188.7 for $C_{12}H_9Cl$ to 498.7 g/mol for $C_{12}Cl_{12}$.

PCBs were manufactured and sold under many names, even if the most common are the "Aroclor" series. Aroclor refers to a mixture of individual chlorinated biphenyl compounds with different degrees of chlorination. A PCB mixture with an Aroclor name usually includes a numerical identifier, which provides additional information on the properties of the mixture. The most common mixtures in the Aroclor series are Aroclor 1016, Aroclor 1242, Aroclor 1248, Aroclor 1254 and Aroclor 1260.

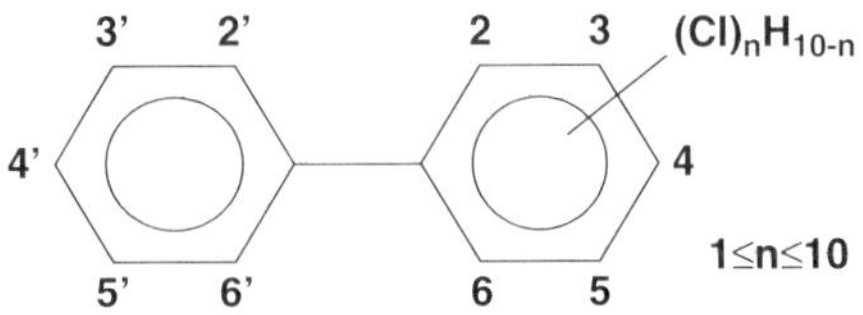

Fig. 25.1. Chemical structure of PCB molecule.

TABLE 25.1

Representation of all the homologues of PCBs

Homologue	Cl substituents	PCB congeners
Monochlorobiphenyl	1	3
Dichlorobiphenyl	2	12
Trichlorobiphenyl	3	24
Tetrachlorobiphenyl	4	42
Pentachlorobiphenyl	5	46
Hexachlorobiphenyl	6	42
Heptachlorobiphenyl	7	24
Octachlorobiphenyl	8	12
Nonachlorobiphenyl	9	3
Decachlorobiphenyl	10	1

25.1.2 Analytical methods for PCBs detection

Conventional methods used for the analysis of PCBs are largely dependent on the use of chromatography and spectrometry, or a combination of both as well as gas-chromatography coupled to an electronic capture detector (GC-ECD) [5–7]. Despite the high degree of specificity, selectivity and accuracy, there are a number of disadvantages associated to these techniques. For example, equipment is expensive, sample preparation and analysis can be complicated and time consuming and the requirement for trained personnel does not permit *in situ* monitoring [8]. Alternative methods as immunoassays, that could provide inexpensive and rapid screening techniques for sample monitoring for laboratory and field analysis, are more and more requested.

Immunoassay methods are simple, sensitive and selective for PCB testing, allow rapid measurements and have also been applied to PCB detection [9–12].

Among the high number of immunoassay techniques, the enzyme-linked immunosorbent assays (ELISAs) combined with a colorimetric end point measurement are the most widely used. These techniques have also been introduced on the market as PCBs ELISA kits by many companies (see Table 25.2).

Immunosensors are affinity biosensors and are defined as analytical devices that detect the binding of an antigen to its specific antibody by coupling the immunochemical reaction to the surface of a device

TABLE 25.2
PCB commercial kits

Producer	Name
EnviroLogix Inc, USA	PCB in soil tube assay
Hach Corporation, USA	PCB immunoassay kit
Strategic Diagnostic Inc. USA	DTECH PCB immunoassay kit
	EnviroGard immunoassay kit
	RaPID assay system

(transducer). Among the different kinds of transducers (piezoelectric, electrochemical and optical), electrochemical immunosensors, based on the electrochemical detection of immunoreaction, have been widely reported. They have been the subject of increasing interest mainly because of their potential application as an alternative immunoassay technique in areas such as clinical diagnostics [13] and environmental and food control [14,15].

Most electrochemical immunosensors use antibodies or antigens labelled with an enzyme that generates an electroactive product which can be detected at the electrochemical transducer surface. The combination of high enzyme activity and selectivity with the sensitive methods of electrochemical detection provides a basis for the development of immunosensors. Horse radish peroxidase (HRP) and alkaline phosphatase (AP) are popular enzyme labels and can be used with a variety of substrates.

Most electrochemical immunosensors use screen-printed electrodes produced by thick-film technology as transducers: the importance of screen-printed electrodes in analytical chemistry is related to the interest for development of disposable and inexpensive immunosensors. A thick-film is based on the layers deposition of inks or pastes sequentially onto an insulating support or substrate; the ink is forced through a screen onto a substrate and the open mesh pattern in the screen defines the pattern of the deposited ink.

To create a thick-film electrode, a conductive or dielectric film is applied to a substrate [16]. The film is applied through a mask contacting the substrate and deposited films are obtained by pattern transfer from the mask.

Screen-printing allows the fast mass production of highly reproducible electrodes at low cost for disposable use. A variety of screen-printed thick-film devices can be produced and in Fig. 25.2 an example of carbon screen-printed electrode is shown.

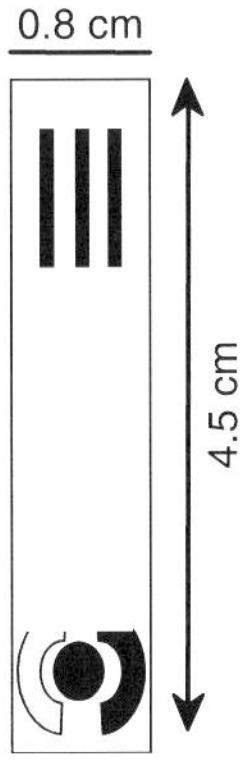

Fig. 25.2. Picture showing a carbon screen-printed electrode.

Electrochemical immunosensors based on screen-printed electrodes have recently been applied to the detection of environmental pollutants such as PCBs, PAHs, pesticides [17–20] and of important molecules in clinical and food field [21–23]. In this case, the screen-printed electrodes are both the solid-phase for the immunoassay and the electrochemical transducers: antibody or antigen molecules are directly immobilised at the sensor surface (transducer) and one of these species is enzyme-labelled in order to generate an electroactive product which can be detected at the screen-printed electrode surface.

The use of the electrode surface as solid phase in immunoassay can present some problems: a shielding of the surface by antibody or antigen molecules can cause hindrance in the electron transfer, resulting in a reduced signal and so a loss of sensitivity. There are different ways that can be used to improve the sensitivity of the system; an interesting approach could involve the screen-printed electrodes are used only for the transduction step, whereas the affinity reaction could be performed using another physical support [24].

The use of micro-particles as solid phase is a relatively new possibility, widely documented in literature [4,25–28]. Various kinds of beads can be used in electrochemical biosystem. The beads range from non-conducting (glass, etc.), conducting (graphite particles) to magnetic materials. Particles are available with a wide variety of surface functional groups and size and have the possibility of reaction kinetics similar to those found in free solution. Graphite or magnetic particles represent the most commonly used beads in bioelectroanalytical systems.

At present, various coated magnetic particles are offered on market [29–32]. Proteins have been coupled covalently to the surface of superparamagnetic particles, with stable and selective groups to bind protein-specific ligands with minimal interference and non-specific binding. These particles respond to an applied magnetic field and re-disperse upon removal of the magnet. They consist of 36–40% magnetite dispersed within a copolymer matrix consisting of styrene and divinyl-benzene. Their binding capacity varies with the bead size, composition and the size of the binding ligand.

The use of magnetic beads as solid support to perform the immunoassay allows obtaining a high yield of solid phase antibody binding and of the affinity reaction: thus the probability that antibodies meet the magnetic beads or the antibody-coated magnetic beads meet the analyte is very high, keeping the solution under stirring. After molecular recognition on the magnetic beads, it is possible to build up the immunosensing surface by localizing the immunomagnetic beads on the working electrode area of a screen-printed electrode with the aid of a magnet and perform the electrochemical measurement.

This approach separates the steps relative to the immunoreaction from the step of electrochemical detection and for this reason the working electrode surface is easily accessible by enzymatic product, which diffuse onto bare electrode surface [28,33] (Fig. 25.3). Using this strategy, finding the optimum conditions for the immunoassay on the magnetic beads and for electrochemical detection on the transducer (carbon screen-printed electrodes) is much easier than in the usual one (electrode) surface systems, because optimum conditions for immunoassay do not conform with those for electrochemical detection and vice versa.

This configuration based on the use of two surfaces, magnetic beads for immunoassay and screen-printed electrodes for electrochemical detection, allows to obtain a faster and a more sensitive detection of the immunoreaction than using a unique surface (screen-printed electrode): in this case it is possible to perform the electrochemical measurement in faster times (less then 30 min) and improve the sensitivity (around two magnitude orders). For this reason, this approach is advised in the development of an electrochemical immunosensor specific to any analyte.

25.2 APPLICATION

25.2.1 Immunoassay scheme for PCBs detection

The choice of the immunoassay scheme is related to the dimensions of the target molecule: when antigen is a small molecule as a PCB molecule, it has got only one epitope and therefore only one antibody can recognise

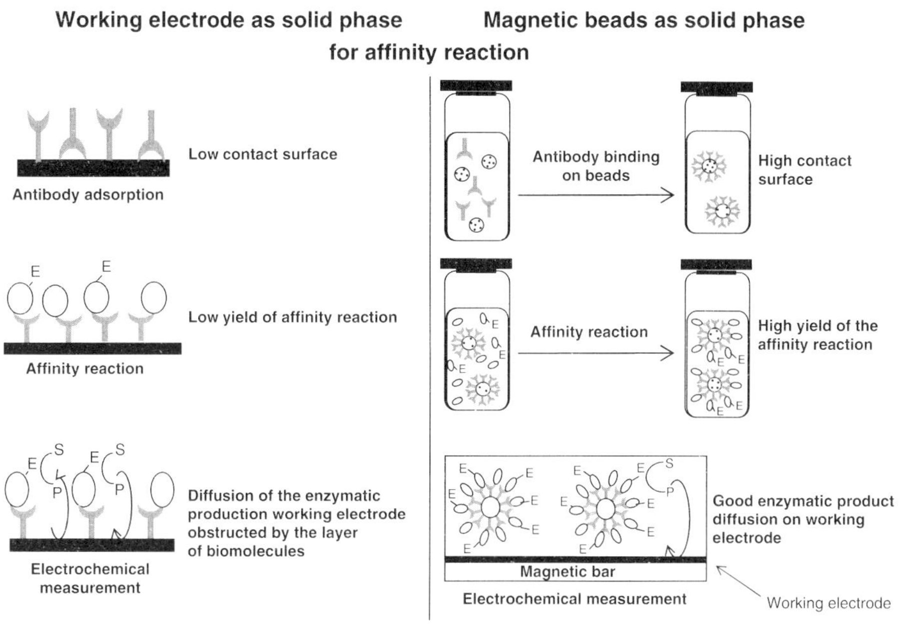

Fig. 25.3. Advantages due to the use of magnetic beads as solid-phase to perform the immunoassay.

it. For this reason, a competitive immunoassay scheme can be developed only when the analyte is a compound with a small molecular weight.

The competitive immunoassays, as the name indicates, are based on a competition reaction between two reagents for a third one. A competitive assay can be carried out in two different ways:

- the specific antibodies can be immobilised on a solid phase and then competition between antigen and an antigen derivate labelled by enzyme can be performed. This scheme is called direct competitive assay.
- a derivate of the antigen is immobilised on a solid phase and then the molecular recognition between the antigen and the specific antibody occurs in a vial. Then the last solution is added to solid phase. If the specific antibody is not labelled by an enzyme, the extent of the affinity reaction can be evaluated by adding a secondary labelled antibody capable of binding the first antibody.

In both cases, a label is used to estimate the extent of the affinity reaction: higher the concentration of analyte in the sample, smaller will be the amount of labelled antigen bound to the solid phase. Therefore,

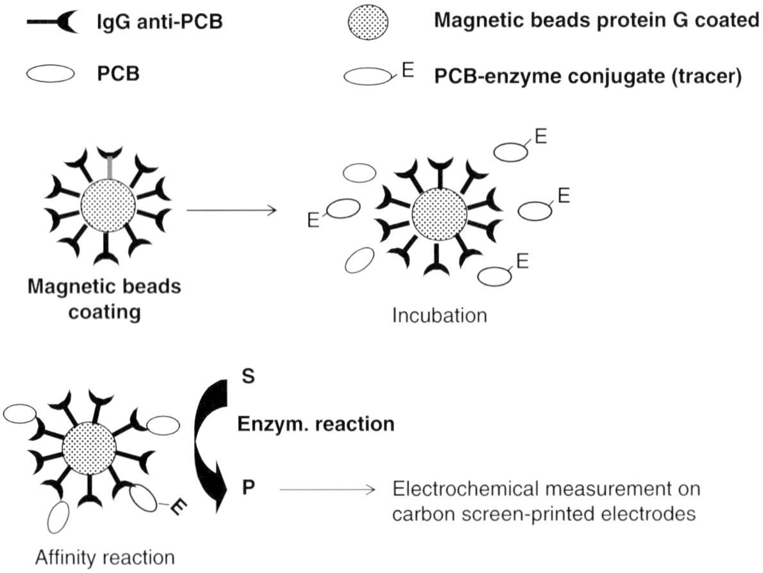

Fig. 25.4. Scheme of a direct competitive immunoassay for PCB detection.

the signal is inversely proportional to the amount of analyte present in the sample.

It is necessary to optimise the following parameters in order to perform a competitive assay:

- Antibody concentration: the sensitivity of a competitive assay is related to the concentration of antibody.
- Antigen labelled concentration: this has to be in limiting amount to saturate the antibodies immobilised on the solid phase.
- Incubation time of the competition solution: it is important that the affinity reaction has completely occurred.

In literature, a direct competitive assay performed using magnetic beads protein G coated as solid phase and carbon screen-printed electrodes as transducers is reported [4]. The main steps of the assay are shown in Fig. 25.4.

25.2.2 Immunochemical reagents: design of immunogen and enzyme labelled compound

Immunoassay specificity and selectivity are strictly dependent on the antibodies used. A good knowledge of the target allows to build up a good immunoassay design.

PCBs are small molecules and they do not have antigenic effects. They are defined as haptens, i.e. they are molecules that need to be linked with a transport protein in order to generate an immunogenic response.

The antibody-PCB recognition mechanism is conformational based, because in this case the antibodies do not recognise a particular chemical group, but the spatial conformation of the molecule [7]. The binding with the protein carrier has to be optimised in order to have a minimal influence on the molecular structure. Spacer arms have an important function too: the distance between the protein and the hapten involves the spatial exposition of the target molecule and the recognition mechanism. In addition, spatial arms must not be constructed using particular chemical groups in order to limit any "cross-reactivity" immunisation effect.

Because of complexity of Aroclor mixtures, production of antibodies able to react with a family of congeners with high specificity is very complicated. However, it is possible to obtain a good affinity using one of the most abundant congeners in industrial formulations as immunisation agent. Aroclor 1016, 1242 and 1248 are constituted by low chlorinated congeners [34], therefore using as immunogen one of more represented low chlorinated congeners it is possible to develop antibodies capable of recognising these mixtures.

In order to obtain an immunogen agent, able to develop polyclonal antibodies against Aroclor mixtures at low chlorination degree, the synthesis strategy developed was to functionalise PCB15 in ortho position in order to produce a final molecule sterically similar to PCB28, which is among the most abundant congeners present in the Aroclor mixtures [9] (Fig. 25.5).

The PCB15 congener was functionalised by nitration in the second position to obtain the azo-derivative and this compound was conjugated with a protein carrier (serum albumine bovine, BSA) using amino-adipate group as spacer arm (Fig. 25.6).

PCBs molecules are unreactive and the insertion step of a chemical group in an aromatic ring can only be carried out under drastic conditions. The nitration step was a critical point in the synthesis and was performed by adding the 4,4′-DB congener in a nitric acid/acetic anhydride mixture at –5°C for 2 h under stirred conditions. From this reaction, the same congener $-NO_2$ ortho-substituted was obtained in a high yield (94%). Another critical point is to obtain the azo derivative; in this case, the previous compound was treated with $SnCl_2$ and HCl for 4 h to obtain an $-NH_2$ group (75% as yield).

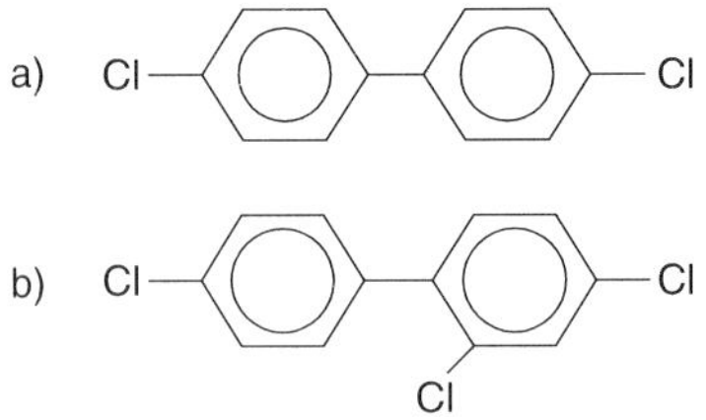

Fig. 25.5. (a) PCB15 (4,4′-dichlorobiphenyl) and (b) PCB28 (2,4,4′-trichlorobiphenyl) structures.

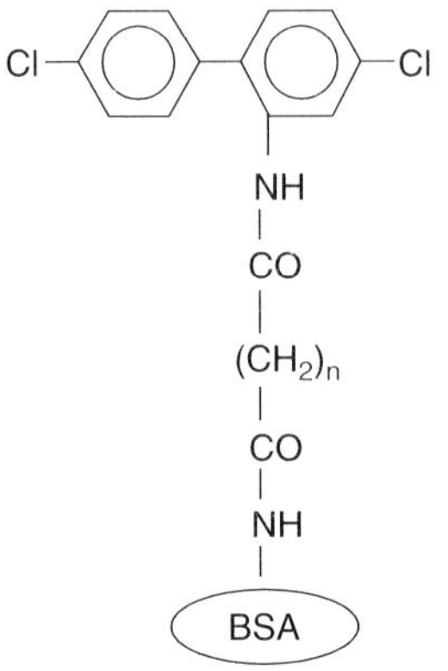

Fig. 25.6. 4,4′-dichlorobiphenyl-BSA functionalised structure.

The enzyme-labelled antigen was obtained using the same procedure.

25.2.3 Dose–response curves for some PCB mixtures

Dose–response curves for PCB standard solutions are carried out. The curves are estimated by nonlinear regression using the following logistic equation [35] by GraphPad Prism 4 program:

$$Y = \text{Bottom} + \frac{(\text{Top} - \text{Bottom})}{1 + 10^{[\log \text{EC}_{50} - X]\text{Hillslope}}}$$

where Top is the Y value at the top plateau of the curve, Bottom is the Y value at the bottom plateau of the curve, EC_{50} is the antigen concentration necessary to halve the current signal, X is $\log_{10}$[free antigen] and Hillslope is the slope of the linear part of the curve.

The calibration curve for the congener PCB28 performed using magnetic beads and carbon screen-printed electrodes is shown in Fig. 25.7. A

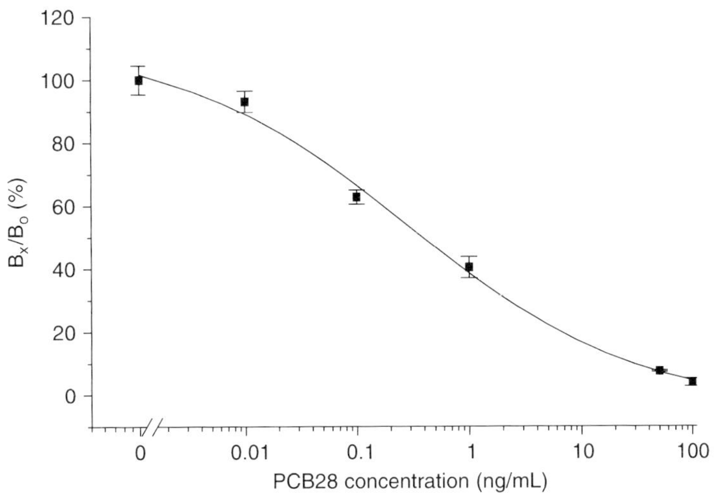

Fig. 25.7. Calibration curve for PCB28 obtained using the electrochemical immunosensor based on magnetic beads and screen-printed electrodes. IgG anti-PCB28 100 μg/mL; PCB28-AP diluted 1:1000.

high sensitivity is expected for this compound, because it had been used to prepare the immunogen for antibody production (see Section 25.2.2). The signal is reported as B_x/B_o percentage units, that is, the percentage of the signal decrease with respect to the blank value (solution containing the tracer only), taken as 100% of the response, versus the logarithm of the congener concentration.

The curve shows the sigmoidal shape typical of a competitive immunoassay; a signal decrease is observed for concentrations greater than 0.01 ng/mL, whereas the lowest current is measured at PCB28 concentration of 100 ng/mL.

The detection limit (DL) of the electrochemical immunosensor based on magnetic beads has been estimated to be equal to 8×10^{-3} ng/mL. This depends on the affinity of antibodies for antigen and is defined as the lowest analyte concentration which can be distinguished and is calculated by evaluation of the mean of the blank solution (containing the tracer only) response minus two times the standard deviations [35].

The EC_{50}, which corresponds to the analyte concentration necessary to displace 50% of the label, has been evaluated to be 0.3 ng/mL.

The immunoassay test has been replicated in order to evaluate its reproducibility; for this purpose, three repetitions of the calibration curves in the concentration range 0–100 ng/mL are carried out, and the average of coefficient of variation (CV) is 9%.

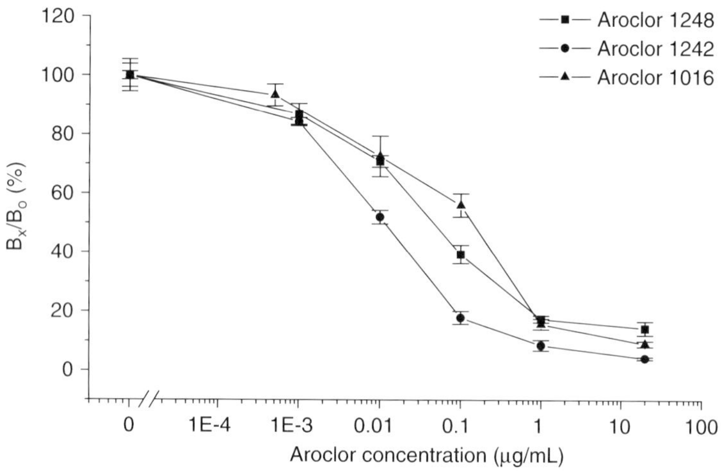

Fig. 25.8. Calibration curves for Aroclor 1242, 1248 and 1016 mixtures. IgG anti-PCB28 100 μg/mL, PCB28-AP diluted 1:1000.

The electrochemical immunosensor based on magnetic beads has been applied to the detection of some Aroclor mixtures too, and calibration curves are shown in Fig. 25.8. Also in this case, the signal is reported as B_x/B_o percentage units versus the logarithm of the mixture concentration. As it can be seen, all the curves exhibit a sigmoidal shape, because the formulations (Aroclor 1242, 1248 and 1016) contain a significant percentage of PCB28 and other structurally similar congeners. Shapes are different because of the different affinities of antibody against all congeners present in the formulations. This consideration is also demonstrated by the EC_{50} values of each mixture and by the different values of DL. The EC_{50} values range between 8 and 94 ng/mL, whereas the DL values are between 0.3 and 0.8 ng/mL.

25.2.4 PCB pollution in environment and food samples

First Soren Jenson, a Swedish researcher at the University of Stockholm, identified PCBs as an environmental problem in 1966: he identified the presence of PCBs in human blood.

The evidence of the acute PCB toxicity came from two industrial incidents—one in Japan in 1968 and the other in Taiwan in 1979: cooking oil was contaminated by big quantities of PCBs and PCDFs [36]. Adults who ingested the contaminated oil showed chloracne and dark brown pigmentation of the skin and lips.

Due to their lipophilic nature, PCBs tend to accumulate or reside in those environmental compartments that are non-polar and are amenable to lipid accumulation, such as the organic components of sediments. PCB presence in polar substances, such as water, is minimal. PCBs are not volatile and thus do not persist in air in any appreciable concentration. Therefore, the major sources of environment exposure to environmental species remain soils and sediments.

Following the ban on PCB production in the United States more than 25 years ago, PCB levels in the environment have declined, particularly where their removal from transformers and capacitors has occurred on a large-scale basis, minimizing the risk of leaks or accidents; anyway we should note that PCB containing equipment is still in use in certain parts of the world [2].

PCBs are deposited atmospherically onto plants and waters consumed by animals and fish. Because of their bioaccumulation in the food web, the major resource of non-occupational exposure to these contaminants ($>90\%$) is via food consumption, particularly fish, meat and dairy products. The half-lives of PCBs are of the order of several years. Since they are apolar and highly lipophilic, the highest concentrations are found in fatty foods, rather than in vegetables cereals and fruits [37].

Analysis of environmental or food samples containing PCBs include matrix preparation, extraction and determination [3]. In this section, some extraction methods, which can be easily coupled to an electrochemical immunosensor, are described. Such extraction techniques have to be simple, fast, useful for *in situ* measurements and do not require trained personnel.

25.2.4.1 Matrix preparation

Samples collected in field are usually preserved by freezing after dissecting into small pieces. Homogenisation and grinding to rupture cell membranes appear to be the most commonly used pre-treatment procedures for tissue matrices (e.g. fish muscle tissue) [38].

25.2.4.2 Extraction

Extractions generally rely on a favourable partition of PCBs from the sample matrix into the extraction matrix. The more favourable the partition coefficient, the higher is the extraction efficiency. Since PCBs are lipophilic, the extraction methods are based on the isolation of lipids from the sample matrix [30]. It should be noted that the concentration of planar or non-ortho-substituted PCBs, which are considered the most toxic PCB congeners, is generally $\approx$1000-fold lower (ng/kg) than those of non-planar or ortho-substituted PCBs (μg/kg). In addition, other compounds, such as pesticides, lipid, biological material or chlorophyll from plants are also extracted and can interfere with the analysis. Sample-recovery measurements can be made by using the method of standard addition.

Many techniques are used for PCBs extraction; among them the most used are:

- Liquid–liquid extraction (LLE). The most common LLE configuration uses about 100 mL of solvent per 5–50 g of sample. Samples are generally frozen prior to extraction to disintegrate the tissues, which are then homogenized by sodium sulphate to bind water present in the sample, followed by overnight drying. The dried powder is then extracted using a suitable solvent. It is essential to match the solvent polarity and generally a combination of non-polar, water-immiscible solvents with solvents of various polarities are used.
- Solid-phase extraction (SPE): it is widely accepted as an alternative extraction/clean up method [39]. SPE has been used for the extraction of PCBs in various types of human milk. Milk powder and evaporated milk were constituted with water prior to extraction. The milk sample was mixed with 5 mL of water and 10 mL of methanol and sonicated, followed by passing the sample through the column.
- Ultrasonic extraction (USE): it is a simple extraction technique, in which the sample is immersed in an appropriate organic solvent in a vessel and placed in an ultrasonic bath. The efficiency of extraction depends on the polarity of the solvent, the homogeneity of the matrix and the ultrasonic time. The mixture of sample and organic solvent is separated by filtration [40,41].
- Microwave-assisted extraction (MAE): it has attracted growing interest, as it allows rapid extraction of solutes from solid matrices by employing microwave energy as a source of heat [42]. The portioning of the analytes from the sample matrix to the extractant depends upon the temperature and the nature of the extractant.

An extraction procedure performed by sonication method for dried marine sediments and soil followed by the analysis of the extracts using an electrochemical immunosensor based on magnetic beads and carbon screen-printed electrodes is described in the protocol (see Procedure 25 in CD accompanying this book).

25.3 CONCLUSIONS

Electrochemical immunosensors are valid tools for the measurement of environmental pollutants such as PCBs. These devices are based on the principles of solid-phase immunoassays and couple the specificity of the immunoassay test to the sensitivity of electrochemical techniques. Screen-printed electrodes are used in the immunosensor development, because they can have a field use and their technology allows mass production at low cost.

In the traditional format, screen-printed electrodes are used as solid phase to carry out the immunoassay and as signal transducers. In recent years, different kinds of beads (glass, graphite particles or magnetic particles) have been used as solid phase to perform the immunoassay. For this reason, it is possible to use a new approach in the immunosensor construction coupling beads to screen-printed electrodes.

Using this strategy, finding the optimum conditions for the immunoassay on the magnetic beads and for electrochemical detection on the transducer (carbon screen-printed electrodes) is much easier than in the traditional format.

Because of their chemical stability and past industrial applications, PCBs are ubiquitous contaminants in the environment; marine sediment, soil and food are the matrices where the presence of PCBs is more documented. For their characteristics electrochemical immunosensors are valid analytical tools to carry out a fast screening analysis on a large number of potentially polluted samples.

ACKNOWLEDGMENT

We wish to thank Prof. M. Fránek of Veterinary Research Institute, Brno (Czech Republic) to have kindly provided the immunoreagents (antibodies anti-PCB28 and PCB28-AP labelled).

REFERENCES

1 M. Fránek, A. Deng, V. Kolář and J. Socha, Direct competitive immunoassays for the coplanar polychlorinated biphenyls, *Anal. Chim. Acta*, 444 (2001) 131–142.
2 G. Ross, The public health implications of polychlorinated biphenyls (PCBs) in the environment, *Ecotoxicol. Environ. Saf.*, 59 (2004) 275–291.
3 F.E. Ahmed, Analysis of polychlorinated biphenyls in food products, *Trend Anal. Chem.*, 22(3) (2003) 170–185.

4 S. Centi, S. Laschi, M. Fránek and M. Mascini, A disposable immunomagnetic electrochemical sensor based on functionalised magnetic beads and carbon-based screen-printed electrodes (SPCEs) for the detection of polychlorinated biphenyls (PCBs), *Anal. Chim. Acta*, 538 (2005) 205–212.
5 S.G. Udai, H.M. Schwartz and B. Wheatley, Congener specific analysis of polychlorinated biphenyls (PCBs) in serum using GC/MSD, *Anal. Chim. Acta*, 30 (1995) 1969–1977.
6 J.D. Berset and R. Holzer, Determination of coplanar and ortho substituted PCBs in some sewage sludges of Switzerland using HRGC/ECD and HRCG/MSD, *Chromosphere*, 32 (1996) 2317–2333.
7 R.G. Luthy, D.A. Dzombak, m.J.R. Shannon, R. Unterman and J.R. Smith, Dissolution of PCB congeners from an Aroclor and an Aroclor/hydraulic oil mixture, *Wat. Res.*, 31 (1997) 561–573.
8 G. Fillmann, T.S. Galloway, R.C. Sanger, M.H. Depledge and J.W. Readman, Relative performance of immunochemical (enzyme-linked immunosorbent assay) and gas chromatography-electron-capture detection techniques to quantify polychlorinated biphenyls in mussel tissues, *Anal. Chim. Acta*, 461 (2002) 75–84.
9 M. Šišak, M. Fránek and K. Hruška, Application of radioimmunoassay in the screening of polychlorinated biphenyls, *Anal. Chim. Acta*, 311 (1995) 415–422.
10 M. Fránek, M.V. Pouzar and V. Kolar, Enzyme-immunoassays for polychlorinated biphenyls. Structural aspects of hapten-antibody binding,, *Anal. Chim. Acta*, 347 (1997) 163–176.
11 N. Lambert, T.S. Fan and J.F. Pilette, Analysis of PCBs in waste oil by enzyme immunoassay, *Sci. Total Environ.*, 196 (1997) 57–61.
12 J.L. Zajicek, D.E. Tillitt, T.R. Schwartz, C.J. Schmitt and R.O. Harrison, Comparison of an enzyme-linked immunosorbent assay (ELISA) to gas chromatography (GC) measurement of polychlorinated biphenyls (PCBs) in selected US fish extracts, *Chemosphere*, 40 (2000) 539–548.
13 P.B. Luppa, L.J. Sokoll and D.W. Chan, Immunosensors—principles and applications to clinical chemistry, *Clin. Chim. Acta*, 314 (2001) 1–26.
14 S.K. Sharma, N. Sehgal and A. Kumar, Biomolecules for development of biosensors and their applications, *Curr. Appl. Phys.*, 3 (2003) 307–316.
15 S. Kröger, S. Piletsky and A.P.F. Turner, Biosensors for marine pollution research, monitoring and control, *Mar. Pollut. Bull.*, 45 (2002) 42–34.
16 C.A. Galán-Vidal, J. Muñoz, C. Domínguez and S. Alegret, Chemical sensors, biosensors and thick film technology, *Trends Anal. Chem.*, 14 (1995) 225–231.
17 S. Laschi, M. Mascini, G. Scortichini, M. Fránek and M. Mascini, Polychlorinated biphenyls (PCBs) detection in food samples using an electrochemical immunosensor, *J. Agric. Food Chem.*, 51 (2003) 1816–1822.
18 K.A. Fähnrich, M. Pravda and G.G. Guilbault, Disposable amperometric immunosensors for the detection of polycyclic aromatic hydrocarbons

(PAHs) using screen-printed electrodes, *Biosens. Bioelectron.*, 18 (2003) 73–82.
19 S. Laschi and M. Mascini, Disposable electrochemical immunosensor for environmental applications, *Ann. Chim.*, 92 (2002) 425–433.
20 S. Hleli, C. Martelet, A. Abdelghani, N. Burais and N. Jaffrezic-Renault, Atrazine analysis using an impedimetric immunosensor based on mixed biotinylated self-assembled monolayer, *Sens. Actuators B Chem.*, 113(2) (2006) 711–717.
21 C. Valat, B. Limoges, D. Huet, J.L. Romette, A disposable protein A-based immunosensor for flow injection assay with electrochemical detection, *Anal. Chim. Acta*, (2000) 187–194.
22 L. Micheli, R. Grecco, M. Badea, D. Moscone and G. Palleschi, An electrochemical immunosensor for aflatoxin M1 determination in milk using screen-printed electrodes, *Biosens. Bioelectron.*, 21(4) (2005) 588–596.
23 L. Micheli, R. Rodoi, R. Guarrina, R. Massaud, C. Bala, D. Moscone and G. Palleschi, Disposable immunosensor for determination of domoic acid in shellfish, *Biosens. Bioelectron.*, 20(2) (2004) 190–196.
24 D.R. Thévenot, K. Toth, R.A. Durst and F.S. Wilson, Electrochemical biosensors: recommended definitions and classifications, *Biosens. Bioelectron.*, 16 (2001) 121–131.
25 S. Solé, A. Merkoçi and S. Alegret, New materials for electrochemical sensing III. Beads, *Trends Anal. Chem.*, 20(2) (2001) 102–110.
26 E. Paleček, R. Kizek, L. Havran, S. Billova and M. Fojta, Electrochemical enzyme-linked immunoassay in a DNA hydridization sensor, *Anal. Chim. Acta*, 469 (2002) 73–83.
27 T. Alefantis, P. Grewal, J. Ashton, A.S. khan, J.J. Valdes and V.G. Del Vecchio, A rapid and sensitive magnetic bead-based immunoassay for the detection of staphylococcal enterotoxin B for high-through put screening, *Mol. Cell. Probes*, 18 (2004) 379–382.
28 M. Dequaire, C. Degrand and B. Limoges, An immunomagnetic electrochemical sensor based on a perfluorosulfonate-coated screen-printed electrode for the determination of 2,4-dichlorophenoxyacetic acid, *Anal. Chem.*, 71 (1999) 2571–2577.
29 Dynal®, http://www.invitrogen.com
30 Promega, http://www.promega.com
31 Bangs Laboratories, http://www.bangslabs.com
32 ProZyme®, http://www.prozyme.com
33 V. Kourilov and M. Steinitz, Magnetic-beads enzyme-linked immunosorbent assay verifies adsorption of ligand and epitope accessibility, *Anal. Biochem.*, 311 (2002) 166–170.
34 A.S. Stack, S. Altaman-Hamamdzic, P.J. Morris, S.D. London and L. London, Polychlorinated biphenyls mixtures (Arocolors) inhibit LPS-induced murine splenocyte proliferation in vitro, *Toxicology*, 139 (1999) 137–154.

35 C. Davies. In: D. Wild (Ed.), *The Immunoassay Handbook*, Stockton Press, New York, 1994, p. 83, Chap. 3.
36 L. Manodori, A. Gambaro, R. Piazza, S. Ferrari, A.M. Stortini, I. Moret and G. Capodaglio, PCBs and PAHs in sea surface microlayer and subsurface water samples of the Venice Lagoon (Italy), *Mar. Pollut. Bull.*, 52 (2005) 184–192.
37 M.D. Erickson, *Analytical Chemistry of PCBs*, 2nd ed, CRC Lewis Publishers, Boca Raton, FL, 1997.
38 C.S. Eskilsson and E. Björklund, Analytical-scale microwave-assisted extraction, *J. Chromatogr. A*, 902 (2000) 227.
39 S. Rodriguez-Mozazm, M.J. López de Alda and D. Barceló, Fast and simultaneous monitoring of organic pollutants in a drinking water treatment plant by a multi-analyte biosensor followed by LC-MS validation, *Talanta*, 69 (2006) 377–384.
40 I. Moret, R. Piazza, M. Benedetti, A. Gambaro, C. Barbante and P. Cescon, Determination of polychlorinated biphenyls in Venice Lagoon sediments, *Chemosphere*, 43 (2001) 559–565.
41 F. Kaštánek and P. Kaštánek, Combined decontamination processes for wastes containing PCBs, *J. Hazard. Mater. B*, 117 (2005) 185–205.
42 M. Ramil Criado, I. Rodriguez Pereiro and R. Cela Torrijos, Determination of polychlorinated biphenyls in ash using dimethylsulfoxidemicrowave assisted extraction followed by solid-phase microextraction, *Talanta*, 63 (2004) 533–540.

Chapter 26

Thick- and thin-film DNA sensors

María Begoña González-García, María Teresa Fernández-Abedul and Agustín Costa-García

26.1 INTRODUCTION

Recently, an impressive number of inventive designs of DNA-based electrochemical sensing are emerging. These types of sensors combine nucleic acid layers with electrochemical transducers to produce a biosensor and promise to provide a simple, accurate and inexpensive platform for patient diagnosis.

A wide variety of electrodes have been used as detectors or supports/detectors (genosensors) in the electrochemical DNA assays, including carbon paste, glassy carbon, graphite, graphite-epoxy composites, gold, mercury or Hg film on silver electrodes. Moreover, thick- and thin-film electrodes have been used and different DNA assays have been carried out on these substrates, either used as genosensors or detectors. The technologies used in their fabrication allow the mass production of reproducible, inexpensive and mechanically robust strip solid electrodes. Other important features that these electrodes exhibit are related to the miniaturisation of the corresponding device along with their ease of handling and manipulation in a disposable manner.

The use of films as electrodes makes possible numerous experiments that would be difficult or impractical to implement with the conventional bulk electrodes. Discussion here emphasises either "thin" ($<5\,\mu m$ thick; usually quite a bit thinner) or "thick" ($>5\,\mu m$; usually quite a bit thicker) film electrode materials, consisting of a conductor, either a continuous or a spatially patterned film, most commonly deposited on a suitably prepared insulating substrate. Films consisting primarily of insulators are not considered here, except to the extent that they may be used to form patterned arrays or electrodes with special geometries. A view of applications and properties of film

Comprehensive Analytical Chemistry 49
Alegret and Merkoçi (Eds)
Volume 49 ISSN: 0166-526X DOI: 10.1016/S0166-526X(06)49026-7

electrodes as well as the methodologies used in their fabrication is given by Anderson and Winograd [1].

26.1.1 Genosensors on thick- and thin-film electrodes

As mentioned above, thick- and thin-film electrodes have been used for different DNA assays, either as genosensors or detectors. Here the discussion will be focused on genosensor devices based on hybridisation event; that is, the electrode is used as support/detector of the hybridisation event.

For designing a genosensor, the crucial steps are the choice of the transducer surface and the immobilisation of the single-stranded DNA (ssDNA) probes onto electrode surface. The immobilisation method will determine the sensitivity and reproducibility of the genosensor. Several strategies for the immobilisation of ssDNA have been carried out and will be discussed later. The ssDNA probe immobilised on the transducer surface recognises its complementary (target) DNA sequence *via* hybridisation. The DNA duplex is then converted into an analytical signal by the transducer. Different strategies for electrochemical detection have been performed. One of them is based on label-free electrochemical detection, *via* the intrinsic electrochemical behaviour of DNA, through guanine or adenine nucleotides. However, most of the strategies are based on the use of indicators or labels. The first ones are based on the differences in the electrochemical behaviour of indicators that interact in a distinct extension with double-strand DNA (dsDNA) and ssDNA. The indicators for hybridisation detection can be anticancer agents, organic dyes or metal complexes. The latter strategies include the use of labels such as ferrocene, enzymes or metal nanoparticles.

Before a more extended explanation of the different strategies for the immobilisation of ssDNA probe on the electrodes surfaces and the different strategies for electrochemical detection of the hybridisation event, some aspects of the fabrication of thick- and thin-film electrodes will be given in the next section.

26.1.1.1 Fabrication of film electrodes

There are numerous film-fabrication methods available, depending on the film material. The two most common materials used for the fabrication of thick- and thin-film electrodes used as support/detector in genosensor devices found in the literature are gold and carbon, although other materials such as platinum, titanium or aluminium are used too but to a minor extent [2–5].

Gold thin-film electrodes (with different thickness ranging between 45 and 500 nm) used in genosensor devices have been mainly prepared by sputtering [6–9] and vacuum evaporation [10–15]. Sputtering method consists of an electrical discharge in a low-pressure gas such as argon at 10–100 mtorr, which is maintained between two electrodes, one of which is the material to be deposited (the target). When the target is made negative, with an accelerating voltage of 1–3 kV, it will be bombarded by energetic argon ions that transfer momentum to the material and cause ejection of atoms or ions, which deposit with relatively high velocity on the substrate.

Vacuum evaporation method is normally carried out in a high-vacuum system (10^{-5}–10^{-8} torr). The high vacuum is necessary to ensure that the mean-free-path of evaporated particles is long enough to reach the substrate, but perhaps more importantly, to maintain cleanliness of the substrate. Evaporation is most commonly initiated by melting the metal onto a conductive support, which may be a filament or boat, usually tungsten or tantalum, followed by further electrical heating of the metal to its melting point. Alternatively, higher-melting metals can be successfully evaporated using electron-beam heating techniques [11,12,14].

The most common substrates used to deposit gold are glass [6,10,12] or various forms of quartz or fused silica [7,8,11,13,14], although an other polyimide substrate named Kapton HN® [9] has been used. In many cases, adhesion of the deposited film to the substrate may be inherently poor. It is extremely desirable in such cases to deposit a thin layer of an intermediate material as titanium [6,8,12] or chromium [10,13] that has better adhesion to the substrate.

Both vacuum evaporation and sputtering methods allow obtaining continuous gold films. To obtain discontinuous films with different geometries, as gold arrays or electrodes with circular geometry with sizes on the order of 10 μm, it is necessary to combine these methods with microlithographic methods. In these methods, the device to be fabricated is coated with a photosensitive polymeric coating, which is baked and then exposed to a light source through a mask or a sequence of masks. The masks are designed using computer-aided design, and are typically produced by covering a metal film (e.g., chromium) on a high-quality glass plate with negative photoresist, exposing and developing the resist and then etching the metal to reveal the desired pattern. Masks may be positive or negative. Positive masks block light whereas negative masks transmit light. The photoresists can also be classified as positive and negative. The positive photoresist is initially polymerised and is then depolymerised and made soluble where light strikes it, whereas the negative photoresist is initially soluble but rendered insoluble at points that have been photopolymerised by exposure to light.

With respect to thick-film electrodes, the screen-printing technology is the most common method used to fabricate genosensor devices. The electrodes obtained with this technology are usually called screen-printed electrodes (SPEs). Most of them are carbon thick film [16–35], although gold thick films are also used [36,37]. The screen-printing technology has become increasingly popular recently for the fabrication of electrodes and complete cells for applications where a disposable, one-time-use electrochemical measurement is desired. Masking patterns are photographically exposed and developed on an emulsion applied to a fine screen whose mesh is open in zones where ink (gold, carbon, silver or Ag/AgCl) is not to be applied. Pastes or inks are then forced by means of a semiflexible squeegee blade through the masking screen onto the substrate to create the film electrode assembly. The finished assembly is then treated according to the type of electrode desired. In some cases (metal electrodes), the inks are fired to form the electrode. In other cases (carbon films), the electrodes are ready as soon as the solvent has evaporated and the film has dried. The most common substrates used to fabricate these SPEs are aluminium ceramic [16–18], polyester flexible film [19–26,36,37] and polycarbonate sheet [27–31].

26.1.1.2 Strategies for immobilisation of ssDNA probe

A crucial step in the design of a genosensor is the immobilisation of the ssDNA probes onto the electrode surface. The immobilisation method will determine the sensitivity and reproducibility of the genosensor. Several strategies for the immobilisation of ssDNA have been carried out and the strategy selected to immobilise the ssDNA probe depends, in many cases, on the nature of the electrode surface. The strategies for the immobilisation of ssDNA probes on solid surfaces include adsorption at controlled potential usually on pretreated screen-printed carbon surfaces [16–22,29]. However, in these cases, the immobilised ssDNA probes are not totally accessible for hybridisation, resulting in poor hybridisation efficiency. Moreover, in these cases, no blocking step is carried out and the background signals obtained in many cases are important.

However, other systems allow obtaining a sensing phase with more strands of DNA than by direct adsorption on the electrode. Moreover, the ssDNA probes are oriented in the genosensing phase, leaving the probes accessible for the reaction with their complementary targets. Actually, in most of the gold thin- [6,7,9,10–15,38–40] as well as thick-film electrodes [36,37], the immobilisation of the ssDNA probe is carried out by the well-known chemisorption of thiol groups on gold surfaces. In these cases, the thiolated probes, thiolated peptide nucleic acids (PNAs) [15] or an adequate thiolated primer (used to carry out PCR amplification on electrode surface [10]) form an ordered self-assembled monolayer on the gold surface. In these cases, a blocking step with a sulphur-containing compound is carried out to avoid the background signals. Other strategies are based on the formation of a polymer, through the electropolymerisation of a monomer-modified ssDNA probe [3] on the electrode surface, through the electropolymerisation of a monomer and subsequently covalent attachment of the amino modified ssDNA probe or through the copolymerisation of the monomer in the presence of ssDNA probe [2,24,25,35]. The avidin–biotin interaction to attach biotinylated ssDNA probes on the electrode surface has also been used to obtain genosensing phases [19,23,27,28,30,31,34].

An other interesting strategy is the modification of the surface of the electrodes with multiwalled carbon nanotubes (MWNTs) or single-walled carbon nanotubes (SWNTs) [13,32]. The MWNTs are grown on the electrodes covered with a nickel catalyst film by plasma-enhanced chemical vapour deposition and encapsulated in SiO_2 dielectrics with only the end exposed at the surface to form an inlaid nanoelectrode array [13]. In the other case, commercial SWNTs are deposited on SPE surface by evaporation [32]. The carbon nanotubes are functionalised with ssDNA probes by covalent attachment. This kind of modification shows a very efficient hybridisation and, moreover, the carbon nanotubes improve the analytical signal.

Finally, there are DNA arrays, whose electrical detection of hybridisation event is based on changes in electrical resistance or in capacitance between neighbouring electrodes, where the immobilisation of the ssDNA probe is carried out on the gaps between electrodes (not on the surface of the electrodes). These arrays are constructed on silicon substrate and the gaps between electrodes are functionalised using glycidosypropyltrimethoxysilane [41] or 7-octenyltrimethoxysilane [5] and then amino modified ssDNA probes are covalently attached. Other authors modified the gap between electrodes using 3-mercaptopropyltrimethoxysilane [42]. One end of this chemical compound is used to silanise the substrate surface while the thiol end is used to bind gold nanoparticles that are bound to thiolate ssDNA capture probes. The presence of nanoparticles allows modifying the gap with ssDNA probes as well as enhancing the current between the electrodes and subsequently the sensitivity of the sensor.

26.1.1.3 Strategies followed for electrochemical detection

Different strategies for electrochemical detection have been performed. One of them is based on label-free electrochemical detection, *via* the intrinsic electrochemical behaviour of DNA, through guanine or adenine nucleotides. However, most of the strategies are based on the use of indicators or labels. The first ones are based on the differences in the electrochemical behaviour of indicators with dsDNA and ssDNA. The indicators for hybridisation detection can be anticancer agents, organic dyes or metal complexes. The latter strategies are the use of labels as ferrocene, enzymes or metal nanoparticles. Table 26.1 shows the most common methods followed to detect the hybridisation event using thick- and thin-film electrodes found in the literature.

Moreover, Fig. 26.1 displays general schemes followed to detect the hybridisation event. The most common schemes used in the detection of hybridisation event are schemes a–c. Schems d_1 and d_2 have been recently developed. The scheme used to detect the hybridisation event depends mainly on the method employed to obtain the analytical signal. Thus, scheme a is used when label-free methods, with or without indicators, are employed to obtain the analytical signal. In this scheme the unlabelled ssDNA target is hybridised with capture probe immobilised on the electrode surface and then the detection is carried out *via* guanine oxidation or *via* the oxidation or reduction of the indicator. Schemes b and c are performed when labels are used to obtain the analytical signal. In the former scheme (scheme b) the labelled ssDNA target is hybridised with capture probe immobilised on the electrode surface and then the detection is carried out in an adequate manner. For this purpose, the labelling of ssDNA target is carried out by PCR amplification using a labelled primer or by the use of labelling kits, e.g., the Universal Linkage System (ULS). In the latter scheme (scheme c), called "sandwich", a first hybridisation step is carried out between the unlabelled ssDNA target and capture probe immobilised on the electrode surface followed by a second hybridisation with a labelled synthetic oligonucleotide (detector probe) complementary to other region of ssDNA target (different to capture probe). Finally, the detection of the label is carried out in an adequate manner. The main advantage of this scheme is that DNA-labelling procedures are avoided.

TABLE 26.1

Methods for detection of hybridisation event based on label-free, with or without indicators, and based on different labels

Methods	Tracers/labels	Strategy	Signal	Detection limit	References
Label-free (without indicators)		Fig. 26.1a	Guanine oxidation	90 ng/mL, 3 μg/mL, 30 μg/mL, 0.15 μg/mL	[16,22,29,32]
	$Os(bpy)_3^{2+}$		Electrocatalytic 8-G or 5-U oxidation	400 fM	[10]
	$Ru(bpy)_3^{2+}$	Fig. 26.1a	Electrocatalytic guanine oxidation	1000 target molecules	[13]
Label-free (with indicators)	$Co(bpy)_3^{3+}$ (intercalator)	Fig. 26.1a	Co^{3+} reduction	50 ng/mL	[18]
	Daunomycin (intercalator)	Fig. 26.1a	Indicator oxidation	1 μg/mL	[19,20]
	FND (ferrocenyl derivate) (intercalator)	Fig. 26.1a	Ferrocene oxidation		[48]
	Methylene blue (interaction with guanines)	Fig. 26.1a	Electrocatalytic		[38]
	Methylene blue (interaction with guanines)	Fig. 26.1a	Indicator reduction	2.4 μg/mL	[21]

TABLE 26.1 (*continued*)

Methods	Tracers/labels	Strategy	Signal	Detection limit	References
	Hoescht 33258 (minor groove binder)	Fig. 26.1a	Indicator oxidation	10^4 copies/mL, 10^{11} copies/mL	[6,15]
Electroactive labels	Pt (II) complex (ULS system)	Fig. 26.1b	Electrocatalytic	0.7 nM	[31]
	Ferrocene	Fig. 26.1c	Ferrocene oxidation		[39,40]
	Ferrocene	Fig. 26.1d_1	Ferrocene oxidation	10 pM	[43]
	Ferrocene		Ferrocene oxidation		[44]
	Colloidal gold	Fig. 26.1b	Silver enhancement and changes in electrical resistance		[41]
	Colloidal gold	Fig. 26.1b	Silver enhancement and changes in capacitance	0.2 nM	[5]
	Colloidal gold	Fig. 26.1b	Changes in electrical resistance	1 fM	[42]
	Colloidal gold	Fig. 26.1c	Redisolution of colloidal gold in HBr/Br_2 and Au^{3+} detection		[33]

Enzymatic labels	Glucose oxidase	Fig. 26.1c	Electrocatalytic using Os complex as mediator	1, 0.5 fM	[11,12]
	Preoxidase	Fig. 26.1b	Substrate OPD	0.3 pM, 0.6 fM	[3,26]
	Preoxidase	Fig. 26.1d$_2$ and c	Electrocatalytic reduction of H_2O_2 using Os complex as mediator	200 fM	[23–25]
	Peroxidase	Fig. 26.1c and b	Substrate TMB	1 pM, 16 copies	[27,28]
	Preoxidase	Fig. 26.1c	Substrate ADPA	16 μg/mL	[34]
	Alkaline phosphatase	Fig. 26.1c	Substrate *p*-aminophenyl phosphate and reduction of Ag^+ to Ag. Oxidation peak of Ag	100 aM	[14]
	Alkaline phosphatase	Fig. 26.1b	Substrate α-naphthyl phosphate	1 nM	[36]
	Alkaline phosphatase	Fig. 26.1c	Substrate BCIP/NBT coupled with $Fe(CN)_6$	1.2 pM	[37]
	Alkaline phosphatase	Fig. 26.1b	Substrate 3-indoxyl phosphate	5, 16 pM	[9,30]

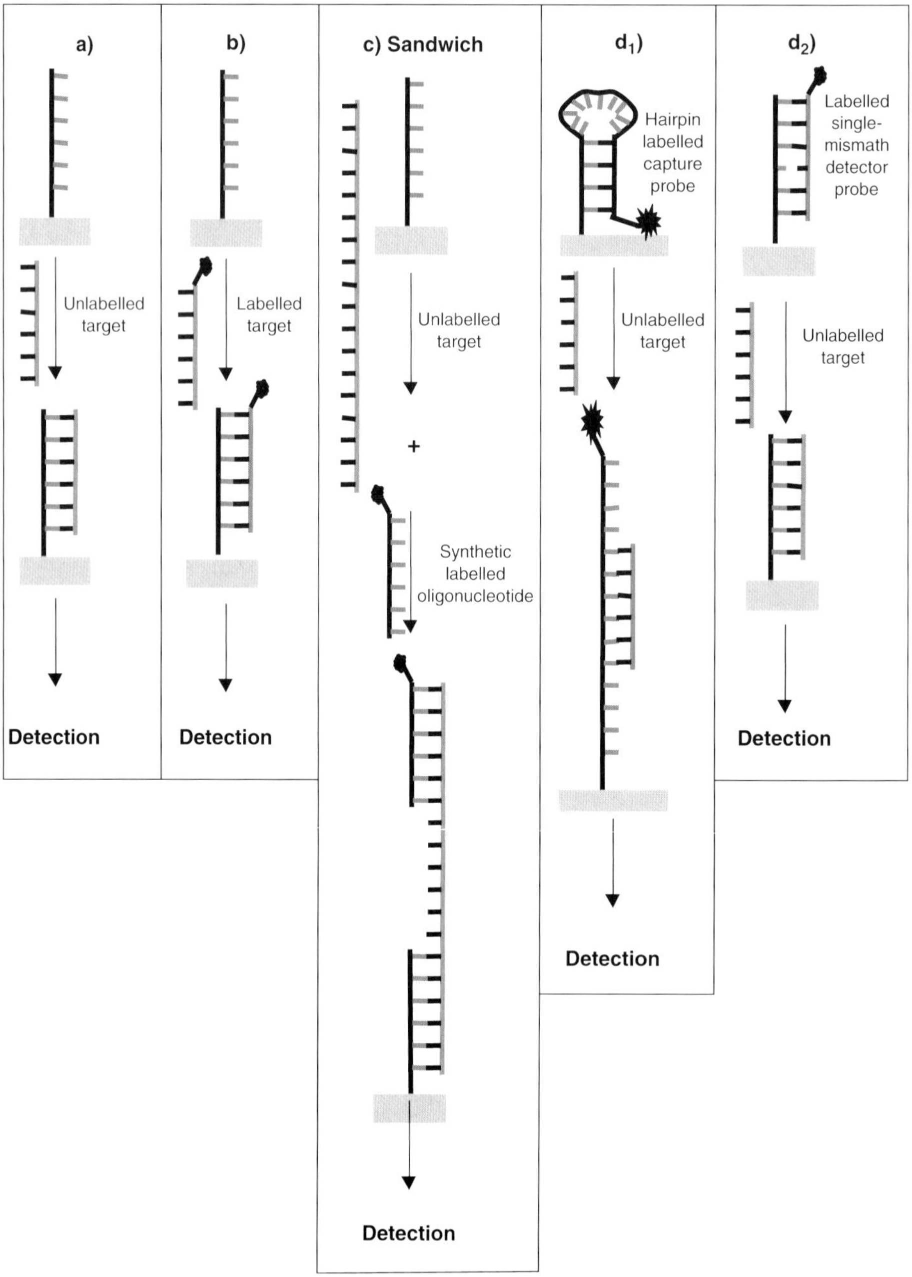

Fig. 26.1. Schemes followed to detect the hybridisation event. Scheme (a) for label-free methods with and without indicators. Schemes (b) and (c) for methods that use labels. Schemes (d_1) reagentless and (d_2) quasi-reagentless.

Finally, in schemes d_1 and d_2 the label is on the electrode surface. The first scheme is called reagentless and the second one is named quasi-reagentless. In the first one, the unlabelled ssDNA target is hybridised with a ferrocene-labelled hairpin capture probe immobilised on the electrode surface [43]. In the hairpin form, the redox-active ferrocene is close to the electrode surface and undergoes oxidation. Hybridisation of the loop region to the target disrupts the hairpin, increasing the distance from the redox-active ferrocene to the electrode surface and thereby lowering the electrochemical signal. The impressive characteristic of the response is the dynamic range, which spans six orders of magnitude of the DNA concentration. Although this broad dynamic range is useful in some applications, the sensitivity of the method (the percentage change per increment of input concentration) is correspondingly less marked, making this scheme more suitable for applications such as the analysis of single nucleotide polymorphisms (SNP) or the detection of pathogens, in which reading small changes in input concentration is less important (than it is in expression analysis).

In the second scheme (scheme d_2), the HRP labelled oligonucleotide containing one base mismatch is suboptimally prehybridised with the capture probe and is displaced upon introduction of the unlabelled fully complementary ssDNA target, producing a decrease in analytical signal [23]. This scheme is called quasi-reagentless by authors because the label used is an enzyme that needs an adequate substrate to obtain an analytical signal.

In label-free systems, the nucleobases inside dsDNA are oxidised to a lesser extent than those in the resulting ssDNA and a decrease in the analytical signal after the hybridisation should be expected. The analytical signal diminishes relative to the signal from the ssDNA probe as the duplex is formed but at the same time as new electroactive bases are added within the target, its contribution to the increase in the analytical signal must be taken into account. Because of these opposite trends, nonlinear calibration curves are often obtained. The alternative is to use guanine-free probes. In these probes guanine is substituted for inosine, which binds preferentially to cytosine and its oxidation process is far removed from that of guanine [22,29]. The signal now appears on hybridisation and background signals are negligible. An other strategy to increase the differences between analytical signals obtained for ssDNA and dsDNA based on guanine oxidation has been developed recently [32]. It consists of the use of a single-strand DNA-binding protein (SSB). This protein binds selectively to ssDNA. The guanine oxidation is hampered when ssDNA probe is bound to SSB but when the hybridisation event is carried out, the SSB does not bind to dsDNA and the guanines oxidation is recorded. Methods based on oxidation of bases are attractive because they are label-free, rapid and the most simple (Fig. 26.1a). Oxidation of guanine at thick- and thin-film electrodes requires high-applied potentials and the associated high background currents must be subtracted to reach competitive detection limits using potentiometric stripping analysis. Another strategy employed to increase signal-to-noise ratio is the use of ruthenium or osmium complexes that act as mediators of the oxidation of guanine, increasing the rate of electron transfer between the nucleic acid and the electrode and enhancing the oxidation signal.

The use of indicators is an other common method to detect the hybridisation event. Electrochemical hybridisation indicators are electroactive compounds with different affinities for ssDNA relative to dsDNA. Three modes of binding of small molecules to DNA are normally described: electrostatic interaction, groove binding and intercalation. Because grooves are formed only within the double helix, groove-binders as Hoescht 33258 have greater affinity for dsDNA than ssDNA and thereby when hybridisation event takes places the oxidation peak of the indicator increases. Other indicators, such as daunomycin, cobalt complex and ferrocenyl-naphtalene diimide (FND), act as intercalators. Shifts in the peak potential or area of the oxidation process of daunomycin on dsDNA when compared to that obtained with ssDNA have been successfully used as electrochemical signals [19,20]. The treading intercalator FND has higher duplex affinity and binding stability and the ferrocene oxidation increases when duplex DNA is formed. With respect to methylene blue (MB), an other indicator, this compound gives lower signals for duplex DNA than those obtained for ssDNA, because the specific interaction of MB with guanine bases is prevented in the duplex and then the reduction peak of MB decreases.

The main advantage of using hybridisation indicators is that DNA-labelling procedures are avoided, but the weak point is that discrimination is not usually high enough and should be improved.

Other methods of detection of hybridisation event consist of the use of labels. These labels can be divided into electroactive and enzymatic labels. The schemes commonly used for detection are scheme b or c. The scheme selected to detect the hybridisation event depends on whether the target is labelled with a marker by PCR amplification or using a labelling kit, or whether a labelled synthetic detector oligonucleotide is used. These methods are usually the most sensitive, but the need of a labelling step makes them more complex, tedious and expensive although the biotechnology companies already offer, at reasonable prices, oligonucleotides labelled with the most commonly used labels, including biotinylated oligonucleotides.

The electroactive labels used in the thick- and thin-film genosensors found in the literature are ferrocene, colloidal gold and platinum (II) complex. The ferrocene moiety is incorporated into a detector probe through the use of a modified adenine residue that has a ferrocene substitution on the ribose ring [39,40] or through the coupling of a ferrocene derivative (ferrocene carboxylic acid) to amino-terminated DNA molecules [44]. The reversible oxidation process of ferrocene is detected using different electrochemical techniques such as cyclic voltammetry, alternating current voltammetry or chronocoulometry.

With respect to colloidal gold, the nanoparticles are usually bound to thiol-terminated DNA molecules [33,41,42], although in other cases [5] the nanoparticles are linked indirectly through biotin–anti-biotin bridges. The detection of gold nanoparticles can be carried out by measuring the changes in the electrical resistance or capacitance between electrodes usually after a silver enhancement procedure or by anodic stripping voltammetry of the gold (III) obtained after a treatment with an oxidant Br_2/HBr mixture.

A platinum (II) complex label has been used by our group [31]. The oligonucleotide targets used are labelled using the Universal Linkage System (ULS®). This kind of labelling consists in the use of a square-planar complex of platinum (II) called BOC-ULS. One of the ends of this complex finishes in a BOC group (*tert*-butoxycarbonyl group) that can be substituted by a molecule (the label, such as fluorescein, digoxigenin, biotin, etc.). The complex "platinum-label", so obtained, is monofunctional, that is, the other end of the complex finishes in one Cl, through which the attachment of the complex to the oligonucleotide takes place. The Cl is substituted by the N7 position of the guanines of the oligonucleotides in a very simple labelling reaction that takes not more than 35 min. The analytical signal is obtained from platinum (II) complex, which is deposited on the electrode surface. In the presence of platinum on the electrode surface and after fixing an adequate potential in acidic medium, the protons are catalytically reduced to hydrogen. The current generated by this catalytic reduction can be measured and increases with platinum concentration and consequently with labelled target concentration following scheme b.

The methods based on the use of enzymatic labels are without doubt among the most sensitive. Moreover, because routine laboratories are usually working with enzymatic labels, this might make them the most attractive labels for implementation of DNA diagnosis in such laboratories. The signal normally arises from a redox process of a product of the enzymatic reaction (Table 26.1). However, in other cases the analytical signal is not based on the redox process of the enzymatic product. Thus, Kwak *et al.* [14] use a biometallisation process where a nonelectroactive substrate (*p*-aminophenyl phosphate) is enzymatically converted into the reducing agent *p*-aminophenol that reduces Ag^+ ions leading to deposition of the metal onto electrode surface. In this case, the oxidation peak of deposited Ag is recorded. Mascini *et al.* [37] use the alkaline phosphatase substrate BCIP/NBT that after the enzymatic reaction generates an insoluble and insulating product on the sensing phase that blocks the electrical communication between the electrode surface and the $Fe(CN)_6^{3-/4-}$ pair. In this case, faradaic impedance spectroscopy is finally used to detect the enhanced electron-transfer resistance.

The enzymes can be linked directly to the DNA strand [3,23–25] or indirectly through biotin–avidin [9,11,12,14,26,36,37], through fluorescein–anti-fluorescein [27,28,30] or digoxigenin–anti-digoxigenin [34] bridges.

Among all enzymes, peroxidases are the most widely used and commercial kits for labelling and are already available [28].

26.1.2 Pretreatments followed with real samples

Before the DNA from real samples are tested with genosensors, the DNA must be isolated from the samples. The DNA-isolation procedures followed depend on the type of sample (tissues, blood, etc.). First of all, the DNA must be extracted from the cells, after a twice centrifugation and lysis procedure. There are several protocols for extraction of DNA. The most common are the salting out [45,46] and phenol–chloroform extraction [47], although commercial extraction kits (e.g., QIAmp Kit from Qiagen or Invisorb kit form Invitek) are used too.

After that, when the amount of DNA is very small and the sensitivity of the genosensor is not enough to detect the DNA contained in the sample, an amplification step using the polymerase chain reaction (PCR) is usually carried out. Although there are several PCR protocols such as normal PCR, asymmetric PCR (A-PCR), long PCR, reverse transcriptase PCR (more used in RNA amplification), etc., the most commonly used is normal PCR. This procedure is performed in thermocyclers, which apply a cycling temperature protocol that comprises: a denaturation step, a primer annealing step and a primer extension step by polymerase enzyme. These steps are cycled several times.

One of the primers used in PCR amplification can be conjugated with a label if the strategy used on the genosensor device is the one explained in scheme b of Fig. 26.1.

After PCR amplification, several authors carry out a purification step of PCR products before they are tested with genosensor devices in order to diminish the PCR blanks [5,15,22,27,36,48]. For this purpose, commercial purification kits are used (e.g., Qiaquick from Qiagen). This purification step removes the primers, dNTPs, Taq polymerase and salts from the PCR products.

As DNA extracted from the sample, with or without PCR amplification step, usually is a double-strand, the hybridisation of the target strand with the capture probe immobilised on genosensor device cannot occur. Then, it is necessary to obtain ssDNA targets previously to test with genosensor device. Several strategies are followed to achieve ssDNA targets.

- *Thermal denaturation*. It is the most commonly used strategy [6,12,13,15,20–22,29,34–36,42]. It consists of denaturing the dsDNA thermally by using a boiling water bath for sometime and then immersing the tube containing the denaturalised DNA in an ice-water bath for a short time in order to retard the re-annealing of the strands. Although the denaturalisation is achieved, one drawback of this procedure is that the sister strand can compete with capture probe by target strand.
- *Asymmetric PCR*. Some authors used this procedure to obtain ssDNA targets [39,40,48]. It consists of a PCR amplification with one of the primers in excess compared with the other one. In this way, the target strand of interest is in excess compared with the sister strand, thus favouring binding at the capture probe as opposed to sister strand reannealing. When the amount of DNA from the sample is enough to detect on the genosensor device without a PCR amplification, this procedure is not used.
- *Alkaline treatment*. Some authors have used this procedure [26,27,48]. It consists of adding to DNA an alkaline solution (e.g., 0.5 M NaOH solution). An increase in pH affects the ionisation of the functional groups of bases implied in the hydrogen bonds, thus decreasing the number of these bonds between the strands. It is a rapid and simple method.

- *Lambda-exonuclease treatment*. This procedure is used by Alderon Biosciences [28]. Exonucleases are enzymes that hydrolyse the terminal phosphodiester bonds of DNA strands obtaining mono-nucleosides monophosphate or biphosphate. In this case a 5′-exonuclease that hydrolyses the 5′-end phosphodiester is used. In this way, the strands are destroyed. To achieve the total hydrolysis of the sister strand, the target strand must be protected, for example, by a label at its 5′-end. Therefore, the PCR amplification is carried out using one of the primers biotinylated at the 5′-end and the other one modified at its 5′-end with a phosphate group. The extension of these primers give rise to the 5′-biotinylated target DNA strand and a 5′-phosphate sister DNA strand, which is hydrolysed by 5′-exonuclease.

26.1.3 Experimental conditions for hybridisation reaction

The hybridisation event is affected by the concentration of DNA target, concentration and size of capture probe, temperature, hybridisation time and hybridisation buffer composition (pH, ionic strength, denaturalising chemical agents, etc.). The control of the experimental variables that affect the hybridisation event is very important in order to obtain an efficient and selective hybridisation. The detection of SNP or genetic mutations requires an efficient discrimination between mismatched and complementary strands. In most of the cases, the selectivity relies on the operating conditions of the assay as hybridisation buffer composition, hybridisation time and hybridisation temperature.

One way to obtain discrimination between mismatched and complementary strands is the use of high temperatures during the hybridisation reaction. The difference in the melting curves of the target from perfectly matched and mismatched capture probes can allow to define a specific temperature to discriminate perfect matches and mismatches. However, the employment of high temperatures during hybridisation step requires precise temperature control, which is difficult to achieve and is expensive [48].

A simple manner to obtain discrimination between mismatched and complementary strands is to use more stringent conditions: varying the saline concentration (ionic strength) of the hybridisation buffer or adding chemical agents that destabilise the DNA duplex as formamide or urea. The presence of ions in DNA solutions stabilises the DNA duplex because they diminish the electrostatic repulsion between phosphate groups of the strands and therefore facilitate the approximation between both strands. If the ionic strength is decreased, only complementary hybrids can be formed. Chemical agents with amino and carbonyl groups in their structure such as formamide or urea present in the hybridisation solution compete with nucleotide bases for hydrogen bonds formation, facilitating the DNA duplex destabilisation and strands separation and achieving discrimination between mismatched and complementary strands. The addition of these chemical agents to obtain more stringent conditions and therefore the discrimination of single-mismatched strand has been used by several authors [9,30,31,36].

However, there are other methods to detect SNP where nonstringent conditions are used. One of them is the one developed by Huang *et al.* [27]. The strategy followed in this case is the use of a hairpin-forming detector probe complementary to the target combined with an unlabelled hairpin-forming competitor probe complementary to mismatched strand in a sandwich format (scheme c of Fig. 26.1). Using the hairpin-forming detector probe, the mismatched target-probe hybrids produce significantly lower signals than that generated by the perfectly matched target-probe hybrids. Moreover, the addition of a hairpin-forming competitor probe containing sequences complementary to mismatched targets increases the sensitivity and specificity of the system by decreasing the formation of mismatched target-probe hybrids.

Other method has been developed by Amano *et al.* [48] to identify genetic mutations on the lipoprotein lipase gene. They combine the use of DNA ligase enzyme and A-PCR with one of the primer (called special primer by authors) containing a Tag sequence at its 5′-end. The Tag sequence is complementary to the genomic region between the mutation point and the special primer. Since the A-PCR products contain self-complementary sequences, they form a self-loop. The capture probes are attached to the electrode surface by their 5′-end. The wild-type probe (W probe) has the same nucleotide as the wild-type genome at the 3′-end and the mutant-type probe (M probe) has the same nucleotide as the mutant-type genome. When the A-PCR product is hybridised with the W probe or the M probe, the ligase enzyme recognises the matched or unmatched nucleotide pair at the 3′-end of the probe. If the 3′-end nucleotide is complementary (or matched), the enzyme repairs the nick between the 3′-end of the probe and the 5′-end of the A-PCR product. As a result, the probe and the A-PCR product become one molecule. However, if the last nucleotide is not complementary (or is mismatched), the enzyme does not repair the nick and the probe and the A-PCR product remain as two different molecules. After a denaturation step, if the enzyme repairs the nick, the probe and the A-PCR product are not separated and then FND (intercalator) is detected; but if the enzyme does not repair the nick, the probe and the A-PCR product are separated and then FND is not detected.

26.2 APPLICATIONS

In this section several examples of genosensors based on hybridisation event, which have been constructed on thick- and thin-film electrodes, will be described. Two of them have been designed to identify the nucleic acids determinants exclusively present on the genome of the pathogen *Streptococcus pneumoniae* [30,31], whereas an other one has been designed to detect a 30-mer SARS (severe acute respiratory syndrome) virus sequence ([9], Procedure 36 at CD accompanying this book). Although in most of them alkaline phosphatase and 3-indoxyl phosphate are used as label and as enzymatic substrate, respectively, other label, a platinum (II) complex, will be presented and its detection discussed [31]. In all cases, synthetic target oligonucleotides as well as three-base mismatch and one-base mismatch strands of the pathogen *Streptococcus pneumoniae* or SARS virus are tested using these genosensor devices. Finally, a genosensor device to detect TNFRSF21 PCR products will be presented (Procedure 37 in CD accompanying of this book).

26.2.1 Enzymatic genosensor on gold thin-films to detect a SARS virus sequence

A DNA hybridisation assay with enzymatic electrochemical detection is carried out on a 100 nm sputtered gold thin film that allows working with small volumes. Reducing the cell volume has several advantages [49]. The first one is the decrease in the diffusion distances required for analytes to reach their surface-bound receptor partners. Moreover, in the case of enzymatic detection, the product dilution, a critical factor in achieving low detection limits, diminishes. A simple, cheap and easy-to-handle homemade device that permits to perform simultaneous hybridisation procedures and sequential detection of more that 20 assay sites is presented.

The sequence chosen as target is included in the 29751-base genome of the SARS (severe acute respiratory syndrome)-associated coronavirus [50]. This is the causative agent of an outbreak of atypical pneumonia, first identified in Guangdong Province, China, that has spread to several countries. The sequence corresponds to a gene that encodes the nucleocapsid protein (422 amino acids), concretely a short lysine rich region that appears to be unique to SARS and suggestive of a nuclear localisation signal. A 30-mer oligonucleotide with bases comprises between numbers 29,218 and 29,247, both included, has been chosen.

26.2.1.1 Fabrication of gold thin-film electrodes and three-electrode potentiostatic system

A kapton slide is cleaned with ethanol and after being dried, it is covered with gold by a sputtering process. Gold atoms are deposited (from the cathode) over kapton (placed on the anode) in a vacuum chamber filled with argon. Gold layer thickness is controlled with the time and the intensity of the discharge. For a 100 nm thick layer a 35 mA discharge is applied over 220 s. Then, a conductor wire is attached to the centre of one of the sides by means of an epoxy resin (CW2400), obtained from RS Components, that is cured at room temperature.

The working area is limited by self-adhesive washers of 5 mm of internal diameter (19.6 mm^2 of internal area). The total area of the gold surface lets approximately 23 washers to stick. The gold film is placed on a support where a crocodile connection is fixed.

Reference and auxiliary electrodes are coupled in a micropipette tip. The reference electrode consists of an anodised silver wire introduced in a tip through a syringe rubber piston. The tip is filled with saturated KCl solution and contains a low-resistance liquid junction. The platinum wire that acts as auxiliary electrode is fixed with insulating tape. For measurement recording the tip is fixed on an electrochemical cell Metrohm support allowing horizontal and vertical movement (see Fig. 36.1 of Procedure 36 in CD accompanying this book).

26.2.1.2 Genosensor design

A complementary strand to the chosen SARS sequence is labelled with a thiol group and immobilised on the gold surface. The target (30-mer oligonucleotide with a sequence included in the SARS-coronavirus) is conjugated to biotin and hybridised with the probe. Addition of AP-labelled streptavidin allows enzymatic detection through the electrochemical signal of the indigo carmine (IC), enzymatic product of 3-indoxyl phosphate (see more details for genosensor construction in Procedure in CD accompanying this book). This substrate was proposed by our group [51] as a suitable substrate based on the favourable processes that IC, a soluble derivative of the product generated (indigo blue), presented. Moreover, compared with other substrates, kinetic constants resulted more favourable. The electrochemical behaviour of IC on gold electrodes has been studied [52].

26.2.1.3 Results

Two steps, the probe immobilisation and the blocking steps, in the design of the genosensor are very important. In the immobilisation step the presence of the thiol group in the probe and the effect of the evaporation during the immobilisation process has been studied. The favourable immobilisation of DNA through its thiol group is observed in Fig. 26.2. Signals were obtained for unmodified DNA and SH-DNA (both biotinylated) following similar procedures. Although ssDNA adsorbs on gold substrates, it adopts a coiled configuration, and therefore cannot form an ordered structure [53]. An almost negligible adsorption of unmodified DNA is presented, that is very favoured by the insertion of a thiol group. Moreover same results were obtained when DNA immobilisation took place at 4°C for 12 h or at 37°C for 20 min. Moreover, the evaporation is a critical condition in the immobilisation of SH-DNA. This effect is shown in Fig. 26.3. It can be observed that the enhancement of the signal with immobilisation time was greater when

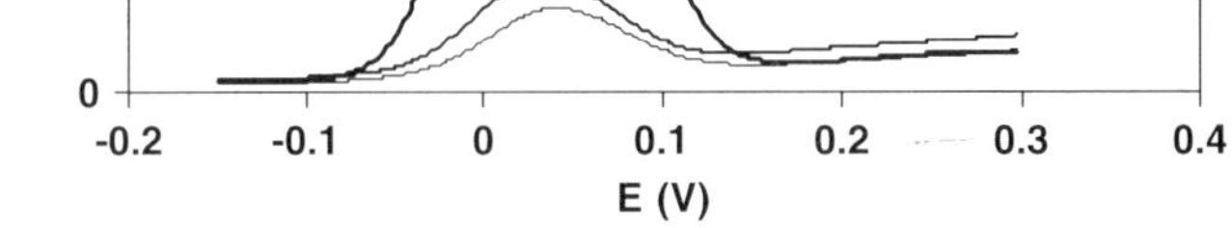

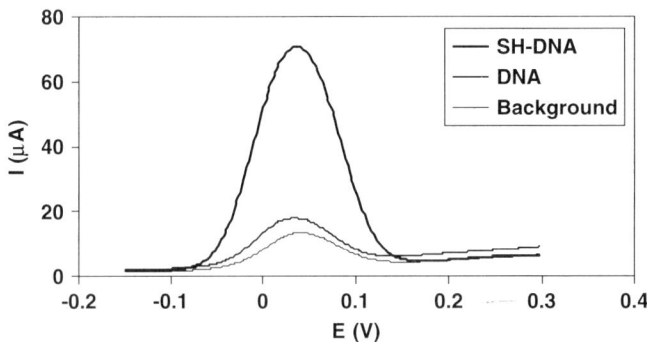

Fig. 26.2. Signals recorded for biotinylated SH-DNA, biotinylated DNA and background. $C = 1\,\mu M$, $V_{drop} = 5\,\mu L$, $T_{immob} = 4°C$, $t_{immob} = 12\,h$. Reprinted from Ref. [9], Copyright 2005, with permission from Elsevier.

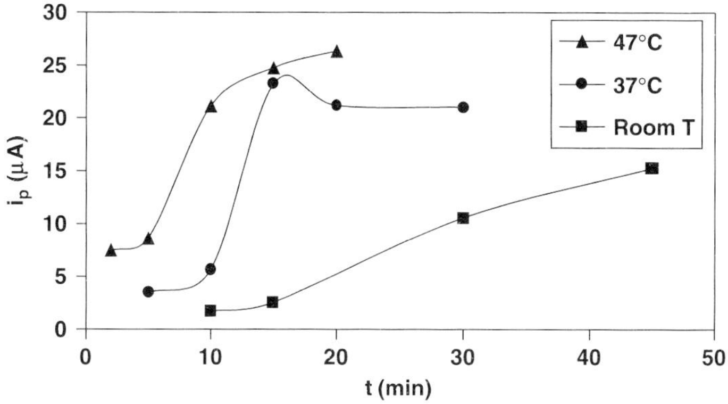

Fig. 26.3. Influence of temperature on DNA immobilisation. $C = 1\,\mu M$, $V_{drop} = 5\,\mu L$. Reprinted from Ref. [9], Copyright 2005, with permission from Elsevier.

evaporation occurred, which corresponds to 30, 15 and 10 min for room temperature, 37 and 47°C, respectively. When the immobilisation took place at 4°C overnight (12 h, attaining evaporation as well) the signal was similar to that obtained at 37°C (27 vs. 29 μA, respectively). Therefore, both methodologies could be employed. The repeatability of the signals obtained for both immobilisation procedures was also similar, obtaining an RSD of 9% and 8% for five measurements when immobilisation was carried out at 37°C for 20 min and 4°C for 12 h, respectively. Considering five different gold films, precision was analogous, with an RSD of 10%. Despite the evaporation that favours DNA adsorption, this effect is the contrary to what occurred when the hybridisation step is carried out, where the evaporation diminishes the hybridisation signal.

Blocking the surface is one of the most important steps to minimise and control non-specific adsorption. Two main types of agents were considered: proteins and sulphur-containing compounds. The signal/background ratio (S/B), and therefore the blocking capacity, was observed for each compound. Among blocking agents assayed, better results were obtained with albumin and 1-hexanethiol. A comparison between the signals obtained for background and immobilised DNA (double labelled with biotin and thiol) when BSA and 1-hexanethiol were employed as blocking agents is shown in Fig. 26.4. It can be observed that the background with albumin is negligible and the S/B ratio is high (22.4), but the DNA signal is better defined for 1-hexanethiol where the capacitively current approaches zero. Moreover, when the

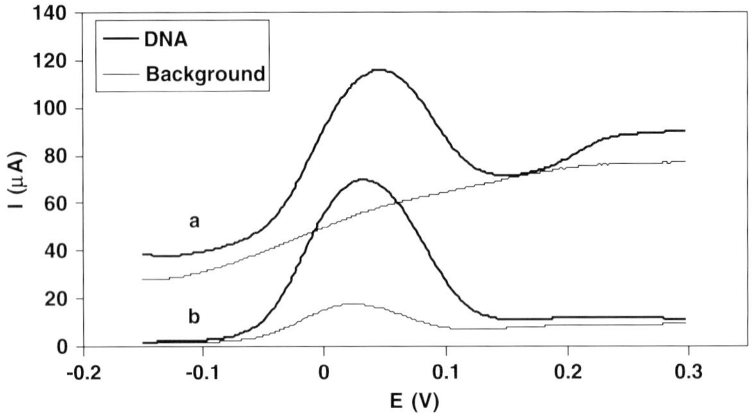

Fig. 26.4. Signals and backgrounds obtained when (a) albumin and (b) 1-hexanethiol are employed as blocking agents. $C_{\text{blocking agent}} = 2\%$, $C_{\text{DNA}} = 1\,\mu\text{M}$, $V_{\text{drop}} = 5\,\mu\text{L}$, $T_{\text{immob}} = 4^{\circ}\text{C}$, $t_{\text{immob}} = 12\,\text{h}$. Reprinted from Ref. [9], Copyright 2005, with permission from Elsevier.

hybridisation step was carried out the S/B ratio obtained for albumin was much lower than that obtained with a double-labelled DNA strand. This is probably caused by the steric hindrance, due to the large size of albumin, which hinders hybridisation.

Repeatability is checked under the experimental conditions explained in Procedure 36 in CD accompanying this book, 1 h of hybridisation and a 2 × SSC buffer containing 50% of formamide. The value of the RSD was 11% for nine measurements.

A complete study on the selectivity of DNA hybridisation has been carried out. It is of relevant interest in the study of SNP as well as for the study of virus mutation. After testing many stringency conditions, it was observed that carrying out the interaction between strands in a medium containing 50% of formamide during 1 h was enough for achieving a high degree of discrimination for all the mutated strands tested.

Studies with mutated (base substitution) 30-mer synthetic oligonucleotides revealed differences for 3, 2 and 1 mismatched strands. The type of interaction that disappears and the possible new interactions that are generated when a base substitution occurs are of importance. One mismatch is detected even if it is located extreme far from the electrode, being the attenuation of the signal lower than when the substitution occurs near the electrode surface (Fig. 26.5). Moreover,

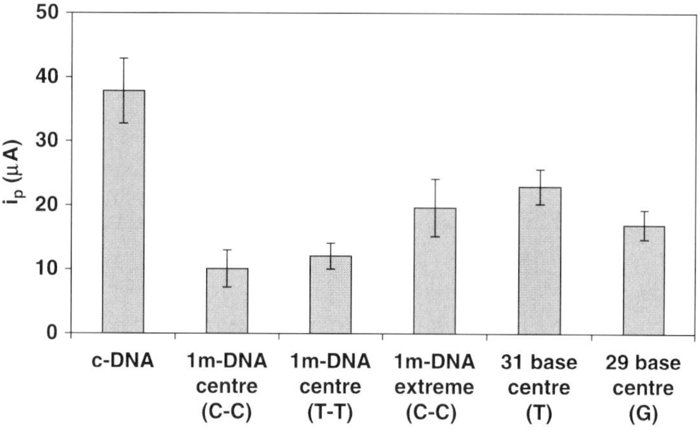

Fig. 26.5. Comparison between signals for the complementary and point mutated strands: substitutions at the centre or 3′-extreme, insertion and deletion. $C_{\text{target}} = 2.5\,\text{nM}$, $V_{\text{target}} = 20\,\mu\text{L}$, $2 \times$ SSC pH 7 buffer with 50% formamide, $t_{\text{hybr}} = 60\,\text{min}$. Data are given as average ± SD ($n = 3$).

mutations that involve deletions and insertions are differentiated from the fully complementary strand. However, although destabilisation of the duplex occurs, lower discrimination than with a substitution mutation is obtained. As a higher number of interactions is present, higher signals are obtained for the base insertion mutation. Hybridisation studies carried out between 30-mer strands were compared with those for 40-mer oligonucleotides. The length of the strand influences the hybridisation in such a way that higher signals are obtained and lower discrimination is achieved when the number of oligonucleotides increases. Experiences with crossed strands (immobilised 40-mer strand that hybridises with a 30-mer target and vice versa, hybrids between immobilised 30-mer strand and a 40-mer one) revealed the importance of an oligonucleotide tail in the immobilised strand for favouring the mobility and therefore the hybridisation. On the other hand, hybridisation is disfavoured when an immobilised strand has to react with another that contains a higher number of oligonucleotides in such a way that a rest near the electrode is present (30–40) (Fig. 26.6).

Discrimination is seen over a wide interval of concentrations (Fig. 26.7). Signals are linear for both complementary and one-base mismatched strands, between 0.01 and 1 nM. Detection limits of 5 and 70 pM were, respectively, obtained.

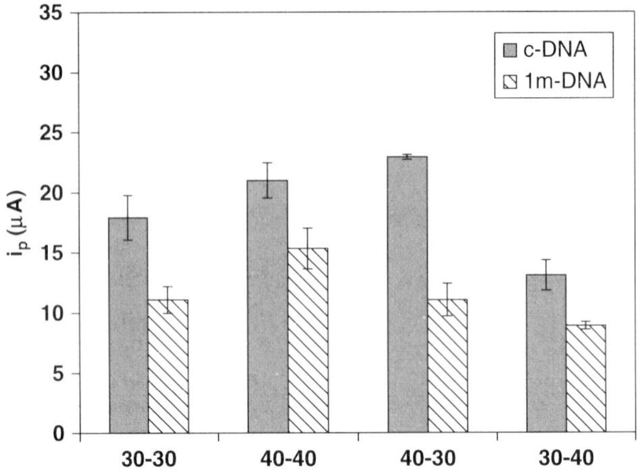

Fig. 26.6. Results obtained for immobilised and complementary (c-DNA) or one-base mismatched (1 m-DNA) target strand hybridisation, combining different lengths. $C_{target} = 2.5$ nM, $V_{target} = 20$ μL, 2 × SSC pH 7 buffer with 50% formamide, $t_{hybr} = 60$ min. Data are given as average ± SD ($n = 3$).

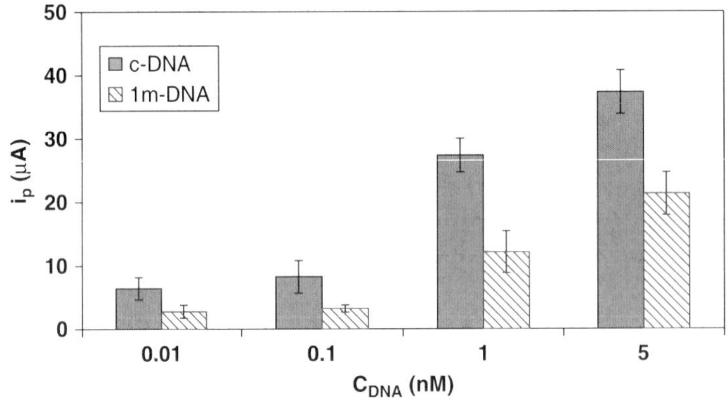

Fig. 26.7. Effect of the concentration on the selectivity. Signals for the complementary (c-DNA) and the one-mismatched strand (C–C at the centre, 1 m-DNA). $V_{target} = 20$ μL, 2 × SSC pH 7 buffer with 50% formamide, $t_{hybr} = 60$ min. Data are given as average ± SD ($n = 3$).

26.2.2 Genosensors on streptavidin-modified thick-film carbon electrodes to detect *Streptococcus pneumoniae* sequences

This section outlines the development of genosensors on screen-printed carbon electrodes (SPCEs) for the identification of nucleic acid determinants exclusively present in the genome of the pathogen *Streptococcus pneumoniae*. Orientation of the strands in the sensing phase is achieved by modifying the surface of the electrode with streptavidin by physical adsorption followed by the immobilisation of biotinylated oligo probes. The physical adsorption of streptavidin must be performed at a constant temperature above the room temperature. Moreover, the electrode surface must be previously electrochemically pretreated at an anodic potential in acidic medium to improve its adsorptive properties. In this way, reproducible, sensitive and stable sensing phases are obtained [54]. The biotinylated oligo nucleic acid probes used in this work target the pneumolysin (ply) gene. This target is randomly labelled with the Universal Linkage System (ULS®). This labelling system consists of the use of a platinum (II) complex that acts as a coupling agent between DNA strands and a label molecule, usually fluorescent. This platinum complex is a monofunctional derivate of cisplatin (a potent anticancer agent used in the treatment of a variety of tumours) that binds to DNA at the N7 position of guanine with release of one Cl^- ion per molecule of the complex. The label molecule used in this study was fluorescein (FITC). Electrochemical detection is achieved using two strategies. One of them is carried out using an anti-FITC alkaline phosphatase-labelled antibody and 3-indoxyl phosphate (3-IP) as enzymatic substrate of AP. The resulting enzymatic product is indigo blue, an aromatic heterocycle insoluble in aqueous solutions. Its sulfonation in acidic medium gives rise to IC, an aqueous soluble compound that shows an electrochemical behaviour similar to indigo blue. Both 3-IP and IC have already been studied on SPCEs [55,56]. However, although these genosensors are stable and sensitive devices for the detection of specific nucleic acid fragments, the need of two additional steps to obtain the analytical signal resulted in a large time-consuming analysis. This fact can be avoided using the second strategy for detection. In this case the analytical signal is directly obtained from platinum (II) complex, which is deposited on the electrode surface. In presence of the platinum on the electrode surface and after fixing an adequate potential in acidic medium, the protons are catalytically reduced to hydrogen. The current generated by this catalytic reduction can be measured and increases with platinum concentration and consequently with labelled target concentration.

Data presented here demonstrate the potential applicability of SPCEs genosensors in the diagnosis of a human infectious pulmonary disease. These electrochemical genosensors are stable and sensitive devices for the detection of specific nucleic acid fragments. Moreover, these devices allow the detection of a one-base mismatch on the targets if adequate experimental conditions are used.

26.2.2.1 Genosensor design

- Electrode pretreatment: 50 μL of 0.1 M H_2SO_4 are dropped on the SPCEs and an anodic current of +3.0 μA is applied for 2 min. Then, the electrodes are washed using 0.1 M Tris buffer pH 7.2.
- Adsorption of streptavidin: an aliquot of 10 μL of a 1×10^{-5} M streptavidin solution is left on the electrode surface overnight at 4 °C. Then, the electrode is washed with 0.1 M Tris buffer pH 7.2 to remove the excess of protein.
- Blocking step: free surface sites are blocked placing a drop of 40 μL of a 2% (w/v) solution of BSA for 15 min followed by a washing step with 0.1 M Tris pH 7.2 buffer containing 1% of BSA.
- Immobilisation of oligonucleotide probes onto the electrode surface: 40 μL of 3′-biotinylated oligonucleotide probes (0.5 ng/μL) is left on the electrode surface for 15 min. Finally, the electrodes are rinsed with 2 × SSC buffer pH 7.2 containing 1% of BSA.

26.2.2.2 Hybridisation step

Hybridisation is performed at room temperature placing 30 μL of FITC labelled oligonucleotide target solutions in 2 × SSC buffer pH 7.2, containing 1% of BSA, on the surface of the genosensor for 45 min and then rinsing with 0.1 M Tris pH 7.2 buffer containing 1% of BSA. The methodology used to detect one-base mismatch strands is similar, but in this case 25% formamide is included in the hybridisation buffer.

26.2.2.3 Analytical signal recording

Two strategies are performed to detect the hybridisation event: enzymatic detection and electrocatalytic detection. The following steps are carried out:

Enzymatic detection

- Reaction with anti-FITC AP conjugate (Ab-AP): an aliquot of 40 μL of Ab-AP solution (1/100 dilution) is dropped on the genosensor device for 60 min. After a washing step with 0.1 M Tris buffer pH 9.8, containing 1% BSA, is carried out.
- Enzymatic reaction: an aliquot of 30 μL of 6 mM 3-IP is deposited on the electrode surface for 20 min. After that, the reaction is stopped

adding 4 μL of fuming sulphuric acid and 10 μL of ultra-pure water. In this step, the corresponding indigo product is converted to its parent hydrosoluble compound IC.
- Analytical signal recording: the SPCEs are held at a potential of −0.25 V for 25 s, and then, a cyclic voltammogram is recorded from −0.25 to +0.20 V at a scan rate of 50 mV/s. The anodic peak current is measured in all experiments.

Electrocatalytic detection
A 50 μL portion of 0.2 M HCl solution is dropped on the electrode surface and the electrode is held at a potential of +1.35 V for 1 min. Then, the chronoamperometric detection is performed at −1.40 V, recording the electric current generated for 5 min.

Figure 26.8 shows the scheme of the genosensor device and the analytical signals obtained with electrocatalytic detection (Fig. 26.8A) and enzymatic detection (Fig. 26.8B).

26.2.2.4 Results
The experimental conditions used to modify the electrode surface with streptavidin by physical adsorption have been studied in detail in a previous work [54], using biotin conjugated with alkaline phosphatase and 3-indoxyl phosphate (3-IP) as enzymatic substrate. The electrode surface must be pretreated by applying an anodic constant current (+3.0 μA) in 0.1 M H_2SO_4 for 2 min to improve its adsorptive properties. The use of this electrochemical pretreatment resulted in an increase in the hydrophilicity of the transducer, allowing the adsorption of streptavidin through hydrophilic and electrostatic attraction. In order to avoid repeatability problems associated to the use of streptavidin, the adsorption of streptavidin on the electrode must be performed at 4°C overnight. Doing this, repeatable signals are obtained and the streptavidin coated SPCEs are stable for months if they are stored at 4°C. Moreover, the significance of the attachment of biotinylated oligonucleotide probes through the streptavidin/biotin interaction has been tested in a previous work [30]. When a double-labelled (biotin and fluorescein) poly-T was attached to the electrode surface through the streptavidin/biotin interaction, the peak currents were much higher than those obtained when it was accumulated on the electrode surface by physical adsorption. This fact means that streptavidin/biotin interaction allows to attach and orient the oligonucleotide strands on electrode surface, whereas the direct adsorption of the oligonucleotide on

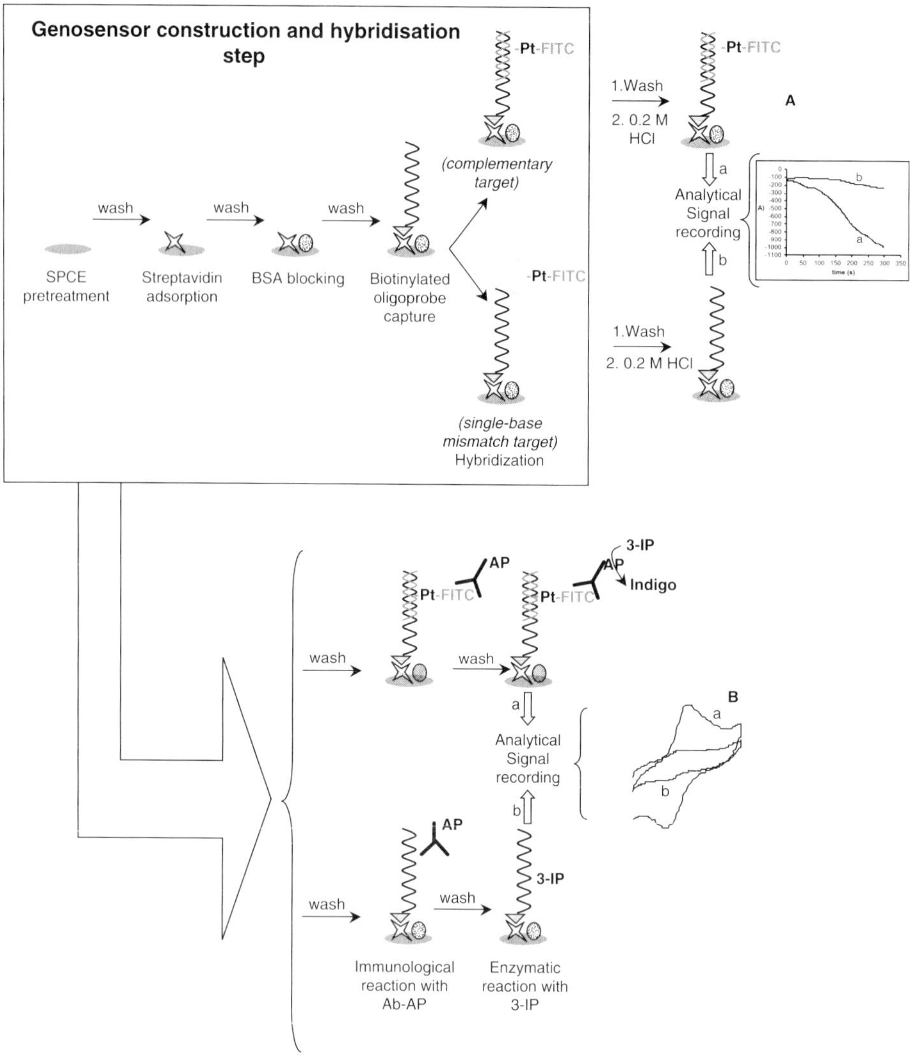

Fig. 26.8. Schematic representation of the analytical procedure followed for the construction of the genosensor and the detection of a complementary target and a single-base mismatch target. (A) Electrocatalytic and (B) enzymatic detection.

the electrode surface results in very poor manner. Using this method of immobilisation of the oligonucleotide probes, the genosensor devices are stable for a year if they are stored at 4°C.

The ply genosensor has been used for detecting oligonucleotide sequences containing a one- or three-base mismatch. Three different concentrations of complementary ply, plymism1 and plymism3 targets were assayed and three genosensors were used for each concentration. Figure 26.9 displays the results obtained with both enzymatic and electrocatalytic detection. For the three concentrations assayed, the analytical signal obtained for the three-base mismatch oligonucleotide sequence is almost the background signal, indicating that three-base mismatch ply targets can be perfectly discriminated from the complementary ply target. For the one-base mismatch oligonucleotide

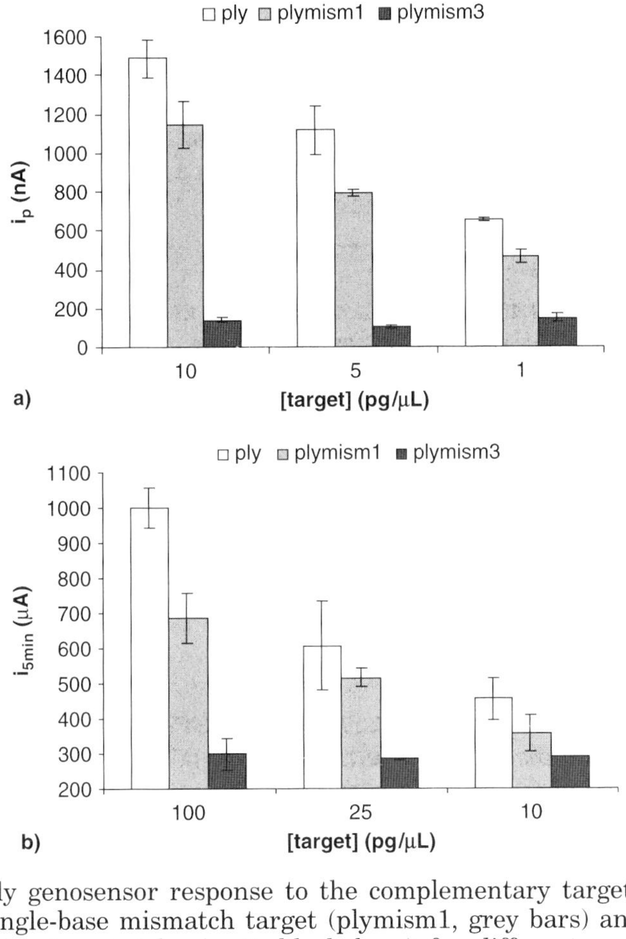

Fig. 26.9. Ply genosensor response to the complementary target (ply, white bars), the single-base mismatch target (plymism1, grey bars) and the three-base mismatch target (plymism3, black bars) for different concentrations. Data are given as average ± SD ($n = 3$). (a) Enzymatic and (b) electrocatalytic detection. Figure (a) reprinted with permission from Ref. [30]. Copyright 2004, American Chemical Society and Figure (b) reprinted with permission from Ref. [31]. Copyright 2005, American Chemical Society.

sequence, the analytical signals obtained only decrease about 30% with respect to those obtained for the complementary target.

In the optimised experimental conditions the ply genosensor has been tested for different concentrations of the complementary oligonucleotide target. In the case of the enzymatic detection, a linear relationship between peak current and concentrations of complementary ply target has been obtained between 0.1 and 5 pg/μL, with a correlation coefficient of 0.9993, according to the following equation:

$$i_p\ (\text{nA}) = 91 + 220[\text{ply target}]\ (\text{pg}/\mu\text{L})$$

Thus, these genosensors can detect 0.1 pg/μL, which is 0.49 fmol of ply target in 30 μL.

In the case of the electrocatalytic detection, a linear relationship between the recorded current and the logarithm of the concentration of ply target is obtained for concentrations between 5 and 100 pg/μL, according to the following equation:

$$i_{5\ \text{min}}(\mu\text{A}) = 510\ \log\{[\text{ply}]\ (\text{pg}/\mu\text{L})\} - 22; \quad r = 0.997\ (n = 5)$$

These genosensors can detect 5 pg/μL (24.5 fmol in 30 μL) of complementary ply target, using the electrocatalytic detection.

To improve the selectivity of the ply genosensor, more stringent experimental conditions are tested. A concentration of 25% formamide is added to the hybridisation buffer. It is well known that this molecule hampers the hybridisation reaction. In these more stringent conditions and using the enzymatic detection, a linear relationship between peak current and concentration of oligonucleotide target is obtained for concentrations between 0.25 and 5 pg/μL, according to the following equation:

$$i_p\ (\text{nA}) = 29 + 175[\text{ply}]\ (\text{pg}/\mu\text{L}); \quad r = 0.9992\ (n = 5)$$

Genosensors can detect about 1.2 fmols of complementary target in 30 μL in these more stringent experimental conditions.

In the case of electrocatalytic detection, a linear relationship between the recorded current and the logarithm of the concentration of oligonucleotide target is obtained for concentrations between 50 and 1000 pg/μL, according to the following equation:

$$i_{5\ \text{min}}\ (\mu\text{A}) = 360\ \log\{[\text{ply}]\ (\text{pg}/\mu\text{L})\} - 275; \quad r = 0.991\ (n = 5)$$

Using this strategy of detection, the genosensors can detect about 245 fmol of complementary target in 30 μL in these more stringent experimental conditions.

As expected, the sensitivity decreases in these stringent experimental conditions for both enzymatic and electrocatalytic detection but the detection of one-base mismatch on an oligonucleotide sequence can be performed for any concentration assayed (Fig. 26.10).

Although the sensitivity of the electrocatalytic detection is 50-fold (under non-stringent conditions) and 200-fold (using 25% formamide in the hybridisation solution) lower than that obtained with the enzymatic

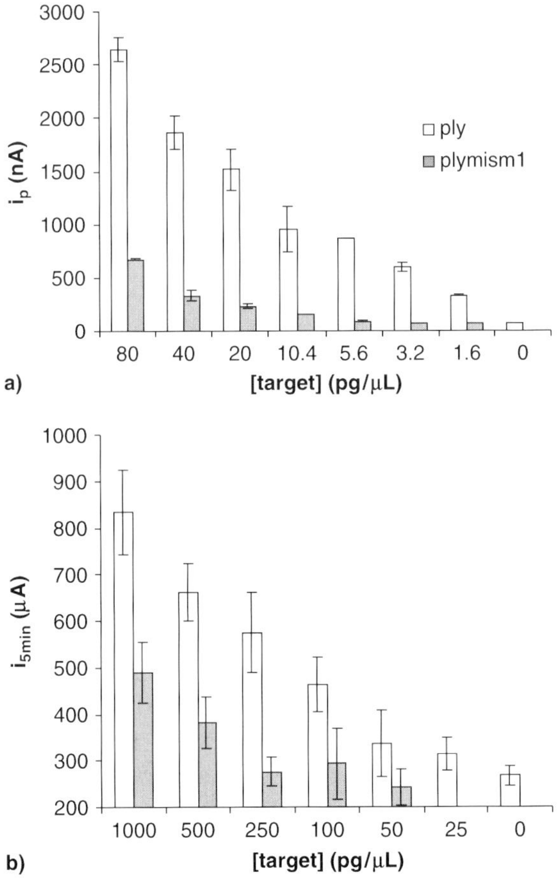

Fig. 26.10. Ply genosensor responses for different concentrations of complementary target (ply, white bars) and the single-base mismatch target (plymisms1, grey bars) when 25% formamide is included in the hybridisation buffer. Data are given as average ± SD ($n = 3$). (a) Enzymatic and (b) electrocatalytic detection. Figure (a) reprinted with permission from Ref. [30]. Copyright 2004, American Chemical Society and Figure (b) reprinted with permission from Ref. [31]. Copyright 2005, American Chemical Society.

detection, the analysis time is considerably shorter, because the analytical signal is achieved directly from the platinum complex whereas in the enzymatic detection two additional steps are necessary to obtain the analytical signal: the reaction with antibody anti-fluorescein and the enzymatic reaction. Thus, the overall analysis time of this chronoamperometric method is about the half than that resulting from the enzymatic method.

26.2.3 Genosensor on streptavidin-modified screen-printed carbon electrodes for detection of PCR products

The genosensor device described in this section can detect a 110 bp region of the TNFRSF21 gene (tumour necrosis factor receptor superfamily, member 21 precursor), bases comprise between 1859 and 1968, both included. This gene is previously inserted in the plasmid expression vector pFLAGCMV1, and amplified by PCR. Then, the PCR product is detected on the genosensor.

The protein encoded by this gene is a member of the TNF-receptor superfamily. This receptor has been shown to induce cell apoptosis. Through its death domain, this receptor interacts with TRADD protein, which is known to serve as an adaptor that mediates signal transduction of TNF receptors. Knockout studies in mice suggested that this gene plays a role in T-helper cell activation, and may be involved in inflammation and immune regulation.

26.2.3.1 Genosensor design

The genosensor device consists of an SPCE where a 20-mer oligonucleotide probe complementary to target is immobilised through the streptavidin/biotin reaction. Orientation of the strands in the sensing phase is achieved by modifying the surface of the electrode with streptavidin by physical adsorption followed by the immobilisation of biotinylated oligo probes. The physical adsorption of streptavidin must be performed at a constant temperature above the room temperature. Moreover, the electrode surface must be previously electrochemically pretreated at an anodic potential in acidic medium to improve its adsorptive properties. The biotinylated oligo nucleic acid probes used in this work target 20-mer of the 110 bp of the PCR product obtained from TNFRSF21 gene. The detection is carried out using an AP labelled antibody anti-FITC and 3-IP as substrate (enzymatic detection explained in last section). A more detailed explanation of the genosensor construction is given in Procedure 37 in CD accompanying this book.

26.2.3.2 PCR amplification and preparation of PCR products for genosensor detection

Two PCR amplifications are carried out. In one of them, an FITC labelled 5′-primer is used for DNA amplification and in the other one the 5′-primer used in the amplification reaction is not labelled. In the latter case the unlabelled PCR products obtained are labelled with fluorescein using a Universal Linkage System, explained in Section 26.2.2. The reactions of amplification are carried out in a thermocycler using a 3′-primer and a 5′-primer with an adequate sequence, Taq polymerase and dNTPs mixture. After an initial denaturation step, 35 cycles of denaturation-annealing-extension are carried out (see more details in Procedure 37 in CD accompanying this book). The unlabelled PCR products obtained, when unlabelled 5′-primer is used, are labelled with Fluorescein ULS® labelling kit before they are checked with the genosensor device (see more details about the labelling procedure in Procedure 37 in CD accompanying this book.

After that, the FITC labelled PCR products must be pretreated before their detection on the genosensor device. The double stranded amplified DNA is denaturated thermally by using a boiling water bath for 6 min at 95°C, then the tube containing the PCR products is immersed in an ice-water bath for 2 min in order to retard strands re-annealing. Finally, an aliquot of 30 μL is placed on the genosensor device.

26.2.3.3 Results

When the PCR products are labelled using the FITC labelled 5′-primer the analytical signals are higher and more reproducible than those obtained using the ULS labelling kit. Moreover, the PCR blank is higher in the latter case (see Fig. 37.2 of Procedure 37 in CD accompanying this book).

For PCR products labelled using the FITC labelled 5′-primer, a linear relationship between the peak current and the logarithm of the number of copies of plasmid DNA is obtained between 2×10^3 and 2×10^5 copies of plasmid DNA, according to the following equation:

$$i_p\,(\text{nA}) = 204 \log\ (\text{copies plasmid DNA}) - 354, \quad r = 0.9998,\ n = 4$$

The limit of detection, calculated as the copies of plasmid DNA corresponding to a signal which is the PCR blank signal plus three times the standard deviation of the PCR blank signal, results to be 360 copies of plasmid DNA.

26.3 CONCLUSIONS

As it has been shown in previous sections, the use of thick- and thin-film electrodes as supports for genosensor devices offers enormous opportunities for their application in molecular diagnosis. The technologies used in the fabrication of both thick- and thin-film electrodes allow the mass production of reproducible, inexpensive and mechanically robust strip solid electrodes. Other important advantages of these electrodes are the possibility of miniaturisation as well as their ease of manipulation in a disposable manner and therefore the use of small volumes. This is an important issue that makes this methodology for detection of DNA more attractive.

Very sensitive methods are always required for DNA sensing. Although enough sensitivity to avoid PCR amplification has been achieved by use of enzymatic labels [12,14,26,28] or metal tags [42], most of the assays routinely start with a PCR or other biochemical amplification. Moreover, although label-free formats are used, most of the strategies followed to obtain the analytical signal involve several washing steps and need the use of labelled reagents (or labelling procedures) or indicators, which complicates the assay performance. Motorola Science has commercialised the eSensor [39,40], which detects the hybridisation of the target by a sandwich assay without clean-up steps. Although this sensor requires the addition of reagents, they are included in the microsystem as dry reagents dissolved by the sample. An important drawback is the need of carrying out DNA isolation and purification before testing the DNA sample with genosensor devices due to the inherent complexity associated with the biological sample. The real goal, for specific pathogen detection, is the development of an assay involving rapid DNA isolation with the detection of few copies in half an hour or less. With these electrochemical genosensors this goal is reliable, but it has not yet been achieved.

Optical gene chips dense arrays of oligonucleotides have been successfully applied to detect transcriptional profiling and SNP discovery, where massively parallel analysis is required. However, the fluorescence-based readout of these chips involves not only highly precise and expensive instrumentation but also sophisticated numerical algorithms to interpret the data, and therefore these methods have been commonly limited to use in research laboratories. In this way, thin-film arrays of 14, 20, 25, 48 and 64 electrodes have already been fabricated [12,15,39,40,44,48], using lithographic techniques. Readout systems for these arrays based on electrical detection have also been developed.

Moreover, a thick-film sensor array suitable for automation combined to readout based on intermittent pulse amperometry (IPA) has been commercialised by Alderon Biosciences [27,28]. These genosensors and the readout instruments provide a simple, accurate and inexpensive platform for patient diagnosis.

It is more than probable that arrays for 50–100 DNA sequences will be needed for some clinical applications. Although it is not difficult to design electrode pads with reproducible dimensions of a micron or less, the electrochemical readout requires mechanical connections to each individual electrode. Therefore, the construction of very large, multiplexed arrays presents a major engineering challenge. Electronic switches in the form of an on-chip electronic multiplexer may provide a possible solution for this problem.

REFERENCES

1 J.L. Anderson and N. Winograd, Film electrodes. In: P.T. Kissinger and W.R. Heineman (Eds.), *Laboratory Techniques in Electroanalytical Chemistry*, Marcel Dekker, New York, 1996, pp. 333–365.

2 J. Cha, J.I. Han, Y. Choi, D.S. Yoon, K.W. Oh and G. Lim, DNA hybridisation electrochemical sensor using conducting polymer, *Biosens. Bioelectron.*, 18 (2003) 1241–1247.

3 G. Marchand, C. Delattre, R. Campagnolo, P. Pouteau and F. Ginot, Electrical detection of DNA hybridisation based on enzymatic accumulation confined in nanodroplets, *Anal. Chem.*, 77 (2005) 5189–5195.

4 S. Hashioka, M. Saito, E. Tamiya and H. Matsumura, Deoxyribonucleic acid sensing device with 40-nm-gap-electrodes fabricated by low-cost conventional techniques, *Appl. Phy. Lett.*, 85 (2004) 687–688.

5 L. Moreno-Hagelsieb, P.E. Lobert, R. Pampin, D. Bourgeois, J. Remacle and D. Flandre, Sensitive DNA electrical detection based on interdigitated Al/Al_2O_3 microelectrodes, *Sens. Actuators B*, 98 (2004) 269–274.

6 K. Hashimoto, K. Ito and Y. Ishimori, Microfabricated disposable DNA sensor for detection of hepatitis B virus DNA, *Sens. Actuators B*, 46 (1998) 220–225.

7 T. Hianik, V. Gajdos, R. Krivanek, T. Oretskaya, V. Metelev, E. Volkov and P. Vadgama, Amperometric detection of DNA hybridisation on a gold surface depends on the orientation of oligonucleotide chains, *Bioelectrochemistry*, 53 (2001) 199–204.

8 X. Li, Y. Zhou, T.C. Sutherland, B. Baker, J.S. Lee and H.B. Kraatz, Chip-based microelectrodes for detection of single-nucleotide mismatch, *Anal. Chem.*, 77 (2005) 5766–5769.

9 P. Abad-Valle, M.T. Fernández-Abedul and A. Costa-García, Genosensor on gold films with enzymatic electrochemical detection of a SARS virus sequence, *Biosens. Bioelectron.*, 20 (2005) 2251–2260.
10 M.R. Gore, V.A. Szalai, P.A. Ropp, I.V. Yang, J.S. Silverman and H.H. Thorp, Detection of attomole quantitites of DNA targets on gold microelectrodes by electrocatalytic nucleobase oxidation, *Anal. Chem.*, 75 (2003) 6586–6592.
11 H. Xie, C. Zhang and Z. Gao, Amperometric detection of nucleic acid at femtomolar levels with a nucleic acid/electrochemical activator bilayer on gold electrode, *Anal. Chem.*, 76 (2004) 1611–1617.
12 H. Xie, Y.H. Yu, F. Xie, Y.Z. Lao and Z. Gao, Breast cancer susceptibility gene mRNAs quantified by microarrays with electrochemical detection, *Clin. Chem.*, 50 (2004) 1231–1233.
13 J.E. Koehne, H. Chen, A.M. Cassell, Q. Ye, J. Han, M. Meyyappan and J. Li, Miniaturised multiplex label-free electronic chip for rapid nucleic acid analysis based on carbon nanotube nanoelectrode arrays, *Clin. Chem.*, 50 (2004) 1886–1893.
14 G. Hwang, E. Kim and J. Kwak, Electrochemical detection of DNA hybridisation using biometallisation, *Anal. Chem.*, 77 (2005) 579–584.
15 K. Hashimoto and Y. Ishimori, Preliminary evaluation of electrochemical PNA array for detection of single base mismatch mutations, *Lab. Chip*, 1 (2001) 61–63.
16 J. Wang, X. Cai, B. Tian and H. Shiraishi, Microfabricated thick-film electrochemical sensor for nucleic acid determination, *Analyst*, 121 (1996) 965–970.
17 J. Wang, X. Cai, G. Rivas, H. Shiraishi and N. Dontha, Nucleic-acid immobilisation recognition and detection at chronopotentiometric DNA chips, *Biosens. Bioelectron.*, 12 (1997) 587–599.
18 J. Wang, G. Rivas and X. Cai, Screen printed electrochemical hybridisation biosensor for the detection of DNA sequences from *Escherichia coli* pathogen, *Electroanalysis*, 9 (1997) 395–398.
19 G. Marrazza, I. Chianella and M. Mascini, Disposable DNA electrochemical sensor for hybridisation detection, *Biosens. Bioelectron.*, 14 (1999) 43–51.
20 G. Marrazza, G. Chiti, M. Mascini and M. Anichini, Detection of human apolipoprotein E genotypes by DNA electrochemical biosensor coupled with PCR, *Clin. Chem.*, 46 (2000) 31–37.
21 B. Meric, K. Kerman, G. Marrazza, I. Palchetti, M. Mascini and M. Ozsoz, Disposable genosensor, a new tool for the detection of NOS-terminator, a genetic element present in GMOs, *Food Control*, 15 (2004) 621–626.
22 F. Lucarelli, G. Marrazza, I. Palchetti, S. Cesaretti and M. Mascini, Coupling of an indicator-free electrochemical DNA biosensor with polymerase chain reaction for the detection of DNA sequences related to the apolipoprotein E, *Anal. Chim. Acta*, 469 (2002) 93–99.

23 I. Katakis, Towards a fast-responding, label-free electrochemical DNA biosensor, *Anal. Bioanal. Chem.*, 381 (2005) 1033–1035.
24 M. Dequaire and A. Heller, Screen printing of nucleic acid detecting carbon electrodes, *Anal. Chem.*, 74 (2002) 4370–4377.
25 Y. Zhang, H.H. Kim, N. Mano, M. Dequaire and A. Heller, Simple enzyme-amplified amperometric detection of a 38-base oligonucleotide at 20 pmol L^{-1} concentration in a 30-μL droplet, *Anal. Bioanal. Chem.*, 374 (2002) 1050–1055.
26 F. Azek, C. Grossiord, M. Joannes, B. Limoges and P. Brossier, Hybridisation assay at a disposable electrochemical biosensor for the attomole detection of amplified human cytomegalovirus DNA, *Anal. Biochem.*, 284 (2000) 107–113.
27 T.J. Huang, M. Liu, L.D. Knight, W.W. Grody, J.F. Miller and C.M. Ho, An electrochemical detection scheme for identification of single nucleotide polymorphisms using hairpin-forming probes, *Nucleic Acids Res.*, 30 (2002) e55.
28 M. Aitichou, R. Henkens, A.M. Sultana, R.G. Ulrich and M.S. Ibrahim, Detection of *Staphylococcus aureus* enterotoxin A and B genes with PCR-EIA and a hand-held electrochemical sensor, *Mol. Cell. Probes*, 18 (2004) 373–377.
29 M. Mascini, M. del Carlo, M. Minunni, B. Chen and D. Compagnone, Identification of mammalian species using genosensors, *Bioelectrochemistry*, 67 (2005) 163–169.
30 D. Hernández-Santos, M. Díaz-González, M.B. González-García and A. Costa-García, Enzymatic genosensor on streptavidin-modified screen-printed carbon electrodes, *Anal. Chem.*, 76 (2004) 6887–6893.
31 D. Hernández-Santos, M.B. González-García and A. Costa-García, Genosensor based on a platinum (II) complex as electrocatalytic label, *Anal. Chem.*, 77 (2005) 2868–2874.
32 K. Kerman, Y. Morita, Y. Takamura and E. Tamiya, *E. coli* single-strand binding protein–DNA interactions on carbon nanotube-modified electrodes from a label-free electrochemical hybridisation sensor, *Anal. Bioanal. Chem.*, 381 (2005) 1114–1121.
33 A. Ruffien, M. Dequaire and P. Brossier, Covalent immobilisation of oligonucleotides on p-aminophenyl-modified carbon screen-printed electrodes for viral DNA sensing, *Chem. Commun.*, 7 (2003) 912–913.
34 K. Metfies, S. Huljic, M. Lange and L.K. Medlin, Electrochemical detection of the toxic dinoflagellate *Alexandrium ostenfeldii* with a DNA-biosensor, *Biosens. Bioelectron.*, 20 (2005) 1349–1357.
35 F. Davis, A.V. Nabok and S.P.J. Higson, Species differentiation by DNA-modified carbon electrodes using an ac impedimetric approach, *Biosens. Bioelectron.*, 20 (2005) 1531–1538.

36 G. Carpini, F. Lucarelli, G. Marrazza and M. Mascini, Oligonucleotide-modified screen-printed gold electrodes for enzyme-amplified sensing of nucleic acids, *Biosens. Bioelectron.*, 20 (2004) 167–175.
37 F. Lucarelli, G. Marrazza and M. Mascini, Enzyme-based impedimetric detection of PCR products using oligonucleotide-modified screen-printed gold electrodes, *Biosens. Bioelectron.*, 20 (2005) 2001–2009.
38 E.M. Boon, D.M. Ceres, T.G. Drummond, M.G. Hill and J.K. Barton, Mutation detection by electrocatalysis at DNA-modified electrodes, *Nat. Biotechnol.*, 18 (2000) 1096–1100.
39 R.M. Umek, S.S. Lin, Y.P. Chen, B. Irvine, G. Paulluconi, V. Chan, Y. Chong, L. Cheung, J. Vielmetter and D.H. Farkas, Bioelectronic detection of point mutations using discrimination of the *H63D* polymorphism of the *Hfe* gene as a model, *Mol. Diagn.*, 5 (2000) 321–328.
40 R.M. Umek, S.W. Lin, J. Vielmetter, R.H. Terbrueggen, B. Irvine, C.J. Yu, J.F. Kayyem, H. Yowanto, G.F. Blackburn, D.H. Farkas and Y.P. Chen, Electronic detection of nucleic acids. A versatile platform for molecular diagnosis, *J. Mol. Diagn.*, 3 (2001) 74–84.
41 M. Urban, R. Möller and W. Fritzsche, A paralleled readout system for an electrical DNA-hybridisation assay based on a microstructured electrode array, *Rev. Sci. Instrum.*, 74 (2003) 1077–1081.
42 C.Y. Tsai, T.L. Chang, C.C. Chen, F.H. Ko and P.H. Chen, An ultra sensitive DNA detection by using gold nanoparticle multilayer in nano-gap electrodes, *Microelectron. Eng.*, 78–79 (2005) 546–555.
43 C.H. Fan, K.W. Plaxco and A.J. Heeger, Electrochemical interrogation of conformational changes as a reagentless method for the sequence-specific detection of DNA, *Proc. Natl. Acad. Sci. USA*, 100 (2003) 9134–9137.
44 P. Liepold, H. Wieder, H. Hillebrandt, A. Friebel and G. Hartwich, DNA-arrays with electrical detection: a label-free low cost technology for routine use in life sciences and diagnostics, *Bioelectrochemistry*, 67 (2005) 143–150.
45 J. Laitinen, J. Samarut and E. Holtta, A nontoxic and versatile protein salting out method for isolation of DNA, *Biotechniques*, 17 (1994) 316–322.
46 S.A. Miller, D.D. Dykes and H.F. Polesky, A simple salting out procedure for extracting DNA from human nucleated cells, *Nucleic Acids Res.*, 16 (1988) 1215–1219.
47 J.E. Sambrook, F. Fritsh and T. Maniatis, *Molecular Cloning: a Laboratory Manual*, Cold Spring Harbor Laboratory Press, New York, 1989.
48 J. Wakai, A. Takagi, M. Nakayama, T. Miya, T. Miyahara, T. Iwanaga, S. Takenaka, Y. Ikeda and M. Amano, A novel method of identifying genetic mutations using an electrochemical DNA array, *Nucleic Acids Res.*, 32 (2004) e141.

49 C.A. Wijayawardhana, H.B. Halsall and W.R. Heineman, Microvolume rotating disk electrode (RDE) amperometric detection for a bead-based immunoassay, *Anal. Chim. Acta*, 399 (1999) 3–11.
50 M.A. Marra, S.J.M. Jones, C.R. Astell, R.A. Holt, A. Brooks-Wilson, Y.S.N. Butterfield, J. Khattra, J.K. Asano, S.A. Barber, S.Y. Chan, A. Cloutier, S.M. Coughlin, D. Freeman, N. Girn, O.L. Griffith, S.R. Leach, M. Mayo, H. McDonald, S.B. Montgomery, P.K. Pandoh, A.S. Petrescu, A.G. Robertson, J.E. Schein, A. Siddiqui, D.E. Smailus, J.E. Stott, G.S. Yang, F. Plummer, A. Andonov, H. Artsob, N. Bastien, K. Bernard, T.F. Booth, D. Bowness, M. Czub, M. Drebot, L. Fernando, R. Flick, M. Garbutt, M. Gray, A. Grolla, S. Jones, H. Feldmann, A. Meyers, A. Kabani, Y. Li, S. Normand, U. Stroher, G.A. Tipples, S. Tyler, R. Vogrig, D. Ward, B. Watson, R.C. Brunham, M. Krajden, M. Petric, D.M. Skowronski, C. Upton and R.L. Roper, The genome sequence of the SARS-associated coronavirus, *Science*, 300 (2003) 1399–1404.
51 C. Fernández-Sánchez and A. Costa-García, 3-Indoxyl phosphate: an alkaline phosphatase substrate for enzyme immunoassays with voltammetric detection, *Electroanalysis*, 10 (1998) 249–255.
52 M. Castaño-Álvarez, M.T. Fernández-Abedul and A. Costa-García, Gold electrodes for detection of enzymes assays with 3-indoxylphosphate as substrate, *Electroanalysis*, 16 (2004) 1487–1496.
53 R.-Y. Zhang, D.-W. Pang, Z.-L. Zhang, J.-W. Yan, J.-L. Yao, Z.-Q. Tian, B.-W. Mao and S.-G. Sun, Investigation of ordered ds-DNA monolayers on gold electrodes, *J. Phys. Chem.*, 106 (2002) 11233–11239.
54 M. Díaz-González, D. Hernández-Santos, M.B. González-García and A. Costa-García, Development of an immunosensor for the determination of rabbit IgG using streptavidin modified screen-printed carbon electrodes, *Talanta*, 65 (2005) 565–573.
55 M. Díaz-González, C. Fernández-Sánchez and A. Costa-García, Comparative voltammetric behaviour of indigo carmine at screen-printed carbon electrodes, *Electroanalysis*, 14 (2002) 665–670.
56 M. Díaz-González, C. Fernández-Sánchez and A. Costa-García, Indirect determination of alkaline phosphatase based on the amperometric detection of indigo carmine at a screen-printed electrode in a flow system, *Anal. Sci.*, 18 (2002) 1209–1213.

Chapter 27

Screen-printed enzyme-free electrochemical sensors for clinical and food analysis

Khiena Z. Brainina, Alisa N. Kositzina and Alla Ivanova

27.1 INTRODUCTION AND APPLICATION

Electrochemical analytical methods, particularly polarography and voltammetry rise in the 1960s was caused by the demand in trace analysis and new technique of preliminary electrochemical concentration of the determined substance on the electrode surface [1,2]. The reason for the new renaissance is the use of screen-printed technologies, which resulted in creation of new electrodes so cheap that they can be easily disposed and there is no need of regenerating the solid electrode surface [3].

Possibilities to manipulate molecular matrix of the electrode or its surface combined with large-scale electrochemical research methods led to the 'explosion' in development and application of biosensors and chemically modified sensors [4]. Biochemically and chemically modified sensors possess high selectivity and sensitivity.

Biosensors include electrodes modified with biomolecules: enzymes, DNA, antibodies, vegetable tissue or microorganisms.

Immobilization of bioactive material on/in the electrode allows combining bio-reaction selectivity with sensitivity of electrochemical detection. Irrespective of reaction in the biosensor, the electrochemical response is measured, in particular, as current at the given potential (amperometric sensor) or electrode potential (potentiometric sensor).

Immobilization methods of bioactive material on the electrode surface are reviewed in Refs. [5,6]. Carbon composites are used as transducers

Comprehensive Analytical Chemistry 49
Alegret and Merkoçi (Eds)
Volume 49 ISSN: 0166-526X DOI: 10.1016/S0166-526X(06)49027-9

in biosensors. Often the composite may include fine-dispersed metal [7] or enzyme [8], thus combining the functions of both transducer and receptor. Recently, electrodes based on screen-printed graphite polymer laid with enzyme have been brought into use [9,10].

In spite of all their advantages, sensitivity and selectivity, bio-sensors, however, do possess disadvantages connected with thermal and timely instability, high cost of bio-receptors and the need to add substrates in the solution under analysis as signal-generating substances. Some attempts to synthesize and use as receptors chemical organic catalytic systems, which will ensure the required selectivity and response rate, have become the basis for developing enzyme-free sensors [11], or biomimetic sensors.

We agree with Baeumner [11] who concludes that very few works on biomimetic-based biosensors have been published. As their recognition element is not a biologically derived molecule, the inclusion of biomimetic sensors into the group of biosensors is questionable. However, it is worth taking into account that they mimic biological activity or any other functions of the biological receptor and they can be applied in the same way as their biological analogues.

Two lines of study are to be considered for the development of chemical sensors based on nanocrystal materials [12]:

- Electrochemical behavior of nanomaterials as transducer for biosensors, immunosensors and chemical sensors.
- Nanomaterials as signal-generating substances to be used as labels for immunosensors.

The use of nanoparticles (metal, carbon, etc.) as transducer material and electrochemical label allows improving analysis sensitivity [12] and broadening its potentiality.

Our work deals with only a few examples: (i) the immunosensor for the forest-spring encephalitis diagnosis (a metal label is used for signal generation), (ii) the enzyme-free urea sensor and (iii) the platinum sensor for antioxidant activity determination. What they all have in common is a screen-printed transducer consisting of graphite or Pt nanoparticles. Transducer configuration is shown in Fig. 27.1.

Linear sweep or preset potential and current measurement providing instrument can be used with immunosensor and enzyme-free sensor. Potential measured instrument can be used with antioxidant activity sensor.

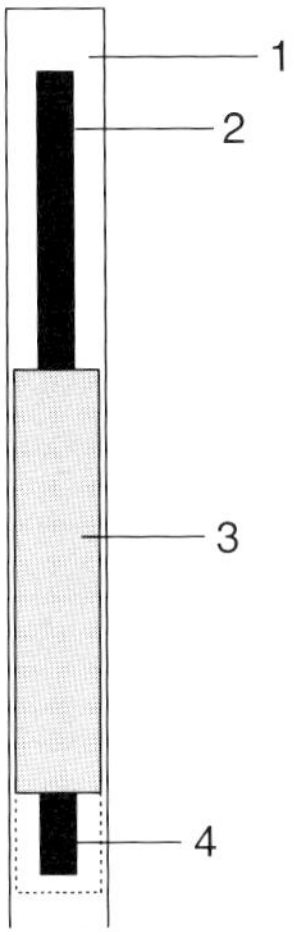

Fig. 27.1. Transducer configuration (45×4.0 mm). (1) Substrate. (2) Graphite or metal containing layer. (3) Insulator. (4) Working area (7.5 mm^2).

27.1.1 Immunosensor

An electrochemical immunosensor includes a transducer (a current-conductive strip deposited on a polymer substrate) and a biorecognition substance (an antigen or antibodies) immobilized on the transducer surface. The formation of an immune complex and localization of non-antigen-specific antibodies (proteins) on this complex, which are usually labeled with enzymes, and a substrate in the solution, generates an active species capable of generating an electrochemical response. The sequential process, the use of enzymes, and occasionally unstable substrates limit and complicate a wide-scale application of such sensors.

Metal labels have been proposed to resolve problems connected with enzymes. Metal ions [13–16], metal-containing organic compounds [17,18], metal complexes [19–21], metalloproteins or colloidal metal particles [22–28] have served as labels. Spectrophotometric [22,25], acoustic [25], surface plasmon resonance, infrared [24] and Raman spectroscopic [28] methods, etc. were used. A few papers have been dealing with electrochemical detection. However, electrochemical methods of metal label detection may be viewed as very promising taking into account their high sensitivity, low detection limit, selectivity, simplicity, low cost and the availability of portable instruments.

Heineman's group [13,14] proposed the method of heterogeneous immunoassay to detect human serum albumin (HSA) labeled with In and Bi ions with the help of the bifunctional chelating agent diethylenetriamine-pentaacetic acid (DTPA). This detection method is based on competition between labeled and unlabeled proteins in the sample for binding with antibodies immobilized on the walls of the polystyrene cell. Bound metal ions then were dissolved in acid and their concentration was measured by the method of anodic stripping voltammetry. This approach of using Bi ions and chelating agent (diethylenetriamine-pentaacetic acid) to detect HSA was applied in Ref. [15]. The only difference was that thick-film screen-printed graphite transducers were used as substrate for enzyme immobilization. Bi ions were detected with stripping voltammetry. An interesting approach of using Cu ions in the competitive immunoassay as catalyst for transforming *o*-phenylendiamine into electroactive 2,3 diaminophenyzine is described in Ref. [16]. To detect HSA chelating agent (diethylenetriamine-pentaacetic acid) was used.

One of the approaches is to use colloidal gold as label [29] in electrochemical immunoassay. In addition to standard stages of the immune complex formation, oxidative release of gold and removal of excessive bromine are included into the analysis procedure.

The electrochemical approaches developed in the aforementioned studies are based on detection of metal ions released from the sandwich. The use of metallic gold as a source of electrochemical data is described in our earlier paper [30], but the stage of metal releasing is omitted in the work mentioned. The present study also deals with the use of a metal label as the signal generating substance.

The proposed electrochemical immunosensor for diagnosis of the forest-spring encephalitis comprises a screen-printed graphite electrode serving as the transducer and a layer of the forest-spring encephalitis antigen immobilized on the transducer and functioning as a bio-recognition substance. The procedure includes formation of an antigen–antibody immune complex, localization of protein A labeled with colloid silver nanoparticles on the complex and recording of silver oxidation voltammograms, which provide information on the presence and concentration of antibodies in blood serum (Fig. 27.2).

27.1.1.1 Selection of the optimum sample of conjugated protein A and colloid silver

Colloid silver was chosen as label for the assay because its standard potential is more negative than the standard potential of gold. Thus, the potential window is more convenient for signal registration than in Ref. [30].

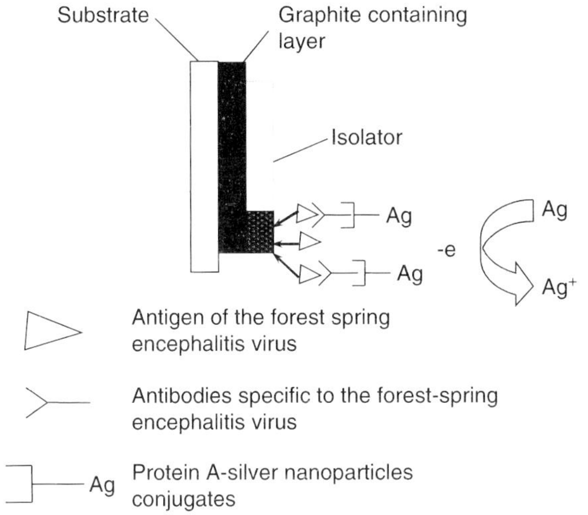

Fig. 27.2. Immune complex formation and its reaction with silver label.

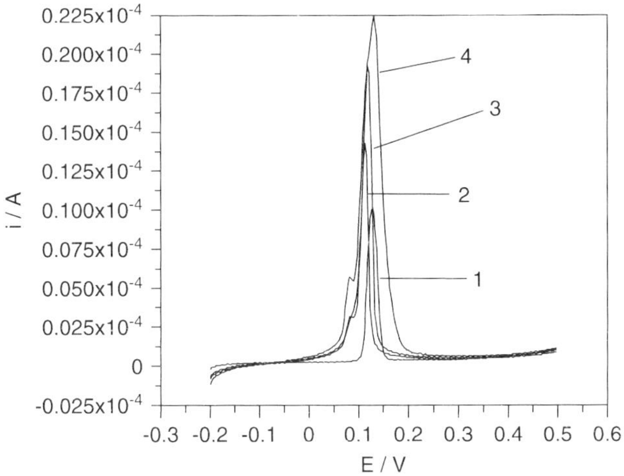

Fig. 27.3. Voltammograms of silver oxidation from conjugate localized on transducer. Background electrolyte: Phosphate buffer solution pH = 7.2. Protein A concentration: 10 and 15 μg/ml (1); 100 μg/ml (2); 70 μg/ml (3); 50 μg/ml (4).

Figure 27.3 shows silver voltammograms recorded after transducer's incubation in conjugates containing protein A in various concentrations. It can be easily observed that current of silver oxidation is the highest one for the sample containing 50 μg/ml of protein. A lower

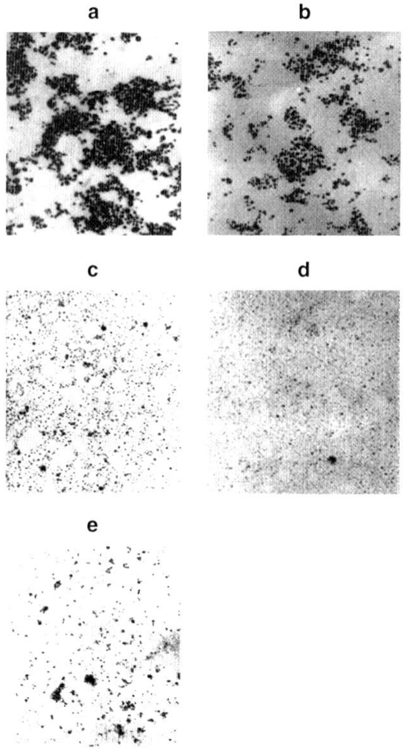

Fig. 27.4. Electron-microscopic photos of protein A-colloid silver conjugates. Starting protein A concentration: (a) 10 μg/ml; (b) 15 μg/ml; (c) 50 μg/ml; (d) 70 μg/ml; (e) 100 μg/ml.

response was registered for the sample containing protein A in higher and lower concentrations.

Figure 27.4 demonstrates the results of electron-microscopic assays. It shows that in conjugates, containing protein in concentrations of 10 and 15 μg/ml, silver particles form big clusters and are allocated unevenly. In the conjugates, containing protein in concentrations of 100 or 70 μg/ml, more fine particles are present than in the samples, containing 50 μg/ml of protein.

The bar charts in Figs. 27.5 and 27.6 illustrate particle size distribution before and after samples centrifugation. After centrifugation an average size of silver particles in the samples, containing 10, 15 and 50 μg/ml of protein A, is close to 5–8 nm. In the samples, containing protein in concentrations 70 or 100 μg/ml the prevailing Ag particle size is 2–3 nm. The results of electrochemical assay demonstrate that the

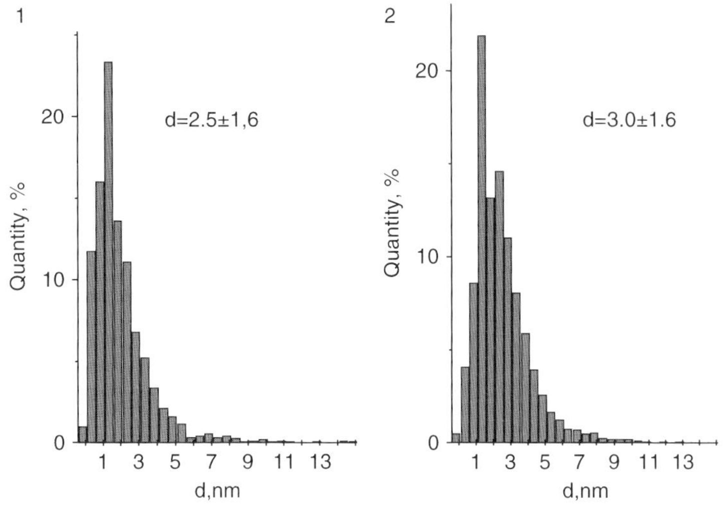

Fig. 27.5. Distribution of silver in order of particle size after centrifugation. Samples contained 70 μg/ml (1) and 100 μg/ml (2) of protein A.

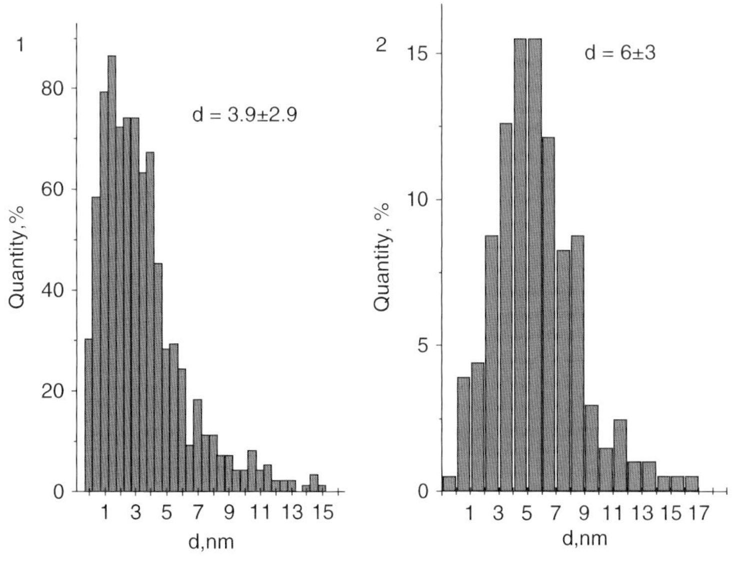

Fig. 27.6. Distribution of silver in order of particle size before (1) and after (2) centrifugation. Samples contained 50 μg/ml of protein A.

optimum particle size is 5–8 nm. Similar results are observed if samples containing 70 μg/ml of protein A are used. It may be concluded that the electrochemical response depends on at least two factors: (1) silver particle concentration in a sample (determined by protein concentration) and (2) allocation by particle size. The sample, containing 50 μg/ml of protein A, can be considered as the optimum one. Higher concentrations of protein A up to 100 μg/ml seems to block the working surface of the sensor.

The conjugate of protein A labeled with silver was used in electrochemical immunosensor for detecting the forest-spring encephalitis antibodies, which is described in Procedure 38 in CD accompanying this book.

27.1.2 Non-enzymatic urea sensor

Determination of urea concentration has acquired great significance in clinical diagnosis of renal abnormalities. The issue of evaluating hemodialysis efficiency (which is judged by urea concentration in dialysis liquid) is of vital importance. The essential parameters are variation dynamics of the urea concentration during the procedure and the level of urea in blood serum before and after dialysis. Typically, its physiological levels range between 2.5 and 8.6 mM in serum depending on the daily protein ingestion [31].

Urea concentration is most frequently determined during a catalytic reaction with urease (Eq. (27.1)) and subsequent measurement of ammonium ions concentration:

$$NH_2CONH_2 + H^+ + 2H_2O \overset{\text{urease}}{\Rightarrow} 2NH_4^+ + HCO_3^- \tag{27.1}$$

For monitoring catalytic (enzymatic) products, various techniques, such as spectrophotometry [32], potentiometry [33,34], coulometry [35,36] and amperometry [37,38], have been proposed. An advantage of these sensors is their high selectivity. However, time and thermal instability of the enzyme, the need of a substrate use and indirect determination of urea (logarithmic dependence of a signal upon concentration while measuring pH) cause difficulties in the use and storage of sensors.

Urease consists of a protein part and metal ions. It is supposed to be a four-coordination complex where amino acids serve as ligands and centered Ni ion serves as co-factor for protein molecules, which by themselves are catalytically passive and activate only in combination with metal ions [27].

For the last few years, some works have been published on electrocatalytic oxidation of phenols, spirits and urea over Ni complexes

[39]. The following reaction serves as a signal generating one:

$$Ni^{2+} \rightarrow Ni^{3+} + e \qquad (27.2)$$

This idea is used in the proposed enzyme-free sensor.

Screen-printed electrodes manufactured with the use of carbon-containing ink were used. (i) Catalytic system was inserted into carbon ink prior to ink immobilization on polymer substrate; (ii) catalyst or polymer matrix (nafion), containing catalyst, was immobilized on the working surface of the transducer. Table 27.1 displays the results of urea determination with the application of catalytic systems inserted into ink or immobilized on the transducer by different techniques. Optimum results (minimum mean square deviation and maximum correlation between the introduced and determined concentrations) are

TABLE 27.1

Urea determination with sensors differently modified with catalysts

Type of catalyst and technique of insertion	Urea		S_r (%)
	Inserted (mM)	Determined (mM)	
Catalyst 1. Nickelhydroxide deposited on transducer surface as a result of reaction between $NiSO_4$ and NaOH; alcoholic solution of nafion is immobilized	10	5.9 ± 0.8	10.4
Catalyst 2. Natriumdiethyldithiocarbaminat dissolved in chloroform and mixed with $NiSO_4$ dissolved in H_2O; solution is extracted; the extract is immobilized on transducer surface	10	8.2 ± 0.6	6.0
Catalyst 3. The extract received in #2 is inserted in carbon-containing ink; the mixture is immobilized on transducer surface	10	9.9 ± 0.1	1.1
Catalyst 4. Natriumdimethilglioximat dissolved in acetone is mixed with $NiSO_4$ dissolved in H_2O; the mixture is immobilized on transducer surface	10	6.3 ± 0.9	11

observed in case 3. The sensor (Fig. 27.7) modified as mentioned in point 3 of Table 27.1 is used at a later time.

Figure 27.8 displays cyclic voltammograms, recorded for both transducer containing and not containing catalyst and urea is either present or absent in the solution. It is seen that oxidation current is observed only in the presence of either a catalyst or both catalyst and urea, which proves the catalytic nature of the process.

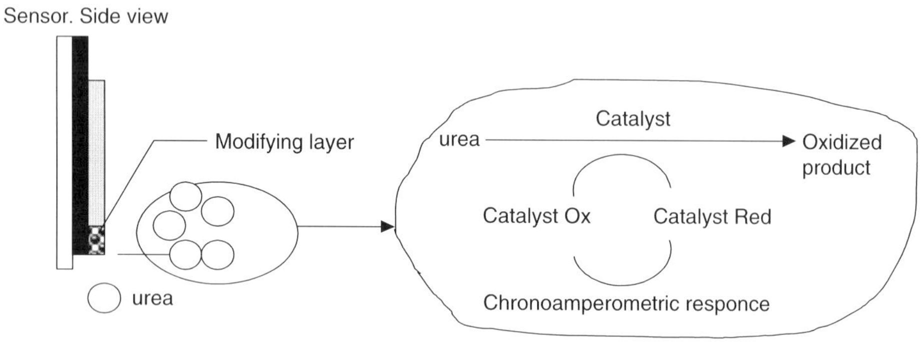

Fig. 27.7. Catalytic process on the sensor/solution boundary.

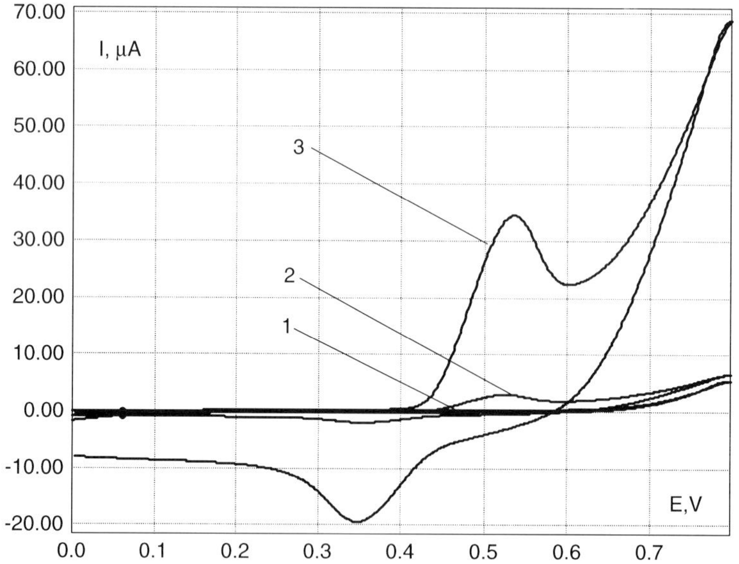

Fig. 27.8. Cyclic voltammograms for (1) non-catalyst-containing transducer; (2) Catalyst 3 containing transducer. Urea: (1, 2) absent in the solution; (3) present in the solution. Background: 0.25 M NaOH.

Table 27.2 illustrates urea determination with the use of sensors, manufactured as described in point 3 of Table 27.1. Initial concentrations of aqueous $NiSO_4$ solution and chloroform NaDDC solution were varied. The optimum results (minimum mean square deviation and maximum correlation between the introduced and determined concentrations) are observed for the case when the initial concentration of $NiSO_4$ exceeds four times the initial chloroform NaDDC concentration. For further assays, the sensor was used meeting the conditions described in point 3 of Table 27.1 and in point 3 of Table 27.2. Solutions contained 0.04 g NaDDC in 2 ml of chloroform and 0.2 g $NiSO_4$ in 2 ml of H_2O.

Figure 27.9 displays an electron microscopic photo (a) and a graph illustrating X-ray microanalysis (b) of the working surface of the

TABLE 27.2

Urea determination with Catalyst 3 containing sensor for different initial concentrations of chloroform NaDDC solution and aqueous $NiSO_4$ solution

Correlation between initial concentrations of NaDDC & $NiSO_4$	Urea		S_r (%)
	Inserted (mM)	Determined (mM)	
(1) 1:1	10	6.3 ± 0.9	11.0
(2) 1:2	10	9.5 ± 0.8	7.0
(3) 1:4	10	9.9 ± 0.1	1.1
(4) 4:1	10	5.5 ± 0.6	8.3
(5) 2:1	10	6.0 ± 0.3	4.5

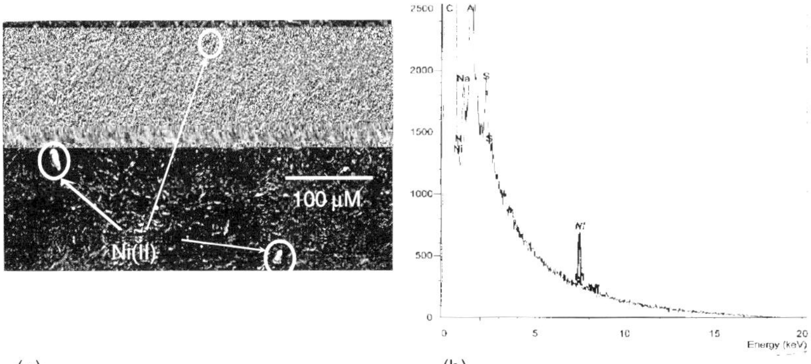

(a) (b)

Fig. 27.9. (a) Electron microscopic photo and (b) energy dispersion spectrum of working surface of transducer with inserted catalyst.

sensor. Small white spots (circled in the photo) of different sizes can be observed on the surface (Fig. 27.9a). The comparison of the photo and the data of X-ray microanalysis shows that the spots contain Ni (Fig. 27.9b). Electrochemical data (Fig. 27.10) obtained with the use of the sensor also proved that 10% of the catalysis is sufficient to form an optimum response.

Besides urea, Ni compounds catalyze oxidation reactions for phenols, alcohols and amines [39]. At the same time, real samples (blood serum and dialysis liquid) contain amino acids, guanidines, uric and ascorbic acids, phenols, cresols, etc. It is reasonable that presence of these compounds may affect the value of an analytical signal. Diagrams given in

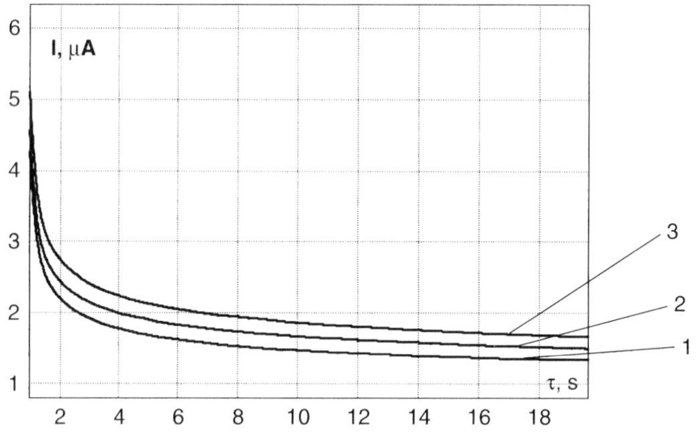

Fig. 27.10. Chronoamperograms for transducer with inserted Catalyst 3. Solution contents: (1) Background: 0.25 M NaOH; (2) 0.25 M NaOH+10^{-4} M urea; (3) 0.25 M NaOH+2×10^{-4} M urea.

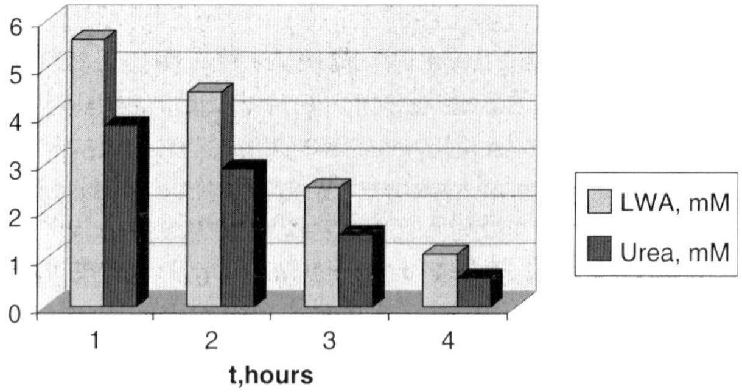

Fig. 27.11. Variation of parameters in dialysis process.

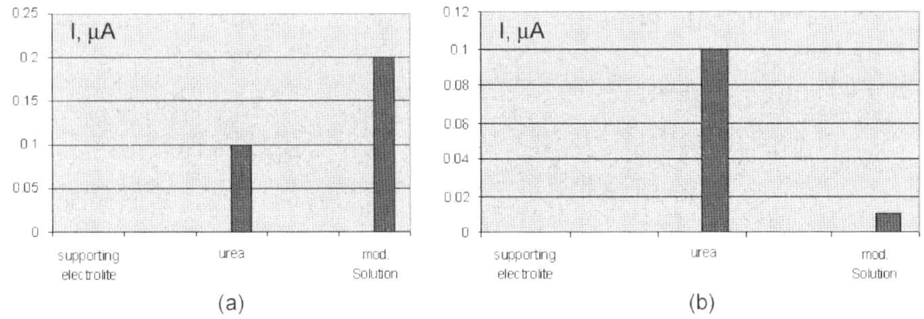

Fig. 27.12. Responses observed in the systems containing model solution and urea (a) before putting the solution through the anion-exchange column; (b) after putting the solution through the anion-exchange column.

Figs. 27.11 and 27.12 demonstrate this effect. It is seen that correlation between the results, obtained by using urease-containing sensor and proposed no enzymatic one, is good. Nevertheless, results of dialysis liquid analysis using no enzymatic sensor are sometimes higher than results given by urease sensor (Fig. 27.11). The reason is low molecular amines are determined together with urea this time. Considering this fact, it is necessary to take into account that this cumulative parameter is not worse diagnostic criteria than urea, because it shows clearance of, for example, protein decomposition products concentration. To estimate the influence of main components of real samples, the solution containing guanidine, uric acid, ascorbic acid, arginine acid, *o*-cresol, indole-3-acetic acid, phenol (concentration 0.1 M each) and urea (0.01 M) was analyzed as the model system. Most of these compounds display acid qualities, which allow them to be separated from urea using an anion-exchange column. Data given in Fig. 27.12 show that after passing the column solution contains not more than 10% of substances interfering with urea determination.

These data and data on blood serum samples analysis given in Procedure 39 in CD accompanying this book confirm that using anion exchange column provides sufficient separation of the interfering compounds from analyte and reliable analysis results.

27.1.3 Sensor and method for AOA measurement

Free radicals are formed in biochemical reactions providing organism cells with 'vital functions'. Negative impact of various environmental factors may lead to excessive free radical formation and as a result to an

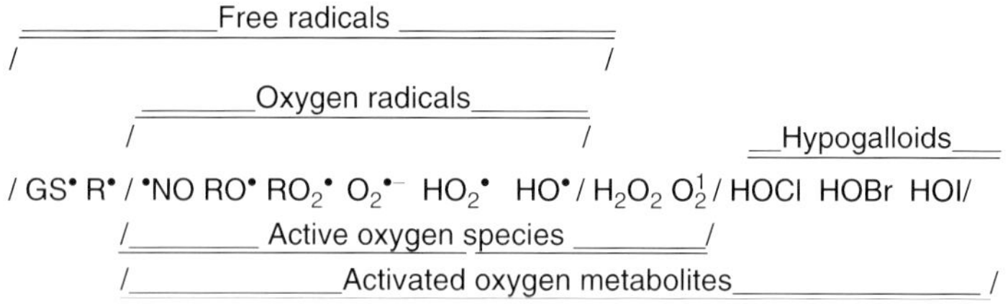

Fig. 27.13. Free radicals and active oxygen compounds. Source: Ref. [40].

increase in oxidative stress. Oxidative stress causes diseases and early organism ageing [40]. On the other hand, activated oxygen metabolites (AOM) containing no unpaired electrons are also formed in living organisms. AOM are not free radicals but possess oxidative properties and are able to generate radicals. The term 'activated oxygen metabolites' comprises compounds of both radical and non-radical nature (Fig. 27.13).

The system protecting organisms from free radical excess comprises enzymes with oxide reductive activity, non-enzyme proteins, polypeptides, water and oil-soluble vitamins, SH-containing amino acids, flavonoids, carotinoids, etc. [40]. Most of these compounds prevent oxidative stress evolution by interrupting chain oxidative reactions. That is why these substances are called substances with antiradical activity as well as antioxidants (AO). Foodstuff, nutrients and some drugs are sources of most antioxidants.

Cumulative antioxidant activity of the sample as the integrated parameter, rather than a simple sum of individual antioxidants, should be taken into consideration. Thus, performance of known and unknown antioxidants and their synergistic interaction will be assessed.

Taking into account the important role of balance in oxidant–antioxidant phenomena for human health and interrelation between internal and external factors, evaluation of oxidant–antioxidant state of human body fluids and natural resources of antioxidants is essential for therapy and prevention of diseases and adverse effects of environment. Thus, analysis of ready-made food and production processes monitoring are of great importance.

The existing analysis techniques, as a rule, consist of a separation stage and the consequent identification of separated substances with spectrophotometric or electrochemical methods [41].

Integrated AOA parameter is based on reduction ability. Consequently, determination methods are based on interaction with long-living radicals or active oxygen compounds. Currently, there exist several methods of measuring antioxidant activity:

- Trolox equivalent capacity (TEAC) assay is based on antioxidants' capability to reduce radical cations 2.2-azinobis(3-etilbenzotiazoline 6-sulfonat) (ABTS) and thus to inhibit absorption in the long wave part of the spectra (600 nm). Water-soluble vitamin E derivation Trolox [42] is used as a standard.
- Oxygen radical absorbency capacity (ORAC) method presumes registration of substrate (B-phycoerythrin or fluorescein) fluorescence [43,44] after reaction with AOC in the sample to be compared with reference template. Trolox is used as a standard to determine peroxyl radicals and gallic acid to determine hydroxyl radicals.
- Total radical trapping parameter (TRAP) assay is widely used in investigations and has various modifications [45–48]. This method presumes antioxidants' capability to react with peroxyl radical 2.2-azobis (2-amidinopropane) dihydrochloride (AAPH). TRAP modifications differ in methods of registering analytical signal. Most often the final stage of analysis include peroxyl radical AAPH reaction with luminescent (luminol), fluorescent (dichlorofluorescein-diacetate, DCFH-DA) or other optically active substrate. Trolox is often used as a standard.
- In ferric reducing/antioxidant power (FRAP) assay reaction of Fe (III)-tripiridyltriazine reduction to Fe (II)-tripiridyltriazine is used. The latter compound is bright blue and its absorption band is 593 nm [49,50].

It is evident that all the described methods are based on reduction reactions. Substrates react either with long-living radicals (TEAC, TRAP), AOC (ORAC) or with the iron complex compound (FRAP), AOM concentration are obtained as a result of enzymatic reactions [51] or their existing level is registered with electron-spin resonance [52]. Various modifications of peroxidation lipid reaction can also be applied [53].

Current methods used to measure antioxidant activity are complicated, expensive, time-consuming, and as a rule, cannot be used for continuous monitoring. Potentiometry with the use of a mediator system and nanoparticles containing electrodes provides a very simple express-method for measuring antioxidant activity of biological liquids, nutrients, drugs and foodstuffs.

The sensor and the mediator system were chosen taking into account the following requirements:

(1) The response of the sensor to Ox/Red potential change must be rapid enough.
(2) The mediator system must meet the following conditions:

$$E^0_{\mathrm{Rad/Rad(red)}} > E^0_{\mathrm{Ox/Red}} > E^0_{\mathrm{AOox/AOred}} \quad (27.3)$$

where $E^0_{\mathrm{Rad/Rad(red)}}$ is the redox potential of the pair oxidized/reduced AOM; $E^0_{\mathrm{Ox/Red}}$ the redox potential of the mediator system; and $E^0_{\mathrm{AOox/AOred}}$ the redox potential of the pair oxidized/reduced antioxidant.
(3) Chemical reaction between antioxidant and oxidized form of the mediator system must be fast enough.

A screen-printed Pt electrode (Fig. 27.1) served as the sensor. Comparison of this sensor with Pt electrodes ('Phoenix', USA; 'Metrohm', Switzerland) showed that their electrochemical behavior (speed of potential establishment and Nernstian slope) differs insignificantly. The advantage of the former is its cost effectiveness, small size and possibility to be used as disposable if the need arises.

Analysis of potentials for selected radical and non-radical systems, participating in metabolism, along with most common antioxidants affects the choice of mediator system (Table 27.3).

Change of ratio between oxidized and reduced components of the mediator in redox chemical reaction with antioxidant provides

TABLE 27.3

Ox-red potential of selected oxidants and antioxidants (vs. N.H.E.)

Oxidants	$E_{ox/red}$ (V)[a]		Antioxidants	$E_{ox/red}$ (V)[b]
$HO_2^{\bullet}/H_2O_2$	+1.44		Cysteine	−0.26
O_2^{-}/HO_2^{-}	+0.20		Glutathione	−0.215
$HO_2^{\bullet}/H_2O$	+1.65		Uric acid	−0.025
O_2^{-}/OH^{-}	+0.65	Mediator	Catechol	−0.115
$H_2O_2/OH^{\bullet}$	+1.44		Ascorbic acid	−0.230
H_2O_2/H_2O	+1.76		Pyrogallol	−0.229
$OH^{\bullet}/H_2O$	+2.38		Caffeic acid	−0.140
$OH^{\bullet}/OH^{-}$	+1.99		Rutin	−0.060

[a] *Source:* Allen J. Bard (Ed.), *Standard Potentials in Aqueous Solution*, Marcel Dekker, New York, 1985.
[b] Our experimental data.

information on 'antioxidant' concentration (activity). Antioxidant activity (concentration) of the solution can be calculated as follows:

$$\mathrm{AOA} = \pm\frac{C_{\mathrm{ox}} - \alpha C_{\mathrm{red}}}{1+\alpha} \tag{27.4}$$

where AOA is the antioxidant activity, mole-eq/l; C_{ox} the concentration of the mediator oxidized form, M; C_{red} the concentration of the mediator reduced form, M; $\alpha = 10^{(E_2-E_1)/b} \cdot C_{\mathrm{ox}}/C_{\mathrm{red}}$, where $b = 2.3RT/nF$ and E_1, E_2 are the system potentials before and after introducing analyzed antioxidant containing sample, V.

Taking into consideration the fact that solution ionic strength remains practically constant and the source of information is potential shift, rather than its absolute value, it is sensible enough not to distinguish between activity and concentration. Concentration value will be used further to characterize antioxidant activity.

The results obtained from the application of $K_3[Fe(CN_6)]$/$K_4[Fe(CN_6)]$ mediator system are described below.

When the sample is introduced into the solution containing the mediator system, the following reaction takes place:

$$n\bullet \mathrm{Fe(III)} + m\bullet \mathrm{X}_{\mathrm{red}} + n\bullet \mathrm{e} \rightarrow n\bullet \mathrm{Fe(II)} + m\bullet X_{\mathrm{ox}} \tag{27.5}$$

Resulting stoichiometric coefficients are displayed in Table 27.4.

Coefficients correlate with the quantity of functional antioxidant groups in the studied compounds (hydroxyl or thiol), which allows the measurement of total AOA in the solutions. The AOA value of the investigated object is expressed in mole-equivalent per liter of mediator system oxidized form, reacted with probe antioxidants.

TABLE 27.4

Stoichiometric coefficients of selected antioxidants in reactions with $K_3[Fe(CN_6)]$

Substance	m:n
Ascorbic acid	1:2
Cysteine	1:1
Glutathione	1:1
Hydroquinone	1:2
Catechol	1:2
Tannin	1:10–12
Uric acid	1:2

Potential/time relation informs about kinetics of antioxidant oxidation. Table 27.5 displays semi-oxidation periods and average constants of oxidation rate for main antioxidants.

It is seen that reaction rates differ significantly. It will be possible in future to obtain information on concentration of selected groups of antioxidants with regard to time intervals of their oxidation.

Table 27.6 shows results of recovery study obtained when proposed method was used. It is seen, that recovery is high enough.

Antioxidant activity values measured with the proposed potentiometric method correlate with antioxidant activity determined by RANDOX methods and photometric method with the use of stable radical 2.2-diphenyl-1-picrylhydrazyl. Correlations are linear.

Figure 27.14 shows the results of AOA determination by potentiometric and RANDOX methods in 10 wine samples. Correlation coefficient is 72%.

TABLE 27.5

Semi-oxidation periods ($\tau_{1/2}$) and average constants of oxidation rate (K) in antioxidant/mediator reaction ($p = 0.95$, $n = 5$)

Substance	$\tau_{1/2}$ (s)	$K \pm \delta$ (1/s)	S_r
Pyrogallol	1.1	0.63 ± 0.08	0.09
Catechol	2.8	0.25 ± 0.07	0.18
Caffeic acid	3.2	0.22 ± 0.02	0.06
Hydroquinone	3.4	0.20 ± 0.05	0.18
Ascorbic acid	3.4	0.20 ± 0.02	0.08
Rutin	3.4	0.20 ± 0.003	0.01
Cysteine	7.3	0.09 ± 0.03	0.12
Uric acid	11.3	0.06 ± 0.01	0.07
Glutathione	155.3	0.0045 ± 0.0002	0.02

TABLE 27.6

Recovery in determination of some antioxidants concentration ($n = 5$, $p = 0.95$)

Substance	Added (μM)	Found ($X \pm \Delta$) (μM)	Recovery (%)
Ascorbic acid	100	110 ± 1.5	10
Caffeic acid	100	120 ± 1.0	20
Cvercetin	100	96 ± 1.2	4
(+)-Catechin	100	98 ± 1.0	2

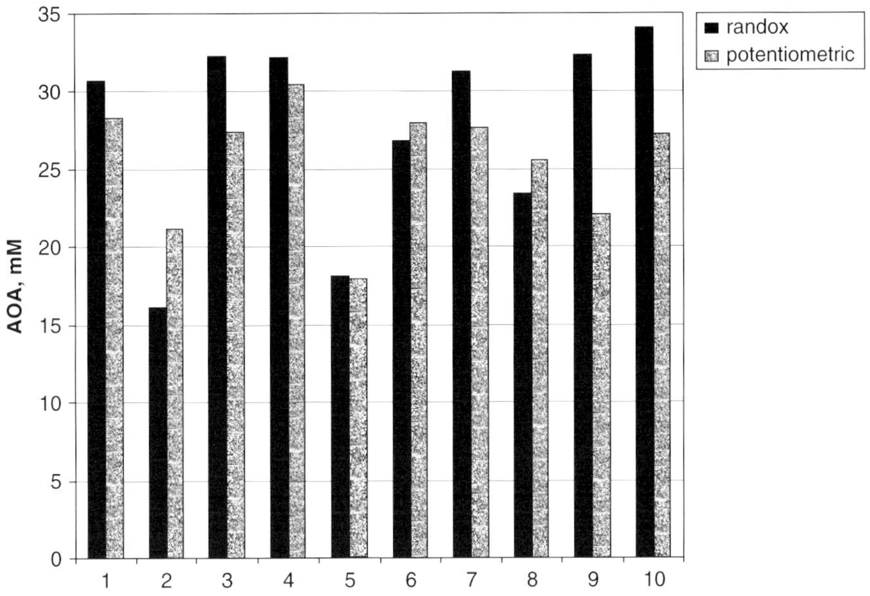

Fig. 27.14. AOA of wine samples obtained by RANDOX and potentiometric methods. Wine samples: (1) Uina Hermina, Rioja 2002, Spain; (2) Vin de Pays d'Oc Jean de Beauvais, Merlot 2002, France; (3) El Condor, Cabernet Sauvignon 2002, Chile; (4) Sunny Ridge, Cabernet Sauvignon 2002, South Africa; (5) Pravova Valley, Sangiovese 1999, Special reserve, Romania; (6) Trivento Bonarda 2003, Argentina; (7) Oltina, La Revedere, Sweden; (8) Cabernet Sauvignon 2001, Bulgaria; (9) Merlot, Bulgaria; (10) Tsar Assen, Cabernet Sauvignon 2002, Bulgaria.

Figure 27.15 displays the results of potentiometric and photometric AOA determination of water and alcohol herb extracts with the use of 2.2-diphenyl-1-picrylhydrazyl stable radical. Correlation coefficients are 94% for water and 99% for alcohol extracts.

The results obtained demonstrate that the suggested potentiometric method provides objective and reliable data on antioxidant activity of various samples. It is a rapid method and does not require expensive instrumentation and additional reagents.

Result validation and reliability, self-descriptiveness and the simple way of carrying out the analysis make the suggested method and the sensor a good alternative to other known methods and sensors for AOA measurement.

The method can be applied in foodstuff production and quality control of finished products, also in nutrient production. Application of the method offered is very simple. It can be realized both in discreet and

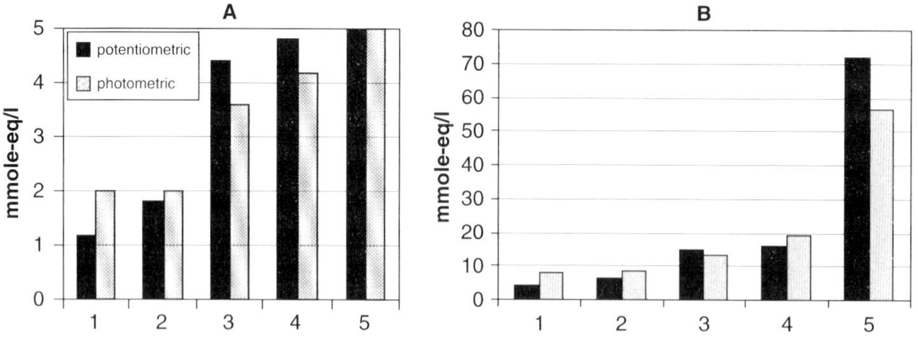

Fig. 27.15. AOA of water (A) and alcohol (B) herb extracts obtained by photometric and potentiometric methods. Herb extracts samples: (1) Haw; (2) Juniper fetus; (3) *Acorus calamus rhizome*; (4) Dandelion roots; (5) Oak bark.

flow-on variants, there is no need of rare and expensive reagents. It should be mentioned that the method provides more complete information on antioxidant activity than commonly used methods of analysis.

27.2 CONCLUSION

Problems with the use of electrochemical methods arise from attempts to create reliable, convenient and user-friendly non-toxic sensors. New technologies and materials enable the development of new sensors with unique properties, which has become possible, in particular, due to application of nanomaterials as (i) transducers, (ii) catalytic constituent of enzyme-free sensors and (iii) labels for immunosensors.

Examples of using nanoparticles in accordance with the points mentioned are given in the paper. They are as follows:

(1) Electrochemical immunosensor: An immunosensor for the spring forest encephalitis diagnosis is based on a carbon-containing transducer and nano silver labeled Protein A used as signal generating substance. Due to easy detection of silver nanoparticles accumulated on the immunocomplex immobilized on the sensor surface and due to reduced quantity of operations in comparison with the case when enzyme-containing sensor applied, the proposed approach allows for shortened analysis duration, and excludes the use of enzymes and associated substrates. At the same time sensitivity and selectivity attributed to the enzyme-containing sensor are preserved. The method can be applied for diagnostics in small- and medium-size hospitals.

(2) Enzyme-free urea sensor: Nano carbon-containing transducer, covered with chemical catalytic system, is used as electrochemical enzyme-free sensor for urea determination in biological materials.

(3) Sensor for AOA measurement: The value of antioxidant activity (AOA) is proposed to be used as an integral parameter for evaluating antioxidant activity of food and nutrients. It can be measured by using a mediator system, which is sensitive to the ox/red potential of test fluids. The paper describes conditions for selecting a mediator system with regard to ox/red potentials of the system and activated oxygen metabolites (AOM); response rate to studied antioxidant. The use of the mediator system and nanomaterial containing transducers allows the solution of problems arising from irreversibility of chemical and electrochemical reactions as well as to reduce the response time.

The proposed potentiometric method of measuring AOA presents a good and prospective alternative to the existing methods, which is confirmed by close correlation between the data obtained with the use of potentiometric and traditional methods. The advantages of the proposed method are its simplicity, cost effectiveness and possibility to be used for *on-site* and *on-line* analysis.

Correlation between the results, obtained by using the known sensors and methods and the proposed sensors, is very good. The approach demonstrates perspectives for creating enzyme-free chemical/biochemical sensors. It also allows the elimination of disadvantages of enzyme-containing sensors, particularly, their time and thermal instability, high cost and necessity to use substrate in the analyzed solution.

Integration of electrochemistry, nanotechnology, mimetic synthesis and bioanalysis makes it possible to create new sensors. It can significantly broaden the area of electroanalysis application with regard to developing smaller as well as quite new instruments and information sources for clinical and food analysis.

It is evident that the range of functions and nature of the nanomaterials used in chemical and biochemical sensors construction will be extended in nearest future. For example, combination of metal/metal oxide, conducting polymers or polymers and specially synthesized molecules-biomimetics are very promising. As a result, new approaches to developing electrochemical sensors will appear. It is not too optimistic to wait for breakthrough in application of electrochemical sensors in medical diagnostics, environmental, food, nutrients and drugs quality monitoring.

ACKNOWLEDGMENT

Authors express their deep gratitude to ISTC for the Project 2132 and to INTAS-00-273 for financial support.

REFERENCES

1 C. Zbinden, *Bull. Soc. Chim. biol.*, 13 (1931) 35–38.
2 G.C. Barker and I.L. Jenkins, *Analyst*, 77 (1952) 685–691.
3 Kh.Z. Brainina and A.M. Bond, *Anal. Chem.*, 67(5) (1995) 2586.
4 A. Turner, *Ann. Chim.*, 87 (1997) 255.
5 J. Labuda. In: P. Frangopol and M. Sandulovicin (Eds.), *Current Topics in Biophysics*, Vol. 5, Al.I. Cuza University Press, Romania, 1996, p. 76.
6 S. Allegret, *Analyst*, 121 (1996) 1751.
7 J. Wang, F. Lu, F. Angnes, J. Liu, H. Sakshend, Q. Chen, M. Pedreno and L. Chen, *Anal. Chim. Acta*, 305 (1995) 3.
8 M. Lutz, E. Burestedt, J. Emneus, H. Liden, S. Gobhadi, L. Gorton and G. Marco-Vorg, *Anal. Chim. Acta*, 305 (1995) 8.
9 A. Cagnin, I. Palchetti, A.P.F. Turner, M. Mascini, *Electroanalysis IX*, Bologna, Italy, 1996, Fr. p. 68
10 G.F. Khan, *Electroanalysis*, 9(4) (1997) 325.
11 A.J. Baeumner, *Anal. Bioanal. Chem.*, 377 (2003) 434–445.
12 D. Hermandes-Santos, M.B. Gonzalez-Garcia and A.C. Garcia, *Electroanalysis*, 14(18) (2002) 1225–1235.
13 M.J. Doyle, H.B. Halsall and W.R. Heineman, *Anal. Chem.*, 54 (1982) 2318.
14 F.J. Hayes, H.B. Halsall and W.R. Heineman, *Anal. Chem.*, 66 (1994) 1860.
15 J. Wang, B. Tian and K.R. Rogers, *Anal. Chem.*, 9 (1998) 1682.
16 W. Guo, J.-F. Song, M.-R. Zhao and J.-X. Wang, *Anal. Biochem.*, 259 (1998) 74–79.
17 A. Varenne, A. Vessière, M. Salmain, S. Durand, P. Brossier and G. Jaouen, *Anal. Biochem.*, 242 (1996) 172.
18 B. Limoges, P. Brossier and C. Degrand, *Anal. Chim. Acta*, 356 (1997) 195–200.
19 E.P. Diamandis and T.K. Christopoulos, *Anal. Chem.*, 62 (1990) 1149A.
20 J. Yuan, K. Matsumoto and H. Kimura, *Anal. Chem.*, 70 (1998) 596–601.
21 G.F. Blackburn, H.P. Shah, J.H. Kenten, J. Leland, R.A. Kamin, J. Link, J. Peterman, M.J. Powell, A. Shah, D.B. Talley, S.K. Tyagi, E. Wilkins, T.G. Wu and R.J. Massey, *Clin. Chem.*, 37 (1991) 1534–1539.
22 J.H.W. Leuvening, B.C. Goverde, P.S.H.M. Thal and A.H.W.M. Schurs, *Immunol. Methods*, 60 (1983) 9–23.
23 K. Sato, M. Tokeshi, T. Odake, H. Kimura, T. Ooi, M. Nakao and T. Kitamori, *Anal. Chem.*, 72 (2000) 1144–1147.

24 L.A. Lyon, M.D. Musik and M.J. Natan, *Anal. Chem.*, 70 (1998) 5177.
25 R. Dahint, M. Grunze, F. Josse and J. Renken, *Anal. Chem.*, 66 (1994) 2888–2892.
26 J. Storhoff, R. Elghanian, R.C. Mucic, C.A. Mirkin and R.L. Letsinger, *Am. Chem. Soc.*, 120 (1998) 1959–1964.
27 A. Perrin, A. Theretz and B. Mandrand, *Anal. Biochem.*, 256 (1998) 200–206.
28 J. Ni, R.J. Lipert, G.B. Dawson and M.D. Porter, *Anal. Chem.*, 71 (1999) 4903–4908.
29 M. Dequaire, Ch. Degrand and B. Limoges, *Anal. Chem.*, 72 (2000) 5521–5528.
30 Kh. Brainina, A. Kozitsina and J. Beikin, *Anal. Bioanal. Chem.*, 376 (2000) 481–485.
31 M.H. Beers and R. Berkow, *The Merck Manual of Diagnosis and Therapy*, 17th ed., Merck Publishing Group, Portland, (1999), 1801.
32 Dissertation for the degree of doctor of philosophy Rebecca Kupcinkas, *A method for optical measurement of urea in effluent hemodialysate*, May 10, 2000, p. 430.
33 R. Koncki, A. Radomska and S. Glab, *Talanta*, 52 (2000) 13–17.
34 A. Schitogullari and A.H. Uslan, *Talanta*, 57 (2002) 1039–1044.
35 W.Y. Lee, K.S. Lee, T.-H. Kim, M.-C. Shin and J.-K. Park, *Electroanalysis*, 12 (2000) 80–82.
36 M.M. Gastillo-Ortega, D.E. Rodrigues, J.C. Encinas, M. Plascencia, F.A. Mendez-Velarde and R. Olayo, *Sens. Actuator B Chem.*, 85 (2002) 19–25.
37 M.D. Osborne and H.H. Girault, *Electroanalysis*, 7 (1995) 714–721.
38 W.J. Cho and H.J. Huang, *Anal. Chem.*, 70 (1998) 3946–3951.
39 S.J. Ferrer, S.G. Granados, F. Bedioui and A.A. Ordaz, *Electroanalysis*, 15 (2003) 70–73.
40 N.K. Zenkov, V.Z. Lankin and E.B. Menschikova, *Oxidative Stress*, MAIK NAUKA/INTERPERIODIKA, 2001.
41 I.F. Abdullin, E.N. Turova and G.K. Budnikov, *Zavod. Lab. Diagn. Mater.*, 6(67) (2001) 3–13.
42 D.D.M. Wayner, G.W. Burton, K.U. Ingold and S. Locke, *FEBS Lett.*, 187 (1985) 33–37.
43 G. Gao, C.P. Verdon, A.H. Wu and R.L. Prior, *Clin. Chem.*, 41 (1995) 1738–1744.
44 R.L. Prior and G. Cao, *Free Radic. Biol. Med.*, 27 (1999) 1173–1181.
45 C. Rice-Evance and N.J. Miller, *Methods Enzymol.*, 234 (1994) 279–293.
46 H. Alho and J. Leinonen, *Methods Enzymol.*, 299 (1999) 14.
47 M. Valkonen and T. Kuusi, *J. Lipid Res.*, 38 (1997) 823–833.
48 A. Gizelli, M. Serafini, G. Maiani, E. Assini and A. Ferro-Luzzi, *Free Radic. Biol. Med.*, 27 (1994) 29–36.
49 I.F.F. Benzie and J.J. Strain, *Methods Enzymol.*, 299 (1999) 15–27.
50 I.F.F. Benzie and J.J. Strain, Patent US6177260, 2001-01-23.

51 V. Lavelly, C. Peri and A. Rizzolo, *J. Agric. Food Chem.*, 48(5) (2000) 1442–1448.
52 G. Bessenderfer and B. Birkenstok, *Talaker. Brauwelt Mir Piva.*, II (2002) 10–18.
53 M. Liebert, U. Licht, V. Bohm and R. Bitsch, *Eur. Food Res. Technol.*, 204(1) (1999) 217–220.

Chapter 28

Analysis of meat, wool and milk for glucose, lactate and organophosphates at industrial point-of-need using electrochemical biosensors

A.L. Hart

28.1 INTRODUCTION

28.1.1 Terms

First, some explanation of terms. By *industry* I mean some situation where analyses are required in a commercial setting or according to some commercial imperative; this will probably mean that the answer is required very quickly according to pre-set specifications. There will be a requirement for large numbers of analyses for long periods of time. The sensors will be manufactured; not made partly or wholly by hand. The cost of the sensors will be compatible with the economics of the industrial enterprise that they are part of. At *point-of-need* is just that; the analyses are required as quickly as possible at the processing line, the distribution centre, the milking parlour, rather than from a central analytical laboratory. There is no real requirement to define *electrochemical biosensors* for readers of this book, although I will use a more relaxed definition than Thevenot *et al.* [1], and refer to electrochemical devices in FIA and other platforms where consideration of these may help us understand how to get a particular analysis done as well as move to the ideal of self-contained devices. '*Thick-film*' is usually regarded as synonymous with '*screen-printing*'. (Occasional reference is made to instances of sensors which are not 'thick-film' where it is felt that they illustrate an important point.)

Comprehensive Analytical Chemistry 49
Alegret and Merkoçi (Eds)
Volume 49 ISSN: 0166-526X DOI: 10.1016/S0166-526X(06)49028-0

28.1.2 Electrochemical biosensors in an industrial context

There are few examples of thick-film sensors being used for fully fledged industrial assays. I have included references to papers from the academic literature where these have gone at least some way in developing sensors that could be used in an industrial setting. Many reviews give good accounts of the nature and production of thick-film biosensors, and the various projects they have been used in [2–5]. This chapter is centred on a review of the use of thick-film electrochemical biosensors for glucose, lactate and organo-phosphates in meat, milk and wool. These analytes are hardy favourites in the biosensor world. This arises from a combination of the availability of suitable enzymes which are not too difficult to put into sensors (in the case of glucose oxidase, all too easy) and the fact that measurement of these substrates is often very useful. Meat, wool and milk are matrices of relevance to the agricultural and food industries. Their 'difficult' nature brings the challenges of using electrochemical sensors in an industrial setting into sharp focus.

Before dealing with the central topic, I would like to raise some further issues pertinent to it, and indeed to the development of thick-film sensors in general. Thick-film sensors are an important part of biosensor research because some blood glucose sensors for use in the home are made this way—if these are successful surely others can be! Further, thick-film technology is not expensive and allows research laboratories to produce quickly, reasonably uniform devices in sufficient numbers for well replicated experiments. At the same time, some insight can be gained into the nature and demands of an industrial production process.

Biosensors with their oft-quoted (ideal) properties would seem to be ready partners for industrial analysts who want information at point-of-need, but as has been pointed out many times, few examples have had the same success as the blood glucose sensors for use in the home (albeit this is an example from medicine rather than industry). The reasons for this have also been pointed out many times, the principal one being that the development and manufacture of the blood glucose sensors is supported by the sadly huge market for diabetic testing and the large amount of investment capital which accrues to that market [6,7]. Further, blood is a sample of reasonably constant composition (in this context), the information is truly useful to the client and the desire for information at home means there is less competition from laboratory-based instruments. This is in contrast to the diverse requirements for analysis in the food industry (for example) which make up a series of relatively small markets for widely differing biosensors expected to cope with complex matrices often under harsh conditions [8–10].

The progression of technology from laboratory to use in the outside world can be described as stages on a Technology Readiness Scale [11]. It has nine stages ranging from basic research to full adoption of the technology. If the truly huge numbers of papers on biosensors in the scientific literature are assessed on this scale, then there would be few describing situations where the upper level of the scale is reached. Most describe detection and transduction at about levels 4 or 5; in other words, at the proof-of-principle stage on the bench-top, with possibly some measurements in the 'real world'. In thinking of how biosensors might be moved to higher levels of technology readiness, the levers for this can be divided into technical and non-technical issues. The latter would include market size, market pull (does someone actually want your sensor?), capital available for further development, protectable intellectual property, cost of sensor manufacture and so on. These factors are perhaps best considered on a case-by-case basis and will not be further discussed here. It is accepted that this is an arbitrary, perhaps even inappropriate, decision as in the development of sensors for industrial use, non-technical issues will determine the final embodiment of technical and scientific phenomena.

An aspiration to use thick-film biosensors for point-of-need analysis in industry means that sample retrieval and preparation will require close attention. The early ideal was that biosensors would rely on the selectivity of the biological recognition element to return an accurate answer about the analyte without sample conditioning, but this has proved an elusive goal. Sample preparation increasingly appears as a topic for discussion in the literature [9,10]. If the sensor can find the analyte untrammelled by electroactive interferents and foulants then it is more likely to return accurate, unbiased results, and to work in a manner which will impress an investor, or the skeptical manager of an industrial process. For samples such as meat, when rapid, point-of-need analyses are desired, the samples are simply not in a convenient liquid phase and sample preparation certainly has a role to play in these instances. In some industrial situations, in-line sensing may be appropriate but in many instances, in-line sensing will be impossible because of concerns about product contamination, the intractable nature of the sample or the harshness of the conditions, e.g. cleaning regimes. At-line sensing where a sample is retrieved from the source will then be necessary. A suitable system for sample retrieval is fundamentally a problem for engineers.

The processes of sample preparation should not be too onerous or slow otherwise there is little incentive to use sensors. Faced with an awkward or slow sample preparation phase for sensor use, the analyst might as well as consider whether instrumental analyses with their often higher sensitivity and specificity (and which, in any case, are regarded as the 'gold standard' for industrial measurements), coupled with on-line sample preparation procedures, might be a more efficient and effective route to chemical information.

A form of sample preparation peculiar to electrochemical biosensors is the removal (or avoidance) of electrochemical interference. Interference is not confined to thick-film sensors but dogs all kinds of electrochemical biosensing in biological matrices. It is less significant where the ratio of analyte to interferent is high (e.g. in determinations of lactate in cultured milk products where the lactate concentration may be 70–100 mM and the effect of interference can be largely negated by dilution) but can become extremely significant when this ratio is lower (e.g. the determination of lactate in fresh milk, where the concentration of lactate is very low but rises with bacterial infection). Ascorbic acid is the stereotypical interferent but it is important to realise that there are other significant interferents in fluids of biological origin, e.g. uric acid in milk.

Developers of thick-film sensors have removed the interference from the sample by chemical oxidation [12,13], by-passed it by selection of applied potentials and electrode materials, including the use of redox mediators [14,15], and kept it from electrode surfaces by inner and outer membranes [16–19].

Above all, it is worth remembering that if insufficient attention is paid to sample preparation then the validity of an analysis will be seriously undermined [20] no matter how sophisticated the measurement instrument is, whether it be a mass spectrometer or sensor. Comments on, and reviews of modern methods of sample preparation such as those of Jinno [21] and Pawliszyn [22,23] are essential reading whether one is a developer of sensors or an instrumental analyst.

Consideration of sample preparation also leads to the question of whether for some applications at least sensors would be more effectively constructed as a collection of modules each with a different function (sample preparation, transduction, detection, data collection) on a platform, large, small or miniaturised, rather than the old ideal of a small, self-contained device with intimately appressed layers which could return a rapid answer from a bucket of slime or some other spectacularly heterogeneous matrix. It is accepted that those with a taxonomic bent would prefer not to call modular systems 'sensors'. 'Large and small' platforms are well represented by flow injection analysers and similar systems, 'miniaturised' platforms are the fashionable 'labs-on-chips'. Complete sensors are sometimes used in chambers, but flow injection analysis also makes it easy to think about the virtues of modularisation, i.e. to take the strain off the sensor by separating, say, sample preparation from detection and transduction or finding modules that are easier to construct than 'sensors' where all functions in the analysis are present in one unitary device. A further important property of flow injection analysis is precise control of fluid flow past the sensor or electrode.

28.2 ELECTROCHEMICAL BIOSENSORS FOR MILK, MEAT AND WOOL

28.2.1 Milk and dairy products

28.2.1.1 Glucose and lactate

Milk is of biological origin but qualifies very much as an industrial matrix as its production and processing underpins the international dairy industry. Point-of-need analysis of milk and its constituent fractions would be of considerable value for farm management and milk processing, e.g. in the diagnosis of mastitis [24], assessment of oestrus [25], detection of antibiotic residues [26], etc. It is a complex matrix—fats, soluble proteins, casein, electrolytes [27]—but at least has the virtue of being an aqueous fluid, so it is amenable to being analysed by electrochemical biosensors placed in platforms where the flow of liquid past the sensor can be controlled. Products such as buttermilk and yoghurt can be diluted if the concentration of analyte allows.

An example of commercial technology sold for point-of-need use which contains thick-film biosensors is that provided by Trace Analytics (www.trace-ag.de). This technology is advertised as being suitable for analyses, including glucose and lactate, in cell cultivations and fermentations. At least one German company uses it to monitor glucose in bioreactors [28]. It is significant that the sensors are provided in the context of sample handling equipment, process controllers and so on. Technology of this kind could well be worth considering for analyses of meat and milk.

Glucose and sucrose in condensed milk (these sugars are not found in raw milk) diluted with phosphate buffer can be measured using analysers made by Yellow Springs Instrument Co. (www.ysi.com/index.html).

The analysers do not contain thick-film electrodes but are mentioned here as a long-standing example of commercial electrochemical sensors available for point-of-need use. Their design, construction and cost per sample are worth considering by those wishing to develop sensors for point-of-need use in the real world. The technology at the core of these analysers is 'first generation', based on platinum electrodes coupled with the appropriate enzyme immobilised on replaceable membranes. The membranes keep out interferents and establish suitable diffusion gradients of the analyte to the enzymes and electrodes. Pumps, reservoirs, mechanical linkages, etc. deliver the samples in a stable, reproducible manner. The significant point about these analysers is that although the core technology is relatively simple, the company is able to manufacture this technology so that it can be successfully used under industrial conditions for a wide range of samples. For some materials, it is still necessary for the analyst to prepare samples for introduction into the analyser but the proper function of the technology surrounding the enzyme electrodes is undoubtedly as important, if not more so, as the electrochemical principle of detection.

A system underpinned by commercially made screen-printed electrochemical cells was described by Palmisano *et al.* [19]. The cells were converted into biosensors for lactate in milk and yoghurt by addition of an electrochemically polymerised barrier to interference and a layer composed of lactate oxidase, glutaraldehyde and BSA. These steps appeared to have been carried out by hand. As there was no outer diffusion-limiting membrane, the linear range of the sensors was quite small (0–0.7 mM). They were incorporated into a FIA with a microdialysis unit based on a planar membrane and a buffer reservoir (earlier work used a microdialysis fibre with a platinum electrode [29]. The concentration of lactate was determined in various milks (0.27–1.64 mM), and in raw milk (c. 0.5–0.9 mM) left to degrade on the laboratory bench. The recovery of the microdialysis unit, 2.6%, implied that the sensor had an ability to return measurable currents for very low concentrations of lactate. A further implication is that the electro-polymerised layer was very effective at preventing interference.

Torriero *et al.* [30] managed to estimate the concentration of lactate in untreated milk without the use of a microdialysis unit. Samples were fed into a reactor consisting of a rotating disc bearing lactate oxidase which by its motion ensured adequate mass transport of hydrogen peroxide to an enzyme electrode. The electrode consisted of horse-radish peroxidase immobilised over osmium on the surface of a glassy carbon electrode. Such an electrode can be poised at 0.0 V so avoiding electrochemical interference. Further matrix effects, thought to arise from micellar casein, were diminished by the addition of EDTA to the milk. The electrodes were not thick-film but this example illustrates again how various steps in an analysis can be spread out over a flow system rather than aggregated into a single sensor.

The concentration of lactate in yoghurt and buttermilk was estimated using sensors made by screen-printing, except for the outer diffusion-limiting membrane which was applied through an air-brush [31]. The core of the sensors was working electrodes containing platinised carbon combined with lactate oxidase in a matrix of hydroxyethyl cellulose. The samples were centrifuged to collect a liquid supernatant which was diluted (75 ×) before being aspirated into a FIA system which housed the sensors in flow cell. Some samples of supernatant were passed through a cation exchange column as it was thought that calcium may be an inhibitor of hydrogen peroxide oxidation [16,32], although this proved to be unnecessary. Seemingly good estimates of concentration were obtained. For example, for six varieties of yoghurt, the mean estimate of lactate concentration was 99.8% (cations present) or 101.4% (cations removed) of spectrophotometric estimates. A more stringent, and sobering, review of the estimates which took into account the statistical distribution of the errors [33] indicated that for all analyses (buttermilk and yoghurt) sensor estimates would on average be 2.1 mM less than spectrophotometric estimates (albeit at total concentrations of 70–100 mM) and 95% of estimates would lie between 9.0 mM above and 4.8 mM below corresponding spectrometric estimates. The results were better than those obtained with similar electrodes (the outer membrane had been applied by screen-printing [16]) used to analyse samples in batch mode. The improved performance was attributed to the better control of hydrodynamic conditions conferred by the FIA system.

Levels of lactate in buttermilk and yoghurt (and blood) were estimated using disposable sensors formed from screen-printed graphite laminated between two polymer sheets [18]. Platinum (deposited by sputter-coating) was the transducing surface. Layers of Nafion were added to reduce interference and were surmounted by lactate oxidase in a mixture of polyethyleneimine and poly(carbamoyl)sulphonate hydrogel. The samples were measured in stirred buffer. A good correlation between biosensor results and those obtained with an enzyme kit was claimed but the data had a considerable amount of scatter—if the enzyme kit is taken as the reference method then a more severe analysis of the biosensor results [33] would not have shown them in a good light. An offset of 0.11 g/100 g sample was found implying that not all electrochemical interference had been removed—or there was some other inhibitor in the sample matrix.

Lamination (used by Patel *et al.* [18] above) is a fabrication technique that could be considered more widely by researchers in laboratories as an adjunct to screen-printing, particularly for deposition of outer diffusion-limiting membranes. Sensors can be constructed entirely by screen-printing technology [16], but it is difficult to maintain control over membrane thickness, porosity, etc. Deposition of pre-cast membranes by lamination may be a way of controlling these parameters more precisely.

28.2.1.2 Organo-phosphates

The level of cholinesterase inhibition in apricot yoghurt was measured with thick-film sensors and an agreeable lack of sample preparation, other than dilution (10 g diluted with 10 ml of 1 M phosphate buffer, pH 7.5) [34]. The sensors were based on mediated (7,7,8,8 tetracyanoquinodimethane) carbon electrodes bearing a layer of enzyme in hydroxyethyl cellulose and BSA fixed with glutaraldehyde. Enzyme inhibition was assessed by the level of acetylcholinesterase activity remaining activity after incubation for 30 min in the diluted yoghurt. The detection limit of sensors, based on electric eel acetylcholinesterase, for paraoxon was lower than the EC limit of 10 μg kg^{-1}. Matrix effects were avoided by soaking the electrodes after incubation in 1 vol.% Tween 20 for 15 min. Success of the assay over a range of samples was judged by comparing sensor results to estimates of residue content by instrumental methods. The diluted apricot yoghurt inhibited electric eel acetylcholinesterase by 12% and human acetylcholinesterase by 6%. Instrumental analysis of the yoghurt revealed 2 μg kg^{-1} of a carbamate and 1–3 μg kg^{-1} of two fungicides and an insecticide synergist.

Sensors of this type were constructed as multi-electrode arrays bearing cholinesterases of differing sensitivity to organo-phosphates, and the assay extended to milk [35]. Both spiked milk and milk from shops inhibited the activity of the electrodes. In two instances, estimates of the level of organo-phosphates in spiked milk made using sensors were very close to those made using GC–MS. This appears to have been a fortunate co-incidence as the response curves used for calibration were not linear. Nine out of ten milk samples from shops inhibited at least some members of the array. In only one case did the GC–MS assay find insecticides in the samples, but the insecticides were not organo-phosphates. The inhibition shown by the enzyme-based arrays was reversible by pyridine-2-aldoxime methochloride indicating that the arrays did appear to have detected organo-phosphates invisible to the GC–MS assay. The attraction of these sensors lies in the relatively straightforward mode of construction and the simple sample preparation but questions remain over the non-linearity in the calibration responses and the utility of a relatively slow assay, subsequent to sample preparation, to industry (assays could of course be done in parallel).

28.2.2 Meat

Analytes in meat are not readily available to the surface of sensors. If the advantages of sensors are to be exploited, something has to be done to 'expose' the analytes. Sample preparation will be the step that determines the performance of the sensor, and the feasibility of using it in an industrial setting.

A procedure in which fluid oozes from a cut made in meat onto a sensor, thus avoiding any other sample preparation, is a tempting prospect. Attempts to build a probe capable of measuring glucose in fluid obtained in this way have been described by Kress-Rogers [36]. A hand-made glucose sensor (graphite–mediator–glucose oxidase) was used to measure glucose in fluid collected from cuts made in pork. Unfortunately, development of a prototype incorporating sensor and knife in the same device does not seem to have proceeded further. This and accounts of other sensors in the book edited by Kress-Rogers and Brimelow [9] are worth reading because of their keen appreciation of the requirements of sensors that are intended for industrial use.

Turning to assays using sensors that require more than a simple cut in the sample, Bergann *et al.* [37] paid considerable attention to sample preparation in attempts to measure lactate in meat with a sensor (not thick film) based on reaction of hydrogen peroxide with a platinum electrode or on the reaction of ferrocene carboxylate with the active site of lactate oxidase. Sample preparation entailed extraction into buffer following grinding. In some cases, ground samples were left to allow the lactate to diffuse into the buffer solution. This was quite effective but slow (up to 90 min). Ultimately, the quality of the assay was dependent on the method of sample preparation.

Extraction of minced meat into buffer was used immediately prior to determination of lactate in meat by screen-printed sensors mounted in a FIA system [17]. The sensors were similar in design to those used for dairy product analysis [31] although they were constructed completely by screen-printing rather than having the outer membrane applied with an air-brush. The outer membrane consisted of a co-polymer of polyvinyl chloride (PVC) and polyvinyl acetate, mixed with cellulose acetate butyrate to improve printing performance. The use of a membrane based on PVC was based on papers published by Vadgama and colleagues [38–41]. Estimates of reasonable quality were obtained with some of the sensors—those with Gafquat and polyethyleneimine added to the enzyme matrix to confer enzyme stability—and it was felt that the two crucial contributions to these were reducing the meat to a liquid slurry and mounting the sensors in a FIA system (previous attempts to use similar sensors in batch mode with juice expressed under pressure from tissue were unsuccessful (unpublished)). No doubt, remaining inaccuracies were due to uncontrolled manufacturing defects.

Lactate in luncheon meat macerated and extracted into water may be determined by the YSI technology mentioned above as an example of sensors developed for point-of-need in food processing [42].

Continuing the theme of matching sample preparation of meat with sensors, an example of an assay which has been incorporated successfully into an industrial setting is the classification of beef carcasses according to meat quality [43,44]. The pH reached by meat some 24 h after slaughter, the 'ultimate' pH, is a guide to meat quality, the best meat having a pH in the range 5.4–5.6. The ultimate pH, whether in this range or higher, is a reflection of the extent to which muscle glycogen has been converted to lactic acid via glycolysis. pH measurement with probe meters at 24 h is widely practised, but where meat is excised from the carcass earlier (and this is increasingly common), the pH values recorded at excision are often meaningless.

Young *et al.* [43] chose an alternative approach to prediction of ultimate pH. They measured the concentration of the muscle glycogen responsible for lactic acid formation. This was done immediately after slaughter. The glycogen concentration, expressed as glucose equivalents per gram of muscle, has a curvilinear relationship to ultimate pH (Fig. 28.1).

About 1.5 g of muscle sample is accurately weighed, then homogenised in acetate buffer at pH 5 in the presence of amyloglucosidase and the resultant slurry is incubated at 55°C for 5 min. Amyloglucosidase completely hydrolyses glycogen to glucose. A Bayer ESPRIT glucose meter and sensor, as routinely used by diabetics, is then used to determine the concentration of glucose. A calibration curve links the glucose value to glycogen and thus to ultimate pH.

The fraction of measured pH values predicted from these glucose determinations is 0.89. A somewhat larger fraction (0.94) can be predicted

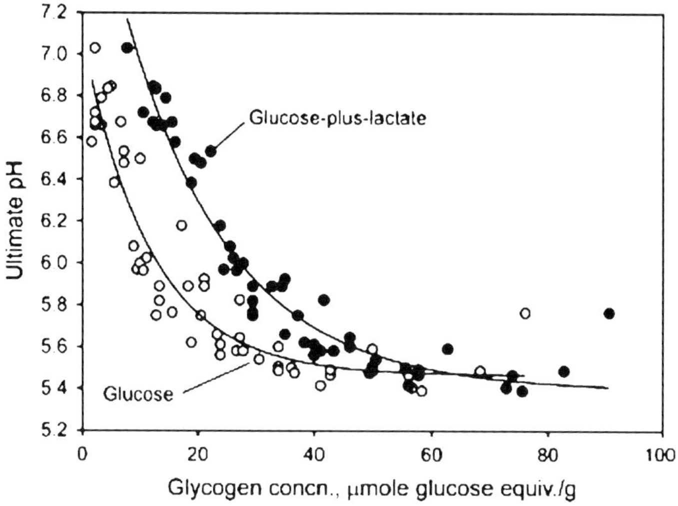

Fig. 28.1. Reproduced from Fig. 5 of Ref. [43], with permission from Elsevier. The relationship between ultimate pH and glycogen concentration in muscle from 51 bulls. The values labelled 'glucose' were derived from glucose estimates returned by the ESPRIT glucometer. Those labelled 'glucose-plus-lactate' include estimates of the lactate levels in the tissue. In practice, decisions about glycogen and pH are based only on glucose estimates.

if the lactic acid is also measured. However, glucose measurements alone are sufficient to predict ultimate pH with enough confidence to satisfy industrial demands.

With regard to using biosensors in industrial settings and the problem of getting biosensors shifted out of the laboratory, the use of ESPRIT sensors in this assay illustrates an important point. The decision to use the ESPRIT sensors was not entirely 'scientific'. The funds available and the funder's expectations about time to develop a functional and effective assay precluded the development of sensors possibly more suited to the matrix. A usable sensor had to be found, and quickly, so an existing technology was selected and adopted. The 'use-once-and-discard' sensors employed in the Bayer ESPRIT system have a major advantage in this industrial application. Their consistent response to analyte (and matrix) is assured because patient use and health depends on it. Relatively unskilled labour can use them without the need for any calibration. This example illustrates that sensors will work when developed to a sufficiently high state. It is ironic that the successful use of sensors in this industrial setting rests on investment of capital and research and development in the medical device sector. There is a further irony in that now that these sensors have been seen to be fit for purpose, there is a substantial barrier to introducing more suitable sensors, designed specifically for the purpose, into this process.

This procedure is now in use at the plants of a major meat company in New Zealand [44]. Trials described in this paper illustrate the demands placed on an analytical process in an industrial setting: cattle were killed at the rate of 90 per hour, samples were taken 10 min after slaughter, and information from the sensors was available 10 min after that—this period included sample preparation time; the sensor reading was returned 30 s after the sample was placed on the sensor. Operationally, the classification of animals was done using glucose values rather than estimating the ultimate pH from the glucose value. As well as being an excellent example of industrial use of biosensors, it pointedly illustrates how 'non-scientific' factors are important in determining sensor use and not always under the control of the sensor scientist.

28.2.3 Wool

The author of this review was asked if electrochemical thick-film sensors could be used in field determinations of the level of organo-phosphate residues in sheep wool. Wool must be the epitome of matrices where sample preparation is absolutely essential before any progress with analysis using electrochemical biosensors is conceivable. Even with meat, one can imagine that analysis by a sensor of the fluid exuded into the space made by an incision might be possible; indeed this has been attempted as noted above [36]. The question as to whether wool could be analysed using a biosensor brought the problems of sensors and difficult samples into sharp and inescapable focus.

A brief sample preparation procedure based on solid-phase extraction was developed [45]. Wool was extracted in acetonitrile for 30 min, the fluid then being filtered and passed through a C18 solid-phase cartridge. The acetonitrile was removed under vacuum and replaced with methanol for application of the putative inhibitors to enzyme-based electrodes.

With regard to assaying the inhibitory activity of extracts electrochemically, one of the problems of assays using sensors based on cholinesterase was that considerable time, e.g. 30–45 min [46,47], could be needed for the activity of the enzyme electrode to fall below control levels. The time increased as the level of inhibition decreased. Such lengthy assays make any number of serial assays impractical. In previous work [48,49], it had been noted that if sensors were exposed to solution containing inhibitors and then allowed to dry, they could be tested very quickly (within half an hour or so) for the current generated on exposure to enzyme substrate. The currents relative to those generated by control electrodes bore a generally regular relationship to dose (Fig. 28.2), and so could be used as a measure of inhibition.

These observations made it possible to think of assays outside the laboratory being carried out in two convenient stages: extraction of wool samples and exposure of electrodes to potential inhibitors using a relatively simple field kit, followed by testing of electrodes, possibly in parallel, in a laboratory equipped with potentiostats, etc. A separation in time between exposure and measurement of inhibition had also been used by Brown *et al.* [50].

Fluid from solid-phase extraction of wool was placed on simple screen-printed electrodes (an outer membrane was applied with an airbrush) [45]. The solvent was allowed to evaporate and, after an overnight incubation, the activity of the electrodes was measured quickly with reference to that of unexposed electrodes. It was possible to detect the presence of organo-phosphates which had been used to contaminate samples of untreated sheep wool (Fig. 28.3) (see Procedure 41 in CD accompanying this book).

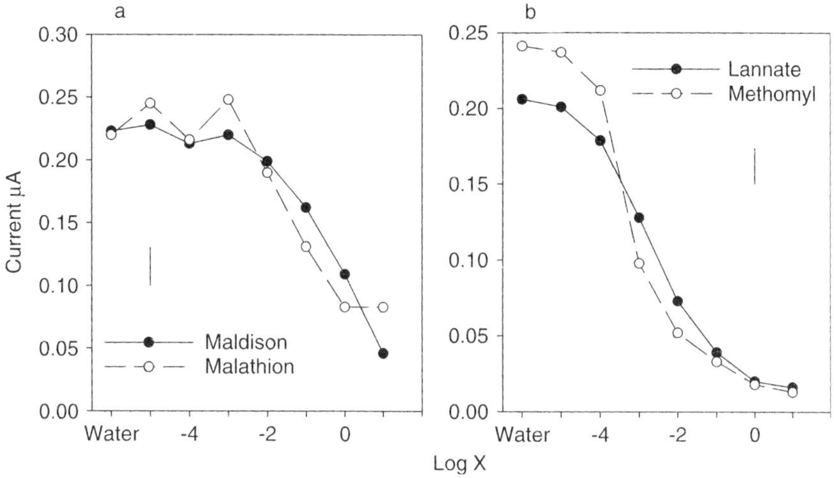

Fig. 28.2. Reproduced from Figs. 1 and 2 of Ref. [49]. Currents generated by electrodes 17 h after exposure to pesticides. (a) Electrodes exposed to an organo-phosphate, malathion and a commercial formulation of it, Maldison. Maldison diluted to the recommended dose (RD) contains 6.1 mM malathion. (b) Electrodes exposed to a carbamate, methomyl and a commercial formulation of it, Lannate. Lannate diluted to the RD contains 2.2. mM methomyl. *X* is a multiple of the RD or 6.1 mM malathion or 2.2 mM methomyl.

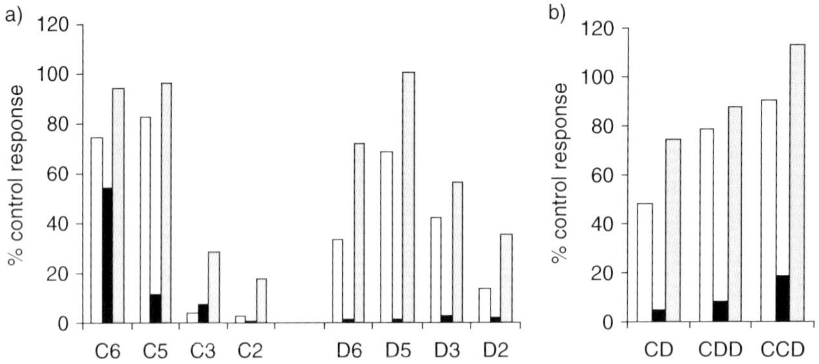

Fig. 28.3. Reproduced from Fig. 3 of Ref. [45], with permission from Elsevier. Detection of organo-phosphates extracted from wool. (a) Extracts containing either chlorfenvinphos (C) or diazinon (D); numbers are X in 10^{-X} M. (b) Extracts containing mixtures of the two; total concentration of organo-phosphates was 10^{-5} M. Electrodes were exposed to the extracts, dried overnight and the currents generated in the presence of butyrylthiocholine measured. White columns, electric eel acetylcholinesterase; grey, bovine erythrocyte acetylcholinesterase; and black, horse serum butyryl cholinesterase.

28.3 TESTING BIOSENSORS: BRIEF COMMENTS ON EXPERIMENTAL DESIGN AND STATISTICS

The intent behind these comments is not to attempt any instruction on this topic but to reiterate, as have many others, that all experiments should have a statistically sound design. This will help ensure that the data gathered will answer the question posed by the experimenter. There are many books and courses available to the sensor researcher. The few remarks that follow are based on the author's experience.

It is particularly important that thought be given to the calculation of the error term in tests of significance. An error term based on repeated measures of a single sensor prototype will not be of any use in establishing the performance of prototypes in general. The sample size should be adequate to support the test of the null hypothesis, i.e. the test is of sufficient power—usually the sample size needs to be bigger than two or three.

Correlations between results of a sensor assay and those from a reference assay are very commonly presented in the literature. These can give an over optimistic view of sensor performance. It is better to use a more rigorous assessment of performance, such as that outlined by Bland and Altman [33] which considers the magnitude and distribution of differences between measurements of the same attribute made by two competing measurement techniques.

28.4 CONCLUSION

My principal conclusion is that electrochemical biosensors will work in an industrial setting if brought to a sufficiently high state of development, at the right price. Like others [51,52], I have suggested that many of the factors required to do this have little to do with electrochemistry *per se*. We know that electrochemical detection works in the sense of electrodes returning a current or potential difference in the presence of analytes; demonstrating this in a laboratory is not very difficult. While electrodes could undoubtedly be made more efficient—less interference, better efficiency of electron transfer between recognition reactions and electrodes—the real challenge is making them work reliably in the real world. Not once, not twice, but thousands of times according to pre-set performance standards. Meeting such a challenge is of course the key difference between science and technology.

We know how to detect things; what we seem to be much less sure of is how to engineer components into a system resilient under industrial conditions at a cost that the market will bear. Resolution of this problem is not made easier by the fact that this task is probably different for every application, even though the central analytical event may not have changed.

The phrase ‘thick-film biosensors’ implies that thick-film technology is suitable for fabrication of complete sensors. Our experience is that while it is certainly possible to make entire sensors this way, screen-printing technology is too extravagant of high-value components such as enzymes (except glucose oxidase) and too uncontrolled for components such as diffusion-limiting membranes. A combination of thick-film printing with automated precision pipetting and lamination of pre-formed membranes (or some similar combination of techniques) may be a better route to functional biosensors.

If the desire is to develop a biosensor for industrial use, then a set of specifications for sensor performance, which includes economic criteria, should be obtained from the end-user before any work begins. Statements such as ‘this surface/binding phenomenon/electrode reaction could be turned into a biosensor for food safety monitoring’ when the work has been done in the absence of specifications set by those responsible for carrying out tests of food safety are highly likely to remain wishful thinking (I am fully appreciative of the reasons why such statements appear in grant applications!).

I would also assert that engineering has essential contributions to make in sensor technology development, engineering skills being as important as the 'science' of detection and transduction, and that drawing up a set of specifications according to the precepts of product development [53] before research and development begins, is highly worthwhile.

As always, in the adoption of technology, science, in the sense of understanding material phenomena, is not enough. It is only in rare cases that the science is so effective and compelling, so obviously meets a need for analysis or diagnosis that it alone drives the development of a market and associated availability of investment capital—and even here, engineering is a key driver of success.

DISCLAIMER

Some examples of commercial technology are mentioned in this review. The reviewer has no relationship with the companies concerned.

REFERENCES

1 D.R. Thevenot, K. Toth, R.A. Durst and G.S. Wilson, Electrochemical biosensors: recommended definitions and classification, *Anal. Lett.*, 34 (2001) 635–639.
2 J. Atkinson, Screen printing: the art of thick-film sensors. In: K.T.V. Grattan (Ed.), *Sensors: Technology, Systems and Applications*, Adam Hilger, Bristol, 1992, pp. 541–546.
3 J.P. Hart and S.A. Wring, Recent developments in the design and application of screen-printed electrochemical sensors for biomedical, environmental and industrial analysis, *Trends Anal. Chem*, 16(2) (1997) 89–103.
4 M. Albareda-Sirvent, A. Merkoci and S. Alegret, Configurations used in the design of screen-printed enzymatic biosensors. A review, *Sens. Actuators B*, 69 (2000) 153–163.
5 J.P. Hart, A. Crew, E. Crouch, K.C. Honeychurch and R.M. Pemberton, Some recent designs and developments of screen-printed carbon electrochemical sensors/biosensors for biomedical, environmental, and industrial analyses, *Anal. Lett.*, 37 (2004) 789–830.
6 E. Magner, Trends in electrochemical biosensors, *Analyst*, 123 (1998) 1967–1970.
7 J.D. Newman and A.P.F. Turner, Home blood glucose biosensors: a commercial perspective, *Biosens. Bioelectron.*, 20 (2005) 2435–2453.

8 A.O. Scott, Biosensors for food analysis: perspectives. In: A.O. Scott (Ed.), *Biosensors for Food Analysis*, The Royal Society of Chemistry, London, 1998, pp. 181–195.
9 E. Kress-Rogers and C.J.B. Brimelow (Eds.), *Instrumentation and Sensors for the Food Industry*, 2nd ed., Woodhead Publishing Ltd., Cambridge, UK, 2001.
10 L.A. Terry, S.F. White and L.J. Tigwell, The application of biosensors to fresh produce and the wider food industry, *J. Agric. Food Chem.*, 53 (2005) 1309–1316.
11 J.C. Mankins, Approaches to strategic research and technology (R&T) analysis and road mapping, *Acta Astron*, 51 (2002) 3–21.
12 G. Cui, S.J. Kim, S.H. Choi, H. Nam and G.S. Cha, A disposable amperometric sensor screen printed on a nitrocellulose strip: a glucose biosensor employing lead oxide as an interference removing agent, *Anal. Chem.*, 72 (2000) 1925–1929.
13 J.H. Shin, Y.S. Choi, H.J. Lee, S.H. Choi, J. Ha, I.J. Yoon, H. Nam and G.S. Cha, A planar amperometric creatinine biosensor employing an insoluble oxidizing agent for removing redox-active interference, *Anal. Chem.*, 73 (2001) 5965–5971.
14 M.J. Green and P.I. Hilditch, Disposable single-use sensors, *Anal. Proc.*, 28 (1991) 374–376.
15 J.D. Newman, S.F. White, I.E. Tothill and A.P.F. Turner, Catalytic materials, membranes, and fabrication technologies suitable for the construction of amperometric biosensors, *Anal. Chem.*, 67 (1995) 4594–4599.
16 W.A. Collier, D. Janssen and A.L. Hart, Measurement of soluble L-lactate in dairy products using screen-printed sensors in batch mode, *Biosens. Bioelectron.*, 11 (1996) 1041–1049.
17 A.L. Hart, C. Matthews and W.A. Collier, Estimation of lactate in meat extracts by screen-printed sensors, *Anal. Chim. Acta*, 386 (1999) 7–12.
18 N.G. Patel, A. Erlenkotter, K. Camman and G.C. Chemnitius, Fabrication and characterization of disposable type lactate oxidase sensors for dairy products and clinical analysis, *Sens. Actuators B*, 67 (2000) 134–141.
19 F. Palmisano, M. Quinto, R. Rizzi and P.G. Zambonin, Flow injection analysis of l-lactate in milk and yoghurt by on-line microdialysis and amperometric detection at a disposable biosensor, *Analyst*, 126 (2001) 866–870.
20 R.M. Smith, Before the injection—modern methods of sample preparation for separation techniques, *J. Chromatogr. A*, 1000 (2003) 3–27.
21 K. Jinno, Modern sample preparation techniques, *Anal. Bioanal. Chem.*, 373 (2002) 1–2.
22 J. Pawliszyn (Ed.), Sampling and sample preparation for field and the laboratory: fundamentals and new directions in sample preparation. In: *Comprehensive Analytical Chemistry*, Vol. XXXVII, Elsevier, Amsterdam, 2002.

23 J. Pawliszyn, Sample preparation: Quo vadis?, *Anal. Chem.*, 75 (2003) 2543–2558.
24 S.R. Davis, V.C. Farr, C.G. Prosser, G.D. Nicholas, S.A. Turner, J. Lee and A.L. Hart, Milk L-lactate concentration is increased during mastitis, *J. Dairy Res.*, 71 (2004) 175–181.
25 R.M. Pemberton, J.P. Hart and T.T. Mottram, An electrochemical immunosensor for milk progesterone using a continuous flow system, *Biosens. Bioelectron.*, 16 (2001) 715–723.
26 E. Gustavsson, J. Degelaen, P. Bjurling and A. Sternesjö, Determination of β-lactams in milk using a surface plasmon resonance-based biosensor, *J. Agric. Food Chem.*, 52 (2004) 2791–2796.
27 N.P. Wong, R. Jenness, M. Keeney and E.H. Marth (Eds.), *Fundamentals of Dairy Chemistry*, Van Nostrand Reinhold Co., New York, 1988.
28 C. Tollnick and M. Beuse, On-line glucose measurements in industrial bioprocess engineering, 2005. www.trace.de/en/onlinemessung/glucose.htm
29 F. Palmisano, D. Centonze, M. Qunito and P.G. Zambonin, A microdialysis fibre based sampler for flow injection analysis: determination of L-lactate in biofluids by an electrochemically synthesised bilayer membrane based biosensor, *Biosens. Bioelectron.*, 11 (1996) 419–425.
30 A.A.J. Torriero, E. Salinas, F. Battaglini and J. Raba, Milk lactate determination with a rotating bioreactor based on an electron transfer mediated by osmium complexes incorporating a continuous-flow/stopped-flow system, *Anal. Chim. Acta*, 498 (2003) 155–163.
31 W.A. Collier, P. Lovejoy and A.L. Hart, Estimation of soluble L-lactate in dairy products using screen-printed sensors in a flow injection analyser, *Biosens. Bioelectron.*, 13 (1997) 219–225.
32 N. Labat-Allietta and D.R. Thevenot, Influence of calcium on glucose biosensor response and on hydrogen peroxide detection, *Biosens. Bioelectron.*, 13 (1998) 19–29.
33 J.M. Bland and D.G. Altman, Statistical methods for assessing agreement between two methods of clinical measurement, *Lancet*, (1986) i: 307–310.
34 H. Schulze, E. Scherbaum, M. Anastassiades, S. Vorlova, R.D. Schmid and T.T. Bachmann, Development, validation, and application of an acetylcholinesterase-biosensor test for the direct detection of insecticide residues in infant food, *Biosens. Bioelectron.*, 17 (2002) 1095–1105.
35 Y. Zhang, S.B. Muench, H. Schulze, R. Perz, B. Yang, R.D. Schmid and T.T. Bachmann, Disposable biosensor test for organophosphate and carbamate insecticides in milk, *J. Agric. Food Chem.*, 53 (2005) 5110–5115.
36 E. Kress-Rogers, Sensors for food flavour and freshness: electronic noses, tongues and testers. In: E. Kress-Rogers and C.J.B. Brimelow (Eds.), *Instrumentation and Sensors for the Food Industry*, 2nd ed., Woodhead Publishing Ltd., Cambridge, UK, 2001, pp. 553–622, Chap. 19.

37 T. Bergann, P. Abel and K. Giffey, Milchsaurekonzentration in Schlachttierkorpen und frischem Fleisch, *Fleischwirtschaft*, 11 (1999) 84–87.
38 Y. Benmakroha, I. Christie, M. Desai and P. Vadgama, Poly(vinyl chloride), polysulfone and sulfonated polyether-ether sulfone composite membranes for glucose and hydrogen peroxide perm-selectivity in amperometric biosensors, *Analyst*, 121 (1996) 521–526.
39 I.M. Christie, P.H. Treloar and P. Vadgama, Plasticised poly(vinyl chloride) as a permselective barrier membrane for high-selectivity amperometric sensors and biosensors, *Anal. Chim. Acta*, 269 (1992) 65–73.
40 M. Kyrolainen, H. Hakanson, B. Mattiasson and P. Vadgama, Minimal fouling enzyme electrode for continuous flow measurement of whole blood lactate, *Biosens. Bioelectron.*, 11 (1997) 1073–1081.
41 M. Kyrolainen, S.M. Reddy and P.M. Vadgama, Blood compatibility and extended linearity of lactate enzyme electrode using poly(vinyl chloride) outer membranes, *Anal. Chim. Acta*, 353 (1997) 281–289.
42 YSI 2700 Application Note 311 L-lactate in lunch meats [www.ysi.com].
43 O.A. Young, J. West, A.L. Hart and F.F.H. Otterdijk, A method for early determination of meat ultimate pH, *Meat Sci.*, 66 (2004) 493–498.
44 O.A. Young, R.D. Thomson, V.G. Mehrtens and M.P.F. Loeffen, Industrial application to cattle of a method for the early determination of meat ultimate pH, *Meat Sci.*, 67 (2004) 107–112.
45 W.A. Collier, M. Clear and A.L. Hart, Convenient and rapid detection of pesticides in extracts of sheep wool, *Biosens. Bioelectron.*, 17 (2002) 815–819.
46 P. Skladal and M. Mascini, Sensitive detection of pesticides using amperometric sensors based on cobalt phthalocyanine-modified composite electrodes and immobilized cholinesterases, *Biosens. Bioelectron.*, 7 (1992) 335–343.
47 A.L. Hart, W.A. Collier and D. Janssen, The response of screen-printed enzyme electrodes containing cholinesterases to organo-phosphates in solution and from commercial formulations, *Biosens. Bioelectron.*, 12 (1997) 645–654.
48 A.L. Hart and W.A. Collier, Stability and function of screen printed electrodes, based on cholinesterase, stabilised by a co-polymer/sugar alcohol mixture, *Sens. Actuators B*, 53 (1998) 111–115.
49 A.L. Hart and W.A. Collier, A contribution to convenience for enzyme-based assays of pesticides in water. In: S.A. Clark, K.C. Thompson, C.W. Keevil and M.S. Smith (Eds.), *Rapid Detection Assays for Food and Water*, Royal Society of Chemistry, Cambridge, 2001, pp. 80–83.
50 W.E. Brown, A.Y. Shamoo, B.L. Hill and M.H. Karol, Immobilized cholinesterase to detect airborne concentrations of hexamethylene diisocyanate (HDI), *Toxicol. Appl. Pharmacol.*, 73 (1984) 105–109.

51 R. Claycomb, C. Kingston and G. Mein, Commercialising biological sensors for agriculture, *AgEng 2004 'Engineering the Future', International Conference on Agricultural Engineering*, Leuven, Belgium, 2004 [available under 'Commercial Aspects of Biosensors' at www.sensortec.co.nz/science/science.htm].
52 P.T. Kissinger, Biosensors—a perspective, *Biosens. Bioelectron.*, 20 (2005) 2512–2516.
53 K.T. Ulrich, S.D. Eppinger, *Product Design and Development*, 2nd ed., McGraw-Hill, Boston, 2000.

Chapter 29

Rapid detection of organophosphates, Ochratoxin A, and Fusarium sp. in durum wheat via screen printed based electrochemical sensors

D. Compagnone, K. van Velzen, M. Del Carlo, Marcello Mascini and A. Visconti

29.1 INTRODUCTION

29.1.1 Biotic and abiotic contaminants in durum wheat

Durum wheat safety is affected by different threats comprising both abiotic and biotic agents. In the first class, pesticides are widely represented including neurotoxic molecules such as organophosphates and carbamates. Organophosphorus (OP) compounds are substances widely used in agricultural practices as pesticides having low environmental persistence and high efficacy. These compounds act by inhibition of acetylcholinesterase (AChE) activity as neurotoxic agents via an excessive stimulation of the cholinergic receptors in both insects and mammals, including humans [1]. This can lead to various clinical implications and high acute toxicity [2]. Among organophosphates dichlorvos (2,2-dichlorovinyl dimethyl phosphate) and pirimiphos methyl (*O*-[2-(diethylamino)-6-methyl-4-pyrimidinyl]*O*,*O*-dimethylphosphorothioate) are important contaminants for the durum wheat industry. Dichlorvos is one of the most widely used pesticides worldwide in the storage of many products, such as corn, rice, and durum wheat, finding widespread use in most European countries [3]. European Union regulation foresees a maximum residue limit (MRL) for dichlorvos in durum wheat at 2.0 μg/g [4]. Dichlorvos is also classified as a probable human carcinogen on the basis of the effects observed in mice and rats. Therefore, the U.S. Environmental Protection Agency (EPA) proposed cancellation of most uses of dichlorvos and proposed

Comprehensive Analytical Chemistry 49
Alegret and Merkoçi (Eds)
Volume 49 ISSN: 0166-526X DOI: 10.1016/S0166-526X(06)49029-2

restrictions on retained uses. In particular, because of dietary cancer risk, the EPA proposed cancelling uses of dichlorvos in processed agricultural commodities that are stored in bulk, packages, or bags including durum wheat and its derivates such as flour and pasta [5]. Pirimiphos methyl is an organothiophosphate pesticide that is active by contact, ingestion, and vapor action, and causes inhibition of AChE of tissues, determining accumulation of acetylcholine at cholinergic neuro-effector junctions (muscarinic effects), and at skeletal muscle myoneural junctions and autonomic ganglia [6]. As well as other thiophosphate molecules, pirimiphos methyl is active in its oxon product that appears to be the active metabolic intermediary. It is mainly used for storage of grain products such as corn, rice, durum wheat, and sorghum where it shows an excellent persistence as reported in the IPCS-INTOX databank of the World Health Organization (WHO) [7].

Among biotic agents fungi that produce toxins are an important and widespread threat that can contaminate cereals.

Fusarium species, including *Fusarium culmorum*, are universal fungal contaminants of maize, wheat, barley, and other small cereals. *F. culmorum* is a pathogen causing "foot rot" and "head blight" diseases and can produce mycotoxins such as zearalenone, deoxynivalenol, and other trichothecenes that can enter the food chain [8–10]. The early identification of this fungal pathogen is therefore crucial in order to avoid crop losses and protect consumer health [11–13]. Other fungi belonging to *Aspergillus* and *Penicillium* species can contaminate cereals including durum wheat resulting in a possible contamination with ochratoxin A (OTA), (7-(L-β-phenylalanylcarbonyl)-carboxyl-5-chloro-8-hydroxy-3,4-dihydro-3*R*-methylisocumarin). OTA is a mycotoxin produced by several *Aspergillus* and *Penicillium* species growing in different agricultural commodities in the field or during storage [14,15]. OTA has been shown to be nephrotoxic, teratogenic, carcinogenic, and immunotoxic to several animal species. The International Agency for Research on Cancer (IARC) has classified OTA as possibly carcinogenic in humans (Group 2B) [16]. The Joint FAO/WHO Expert Committee on Food Additives (JECFA), after evaluation of OTA nephrotoxicity, proposed for this mycotoxin a provisional tolerable weekly intake (PTWI) of 0.1 μg/kg body mass (equivalent to 14 ng/kg body mass/day) [17].

Recently, the European Commission fixed the maximum levels for OTA at 5 μg/kg in cereals, derivative products, and roasted coffee, 3 μg/kg in all cereal products intended for direct human consumption, 10 μg/kg in dried vine fruits and soluble coffee, and 2 μg/kg in wine, grape juice, and must [18]. Lower levels (0.5 μg/kg) have been established in foods for infants and young children [19].

29.1.2 Screen-printed electrochemical sensors for the detection of acetylcholinesterase inhibitors

OP compounds and carbamate are widely used as insecticides, pesticides, and warfare agents [20,21]. Detection of pesticides is usually carried out by multiresidue methods (MRMs) of analysis, which are able to detect simultaneously more than one residue and have been developed mainly based on chromatographic techniques. Two groups of MRMs are used: (i) multiclass MRMs that involve coverage of residues of various classes of pesticides, and (ii) selective MRMs, which concern multiple residues of chemically related pesticides (e.g., *N*-methyl carbamate pesticides (NMCs), carboxylic acids, phenols, etc.). As foods are usually complex matrices all of the pre-analytical steps (matrix modification, extraction, and clean-up) are often necessary.

The anticholinesterase activity of organophosphate and carbamate has been used as the basis to build up a number of detection schemes for these classes of compounds. A large number of applications in pure standard solution or in environmental samples have been published [22] but only few applications on real food samples have been reported. In spite of the significant demand for sensing pesticides in food samples, it seems that the use of biosensors penetrate very slowly in the food industry. Since the inhibition of AChE by OP is irreversible and because of the matrix fouling of the sensing devices, a cheap, disposable electrode is foreseen as a simplification in the analytical procedure. Single-use biosensors, based on thick- or thin-film technology, are of high interest for real sample analysis due to their low cost, which make them disposable, and their compatibility with portable instrumentation for in-loco measurements.

Most of the inhibition bioassays or biosensors for organophosphate and carbamate pesticides are based on the amperometric detection of the enzymatic product of the reaction.

Some applications of amperometric biosensing strategies for pesticide detection in real or spiked food samples have been recently reported. Most of the applications have been developed for vegetable matrices. Different formats of biosensors have been used: disposable screen-printed choline oxidase biosensors [23] using AChE in solution were utilized to detect pesticides in real samples of fruit and vegetables.

Potato, carrot, and sweet pepper samples spiked with aldicarb, propoxur, carbaryl, carbofuran, and methomyl tested with this AChE biosensor, exhibited acceptable recoveries (79–96%) [24].

An interesting application on spiked (aldicarb, carbaryl, carbofuran, methomyl, and propoxur) fruit and vegetable samples was based on screen-printed electrodes (SPEs) chemically modified with a carbon paste mix of cobalt(II)phtalocyanine and acetylcellulose [25]. In this paper, an appealing solvent-free extraction procedure of the spiked samples is reported. This was performed by mixing the fruits and passing the resulting juice through a sieve. No effect of the matrices pH on the biosensor performance is reported. Screen-printed sensor developed using photolithographic conducting copper track, graphite–epoxy composite, and either AChE or butirrylcholinesterase was also used in the analysis of spiked (paraoxon and carbofuran) samples of tap water and fruit juices at sub-nanomolar concentration [26].

A recent application of a tetracianoquinodimethane (TCNQ)-modified SPE for the development of a biosensing device for chlorpyrifos methyl was also reported. This method was demonstrated to detect the active molecule both in standard solution and in commercial products (Reldan® 22) with comparable sensitivity. The analytical protocol was then applied to grapes and vine leaf samples in order to improve safety in wine-making process [27].

Other food samples that have been recently investigated with electrochemical biosensor based on AChE inhibition are infant foods. The European Union has set a very low limit (10 μg/kg) for pesticides in infant food [28]. An amperometric biosensor that met these limits both for infant food and orange juice has been reported. The method included an oxidation step of phosphorothionates pastiches to produce the oxygenated derivative, which represent the active pesticide molecule. Moreover, the biosensors could be regenerated via the recovery of AChE activity through a chemical activation with pyridine-2-aldoxime methochloride (PAM). The biosensors performed well in solvent extract containing water, though they exhibited a reduced recovery in food with a lower water content [29].

Following work by the same group addressed some of the major problems arising when electrochemical biosensors are in contact with food matrices: pH effect and particle effect. Both problems were solved treating the biosensor surface with a Tween20®/phosphate buffer solution (pH 7.5) after the incubation with pesticide. The treatment was successful in removing the particulate, the correct pH for enzyme activity measurement was attained and the pesticide enzyme inhibition retained. A large number of samples were analyzed and the results were in agreement with reference standard methods [30].

To our knowledge, there are only few reports on application of screen-printed cholinesterase-based biosensors to food samples. Many other amperometric detection schemes and chemistries have been recently investigated that could enhance the overall analytical performance of screen-printed devices. Nevertheless, the possibility to successfully apply these approaches to real food samples is strictly linked to their ability to surmount the matrix effect that leads either to complicated multi-step sample preparation or to poor recovery.

Carbon nanotubes (CNTs) are an emerging class of material due to their exceptional structural and electronic characteristics [31]. The possibility of the promotion of electron transfer at a lower overpotential and their high surface area provide the basis for improving biosensing systems [32]. They catalyze the electrochemical reaction of important analytes involved in biosensor development such as NADH [33], thiols [34], and hydrogen peroxide [35]. Moreover, CNTs appear to be an interesting material for the immobilization of biological molecules [36–38] in biosensor applications [39–41]. The fabrication and evaluation of CNT-derived screen-printed disposable electrochemical sensors based on a CNT ink has been reported [42]. CNTs have recently been used in the development of a disposable biosensor based on thick film-strips for the sensitive detection of OP pesticides. In this work, the dual role of CNT, electrocatalytic activity toward thiocholine and immobilization matrix for the enzyme, is demonstrated leading to the development of a redox mediator-free, simple, and robust single-enzyme biosensor able to detect 0.5 nM of paraoxon in solution [43].

An interesting and emerging technology that has led to astonishing results in OP determination is the use of sonochemical fabrication of SPE to obtain micro-electrode array. This technology result in the ablation of non-conductive polymer films that coat, and so insulate, underlying conductive surfaces. Sensors were fabricated via the electropolymerization of an insulating film at planar electrode surfaces and then ablated with ultra-sound to expose microdiameter scale areas of underlying conductor. A second polymer of polyaniline with conducting properties, carrying co-entrapped AChE, may then be electrodeposited *in situ* at the micro-electrode cavities forming an enzyme micro-electrode array. With this device the authors obtained a stir-independent response, as expected for micro-electrodes able to detect as low as 10^{-20} M of dichlorvos. The authors claim that this astonishing result is due to the combined effect of enzyme inhibition amplification because of the biosensor geometry and the use of recombinant AChE selective for the particular pesticide used [44].

To complete this overview on the strategies that are pursued to enhance the sensitivity and stability of screen-printed biosensors for the detection of pesticides new immobilization approaches have to be mentioned. SPE modified with concanavalin A have been used to bind, via a high affinity interaction, AChE. The obtained biosensors, optimized for manufacturing conditions, were able to detect 10^{-8} M of clorpyrifos [45].

Vakurov *et al.* [46] evaluated a strategy to improve the covalent binding of AChE to screen-printed carbon electrodes modified with polyamines. To improve the extent of dialdehyde modification, electrodes were aminated. Initially, this was performed by electrochemical reduction of 4-nitrobenzenediazonium to a nitroaryl radical permitting attachment to the carbon surface; subsequent reduction of the 4-nitrobenzene yielded a 4-aminobenzene-modified carbon surface. The obtained biosensors resulted in very sensitive devices measuring as low as 10^{-10} M of OPs.

The application of such interesting technologies to real food samples has to overcome the difficulties of extraction strategies and the problems arising from matrix components that can affect the biosensor performance. In the recent years, our group have been involved in the realization of biosensing strategies based on SPEs for the detection of dichlorvos and pirimiphos methyl in durum wheat. The proposed methods were all based on a choline oxidase biosensor obtained by modification of carbon SPEs able to determine AChE activity in solution. In order to avoid cumbersome regeneration steps, we decide to use free cholinesterases rather than immobilized on the choline oxidase biosensor surface.

We successfully applied an AChE inhibition assay to the detection of dichlorvos in durum wheat samples using a simplified extraction procedure. The aqueous extraction solvent (phosphate buffer) was used as the assay buffer, thus simplifying the overall procedure and limiting the negative effects of organic solvents on enzyme activity. The total assay time, including the extraction step, was 30 min.The choline oxidase biosensor exhibited an excellent stability: after 20 days from preparation, the blank measurement lost only 10% of the signal intensity. The calculated limits of detection (LODs) in buffer solution were 10 and 0.05 ng/mL, respectively, using either electric eel AChE (eeAChE) or a recombinant AChE specifically engineered to be inhibited by dichlorvos (rAChE) [47].

The developed assay was also applied to milled samples. An optimized extraction protocol using exane as extraction solvent exhibited recoveries of dichlorvos at 0.25–1.50 μg/g from 96.5% to 100.9%, with a LOD of 0.02 μg/g. An aliquot of the filtered hexane extract was partitioned with phosphate buffer solution, and the organic layer was evaporated prior to electrochemical analysis. An LOD of 0.05 μg/g of dichlorvos was obtained with mean recoveries of 97–103% at spiking levels of 0.25–1.50 μg/g. A good correlation (0.9919) was found between the results obtained with the electrochemical and those obtained with the gas chromatographic methods. The electrochemical method was peer-validated in two laboratories that analyzed 10 blind samples (five duplicates), including a blank and four spiked samples with dichlorvos at levels of 0.25, 0.60, 1.00, and 1.50 μg/g. Within-laboratory repeatability (RSD_r) and between-laboratory reproducibility (RSD_R) ranged from 5.5% to 7.8% and from 9.9% to 17.6%, respectively [48].

The possibility of detecting a phosphotionate pesticide in durum wheat has been also investigated. The determination was accomplished via chemical oxidation of the phosphothionate molecule both in buffer and in matrix extract solution optimizing the experimental parameters (reagents concentration and reaction time). The procedure was then applied to standardize the pirimiphos methyl detection obtaining the calibration curves under different conditions. The LOD with matrix extract was 50 or 100 ng/mL, depending on the extract % addition, which allowed the detection of samples contaminated at the MRL = 5 mg/kg. The samples mean recovery was 70.3% and no false positive samples were detected [49].

29.1.3 Screen-printed electrochemical DNA sensors for identification of microorganisms

In the last 20 years, there has been a continuous increase in the use of nucleic acid combined with electrochemical transducers to produce a new kind of affinity biosensor. Among the different kind of electrochemical sensor formats available, SPE based on thick and thin film technology have played an important role. This is surely due to their recognized advantages in terms of cost that allow their disposable use.

As well as other molecular-based biosensors DNA biosensors rely on highly specific recognition events to detect their target analytes. The role of the transducer is to provide a suitable platform that facilitates formation of the probe-target complex in such a way that the binding event triggers a usable signal for electronic readout. According to IUPAC definition, a biosensor includes a molecular recognition layer and a signal transducer that can be coupled to an appropriate readout device. DNA is especially well suited for biosensing applications, because the base-pairing interactions between complementary sequences are both specific and robust.

Different strategies of electrochemical sensing have been used for DNA electrochemical detection: (1) direct DNA electrochemistry, (2) indirect DNA electrochemistry, (3) DNA-specific redox indicator detection, (4) DNA-mediated charge transport, and (5) nanoparticle-based electrochemistry amplification.

Direct DNA electrochemistry allows highly sensitive detection of DNA, down to femtomoles of target sequence without a labelling step, which simplifies the overall detection protocol, the main limitation of this approach being the high background signals that often limits a simple readout. This strategy has been applied to different kinds of electrodes in terms of geometry and material ranging from the hanging mercury drop electrode [50], carbon paste electrode [51], to screen-printed carbon transducers [52]. Methods to oxidize target DNA indirectly through the use of electrochemical mediators have also been explored. An especially attractive approach uses polypyridyl complexes of Ru(II) and Os(II) to mediate the electrochemical oxidation of guanine [53]. This detection approach provide high sensitivity without complex instrumentation through redox-mediated indirect DNA oxidation, the main limitation consisting in the electrode preparation that can be difficult to handle. Several applications have been described in which target DNA sequences are labelled with redox-active reporter molecules. Appearance of the characteristic electrochemical response of the redox reporter signals the occurrence of the hybridization event. Using physical separation methods to isolate the labelled sequences, LODs of the order of $\sim 10^{-10}$ molecules have been reported [54,55]. As an alternative to chemical labelling schemes, DNA-mediated charge transport electrochemistry has been exploited as detection scheme using redox-active molecules that non-covalently associate with the double helix. In these analyzes, rather than serving as a reactant, the DNA is the mediator. These assays can provide high sensitivity and simplicity. Nanoparticle-based electrochemistry amplification is another detection scheme that has been exploited. In this case, intercalative probe molecules are used taking advantage of the electronic structure of double-helical DNA, using intercalated redox probe molecules to report the perturbations occurring in base stacking [56]. In this format the intercalator is not used to report the amount of DNA or whether it is double stranded versus single stranded; here the DNA base pair stack mediates charge transport to the intercalator bound to the DNA. The method relies on the switch of inherent characteristic of duplex DNA and its ability to mediate charge transport is used for the detection. Assays of DNA-mediated electrochemistry, using both DNA-mediated charge transport, and nanoparticle-based electrochemistry amplification, are therefore uniquely suited to sense changes in DNA: damage, mistakes, mismatches, and even protein binding [57,58].

DNA biosensors, based on the summarized schemes, have been proposed in different area of analysis: environmental, clinical, food control, biotechnology, and pharmaceutical.

In this report, we review some recent applications of electrochemical DNA screen printed based biosensor for the detection of microorganisms including, viruses, bacteria, and fungi. An electrochemical hybridization biosensor based on the intrinsic oxidation signals of nucleic acids (guanine oxidation) and proteins (tyrosine and tryptophan) has been designed, that makes use of the unique binding event between *Escherichia coli* single-strand binding (SSB) protein and single-stranded DNA (ssDNA). The voltammetric signal from guanine oxidation significantly decreased upon binding of SSB to single-stranded oligonucleotides (probe), anchored on a single-walled CNT-modified screen-printed carbon electrode. Simultaneously, oxidation of the tyrosine and tryptophan residues of the SSB protein increased upon binding of the SSB protein to ssDNA and ss-oligonucleotides. The LOD of 0.15 μg/mL of target DNA can be applied to genetic assays [59]. Some work devoted to food safety resulted in the realization of an electrochemical biochip able to specifically detect *Bacillus cereus*. *B. cereus* constitutes a significant cause of acute food poisoning in humans. The DNA biosensor was specifically designed using a capture probe for the toxin-encoding genes. A bead-based sandwich hybridization system was obtained in conjugation with electric chips for detection of both vegetative cells and spores of *Bacillus* strains. The system consisted of a silicon chip-based potentiometric cell, and utilizes paramagnetic beads as solid carriers of the DNA probes. The specific signals from 20 amol of bacterial cell or spore DNA were achieved in less than 4 h. The method could be also applied directly to unpurified spore and cell extract samples. The developed method can offer a contribution in the rapid identification of *Bacillus* strains without preceding nucleic acid amplification [60]. A recent advancement in DNA biosensor consisted in the realization of an electrochemical microfluidic biosensor with an integrated minipotentiostat for the quantification of RNA based on nucleic acid hybridization and liposome signal amplification for dengue virus detection. The detection scheme was ensured by short DNA probes that hybridize with the target RNA or DNA sequence. A reporter probe, coupled to liposomes entrapping the electrochemically active redox couple potassium ferri/ferrohexacyanide, was used to amplify the recognition reaction. The capture probes were bound to superparamagnetic beads that were isolated on a magnet in the biosensor. Upon hybridization reaction, the liposomes were lysed to release the electrochemical markers. The detection occurred at interdigitated ultramicroelectrode array. The system was completed by a miniaturized instrumentation (miniEC). The functionality of the miniEC was successfully demonstrated with the detection of dengue virus RNA [61]. A biosensor for virus detection, particularly for SARS (severe acute respiratory syndrome) virus, has been described by Abad-Valle *et al.* [62]. The functioning scheme was a typical hybridization biosensor constructed on a gold thin-film electrode. The selected target, a 30-mer sequence, was chosen in a lysine-rich region, unique to SARS virus and the complementary strand, the probe, was immobilized to the gold surface. Hybridization reaction with the biotin-conjugated SARS strand (at the 3′-end) was obtained and monitored using alkaline phosphatase (AP)-labelled streptavidin that permitted amplified indirect electrochemical detection. The analytical signal is constituted by an electrochemical process of indigo carmine, the soluble product of the enzymatic hydrolysis of 3-indoxyl phosphate. The use of a sensitive electrochemical technique such as square wave voltammetry allowed an LOD of 6 pM to be obtained for this DNA sequence [62]. A multiple pathogen detection system based on hybridization coupled with bioelectronic detection was described by Vernon *et al.* [63]. The method was developed using papillomaviruses as a model system. In this case, two chips were spotted with capture probes consisting of DNA oligonucleotide sequences specific for HPV types and two electrically conductive signal probes were synthesized to be complementary to a distinct region of the amplified HPV target DNA. The measuring system was able to successfully detect 86% of the HPV types contained in clinical samples demonstrating the feasibility of integrated detection of multiple pathogens.

Recently, our group has devoted some research activity to the realization of DNA screen-printed biosensors for the detection of fungi [64]. To our knowledge there is no evidence in the literature of analysis of fungi via electrochemical detection of DNA or RNA.

In our research, we proposed a post amplification analytical method to detect *F. culmorum*, a pathogen causing ''foot rot'' and ''head blight'' diseases in cereals, and can produce mycotoxins such as zearalenone, deoxynivalenol, and other trichothecenes that can enter the food chain. The early identification of this fungal pathogen is therefore recommended in order to avoid crop losses and protect consumer health [65].

The sensing principle of this work is based on the hybridization reaction between a probe and the complementary target sequence. The hybridization is followed by the electrochemical detection using a square wave voltammetry (SWV). The direct label-free detection was accomplished by monitoring the guanine oxidation peak of the target sequence relying on the use of inosine-modified (guanine-free) probes [66–68]. The specificity and selectivity of the methods aroused from the selection of four different probes immobilized on sensor surface, complementary to different regions of the amplified sequence. A sequence was identified that is able to grant selectivity and specificity to the sensing system.

29.1.4 Screen-printed electrochemical immunosensors for the detection of toxins

Immunosensors are analytical devices, which selectively detect analytes and provide a concentration-dependent signal [69]. Electrochemical immunosensors employ either antibodies or their complementary binding partners (antigens) as biorecognition elements in combination with electrochemical transducers [70]. Among electrochemical transducers, SPE appears as the most useful due to their characteristics [71]. SPEs are mass produced and therefore the single device has a low cost as compared to conventional electrodes, hence they can be used in a disposable manner, avoiding cumbersome regeneration steps following the sample measurement. They have a small size thus reducing the volume of sample that can be analyzed (20–100 μL).

Some work has been carried out from Prof. Palleschi's group in Rome leading to the development of disposable immunosensor for the determination of domoic acid (DA) in shellfish [72]. DA is neurotoxic amino acid responsible for the human syndrome known as "amnesic shellfish poisoning" (ASP). The method involves the use of disposable SPEs for the immunosensor development based on a "competitive indirect test". DA conjugated to bovine serum albumin (BSA–DA) was coated onto the working electrode of the SPE, followed by incubation with sample (or standard toxin) and anti-DA antibody. An anti-goat IgG–alkaline phosphatase (AP) conjugate was used for signal generation via the use of an electrochemical substrate, naphtyl phosphate. The enzyme product, naphtol was detected by differential pulse voltammetry (DPV). The immunosensors allowed the detection of 20 μg/g of DA in mussel tissue. Moving toward toxins produced by fungi, the same group has devised an electrochemical immunosensor for aflatoxin M1. The analyzed matrix was milk [73]. The immunosensors were fabricated by immobilizing AFM1 antibodies directly on the surface of SPEs, and allowing the competition to occur between free AFM1 and that conjugated with peroxidase (HRP) enzyme. The electrochemical technique used was the chronoamperometry, performed at −100 mV. Results have shown that using SPEs aflatoxin M1 can be measured with an LOD of 25 ppt and with a working range between 30 and 160 ppt. Moreover, the authors claim that interference problems have been addressed for the direct analysis of aflatoxin M1 in milk.

Another micotoxin that is widely present in food commodities is OTA, 7-(L-b-phenylalanylcarbonyl)-carboxyl-5-chloro-8-hydroxy-3,4-dihydro-3R-methylisocumarin. It is produced by several *Aspergillus* and *Penicillium* species. It is considered as secondary toxic metabolite with nephrotoxic, teratogenic, carcinogenic, and immunotoxic activity in several animal species. Immunosensing strategies have been proposed to be able to detect OTA in different matrix, such as wine [74] and durum wheat [75]. In the former matrix, the assay was carried out on carbon-based SPEs. The immunosensors were developed using polyclonal antibodies. The assay gave an LOD of 180 pg/mL and sensitivity of 6.1 ± 0.1 ng/mL. The immunosensors were challenged with wine to assess a matrix effect. Recoveries obtained were in the 70–118% range.

A more challenging matrix was faced in the latter research [75] carried out in collaborations with authors' groups. In this case, the immunosensors were realized using monoclonal antibodies, and two formats of immunosensors compared. The immunosensor in the direct format was then used for the determination of OTA in wheat. Samples were extracted with aqueous acetonitrile and the extract analyzed directly by the assay without clean up. The I_{50} in real samples was 0.2 μg/L corresponding to 1.6 μg/kg in the wheat sample with a LOD of 0.4 μg/kg (calculated as blank signal—3σ). Within- and between-assay variability were less than 5% and 10%, respectively. Results obtained on naturally contaminated wheat samples were compared with a reference method with a good correlation ($r = 0.9992$).

29.2 APPLICATION

In this report, we describe the application of SPEs to the determination of different contaminants in durum wheat samples. All the reported applications have been developed using PalmSens hand-held potentiostat equipped with PalmSens PC software for the elaboration of current data (PalmSens, Amsterdam, The Netherlands) (Fig. 29.1).

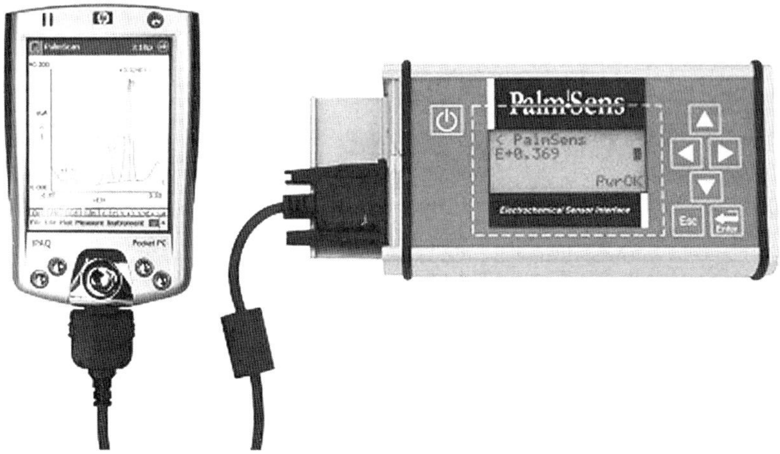

Fig. 29.1. Hand-held potentiostat by Palmsens used in all the optimization protocols described in the text.

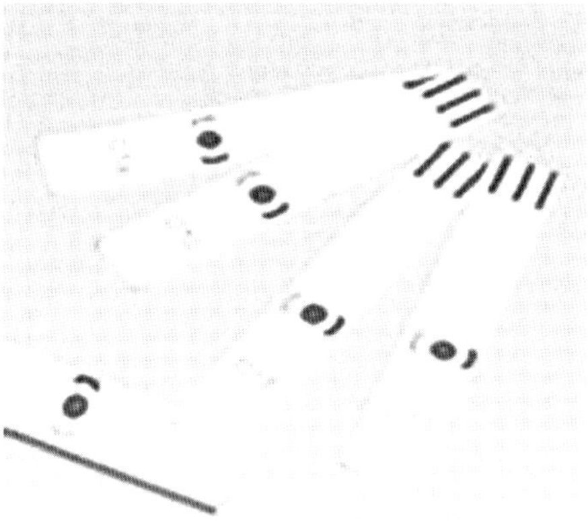

Fig. 29.2. Screen-printed electrodes used in all the proposed applications.

The different biosensors were obtained using thick-film SPEs produced by Biosensor Laboratory, University of Florence and commercialized by PalmSens. The electrochemical cell, consisting of a graphite working electrode and silver counter and pseudo-reference electrodes, was printed on a planar polyester substrate (Fig. 29.2).

The applications (Screen-printed electrochemical sensors for the detection of AChE inhibitor, Screen-printed electrochemical DNA sensors for identification of microorganisms, and Screen-printed electrochemical immunosensors for the detection of toxins) have been finalized to

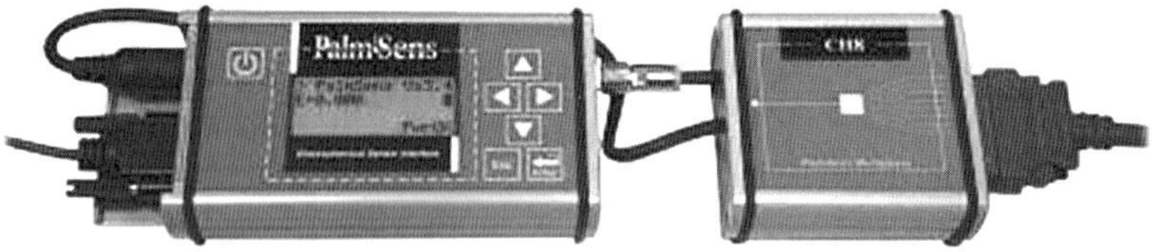

Fig. 29.3. The Palmsens hand-held potentiostat equipped with the dedicated multiplexer for eight-channel measurements.

the assembly of a dedicated instrumentation and software able to handle delocalized analysis of OPs, *Fusarium* sp. DNA, and OTA.

The instrumentation consisted of the hand-held potentiostat interfaced with a CH8 multiplexer (PalmSens) (Fig. 29.3) that allow different sensor configuration:

a. Sensorarrays with eight working and eight combined reference/counter electrodes
b. Sensorarrays with eight working electrodes sharing a reference and a counter electrode
c. Sensorarrays with eight working electrodes sharing a combined reference/counter electrode

The software front page reported in Fig. 29.4 enables the choice of the analytical application among pesticides, *Fusarium* sp., and ochratoxin.

Each application allows the use of eight independent channels that depending on the procedure can be used to perform individual or duplicate analysis. For instance, in the OTA application the default setting of the software permit to dedicate two channels for the blank measurement and two channels for each of the three calibrators or samples whereas in the OPs protocol the user can utilize single channel or multiple channels. In each application, a check biosensor option is available to test the correct functioning and positioning of the sensors.

In OTA and OPs protocols calibration and determination windows are present. The user can either create his/her own three-point experimental calibration, which can be saved for further use, or load external data to obtain the calibration curve. The concentration levels of the standard solutions are then used by the software to classify the sample result according to a "traffic light scheme" (red light for sample exceeding the legal limit, yellow for samples that need analytical confirmation, and green light for negative samples).

The *Fusarium* sp. DNA application consists of a check biosensor tool and a PCR sample window. The check biosensor enables the control of

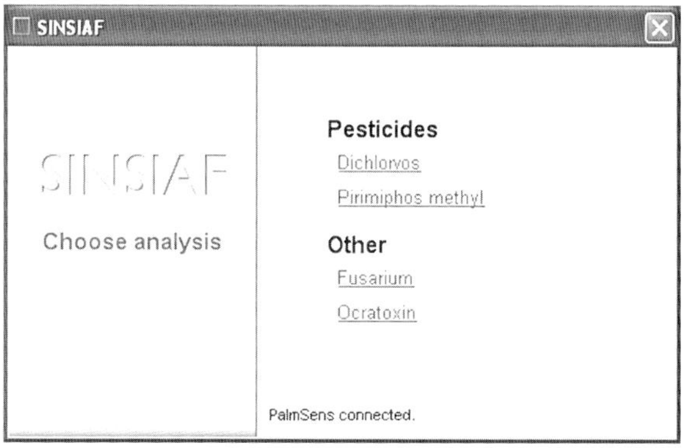

Fig. 29.4. Dedicated software for durum wheat safety control, the front page enables the choice of the experimental procedure to be carried out.

the probe immobilization procedure, then in the PCR sample window the amplified samples are measured. Four out of the eight channels are used as "negative" control and four for sample measurement. The software, using an internal algorithm, allows the discrimination between positive and negative samples.

29.2.1 Screen-printed electrochemical sensors for the detection of dichlorvos and pirimiphos-methyl

Experimental details are reported in Procedure 42 (CD accompanying this book).

Both the determinations reported here rely on the inhibition activity of OPs pesticides toward AChE combined with the detection of the AChE enzyme product choline at the surface of a mediator-modified screen-printed choline oxidase electrochemical biosensor.

The biochemical–electrochemical pathway used to determine the inhibition consisted of two enzymatic reactions (Eqs. (29.1) and (29.2)) generating a chemical oxidation (Eq. (29.3)) that was determined by cathodic chronoamperometry (Eq. (29.4)). The used iron containing electrochemical mediator was the widely used Prussian blue.

$$\begin{array}{l}\text{Acetylcholine} + H_2O \rightarrow \text{acetic acid} + \text{choline} \\ (\text{enzyme I}: \text{ AChE})\end{array} \tag{29.1}$$

$$\begin{array}{l}\text{Choline} + 2O_2 + H_2O \rightarrow 2H_2O_2 + \text{betaine} \\ (\text{enzyme II}: \text{ choline oxidase})\end{array} \tag{29.2}$$

$$Fe(II) + H_2O_2 \rightarrow Fe(III) + 2OH^-(\text{chemical oxidation}) \quad (29.3)$$

$$Fe(III) + e^- \rightarrow Fe(II)(\text{electrochemical reduction}) \quad (29.4)$$

Standard and sample extract solutions were analyzed according to the following experimental scheme: first, the current intensity of a blank sample extract was measured, and then the current intensity of either the standard or the sample extract was measured. The current intensity of the blank sample ($I0$) and of the contaminated sample or standard solution (II) was used to calculate the percent inhibition according to the following formula:

$$I(\%) = 100\frac{(I0 - II)}{I0} \quad (29.5)$$

Electrochemical experiments were carried out using a hand-held potentiostat equipped with dedicated software for the elaboration of current data (Fig. 29.5). The current was sampled 2 min after the reaction started. Figure 29.5 shows the chronoamperogram where current versus time is plotted. The current shape was reproducible from time to time and hence the precision assured sampling the current, 2 min after the start of the measurement. After 2 min the current was continuously increasing due to the residual activity of AChE in solution. The biosensor

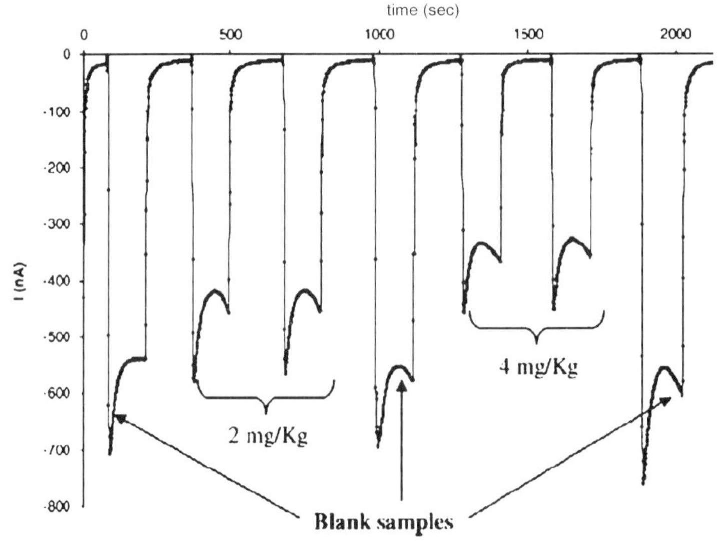

Fig. 29.5. Typical amperogram recorded for durum wheat extract at 2 and 4 mg/kg using the hand-held potentiostat. Reprinted with permission from Ref. [47].

used to detect the extent of the AChE inhibition was a choline oxidase biosensor assembled on thick-film electrode.

29.2.1.1 Dichlorvos determination in durum wheat samples

Here we report experimental data obtained using the proposed electrochemical assay with screen-printed choline oxidase biosensor for the detection of dichlorvos in durum wheat samples. As described in P1E9, two different extraction approaches have been optimized: the former using whole wheat kernels and aqueous extraction and the latter using grounded samples and hexane extraction. Both studies were carried out using free cholinesterases because the aim of the protocol was to devise a rapid procedure that did not include cumbersome regeneration steps, which were unavoidable with immobilized AChE.

The former approach led to a simplified extraction protocol where the extraction solvent was then used as assay buffer.

In this application, the use of wild-type electric eel AChE and a recombinant AChE, specifically selected as very sensitive to dichlorvos, was compared. The effect of the matrix extract was determined by using various sample: solvent ratios, 1:2.5, 1:5, 1:10, and 1:20. The optimal extraction ratio, considering the electrochemical interferences and the effect on enzyme activity and bioavailability of the pesticide, was 1:10.

The method was calibrated both in buffer and durum wheat extract. The LODs in durum wheat samples were 0.45 mg/kg for the wild-type AChE and 0.07 mg/kg for rAChE. These characteristics allowed the detection of contaminated samples at the legal MRL, which is 2 mg/kg [4]. Moreover, fortified samples of durum wheat were obtained with both dichlorvos and the commercial product Didivane, which contains dichlorvos as active molecule. At all the tested levels, the occurrence of contaminant was detected with an average recovery of 75%. The total assay time, including the extraction step, was 30 min. Because several extractions as well as most of the assay steps can be run simultaneously, the throughput for one operator is 12 determinations per hour. In Table 29.1, we summarize the results obtained for the fortified samples.

Samples ($n = 55$) fortified at different levels (4, 2, 1, and 0.5 mg/kg) were extracted and analyzed with the proposed electrochemical assay using both eeAChE and rAChE. For eeAChE, the recovery for dichlorvos-spiked samples ranged between 74 and 78%. The incomplete recovery may be due to concurrent causes, such as the use of whole kernels as sample, water-based extraction solvent, and adsorption

TABLE 29.1

Recovery study on wheat samples fortified at various levels with dichlorvos and Didivane using both eeAChE and rAChE

Fortification level (mg/kg)	eeAChE			rAChE		
	Inhibition (%)[c]	Calc. (DDPV) (mg/kg)	Recovery (%)	Inhibition (%)[c]	Calc. (DDPV) (mg/kg)	Recovery (%)
4.0 ($n = 15$)[a]	42 ± 3	3.1 ± 0.1	78 ± 3	NA[d]	NA	NA
2.0 ($n = 15$)[a]	27 ± 1	1.5 ± 0.1	77 ± 5	70 ± 1	1.5 ± 0.2	74 ± 2
2.0 ($n = 15$)[b]	26 ± 1	1.5 ± 0.1	75 ± 5	68 ± 1	1.4 ± 0.1	69 ± 2
1.0 ($n = 5$)[a]	16 ± 1	0.8 ± 0.1	77 ± 10	46 ± 2	0.7 ± 0.1	74 ± 4
0.5 ($n = 5$)[a]	8 ± 2	0.4 ± 0.1	74 ± 20	26 ± 1	0.4 ± 0.1	72 ± 3

Reprinted with permission from Ref. [47].
[a]Fortified with dichlorvos.
[b]Fortified with Didivane.
[c]Inhibition ($I\%$) is the analytical signal.
[d]NA, not available.

exerted by the matrix. Although the advantage of using the extraction solvent as measuring buffer has an important influence on the speed of the assay, in order to detect dichlorvos at 0.5 mg/kg with better accuracy, it was necessary to use the rAChE; in that case, an *I*% of 26.3% was obtained, which is well above the minimum detectable inhibition of the method (8%). Moreover, the recovery of the samples spiked with Didivane was comparable to that obtained with the pure molecule, dichlorvos. The recovery data suggest that the buffer used in the extraction protocol allowed a repeatable recovery at an applicable contamination level of both dichlorvos in pure form and dichlorvos in commercial formulation (Didivane), thus avoiding the use of a polar organic solvent such as methanol or acetonitrile, which may influence enzyme activity and sensor stability.

Despite aqueous-based extraction using whole wheat kernels has been demonstrated as effective to detect dichlorvos, the procedure might lead to inadequate homogeneity of testing samples. Moreover, aqueous solvents cannot be used for extraction of ground wheat samples due to the formation of a slurry, preventing filtration of adequate amounts of extract. To apply the electrochemical method to ground wheat samples, as generally required in food analysis, a non-aqueous extraction solvent was required.

We have hence tested a number of extraction solvents that could be used with the grounded samples for dichorvos extraction and then easily coupled, via liquid–liquid partitioning with the assay procedure.

To perform electrochemical analysis with the choline oxidase biosensor, dichlorvos needed to be transferred to PBS solution, thus avoiding any electrochemical interference by organic solvents.

Therefore, the filtered hexane extract was submitted to liquid–liquid partitioning with PBS and the upper organic layer was removed by evaporation. Finally, the buffer solution containing dichlorvos was analyzed by the biosensor. Dichlorvos was easily measured in ground wheat by electrochemical bioassay at levels as low as 0.05 μg/g. The method was peer-validated by two laboratories, and the results of the validation test are reported in Table 29.2.

As shown in Table 29.2, mean recoveries of dichlorvos ranged from 97% to 108%, RSD_r values ranged from 5.5% to 7.8%, and RSD_R values ranged from 9.9% to 17.6%. The dichlorvos mean recovery was calculated for each spiking level as the mean of four measurements, $n = 4$ (Table 29.2). No false negative or false positive results were obtained by the electrochemical assay. A good correlation between dichlorvos concentrations obtained by electrochemical biosensor and GC analysis was

TABLE 29.2

Results of the validation test performed by two laboratories for determination of dichlorvos in ground durum wheat by electrochemical bioassay

	Spiked level				
	Blank	0.25 μg/g	0.60 μg/g	1.00 μg/g	1.50 μg/g
Lab 1, rep 1	Nd[a]	0.27	0.54	1.03[b]	1.71[b]
Lab 1, rep 2	Nd	0.24	0.48	1.18[b]	1.64[b]
Lab 2, rep 1	Nd	0.29	0.67	0.99[b]	1.38[b]
Lab 2, rep 2	Nd	0.29	0.63	0.93[b]	1.48[b]
Mean recovery (%)	–[c]	108	97	103	103
S_r	–	0.015	0.036	0.081	0.061
RSD_r (%)	–	5.5	6.2	7.8	3.9
S_R	–	0.027	0.102	0.117	0.179
RSD_R (%)	–	9.9	17.6	11.4	11.5

Reprinted with permission from Ref. [48].

[a]Not detected, LOD = 0.05 μg/g.

[b]Data calculated with calibration curve obtained in diluted (1:3) blank extract.

[c]Not applicable.

also found as shown in Fig. 29.6. The correlation coefficient (r) was 0.9919.

29.2.1.2 Pirimiphos-methyl determination in durum wheat samples

The electrochemical assay for the detection of AChE inhibitors outlined above has been optimized for the detection of pirimiphos methyl-pirimiphos-methyl in durum wheat. Pirimiphos methyl is a phosphotionate insecticide and therefore it requires to be transformed in the corresponding oxo-form to act as an effective AChE inhibitor, in fact it did not show inhibition of AChE in a concentration range 50–5000 ng/mL (data not shown).

The procedure for the oxidation of pirymiphos methyl via *N*-bromosuccinimide (NBS) and AChE inhibition was optimized in buffer solution for reagents, concentration, and inhibition time (see Procedure 43 in CD accompanying this book for more details).

As a compromise, between analytical performance and overall assay length, a 10-min incubation of NBS and ascorbic acid (AA) and 30-min AChE incubation was selected. A statistical analysis, using stepwise general least squares analysis (SGLSA), of the data demonstrated that only AChE incubation time had a significant and positive influence on the inhibition of AChE.

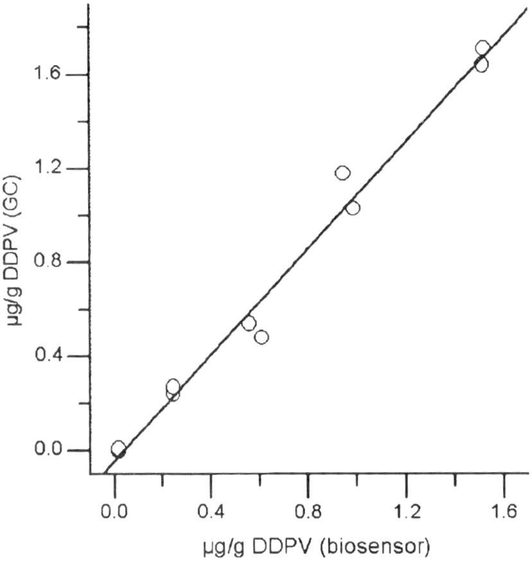

Fig. 29.6. Comparison of dichlorvos (DDPV) contents in spiked durum wheat samples analyzed by GC and electrochemical biosensor. Reprinted with permission from Ref. [48].

The method was calibrated both in buffer solution and in durum wheat extract. The intra-electrode RSD (%) ranged between 1.6 and 15.0, whereas the inter-electrode RSD (%) was comprised between 4.6 and 15. The LOD was 38 ng/mL, and the *I*50% was 360 ng/mL. The assay conditions were then re-optimized to work with durum wheat extracts and calibrations were obtained under different experimental conditions such as sample pretreatment (milled or whole grains) and extract concentration (2% or 4%). The calibrations were slightly affected by the sample matrix resulting in an increased LOD (65–133 ng/mL) and *I*50% (640–1650 ng/mL). The LOD referred to the sample, determined using the best operational condition, was 3 mg/kg.

Spiked samples were prepared at the EU regulated level (5 mg/kg) and analyzed with the optimized protocol resulting in an average recovery of 70.3%.

Acetone and methanol were tested for the extraction and the effect on the assay compared (Fig. 29.7).

Acetone appeared to strongly affect the enzyme activity and was therefore discarded, whereas methanol did not strongly influence the enzyme and hence was used for sample extraction.

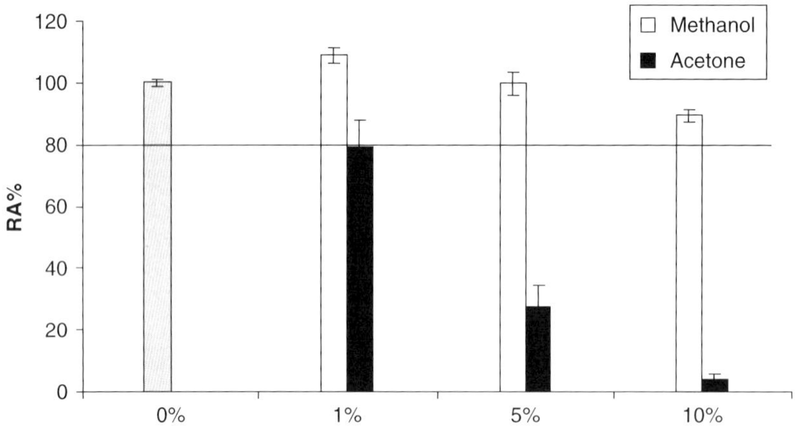

Fig. 29.7. Effect on the residual enzyme activity of methanol and acetone at different concentration (0–10%) in phosphate buffer. Redrawn with permission from Ref. [49].

TABLE 29.3

Recovery data of spiked samples ($n = 9$)

	I%	Recovery (%)
S1	15.6	80.0
S2	12.7	69.0
S3	14.3	75.0
S4	12.7	69.5
S5	13.5	72.5
S6	11.3	63.5
S7	14.8	76.5
S8	12.7	69.0
S9	9.9	58.5
Mean	13.1 ± 1.7	70.3 ± 7.0

I% values are the mean of three measurements for each sample, $8 < CV < 15$. The recovery has been calculated with respect to the spiked level (5 mg/kg). Reprinted with permission from Ref. [49].

Once the oxidation of the phosphothionate pesticide was optimized, the assay was applied to nine spiked samples resulting in acceptable recovery (Table 29.3). The spiking concentration was 5 mg/kg for all the samples, which correspond to the legal limit settled by the European Union.

In this application, the samples were extracted using pure methanol, which was then directly diluted in the assay buffer.

The assay scheme used to detect AChE inhibitors is not selective, as a great number of molecules inhibit AChE.

The methods here proposed are "target oriented"; that is the sought for analyte is known to the analyst. This occurs in the situations where the grower must use a certain production protocol, including pesticide treatments, as stated by contract. In the case where the analyte is not known, the assay can be used as a toxicity test able to detect the presence of a total anticholinesterase activity (TAA).

29.2.2 Screen-printed electrochemical sensors for the detection of ochratoxin in durum wheat

Competitive electrochemical enzyme-linked immunosorbent assays based on disposable SPEs have been developed for quantitative determination of OTA.

Indirect and direct formats of immunosensors-based assay were developed. A total of 6 μL of OTA–BSA in buffer (indirect format) and 6 μL of goat IgG (anti-mouse IgG) (direct format) in buffer were dispensed on the graphite-based screen-printed working electrodes and kept overnight at 4°C. Another 6 μL of 1% PVA solution was used to block the surface for 30 min at room temperature. Also, 6 μL of OTA monoclonal antibody were added to the electrode surface for 30 min at room temperature. Binding or competition was run with 6 μL of OTA–AP conjugate or conjugate+standard for 30 min at room temperature. Washing was then carried out. The activity of the label enzyme was measured electrochemically by the addition of 100 μL of substrate solution (5 mg/mL 1-naphthyl phosphate in DEA buffer; prepared daily), for 2 min at room temperature. The enzymatic product, 1-naphthol, was detected by differential pulse voltammetry (DPV) using the following conditions: potential range 0–600 mV, pulse width 60 ms, pulse amplitude 50 mV, and scan speed 50 mV/s (Fig. 29.8).

The assays were carried out using monoclonal antibodies in the direct and indirect formats. OTA working range, I_{50} and LODs were 0.05–2.5 μg/L and 0.1–7.5 μg/L, 0.35 (±0.04) μg/L and 0.93 (±0.10) μg/L, 60 μg/L and 120 μg/L in the direct and indirect assay formats, respectively. The immunosensor in the direct format was selected for the determination of OTA in wheat. Samples were extracted with aqueous acetonitrile and the extract analyzed directly by the assay without clean-up. The I_{50} in real samples was 0.2 μg/L corresponding to 1.6 μg/kg in the wheat sample with an LOD of 0.4 μg/kg (calculated as blank signal—3σ). Within- and between-assay variability were less than 5%

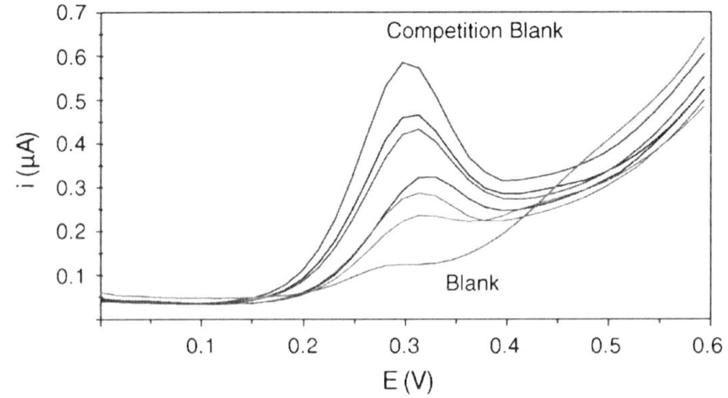

Fig. 29.8. Differential pulse voltammograms obtained for the samples assayed (one representative voltammogram for each of the spiked levels, from 0.5 to 15 μg/kg). Competition blank signal is relative to OTA level = 0, blank signal is OTA-free wheat extract with no OTA–AP conjugate used in the assay. Reprinted with permission from Ref. [75].

TABLE 29.4

Comparison of wheat samples contaminated with OTA determined by electrochemical immunosensor and HPLC

Theoretical OTA level (μg/kg)	HPLC		Immunosensor
	OTA ± SD[a] (μg/kg)	RSD[b] (%)	OTA ± SD (μg/kg)
0.5	0.44 ± 0.08	18.2	0.6 ± 0.1
1	0.9 ± 0.1	11.1	1.0 ± 0.1
4	3.8 ± 0.4	10.5	3.1 ± 0.4
5	4.5 ± 0.5	11.1	4.2 ± 0.6
15	13.1 ± 0.9	6.9	12 ± 1

Reprinted with permission from Ref. [75].
[a]Mean value ± SD ($n = 3$, SD = standard deviation).
[b]RSD = relative standard deviation.

and 10%, respectively. A good correlation ($r = 0.9992$) was found by comparative analysis of naturally contaminated wheat samples using this assay and an HPLC/immunoaffinity clean-up method based on the AOAC Official Method 2000.03 for the determination of OTA in barley as reported in Table 29.4.

29.2.3 Screen-printed electrochemical sensors for the detection of *Fusarium* sp. DNA

DNA electrochemical sensors were developed for the detection of specific sequences of *F. culmorum*. The sensing principle of the proposed application was based on the affinity interaction between complementary strand of nucleic acids: the probe was immobilized on the sensor surface and the target analyte was free in solution. The hybridization is followed using an SWV. The direct label-free detection was accomplished by monitoring the guanine oxidation peak of the target sequence. This approach rely on the use of inosine-modified (guanine-free) probes. In fact, the non-electroactive base inosine still forms a specific base pair with the cytosine residue. This results in a flat baseline (in the potential region +0.9–1 V versus pseudo Ag/AgCl reference electrode) for the probe-modified electrode. The duplex formation was thus detected through the appearance of the guanine oxidation peak of the target sequence, following hybridization. The use of such a label-free biosensor for the detection of post PCR samples can improve the safety, since the indicators are usually toxic or carcinogenic compounds, and save time. The sensors were first applied to synthetic complementary oligonucleotides, and then to PCR-amplified samples of *F. culmorum*. The specificity of the genosensors was tested in the presence of different amounts of non-complementary PCR sample. The crucial part of this work was to understand how to increase the occurrence of the hybridization reaction between the probe and the denatured single strand of PCR product. In this application, the probe selection was carried out using a computational approach. Computational tools can greatly improve the rational design of the ligands by exploiting different knowledge about the structure and the chemical properties of the target molecule. The computational run was performed by fixing the folding temperature at 37°C with a concentration of sodium of 330 mM to fit the experimental conditions. The other parameters were left as default. An important purpose of this work was to evaluate the relation between experimental and computational data. Therefore the probe selection was focused on the study of the steric hindrance of the simulated secondary structure by emphasizing as primary parameter the intra-strand hydrogen bonds within the PCR single strand. For that reason, we selected the probes complementary to the PCR single-strand regions with different intra-strand hydrogen bonds as reported below (Fig. 29.9):

- probe 1 complementary to a region with six intra-strand hydrogen bonds (3 G-C; 3 A-T)

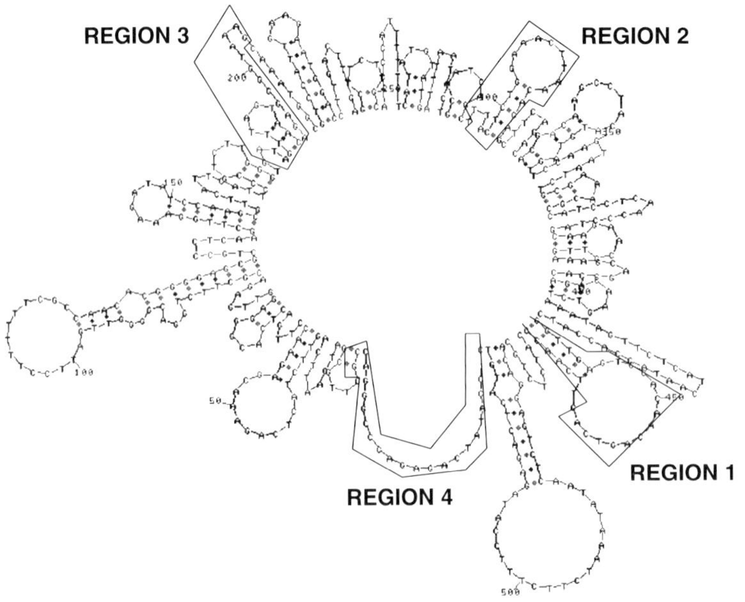

Fig. 29.9. Dots indicate intra-strand hydrogen bonds. Folding conditions: $T = 37^{\circ}C$, $[Na^{+}] = 330\,mM$.

- probe 2 complementary to a region with five intra-strand hydrogen bonds (3 G-C; 2 A-T)
- probe 3 complementary to a region with three intra-strand hydrogen bonds (1 G-C; 2 A-T)
- probe 4 complementary to a region with two intra-strand hydrogen bonds (2 G-C; 0 A-T)

The different probes were used to obtain four independent sensors that were then tested for their selectivity and specificity toward synthetic complementary sequences and PCR samples. In Fig. 29.10 we report the response curve obtained for probe 3 DNA sensor versus different concentration of the complementary strand and versus three non-complementary sequences.

Thereafter, the different DNA biosensors were used to test PCR samples obtained from *F. culmorum*. The same DNA biosensor was also challenged with completely non-complementary DNA (Table 29.5). The results exhibited a dose-dependent response up to 7 μg/mL; for higher concentrations a hook effect was observed.

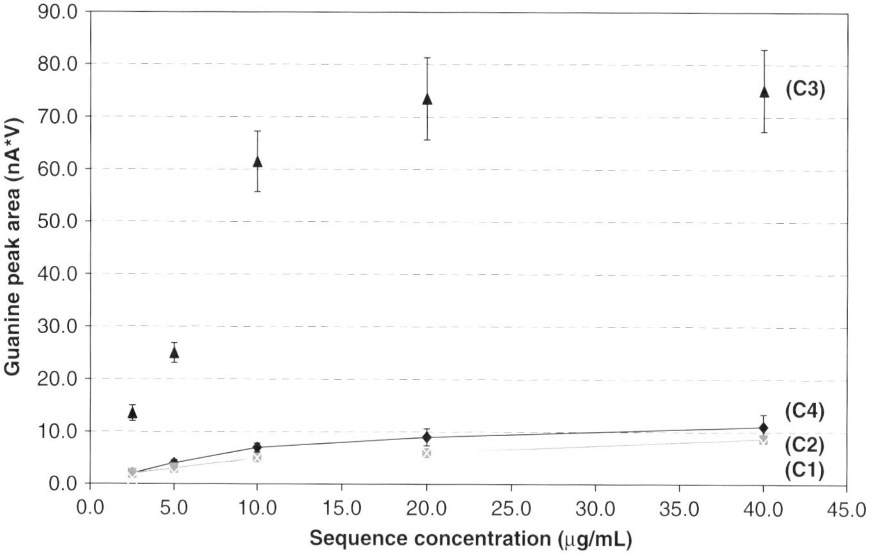

Fig. 29.10. Correlation between analytical signals observed with probe 3 genosensor and concentrations of the oligonucleotide complementary to the probe 3 (C3) and of the oligonucleotides complementary to the probes 1, 2, and 4 (C1, C2, and C4, respectively).

TABLE 29.5

Analytical signals observed with different concentrations of *F. culmorum* PCR product and of a non-complementary DNA fragment

PCR concentration (μg/mL)	*F. culmorum* PCR product (±SD) (nA V)	Non-complementary PCR product (±SD) (nA V)
30	14.03±1.96	0.32±0.04
15	24.16±3.14	0.05±0.02
7	36.40±6.55	0.03±0.01
3	10.10±2.12	0.10±0.02

PCR products were thermally denatured prior to analysis.

In order to avoid false negatives, a control procedure was introduced. The PCR samples were split into two aliquots, one of which was thermally denatured, for the analysis.

It was experimentally evaluated that if the denaturated/non-denaturated signal was higher than 3, the sample could be considered as positive, while for lower ratio the sample has to be classified as negative. Particularly, interesting results were obtained using the probes complementary to the regions 3 and 4. The SPE modified with the probe complementary to region 3 exhibited acceptable repeatability (CV within 20% on three consecutive measurements) with no measurable analytical signal in the presence of different amounts of non-complementary PCR sample.

29.3 CONCLUSIONS

The applications described above, coupled with the realization of a dedicated portable instrumentation and software, represent a user-friendly analytical tool dedicated to durum wheat safety. Moreover, all the applications are based on the use of one single type of thick-film SPE facilitating the overall procedure for the final user that has to store and handle one single type of transducer. The developed device, which consists of the hand-held potentiostat, the multiplexer for eight-channel control and a dedicated software, can be used to detect OPs pesticides, such as dichlorvos and pirimiphos methyl at contamination level below the MRL settled by the European Union, OTA, and also amplified DNA of *F. culmorum*.

The software has a user-friendly interface that informs the user on the procedural steps that must be performed (e.g., “add 20 μL of Vial A to Vial B, mix and transfer on the sensor surface”) allowing the realization of the measurement even for non-trained personnel. All the internal controls result in a “pass or fail” message as well as all the incubation times are controlled by the software. At the end of every protocol the user receives a screen message that contains a self-explanative result via a color code according to a “traffic light scheme”.

For dichlorvos detection, the major achievements were the development of two alternative extraction methods using alternatively the assay buffer or organic solvent as extraction mean. The former procedure allowed a simplified protocol, using whole kernels, that could readily be used in field to measure as low as 0.45 mg/kg using wild-type eeAChE and 0.07 mg/kg using a specifically designed engineered rAChE. The latter procedure performed using solvent extraction allowed a complete recovery of the pesticide from grounded kernels. The procedure that can be used in a laboratory environment as screening method allowed the detection at one order of magnitude lower (0.05 mg/kg) than in the former protocol using wild-type eeAChE. The method has been peer-validated in an inter-laboratory experiments. A second challenging pesticide was investigated for method development resulting in a procedure, finally transferred on the developed software, for the analysis of pirimiphos methyl in durum wheat. The phosphotionate pesticide was oxidized and transformed in the active oxo form using a dedicated optimized procedure able to function in durum wheat extract. An extraction protocol carried out in methanol was used as this solvent could then be used in the electrochemical assay protocol. The LOD referred to the sample, determined using the best operational condition, was 3 mg/kg. Spiked samples were prepared at the EU-regulated level (5 mg/kg) and analyzed with the optimized protocol resulting in an average recovery of 70.3%.

For OTA detection, the optimized immunosensors and the protocol that was implemented on the electrochemical device allowed the detection of 0.4 μg/kg, with within- and between-assay variability less than 5% and 10%, respectively. The method was evaluated with respect to a reference instrumental method (HPLC/immunoaffinity clean-up method based on the AOAC Official Method 2000.03) obtaining good agreement ($r = 0.9992$).

Finally, an electrochemical DNA biosensor able to discriminate between positive and negative PCR-amplified samples of *F. culmorum* was shown.

In conclusion, the overall device enables a thorough control on durum wheat samples in a user-friendly manner, at detection levels that allow an improvement in the control protocols that have to be carried in field by non-specialized personnel.

ACKNOWLEDGMENTS

The authors wish to thank the MIUR (DL 297/27 July1999), SINSIAF Project for funding.

REFERENCES

1 U.S. FDA. Pesticide Monitoring Database 1996–1999, http://www.cfsan.fda.gov/
2 Council Directive 80/778/EEC (OJ L229, p 11, 30/08/1980) of 15 July, 1980.
3 H. Kidd, D. Hartley and J. Kennedy. In: Insecticides and Acaricides, *European Directory of Agrochemical Products*, 2nd ed., Vol. 3, The Royal Society of Chemistry, London, UK, 1986, pp. 99–247.
4 Commission Directive 2001/57/EC of 25 July, 2001.
5 U.S. Environmental Protection Agency (EPA). Federal Register (1995), 60 FR 50338 (September 28).

6 W.J. Donarsky, D.P. Dumas, D.P. Heitmeyer, V.E. Lewis and F.M. Raushel, *Arch. Biochem. Biophys.*, 227 (1989) 4650–4655.
7 www.intox.org/databank/documents/chemical/pirimet/pest49_e.htm.
8 A. Bottalico and G. Perrone, *Eur. J. Plant Pathol.*, 108 (2002) 611–624.
9 H. Hestbjerg, K.F. Nielsen, U. Thrane and S. Elmholt, *J. Agric. Food Chem.*, 50 (2002) 7593–7599.
10 A. Logrieco, C. Altomare, A. Moretti and A. Bottalico, *Mycol. Res.*, 96 (1992) 518–523.
11 Joint FAO/WHO Expert Committee on Food Additives (JECFA), 2000. Safety evaluation of certain food additives and contaminants. Zearalenone. WHO Food Additives Series 44. http://www.inchem.org/documents/jecfa/jecmono/v44jec14.htm
12 Joint FAO/WHO Expert Committee on Food Additives (JECFA), 2001. Safety evaluation of certain mycotoxins in food. Deoxynivalenol, HT-2 and T-2 toxin. FAO Food and Nutrition Paper 74. http://www.inchem.org/documents/jecfa/jecmono/v47je01.htm
13 M. Schollenberger, H.M. Muller, M. Rufle, S. Suchy, S. Planck and W. Drochner, *J. Food Microbiol.*, 97 (2005) 317–326.
14 P. Krogh, *Mycotoxins in Food*, Academic Press, London, 1987.
15 J.D. Miller and H.L. Trenholm, *Mycotoxins in Grain. Compounds Other than Aflatoxin*, Eagan press, St. Paul, USA, 1994.
16 *Some Naturally Occurring Substances. Food Items and Constituents, Heterocyclic Aromatic Amines and Mycotoxins* (IARC Monographs on the Evaluations of Carcinogenic Risks to Humans), No. 56, International Agency for Research on Cancer, Lyon, 1993, pp. 489–521.
17 *Ochratoxin A—Toxicological Evaluation of Certain Food Additives and Contaminants*, WHO Food Additives Series 35, World Health Organization (WHO), Geneva, 1996, pp. 363–376.
18 Commission Regulation No. 472/2002 of 12 March 2002 and No. 123/2005 of 26 January 2005.
19 Commission Regulation No. 683/2004 of 13 April 2004.
20 J.A. Compton, *Military Chemical and Biological Agents*, Telford Press, NJ, 1988, p. 135.
21 Food and Agricultural Organization of the United Nations Product Yearbook 43, 1989, pp. 320.
22 A. Mulchandani, W. Chen, P. Mulchandani, J. Wang and K.R. Rogers, *Biosens. Bioelectron.*, 16 (2001) 225–230.
23 I. Palchetti, A. Cagnini, M. Del Carlo, C. Coppi, M. Mascini and A.P.F. Turner, *Anal. Chim. Acta*, 337 (1997) 315–321.
24 G.S. Nunes, P. Skládal, H. Yamanaka and D. Barceló, *Anal. Chim. Acta*, 362(1) (1998) 59–68.
25 G.S. Nunes, D. Barceló, B.S. Grabaric, J.M. Díaz-Cruz and M.L. Ribeiro, *Anal. Chim. Acta*, 399(1,2) (1999) 37–49.
26 M. Albareda-Sirvent, A. Merkoçi and S. Alegret, *Anal. Chim. Acta*, 442(1) (2001) 35–44.

27 M. Del Carlo, M. Mascini, A. Pepe, D. Compagnone and M. Mascini, *J. Agric. Food Chem.*, 50 (2002) 7206–7210.
28 Commission Directive 1999/50/EC of 25 May 1999 amending Directive 91/321/EEC on infant formulae and follow-on formulae, Official Journal of the European Communities No L 139, 02/06/1999, p. 29.
29 H. Schulze, R.D. Schmid and T.T. Bachmann, *Anal. Bioanal. Chem.*, 372 (2002) 268–272.
30 H. Schulze, E. Scherbaum, M. Anastassiades, S. Vorlova, R.D. Schmid and T.T. Bachmann, *Biosens. Bioelectron.*, 17 (2002) 1095–1105.
31 H.B. Ray, A.A. Zakhidov and W.A. de Heer, *Science*, 297 (2002) 787.
32 S. Sotiropoulou and N.A. Chaniotakis, *Anal. Bioanal. Chem.*, 375 (2003) 103.
33 M. Musameh, J. Wang, A. Merkoci and Y. Lin, *Electrochem. Commun.*, 4 (2002) 743.
34 X.N. Cao, L. Lin, Y.Z. Xian, W. Zhang, Y.F. Xie and L.T. Jin, *Electroanalysis*, 15 (2003) 892.
35 J. Wang, M. Musameh and Y. Lin, *J. Am. Chem. Soc.*, 125 (2003) 2408.
36 R.J. Chen, Y. Zhang, D. Wang and H. Dai, *J. Am. Chem. Soc.*, 123 (2001) 3838.
37 J.J. Davis, M.L.H. Green, H.A.O. Hill, Y.C. Leung, P.J. Sadler, J. Sloan, A.V. Xavier and S.C. Tsang, *Inorg. Chim. Acta*, 261 (1998) 1272.
38 F. Balavoine, P. Schultz, C. Richard, V. Mallouh, T.W. Ebbesen and C. Mioskowski, *Angew. Chem. Int. Ed.*, 38 (1999) 1912.
39 Y. Lin, F. Lu and J. Wang, *Electroanalysis*, 16 (2004) 145.
40 A. Guiseppi-Elie, C. Lei and R.H. Baughman, *Nanotechnology*, 13 (2002) 559.
41 J.Z. Xu, J.L. Zhu, Q. Wu, Z. Hu and H.Y. Chen, *Electroanalysis*, 15 (2003) 219.
42 J. Wang and M. Musameh, *Analyst*, 129 (2004) 1.
43 A.J. Kanchan, J. Tang, R. Haddon, J. Wang, W. Chen and A. Mulchandania, *Electroanalysis*, 17(1) (2005) 54–58.
44 A. Karen Law and P.J. Seamus Higson, *Biosens. Bioelectron.*, 20 (2005) 1914–1924.
45 B. Bucur, A.F. Danet and J.L. Marty, *Biosens. Bioelectron.*, 20(2) (2004) 217–225.
46 A. Vakurov, C.E. Simpson, C.L. Daly, T.D. Gibson and P.A. Millner, *Biosens. Bioelectron.*, 20(11) (2005) 2324–2329.
47 M. Del Carlo, A. Pepe, M. De Gregorio, M. Mascini, J.L. Marty, D. Fournier, A. Visconti and D. Compagnone, *J. Food Prot.*, 2005, Accepted for publication.
48 F. Longobardi, M. Solfrizzo, D. Compagnone, M. Del Carlo and A. Visconti, *J. Agric. Food Chem.*, 53 (2005) 9389–9394.
49 M. Del Carlo, A. Pepe, M. Mascini, M. De Gregorio, A. Visconti and D. Compagnone, *Anal. Bioanal. Chem.*, 381 (2005) 1367–1372.
50 E. Palecek, *Anal. Biochem.*, 170 (1988) 421–431.
51 K. Kerman, D. Ozkan, P. Kara, A. Erdem, B. Meric, P.E. Nielsen and M. Ozsoz, *Electroanalysis*, 15 (2003) 667–670.

52 J. Wang and A.B. Kawde, *Analyst*, 127 (2002) 383–386.
53 I.V. Yang and H.H. Thorp, *Anal. Chem.*, 73 (2001) 5316–5322.
54 E. Palecek, M. Fojta and F. Jelen, *Bioelectrochemistry*, 56 (2002) 85–90.
55 M. Fojta, L. Havran, S. Billova, P. Kostecka, M. Masarik and R. Kizek, *Electroanalysis*, 15 (2003) 431–440.
56 E.M. Boon and J.K. Barton, *Curr. Opin. Struct. Biol.*, 12 (2002) 320–329.
57 E.M. Boon, D.M. Ceres, T.G. Drummond, M.G. Hill and J.K. Barton, *Nat. Biotechnol.*, 18 (2000) 1096–1100.
58 S.O. Kelley, E.M. Boon, J.K. Barton, N.M. Jackson and M.G. Hill, *Nucleic Acids Res.*, 27 (1999) 4830–4837.
59 K. Kerman, Y. Morita, Y. Takamura and E. Tamiya, *Anal. Bioanal. Chem.*, 381(6) (2005) 1114–1121.
60 M. Gabig-Ciminska, H. Andresen, J. Albers, R. Hintsche and S.O. Enfors, *Microbiol. Cell Fact*, 3(1) (2004) 2.
61 S. Kwakye, V.N. Goral and A.J. Baeumner, *Biosens. Bioelectron.*, 21 (2006) 2217.
62 P. Abad-Valle, M.T. Fernandez-Abedul and A. Costa-Garca, *Biosens. Bioelectron.*, 20(11) (2005) 2251–2260.
63 S.D. Vernon, D.H. Farkas, E.R. Unger, V. Chan, D.L. Miller, Y.P. Chen, G.F. Blackburn and W.C. Reeves, *BMC Infect. Dis.*, 3(12) (2003) 56–60.
64 M. Mascini, M. Del Carlo, I. Cozzani and D. Compagnone, *Anal. Lett.*, (2006), submitted for publication.
65 M. Schollenberger, H.M. Muller, M. Rufle, S. Suchy, S. Planck and W. Drochner, *Int. J. Food Microbiol.*, 97 (2005) 317–326.
66 J. Wang, G. Rivas, J.R. Fernandes, J.L.L. Paz, M. Jiang and R. Waymire, *Anal. Chim. Acta*, 375 (1998) 197–203.
67 J. Wang and A.N. Kawde, *Anal. Chim. Acta*, 431 (2001) 219–224.
68 F. Lucarelli, G. Marrazza, I. Palchetti, S. Cesaretti and M. Mascini, *Anal. Chim. Acta*, 469 (2002) 93–99.
69 P. Skládal, *Electroanalysis*, 9 (1997) 737–745.
70 A.L. Ghindilis, P. Atanasov, M. Wilkins and E. Wilkins, *Biosens. Bioelectron.*, 13(1) (1998) 113–131.
71 M. Del Carlo, I. Lionti, M. Taccini, A. Cagnini and M. Mascini, *Anal. Chim. Acta*, 342 (1997) 189–197.
72 L. Micheli, A. Radoi, R. Guarrina, R. Massaud, C. Bala, D. Moscone and G. Palleschi, *Biosens. Bioelectron.*, 20(2) (2004) 190–196.
73 L. Micheli, R. Grecco, M. Badea, D. Moscone and G. Palleschi, *Biosens. Bioelectron.*, 21(4) (2005) 588–596.
74 S. Hugo Alarcon, L. Micheli, G. Palleschi and D. Compagnone, *Anal. Lett.*, 37(8) (2004) 1545–1558.
75 S. Hugo Alarcon, G. Palleschi, D. Compagnone, M. Pascale, A. Visconti and I. Barna-Vetrò, *Talanta*, 69 (2006) 1031–1037.

NOVEL TRENDS

Chapter 30

Potentiometric electronic tongues applied in ion multidetermination

Manel del Valle

30.1 INTRODUCTION

Ion selective electrodes (ISEs) or, in a wider sense, potentiometric sensors have demonstrated its usefulness to yield information of chemical species in automated and autonomous operation. This feature has fostered their use in the monitoring of numerous processes, in the industrial, clinical and environmental fields, among others. Current practice with these devices relies on sensors with high selectivity; only in this way, a simple determination of a single ion is possible in presence of its interferents. Some reluctances on the broadening of their use are surely due to the fact that ISEs are not specific but show high selectivity towards a reduced number of ions.

One recent proposal to cope with this problem is to use an array of non-specific sensors and to let the obtained data undergo a multivariate chemometric treatment [1]. The purpose of the latter will be to identify the chemical species present or to determine their concentrations without having to eliminate interferences, quantifying them as well [2]. The key point is that multidimensional data generated contain all the information about the system the array is designed for, and that information concerning single components can be extracted [3]. Maybe the greatest advantage using a sensor array is its ability to generate multivariate analytical data in real time and simultaneously to permit the identification of matrix effects.

This strategy was first proposed for analysis in the gas phase employing an array of gas sensors [4], where it is known as "electronic nose" [5]. The approach receives the biomimetic qualifier, given it is inspired in the physiological basis of animal olfaction. Information obtained from

Comprehensive Analytical Chemistry 49
Alegret and Merkoçi (Eds)
Volume 49 ISSN: 0166-526X DOI: 10.1016/S0166-526X(06)49030-9

chemical sensor arrays of complex gaseous samples has allowed identification of species as well as their quantitative determination [6].

The application of this concept to liquid samples is what we already refer to "electronic tongue". It entails the use of multidimensional information coming from an array of chemical sensors, mimicking the animal sense of taste. As several possibilities exist on the side of which sensors form the array, the general response shown by the different sensors used is of paramount importance; that is, cross-selectivity features are needed in order to profit from the multidimensional aspects of the information [7]. The performance of electronic tongues can be suited not only to qualitative purposes like identification of species and classification of sample varieties, but also to quantitative uses, normally the multidetermination of a set of chemical species, an interesting objective for process control. A more bioinspired trend is the "artificial taste" [8] in order to perform automated taste perception, especially in the industrial field.

Let us recall the agreed definition [9] of the "electronic tongue", defined by significant research groups working in this topic as: "an analytical instrument comprising an array of non-specific, poorly selective chemical sensors with partial specificity (cross sensitivity) to different compounds in a solution, and an appropriate chemometric tool for data processing", or by the recent IUPAC report [10]: "a multisensor system, which consists of a number of low-selective sensors and uses advanced mathematical procedures for signal processing based on pattern recognition and/or multivariate data analysis—artificial neural networks (ANNs), principal component analysis (PCA), etc".

With respect to the type of sensors that can be used in an electronic tongue, practically all the main families of chemical sensors have been used to form the sensor array, viz. potentiometric, voltammetric, resistive, gravimetric and optical, if main sensor families have to be quoted [11]. Table 30.1 sketches a survey of different approaches that can be recorded when the specialized literature is inspected. Even hybrid systems have been proposed, mainly those combining potentiometric and voltammetric sensors [3,12]. The combination of electronic noses and electronic tongues to improve detection or identification capabilities, in a sensor fusion approach, has also been proposed [13,14].

Since all the sensors may respond to all the analytes, a great amount of complex data are generated that must be processed using a multivariable calibration approach [21]. Chemometrics is in charge then, for extraction of the information sought by appropriate processing [22]. This coupling, which represents one of the more clear benefits

TABLE 30.1
Examples of different families of sensors used to form the sensor array in an electronic tongue

Sensor family	Research group	Example	Ref.
Electrochemical, potentiometric	Legin (Russia)	Determination of heavy metals with an array of chalcogenide membrane sensors.	[15]
Electrochemical, ISFETs	Bratov (Spain)	Determination of several ions for water characterization	[16]
Electrochemical, voltammetric	Winquist (Sweden)	Characterization of waters with an array of noble metals	[17]
Electrical, resistive	Mattoso (Brasil)	Conducting polymer sensors	[18]
Optical	Mc.Devitt (USA)	Microspheres with immobilized dyes	[19]
Gravimetric	Gardner (UK)	Surface acoustic wave (SAW) sensors	[20]

accounted for in the combination of chemometrics and electrochemical sensors [1], was recently stated as one possible way to improve sensor performance [23]. Different methods exist which allow processing of this data depending on the type of application; two of these are ANNs and PCA, as it was mentioned. Curiously, if the system employs ANNs, a double biomimicry circumstance occurs, given ANNs are also inspired in the physiology of the animal nervous system. One can classify the different available techniques according to the type of application. If this is a qualitative goal, PCA is the first step as it will visualize if the samples can be separated in classes (classified) or will identify a specific variety. After this, some pattern recognition or mean to predict the membership to any of the classes will be needed; for this, tools like linear discriminant analysis, nearest neighbour, soft independent modelling of class analogy (SIMCA), Support Vector machines or ANNs can be used [24]. When the purpose is quantitative, different tools are used, given the numeric information is the end result. Some of these are principal component regression (PCR), which departs from a first PCA transformation to build a multivariate regression, partial least squares regression (PLS) or ANNs. Even, some expert system approaches, like case-based reasoning, have been used to process the multivariate response from sensor arrays [25]. When the departure information is extremely complex, a previous feature extraction is needed, useful to suppress redundant, non-significant information, and to retain key data. Some examples of procedures used for feature extraction are PCA, fitting of splines or other functions, for example Legendre polynomials [26], or Fourier or wavelet transform [27].

When one pays a close look at the analytical literature in search of the different works related with electronic tongues, although there are several variants in the type of sensors used, practically half share corresponds to the use of potentiometric sensors, almost all using ISEs in front of one case using ISFETs [16]. Therefore, the most significant groups working with potentiometric electronic tongues are the group of A. Legin in St. Petersburg (Russia) [28,29], which can be considered founders of the field; noteworthy is also the work of K. Toko in Japan [30,31] most dedicated to the "artificial taste" concept. Once collaborators of the Russian laboratories and now independent research teams are those of D'Amico in Italy [32], Mortensen in Denmark [33] and Nam in South Korea [34]. Apart from these, two other significant groups have appeared; one is that of Wróblewski, in Warsaw (Poland) [35], together with the research group represented by this chapter in Barcelona (Spain) [36].

To complete this scenery it is also necessary to mention the commercial offer of three systems, available for any laboratory to start any specific application at once. These are: the Astree system [37], commercialized by Alpha-MOS, a French company with large experience in electronic noses that also presents an electronic tongue; the "artificial taste" from Intelligent Sensor Technology Inc. (Japan, formerly part of Anritsu Corporation) [38], a company linked to Toko's research; and different components available from Electronic Tongue [39], a company linked to Legin's laboratory. All these supply sensors, hardware for operation and also the software part. Support is normally offered to develop user's specific applications, as the training and building of quantification or identification models can be a delicate task.

Antecedents of the treated topic can be traced to the building of response models for arrays of ISEs, which considers the case of cross-response terms. This has been historically addressed by the application of different chemometric tools. The first attempt was by Otto and Thomas [40] in the 1980s, who employed an eight-sensor array and their Nicolsky–Eisenman response model fitted by multiple linear regression. Beebe and Kowalski retook this approach, in this case employing a five-sensor array with a nonlinear regression method (SIMPLEX) [41] or projection pursuit regression [42], a multivariate chemometric technique. A further approach was that of Forster *et al.* who employed a four-electrode array and again SIMPLEX nonlinear regression [43]. All these examples presupposed a sensor array with at least as many ISEs as the number of the different ions present plus some with partial cross-selective response [44,45], in order to obtain an overdetermined system.

A convenient approach to the case is the use of ANNs, as first demonstrated by the seminal work by Bos and Van der Linden [46]. ANNs generate ''black-box'' models, which have shown special abilities to describe nonlinear responses obtained with sensors of different families. Unfortunately, these tools create models only from a large amount of departure information, the training set, which must be carefully obtained [47]. The extra information is in account of the absence of a thermodynamical or physical model, for example the Nicolsky–Eisenman equation.

An interesting additional feature of using potentiometric sensors in array mode has been pointed out, which is to profit from the extra information to gain confidence in measurements, that is to perform redundant analysis. This strategy can be the basis for automated fault detection of the sensory elements [48] and also be an aid for more robust calibrations [49].

Recently in our laboratory, we have started a research line dealing with the use of electronic tongues, mainly for quantitative multidetermination applications [36], taking profit of accumulated experience in potentiometric sensors. In this line, soon we realized the high experimental effort involved in the generation of the departure information, i.e. appropriate standards and their corresponding measured responses, needed for the building of the quantitative model, normally the training process of an ANN. The proposed strategy to overcome this disadvantage has been the use of automated analytical systems to generate mixtures of analytes as standards, and to simultaneously acquire the complex responses to be processed by the chemometric modelling tools [50,51]. Hence, the acquisition of the large number of empirical information needed to train the system, *ca.* one hundred experimental points, is facilitated by the use of a sequential injection analysis (SIA) system. Its operation, employing in-lab developed software, generates individual mixtures of considered species, which are brought to the sensor array

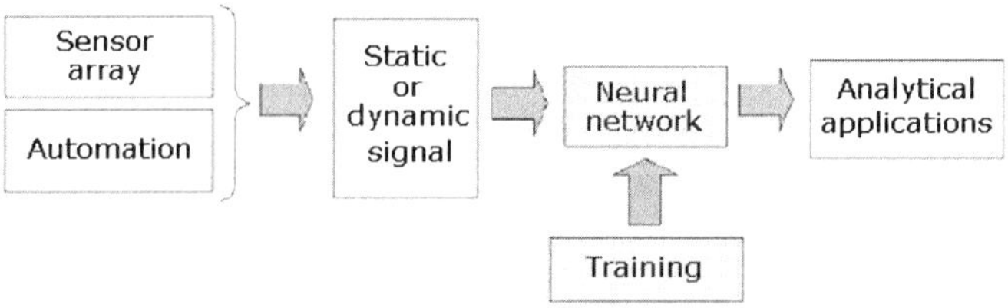

Fig. 30.1. Schematic representation of the combination of the used concepts.

and permits to perform the measurements. Also, samples are processed in the same manner, in order to perform equivalent operation to standards. The general approach described above is schematized in Fig. 30.1. The goal is to develop advanced analytical systems based on the treatment of the multidimensional data generated by a sensor array in combination with an automated liquid handling system. The sort of measurement performed, which can be a steady-state reading but also a dynamic profile recording, will allow to exploit different possibilities in the data treatment.

The remaining part of the chapter is then to report some of the accumulated experience on the different applications on electronic tongue systems, mainly with those employing arrays of potentiometric sensors in batch conditions and ANNs. This combination has been predictably aimed to quantitative purposes, namely the multidetermination of mixtures of species or the determination of specific ions in the presence of their typical interfering substances. In the next section, the ANN concepts and know-how for readers not introduced in the field are summarized. After that, a description of the different potentiometric sensor arrays employed together with some of our experience in batch systems and flow systems will be presented. A complete application yet simple enough for beginners in this field might be reproduced following Procedure 45 in this book.

30.1.1 Artificial neural networks

An ANN is a systematic procedure for data processing inspired by the nervous system function in animals. It tries to reproduce the brain logical operation using a collection of neuron-like entities to perform processing of input data. Furthermore, it is an example of a connectionist model, in which many small logical units are connected in a network [52]. Let us consider the proposed case, a multidetermination application employing an array of ISEs as input data. In a rather complex system, where a number of interferents may be present for a given primary ion, a thermodynamical relation can be established using the Nicolsky–Eisenman expression that defines the response of a sensor (i) towards the activities of the intervening ions, as:

$$E_i = K_i + s_i \log\left[a_1 + k^{\text{pot}}_{1,j} \cdot (a_j)^{z_1/z_j} + k^{\text{pot}}_{1,k} \cdot (a_k)^{z_1/z_k} + \ldots\right] \tag{30.1}$$

where a_1 is the primary ion of this sensor, and a_j, a_k, etc. the interfering ions of charge z_j and z_k, K_i and s_i the electrode constant and sensitivity, respectively. The cross-sensitivity is expressed through the potentiometric selectivity coefficient, $k^{\text{pot}}_{1,h}$, a measure on how an interfering ion h generates a distorting response when measuring the species (Eq. (30.1)). Given the number of ISEs will be larger than the number of species, for any cross-response situation considering more than two ions, the subsequent equation system that is obtained represents a challenge for any chemometrics specialist. Instead of the thermodynamical model, various authors propose the use of an ANN for the processing of data obtained through an array of ISEs [44,46]. Since ANNs are powerful in the modelling of nonlinear systems, and the subjacent phenomena are highly nonlinear, preliminary results were very promising, permitting both qualitative and quantitative analysis. Other authors support the use of PLS regression in front of ANN [21,45].

The basic processing unit of an ANN (or its building block) is called perceptron [53]. This is a crude approximation to the biological neuron, the cell in the nervous system. It is a decision-making unit with several input connections and a single output, as sketched in Fig. 30.2. A signal x_j which is delivered from input j is multiplied on arrival by a connection

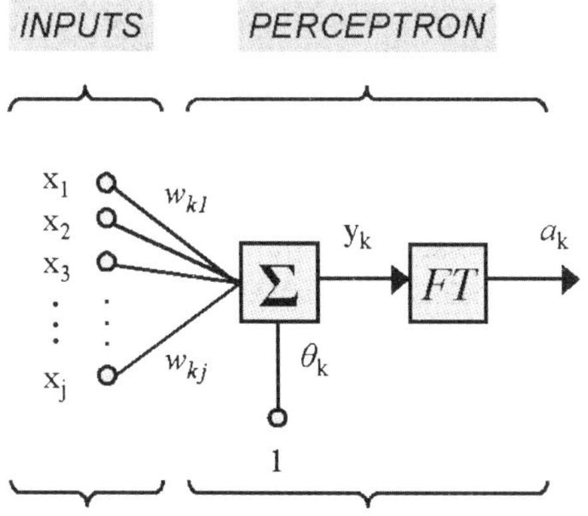

Fig. 30.2. The basic ANN unit, the perceptron K.

weight w_{kj}, so that each signal appears at the k perceptron as the weighted value $w_{kj}.x_j$. The perceptron sums the incoming signals and adds a bias θ_k to give a total signal y_k. To this sum, a transfer function, normally a step-function, is applied to produce the output a_k. Inspired on its physiology, if the sum of inputs is below a threshold value, the neuron is quiescent and remains "off". If the sum reaches the threshold level, the neuron is turned "on" and a message is sent out.

Therefore, the behaviour of the perceptron to certain input information is determined by the weights of its input connections and by the level at which the threshold is set. The transfer function used will also define the shape of the transition step. Knowledge is stored as the values of adjustable parameters (w_{kj}), whereas initially, the connection weights are set to small random values. Learning is then the process of adjusting the values in a way that roughly parallels the training of a biological system [54]. A unique condition must be fulfilled: the problem has to be linearly separable—but most significant scientific problems are not.

Putting it into mathematics, the cumulative input is calculated as

$$y_k = \sum w_{kj} \cdot x_j + \theta_k \tag{30.2}$$

And the firing of the neuron happens if the threshold, as defined by a transference function used, is surpassed:

$$a_k = FT\{y_k\} \tag{30.3}$$

The failure of the perceptron to handle real-world scientific problems first suggested a wrong model for the brain, and the use of the perceptron was abandoned. Years later, someone realized that the brain can process several independent streams of information simultaneously—it is referred to as a parallel device. Therefore, more than one perceptron may be used in order to accomplish a similar effect. This can be done in two different ways: first giving the perceptrons neighbours to form a layer of units which share inputs from the environment; and secondly by introducing further layers, each taking as their input, the output from the previous layer.

In this way, the most common ANN used for numerical models is known as the multilayer feedforward network, and is sketched in Fig. 30.3; the scheme represents the approach for a simultaneous calibration model of two species, A and B, departing from the readings of four ISE sensors.

The path of the departure information (potential readings) begins entering an input layer, whose purpose is just to distribute incoming

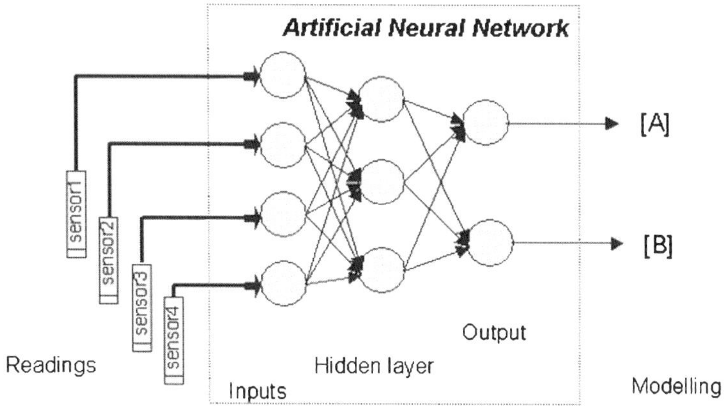

Fig. 30.3. The electronic tongue concept applied for multidetermination employing a neural network model.

signals to the next layer; it does not perform any thresholding, thus the units are not perceptrons in the right sense. The perceptrons in the second layer constitute a hidden layer, as they communicate with the environment only by sending or receiving messages to units in the layers to which they are connected. The output layer provides a link between the artificial network and the outside world, submitting the processed information (the sought concentrations). Every perceptron is connected to all units in the adjoining layers, but there are no connections between units in the same layer. That is why it is called a fully connected layered feedforward network—messages flow in the forward direction only.

Mathematically, each output represents just a specific linear combination with specific weights from each preceding perceptron, though passage from layer to layer is also modulated by the transfer function used (usually a given transfer function for all nodes in a given layer):

$$[X] = FT^{\text{output}}\left\{\theta_j^{\text{o}} + \sum w_{ij}^{\text{o}} \cdot FT^{\text{hidden}}\left(\sum w_{ik}^{h} \cdot E_k + \theta_k^{h}\right)\right\} \tag{30.4}$$

where the superscripts o and h indicate output and hidden layers, respectively, and the input information (E_k) are the different readings from the ISEs. The operation, then, is formed by two stages: first, the ANN response model is built, using some training data, and next the model can be used for prediction of unknown samples.

In this approach, there are a large number of conditions to fix before the training of the ANN can be started [55]. In fact, this is a delicate part of the modelling with neural networks, as there are too many possibilities, and experience is the way to arrive quickly to a good configuration. First, there is to decide if the input data must suffer any kind of pretreatment. A recommended step here is to normalize the range of the different input channels, to avoid any imbalance between them. A second point is to decide the topology of the network; although the number of input and output neurons is fixed by the nature of input and output information, there is no guide to infer how many neurons in the hidden layer will yield better ANN models. In fact, the architecture of the multilayer network can still be more complex, as several hidden layers can be used. Nevertheless, in almost all the recorded chemistry situations, single hidden layers are sufficient [55]. The recommendation to deduce the best number of neurons is then trial and error, beginning by large number, and decreasing it till the model gets worse. Beginning in this way, like when fitting polynomials, some degree of adjustment is first obtained, subsequently improved reducing the number of neurons, and once it deteriorates due to a manifestly simple network, the optimal configuration is met. Another factor is the kind of transfer functions used, normally the same for the whole layer. The input layer, playing only a distributive role, uses a linear transfer function; meanwhile, the hidden and output layers can use one among around one dozen possible functions. Some of the transfer functions that can be used are shown in Fig. 30.4. It is common to use sigmoidal transfer functions in the hidden layer. Some used sigmoidal transfer functions are the log-sigmoidal (*logsig*) and the tan-sigmoidal (*tansig*), whereas the linear function is called *purelin*. The *logsig* function generates outputs between 0 and 1, at which the sum of the outputs goes towards infinite. The *tansig*-function is very similar to the *logsig*, but it generates outputs between –1 and 1. A saturated-linear function (*satlins*) represents a linear correspondence but with a way to avoid saturation at its output. The type of transfer function that will be best

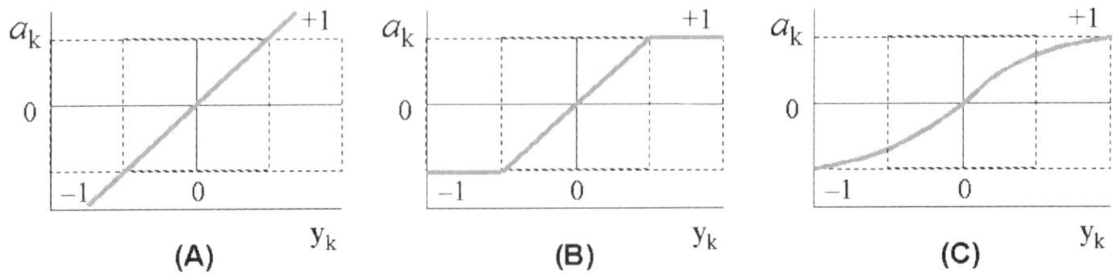

Fig. 30.4. Representation of three commonly used transfer functions: (A) *purelin*, (B) *satlins* and (C) *tansig*.

for our problem will depend on the nature of the relationship considered, and again, different possibilities may be checked for a given case. Other factors to consider are how to accomplish the "learning" of the network, how to check the progress of this learning process and how to avoid some defects that can arise during learning.

Now let us consider the procedure of learning, that is, the attainment of the numerical model for our data. In the elementary perceptron, learning is simple and the learning rules unambiguous: for a set of known samples, the weights w_{kj} are adjusted till the obtained outputs agree with the expected values. Matters are more complicated in a network, because it must be established how changes in the connection weights should be allocated to connections between different layers to promote learning.

The most widely used solution to this problem is backpropagation of errors. Let us represent as o_{pk} the certain instant output of perceptron k for a sample p in the training set, calculated with an expression like Eq. (30.4); analogously, let us represent as t_{pk} the target value. The error signal $(t_{pk} - o_{pk})$ is calculated, and a proportion of the error signal is allocated to the various connections in the network (backpropagated), tuning the values of connection weights (w_{kj}). The goal of this procedure is to reduce the error signal. Using standard backpropagation, the error signals are collected for all output units and all training targets, and the connection weights are adjusted at the end of every epoch, that is, after all samples in the training set have been shown to the network once.

The objective of backpropagation is to adjust the weights in order to minimize the error function F_{p}, defined as the semisum of the individual output unit errors for all samples (k) in the training set [56]:

$$F_{\mathrm{p}} = \frac{1}{2}\sum_{k}(t_{pk} - o_{pk})^2 \qquad (30.5)$$

In this way, the learning process of an ANN is equivalent to a minimization in a multidimensional space (the space of connection weights). A way to accomplish this is by using the gradient-descent algorithm, an iterative optimization procedure in which the connection weights are adjusted in a fashion which reduces the error most rapidly, by moving the system downwards in the direction of maximum gradient [57]. The weight of a connection at stage $(t+1)$ of the training is related to its weight at stage (t) by the following equation:

$$w_{kj}(t+1) = w_{kj}(t) - \eta\frac{\partial F_{\mathrm{p}}}{\partial w_{kj}} \qquad (30.6)$$

where η is a gain term, known as the training rate factor; next, it is re-expressed as

$$w_{kj}(t+1) = w_{kj}(t) + \eta\delta_{pk}o_{pk} \tag{30.7}$$

where δ is the size of change; the product $\delta_{pk} \cdot o_{pk}$represents the gradient contribution. The training rate factor varies between 0 and 1 and accelerates or slows down the descent towards the global minimum of the system; numbers around 0.5 are of typical use here. It is possible to derive expressions prescribing the size of the changes that must be made at the connection weights to reduce the error signal [58]:

For the output layer:

$$\delta_{pk} = ko_{pk}(1 - o_{pk})(t_{pk} - o_{pk}) \tag{30.8}$$

For the hidden layer:

$$\delta_{pk} = ko_{pk}(1 - o_{pk})\sum_{i} \delta_{pi}w_{ki} \tag{30.9}$$

These expressions, which are known as the *generalized delta rule*, show that the extent of the adjustment of connection weights to hidden layers depends upon errors in the subsequent layers, so modifications are made first to the output layer weights, and the error is then propagated successively back through the hidden layers—this is referred to as backpropagation (of error). Each unit receives an amount of the error signal which is in proportion to its contribution to the output signal, and the connection weights are adjusted by an amount proportional to this error.

Backpropagation by gradient-descent is generally a reliable procedure; nevertheless, it has its limitations: it is not a fast training method and it can be trapped in local minima. To avoid the latter, a variant of the above algorithm called gradient-descent with momentum (GDM) introduces a third term, β:

$$w_{kj}(t+1) = w_{kj}(t) + \eta \cdot \delta_{pk}o_{pk} + \beta\Delta w_{kj}(t) \tag{30.10}$$

The term β, referred to as the *momentum*, takes a fixed value between 0 and 1 and serves to reduce to the probability of the system being trapped in a local minimum.

The above-related situation can also be improved by using a different algorithm for training. One of the most efficient minimization algorithms is the Levenberg–Marquardt (LM) [56,59]. It is between 10 and 100 times faster than gradient-descent, given it employs a second-derivative approach, while GDM employs only first-derivative terms. As the calculation of the Hessian matrix (matrix of the second derivatives of the error in respect of the weights) is a very cumbersome task, LM algorithm employs an approximation starting with a Jacobian matrix (matrix of the first derivatives of the error in respect of the weights), since it is much easier to calculate the Jacobian than the Hessian matrix [60,61]. Therefore, the weights can be calculated as

$$w_{kj}(t+1) = w_{kj}(t) - [\mathbf{J}^{\mathrm{T}} \cdot \mathbf{J} + \mu \cdot \mathbf{I}]^{-1}\mathbf{J}^{\mathrm{T}} \cdot e \qquad (30.11)$$

where $\mathbf{J}$ stands for the Jacobian matrix, μ for an adjustment factor, $\mathbf{I}$ for the identity matrix and e for a vector of network errors. When μ is large, this becomes gradient-descent with a small step size. Thus, the aim is to keep μ as small as possible. This way, if μ is decreased after every epoch, this becomes a very effective algorithm [56].

One of the problems that may occur during neural network training is called overfitting [53,62]. This situation occurs when the error on the training set is driven to a very small value, but when new data are presented to the network the error is large. Figure 30.5 illustrates the overfitting concept: the network has memorized the training examples, but the generalization to new situations is incorrect. Two methods can be used to avoid overfitting.

1. *Bayesian regularization* (BR): This technique searches for the simplest network which adjusts itself to the function to be approximated, but which also is able to predict most efficiently the points that did not participate in the training [63]. In contrast to gradient-descent, in this case not only the global error of the ANN is taken

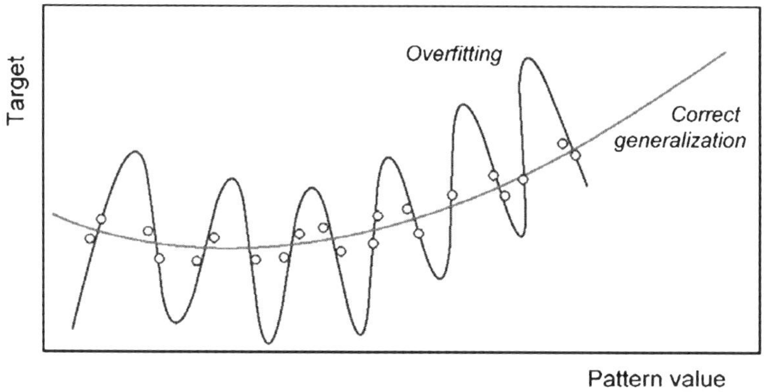

Fig. 30.5. Illustration of the overfitting situation in a case with noisy data. The ANN model memorizes non-significant details yielding increased errors in data not used for training.

into consideration, but also the value of every single weight of the network. Therefore, the values of the weights are minimized, and the network's complexity is reduced, the responses are smoothened and overfitting is avoided. Furthermore, certain neurons are pruned if all its weights are equal to zero:

$$F_{\text{tot}} = \gamma \cdot F_{\text{p}} + (1-\gamma)\frac{1}{n}\sum w_{kj}^2 \qquad (30.12)$$

2. *Early stopping:* Another method for improving generalization is called early stopping, and employs additional data to avoid the undesired circumstance mentioned above. In this technique, the available data are divided into three subsets, as schematized in Fig. 30.6. The first subset is the training set, which is used for computing the gradient and updating the network weights and biases, viz. to accomplish learning of the ANN. The second subset is the validation set, which is used during the training process to check the trend presented by this error from data not used for training. The validation error will normally decrease during the initial phase of training, as does the training set error. However, when the network begins to overfit the data, the error on the validation set will typically begin to rise. When the validation error increases for a specified number of iterations, the training is stopped, and the weights and biases at the minimum of the validation error are returned. The test set error is a third subset, not used at all during the training process or its internal monitoring, but it is used to compare performance of different models. If the

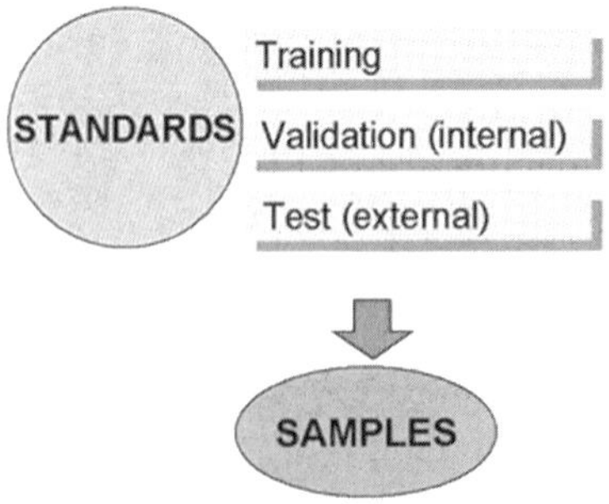

Fig. 30.6. Subdivision of the sample set in the different subsets recommended to prepare the training process of an ANN model.

error in this external test set reaches a minimum at a significantly different iteration number than the validation set error, this may indicate a poor division of the dataset.

All these precautions represent an increased experimental effort in generating data for the numerical model, which can be one of the inconveniences when working with ANNs. In order to achieve convergence with a proper modelling ability and a demanded level of accuracy and precision, two independent dataset are used at least, a training set and an (internal) test or validation set, although the third set is also suggested [56]. Each set contains two kinds of information that interrelates. The first type is formed by the responses of the sensor array (patterns); the second is their corresponding searched information (targets), which in a quantitative application case are the concentration values of the analytes. This training set must be large enough and contain sufficient variability to yield a proper modelling of the response. In order to assure the quality of the final results, some check is needed. For this purpose, the second set of data, the validation set, is used. From the total of generated cases, one can equally distribute the available cases between the different subsets, or reserve the largest part of the data (up to 50%) for training. Reference authors in the field recommend the use of 5–10 training samples per connection weight in the ANN model, which can reach hundreds or thousands of cases for complex network structures [55]. Ways to generate the complete set is through experimental design schemes, for example a factorial design or others. When working with electronic tongues employing ISEs, two other situations are also typical, the use of a collection of sample varieties (juices, wines, an intermediate industrial product, etc.) that need to bring some reference value (the target, for example a parallel analytical determination) in order to build the training set, or a set of examples generated by accumulated microadditions of certain standard in a predefined background [64].

Now, from a practical point of view, the best way to incorporate all these variants and recommendations is to use MATLAB [65] as the programming environment to generate electronic tongue software code. When the basic core is combined with its neural network toolbox [56] (an added option), dozens of types of ANNs and learning algorithms can be easily implemented through preprogrammed instructions; after that, just the selection of the various parameters and details that define an ANN configuration (for example number of neurons in the hidden layers, or learning rate values) is the whole lot needed to start an application.

Much commercial software is limited in comparison to MATLAB, as the latter permits to alter every single factor intervening in an ANN; this feature, which can upset beginners for the large number of factors to control, pays off thanks to the optimized code that furthermore offers an astounding execution speed. If the programming abilities of the electronic tongue practitioner are far away from those desirable, the left option is to use integrated packages, which are of much more friendly use but much more limited in available options. Some of these packages are Trajan Neural Network Simulator (Trajan software Ltd., UK) [66] or Statistica Neural Network (StatSoft Inc., USA) [67]. Another option is to use shareware software (a list of different offered packages can be found at exchange software sites, like Simtelnet, under the entry "neural"); a very convenient package is EasyNN, a program that can be used to create, train, validate and test ANN models within a few mouse clicks, and, to the author's surprise, to fulfil the training of one case study (enter pattern and target data, define ANN configuration, and complete the learning process), just a few minutes after installation [68], which speaks about its user-friendliness.

30.2 APPLICATION

When one begins to consider which sensors to incorporate in an electronic tongue system, even if these have been chosen to be of the potentiometric type, there are several variants that can be identified in the literature. The most used type of ISE, given it is favourite among the most active research group (Legin in Russia), is that formed by chalcogenide glass. This is a vitreous material that can incorporate different heavy metal oxides, and, as a result, provides potentiometric response to heavy metal cations and counteranions [15,21]. The second most typical variant used is the PVC membrane ISE, a most versatile variant, given it can incorporate hundreds of different ionophores to induce response to cations, anions and even neutral species [34–36,69]. The variant related with the "artificial taste" concept usually employs what the authors recall sensors employing lipid/polymer membrane [30,70], which in fact corresponds to the plasticized PVC membrane but without using any ionophore. Simple metallic elements have also been used in potentiometric mode, as wires [71] or as screen-printed inks [72]. Obviously, commercial ISEs can also be used to construct the ISE array [25,44–46]. To complete the scenery, sensor arrays employing electropolymerized layers of different metallic porphyrins, a typical material used for electronic noses, have also been described [32]. Finally, the use of biosensors of the potentiometric type for certain specific applications must also be mentioned [73–75].

A convenient technology to get an array of potentiometric sensors is to use the classical PVC membrane, as a wide experience exists on their formulation and response characteristics, together with the commercial availability of membrane components. Figure 30.7 depicts the composition of a typical PVC membrane used to develop potentiometric sensors [76]. The components are normally dissolved with the aid of a volatile solvent (for example, tetrahydrofuran) and after its evaporation, a plasticized membrane results.

This membrane is normally employed in what is called a "symmetrical configuration", to form the classical ISE which employs an internal reference solution, as sketched in Figure 30.8. This membrane can also be used directly fixed on a solid contact, per example a metallic wire or a screen-printed substrate, in the coated-wire configuration [77]. The proposed Procedure 45 (see in CD accompanying this book) suggests the use of coated wire sensors, as these are the easiest to construct.

The experience from our laboratory in electronic tongues departs from *all-solid-state* ISEs. These are based on a solid electrical contact made from a conductive composite formed by epoxy resin and conducting graphite [64]. Figure 30.9 schematically shows the construction procedure of one of these sensors. It is formed filling a plastic cylinder (8 mm i.d.) with a homogeneous mixture of 35.7% Araldite M, 14.3% Hardener HR and 50% graphite powder (50-μm particle size), and cured for 6 h at 50°C. When hardened, a 0.5-mm depth cavity is formed at the top of the constructed body. This space is used to solvent-cast a plasticized PVC membrane incorporating an appropriate ionophore. Once formed, membranes need to be conditioned on a 0.1-M solution of its primary ion for 24 h.

One of the applications first developed in our laboratory was aimed at the determination of alkaline ions. This case was originally selected, given there are many ionophores for these ions with low and also high

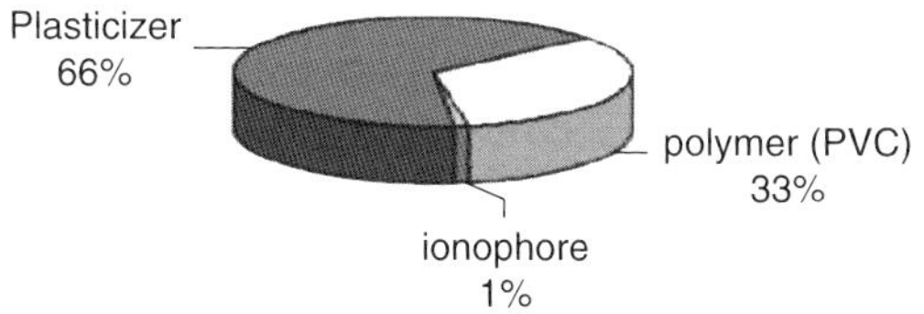

Fig. 30.7. Typical composition of a potentiometric PVC membrane.

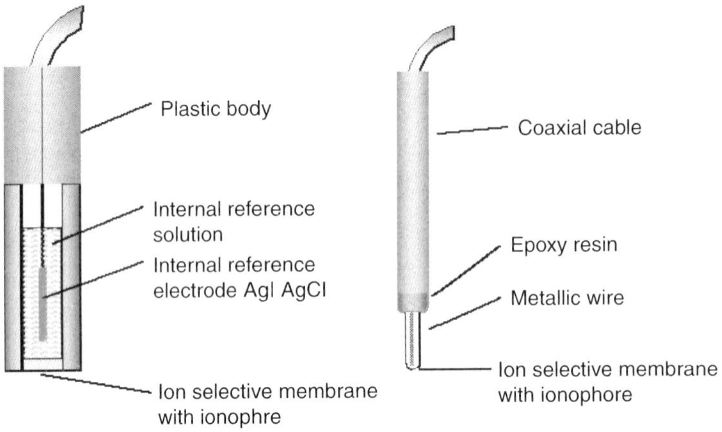

Fig. 30.8. (A) Classical ion-selective electrode. (B) Coated-wire electrode.

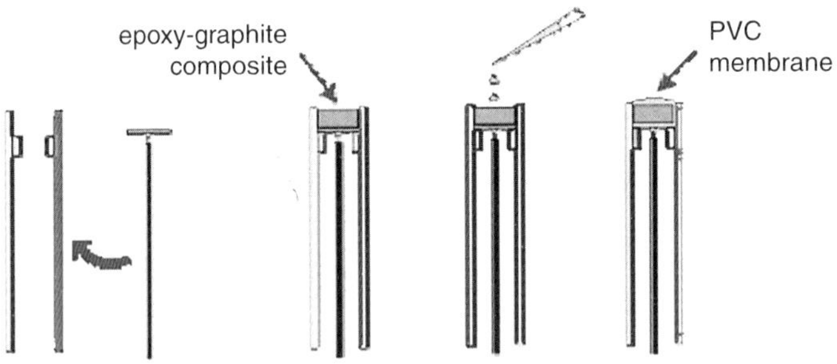

Fig. 30.9. Manufacture of the solid-contact potentiometric sensors.

selectivity, and because there are many interesting applications to be derived. Among others, the determination of ammonium together with its potassium and sodium interferences in different fields, or the determination of electrolytes in clinical samples may be interesting. For example, to attempt the determination of ammonium ions in presence of interfering substances, eight ISEs were initially chosen, using ionophores of two types, specific and of generic response. Nonactin for ammonium, or valynomicin for potassium were two of the first type. Although thought of high selectivity, in fact, this is the case only for certain application fields, given nonactin has a known interference of potassium and sodium, or valinomycin shows a high interference for ammonium, a cation-absent in plasma [78]. For sodium, we used monensin, an ionophore of the same family, the ionophoric antibiotics. Lasalocide, also an ionophoric antibiotic was used as the component in a

TABLE 30.2

Formulation of the ion-selective membranes employed in the construction of the potentiometric sensor array

Sensor	PVC (%)	Plasticizer	Ionophore	Ref.
NH_4^+	33	Bis(1-butyl-pentyl)adipate (66%)	Nonactin (1%)	[79]
K^+	30	Dioctylsebacate (DOS) (66%)	Valinomicyn (3%)	[80]
Na^+	27	Dibutylsebacate (DBS) (70%)	Monensin (3%)	[81]
Generic 1	29	DOS (67%)	Dibenzo-18-crown-6 (4%)	[78]
Generic 2	27	DBS (70%)	Lasalocide (3%)	[79]

generic response sensor, together with dibenzo-18-crown-6, a complexing agent that can act as neutral carrier. Table 30.2 summarizes the formulation of the different membranes, specifying the electroactive components used. The sensor array comprised duplicated sensors for ammonium, potassium and sodium, plus two generic formulations to alkali ions, one employing the crown ether and the other the generic antibiotic.

The formulations shown are usual recipes used in laboratories working with potentiometric sensors, not forgetting that some degree of generic response sensors is needed for the proper operation of the electronic tongue system. An additional trend, not used here, can be picked out when the literature is examined for the formulations of PVC membrane sensors used. This special fashion accomplishes the cross-response characteristics through the use of mixtures of ionophores in the formulation of membranes [34,35], per example using an alkaline-ion ionophore plus and alkaline-earth one, in order to get overall response to cations of different charge.

Coming back to the application developed, the eight ISE arrays were applied to the determination of ammonium ion plus its alkaline interfering ions, in different ways and in different types of samples. Table 30.3 summarizes some of the developed applications in our laboratories with potentiometric electronic tongues. The ammonium determination was first attempted in ammonium/potassium mixtures [36,64], and later, also with sodium, in river waters, wastewaters and fertilizer solutions [82]. This versatility encouraged us to use the electronic tongue for monitoring purposes in fertigation strategy in

TABLE 30.3

Some multidetermination cases employing electronic tongues with ISE arrays and ANNs developed in our laboratories

Species determined	Sensor array used	Special technique	Application to samples	Ref.
NH_4^+ and K^+	8 Potentiometric sensors	–	–	[36,64]
NH_4^+ plus alkaline ions	8 Potentiometric sensors	–	River waters	[82]
NH_4^+ plus alkaline ions	5 Tubular sensors	SIA system	Fertilizers	[51]
Plant nutrients	8 Potentiometric sensors	–	Greenhouse	[83]
NO_3^- and Cl^-	4 Tubular sensors	FIA system	Synthetic samples	[84]
Alkaline-earth ions	5 Tubular sensors	SIA system	Mineral waters	[85]
Mixture of anions	5 Tubular sensors	SIA Dynamic system	Mineral waters	[86]
Mixture of cations	5 Tubular sensors	SIA dynamic system	Mineral waters	[87]
Urea plus alkaline ions	Array of potentiometric sensors and biosensors	–	Urine	[75]

greenhouse cultivation [83]. In this case, a different sensor array of eight ISEs was designed for the monitoring of nutrients (NH_4^+, K^+ and NO_3^-) plus undesirable compounds (Na^+ and Cl^-) that accumulate during recirculation of feed solutions.

In the different cases, the response model of the electronic tongue was built employing ANNs. For this purpose, a large number of samples (mixtures of the ions considered) were generated by cumulative micro-additions of standards over a defined initial volume. Per example, in the three ions case applied to river waters, series of additions employing NH_4^+, K^+, Na^+ or combinations of two or three ions were alternated, generating 174 different mixtures of the three considered ions, and for each, the responses of the ISE array were recorded. The net laboratory effort implied corresponded to two work days. This initial information was next subdivided into the training, validation and test subsets, in order to perform the different stages for the building of the model. The subdivision was done randomly, with 50% of the data for training, 25% for internal validation and 25% for testing.

Next, different configurations of the ANN model were evaluated, in order to attain the best model. Factors to vary were, for example, the transference functions used, pretreatment of initial data, training algorithm used, number of neurons in hidden layers, convergence criteria, or previous initialization of weights. Normally, we use a single hidden layer, given in our experience, two or more hidden layers are rarely needed. Intrinsic parameters of the training process, as learning rate or momentum, should also take reliable values. A recommendable practice is to perform the training process several times, and to check coherence of final results, as with any multivariate optimization process. On every restart, a different set of initialization weights should be used to give consistence. In our case, when fixing details from previous experience, the number of different configurations that were tested could be reduced to *ca.* 50. For each, the goodness of the model was assessed by two indicators. First, a sufficiently reduced sum of residuals, as a measure of the model-fitting degree was reached; therefore, the root mean squares of errors (RMSE) was computed for each ion, according to the following equation:

$$\mathrm{RMSE} = \sqrt{\frac{\sum_j (c_j - \hat{c}_j)^2}{n-1}} \qquad (30.13)$$

where c_j is the predicted concentration, $\hat{c}_j$ its expected value and n the number of samples in the considered subset. Secondly, comparison graphs of predicted versus expected concentration were constructed for each ion. The objective here is to obtain distributions of points the closest to ideality, *viz.*, the $y = x$ identity line, assessed through its least squares regression line $y = m \cdot x + b$. Then, obtained slopes close to 1 and intercepts close to 0 will indicate the proper prediction ability for the developed model. This result is easily attainable when the considered data are that used for training, as ANNs are very efficient in modelling, but the significant result is to check this comparison data for the external test set, the samples not participating in the process. This generalization ability, in the case of the simultaneous multidetermination of ammonium, potassium and sodium yielded (for the test set) the comparison lines $y = 1.07(\pm 0.07) \cdot x + 0.0005(\pm 0.0005)$, $y = 0.98 (\pm\ 0.03) \cdot x + 0.0003(\pm 0.0005)$ and $y = 1.01(\pm 0.05) \cdot x + 0.000(\pm 0.009)$, respectively. The correlation coefficients were 0.995, 0.992 and 0.975 in the same order, showing a low scattered distribution, especially for NH_4^+. As can be seen in all three cases, the uncertainty intervals (calculated at the 95% confidence level) included the theoretical slopes of 1.0 as well as the intercepts of 0.0. These results validated the correct prediction ability for the generated model. Figure 30.10 shows the obtained comparison lines for the considered case.

Automation is one of the trends present in modern Analytical Chemistry. In this way, electronic tongues have been implemented with automated systems, normally employing the flow injection analysis (FIA) technique [33,34,84,88]. When the electronic tongue is devised to be implemented in automated systems, per example an FIA system, or an SIA system, a special design of the ISE is needed to allow integration in the flow path. This design may be of tubular shape as shown in Fig. 30.11 [89], very appropriate to place the sensors in series, as very low distortion develops from one sensor to the next. The construction departs from a Perspex tube to form the outside of the body, into which an electrical connector is plugged. The tube is filled with a conducting epoxy–graphite paste, and, when hardened, a small hole of 1.5 mm in diameter is drilled longitudinally. The inner wall of the hole is covered with the PVC membrane and once formed, letting a flow path of *ca.* 1 mm, conditioned as above.

The existing experience with the FIA technique consisted in the implementation of an electronic tongue in an FIA system [84]. This system employed four tubular ISEs with cross-term response, and permitted the direct determination of nitrate in surface waters with presence of chloride, its typical interferent. Figure 30.12 sketches the concepts in the automated FIA electronic tongue: the injection of the sample generates a peak profile on each sensor used, being the peak height the information fed into the ANN model. The FIA system is shown in Fig. 30.13, a two-line manifold to avoid any distortion in ionic strength due to sample injection. It employed three different sensors to nitrate ion, with different selectivity to chloride, plus a fourth sensor sensitive to chloride. The performance of the system was specially noticeable at the lower concentrations of nitrate ion (*ca.* 1 mg L^{-1} of NO_3^-), where interference of levels up to 800 mg L^{-1} of Cl^- could be compensated.

When using ANNs, lots of information about the system has to be generated for its proper training and validation. In order to generate this information, samples with different concentrations of the analytes have to be prepared and measured. Doing this by hand is a laborious

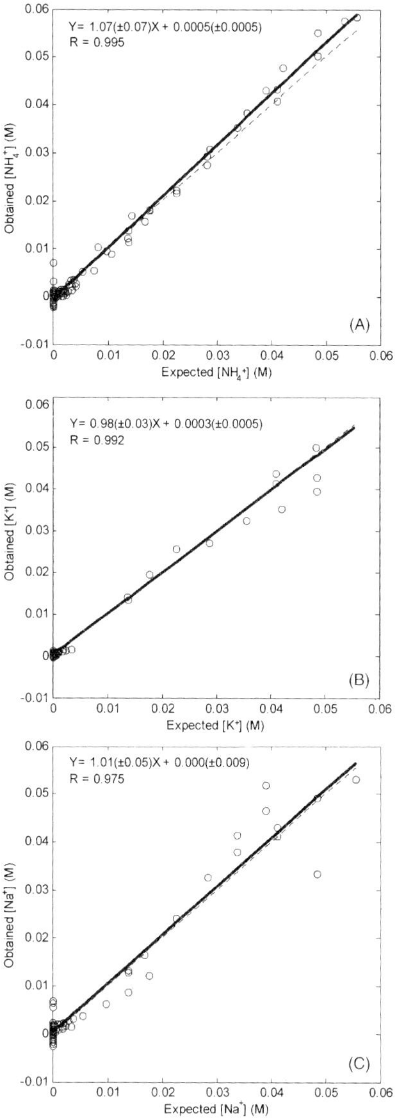

Fig. 30.10. Modelling performance achieved with the ANN model with the samples of the external test set, those not participating in training; (A) Ammonium, (B) Potassium, (C) Sodium. The dashed line corresponds to the theoretical comparison line $y=x$ and the solid line is the regression of the comparison data.

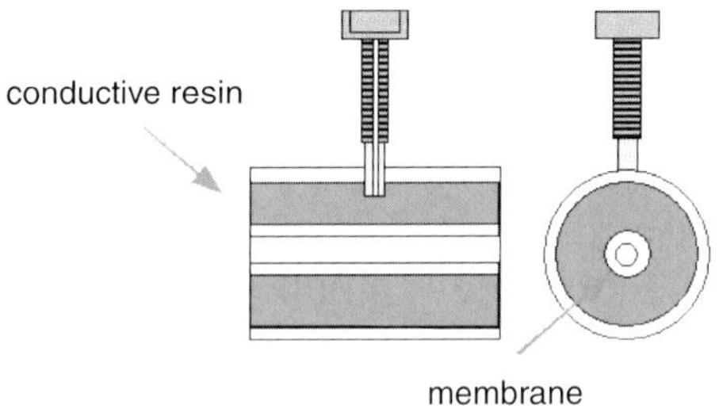

Fig. 30.11. Design of the flow-through solid-contact potentiometric sensors to be used in FIA or SIA systems.

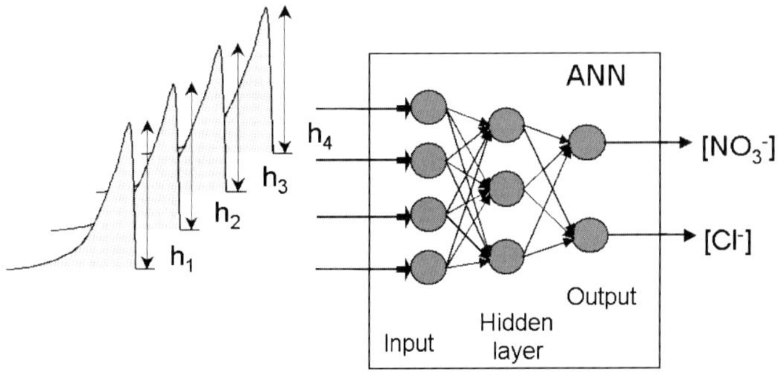

Fig. 30.12. Conceptual scheme of the FIA electronic tongue for the simultaneous determination of nitrate and chloride.

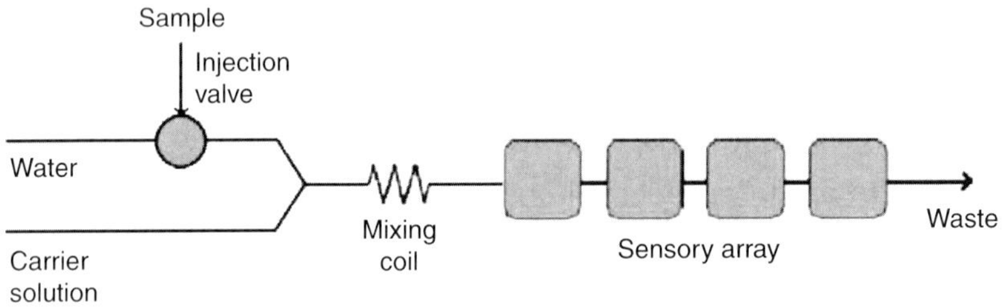

Fig. 30.13. Simple manifold to implement a potentiometric electronic tongue in a flow injection system.

manual procedure, which is susceptible to variability and human error; moreover, it takes by far more time than using an automated liquid handling system. Using automated systems each individual sample can be produced in a random order to negate systematic errors caused by temperature variation or electrode drift; also, the human-induced error may be reduced [90]. SIA systems represent a further evolution of FIA systems [91]. FIA, being well established as a powerful sample handling procedure for automated laboratory and process analytical chemistry, turns up to difficulties in monitoring applications when the required manifolds are too complex, particularly for multicomponent analysis. In certain circumstances, multicomponent process analytical chemistry is only feasible by utilizing several independent FIA systems [92]. In fact, the coupling of FIA systems with electronic tongues does not facilitate its training, but for the speed in the measurement process. The manual preparation of dozens of standards to generate the response model is not prevented anyway. In contrast, SIA systems, featuring increased versatility in handling of solutions, can be employed to generate the mixed standards needed for the training process of an electronic tongue. This idea has already been accomplished in our laboratories, where a specific manifold [50,51] and control software based in Labview environment was developed [93]. The scheme of the SIA system is shown in Fig. 30.14.

The key part of the system is the multiport valve, which interconnects the different parts and solutions used by the system. The common port is connected to a reversible pump with the retention coil placed in between. The pump is connected to the carrier solution reservoir. The common port can access any of the other ports, which lead to sample, standard solutions or reagents, mixing chamber and sensor array, by electrical rotation of the valve. Since the system is bidirectional, volumes can not only be propelled directly to the detector, but also be injected into the retention coil, therefore merging accurate aliquots of different solutions. In order to assure proper mixing of the solutions not only via diffusion

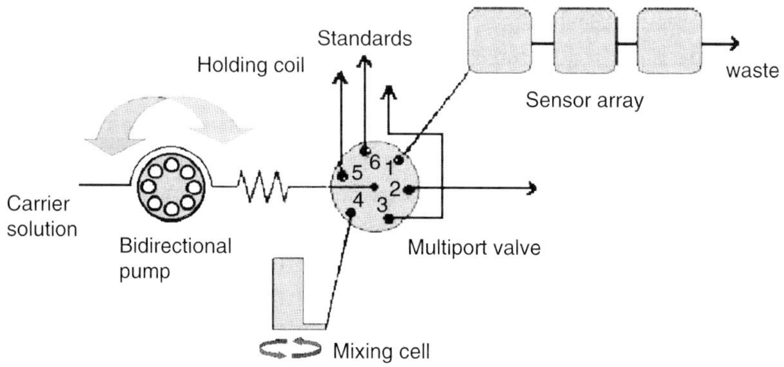

Fig. 30.14. Scheme of the automated SIA electronic tongue.

and secondary forces, a mixing chamber is also used in our application, into which the volume of the retention coil can be injected and homogenized. After that, the solution is propelled to the detector for measurement, which is completed on-line. In this way, the SIA electronic tongue makes possible the automated unattended preparation of *ca.* 100 standards, used for the building of the ANN response model. The system takes controlled volumes of different stock solutions, mixes and dilutes them, and finally, brings the mixture to the sensor array and takes the measurement. The complete process is completed in less than 5 min per standard. Training, validation and test samples are generated alternatively, and external samples are also taken and brought to the sensors in an equivalent way. The determination of alkaline ions, as already studied with batch sensors, was implemented with the SIA technique, now employing an array of five tubular ISEs, yielding satisfactory results [51]. More recently, the same concept has been extended to alkaline-earth cations, where the satisfactory simultaneous determination of mixtures of calcium, magnesium and barium was feasible [85].

When using automated systems, a new possibility can be exploited; this comes from the dynamic nature of the measurements that occur in these. In this way, we have adapted our procedures to record the transient response corresponding to the introduction of a sample. The idea is that the kinetic resolution may improve the modelling ability of electronic tongue systems. This concept has already been presented by the Legin group with an FIA electronic tongue [94]; in our case, we have used the SIA electronic tongue, as it is capable of preparing our training standards. Figure 30.15 depicts the mentioned concept. The novelty is the use of the dynamic components of the signal in order to better discriminate or differentiate a sample taking profit of the different response rate of primary ion and interferents [77]. Two studies have been completed by now, one with alkaline–alkaline earth cation mixtures [86], and the other with anion mixtures [87]. In both cases, results are improved when compared to using steady-state signals as input data for the electronic tongue.

To illustrate similar works present in the literature, of multideter-mination of ions employing potentiometric electronic tongues with ANNs as the chemometric tool, an inspection of the existing contributions was done, which permitted the construction of Table 30.4. Many of the examples come from the group of Legin in Russia, as the most active research group in this field, even when most of their work is not included, given their preference for PCA or PLS as the data treatment procedure. As can be seen, the fruitful coupling of ANNs and ISE arrays has permitted the attainment of a new trend in potentiometric analysis. The

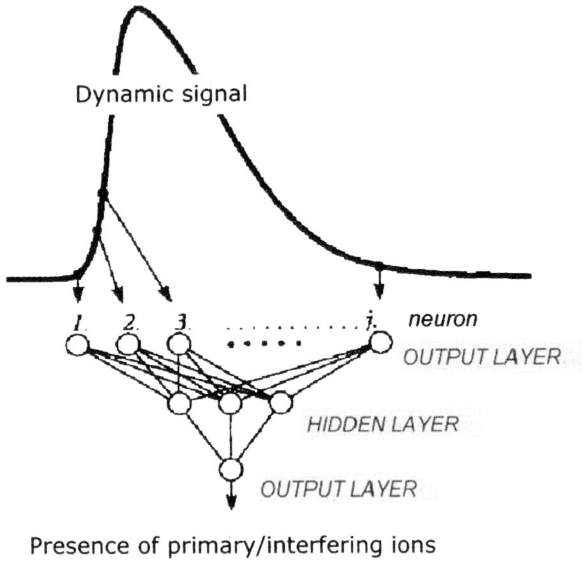

Fig. 30.15. Conceptual scheme of an electronic tongue employing the dynamic component of the FIA/SIA signal.

diagnostic may be that, for its proper dissemination over world laboratories, it only lacks a proper commercial consolidation of the product.

30.3 CONCLUSIONS

This chapter has presented the detailed application of electronic tongues as analytical systems for multidetermination of species, in batch conditions, and also integrated in FIA and SIA systems. The preferred chemometrics tool to model its response, ANNs, has been presented in detail. The different precautions for its use together with options and configurations have been described. The ways employed to check the goodness of fit of a developed response model have also been explained. In our applications, we usually employ a multiple output ANN, but nobody impedes the use of a different ANN with a single output for each species considered, except for the training effort which is multiplied.

The electronic tongue concept has been successfully proved to be a feasible alternative for multiple analyte determinations, specially suited for monitoring applications, where it can provide real-time multicomponent information, simultaneously compensating matrix effects when used in a dynamic manner. For its proper set-up, the main effort is related with the large amounts of training information needed. In this

TABLE 30.4

Some multidetermination cases employing electronic tongues with ISE arrays and ANNs present in the literature

Species determined	Sensor array used	Special technique	Application to samples	Ref.
Cl^-/Br^- mixtures	4 Commercial ISEs	–	–	[44]
K^+, Ca^{2+}, NO_3^-, Cl^-	4 Commercial ISEs	–	–	[46]
Lysine aminoacid+interferents	7 PVC sensors+biosensor	–	Animal feeds	[74]
Zn^{2+}, Pb^{2+}	Chalcogenide ISEs	FIA	–	[94]
Ca^{2+}, Mg^{2+}, Na^+, $HCO_3^-Cl^-$, H^+, HPO_4^{2-}	30 PVC membrane ISEs	–	Artificial blood plasma	[95]
Pb^{2+}, Cd^{2+}, Zn^{2+}, Fe^{3+}	12 Chalcogenide ISEs	Solid-state thin-film sensors	–	[96]
Total solid content	12 PVC membrane sensors	Screen-printed sensors, FIA	Water, beverages	[34]
Catechines, sugars+arginine aminoacid	30 sensors+biosensors	Screen-printed sensors, FIA	Teas	[98]
Ammonium, citrate and oxalate	8 PVC sensors	–	Fermentation broth	[97]
Apparent+perceived quinine content	30 ISEs, PVC+chalcogenide	Bitterness assessment	Pharmaceutical products	[99]
Mixtures of alkylsulphate surfactants	10 PVC membrane ISEs for anionic surfactants	–	–	[100]

aspect, automated systems can help in the generation of information, thus providing simple, low-maintenance, and green analytical systems, where the complexity has moved from special function parts to the software aspect. On the other hand, the use of automated systems adds new possibilities to electronic tongues, as the dynamic component of sensor signals can be used, thus permitting a better resolution among analytes and interfering substances.

As in other sensor applications, the stability of the response if the model is to be applied for monitoring purposes is of paramount importance. This aspect however is one of the circumvented points in current research, which needs to be studied and compensated for the success of their application.

Multicomponent determination is one of the possibilities brought by electronic tongues. Other applications are in its qualitative application, where a classification of sample varieties is one of the possibilities [101], but where assessment of non-directly or non-easily quantifiable aspects, such as aging or perception can be also attempted. Although the chapter has been limited to electronic tongues employing potentiometric sensors, the field is completed with electronic tongues employing other variants, for example voltammetric sensors. In fact, our group has been doing significant contributions also with voltammetric electronic tongues, either using arrays of sensors or even from the complex data supplied by a single sensor. In this case, the multivariate information can be considered to come from the voltammogram. Especially, key contributions have been the incorporation of biosensors to the electronic tongue concept [102,103] or the development of specific variants for the data processing, namely, the use of the wavelet transform for feature extraction prior to modelling using ANNs [104].

ACKNOWLEDGMENTS

The work presented here has been supported by the MECD (Madrid Spain) through project CTQ2004-08134, and by the Department of Innovations, Universities and Enterprise from the Generalitat de Catalunya.

REFERENCES

1 V. Pravdová, M. Pravda and G.G. Guilbault, *Anal. Lett.*, 35 (2002) 2389.
2 J. Workman, K.E. Creasy, S. Doherty, L. Bond, M. Koch, A. Ullman and D.J. Veltkam, *Anal. Chem.*, 73 (2001) 2705.

3 F. Winquist, S. Holmin, D. Krantz-Rülcker, P. Wide and I. Lundström, *Anal. Chim. Acta*, 406 (2000) 147.
4 K.C. Persaud, *Trends Anal. Chem.*, 11 (1992) 61.
5 J.W. Gardner and P.N. Bartlett, *Electronic Noses. Principles and Applications*, Oxford University Press, Oxford, 1999.
6 T.C. Pearce, S.S. Schiffman, T.H. Nagle and J.W. Gardner (Eds.), *Handbook of Machine Olfaction—Electronic Nose Technology*, Wiley-VCH, Weinheim, 2002.
7 A. Legin, A. Rudnistskaya and Y. Vlasov, Electronic tongues new analytical perspective for chemical sensors. In: S. Alegret (Ed.), *Integrated Analytical Systems, Comprehensive Analytical Chemistry series*, Vol. 39, Elsevier, Amsterdam, 2003, pp. 437–486.
8 K. Toko, *Biomimetic Sensor Technology*, Cambridge University Press, Cambridge, 2000.
9 M. Holmberg, M. Eriksson, C. Krantz-Rülcker, T. Artujrsson, F. Winquist, A. Lloyd-Spetz and I. Lundström, *Sens. Actuators B*, 101 (2004) 213.
10 Y. Vlasov, A. Legin, A. Rudnitskaya, C. Di Natale and A. D'Amico, *Pure Appl. Chem.*, 77 (2005) 1965.
11 A. Hulanicki, S. Glab and F. Ingman, *Pure Appl. Chem.*, 63 (1991) 1247.
12 A. Söderström, A. Rudnitskaya, A. Legin and C. Krantz-Rülcker, *J. Biotech.*, 119 (2005) 300.
13 F. Winquist, I. Lundström and P. Wide, *Sens. Actuators B*, 58 (1999) 512.
14 C. Di Natale, R. Paolesse, A. Macagnano, A. Mantini, A. D'Amico, A. Legin, L. Lvova, A. Rudnitskaya and Y. Vlasov, *Sens. Actuators B*, 64 (2000) 15.
15 Y. Vlasov, A. Legin and A. Rudnitskaya, *Sens. Actuators B*, 44 (1997) 532.
16 L. Moreno, A. Merlos, N. Abramova, C. Jimenez and A. Bratov, *Sens. Actuators B*, 116 (2006) 130.
17 C. Krantz-Rülcker, M. Stenberg, F. Winquist and I. Lundström, *Anal. Chim. Acta*, 426 (2001) 217.
18 M. Ferreira, A. Riul, K. Wohnrath, F.J. Fonseca, O.N. Oliveira and L.H.C. Mattoso, *Anal. Chem.*, 75 (2003) 953.
19 J.J. Lavigne, S. Savoy, M.B. Clevenger, J.E. Ritchie, B. McDoniel, S.J. Yoo, E.V. Anslyn, J.T. McDevitt, J.B. Shear and D. Neikirk, *J. Am. Chem. Soc.*, 120 (1998) 6429.
20 G. Sehra, M. Cole and J.W. Gardner, *Sens. Actuators B*, 103 (2004) 233.
21 C. Di Natale, A. Macagnano, F. Davide, A. D'Amico, A. Legin, Y. Vlasov, A. Rudnitskaya and B. Selezenev, *Sens. Actuators B*, 44 (1997) 423.
22 D.L. Massart, B.G.M. Vandeginste, L.M.C. Buydens, S. de Jong, P.J. Lewi and J. Smeyers-Verbeke, *Handbook of Chemometrics and Qualimetrics*, Vol. A and B, Elsevier, Amsterdam, 1997.
23 B.K. Lavine and J. Workman, *Anal. Chem.*, 74 (2002) 2763.
24 P. Ciosek and W. Wróblewski, *Sens. Actuators B*, 114 (2006) 85.

25 M. Colilla, C.J. Fernandez and E. Ruiz-Hitzky, *Analyst*, 127 (2002) 1580.
26 J. Simons, M. Bos and W.E. Van der Linden, *Analyst*, 120 (1995) 1009.
27 S. Holmin, P. Spångeus, C. Krantz-Rülcker and F. Winquist, *Sens. Actuators B*, 76 (2001) 455.
28 Y. Vlasov and A. Legin, *Fresenius J. Anal. Chem.*, 361 (1998) 255.
29 Y. Vlasov, A. Legin, A. Rudnitskaya, A. D'Amico and C. Di Natale, *Sens. Actuators B*, 65 (2000) 235.
30 K. Toko, *Sens. Actuators B*, 64 (2000) 205.
31 K. Toko, *Biosens. Bioelectron.*, 13 (1998) 701.
32 R. Paolesse, C. Di Natale, M. Burgio, E. Martinelli, E. Mazzone, G. Palleschi and A. D'Amico, *Sens. Actuators B*, 95 (2003) 400.
33 J. Mortensen, A. Legin, A. Ipatov, A. Rudnitskaya, Y. Vlasov and K. Hjuler, *Anal. Chim. Acta*, 403 (2000) 273.
34 L. Lvova, S.S. Kim, A. Legin, Y. Vlasov, J.S. Yang, G.S. Cha and H. Nam, *Anal. Chim. Acta*, 468 (2002) 303.
35 P. Ciosek, E. Augustyniak and W. Wróblewski, *Analyst*, 129 (2004) 639.
36 J. Gallardo, S. Alegret, R. Muñoz, M. De Roman, L. Leija, P.R. Hernandez and M. del Valle, *Anal. Bioanal. Chem.*, 377 (2003) 248.
37 http://www.alpha-mos.com/en/products/proast.php.
38 http://www.insent.co.jp/english/top.htm.
39 http://www.electronictongue.com.
40 M. Otto and J.D.R. Thomas, *Anal. Chem.*, 57 (1985) 2647.
41 K. Beebe, D. Uerz, J. Sandifer and B. Kowalski, *Anal. Chem.*, 60 (1988) 66.
42 K. Beebe and B. Kowalski, *Anal. Chem.*, 60 (1988) 2273.
43 R.J. Forster, F. Regan and D. Diamond, *Anal. Chem.*, 63 (1991) 876.
44 M. Baret, D.L. Massart, P. Fabry, C. Menardo and F. Conesa, *Talanta*, 50 (1999) 541.
45 J. Saurina, E. López-Aviles, A. Le Moal and S. Hernandez-Cassou, *Anal. Chim. Acta*, 464 (2002) 89.
46 M. Bos, A. Bos and W.E. Van der Linden, *Anal. Chim. Acta*, 233 (1990) 31.
47 M. Bos, A. Bos and W.E. Van der Linden, *Analyst*, 118 (1993) 323.
48 C.L. Stork and B.R. Kowalski, *Chemometr. Intell. Lab. Syst.*, 46 (1999) 117.
49 J.C. Seiterand and M.D. DeGrandpre, *Talanta*, 54 (2001) 99.
50 A. Gutés, F. Céspedes, S. Alegret and M. del Valle, *Talanta*, 66 (2005) 1187.
51 M. Cortina, A. Gutés, S. Alegret and M. del Valle, *Talanta*, 66 (2005) 1197.
52 H.M. Cartwright, *Applications of Artificial Intelligence in Chemistry*, Oxford University Press, Oxford, 1993.
53 D. Svozil, V. Kvasnicka and J. Pospichal, *Chemometr. Intell. Lab. Syst.*, 39 (1997) 43.
54 J. Zupan and J. Gasteiger, *Neural Networks for Chemists*, Wiley-VCH, Weinheim, 1993.

55 F. Despagne and D.L. Massart, *Analyst*, 123 (1998) 157R.
56 H. Demuth and M. Beale, *Neural Network Toolbox User's Guide*, Version 4, Mathworks Inc., Natick, MA, 2002.
57 C.M. Bishop, *Neural Networks for Pattern Recognition*, Oxford University Press, Oxford, 1995.
58 D.E. Rumelhart and J.L. McClelland, *Parallel Distributed Processing: Explorations in the Microstructure of Cognition*, Vol. 1, The MIT Press, Cambridge, MA, 1986, pp. 318–362.
59 S.S. Rao, *Optimisation: Theory and Applications*, Wiley Eastern, New Delhi, 1978.
60 K. Levenberg, *Quart. Appl. Math.*, 2 (1944) 164.
61 D. Marquardt, *SIAM J. Appl. Math.*, 11 (1963) 431.
62 J.A. Freeman and D.M. Skapura, *Neural Networks: Algorithms, Applications, and Programming Techniques*, Addison-Wesley, Reading, MA, 1991.
63 J.C. Mackay, *Probable networks and plausible predictions a review of practical Bayesian methods for supervised neural networks, Technical Report*, Cavendish Laboratory, Cambridge, UK, 1995.
64 J. Gallardo, S. Alegret, M.A. De Roman, R. Muñoz, P.R. Hernández, L. Leija and M. del Valle, *Anal. Lett.*, 36 (2003) 2893.
65 http://www.mathworks.com.
66 http://www.trajan-software.demon.co.uk.
67 http://www.statsoft.com.
68 http://www.easynn.com.
69 A. Legin, A. Rudnitskaya, Y. Vlasov, C. Di Natale, F. Davide and A. D'Amico, *Sens. Actuators B*, 44 (1997) 291.
70 K. Toko, A taste sensor. In: S. Alegret (Ed.), *Integrated Analytical Systems, Comprehensive Analytical Chemistry Series*, Vol. 39, Elsevier, Amsterdam, 2003, pp. 487–511.
71 L. Lvova, E. Martinelli, E. Mazzone, A. Pede, R. Paolese, C. Di Natale and A. D'Amico, *Talanta*, 70 (2006) 833.
72 R. Martínez-Máñez, J. Soto, E. Garcia-Breijo, L. Gil, J. Ibáñez and E. Llobet, *Sens. Actuators B*, 104 (2005) 302.
73 J.M.C.S. Magalhaes and A.A.S.C. Machado, *Analyst*, 127 (2002) 1069.
74 N. Garcia-Villar, J. Saurina and S. Hernandez-Cassou, *Fresenius J. Anal. Chem.*, 371 (2001) 1001.
75 M. Gutiérrez, S. Alegret and M. del Valle, *Biosens. Bioelectron.*, 22 (2007) 2171.
76 R.W. Cattrall, *Chemical Sensors*, Oxford University Press, Oxford, 1997.
77 W.D. Diamond (Ed.), *Principles of Chemical and Biological Sensors*, Wiley, New York, 1998.
78 Y. Umezawa (Ed.), *Handbook of Ion-Selective Electrodes Selectivity Coefficients*, CRC Press, Boca Raton, FL, 1990.
79 O.G. Davies, G.J. Moody and J.D.R. Thomas, *Analyst*, 113 (1998) 497.

80 H. Sheen, T.J. Cardwell and R.W. Cattrall, *Analyst*, 123 (1998) 2181.
81 K. Tohda, K. Suzuki, H. Aruga, M. Matsuzoe, H. Inove and T. Shirai, *Anal. Chem.*, 60 (1988) 1714.
82 J. Gallardo, S. Alegret, R. Muñoz, M. De Román, L. Leija, P.R. Hernández and M. del Valle, *Electroanalysis*, 17 (2005) 348.
83 M. Gutiérrez, S. Alegret, R. Cáceres, J. Casadesús, O. Marfà and M. del Valle, *Comput. Electron. Agric.*, (2007), doi:10.1016/j.compag.2007.01.012.
84 J. Gallardo, S. Alegret and M. del Valle, *Sens. Actuators B*, 101 (2004) 72.
85 D. Calvo, M. Größl, M. Cortina and M. del Valle, *Electroanal*, 19 (2007) 644.
86 M. Cortina, A. Duran, S. Alegret and M. del Valle, *Anal. Bioanal. Chem.*, 385 (2006) 1186.
87 D. Calvo, A. Duran and M. del Valle, *Anal. Chim. Acta*, (2007), doi:10.1016/j.aca.2006.11.079.
88 P. Ciosek and W. Wróblewski, *Talanta*, 69 (2006) 1156.
89 J. Alonso, J. Baró, J. Bartrolí, J. Sànchez and M. del Valle, *Anal. Chim. Acta*, 308 (1995) 115.
90 E. Richards, C. Bessant and S. Saini, *Sens. Actuators B*, 88 (2003) 149.
91 G.D. Christian, *Analyst*, 119 (1994) 2309.
92 N.W. Barnett, C.E. Lenehan and S.W. Lewis, *Trends Anal. Chem.*, 5 (1999) 346.
93 A. Duran, M. Cortina, L. Velasco, J.A. Rodriguez, S. Alegret and M. del Valle, *Sensors*, 6 (2006) 19.
94 A. Legin, A. Rudnitskaya, A. Legin, A. Ipatov and Y. Vlasov, *Russ. J. Appl. Chem.*, 78 (2005) 89.
95 A. Legin, A. Smirnova, A. Rudnitskaya, L. Lvova, E. Suglobova and Y. Vlasov, *Anal. Chim. Acta*, 385 (1999) 131.
96 Y. Mourzina, J. Schubert, W. Zander, A. Legin, Y. Vlasov, H. Lüth and M.J. Schöning, *Electrochim. Acta*, 47 (2001) 251.
97 A. Legin, D. Kirsanov, A. Rudnitskaya, J.J.L. Iversen, B. Seleznev, K.H. Esbensen, J. Mortensen, L.P. Houmøller and Y. Vlasov, *Talanta*, 64 (2004) 766.
98 L. Lvova, A. Legin, Y. Vlasov, G.S. Cha and H. Nam, *Sens. Actuators B*, 95 (2003) 391.
99 A. Legin, A. Rudnitskaya, D. Clapham, B. Seleznev, K. Lord and Y. Vlasov, *Anal. Bioanal. Chem.*, 380 (2004) 36.
100 A.I. Kulapin, R.K. Chernova, E.G. Kulapina and N.M. Mikhaleva, *Talanta*, 66 (2006) 619.
101 J. Gallardo, S. Alegret and M. del Valle, *Talanta*, 66 (2005) 1303.
102 A. Gutés, F. Céspedes, S. Alegret and M. del Valle, *Biosens. Bioelectron.*, 20 (2005) 1668.
103 A. Gutés, A.B. Ibáñez, M. del Valle and F. Céspedes, *Electroanalysis*, 18 (2006) 82.
104 L. Moreno-Barón, A. Merkoçi, S. Alegret, M. del Valle, L. Leija, P.R. Hernandez, R. Muñoz and R. Cartas, *Sens. Actuators B*, 113 (2006) 487.

Chapter 31

Electrochemical sensors for food authentication

Saverio Mannino, S. Benedetti, S. Buratti, M.S. Cosio and Matteo Scampicchio

31.1 INTRODUCTION

Food authentication is the process by which a food is verified as complying with its label description. Authenticity problems are directly associated with the labelling of a product and compositional regulations, which may differ from country to country and that have a fundamental place in determining which scientific tests are appropriate for a particular issue.

In the case of wine, for instance, the denomination of origin is granted only to wines manufactured in particular geographical areas, with typical characteristics linked to the environment, to natural factors and to traditions of the area. Also for other food products, such as cheese and extra virgin olive oil, the Denomination of Protected Origin (DPO) indicates the place of origin of the product. In order to have the DPO mark, the two conditions are required: (1) raw materials and the transformation into the final product have to take place in the delimited area indicated by the "Disciplinary of Production" and (2) the quality and characteristic of the product have to be related to the geographical environment and the traditions of the place of origin.

Considering the importance to characterise food products, the analytical quality control needs methods that are able to extract useful information from the food matrix. Infact, each food product on the market can be viewed as "a typical piece" that can be characterised by some precise analytical indices. It is not important to know all the quantitative aspects that characterise the products, but it is sufficient to create a map of analytical semi-quali-quantitative signals that constitutes the product fingerprint. Since the majority of the methods reported for food analysis

Comprehensive Analytical Chemistry 49
Alegret and Merkoçi (Eds)
Volume 49 ISSN: 0166-526X DOI: 10.1016/S0166-526X(06)49031-0

are time-consuming, complex, require specialised personnel and cannot be completely automatic, the development of rapid, low-cost, multi-elementary, simple and objective methods allowing to have the signals that are in relation with the product characteristics is today of great interest for food industry and laboratories.

One of the promising directions for the development of innovative analytical method is the use of electrochemical methods whose speed and on-line capabilities nicely address the trends of automation and continuous processing in the food industry. Recently, devices such as electronic nose and electronic tongue have been proposed for the characterisation and authentication of different type of food products, and also for medical and environmental application.

The electronic nose and the electronic tongue are the common names of electrochemical sensor systems responding to flavour/odour (volatiles) or taste (solubles), using an array of simple and non-specific sensors and a pattern recognition software system [1]. Essentially, each odour or taste leaves a characteristic pattern or fingerprint on the sensor array, and a pattern recognition able to distinguish and recognise the odours and the tastes.

The key principle involved in the electronic nose concept is the transfer of the total headspace of a sample to a sensor array that detects the presence of volatile compounds in the headspace and a pattern of signals is provided that are dependent on the selectivity and sensitivity of sensors and the characteristics of the volatile compounds in the headspace [2].

The electronic nose consists of an array of gas sensors with different selectivity, a signal collecting unit and a pattern recognition software. It is particularly useful for the analysis of headspace of liquid or solid food samples [3], and numerous attempts of using the electronic nose have been reported [4,5].

Various kinds of gas sensors are available, but only four technologies are currently used in commercialised electronic noses: metal oxide semiconductors (MOSs), metal-oxide semiconductor field-effect transistors (MOSFETs), conducting organic polymers (CPs) and piezoelectric crystals (bulk acoustic wave (BAW)). Others, such as fibreoptic [6], electrochemical [7] and bi-metal sensors, are still in developmental stage and may be integrated in the next generation of the electronic noses. In all cases, the goal is to create an array of differentially sensitive sensing elements.

Metal oxide semiconductor (MOS) sensors were first used commercially in the 1960s as household gas alarms in Japan under names of

Taguchi (the inventor) or Figaro (the company's name). These sensors rely on changes of conductivity induced by the adsorption of gases and subsequent surface reactions [8]. They consist of a ceramic substrate (round or flat) heated by wire and coated by a metal oxide semiconducting film. The metal oxide coating may be either of the n-type (mainly tin dioxide, zinc oxide, titanium dioxide or iron (III) oxide), which responds to oxidising compounds, or of the p-type (mainly cobalt oxide or nickel oxide), which responds to reducing compounds [9]. The film deposition technique divides each sensor type into thin- (6–1000 nm) or thick- (10–300 μm) film MOS sensors. The first one offer a faster response and significantly higher sensitivity but are much more difficult to manufacture in terms of reproducibility. Therefore, commercially available MOS sensors are often based on thick-film technologies. Due to the high operating temperature (200–650°C), the organic volatiles transferred to the surface of the sensors are totally oxidised to carbon dioxide and water, leading to the change in the resistance. MOS sensors are extremely sensitive to ethanol which blinds them to any other volatile compound of interest.

Metal-oxide semiconductor field-effect transistor (MOSFET) sensors rely on a change of electrostatic potential. A MOSFET sensor comprises three layers, a silicon semiconductor, a silicon oxide insulator and a catalytic metal (usually palladium, platinum, iridium or rhodium), also called the gate. When polar compounds interact with this metal gate, the electric field, and thus the current flowing through the sensor, are modified. The recorded response corresponds to the change of voltage necessary to keep a constant pre-set drain current [10]. The selectivity and sensitivity of MOSFET sensors may be influenced by the operating temperature (50–200°C), the composition of the metal gate and the microstructure of the catalytic metal. MOSFET sensors have a relatively low sensitivity to moisture and are thought to be very robust.

Conducting organic polymer (CP) sensors like MOS sensors rely on changes in resistance by adsorption of gas. These sensors comprise a substrate (such as a fibre-glass or silicon), a pair of gold-plated electrodes and a conducting organic polymer such as polypyrrol, polyaniline or polythiophene as a sensing element. The polymer film is deposed by electrochemical deposition between both the electrodes previously fixed to a substrate. When a voltage is passed across the electrodes, a current passes through the conducting polymer [11]. The addition of volatile compounds to the surface of the sensor alters the electron flow in the system and therefore the resistance of the sensor. In general, CP sensors show good sensitivities especially for polar compounds. However, their low operating temperature ($<50^{\circ}C$) makes them extremely sensitive to moisture. Although such sensors are resistant to poisoning, they have a lifetime of only about 9–18 months.

Piezoelectric crystal sensors are based on the change of the mass, which may be measured as a change in resonance frequency. These sensors are made of tiny discs, usually quartz, lithium niobate ($LiNbO_3$) or lithium tantalate ($LiTaO_3$), coated with materials such as chromatographic stationary phase, lipids or any non-volatile compounds that are chemically and thermally stable [12]. When an alternating electrical potential is applied at room temperature, the crystal vibrates at a very stable frequency, defined by its mechanical properties. Upon exposure to a vapour, the coating adsorbs certain molecules, which increase the mass of the sensing layer and hence decrease the resonance frequency of the crystal. This change may be monitored and related to the volatile present. The crystal may be made to vibrate in a BAW or in a surface acoustic wave (SAW) mode by selecting the appropriate combination of crystal cut and type of electrode configuration. BAW and SAW sensors differ in their structure: BAW are three-dimensional waves travelling through the crystal, while SAW are two-dimensional waves that propagate along the surface of the crystal at a depth of approximately one wavelength. These devices are also called "quartz crystal microbalance" (QCM or QMB) because, similar to a balance, their responses change in proportion to the amount of mass adsorbed. Since piezoelectric sensors may be coated with an unlimited number of materials, they present the best selectivity. However, the coating technology is not yet well controlled, which induces poor batch-to-batch reproducibility.

The principle of the electronic tongue is similar to that of electronic nose, but the array of sensors is designed for liquids. There are various techniques that can be used for an electronic tongue, such as conductimetric, potentiometric or voltammetric techniques [13]. In literature, studies using potentiometric electronic tongue for beverage analysis and wine discrimination have been reported [14,15]. An electronic tongue based on voltammetry was used to classify fruit juice, still drink and milk and also to follow the aging process of milk and orange juice when stored at room temperature [16,17], while information about the use of an electronic tongue based on flow injection system operating in amperometry are limited to our lab. There are two main trends in the work with electronic tongues, those that try to mimic tastes like sourness, sweetness, saltiness, bitterness, etc. and those more related to conventional analytical applications both qualitative and quantitative.

In literature there are several works about the combination of an electronic nose and an electronic tongue for clinical and food analysis [18–22]. In all these works, it has been demonstrated that the discrimination and the classification properties are improved when information from both electronic nose and electronic tongue are combined.

It must be pointed out that the electronic nose and the electronic tongue methodology are able to show the authenticity of the product only if supported by the multivariate statistical techniques useful not only to classify but also to validate and predict unknown samples.

The data processing of the multivariate output data generated by the gas sensor array signals represents another essential part of the electronic nose concept. The statistical techniques used are based on commercial or specially designed software using pattern recognition routines like principal component analysis (PCA), cluster analysis (CA), partial least squares (PLSs) and linear discriminant analysis (LDA).

Principal components analysis (PCA) is a procedure that permits to extract useful information from the data, to explore the data structure, the relationship between objects, the relationship between objects and variables and the global correlation of the variables. It was used for explorative data analysis, as it identifies orthogonal directions of maximum variance in the original data, in decreasing order, and projects the data into a lower-dimensionality space formed of a subset of the highest-variance components. The orthogonal directions are linear combinations (principal components (PCs)) of the original variables and each component explains in turn a part of the total variance of the data; in particular, the first significant component explains the largest percentage of the total variance and the second one, the second-largest percentage, and so forth [23].

Cluster analysis (CA) performs agglomerative hierarchical clustering of objects based on distance measures of dissimilarity or similarity. The hierarchy of clusters can be represented by a binary tree, called a dendrogram. A final partition, i.e. the cluster assignment of each object, is obtained by cutting the tree at a specified level [24].

Partial least square (PLS) regression model describes the dependences between two variables blocks, e.g. sensor responses and time variables. Let the **X** matrix represent the sensor responses and the **Y** matrix represent time, the **X** and **Y** matrices could be approximated to few orthogonal score vectors, respectively. These components are then rotated in order to get as good a prediction of y variables as possible [25].

Linear discriminant analysis (LDA) is among the most used classification techniques. The method maximises the variance between categories and minimises the variance within categories. This method renders a number of orthogonal linear discriminant functions equal to the number of categories minus one [26].

Innovative multivariate statistical analyses, such as artificial neural network (ANN) and genetic algorithms (GAs), were also used in order build regression models with real predictive capability and applicable to unknown samples.

The artificial neural networks (ANN) are very sophisticated modelling techniques capable of modelling extremely complex functions [27]. The basic unit of an ANN is the neuron. A neural network consists of a set of interconnected network of neurons. The input layer has one neuron for each of the sensor signals, while the output layer has one neuron for each of different sample properties that should be predicted. Usually, one hidden layer with a variable number of neurons is placed between the input and the output layer. During the ANN training phase, the weights and transfer function parameters are adjusted such that the calculated output value for a set of input values is as close as possible to the unknown true value of the sample properties. The model estimation is more complex than for a linear regression model due to the non-linearity of the model. Neural networks learn from examples through iteration, without requiring *a priori* knowledge of the relationship among variables under investigation. The mostly used neural network was a multilayer perceptron (MLP) that is based on an algorithm called backpropagation. Since a neural network can arrive at different solutions for the same data, if different values of the initial network weights are provided, the network was trained several times. The goal was to try and find a neural network model for which multiple training approaches the same final mean squared error (MSE).

Genetic algorithms (GAs) were employed to select variables and build predictive regression models [28]. GAs select subsets of variables that maximise the predictive power of regression models and perform this selection by considering populations of models generated with an evolution process and optimised according to an objective function (in this case Q^2 leave-one-out calculated with the ordinary least square regression) [29]. Q^2 leave-one-out cross-validation involves the use of a single observation from the original matrix as the validation data, and the remaining observations as the training data. This is repeated such that each observation in the sample is used once as the validation data.

A population is made of a series of chromosomes. Each chromosome is a binary vector, where each position (a gene) corresponds to a variable (i.e. a chromosome represents a model made up of a subset of selected variables). The evolution process is based on three main steps: initially, the model population is randomly built. The value of the objective function of each model is calculated and the models are then ordered with respect to this objective function. After that, the reproduction step selects pairs of models (parents) and from each pair of models a new model (son) is generated preserving the common characteristics of the parents (i.e. variables excluded in both models remain excluded; variables included in both models remain included) and mixing the opposite characteristics. If the generated son coincides with one of the individuals already present in the actual population, it is rejected; otherwise, it is evaluated. If the objective function value is better than the worst value in the population, the model is included in the population, in the place corresponding to its rank; otherwise, it is no longer considered. This procedure is repeated for several pairs. The mutation step instead changes every gene of each chromosome into its opposite according to a defined probability. If the objective function of each mutated model is better than the worst value in the population, the model is included in the population. Reproduction and mutation steps are alternatively repeated until a stop condition has occurred or the evolution process is ended arbitrarily.

In the last years, the aim of our research was to investigate the potential of the gas and liquid sensor arrays for food authentication and quality control: investigation on the freshness evolution of Italian Crescenza cheeses, characterisation and classification of honey of different botanical and geographical origin, and characterisation and classification of Italian Barbera wines.

31.2 APPLICATION

31.2.1 Crescenza cheese

The aim of this work was to assess the suitability of the electronic nose to define the shelf-life of Crescenza cheese. Crescenza is a typical soft Lombardy (Italy) cheese. Two lots of Crescenza cheese packets (250 g), directly supplied by a manufacturer at the beginning of their commercial life, were stored under constant temperature conditions at 8 and 15°C in the original paper packaging and were analysed every 3 days in a 31-day period and every day for a 10-day period, respectively. Furthermore, 14 Crescenza packets of the same brand and purchased from a local market at the beginning of their commercial life were used for the validation of the electronic nose (EN) tests at 0, 6, 13 and 20 day storage at 8 and 15°C. All samples were kept at room temperature (20+1°C) for 30 min before analyses. Each sample was evaluated two times and the average of the results was used for statistical analysis.

Analyses were performed with an electronic nose (model 3320 Applied Sensor Lab Emission Analyser; Applied Sensor Co., Linkoping, Sweden), which consisted of three parts: automatic sampling apparatus, detector unit containing the array of sensors (10 MOSFETs and 12 MOS) and software for pattern recognition. The MOSFET sensors were divided into two arrays of five sensors each, one array operating at 140°C and the other at 170°C, while the MOSs, mounted in a separate chamber, were kept at 400–500°C during all the process phases. The automatic sampling system supported a carousel of 12 loading sites for the samples under controlled temperature.

Two grams of each sample were placed in 40 mL Pyrex® vials with silicone caps and then settled into the automatic sampling carousel. Preliminary trials indicated that larger sample volumes did not significantly enhance signal intensity and reproducibility. After 20 min equilibration at room temperature, the measurement run started with a 10 min incubation at 25°C. An air flow (room air filtered through active carbon) was conveyed over the sensors at a constant rate (1 cm^3 s^{-1}) for 10 s to stabilise the baseline. An automatic syringe then suckled Crescenza cheese headspace and conveyed it over the sensor surfaces for 30 s. The total cycle time for each measurement was 5 min.No sensor drift during the measurement period was experienced. Each sample was evaluated three times and the average of the results was used for the analysis.

The sensor responses were collected and elaborated by PCA performed on scaled data to achieve a partial visualisation of the set in a reduced dimension. The two first principal components represented 98% of the total variance and their score plot (Fig. 31.1) allowed a separation of the samples according to the storage conditions. Samples were distributed along PC1 and PC2 according to the storage time and storage temperature, respectively.

In order to characterise Crescenza cheese samples, a supervised pattern recognition method (LDA) was applied. The results of LDA and leave-one-out cross-validation are reported. LDA applied to the sensor responses gave a 100% recognition for all Crescenza cheese (error rate 0%), and also during the leave-one-out cross-validation, all samples were correctly classified (cross-validation error rate 0%) in three groups: "fresh" corresponding to 0–7 days at 8°C and 0–4 days at 15°C, "aged"

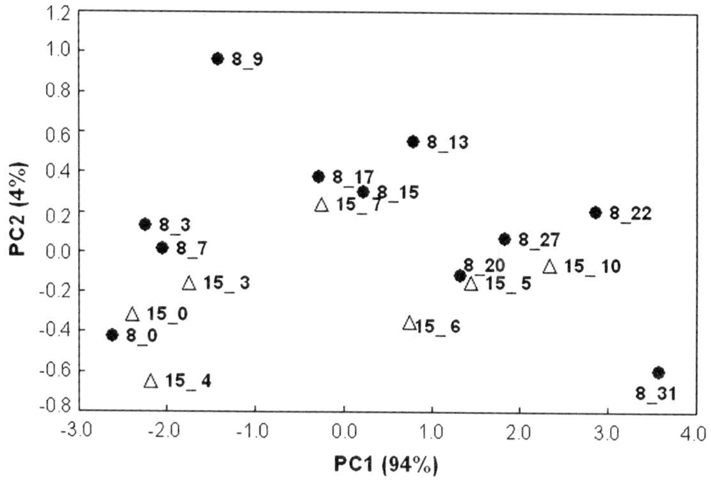

Fig. 31.1. PCA score plot of Crescenza cheese samples in the plane defined by the first two principal components: (•) samples stored at 8°C and (Δ) samples stored at 15°C.

relating to 9–20 days at 8°C and 5–7 days at 15°C and "very aged" for samples from 22 to 31 days at 8°C and after 10 days at 15°C. The classification model was applied to a new set of electronic nose data, i.e. the seven Crescenza packets purchased from a local market and analysed at the beginning of the storage and after 6, 13 and 20 days at 8 and 15°C. Figure 31.2 shows the predictive ability of LDA model.

It can be seen that three samples stored at time 0 and 6 days at 8 and 15°C, respectively, were grouped in the "fresh" class, while four samples stored for 13 and 20 days at 8 and 15°C, respectively, were classified in the "aged" class.

From the results illustrated in this work, it is clear that the sensor responses of the electronic nose can be used to define the threshold of the shelf-life of Crescenza samples stored at different temperatures.

31.2.2 Honey

Seventy different unifloral honey samples of specific botanical and geographical origins were analysed by an electronic nose. Fourteen samples from *Robinia pseudoacacia* L., 30 from *Rhododendron* spp. and 20 from *Citrus* spp. were of Italian origin, and six samples from *Robinia pseudoacacia* honey were of Hungarian origin. All samples were provided by the Istituto Nazionale di Apicoltura (Bologna, Italy) and their

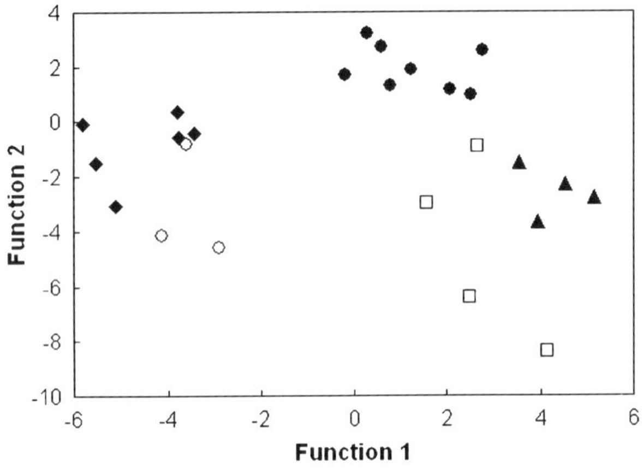

Fig. 31.2. Projection of Crescenza cheese samples predicted by the LDA model: (♦) "fresh" samples, (•) "aged" samples, (▲) "very aged" samples, (○) unknown samples from 0 to 6 days and (□) unknown samples from 13 to 20 days.

unifloral authenticity was verified in the accredited laboratory by pollen, organoleptic and chemical analyses. Analyses were performed with the same commercial electronic nose (model 3320 Applied Sensor Lab Emission Analyser; Applied Sensor Co.) used for Crescenza cheese samples.

Three grams of each sample were placed in 40 mL Pyrex® vials with silicone caps and then introduced inside the automatic sampling carousel of the electronic nose. After an equilibration time of 20 min at room temperature, the measurement sequence started with an incubation of 40°C for 10 min. Each sample was evaluated three times and the average of the results was used for subsequent statistical analysis (PCA and ANN).

The cumulative variance explained by the first two principal components was 79.3%, with the first component providing 48.0% of the total (Fig. 31.3A).

The first two principal components on electronic nose data clearly separated the *Rhododendron* honey from the three others. The third component explained 13.2% of the total variance, and the fourth 4.4% (Fig. 31.3B). The third and fourth components separated the other three types of honey: the *Citrus* honey was located in the right part of the plot, the Hungarian *Robinia* in the upper central part and the Italian *Robinia* in the lower left part of the plot.

Once explored the sensor data by PCA and determined that the four types of honey could be separated, a classification model was performed by the ANN technique. The results of ANN analysis can be visualised

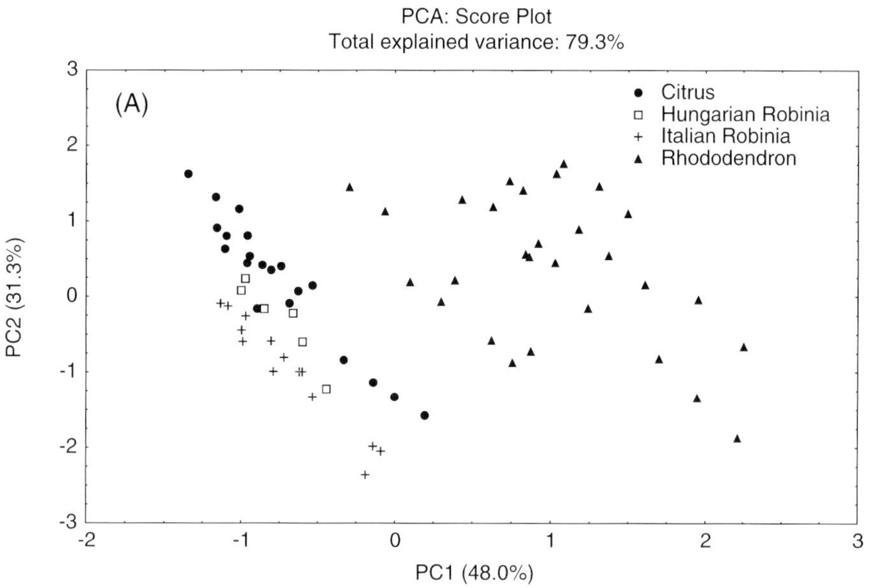

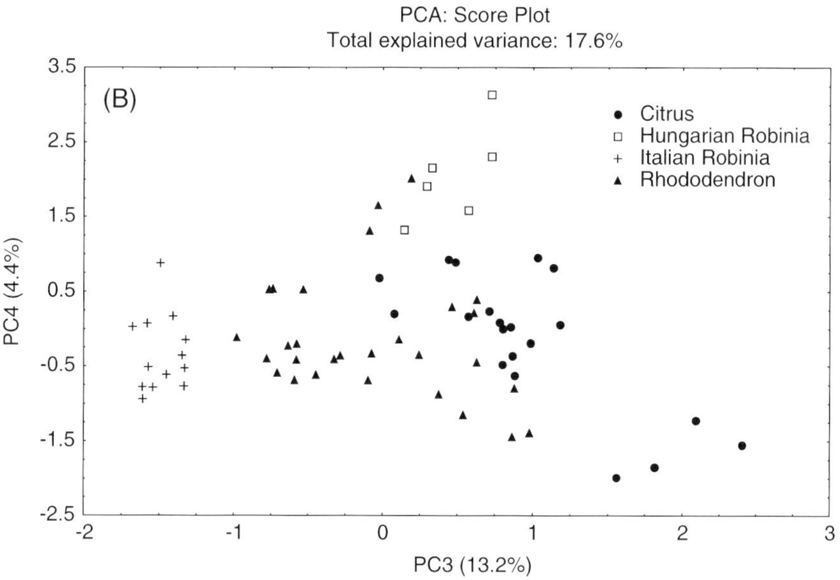

Fig. 31.3. PCA plot of the scores produced by the 22 sensor responses along (A) the first two principal components and (B) the third and fourth principal components.

TABLE 31.1

Confusion matrix for the 17 honey samples in the testing set analysed with the ANN

Sample honey type	Output of the ANN			
	Rhododendron	Italian *Robinia*	Hungarian *Robinia*	*Citrus*
Rhododendron	6	0	0	0
Italian *Robinia*	0	2	0	0
Hungarian *Robinia*	0	0	1	0
Citrus	0	0	0	8

by a confusion matrix of the desired and actual network output in Table 31.1. A confusion matrix displays the desired classification on the rows and the predicted classification on the columns. The ideal situation is to have all the exemplars end up on the diagonal cells of the matrix. The confusion matrix in Table 31.1 shows perfect classification. Every honey type was correctly classified by the network as *Rhododendron*, Italian *Robinia*, Hungarian *Robinia* and *Citrus*.

However, for each set of samples, a two-way binomial test (Engels, 1988) was used to determine the lowest probability of correct identification that would have given a 95% chance of assigning all samples in this set to the correct category. Thus, for the 6 samples of *Rhododendron* honey, the lowest probability of correct identification was 0.616; for the 8 samples of *Citrus* honey, this probability was 0.679; and for all 17 samples, the lowest probability was 0.835, indicating that there was a 95% chance that the probability of correct identification of the ANN used was 0.835 or better.

31.2.3 Wine

In the European Union quality wine produced in determined regions (VQPRD—Vins de Qualité Produits dans des Régions Déterminées) is normally used to indicate the denomination of origin of wine. The VQPRD corresponds to the Italian denomination of controlled and guaranteed origin (DOCG) and denomination of controlled origin (DOC) denomination (Italian law no. 164/1992 and Reg. CEE 823/87). In order to have the DOC mark, two conditions are required: (1) the supply of raw materials and their transformation into the final product have to take place in a delimited area as indicated by the "Disciplinary of Production" and (2) the quality and the characteristic of wine have to be exclusively related to the geographical area so that this wine must be unique and not reproducible outside the area of origin. DOCG is exclusively used for wines known and appreciated in the world, which have held the DOC denomination for at least 5 years.

Fifty-three different Barbera wines, chosen from two northern Italian regions (Piemonte and Lombardia), were analysed by electronic nose and electronic tongue. All samples were commercially available DOC wines. Twenty-three of them were Barbera Oltrepò produced in the Province of Pavia (Lombardia), 13 Barbera Piemonte produced in the province of Alessandria, Asti, Cuneo (Piemonte), 12 Barbera Asti produced in the province of Asti and Alessandria (Piemonte) and 5 Barbera Alba produced in the province of Cuneo (Piemonte). All Barbera wines were of the same vintage (2002) and of different wine-growers.

Analyses were performed with a portable electronic nose (PEN2) operating with the enrichment and desorbtion unit (EDU). The system was from WMA (Win Muster Airsense) Analytics Inc. (Germany). PEN2 consists of a sampling apparatus, a detector unit containing the array of sensors and a pattern recognition software (Win Muster v.3.0) for data recording. The sensor array is composed of 10 MOS-type chemical sensors. The sensor response is expressed as resistivity (Ohms).

EDU is a microprocessor-controlled device capable of automatically trapping and thermally desorbing the samples. The adsorbent material is Tenax-TA® polymer, 150 mg.Three samples, each of 1 mL, were taken from the same wine bottle immediately after opening and placed in 30 mL Pyrex® vials provided with a pierceable silicon Teflon disk in the cap. After 10 min headspace equilibration time, the measurement sequence started. Ambient air cleaned by active charcoal was used as reference gas. In EDU, four analytical steps are particularly important and are performed automatically: sampling at 20°C for 60 s, desorption at 80°C for 60 s, injection at 80–160°C for 80 s in order to have the gradual release of all volatile compounds and cleaning and cooling at 300°C for 240 s. Each sample was evaluated three times and the average of the sensor responses was used for subsequent statistical analysis.

The electronic tongue system based on flow injection analysis (FIA) with two amperometric detectors was set up. The FIA apparatus consisted of a Jasco (Tokyo, Japan) model 880 PU pump and two EG&G Princeton Applied Research (Princeton, NJ, USA) Model 400 thin-layer electrochemical detector connected in series. Each detector was equipped with a working electrode (a dual glassy carbon electrode and a gold electrode), a reference (Ag/AgCl saturated) electrode and a platinum counter electrode. The connecting tubes were of PEEK (1.5 mm o.d. × 0.5 mm i.d.). Data were recorded using a Philips (Eindhoven, Netherlands) PM 8252 recorder. In the flow system, a carrier solution is continuously pumped through the amperometric detectors and the samples are injected into the flow stream. Data are obtained by measuring the current resulting from the oxidation or reduction of the electroactive compounds present in the samples. Analyses were performed at room temperature using a carrier solution composed of 70% methanol and 30% acetate buffer (0.1 M, pH 4) to which 2% sodium chloride was added. A 1 mL min^{-1} flow rate and a 20 mL injection volume were employed. No sample preparation was needed except 1:20 dilution with the carrier solution before injection. For each sample, a double sequence of injections was set. In the first injection sequence, the dual glassy carbon electrode and the gold electrode operated in a parallel configuration at potentials of 0.4 and 0.6 V for carbon and 0.4 V for gold. In the second injection sequence, the dual glassy carbon working electrode operated at 0.8 and –0.2 V in a series configuration. Each sample was evaluated twice and the average of the results was used for subsequent statistical analysis.

All data collected from the electronic nose (10 variables) and the electronic tongue (5 variables) were compared and elaborated by PCA.

The four first principal components represent 69% of the total variance. On examining the score plot (Fig. 31.4) in the area defined by the first two principal components (51.7% of the total variance), a clear

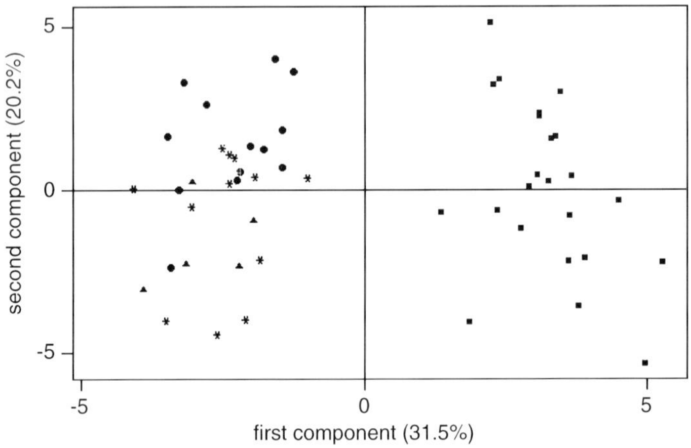

Fig. 31.4. Score plot of wine samples in the plane defined by the first two principal components: (■) Barbera Oltrepò Pavese, (●) Barbera Piemonte, (*) Barbera Asti and (▲) Barbera Alba.

separation of the samples into two groups was found according to the region of origin.

The first group, in the positive part of the plot, was completely composed of Barbera Oltrepò samples; the second group, in the negative part, was formed of samples from Piemonte. Within the second group, Barbera Piemonte samples tended to be in the upper part of second principal component, while Barbera Asti and Barbera Alba samples were in the bottom part of the second principal component and the differentiation of these wines was not possible.

In order to characterise wine samples into the mentioned four classes, a supervised pattern recognition method (LDA) was applied. The results obtained gave 100% correct classification for the three classes (Barbera Oltrepò, Barbera Piemonte and Barbera Alba) and only one Barbera Asti sample was not correctly classified (cross-validation error rate 1.89%).

31.3 CONCLUSIONS

Electronic nose and electronic tongue are particularly suitable for carrying out rapid and objective sensory measurements, which are important in food industries.

Given their non-specific nature, the electronic nose and electronic tongue sensor arrays can only perform "yes or no" tests inside the set of product. Contrary to traditional analytical methods, the electrochemical sensor responses do not need and do not provide information on the nature of the compounds under investigation, but only on digital fingerprint of the typical food products.

REFERENCES

1 A.K. Deisingh, D.C. Stone and M. Thompson, *Int. J. Food Sci. Technol.*, 39 (2004) 587–604.
2 J.W. Gardner and P.N. Bartlett, *Sens. Actuators B*, 18 (1993) 211–220.
3 E. Schaller, J.O. Bosset and F. Escher, *Lebensm. Technol.*, 31 (1998) 305–303.
4 S. Ampuero and J.O. Bosset, *Sens. Actuators B*, 94 (2003) 1–12.
5 P.N. Bartlett, J.M. Elliot and J.W. Gardner, *Food Technol.*, 57 (1997) 44–48.
6 T.A. Dickinson, J. White, J.S. Kauer and D.R. Walt, *Nature*, 382 (1996) 697–700.

7 H. Baltruschat, I. Kamphauser, R. Oelgeklaus, J. Rose and M. Wahlkamp, *Anal. Chem.*, 69 (1997) 743–748.
8 D. Kohl. In: G. Sberveglieri (Ed.), *Gas Sensors*, Kluwer Academic Publishers, Dordrecht, 1992, pp. 43–88.
9 P. Mielle, *Trends Food Sci. Technol.*, 7 (1996) 432–438.
10 I. Lundstrom, A. Spetz, F. Winquist, U. Ackelid and H. Sundgren, *Sens. Actuators B*, 1 (1990) 15–20.
11 M.E.H. Amrani, K.C. Persaud and P.A. Payne, *Meas. Sci. Technol.*, 6 (1995) 1500–1507.
12 G. Guilbault and J.M. Jordan, *Crit. Rev. Anal. Chem.*, 19 (1988) 1–28.
13 F. Winquist, S. Holmin, C. Krantz-Rulcker, P. Wide and I. Lundstrom, *Anal. Chim. Acta*, 406 (2000) 147–157.
14 A. Legin, A. Rudnitskaya, Y. Vlasov, C. Di Natale, E. Mazzone and A. D'Amico, *Electroanalysis*, 11 (1999) 814–820.
15 A. Legin, A. Rudnitskaya, L. Lvova, Y. Vlasov, C. Di Natale and A. D'Amico, *Anal. Chim. Acta*, 484 (2003) 33–44.
16 F. Winquist, E. Rydberg, S. Homlin, C. Krantz-Rulcker and I. Lundstrom, *Anal. Chim. Acta*, 471 (2002) 159–172.
17 F. Winquist, C. Krantz-Rulcker, P. Wide and I. Lundstrom, *Meas. Sci. Technol.*, 9 (1998) 1937–1946.
18 C. Di Natale, R. Paolesse, A. Magagnano, A. Martini, A. D'Amico, A. Legin, L. Lvova, A. Rudnitskaya and Y. Vlasov, *Sens. Actuators B*, 64 (2000) 15–21.
19 R.N. Bleibaum, H. Stone, T. Tan, S. Labreche, E. Saint-Martin and S. Isz, *Food Qual. Pref.*, 13 (2002) 409–422.
20 S. Buratti, S. Benedetti, M. Scampicchio and E.C. Pangerod, *Anal. Chim. Acta*, 525 (2004) 133–139.
21 F. Winquist, I. Lundstrom and P. Wide, *Sens. Actuators B*, 58 (1999) 512–517.
22 V. Parra, A.A. Arrieta, J.A. Fernandez-Escudero, M. Iniguez, J.A. de Saja and M.L. Rodriguez-Mendez, *Anal. Chim. Acta*, 563 (2006) 229–237.
23 K.R. Beebe, R.J. Pell and M.B. Seasholtz, *Chemometrics, A Practical Guide*, Wiley, New York, USA, 1998.
24 J.W. Gardner and P.N. Bartlett, Pattern recognition in odour sensing. In: *Sensors and Sensory Systems for an Electronic Nose*, Vol. 212, Kluwer Academic Publishers, Dordrecht, 1992, pp. 161–179.
25 K.G. Joreskogand and H. Wold (Eds.), *Systems Under Indirect Observation, Parts I and II*, North Holland, Amsterdam, 1982.
26 M. Meloun, J. Militky and M. Forina, *Chemometrics for Analytical Chemistry*, Ellis Horwood, New York, USA, 1992.
27 C.M. Bishop, *Neural Network for Pattern Recognition*, Oxford University Press, UK, 2002.
28 D.E. Goldberg, *Genetic Algorithms in Search, Optimization and Machine Learning*, Addison-Wesley Pub. Co., MA, USA, 1989.
29 R. Leardi, R. Boggia and M. Terrile, *J. Chemomet.*, 6 (1992) 267–281.

Chapter 32

From microelectrodes to nanoelectrodes

Pedro Jose Lamas, Maria Begoña González and Agustin Costa

32.1 INTRODUCTION

32.1.1 General considerations about electrodes with reduced dimensions (ERD)

Electrodes have a dimension which controls the electrochemical response i.e. the critical dimension. That is a radius (for cylindrical, spherical, hemispherical and disk geometry) or a width (for ring and band geometry). Based on the scale of the critical dimension, the electrodes can be classified into three groups [1]:

- Higher than 25 μm: conventional electrodes or macroelectrodes.
- Below 25 μm: microelectrodes or ultramicroelectrodes (UME).
- Below 10 nm: nanodes or nanoelectrodes (NE).

The mass transport at electrodes is a very complex issue that is the key to understand the electrode behaviour when its critical dimension is reduced [2]. Briefly, if diffusion is the only mass transport, two mass fluxes will be present in an electrolysis: consumption flux (at the electrode surface) and diffusive flux (across the diffusion layer). For equilibrating fluxes, the bigger the electrode surface the thicker is the diffusion layer and longer the distance across it (d_{Eq}). Nevertheless, the distance that the electroactive species cover by diffusion transport is limited (d_{D}). At macroelectrodes, the fluxes equilibrium never takes place because d_{Eq} is too long for the d_{D}. Therefore, a transient state is always present. As the critical dimension is decreased, the consumption flux and d_{Eq} are reduced. There is a value for the critical dimension at which d_{Eq} approaches d_{D}. Then, fluxes can be equilibrated and a steady state is achievable. Apart from other questions (like electrode geometry

Comprehensive Analytical Chemistry 49
Alegret and Merkoçi (Eds)
Volume 49 ISSN: 0166-526X DOI: 10.1016/S0166-526X(06)49032-2

and experimental time-scale [3–5]), the capacity to establish this effective diffusive mass transport can be considered the basis of microelectrodic behaviour. Generally, the decrease in the electrode dimensions implies other important changes like the reduction of capacitance and ohmic drop [6].

What happens when the dimensions are furthermore reduced? Initially, an enhanced diffusive mass transport would be expected. That is true, until the critical dimension is comparable to the thickness of the electrical double layer or the molecular size (a few nanometers) [7,8]. In this case, diffusive mass transport occurs mainly across the electrical double layer where the characteristics (electrical field, ion solvent interaction, viscosity, density, etc.) are different from those of the bulk solution. An important change is that the assumption of electroneutrality and lack of electromigration mass transport is not appropriate, regardless of the electrolyte concentration [9]. Therefore, there are subtle differences between the microelectrodic and nanoelectrodic behaviour.

The above-mentioned values for the critical dimensions should be considered like approximations. In fact, a UME with a critical dimension of 25 μm is not totally different from another of 30 μm. Nevertheless, sometimes the terms UME and, specially, NE are erroneously employed. Today, in keeping with other aspects of nanotechnology and nanoscience, the electrochemical scale of interest is around 100 nm (far from the scale of the nanoelectrodic behaviour). Perhaps, the term ERD (electrodes with reduced dimensions) is more adequate for UMEs and those "NEs" (i.e. electrodes whose critical dimension is between 25 μm and 10 nm), because their electrochemical behaviours are the same.

Finally, another relevant question is the possibility of grouping several ERDs (electrode arrays or ensembles). In those systems, the behaviour of each electrode can be affected by neighbouring ones [10]. Depending on the inter-electrode distance and the experimental time-scale, five cases can be differentiated (Fig. 32.1). The main interest is focused on designs where each ERD works independently because the properties of the ERDs are maintained while the currents are summed up (Figs. 32.1a and b). Nevertheless, the case of the "overlapping" diffusion layers (Fig. 32.1d) is also interesting because although the system responses as a macroelectrode, the limit detection is improved [11].

32.1.2 Methods for the construction of ERDs and NEs

A wide variety of electrodes can be classified like ERDs. Therefore, the bibliography about their construction is vast. Table 32.1 shows some

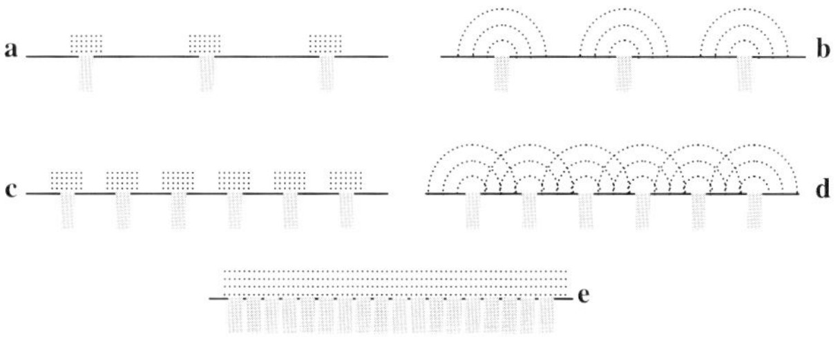

Fig. 32.1. Diffusion profiles of grouped ERDs; (a) high interelectrode distance (i.d.) and short experimental time-scale (e.t.), (b) high i.d. and long e.t., (c) low i.d. and short e.t., (d) low i.d. and long e.t., (e) very low i.d., the electrode works as macroelectrode independently of e.t.

general and simple methods. The sealing of micrometric conductive materials (fibres or films) into insulator matrices (curing epoxy resins or melting soft glass) is the easier way to achieve UMEs (see more details in Section 32.2.1). Micrometric fibres and films are commercially available. For thinner films, plane or cylindrical surfaces are covered with atomic depositions but a more specific instrumentation is required. The pipette puller method is based on the local heating of a glass capillary together with an inserted fibre. Both are simultaneously pulled and disk UMEs of very small dimensions (electrode and surrounding insulator) can be obtained. Recently, local heating produced by a laser source allows to obtain reproducible NEs. The etching of UMEs is an easy way to diminish the electrode dimensions or to fabricate other UMEs. By flame etching of carbon fibres sealed in soft glass, minute needle carbon ERDs with low noise are obtained. Recessed or pore electrodes can be generated by electrochemical etching of disk electrodes, which can be filled with other electrode material like carbon paste or epoxy–graphite composite. The so-called "etching/insulator" method is based on the sequential electrochemical etching of a cylinder UME (to a sharp tip) and further insulation of all (except the end of the tip). Waxes, varnishes or electrophoretic paints are insulation materials commonly employed in these cases. In this way, true NE could be fabricated. The template method [24] is a low-cost methodology for manufacturing randomly grouped ERDs, i.e. electrode ensembles. This method involves the filling or synthesis of a desired electrode material within the micro or nanopores of a membrane. Usually, alumina and "track-etch" polycarbonate membranes are used like templates. The second ones are more common and membranes with different porosities are commercially available. The process of filling

TABLE 32.1
General methods for construction of ERDs and NEs

Method	Conductive material	Geometry	Critical dimension	Additional comment	Ref.
Sealing into an insulator material	Metal or carbon fibre	Cylindrical	−5 μm	The fibre is cut to the required length.	[12]
		Disk		Finally, the surface is polished.	
		Hemispherical		Hg electrodeposition on a disk (Pt/Ir)UME.	[13]
		Spherical			
	Metal foils	Band		Finally, the surface is polished.	[12]
	Organometallic paints	Ring		The paint became a metallic film by thermochemical reduction.	[14]
	Metal films	Band	−5 nm	Metal deposition by sputtering or atomic vapour.	[7]
		Ring			[15]
	Carbon films		−50 nm	Pyrolizing a methane flow into a capillary.	[16]

Pipette puller	Metal or carbon fibre	Disk	–1 μm	Tiny tip electrode is achieved.	[17]
			–10 nm	A laser is required.	[18]
Etching of UMEs	Carbon fibres	Needle	–100 nm	Flame etching of carbon fibre sealed into glass.	[19]
	Metal or carbon fibre		–1 nm	Specific electrochemical etching for each case.	[1]
		Recessed		From disk electrodes.	[20]
Etching/insulator	Metal or carbon fibre	Conic		Adequate sealing of needle electrodes.	[1]
Template	Carbon paste	Disk	–1 μm	Rubbing the membrane onto the paste.	[21]
	Metal		–200 nm	Electrodeposition.	[22]
			200–10 nm	Electroless deposition.	[11]
		Cylindrical	–10 nm	Etching metal disk ensembles.	[23]

depends on pore size. At the micrometer range, the pores can be filled by metal electrodeposition or by rubbing on carbon pastes. For smaller pores (diameter bellow 200 nm), filling by electroless deposition is more suitable. Finally, another approach consists of insulating a planar electrode and further drilling of holes. They are created in the thin insulator layer through the underlying electrode. In this way, mixed alkene thiol self-assembled monolayers [25] and block self-assembled copolymers [26] are being employed to fabricated NE ensembles.

Moreover, nanostructured materials are useful for the construction of NE. For example, a carbon NE has been reported by sealing of a carbon nanotube under an insulator layer [27]. NE ensembles have been obtained through self-assembling of gold nanoparticles [28] and carbon nanotubes [29] at derivatized substrates. Another interesting approach is the direct growth of carbon nanotubes on electrodes with dispersed catalytic nickel nanoparticles. In this case, highly dispersed carbon NE ensembles can be constructed [30].

Although some low-cost methodologies have been reported [31,32], conventional lithographic techniques are the main ways for the construction of UME arrays, i.e. orderly grouped electrodes. The equipment is not cheap but a wide variety of designs and a final inexpensive manufacturing cost can be achieved. Carbon electrode arrays with suitable properties are not obtained by these techniques. Nevertheless, metallic arrays can be modified by pyrolizing organic molecules previously adsorbed on their surfaces [33]. Advances in photo- and e-beam lithographic techniques continue to enable the fabrication of ERD arrays on the nanometer scale [34]. Recently, carbon NE arrays were obtained by direct growth of carbon nanotubes from a catalytic nickel nanopattern generated by lithographic techniques [35].

The main inconvenience of the ERDs construction is the lack of reproducibility. Due to the tiny electrode surfaces, small variations imply big changes. The sealing between the electrode surface and the insulator material is very crucial for obtaining a well-defined electrode surface and low noise. Their characterization can be achieved by different techniques [17]. Scanning electron microscopy (SEM) is suitable for UMEs but not for smaller ERDs. Information about ERD dimensions can be obtained from the experimental (by chronoamperometry or cyclic voltammetry) and theoretical response in well-defined electrochemical systems [5]. Moreover, this electrochemical characterization shows several limitations when ERDs approach the low nanometric scale [8,14,36].

32.1.3 Applications of ERDs

Single ERDs could replace macroelectrodes for electroanalytical applications. Moreover, special properties of ERDs have made possible the use of different electroanalytical techniques where macroelectrodes were not suitable. Some of these special applications are consequence of their low ohmic drop (voltammetric experiments in very high resistive mediums such as organic, low temperature systems, solids, gases, supercritic fluids or mediums without supporting electrolyte [6]). As consequence of their reduced capacitance and ohmic drop, other important application of ERDs is high-speed cyclic voltammetry in static or hydrodynamic mediums [14,37]. This is a useful tool for studies of fast electrokinetic processes and for use as detectors in flow systems (FIA and HPLC) [38]. Other specific applications of ERDs are derived from their tiny dimensions. In this way, ERDs are employed for electroanalysis with small sample volumes [39] or high spatial resolution. Moreover, the use of NE in scanning electrochemical microscopy (SECM) allows the approaching to single molecular electrochemistry [40] and the improvement of the spatial resolution [1]. However, the most characteristic application of ERDs is the in vivo analysis and the monitoring of biological analytes.

ERDs and the (bio)sensors based on them are largely exploited in biology and medicine (especially in neuroscience) because these devices are very suitable for implants [41,42]. Apart from their analytical characteristics (specificity, time response, sensitivity and limit of detection), the biocompatibility of the ERDs as implants is higher than others (like devices based on mirodialysis probes) and, therefore, their results are more representatives [43]. The use of ERDs for studies of neurotransmitters in brains has been achieved in three ways [44]. First, bare (or with a Nafion film to avoid the ascorbic acid interference) carbon fibre or disk UMEs have been widely employed to monitoring catecholamines [45,46]. Second, modified ERDs and enzymatic sensors based on them allow a selective detection of electroactive and not electroactive neurotransmitters, respectively [47–49] (see Table 32.2). Finally, potentiometric detection of electrolytes by capillary microelectrodes with adequate membranes [67–69] permits following the influx or efflux of ions (like Ca^{2+}, Na^{+} and K^{+}) that are involved on the neuronal mechanisms. The practical applications of ERDs in the nanometer range are scarce. On the one hand, the most relevant examples of biosensors are two glucose sensors based on a Pt NE [70] and a needle carbon ERD [71]. On the other, the greater spatial resolution of the

TABLE 32.2

Modified ERDs or enzymatic microsensors for analysis in vivo of analytes with biological interest (DME and CME are disk and cylindrical UME, respectively)

Analyte	Matrix	ERD	Description	Ref.
Glutamate	Brain	C CME	Multienzyme redox-hydrogel under Nafion	[50]
	Brain	Pt CME	Enzyme attached at electropolymer	[51]
	Neuron culture	Pt DME	Bienzyme redox-hydrogel coating	[52]
Acetylcholine and choline	Brain	C CME	Multienzyme redox-hydrogel under Nafion	[53]
	Brain	Pt/Ir CME	Multienzyme onto an electropolymer	[54]
Lactate	Brain	C CME	Enzyme cross-linked under cellulose acetate	[55]
Serotonin	Urine	C CME	Electrodeposited polymer cover with Nafion	[56]
	Blood serum	C CME	Electrodeposited polymer modified with DNA	[57]
Adenosine	Spinal cord	Pt CME	Multienzyme in electrogenerated polymer	[58]
Dopamine	Brain	Pt/Ir CME	Enzyme trapped in electropolymer	[59]
ATP	Spinal cord	Pt/Ir CME	Bienzyme in sol–gel	[60]
	Artificial membrane	Pt DME	Bienzyme trapped in electropolymer	[61]
NO	Brain slice	C CME	Electropolymer over Nafion film	[62]
	Neuron culture	C CME	Electropolymer covered with Nafion	[63]
Glucose	Brain	C CME	Enzyme trapped in electropolymer	[64]
Insuline	β-Cells	C DME	Electrodeposited ruthenium oxide	[65]
O_2	Cells culture	Recessed Pt	Holes filled with cellulose acetate	[66]

needle carbon ERD (with respect to the carbon disk UME) allowed to demonstrate that the dopamine release by PC12 single cells take place only from dispersed superficial vesicles [72].

The small currents generated at single ERDs can be an obstacle for conventional instrumentation. Therefore, the use of grouped ERDs is an easy way to avoid this problem. Some UME ensembles are commercially available but their practical applications are few [73,74]. Recently, the interest about disk ERD ensembles (generated by the template method) have been increased. Apart from their low detection limits for diffusive electrochemical species, they show several advantages for cell culture studies [75] and for DNA [76] and enzymatic sensor performance [77–80]. Furthermore, gold nanotube electrodes ensembled in polycarbonate membranes have demonstrated an improvement in the characteristics of glucose biosensors although those systems do not present the low detection limits of the disk ensembles [81–84].

Arrays can be divided into three groups depending on their design (Fig. 32.2). In the first design, each electrode is independently addressed (Fig. 32.2a). This type of array allows several practical applications. On one hand, these devices can be fabricated for obtaining the same information at different sites of a medium. For example, studies about the responses of neuronal cell cultures under different stimulus [85–87]. On the other hand, these devices can be designed to obtain different information at reduced spaces, i.e. multianalyte determination. Those arrays can be employed directly like detectors in flow systems [88] (where each electrode is maintained at a different potential) or they can be individually modified for specific analytes [89].

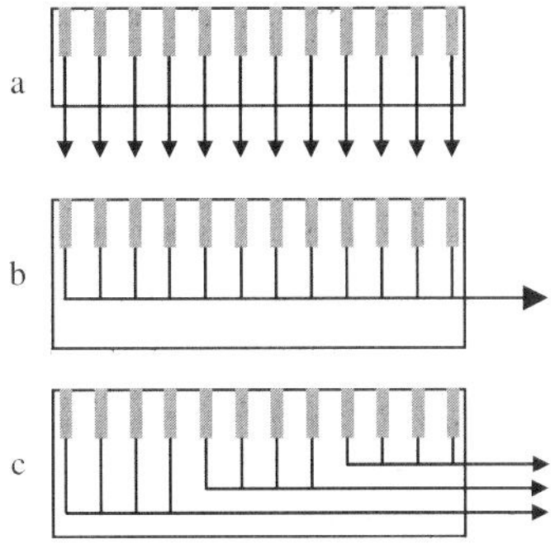

Fig. 32.2. Designs of electrode arrays: (a) each electrode is independently addressed, (b) all electrodes work in parallel fashion and (c) groups of electrodes work in parallel fashion but each group is independently addressed (for more details see text).

In this way, recent ceramic-based ERDs arrays fabricated by lithographic techniques were employed for multineurochemical determination in the brain of rats [90]. These devices are a suitable alternative to microdialysis systems because they allow monitoring several neurochemicals with high biocompatibility and spatial resolution. In the second design, all electrodes work in parallel fashion (Fig. 32.2b). Usually, ERD distribution in these arrays is high enough to avoid overlapping of the diffusion layers. Heavy metal determination in environmental media [91–93] and development of microbiosensor devices [94] are some of their practical applications. Nowadays, low-noise carbon NE arrays are becoming one of the most important transducers for electrochemical genosensor devices because ultrasensitive DNA determinations that can be achieved [35]. Finally, in the third design (Fig. 32.2c), ERD arrays are grouped and independently addressed, i.e. the first and second designs at the same time. In this way, small devices with high performance for multianalyte determination are obtained (for example inmunosensor, genosensor or heavy metals determination [95,96]). Interdigitated electrode arrays (IDA) are a special case of the abovementioned ‘‘third’’ design.

The specific electrochemical behaviour of IDAs is result of its design [97], i.e. two arrays intercalated and individually addressed in a bipotentiostatic system where reversible redox species can be cycled between one array (generator) and the other array (collector) (Fig. 32.3). The feedback obtained, greatly enhances the current and high sensitive detection can be achieved. An important application of IDAs is the electrochemical detection of *p*-aminophenol when it is generated from *p*-aminophenyl phosphate, by enzymatic reaction with alkaline phosphatase (like enzymatic label), in geno- [98–100] and immunoassays [101–103]. Another interesting feature of IDAs is the possibility of

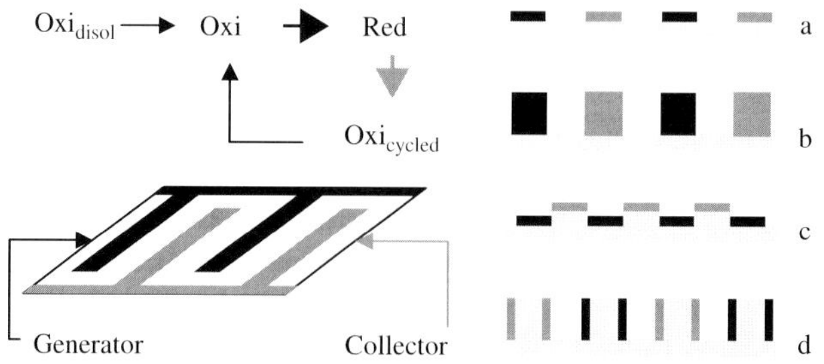

Fig. 32.3. Left, Scheme of redox cycling at IDAs. Right, different design of IDAs; (a) flat [107], (b) 3D comb [110], (c) step [111] and (d) comb [112].

avoiding electrochemical interferences, for example, the dopamine detection (a reversible redox specie) without interference of ascorbic acid (irreversible redox specie) in brain fluids [104–106].

Long time ago, theoretical works predicted that the feedback at IDAs is directly linked to the dimensions of the electrodes and the interelectrode distance (gap) [107,108] and, recently, the height of the electrodes was found to be another crucial parameter [109]. Different models have been developed to enhance the efficiency collection (Fig. 32.3) but the greatest advances were with nanometric dimensions [113]. In this way, IDAs have improved the analytical characteristics of current methodologies (as higher sensitive *p*-aminophenol detection [114] or capacitive biosensors [115,116]) and other applications have emerged [117].

32.2 APPLICATION

32.2.1 Construction of carbon UMEs

Before explaining the construction procedure, we have to warn about the importance of the carbon fibre choice. These fibres are produced by high temperature pyrolysis of polymeric materials. Nevertheless, it has been reported that their physical and chemical properties depend on the manufacturing procedure. In our works, the carbon fibres (nominal diameter of 7.5 μm) used were supplied by Donnay (Belgium). Therefore, our procedures and results would be unsuitable for other carbon fibres.

UMEs used in our laboratory were constructed by sealing of carbon fibre into low viscosity epoxy resin (see Fig. 32.4) [118]. This method is simple, rapid and no specialised instrumentation is required. Firstly, the fibres are cleaned with this aim. They are immersed in dilute nitric acid (10%), rinsed with distilled water, soaked in acetone, rinsed again with distilled water and dried in an oven at 70°C. A single fibre is then inserted into a 100-μL standard micropipette tip to a distance of 2 cm. A small drop of low-viscosity epoxy resin (A. R. Spurr, California) is carefully applied to the tip of the micropipette. Capillary action pulls the epoxy resin, producing an adequate sealing. The assembly is placed horizontally in a rack and cured at 70°C for 8 h to ensure complete polymerization of the resin. After that, the electric contact between the carbon fibre and a metallic wire or rod is made by back-filling the pipette with mercury or conductive epoxy resin. Finally, the micropipette tip is totally filled with epoxy resin to avoid the mobility of the external connection. Then, the carbon fibre UME is ready. An optional protective sheath can be incorporated to prevent electrode damage.

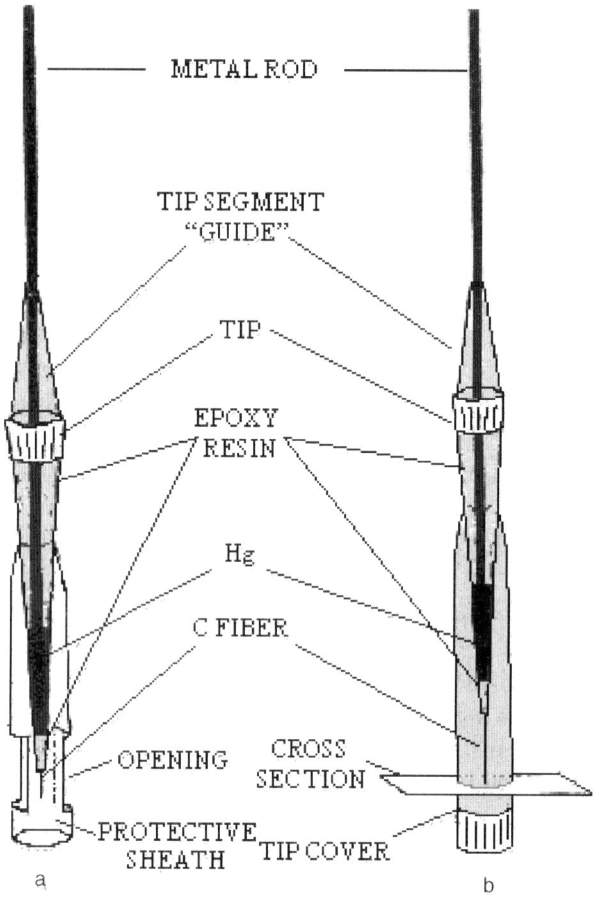

Fig. 32.4. Schematic drawing of (a) cylindrical and (b) disk carbon UMEs.

The preparation of carbon disk-shape UMEs is similar to that of the carbon fibre UME. The only difference is that the active part is sealed on epoxy resin. With this purpose, a micropipette tip (1 mL) is glued to the head-tip of a carbon fibre UME. The carbon fibre is maintained in vertical position with a metal hook and the micropipette tip (1 mL) is filled with epoxy resin. Once the resin is cured, the tip is cross-sectioned with a microtome or a blade. Then, the disk carbon UME is ready.

32.2.2 Pretreatment of the UMEs

When UMEs are used as analytical tools, the main disadvantage is the lack of reproducible measurements. The electrode surface changes often result in variations in sensitivity or reversibility and in extreme cases may lead to complete inhibition of charge transfer. Therefore, it is widely recognised that pretreatment of solid electrodes has a marked effect on their response to many species and it constitutes a critical and essential step in order to get well-defined and reproducible voltammetric signals. In our experience, the choice of the activation procedure is not a simple and easily answered question since firstly, it always depends on the analyte assayed and secondly, the activation procedures could damage the UME.

The pretreatment of the carbon disk UMEs (like a greater part of plane electrodes) is commonly based on mechanical procedures, i.e. polishing [118]. Initially, carbon disk UMEs are grounded down on emery paper (600 grit). Then, the electrode surface is polished with successively finer grades of alumina slurries (0.1- and 0.05-μm diameter particles) on respective cloths. Usually, that is enough to obtain a clean and adequate electroactive electrode surface.

Pretreatment of the carbon fibre UMEs is more complicated. An effective mechanical procedure is not possible due to the electrode geometry. Therefore, a chemical and/or electrochemical pretreatment must be optimized for each case. In some of our works, a chemical pretreatment had been employed. This pretreatment is an oxidative procedure that consists of placing the carbon fibre UME for 5 min in a test-tube containing chromic acid mixture and then, after washing with distilled water, it is dipped in concentrated nitric acid for 2 min. Finally, it is washed again [119–121]. In other works, electrochemical pretreatments have been employed. For folic acid determination [122], the pretreatment selected as optimum was a potential cycle from 0 to 2 V at a frequency of 10 Hz for a period of 1 min in a Britton-Robinson buffer (pH 2). A more general procedure is achieved in acetonitrile solution (tetraethylamonium perchlorate [118,123] or tetra-*n*-butylammonium hexafluorophosphate (TBAHFP) 5.0×10^{-3} M [124,125]). A cathodic potential of −3.5 V for 15 s is applied to the electrode. Then, 10 cycles of an alternating potential using a triangular waveform starting at 0.0 V and following the sequence ($0 \rightarrow -1 \rightarrow +2 \rightarrow 0$ V) at 10 V/s is applied to the fibre. After that, the electrode is checked with ferrocene in acetonitrile (and electrolyte support). The pretreatment is applied several times until well-defined waves of ferrocene are obtained. This procedure is more general, it does not damage the electrode and it is suitable for reproducibility measurements of different analytes and media.

Usually, the effect of pretreatment results in an increment of faradaic and charging currents. The increase in charging currents would be interpreted as an increase in the surface of the electrode. However, the enhancement of faradic currents would not be explained by this relatively small increase in electrode surface [122]. The increased sensitivity to faradaic process is attributed to the generation of a surface quinone– hydroquinone system as a consequence of the oxidative pretreatments. After the pretreatment, the electrode sensitivity decreases progressively. This can be due to an electrode surface passivation (by-products of electrodic reactions) or loss of active quinoidal functions. In these cases, the pretreatment must be repeated for each measurement.

32.2.3 Bare carbon UMEs

The well-known equation "$i = 4nFDCr$" (i limiting current, n number of electrons implied in the electrochemical process, F Faraday constant, D diffusion coefficient, C electroactive specie concentration and r radius of the disk) describes the theoretical steady-state limiting currents of the disk UMEs. This equation is useful to determine the effective radius of a disk UME and to estimate diffusion coefficients. In this sense, the above-mentioned polished carbon disk UMEs have been characterised through the limiting currents obtained in solution with known parameters, i.e. ferrocyanide aqueous solutions (0.05 M and 2 M KCl) [118]. The experimental limiting currents were fairly accurately described by this equation ($\pm 10\%$). When the effective radius is determined, this equation can be employed to obtain unknown diffusion coefficients. In this way, we have estimated the diffusion coefficients for β-carotene in several aprotic solvents with different electrolytic concentrations [123].

β-Carotene is an important source of vitamin A. The oxidative cleavage at the 15,15′-double bond in the liver generates two molecules of vitamin A_1 retinal, which is eventually reduced to vitamin A_1 retinol. Studies about the kinetic and reaction mechanism of this first oxidation step were carried out in our laboratory. β-Carotene is highly insoluble in water, and consequently, its redox characteristics should be studied in nonaqueous media. In these cases, bare carbon fibre UMEs (BCFMEs) are a suitable analytical tool [126]. The goal of our study was not the electroanalytical determination of β-carotene but this is a very good example of the versatility and advantages of the UMEs. BCFMEs allowed to perform measurements in aprotic solvents (dichloromethane, chlorobenzene, tetrahydrofuran and acetonitrile) and to achieve high scan rates (even 10 V/s). The voltammetric signals suggested that the electrooxidation of β-carotene in inert solvents consists of a fast (quasi-reversible) electrochemical reaction. The same process in non-inert aprotic solvent (with basic character) is of more complex nature because the initial charge transfer is affected by a coupled chemical reaction that follows.

A practical application of BCFMEs is the determination of gold by anodic stripping voltammetry [124]. In this work, the BCFMEs were employed for the determination of tetrachloroaurate(III) complex in spiked samples of tap water. The methodology proposed is very easy and the details can be found in Procedure 46 in CD accompanying this book.

Far from the metal trace analysis, our initial studies with BCFMEs were focused on the determination of folic acid [122]. In this case, the main goal was the optimisation of the electrode pretreatment for this analyte. An acidic medium (0.1 M perchloric acid) was considered optimum for folic acid determination by differential pulse voltammetry. A linear range between 2.0×10^{-8} and 1.0×10^{-6} M with a detection limit of 1.0×10^{-8} M was obtained. Nevertheless, in this work, the adsorptive properties of the folic acid on mercury were noted and the employment of mercury-coated carbon fibre UMEs for folic acid determination has been targeted as a future goal.

32.2.4 Mercury thin films on carbon fibre UMES (HgCFMEs)

On one hand, mercury-film electrodes give increased resolution when compared to the hanging mercury drop electrode (HMDE). On the other hand, BCFMEs have the inherent characteristics of an ERD. Hence, it would be extremely desirable to combine all properties. Furthermore, this combination may provide additional advantages such as easy handling, low cost, and other well-known analytical advantages associated with the use of UMEs itself, for example, the elimination of convective hydrodynamics along the accumulation step in stripping voltammetric techniques.

The majority of the works with HgCFMEs focused on the anodic stripping of metals [127]. However, our works have been mainly focused on the determination of organic molecules that show the ability of adsorption on mercury [119–121], for example pteridines. Therefore, those molecules can be primarily preconcentrated on HgCFMEs and, subsequently, determined by cathodic stripping techniques. This procedure is described like adsorptive stripping voltammetry and it is a very interesting technique for trace and ultratrace analysis due to its excellent sensitivity, accuracy, precision and the low cost of instrumentation.

32.2.4.1 Mercury thin film formation

For mercury electrodeposition on carbon UMEs, several parameters must be considered. Firstly, unlike metals, carbon materials are completely inert against mercury and its adherence on lateral surfaces of carbon fibres is much better than on other carbon surfaces Therefore, BCFMEs are a more suitable support for mercury electrodeposition than disk carbon UMEs. Secondly, the BCFMEs must be conditioned to obtain a suitable surface for the mercury electrodeposition. The abovementioned chemical pretreatment (dipping in chromic acid mixture and concentrated nitric) has been found to be an adequate procedure for obtaining consistent mercury films. Thirdly, the compatibility between analyte and mercury salts is desirable. In our studies, mitoxantrene was the only analyte compatible with mercury salts while folic acid, edatrexate, methotrexate and aminopterin had to be determined by *ex situ* mercury film generation because those analytes precipitate in the presence of mercury salts. Fourthly, the choice of the supporting electrolyte is important for the film formation and the adsorptive stripping step. For example, hydrochloric acid allows good film deposition and consequently, good stripping responses are obtained when pteridines are assayed. However, perchloric acid was more suitable when mitoxantrone was studied. Figure 32.5 shows a typical study of the influence of the supporting electrolyte composition on the a.c. stripping signal of 4.0×10^{-8} M folic acid using the first reduction process common to many pteridines, in which the pteridine ring is reversibly reduced to the 5,8-dihydro corresponding derivative. Fifthly, deposition of larger amounts of mercury facilitates adsorption of greater amounts of the analyte but then the electrode would not possess the stability of a mercury thin film. Thus, the mercury plating conditions on BCFMES are very critical and need to be studied carefully. The main parameters that influence the mercury electrodeposition are the mercury salt concentration, the applied preplating potential (E_{film}) and the preplating time (t_{film}). The films generated can be characterised by the anodic stripping process of the mercury.

Our studies indicate that a 10^{-3} M solution of $Hg(NO_3)_2$ in 5 M HCl is suitable for the *ex situ* formation of the mercury film. Films formed using these conditions are thought to exist as mercury microdroplets of high surface area facilitating efficient adsorption of the analyte compound. Figure 32.6a–d shows the adsorptive stripping voltammograms

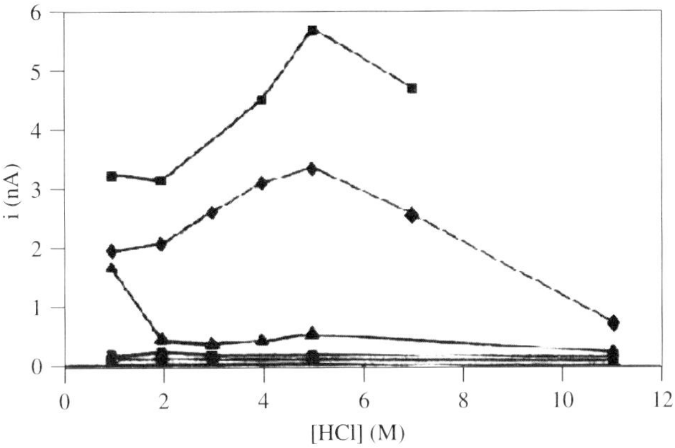

Fig. 32.5. Optimisation of thin mercury film conditions: influence of the supporting electrolyte composition on the a.c. stripping voltammetric response of 4.0×10^{-8} M folic acid. $Hg(NO_3)_2$ concentration • = 1.0×10^{-5}; ▼ = 1.0×10^{-4}; ▲ = 1.0×10^{-3}; ♦ = 0.01; ■ = 0.1 M, $t_{film} = 30$ s; $E_{film} = -0.2$ V. Stripping peak in acetate buffer (pH 5.0).

recorded in a 5.0×10^{-9} M aminopterin solution for the optimisation of the mercury thin film, formed *ex situ*, with different values of electroplating parameters. Similar behaviour has been shown for other pterines studies. The mercury preplating conditions were optimised in two steps. The results were evaluated by the peak current obtained with the anodic stripping of each film. First, E_{film} was varied while t_{film} was maintained for 60 s. As the E_{film} becomes more negative, the number of mercury droplets increases. At E_{film} above –0.8 V, hydrogen gas is generated causing droplet detachment. Therefore, an E_{film} of –0.8 V was chosen to ensure the stability of the film. Second, t_{film} was varied for a fixed E_{film} i.e –0.8 V. As shown in Figs. 32.6b and c, the stripping currents began to decrease when higher t_{film} were employed. An optimum t_{film} of 90 s was chosen because it seems that, at greater times, a growth in droplet size rather than in droplet number occurs, thus causing an overall decrease in the surface area of mercury available for analyte adsorption, and also producing a less stable film. For *ex situ* experiments, HgCMFEs were quickly transferred (10 s) to the analytical cell. A –0.8 V closed circuit was maintained throughout this procedure. Once the HgCMFEs were transferred, the potential was moved to –1.4 V for 30 s to ensure that a clean film was present. Then, it is ready for any analytical purposes and accumulation studies of the selected molecules could be made. By careful selection of the accumulation time, different concentration ranges could be studied. Table 32.3 shows some of the

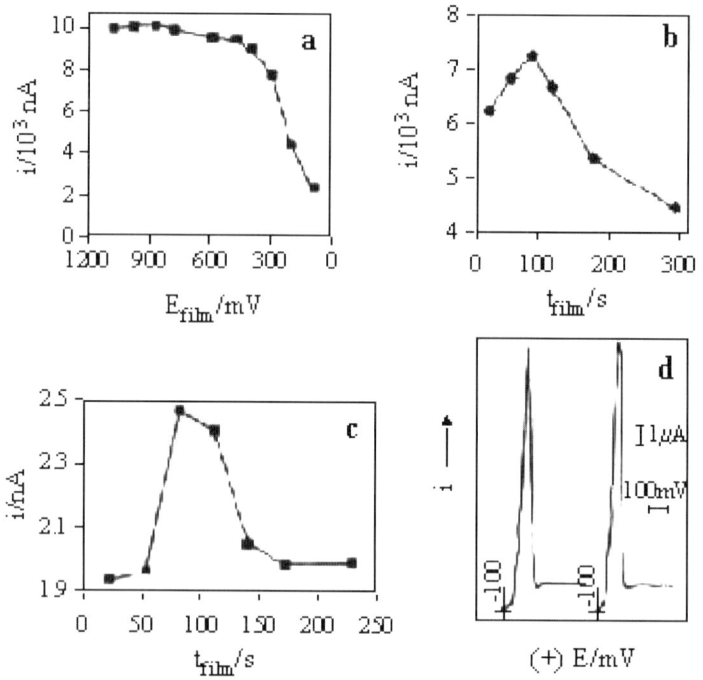

Fig. 32.6. (a) Optimisation of the thin mercury film in terms of $E_{\rm film}$ using d.c. adsorptive stripping voltammetry of 5.0×10^{-8} M aminopterin in pH 5 acetate buffer. Mercury salt solution = 1.0×10^{-3} M in 0.5 M HCl; $t_{\rm film} = 60$ s; $E_{\rm acc.} = 0.0$ V, $t_{\rm acc.} = 60$ s; film was stripped anodically at 100 mV/s. (b) Optimisation of the thin mercury film in terms of $t_{\rm film}$ using d.c. adsorptive stripping voltammetry of aminopterin. $E_{\rm film} = -0.8$ V. Rest of conditions as in (a). (c) Optimisation of the thin mercury film in terms of $t_{\rm film}$ versus a.c. stripping voltammetry of aminopterin 5×10^{-9} M. Rest of the conditions as above. (d) Two anodic stripping peaks of the thin mercury film under optimum deposition conditions; $t_{\rm film} = 90$ s; $E_{\rm film} = -0.8$ V; scan rate = 100 mV/s.

main analytical characteristics of the compounds under examination by a.c. adsorptive stripping voltammetry with HgCFMEs.

For the optimisation of the film generation, the BCFME was regenerated between each film formation. The optimum conditions for this regeneration were found to be the application of a potential of +0.79 V for 40 s. Figure 32.6d shows two typical anodic stripping signals of this procedure. The reproducibility of the formed film was excellent (RSD of 1.1%, $n = 6$). However, when the HgCFMEs were transferred, the reproducibility yielded only an RSD of 13.3%. Therefore, the necessity to use the same film for *ex situ* experiments is suggested for analytical

TABLE 32.3

Analytical characteristics of selected pteridines/antitumoural drugs using a.c. stripping voltammetry on HgCFMEs

Analyte	Calibration range (M)	t_{acc}	L.D. (S/N = 3)	Reproducibility (RSD, $n = 10$)	Ref.
Folic acid	1.0×10^{-9}–2.5×10^{-8}	60	9.0×10^{-10}	1.44% (at 10^{-8} M)	[119]
Mitoxantrone	5.0×10^{-10}–2.0×10^{-8}	300	5.0×10^{-10}	5.05% (at 10^{-9} M)	[119]
Aminopterin	2.0×10^{-10}–8.0×10^{-8}	180	1.0×10^{-10}	3.57% (at 10^{-10} M)	[120]
Edatrexate	1.0×10^{-10}–5.0×10^{-8}	240	5.0×10^{-14a}	1.39% (at 10^{-8} M)[b]	[121]
Metotrexate	1.0×10^{-10}–3.0×10^{-8}	180	3.0×10^{-13a}	–	[121]

[a] $t_{acc} = 350$ s.
[b] $t_{acc} = 10$ s.

purposes. This is not necessary for mitoxantrone because the mercury film can be regenerated *in situ* and different films can be used for recording different voltammograms. In any case, after each adsorptive stripping, an electrochemical activation is necessary in order to remove the adsorbed reduction products. It is usually carried out by holding the potential at a rather negative value for a short time (if the same mercury film has to be used for the analysis run) or at a more positive value (if the mercury film is renewed after each measurement).

When the performance of HgCFME was compared to that reported using classical mercury electrodes, several advantages were evident. First of all, the charging current associated to these electrodes is minimal compared to that of the conventional. This fact, combined with the greater mass transport characteristic of UMEs, allowed the employment of much lower accumulation times and the possibility of accumulation from quiescent solutions. These factors are particularly advantageous for analysis of biological fluids since longer accumulation times and stirring of the solution enhance the diffusion of interfering large compounds which normally diffuse very slowly to the electrode surface in quiescent solutions. Much lower limits of detection were achieved using the HgCFMEs and the precision of the signal compared favourably with that obtained using classical electrodes. The same fibre could be used for a period of at least 8 weeks with no significant diminution of performance.

32.2.4.2 Applications of the HgCFMEs

The analysis of real (complex) samples was carried out by first evaluating the adsorptive preconcentration/stripping responses of various analytes in the presence of surfactants. Once it was shown that this interference was lower than when other conventional electrodes were used, the next step involved assessing the performance of the HgCFMEs for direct analysis in real biological samples. The results obtained were quite satisfactory depending on the sample matrices implicated. Compounds such as mitoxantrene, aminopterin or edatrexate could be determined directly in urine samples. In all cases, a dilution of the spiked urine sample was necessary in order to minimize interferences arising from the medium and better limits of detection could be reached after sample clean-up using solid-phase extraction methodology. After introduction of the sample into the cell, a standard addition method was followed and the initial concentration of the urine was evaluated by extrapolation (see Fig. 32.7). Table 32.4 shows some of the results obtained when edatrexate is determined in urine. Although the

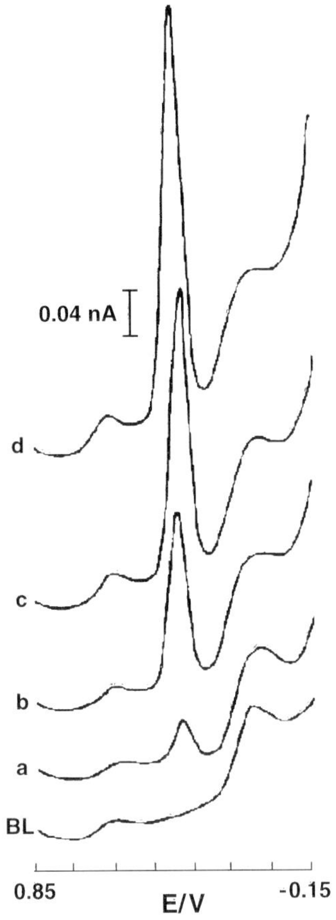

Fig. 32.7. Phase-selective a.c. adsorptive stripping voltammograms of: (BL) urine blank, (a) direct injection of 40 μL of urine sample 1×10^{-6} M in edatrexate; followed by (b) 10 μL; (c) 20 μL; (d) 30 μL of a 1.0×10^{-5} M edatrexate stock solution. Procedure and a.c. voltammetric conditions are in Ref. [121].

TABLE 32.4

Direct analysis of edatrexate in urine

True concentration (M)	Concentration determined (M)	Relative error (%)	RSD (%)
5.0×10^{-6}	5.26×10^{-6}	5.09	1.48
1.0×10^{-6}	1.04×10^{-6}	7.24	7.07
5.0×10^{-7}	5.65×10^{-6}	12.87	4.17

results are acceptable for normal therapeutic levels, as expected, the relative error increases at lower analyte concentration. This was due to the modification of the mercury film or a slight electrode passivation by compounds naturally present in urine. When the same samples were submitted to the above-mentioned previous extraction clean-up procedure, a detection limit of 1.0×10^{-8} M edatrexate in urine was recorded.

The main inconvenience was the lifetime of the electrode, which was lowered by gradual passivation with compounds present in urine. However, the use of an optimised mercury film, high urine dilutions and short accumulation times minimised this effect. Surface fouling was also alleviated by fibre regeneration along 1 min in concentrated chromic acid. The same fibre could normally be used for five different urine analysis runs involving approximately 35 measurements before it had to be replaced. This problem had not been noticed when the mercury film was generated *in situ*.

Improvements of the analytical performance have been achieved by the incorporation of the HgCFMEs into a *T* shape flow cell [128]. In this work, a.c. was used for adsorptive stripping analysis of mitoxantrone in flowing systems and better reproducibility of the measurements was obtained. The method is more versatile and lower detection limits (9.0×10^{-9} M mitoxantrone) were obtained. Due to the use of a carrier, the mercury film seemed to be more resistant to interferences arising from the matrix when real samples (serum) were analysed. Again, the main inconvenience was the impossibility of using the same fibre for making more than one calibration graph when serum samples were analysed.

Finally, we have reported the application of the HgCFMEs to the voltammetric study of vitamin K_1 [125]. In this study, the advantages of both BCFMEs and HgCFMEs in non-aqueous medium (acetonitrile) were exploited. Initially, the carbon fibre UMEs were pretreated in acetonitrile/TBHFP and tested with ferrocene in this medium (as mentioned above). In order to avoid the difficulties associated to the working electrode transference, the formation of the mercury thin films *in situ* was optimised in acetonitrile. Following a procedure similar to the one previously described, the optimal experimental conditions were determined as deoxygenated quiescent 1.0×10^{-4} M $HgCl_2$/acetonitrile/ 5.0×10^{-2} M TBHFP medium, deposition potential of −2 V and accumulation time of 60 s. The rate constant of the heterogeneous charge transfer reaction ($k^{\circ\prime}$) for the first electroreduction diffusive process of vitamin K_1 in acetonitrile medium was determined from data obtained by cyclic voltammetry at different scan rates. For BCFMEs and HgCFMEs, $k^{\circ\prime}$ were found to be affected by electrolyte concentration but HgCFMEs

TABLE 32.5
Linearity ranges and detection limits for calibrations of vitamin K_1

Electrode (medium)	Voltammetric technique	Linearity range (M)	Detection limit (M)
BCFME (MeCN/ TBAHFP 5 mM)	CV; 2 V/s	5×10^{-6}–6×10^{-4}	3.7×10^{-6}
	SWV; $E_{sw} = 50$ mV, $f = 100$ Hz	2×10^{-5}–4×10^{-4}	4.0×10^{-6}
	SWV; $E_{sw} = 100$ mV, $f = 1000$ Hz	2×10^{-5}–4×10^{-4}	4.0×10^{-6}
HgCFMEs ($HgCl_2$ 0.1 mM/MeCN/ TBAHFP 5 mM)	CV; 2 V/s	5×10^{-6}–1×10^{-4}	3.4×10^{-7}
	SWV; $E_{sw} = 25$mV, $f = 200$ Hz	5×10^{-6}–4×10^{-4}	1.8×10^{-7}
HMDE (MeCN/ TBAHFP 5 mM)	CV; 100 mV/s	3×10^{-5}–1×10^{-3}	1.8×10^{-5}
	SWV; $E_{sw} = 25$ mV, $f = 200$ Hz	1×10^{-5}–5×10^{-4}	3.3×10^{-6}
	DPV; PH = 100 mV, PW = 50 ms, 50 mV/s	1×10^{-5}–1×10^{-3}	4.3×10^{-6}

MeCN = acetonitrile, TBAHFP = tetra-*n*-butylammonium hexafluorophosphate, CV = cyclic voltammetry, SWV = square wave voltammetry and DPV = differential pulse voltammetry.

always showed faster kinetic (usually, one order of magnitude higher). Calibrations of vitamin K_1 were achieved using different electrodes (HDME, BCFMEs and HgCFMEs) and voltammetric techniques. The results are shown in Table 32.5. The best reproducibility (RSD of 0.86% ($n = 10$) for 4.0×10^{-4} M) and the lowest detection limit (1.8×10^{-7} M) were obtained by square wave voltammetry with HgCFMEs.

32.3 CONCLUSIONS

Initially, little attention was paid to the influence that size and geometry of electrode surface have on electrochemical processes. However, today it is known that both parameters are decisive in the behaviour and electrochemical responses of the electrode. The reduction of the electrode dimensions should be considered from two points of view: theoretical and practical.

Some time ago, measurements with UMEs did not fit into the classical theories. Due to this, more general theories were developed, for explaining the relation between the geometry, size and electrode response. Today, these theories show several limitations when electrodes with dimensions in the low range of the nanometric scale (NEs) are employed. In this way, a better knowledge about the fundamental performance of the electrodes can be achieved by studies with NEs.

From a practical point of view, the reduction of the electrode dimensions shows advantages and drawbacks. On one hand, minimal dimensions are desired when the application exploits some of the characteristics inherent to ERD (high diffusive mass transport, diminished capacitance, reduced ohmic drop, high spatial resolution, high biocompatibility, etc.). On the other hand, the smaller the electrode, the more difficult is the measurement of its response and worse is the reproducibility of its construction. Therefore, a question can be proposed: "what is the adequate critical dimension of an electrode?"

For single electrodes, UMEs have shown important advantages in a wide variety of applications although improvements can be obtained with ERDs slightly smaller. In this context, ERDs in the 100-nm range promise a successful future. Apart from theoretical and instrumental limitations, the use of single NEs is still in its infancy and the enhancement in the spatial resolution of the SECM is the more noteworthy of their applications. However, experimental results (like the low detection limits achieved with NE ensembles) have proved that grouped NEs are suitable devices for several goals. Other grouped ERDs present a high electrochemical performance with bigger electrodes, as IDAs of electrodes whose critical dimension is around 100 nm. Therefore, for practical applications, the reduction of the electrode dimensions is interesting but the lowest scale is not always the best.

Finally, it has to be indicated that the actual trend for miniaturising analytical systems implies the use of ERDs or NEs in electroanalytical applications. With this aim, highly packed ERD or NE arrays have demonstrated to be suitable candidates from the analytical and economical criterion.

REFERENCES

1 D. Arrigan, *Analyst*, 129 (2004) 1157–1165.
2 R.M. Wightman and D.O. Wipf, Voltammetry at ultramicroelectrodes. In: A.J. Bard (Ed.), *Electroanalytical Chemistry*, Dekker, New York, 1989.

3 J. Heinze, *Angew. Chem. Int. Ed. Engl.*, 32 (1993) 1268–1288.
4 R.C. Engstrom, C.M. Pharr and M.D. Koppang, *J. Electroanal. Chem.*, 221 (1987) 251–255.
5 K. Oaki, *Electroanalysis*, 5 (1993) 627–639.
6 R.J. Foster, *Chem. Soc. Rev.*, 23 (1994) 289–297.
7 R.B. Morris, D.J. Franta and H.S. White, *J. Phys. Chem.*, 91 (1987) 3559–3564.
8 J.L. Conyers Jr. and H.S. White, *Anal. Chem.*, 72 (2000) 4441–4446.
9 C.P. Smith and H.S. White, *Anal. Chem.*, 65 (1993) 3343–3353.
10 T.J. Davies and R.G. Compton, *J. Electroanal. Chem.*, 585 (2005) 63–82.
11 V.P. Menon and C.R. Martin, *Anal. Chem.*, 67 (1995) 1920–1928.
12 A.C. Michael and R.M. Wightman, Microelectrodes. In: P.T. Kissinger and W.R. Heinemann (Eds.), *Laboratory Techniques in Electroanalytical Chemistry*, Dekker, New York, 1996.
13 K.R. Wehmeyer and R.M. Wightman, *Anal. Chem.*, 57 (1985) 1989–1993.
14 J.V. Macpherson, N. Simjee and P.R. Unwin, *Electrochim. Acta*, 47 (2001) 29–45.
15 D.R. MacFarlane and D.K.Y. Wong, *J. Electroanal. Chem.*, 185 (1985) 197–202.
16 Y.T. Kim, D.M. Scarnulis and A.G. Ewing, *Anal. Chem.*, 58 (1986) 1782–1786.
17 C.G. Zoski, *Electroanalysis*, 14 (2002) 1041–1051.
18 B.B. Katemann and W. Schuhmann, *Electroanalysis*, 14 (2002) 22–28.
19 W.H. Huang, D.W. Pang, H. Tong, Z.L. Wang and J.K. Cheng, *Anal. Chem.*, 73 (2001) 1048–1052.
20 S. Ramírez García, S. Alegret, F. Céspedes and R.J. Foster, *Anal. Chem.*, 76 (2004) 503–512.
21 I.F. Cheng, L.D. Whiteley and C.R. Martin, *Anal. Chem.*, 61 (1989) 762–766.
22 R.M. Penner and C.R. Martin, *Anal. Chem.*, 59 (1987) 2625–2630.
23 K. Krihnamoorthy and C.G. Zoski, *Anal. Chem.*, 77 (2005) 5068–5071.
24 J.C. Hulteen and C.R. Martin, *J. Mater. Chem.*, 7 (1997) 1075–1087.
25 E. Sabatini and I. Rubinstein, *J. Phys. Chem.*, 91 (1987) 6663–6669.
26 E. Jeoung, T.H. Galow, J. Schotter, M. Bal, A. Ursache, M.T. Tuominen, C.M. Stafford, T.P. Russell and V.M. Rotello, *Langmuir*, 17 (2001) 6396–6398.
27 J.K. Campbell, L. Sun and R.M. Crooks, *J. Am. Chem. Soc.*, 121 (1999) 3779–3780.
28 W. Cheng, S. Dong and E. Wang, *Anal. Chem.*, 74 (2002) 3599–3604.
29 J. Liu, A. Chou, W. Rahmat, M.N. Paddon-Row and J.J. Gooding, *Electroanalysis*, 17 (2005) 38–46.
30 Y. Tu, Y. Lin, W. Yantasee and Z. Ren, *Electroanalysis*, 17 (2005) 79–84.
31 S. Szunerits, P. Garrigue, J.L. Bruneel, L. Servant and N. Sojic, *Electroanalysis*, 15 (2003) 548–555.

32 H.P. Wu, *Anal. Chem.*, 65 (1993) 1643–1646.
33 O. Niwa and H. Tabei, *Anal. Chem.*, 66 (1994) 285–289.
34 C. Vieu, F. Carcenac, A. Pépin, Y. Chen, M. Mejias, A. Lebib, L. Manin-Ferlazzo, L. Couraud and H. Launois, *Appl. Surf. Sci.*, 164 (2000) 111–117.
35 J. Koehne, J. Li, A.M. Cassell, H. Chen, Q. Ye, H. Tee Ng, J. Han and M. Meyyappan, *J. Mater. Chem.*, 14 (2004) 676–684.
36 Y. Shao and M.V. Mirkin, *Anal. Chem.*, 69 (1997) 1627–1634.
37 C. Amatore and E. Maisonhaute, *Anal. Chem.*, 77 (2005) 303–311A.
38 B. Soucaze-Guillous, W. Kutner and K.M. Kadish, *Anal. Chem.*, 65 (1993) 669–672.
39 F.M. Matysik, *Anal. Bioanal. Chem.*, 375 (2003) 33–35.
40 F.F. Fan, J. Kwak and A.J. Bard, *J. Am. Chem. Soc.*, 118 (1996) 9669–9675.
41 N.S. Lawrence, E.L. Beckett, J. Davis and R.G. Compton, *Anal. Biochem*, 303 (2002) 1–16.
42 G.S. Wilson and R. Gifford, *Biosens. Bioelectron.*, 20 (2005) 2388–2403.
43 A.S. Khan and A.C. Michael, *Trends Anal. Chem.*, 22 (2003) 503–508.
44 M. Koudelka-Hep and P.D. Van der Wal, *Electrochim. Acta*, 45 (2000) 2437–2441.
45 M.A. Dayton, A.G. Ewing and R.M. Wightman, *Anal. Chem.*, 52 (1980) 2392–2396.
46 K.T. Kawagoe and M. Wightman, *Talanta*, 41 (1994) 865–874.
47 N. Dale, S. Hatz, F. Tian and E. Llaudet, *Trends Biotechnol*, 23 (2005) 420–428.
48 P. Pantano and W. Kuhr, *Electroanalysis*, 7 (1995) 405–416.
49 G.S. Wilson and Y. Hu, *Chem. Rev.*, 100 (2000) 2693–2704.
50 W.H. Oldenziel and B.H.C. Westerink, *Anal. Chem.*, 77 (2005) 5520–5528.
51 Md.A. Rahman, N.H. Kwon, M.S. Won, E.S. Choe and Y.B. Shim, *Anal. Chem.*, 77 (2005) 4854–4860.
52 E. Mikeladze, A. Schulte, M. Mosbach, A. Blöchl, E. Csöregi, R. Solomonia and W. Schuhmann, *Electroanalysis*, 14 (2002) 393–399.
53 O.N. Schuvailo, S.V. Dzyadevych, A.V. El′skaya, S. Gautier Sauvigne, E. Csöregi, R. Cespuglio and A.P. Soldatkin, *Biosens. Bioelectron.*, 21 (2005) 87–94.
54 K.M. Mitchell, *Anal. Chem.*, 76 (2004) 1098–1106.
55 N.F. Sharam, L.I. Netchiporouk, C. Martelet, N. Jaffrezic Renault, C. Bonnet and R. Cespuglio, *Anal. Chem.*, 70 (1998) 2618–2622.
56 S. de Irazu, N. Unceta, M.C. Sampedro, M.A. Goicolea and R.J. Barrio, *Analyst*, 126 (2001) 495–500.
57 X. Jiang and X. Lin, *Anal. Chim. Acta*, 537 (2005) 145–151.
58 E. Llaudet, N.P. Botting, J.A. Crayston and N. Dale, *Biosens. Bioelectron.*, 18 (2003) 43–52.

59 S. Cosnier, C. Innocent, L. Allien, S. Poitry and M. Tsacopoulos, *Anal. Chem.*, 69 (1997) 968–971.

60 E. Llaudet, S. Hatz, M. Droniou and N. Dale, *Anal. Chem.*, 77 (2005) 3267–3273.

61 A. Kueng, C. Kranz and B. Mizaikoff, *Biosens. Bioelectron.*, 21 (2005) 346–353.

62 N.R. Ferreira, A. Ledo, J.G. Frade, G.A. Gerhardt, J. Laranjinha and R.M. Barbosa, *Anal. Chim. Acta*, 535 (2005) 1–7.

63 J. Pei, N.T. Yu and X.Y. Li, *Anal. Chim. Acta*, 402 (1999) 145–155.

64 L.I. Netchiporouk, N.F. Sharam, N. Jaffrezic Renault, C. Martelet and R. Cespuglio, *Anal. Chem.*, 68 (1996) 4358–4364.

65 W. Gorski, C.A. Aspinwass, J.R.T. Lakey and R.T. Kennedy, *J. Electroanal. Chem.*, 425 (1997) 191–199.

66 S.K. Jung, W. Gorski, C.A. Aspinwass, L.M. Kauri and R.T. Kennedy, *Anal. Chem.*, 71 (1999) 3642–3649.

67 U.E. Spichinger-Keller, *Chemical Sensors and Biosensors for Medical and Biomedical Applications*, Wiley-VCH, Weinheim, 1998.

68 D. Ammann, *Ion Selective Microelectrodes*, Springer-Verlag, Berlin, 1986.

69 R.Q. Yu, Z.R. Zhang and G.L. Shen, *Sens. Actuat B*, 65 (2000) 150–153.

70 S. Hrapovic and J.H.T. Luong, *Anal. Chem.*, 75 (2003) 3308–3315.

71 J. Fei, K. Wu, F. Wang and S. Hu, *Talanta*, 65 (2005) 918–924.

72 W.Z. Wu, W.H. Huang, W. Wang, Z.L. Wang, J.K. Cheng, T. Xu, R.Y. Zhang. Y. Chen and J. Liu, *J. Am. Chem. Soc.*, 127 (2005) 8914–8915.

73 M. Lacroix, P. Bianco and E. Lojou, *Electroanalysis*, 11 (1999) 1068–1076.

74 S. Zhang, H. Zhao and R. John, *Anal. Chim. Acta*, 421 (2000) 175–187.

75 Y. Xian, M. Liu, Q. Cai, H. Li, J. Lu and L. Jin, *Analyst*, 126 (2001) 871–876.

76 R. Gasparac, B.J. Taft, M.A. Lapierre Devlin, A.D. Lazarek, J.M. Xu and S.O. Kelley, *J. Am. Chem. Soc.*, 126 (2004) 12270–12271.

77 T.H. Hsia, K.T. Liao and H.J. Huang, *Anal. Chim. Acta*, 537 (2005) 315–319.

78 B. Brunetti, P. Ugo, L.M. Moretto and C.R. Martin, *J. Electroanal. Chem.*, 491 (2000) 166–174.

79 P. Ugo, N. Pepe, L.M. Moretto and M. Battagliarin, *J. Electroanal. Chem.*, 560 (2003) 51–58.

80 L.M. Moretto, N. Pepe and P. Ugo, *Talanta*, 62 (2004) 1055–1060.

81 M. Delvaux and S. Demoutier-Champagne, *Biosens. Bioelectron.*, 18 (2003) 943–951.

82 M. Delvaux, A. Walcarius and S. Demoutier-Champagne, *Anal. Chim. Acta*, 525 (2004) 221–230.

83 M. Delvaux, A. Walcarius and S. Demoutier-Champagne, *Electroanalysis*, 16 (2004) 190–198.

84 M. Delvaux, A. Walcarius and S. Demoutier-Champagne, *Biosens. Bioelectron.*, 20 (2005) 1587–1594.
85 S. Martinoia, L. Bonzano, M. Chiappalone and M. Tedesco, *Sens. Actuators B*, 108 (2005) 589–596.
86 H. Ecken, S. Ingebrandt, M. Krause, D. Richter, M. Hara and A. Offenhäusser, *Electrochim. Acta*, 48 (2003) 3355–3362.
87 K. Hayashi, T. Horiuchi, R. Kurita, K. Torimitsu and O. Niwa, *Biosens. Bioelectron.*, 15 (2000) 523–529.
88 T. Matsue, A. Aoki, E. Ando and I. Uchida, *Anal. Chem.*, 62 (1990) 407–409.
89 T. Livache, B. Fouque, A. Roget, J. Marchand, G. Bidan, R. Téoule and G. Mathis, *Anal. Biochem.*, 255 (1998) 188–194.
90 J.J. Burmeister and G.A. Gerhardt, *Trends Anal. Chem.*, 22 (2003) 498–502.
91 R. Feeney and S.P. Kounaves, *Electroanalysis*, 12 (2000) 677–684.
92 R. Feeney and S.P. Kounaves, *Anal. Chem.*, 72 (2000) 2222–2228.
93 P.R.M. Silva, M.A. El Khakani, M. Chaker, A. Dufresne and F. Courchesne, *Sens. Actuators B*, 76 (2001) 250–257.
94 J.H. Kim, B.G. Kim, J.B. Yoon, E. Yoon and C.H. Han, *Sens. Actuators A*, 95 (2002) 108–113.
95 K. Dill, D.D. Montgomery, A.L. Ghindilis, K.R. Schwarzkopf, S.R. Ragsdale and A.V. Oleinikov, *Biosens. Bioelectron.*, 20 (2004) 736–742.
96 X. Xie, D. Stüben, Z. Berner, J. Albers, R. Hintsche and E. Jantzen, *Sens. Actuators B*, 97 (2004) 168–173.
97 O. Niwa, *Electroanalysis*, 7 (1995) 606–613.
98 J. Albers, T. Grunwald, E. Nebling, G. Piechotta and R. Hintsche, *Anal. Bioanal. Chem.*, 377 (2003) 521–527.
99 E. Nebling, T. Grunwald, J. Albers, P. Schafer and R. Hintsche, *Anal. Chem.*, 76 (2004) 689–696.
100 D. Liu, R.K. Perdue, L. Sun and R.M. Crooks, *Langmuir*, 20 (2004) 5905–5910.
101 O. Niwa, Y. Xu, H.B. Halsall and W.R. Heineman, *Anal. Chem.*, 65 (1993) 1559–1563.
102 J.H. Thomas, S.K. Kim, P.J. Hesketh, H.B. Halsall and W.R. Heineman, *Anal. Chem.*, 76 (2004) 2700–2707.
103 J.H. Thomas, S.K. Kim, P.J. Hesketh, H.B. Halsall and W.R. Heineman, *Anal. Biochem.*, 328 (2004) 113–122.
104 O. Niwa, M. Morita and H. Tabei, *Electroanalysis*, 3 (1991) 163–168.
105 M. Morita, O. Niwa and T. Horiuchi, *Electrochim. Acta*, 42 (1997) 3177–3183.
106 O. Niwa, R. Kurita, Z. Liu, T. Horiuchi and K. Torimitsu, *Anal. Chem.*, 72 (2000) 949–955.
107 D.G. Sanderson and B. Anderson, *Anal. Chem.*, 57 (1985) 2388–2393.

108 A.J. Bard, J.A. Crayston, G.P. Kittlesen, T.V. Shea and M.S. Wrighton, *Anal. Chem.*, 58 (1986) 2321–2331.
109 J. Min and A.J. Baeumner, *Electroanalysis*, 16 (2004) 724–729.
110 N. Honda, M. Inaba, T. Katagiri, S. Shoji, H. Sato, T. Homma, T. Osaka, M. Saito, J. Mizuno and Y. Wada, *Biosens. Bioelectron.*, 20 (2005) 2306–2309.
111 K. Aoki, *J. Electroanal. Chem.*, 270 (1989) 35–41.
112 S.K. Kim, P.J. Hesketh, C. Li, J.H. Thomas, H.B. Halsall and W.R. Heineman, *Biosens. Bioelectron.*, 20 (2004) 887–894.
113 K. Ueno, M. Hayashida, J.Y. Ye and H. Misawa, *Electrochem. Commun.*, 7 (2005) 161–165.
114 X. Zhu and C.H. Ahn, *IEEE Trans. Nanobiosci.*, 4 (2005) 164–169.
115 C. Berggren, B. Bjarnason and G. Johansson, *Electroanalysis*, 13 (2001) 173–180.
116 P.V. Gerwen, W. Laureyn, W. Laureys, G. Huyberechts, M.O. De Beeck, K. Baert, J. Suls, W. Sansen, P. Jacobs, L. Hermans and R. Mertens, *Sens. Actuators B*, 49 (1998) 73–80.
117 L. Malaquin, C. Vieu, M. Geneviéve, Y. Tauran, F. Carcenac, M.L. Pourciel, V. Leberre and E. Trévisiol, *Microelectron. Eng.*, 73–74 (2004) 887–892.
118 A.L. Suarez, J.A. García, A. Costa and P. Tuñon, *Electroanalysis*, 3 (1991) 413–417.
119 J.A. Pozo, A. Costa and P. Tuñon, *Anal. Chim. Acta*, 273 (1993) 101–109.
120 M.A. Malone, A. Costa, P. Tuñon and M.R. Smyth, *Analyst*, 118 (1993) 649–655.
121 M.A. Malone, A. Costa, P. Tuñon and M.R. Smyth, *Anal. Methods Instrum.*, 1 (1993) 164–171.
122 T.J. O'Shea, A. Costa, P. Tuñon and M.R. Smyth, *J. Electroanal. Chem.*, 307 (1991) 63–71.
123 A.L. Suarez, G. Alarnes and A. Costa, *Electrochim. Acta*, 44 (1999) 4489–4498.
124 G. Alanes and A. Costa, *Electroanalysis*, 9 (1997) 1262–1266.
125 G. Alarnes, A.L. Suarez and A. Costa, *Electrochim. Acta*, 44 (1998) 763–772.
126 L. Agüí, J.E. López, Guzmán A. González Cortés, P. Yañéz Sedeño and J.M. Pingarrón, *Anal. Chim. Acta*, 385 (1998) 241–248.
127 A. Economou and P.R. Fielden, *Analyst*, 128 (2003) 205–212.
128 J. Amez del Pozo, A. Costa and P. Tuñon, *Anal. Chim. Acta*, 289 (1994) 169–176.

Chapter 33

DNA/RNA aptamers: novel recognition structures in biosensing

Tibor Hianik

33.1 INTRODUCTION

Biosensors based on DNA or RNA aptamers (aptasensors) are of considerable interest as an alternative to the biosensors based on antibodies. Aptamers are artificial oligonucleotides (DNA or RNA) that at certain conditions (ionic composition, pH, temperature) adopt three-dimensional structure with binding site specific for certain proteins or even for low-molecular-weight compounds [1]. Advantage of aptasensors in comparison with those based on antibodies consists in the possibility to develop aptamers with high specificity to large variety of compounds by means of chemical synthesis without using experimental animals. DNA aptamers are more stable than antibodies and considerably less expensive. Aptamers can be advantageously used in the development of affinity biosensors. They can be easily chemically modified and immobilized on a solid support. In contrast with antibodies, which at certain conditions can be irreversibly denaturated, aptamers-based biosensors can be regenerated without loss of integrity and selectivity [2–5].

The method of selection of aptamers was discovered independently by three groups of investigators. Robertson and Joyce [6] described the method of selection of RNA with improved enzymatic activity to cleave DNA. Tuerk and Gold [7] patented the process of selection of DNA ligands as a target for T4 RNA polymerase. This method is known as SELEX (systematic evolution of ligands by exponential enrichment). Ellington and Szostak [1] reported method of in vitro selection of RNA that specifically binds organic dyes. In this paper also the term "aptamer" was introduced. The identification of aptamers is based on a combinatorial approach. Specific oligonucleotides are

Comprehensive Analytical Chemistry 49
Alegret and Merkoçi (Eds)
Volume 49 ISSN: 0166-526X DOI: 10.1016/S0166-526X(06)49033-4

isolated from complex libraries of synthetic nucleic acids. For this purpose, random-sequence DNA libraries are obtained by automated DNA synthesis. The size of a randomized region can vary from 30 to 60 nucleotides, flanked on both sides with a specific unique DNA sequence for polymerase chain reaction (PCR) amplification. The theoretical diversity of individual oligonucleotides in these random DNA libraries is rather large. For example, in the case of oligonucleotides composed of 40 bases, it is $4^{40} = 1.2 \times 10^{24}$. In practice, however, a considerably smaller library of approximately 10^{13}–10^{15} molecules is used [8]. The selection consists of DNA binding with immobilized ligands, e.g., proteins or haptens. The stability of complexes is characterized by apparent dissociation constant, K_d. For aptamer–protein complexes, K_d vary within the 1–100 nmol/L range, which is similar to affinity range of antibody–antigen complexes. Unbound DNA/RNA molecules are eluted from the column, while bound aptamers are isolated from the complex and then amplified by PCR. The SELEX technique is discussed in detail in a number of papers and reviews (see, e.g., Ref. [9]).

Although the first SELEX-related patent was filed in 1989 [7], the potentialities of the aptamer-based biosensors have not been realized in full scale due to the problems with aptamer stability during immobilization and signal registration. Several problems related to the practical application of aptamers are still under study, for example how immobilization of aptamers to the supported films and their microenvironment will affect the aptamer structure and aptamer–ligand interactions. Problems are connected with application of aptamers in a complex biological systems, where interferences with other molecules could take place, and especially RNA aptamers are unstable due to cleavage by nucleases. So far mostly radiolabeled aptamers were used, e.g., for quantification of protein kinase [10] or in vivo detection of clots [11]. However, to be widely employed in clinical practice, aptamers must be detected via a non-radioisotope method with a comparable sensitivity, e.g., aptamers can be covalently linked to an enzyme [12], or fluorescently labeled aptamers can be exploited [13]. Moreover, the most reliable and cost-effective way would be exploitation of the direct physical methods that do not require labeling of aptamers by additional chemical ligands. This highly promising direction route has not been exploited so far in sufficient detail. It is highly advantageous to explore the possibility of immobilization of aptamers onto novel materials, e.g., nanotubes or dendrimers, and using nanoparticles in detection of the ligand–aptamer interactions.

In this chapter we will focus on the methods of immobilization of aptamers and the detection of aptamer–ligand interactions by various methods. The application of aptamers in therapy and in conventional analytical assay has been reviewed elsewhere [2,4,5,8]. A comparison of aptamer- and antibody-based assays was recently published in a review by Leca-Bouvier and Blum [14].

33.1.1 Structure of DNA/RNA aptamers

Currently, the SELEX is a highly automatized procedure and only several weeks are necessary for development of aptamers for certain ligands. This is much shorter in comparison with the selection of antibodies, where usually several months are required. Due to effectivity of the SELEX, the library of aptamers against various ligands has become wider. On the other hand, the primary procedure does not results in all cases in aptamers with desired affinity. Therefore, optimization of aptamer structure is required. This optimization is performed through biased library generation [8]. As a result, it is possible to select aptamers with sensitivity to small ligand modification. Aptamers can even distinguish the chirality of molecules and their secondary structure. In principle, there is no restriction in the type of target for which the aptamer can be selected. To date, aptamers with affinity to various ligands were synthesized, including metal ions, organic dyes, drugs, amino acids, co-factors, antibiotics and nucleotide base analogs. Special interest was focused on the development of aptamers for detection of proteins including enzymes, antibodies, prions, growth factors, gene regulatory factors, cell adhesion molecules and lectins. Also, aptamers for viral particles and pathogenic bacteria were developed (see Refs. [2,8] for review). The aptamers were first used as therapeutic agents. For example, antithrombin aptamer was developed with the purpose of application as an anticoagulant [15]. Only recently the aptamers have been used as recognition elements in biosensing (see, e.g., Ref. [3]).

To date, one of the best investigated aptamers are that for thrombin. Thrombin is a multifunctional serine protease that plays important role in procoagulant and anticoagulant functions. Thrombin converts soluble fibrinogen to insoluble fibrin that forms the fibrin gel, which is responsible either for a physiological plug or for pathological thrombus [16]. This process is catalyzed by positively charged fibrinogen binding site at the thrombin molecule. The more positive heparin binding site is responsible for anticoagulant function of thrombin. These binding sites are spatially separated and localized at opposite poles of thrombin molecule [17]. Initially, Bock et al. [15] developed 15-mer DNA aptamer, which selectively binds fibrinogen site of the thrombin (Fig. 33.1A). This aptamer inhibits the thrombin-catalyzed fibrin clot formation [15]. The binding constant for this structure determined by Bock et al. [15] was $K_d = 200$ nmol/L, while Tasset et al. [17] reported $K_d = 100$ nmol/L, which agreed with that reported by Macaya et al. [18] ($K_d = 75$–100 nmol/L). However, Tsiang et al. [19] reported much lower K_d value, 1.6–6.2 nmol/L, for this aptamer. It should be, however, noted that the dissociation constant depends both on thrombin activity as well as on experimental conditions, e.g., temperature, ionic strength, etc., and method of detection. In the aptamer developed by Bock et al. [15], the intramolecular G-quadruplex plays the crucial role. According to NMR studies [20] and X-ray crystallography [21], the eight guanine

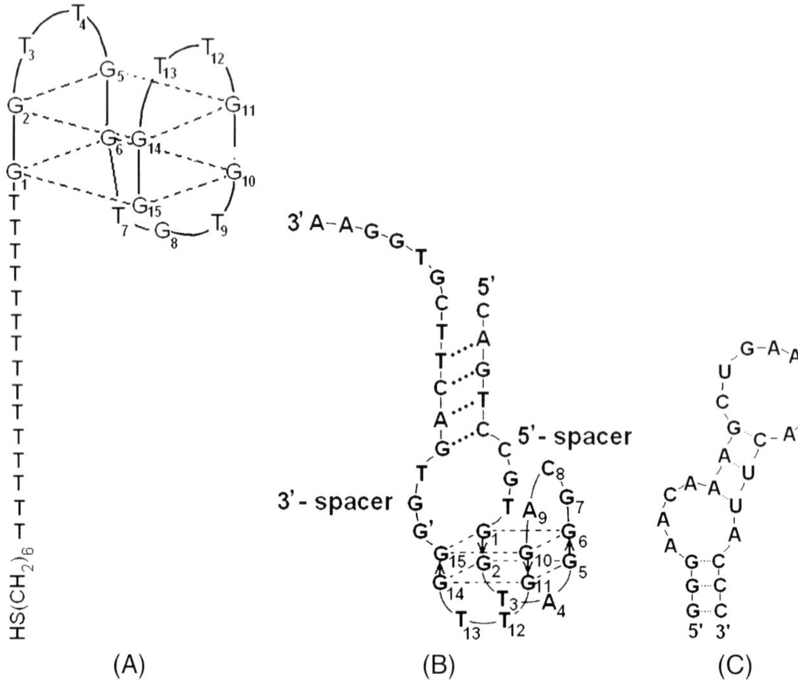

Fig. 33.1. Structure of DNA aptamers for recognition fibrinogen (A) [15,29] and heparin binding sites (B) [17] of thrombin. The non-canonical hydrogen bonds between guanines are shown by dotted lines. In the case of structure (A), the spacer composed of 15 T chain terminated by thiol group at the end of hydrophobic spacer is shown [34]. (C) Structure of RNA aptamer against fibrinogen binding site of thrombin [23].

residues form G tetrads that are connected at one end by TT loop (T_3–T_4, see Fig. 33.1A) and at the other end by the TGT loop ($T_7G_8T_9$, Fig. 33.1A). The non-canonical Watson–Crick pair T_4–T_{13} is important for stabilization of the aptamer in conformation that provides binding of T_3–T_4 side of quadruplex to fibrinogen binding site of the thrombin. Substitution of T_4 to adenine (A) resulted in loss of binding affinity. The structure of this aptamer is stabilized by K^+ ions. Later Tasset et al. [17] developed 38-mer aptamer that selectively recognizes heparin binding site with considerably higher affinity, $K_d = 0.5\,nmol/L$ (Fig. 33.1B). This aptamer had certain similarities with that developed by Bock et al. [15]. The substantial difference consisted in replacement of T_4 with A_4 in a quadruplex loop (Fig. 33.1B). However, the quadruplex structure remained stable due to extension of aptamer, in which four Watson–Crick pairs stabilized the aptamer conformation. It is interesting that for this aptamer K^+ ions had no influence on binding properties. Special investigations by means of photocrosslinking of 5-iodo-2′-deoxyuridine (5-IdU) performed in this work showed that exclusively one nucleotide T_{12} is responsible for binding to thrombin through Phe245 in heparin binding site. However, other nucleotides also play crucial role in providing high affinity to this aptamer. The adenine at position 4 of the quadruplex is important for determination of binding site of this DNA aptamer. The first guanine (G′) of the 3′ spacer (Fig. 33.1B) contributes to high affinity of the aptamer to heparin binding site of thrombin (see Ref. [17] for structure of the aptamer–thrombin complex). Apart from DNA aptamers, RNA aptamers that selectively bind thrombin were also developed. The RNA aptamer developed by Kubik et al. [22] was selective to heparin binding site of thrombin. The K_d for this RNA aptamer was 2–5 nmol/L [22], which was 10 time higher than that for DNA aptamer of high affinity to heparin binding site, i.e., DNA aptamer revealed higher affinity to thrombin. Later, White et al. [23] developed RNA aptamer specific for fibrinogen binding site of thrombin. The structure of this aptamer is shown on Fig. 33.1C. The binding affinity of this aptamer to thrombin on the surface was determined by Gronewold et al. [24] using filter binding assay, surface plasmon resonance (SPR) and acoustic methods. The K_d values determined by these methods were 294, 113 and 181 nmol/L, respectively.

Typical two-stacked guanine tetramers in the active binding site of the DNA aptamers against thrombin were typical also for other aptamers subsequently developed, including RNA aptamer (see, *e.g.*, aptamer against cellular prion [25–27]). Apart from linear aptamers shown in Fig. 33.1, so-called aptamer beacons were also developed. These aptamers contain special complementary spacers at both 3′ and 5′ terminals. In absence of target (*e.g.*, thrombin), these terminals form double helix that stabilizes the molecular beacon structure. However, in presence of target, the aptamer is transformed into a typical three-dimensional structure allowing specific interaction with the ligand. The advantage of aptamer beacon is that it allows sensitive detection of ligands, using, *e.g.*, fluorescence method (see below and Ref. [28]).

33.1.2 Folding of aptamers into three-dimensional structure

Folding of aptamer into a functional three-dimensional structure depends on the properties of aqueous environment and requires certain ions. The influence of various ions on the stability of the G-quadruplexes was studied in detail for thrombin-sensitive DNA aptamer developed by Bock et al. [15]. We have already mentioned above that this aptamer, with oligonucleotide sequence $d(G_2T_2G_2TGTG_2T_2G_2)$, specifically binds to fibrinogen binding site of the thrombin. Formation of a three-dimensional structure (Fig. 33.1A) requires the presence of K^+ ions. By means of circular dichroism (CD) method it was shown that stable intramolecular G-quadruplexes are formed also in presence of Rb^+, NH_4^+, Sr^{2+} or Ba^{2+}. Using differential scanning calorimetry it has been shown that the G-quadruplex is most stable in presence of Sr^{2+} ions. The transition of G-quadruplex from folded to unfolded conformation takes place at approximately 70°C, while for K^+ it was approximately 50°C and in presence of Cs^+ ions, when aptamer was less stable, the transition temperature was around 15°C [29]. Rather high stability of the G-quadruplex is provided also in presence of Pb^{2+} ions [30]. On the other hand, the cations Li^+, Na^+, Cs^+, Mg^{2+} and Ca^{2+} form weaker complexes with quadruples. The results have been explained by ionic radius. The cations with ionic radius in the range 1.3–1.5 Å, which form stable complexes with quadruplex, fit well within the two G-quarters, while other cations cannot. Analysis of the quadruplex stability using also densitometry and ultrasound velocimetry methods revealed that in presence of K^+ and Sr^{2+} ions, the specific volume of G-quadruplex decreases. This has been explained by increase in hydration of the quadruplex (the specific volume of hydrated water is lower in comparison with bulk water molecules). On the other hand, ultrasound velocimetry together with density data allowed to estimate apparent compressibility of aptamers, which was negative. This indicates the opposite effect—release of water molecules from quadruplex.

The results were explained by two effects: (1) dehydration of cations and guanine O6 atomic group and (2) the water uptake upon folding of a single-stranded DNA into a G-quadruplex [29]. It is, however, interesting that DNA aptamer with high affinity to heparin binding site is insensitive to K^+ ions [17].

Keeping pH close to neutral values (pH $\sim$ 7) is important for stability of aptamers and for their binding effectivity. For example, increase in pH to 7.5–8.5 or decrease to 4.5 resulted in loss of binding affinity of the aptamer to the thrombin [31]. The influence of increased pH (up to 8.5) on aptasensor response has also been reported by Xiao et al. [32]. The effect of pH is substantial also for RNA aptamers because hydroxyl at the 2′ position is reactive especially at higher pH. At this condition, the hydroxyl will attack the neighboring phosphodiester bond to produce cyclic 2′,3′-phosphate and will break the nucleic acid backbone. In order to avoid this damage, the hydroxyl groups at 2′ position is substituted by amino group or by fluorine atom [2].

33.2 APPLICATIONS OF APTAMERS IN BIOSENSING

In this part we will describe recent achievements in the development of biosensors based on DNA/RNA aptamers. These biosensors are usually prepared by immobilization of aptamer onto a solid support by various methods using chemisorption (aptamer is modified by thiol group) or by avidin–biotin technology (aptamer is modified by biotin) or by covalent attachment of amino group-labeled aptamer to a surface of self-assembly monolayer of 11-mercaptoundecanoic acid (11-MUA). Apart from the method of aptamer immobilization, the biosensors differ in the signal generation. To date, most extensively studied were the biosensors based on optical methods (fluorescence, SPR) and acoustic sensors based mostly on thickness shear mode (TSM) method. However, recently several investigators reported electrochemical sensors based on enzyme-labeled aptamers, electrochemical indicators and impedance spectroscopy methods of detection.

33.2.1 Immobilization of aptamers onto a solid support

As we have already mentioned, various methods of immobilization of aptamers onto a solid support are used. In principle these methods are similar to those applied previously for immobilization of single- or double-stranded DNA in genosensors or DNA biosensors for detection of DNA damage (see Pividori *et al.* [33] for review). The methods of immobilization based on physical adsorption of DNA by means of electrostatic interactions are not suitable due to low stability caused by aptamers desorption from the surface. The most effective methods are based on chemisorption of thiol-labeled aptamers onto a gold surface [33,34] or on strong affinity of biotin to avidin, streptavidin or neutravidin. In latter case, one end of DNA or RNA aptamer is modified by biotin. The solid support is covered by avidin [35], streptavidin [36] or neutravidin [37]. In the case of avidin or streptavidin, the proteins are usually chemically linked to the organic layer formed by, e.g., 3,3′-dithiopropionic acid di(*N*-succinimidylester) (DSP) [35]. Neutravidin is very convenient because it can be directly chemisorbed on gold and does not require additional chemical modification of the surface [37]. Recently, poly(amidoamine) dendrimers (PAMAM) were also used for aptamer immobilization. PAMAM dendrimers are globular macromolecules with amino terminal groups. Using glutaraldehyde, cross-linking of avidin to a dendrimer surface is possible, so it can be then used for immobilization of biotinylated aptamers. The advantage of dendrimers is their high stability and relatively large surface in comparison with flat electrode [38]. So *et al.* [39] reported immobilization of thrombin aptamer onto the surface of carbon nanotubes. The nanotubes were pretreated with carbodiimidazole-activated Tween 20 (CDI-Tween). 3′-End of thrombin aptamer was modified by $-NH_2$ groups, which allowed covalent binding of aptamer to CDI-Tween. Application of conducting nanotubes allowed to fabricate field-effect transistor-based biosensor for detection of thrombin. For choosing the method of immobilization, it is important to provide sufficient conformational freedom to the aptamer. Particularly this can be achieved also by using sufficiently long spacer. For example, Liss *et al.* [35] showed that substantial improvement in the binding properties of anti-IgE aptamer can be reached by extension of the aptamer length. Aptamer can be immobilized also on glass slides [40] or on a Si surface using UV radiation [41]. Some examples of immobilization of aptamers onto a solid support are schematically shown in Fig. 33.2.

33.2.2 Detection of aptamer–ligand interactions

Aptamer–ligand interactions can be detected by conventional assay based on, e.g., radioactive labeling, chromatography, capillary electrophoresis and mass spectrometry. These methods have been reviewed recently by Tombelli *et al.* [5]. Novel approach in detection of aptamer–ligand interaction is connected with aptasensors. In aptasensors,

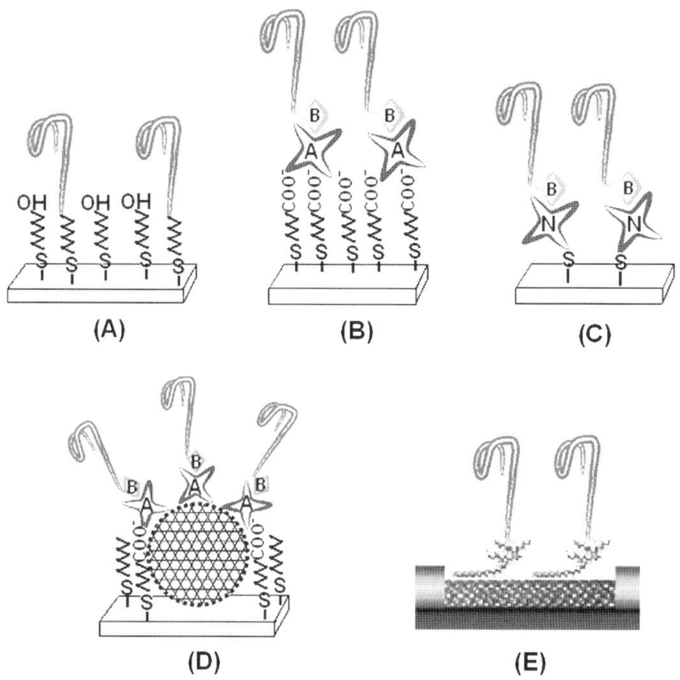

Fig. 33.2. Schematic picture of the methods of immobilization of the aptamers: (A) chemisorption. Biotinylated aptamer is immobilized on a surface covered by (B) avidin, (C) neutravidin and (D) avidin on a surface of poly(amidoamine) dendrimers (PAMAM). (E) Aptamer is immobilized on carbon nanotubes activated by Tween 20 (it is partially adapted from Ref. [39] with permission of American Chemical Society).

the aptamer is immobilized onto a solid support and the signal following binding of the ligand is detected by various methods, mostly electrochemical, acoustical and optical. Recent progress in aptasensors has been reviewed [5,14]. However, this field of research is progressing so strongly that already new achievements have appeared in literature, especially focused on electrochemical methods of detection and on application of nanotubes and nanoparticles in aptasensors. Therefore, we will describe these new directions in more detail.

33.2.2.1 Electrochemical methods

Electrochemical methods of detection affinity interactions at the surfaces are rather effective due to their relative simplicity and low cost. Amperometric aptasensor based on sandwich assay was proposed by Ikebukuro *et al.* [42]. They used two aptamers selective to thrombin.

A 15-mer thiol-labeled aptamer selective to fibrinogen binding site of thrombin was immobilized onto a gold surface, while second 29-mer biotinylated aptamer sensitive to heparin binding site of thrombin was labeled by enzyme glucose dehydrogenase from *Burkholdelia cepacia* (GDHPc) [42] or by oxygen-insensitive pyrroquinoline quinone glucose dehydrogenase from *Acinobacter calcoaceticus* ((PQQ)GDH) [43]. First the thrombin was added into the measuring cell containing aptasensor immobilized onto the surface of a gold electrode. Then the enzyme-labeled aptamer was added and attached to the sensor surface due to existence of second binding site at thrombin (Fig. 33.3A). In presence of glucose, the substrate for the enzyme, an electrochemical signal

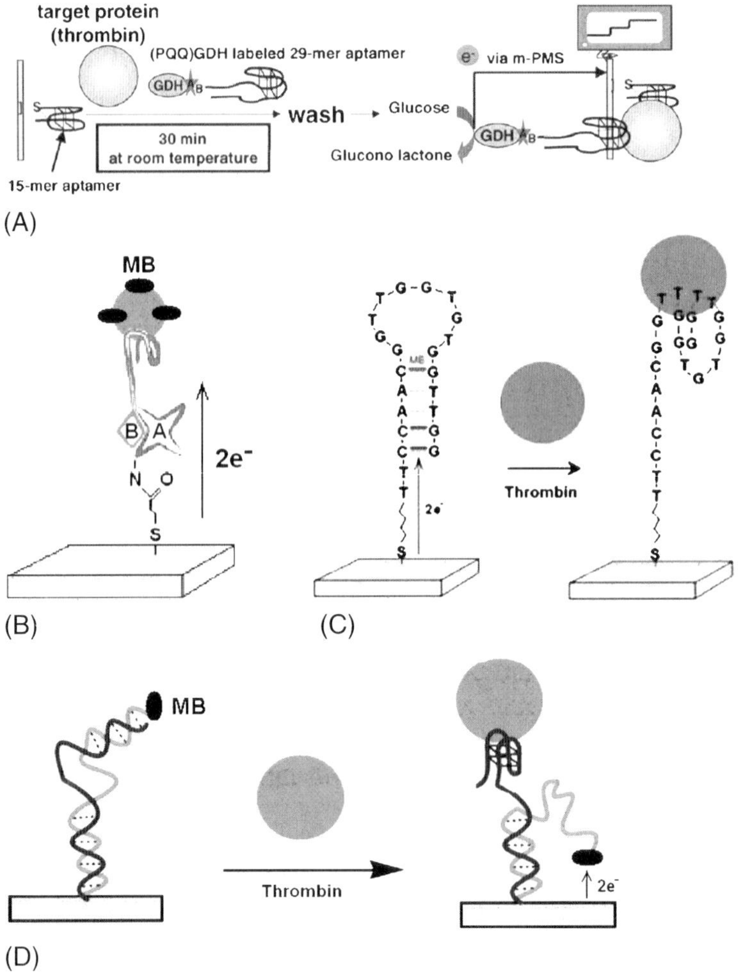

(current) appeared. The current was proportional to the amount of thrombin on the sensor surface. The detection limit was 1 μmol/L for GDHPc and 10 nmol/L for (PQQ)GDH aptasensor. Using fluorescence polarization, the authors also proved binding of aptamers to different binding sites of the thrombin. The sensor was selective to thrombin and only negligible response took place in presence of bovine serum albumin. In this work there were also reported difficulties with sensor regeneration due to strong binding of aptamers to thrombin.

Electrochemical indicator methods are based on the application of redox probe that undergoes oxidation and reduction transition due to electron transfer from electrode surface to a probe. In 2005, several studies that used methylene blue (MB) as an electrochemical indicator were published. MB is positively charged low-molecular-weight compound that can be reduced by two electrons to a leucomethylene blue (LB). The reduction process can be effectively monitored, e.g., by differential pulse voltammetry or coulometry. In presence of redox probe $Fe(CN)_6^-$, the LB is oxidized to MB and the system is regenerated [44,45]. In papers by Hianik *et al.* [31,46], MB was used as the indicator of detection of interaction of human thrombin with DNA aptamer. The method of detection is schematically shown in Fig. 33.3B. MB binds both to DNA and to the protein. For charge transfer from electrode to MB, i.e., for MB reduction, it is important that MB should be close to the electrode surface. Therefore, the charge transfer from the electrode

Fig. 33.3. Schematic representation of detection of thrombin–aptamer interactions by electrochemical methods. (A) Sandwiched assay according to Ref. [43] (reproduced by permission of Elsevier). Thiol-modified 15-mer aptamer sensitive to fibrinogen binding site of thrombin was immobilized onto a surface of gold electrode. After addition of thrombin, the second, 29-mer enzyme-labeled aptamer sensitive to heparin binding site of thrombin was added. The thrombin was detected be means of measurement of current due to glucose degradation. (B–D) Detection of thrombin using electrochemical indicator MB: (B) MB binds to both aptamer and thrombin. Binding of thrombin with adsorbed MB to aptamer resulted in an increase in charge transfer from electrode to MB [46]. (C) MB is intercalated between base pairs of aptamer beacon, and electron transfer from electrode to MB is favorable. Addition of thrombin resulted in changes in conformation, release of MB from double helix and a decrease in electron transport [48]. (D) (MB)-tagged oligonucleotide with aptamer forms rigid duplex that prevents the MB tag from approaching the electrode surface. The addition of thrombin resulted in quadruplex formation and approach of MB to the surface of electrode. In this configuration, reduction of MB is favorable (adapted from Ref. [49] with permission of American Chemical Society).

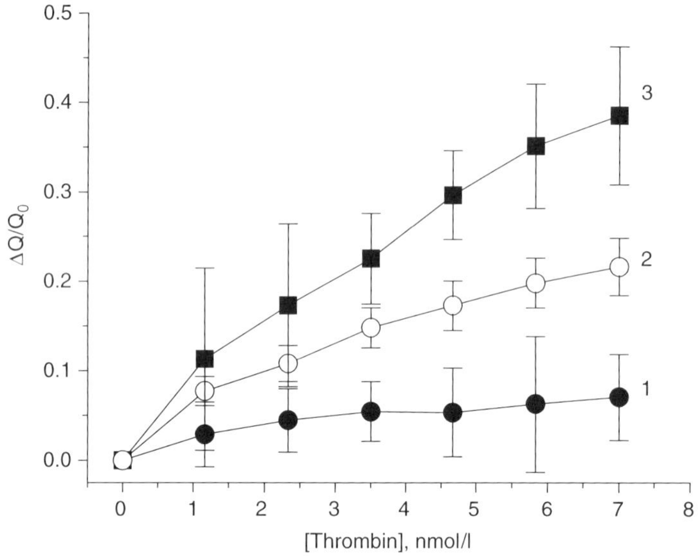

Fig. 33.4. The plot of the relative changes in charge transfer $\Delta Q/Q_0$ ($\Delta Q = Q - Q_0$, where Q_0 is the charge transfer without thrombin and Q that at certain thrombin concentration) as a function of thrombin concentration and for different methods of aptamer immobilization: (1) chemisorption of thiolated aptamer, (2) dendrimer surface covered by avidin and (3) surface covered by avidin. Results represent mean ± S.D. obtained in three experiments for each system. 32-mer DNA aptamer 3′-GGG TTT TCA CTT TTG TGG GTT GGA CGG GAT GG-5′ was modified by either thiol group or biotin at 3′ end. This aptamer was sensitive to heparin binding site of thrombin. Aptamer was immobilized onto the surface of gold electrode (CH Instruments, USA) of 2 mm diameter. As an electrolyte 140 mmol/L NaCl, 5 mmol/L KCl, 1 mmol/L $CaCl_2$, 1 mmol/L $MgCl_2$, 20 mM TRIS, pH 7.4 containing 2 μmol/L MB was used. The charge transfer was determined from differential pulse voltammetry measurements performed by means of potentiostat CHI 410 (CH Instruments, USA).

to MB should be more intensive to the proteins that specifically bind to the aptamer in comparison with that for unbounded protein in a solution. As an example, Fig. 33.4 shows the plot of relative changes in charge transfer as a function of the thrombin concentration and for different methods of immobilization of aptamers onto a solid support by means of chemisorption (curve 1), or using avidin–biotin technology, when aptamer was immobilized onto a gold support covered with G1 PAMAM dendrimers conjugated with avidin (curve 2) or on a gold support modified by avidin (curve 3) [31]. We can see that in all cases, the charge transfer increases on increasing the concentration of thrombin. The sensitivity of the sensor was highest for the method of immobilization of aptamer onto a gold surface covered by avidin, medium for that using dendrimer layers and minimal for aptamer chemisorbed directly onto a gold surface. The minimal sensitivity in the latter case can be probably due to higher density of aptamers on a surface that restricts maintaining optimal configuration of aptamer binding site to the thrombin. The differences between immobilization of aptamer onto G1 and onto avidin-covered surface could be connected with higher flexibility of aptamers immobilized on avidin, which is in addition linked to the gold support through a flexible linker. In contrast, when G1 is used as a support, avidin is relatively rigidly attached to this surface by glutaraldehyde. We showed [31] that this sensor has comparable sensitivity and detection limit (around 5 nmol/L) like mass detection (QCM) and SPR methods [36] and is also sufficiently selective in comparison with non-specific binding of, e.g., human serum albumin (HSA) or human immunoglobulin (IgG) (see Procedure 47 in CD accompanying this book). The obtained detection limit is sufficient for detection of thrombin in real blood samples (the physiological concentration of thrombin is in a range of low nmol/L to low μmol/L) [47].

Another method of detection of the thrombin–aptamer interaction was proposed by Bang *et al.* [48]. They immobilized aptamer beacon modified by amino group to a self-assembled monolayer onto a gold surface formed by 11-MUA. Aptamer beacon without thrombin is composed of short double helix. In presence of thrombin, the double helix is not stable and the aptamer changes its conformation. The detection method is based on measurement of the electrochemical reduction of MB. At aptamer hairpin configuration, the MB is intercalated into the double helix and possesses a well-resolved reduction peak at −197 mV vs. Ag/AgCl reference electrode. The conformational changes of aptamer due to binding of thrombin resulted in the release of MB and a decrease in reduction peak. The detection limit of this sensor was 11 nmol/L and the linear range of the signal was observed between 0 and 50.8 nmol/L of thrombin. This method of detection is schematically shown in Fig. 33.3C.

Two different formats of electrochemical detection of thrombin–aptamer interactions using MB were recently developed [32,49]. The aptamer was modified at one end by thiol group and at the other end by MB. Without thrombin, the MB possesses the reduction signal. Addition of thrombin shifted equilibrium from unfolded to folded aptamer conformation. This resulted in an increase in distance of MB from the electrode. As a result the reduction signal decreased.

The limitation of this "signal-off" architecture was improved in another work [49] in which the aptamer was first allowed to interact with short partially complementary nucleotide modified by MB. (MB)-tagged oligonucleotide with aptamer formed rigid duplex that prevents the MB tag from approaching the electrode surface, so the reduction current is suppressed. The addition of thrombin resulted in quadruplex formation and approach of MB to the surface of electrode. Thus, the amplitude of reduction signal increased (Fig. 33.3D). This method allowed to detect thrombin at concentrations as low as 3 nmol/L.

MB has been shown useful also for detection of cocaine by means of specific DNA aptamer [50]. The MB-tagged aptamer has been immobilized via thiol group onto a gold support. In absence of cocaine, the aptamer was partially unfolded. Addition of cocaine resulted in folding of aptamer into three-way junction, moving MB to a close proximity with the electrode surface. This resulted in an increase in reduction peak measured by AC voltammetry. Sensor was regenerable and allowed to detect cocaine within several seconds with sensitivity below 10 μmol/L.

Direct detection of thrombin using single-walled carbon nanotube field-effect transistor (SWNT-FET) was reported by So *et al.* [39]. In this work, aptamer was covalently immobilized onto carbon nanotubes modified by Tween 20. Addition of thrombin resulted in a drop in the conductance of the device. This has been explained by screening of negative charge of DNA by thrombin, which was positively charged under experimental conditions (pH 5.4). Certainly, the isoelectric point of thrombin is rather high: 7–7.6 [51]. The lowest detection limit (LOD) of the sensor was 10 nmol/L. The sensor was selective to thrombin. This was proved by an experiment in which elastase instead of thrombin was used. Elastase is a serine protease with isoelectric point and molecular weight similar to those of thrombin. Addition of elastase resulted in considerably lower changes in the current in comparison with thrombin. Sensor was regenerable by treatment with 6 mol/L guanidine hydrochloride solution, which removed bounded thrombin molecules.

Label-free detection of ligand–aptamer interaction was also demonstrated by means of impedance spectroscopy technique [52,53]. Simultaneously, Radi *et al.* [52] and Rodriguez *et al.* [53] reported application of Faradaic impedance spectroscopy (FIS) in detection of interaction of proteins with DNA aptamers. The detection method is based on the measurement of resistance in presence of redox mediator $Fe(CN)_6$. In absence of target protein, the negatively charged aptamer repulse the redox mediator molecules from the sensor surface. In a paper by

Radi *et al.* [52], thiol-modified antithrombin aptamer was immobilized onto a gold electrode together with 2-mercaptoethanol. Thus, the surface was blocked against non-specific adsorption. This sensor allowed to detect thrombin already at 2 nmol/L with linear range 5–35 nmol/L. The sensor was regenerable using 2 mol/L NaCl. Rodriguez *et al.* [53] detected lysozyme with biotinylated DNA aptamer immobilized onto an indium oxide layer coated with a polymer and streptavidin. Binding of positively charged protein (at $pH < pI$) resulted in enhancement of electron transfer and, as a result, a decrease in the resistance takes place. The method allowed to detect the lysozyme with high sensitivity. It was possible to reproducibly detect 14 nmol/L of the protein, while sensor was considerably less sensitive to other proteins: bovine serum albumin, cytochrome *c* and thrombin.

Recently, a new electrochemical method of detection of the aptamer–protein interactions based on quantum-dot (QD) semiconductor nanocrystals was reported [54]. In a competitive binding assay, the proteins (thrombin or lysozyme) were conjugated with Cd or Pb nanocrystals. The sensor was prepared by immobilization onto a gold support of thiolated DNA aptamers that selectively bind thrombin and/or lysozyme. After competitive binding with the sensor surface and after the washing step, the conjugated proteins were detected on a surface of mercury-coated glassy carbon electrode using square-wave stripping voltammetry method. This method allowed to reach extremely low detection limit of approximately 0.5 pM of the protein.

The electrochemical methods have great potential due to existing wide market of relatively not expensive instruments produced by, e.g., CH Instruments (USA), Eco Chemie (The Netherlands), BAS (USA) and others. Advantage of this approach consists also in simple and low-cost sensor fabrication as well as in easy-to-use and simple evaluation of the binding processes.

33.2.2.2 Acoustical methods

Acoustical methods belong to label-free detection tolls in biosensing. In TSM method the sensor is fabricated on one side of AT-cut quartz covered by thin gold layer, while the other side is free and in contact with air or nitrogen gas. When high-frequency voltage (few MHz) is applied to quartz, it starts oscillating in direction parallel to the surface and propagates transverse shear waves into the surrounding medium with decay length of the order of micrometres. The response of the TSM sensor in liquid depends on a number of factors such as viscoelasticity, acoustoelectrical and interfacial acoustic phenomena. When analysis of crystal oscillations is performed by network analyzer, several parameters can be obtained, which characterize the properties of the crystal and interaction of acoustic wave with the surrounding liquid: series resonant frequency (f_s), motional resistance (R_m), inductance (L) and static and motional capacitances (C_s and C_m, respectively) [55]. In simplest format usually only series resonant frequency is measured. This particular method is known as quartz crystal microbalance (QCM) and has been developed by Sauerbrey [56], who derived relation between changes in resonant frequency, Δf_s, and changes in the surface mass density, ρ_S:

$$\Delta f_s = -\frac{2f_0^2\rho_s}{\sqrt{\mu_q\rho_q}} \tag{33.1}$$

where f_0 is the fundamental frequency of the crystal (typically 5–10 MHz), and $\mu_q = 2.947 \times 10^{10}$ Pa and $\rho_q = 2648\,\text{kg}\,\text{m}^{-3}$ are the shear stiffness and mass density of the quartz, respectively. For practical purposes it is convenient to use this equation in a format:

$$\Delta f_s = -2.26 \times 10^{-6} f_0^2 \left(\frac{\Delta m}{A}\right) \tag{33.2}$$

where Δm signifies the change in the mass (in ng) and A is the area of working electrode (in cm^2) (typically $A{\sim}0.3\,\text{cm}^2$) [57].

Equations (33.1) and (33.2) are exactly valid only for dry crystals. In liquid, however, the viscous forces cause damping of acoustic waves. As a result, observed changes in series resonant frequency are connected not only with the changes in the mass but also with shearing viscosity. This has been demonstrated in a work by Thompson and coworkers, which applied TSM method to study protein–RNA interactions [58]. A detailed study of the interaction of short peptides that model the HIV-1 Tat protein with the HIV-1 mRNA (TAR) showed that for relatively short peptides (12–22 amino acid residues) an increase in f_s has been observed. Almost no changes in f_s took place when peptide composed of 27 amino acid residues interacted with TAR. This was in contrast with mass-response model, which predicts a decrease in f_s (see Eq. (33.1)). The f_s value decreased only for longer peptides [59]. The decrease in the frequency was observed also when native Tat protein (101 amino acid residues) interacted with TAR [60,61]. The observed discrepancy has been connected with structural changes in RNA due to binding processes. Thus, various layers move with different velocities which induces changes in acoustic coupling and energy dissipation. This is reflected in changes in series resonant frequency and in a motional resistance (see Ref. [59] for more detailed explanation). The TSM method has been further used for studying the interaction of antibiotics neomycin [59] and streptomycin [62] with TAR in the presence of Tat. It has been confirmed that neomycin effectively blocked interaction of Tat with TAR. TSM method is rather effective also for determination of kinetics of binding of proteins to RNA [55] and DNA [31] aptamers.

In simplest, QCM, format the protein–aptamer interactions were analyzed by Liss *et al.* [35]. They compared the interaction of IgE with DNA aptamer as well as with anti-IgE antibodies. While the detection limit was similar in both cases, the advantage of aptasensors was in the possibility of surface regeneration, which was impossible for antigen-based biosensor. Mascini and co-workers showed that similar results for sensitivity and selectivity in detection of Tat peptide with RNA aptamer can be obtained by QCM and SPR methods [61]. Comparison of QCM method with electrochemical biosensor assay of thrombin detection was recently performed [31,46]. In our recent work, we used neutravidin for immobilization of biotinylated DNA aptamers sensitive to thrombin and studied their binding properties and determined the binding and dissociation constants of thrombin to aptamers [31]. Fig. 33.5 shows the changes in series resonant frequency of the AT-cut quartz covered by neutravidin with immobilized 32-mer DNA aptamer selective to heparin binding site of thrombin (3′-BIOTIN-GGG TTT TCA CTT TTG TGG GTT GGA CGG GAT GG-5′) following addition of human thrombin or HSA into the flow-through cell. The decrease in frequency following addition of thrombin was observed, while a considerably smaller response took place when HSA was added. Please note that concentration of HSA was approximately 1000 times higher in comparison with that of thrombin. The detection limit for thrombin was 1 nmol/L.

Sensitivity of QCM for detection of thrombin by DNA aptamer was improved using gold nanoparticles [34]. In this work, thiolated DNA aptamer specific to fibrinogen binding site was immobilized onto AT-cut quartz surface. Addition of thrombin resulted in a decrease in oscillation frequency. The aptamers of the same structure conjugated with gold nanoparticles were added into the measuring cell and a considerably larger decrease in frequency was observed due to larger mass of aptamer–nanoparticle complex. However, in contrast with the paper by Ikebukuro *et al.* [43], Pavlov *et al.* [34] used the aptamer of the same structure for immobilization onto electrode surface and the

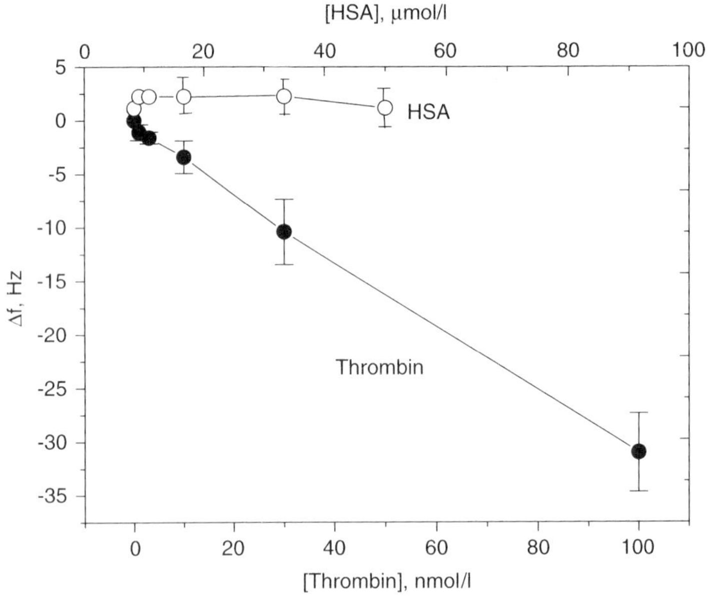

Fig. 33.5. The plot of the series resonance frequency of AT-cut crystal covered by neutravidin and with immobilized biotinylated 32-mer DNA aptamer selective to heparin binding site of thrombin as a function of thrombin and HSA concentration, respectively. The fundamental frequency of the crystal was 9 MHz. The frequency was determined by HP4395A Network-Spectrum analyzer (Hewlet Packard, Colorado Springs, CO, USA). The crystal was placed in a flow cell developed by Thompson *et al.* [37] (experiment was performed by I. Grman in M. Thompson laboratory [75]).

nanoparticles. It should be therefore surprising that they observed changes in the mass following addition of aptamer conjugated with nanoparticles, because already the binding site for fibrinogen was occupied by aptamers immobilized onto the crystal. However, as it was shown by Tasset *et al.* [17] in competitive binding studies, the aptamer specific for fibrinogen developed by Bock *et al.* [15] and used in the paper by Pavlov *et al.* [34] has certain, although low, affinity to heparin binding site. Therefore, one can assume that when another aptamer specific to heparin binding site of thrombin is used for conjugation with nanoparticles, the sensitivity of the detection can be additionally improved.

Low-wave acoustic sensor [63] was used to detect interaction of thrombin with RNA and DNA aptamers [24]. The authors compared the binding of thrombin to RNA aptamer also by using filter binding method utilizing radiolabeling of RNA aptamer by $3'$-P^{32} and also by applying SPR method. They determined K_d values of 294 ± 30 nmol/L for filter binding, 113 ± 20 nmol/L for SPR and 181 ± 20 nmol/L for acoustic method. DNA aptamers, which bind thrombin to fibrinogen binding site, revealed about twice lower effectivity in comparison with RNA aptamers. They also studied the effect of heparin, human antithrombin III and their complexes on the binding of DNA aptamer to thrombin. The results suggest that binding of thrombin to DNA aptamer affects the binding of heparin–antithrombin III complex to heparin binding site. This has been explained by conformational changes of thrombin following binding of DNA aptamers to fibrinogen binding site (exocite II), which affect the heparin binding site (exocite I) and vice versa.

Thus, acoustic sensors represent a rather useful tool for studying the protein–aptamer interactions without additional labeling. Using acoustic sensor in TSM format, it is possible to study also subtle structural effects on the sensor surface connected with conformational changes of aptamers induced by protein or low-molecular-weight ligands. These sensors can be advantageously used also for studying kinetics of the ligand–aptamer interactions. For a simple QCM format, a wide range of low-cost instruments are available (see Ref. [64] for review). The TSM method, however, requires application of network analyzer and special analysis of the results. However, Agilent Technology (USA) also produced an economic series of vector network analyzers, e.g., 5061A, that are suitable for detailed analysis of affinity interactions on surfaces using AT-cut quartz resonators. It should be noted that the cell design is very important in QCM and TSM studies. Most suitable is a flow-through cell that provides sufficient signal stability, but fixing of the crystal in a cell in a way that minimizes mechanical stress is also very important [55,65].

33.2.2.3 Optical methods

Optical methods are based on fluorescence probe-labeled aptamers (fluorescence intensity, fluorescence anisotropy), or label-free aptamers can be used for detection of analyte using SPR or Fourier transform infrared attenuated total reflection (FTIR-ATR).

The first aptasensor reported was particularly based on optical detection [66]. The 58-mer RNA aptamer selective to L-adenosine was immobilized onto the core of multimode fiber using avidin–biotin method. The detection was based on competitive binding of FITC-labeled L-adenosine with unlabeled analyte. This sensor also allowed to study the kinetics of binding and determine equilibrium constants.

The sensitive detection of thrombin with detection limit of 5 nmol/L was obtained using polarization fluorescence method [40]. In this work fluorescein-modified aptamer was immobilized onto a glass slide that was optically coupled to the prism. Fluorescence was excited by polarized evanescent wave at 488 nm. The fluorescence emission was collected from the top of the prism. The sensor was highly selective (a comparative study of the binding of elastase was performed with practically no interference from thrombin) and regenerable. A similar approach was used by McCauley *et al.* [13] for fabrication of aptamer sensor array composed of four different aptamers, which was able to detect simultaneously thrombin, inosine monophosphate dehydrogenase (IMPDH), vascular endothelial growth factor (VEGF) and basic fibroblast growth factor (bFGF). The fiber-optic microarray system for aptamer-based detection of thrombin was developed by Lee and Walt [47]. They were able to detect thrombin with a detection limit of 1 nmol/L.

Original method of detection of aptamer–thrombin interaction was proposed by Hamaguchi *et al.* [28]. They developed aptasensor using aptamer beacon based on known DNA aptamer against thrombin [15], but extended the oligonucleotide chain at the 5′ end by addition of the sequence of nucleotides complementary to that at 3′ end. Thus, without thrombin the aptamer was in hairpin configuration with five base pairs close to the solid support. The 3′ end of the aptamer was modified by fluorescence quencher DABCYL, while 5′ terminal was modified by fluorescein. Without thrombin the fluorescence was depressed due to close distance between fluorescence probe and quencher. However, addition of thrombin resulted in changes in aptamer conformation and in movement of the fluorescently labeled terminal away from the quencher. This has been detected as an increase in fluorescence intensity. This method allowed to detect thrombin with limit of detection (LOD) of 10 nmol/L. Authors also analyzed the aptamer beacon properties using circular dichroism method that has been applied to three aptamer beacon structures that differ in the length of complementary chain. It has been shown that for optimal detection of thrombin, it is crucial that the free energy of the complex aptamer–thrombin is lower in comparison with aptamer beacon. For example, when aptamer beacon contained six instead of five complementary base pairs, the beacon structure was so stable that thrombin did not cause folding of aptamer into a configuration with typical binding motif. In contrast, when a shorter complementary chain was used (four base pairs), the aptamer was always in extended configuration.

Yamana *et al.* [67] used bis-pyrene-labeled DNA aptamer for detection of ATP. The pyrene excimer was incorporated into several nucleotide positions. Addition of ATP resulted in an increase in fluorescence only for aptamers labeled by fluorescence probe between residues that were responsible for ATP binding. Using this sensor it was possible to detect ATP with mmol/L sensitivity.

SPR method is based on the generation of surface plasmons by light evanescent waves. The intensity of totally reflected light is measured. The intensity of reflected light depends particularly on changes in thickness of the sensing layer (see, e.g., Ref. [68]). The advantage of SPR is that the sensing layer composed of aptamers or other receptors does not require additional chemical modification by fluorescence probes. This method has been used for development of RNA aptamer biosensor for detection of SD4 antigen [69], HIV-Rev protein [70] and TTF1 [71]. HIV-Tat protein [60,61] was detected using RNA aptamer with 0.25 ppm detection limit. In contrast with DNA aptamers, which are rather stable also in complex biological liquids, the RNA aptamers are unstable due to degradation by nucleases. Even in laboratory conditions special care is required for work with RNA aptamers [61].

SPR method is currently well established and used in analysis of affinity interactions. The commercial instruments produced by Biacore AB or Texas Instruments are available in the market. Substantial progress in SPR instrumentation is connected also with the possibility to detect simultaneously the affinity interaction from several surfaces [72]. This approach has been proved equally effective for detection of thrombin–DNA aptamer interactions [36].

Another label-free optical detection method—FTIR-ATR—has been applied for detection of thrombin by means of DNA aptamers [73]. The antithrombin DNA aptamer previously developed by Tasset *et al.* [17] was immobilized covalently onto Si surface using UV irradiation method. As a quantitative measure, the area of N–H and CH_2 bands was used. This method allowed to detect thrombin with a sensitivity around 10 nmol/L. The specificity of binding of protein to aptamer was also investigated using DNA with no binding site for thrombin. It has been noted that for effective binding study by FTIR-ATR method, the concentration of protein should be kept lower than 100 nmol/L.

Applications of cantilevers also allows label-free detection of aptamer–ligand interactions. Although it is not an optical sensor, the interferometry is used for detection of the differential bending between cantilever and reference modified with non-specific aptamer [74].

33.3 CONCLUSIONS

Aptamer-based biosensors are rather promising. They can be used for detection of proteins or low-molecular-weight compounds with high sensitivity and selectivity comparable and even higher than those available so far for antibody-based assay. However, major advantage of the aptamers in comparison with antibodies is the possibility of their synthesis in vitro without using experimental animals. This allows to prepare aptamers even for toxic compounds, which is impossible in the case of antibodies. Aptamers can be chemically modified by various ligands and indicator groups. Thus, they can be immobilized onto solid supports of different types. Various indicators (optical or electrical) allow to use a variety of methods for detection of aptamer–ligand interactions, including electrochemical and optical methods. Certain methods, e.g., SPR, QCM, TSM or FTIR, even do not require aptamer labeling. This detection, especially in the case of acoustic methods, can be substantially amplified using nanoparticles. Recent achievements in electrochemical detection of aptamers–ligand interactions are rather promising because they open new routes for easy-to-use and low-cost assays that may be rather useful in medical and environmental applications of aptasensors. However, further efforts are required to optimize sensor properties, especially for application in real samples. For example, the application of RNA aptamers is still in premature stage due to problems with their stability in real samples. The limitations also consist in difficulties in synthesizing longer RNA aptamers (more than 50 bases). DNA aptamers are more stable, but further works are required for optimization of aptamer immobilization, sensor regeneration and detection of ligand–aptamer interactions, especially in the case of small molecules.

ACKNOWLEDGMENTS

This work was financially supported by Slovak Grant Agency (Projects No. 1/1015/04), by Agency for Promotion Research and Development under the contract No. APVV-20-P01705 and by European Commission under the Noe NeuroPrion contract No. FOOD-CT-2004-506579. I would like to thank Igor Grman for technical assistance.

REFERENCES

1 A.D. Ellington and J.W. Szostak, *Nature*, 346 (1990) 818–822.
2 W. James, Aptamers. In: R.A. Meyers (Ed.), *Encyclopedia of Analytical Chemistry*, J. Wiley & Sons Ltd., Chichester, 2000, pp. 4671–4848.
3 F.W. Scheller, U. Wollenberg, A. Warsinke and F. Lisdat, *Curr. Opin. Biotechnol.*, 12 (2001) 15–40.
4 C.K. O'Sullivan, *Anal. Bioanal. Chem.*, 372 (2002) 44–48.
5 S. Tombelli, M. Minunni and M. Mascini, *Biosens. Bioelectron.*, 20 (2005) 2424–2434.
6 D.L. Robertson and G.F. Joyce, *Nature*, 344 (1990) 467–468.
7 C. Tuerk and L. Gold, *Science*, 249 (1990) 505–510.
8 S.D. Jayasena, *Clin. Chem.*, 45 (1999) 1628–1650.
9 L. Gold, B. Polisky, O. Uhlenbeck and M. Yarus, *Annu. Rev. Biochem.*, 64 (1995) 763–797.
10 R. Conrad and A.D. Ellington, *Anal. Biochem.*, 242 (1996) 261–265.
11 H. Dougan, J.B. Hobbs, J.I. Weitz and D.M. Lyster, *Nucleic Acids Res.*, 25 (1997) 2897–2901.
12 S.E. Osborne, I. Matsumura and A.D. Ellington, *Curr. Opin. Chem. Biol.*, 1 (1997) 5–9.
13 G.T. McCauley, N. Hamaguchi and M. Stanton, *Anal. Biochem.*, 319 (2003) 244–250.
14 B. Leca-Bouvier and J. Blum, *Anal. Lett.*, 38 (2005) 1491–1517.
15 L.C. Bock, L.C. Griffin, J.A. Latham, E.H. Vermaas and J.J. Toole, *Nature (London)*, 355 (1992) 564–566.
16 C.A. Holland, A.T. Henry, H.C. Whinna and F.C. Church, *FEBS Lett.*, 484 (2000) 87–91.
17 D.M. Tasset, M.F. Kubik and W.J. Steiner, *J. Mol. Biol.*, 272 (1997) 688–699.
18 R.F. Macaya, P. Schultze, F.W. Smith, J.A. Roe and J. Feigon, *Proc. Natl. Acad. Sci. USA*, 90 (1993) 3745–3749.
19 M. Tsiang, C.S. Gibbs, L.C. Griffin, K.E. Dunn and L.K. Leung, *J. Biol. Chem.*, 270 (1995) 19370–19376.
20 P. Schultze, R.F. Macaya and J. Feigon, *J. Mol. Biol.*, 235 (1994) 1532–1547.
21 K. Padmanabhan, K.P. Padmanabhan, J.D. Ferrara, J.E. Sadler and A. Tulinsky, *J. Biol. Chem.*, 268 (1993) 17651–17654.
22 M.F. Kubik, A.W. Stephens, D. Schneider, R.A. Marlar and D. Tasset, *Nucleic Acids Res.*, 22 (1994) 2619–2626.
23 R. White, C. Rusconi, E. Scantino, A. Wolberg, J. Lawson, M. Hoffman and B. Sullenger, *Mol. Ther.*, 4 (2001) 567–573.
24 T.M.A. Gronewold, S. Glass, E. Quantdt and M. Famulok, *Biosens. Bioelectron.*, 20 (2005) 2044–2052.

25 S. Weiss, D. Proske, M. Neumann, M.H. Groschup, H.A. Kretzschmar, M. Famulok and E.-L. Winnacker, *J. Virol.*, 71 (1997) 8790–8797.
26 D. Proske, S. Gilch, F. Wopfner, H.M. Schaetzl, E.-L. Winnacker and M. Famulok, *Chembiochem.*, 3 (2002) 717–725.
27 A. Rhie, L. Kirby, N. Sayer, R. Wellesley, P. Disterer, I. Sylvester, A. Gill, J. Hope, W. James and A. Tahiri-Alaouit, *J. Biol. Chem.*, 278 (2003) 39697–39705.
28 N. Hamaguchi, A. Ellington and M. Stanton, *Anal. Biochem.*, 294 (2001) 126–131.
29 B.I. Kankia and L.A. Marky, *J. Am. Chem. Soc.*, 123 (2001) 10799–10804.
30 I. Smirnov and R.H. Shafer, *J. Mol. Biol.*, 296 (2000) 1–5.
31 T. Hianik, V. Ostatná, M. Sonlajtnerova and I. Grman, *Bioelectrochemistry* (2006), 70 (2007) 127–133.
32 Y. Xiao, A.A. Lubin, A.J. Heeger and K.W. Plaxco, *Angew. Chem. Int. Ed.*, 44 (2005) 5456–5459.
33 M.I. Pividori, A. Merkoci and S. Alegret, *Biosens. Bioelectron.*, 15 (2000) 291–303.
34 V. Pavlov, Y. Xiao, B. Shlyahovsky and I. Willner, *J. Am. Chem. Soc.*, 126 (2004) 11768–11769.
35 M. Liss, B. Petersen, H. Wolf and E. Prohaska, *Anal. Chem.*, 74 (2002) 4488–4495.
36 V. Ostatná, M. Sonlajtnerová, I. Grman, H. Vaisocherová, J. Homola and T. Hianik, *Book of Abstracts, EuroConference Biomaterials and Analytical Techniques*, June 18–23, 2005, Sant Feliu, Spain.
37 B.A. Cavic and M. Thompson, *Anal. Chim. Acta*, 469 (2002) 101–113.
38 L. Svobodová, M. Šnejdárková, K. Tóth, R.E. Gyurcsanyi and T. Hianik, *Bioelectrochemistry*, 63 (2004) 285–289.
39 H.-M. So, K. Won, Y.H. Kim, B.-K. Kim, B.H. Ryu, P.S. Na, H. Kim and J.-O. Lee, *J. Am. Chem. Soc.*, 127 (2005) 11906–11907.
40 R.A. Potyrailo, R.C. Conrad, A.D. Ellington and G.M. Hieftje, *Anal. Chem.*, 70 (1998) 3419–3425.
41 F. Wei, B. Sun, W. Liao, J.H. Ouyang and X.S. Zhao, *Biosens. Bioelectron.*, 18 (2003) 1149–1155.
42 K. Ikebukuro, C. Kiyohara and K. Sode, *Anal. Lett.*, 27 (2004) 2901–2909.
43 K. Ikebukuro, C. Kiyohara and K. Sode, *Biosens. Bioelectron.*, 20 (2005) 2168–2172.
44 E.M. Boon, D.M. Ceres, T.G. Drummond, M.G. Hill and J.K. Barton, *Nat. Biotechnol.*, 18 (2000) 1096–1100.
45 V. Ostatná, N. Dolinnaya, S. Andreev, T. Oretskaya, J. Wang and T. Hianik, *Bioelectrochemistry*, 67 (2005) 205–210.
46 T. Hianik, V. Ostatná, Z. Zajacová, E. Stoikova and G. Evtugyn, *Bioorg. Med. Chem. Lett.*, 15 (2005) 291–295.
47 M. Lee and D. Walt, *Anal. Biochem.*, 282 (2000) 142–146.
48 G.S. Bang, S. Cho and B.-G. Kim, *Biosens. Bioelectron.*, 21 (2005) 863–870.

49 Y. Xiao, B.D. Pores, K.W. Plaxco and A.J. Heeger, *J. Am. Chem. Soc.*, 127 (2005) 17990–17991.
50 B.R. Baker, R.Y. Lai, M.S. Wood, E.H. Doctor, A.J. Heeger and K.W. Plaxco, *J. Am. Chem. Soc.*, 128 (2006) 3138–3139.
51 J.W. Fenton, B.H. Landis, D.A. Walz and J.S. Finalyson. In: R.L. Lundblat, J.W. Fenton and K.G. Mann (Eds.), *Chemistry and Biology of Thrombin*, Ann Arbor Science Publishers, Ann Arbor, MI, 1977, pp. 43–70.
52 A.E. Radi, J.L.A. Sanchez, E. Baldrich and C.K. O'Sullivan, *Anal. Chem.*, 77 (2005) 6320–6323.
53 M.C. Rodriguez, A.-N. Kawde and J. Wang, *Chem. Commun.*, 34 (2005) 4267–4269.
54 J.A. Hansen, J. Wang, A.-N. Kawde, Y. Xiang, K.V. Gothelf and G. Collins, *J. Am. Chem. Soc.*, 128 (2006) 2228–2229.
55 N. Tassew and M. Thompson, *Biophys. Chem.*, 106 (2003) 241–252.
56 G. Sauerbrey, *Z. Phys.*, 155 (1959) 206–210.
57 G.L. Hayward and G.Z. Chu, *Anal. Chim. Acta*, 288 (1994) 179–185.
58 L.M. Furtado, H. Su and M. Thompson, *Anal. Chem.*, 71 (1999) 1167–1175.
59 N. Tassew and M. Thompson, *Anal. Chem.*, 74 (2002) 5313–5320.
60 M. Minunni, S. Tombelli, A. Guilotto, E. Luzi and M. Mascini, *Biosens. Bioelectron.*, 20 (2004) 1149–1156.
61 S. Tombelli, M. Minunni, E. Luzi and M. Mascini, *Bioelectrochemistry*, 67 (2005) 135–141.
62 N. Tassew and M. Thompson, *Org. Biomol. Chem.*, 1 (2003) 3268–3270.
63 M. Schlensog, M.D. Schlensog, T. Gronewold, M. Tewes, M. Famulok and E. Quandt, *Sens. Actuators B*, 101 (2004) 308–315.
64 C.K. O'Sullivan and G.G. Guilbault, *Biosens. Bioelectron.*, 14 (1999) 663–670.
65 A. Saluja and D.S. Kalonia, *AAPS PharmSciTech.*, 5 (2004) 1–14.
66 F. Kleinjung, S. Klusmann, V.A. Erdmann, F.W. Scheller, J.P. Fürste and F.F. Bier, *Anal. Chem.*, 70 (1998) 328–331.
67 K. Yamana, Y. Ohtani, H. Nakano and I. Saito, *Bioorg. Med. Chem. Lett.*, 13 (2003) 3429–3431.
68 B.R. Eggins, *Chemical Sensors and Biosensors*, Wiley, Chichester, 2003.
69 E. Kraus, W. Janes and A.N. Barclay, *J. Immunol.*, 160 (1998) 5209–5212.
70 D.I. Van Ryk and S. Venkatesan, *J. Biol. Chem.*, 274 (1999) 17452–17463.
71 M.B. Murphy, S.T. Fuller, P.M. Richardson and S.A. Doyle, *Nucleic Acids Res.*, 31 (2003) e110.
72 M. Piliarik, H. Vaisocherová and J. Homola, *Biosens. Bioelectron.*, 20 (2005) 2104–2110.
73 W. Liao, F. Wei, D. Liu, M.X. Qian, G. Yuan and X.S. Zhao, *Sens. Actuators B*, 114 (2006) 445–450.
74 C.A. Savran, S.M. Knudsen, A.D. Ellington and S.R. Manalis, *Anal. Chem.*, 76 (2004) 3194–3198.
75 T. Hianik, I. Grman and M. Thompson, *Book of Abstracts. 2nd Symposium of Slovak Biophysical Society*, Herl'any, Slovakia, March 26–29, 2006.

Chapter 34

Miniaturised devices: electrochemical capillary electrophoresis microchips for clinical application

Mario Castaño-Álvarez, María Teresa Fernández-Abedul and Agustín Costa-García

34.1 INTRODUCTION

Miniaturisation and automatisation of analytical and bioanalytical systems have had a great development in the last decade. The final goal of miniaturised devices is represented by micro-total analysis systems (μ-TAS), also called "lab-on-a-chip" (LOC). These devices bring advantages in terms of mass production, low cost and disposability as well as speed, high performance, portability and low energy requirements.

Many of these microsystems have been described, but capillary electrophoresis (CE) microchips seem to be the most widely developed. The concept of μ-TAS was proposed by Manz *et al.* in 1990 [1] and 2 years later, electrophoresis in planar chips was successfully integrated and its use demonstrated using silicon and glass substrates [2,3]. This area of miniaturised analysis systems has grown rapidly, as is shown in two complete revisions by Manz *et al.* [4,5], the first dealing with theory and technology and the second one with analytical standard operations and applications, and the development continues exponentially.

As stated by Manz, the μTAS was envisioned as a new concept for chemical sensing, needed since sensors at that time were not providing the best results in terms of selectivity and lifetime. Initially, the main reason for miniaturisation was to enhance the analytical performance of the device rather than to reduce its size. However, it was also recognised that a small scale presented the advantage of a smaller consumption of sample and reagents. Moreover, the total chemical analysis system scheme could provide an integration of several laboratory procedures such as sample preparation, filtration, preconcentration,

Comprehensive Analytical Chemistry 49
Alegret and Merkoçi (Eds)
Volume 49 ISSN: 0166-526X DOI: 10.1016/S0166-526X(06)49034-6

derivatisation and separation that could enable the monitoring of many components within a single device. In essence, a μTAS is a device that improves the performance of an analysis by virtue of its reduced size. Thus, the sentence "the biggest changes often depend on the smallest things" could be perfectly applied here.

As stated above, small volume of the sample and reagent are representative of most miniaturised systems. This characteristic has clear advantages associated with cost and analytical throughput, but also has disadvantages, such as the suitability of detection techniques. Consequently, much research effort has focused on developing miniaturised and sensitive detection units [6], and detection improvements are still one of the most important focuses of research.

The aim of this chapter is to show through the subsequent sections the present trends in capillary electrophoresis microchips paying special attention to the manufacturing and designs employed in those combined to electrochemical (EC) detection. Conductimetric and amperometric detection are considered. Clinical application of these devices is further revised.

34.1.1 Microchip fabrication

Microchip fabrication clearly depends on the material employed. Although these devices debuted in silicon and glass as conceived at the beginning [2,3] in the early 1990s, plastic fabrication is being common nowadays. Various glass substrates can be employed, from inexpensive soda-lime glass to high-quality quartz, including borofloat (pyrex, BK7, etc.) between them. Good optical properties, efficiency in dissipating heat, well-understood surface characteristics and well-developed microfabrication methods, adapted from the microelectronics industry, made them the first option. Moreover, their behaviour was quite similar to traditional fused silica electrophoresis capillaries. Structures on glass substrates are usually generated using standard photolithographic technologies and wet etching (Fig. 34.1A) [7]. Ideal geometries involve high aspect ratio channels, that is to say, deep channels with parallel sides. Non-parallel walls occur in glass with wet-etching procedures because this process occurs on the exposed glass surface; hence, as the channel etches deeper, the walls are also etched. The result is channels are wider at the top than at the base. An alternative to produce very deep channels with parallel sides is the use of dry etching such as powder blasting and plasma or deep reactive ion etching (DRIE).

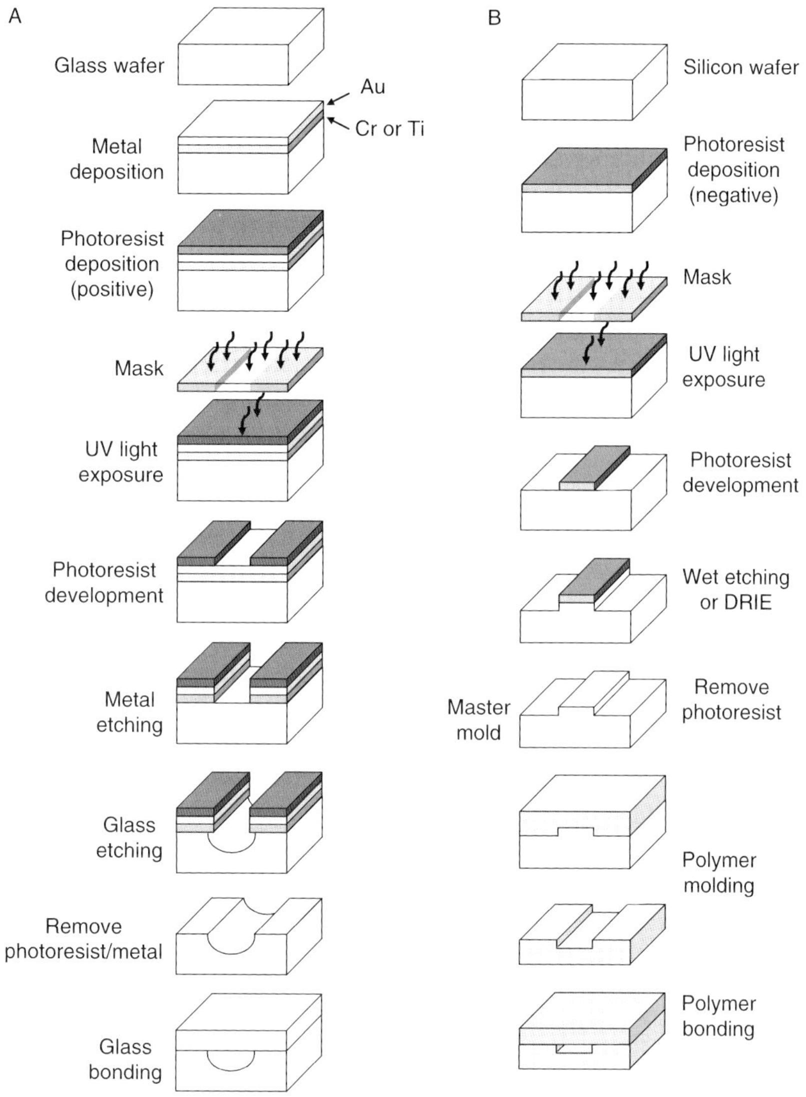

Fig. 34.1. Schematic representation of fabrication of glass (A) and polymer (B) microchip.

With the aim of developing microchip CE systems from materials that have properties similar to those of glass but are easier and less expensive to fabricate, a ceramic material, aluminium borosilicate, has been employed [8]. Although electroosmotic properties are similar to those of fused silica and fabrication can be accomplished using either milling or laser ablation, its use has not been widespread.

Polymer microchips are of increasing interest because their potentially low manufacturing costs may allow them to be disposable. Moreover, clean room and etching facilities are not required. The numerous methods for manufacturing systems of reservoirs and capillary-size channels in polymeric substrates can be divided into two areas: direct fabrication (mechanical machining, wire imprinting, laser ablation, etc.) and replication. In the last case, an accurate template or master is formed, usually by LIGA (a German acronym for lithography, electroplating and moulding) and then it is replicated faithfully again and again by methods such as injection or compression moulding, embossing or casting [9] to transfer the channel pattern to the polymer (Fig. 34.1B). Simplified procedures such as those including UV-initiated atmospheric polymerisation [10,11] or the employment of laser printers [12–14] have been reported.

A polymeric material must fulfil several requirements [15]: the material must be machinable, inert to the conditions of the assay, have acceptable thermal and electrical properties (since high electric fields and Joule heating are produced) and be annealable (because the fabricated features must be enclosed to allow fluid flow). Depending on the methodology, the surface of the material must be modifiable or optically transparent. Many are opaque in the UV and they may generate background fluorescence with laser-induced excitation, and therefore an EC detection could be required. Materials such as PMMA (polymethylmethacrylate), PDMS (poly(dimethylsiloxane)), PC (polycarbonate), PE (polyethylene), PET (poly(ethylene terephthalate), PVC (polyvinylchloride), PS (polysterene), PU (polyurethane) and PI (polyimide) have been used, the first two being the most usual. Recently, a new polymer composition that includes cyclic olefin polymers or copolymers (COP or COC) is being applied due to advantageous characteristics such as good machinability and high chemical resistance. Studies on the zeta potential of microchannels [16] and on channel-surface modification for carrying out protein analysis [17] in Zeonor® microchips have been reported. A COP CE microchip has been employed in connection with MS detection due to its high solvent resistance [18]. A Topas® microchip has been proposed for the first time in our lab with analytical purposes employing EC detection [19].

Although most photoresists are generally considered to be sacrificial materials, liquid-type negative photoresists, such as SU-8, can be used to create microchannels within microfluidic chips [20]. The photoresist then becomes a structural material, in such a way that its thickness determines the depth of the microchannel. A negative dry photoresist (acrylate-based photopolymer) laminated on PMMA can also be employed for the fabrication of CE–EC microchips [21].

Last step of the fabrication is the system assembly, enclosing the channel networks or microstructures to allow fluids to flow through the device. In most cases, a cover plate made of the same material as the base is sealed to the device. When a thermal bonding is employed, it is necessary to heat glass structures near the glass transition temperature (T_g), around 600°C. Because polymers have a T_g comprised between 120 and 180°C, relatively low-temperature annealing is needed. Care must be taken near T_g, because the material can severely deform and any high aspect ratio microstructure (HARM) will likely deform. Moreover, gas can be formed creating bubbles. Low-temperature annealing is an attractive feature of polymer-based devices because it facilitates modification of the channels before enclosure. Moreover, when electrodes are included in one of the plates, harmless procedures have to be employed. An alternative is the use of adhesive materials that are cured at lower temperatures. Oxidation of the surface with a plasma of oxygen activates the surface and makes it more hydrophilic, which favors annealing. Low-temperature bonding is also obtained by *in situ* polymerisation procedures [11].

Although more robust and handy devices are obtained when both plates are sealed together, reversible enclosing of the channels that permit the easy disassembly for cleaning is obtained by covering the substrate containing microstructures with a thin slab [22] or sealing it against adhesive tapes [23]. Cell-coated PDMS microchannels were reversibly sealed to a glass plate containing electrodes for amperometric detection [24], resulting in an immobilised cell reactor with integrated electrodes.

Hybrid devices can be obtained when different material is used for practicing the microchannels and for covering these microstructures. One of the reasons is that the employment of a cover plate with different composition allows a system assembly at a temperature significantly below the T_g of the plate that contains the microstructures [25]. These devices are also currently employed when detection and/or high voltage electrodes and/or decouplers (see Section 34.1.3) are integrated into the chip. For instance, microchannels can be fabricated in a PDMS layer and meanwhile the working electrode and/or decouplers are fabricated on a glass substrate [26–29]. This possibility has also been employed for conductivity measurements, in such a way that several PDMS replicas can be used with a glass gold electrode plate [30]. Reversible sealing also permit modification or cleaning of the channels [24].

34.1.2 Microchip designs

The design of a microchip is one of the most important characteristics since it depends on the operations that want to be integrated. Therefore, the design defines its function and the sequence of actions taking place in the device. As commented by E. Zubritsky [31], one sure way to make a good electrophoretic separation go bad has been to add a turn to the separation channel. The effect of geometrical dispersion, due to variations in the distance travelled by molecules and in the strength of the electric field, can drastically reduce the number of theoretical plates in a separation. In order to solve the problem, spirals, tapers and complex geometries were proposed. A 10-cm serpentine chip was employed for the separation and detection of carbohydrates since in spite of lengthening the migration time it provided better efficiency [32]. Two different designs that also include serpentine flow channels and allow high sample loading and reducing fabrication costs were employed for isotachophoretic separation with conductivity detection [33,34]. A meander design was also designed for the separation channel of a CE microchip with integrated contactless conductivity [35].

The most common design is the crossing of injection and separation channels. When reactions are performed, particular schemes are employed. This is the case of bioanalytical reactions or others such as derivatisations. Additional reservoirs are added for carrying out pre-column enzymatic reactions [36]. A microchip platform in which reagent (enzymes/labelled antibody) and sample (substrate/antigen) reservoirs are connected through a reaction chamber to a four-way injection cross allows complete integration of the multiple steps of EC enzyme [37]/immuno [38] assays. In case the antibody for the immunoassay is enzymatically labelled, a postcolumn channel permits the introduction of the substrate [39]. The same scheme, two reservoirs connected through a reactor before the injection cross, is also employed for performing derivatisation reactions prior to separation [40]. A gated injector with reservoirs for sample and sample plus a reduction reaction mixture has been designed for the determination of nitrite and nitrate [41].

To overcome limitations of end-channel detection systems (see Section 34.1.3), a design with sheath-flow channels (1 cm long) that flank the separation channel and join it just before the channel exit has been reported [42]. At the end of these auxiliary channels, buffer reservoirs are created by inserting pipette tips into the drilled access holes for generating a gravity constant flow into the detection reservoir.

Positioning of the working electrode has also led to different designs. Since they refer mainly to the EC detection they will be commented in Section 34.1.3.2.

34.1.3 Electrochemical detection

EC detection is a promising alternative for capillary electrophoresis microchips due to its inherent characteristics, allowing a proper miniaturisation of the devices and compatibility with the fabrication processes, in case of an integrated detection. Moreover, the low cost associated permit the employment of disposable elements. As the EC event occurs on the surface of electrodes and the decrease in size usually results in new advantages (see Chapter 32), the possibilities of incorporating EC detectors are broad. The simplicity of the required instrumentation, portable in many cases, suit well with the scaling-down trend. Moreover, as the sample volume in conventional microchannel devices is less than 1 nL, a very highly sensitive detector should be constructed to analyse even modest concentrations of sample solutions. Since sensitivity is one of the accepted characteristics of EC detection EC–CE microchips approach to the ideal analytical devices.

Usual detection methods for CE microchips are reviewed in Refs. [4,43–45]. A common detection scheme for microchannel electrophoresis is laser-induced fluorescence (LIF), especially with the development of light-emitting diodes (LED). Among the inconveniences are the small dimensions of the optical path length and the difficulties for filtering the LED light from fluorescence emission. Moreover, derivatisation with a fluorophore is almost always necessary. The increased interest in proteomics and the development of CE with mass spectrometry (MS) detection has promoted the study of possible interfaces between chip and MS, especially with electrospray ionisation. However, the high cost and large size of the instrumental setup are sometimes incompatible with the concept of μ-TAS. A simpler and highly sensitive technique is chemical luminescence that does not need a light source, although the method requires optimisation of the interface and as with fluorescence, labelling of the analyte is usually needed.

Some cases can be found of combination between both, optical and EC principles, with the aim of enhancing characterisation ability and providing more information. An example is the fluorescence and EC dual detection performed on an electrophoresis microchip [46] that enables the analysis of a wider range of compounds, including the resolution of two neutral analytes, NBD-arginine and catechol (see Fig. 34.2). Simultaneous EC and electrochemiluminescent (ECL) detection has been proposed for microchip capillary electrophoresis [47]. In the ECL process, $Ru(bpy)_3^{3+}$ is generated from $Ru(bpy)_3^{2+}$ and reaction with analytes results in an ECL emission and a great current enhancement in EC detection.

EC detectors have already proven to be well suited for CE microchip systems, and therefore as commented in a recent review of EC detection for capillary electrophoresis microchips by Wang [48], "the rapid progress of EC detection for CE microchips over the past six years suggests the major impact it may have in the near future". Other revisions on developments in EC detection for microchip capillary electrophoresis can be found in Refs. [49–51].

EC detection methods have often been considered incompatible with electrophoresis because the combination of high voltages applied for electrophoretic separation and sensitive electrodes is seen as a conflict. However, traditional capillary electrophoresis takes advantage of many of them and with appropriate designs of the detector cell the separation voltage does not interfere with the EC measurement. On the other hand, there are several reports in which this interference is taken as a benefit for generating new detection approaches (see Sections 34.1.3.1 and 34.1.3.2) [52–54].

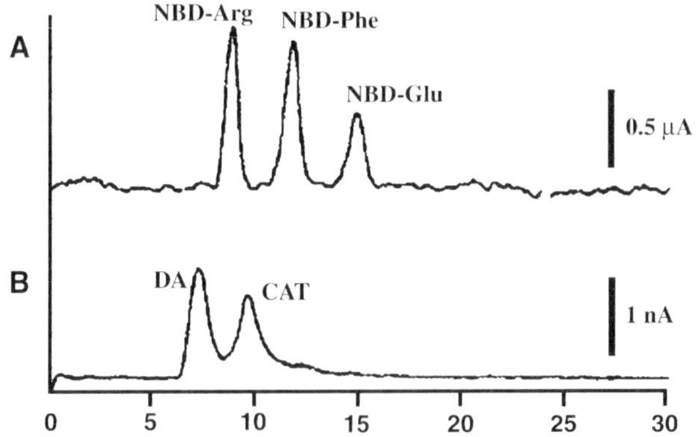

Fig. 34.2. Electropherograms using (A) LIF and (B) EC for simultaneous dual detection. Detection of dopamine (DA 50 μM), catechol (CA 110 μM), and fluorescently labelled arginine (Arg 140 μM), phenylalanine (Phe 180 μM), and glutamic acid (Glu 220 μM). Reprinted with permission from Ref. [46]. Copyright (2002) American Chemical Society.

Among the various possibilities that offer the EC detection, amperometry and conductimetry are, in this order, the most common. Although potentiometry results are a very interesting technique in many fields of Analytical Chemistry, it has not found enough echo in the microchip technology. Its incursion in microchips is related with the employment of ion-selective electrodes for Ba^{2+} determination [55] or potentiometric titration of iron ferrocyanide [56], but it has not yet been associated with CE microchips.

34.1.3.1 Conductimetry

This type of detection has achieved much development in the last few years due to its simplicity. A specific revision on conductimetric (and potentiometric) detection in conventional and microchip capillary electrophoresis can be found in Ref. [57]. It is considered a universal detection method, because the conductivity of the sample plug is compared with that of the solution and no electroactivity of the analytes is required. Two electrodes are either kept in galvanic contact with the electrolyte (contact conductivity) or are external and coupled capacitively to the electrolyte (contactless mode). An alternating current potential is applied across the electrodes and the current due to the conductivity of the bulk solution is measured. As the signal depends on the difference in conductivity between solution and analyte zones, the choice of the electrolyte is crucial. It is necessary that it presents different conductivity without affecting sensitivity.

In the contact mode, which has also been combined with isotachophoretic separations [33,34], although an end-channel configuration (with both electrodes in the outlet reservoir [30]) can be adopted, the more common one is the in-channel format [58]. Apart from being of easy fabrication, dilution of analytes before reaching the electrodes does not occur, which minimises peak broadening. Moreover, several detectors can be located along the separation channel, so that sample plug can be monitored at various locations [59]. The geometry of the electrodes influences the performance of the detector, in such a way that smaller electrodes, situated parallel to the analyte flow, produced the most sensitive response with higher resolution [60]. On the other hand, the separation electrical field can be used to generate a potential difference between two electrodes located along the channel [52]. Decoupling is not necessary and the signal is simply measured by a high-impedance voltmeter.

Reduction in the noise and avoiding the creation of bubbles make the contactless mode (CCD—contactless conductivity detection) more sensitive when compared with the contact conductivity detection, also named capacitively coupled contactless conductivity detection (C^4D). In the first C^4D design, proposed by Guijt *et al.* [61], silicon carbide was deposited as an insulating layer on electrodes located in the separation channel but in the majority of the references the electrodes are outside the chip. It is important to consider not only the electrode system geometry [62,63] but also operational parameters [64–66]. Wang *et al.* [67] have designed a detector with two planar sensing aluminium film electrodes situated perpendicularly to the channel and separated by 700 μm [68]. By mounting these electrodes on a thin polymer plate that is clipped on the microchip, a movable CCD is obtained [69]. The use of a high excitation a.c. voltage (HV-CCD) [70] and the decrease in the distance from adhesive copper tape electrodes to the solution (Fig. 34.3) [71] allow an improvement in the sensitivity of the assays. With this aim, recesses has been created at the same time as microchannels in a glass substrate for locating Pt detection electrodes at 15–20 μm from the separation channel [72].

A dual EC detection employing amperometry/conductivity allows measurements of a wider range of analytes *per* separation (ionic *via* CCD mode and electroactive compounds *via* amperometric detection) has also been reported [73]. This scheme also gave two simultaneous signals for the same component that is both charged and electroactive. This is the only report in which both principles are combined. Most of the CE–EC microchips are based in the amperometric detection mode that is more deeply considered in next section.

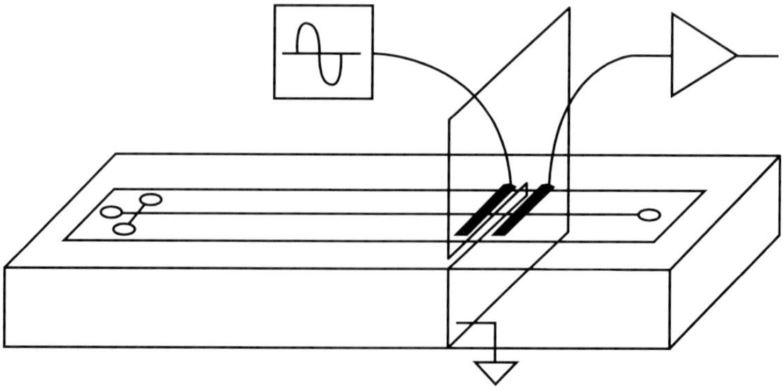

Fig. 34.3. Schematic drawing of the cell arrangement for contactless conductivity detection in CE microchip. Reprinted in part with permission from Ref. [158]. Copyright (2004) American Chemical Society.

34.1.3.2 Amperometry

Amperometry is the most widely reported EC detection mode for CE microchips, which primarily relies on oxidation or reduction of electrochemically active species by applying a constant potential to a working electrode. The current is then monitored as a function of time. Since it is based on the redox reaction that occurs at the electrode surface, electrodes can be miniaturised without loss in sensitivity. The relevance of this simple technique is reported in several reviews [48,74]. In this section, a general overview of the combination of this detection technique to CE microchips together with special sections for different amperometric techniques and electrode materials and types are considered.

The techniques that are already used to make the microchips can be employed to fabricate miniaturised detection electrodes reproducibly and inexpensively. A simple way of classifying the detectors relies on the position of the EC detector (usually referring to the working electrode, where the EC reaction takes place). Obvious separation in the "in-channel" and "end-channel" type is then produced (Fig. 34.4). In the first case (*in-channel*), the electrode is situated inside the separation channel. Although it is very advantageous because analytes migrate over the electrode when they are confined to the channel, the major inconvenience is the isolation of the detector from the high separation voltage.

Actually, there are reports on transforming the negative effect of separation electric field into new detection approaches. A potentiostatless detection scheme for amperometric detection in CE based on the use of microband array electrodes situated in the CE electric field has been proposed [53] as well as the use of an indirect amperometric detection with a carbon fibre in-channel configuration [54]. In this case, the potential difference induced by the CE separation electric field produces a change in the reduction potential of oxygen, which can be used to determine nonelectroactive analytes.

Apart from these reports, in most of the approaches, interferences between separation voltage and detection potential are usually avoided. With the aim of minimising these interferences, the electrode is aligned at the exit of the separation channel (*end-channel*). As there is a small distance (10–20 μm approximately) between the channel outlet and the working electrode, the separation voltage is grounded before attaining this electrode, and therefore, its influence is much lower. However, a loss of separation efficiency and decreased sensitivity due to the dilution of analytes from the end of the channel to the large reservoir

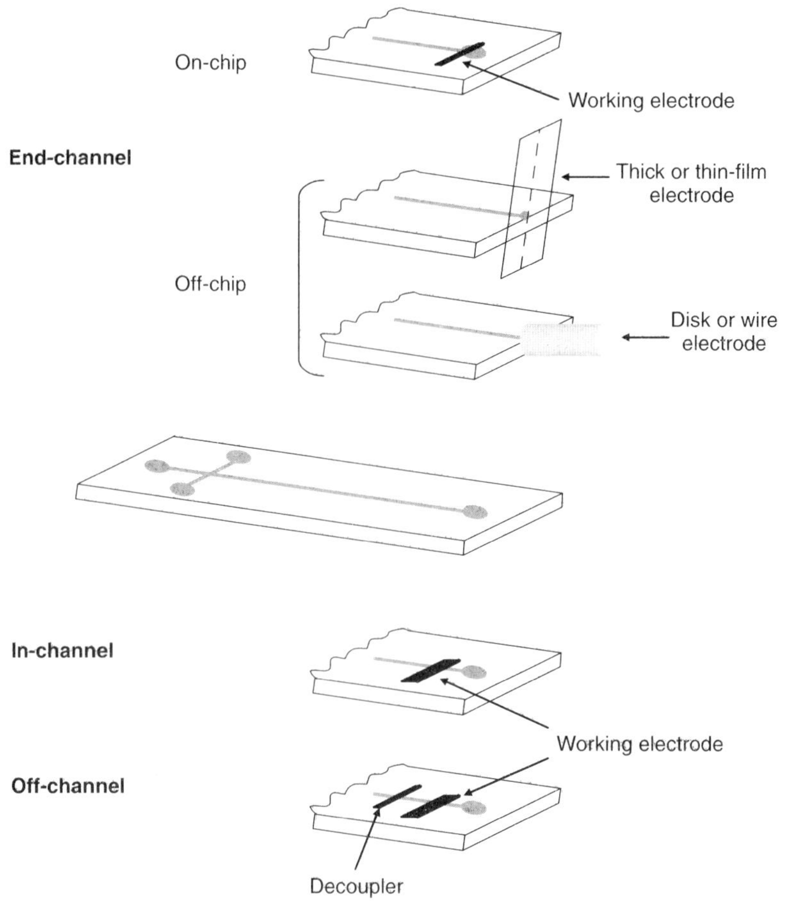

Fig. 34.4. Configurations for aligning the working electrode in CE microchip with amperometric detection.

can be produced, as it was demonstrated by using an electrically isolated potentiostat [75] as well as scanning EC microscopy [76], useful also for imaging the outlet of the microchannel and the correct alignment of the electrode. In order to isolate working and reference electrodes situated in an end-channel configuration, a passivating layer of silicon oxide was deposited on the platinum leads addressing working and reference electrodes and sheath-flow channels that join the end of the separation channel [42] allow the location of the working electrode up to 250 μm from the channel end minimising the influence of separation voltages. Due to these interferences, hydrodynamic voltammograms (HDVs) at the separation voltage employed are usually carried out to fix the optimal detection potential.

Alternatively, new strategies have been designed to take advantage of the "in-channel" format. Although the use of lower separation voltages [77] allows working with this configuration without interferences, sensitivity can be improved with the employment of higher voltages. In this case, the use of decouplers is necessary. As the working electrode is situated downstream the decoupler, electric field is established between the high-voltage electrode situated in the buffer reservoir and the decoupler. Therefore, this configuration is commonly named "*off-channel*", even when the working electrode is situated inside the "physical" channel. Different decouplers have been described and they can be classified into two main types. A path for grounding the separation voltage can be obtained by practicing small fracture/s in the channel just prior to the working electrode: a microhole (10 μm) array [78] and holes practiced with a CO_2 laser and covered with cellulose acetate [79] were reported. Electrodes situated before the working electrode can also be used for grounding the separation voltage and those constitute the second type of decouplers. Metals that are capable of absorbing H_2 gas, which is produced by water electrolysis at the cathode, were employed: Pt (Pt wire [21] or nanoparticles electrically deposited on a gold thin layer [80,81]) and Pd (thin-film [26,27,82,83] or microwire [84]) cathodes, the last one with higher H_2 absorption ability, have been employed for the isolation of detection and separation currents. The size and the distance to the working electrode are crucial for an optimal performance. An indium–tin oxide electrode was positioned in front of the three-electrode EC system in order to avoid the interference of a high-electric field in EC detection [85]. In this case, the reference electrode was also designed to envelope the working electrode so that the region around the working electrode is maintained at a stable potential.

Possibilities for *integration* of the electrodes in the microchip are multiple, from total to partial [6]. They include integration of detection electrodes (working, reference and/or auxiliary), decoupler (if it is the case) and high-voltage electrodes. Most of the designs attempt to integrate the working electrode (and decoupler if it is the case) and high-voltage electrodes are usually placed in the corresponding reservoirs. Integration of EC detection with capillary electrophoresis chips was already proposed by Woolley *et al.* in 1998 [86]. Inclusion of all electrodes directly onto the chip surface is required to take full advantage of the microfabrication approach. Without this, additional manual operations are needed to complete construction of the device. A total of six gold electrodes [87], seven platinum electrodes [88] or metal layers of platinum, titanium and palladium that were the basis of thirteen electrodes [27] were integrated in different CE microchips. Crucial for the success of integrating electrodes is the process used for bonding. For instance, a thermal bonding of glass substrates cannot be used with sputtered gold layers as they cannot withstand temperatures above 400°C.

A potentiostatic *mode* system of three electrodes (working, reference and auxiliary) is commonly employed. However, a two-electrode system is currently found in the literature. A gold working and a reference electrode system have been reported [87]. A two-electrode configuration in which the working electrode was aligned at the end of the separation channel and the electrophoretic ground electrode acts both as a pseudoreference and as the counter electrode was also proposed [89]. Electrolysis of water takes place at the electrophoretic ground, yielding a stable potential for a given set of separation conditions.

Dual-electrode detection for CE is much simpler to implement in the microchip format and offers additional sensitivity and selectivity due to the ability to monitor reversible redox reactions as well as help for identifying peaks in complex mixtures. Dual detection with serial gold [90,91] and carbon paste [92], fibre [93] and pyrolysed photoresist [94] working electrodes was reported.

Amperometric techniques

Even when the simple oxidation or reduction of electrochemically active species by applying a constant potential to a working electrode and monitoring of the resulting current as a function of time is the more common approach, particularities can be found. Pulsed electrochemical detection (PED), in which *pulsed amperometric detection* (PAD) is included as the more common format, has many advantages over a simple amperometric technique. A potential waveform consisting of a high positive potential for oxidising and cleaning the surface and a negative potential step to reactivate the electrode surface is followed by the application of a third potential for detection. Usually combined with noble metal electrodes, it is very useful for overcoming problems associated with electrode fouling and as a cleaning cycle provides a fresh electrode surface, analytes can be adsorbed and the electron transfer process can be favoured. Therefore, numerous microelectrode applications have been recently reviewed [95] as well as the application of PED to CE, including microchip CE [96].

Indirect amperometry has been combined with in-channel detection without the use of any decoupler for the detection of both electroactive and nonelectroactive analytes by using dissolved oxygen as electroactive indicator [54]. The potential of the working electrode in the case of a reduction reaction is coupled by the separation electric field, while the potential of the working electrode in the case of EC oxidation is not coupled. Therefore, electroactive analytes can be detected based on their own amperometric response and nonelectroactive ones can be detected through the indirect amperometric response of dissolved oxygen in solution.

Microfluidic chip devices are also shown to be attractive platforms for performing microscale *voltammetric* analysis and for integrating voltammetric procedures (linear-sweep, square-wave and adsorptive-stripping voltammetry) with on-chip chemical reactions and fluid manipulations [97].

Sinusoidal voltammetry (SV) is an EC detection technique that is very similar to fast-scan cyclic voltammetry, differing only in the use of a large-amplitude sine wave as the excitation waveform and analysis performed in the frequency domain. Selectivity is then improved by using not only the applied potential window but also the frequency spectrum generated [28]. Brazill's group has performed a comparison between both constant potential amperometry and sinusoidal voltammetry [98].

Electrode types and materials

Metal-based electrodes have been made using conventional electrode materials, gold and platinum being the most common. Wires or thin-layer electrodes are the most usual configurations. *Gold* thin layers are in combination with in-channel [72,80,87,91,99] or end-channel configurations [42,90]. Gold wires or disks can be included in predesignated electrode channels [84] or aligned in front of the outlet of the separation channel [77,100]. Since the waste reservoir is cut off for aligning the electrode, this detection is named "*off-chip*", for distinguishing it from this where the detector is situated "*on-chip*" (Fig. 34.4). This is a "flow-onto" configuration, which can be differentiated from the "flow-through" one, obtained for instance when a thin gold film is sputtered around the outlet of the separation channel [101], or other such as "flow-over" or "flow-onto/flow-over hybrid" configurations [49]. Other gold working electrodes such as this fabricated by coating a carbon screen-printed electrode (see below) through the application of a square-wave potential pulse [37] or the "Au-CDtrodes" produced by a technology based on laser-printed toner mask process that utilises the thin gold film from CD-R have been reported in

Ref. [14]. From one CD-R, it is possible to produce more than 50 pairs of such electrodes, which reduces the cost *per* electrode and suggests the disposability of the working electrode. A single-wall carbon nanotube (SWCNT) modified gold electrode was also employed in connection with CE microchips and displayed improved sensitivity and separation resolution compared to bare gold electrode, reflecting the electrocatalytic activity of SWCNT [102].

Platinum films [27,88], microwires [19,84] and microdisks [12,103] were also employed. Characterisation of electrode fouling and surface regeneration for platinum electrode on an electrophoresis microchip was reported [104]. The platinum tip of a scanning electron microscope has also been used for carrying out EC measurements combined in an end-configuration to a CE microchip [76].

A *silver* electrode was also constructed by EC plating on a gold layer made through a micromoulding in capillary and electroless deposition [29]. A *mercury–gold amalgam* electrode [105] was fabricated using 60-μm diameter gold wire that was amalgamated by dipping it into mercury for 5 s. *Palladium*, also used for decoupling separation voltage from detection has also been employed [82,83,106], as well as *nickel* fabricated by electroless deposition [107].

Copper has been employed commonly in relation with carbohydrates and aminoacids but other analytes can also be determined. Microalignment of the electrode with the channel outlet [89], deposition of a thin layer [81,108] or incorporation of a wire [21] are examples of in-channel formats. End-channel configurations such as those obtained by using a guide tube [11] or attaching a Cu wire perpendicularly to the outlet [32] after chip bonding were tested. Carbon nanotube (CNT)/copper composite electrodes have proved to be more sensitive, compared to the Cu and CNT alone, in the determination of carbohydrates [109].

Carbon is another material widely used for electroanalytical applications due to its low cost, large potential window and low background. Carbon paste (unmodified and cobalt phthalocianine modified) filled electrode channels (one or two in case of dual detection) fabricated in a PDMS plate and situated perpendicular to the separation channel of another PDMS plate [92]. Micromoulding of carbon inks has been used to pattern carbon microelectrodes in such a way that PDMS microchannels are sealed to a glass substrate to define the size of the microelectrode [26]. After getting filled with carbon ink and cured, the PDMS is removed, leaving the carbon microelectrode that can be integrated in a rather large microchannel. Pyrolised photoresist carbon films (carbon PPF) that possess an exceptionally smooth surface with a low oxygen/carbon atomic ratio, which contribute to low background levels, can be photolithographically included in the chip [28,94,108].

Carbon fibres microdisk electrodes have been included in a perpendicular electrode channel [93] and also aligned at the outlet of the channel [110,111]. Although the employment of thick-film electrodes (planar screen-printed electrodes, SPEs) has become widespread due to their inherent advantages (see Chapter 26), true integration within the microchannel is not possible due to the thickness of the resulting electrode. They are then situated at the end of the microchip using micropositioners and spacers in an "off-chip" format and perpendicular to the flow direction. Wang *et al.* produced the first design in 1999 [112] (Fig. 34.5) and many related articles have been reported since then [113–115]. Boron-doped diamond (BDD) electrodes glued onto an alumina ceramic plate were employed with this same configuration [116]. In these cases, a plastic screw is used for fixing the distance to the outlet. A conventional glassy carbon electrode is fixed with the same system [117] and alternatively, alignment of 300-μm pencil lead [118] and graphite–epoxy composite [119] was performed with the aid of a three-dimensional adjustor and a guide tube, respectively.

Referring to the designs of the microchips they can vary in order to correctly position the working electrode. In this context, electrode channels were constructed in a PDMS layer that is reversibly [41] or irreversibly [120] bonded to another, once the separation channel has been correctly aligned (perpendicularly) with the working electrode. A guide tube for precise parallel alignment of the working electrode in glass [121] and PDMS [122] microchips was also proposed.

34.2 APPLICATIONS

Capillary electrophoresis microchips have been employed for a variety of applications. Reviews on different applications such as environmental [123], security [124], forensic [125] and biochemical [126–128] among others are reported. Clinical diagnostics is one of the fields that is having a more rapid development. This rapid growth needs an urgent development of new point-of-care and high-throughput clinical devices. In addition, traditional assays for the determination of clinically important analytes require large amount of time and money for reagents and instruments that is not compatible with the enormous number of analysis will be needed in our future daily life. Capillary electrophoresis microchips application for clinical assays can be a solution to this

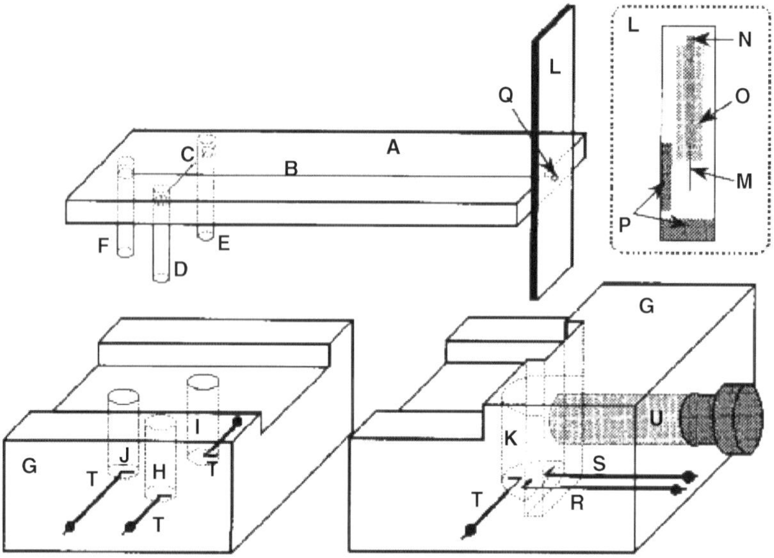

Fig. 34.5. Capillary electrophoretic system with electrochemical detection. (A) Glass microchip, (B) separation channel, (C) injection channel, (D) pipette tip for buffer reservoir, (E) pipette tip for sample reservoir, (F) pipette tip for reservoir not used, (G) Plexiglass body, (H) buffer reservoir, (I) sample reservoir, (J) blocked (unused) reservoir, (K) detection reservoir, (L) screen-printed working-electrode strip, (M) screen-printed working electrode, (N) silver ink contact, (O) insulator, (P) tape (spacer), (Q) channel outlet, (R) counter electrode, (S) reference electrode, (T) high-voltage power electrodes, (U) plastic screw. For clarity, the chip, its holder, and the screen-printed electrode strip are separated, and dimensions are not in scale. Reprinted with permission from Ref. [112]. Copyright (1999) American Chemical Society.

problem [129]. In this section, an overview of the most important clinical applications performed with the most common EC principles found in CE microchip, amperometry and conductimetry is given. Further, the steps for designing an amperometric detector (in-channel, end-channel or dual configuration) together with the parameters that have to be taken into account for an optimal performance of a CE–EC microchip are commented on in more detail. Studies are carried out for analytes of clinical interest such as ascorbic acid, p-aminophenol and hydrogen peroxide. They are important not only as proper analytes but also as enzymatic products, what can be employed for the indirect determination of molecules through enzymatic assays (including immuno and DNA assays).

Different compounds of clinical interest have been analysed in CE microchips with EC detection. Due to their favourable inherent characteristics, *neurotransmitters* [24,27,94,121] and *ions* [59,60,67,72,130] have been the class of choice for testing amperometric and conductimetric detectors, respectively. Much research has focused on the separation and detection of these analytes and the number of references that can be found is broad [50]. Applications of amperometric or conductimetric detection for other analytes of clinical significance can be found in Table 34.1, where information about the microchip material, EC technique, electrode configuration as well as electrode material employed is included.

Amperometric detection can be then applied for the determination of analytes such as glucose [131–137], ascorbic acid [138–141], uric acid [142,143], amino acids [144–149], peptides or DNA [150,151]. Conductimetric detection has been also employed in combination with CE microchip for amino acids [152,153], peptides/proteins [152,154,155], carbohydrates [153] or DNA [152].

Glucose has been determined either directly using its oxidation process [84,109,122,131–133] or through different enzyme assays [36,37,87,134–136]. Thus, glucose–oxidase (GOD) or glucose–dehydrogenase (GDH) can be used in combination with single glucose or glucose/NAD^+ for producing hydrogen peroxide or NADH, respectively that can be detected amperometrically. On the other hand, a simultaneous enzymatic and immunochemical assay has been integrated in a microchip for measurement of glucose and insulin [137].

Different possibilities for conducting enzymatic assays on microchip platforms including pre-, on- or post-column reactions have been reviewed [156]. An enzymatic assay (employing creatininase, creatinase and sarcosine oxidase) has also been developed in a microchip for analysing *renal marks* such as creatinine and creatine as well as *p*-aminohippuric acid and uric acid [157].

Amino acids (isoleucine, phenylalanine, arginine and alanine) have been analysed on a microchip with a post-channel reaction with amino acid oxidase reaction [144]. Pre-channel derivatisation of amino acids with naphthalene-2,3-dicarboxyaldehyde (NDA) has been described for facilitating its amperometric detection [145]. Separation and direct detection of amino acids without derivatisation have also been achieved in microchips [89,109,122,132,146–148].

Different *immunoassays* have been developed in CE microchips for detecting immunoglobulins, such as mouse IgG. Enzymatic (alkaline phosphatase) [39] or electroactive (ferrocene) [38] labelled anti-mouse

TABLE 34.1

Recent clinical applications of CE microchip with electrochemical detection

Analyte	Chip material	EC technique	Electrode configuration	Electrode type	Ref.
Arg, Met-Gly, glucose	PDMS	DC amperometry	End-channel (off-chip)	Copper	[122]
Carbohydrates	PDMS/Glass	PAD	End-channel (off-chip)	Platinum	[131]
Carbohydrates, aminoacids, antibiotics	PDMS	PAD	End-channel (on-chip)	Gold	[132]
Glucose	PDMS	PAD	End-channel (on-chip)	Gold	[133]
Carbohydrates, aminoacids	Glass	DC amperometry	End-channel (off-chip)	CNT/copper composite	[109]
Glucose	PDMS	PAD	Off-channel	Gold and platinum	[84]
Ascorbic acid	PDMS/Glass	DC amperometry	End-channel (on-chip)	Platinum and carbon microdisk	[103]
Ascorbic acid	Glass	DC amperometry	End-channel (off-chip)	Carbon fiber	[138]
Ascorbic acid, uric acid	Glass	DC amperometry	In-channel	Gold	[87]
Ascorbic acid	PDMS/Glass	Sinusoidal voltammetry	In-channel	PPF carbon	[139]
Ascorbic acid	PMMA	DC amperometry	End-channel (on-chip)	Platinum	[19]
Ascorbic acid	Topas	DC amperometry	End-channel (on-chip)	Platinum and gold	[19,77]
Ascorbic acid	Polyester	DC amperometry	End-channel (off-chip)	Gold	[14]
Ascorbic acid	PDMS	DC amperometry	End-channel (on-chip)	Carbon ink	[140]
Ascorbic acid	Glass	DC amperometry	End-channel (off-chip)	Glassy carbon	[141]
Uric acid	PDMS/Glass	DC amperometry	End-channel (off-chip)	Platinum	[142]
Creatine, creatinine, uric acid	PDMS	PAD	End-channel (on-chip)	Gold	[143]
Amino acids (Arg, Lys, Gly, Cys, PhenA)	Glass	DC amperometry	End-channel (off-chip)	Screen-printed carbon	[145]
Aromatic aminoacids	Glass	DC amperometry	End-channel (off-chip)	Screen-printed carbon	[146]
Carbohydrates, aminoacids	Glass	DC amperometry	End-channel (on-chip)	Copper-coated platinum	[89]
Aminoacids	PDMS	DC amperometry	End-channel (off-chip)	Copper	[118]
Aminoacids	PDMS/Glass	DC amperometry	End-channel (off-chip)	Carbon fibre	[147]
Seleno aminoacids	Glass	DC amperometry	End-channel (off-chip)	Screen-printed carbon	[148]
Aminoacids	Glass	DC amperometry	End-channel (off-chip)	CNT modified screen-printed carbon	[149]
Proteins, peptides	PMMA	Conductimetry	Contact conductivity	Platinum	[152,154,155]
Aminoacids	PMMA	Conductimetry	Contact conductivity	Platinum	[152]

Aminoacids	PMMA	Conductimetry	Contactless conductivity	Copper	[70,153]
Carbohydrates	PMMA	Conductimetry	Contactless conductivity	Copper	[153]
Enzymeassays					
Glucose, creatinine, uric and ascorbic acid	Glass	DC amperometry	End-channel (on-chip)	Platinum	[36]
Glucose, uric and ascorbic acid	Glass	DC amperometry	End-channel (off-chip)	Gold coated screen-printed carbon	[136]
Glucose, ethanol	Glass	DC amperometry	End-channel (off-chip)	Gold coated screen-printed carbon	[134]
Glucose	Glass	DC amperometry	End-channel (off-chip)	Gold coated screen-printed carbon	[37]
Glucose	Glass	DC amperometry	In-channel	Gold	[87]
Glucose, ascorbic acid	Glass	DC amperometry	End-channel (off-chip)	Gold coated screen-printed carbon	[135]
Creatine, creatinine, *p*-aminohippuric acid, uric acid	Glass	DC amperometry	End-channel (off-chip)	Gold coated screen-printed carbon	[157]
Amino acids	Glass	DC amperometry	End-channel (off-chip)	Gold coated screen-printed carbon	[144]
Immunoassays					
Insulin, glucose	Glass	DC amperometry	End-channel (off-chip)	Gold	[137]
Mouse IgG	Glass	DC amperometry	End-channel (off-chip)	Screen-printed carbon	[39]
Mouse IgG, T_3	Glass	DC amperometry	End-channel (off-chip)	Gold coated screen-printed carbon	[38]
Human IgM	PMMA	Conductimetry	Contactless conductivity	Copper	[158]
DNA assays					
ΦX174 *Hae*III, *Salmonella* PCR product	Glass	DC amperometry	End-channel (on-chip)	Platinum	[86]
dsDNA, SNP	Glass	DC amperometry	End-channel (on-chip)	Gold	[42]
DNA ladder	PDMS/Glass	DC amperometry	End-channel (on-chip)	ITO	[151]
SNP	PDMS	Sinusoidal voltammetry	In-channel	PPF carbon	[150]
DNA ladder	PDMS/Glass	DC amperometry	Off-channel	ITO	[85]
dsDNA fragments	PMMA	Conductimetry	Contact conductivity	Platinum	[152]
PCR products	PMMA	Conductimetry	Contact conductivity	Platinum	[58]

IgG was employed for monitoring the immunological reaction through amperometric detection. Human immunoglobulin M has been determined on a CE microchip using high-voltage contactless conductivity detection (HV-CCD) [158]. The immunological interaction between IgM and anti-IgM was also studied. This device avoids the use of redox or enzymatic labels and pre/post-channel reactions.

Bioanalysis in microchips not only includes enzymatic or immunoassays but also *DNA* analysis. Different modes of capillary electrophoresis such as capillary gel electrophoresis, capillary electrochromatography (CEC), micellar electrokinetic chromatography or isotachophoresis have been employed for DNA separation. CE microchips have been employed to separate and detect DNA restriction fragments and PCR products using an indirect EC detection through the complex between, iron (III) and phenanthroline ($Fe(phen)_3^{2+}$) that acts as intercalation agent [42,86]. In addition, an allele-specific, PCR-based single-nucleotide polymorphism typing assay for the C282Y substitution diagnostic for hereditary hemochromatosis has been developed and evaluated using ferrocene-labeled primers [42].

Single-nucleotide polymorphisms (SNPs) have been analysed on a capillary gel electrophoresis (CGE) microchip with EC detection [150]. The genetic section that contained the SNP was amplified by PCR and purified. Then, it was used in a single-base extension (SBE) reaction with a redox-labeled chain terminator, ferrocene-acycloATP. Products of the SBE, ferrocene-labeled SNP and free ferrocene-acycloATP, were separated employing CGE on microchip and detected using sinusoidal voltammetric detection at a pyrolysed photoresist film (PPF) electrode.

CEC was used to separate double-stranded DNA (dsDNA) on microchips [58,152]. In these cases, C_{18}-modified PMMA microchips were employed with ion-pair reverse phase chromatography and DNA fragments with different sizes were analysed by contact conductivity detection.

Ascorbic acid is an important vitamin and its determination is needed not only in foods but also in drugs, blood and urine. Capillary electrophoresis microchips with EC detection allow a simple and rapid detection of ascorbic acid as well as its separation from interferences.

In our laboratory, polymer CE microchips in combination with EC detection have been successfully used as miniaturised devices for determination of clinically important analytes. As commented in Section 34.1.2, poly(methylmethacrylate) (PMMA) is one of the most used polymers for manufacturing microchips. Recently, cyclic olefin copolymers (COCs) such as Topas (thermoplastic olefin polymer of amorphous structure) have also received attention due to its high chemical resistance, good machinability and optical transparency.

PMMA and Topas microchips with EC detection can be used for determination of vitamins such as L-ascorbic acid (vitamin C) and other compounds such as *p*-aminophenol and hydrogen peroxide.

Aspects such as the detector design, pretreatment, injection/separation control and running buffer are the basis of a good performance of the microchips. All these points are treated in the next sections in order to obtain a standard methodology for analysing ascorbic acid and other important compounds.

34.2.1 Amperometric detector design

Different detector configurations and working electrodes can be used in combination with polymer CE microchips.

34.2.1.1 End-channel detector

Integration of an end-channel working electrode is possible employing wires of small diameter or thin and thick films.

The integration and alignment of the metal-wire working electrode is easy when the diameter of the electrode is greater than that of the separation channel and high enough for not being malleable. Thus, gold or platinum wire can be manually aligned at the outlet of the separation channel (see more details in Procedure 48 in CD accompanying this book) in a flow-onto configuration.

On the other hand, thin- and thick-film electrodes can be perpendicularly aligned at the outlet of the channel. In this case, a methacrylate detection cell based on Wang design [112] is made in order to integrate the working electrode film [77]. This design allows an easy alignment of a screen-printed carbon electrode (SPE) or a gold film, as well as a rapid replacement of the working electrode when electrode surface fouling happens.

34.2.1.2 In-channel detector

In the in-channel detector, the working electrode is directly placed inside the separation channel. Thus, analytes migrate over the electrode while they are still confined to the channel. The design of the working electrode would have to minimise the coupling between the high voltage and the detection potential. As commented in Section 34.1.3, this is an inconvenience in most of the cases and the electrode width and the distance to the channel outlet have to be considered in order to minimise this effect.

A gold thin layer (100 nm) can be deposited on the cover plate that is used for closing microchip channels. Thus, a gold film with 0.4 or 1.0 mm width is situated at 10 or 100 μm from the channel outlet, respectively [77,99]. These configurations were provided by Microfluidic-ChipShop (Jena, Germany). As the cost related with modifications of the standard designs was relatively high, no more designs were tested.

34.2.1.3 Dual detector

The last detector considered is based on the combination of end- and in-channel configuration [77]. In this detector, two gold working electrodes are used: a film inside the channel and a wire aligned at the end of the separation channel.

34.2.2 Microchip pretreatment

The state of the channel surface is very important in order to obtain a good performance of the microchips. This depends on the chip material and pretreatment, mainly. Hence, prior to using the microchip, channels have to be pretreated adequately, which is needed in order to clean the channels and obtain an appropriate electroosmotic flow (EOF) [159]. It can also be useful for performing a chemical cleaning of the working electrode.

EOF is a wall-generated phenomenon with an important role in separation. EOF in microchips is due to the negative charge density in the capillary surface. Therefore, the generation of EOF in glass microchips is due to dissociation of silanol groups present on the channel surface; meanwhile in polymer microchips the factor responsible for the surface net charge is not clear. It is very likely that the carboxyl groups are the main cause for EOF formation, while some ions (OH^-) adsorbed also on the wall cannot be excluded. Topas is a COC that combines chains of ethylene and norbornene. The chemical structure of Topas neither has carboxyl groups nor chemical sites expected to ionise, so the source of charge on its surface remains unclear. Although adsorption of ions may be postulated as the source of charge, the nature of such adsorption processes is difficult to ascertain.

Different solutions can be used for conditioning polymer channels. Strong acid or alkaline solutions as well as most of the common organic solvents should not be used in order to avoid polymer degradation. Slightly acid solutions allow obtaining a stable EOF extending the lifetime of the polymer microchip. Thus, 0.1 M HCl and $HClO_4$ enable improving the performance of PMMA and Topas microchips, respectively. When PMMA microchips are employed with a gold working electrode, HCl cannot be used because Cl^- anions can form a complex with gold [160]. In this case, therefore, the channels are only rinsed with Milli-Q water.

The value of EOF in PMMA and Topas microchips is quite similar but slightly higher for Topas under the same conditions [19]. The electroosmotic mobility due to the EOF in PMMA and Topas microchips comprises between 2 and $5 \times 10^{-4}\,cm^2/V\,s$.

34.2.3 Amperometric detector performance

Different parameters such as electrode alignment, detection potential and separation/injection voltage can affect the good performance of the amperometric detection in microchips.

34.2.3.1 Effect of electrode alignment and detection potential

In the end-channel metal-wire detector, the working electrode must suitably be aligned with the separation channel [159]. In this context, if the electrode is too far from the channel outlet or is not parallel to the separation channel a decrease in sensitivity would be observed due to analyte dilution in the reservoir (Fig. 34.6).

In the in-channel detector, the working electrode is situated within the separation channel and analytes always migrate over the electrode while they are still confined to the channel. Therefore, the in-channel electrode does not need alignment. In this case, the layer thickness, the width and the distance to the outlet of the channel will be the parameters that influence the good performance of the detector.

A dual-electrode detector, based on a film inside the channel and a wire at the end of the channel, can be used in order to evaluate the accurate alignment of the end-channel electrode. Thus, when the end-channel wire is properly aligned, analytical signals are quite similar for the in- and end-channel detector [159].

The end-channel detector can also be based on thin- or thick-film electrodes. In this case, the original waste/detection reservoir is cut off, leaving the channel outlet at the end side of the microchip, in order to make the alignment of the film electrode easier. The electrode is perpendicularly aligned and fixed at the end of the separation channel

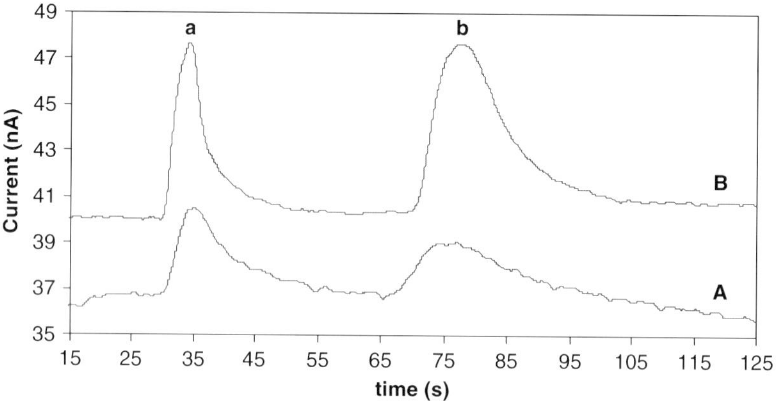

Fig. 34.6. Electropherograms for (a) *p*AP and (b) AsA using a polymer-microchip with an end-channel Au wire detector inadequately (A) and properly (B) aligned. Conditions: $V_{sep} = +2000\,V$; injection: 5 s at +2000 V; $E_d = +0.8\,V$ (vs. Ag/AgCl), running buffer: 50 mM Tris-Gly pH 9.0. Copyright (2006) from Instrumentation Science and Technology by Ref. [159]. Reproduced by permission of Taylor & Francis Group, LLC., http://www.taylorandfrancis.com.

using a plastic screw. The screw allows to control the distance between the outlet and the working electrode [77,159].

The use of films (SPE or gold) compared to that of wires (platinum or gold) allows the rapid replacement of the working electrode when electrode surface fouling happens. This can be made by changing the whole film or due to the small area involved, by slightly moving the position of the film, which in turn, proves simpler.

Although the detector design is very important for a good performance of the microchip, the material of the working electrode also affects the electron transfer that ultimately depends on the compounds that are detected.

Amperometric detection is based on applying a constant potential to the working electrode and measuring the resulting current. The selection of the optimum detection potential relies on the construction of HDVs. Since the detection potential can shift slightly depending on the electrode material and the separation voltage, HDVs must be recorded under the same separation conditions rather than those of the final sample analysis (Fig. 34.7).

34.2.3.2 Effect of separation/injection voltage

Separation and injection in electrophoresis microchips is electrokinetically driven using high voltages. The high voltage applied for the

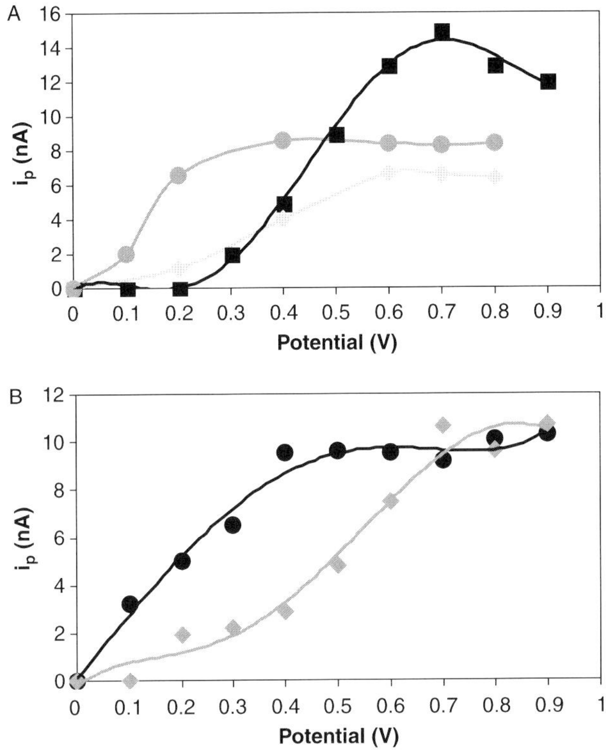

Fig. 34.7. Hydrodynamic voltammograms for AsA (◆), *p*AP (●) and H_2O_2 (■) using a polymer-microchip with an end-channel Pt wire (A) or Au wire (B) detector.

separation, as was mentioned above, can affect the amperometric detector response.

The influence of the separation voltage on the analytical signals has been studied in a PMMA microchip with an end-channel Pt-wire detector by varying this parameter from +500 to +4000 V (Fig. 34.8A) [19]. Thus, when a higher separation voltage is applied, the peak current increases reaching a plateau for +2000 V. Since the migration time of the sample plug in the separation channel decreases with separation voltage, smaller sample dispersion is produced, which means an increase in the peak current and a decrease in the peak width. The inconvenience of a higher separation voltage is an increase in the baseline noise level as well as higher Joule heating and bubble formation in the separation channel. Normally, a good performance is obtained working with a separation voltage around +2000 V.

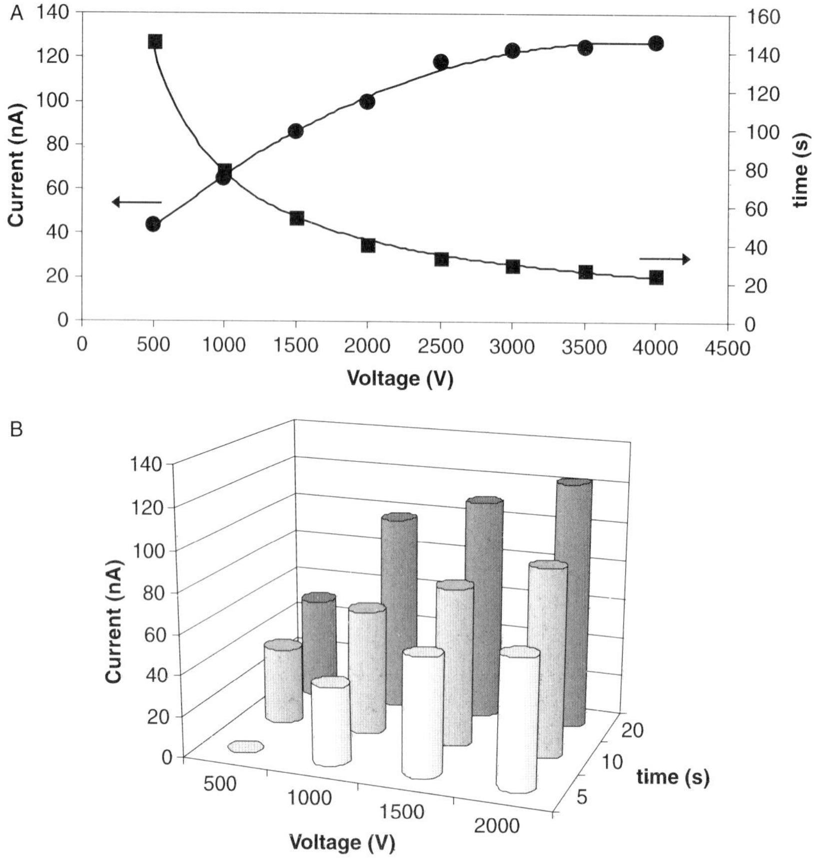

Fig. 34.8. Influence of (A) separation voltage and (B) injection time/voltage for H_2O_2 using a microchip with an end-channel Pt-wire detector. Reprinted with permission from Ref. [19].

The influence of the separation voltage has also been studied in a PMMA microchip with an in-channel gold-film detector [77]. In this case, separation voltages higher than +500 V should not be applied in order to avoid an excessive baseline noise level. Thus, when +1000 V are applied the noise level is too high, bubbles are formed within the channel and therefore, analytical signals cannot be recorded. However, the use of low voltages reduces Joule heating, increasing the lifetime of the microchip.

On the other hand, although injection voltage does not directly influence the amperometric detection, it affects the analytical signals recorded. Thus, the influence of the injection voltage on the analytical signals has been studied in a PMMA microchip with an end-channel

Pt-wire detector (Fig. 34.8B) [19]. When an injection time is fixed, the peak current increases with the injection voltage. This could be due to the fact that a higher flow is obtained for higher voltages and a better-defined sample plug (with smaller dispersion) is obtained. When an injection voltage is fixed, the peak current also increases with the injection time. However, peaks are widened. Since an unpinched injection is performed, this peak current and width increase could be due to a small increase in the injected volume *via* dispersion at the intersection.

34.2.4 Separation performance

In capillary zone electrophoresis microchips, where the background electrolyte consists only of aqueous buffer, analytes are separated based on a size-to-charge ratio, and neutral analytes are not resolved from each other.

In this way, the running buffer solution plays an important role for electrophoresis separations in microchips. The main properties that have to be taken into account are a good buffering capacity in the pH range of choice and low mobility, which means minimally charged ions and low ionic strength in order to minimise current generation.

Separation of hydrogen peroxide and ascorbic acid (AsA) has been evaluated in different buffer systems using a PMMA microchip with an end-channel Pt-wire detector [19]. Zwitterionic buffers based on Tris, TAPS and EPPS have been tested. Buffers are adjusted to pH to 9.0 using different alkaline solutions (NaOH, AMPD and Tris), inorganic acids (boric and hydrochloric acid) and organic acids (acetic acid, glycine, TAPS, EPPS and TES). Electrophoretic current due to the separation and injection high voltage is sufficiently low when zwitterionic buffers are used. The low conductivity of these buffers, especially Tris-based buffer, minimises Joule heating allowing the use of high voltage for rapid separations. Buffers based on mixtures of TAPS with Tris or AMPD seem to be more appropriate considering the resolution, efficiency and bandwidth. The effect of the buffer concentration has also been tested. Thus, an increase in the concentration implies lower EOF and higher ionic strength obtaining longer migration time and higher number of theoretical plates.

Separation of hydrogen peroxide and ascorbic acid (AsA) has also been evaluated in a Topas microchip with an end-channel Pt-wire detector [19]. In this case, better results in the separation efficiency and sensitivity were obtained for the same conditions than in PMMA microchip (see more details about comparison of PMMA and Topas microchip in

Section 48.6 of Procedure 48 in CD accompanying this book). The Topas microchip has demonstrated a good performance for separation of different analytes in combination with amperometric detection.

Separation and detection of *p*-aminophenol and ascorbic acid has also been evaluated in Topas microchips using different end-channel amperometric detectors. Thus, platinum- and gold-wire, screen-printed carbon electrode and gold film have been used as working electrodes [77].

In this case, the precision of the platinum and gold-wire detectors was evaluated from a series of repetitive injections of a sample mixture containing *p*AP and AsA using 50 mM Tris-based buffer $pH = 9.0$ as background electrolyte. Data of the standard deviation of peak current and migration time for 15 successive signals for *p*AP and AsA are given in Table 34.2. Electropherograms corresponding to the separation of *p*AP and AsA with a gold-wire detector are displayed in Fig. 34.9. The separation efficiency of the Topas microchip with the platinum and gold-wire detectors was determined through the theoretical plate number (N). This parameter as well as the half-peak width ($w_{1/2}$), the peak current and resolution (R) are shown in Table 34.2. The separation parameters obtained with end-channel gold-wire detector resulted better than those at platinum-wire detector, as well as the RSD values obtained.

Screen-printed carbon electrodes were also evaluated in the Topas microchip for separation and detection of *p*AP and AsA. The theoretical plate number (N), half-peak width ($w_{1/2}$), peak current and resolution are also shown in Table 34.2. Peak current at SPEs was higher than those obtained with gold and platinum wires.

A gold film was also evaluated using *p*AP as analyte model. The theoretical plate number (N) was 5990 m^{-1} with a half-peak width ($w_{1/2}$) of 4.1 s for *p*AP. Peak current at gold-film electrode was lower than at SPEs but higher than at gold and platinum wire.

As it has been previously said, only low separation voltages can be used with the in-channel detector when a decoupler is not combined, which increases the migration time. The performance of a PMMA microchip with an in-channel gold-film detector, working at low separation voltages, has been improved using the surfactant sodium dodecyl sulfate (SDS) [77]. When SDS (negatively charged) is used in the background electrolyte with a concentration lower than the critical micellar concentration (CMC ~8 mM), it can interact with the capillary surface. Thus, the negative charge density in the capillary is increased improving the EOF [161]. The employment of this dynamic modifier decreases

TABLE 34.2

Separation parameters for *p*AP and AsA using Topas microchip with different end-channel detectors

End-channel detectors	E_d (V)	*P*AP				AsA				*R*
		i_p (nA)	t_m (s)	$w_{1/2}$ (s)	N (m^{-1})	i_p (nA)	t_m (s)	$w_{1/2}$ (s)	N (m^{-1})	
Platinum wire	+0.7	10.5 ± 0.5	39 ± 1	7.5 ± 0.4	3000 ± 250	6.2 ± 0.7	98 ± 2	16.5 ± 0.8	3900 ± 400	1.3
Gold wire	+0.8	8.6 ± 0.4	34.0 ± 0.5	4.9 ± 0.2	5400 ± 400	6.8 ± 0.7	78 ± 1	11.9 ± 0.6	4800 ± 550	2.6
SPE	+0.8	22.5 ± 0.6	29.0 ± 0.5	3.7 ± 0.2	6900 ± 300	25.5 ± 0.8	76 ± 1	12.6 ± 0.7	4100 ± 250	2.9

Reprinted from Ref. [77], Copyright (2006), with permission from Elsevier.

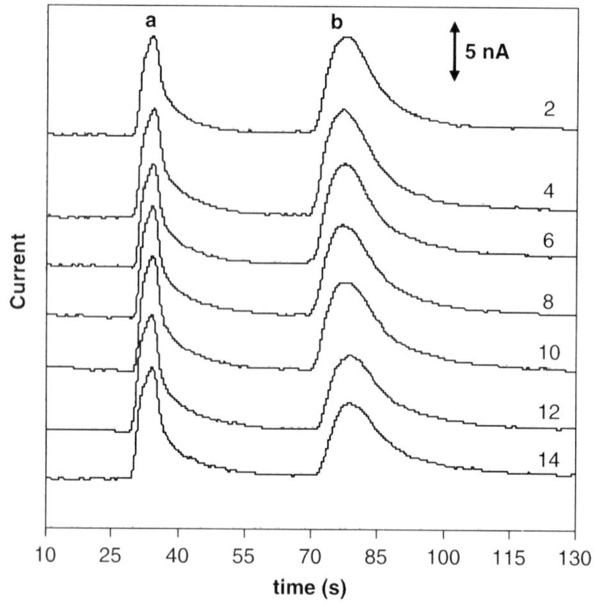

Fig. 34.9. Even electropherograms for succesive records of (a) *p*AP and (b) AsA using a microchip with an end-channel Au wire detector. (Running buffer: 50 mM Tris-Gly pH 9.0; $V_{sep} = +2000\,V$; injection: 5 s at +2000 V; $E_d = +0.8\,V$). Reprinted from Ref. [77], Copyright (2006), with permission from Elsevier.

the migration time and the peak width, and enhances the peak current and the efficiency for the separation of *p*-aminophenol and ascorbic acid (Fig. 34.10).

The selectivity of the microchip with EC detection can be improved by using a dual electrode detector (see the end of Section 34.1.3.2). A dual detector based on a gold film inside the separation channel and a gold wire aligned at the end of the channel has been used in combination with a PMMA microchip [77]. In the dual detector, different detection potentials can be applied to each working electrode. This system is useful for monitoring species, such as *p*-aminophenol, which undergo a chemical reversible redox reaction. In this case, two signals are simultaneously recorded for the same analyte (Fig. 34.11), meanwhile, species with only oxidation or reduction process, a unique signal is recorded. This allows the identification and discrimination of species as well as the improvement in the precision of the determinations or the elimination of interferences enhancing the selectivity.

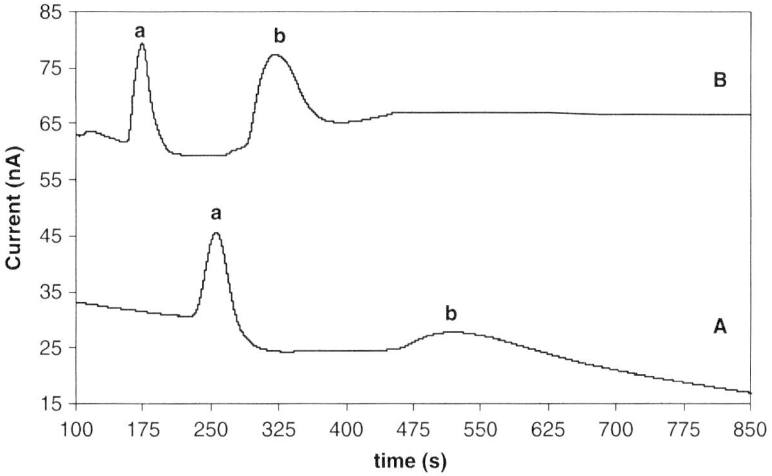

Fig. 34.10. Electropherograms for (a) *p*AP and (b) AsA using 50 mM Tris-Gly pH 9.0 without (A) and with (B) 1 mM SDS in a microchip with an in-channel gold-film detector. Conditions: $V_{sep} = +250\,V$; injection: 5 s at +2000 V; $E_d = +0.8\,V$ (vs. Ag/AgCl). Reprinted from Ref. [77], Copyright (2006), with permission from Elsevier.

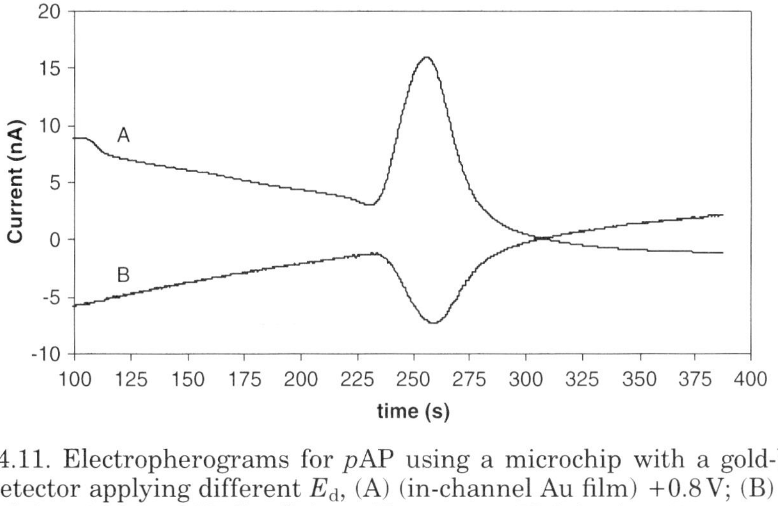

Fig. 34.11. Electropherograms for *p*AP using a microchip with a gold-based dual detector applying different E_d, (A) (in-channel Au film) +0.8 V; (B) (end-channel Au wire) −0.1 V. Conditions: $V_{sep} = +250\,V$; injection: 5 s at +2000 V; running buffer: 50 mM Tris-Gly pH 9.0. Reprinted from Ref. [77], Copyright (2006), with permission from Elsevier.

34.3 CONCLUSIONS

Miniaturised analytical systems, especially capillary electrophoresis microchips, have demonstrated to be a promising tool for analytical purposes.

These devices require a sensitive detection system such as the EC detection. There is no doubt that EC detection is becoming a powerful tool for microscale analytical systems. The ease of miniaturisation and integration of the electrodes in the microchips has been described. Among all the EC techniques, amperometry and conductimetry have been noted due to their simplicity and analytical possibilities. Although capillary electrophoresis seemed initially incompatible with amperometric detection, different electrode configurations have been successfully employed for coupling EC detection with CE microchips. The relevance of the working electrode alignment has also been demonstrated.

The feasibility of constructing a miniaturised system including reaction, separation and detection units integrated directly onto the EC–CE microchip device has also been shown and approximates the concept of "total analysis system". Furthermore, the low cost of the EC detection in connection with new polymer material can bring a real disposable device.

These versatile microfluidic systems will provide new tools for clinical diagnostics and other important fields of Analytical Chemistry. Future directions point toward the development and refinement of truly self-contained portable μ-TAS devices that can be used for point-of-care or on-site analysis. It is foreseeable that in the near future these devices could be routinely employed for the detection of numerous clinically relevant compounds.

ACKNOWLEDGMENTS

Part of the work has been supported by the FICYT under project no. IB05-151C2. M. Castaño-Álvarez thanks FICYT—Principado de Asturias for the award of a PhD grant.

REFERENCES

1 A. Manz, N. Graber and H.M. Widmer, Miniaturized total chemical analysis systems: A novel concept for chemical sensing, *Sens. Actuator B*, 1 (1990) 244–248.

2 A. Manz, D.J. Harrison, E.M.J. Verpoorte, J.C. Fettinger, A. Paulus, H. Ludi and H.M. Widmer, Planar chips technology for miniaturization and integration of separation techniques into monitoring systems: Capillary electrophoresis on a chip, *J. Chromatogr. A*, 593 (1992) 253–258.
3 D.J. Harrison, A. Manz, Z.H. Gan, H. Ludi and H.M. Widmer, Capillary electrophoresis and sample injection systems integrated on a planar glass chip, *Anal. Chem.*, 64 (1992) 1926–1932.
4 P.-A. Auroux, D. Iossifidis, D.R. Reyes and A. Manz, Micro total analysis systems. 1. Introduction, theory, and technology, *Anal. Chem.*, 74 (2002) 2623–2636.
5 D.R. Reyes, D. Iossifidis, P.-A. Auroux and A. Manz, Micro total analysis systems. 2. Analytical standard operations and applications, *Anal. Chem.*, 74 (2002) 2637–2652.
6 A. Ríos, A. Escarpa, M.C. González and A.C. Crevillén, Challenges of analytical microsystems, *Trends Anal. Chem.*, 25 (2006) 467–479.
7 T. McCreedy, Fabrication techniques and materials commonly used for the production of microreactors and micro total analytical systems, *Trends Anal. Chem.*, 19 (2000) 396–401.
8 C.S. Henry, M. Zhong, S.M. Lunte, M. Kim, H. Bau and J.J. Santiago, Ceramic microchips for capillary electrophoresis-electrochemistry, *Anal. Commun.*, 36 (1999) 305–307.
9 T.D. Boone, Z.H. Fan, H.H. Hooper, A.J. Ricco, H. Tan and S.J. Williams, Plastic advances microfluidic devices, *Anal. Chem.*, 74 (2002) 78A–86A.
10 A. Muck Jr., J. Wang, M. Jacobs, G. Chen, M.P. Chatrathi, V. Jurka, Z. Výborný, S.D. Spillman, G. Sridharan and M.J. Schöning, Fabrication of poly(methylmethacrylate) microfluidic chips by atmospheric molding, *Anal. Chem.*, 76 (2004) 2290–2297.
11 G. Chen, J. Li, S. Qu, D. Chen and P. Yang, Low temperature bonding of poly(methylmethacrylate) electrophoresis microchips by *in situ* polymerisation, *J. Chromatogr. A*, 1094 (2005) 138–147.
12 A.-L. Liu, F.-Y. He, Y.-L. Hu and X.-H. Xia, Plastified poly(ethylene terephthalate) (PET)-toner microfluidic chip by direct-printing integrated with electrochemical detection for pharmaceutical analysis, *Talanta*, 68 (2006) 1303–1308.
13 N. Bao, Q. Zhang, J.-J. Xu and H.-Y. Chen, Fabrication of poly(dimethylsiloxane) microfluidic system based on masters directly printed with an office laser printer, *J. Chromatogr. A*, 1089 (2005) 270–275.
14 W.K.T. Coltro, J.A.F. da Silva, H.D.T. da Silva, E.M. Richter, R. Furlan, L. Angnes, C.L. do Lago, L.H. Mazo and E. Carrilho, Electrophoresis microchip fabricated by a direct-printing process with end-channel amperometric detection, *Electrophoresis*, 25 (2004) 3832–3839.
15 S.A. Soper, S.M. Ford, S. Qi, R.L. McCarley, K. Kelly and M.C. Murphy, Polymeric microelectromechanical systems, *Anal. Chem.*, 72 (2000) 643A–651A.

16 P. Mela, A. van der Berg, Y. Fintschenko, E.B. Cummings, B.A. Simmons and B.J. Kirby, The zeta potential of cyclo-olefin polymer microchannels and its effects on insulative (electrodeless) dielectrophoresis particle trapping devices, *Electrophoresis*, 26 (2005) 1792–1799.
17 C. Li, Y. Yang, H.G. Craighead and K.H. Lee, Isoelectric focusing in cyclic olefin copolymer microfluidic channels coated by polyacrylamide using a UV photografting meted, *Electrophoresis*, 26 (2005) 1800–1806.
18 Y. Yang, J. Kameoka, T. Wachs, J.D. Herion and H.G. Craighead, Quantitative mass spectrometric determination of methylphenidate concentration in urine using an electrospray ionization source integrated with a polymer microchip, *Anal. Chem.*, 76 (2004) 2568–2574.
19 M. Castaño-Álvarez, M.T. Fernández-Abedul and A. Costa-García, Poly(methylmethacrylate) and Topas capillary electrophoresis microchip performance with electrochemical detection, *Electrophoresis*, 26 (2005) 3160–3168.
20 S. Metz, S. Jiquet, A. Bertsch and P. Renaud, Polyimide and SU-8 microfluidic devices manufactured by heat-depolymerizable sacrificial material technique, *Lab Chip*, 4 (2004) 114–120.
21 Y.-C. Tsai, H.-P. Jen, K.-W. Lin and Y.-Z. Hsieh, Fabrication of microfluidic devices using dry film photoresist for microchip capillary electrophoresis, *J. Chromatogr. A*, 1111 (2006) 267–271.
22 C.S. Effenhauser, G.J.M. Bruin, A. Paulus and M. Ehrat, Integrated capillary electrophoresis on flexible silicone microdevices: Analysis of DNA restriction fragments and detection of single DNA molecules on microchips, *Anal. Chem.*, 69 (1997) 3451–3457.
23 J.C. McDonald and G.M. Whitesides, Poly(dimethylsiloxane) as a material for fabricating microfluidic devices, *Acc. Chem. Res.*, 35 (2002) 491–499.
24 M.W. Li, D.M. Spence and R.S. Martin, A microchip-based system for immobilizing PC 12 cells and amperometrically detecting catecholamines released after stimulation with calcium, *Electroanalysis*, 17 (2005) 1171–1180.
25 L.E. Locascio, C.E. Perso and C.S. Lee, Measurement of electroosmotic flow in plastic imprinted microfluid devices and the effect of protein adsorption on flow rate, *J. Chromatogr. A*, 857 (1999) 275–284.
26 M.L. Kovarik, M.W. Li and R.S. Martin, Integration of a carbon microelectrode with a microfabricated palladium decoupler for use in microchip capillary electrophoresis/ electrochemistry, *Electrophoresis*, 26 (2005) 202–210.
27 M.J. Schöning, M. Jacobs, A. Muck, D.-T. Knobbe, J. Wang, M. Chatrathi and S. Spillmann, Amperometric PDMS/glass capillary electrophoresis-based biosensor microchip for catechol and dopamine detection, *Sens. Actuator B*, 108 (2005) 688–694.

28 N.E. Hebert, B. Snyder, R.L. McCreery, W.G. Kuhr and S.A. Brazill, Performance of pyrolyzed photoresist carbon films in a microchip capillary electrophoresis device with sinusoidal voltammetric detection, *Anal. Chem.*, 75 (2003) 4256–4271.
29 J. Yan, X. Yang and E. Wang, Electrochemical detection of anions on an electrophoresis microchip with integrated silver electrode, *Electroanalysis*, 17 (2005) 1222–1226.
30 Y. Liu, D.O. Wipf and C.S. Henry, Conductivity detection for monitoring mixing reactions in microfluidic devices, *Analyst*, 126 (2001) 1248–1251.
31 E. Zubritsky, Taming turns in microchannels, *Anal. Chem.*, 72 (2000) 687A–690A.
32 H.-L. Lee and S.-C. Chen, Microchip capillary electrophoresis with amperometric detection for several carbohydrates, *Talanta*, 64 (2004) 210–216.
33 J.E. Prest, S.J. Baldock, P.R. Fielden and B.J.T. Brown, Determination of metal cations on miniaturised planar polymeric separation devices using isotachophoresis with integrated conductivity detection, *Analyst*, 126 (2001) 433–437.
34 B. Grass, A. Neyer, M. Höhnck, D. Siepe, F. Eisenheiss, G. Weber and R. Hergenröder, A new PMMA-microchip device for isotachophoresis with integrated conductivity detector, *Sens. Actuator B*, 72 (2001) 249–258.
35 A. Berthold, F. Laugere, H. Schellevis, C.R. de Boer, M. Laros, R.M. Guijt, P.M. Sarro and M.J. Vellekoop, Fabrication of a glass-implemented microcapillary electrophoresis device with integrated contactless conductivity detection, *Electrophoresis*, 23 (2002) 3511–3519.
36 H.-L. Lee and S.-C. Chen, Microchip capillary electrophoresis with electrochemical detector for precolumn enzymatic analysis of glucose, creatinine, uric acid and ascorbic acid in urine and serum, *Talanta*, 64 (2004) 750–757.
37 J. Wang, M.P. Chatrathi and A. Ibañez, Glucose biochip: dual analyte response in connection to two pre-column enzymatic reactions, *Analyst*, 126 (2001) 1203–1206.
38 J. Wang, A. Ibáñez and M.P. Chatrathi, Microchip-based amperometric immunoassays using redox tracers, *Electrophoresis*, 23 (2002) 3744–3749.
39 J. Wang, A. Ibáñez, M.P. Chatrathi and A. Escarpa, Electrochemical enzyme immunoassays on microchip platforms, *Anal. Chem.*, 73 (2001) 5323–5327.
40 J. Wang, J. Zima, N.S. Lawrence and M.P. Chatrathi, Microchip capillary electrophoresis with electrochemical detection of thiol-containing degradation products of V-type nerve agents, *Anal. Chem.*, 76 (2004) 4721–4726.
41 R. Kikura-Hanajiri, R.S. Martin and S.M. Lunte, Indirect measurement of nitric oxide production by monitoring nitrate and nitrite using

microchip electrophoresis with electrochemical detection, *Anal. Chem.*, 74 (2002) 6370–6377.
42 P. Ertl, C.A. Emrich, P. Singhal and R.A. Mathies, Capillary electrophoresis chips with a sheath-flow supported electrochemical detection system, *Anal. Chem.*, 76 (2004) 3749–3755.
43 M.A. Schwarz and P.C. Hauser, Recent developments in detection methods for microfabricated analytical devices, *Lab Chip*, 1 (2001) 1–6.
44 K. Uchiyama, H. Nakajima and T. Hobo, Detection methods for microchip separations, *Anal. Bioanal. Chem.*, 379 (2004) 375–382.
45 K.B. Mogensen, H. Klank and J.P. Kutter, Recent developments in detection for microfluidic systems, *Electrophoresis*, 25 (2004) 3498–3512.
46 J.A. Lapos, D.P. Manica and A.G. Ewing, Dual fluorescence and electrochemical detection on an electrophoresis microchip, *Anal. Chem.*, 74 (2002) 3348–3353.
47 H. Qiu, X.-B. Yin, J. Yan, X. Zhao, X. Yang and W. Wang, Simultaneous electrochemical and electrochemiluminescence detection for microchip and conventional capillary electrophoresis, *Electrophoresis*, 26 (2005) 687–693.
48 J. Wang, Electrochemical detection for capillary electrophoresis microchips: A review, *Electroanalysis*, 17 (2005) 1133–1140.
49 M. Pumera, A. Merkoçi and S. Alegret, New materials for electrochemical sensing VII. Microfluidic chip platforms, *Trends Anal. Chem.*, 25 (2006) 219–235.
50 W.R. Vandaveer IV, S.A. Pasas-Farmer, D.J. Fischer, C.N. Frankenfeld and S.M. Lunte, Recent developments in electrochemical detection for microchip capillary electrophoresis, *Electrophoresis*, 25 (2004) 3528–3549.
51 J. Rossier, F. Reymond and P.E. Michel, Polymer microfluidic chips for electrochemical and biochemical analysis, *Electrophoresis*, 23 (2002) 858–867.
52 X. Bai, Z. Wu, J. Josserand, H. Jensen, H. Schafer and H.H. Girault, Passive conductivity detection for capillary electrophoresis, *Anal. Chem.*, 76 (2004) 3126–3131.
53 O. Klett and L. Nyholm, Separation high voltage field driven on-chip amperometric detection in capillary electrophoresis, *Anal. Chem.*, 75 (2003) 1245–1250.
54 J.-J. Xu, N. Bao, X.-H. Xia, Y. Peng and H.-Y. Chen, Electrochemical detection method for nonelectroactive and electroactive analytes in microchip electrophoresis, *Anal. Chem.*, 76 (2004) 6902–6907.
55 R. Tantra and A. Manz, Integrated potentiometric detector for use in chip-based flow-cells, *Anal. Chem.*, 72 (2000) 2875–2878.
56 R. Ferrigno, J.N. Lee, X. Jiang and G.M. Whitesides, Potentiometric titrations in a poly(dimethylsiloxane)-based microfluidic device, *Anal. Chem.*, 76 (2004) 2273–2280.

57 J. Tanyanyiwa, S. Leuthardt and P.C. Hauser, Conductimetric and potentiometric detection in conventional and microchip capillary electrophoresis, *Electrophoresis*, 23 (2002) 3659–3666.
58 M. Galloway and S.A. Soper, Contact conductivity detection of polymerase chain reaction products analyzed by reverse-phase ion pair microcapillary electrochromatography, *Electrophoresis*, 23 (2002) 3760–3768.
59 M. Masar, M. Dankova, E. Olvecka, A. Stachurova, D. Kaniansky and B. Stanislawski, Determination of free sulfite in wine by zone electrophoresis with isotachophoresis sample pretreatment on a column-coupling chip, *J. Chromatogr. A*, 1026 (2004) 31–39.
60 B. Grass, D. Siepe, A. Neyer and R. Hergenroder, Comparison of different conductivity detector geometries on an isotachophoresis PMMA-microchip, *Fresenius' J. Anal. Chem.*, 371 (2001) 228–233.
61 R.M. Guijt, E. Baltussen, G. van der Steen, H. Frank, H. Billiet, T. Schalkhammer, F. Laugere, M. Vellekoop, A. Berthold, L. Sarro and G.W.K. van Dedem, Capillary electrophoresis with on-chip four-electrode capacitively coupled conductivity detection for application in bioanalysis, *Electrophoresis*, 22 (2001) 2537–2541.
62 J.G.A. Brito-Neto, J.A.F. da Silva, L. Blanes and C.L. do Lago, Understanding capacitively coupled contactless conductivity detection in capillary and microchip electrophoresis. Part 1. Fundamentals, *Electroanalysis*, 17 (2005) 1198–1206.
63 P. Kubán and P.C. Hauser, Effects of the cell geometry and operating parameters on the performance of an external contactless conductivity detector for microchip electrophoresis, *Lab Chip*, 5 (2005) 407–415.
64 J.G.A. Brito-Neto, J.A.F. da Silva, L. Blanes and C.L. do Lago, Understanding capacitively coupled contactless conductivity detection in capillary and microchip electrophoresis. Part 2. Peak shape, stray capacitance, noise, and actual electronics, *Electroanalysis*, 17 (2005) 1207–1214.
65 P. Kubán and P.C. Hauser, Fundamental aspects of contactless conductivity detection for capillary electrophoresis. Part I: Frequency behavior and cell geometry,, *Electrophoresis*, 25 (2004) 3387–3397.
66 P. Kubán and P.C. Hauser, Fundamental aspects of contactless conductivity detection for capillary electrophoresis. Part II: Signal-to-noise ratio and stray capacitance,, *Electrophoresis*, 25 (2004) 3398–3407.
67 M. Pumera, J. Wang, F. Opekar, I. Jelínek, J. Feldman, H. Löwe and S. Hardt, Contactless conductivity detector for microchip capillary electrophoresis, *Anal. Chem.*, 74 (2002) 1968–1971.
68 J. Wang, G. Chen, A. Muck, Jr, M.P. Chatrathi, A. Mulchandani, W. Chen, Microchip enzymatic assay of organophosphate nerve agents, *Anal. Chim. Acta*, 505 (2004) 183–187
69 J. Wang, G. Chen and A. Muck Jr., Movable contactless-conductivity detector for microchip capillary electrophoresis, *Anal. Chem.*, 75 (2003) 4475–4479.

70 J. Tanyanyiwa, E.M. Abad-Villar, M.T. Fernández-Abedul, A. Costa-García, W. Hoffmann, A.E. Guber, D. Herrmann, A. Gerlach, N. Gottschlich and P.C. Hauser, High-voltage contactless conductivity-detection for lab-on-chip devices using external electrodes on the holder, *Analyst*, 128 (2003) 1019–1022.
71 J. Tanyanyiwa and P.C. Hauser, High-voltage contactless conductivity detection of metal ions in capillary electrophoresis, *Electrophoresis*, 23 (2002) 3781–3786.
72 J. Lichtenberg, N.F. de Rooij and E. Verpoorte, A microchip electrophoresis system with integrated in-plane electrodes for contactless conductivity detection, *Electrophoresis*, 23 (2002) 3769–3780.
73 J. Wang and M. Pumera, Dual conductivity/amperometric detection system for microchip capillary electrophoresis, *Anal. Chem.*, 74 (2002) 5919–5923.
74 W.R. Vandaveer IV, S.A. Pasas, R.S. Martin and S.M. Lunte, Recent developments in amperometric detection for microchip capillary electrophoresis, *Electrophoresis*, 23 (2002) 3667–3677.
75 R.S. Martin, K.L. Ratzlaff, B.H. Huyng and S.M. Lunte, In-channel electrochemical detection for microchip capillary electrophoresis using an electrically isolated potentiostat, *Anal. Chem.*, 74 (2002) 1136–1143.
76 K. Wang and X.-H. Xia, Microchannel-electrode alignment and separation parameters comparison in microchip capillary electrophoresis by scanning electrochemical microscopy, *J. Chromatogr. A*, 1110 (2006) 222–226.
77 M. Castaño-Álvarez, M.T. Fernández-Abedul and A. Costa-García, Amperometric detector designs for capillary electrophoresis microchips, *J. Chromatogr. A*, 1109 (2006) 291–299.
78 J.S. Rossier, R. Ferrigno and H.H. Girault, Electrophoresis with electrochemical detection in a polymer microdevice, *J. Electroanal. Chem.*, 492 (2000) 15–22.
79 D.M. Osbourn and C.E. Lunte, On-Column electrochemical detection for microchip capillary electrophoresis, *Anal. Chem.*, 75 (2003) 2710–2714.
80 C.-C. Wu, R.-G. Wu, H.-G. Huang, Y.-C. Lin and H.-C. Chang, Three-electrode electrochemical detector and platinum film decoupler integrated with a capillary electrophoresis microchip for amperometric detection., *Anal. Chem.*, 75 (2003) 947–952.
81 Y. Du, J. Yan, W. Zhou, X. Yang and E. Wang, Direct electrochemical detection of glucose in human plasma on capillary electrophoresis microchips, *Electrophoresis*, 25 (2004) 3853–3859.
82 D.-C. Chen, F.-L. Hsu, D.-Z. Zhan and C.-H. Chen, Palladium film decoupler for amperometric detection in electrophoresis chips, *Anal. Chem.*, 73 (2001) 758–762.
83 N.A. Lacher, S.M. Lunte and R.S. Martin, Development of a microfabricated palladium decoupler/electrochemical detector for microchip

capillary electrophoresis using a hybrid glass/poly(dimethylsiloxane) device, *Anal. Chem.*, 76 (2004) 2482–2491.
84 J. Vickers and C.S. Henry, Simplified current decoupler for microchip capillary electrophoresis with electrochemical and pulsed amperometric detection, *Electrophoresis*, 26 (2005) 4641–4647.
85 J.-H. Kim, C.J. Kang and Y.-S. Kim, A disposable capillary electrophoresis microchip with an indium tin oxide decoupler/amperometric detector, *Microelectr. Eng.*, 78–79 (2005) 563–567.
86 A.T. Woolley, K. Lao, A.N. Glazer and R.A. Mathies, Capillary electrophoresis chips with integrated electrochemical detection, *Anal. Chem.*, 70 (1998) 684–688.
87 R. Wilke and S. Büttgenbach, A micromachined capillary electrophoresis chip with fully integrated electrodes for separation and electrochemical detection, *Biosens. Bioelectron.*, 19 (2003) 149–153.
88 R.S. Keynton, T.J. Roussel Jr., M.M. Crain, D.J. Jackson, D.B. Franco, J.F. Naber, K. Walsh and R.P. Baldwin, Design and development of microfabricated capillary electrophoresis devices with electrochemical detection, *Anal. Chim. Acta*, 507 (2004) 95–105.
89 M.A. Schwarz, B. Galliker, K. Fluri, T. Kappes and P.C. Hauser, A two-electrode configuration for simplified amperometric detection in a microfabricated electrophoretic separation device, *Analyst*, 126 (2001) 147–151.
90 R.S. Martin, A.J. Gawron, S.M. Lunte and C.S. Henry, Dual-electrode electrochemical detection for poly(dimethylsiloxane)-fabricated capillary electrophoresis microchips, *Anal. Chem.*, 72 (2000) 3196–3202.
91 C.-Ch.J. Lai, Ch.-h. Chen and F.-H. Ko, In-channel dual-electrode amperometric detection in electrophoretic chips with a palladium film decoupler, *J. Chromatogr. A*, 1023 (2004) 143–150.
92 R.S. Martin, A.J. Gawron, B.A. Fogarty, F.B. Regan, E. Dempsey and S.M. Lunte, Carbon paste-based electrochemical detectors for microchip capillary electrophoresis/electrochemistry, *Analyst*, 126 (2001) 277–280.
93 A.J. Gawron, R.S. Martin and S.M. Lunte, Fabrication and evaluation of a carbon-based dual-electrode detector for poly(dimethylsiloxane) electrophoresis chips, *Electrophoresis*, 22 (2001) 242–248.
94 D.J. Fischer, W.R. Vandaveer IV, R.J. Grigsby and S.M. Lunte, Pyrolyzed photoresist carbon electrodes for microchip electrophoresis with dual-electrode amperometric detection, *Electroanalysis*, 17 (2005) 1153–1159.
95 W.R. LaCourse and S.J. Modi, Microelectrode applications of pulsed electrochemical detection, *Electroanalysis*, 17 (2005) 1141–1152.
96 C.D. García and C.S. Henry, Coupling capillary electrophoresis and pulsed electrochemical detection, *Electroanalysis*, 17 (2005) 1125–1131.
97 J. Wang, R. Polsky, B. Tian and M.P. Chatrathi, Voltammetry on microfluidic chip platforms, *Anal. Chem.*, 72 (2000) 5285–5289.

98 N.E. Hebert, W.G. Kuhr and S.A. Brazill, A microchip electrophoresis device with integrated electrochemical detection: A direct comparison of constant potential amperometry and sinusoidal voltammetry, *Anal. Chem.*, 75 (2003) 3301–3307.
99 M. Castaño-Álvarez, M.T. Fernández-Abedul and A. Costa-García, Amperometric PMMA-microchip with integrated gold working electrode for enzyme assays, *Anal. Bioanal. Chem.*, 382 (2005) 303–310.
100 J. Wang, A. Escarpa, M. Pumera and J. Feldman, Capillary electrophoresis-electrochemistry microfluidic system for the determination of organic peroxides, *J. Chromatogr. A*, 952 (2002) 249–254.
101 J. Wang, B. Tian and W. Sahlin, Integrated electrophoresis chips/amperometric detection with sputtered gold working electrodes, *Anal. Chem.*, 71 (1999) 3901–3904.
102 M. Pumera, X. Llopis, A. Merkoci and S. Alegret, Microchip capillary electrophoresis with a single-wall carbon nanotube/gold electrochemical detector for determination of amino phenols and neurotransmitters, *Microchim. Acta*, 152 (2006) 261–265.
103 U. Backofen, F.M. Matysik and C.E. Lunte, A chip-based electrophoresis system with electrochemical detection and hydrodynamic injection, *Anal. Chem.*, 74 (2002) 4054–4059.
104 D.P. Manica, Y. Mitsumori and A.G. Ewing, Characterization of electrode fouling and surface regeneration for a platinum electrode on an electrophoresis microchip, *Anal. Chem.*, 75 (2003) 4572–4577.
105 S.A. Pasas, N.A. Lacher, M.I. Davies and S.M. Lunte, Detection of homocysteine by conventional and microchip capillary electrophoresis/electrochemistry, *Electrophoresis*, 23 (2002) 759–766.
106 J. Wang, M.P. Chatrathi and B. Tian, Capillary electrophoresis chips with thick-film amperometric detectors: Separation and detection of hydrazine compounds, *Electroanalysis*, 12 (2000) 691–694.
107 J. Wang, G. Chen and M.P. Chatrathi, Nickel amperometric detector prepared by electroless deposition for microchip electrophoretic measurement of alcohols and sugars, *Electroanalysis*, 16 (2004) 1603–1608.
108 N.E. Hebert, W.G. Kuhr and S.A. Brazill, Microchip capillary electrophoresis coupled to sinusoidal voltammetry for the detection of native carbohydrates, *Electrophoresis*, 23 (2002) 3750–3759.
109 J. Wang, G. Chen, M. Wang and M.P. Chatrathi, Carbon-nanotube/copper composite electrodes for capillary electrophoresis microchip detection of carbohydrates, *Analyst*, 29 (2004) 512–515.
110 A.-J. Wang, J.-J. Xu, Q. Zhang and H.-Y. Chen, The use of poly(dimethylsiloxane) surface modification with gold nanoparticles for the microchip electrophoresis, *Talanta*, 69 (2006) 210–215.
111 A.-J. Wang, J.-J. Xu and H.-Y. Chen, Proteins modification of poly(dimethylsiloxane) microfluidic channels for the enhanced microchip electrophoresis, *J. Chromatogr. A*, 1107 (2006) 257–264.

112 J. Wang, B. Tian and E. Sahlin, Micromachined electrophoresis chips with thick-film electrochemical detectors, *Anal. Chem.*, 71 (1999) 5436–5440.
113 J. Wang, M. Pumera, M.P. Chatrathi, A. Rodriguez, S. Spillman, R.S. Martin and S.M. Lunte, Thick-film electrochemical detectors for poly(dimethylsiloxane)-based microchip capillary electrophoresis, *Electroanalysis*, 14 (2002) 1251–1255.
114 J. Wang, M. Pumera, M.P. Chatrathi, A. Escarpa, R. Koonrad, A. Griebel, W. Dörner and H. Löwe, Towards disposable lab-on-a-chip: Poly(methylmethacrylate) microchip electrophoresis device with electrochemical detection, *Electrophoresis*, 23 (2002) 596–601.
115 J. Wang, M. Pumera, M.P. Chatrathi, A. Escarpa and M. Musameh, Single-channel microchip for fast screening and detailed identification of nitroaromatic explosives or organophosphate nerve agents, *Anal. Chem.*, 74 (2002) 1187–1191.
116 J. Wang, G. Chen, A. Muck, Jr, D. Shin, A. Fujishima, Microchip capillary electrophoresis with a boron-doped diamond electrode for rapid separation and detection of purines, *J. Chromatogr. A*, 1022 (2004) 207–212
117 A. Collier, J. Wang, D. Diamond and E. Dempsey, Microchip micellar electrokinetic chromatography coupled with electrochemical detection for analysis of synthetic oestrogen mimicking compounds, *Anal. Chim. Acta*, 550 (2005) 107–115.
118 N. Bao, J.-J. Xu, Y.-H. Dou, Y. Cai, H.-Y. Chen and X.-H. Xia, Electrochemical detector for microchip electrophoresis of poly(dimethylsiloxane) with a three-dimensional adjustor, *J. Chromatogr. A*, 1041 (2004) 245–248.
119 G. Chen, L. Zhang and J. Wang, Miniaturized capillary electrophoresis system with a carbon nanotube microelectrode for rapid separation and detection of thiols, *Talanta*, 64 (2004) 1018–1023.
120 Y. Liu, J.A. Vickers and C.S. Henry, Simple and sensitive electrode design for microchip electrophoresis/electrochemistry, *Anal. Chem.*, 76 (2004) 1513–1517.
121 Y. Wu, J.-M. Lin, R. Su, F. Qu and Z. Cai, An end-channel amperometric detector for microchip capillary electrophoresis, *Talanta*, 64 (2004) 338–344.
122 Y.-H. Dou, N. Bao, J.-J. Xu and H.-Y. Chen, A dynamically modified microfluidic poly(dimetilsiloxane) chip with electrochemical detection for biological analysis, *Electrophoresis*, 23 (2002) 3558–3566.
123 G. Chen, Y. Lin and J. Wang, Monitoring environmental pollutants by microchip capillary electrophoresis with electrochemical detection, *Talanta*, 68 (2006) 497–503.
124 J. Wang, Microchip devices for detecting terrorist weapons, *Anal. Chim. Acta*, 507 (2004) 3–10.

125 E. Verpoorte, Microfluidic chips for clinical and forensic analysis, *Electrophoresis*, 23 (2002) 677–712.
126 J. Khandurina and A. Guttman, Bioanalysis in microfluidic devices, *J. Chromatogr. A*, 943 (2002) 159–183.
127 U. Bilitewski, M. Genrich, S. Kadow and G. Mersal, Biochemical analysis with microfluidic systems, *Anal. Bioanal. Chem.*, 377 (2003) 556–569.
128 K. Sato, A. Hibara, M. Tokeshi, H. Hisamoto and T. Kitamori, Microchip-based chemical and biochemical analysis systems, *Adv. Drug Deliv. Rev.*, 55 (2003) 379–391.
129 S.F.Y. Li and L.J. Kricka, Clinical analysis by microchip capillary electrophoresis, *Clin. Chem.*, 52 (2006) 37–45.
130 E.X. Vrouwe, R. Luttge, W. Olthuis and A. van den Berg, Microchip analysis of lithium in blood using moving boundary electrophoresis and zone electrophoresis, *Electrophoresis*, 26 (2005) 3032–3042.
131 J.C. Fanguy and C.S. Henry, Pulsed amperometric detection of carbohydrates on an electrophoretic microchip, *Analyst*, 127 (2002) 1021–1023.
132 C.D. García and C.S. Henry, Direct determination of carbohydrates, amino acids, and antibiotics by microchip electrophoresis with pulsed amperometric detection, *Anal. Chem.*, 75 (2003) 4778–4783.
133 C.D. García and C.S. Henry, Enhanced determination of glucose by microchip electrophoresis with pulsed amperometric detection, *Anal. Chim. Acta*, 508 (2004) 1–9.
134 J. Wang, M.P. Chatrathi and B. Tian, Microseparation chips for performing multienzymatic dehydrogenase/oxidase assays: Simultaneous electrochemical measurement of ethanol and glucose, *Anal. Chem.*, 73 (2001) 1296–1300.
135 G. Chen and J. Wang, Fast and simple sample introduction for capillary electrophoresis microsystems, *Analyst*, 129 (2004) 507–511.
136 J. Wang, M.P. Chatrathi, B. Tian and R. Polsky, Microfabricated electrophoresis chips for simultaneous bioassays of glucose, uric acid, ascorbic acid, and acetaminophen, *Anal. Chem.*, 72 (2000) 2514–2518.
137 J. Wang, A. Ibañez and M.P. Chatrathi, On-Chip integration of enzyme and immunoassays: Simultaneous measurements of insulin and glucose, *J. Am. Chem. Soc.*, 125 (2003) 8444–8445.
138 F. Xia, W. Jin, X. Yin and Z. Fang, Single-cell analysis by electrochemical detection with a microfluidic device, *J. Chromatogr. A*, 1063 (2005) 227–233.
139 N.E. Hebert, B. Zinder, R.L. McCreery, W.G. Kuhr and S.A. Brazill, Performance of pyrolyzed photoresist carbon films in a microchip capillary electrophoresis device with sinusoidal voltammetric detection, *Anal. Chem.*, 75 (2003) 4265–4271.

140 M.L. Kovarik, N.J. Torrence, D.M. Spence and R.S. Martin, Fabrication of carbon microelectrodes with a micromolding technique and their use in microchip-based flow analyses, *Analyst*, 129 (2004) 400–405.
141 A.J. Blasco, I. Barrigas, M.C. González and A. Escarpa, Fast and simultaneous detection of prominent natural antioxidants using analytical microsystems for capillary electrophoresis with a glassy carbon electrode: A new gateway to food environments, *Electrophoresis*, 26 (2005) 4664–4673.
142 J.C. Fanguy and C.S. Henry, The analysis of uric acid in urine using microchip capillary electrophoresis with electrochemical detection, *Electrophoresis*, 23 (2002) 767–773.
143 C.D. Garcia and C.S. Henry, Direct detection of renal function markers using microchip CE with pulsed electrochemical detection, *Analyst*, 129 (2004) 579–584.
144 J. Wang, M.P. Chatrathi, A. Ibañez and A. Escarpa, Micromachined separation chips with post-column enzymatic reactions of class enzymes and end-column electrochemical detection: assays of amino acids, *Electroanalysis*, 14 (2002) 400–404.
145 J. Wang, G. Chen and M. Pumera, Microchip separation and electrochemical detection of amino acids and peptides following precolumn derivatization with naphthalene-2,3-dicarboxyaldehyde, *Electroanalysis*, 15 (2003) 862–865.
146 J. Wang and G. Chen, Microchip capillary electrophoresis with electrochemical detector for fast measurements of aromatic amino acids, *Talanta*, 60 (2003) 1239–1244.
147 J.-J. Xu, Y. Peng, N. Bao, X.-H. Xia and H.-Y. Chen, Simple method for the separation and detection of native amino acids and the identification of electroactive and non-electroactive analytes, *J. Chromatogr. A*, 1095 (2005) 193–196.
148 J. Wang, S. Mannino, C. Camera, M.P. Chatrathi, M. Scampicchio and J. Zima, Microchip capillary electrophoresis with amperometric detection for rapid separation and detection of seleno amino acids, *J. Chromatogr. A*, 1091 (2005) 118–177.
149 J. Wang, G. Chen, M.P. Chatrathi and M. Musameh, Capillary electrophoresis microchip with a carbon nanotube-modified electrochemical detector, *Anal. Chem.*, 76 (2004) 298–302.
150 N.R. Hebert and S.A. Brazill, Microchip capillary gel electrophoresis with electrochemical detection for the analysis of known SNPs, *Lab Chip*, 3 (2003) 241–247.
151 J.-H. Kim, C.J. Kang and Y.-S. Kim, Development of a microfabricated disposable microchip with a capillary electrophoresis and integrated three-electrode electrochemical detection, *Biosens. Bioelectron*, 20 (2005) 2314–2317.

152 M. Galloway, W. Stryjewski, A. Henry, S.M. Ford, S. Llopis, R.L. McCarley and S.A. Soper, Contact conductivity detection in poly(methyl methacylate)-based microfluidic devices for analysis of mono- and polyanionic molecules, *Anal. Chem.*, 74 (2002) 24072415.
153 J. Tanyanyiwa, E.M. Abad-Villar and P.C. Hauser, Contactless conductivity detection of selected organic ions in on-chip electrophoresis, *Electrophoresis*, 25 (2004) 903–908.
154 M. Zuborova, Z. Demianova, D. Kaniansky, M. Masar and B. Stanislawski, Zone electrophoresis of proteins on a poly(methyl methacrylate) chip with conductivity detection, *J. Chromatogr. A*, 990 (2003) 179–188.
155 Z. Deyl, I. Miksik and A. Eckhardt, Comparison of standard capillary and chip separations of sodium dodecylsulfate protein complexes, *J. Chromatogr. A*, 990 (2003) 153–158.
156 J. Wang, On-chip enzymatic assays, *Electrophoresis*, 23 (2002) 713–718.
157 J. Wang and M.P. Chatrathi, Microfabricated electrophoresis chip for bioassay of renal markers, *Anal. Chem.*, 75 (2003) 525–529.
158 E.M. Abad-Villar, J. Tanyanyiwa, M.T. Fernandez-Abedul, A. Costa-Garcia and P.C. Hauser, detection of human immunoglobulin in microchip and conventional capillary electrophoresis with contactless conductivity measurements, *Anal. Chem.*, 76 (2004) 1282–1288.
159 M. Castaño-Álvarez, M.T. Fernández-Abedul and A. Costa-García, Analytical performace of CE-microchips with amperometric detection, *Instrum. Sci. Technol.*, 34 (2006) 697–710.
160 E.M. Abad-Villar, M.T. Fernández-Abedul and A. Costa-García, Flow injection electrochemical enzyme immunoassay based on the use of gold bands, *Anal. Chim. Acta*, 409 (2000) 149–158.
161 C.D. García, B.M. Dressen, A. Henderson and C.S. Henry, Comparison of surfactants for dynamic surface modification of poly(dimethylsiloxane) microchips, *Electrophoresis*, 26 (2005) 703–709.

Chapter 35

Microchip electrophoresis/ electrochemistry systems for analysis of nitroaromatic explosives

Martin Pumera, Arben Merkoçi and Salvador Alegret

35.1 INTRODUCTION

Recent terrorist attacks on October 12, 2002, in Bali, Indonesia, showed that the homemade and industrial explosive bombs present a serious threat to the world community. These terrorist activities generated enormous demand for rapid identification of nitroaromatic explosives at the site of terrorism. The decentralized detection of explosives at low concentrations is fundamental for safety of civilized people. The ideal counter-terrorism detection device should allow the security forces to make the important decision concerning evacuating, barricading, effective decontamination of particular site or efficient pursuing of suspects.

There are several hundreds of explosive materials officially listed [1]. They are usually based on (i) nitrated organic compounds (i.e. 2,4,6-trinitrotoluene (TNT), hexahydro-1,3,5-trinitro-1,3,5-triazine (RDX), tetryl or nitroglycerin) or (ii) inorganic nitrate, chlorate or perchlorate salts (i.e. NH_4NO_3, KNO_3 or NH_4ClO_4). Nitrated organic explosives are usually used for military purposes; TNT and DNB have been in use for more than 100 years, mostly during World War I and II. RDX and HMX are very powerful explosives widely used in present days in plastic explosives or warheads. The current military explosives are usually mixtures of explosive component with other organic compounds, such as stabilizers and plasticizers (see Table 35.1). Military and industrial explosives are highly effective and therefore attract the attentions of terrorists.

Comprehensive Analytical Chemistry 49
Alegret and Merkoçi (Eds)
Volume 49 ISSN: 0166-526X DOI: 10.1016/S0166-526X(06)49035-8

TABLE 35.1

Examples of mixtures of explosives

Explosive name	Composition
Amatol	NH_4NO_3 and TNT
A-3	RDX and wax
B	RDX (60%) and TNT (40%)
C-4	RDX and wax/oils
Cyclotol	RDX (75%) and TNT (25%)
H-6	RDX and TNT with aluminum particles
Octol	HMX (75%) and TNT (25%)
Semtex-H	RDX, PETN, oils
Torpex	TNT (42%), RDX (40%) and aluminum particles (18%)
Tritonal	TNT with aluminum particles

The only official method for analysis of nitrated explosives is specified by United States Environmental Protection Agency (US EPA) under EPA method 8330 for monitoring of nitroaromatic explosives in soils and ground water [2]. The method describes HPLC separation based on the use of two reverse-phase columns and it requires 60 min for full resolution of 14 nitrated explosives. It is obvious that this method does not fulfill the urgent need of security forces for the fast detection of explosives. Recent Bali terrorist bombing case study demonstrated the advantages of the mobile and portable laboratory for study of post-blast explosive residues at the terrorist attack site [3]. It is clear that the laboratory-based equipment is too bulky and slow to meet the requirements of security forces for rapid and decentralized detection of explosives. Microchip electrophoresis devices can offer the ability to monitor nitrated explosives at the sample source (before or after terrorist attack) with significant advantages in terms of high-throughput, efficiency, low-cost and sample size.

There can be found good reviews on conventional and microchip capillary electrophoresis in forensic/security analysis [4–7] in the literature. The aim of this chapter is to overview the progress which has been made towards the development of portable microfluidic device for on-site and fast detection of nitrated explosives and to describe the major developments in this field (summarized details on analytical methods for microchip determination of nitroaromatic explosives can be found in Table 35.2). The corresponding practical protocol for measurements of explosives on microfluidic device with amperometric detector is described in Procedure 49 (see CD accompanying this book).

TABLE 35.2

Analysis of nitrated organic explosives by microchip electrophoresis

Analyte	Detection technique	Limit of detection	Reference
TNT; DNB; 2,4-DNT; 2,6-DNT; 4-NT	Amperometry, screen-printed carbon electrode, at −0.5 V	~600 ppb (TNT)	[19]
TNT; RDX; 2,4-DNT; 2,6-DNT; 2,3-DNT	Amperometry, gold wire electrode, at −0.7 V	110 ppb	[20]
TNT; 2,4-DNT; 2,6-DNT; 2,3-DNT	Amperometry, gold electrode deposited onto channel outlet, at −0.8 V	24 ppb (TNT)	[21]
TNT	Amperometry, mercury/gold amalgam electrode, at −0.6 V	7 ppb (TNT)	[22]
TNT; 1,3-DNB; 2,4-DNT	Amperometry, boron-doped diamond electrode, at −0.7 V	70 ppb (1,3-DNB); 110 ppb (2,4-DNT)	[23]
TNT; TNB; DNB; 2,4-DNT; 2-Am-4,6-DNT; 4-Am-2,6-DNT	Amperometry, screen-printed carbon electrode, at −0.5 V	60 ppb (TNT and DNB)	[24]
TNT, DNT, TNB	Amperometry, screen-printed carbon electrode, at −0.4 V	800 ppb (TNT); 450 ppb (TNB)	[25]
TNT; DNB; TNB; NB; tetryl; 2,4-DNT; 2,6-DNT; NT; 2-Am-4,6-DNT; 4-Am-2,6-DNT	Indirect LIF, visualizing agent Cy7	Around 1 ppm	[26]
TNB; TNT; 2,4-DNB; 2-Am-4,6-DNB	Amperometry, screen-printed carbon electrode, at −0.5 V	80 ppb (TNT)	[30]
TNT; TNB; 2,4-DNT; 1,3-DNB; 2,4-DNP	Direct LIF	Around 1 ppb	[11]

TABLE 35.2 (*continued*)

Analyte	Detection technique	Limit of detection	Reference
TNT; TNB; tetryl	UV–Vis at 505 nm	160 ppb (TNT); 60 ppb (TNB); 200 ppb (tetryl) without *ex situ* preconcentration; 340 ppt (TNT); 250 ppt (TNB); 190 ppt (tetryl) with *ex situ* preconcentration	[18]

35.1.1 Detection techniques

Nitroaromatic compounds (see Fig. 35.1) can be directly detected by UV–Vis absorbance [8]. This detection approach is widely used in conventional capillary electrophoretic analysis of nitroaromatic explosives; however, to less extent in microfluidic chips. The major problem of the UV–Vis detection is the linear dependence of absorbency on optical path length. To address this, many commercial conventional CE instruments use an UV–Vis detection cell with an increased path length, i.e. bubble cell, Z or U cell. In the chip configuration, this is not easy to develop, although the use of a micromachined U cell for integrated on-chip UV detection was described [9]. An additional difficulty with UV detection on microchip format is that the mostly used material for fabrication of microchmachined chips, borosilicate glass, absorbs light with wavelengths shorter than 380 nm. Fused silica chip substrates can be used to overcome these problems but have not been investigated due to the higher cost. Laser-induced fluorescence in its direct or indirect mode can also be used on microfluidic devices for detection of explosives [10,11].

Amperometric detection is the preferred method for the analysis of nitroaromatic explosives on microchip devices since it offers up to three orders of magnitude higher sensitivity than indirect LIF and it has a great potential for miniaturization and integration on microchip platform. Presence of nitrogroup allows its cathodic reduction to form alkylhydroxyamines. The reduction mechanism of polynitroaromatic compounds is complex and depends on the number of nitro groups,

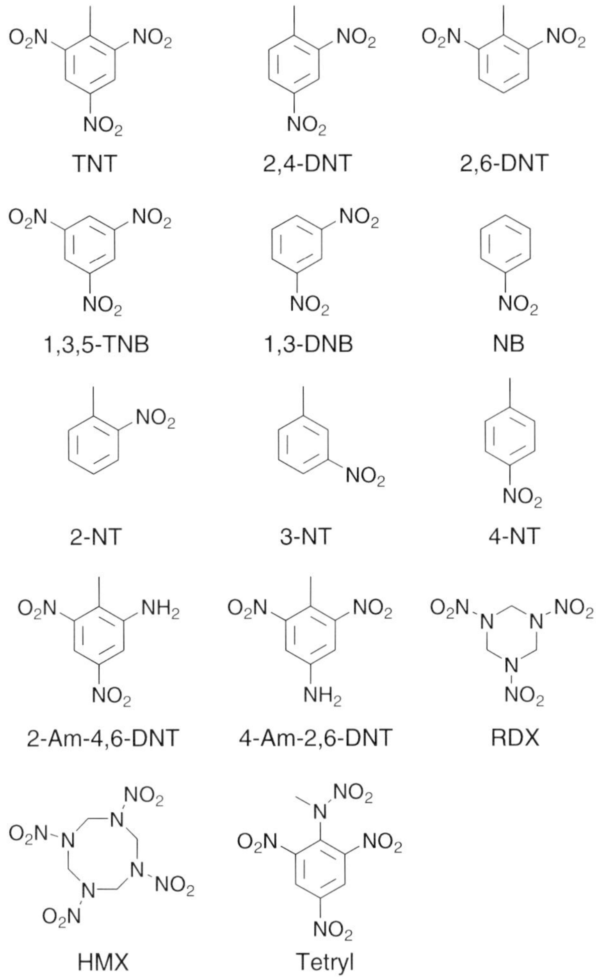

Fig. 35.1. Structures of nitroaromatic and nitroamine explosives and explosive residues described in US EPA method 8330.

on their relative positions on the rings, on the nature of the other substituents in the aromatic system and on pH of the solution [12–15]. Trinitrocompounds, such as TNT, are more readily reduced than dinitro and mononitro compounds. The reduction potentials of nitroaromatic compounds occur at more negative potentials on glassy carbon than on mercury film electrode [16]. Reduction of nitrobenzenes and nitrotoulenes occurs in one single four-electron step to form arylhydroxyamines, which is followed by two-electron step forming arylamines [16,17]. Two-electron reduction of aromatic nitro group to a nitroso group has never been reported. This is attributed to relatively higher reduction potential of the nitroso group in comparison to the corresponding nitro group [16].

35.1.2 Separation techniques

Nitroaromatic explosives and other nitrated organic explosives are under the normal conditions neutral compounds and therefore cannot be separated directly by capillary zone electrophoresis (CZE) technique. Another separation vector must be introduced in order to achieve the resolution between the solutes. Micellar electrokinetic chromatography (MEKC) is typically employed on microchip scene for separation of nitroaromatic explosives.

Original separation method was reported by Collins' group which used strongly basic nonaqueous (acetonitrile/methanol) medium for ionization and consequent μCZE separation of trinitroaromatic explosives [18].

35.2 APPLICATIONS OF MICROFLUIDIC DEVICES FOR MONITORING OF NITRATED ORGANIC EXPLOSIVES

Microfluidic device for detection of five TNT-related explosive compounds with exchangeable carbon thick-film screen-printed amperometric detector was described by Wang *et al.* [19]. This detection design permitted convenient and rapid replacement of the detector. The limit of detection of this method was 600 ppb. The need to improve detection limits of nitroaromatic explosives on microfluidic platform led several researchers to explore other electrode materials, offering better sensitivity. Luong's group employed gold-wire electrode in end-column wall-jet configuration, which resulted into detection limit of explosives around 150 ppb [20]. The same group exploited further the use of gold electrode material. The microchip with gold electrode fabricated on the channel outlet by electroless deposition was prepared [21]. Such a deposited gold detector showed limit of detection of 24 ppb for TNT. So far lowest detection limit for microchip electrophoresis–amperometry was found to be 7 ppb for TNT using mercury/gold amalgam electrode [22]. However, such a low detection limit was coupled only to μ-FIA mode.

Electrode surface fouling is a problem in amperometric detection on microfluidic platform. Wang *et al.* introduced boron-doped diamond electrode with highly stable response towards detection of explosives [23]. While thick-film carbon detector displayed a gradual decrease in the TNT response (with a 30% decrease and an RSD of 10.8%; $n = 60$), a highly stable signal was observed upon using the diamond electrode (RSD of 0.8%; $n = 60$). Such resistance to surface fouling reflected the negligible adsorption of explosive reduction products at the boron-doped diamond electrode surface.

The majority of the developed techniques for detection of explosives on a lab-on-a-chip platform show analysis times around 120 s, insufficiently long for fast detection of terrorist weapons. To solve this problem, a single-channel chip-based analytical microsystem that allowed rapid flow injection measurements of the total content of organic explosive compounds, as well as detailed micellar chromatographic identification of the individual ones, was described [24]. The protocol involved repetitive rapid flow injection (screening) assays to provide a timely warning and switching to the separation (fingerprint identification) mode only when explosive compounds were detected (for protocol scheme, see Fig. 35.2). While MEKC was used for separating the neutral nitroaromatic explosives, an operation without sodium dodecyl sulfate (SDS) led to high-speed measurements of the "total" explosives content. Switching between the "flow injection" and "separation" modes was accomplished by rapidly exchanging the SDS-free and SDS-containing buffers in the separation channel. Amperometric detection was used for monitoring the separation. Assay rates of about 360 and 30 h^{-1} were thus realized for the "total" screening and "individual" measurements, respectively. Method for fast FIA microchip screening of TNT with novel world-to-chip interface was described later by the same authors [25].

Indirect laser-induced fluorescence was used to detect explosive compounds after their separation by MEKC [26]. To achieve indirect detection, a low concentration of a dye (5 μM Cy7) was added to the running buffer as a visualizing agent. Using this methodology, a sample containing 14 explosives (US EPA 8330 mixture) was examined, however with poor detection limits for nitroaromatic explosives (around 1 ppm) and nitramine explosives (RDX, HMX; LOD around 2000 ppm). Such a huge difference in LOD was attributed to the low fluorescence quenching efficiencies of nitramines compared to the nitroaromatic explosives. The significantly improved detection limits of explosives on lab-on-a-chip/LIF platform were demonstrated by Bromberg and Mathies [11]. A homogenous immunoassay of TNT was based on the rapid microchip electrophoretic separation of an equilibrated mixture of an anti-TNT antibody, fluorescein-labeled TNT and unlabeled TNT or its analogue. A TNT immunoassay was sensitive (LOD of 1 ppb), having a wide dynamic range (1–300 ppb).

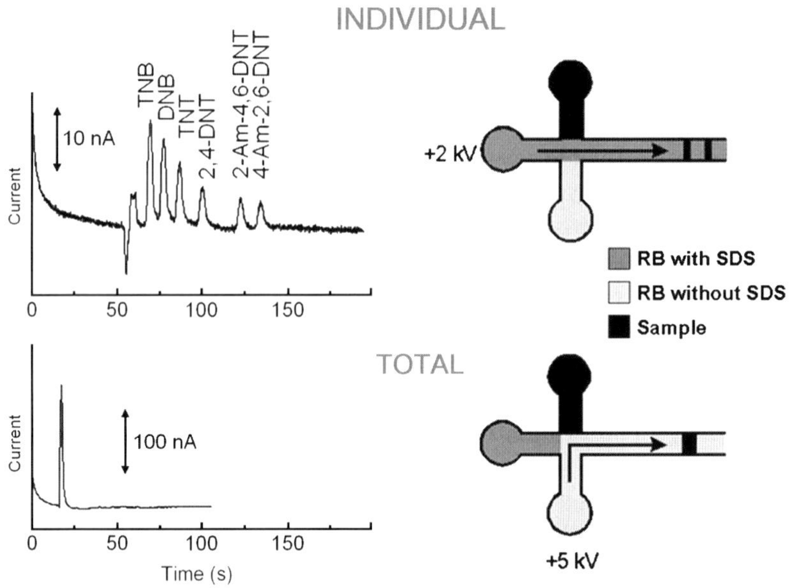

Fig. 35.2. Amperometric chip for rapid screening (total, bottom) and detailed fingerprint characterization (individual, top) of nitroaromatic explosives. The protocol involves repetitive rapid flow injection (screening) assays to provide a timely warning and alarms and switching to the separation (fingerprint identification) mode only when harmful compounds are detected. While micellar electrokinetic chromatography, in the presence of sodium dodecyl sulfate (SDS) (right, top), is used for separating the neutral nitroaromatic explosive (left, top), an operation without SDS (right, bottom) leads to high-speed measurements of the ''total'' explosives content (left, bottom). Switching between the ''flow injection'' and ''separation'' modes is accomplished by rapidly exchanging the SDS-free and SDS-containing buffers in the separation channel. Adapted with permission from Ref. [24].

Pushing detection limits of nitroaromatic explosives into the parts per trillion (ppt) level requires sample preconcentration. Collins and coworkers used solid-phase extraction (SPE) of explosives from sea water which was followed by rapid on-chip separation and detection [18]. Explosives were eluted from SPE column by acetonitrile and were injected in the microchip separation channel. Lab-on-a-chip analysis was carried out in nonaqueous medium. The mixed acetonitrile/methanol separation buffer was used to produce the ionized red-colored products of TNT, TNB and tetryl [27,28]. The chemical reaction of the bases (hydroxide and methoxide anions) with trinitroaromatic explosives resulted in negatively charged products, which were readily separated by microchip zone electrophoresis (for reaction scheme, see Fig. 35.3) [18]. It is expected that nonaqueous microchip electrophoresis will play more important role in the future in the detection of explosives for its inherent compatibility with organic solvents used in preconcentration techniques [18,29].

Several explosive devices contain mixtures of organic and inorganic explosives, i.e. Amatol or Tritonal (containing TNT and NH_4NO_3 or TNT and aluminum particles, respectively) and analytical method for their fast detection is needed. The dual electrochemical microchip detection system containing two orthogonal detection modes (conductivity and amperometry) facilitated the measurements of inorganic explosives and nitroaromatic explosive components in one analytical run on single-channel microchip [30]. The conductivity detector profiled only the ionic species, the amperometric one responded to the redox-active nitroaromatic components. Total assay of explosive mixture related to Amatol was performed within 2 min (see Fig. 35.4).

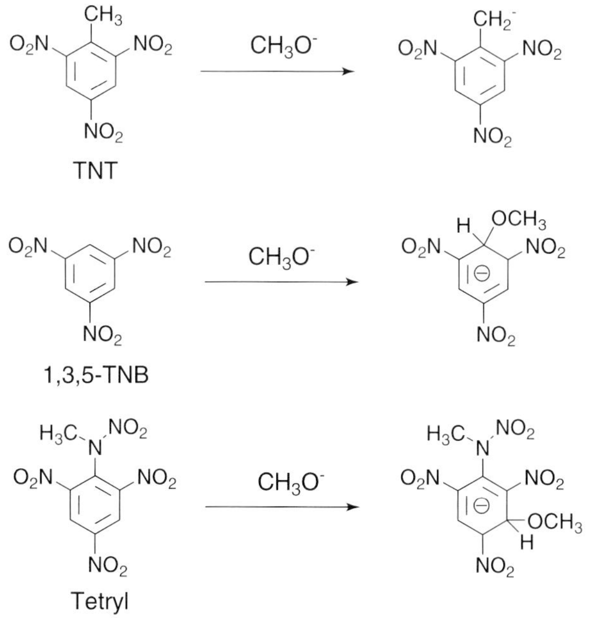

Fig. 35.3. Chemical reaction of TNT, TNB and tetryl in basic acetonitrile/methanol. Reprinted with permission from Ref. [18].

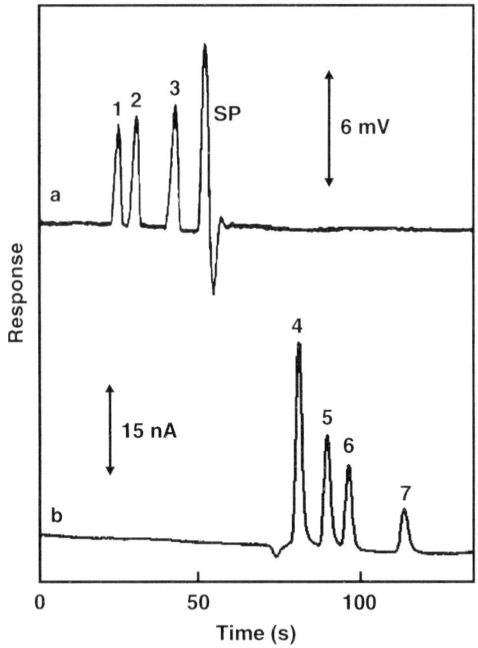

Fig. 35.4. Electrophoregrams showing the simultaneous measurement of inorganic and nitroaromatic explosives, as recorded with (a) the contactless conductivity and (b) amperometric detectors. Analytes: (1) ammonium, (2) methylammonium, (3) sodium, (4) TNB, (5) TNT, (6) 2,4-DNB and (7) 2-Am-4,6-DNB; SP: system peak. Reprinted with permission from Ref. [30].

35.3 CONCLUSION

This chapter demonstrated that microchip electrophoresis reached maturity and is appropriate for analysis of nitrated explosives. However, to create easy-to-operate field portable instruments for pre-blast explosive analysis would require incorporation of world-to-chip interface, which would be able to continuously sample from the environment. Significant progress towards this goal was made and integrated on-chip devices which allow microfluidic chips to sample from virtually any liquid reservoir were demonstrated [25,31].

ABBREVIATIONS

2-Am-4,6-DNT: 2-amino-4,6-dinitrotoluene,
4-Am-2,6-DNT: 4-amino-2,6-dinitrotoluene,

2,4-DAm-NT:	2,4-diaminonitrotoluene,
2,6-DAm-NT:	2,6-diaminonitrotoluene,
1,3-DNB:	1,3-dinitrobenzene,
2,4-DNP:	2,4-dinitrophenol,
2,3-DNT:	2,3-dinitrotoluene,
2,4-DNT:	2,4-dinitrotoluene,
2,6-DNT:	2,6-dinitrotoluene,
3,4-DNT:	3,4-dinitrotoluene,
EPA:	U.S. Environmental Protection Agency,
2-HADNT:	2-hydroxylamino-4,6-dinitrotoluene,
4-HADNT:	4-hydroxylamino-2,6-dinitrotoluene,
HMX:	octahydro-1,3,5,7-tetranitro-1,3,5,7-tetrazocine,
MMA:	monomethylammonium,
NB:	nitrobenzene,
2-NT:	2-nitrotoluene,
3-NT:	3-nitrotoluene,
4-NT:	4-nitrotoluene,
PETN:	pentaerythritol tetranitrate,
PMMA:	poly(methylmethacrylate),
RDX:	hexahydro-1,3,5-trinitro-1,3,5-triazine,
SDS:	sodium dodecyl sulfate,
Tetryl:	methyl-2,4,6-trinitrophenylnitramine,
1,3,5-TNB:	1,3,5-trinitrobenzene,
TNT:	2,4,6-trinitrotoluene.

ACKNOWLEDGMENTS

Martin Pumera is grateful for the financial support from the Japanese Ministry for Education, Culture, Sports, Science and Technology (MEXT) through ICYS program.

REFERENCES

1 U.S. Department of the Treasury, Bureau of Alcohol, Tobacco and Firearms, *Commerce in Explosives; List of Explosive Materials*, http://www.atf.treas.gov/pub/fire-explo_pub/listofexp.htm
2 *Test Methods for Evaluating Solid Waste*, Proposed Update II, Method 8330, U.S. Environmental Protection Agency, Washington, DC, EPA Report SW846, 3rd ed., November 1992.
3 D. Royds, S.W. Lewis and A.M. Taylor, *Talanta*, 67 (2005) 262–268.

4 W. Thormann, I.S. Lurie, B. McCord, U. Marti, B. Cenni and N. Malik, *Electrophoresis*, 22 (2001) 4216–4243.
5 E. Verpoorte, *Electrophoresis*, 23 (2002) 677–712.
6 J. Wang, *Anal. Chim. Acta*, 507 (2004) 3–10.
7 M. Pumera, *Electrophoresis*, 27 (2006) 244–256.
8 S.A. Oehrle, *J. Chromatogr. A*, 745 (1996) 233.
9 N.J. Petersen, K.B. Mogensen and J.P. Kutter, *Electrophoresis*, 23 (2002) 3528–3536.
10 S.R. Wallenborg and C. Bailey, *Anal. Chem.*, 72 (2000) 1872–1878.
11 A. Bromberg and R.A. Mathies, *Anal. Chem.*, 75 (2003) 1188–1195.
12 A. Tallec, *Ann. Chim.*, 3 (1968) 345–349.
13 D.C. Schmelling, K.A. Gray and P.V. Kamat, *Environ. Sci. Technol.*, 30 (1996) 2547–2555.
14 J. Barek, M. Pumera, A. Muck, M. Kadeřábková and J. Zima, *Anal. Chim. Acta*, 292 (1999) 141–146.
15 J. Wang, F. Lu, D. MacDonald, J. Lu, M.E.S. Ozsoz and K.R. Rogers, *Talanta*, 46 (1998) 1405–1412.
16 K. Bratin, P.T. Kissinger, R.C. Briner and C.S. Bruntlett, *Anal. Chim. Acta*, 130 (1981) 295–311.
17 I.M. Kolthoff, J.J. Lingane, Polarography, Vol. 2, Interscience Publishers, New York, 1952.
18 Q. Lu, G.E. Collins, M. Smith and J. Wang, *Anal. Chim. Acta*, 469 (2002) 253–260.
19 J. Wang, B. Tian and E. Sahlin, *Anal. Chem.*, 71 (1999) 5436–5440.
20 A. Hilmi and J.H.T. Luong, *Environ. Sci. Technol.*, 34 (2000) 3046–3050.
21 A. Hilmi and J.H.T. Luong, *Anal. Chem.*, 72 (2000) 4677–4682.
22 J. Wang and M. Pumera, *Talanta*, 69 (2006) 984–987.
23 J. Wang, G. Chen, M.P. Chatrathi, A. Fujishima, D.A. Tryk and D. Shin, *Anal. Chem.*, 75 (2003) 935–939.
24 J. Wang, M. Pumera, M.P. Chatrathi, A. Escarpa, M. Musameh, G. Collins, A. Mulchandai, Y. Lin and K. Olsen, *Anal. Chem.*, 74 (2002) 1187–1191.
25 G. Chen and J. Wang, *Analyst*, 129 (2004) 507–511.
26 S.R. Wallenborg and C.G. Bailey, *Anal. Chem.*, 72 (2000) 1872–1878.
27 C.A. Fyfe, C.D. Malkiewich, S.W.H. Damji and A.R. Norris, *J. Am. Chem. Soc.*, 98 (1976) 6983–6988.
28 C.F. Bernasconi, *J. Am. Chem. Soc.*, 92 (1970) 129–137.
29 J. Wang and M. Pumera, *Anal. Chem.*, 75 (2003) 341–345.
30 J. Wang and M. Pumera, *Anal. Chem.*, 74 (2002) 5919–5923.
31 S. Attiya, A.B. Jereme, T. Tang, G. Fitzpatrick, K. Seiler, N. Chiem and D.J. Harrison, *Electrophoresis*, 21 (2001) 318–327.

Chapter 36

Microfluidic-based electrochemical platform for rapid immunological analysis in small volumes

Joël S. Rossier and Frédéric Reymond

36.1 INTRODUCTION

This chapter presents an approach to perform enzyme linked immunosorbent assays (ELISA) in a microfluidic format with electrochemical detection. This field of analytical chemistry has shown a strong activity in recent years, and many reports have presented the use of capillary-sized reactors for running immunoassays either in homogeneous format (where the antigen–antibody complex and the labelled revelation reagents are separated prior to detection, as for instance by capillary electrophoresis [1–3]) or in heterogeneous format (where the antibody is immobilised on the inner surface of the microsensor device [4] or on microbeads [5,6]).

Despite the numerous interesting activities in this field, this chapter is not intended to provide a systematic review; various articles describe the most recent developments in a systematic way, and the reader is invited to complete the present overview by reading some recent reports such as the one of Diaz-Gonzalez *et al.* [7] or from Heineman's group [8].

This chapter focuses on the approach we followed for developing a novel electrochemical sensor platform based on disposable polymer microchips with integrated microelectrodes for signal transduction. It presents the development of the so-called *Immuspeed*™ technology, which is dedicated to quantitative immunoassays with reduced time-to-results as well as sample and reagent volumes. Prior to presenting the specific characteristics of *Immuspeed*, the basic principles integrated in this platform are first presented and illustrated with reference to

Comprehensive Analytical Chemistry 49
Alegret and Merkoçi (Eds)
Volume 49 ISSN: 0166-526X DOI: 10.1016/S0166-526X(06)49036-X

examples from the literature reviewing the features enabling reduced incubation times, easy handling devices providing a quantitative answer in a few minutes, while, generally speaking, standard ELISA techniques require hours of incubation and cumbersome manipulation.

36.1.1 Basic principle of the standard ELISA technique

Immunosorbent affinity assays were invented more than 30 years ago by Diamandis and Christopoulos [9], and they rely on the immobilisation of a biological recognition entity (generally an antibody or an antigen) in order to capture an analyte of interest. This immunosorbent assay format has become very robust and it is strongly appreciated because of its outstanding degree of selectivity and sensitivity. The ELISA format indeed takes advantage of the fantastic ability of nature to develop very specific molecules (antibodies) in higher living organisms. This property has been used to engineer antibodies with high affinity to molecules that have to be detected selectively and quantitatively. Obtaining molecules with higher specificity has been one of the most successful research areas in biology during the last decades. Various bio-engineered molecules (among which polyclonal and monoclonal antibodies or antibody fragments) are now widely used and recognised as the best way to diagnose most of the current diseases in modern medical diagnostics.

Affinity assays are not limited to the use of antibodies. Indeed, other kinds of molecules such as oligonucleotides or imprinted polymers can serve for the specific capture of a given marker. All these molecular moieties can be immobilised on a surface that is then placed in contact with the molecule to be detected.

Various strategies have been developed for the immobilisation of the affinity partner, among which physical adsorption as well as covalent binding are used, either to coat the walls of the reaction chamber, or to functionalise beads that are put in contact with the analyte solution or even in some cases to create a biologically modified membrane or electrode.

The methodology and assay protocols described hereinafter will be limited to the primary applications of the present microfluidic platform and will thus concentrate mainly on the ELISA technique, where the affinity reaction incorporates an antibody to recognise the diagnostic molecule.

36.1.2 Specific feature of ELISA in microtitre plates

In general, ELISA tests have been implemented in meso-scale instrumentation based on the microtitre plate format, which has become a standard, very widely spread configuration. The analyses are usually performed with a protocol that enables the thermodynamic equilibrium of the immunoreaction to be reached at each step of the assay. In this manner, the capture efficiency is optimised and the obtained results are often very satisfactory in terms of sensitivity and reproducibility. In order to further increase the performances and throughput of these tests, fully automatic robotised stations have been developed, thereby reducing manipulation errors such as dilution or pipetting imprecision, for instance.

The ELISA tests are performed using different incubation steps, depending on the assay format, namely sandwich, competitive or inhibition immunoassays that correspond to the way the analyte is captured and further revealed, which is generally dictated by the nature of the analyte.

What characterises the different incubation steps is the time required to reach thermodynamic equilibrium between an antibody and an antigen in the standard format of microtitre plates. In fact the volume used in each of the incubation steps has been fixed between 100 and 200 μL to be in contact with a surface area of approx. 1 cm^2 where the affinity partner is immobilised. The dimensions of the wells are such that the travel of the molecule from the bulk solution to the wall (where the affinity partner is immobilised) is in the order of 1 mm. It must be taken into account that the generation of forced convection or even of turbulence in the wells of a microtitre plate is rather difficult due to the intrinsic dimensions of the wells [10]. Indeed, even if some temperature or shaking effects can help the mass transport from the solution to the wall, the main mass transport phenomenon in these dimensions is ensured by diffusion.

36.1.3 Analysis time in diffusion-controlled assays (Nernst–Einstein diffusion rule)

When taking into account the typical diffusion coefficient of a protein (typically about 10^{-10} m^2/s) and the average distance along which it has to diffuse to reach the wall where the affinity partner is immobilised (i.e. about 1 mm in a 96-well plate), the time required for capturing the analyte can be easily approximated to ~3600 s using the Nernst–Einstein relation given by the following equation [11]:

$$L = 2\ \mathrm{sqrt}(Dt) \quad (36.1)$$

This long time is therefore necessary to statistically bring all molecules from the bulk solution towards the surface in a standard microtitre plate well. In case of rapid thermodynamic reaction, which characterises most of the tests run in routine *in vitro* diagnostics (IVD), either the reduction of the diffusion distance or the increase of the mass transport efficiency through turbulent flow with micro- or nanoparticles enable a dramatic reduction of the assay time from typically 3 h to a few tens of minutes. Many clinical analysers in hospitals indeed work with microbeads, for instance the Elecsys™ from Roche, the Vidia™ from BioMerieux or the Luminex instruments. Other examples of rapid immunological reactions favoured by an efficient mass transport are to be seen in the lateral flow assays where rapid assays can be obtained even by unskilled persons, such as pregnancy testing kits. (Many rapid diagnostic lateral flow devices are commercially available, for instance, at Unipath.) It should be noted however that the large majority of lateral flow assays are not quantitative but only provide a yes/no answer.

Our approach has been to keep the standard ELISA chemistry, with the specificity to keep each assay step at equilibrium while reducing dramatically the time of diffusion of the molecules from the bulk to the wall where the affinity partner is immobilised. When the Nernst–Einstein diffusion distance relation is drawn for a protein of medium size (14 kDa) as in Fig. 36.1, it is interesting to see that such a molecule will need as long as 1 h to diffuse over a distance of 1 mm, whereas only a few minutes are required to move along a distance of 0.1 mm and only 5 s for 0.02 mm. The diffusion time thus dramatically decreases with the distance between the affinity partners. When the reaction of the antigen–antibody complex formation is diffusion-controlled as in many diagnostic kits, it is expected that the immunoassay incubation can be strongly reduced as well.

The Nernst–Einstein relationship plotted in Fig. 36.1 motivated the development of DiagnoSwiss' technology, namely reducing the analysis time and reaction volumes by replacing the commonly used microtitre plate wells with typical dimensions of 100 μL volume and 1 cm^2 surface area by a microchannel with a volume of 60 nL and a surface area of 0.03 cm^2. The microchannel is an isotropically etched microstructure (see Section 36.2) with a minimal cross-section dimension of 40 μm,

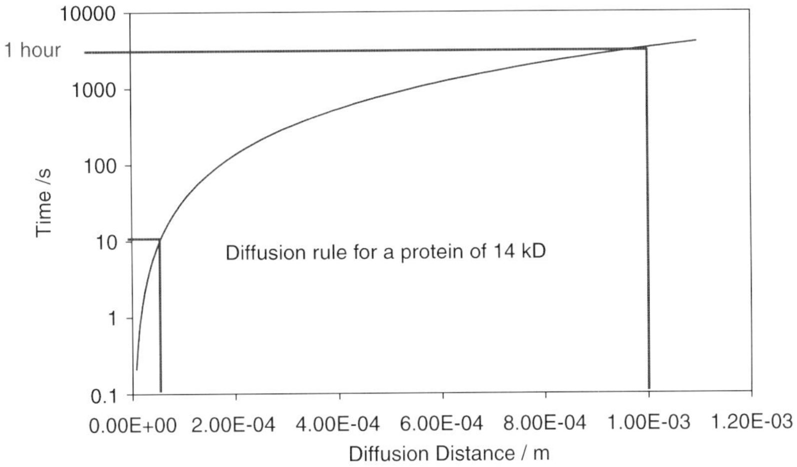

Figure 36.1. Nernst–Einstein relationship representing the time for a protein of 14 kD to diffuse through different distances (diffusion coefficient $10^{-10}\,m^2/s$).

meaning that the largest distance between a molecule in the bulk solution and the wall where the affinity partner is immobilised is no more than 20 μm. In our typical example, a molecule of 14 kDa would take no more than a few seconds to reach the surface coated with the immobilised partner. This rule shows that when the recognition of the antigen by the antibody is diffusion controlled, the theoretical time gain from a microtitre plate to such a microchannel ELISA would be a factor of more than 100. Even if only a part of this gain can be realised in practice, this is of great interest for implementation in an immunoassay analyser.

36.1.4 Capillary immunoassays

The use of capillaries to run immunoassay reactions was presented in the 1980s and 1990s by Halsall group [8] for the detection of atrazine. The system was based on a flow injection analysis protocol where different solutions (sample, conjugate) were injected and reacted in the capillary itself; the affinity-captured enzyme conjugate then reacted with a substrate solution injected in the capillary prior to detection. The reacted substrate solution was then pushed towards an electrode for amperometric detection. The authors showed that the incubation time reached equilibrium after about 20 min, but they worked with a capillary of 360 μm diameter. In order to really reduce the incubation time below 1 min, we developed disposable polymer chips with an internal diameter of 40 μm (almost one order of magnitude smaller than that in the pioneering work of Haslall *et al.* [12]) and integrated the electrodes inside the chip, which enables detection at the location of the enzyme reaction and therefore allows us to follow the enzyme kinetics. The electrodes can also be used for monitoring the flow rate inside the channel and serve as a foolproof system as will be described hereafter.

36.2 POLYMER MICROFLUIDIC-BASED ELISAs WITH ELECTROCHEMICAL DETECTION

The system presented here consists of a disposable cartridge called Immuchip^{TM}, an instrument called Immuspeed^{TM} and its associated software Immusoft^{TM}.

36.2.1 Immuchip^{TM}: a disposable cartridge with polymer microfluidic electrochemical cells

The cartridge where the immunoassay takes place is composed of a polyimide flex material in which a microchannel is etched and in which electrodes as well as inlet and outlet access holes for the liquid are integrated using a proprietary fabrication process [13] (Fig. 36.2). The microchannel is etched by successive fabrication steps including photo-patterning, copper etching, plasma etching and finally gold electroplating to give a favourable electrode material. A detailed description of the chip fabrication has been reported previously [14]. The microchannel is then sealed by means of a lamination and assembled

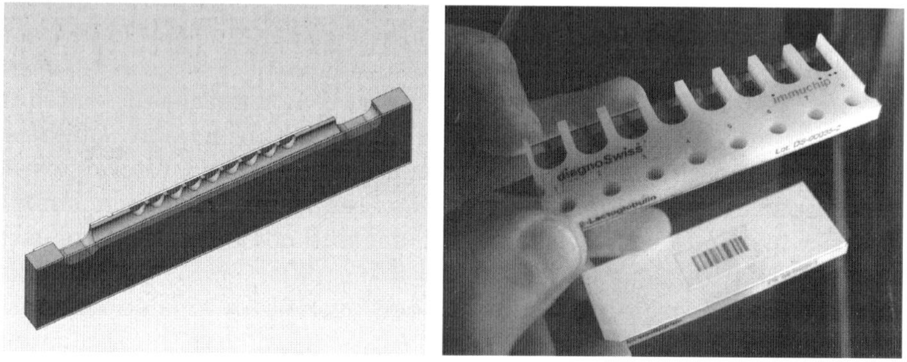

Figure 36.2. Left: cross section of a microchannel with an inlet port, a channel, an outlet port and integrated gold electrodes. Right: a cartridge containing eight parallel microchannels, embedded in a white polymer support, serving as rigidifier, guide and reservoir.

with a holder cartridge for easy handling, registration and to have a reservoir for the solution. Each of the eight parallel microchannels is a true electrochemical cell composed of an array of working electrodes, a reference and a counter electrode [15].

An application example of Immuchip™ for the analysis of Interleukin 1B by enzyme-linked immunosorbent assay is described in Procedure 50 (see in CD accompanying this book).

36.2.2 Immuspeed™: a bench-top instrument for microfluidic assays

The instrument is composed of an interface to plug the disposable cartridge, a temperature controller, a multichannel pumping device and valves as well as a multiplexer electrochemical detector for sequential or parallel detection of the amperometric events occurring in each of the eight channels; for some protocols, it can also serve to deliver different reagents such as the secondary antibody or the substrate and thus provide fully automatic assays.

36.2.3 Principles of microfluidic ELISAs with electrochemical detection

The microchannels are coated with antibodies that are ready to capture antigen entering the channels. For a good depletion of the volume of solution that passes through the channel, we use a multi-loading procedure where each aliquot of solution present in the chip has time to be depleted by diffusion. Thanks to the Nernst–Einstein diffusion law, it is possible to calculate that the depletion of a protein of 14 kDa would take less than 10 s by diffusion, taking into account that the antibody–antigen reaction is fast. After 10 s, the pump will thus aspirate another aliquot of about 85 nL of new solution through the chip at 5 μL/min during 1 s and the solution is stopped again for 10 s. Such multi-loadings enable pre-concentration of the antigen in the chip and therefore decreases the limit of detection of the system. The signal is then further amplified by the enzyme and the reaction is followed by amperometry as will be explained later (Fig. 36.3).

36.2.4 Microfluidics control thanks to integrated electrochemical flow sensors

In order to obtain a foolproof system, it is necessary to have indicators for each event occurring in the sensor during an assay. To this end, it is

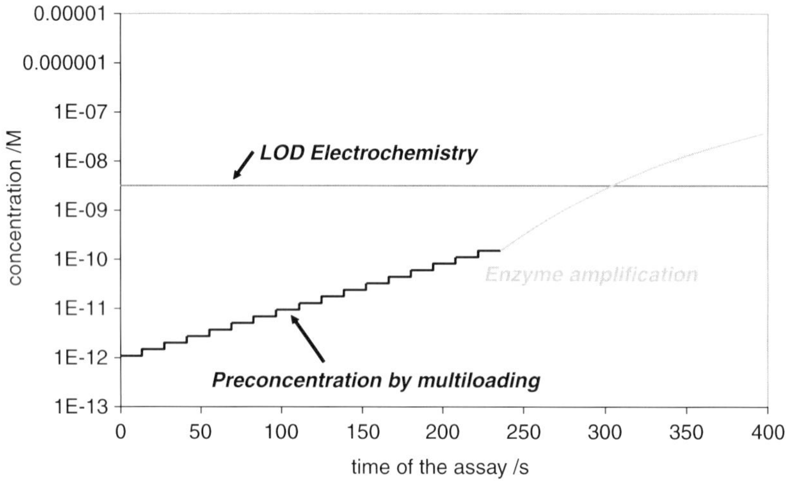

Figure 36.3. Typical behaviour of the concentration of the antigen by multi-loading and the enzyme amplification that enables to obtain a signal detectable by electrochemistry even for a starting antigen concentration of 1 pM.

for instance important to know whether the microchannel has been properly filled at each loading step and hence to control that it had the same amount of solution as expected. If a particle blocks the flow regeneration, this would block the channel and could lead to false-positive results. In order to identify such events, each channel is equipped with its own embedded flow sensor. The electrode integrated within the microchannels can be used as a flow sensor, which enables each of the fluidic events occurring in the chip to be followed, for instance by detecting a tracer redox molecule [16,17]. If a fresh solution containing oxygen enters the channel, a peak of current will be detected, proportional to the flow rate at the power one third as already presented elsewhere [18]. At each load the peak will appear, acting therefore as a quality control that the number of the pre-concentration runs is the one set in the assay protocol (Fig. 36.4).

36.2.5 Enzymatic detection by means of amperometry

Similar to many conventional immunoassays, the detection of the analyte of interest is performed indirectly by formation of a complex using a secondary affinity partner. In order to increase the sensitivity of the detection, this secondary affinity partner can be labelled with an enzyme. This enzyme thus serves as catalyst for the conversion of a

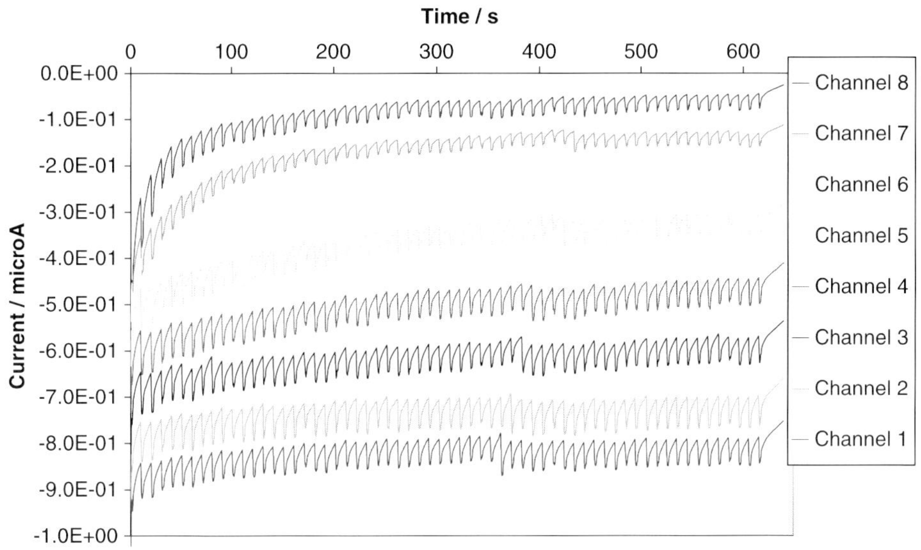

Figure 36.4. Results of flow monitoring during multiple sample loadings multi-loadings in eight parallel microchannels by means of amperometric detection of dissolved oxygen. It is interesting to note that each time the pump aspirates a new aliquot of antigen-containing solution, a negative (reduction) peak appears on the graph. If one channel is blocked, the problem will be directly identified and the channel results excluded from the final detection.

substrate into a detectable product. For electrochemical detection purposes, the substrate is chosen in such a manner that an electroactive product is created by the enzymatic reaction, so that a current can be generated and measured by oxidation or reduction of this product. In the general scheme of the assay, this substrate selection is the single difference compared to conventional immunoassays where luminescence is used for detection of the analyte, since the affinity partners as well as the enzymatic label can remain the same in both cases.

In order to detect the analyte specifically, a complex has to be formed first. To this end, the revelation moiety (e.g. an enzyme-labelled antigen or antibody) is for instance incubated in the chip so as to bind to the analyte that has previously been captured within the microchannel. In another scheme, the analyte solution is first mixed with the revelation moiety, and the formed complex is then incubated in the chip in order to be captured on the bed of antibodies coating the walls of the microchannel. After a washing step (to remove the excess affinity partner), the microchannel is filled with the substrate which shall thus react with the enzyme. In the case where the enzyme is alkaline phosphatase (ALP), the most commonly used substrates are *p*-aminophenyl phosphate [4,15,19–21], *p*-nithophenyl phosphate [20] and 4-methylumbelliferyl phosphate [22] that are transformed into *p*-aminophenol, *p*-nitrophenol and 7-hydroxy-4-methylcoumarin, respectively, which are all detectable by electrochemistry.

In electrochemical enzyme immunosensors, amperometry is the most widely used technique for determining the concentration of the product of the enzymatic reaction and hence of the analyte of interest. In our detection scheme, this product concentration is measured at several time intervals, so as to follow the catalytic activity of the enzyme label. Depending on the enzyme concentration, the current will increase faster or slower, according to the Michaelis–Menten kinetics. For small enzyme concentrations, the product generation follows a pseudo first-order kinetics (so that the enzyme and product concentrations are related by a linear relationship), whereas a second-order reaction takes place at larger enzyme concentrations. In this manner, the final signal determination is obtained by extracting the value of the slope at the origin of the current versus time curve.

As an example, Fig. 36.5 presents a calibration curve for the determination of Interleukin 1B in plasma sample. The insert in Fig. 36.5 shows the row data obtained for the parallel detection of *p*-aminophenol in a series of eight individually addressable microchannels as a function of time, and the resulting calibration points are given by the slope at the origin of these different curves.

36.3 IMMUSOFT^{TM}: A PROGRAM FOR COMPUTER-DRIVEN MICROFLUIDIC ASSAYS

Immusoft^{TM} is a software that has been developed to perform computer-driven assays in our microchips. This software has a user-friendly graphical user interface, and it enables control of the pump, the valves and the electrochemical detection system, as well as the development of specific assay protocols, the running of simultaneous or sequential experiments in eight parallel microchannels, the automatic read-out of the results and the processing of the obtained data. These different functions are managed by way of three main menus, named Method, Analysis and Results, and the software also comprises two additional items dedicated to the setting of the computing parameters and to the maintenance of the instrumentation.

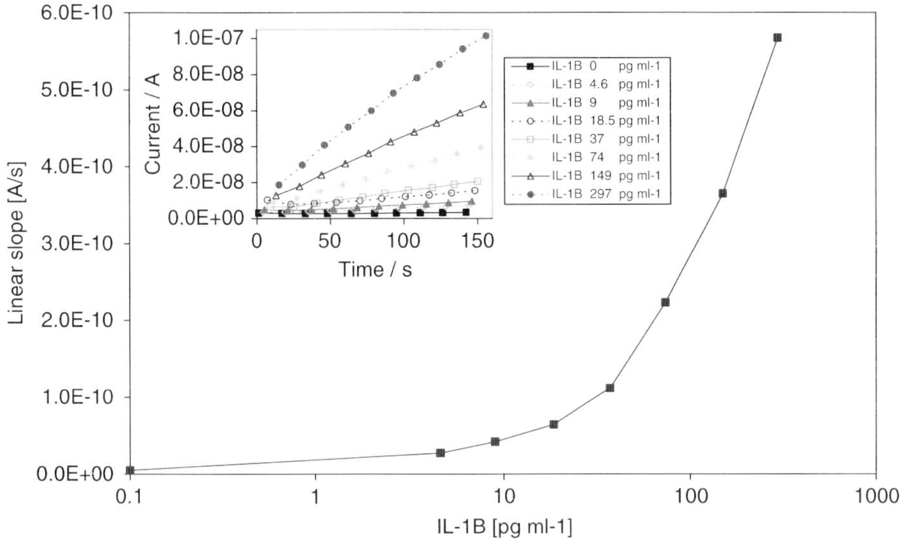

Figure 36.5. Calibration curve obtained for the determination of Interleukin 1B in eight parallel microchannels by sandwich immunoassay. The insert shows the currents resulting from the oxidation of *p*-aminophenol as a function of time.

In order to provide a deeper insight into the functions and possibilities of this integrated microchip platform, this section describes the three main menus in more detail.

36.3.1 Method creator: establishment of assay protocols

The method part of Immusoft^{TM} enables the design of the various assay steps in order to further perform the analysis automatically. Actually, the user can design an assay by adding fluidic and detection steps according to the desired assay protocol. For instance, if the assay comprises only two steps, such as the incubation of the sample mixed with a conjugate and the substrate addition, two fluidic steps are programmed, namely: one for the incubation of the sample/conjugate with a given number of sample loadings, thereby enabling pre-concentration of the enzyme on the surface of the chip, and a second fluidic step (for instance induced as a back-flow from the instrument) which serves both to wash out the non-reacted sample/conjugate solution and to introduce the enzyme substrate, which is then comprised in the washing buffer solution. Further to these fluidic steps, the protocol shall then be completed by adding a measurement step, in which the detection parameters are introduced. For chrono-amperometric detection for instance, the detection parameters include the value of the potential to apply, the duration of each measurement and, in order to follow the kinetics of the enzyme reaction, the time interval between two chrono-amperometric measurements as well as the number of such measurement cycles. For the development of new assays or of new protocols, it may be useful to check the quality of the assay by repeating the detection. In the Method menu, it is thus possible to ask the instrument to fill the microchannels with fresh substrate solution after a first detection cycle and to perform a second detection. In the example shown in Fig. 36.6, the assay protocol comprises a first detection where 10 chrono-amperometric measurement points are performed, and this first detection is followed by the introduction of fresh substrate and by a second detection for which five measurement points are taken.

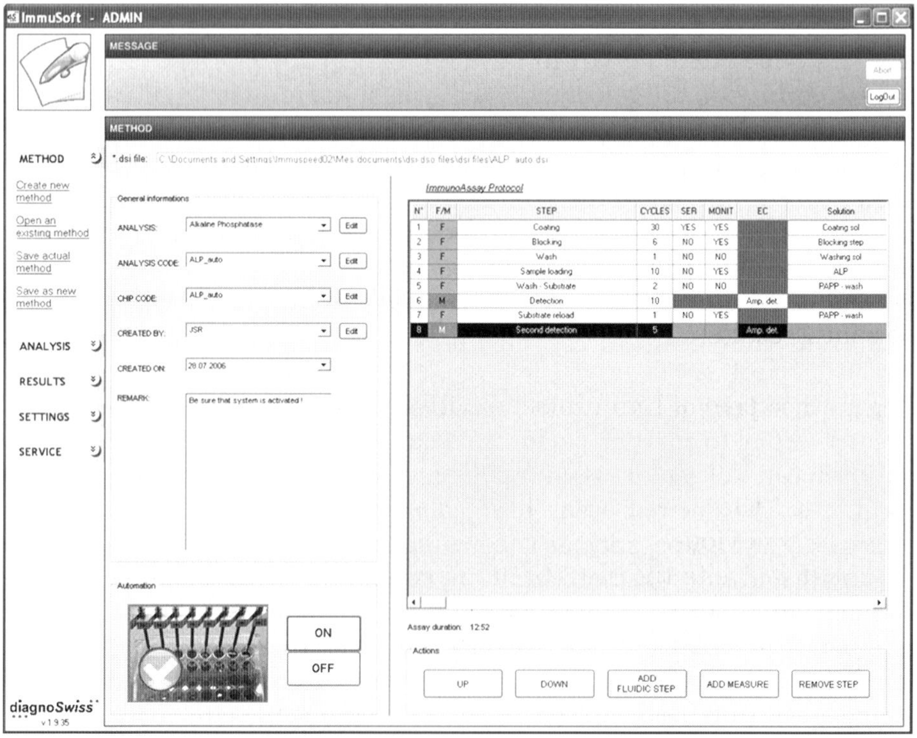

Figure 36.6. Snapshot of the Method creator of the Immusoft computer programme serving for the establishment of assay protocols. The figure shows on the right-hand side the various steps of the protocol developed for the assay of alkaline phosphatase (ALP), in which the functionalisation of the microchannels (coating and blocking steps) is directly integrated in the assay progress.

If desired, the functionalisation of the microchip with the capture antigen/antibody can be directly integrated within the assay protocol. As illustrated in Fig. 36.6, which shows an example of the Method creation window used for ALP assays, the protocol is designed such as to integrate the coating of the microchannels with avidin and biotin-anti-ALP, as well as the blocking against non-specific adsorption with bovine serum albumin, followed by a washing with a PBS buffer solution. The proper assay thus starts after these three chip preparation steps, and Immusoft^{TM} automatically indicates the actions to be performed by the experimenter as well as the volumes of reagents to dispense at each step of the chip functionalisation and of the subsequent analysis.

All the functions set during the establishment of the assay protocol are then recorded as a given method and, as will be presented in the next section, this method is called by the Analysis menu when the assay procedure is launched so that the assays are carried out in an interactive manner for the user.

It should also be noted that the Method part of the software integrates all the data concerning the nature of the test to perform, the reagent used, the references of the microchip used, the experimenter's name as well as the date and time of the assay. All these data are then automatically stored in the final report concerning the experiments run with this protocol, which greatly facilitates the discovery of possible experimental errors as well as the traceability of the obtained results.

36.3.2 Analysis menu: computerised assay realisation and control

In order to run the assays, the Analysis menu of Immusoft^{TM} is selected, and the user has to choose the correct protocol for the experiment as previously programmed and stored in the Method menu. The user deposits the sample/conjugate in the chip reservoirs and the instrument proceeds to the reagent introduction and sample loading following the desired protocol. When an experiment is launched, the Analysis part of the software directly calculates all the required volumes of solution (coating agents, sample, washing buffer and enzymatic substrate), and it directly informs the user about the status of the experiment and about the manipulations to undertake. Each manipulation has to be confirmed by the user, who also has the possibility to follow all the fluidic steps so as to make sure that they have been properly fulfilled. During the entire assay duration, the software displays a window which shows on-line the monitoring of the fluid flows within the microchannels, so as to check that the microfluidics runs correctly in each microchannel and hence that they are filled with the appropriate volume of reagent, washing and/or sample solution as presented above in Section 36.2.4.

As an example, Fig. 36.7 shows the current measured during the coating step of the ALP assay described above, in which 30 loadings have been selected to functionalise the microchannels. As can be seen in Fig. 36.7, the measured current exhibits a kind of peak shape, which results from the introduction of new sample solution at given time intervals. Indeed, each time the pump is activated, the current drops due to the induced fluid flux and the concomitant renewal of the reporter molecule (dissolved oxygen in the present case) at the vicinity of the electrodes. Once the pump is stopped, the current increases rapidly

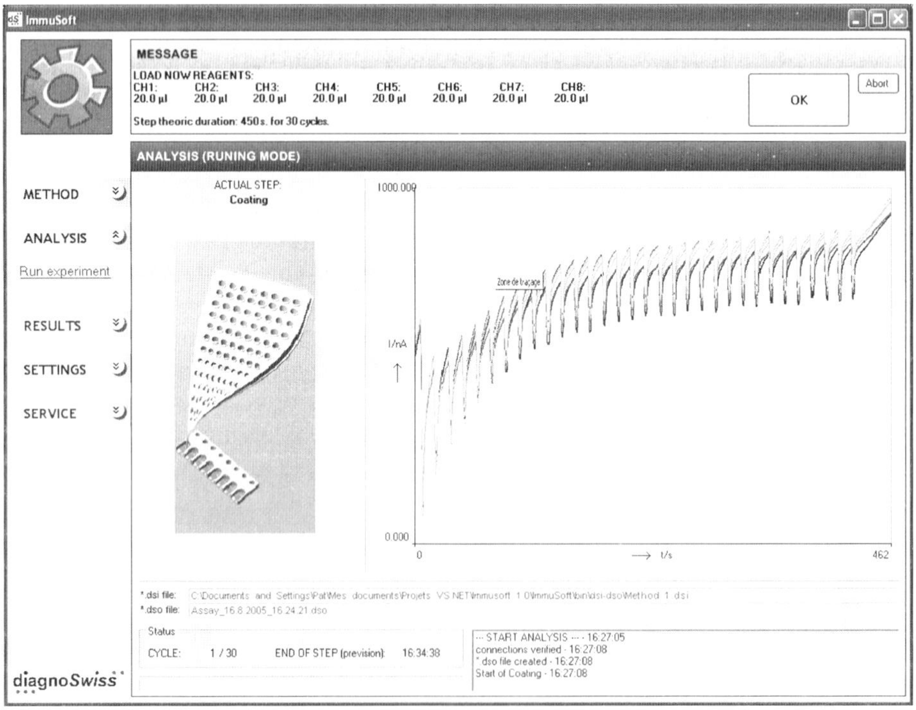

Figure 36.7. Example of Immusoft^{TM} window appearing in the Analysis menu during the processing of the fluidic steps of an assay. The figure shows the electrochemical monitoring of the fluid flux within the microchannels obtained during the 30 loading steps of the microchannel coating procedure chosen for alkaline phosphatase assays.

and tends to approach a saturation which is indicative of the limitation of the measured current due to the diffusion towards the electrodes in a confined environment. The peak shape of the signal still allows identification of the number of loads in each channel and verification that they have been properly filled.

In the last step of an assay, i.e. during the electrochemical detection of the electroactive species of interest, a new window appears in order to let the experimenter follow the detection on-line. As presented in Fig. 36.8, the software has been designed to perform chrono-amperometric measurements of the desired analyte. On the left part of the window the raw measurement data that provide the evolution of the current for each microchannel that is measured during a time period of generally 2 s appears. With the multi-potentiostat used, the eight channels are

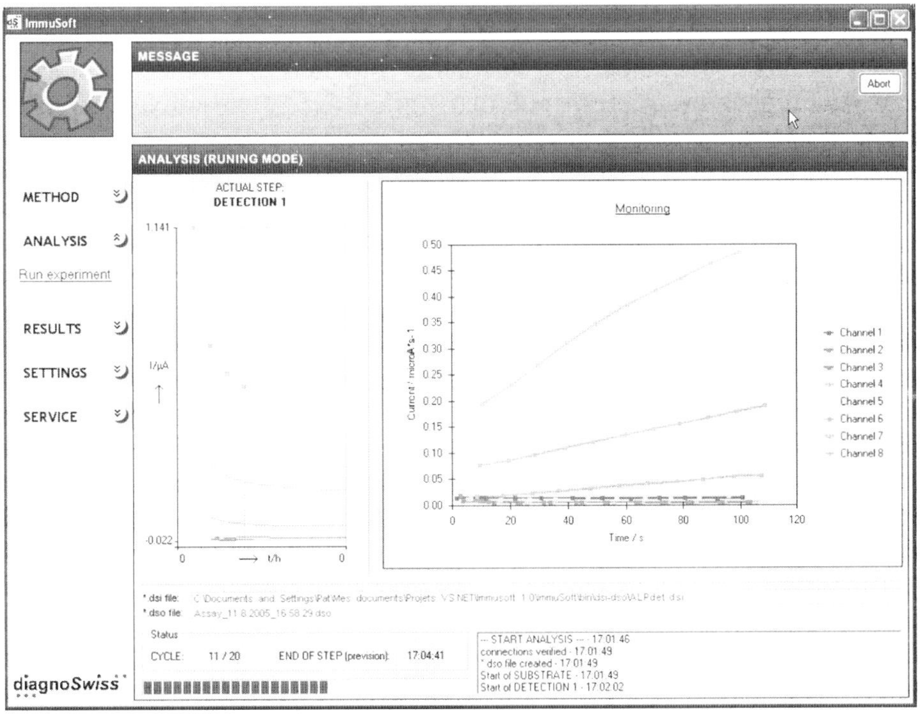

Figure 36.8. Example of Immusoft^{TM} window appearing during the electrochemical detection step of an assay. On the left-hand side, the figure shows the row amperometric data, from which the deduced faradaic current is automatically displayed for each microchannel in the current versus time plot shown on the left-hand side of the window.

probed sequentially, and the measurement is repeated over the number of cycles selected by the user. On the right-hand side of the window, the value of the faradaic current measured at each cycle is directly plotted as a function of time. In applications relying on enzymatic amplification of the analyte to be detected, the concentration of the electroactive substrate increases with time, as reflected in the example of Fig. 36.8 in which *p*-aminophenyl-phosphate is hydrolysed by ALP into *p*-aminophenol, which is then oxidised at the electrodes during the detection.

36.3.3 Results menu: measurement display and data processing

At the end of the measurement, the Results part of the Immusoft^{TM} programme recalls the time evolution of the measured currents, which follows the kinetics of the enzymatic reaction. As mentioned above, the slope at origin of these current versus time curves (noted dI/dt) is directly proportional to the enzyme concentration and hence to the concentration of the captured analyte(s). This part of the software is thus designed to calculate these slopes, and it enables automatic plotting of these values as a function of the effective analyte concentrations. As illustrated in the example of Results window of Fig. 36.9, which shows the data obtained for β-lactoglobulin tests in an eight-channel chip, the left-hand-side plot displays the values of the faradaic currents measured during the various detection cycles chosen in the assay protocol, while the right-hand-side diagram provides the final results. In this example, four data points in the pM range have been assayed in duplicate, and the values of the slopes of the measured current–time curves determine the effective sample concentrations.

With known analyte concentrations, the processed data provide calibration points, and Immusoft^{TM} comprises fitting procedures to deduce the corresponding calibration curve that can be stored in the program for further experiments. Different calibration curves can be stored depending for instance on the assay protocol, on the specific features of the chip used or on the medium in which the assay is performed. In most cases, the calibration is performed with six independent chips of eight channels and cumulated in order to get a stable batch calibration. Then the results can be referred to this internal batch calibration. For routine control, one calibration each week is recommended to be sure that the chemistry is still in the specifications (e.g. $\pm 10\%$ of inter-assay standard deviation).

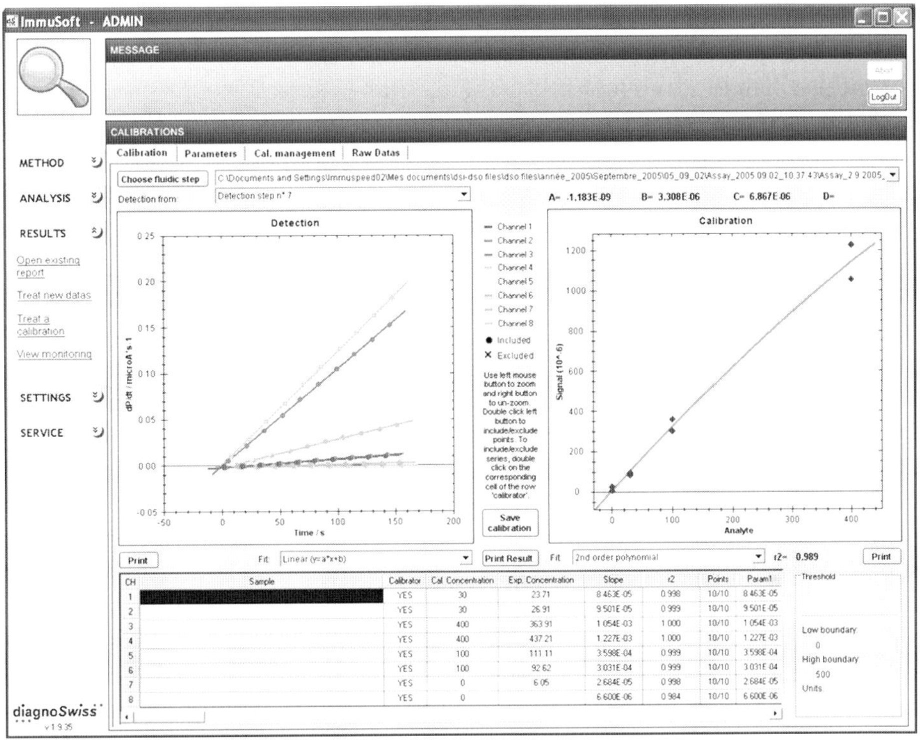

Figure 36.9. Example of Results window for the display of the current versus time curves measured in duplicates for β-lactoglobulin assays (left plot) and used for the automatic calculation of the corresponding slopes at origin and for the delivery of the final assay results including curve fitting (right plot).

36.4 PERFORMANCES EXEMPLIFIED WITH THE IMMUNOASSAY OF ALKALINE PHOSPHATASE

In order to demonstrate the performance of this electrochemical microimmunoassay platform in terms of limits of detection and dynamic range, a series of ALP tests has been conducted in 100 nL polyimide microchips. To this end, anti-phosphatase antibodies have first been immobilised on the surface of the microchannels at a concentration of 10 µg/mL in a flow-through mode (4 mL of anti-ALP solution pumped at 0.4 mL/min during 10 min) so as to saturate the microchannel surface by physical adsorption. Then, the surface was blocked with a 5% BSA in phosphate buffer in order to block the free sites remaining on the surface. Solutions of ALP at various concentrations (namely 0, 0.1, 1, 10 and 100 pM) were then injected and incubated during 9 min in the

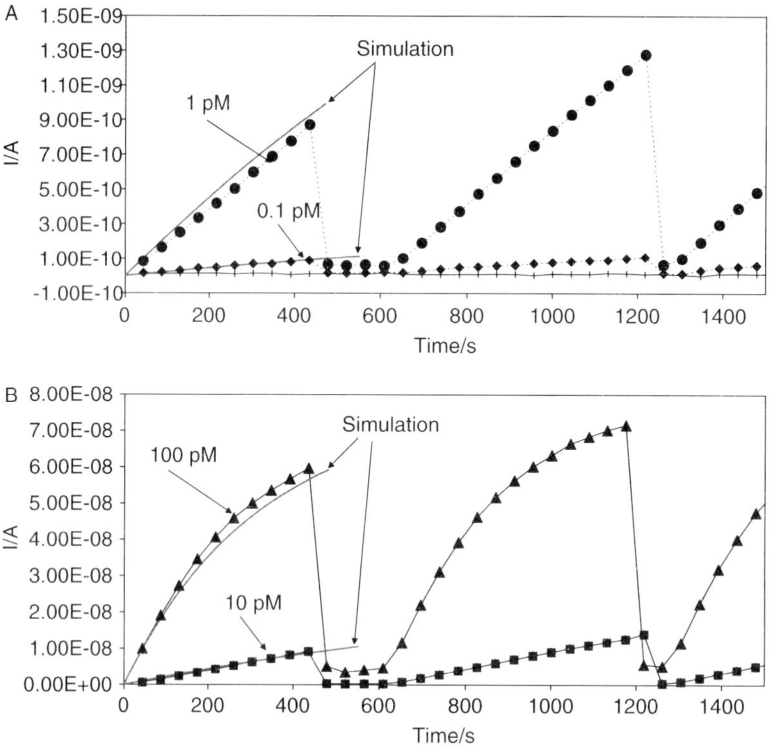

Figure 36.10. Comparison between calculated and experimental results for the detection of alkaline phosphatase in polyimide microchannels at various concentrations, namely: (A) 0, 0.1 and 1 pM and (B) 10 and 100 pM

eight-microchannel array. Finally, the microchannels were filled with the enzyme substrate solution (namely 10 mM *p*-aminophenyl phosphate in 0.5% Tween Tris buffer at pH 9), and the resulting product concentration detected by chrono-amperometric measurements at a potential of 250 mV versus Ag/AgCl. The time evolution of the currents obtained for these five concentrations has been measured in duplicate, and Fig. 36.10 shows the results obtained for one of these measurement series.

In order to ensure that the signal is effectively coming from the immobilised enzymes and not from possible enzymes remaining in the channel by lack of washing, it is possible to restart a detection after a given time (here once after 500 s and another one after 1250 s) by pumping fresh substrate through the channel, thereby reducing the product concentration to almost zero and to restart the detection. It can clearly be seen in Fig. 36.10 that at ALP concentrations as high as

100 pM, the signal reaches the limit of Michaelis–Menten kinetics, so that it is not growing linearly as for a pseudo first-order kinetics but follows a second-order reaction scheme. At lower ALP concentrations (namely below 10 pM), the signal evolves linearly over the entire duration of the detection, so that the reaction follows a pseudo first-order kinetics, as expected. It is interesting to note at this point that a small starting concentration of 0.1 pM of ALP can be detected in the chip with a signal which is significantly different from that measured for the blank point at 0 pM of ALP. It should also be noted that the current increase for this concentration after 500 s of detection is only 100 pA, but the detection is sufficiently sensitive to distinguish such low currents.

The results obtained for this ALP test demonstrate that this amperometric microchip platform has a good sensitivity (limit of detection lower than 0.1 pM for ALP) and that it works over a large dynamic range (at least four orders of magnitude in the present case).

In order to further validate these results, a theoretical model has also been developed in-house in order to mathematically describe the physicochemical processes occurring in the chip and hence to calculate the current that can be expected for the detection of ALP in microchip-based immunoassays with amperometric detection. For such multi-step assays, the analytical expressions governing the three basic phenomena involved for the generation of the final signal have been established by taking account of: (a) the Levich equation for forced convection of solution during in-flow incubations which enables the determination of the surface concentration of captured molecules; (b) the Michaelis–Menten expression for the calculation of the concentration of the generated product as a function of the duration of the enzymatic reaction; and (c) Fick's second law of diffusion, adapted for recessed electrodes in a finite environment as described by Bartlett and Taylor [23], in order to calculate the electrical current generated by the oxido-reduction of the product of the enzymatic reaction.

In order to illustrate the quality of this analytical model, the parameters have been optimised for the detection of ALP in our microchannels. As can be deduced from Fig. 36.10, the currents obtained from these analytical expressions are in very good agreement with the experimental data, and revealed to be valid over a large range of analyte concentrations (here from 0.1 to 100 pM). This model confirms that the measured signals correspond very well to the currents that can be expected for ALP determination in such an amperometric microsensor, and it constitutes a very useful tool for the optimisation of both the microchip features and the parameters of the assay protocols.

36.5 CONCLUSION AND PERSPECTIVES

The use of ELISA is broad and it finds applications in many biological laboratories; over the last 30 years many tests have been developed and validated in different domains such as clinical diagnostics, pharmaceutical research, industrial control or food and feed analytics for instance. Our work has been to redesign the standard ELISA test to fit in a microfluidic system with disposable electrochemical chips. Many applications are foreseen since the biochemical reagents are directly amenable from a conventional microtitre plate to our microfluidic system. For instance, in the last 5 years, we have reported previous works with this concept of microchannel ELISA for the detection of thromboembolic event marker (D-Dimer) [4], hormones (TSH) [18], or vitamin (folic acid) [24]. It is expected that similar technical developments in the future may broaden the use of electroanalytical chemistry in the field of clinical tests as has been the case for glucose monitoring. This work also contributes to the novel analytical trend to reduce the volume and time consumption in analytical labs using lab-on-a-chip devices. Not only can an electrophoretic-driven system benefit from the miniaturisation but also affinity assays and in particularly immunoassays with electrochemical detection.

REFERENCES

1 J. Wang, A. Ibanez and M.P. Chatrathi, *Electrophoresis*, 23 (2002) 3744–3749.
2 N. Chiem and D.J. Harrison, *Anal. Chem.*, 69 (1997) 373–378.
3 N.H. Chiem and D.J. Harrison, *Clin. Chem.*, 44 (1998) 591–598.
4 J.S. Rossier and H.H. Girault, *Lab Chip*, 1 (2001) 153–157.
5 K. Sato and T. Kitamori, *J. Nanosci. Nanotechnol.*, 4 (2004) 575–579.
6 K. Sato, M. Yamanaka, H. Takahashi, M. Tokeshi, H. Kimura and T. Kitamori, *Electrophoresis*, 23 (2002) 734–739.
7 M. Diaz-Gonzalez, M.B. Gonzalez-Garcia and A. Costa-Garcia, *Electroanalysis*, 17 (1995) 1901–1918.
8 A. Bange, H.B. Halsall and W.R. Heineman, *Biosens. Bioelectron.*, 20 (2005) 2488–2503.
9 E.P. Diamandis and T.K. Christopoulos, *Immunoassay*, Academic Press, San Diego, 1996, p. 579.
10 T. Beumer, P. Haarbosch and W. Carpay, *Anal. Chem.*, 68 (1996) 1375–1380.
11 J.S. Rossier, G. Gokulrangan, H.H. Girault, S. Svojanovsky and G.S. Wilson, *Langmuir*, 16 (2000) 8489–8494.

12 H.B. Halsall, W.R. Heineman and S.H. Jenkins, *Clin. Chem.*, 34(9) (1988) 1701–1702.

13 J.S. Rossier, F. Reymond and W. Schmidt, Patent: Method for Fabricating Micro-Structures with Various Surface Properties in Multilayer Body by Plasma Etching, 2001, WO 2001/056771.

14 J.S. Rossier, C. Vollet, A. Carnal, G. Lagger, V. Gobry, H.H. Girault, P. Michel and F. Reymond, *Lab Chip*, 2 (2002) 145–150.

15 J.S. Rossier, F. Reymond and P.E. Michel, *Electrophoresis*, 23 (2002) 858–867.

16 J.S. Rossier, P. Michel and F. Reymond, *Patent: Microfluidic Chemical Assay Apparatus and Method*, 2003, WO 2003/004160 A004161.

17 J.S. Rossier, P. Morier and F. Reymond, *Patent: Microfluidic Flow Monitoring Device*, 2003, WO 2003/004160 A004161.

18 P. Morier, C. Vollet, P.E. Michel, F. Reymond and J.S. Rossier, *Electrophoresis*, 25 (2004) 3761–3768.

19 J.-W. Choi, K.W. Oh, J.H. Thomas, W.R. Heineman, H.B. Halsall, J.H. Nevin, A.J. Helmicki, H.T. Henderson and C.H. Ahn, *Lab Chip*, 2 (2002) 27–30.

20 J. Wang, A. AIbáñez and M.P. Chatrathi, *J. Am. Chem. Soc.*, 125 (2003) 8444–8445.

21 J. Kulys, V. Razumas and A. Malinauskas, *Bioelectrochem. Bioenerg.*, 7 (1980) 11–24.

22 H. Mao, T. Yang and P.S. Cremer, *Anal. Chem.*, 74 (2002) 379–385.

23 P.N. Bartlett and S.L. Taylor, *Electroanal. Chem.*, 453 (1998) 49–60.

24 D. Hoegger, P. Morier, C. Vollet, D. Heini, F. Reymond and J.S. Rossier, Disposable Microfluidic ELISA for the Rapid Determination of Folic Acid Content in Food Products, *Anal. Bioanal. Chem.*, 387 (2007) 267–275.

Chapter 37

Scanning electrochemical microscopy in biosensor research

Gunther Wittstock, Malte Burchardt and Carolina Nunes Kirchner

37.1 INTRODUCTION

Scanning electrochemical microscopy (SECM) is a technique that allows to record spatially resolved maps of chemical reactivities, i.e. images that reflect the *rate of heterogeneous chemical reactions*. The acronym is used for both, the method and the instrument. An amperometric or, in rare cases, a potentiometric microsensor is immersed into an electrolyte solution and scanned over the sample surface. By detecting dissolved chemical species in the immediate vicinity of the sample surface, this technique lends itself to the characterization of surfaces at which substances are locally released into the solution. It can be applied to a large variety of interfaces including solid–liquid, liquid–liquid, and liquid–gas interfaces [1–5]. The sample can be conductive, semiconductive or insulating. SECM is historically rooted in two experimental areas. As a scanning probe, it shares many similarities with related techniques such as scanning force microscopy in an electrochemical cell (ECSFM) or electrochemical scanning tunneling microscopy (ECSTM). However, the signal in SECM is based on an electrochemical signal specific for a certain chemical compound. In this respect, the scanning probe can also be regarded as a positionable chemical microsensor. Looking back on almost two decades of SECM development, it seems logical to use an amperometric ultramicroelectrode (UME) in order to measure local concentrations of reactants and products close to a macroscopic sample electrode. This experiment was performed by Engstrom *et al.* [6] and can be considered the first SECM experiment. At about the same time, Bard *et al.* [7] reported currents at unusually large sample–tip distances in an

Comprehensive Analytical Chemistry 49
Alegret and Merkoçi (Eds)
Volume 49 ISSN: 0166-526X DOI: 10.1016/S0166-526X(06)49037-1

ECSTM. The signals were considered to be Faradaic currents, i.e. currents that result from the heterogeneous electrochemical conversion of a chemical species at the metal tip. Bard *et al.* [7] realized that a Faradaic current at the tip can be modulated by the presence of the sample in various ways. By 1989 Bard *et al.* [8,9] had worked out an experiment in an analogous micrometer-sized system where all the effects could clearly be ascribed to well-understood physical processes such as diffusion or heterogeneous electrochemical reactions. Very soon it became clear that SECM is not just suitable to measure local solute concentrations but also, and more importantly, represents a tool to map local (electro)chemical reactivities, to induce localized electrochemical surface modifications, or to investigate heterogeneous and homogeneous kinetics. Since SECM instruments became commercially available[1], the number of groups using this technique is rapidly increasing. Although the potential of SECM to contribute to sensor research was recognized very early by Wang *et al.* [10], it can be noted that till now most successful SECM groups are rooted in fundamental electrochemistry. This chapter aims to summarize the progress made so far and tries to encourage a broader application of the technique by the sensor community. This should be attractive because SECM offers unique possibilities to prepare and investigate advanced sensing concepts. Moreover, chemical microsensors might become very attractive new probe electrodes for SECM that can be applied in the context of cell biology or materials research. Finally, the electrochemical sensor community is well familiar with the concepts of diffusion and heterogeneous reactions underlying each SECM experiment, which should considerably ease the first steps in SECM application.

37.1.1 Instrument and probes

The general experimental setup is shown in Fig. 37.1. The probe is an amperometric disk-shaped UME that is embedded in an insulating sheath, typically made from glass. Most often the electrode is made from Pt but electrodes from Au and carbon fibers have been used as well. Typical probe diameters are 10 or 25 μm. Of course, smaller electrodes may be used and this area is currently extensively explored. As it will be evident later, it is convenient to characterize the probe by two important radii: the radius r_T of the active electrode area and

[1]Details may be requested from the authors, as they are constantly changing.

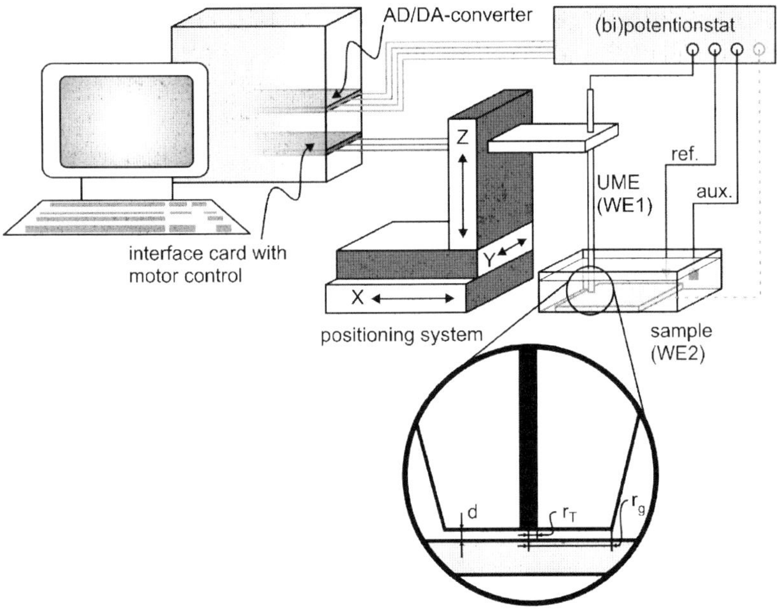

Fig. 37.1. Schematic setup of the SECM. Depending on the experiment, the sample may or may not be connected to the second channel of a bipotentiostat.

the radius r_g of the insulating glass sheath. Often the ratio $RG = r_g/r_T$ is also important. The potential of the UME E_T relative to a thermodynamically defined reference electrode is controlled by a bipotentiostat. The probe is scanned in a distance d of 0.5–3r_T over the sample.

37.1.2 The feedback mode: hindered diffusion and mediator recycling

For measurements in the feedback mode, the working solution contains one redox form of a quasi-reversible redox couple ($R \rightarrow O + ne^-$). For the discussion of the working principle, it is assumed that initially only the reduced form R is present. This compound serves as electron mediator and is added typically in millimolar concentrations to an excess of an inert electrolyte[2]. The UME is poised at a potential sufficiently large to cause the diffusion-controlled oxidation of R. In solution bulk the

[2]Of course, analogous experiments may be carried out if the oxidized form of the mediator is provided. Reaction directions are reversed in this case.

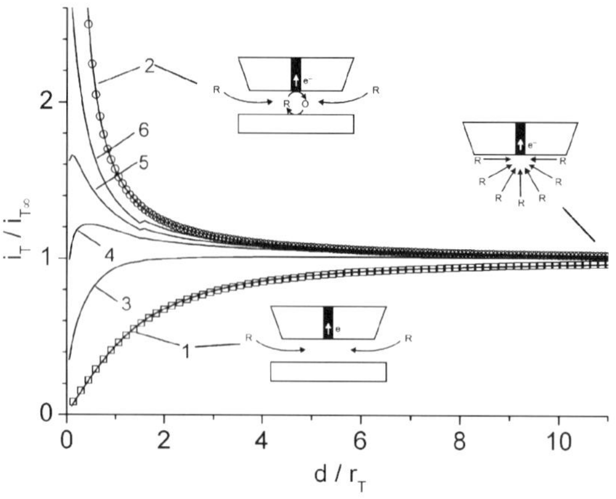

Fig. 37.2. Normalized current as a function of normalized UME–sample distance and schematic representation of the diffusion at the UME; experimental approach to glass (hindered diffusion, □) and to a gold sample (mediator regeneration by diffusion-controlled reaction at the sample, ○). Calculated curves (—) for hindered diffusion (1), diffusion-controlled reaction at the sample (2) and normalized rate constants $\kappa = k_{\mathrm{eff}} r_{\mathrm{T}}/D$: 0.3 (3), 1.0 (4), 1.8 (5), and 3.6 (6) are given.

current, $i_{\mathrm{T}\infty}$, can be calculated according to

$$i_{\mathrm{T}\infty} = gnFD_{\mathrm{R}}c_{\mathrm{R}}^{*}r_{\mathrm{T}} \tag{37.1}$$

where n is the number of electrons transferred per molecule, F the Faraday constant, D_{R} the diffusion coefficient of R and c_{R}^{*} the concentration of R in the bulk solution. The geometry-dependent factor g assumes the value 4 for an infinitely large insulating sheath [11]. The value $g = 4$ is also a good approximation if RG$\geqslant$10. For quantitative work with smaller RGs, slightly different values should be preferred (e.g. $g = 4.43$ for RG = 2) [12]. In analogy to other scanning probe techniques, the UME is frequently called "tip" from which the subscript "T" was taken to specify quantities of the UME. Correspondingly, the subscript "S" is used for quantities of the specimen electrode[3]. The subscript "∞" indicates that the UME has a quasi-infinite distance to the surface. Practically, $i_{\mathrm{T}\infty}$ is observed if the UME–sample distance d is larger than $20\,r_{\mathrm{T}}$.

Figure 37.2, curve 1 shows the decrease in the steady-state current, $i_{\mathrm{T}}(d)$ if the UME approaches an insulating and inert specimen

[3]Synonyms are "sample" and "substrate". To avoid confusion with the biochemical term "substrate" as one of the reagents in an enzymatic reaction, "specimen", "sample", or "support" are preferred in this paper.

surface, e.g. glass. For a unified description, the UME current is normalized to $i_{T\infty}$ and d is given in units of r_T. In order to reach the active area of the UME, R has to diffuse through the gap formed by the specimen surface and the insulating sheath of the UME. The resulting mass-transfer resistance will increase (and i_T will decrease) as the interelectrode space narrows (decreasing d) and the thickness of the insulating shielding (RG) increases. While many schematics in this review as well as elsewhere highlight the different reactions occurring at the UME and at the sample, Fig. 37.1 gives a realistic picture of the relative sizes of the active electrode area, the UME–sample gap and the insulating sheath. The hindered diffusion of the mediator provides a background signal for activity imaging and has been computed in an increasingly sophisticated manner [12,13]. From such simulations for specific d,approximate analytical functions have been derived, e.g. curve 1 in Fig. 37.2 can be described by the following equation:

$$I_T^{\mathrm{ins}}(L) = \frac{i_T}{i_{T\infty}} = \frac{1}{k_1 + (k_2/L) + k_3 \exp(k_4/L)} \tag{37.2}$$

The numerical values for k_1 ... k_4 vary with RG. For instance, for RG = 10, the following values provide the analytical function: $k_1 = 0.40472$, $k_2 = 1.60185$, $k_3 = 0.58819$, and $k_4 = -2.37294$ [12]. The analytical approximations for hindered diffusion provide a way to determine d from experimental approach curves. For this purpose, one can use an irreversible reaction at the UME (often O_2 reduction). In such a case, Fig. 37.2, curve 1 is obtained irrespective of the nature of the sample. Besides the mediator flux from the solution bulk, there might be a heterogeneous reaction at the sample surface during which the UME-generated species O is recycled to the mediator R. The regeneration process of the mediator might be (i) an electrochemical reaction (if the sample is an electrode itself) [9], (ii) an oxidation of the sample surface (if the sample is an insulator or semiconductor) [14], or (iii) the consumption of O as an electron acceptor in a reaction catalyzed by enzymes or other catalysts immobilized at the sample surface [15]. All these processes will increase i_T above the values in curve 1 of Fig. 37.2. How much i_T increases, depends on the kinetics of the reaction at the sample. If the reaction of the sample occurs with a rate that is controlled by the diffusion of O towards the sample, Fig. 37.2, curve 2 is recorded. If the sample is an electrode itself, such a curve is experimentally obtained if the sample potential is sufficiently negative to the formal potential of the mediator. The curve is described by the following equation:

$$I_{\mathrm{T}}^{\mathrm{cond}}(L) = \frac{i_{\mathrm{T}}}{i_{\mathrm{T}\infty}} = k_1 + \frac{k_2}{L} + k_3 \exp\left(\frac{k_4}{L}\right) \tag{37.3}$$

The curve with $k_1 = 0.72627$, $k_2 = 0.76651$, $k_3 = 0.26015$, and $k_4 = -1.4132$ [12] is the upper limit of the response for any given distance d of an UME with RG 10. The term "positive feedback" was coined by Bard *et al.* [9] for the communication between the UME and the sample by diffusing redox mediators[4]. Correspondingly, the term "negative feedback" gained popularity for the hindered diffusion of the mediator to the UME located above an inert, insulating surface following a suggestion of Bard *et al.* [16].

Curves for selected values of the normalized first-order rate constant $\kappa = k_{\mathrm{eff}} r_{\mathrm{T}}/D$ are given in Fig. 37.2, curves 3–6. They can be described by an analytical approximation under three conditions: (i) the distance range is $0.1 < L < 1.6$, (ii) RG ≈ 10, and (iii) the reaction at the sample is of first order with respect to the mediator (Eq. (37.4)).

$$I_{\mathrm{T}}(L) = \frac{i_{\mathrm{T}}}{i_{\mathrm{T}\infty}} = I_{\mathrm{T}}^{\mathrm{ins}}(L) + I_{\mathrm{S}}^{\mathrm{kin}}(L)\left(1 - \frac{I_{\mathrm{T}}^{\mathrm{ins}}(L)}{I_{\mathrm{T}}^{\mathrm{cond}}(L)}\right) \tag{37.4}$$

The normalized substrate current $I_{\mathrm{S}}^{\mathrm{kin}} = i_{\mathrm{S}}/i_{\mathrm{T}\infty}$ is the current equivalent at the sample i_{S} normalized by $i_{\mathrm{T}\infty}$. It can be estimated for RG = 10 and $0.1 < L < 1.6$ by

$$I_{\mathrm{S}}^{\mathrm{kin}}(L, k_{\mathrm{eff}}) = \frac{0.78377}{L(1 + (1/\Lambda))} + \frac{0.68 + 0.3315\exp(-1.0672/L)}{1 + F(L, \Lambda)} \tag{37.5}$$

where $\Lambda = \kappa L$ and $F(L,\Lambda) = (11/\Lambda + 7.3)/(110 - 40L)$ [17].

This situation allows making two basic experiments: scanning the UME at constant distance d provides an image that reflects the distribution of heterogeneous reaction rates on the sample (reaction rate imaging). Moving the UME vertically towards the sample allows a more detailed kinetic investigation of the reaction $\mathrm{O} + n\mathrm{e}^- \rightarrow \mathrm{R}$ at

[4]The term "feedback mode" refers to the coupling of heterogeneous reactions at the specimen and the UME and not to an electronic control principle as commonly used in other scanning probe techniques to maintain a constant sample–probe distance.

a specific location of the sample. This experiment is described in detail in Procedure 51 (see in CD accompanying this book). Both experiments can give valuable information for the optimization of electrochemical sensors.

37.1.3 Generation/collection mode

The generation/collection mode (GC mode), and specifically the substrate-generation/tip-collection mode (Fig. 37.3a), refers to an experiment in which the UME is used to detect species that are generated or released from the sample. It can be performed with amperometric and potentiometric probes. Usually the detected compound is initially not present in solution and therefore, the concentration far away from the surface approaches zero. The experiment may at first appear straightforward. However, there are several problems associated with the interpretation of the results: (i) only if the active regions are well-separated microstructures, there will be a steady-state concentration profile. Above macroscopic active regions of sensors, the local concentrations depend on the time that has passed after the onset of the reaction at the sample. (ii) The diffusion layer of the specimen is disturbed by the presence of the UME (stirring, blocking of reactant diffusion to the sample, overlap of the diffusion layers of the UME and

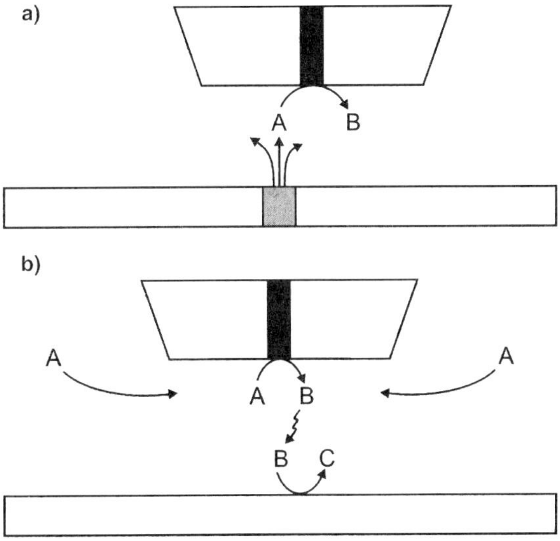

Fig. 37.3. Schematics of the GC mode. (a) Substrate-generation/tip-collection mode and (b) tip-generation/substrate-collection mode.

the sample in case of amperometric UMEs). (iii) In the GC mode there might be a current enhancement due to recycling if the reaction at the microelectrode is a reversible reaction and $d \leqslant 5r_T$. GC experiments have been performed with potentiometric microelectrodes as local probes (see Section 37.2.5). Because potentiometric microelectrodes do not convert the detected species[5], feedback effects are not possible, and disturbance of the sample diffusion layer is reduced. The lateral resolution of GC experiments is always inferior compared to corresponding feedback experiments. A number of applications have emerged for GC experiments, where feedback experiments are either not possible or not sensitive enough. The question of sensitivity is particularly important when mapping enzyme activities. If the UME behaves approximately as a passive probe, i_T depends on the radius of the active region of the sample r_S, the concentration of the detected species c_S at the surface of the sample, and a dilution factor θ:

$$i_T = gnFDr_Tc_S\theta \qquad (37.6)$$

The factor θ depends on d and the lateral distance r of the UME from the center of the active region. For an x line scan over the center of the spot ($r = \Delta x$), θ assumes the following simple form:

$$\theta = \frac{2}{\pi}\arctan\frac{\sqrt{2}r_S}{\sqrt{(\Delta x^2 + d^2 - r_S^2) + \sqrt{(\Delta x^2 + d^2 - r_S^2)^2 + 4d^2r_S^2}}} \qquad (37.7)$$

A fit of an experimental line scan to Eqs. (37.6) and (37.7) provides c_S. The total flux Ω from the active region can be calculated by the following equation [11]:

$$\Omega = 4Dc_Sr_S \qquad (37.8)$$

Alternatively, the UME can be used to produce locally a reagent that is converted at the sample with a limited reaction rate (Fig. 37.3b, tip-generation/sample-collection mode). Recording the sample current i_S as a function of the UME position can provide a mapping of the local sample reactivity. This has been used for irreversible reactions such as O_2 reduction (vide infra).

[5]This assumption usually holds because the concentration is approximately constant over the active area of the micrometer-sized potentiometric microelectrode. Only if there is a very steep concentration gradient, an exchange current occurs at the potentiometric microelectrode altering the local solution composition.

37.2 APPLICATION OF SECM IN CHEMICAL AND BIOCHEMICAL SENSOR RESEARCH

37.2.1 Investigation of electrochemical sensor surfaces

The most obvious application of SECM in the sensor area is the inspection of UME arrays and a number of studies have been reported [18–24]. Very early SECM has been used to investigate the heterogeneous distribution of electrochemical reaction rates on composite materials such as graphite spray by Wittstock *et al.* [25] (Fig. 37.4). Figure 37.4a shows an SECM image of the graphite spray after coating the surface. Only a small number of very active regions can be seen. After thermal activation, the number of active grains is increased in the SECM image of Fig. 37.4b, while scanning electron microscopic pictures showed no change in the topography. This was interpreted as a partial decomposition of the binding polymer by which more grains are exposed at the outer surface of the carbon spray-coated areas. Mechanisms of thermal activation procedures were elucidated. More recently, a number of applications appeared in which SECM was used as a screening tool for electrocatalysts. While most of these studies were prompted by the search for new fuel cell catalysts, similar experiments may be used to optimize, for example, amperometric oxygen electrodes. Jambunathan *et al.* [26] investigated the potential-dependent rate of hydrogen oxidation by analyzing approach curves recorded at different sample potentials similar to the approach explained in Section 37.1 and in the Protocol #P1 H37. Fernandez *et al.* [27] investigated arrays of noble metal catalysts for oxygen reduction. The author employed the tip-generation/sample-collection mode in which a constant flux of

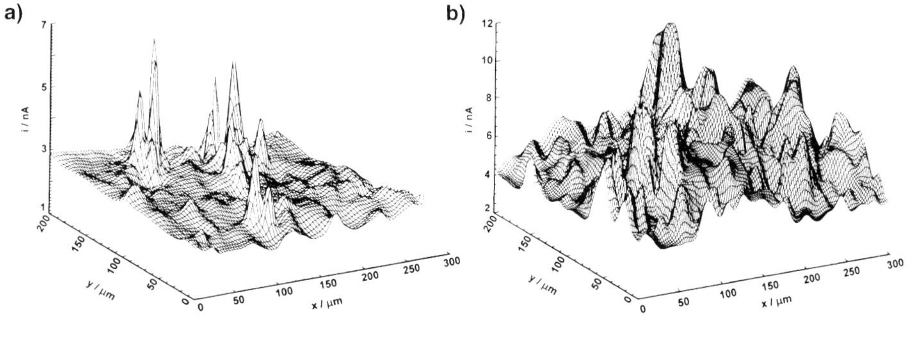

Fig. 37.4. Image of carbon spray (a) before and (b) after thermal activation. Reproduced with permission from Ref. [25]. Copyright 1994, Springer-Verlag.

molecular oxygen was generated at the UME under galavanostatic condition. The sample reduction current reflects the ability of the sample region underneath the UME to reduce the molecular oxygen. Fernandez *et al.* [28] used a similar approach to optimize the weight fraction of wired enzymes for oxygen reduction electrodes.

SECM is a very suitable tool for corrosion research including corrosion of sensor materials. Toth *et al.* [29] investigated the corrosion of AgI electrodes. Gründig *et al.* [30] investigated composites of a carbon paste and *N*-methyl phenazine (NMP^+) methosulfate that showed favorable characteristics as electrodes for NADH oxidation. It could be proven that these electrodes function by slow release of NMP^+ that then acts as a dissolved redox mediator. The remarkable long-term operational stability could be explained by the large reservoir of NMP^+ in the composite material. Janotta *et al.* [31] investigated the corrosion of diamond-like carbon-coated optical waveguides from zinc selenide used for optical sensors. Details about the corrosion mechanism could be elucidated.

37.2.2 Investigation of immobilized enzymes

Immobilized enzymes can be investigated in the feedback and in the GC mode. In the feedback mode oxidoreductases can be imaged. They use the SECM mediator as electron donor (or acceptor). Table 37.1 gives an overview about the enzymes investigated.

The GC mode can also be used to image some enzymes that are not oxidoreductases. For the important enzymes alkaline phosphatase (ALP) and galactosidase, this has been achieved by using an enzyme substrate that is not redox active at the UME potential, whereas one of the products (*p*-aminophenol (PAP)) can be oxidized. The experiment is detailed in the Protocol #P2 H37. The use of potentiometric probes is also possible. Table 37.2 provides an overview about the investigated enzymes.

One of the promising potentials of SECM for biosensor research is the possibility to investigate immobilized enzymes independent of the communication to the electrode onto which they are immobilized. In fact, not too seldom, the immobilization of proteins onto electrode surfaces inhibits fast electron transfer reactions. SECM can be used to probe the enzymatic activity from the solution side of an immobilized enzyme film with a UME that is free of any cover layer. When designing an SECM experiment for the investigation of immobilized enzymes, one should consider the following guidelines.

TABLE 37.1

SECM feedback imaging of local activity of immobilized oxidoreductases

Enzyme imaged	Substrate	Mediator/ reaction at the UME	References
Glucose oxidase, EC 1.1.3.4	50–100 mM glucose	0.05–2 mM ferrocene monocarboxylic acid, dimethylaminomethyl ferrocene oxidation	[15,32,33]
		0.05–2 mM $K_4[Fe(CN)_6]$ oxidation	[15]
		0.02–2 mM hydroquinone oxidation	[15,34]
		0.5 mM [Os fpy $(bpy)_2Cl]Cl$, fpy = formylpyridine, bpy = bipyridine oxidation	[33]
PQQ-dependent glucose dehydrogenase, EC 1.1.99.17	50 mM glucose	0.05–2 mM ferrocene monocarboxylic acid, 0.05–2 mM ferrocene methanol, 0.05–2 mM *p*-amniophenol oxidation	[35]
NADH-cytochrome *c* reductase, EC 1.6.99.3 within mitochondria	50 mM NADH	0.5 mM *N,N,N′,N′*-tetramethyl-*p*-phenylenediamine oxidation	[34]
Diaphorase (NADH acceptor oxidoreductase, EC 1.6.99.–)	5.0 mM NADH	0.5 mM hydroxymethyl ferrocene oxidation	[36]
Horseradish peroxidase, EC 1.11.1.7	0.5 mM H_2O_2	1 mM hydroxymethyl ferrocenium reduction	[37]
Nitrate reductase, EC 1.7.99.4	23–65 mM NO_3^-	0.25 mM methylviologen reduction	[38,39]

TABLE 37.2

SECM imaging of local enzyme activities in GC mode

Enzyme	Species detected at the UME	Probe	References
Glucose oxidase, EC 1.1.3.4	H_2O_2	Amperometric Pt UME	[40,41]
		Amperometric enzyme electrode	[42]
PQQ-dependent glucose dehydrogenase, EC 1.1.99.17	$[Fe(CN)_6]^{4-}$	Amperometric Pt UME	[35]
Urease, EC 3.5.1.5	H^+	Potentiometric Sb UME	[43]
	NH_4^+	Liquid membrane ISE	[44]
Horseradish peroxidase, EC 1.11.1.7	Ferrocene derivatives	Amperometric Pt UME	[37,45,46]
Alkaline phosphatase, EC 3.1.3.1	4-Aminophenol	Amperometric Pt UME	[32,47–50]
Galactosidase, EC 3.2.1.23	4-Aminophenol	Amperometric Pt UME	[51,52]
Alcohol dehydrogenase, EC 1.1.1.1	H^+	Potentiometric Sb UME	[53]
NADPH-dependent oxidase in osteoclasts	$O_2{\bullet}^-$	Cytochrome *c*-modified Au UME (amperometric)	[54]

(1) If the enzymes are immobilized on an electrode surface, generally only the GC mode can be used for their investigation, because the signal is independent of the nature of the support (Fig. 37.5c). This was demonstrated by Wittstock and Schuhmann [41]. If oxidoreductases are immobilized on conducting surfaces, the feedback can result from a heterogeneous electron transfer at the electrode or the enzymatic reaction (Fig. 37.5b). Kranz *et al.* [33] isolated the contribution of the enzymes by carefully designed control experiments. In most cases such an approach is extremely

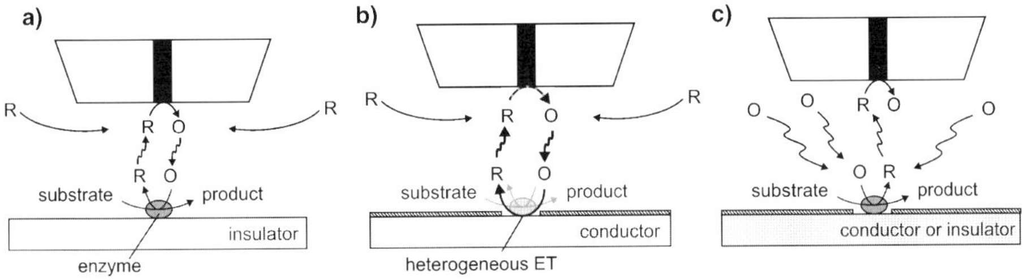

Fig. 37.5. Imaging of enzymes immobilized on electrode surfaces. (a) Oxidoreductase immobilized on insulators (imaging and quantification possible); (b) oxidoreductase immobilized on conductor; the feedback may be caused by the electrode or the enzymes; and (c) GC mode experiment does not depend on the nature of the support surface.

difficult because the electrode surface will contribute much more efficiently to the mediator recycling.

(2) When the feedback mode is possible (i.e. the enzymes is immobilized on an insulating support, Fig. 37.5a), it provides a much better lateral resolution but has only a very limited sensitivity. Therefore, it can only be applied for very active enzymes or if enzymes are bound in high surface concentrations. Bard *et al.* [15] gave a quantitative detection limit for this situation:

$$k_{\mathrm{cat}}\Gamma_{\mathrm{enz}} \geq \frac{10^{-3}Dc}{r_{\mathrm{T}}} \qquad (37.9)$$

The left side summarizes the enzyme-dependent terms: turnover number k_{cat} and surface concentration Γ_{enz}. In case of enzyme-loaded films, Γ_{enz} should be replaced by the product of enzyme volume concentration in the film and the film thickness. The right side summarizes the experimental conditions: diffusion coefficient D, concentration c of the mediator, and UME radius r_{T}. The feedback mode always requires an as small as possible working distance d. The smaller the UME the more difficult it will be to detect the activity of the immobilized enzyme.

The generation-collection experiments have a much higher sensitivity, which was brought into an analogous formula by Horrocks and Wittstock [55].

$$k_{\mathrm{cat}}\Gamma_{\mathrm{enz}} \geq \frac{Dc'}{r_{\mathrm{S}}} \qquad (37.10)$$

where r_S is the radius of the enzymatically active region on the *sample*, c' and D denote the detection limit for the species observed at the UME and its diffusion coefficient, respectively. Assuming $c' = 1 \times 10^{-6}$ M, $D = 5 \times 10^{-6}$ cm^2 s^{-1}, $r_S = 50$ μm, and $\Gamma_{enz} = 1 \times 10^{-12}$ mol cm^{-12} (about one monolayer), enzymes can be detected with a specific activity $k_{cat} \geqslant 1$ s^{-1} in GC imaging [55].

(3) In order to quantify the results in the GC mode, the sample must be a microstructure by itself, so that it develops a steady-state diffusion layer. This restriction does not apply to feedback mode experiments.

(4) The GC image is less critical dependent on d. Therefore, larger d can be used that are more forgiving to protruding samples.

(5) The concentrations of the enzyme substrate have to be selected with some care. In the GC experiment, it should be well above the Michaelis–Menten constant of the enzyme for this substrate. The presence of the UME with its sheath will also limit the diffusion of the enzyme substrate towards the active regions. This may lead to an underestimation of enzyme activity on the surface or to great distorsion in recorded images. Using electrodes with a small RG is a good idea for GC experiments. This question is explained in Procedure 52 (see in CD accompanying this book).

(6) Studies of immobilized enzymes should be done with control experiments. Figure 37.6 illustrates how misleading SECM experiments can be without properly designed controls. The line scan in Fig. 37.6b was obtained from a surface which consists of a patterned self-assembled monolayer of mercaptoundecanoic acid (MUA) and (1-mercapto-undec-11-yl)hexaethyleneglycol (EG_6). The MUA pattern was further modified through layer-by-layer adsorption during which horseradish peroxidase (HRP) was incorporated into a polyionic gel. In an attempt to image the HRP activity in the GC mode (according to Fig. 37.6a, scheme 1), H_2O_2 and ferrocene methanol were provided and ferrocinium ions were detected at the UME. A clear difference can be seen in Fig. 37.6a between the HRP-containing and HRP-free film. However, a control experiment without H_2O_2 shows that the same current difference is observed (Fig. 37.6c). Therefore, the pattern in the line scan cannot originate from the enzymatic activity of the film! Most likely, the solution contained a minute fraction of oxidized ferrocene molecules. They acted as a mediator for a feedback image of the patterned monolayer and probed the different permeability of the monolayers formed from MUA and

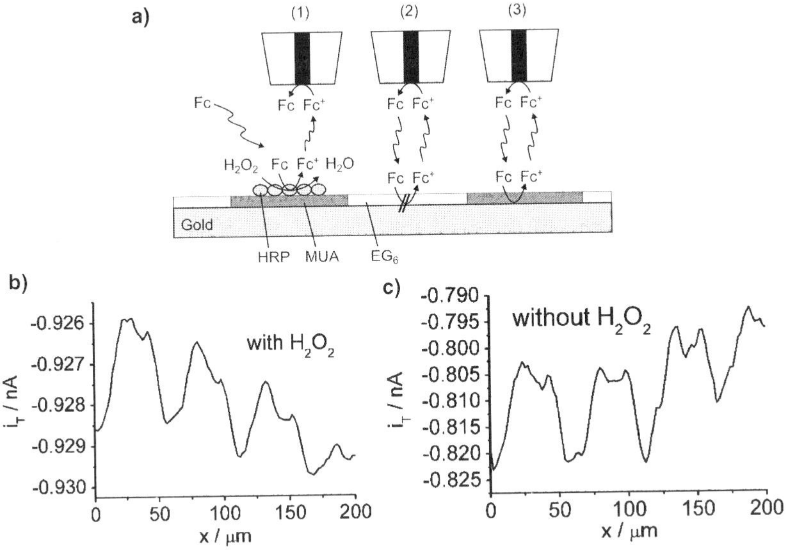

Fig. 37.6. Experimental example of an unintended feedback imaging. (a) Schematic showing (1) the intended GC imaging of HRP activity; (2) hindered diffusion above a densely packed monolayer using the ferrocinium derivative (Fc^+) as mediator. Fc^+ is present in low concentration possibly due to partial chemical oxidation by O_2 or H_2O_2; (3) the regeneration of Fc^+ at the gold electrode surface under a less-organized monolayer. (b) Experimental line scan in the presence of ferrocene methanol (and traces of Fc^+) and H_2O_2. (c) Same as (b) but in absence of H_2O_2.

polyelectrolyte gel and EG_6 (Fig. 37.6, schemes 2 and 3).

Diffusional spreading in GC imaging of enzymes can be limited by a coupled homogenous reaction that consumes the reagent generated at the sample. This was first demonstrated for imaging of glucose oxidase activity by Wittstock and Schuhmann [41] and consequently found further application [56]. The comparison of the detection modes for the field of biochip detection was experimentally verified by Zhao and Wittstock [47] using PQQ-dependent GDH as example.

37.2.3 Advanced interfacial architectures for sensors

Microstructured biosensor surfaces have been investigated by SECM [10,57]. The investigations were mainly prompted by the idea to produce compartmentalized surfaces, where some regions are surface-modified in order to provide an ideal environment for the biochemical component and other regions are optimized for electrochemical detection [58,59]. SECM can also be used to locally modify surfaces [36,46,60–69] by different defined electrochemical mechanisms. The combination of the imaging capabilities for specific enzymatic reactions and the possibility to modify the surfaces in a buffer solution make SECM an ideal tool to explore the potential of such micropatterned surfaces for sensing applications.

The group of Kuhr pursued an approach where enzymes were only immobilized on specific areas of the electrode. The electrochemical detection was performed on the unmodified regions of the same electrode leading to faster response times of the sensor [58]. The authors used SECM to show the different kinetics at the modified and unmodified regions of the sensor surface [57]. SECM was (among other techniques [58,70]) used for microderivatization of the surface [63].

Wilhelm and Wittstock [65] created a dual-enzyme surface in which a reaction sequence was performed. HRP was immobilized by a combination of microcontact printing and further covalent modification of the layers (Fig. 37.7a) [71]. The periodic pattern was then locally modified by electrochemical desorption of the alkanethiolate layer [46]. The bare gold surface was then refilled with modified glucose oxidase (GOx, Fig. 37.7a). An image of the combined reactivity of both enzymes is shown in Fig. 37.7c. One can see that the diffusional spreading of H_2O_2 is prevented by the HRP pattern. The individual reactivity of GOx is imaged after exchanging the solution (glucose and O_2) and detecting H_2O_2 at the UME (Fig. 37.7d and e). The HRP activity can be imaged when the solution bulk contains H_2O_2 and ferrocene methanol (Fig. 37.7f and g). The larger reduction currents mark the position of the HRP pattern. One can see that the activity is periodically changed on the stamped pattern and the different conversion rates in Fig. 37.7c must be caused by the locally different supply of H_2O_2.

Fig. 37.7. Investigation of patterned bienzymatic surface containing HRP and GOx. (a) Scheme of the patterned surface after electrochemical desorption and attachment of GOx. (b) Schematic representation of the SECM GC experiment. (c) Experimental results in the presence of O_2, glucose and Fc, image frame $500 \times 500\,\mu m^2$. (d) Schematic representation of the SECM GC experiment for recording the GOx activity. (e) SECM line scan along the line in (c) while detecting the formation of H_2O_2 in a solution initially containing glucose and O_2. (f) Schematic representation of the SECM GC experiment for recording the HRP activity. (g) SECM line scan along the line in (c) while detecting the formation of Fc^+ in a solution initially containing H_2O_2 and Fc. Reproduced with permission from Ref. [65], Copyright 2003, Wiley-VCH.

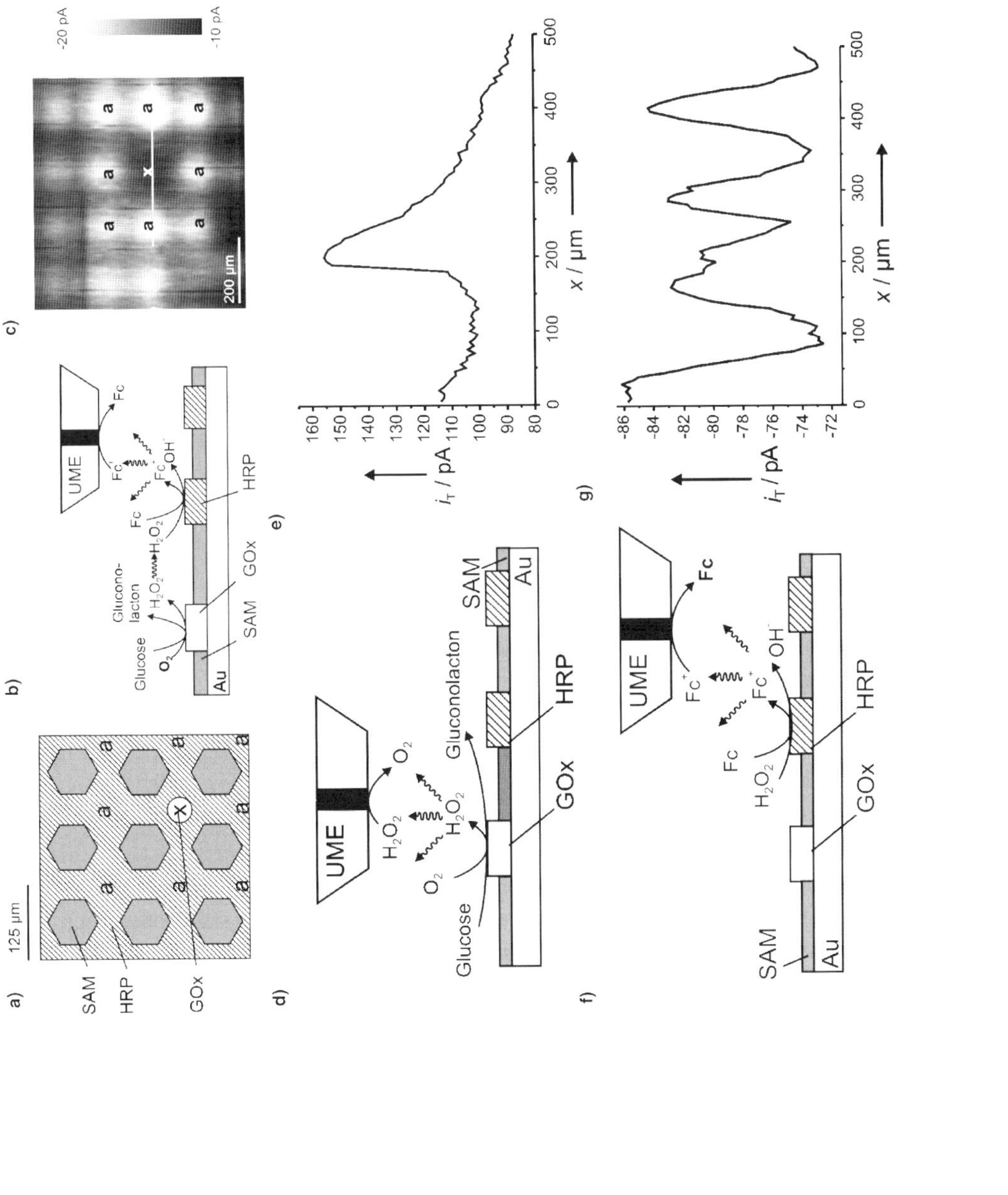
a)
b)
c)
d)
e)
f)
g)
125 µm
SAM
HRP
GOx
UME
Glucose
Gluconolacton
Au
Fc
200 µm
-20 pA
-10 pA
i_T / pA
x / µm

Niculescu *et al.* [72] formed microstructures of enzyme-loaded hydrogels and imaged the topography and the enzymatic activity of the entrapped enzymes. Turcu *et al.* [73] prepared gradient materials with a piezo-electric dispenser. They investigated the SECM response of the material in one and the same solution. The authors envision that such approaches can be used to enhance the reliability of sensor data, because an image represents the response of many slightly different glucose sensors that are read out by SECM. Furthermore, such approaches may hold a considerable potential for combinatorial optimization of sensor preparations. Instead of preparing many individual sensors with slightly varying protocols, the same variation in the recognition layer is generated onto *one support electrode*. The gradient material is then investigated by SECM and the optimum composition is found by the image region that gives the best SECM signal. As with any combinatorial experiment, such an approach shall be regarded as a screening tool. The SECM experiment will not be sufficient to provide an estimate of all relevant sensor characteristics.

Kurzawa *et al.* [74] used SECM in the characterization of enzyme microstructures that were obtained by a new non-manual immobilization technique based on the site-directed deposition of polyelectrolytes induced by a local pH-shift. This method has already been used to modify microscopy sensor regions [75].

37.2.4 SECM as a readout for protein and DNA chip as well as for electrophoresis gels

SECM can be used to detect enzyme labels in heterogeneous immunoassays as first demonstrated by Wittstock *et al.* [50]. In this application, SECM competes with fluorescence detection. The main disadvantage of SECM in this respect can be seen in long experimental time, lack of high throughput and required skills of the experimenter. On the other hand, the sensitivity of the detection is comparable, the required instrumentation is less expensive and common problems in optical detection such as background fluorescence are of no concern in SECM. Given these circumstances, one can expect that electrochemical techniques will become important in the area of protein arrays, where a limited number of analytes will be detected (and therefore reading time is less important than in high-density arrays). Due to its flexibility, SECM is ideally suited for the optimization of such protocols. A readout in a real mass-produced assay can then be made in microstructured electrochemical cells that exploit the SECM working modes but avoid the mechanically demanding and time-consuming scanning as it has been demonstrated by Kaya *et al.* [76] and Ogasawara *et al.* [77].

Along this line Matsue *et al.* [37,45,78,79] have developed a number of biochips. Among them are multi-analyte assays for human placental lactogen (HPL) and human chorionic gonadotropin (HCG) [45] and leukocidin, a toxic protein produced by methicillin-resistant *Staphylococcus aureus* [79]. Figure 37.8 shows an example of a dual immunoassay with SECM detection. The analyte is defined by the position on the chip and the amount of analyte is quantified via the collection current at the UME. The current originates from the reduction of ferrocinium methanol (Fc^+) at the UME. Fc^+ is produced locally at the chip surface by the enzyme HRP under consumption of H_2O_2.

Microbeads can be used as a new platform for immunoassays and a model assay for immunoglobulin G has been demonstrated by Wijayawardhana *et al.* [48,80] with ALP as the labeling enzyme. Quite reasonable detection limits of 6.4×10^{-11} mol L^{-1} or 1.4×10^{-15} mol

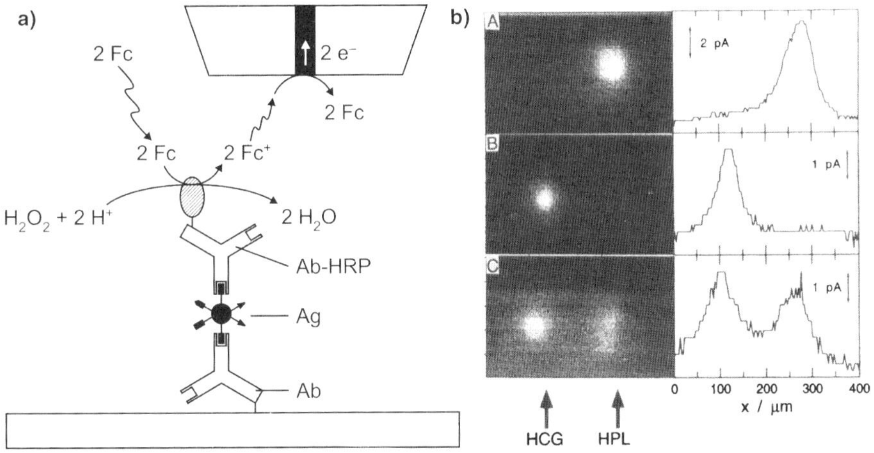

Fig. 37.8. (a) Scheme of the SECM detection in a sandwich immunoassay. (b) SECM images and cross-sectional profiles of microstructured glass supports with (anti-HCG)-Ab and (anti-HPL)-Ab immobilized at two distinct regions, treated with (A) 56 ng mL^{-1} HPL; (B) 2.0 IU mL^{-1} HCG; and (C) a mixture containing 31 ng mL^{-1} HPL+0.63 IU mL^{-1} HCG. After rinsing, the glass support was dipped into a solution of 20 µg mL^{-1} (anti-HCG)-Ab-HRP+7 µg mL^{-1} (anti-HPL)-Ab-HPL. Working solution, 1.0 mM Fc+0.5 mM H_2O_2+0.1 M KCl+0.1 M phosphate buffer, pH 7.0; $v_t = 9.8$ µm s^{-1}; $E_T = +0.05$ V (Ag/AgCl). Reprinted with permission from Ref. [45]. Copyright 1997, Elsevier Science.

mouse immunoglobulin G could be demonstrated (Fig. 37.9). Since PAP produced by the enzyme was detected close to the point of origin, an incubation step for enzymatic amplification could be avoided. Probably more advantageous are assays using a similar principle but employing galactosidase as an enzyme label. In this way, the oxygen-sensitive substrate for ALP, *p*-aminopenylphosphate, is avoided [52]. Zhao and Wittstock [51] generated a dual enzyme structure which could be used for signal amplification of an assay that uses a galactosidase label. PAP generated at the galactosidase label was converted to *p*-quinone-imine (PQI) at the UME. PQI was used as an electron acceptor for glucose

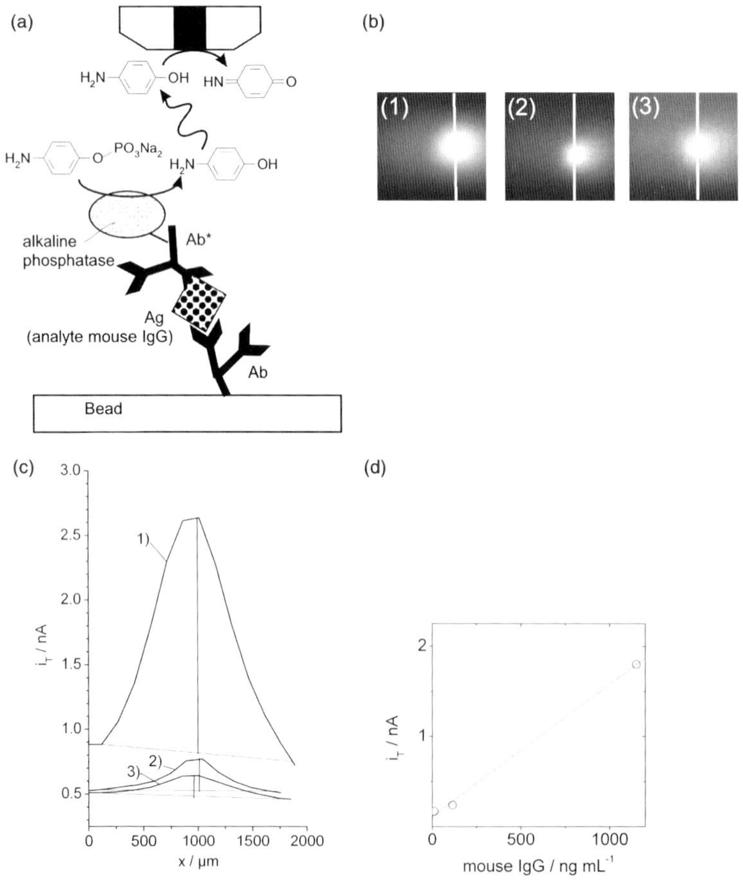

Fig. 37.9. SECM detection of a sandwich immunoassays for IgG. (a) Schematic showing the detection principle. (b) Images obtained from three spots; (c) extracted line scans; and (d) calibration curve. Reprinted with permission from Ref. [48]. Copyright 2000, Wiley-VCH.

dehydrogenase in the presence of glucose. In this process, PAP was regenerated and could be detected again at the UME leading to a significant signal amplification and an improvement in lateral resolution (Fig. 37.10).

More recently, another application of SECM detection in DNA and protein chips and in electrophoresis gels has emerged with different detection principles. Wang *et al.* [81] labeled single-stranded DNA (ssDNA) with gold nanoparticles. After binding to their complementary strand at the chip surface, silver was electroless deposited at the

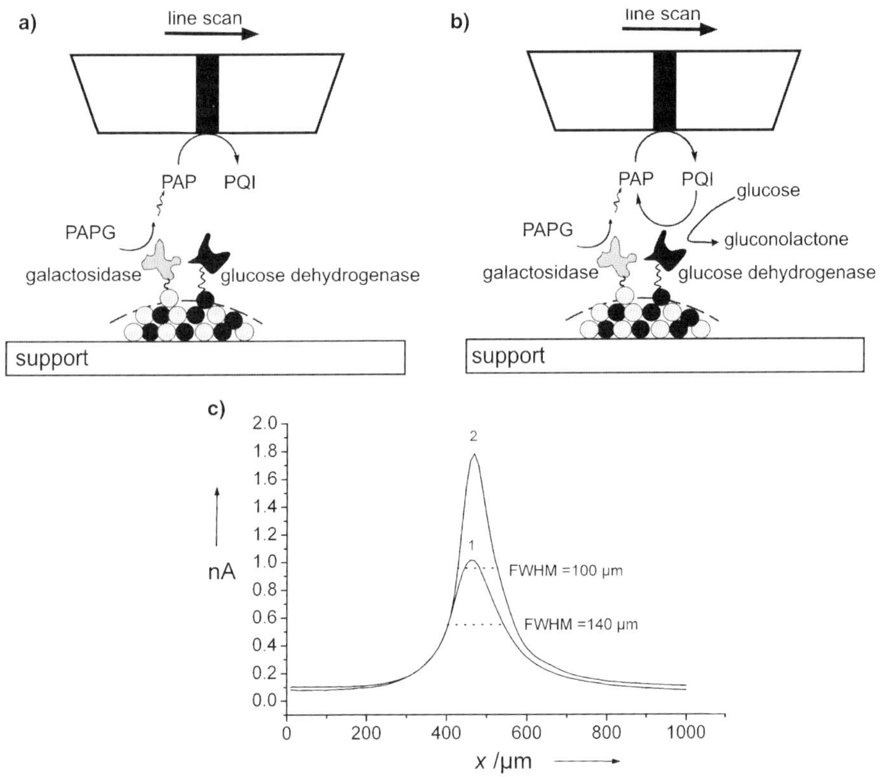

Fig. 37.10. (a) Schematic of conventional GC mode of Gal in the absence of glucose; (b) schematic of the signal amplification in the presence of glucose. PAPG, *p*-aminophenyl-β-D-galactopyranoside; PAP, *p*-aminophenol; PQI, *p*-quinone imine. (c) SECM line scans across the center of the microspot in the conventional GC mode (curve 1) and in the combined mode (curve 2). Curve 1 was obtained in the assay solution without glucose and curve 2 was obtained after adding 50 mM glucose to the solution. The schematics are not in scale. Reprinted with permission from Ref. [51]. Copyright 2004, Wiley-VCH.

metallic particles. The formed metal features were imaged in the feedback mode with $[Ru(NH_3)_6]^{3+}$ as mediator, in which the mediator is regenerated by an electron transfer from silver metal at those regions of the chip surface where a thin silver film was formed. A similar approach was used by Carano *et al.* [82]. Proteins were blotted on poly(vinylidene difluoride) membranes. The model protein bovine serum albumin was then tagged with previously prepared silver nanoparticles. The position of the protein bands was detected by SECM. The mediator was $[Os(bpy)_3]^{2+}$, which was oxidized at the UME. The oxidized mediator caused the oxidation and dissolution of the deposited silver nanoparticles that were not interconnected to each other. In these applications, the exact quantitative relation between the SECM signal and the amount of analyte is not well understood and it is empirically established by calibration curves that might be sensitive to a number of factors that are difficult to control.

Wang and Zhou [83] detected hybridized DNA on a chip by electrogenerating $[Ru(bpy)_3]^{3+}$ at the UME in an SECM experiment. $[Ru(bpy)_3]^{3+}$ is a strong oxidizer and can oxidize guanine, one of the bases in DNA. Where DNA was located on the chip, this led to an enhanced current at the UME. Turcu *et al.* [84,85] immobilized ssDNA on microspots at a gold surface. The location of the ssDNA could be imaged using a negatively charged redox mediator ($[Fe(CN)_6]^{3-/4-}$) in the SECM feedback mode. The mediator regeneration proceeded by an heterogeneous electron transfer to the gold surface of the chip. At the spots of the negatively charged ssDNA, the diffusion of the negatively charged mediator to the gold surface was blocked and a reduced feedback current was observed. The observed current was further reduced, when hybridization with the complementary strand occurred and the negative charge at the surface was increased.

Fortin *et al.* [86] reported an alternative detection scheme based on an enzymatic amplification without imaging the enzyme activity in the feedback mode. They first produced the DNA microarray by depositing a polypyrrole pattern with a procedure developed earlier by Kranz *et al.* [87]. By using a mixture of pyrrole and functionalized monomers covalently attached to a 15mer ssDNA oligomer, polypyrrole spots were formed carrying the 15mer oligomer as single DNA strand [88]. Hybridization was performed with biotinylated complementary ssDNA strands. After reaction with streptavidin and biotinylated HRP, the enzymatically catalyzed oxidation of soluble 4-chloro-1-naphthol to insoluble 4-chloro-1-naphthon was used to create an insoluble, insulating film at the polypyrrole spots. SECM feedback imaging with the mediator $[Ru(NH_3)_6]^{3+}$ visualized these regions by reduced currents compared to the bare gold surface [86].

37.2.5 Application of chemo and biosensors as SECM probes

In principle, any amperometric UME can be regarded as a chemical sensor. In this section, we want to review the use of electrodes that are further modified in order to achieve a particular selectivity or that cannot be easily positioned using Eqs. (37.2) and (37.3). Daniele *et al.* [89,90] used mercury electrodes deposited onto platinum to perform a stripping analysis of heavy metals released from sediments. In this case, the positioning could be achieved with a conventional redox mediator, because the theory for approach curves of a sphere-capped electrode is known. For potentiometric probes such an approach will clearly not work [29,43,44,91–98]. Also, enzyme electrodes [42,75,99–102] and modified amperometric electrodes [103–105] have been applied as SECM probes although Eqs. (37.2) and (37.3) cannot be used to position them. The proper positioning of the probe represents the main problem in using these electrodes. In the case of potentiometric probes also the long response time might be an important factor, in particular, when quantitative results shall be obtained. Occasionally sensitivity may become a problem and then slightly larger electrodes can be used.

37.2.5.1 Dual probes

Several approaches have been used successfully for the positioning of these electrodes. In initial attempts, Horrocks *et al.* [43] prepared potentiometric microelectrodes from Sb that could also serve as amperometric electrodes at other potentials. Positioning was performed by exploiting the amperometric function, while the imaging in the GC mode was done using the potentiometric function. Other groups prepared dual probes for instance in θ-shaped capillaries. Wei *et al.* [92] filled one capillary with Ga and used this electrode for positioning the probe in the feedback mode. The other capillary was used to form a liquid membrane ion-selective electrode that was used for imaging in the GC mode. Alternatively, Wei *et al.* [93] filled the second barrel with an electrolyte and used it in the same manner as in scanning ion-conductance microscopy. The positioning could be made via a conductivity measurement which follows a relation very similar to Eq. (37.2). Horrocks *et al.* [42] applied a small ac potential perturbation to amperometric enzyme UME in order to position them via the solution resistance. More recently, Yasukawa *et al.* [105] proposed probes with two amperometric electrodes. One is used for positioning, and the other for recording the signal of interest. In order to avoid feedback effects between the two electrodes of the probe, the glass body had a particular shape separating the two electrodes. Pailleret *et al.* [103] and Isik *et al.* [104] also used dual probes. One was an amperometric NO sensor with a diameter of 50 μm, and the other was an amperometric electrode used for positioning in the feedback mode. This probe has been used to measure topography and NO release from endothelial cells (Fig. 37.11). Kueng *et al.* [99] detected adenosine triphosphate (ATP) released from a membrane with a dual electrode probe. One electrode was an unmodified amperometric electrode and allowed positioning of the probe. The second electrode was a dual enzyme electrode operating on competitive consumption of the substrate glucose between the immobilized enzymes glucose oxidase and hexokinase involving ATP as a co-substrate.

37.2.5.2 Use of shear force distance control systems

Several attempts exist to design a setup by means of feeding back a *current*-independent signal into an actuator that keeps d constant. Most convincing results could be obtained by systems that detect mechanical shear forces between a UME that vibrates parallel to the sample surface with amplitude in the nanometer range ($\ll r_T$). Since the electrodes are immersed in a viscous solution, stable detection of the shear forces appears to be challenging. The vibrations can be excited by a tuning fork similar to the concept in scanning near-field optical microscopy [106–109]. However, since the resonance frequency is determined by the tuning fork but the UME represents usually a much bigger mass than the legs of the tuning fork, the mechanical resonator tends to be ill-defined and the number of published practical examples is rather limited despite the high expectations which have been raised earlier. Another approach excites the microelectrode with a frequency tuned to the mechanical properties of the microelectrode itself. The vibration amplitude is detected either by projecting a diffraction pattern produced by a light emitting diode onto a split photo diode [110] or by detecting the vibration amplitude with a second piezo attached to the UME [111]. Both approaches have been largely popularized in the SECM community by the work of Schuhmann *et al.* [19,64,95,112–115]. Meanwhile, a commercially self-standing shear force system is available that can be attached to any well-designed

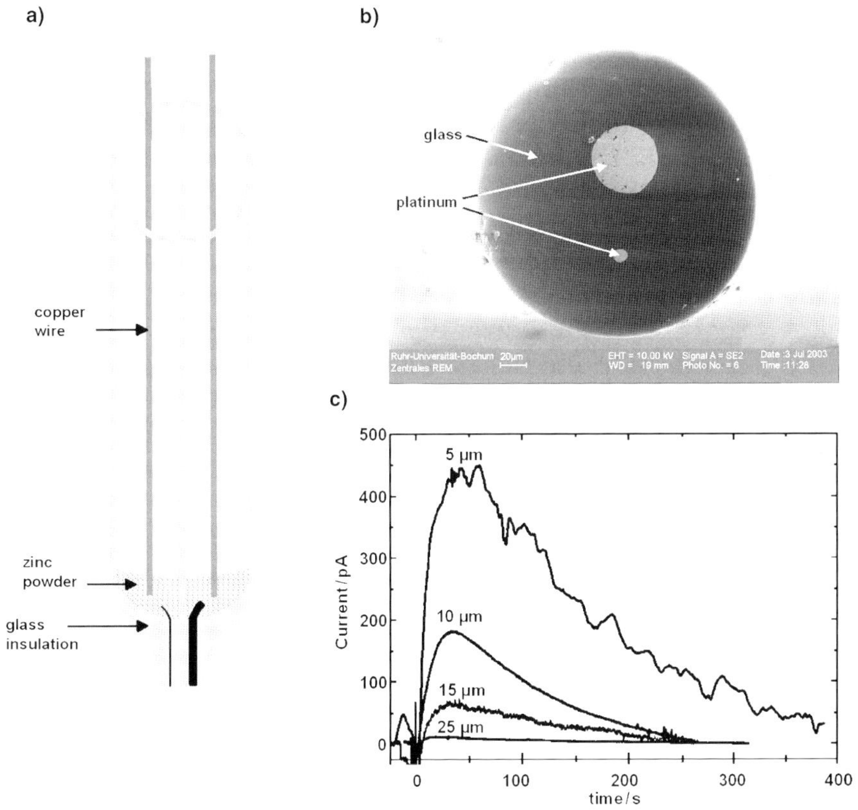

Fig. 37.11. Use of an NO microsensor for detection of the NO release from cultured endothelial cells. The sensor is a dual probe microsensor. The small sensor is a bare Pt UME used to position the sensor in the feedback mode. Onto the larger Pt electrode a polymer was deposited from an acrylic resin containing Ni(4-*N*-tetramethyl) pyridyl porphyrin and served as amperometric NO sensor. (a) Schematic of the sensor, (b) optical microphotograph of the sensor surface. (c) Response of the NO sensor to the stimulation of the cells with bradykinin at different distances of the sensor to the surface of the cells. Reprinted with permission from Ref. [104]. Copyright 2004, American Chemical Society.

SECM instrumentation (see footnote 1). Nevertheless, the use of a shear force system adds another level of complexity to the experiment because the mechanical properties of the UME, the sample, and the entire setup become important and have to be optimized. Shear force systems have been used to investigate electrode arrays [22,114,116] and detect metabolic activity of living cells [115] or following cell–cell communication processes [112] by the group of Schuhmann.

37.2.5.3 Integration of electrodes into other scanning probes

A number of combined scanning probe techniques have been realized, including combinations of SECM and ECSTM [117,118]. However, the biggest promise holds the combination of SECM and scanning force microscopy (SFM). This area was pioneered by Kranz *et al.* [119] and Macpherson and Unwin [120]. Both groups have used different integration concepts. Kranz *et al.* [121] integrated a frame-shaped UME into an SFM probe by coating the SFM probe with gold and an insulating layer and later cut the apex of the probe by fast ion bombardment (FIB). This produces an insulating thorn that enables a constant working distance of the recessed frame electrode. This idea has proven to be extensible to a number of different practical problems. The signals can be modeled and quantitative agreement between simulations and experiments was observed [122]. The cantilevers show good imaging properties and can also be used for simultaneous measurements in SFM tapping mode and SECM [123]. This enabled topographic and activity imaging of enzyme-loaded gel pads [123]. After electrolytic deposition of mercury on the gold electrode, the application to geochemical problems could be demonstrated [124]. Based on the integrated electrodes, Kueng *et al.* [75,102] formed miniaturized biosensors by entrapping the enzymes into a polyelectrolyte gel and used them to monitor the transport of ATP through a model membrane. The development by Macpherson *et al.* [125] was continued and resulted in disk-shaped microdisk electrodes of 80 nm diameter based on metalized, insulated and FIB-milled carbon nanotubes. Such probes have been used in taping mode and contact mode to image a 2 μm diameter Pt disk electrode [125]. The electrochemical activity was recorded in the GC mode using a lift-off scanning regime.

37.3 CONCLUSION

SECM has been applied to study a large variety of heterogeneous reactions at solid–liquid interfaces. The examples of high practical relevance include studies of corrosion mechanisms of passivated metals, heterogeneous catalysis for fuel cell materials, catalysis by immobilized enzymes used in biosensors, and metabolic activity of single cells and intact organs. The SECM image provides a direct representation of interfacial reactivity even in those cases where the topography of the interface does not change during the reaction, e.g. during an electron transfer from an electrode to a dissolved compound without accompanying deposition or dissolution processes.

Furthermore, detailed kinetic analysis can be obtained by recording approach curves and by fitting the data to models of the interfacial reaction and mass transfer processes. This analysis can provide quantitative results because the mass transport between the UME and the sample is usually controlled by diffusion, a theoretically well-understood phenomenon that can be described by continuum models. This makes SECM a very interesting tool to analyze a large variety of interfacial *processes*—many of them of high relevance for electrochemical and biochemical sensors. Among the processes studied are the permeability of passivating layers, the kinetics of heterogeneous electron transfer, the kinetics of immobilized enzymes, the distribution of reactive regions on sensor surfaces, the local activity of immobilized enzymes in particular on microstructured sensor surfaces, or the overlapping of diffusion layers in microstructured sensors.

The possibility to modify surfaces by a large variety of chemically *well-defined* reactions holds a great potential for prototyping of advanced sensor surfaces.

SECM can also be used as a read-out tool for protein and DNA chips. While any scanning technique is perhaps too slow for a routine application, SECM is ideally suited for testing concepts in prototype applications which are later transferred to microstructured devices without moving parts.

Finally, SECM offers a new application area for miniaturized electrochemical and biochemical sensors. They can be used in connection with a positioning system to solve, for instance, problems of cell biology, material science, and interfacial geochemistry. Since SECM instruments are now available from different commercial sources, a much broader application in the electrochemical sensor community is expected within the next years.

The lateral resolution in routine SECM experiments does not reach the resolution easily achieved with SFM or STM. Therefore, it should not be used for pure topographic imaging. The importance of SECM derives from its unique ability to analyze local chemical fluxes.

ACKNOWLEDGMENTS

The work on SECM in the authors' laboratory has been supported over the years by the Deutsche Forschungsgemeinschaft, The State of Lower Saxony, the VW Foundation, the Alexander von Humboldt Foundation, The Fonds of the Chemical Industries and the Hanse Institute of

Advanced Studies. G.W. would like to express sincere thanks to the University of Oldenburg for granting a sabbatical and in particular to Carl H. Hamann for taking up the teaching load during that time!

REFERENCES

1 A.J. Bard and M.V. Mirkin (Eds.), *Scanning Electrochemical Microscopy*, Marcel Dekker, New York, Basel, 2001.
2 A.L. Barker, M. Gonsalves, J.V. Macpherson, C.J. Slevin and P.R. Unwin, *Anal. Chim. Acta*, 385 (1999) 223–240.
3 G. Wittstock, Imaging localized reactivities of surfaces by scanning electrochemical microscopy. In: K. Wandelt and S. Thurgate (Eds.), *Solid–Liquid Interfaces, Macroscopic Phenomena—Microscopic Understanding*, Springer-Verlag, Berlin, Heidelberg, 2003, pp. 335–364.
4 G. Wittstock, *Fresenius J. Anal. Chem.*, 370 (2001) 303–315.
5 B.R. Horrocks, Scanning electrochemical microscopy. In: P.R. Unwin, A.J. Bard and M. Stratmann (Eds.), *Instrumentation and Electroanalytical Chemistry*, Wiley-VCH, Weinheim, 2003, pp. 444–490.
6 R.C. Engstrom, M. Weber, D.J. Wunder, R. Burges and S. Winquist, *Anal. Chem.*, 58 (1986) 844–848.
7 H.-Y. Liu, F.-R.F. Fan, C.W. Lin and A.J. Bard, *J. Am. Chem. Soc.*, 108 (1986) 3838–3839.
8 J. Kwak and A.J. Bard, *Anal. Chem.*, 61 (1989) 1794–1799.
9 A.J. Bard, F.-R.F. Fan, J. Kwak and O. Lev, *Anal. Chem.*, 61 (1989) 132–138.
10 J. Wang, L.-H. Wu and R. Li, *J. Electroanal. Chem.*, 272 (1989) 285–292.
11 Y. Saito, *Rev. Polarogr.*, 15 (1968) 177–187.
12 J.L. Amphlett and G. Denuault, *J. Phys. Chem. B*, 102 (1998) 9946–9951.
13 J. Kwak and A.J. Bard, *Anal. Chem.*, 61 (1989) 1221–1227.
14 D. Mandler and A.J. Bard, *Langmuir*, 6 (1990) 1489–1494.
15 D.T. Pierce, P.R. Unwin and A.J. Bard, *Anal. Chem.*, 64 (1992) 1795–1804.
16 A.J. Bard, F.-R.F. Fan, D.T. Pierce, P.R. Unwin, D.O. Wipf and F. Zhou, *Science*, 254 (1991) 68–74.
17 C. Wei, A.J. Bard and M.V. Mirkin, *J. Phys. Chem.*, 99 (1995) 16033–16042.
18 C.G. Zoski, N. Simjee, O. Guenat and M. Koudelka-Hep, *Anal. Chem.*, 76 (2004) 62–72.
19 M. Etienne, A. Schulte and W. Schuhmann, *Electrochem. Commun.*, 6 (2004) 288–293.
20 J. Ufheil, K. Borgwarth and J. Heinze, *Anal. Chem.*, 74 (2002) 1316–1321.

21 O. Köster, W. Schuhmann, H. Vogt and W. Mokwa, *Sens. Actuators B*, 76 (2001) 573–581.

22 D.J. Strike, A. Hengstenberg, M. Quinto, C. Kurzawa, M. Koudelka-Hep and W. Schuhmann, *Mikrochim. Acta*, 131 (1999) 47–55.

23 G. Wittstock, B. Gründig, B. Strehlitz and K. Zimmer, *Electroanalysis*, 10 (1998) 526–531.

24 G. Wittstock, H. Emons, T.H. Ridgway, E.A. Blubaugh and W.R. Heineman, *Anal. Chim. Acta*, 298 (1994) 285–302.

25 G. Wittstock, H. Emons, M. Kummer, J.R. Kirchhoff and W.R. Heineman, *Fresenius J. Anal. Chem.*, 348 (1994) 712–718.

26 K. Jambunathan, B.C. Shah, J.L. Hudson and A.C. Hillier, *J. Electroanal. Chem.*, 500 (2001) 279–289.

27 J.L. Fernandez, D.A. Walsh and A.J. Bard, *J. Am. Chem. Soc.*, 127 (2005) 357–365.

28 J.L. Fernandez, N. Mano, A. Heller and A.J. Bard, *Angew. Chem. Int. Ed. Engl.*, 43 (2004) 6355–6357.

29 K. Toth, G. Nagy, B.R. Horrocks and A.J. Bard, *Anal. Chim. Acta*, 282 (1993) 239–246.

30 B. Gründig, G. Wittstock, U. Rüdel and B. Strehlitz, *J. Electroanal. Chem.*, 395 (1995) 143–157.

31 M. Janotta, D. Rudolph, A. Kueng, C. Kranz, H.-S. Voraberger, W. Waldhauser and B. Mizaikoff, *Langmuir*, 20 (2004) 8634–8640.

32 C.A. Wijayawardhana, G. Wittstock, H.B. Halsall and W.R. Heineman, *Anal. Chem.*, 72 (2000) 333–338.

33 C. Kranz, G. Wittstock, H. Wohlschläger and W. Schuhmann, *Electrochim. Acta*, 42 (1997) 3105–3111.

34 D.T. Pierce and A.J. Bard, *Anal. Chem.*, 65 (1993) 3598–3604.

35 C. Zhao and G. Wittstock, *Anal. Chem.*, 76 (2004) 3145–3154.

36 H. Shiku, T. Takeda, H. Yamada, T. Matsue and I. Uchida, *Anal. Chem.*, 67 (1995) 312–317.

37 H. Shiku, T. Matsue and I. Uchida, *Anal. Chem.*, 68 (1996) 1276–1278.

38 G. Wittstock, T. Wilhelm, S. Bahrs and P. Steinrücke, *Electroanalysis*, 13 (2001) 669–675.

39 J. Zaumseil, G. Wittstock, S. Bahrs and P. Steinrücke, *Fresenius J. Anal. Chem.*, 367 (2000) 352–355.

40 T. Wilhelm and G. Wittstock, *Mikrochim. Acta*, 133 (2000) 1–9.

41 G. Wittstock and W. Schuhmann, *Anal. Chem.*, 69 (1997) 5059–5066.

42 B.R. Horrocks, D. Schmidtke, A. Heller and A.J. Bard, *Anal. Chem.*, 65 (1993) 3605–3614.

43 B.R. Horrocks, M.V. Mirkin, D.T. Pierce, A.J. Bard, G. Nagy and K. Toth, *Anal. Chem.*, 65 (1993) 1213–1224.

44 B.R. Horrocks and M.V. Mirkin, *J. Chem. Soc.*, 94 (1998) 1115–1118.

45 H. Shiku, Y. Hara, T. Matsue, I. Uchida and T. Yamauchi, *J. Electroanal. Chem.*, 438 (1997) 187–190.

46 T. Wilhelm and G. Wittstock, *Electrochim. Acta*, 47 (2001) 275–281.
47 C. Zhao and G. Wittstock, *Biosens. Bioelectron.*, 20 (2005) 1277–1284.
48 C.A. Wijayawardhana, G. Wittstock, H.B. Halsall and W.R. Heineman, *Electroanalysis*, 12 (2000) 640–644.
49 G. Wittstock, S.H. Jenkins, H.B. Halsall and W.R. Heineman, *Nanobiology*, 4 (1998) 153–162.
50 G. Wittstock, K.-j. Yu, H.B. Halsall, T.H. Ridgway and W.R. Heineman, *Anal. Chem.*, 67 (1995) 3578–3582.
51 C. Zhao and G. Wittstock, *Angew. Chem. Int. Ed. Engl.*, 43 (2004) 4170–4172.
52 C. Zhao, J.K. Sinha, C.A. Wijayawardhana and G. Wittstock, *J. Electroanal. Chem.*, 561 (2004) 83–91.
53 Y.N. Antonenko, P. Pohl and E. Rosenfeld, *Arch. Biochem. Biophys.*, 333 (1996) 225–232.
54 C.E.M. Berger, B.R. Horrocks and H.K. Datta, *J. Endocrinol.*, 158 (1998) 311–318.
55 B.R. Horrocks and G. Wittstock, Biological systems. In: A.J. Bard and M.V. Mirkin (Eds.), *Scanning Electrochemical Microscopy*, Marcel Dekker, New York, Basel, 2001, pp. 445–519.
56 I. Turyan, T. Matsue and D. Mandler, *Anal. Chem.*, 72 (2000) 3431–3435.
57 W.B. Nowall, N. Dontha and W.G. Kuhr, *Biosens. Bioelectron.*, 13 (1998) 1237–1244.
58 S.E. Rosenwald, N. Dontha and W.G. Kuhr, *Anal. Chem.*, 70 (1998) 1133–1140.
59 L.X. Tiefenauer and C. Padeste, *Chimia*, 53 (1999) 62–66.
60 D. Mandler, S. Meltzer and I. Shohat, *Isr. J. Chem.*, 36 (1996) 73–80.
61 D. Mandler, Micro- and nanopatterning using the scanning electrochemical microscope. In: A.J. Bard and M.V. Mirkin (Eds.), *Scanning Electrochemical Microscopy*, Marcel Dekker, New York, Basel, 2001, pp. 593–627.
62 R.C. Tenent and D.O. Wipf, *J. Electrochem. Soc.*, 150 (2003) E131–E139.
63 W.B. Nowall, D.O. Wipf and W.G. Kuhr, *Anal. Chem.*, 70 (1998) 2601–2606.
64 C. Kranz, H.E. Gaub and W. Schuhmann, *Adv. Mater.*, 8 (1996) 634–637.
65 T. Wilhelm and G. Wittstock, *Angew. Chem. Int. Ed.*, 42 (2003) 2247–2250.
66 S. Sauter and G. Wittstock, *J. Solid State Electrochem.*, 5 (2001) 205–211.
67 Y.-S. Torisawa, H. Shiku, T. Yasukawa, M. Nishizawa and T. Matsue, *Biomaterials*, 26 (2005) 2165–2172.
68 H. Shiku, I. Uchida and T. Matsue, *Langmuir*, 13 (1997) 7239–7244.
69 D. Oyamatsu, N. Kanaya, H. Shiku, M. Nishizawa and T. Matsue, *Sens. Actuators B*, B91 (2003) 199–204.
70 N. Dontha, W.B. Nowall and W.G. Kuhr, *Anal. Chem.*, 69 (1997) 2619–2625.

71 T. Wilhelm and G. Wittstock, *Langmuir*, 18 (2002) 9485–9493.
72 M. Niculescu, S. Gaspar, A. Schulte, E. Csoregi and W. Schuhmann, *Biosens. Bioelectron.*, 19 (2004) 1175–1184.
73 F. Turcu, G. Hartwich, D. Schaefer and W. Schuhmann, *Macromol. Rapid Commun.*, 26 (2005) 325–330.
74 C. Kurzawa, A. Hengstenberg and W. Schuhmann, *Anal. Chem.*, 74 (2002) 355–361.
75 A. Kueng, C. Kranz, A. Lugstein, E. Bertagnolli and B. Mizaikoff, *Angew. Chem. Int. Ed. Engl.*, 44 (2005) 3419–3422.
76 T. Kaya, K. Nagamine, N. Matsui, T. Yasukawa, H. Shiku and T. Matsue, *Chem. Commun.* (2004) 248–249.
77 D. Ogasawara, Y. Hirano, T. Yasukawa, H. Shiku, T. Matsue, K. Kobori, K. Ushizawa and S. Kawabata, *Chem. Sens.*, 20 (2004) 139–141.
78 N. Motochi, Y. Hirano, Y. Abiko, D. Oyamatsu, M. Nishizawa, T. Matsue, K. Ushizawa and S. Kawabata, *Chem. Sens.*, 18 (2002) 172–174.
79 S. Kassai, A. Yokota, H. Zhou, M. Nishizawa, T. Onouchi, K. Niwa and T. Matsue, *Anal. Chem.*, 72 (2000) 5761–5765.
80 C.A. Wijayawardhana, N.J. Ronkainen-Matsuno, S.M. Farrel, G. Wittstock, H.B. Halsall and W.R. Heineman, *Anal. Sci.*, 17 (2001) 535–538.
81 J. Wang, F. Song and F. Zhou, *Langmuir*, 18 (2002) 6653–6658.
82 M. Carano, N. Lion, J.-P. Abid and H.H. Girault, *Electrochem. Commun.*, 6 (2004) 1217–1221.
83 J. Wang and F. Zhou, *J. Electroanal. Chem.*, 537 (2002) 95–102.
84 F. Turcu, A. Schulte, G. Hartwich and W. Schuhmann, *Angew. Chem. Int. Ed. Engl.*, 43 (2004) 3482–3485.
85 F. Turcu, A. Schulte, G. Hartwich and W. Schuhmann, *Biosens. Bioelectron.*, 20 (2004) 925–932.
86 E. Fortin, P. Mailley, L. Lacroix and S. Szunerits, *Analyst*, 131 (2006) 186–193.
87 C. Kranz, M. Ludwig, H.E. Gaub and W. Schuhmann, *Adv. Mater.*, 7 (1995) 38–40.
88 S. Szunerits, N. Knorr, R. Calemczuk and T. Livache, *Langmuir*, 20 (2004) 9236–9241.
89 S. Daniele, I. Ciani, C. Bragato and M.A. Baldo, *J. Phys. IV*, 107 (2003) 353–356.
90 S. Daniele, C. Bragato, I. Ciani and M.A. Baldo, *Electroanalysis*, 15 (2003) 621–628.
91 G. Denuault, G. Nagy and K. Toth, Potentiometric probes. In: A.J. Bard and M.V. Mirkin (Eds.), *Scanning Electrochemical Microscopy*, Marcel Dekker, New York, Basel, 2001, pp. 397–444.
92 C. Wei, A.J. Bard, I. Kapui, G. Nagy and K. Toth, *Anal. Chem.*, 68 (1996) 2651–2655.
93 C. Wei, A.J. Bard, G. Nagy and K. Toth, *Anal. Chem.*, 67 (1995) 1346–1356.

94 K. Toth, G. Nagy, C. Wei and A.J. Bard, *Electroanalysis*, 7 (1995) 801–810.
95 M. Etienne, A. Schulte, S. Mann, G. Jordan, I.D. Dietzel and W. Schuhmann, *Anal. Chem.*, 76 (2004) 3682–3688.
96 A. Schulte, S. Belger and W. Schuhmann, *Mater. Sci. Forum*, 394–395 (2002) 145–148.
97 C.E.M. Berger, B.R. Horrocks and H.K. Datta, *Electrochim. Acta*, 44 (1999) 2677–2683.
98 E. Klusmann and J.W. Schultze, *Electrochim. Acta*, 42 (1997) 3123–3134.
99 A. Kueng, C. Kranz and B. Mizaikoff, *Biosens. Bioelectron.*, 21 (2005) 346–353.
100 A. Kueng, C. Kranz and B. Mizaikoff, *Biosens. Bioelectron.*, 19 (2004) 1301–1307.
101 A. Kueng, C. Kranz and B. Mizaikoff, *Sensor Lett.*, 1 (2003) 2–15.
102 A. Kueng, C. Kranz, A. Lugstein, E. Bertagnolli and B. Mizaikoff, *Methods Mol. Biol. (Totowa, NJ)*, 300 (2005) 403–415.
103 A. Pailleret, J. Oni, S. Reiter, S. Isik, M. Etienne, F. Bedioui and W. Schuhmann, *Electrochem. Commun.*, 5 (2003) 847–852.
104 S. Isik, M. Etienne, J. Oni, A. Bloechl, S. Reiter and W. Schuhmann, *Anal. Chem.*, 76 (2004) 6389–6394.
105 T. Yasukawa, T. Kaya and T. Matsue, *Anal. Chem.*, 71 (1999) 4631–4641.
106 P. James, L.F. Garfias-Mesias, P.J. Moyer and W.H. Smyrl, *J. Electrochem. Soc.*, 145 (1998) 64–66.
107 Y. Hirano, D. Oyamatsu, T. Yasukawa, H. Shiku and T. Matsue, *Electrochemistry (Tokyo)*, 72 (2004) 137–142.
108 Y. Lee and A.J. Bard, *Anal. Chem.*, 74 (2002) 3626–3633.
109 H. Yamada, H. Fukumoto, T. Yokoyama and T. Koike, *Anal. Chem.*, 77 (2005) 1785–1790.
110 M. Ludwig, C. Kranz, W. Schuhmann and H.E. Gaub, *Rev. Sci. Instrum.*, 66 (1995) 2857–2860.
111 R. Brunner, A. Bietsch, O. Hollricher and O. Marti, *Rev. Sci. Instrum.*, 68 (1997) 1769–1772.
112 L.P. Bauermann, W. Schuhmann and A. Schulte, *Phys. Chem. Chem. Phys.*, 6 (2004) 4003–4008.
113 B. Ballesteros Katemann, A. Schulte and W. Schuhmann, *Electroanalysis*, 16 (2004) 60–65.
114 B. Ballesteros Katemann, A. Schulte and W. Schuhmann, *Chem. Eur. J.*, 9 (2003) 2025–2033.
115 A. Hengstenberg, A. Blöchl, I.D. Dietzel and W. Schuhmann, *Angew. Chem. Int. Ed. Engl.*, 40 (2001) 905–907.
116 B. Ballesteros Katemann, A. Schulte, E.J. Calvo, M. Koudelka-Hep and W. Schuhmann, *Electrochem. Commun.*, 4 (2002) 134–138.
117 O. Sklyar, T.H. Treutler, N. Vlachopoulos and G. Wittstock, *Surf. Sci.*, 597 (2005) 181–195.

118 T.H. Treutler and G. Wittstock, *Electrochim. Acta*, 48 (2003) 2923–2932.
119 C. Kranz, G. Friedbacher, B. Mizaikoff, A. Lugstein, J. Smolier and E. Bertagnolli, *Anal. Chem.*, 73 (2001) 2491–2500.
120 J.V. Macpherson and P.R. Unwin, *Anal. Chem.*, 72 (2000) 276–285.
121 A. Lugstein, E. Bertagnolli, C. Kranz, A. Kueng and B. Mizaikoff, *Appl. Phys. Lett.*, 81 (2002) 349–351.
122 O. Sklyar, A. Kueng, C. Kranz, B. Mizaikoff, A. Lugstein, E. Bertagnolli and G. Wittstock, *Anal. Chem.*, 77 (2005) 764–771.
123 A. Kueng, C. Kranz, A. Lugstein, E. Bertagnolli and B. Mizaikoff, *Angew. Chem. Int. Ed. Engl.*, 42 (2003) 3238–3240.
124 D. Rudolph, S. Neuhuber, C. Kranz, M. Taillefert and B. Mizaikoff, *Analyst*, 129 (2004) 443–448.
125 D.P. Burt, N.R. Wilson, J.M.R. Weaver, P.S. Dobson and J.V. Macpherson, *Nano Lett.*, 5 (2005) 639–643.

Chapter 38

Gold nanoparticles in DNA and protein analysis

María Terra Castañeda, Salvador Alegret and Arben Merkoçi

38.1 INTRODUCTION

According to IUPAC recommendations, a biosensor is a self-contained integrated receptor–transducer device, which is capable of providing selective quantitative or semi-quantitative analytical information using a biological recognition element [1]. Biosensors convert chemical information to an electrical signal through a molecular recognition reaction on a physical transducer. The amount of signal generated is proportional to the concentration of the analyte, allowing for both quantitative and qualitative measurements in time [2].

A biosensor consists of three main components: a biological recognition element or bioreceptor for detection, the transducer, component for readout and an output system. The bioreceptor is a biomolecule, such as enzymes, antibodies, receptors proteins, nucleic acids, cells or tissue sections, which recognises the target analyte. Generally, there are three principal classes of biosensors in terms of their biological component: (1) biocatalytic, depending on the use of pure or crude enzymes to moderate a biochemical reaction and using the chemical transformation of the biomarker as the source of signal; (2) bioaffinity, relying on the use of proteins or DNA to recognise and bind a particular target; and (3) microbe based that use microorganisms as the biological recognition element. These generally involve the measurement of microbial respiration, or its inhibition, by the analyte of interest. Biosensors have also been developed using genetically modified microorganisms (GMOs) that recognise and report the presence of specific environmental pollutants [3–5]. Alternative (bio)recognition molecules include RNA and DNA aptamers [6], molecularly imprinted polymers and templated surfaces [7,8]. The aptamers are functional nucleic acids selected from combinatorial oligonucleotide

Comprehensive Analytical Chemistry 49
Alegret and Merkoçi (Eds)
Volume 49 ISSN: 0166-526X DOI: 10.1016/S0166-526X(06)49038-3

libraries by *in vitro* selection against a variety of targets, such as small organic molecules, peptides, proteins and even whole cells [9].

Recently, introduction of DNA analogous detection systems such as the use of peptide nucleic acid (PNA) technology has attracted considerable attention [10]. The PNAs are synthetic analogues of DNA that hybridise with complementary DNAs or RNAs with high affinity and specificity, essentially because of an uncharged and flexible polyamide backbone. The unique physico-chemical properties of PNAs have led to the development of a variety of research and diagnostic assays where these are used as molecular hybridisation probes [11].

The transducers most commonly employed in biosensors are: (a) Electrochemical: amperometric, potentiometric and impedimetric; (b) Optical: vibrational (IR, Raman), luminescence (fluorescence, chemiluminescence); (c) Integrated optics: (surface plasmon resonance (SPR), interferometery) and (d) Mechanical: surface acoustic wave (SAW) and quartz crystal microbalance (QCM) [4,12].

The biosensors can be divided into: nonlabelled or label-free types, which are based on the direct measurement of a phenomenon occurring during the biochemical reactions on a transducer surface; and labelled, which relies on the detection of a specific label. Research into 'label-free' biosensors continues to grow [13]; however 'labelled' ones are more common and are extremely successful in a multitude of platforms.

Recently, the field of biosensors for diagnostic purposes has acquired a great interest regarding the use of nanomaterials such as DNA and protein markers. Some biosensing assays based upon bioanalytical application of nanomaterials have offered significant advantages over conventional diagnostic systems with regard to assay sensitivity, selectivity and practicality [14–16]. Nanoparticles (NPs) in general [2,17] and quantum dots (QDs) [18] have been used particularly successfully as DNA tags.

Although the spectrum of NPs for labelling applications is relatively broad, this chapter discusses only the application of gold NPs (AuNPs) in electrochemical genosensors and immunosensors and some of the trends in their use for environmental and biomedical diagnostics between other application fields.

38.1.1 Current labelling technologies for affinity biosensors

It is well known that affinity biosensors, usually DNA sensors or immunosensors, require a biorecognition molecule that demonstrates a high affinity and specificity for the target biomarker.

The formation of double-stranded DNA upon hybridisation is commonly detected in connection with the use of an appropriate electroactive hybridisation intercalator or labelling DNA by a simple electroactive molecule or an adequate NP.

Electrochemical detection of hybridisation is mainly based on the differences in the electrochemical behaviour of the labels connected with double-stranded DNA (dsDNA) or single-stranded DNA (ssDNA). The labels for hybridisation detection can be anticancer agents, organic dyes, metal complexes, enzymes or metal NPs. There are basically four different pathways for electrochemical detection of DNA hybridisation: (1) A decrease/increase in the oxidation/reduction peak current of the label, which selectively binds with dsDNA/ssDNA, is monitored. (2) A decrease/increase in the oxidation/reduction peak current of electroactive DNA bases such as guanine or adenine is monitored. (3) The electrochemical signal of the substrate after hybridisation with an enzyme-tagged probe is monitored. (4) The electrochemical signal of a metal NP probe attached after hybridisation with the target is monitored [19].

On the other hand, the main types of immunoassays that can be performed by using labelled antibodies or antigens are: direct sandwich, competitive and indirect assays. The labels can be: enzymes (alkaline phosphatase, peroxidise or glucose oxidase); metal NPs (gold); fluorescent or electrochemiluminescent probes.

38.1.2 Nanoparticles as labels

The use of NPs as labels of DNA molecules [20,21] opened a new alternative for the detection of hybridisation events. NPs labels offer a number of advantages for DNA detection platforms as well as for immunoassays. The NP-based DNA and immunoassays are easy to use, offer good sequence selectivity and sensitivity. Moreover, the NP labelling technology is compatible with chip techniques [22].

Metal NPs have received tremendous attention in the field of bioanalytical science, in particular the sequence-specific DNA detection [23,24]. This is attributed to their unique properties in the conjugation with biological recognition elements (e.g., DNA oligonucleotide probe) as well as in the signal transduction with optical [22,25], electrical [26], microgravimetric [27] and electrochemical [23,28–30] methods.

The commonly used fluorescence labels provide good sensitivity but have various disadvantages. Fluorescent dye labels are expensive, they photobleach rapidly, and the records are not permanent. With metal NPs it seems to be possible to overcome most of the disadvantages described for fluorescence labelling. Metal NPs in general and particularly gold NPs (AuNPs) can be linked to DNA and other biomolecules without changing their ability to bind to their complementary biomolecule [20,21].

With regard to immunosensors, a number of different reporter groups are used, including enzymes which convert a substrate into a highly coloured product (enzyme-linked immunosorbent assay, 'ELISA') or which digest a substrate to give a photon of light to expose a film (chemiluminescence).

The current research in the field of electrochemical indicators is mainly to find new labels that have powerful electrochemical signals. Metal NPs with well-defined redox properties that can be followed by electrochemical stripping techniques are of great interest [28,31].

38.2 DNA ANALYSIS

Much interest in the development of different non-radioactive DNA sensing techniques for genomics analysis with respect to environmental applications [32] or diagnostics has been shown in the last decade.

The DNA biosensors or genosensors are based on the immobilisation of a single-stranded oligonucleotide on a transducer surface that recognises its complementary DNA sequence via the hybridisation reaction. The immobilisation of oligonucleotides (bioreceptors) onto transducer surfaces plays a crucial role in the performance of the genosensors or the bioanalytical device. Indeed, the bioreceptor must be readily accessible to the analyte [33]. The techniques for the immobilisation of purified oligonucleotides on an electrochemical transducer in the design of genosensors and their corresponding detection methods using these sensors were reviewed in detail by Pividori *et al.* [34]. The most generally employed are the following: (1) entrapment of the bioreceptor within a polymeric matrix such as agar gel, polyacrylamide, polypyrolle matrix or sol–gel matrices [35]; (2) covalent bonding onto surfaces (glassy carbon or carbon paste modified electrodes and polypyrolle, platinum or gold surfaces) by means of bifunctional groups or spacers such as glutaraldehyde, carbodiimide or a self-assembled monolayer of bifunctional silanes; and (3) adsorption of a thiol-modified bioreceptor for self-assembly on a gold surface [33,34].

In a typical configuration of a DNA biosensor, the bioreceptor is an ssDNA called the 'capture probe' that is immobilised by one of the methods described above. The analyte, a complementary ssDNA called the 'target DNA', is recruited to the surface via base-pairing interactions with the capture probe. In most current applications, the target DNA is the product of an amplification reaction (e.g., PCR or related method). This amplification step has two goals: (1) Allow reaching the lower detection limit of the device. (2) Allow incorporating a label into the target DNA that can then be used in the detection. The labels are generally fluorophores or a biotin molecule. When using biotin labelling, additional steps are required for further detection by use of an antibiotin antibody coupled to a fluorophore. The labelling efficiency of the target DNA depends on the polymerase's ability to incorporate the labelled nucleotides into the growing nucleic acid chain. Since this labelled nucleotide is not the natural substrate for the enzyme, the efficiency of the incorporation is low. An alternative to this approach involves a three-component 'sandwich' assay, in which the label is associated with a third DNA sequence (the signalling probe) designed to be complementary to an overhanging portion of the target. This dual hybridisation then eliminates the need to modify the target strand [33].

To date, the detection of the target DNA has been accomplished using fluorescent labels. In addition to their sensitivity, the great diversity of fluorophores allows as to compare gene expressions (e.g., in normal as compared to pathological cells). However, fluorescent labelling has some drawbacks. Fluorescent dyes and the equipment required to image them are expensive; the dyes rapidly photobleach and their manipulation requires special care to avoid interfering signals. As an alternative, NP labels offer excellent prospects for biological sensing due to their low cost, stability and unique physico-chemical properties [33]. The AuNPs functionalised with oligonucleotides are extensively used as tags in many highly sensitive and selective DNA recognition schemes by means of electrochemical sensing. The integration of NPs DNA labelling technology in chips has been also been studied in DNA analysis [23].

The hybridisation of a nucleic acid to its complementary target is one of the most definite and well-known molecular recognition events. Therefore, the hybridisation of a nucleic acid probe to its DNA target can provide a very high degree of accuracy for identifying complementary DNA sequences [32–36].

Diverse techniques for detection of DNA hybridisation have been developed. In most of them, the hybridisation event and electrochemical detection are carried out on the electrode surface [37–41]. In other cases, the electrode only acts as a detector of the hybridisation event

[42–48], which occurs in a separate step, either because it takes place in a microwell [42] or because the hybridisation event occurs on the surface of magnetic beads, which are separated from the hybridisation solution and then redissolved [43,44].

Electrochemical transduction of the hybridisation event can be classified into two categories: label-based and label-free approaches. The label-based approach can be further subdivided into intercalator/groove binder, non-intercalating marker, and NP. The label-free approach is based on the intrinsic electroactivity of the DNA purine bases or the change in interfacial properties (e.g., capacitance and electron transfer resistance) upon hybridisation [49].

The sensitivities of the label-based approach depend mainly on the specific activity of the labels linked to the oligonucleotide probe. Radioisotopic [48], fluorescent [50] and enzymatic [38,40,47] labels have been commonly used. Besides the above labels, NPs have attractive properties to act as DNA tags [18,51]. The fact that NPs present an excellent biocompatibility with biomolecules and display unique structural, electronic, magnetic, optical and catalytic properties have made them a very attractive material to be used as label [52,53].

Cai *et al.* [28] have synthesised AuNPs-labelled ssDNA as a probe to be hybridised with their complementary strand on chitosan-modified glassy carbon electrode (GCE). In their experiments, SH-ssDNA was self-assembled onto an AuNP (16 nm in diameter) (details given previously in Ref. [54]). After hybridisation, the AuNPs tags were detected by differential pulse voltammetry (DPV). The detection limit was 1.0×10^{-9} mol L^{-1} of 32-base synthesised complementary oligonucleotide.

A novel NP-based detection of DNA hybridisation based on magnetically induced direct electrochemical detection of the 1.4 nm Au_{67} QD tag linked to the target DNA was reported by Pumera *et al.* [55]. The Au_{67} NP tag was directly detected after the DNA hybridisation event, without need of acidic (i.e., HBr/Br_2) dissolution [55].

A triple-amplification bioassay that couples the carrier-sphere amplifying units (loaded with numerous AuNP tags) with the 'built-in' preconcentration feature of the electrochemical stripping detection and the catalytic enlargement of the multiple gold particle tags was demonstrated [56]. The gold-tagged beads were prepared by binding biotinylated AuNPs to streptavidin-coated polystyrene spheres. These beads were functionalised with a single-stranded oligonucleotide, which was further hybridised with a complementary oligonucleotide that was linked to a magnetic particle. The numerous AuNP labels associated with one ds-oligonucleotide pair were enlarged by the electroless

TABLE 38.1

Electrochemical genosensors using AuNPs tags, with potential biomedical and environmental applications

Transducer	AuNPs	Detection method	Sample/target DNA	Detection limit	Ref.
SPMBE	20-nm (Sigma)	ASV	Cell culture/ 406-bp HCMV-amplified	5 pM	[42]
PGE	5 ± 1.3 nm (prepared as described in Ref. [61]	DPV	Real PCR amplicons/ factor V Leiden Mutation	0.78 fmol	[29]
SPEs	10 nm (Sigma)	PSA	Synthetic/ breast cancer	4×10^{-9} M	[62]
M-GECE	10 nm (Sigma)	DPV	Synthetic/ breast cancer	33 pmols	[64]
M-GECE	10 nm (Sigma)	DPV	Synthetic/ cystic fibrosis	–	[64]

SPMBE = screen-printed microband electrode, ASV = anodic stripping voltammetry, HCMV = *human cytomegalovirus*, PGE = pencil-graphite electrode, DPV = differential pulse voltammetry, SPEs = screen-printed electrodes, PSA = potentiometric stripping analysis, M-GECE = magnetic graphite–epoxy composite electrode.

deposition of gold and transported to the electrode array with the use of the magnetic particle. Then, the Au assembly was dissolved upon treatment with HBr/Br_2 dissolution and electrochemically analysed by using electrochemical deposition/stripping voltammetry. Such a triple-amplification route offered a dramatic enhancement of the sensitivity.

Several protocols with potential biomedical and environmental application have been developed. Table 38.1 summarises information of some typical electrochemical genosensors based on DNA hybridisation detection using AuNPs tags.

38.2.1 Clinical

Electrochemical assays of nucleic acids based on DNA hybridisation have received considerable attention [34,57–60]. DNA hybridisation biosensors are a very attractive topic in the clinical diagnostics of inherited diseases and the rapid detection of infectious microorganisms.

Authier *et al.* [42] developed an electrochemical DNA detection method for the sensitive quantification of an amplified 406-base pair human cytomegalovirus DNA sequence (HCMV DNA). The HCMV DNA was extracted from cell culture, amplified by polymerase chain reaction (PCR), and then quantified by agarose gel electrophoresis. The HCMV DNA was immobilised on a microwell surface and hybridised with the complementary oligonucleotide-modified AuNPs (20 nm) followed by the release of Au(III) by treatment with acidic bromine–bromide solution, and the indirect determination of the solubilised Au(III) ions by anodic stripping voltammetry (ASV) at a sandwich-type screen-printed microband electrode (SPMBE). The combination of the sensitive Au(III) determination at a SPMBE with the large number of Au(III) released from each gold NP probe allows detection of as low as 5 pM amplified HCMV DNA fragment.

Ozsoz *et al.* [29] described an electrochemical genosensor based on AuNPs for detection of Factor V Leiden Mutation from PCR amplicons, which were obtained from real samples. The covalently bound amplicons onto a pencil graphite electrode (PGE) were then hybridised with oligonucleotide-AuNPs (the AuNPs, with average diameter of 5 ± 1.3 nm were prepared as reported in the literature [61]). The oxidation signal of AuNPs was measured directly by using DPV at PGE. Direct electrochemical oxidation of the AuNPs was observed at a stripping potential of approximately +1.2 V. The response is greatly enhanced due to the large electrode surface area and the availability of many oxidisable gold atoms in each NP label. The detection limit for PCR amplicons was as low as 0.78 fmol.

Wang *et al.* [62] developed an AuNPs based protocol for the detection of DNA segments related to the breast cancer *BRCA1* gene. This bioassay consisted in the hybridisation of a biotinylated target DNA to streptavidin-coated magnetic bead-binding biotinylated probe and followed by binding of streptavidin-coated AuNPs (5 nm) to the target DNA, dissolution of the AuNPs and electrochemical detection using potentiometric stripping analysis (PSA) of the dissolved gold tag at single-use thick-film carbon electrodes, obtaining a detection limit of 4×10^{-9} M.

The sensitivity of the detection is usually improved by the silver enhancement method. A better detection limit was reported when a silver enhancement method was employed, based on the precipitation of silver on AuNPs tags and its dissolution (in HNO_3) and subsequent electrochemical potentiometric stripping detection [43]. The new silver-enhanced colloidal gold stripping detection strategy represented an attractive alternative to indirect optical affinity assays of nucleic acids and other biomolecules.

Wang *et al.* [63] also reported a new NP-based protocol for detecting DNA hybridisation based on a magnetically induced solid-state electrochemical stripping detection of metal tags. The new bioassay involves the hybridisation of a target oligonucleotide to probe-coated magnetic beads, followed by binding of the streptavidin-coated AuNPs (5 nm) to the captured target, catalytic silver precipitation on the AuNPs tags, a magnetic 'collection' of the DNA-linked particle assembly and solid-state stripping detection (PSA) at a thick-film carbon electrode with a magnet placed below the working electrode. The high sensitivity and selectivity of the new protocol was illustrated for the detection of DNA segments related to the *BRCA1* breast cancer gene. A detection limit of around 150 $pg\,mL^{-1}$ (i.e., 1.2 fmol) was obtained.

The application of AuNPs as oligonucleotide labels in DNA hybridisation detection assays using a magnetic graphite–epoxy composite electrode (M-GECE) has been reported by Pumera *et al.* [55] (a detailed description of DNA detection using AuNPs as labels is given in Procedure 53–see the accompanying CD-Rom).

Recently, Castañeda *et al.* [64] developed two AuNPs based genosensor designs, for detection of DNA hybridisation. Both assay formats were based on a magnetically induced direct electrochemical detection of the AuNPs tags on M-GECE. The AuNPs tags are also directly detected after the DNA hybridisation event without the need of acidic dissolution.

The first assay is based on the hybridisation between two single-strand biotin-modified DNA probes: a capture DNA probe and a target DNA related to the *BRCA1* breast cancer gene, which is coupled with streptavidin–AuNPs (10 nm). The second assay (Fig. 38.1A–F) is based on hybridisation between three single-strand DNA probes: a biotin-modified capture DNA probe (CF-A), a target DNA, related to cystic fibrosis gene (CF-T) and DNA signaling probe (CF-B) modified with AuNPs via biotin–streptavidin complexation reactions. In this assay, the target is 'sandwiched' between the other two probes. The electrochemical detection of AuNPs by DPV was performed in both cases (a detailed description of DNA detection using AuNPs as labels is given in Procedure 53 in the accompanying CD-Rom).

38.2.2 Environmental

Nucleic acid based affinity and electrochemical biosensors for potential environmental applications have recently been reported. Application areas for these include the detection of chemically induced DNA

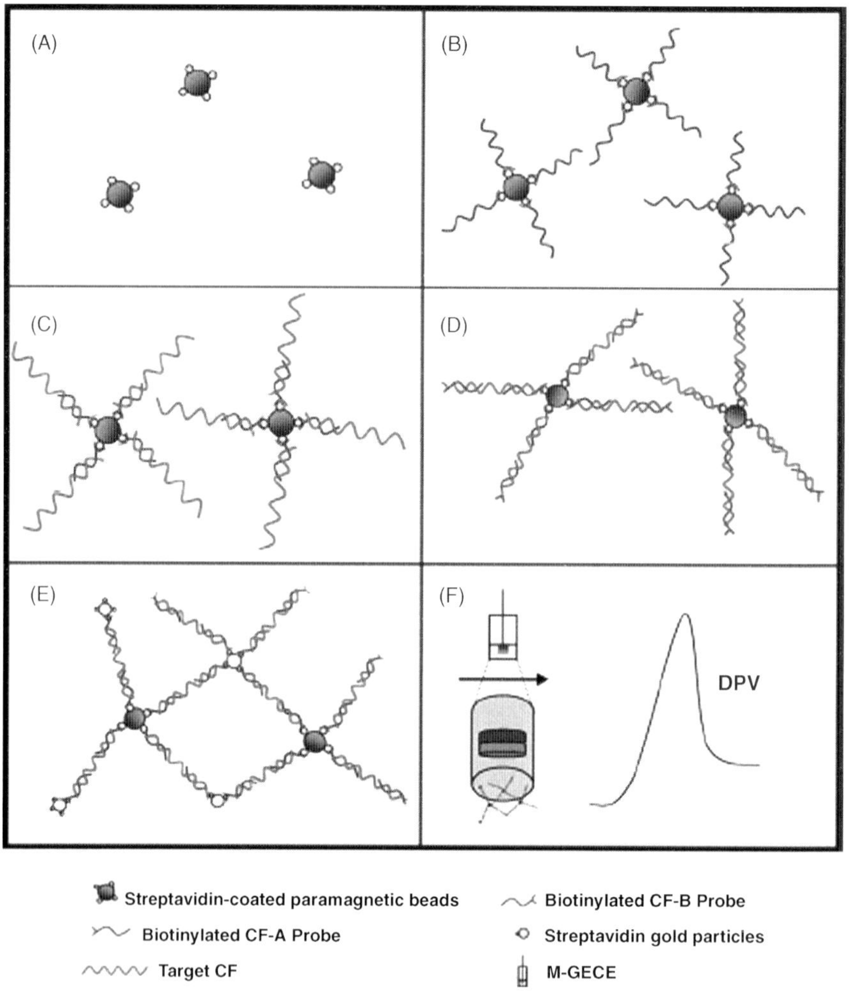

Fig. 38.1. Schematic representation of the sandwich system analytical protocol (not in scale): (A) streptavidin-coated magnetic beads; (B) immobilisation of the biotinylated CF-A probe onto the magnetic beads; (C) addition of CF-T (first hybridisation event); (D) addition of biotinylated CF-B probe (second hybridisation event); (E) tagging by using the streptavidin–gold nanoparticles; (F) accumulation of AuNPs–DNA–magnetic bead conjugate on the surface of M-GECE and magnetically trigged direct DPV electrochemical detection of AuNPs tag in the conjugate. Reprinted from Ref. [64]. Copyright 2006; with permission from Elsevier Science.

damage and the detection of microorganisms through the hybridisation of species-specific sequences of DNA [36].

In the case of DNA biosensors, two strategies are applied to detect pollutants: one is the hybridisation detection of nucleic acid sequences from infectious microorganisms, and the other the monitoring of small pollutants interacting with the immobilised DNA layer (drugs, mutagenic pollutants, etc.) [65].

38.3 PROTEINS ANALYSIS

Proteins are present at various concentrations in samples from very different origins and the determination of their concentration is of particular interest. Biosensors offer an alternative to the classical analytical methods due to their inherent specificity, simplicity, relative low cost and rapid response.

Immunosensors are affinity ligand-based biosensors in which the immunochemical reaction is coupled to a transducer [66]. These biosensors use antibodies as the biospecific sensing element, and are based on the ability of an antibody to form complexes with the corresponding antigen [16]. The fundamental basis of all immunosensors is the specificity of the molecular recognition of antigens by antibodies to form a stable complex. Immunosensors can be categorised based on the detection principle applied. The main developments are electrochemical, optical and microgravimetric immunosensors [66].

The interactions between an antibody and an antigen are highly specific. Such a specific molecular recognition has been exploited in immunosensors to develop highly selective detection of proteins.

Several protocols with potential biomedical and environmental application have been developed. Table 38.2 summarises information of some typical electrochemical immunosensors using AuNPs tags.

38.3.1 Clinical

The development of immunosensors for the detection of diseases has received much attention lately and this has largely been driven by the need to develop hand-held devices for point-of-care measurements [67,68]. Immunosensors can incorporate either the antigen or the antibody onto the sensor surface, although the latter approach has been used most often [67]. Optical [69,70] and electrochemical [70] detection methods are most frequently used in immunosensors [67]. Detection by

TABLE 38.2

Electrochemical immunosensors using AuNPs tags, with potential biomedical applications

Transducer	AuNPs	Detection method	Sample/ antigen	Detection limit	Ref.
MCPE	10 nm (Sigma)	SWASV	–/Mouse IgG	$0.02\,\mu g\,mL^{-1}$	[72]
SPE	18 nm	ASV	–/IgG	$3\times10^{-12}\,M$	[73]
Platinum electrode	–	PSA	–/Diptheria antigen	$2.4\,ng\,mL^{-1}$	[76]
GCE	–	ASV	–/Human IgG	$6\times10^{-12}\,M$	[75]
TFGE	Protein A labelled with colloidal gold (Sigma)	PSA	Blood serum/ Forest-Spring encephalitis virus	$10^{-7}\,mg\,mL^{-1}$	[77]

MCPE = magnet carbon paste electrode, SWASV = square wave anodic stripping voltammetry, ASV = anodic stripping voltammetry, PSA = potentiometric stripping analysis, SPEs = screen-printed electrodes, TFGE = thick-film graphite electrode, GCE = glassy carbon electrode.

electrochemical immunosensors is generally achieved by using either electroactive labels or enzyme labelling [71].

Liu and Lin [72] have developed a renewable electrochemical magnetic immunosensor by using magnetic beads and AuNPs labels. Anti-IgG antibody-modified magnetic beads were attached to a renewable carbon paste transducer surface by a magnet that was fixed inside the sensor. The magnet carbon paste electrode (MCPE) offers a convenient immunoreaction and electrochemical sensing platform. AuNP (10 nm) labels were capsulated to the surface of magnetic beads by sandwich immunoassay. A highly sensitive electrochemical stripping analysis using square wave anodic stripping voltammetry (SWASV) that offers a simple and fast method to quantify the captured AuNPs tags and avoid the use of an enzyme label and substrate was used. The stripping signal of AuNPs is related to the concentration of target mouse IgG in the sample solution. The detection limit of $0.02\,\mu g\,mL^{-1}$ of IgG was obtained under optimum experimental conditions. Such particle-based electrochemical magnetic immunosensors could be readily used for simultaneous parallel detection of multiple proteins by using multiple inorganic metal NP tracers and are expected to open new opportunities for disease diagnostics and biosecurity. The illustrative representation of the main steps of protocol is shown in Fig. 38.2.

An electrochemical immunoassay using a colloidal gold label (18 nm) that after oxidative gold metal was indirectly determined by ASV at a

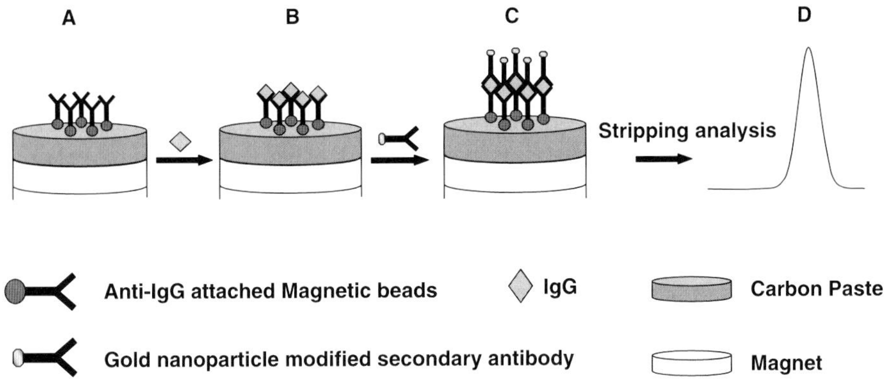

Fig. 38.2. Particle-based electrochemical immunoassay protocol. (A) Introduction of antibody-modified magnetic beads to magnet/carbon paste electrochemical transducer surface; (B) binding of the IgG antigen to the antibodies on the magnetic beads; (C) capture of the gold nanoparticle labelled secondary antibodies; (D) electrochemical stripping detection of AuNPs. Reprinted with permission from Ref. [72].

single-use carbon-based SPE was developed by Dequaire *et al.* [73]. A noncompetitive heterogeneous immunoassay of an IgG was carried out. Primary antibodies specific to goat IgG were adsorbed passively on the bottom of a polystyrene microwell. The goat IgG analyte was first captured by a primary antibody and was then sandwiched by a secondary colloidal gold-labelled antibody. The unbound labelled antibody was removed. To perform the detection, the colloidal gold present in the bound phase was dissolved in an acidic bromine–bromide solution and the Au(III) ions were measured by ASV. A limit of detection in the picomolar range (3×10^{-12} M) was obtained, which is comparable with colorimetric ELISA or with immunoassays based on fluorescent europium chelate labels.

Tang *et al.* [74] developed a potentiometric immunosensor by means of self-assembling AuNPs, polyvinyl butyral and diphtheria antibody to the surface of platinum electrodes. Diphtheria antigen was detected via the change in the electric potential upon antigen–antibody interaction. The detection limit was $2.4\,\mathrm{ng\,mL^{-1}}$ of diphtheria antigen and the linear semilogarithmic range extended from 4.4 to $960\,\mathrm{ng\,mL^{-1}}$.

An electrochemical immunoassay has been developed by Chu *et al.* [75], based on the precipitation of silver on colloidal gold labels which, after silver metal dissolution in an acidic solution, was indirectly determined by ASV at a glassy carbon electrode. The method was evaluated for a noncompetitive heterogeneous immunoassay of an IgG as a model. The anodic stripping peak current depended linearly on the IgG concentration over the range of $1.66\,ng\,mL^{-1}$–$27.25\,\mu g\,mL^{-1}$ in a logarithmic plot. A detection limit as low as $1\,ng\,mL^{-1}$ (i.e., $6\times10^{-12}\,M$) human IgG was achieved, which is competitive with colorimetric ELISA or with immunoassays based on fluorescent europium chelate labels.

A direct electrochemical immunoassay system based on the immobilisation of α-1-fetoprotein antibody (anti-AFP), as a model system, on the surface of core-shell Fe_2O_3/Au magnetic nanoparticles (MNP) has been demonstrated. To fabricate such an assay system, anti-AFP was initially covalently immobilised onto the surface of core-shell Fe_2O_3/Au MNP. Anti-AFP-modified MNP (bio-NPs) were then attached to the surface of carbon paste electrode with the aid of a permanent magnet. The performance and factors influencing the performance of the resulting immunosensor were studied. α-1-Fetoprotein antigen was directly determined by the change in current or potential before and after the antigen–antibody reaction versus saturated calomel electrode. The electrochemical immunoassay system reached 95% of steady-state potential within 2 min and had a sensitivity of 25.8 mV. The linear range for AFP determination was from 1 to $80\,ng\,AFP\,mL^{-1}$ with a detection limit of $0.5\,ng\,AFP\,mL^{-1}$. Moreover, the direct electrochemical immunoassay system, based on a functional MNP, can be developed further for DNA and enzyme biosensors [76].

Another interesting electrochemical immunosensor including AuNPs labels for diagnosis of Forest-Spring encephalitis has been proposed by Brainina *et al.* [77]. It comprises a screen-printed thick-film graphite electrode (the transducer) and a layer of the Forest-Spring encephalitis antigen (the biorecognition substance) immobilised on the electrode surface. The procedure includes formation of an antigen–antibody immune complex, localisation of colloidal gold-labelled protein A on the complex, and recording of gold oxidation voltammogram, which provides information about the presence and the concentration of antibodies in blood serum. The response is proportional to the concentration of antibodies over the interval from 10^{-7} to $10^{-2}\,mg\,mL^{-1}$. The detection limit is $10^{-7}\,mg\,mL^{-1}$. A sandwich comprising the antigen, an antibody, and protein A labelled with colloidal gold was formed on the sensor surface during the analysis. The gold oxidation current provided information about the concentration of antibodies in test samples. Taking Forest-Spring encephalitis as an example, it was shown that the use of metal-labelled protein A is promising for diagnosis of infectious diseases [77].

38.4 CONCLUSIONS

The development of electrochemical genosensors and immunosensors based on labelling with NPs has registered an important growth, principally for clinical and environmental applications. The electrochemical detection of NP labels in affinity biosensors using stripping methods allows the detailed study of DNA hybridisation as well as immunoreactions with interest in genosensor or immunosensor applications.

Beside different kinds of nanocrystals (or QDs) AuNPs are showing a special interest in several applications. Electrochemical methods used for AuNPs label detection may be very promising taking into account their high sensitivity, low detection limit, selectivity, simplicity, low cost and availability of portable instruments.

The sensitivity of AuNPs detection is usually improved by the silver enhancement method. In this procedure, silver ions are reduced to silver metal by a reducing agent, at the surface of a gold NP, causing it to grow and so facilitating the detection [36].

Many strategies based on DNA hybridisation assays using AuNPs have been developed. Most of them rely on capturing the NP to the hybridised target in a three-component 'sandwich' format.

Actually, special attention is paid in the application of biosensors in environmental analysis. The high cost and slow turnaround times typically associated with the measurement of regulated pollutants clearly indicates a need for environmental screening and monitoring methods that are fast, portable and cost-effective. To meet this need, a variety of field analytical methods have been introduced. Because of their unique characteristics, however, technologies such as DNA sensors and immunosensors based on AuNPs might be exploited to fill specific niche applications in the environmental monitoring area.

The analysis of trace substances in environmental science, pharmaceutical and food industries is a challenge since many of these applications demand a continuous monitoring mode. The use of immunosensors based on AuNPs in these applications should also be appropriate. Although there are many recent developments in the immunosensor field, which have potential impacts [36], nevertheless there are few papers concerning environmental analysis with electrochemical detection based on AuNPs. The application of some developed clinical immunosensors can also be extended to the environmental field.

ACKNOWLEDGEMENTS

Authors thank the Spanish "Ramón Areces" foundation (project 'Bionanosensores') and MEC (Madrid) (Projects MAT2005-03553, and the Consolider-Ingenio 2010 CSD2006-00012).

REFERENCES

1 D.R. Thevenot, K. Toth, R.A. Durst and G.S. Wilson, *Pure Appl. Chem.*, 71 (1999) 2333–2348.
2 C.M. Niemeyer, *Angew. Chem. Int. Ed.*, 40 (2001) 4128–4158.
3 S.F. D'Souza, *Biosens. Bioelectron.*, 16 (2001) 337–353.
4 J. Wang, *Anal. Electrochem.*, 2nd ed, Wiley-VCH, New York, 2000.
5 T. Vo-Dinh and B. Cullum, *Fresenius J. Anal. Chem.*, 366 (2000) 540–551.
6 T.S. Misono and P.K.R. Kumar, *Anal. Biochem.*, 342 (2005) 312–317.
7 O. Hayden, R. Bindeus, C. Haderspock, K.J. Mann, B. Wirl and F.L. Dickert, *Sens. Actuators B-Chem.*, 91 (2003) 316–319.
8 O. Hayden and F.L. Dickert, *Adv. Mater.*, 13 (2001) 1480–1483.
9 E.J. Cho, J.R. Collett, A.E. Szafranska and A.D. Ellington, *Anal. Chim. Acta*, 564 (2006) 82–90.
10 G.L. Igloi, *Expert Rev. Mol. Diagn.*, 3 (2003) 17–26.
11 F. Pellestor and P. Paulasova, *Int. J. Mol. Med.*, 13 (2004) 521–525.
12 A.F. Collings and F. Caruso, *Rep. Prog. Phys.*, 60 (1997) 1397–1445.
13 M.A. Cooper, *Anal. Bioanal. Chem.*, 377 (2003) 834–842.
14 P. Alivisatos, *Nanotechnology*, 14 (2003) R15–R27.
15 P. Alivisatos, *Nat. Biotechnol.*, 22 (2004) 47–52.
16 W.C.W. Chan, D.J. Maxwell, X.H. Gao, R.E. Bailey, M.Y. Han and S.A. Nie, *Curr. Opin. Biotechnol.*, 13 (2002) 40–46.
17 M. Huber, T.F. Wei, U.R. Müller, P.A. Lefebvre, S.S. Marla and Y.P. Bao, *Nucleic Acids Res*, 32 (2004) e137.
18 A. Merkoçi, M. Aldavert, S. Marin and S. Alegret, *Trends Anal. Chem.*, 24 (2005) 341–349.
19 K. Kerman, M. Kobayashi and E. Tamiya, *Meas. Sci. Technol.*, 15 (2004) R1–R11.
20 C.A. Mirkin, R.L. Letsinger, R.C. Mucic and J.J. Storhoff, *Nature*, 382 (1996) 607–609.
21 P. Alivisatos, K.P. Johnsson, X. Peng, T.E. Wilson, C.J. Loweth, M.P. Bruchez Jr. and P.G. Schultz, *Nature*, 382 (1996) 609–611.
22 T.A. Taton, C.A. Mirkin and R.L. Letsinger, *Science*, 289 (2000) 1757–1760.
23 W. Fritzsche and T.A. Taton, *Nanotechnology*, 14 (2003) R63–R73.
24 J. Wang, *Anal. Chim. Acta*, 500 (2003) 247–257.
25 Y.C. Cao, R. Jin and C.A. Mirkin, *Science*, 297 (2002) 1536–1540.

26 S.J. Park, T.A. Taton and C.A. Mirkin, *Science*, 295 (2002) 1503–1506.
27 F. Patolsky, K.T. Ranjit, A. Lichtenstein and I. Willner, *Chem. Commun.*, 12 (2000) 1025–1026.
28 H. Cai, Y. Xu, N. Zhu, P. He and Y. Fang, *Analyst*, 127 (2002) 803–808.
29 M. Ozsoz, A. Erdem, K. Kerman, D. Ozkan, B. Tugrul, N. Topcuoglu, H. Ekren and M. Taylan, *Anal. Chem.*, 75 (2003) 2181–2187.
30 T.M.H. Lee, L.L. Li and I.M. Hsing, *Langmuir*, 19 (2003) 4338–4343.
31 H. Cai, Y.Q. Wang, P.G. He and Y.Z. Fang, *Chem. J. Chinese Universities*, 24 (2003) 1390–1394.
32 K. Shanmugam, S. Subramanayam, S.V. Tarakad, N. Kodandapani and S.F. D'Souza, *Anal. Sci.*, 17 (2001) 1369–1374.
33 B. Foultier, L. Moreno-Hagelsieb, D. Flandre and J. Remacle, *IEE Proc. Nanobiotechnol.*, 152 (2005) 3–12.
34 M.I. Pividori, A. Merkoci and S. Alegret, *Biosens. Bioelectron.*, 15 (2000) 291–303.
35 K. Vivek, T. Vijay and J. Huangxian, *Crit. Rev. Anal. Chem.*, 36 (2006) 73–106.
36 L.D. Mello and L.T. Kubota, *Food Chem*, 77 (2002) 237–256.
37 E. Williams, M.I. Pividori, A. Merkoçi, R.J. Forster and S. Alegret, *Biosens. Bioelectron.*, 19 (2003) 165–175.
38 M.I. Pividori, A. Merkoçi and S. Alegret, *Biosens. Bioelectron.*, 19 (2003) 473–484.
39 A. Erdem, K. Kerman, B. Meric, U.S. Akarca and M. Ozsoz, *Anal. Chim. Acta*, 422 (2000) 139–149.
40 M.I. Pividori, A. Merkoçi, J. Barbé and S. Alegret, *Electroanalysis*, 15 (2003) 1815–1823.
41 D. Hernández-Santos, M. Díaz-González, M.B. González-García and A. Costa-García, *Anal. Chem.*, 76 (2004) 6887–6893.
42 L. Authier, C. Grossiord and P. Brossier, *Anal. Chem.*, 73 (2001) 4450–4456.
43 J. Wang, R. Polsky and D. Xu, *Langmuir*, 17 (2001) 5739–5741.
44 J. Wang, G. Liu and A. Merkoçi, *Anal. Chim. Acta*, 482 (2003) 149–155.
45 M. Fojta, L. Havran, M. Vojtiskova and E. Palecek, *J. Am. Chem. Soc.*, 126 (2004) 6532–6533.
46 E. Palecek, R. Kizek, L. Havran, S. Billova and M. Fojta, *Anal. Chim. Acta*, 469 (2002) 73–83.
47 J. Wang, D. Xu, A. Erdem, R. Polsky and M.A. Salazar, *Talanta*, 56 (2002) 931–938.
48 S.F. Wolf, L. Haines, J. Fisch, J.N. Kremsky, J.P. Dougherty and K. Jacobs, *Nucleic Acids Res*, 15 (1987) 2911–2926.
49 T.M. Lee and I.-M. Hsing, *Anal. Chem. Acta*, 556 (2006) 26–37.
50 P.O. Part, E. López and G. Mathis, *Anal. Biochem.*, 195 (1991) 283–289.
51 S.G. Penn, L. He and M. Natan, Nanoparticles for bioanalysis, *Curr. Opin. Chem. Biol.*, 7 (2003) 609–615.

52 Z. Zhong, K.B. Male and J.H.T. Luong, *Anal. Lett.*, 36 (2003) 3097–3111.
53 D. Hernández-Santos, M.B. González-García and A. Costa-García, *Electroanalysis*, 14 (2002) 1225–1235.
54 A. Doron, E. Katz and I. Willner, *Langmuir*, 11 (1995) 1313–1317.
55 M. Pumera, M.T. Castañeda, M.I. Pividori, R. Eritja, A. Merkoçi and S. Alegret, *Langmuir*, 21 (2005) 9625–9629.
56 A. Kawde and J. Wang, *Electroanalysis*, 16 (2004) 101–107.
57 J. Wang, *Anal. Chim. Acta*, 469 (2002) 63–71.
58 F. Lucarelli, G. Marrazza, A.P. Turner and M. Mascini, *Biosens. Bioelectron.*, 19 (2004) 515–530.
59 J.J. Gooding, *Electroanalysis*, 14 (2002) 1149–1156.
60 A. Kouřilová, S. Babkina, K. Cahová, L. Havran, F. Jelen, E. Paleček and M. Fojta, *Analytical Letters*, 38 (2005) 2493–2507.
61 L.M. Demers, C.A. Mirkin, R.C. Mucic, R.A. Reynolds, R.L. Letsinger, R. Elghanian and G. Viswanadham, *Anal. Chem.*, 72 (2000) 5535–5541.
62 J. Wang, D. Xu, A.-N. Kawde and R. Polsky, *Anal. Chem.*, 73 (2001) 5576–5581.
63 J. Wang, D. Xu and R. Polsky, *J. Am. Chem. Soc.*, 124 (2002) 4208–4209.
64 M.T. Castañeda, A. Merkoçi, M. Pumera and S. Alegret, *Biosens. Bioelectron.*, 22 (2007) 1961–1967.
65 J. Wang, G. Rivas, X. Cai, E. Palecek, P. Nielsen, H. Shiraishi, N. Dontha, D. Luo, C. Parrado and M. Chicharro, *Anal. Chim. Acta*, 347 (1997) 1–8.
66 P.B. Luppa, L.J. Sokoll and D.W. Chan, *Clin. chim. Acta*, 314 (2001) 1–26.
67 R.I. Stefan, J.F. van Staden and H.Y. Aboul-Enein, *Fresenius J. Anal. Chem.*, 366 (2000) 659–668.
68 A. Warsinke, A. Benkert and F.W. Scheller, *Fresenius J. Anal. Chem.*, 366 (2000) 622–634.
69 M.J. Gomara, G. Ercilla, M.A. Alsina and I. Haro, *J. Immunol. Methods*, 246 (2000) 13–24.
70 V. Koubova, E. Brynda, L. Karasova, J. Skvor, J. Homola, J. Dostalek, P. Tobiska and J. Rosicky, *Sens. Actuators B Chem*, 74 (2001) 100–105.
71 I.E. Tothill, *Comput. Electron. Agric.*, 30 (2001) 205–218.
72 G. Liu and Y. Lin, *J. Nanosci. Nanotechnol.*, 5 (2005) 1060–1065.
73 M. Dequaire, C. Degrand and B. Limoges, *Anal. Chem.*, 72 (2000) 5521–5528.
74 D. Tang, R. Yuan, Y. Chai, X. Zhong, Y. Liu and J. Dai, *Biochem. Eng.*, 22 (2004) 43–49.
75 X. Chu, X. Fu, K. Chen, G.L. Shen and R.Q. Yu, *Biosens. Bioelectron.*, 20 (2005) 1805–1812.
76 D. Tang, R. Yuan and Y. Chai, *Biotechnol. Lett.*, 28 (2006) 559–565.
77 K. Brainina, A. Kozitsina and J. Beikin, *Anal. Bioanal. Chem.*, 376 (2003) 481–485.

Subject Index